Numerical Methods in Laminar and Turbulent Flow

Editors:

C. Taylor

J. A. Johnson

W. R. Smith

Proceedings of the Third International Conference held in Seattle, 8th-11th August, 1983

PINERIDGE PRESS

Swansea, U.K.

First Published 1983 by
Pineridge Press Limited
91, West Cross Lane, West Cross, Swansea, U.K.

ISBN 0-906674-22-0

International Conference on Numerical Methods
in Laminar and Turbulent Flow: *3rd: 1983: University of Washington, Seattle*

Numerical methods in laminar and turbulent flow.
1. Fluid dynamics — Mathematics — Congresses
I. Title II. Taylor, C. III. Johnson, J. A.
IV. Smith, R.
532'.51015'117 QA911
ISBN 0-906674-22-0

Printed and bound in Great Britain by
Dotesios (Printers) Ltd., Bradford-on-Avon, Wiltshire

PREFACE

The proceedings contain the papers presented at the Third International Conference on Numerical Methods in Laminar and Turbulent Flow held at Seattle during the period August 8th-11th 1983. A large number of abstracts, approximately 165, were received. Of these more than forty had to be rejected and the organisers apologise to those whose papers could not be included for presentation.

One of the most active areas of research in fluid mechanics currently being investigated is that where non-linearity predominates. The current proceedings contains innovative numerical techniques and application to solve a wide range of engineering problems. Again, as in both previous proceedings with the same title, accompanying such advances and applications are indicators of further areas of study where research effort should be concentrated. Indeed, some papers act as pointers to areas of research that would be of considerable current benefit to industry.

The papers have been produced directly from lithographs of the respective authors' manuscripts and the editors do not accept responsibility for any erroneous comments or opinions expressed.

C. TAYLOR — University College of Swansea, U.K.

J.A. JOHNSON — Weyerhaeuser Company, Tacoma, USA

R. SMITH — College of Forest Resources, University of Washington, Seattle, U.S.A.

CONTENTS

PREFACE

SECTION 1 LAMINAR FLOW AND LUBRICATION

SECTION 2 TURBULENT FLOW

SECTION 3 BOUNDARY LAYERS

SECTION 4 FLOW WITH SEPARATION

SECTION 5 ESTUARY AND COASTLINE HYDRODYNAMICS

SECTION 6 FREE SURFACE FLOW

SECTION 7 TURBO MACHINERY

SECTION 8 DRIVEN CAVITY FLOW

SECTION 9 NON-NEWTONIAN FLOW

SECTION 11 CONVECTION/DIFFUSION

SECTION 12 TWO/MULTI PHASE FLOW

SECTION 13 COMBUSTION

SECTION 14 MATHEMATICAL CONCEPTS AND GENERAL APPLICATIONS

SECTION 1

LAMINAR FLOW

AND

LUBRICATION

FLOW-THROUGH BOUNDARY CONDITIONS FOR TIME-DEPENDENT, BUOYANCY-INFLUENCED FLOW SIMULATIONS USING LOW ORDER FINITE ELEMENTS

J. M. Leone, Jr., P. M. Gresho, R. L. Lee, and R. L. Sani[(1)]

Lawrence Livermore National Laboratory, University of California, Livermore, CA 94550

(1) CIRES/NOAA, University of Colorado, Boulder, CO 80309

SUMMARY. Two methods are presented to extend the utility of outflow natural boundary conditions from isothermal flows to those in which an additional vertical pressure gradient, associated with buoyancy-influenced flows, is important. Each of these methods approximates the pressure contribution to the natural boundary conditions in a time-dependent way. The methods are derived and demonstrated in two-dimensions and the extension to three-dimensions is discussed.

1. INTRODUCTION. One of the problems that arises when modeling fluid flow in a physically unbounded domain is the specification of boundary conditions at the computational boundary. When modeling isothermal flows with the finite element method (FEM), this has usually been successfully accomplished using the so called natural boundary conditions. The use of natural boundary conditions is more complicated, however, in buoyancy-influenced flows because of the vertical pressure gradient generated by the buoyancy terms. Herein we discuss the additional difficulties and describe our current techniques for surmounting them.

2. THEORY. The weak form of the equations governing the flow of a Boussinesq fluid are:

$$\int_\Omega \phi_i(\partial \underline{u}/\partial t + \underline{u}\cdot\nabla\underline{u})d\Omega + \int_\Omega \nabla\phi_i\cdot\underline{\underline{\tau}}d\Omega + \int_e \phi_i\,\underline{g}\beta T d\Omega = \int_{\partial\Omega} \phi_i\underline{\underline{\tau}}\cdot\underline{n}d(\partial\Omega), \quad (1a)$$

$$\int_\Omega \phi_i(\partial T/\partial t + \underline{u}\cdot\nabla T)d\Omega + \int_\Omega \nabla\phi_i\cdot K\nabla T d\Omega = \int_{\partial\Omega} \phi_i K\nabla T\cdot\underline{n}d(\partial\Omega), \quad (1b)$$

and $$\int_{\Omega} \psi_j \nabla \cdot \underline{u} d\Omega = 0 , \tag{1c}$$

where

$$\underline{\underline{\tau}} = \tau_{ij} = - P\delta_{ij} + \nu(\partial u_i/\partial x_j + \partial u_j/\partial x_i) \quad \text{or}$$

$$\tau_{ij} = (- P + \partial u_i/\partial x_j)\delta_{ij},$$

where ψ_j and ϕ_i are appropriate weighting functions, $\underline{u}$ is the velocity vector, Ω is the computational domain, $\partial\Omega$ is the boundary of Ω, $\underline{n}$ is the outward pointing unit normal vector, $P=p/\rho$, and all other symbols are standard. For convenience, this paper utilizes the ∇^2 form of the viscous terms (leading to the second form for τ_{ij}), but the ideas presented are independent of this aspect.

The boundary integrals on the right side of (1a) and (1b) represent the FEM natural boundary conditions. There is a fundamental difference between the normal and tangential components of the natural boundary conditions for the momentum equations; the pressure occurs in the former but not in the latter, a fact which can complicate the use of the normal natural boundary condition. In this paper, we address the treatment of the pressure contribution to the normal component of the natural boundary condition in stratified flows applied to a vertical outflow boundary. For purposes of discussion, let us assume that the outflow boundary is normal to the positive x direction. Then the i-th nodal value of the normal component of the natural boundary condition, G_i, is ;

$$G_i = \int_{\partial\Omega} \phi_i f_n d(\partial\Omega) \qquad \text{where } f_n = - P + \nu\, \partial u/\partial x . \tag{2}$$

In isothermal flows, setting both the normal and tangential components of the natural boundary condition to zero, particularly for large Re, has been used with success in many simulations. For example, Leone and Gresho [1] used this boundary condition to model flow over a step and Gresho, Lee, and Sani [2] modeled vortex shedding from a circular cylinder where discrete vorticies left the outflow boundary without apparent distortion, again using $f_n=0$.

In stratified flows, however, there is an additional (vertical) pressure gradient generated by the buoyancy terms which is not present in isothermal flows. Thus, in simulating such a flow, it is usually important that the normal component of the natural boundary condition account for this pressure variation at an outflow boundary, and it is no longer appropriate to set it equal to zero. This can be demonstrated by a simple example, Poiseuille flow of a stably stratified fluid. For this simulation, the top and bottom of the domain are no slip surfaces, inflow velocities are fixed at the analytic solution, the temperature is fixed at 1 on the top and 0 on the bottom, and $\partial T/\partial x = 0$ on the right and left. On the right boundary the natural boundary conditions on velocity were set to zero.

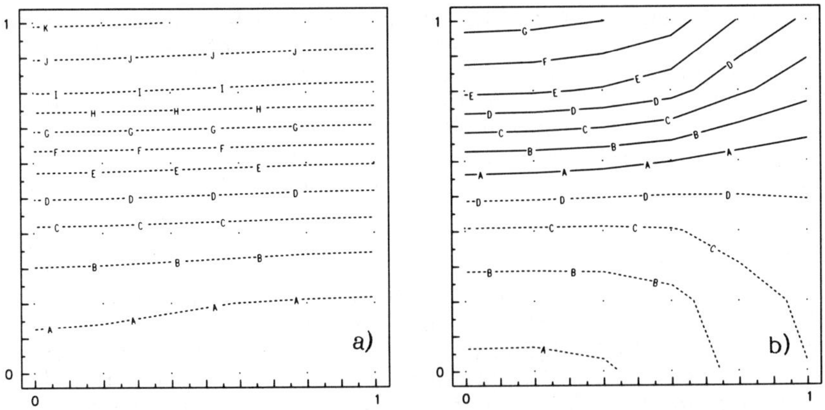

FIGURE 1. Pressure contours for stratified Poiseuille flow: a) analytic solution, b) $f_n = 0$ results.

The analytic solution was used for initial conditions. Figure 1a shows the analytic pressure field while Figure 1b shows the derived pressure field (at t = 0) from the simulation above. While there is a small distortion in both fields due to extrapolation errors within the contour package, the distortion of the pressure field near the right boundary in Figure 1b is real and is caused by the inappropriate boundary condition at the outflow. (If the equations are integrated forward in time, the distorted pressure field generates a two-dimensional flow field which deviates significantly from the desired solution). One way to remedy this problem is to allow f_n to vary with y in the proper manner; herein we shall present two techniques for doing so.

Another way to ameliorate this problem would be to recast the equations without using the weak form of the pressure gradient term. This eliminates the pressure in the boundary integrals, but at the cost of increasing the required continuity of the pressure shape functions from C^{-1} functions to C^0 functions. (See [3] for a discussion of this approach using higher order elements). However, for the simple element of interest here, using bilinear velocity (and temperature) and piecewise constant pressure, this requirement is not met. Therefore, the pressure gradient term must be written in the weak form and the pressure term included in the boundary integrals whenever this element is used.

With this in mind we have developed two methods to treat flow-through boundaries in time-dependent stratified flow simulations. Both of these methods are based on the assumption that the pressure contribution dominates the boundary integral (this is generally true for all but very low Reynolds numbers), i.e. $f_n \simeq -P$ in (2). The two methods differ in how the boundary pressures are estimated.

The first method, which we call the hydrostatic natural boundary condition (HYNBC) involves a second assumption: the boundary pressure can be well-approximated by its hydrostatic component, i.e. $f \simeq -P_h$ where P_h is calculated from the appropriate discrete analog of $P_h = \int g\beta T dy$. The hydrostatic assumption is usually reasonable in those cases where buoyancy effects are important. Thus, at time t, P_h is calculated along the outflow boundary using the current temperature T and is used to evaluate the integral in (2) necessary to advance the velocities to time $t+\Delta t$.

The proper way to compute P_h is described below and is based on the requirement that the specified values of $f_n(y)$ assure that both horizontal and vertical momentum equations are precisely satisfied for the special case of "no-flow"; i.e. for $\underline{u}=\underline{0}$ and $\partial T/\partial x = 0$, the discretized versions of $\partial P/\partial x = 0$ and $\partial P/\partial y = \beta g T(y)$ must be satisfied at the "outflow" boundary. Since $f_n(y) = -P_h(y)$ at the outflow, and $\partial P_h/\partial y = \beta g T$, the first requirement is that $f_n(y)$ in (2) be expressed using the same type of basis functions as those for pressure; e.g. if P is expressed using (2D) piecewise constant functions over the area of each element, then f_n must be expressed using (1D) functions which are piecewise constant over each element side. This assures that the "hydrostatic" momentum equations at the outlet are of precisely the same form as those in the interior of the domain. We will demonstrate this construction for the bilinear velocity (and temperature), piecewise constant pressure element, but an equivalent procedure can be used for (some) higher-order elements. The appropriate (weak) form of the 1D hydrostatic equation is

$$\phi_i P \Big|_0^H - \sum_{j=1}^{M} P_j \int_0^H d\phi_i/dy \, \psi_j \, dy = \beta g \sum_{j=1}^{N} T_j \int_0^H \phi_i \phi_j \, dy , \tag{3}$$

where $\phi_i(y)$ is a linear basis function, ψ_j is piecewise constant, H is the domain height at the outlet, and there are M elements and N nodes in $0 \le y \le H$. The nodal values of temperature, $T_j(t)$ are known and (3) is used to compute the M values of $P_j(t)$ (which are then used as f_{n_j} in (2) for the next time step; i.e.

$$f_n(y,t) = \sum_{j=1}^{M} f_{n_j}(t)\psi_j(y) . \tag{4}$$

Eqn (3) is solvable only in certain cases; there must be a proper balance between the number of nodes (N) and the number of elements (M). When the system is solvable, it leads to a simple marching technique; this is the case for the element considered herein.

Consider the column of outflow elements in Figure 2. Application of (3) to this grid yields

$$P_1 = P_o - H_1(\beta g/6)(2T_1 + T_2),$$

$$P_2 = P_1 - (\beta g/6)[H_1(T_1+2T_2)+H_2(2T_2+T_3)] ,$$

$$\vdots$$

$$P_M = P_{M-1} - (\beta g/6)[H_{M-1}(T_{N-2} + 2T_{N-1}) + H_M(2T_{N-1}+T_N] ,$$

where P_o (the pressure at node 1) is an arbitrary constant. If the boundary mass matrix for temperature is lumped in (3), the results are

$$P_1 = P_o - (\beta g H_1/2)T_1 ,$$

$$P_2 = P_1 - (\beta g/2)(H_1+H_2)T_2 ,$$

$$\vdots$$

$$P_M = P_{M-1} - (\beta g/2)(H_{M-1}+H_M)T_{N-1} ; \qquad (5)$$

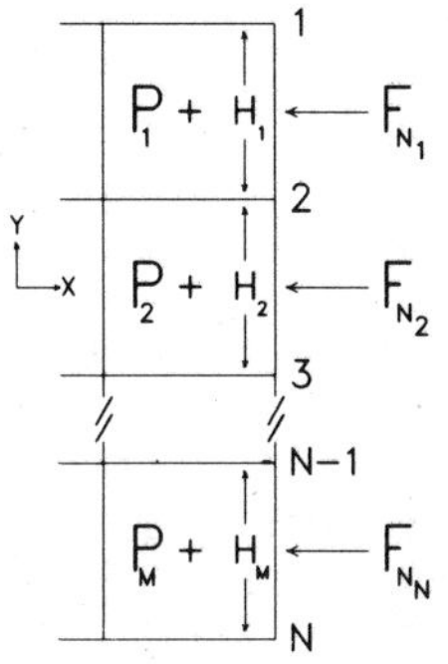

FIGURE 2.

this is the form used in our codes. These pressure values are computed at the end of each time step and, replacing P_j by $-f_{n_j}$, are used in (2) and (4) to update the "normal force" boundary condition for the next time step. This yields

$$G_1 = H_1 f_{n_1}/2 ,$$

$$G_2 = (H_1 f_{n_1} + H_2 f_{n_2})/2, \text{ etc.} \qquad (6)$$

A similar technique can be applied to some higher-order elements; e.g. the element with biquadratic velocity and either discontinuous (C^{-1}) bilinear pressure (e.g. nodes at the 2x2 Gauss points) or linear (3-node) pressure. In these cases the "marching" solution of (3) now requires the intermediate step of solving a 2x2 linear system for each element (a modified marching scheme) since there are two values of f_{n_i} on each element face. However, if an element with biquadratic velocity (or 8-node serendipity) and C^0 bilinear pressure is employed, the technique fails since (3) has, in general, no solution in these cases; there are more momentum equations than continuity equations ($N \simeq 2M$) for these elements. This failure is directly related to the problems discussed in [4] when these elements are employed for stratified flows.

Returning to the piecewise constant pressure element, we briefly address the analogous problem in 3D. It appears possible to extend the 2D scheme to a plane of outflow elements in such a way that a 1D marching scheme for each column of elements can be employed. However, we have not yet implemented such a scheme, nor even studied it in much detail; it may lead to additional problems relating to reference pressure specification.

The second method, which we call the extrapolated pressure natural boundary condition (EPNBC), does not invoke the hydrostatic assumption but rather calculates boundary pressures by linear extrapolation of the neighboring element pressures (see Fig. 3): e.g.,

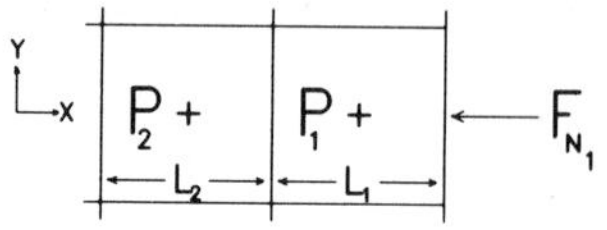

$$F_{N_1} = - P_1 - [L_1/(L_1+L_2)] (P_1-P_2) \, . \quad (6)$$

FIGURE 3.

Thus, at time t the boundary pressures are extrapolated from the calculated pressure field and are used to evaluate the boundary integral (2) for advancing the solution to time $t+\Delta t$. When using this process one must first assure that there will be no checkerboard (CB) pressure mode [5]. (In an attempt to overcome this restriction, we tested the idea of extrapolating the boundary pressure from nodal pressures which had been filtered to eliminate any CB pressure modes. However, this proved unsatisfactory because the bilinear extrapolation scheme generated a spurious horizontal pressure gradient at the boundary when the vertical pressure gradient was higher order than linear). Another difficulty is that this method is not self starting because the velocity (natural) boundary conditions are required in the solution of the initial pressure field. Currently we use HYNBC to begin the simulation - in a mix of the two methods.

3. NUMERICAL RESULTS. The first test of these new boundary conditions was the Poiseuille flow problem discussed earlier. Not surprisingly, both of the new boundary conditions reproduced the analytic velocity field. While similar results can be obtained for this particular flow by specifying the appropriate values of $f_n(y)$, since they are known and time independent, this is impossible for general stratified flows because the "proper" f_n is time-dependent and varies in an unknown manner.

The initialization of an atmospheric flow over a slope is a problem we are often required to model which has such characteristics. For this problem a geostrophic wind (large scale pressure gradient) and an inflow vertical temperature profile are known. We desire a "balanced state" to be used as initial conditions in other simulations; e.g. drainage flows caused by the cooling of the surface. (Since many of our simulations involve atmospheric flows, the equations we actually solve use slightly different variables than in (1) and contain additional terms due to the coriolis force caused by the earth's rotation; see [6]). Figure 4 shows the domain and initial conditions used in this example. A balanced state was achieved after approximately 9 hours of integration of the governing equations using each of the proposed boundary conditions at the outflow boundary. These balanced states agreed with each other to within 3%. During the integration, warm air was brought into the domain through the

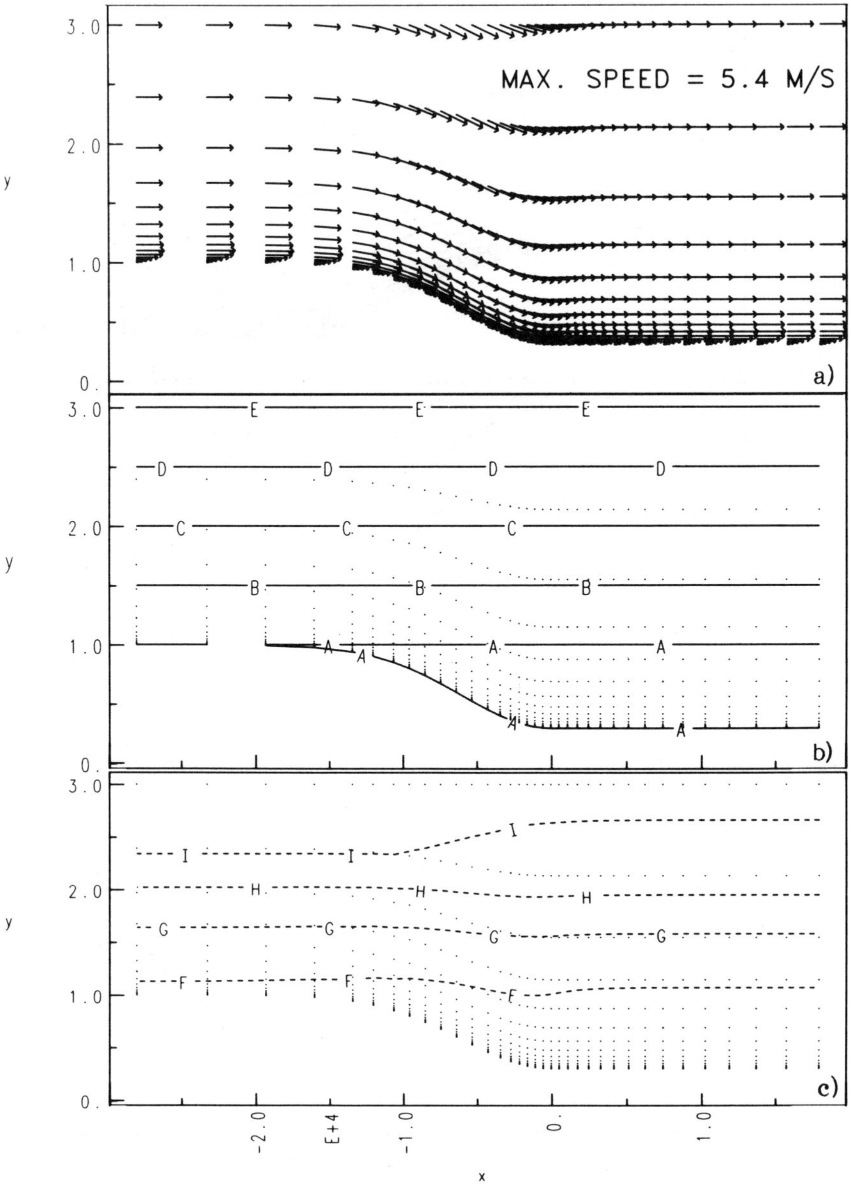

FIGURE 4. Initial conditions for slope flow initialization: a) velocity field, b) potential temperature field, $\partial T/\partial y$ = 0.2° K/km, c) pressure field.

top, in the region above the slope. This warm air displaced the cooler air initially present by gradually forcing much of it out through the right (outflow) boundary. This replacement caused significant time-dependent changes in the temperature structure, and thus in the pressure structure, at the outflow boundary. These changes

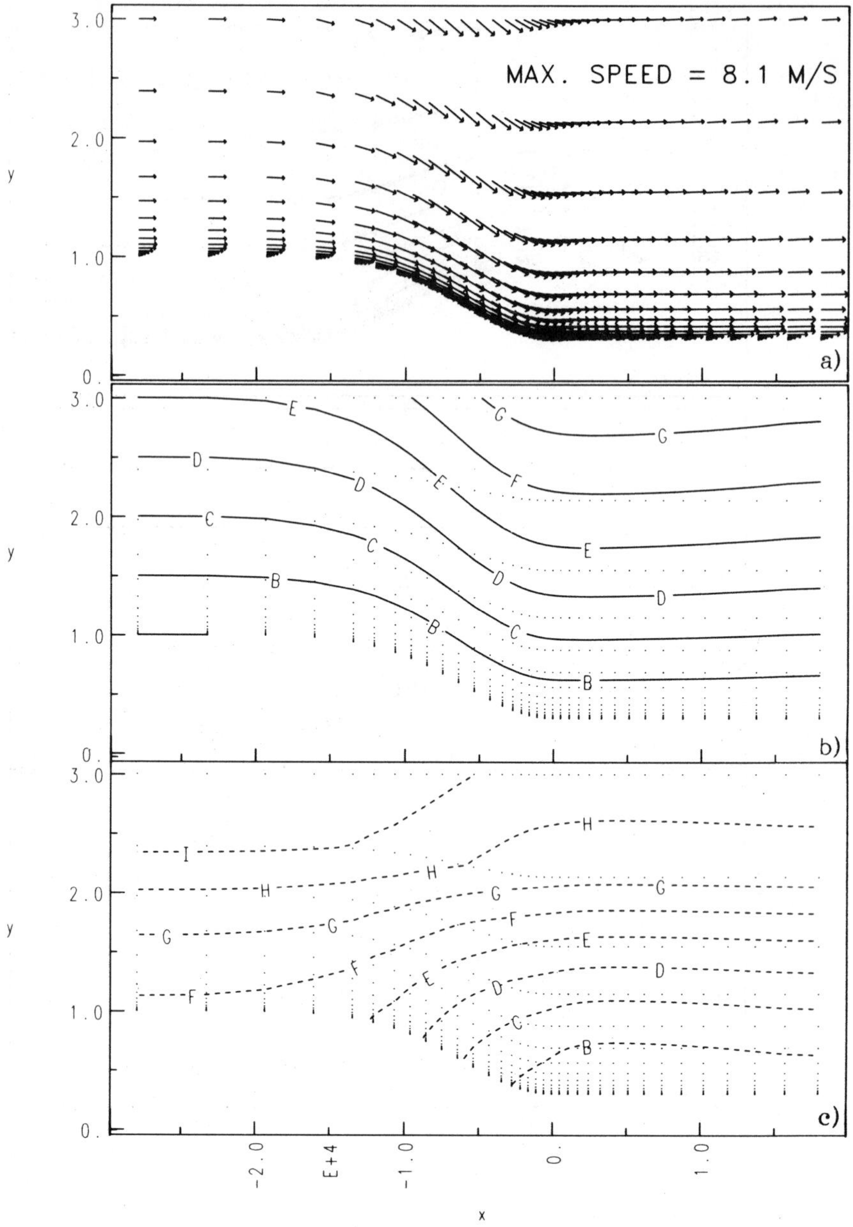

FIGURE 5. Balanced field from slope flow initialization: a) velocity field, b) potential temperature field, c) pressure field.

were accounted for in a very reasonable way by each of the new boundary conditions. The balanced state achieved is presented in Figure 5.

When the same simulation was attempted with natural boundary conditions on the right boundary which were prescribed for all time

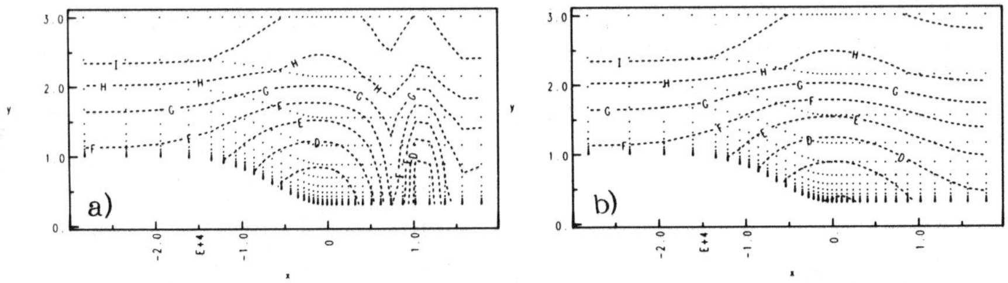

FIGURE 6. Pressure field at 3 hrs from slope flow initialization: a) time-independent outflow f_n, b) time-dependent outflow f_n.

at their initial values, the results were physically unreasonable. Figure 6a shows the pressure field at 3 hours, when the velocity and temperature fields began showing obviously unrealistic behavior. For comparison, Figure 6b shows the 3 hour pressure field from the simulation using HYNBC. The time-independent natural boundary condition severely limits the ability of the pressure field to respond to changes in the thermal field near the boundary. This in turn caused unrealistic pressure gradients which distorted the velocity field. This anomalous behavior was absent when either of the proposed boundary conditions was used.

A final example of the utility of these boundary conditions involves the gravitational spread of a cold (heavy) gas in a neutrally stable atmosphere. Figure 7 shows a "short grid," 134m, and the initial temperature field. A long grid simulation, 240m, was also performed to establish a base solution. Then three simulations were run on the short grid of Figure 7 and compared to the long grid solution. When the outflow natural boundary condition was fixed at zero (which is reasonable only prior to t=12s, at which time the dense fluid reaches the outflow boundary) the cold gas was forced too close to the ground as it left the grid and the velocity near the ground became quite large. The simulations using the new boundary conditions compared fairly well with the long grid simulation. Figure 8 is the time history of the horizontal velocity at the node marked (see arrow) on Figure 7. The excessive velocity of the $f_n = 0$ calculation is clearly evident while the other two runs compare much better. (The oscillations in these results seem to be caused by a Kelvin-Helmholtz rollup-like instability which then leads to internal gravity waves).

4. CONCLUDING REMARKS. We have developed two methods of modeling outflow boundary conditions for stratified flow simulations, HYNBC and EPNBC, and demonstrated their utility in 2D simulations. However, further research in this area is still necessary; e.g., we do not yet have sufficient experience to judge the relative merits of the two schemes. There is some indication that HYNBC is somewhat more robust, but this remains to be verified. Also, while the generalization of the methods to 3D is fairly obvious, it is not clear how they will actually perform. HYNBC may have problems

related to reference pressure specification and EPNBC is sensitive to CB pressure modes which are more numerous in 3D than in 2D.

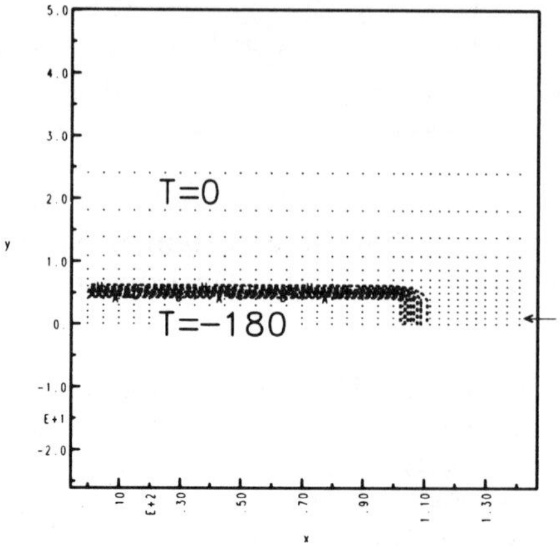

FIGURE 7. Initial temperature field for heavy gas spread (Isotherm ΔT = 20° C).

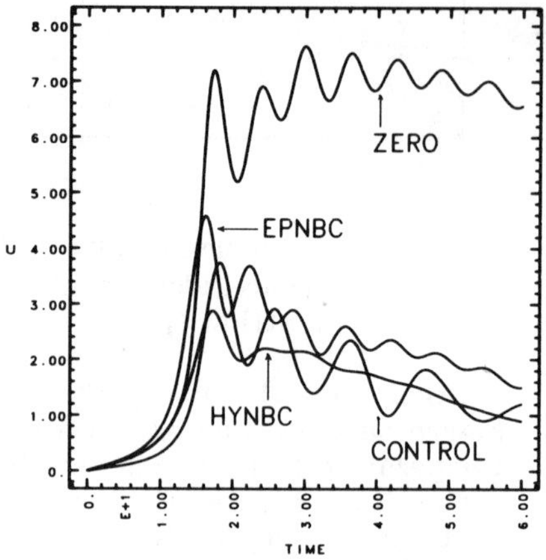

FIGURE 8. Time histories of u velocity at node marked by arrow in Figure 7. ZERO is $f_n = 0$ result and CONTROL represents the long grid results.

5. ACKNOWLEDGMENTS. This work was performed under the auspices of the U. S. Department of Energy by the Lawrence Livermore National Laboratory under contract No. W-7405-Eng-48.

6. REFERENCES.

1. LEONE, J. M., JR. and P. M. GRESHO - Finite Element Simulations of Steady, Two-Dimensional Viscous Incompressible Flow over a Step. J. Comput. Phys., Vol. 41, pp. 167-181, 1981.

2. GRESHO, P. M., R. L. LEE, and R. L. SANI - On the Time Dependent Solution of the Incompressible Navier-Stokes Equations in Two and Three Dimensions, Recent Advances in Numerical Methods in Fluids, Volume I, Ed. Taylor, C. and K. Morgan, Pineridge Press, 1980.

3. SANI, R. L., M. S. ENGELMAN, P. M. GRESHO, and C. D. UPSON - On Flow-through Boundary Conditions for Buoyancy Influenced and/or Rotating Flows Using Higher-Order Elements, to be presented at The Fifth International Symposium on Finite Element Methods in Flow Problems, Austin, TX, January 23-26, 1984.

4. GRESHO, P. M., R. L. LEE, S. T. CHAN, J. M. LEONE, JR., - A New Finite Element for Boussinesq Fluids, Proceedings of the Third Internation Conference on Finite Elements in Flow Problems, Ed. D. H. Norrie, Banff, Alberta, Canada, 1980.

5. SANI, R. L., P. M. GRESHO, R. L. LEE, and D. R. GRIFFITHS - The Cause and Cure (?) of The Spurious Pressures Generated by Certain FEM Solutions of the Incompressible Navier-Stokes Equations: Part 1. Int. J. Num. Meth. Fluids, Vol. 1, pp. 17-43, 1981.

6. LEE, R. L. and J. M. LEONE, JR. - A Modified Finite Element Model for Application to Terrain - Induced Meso-scale FLows. Presented at the Sixth Symposium on Turbulence and Diffusion, Boston, MA, March 22-25, 1983.

FINITE ELEMENT ANALYSIS OF VISCOUS FLOW - SOLID BODY INTERACTION

M.D. Olson and M.B. Irani*

ABSTRACT

This paper describes preliminary development work on extending the finite element method to cover interaction between viscous flow and a moving solid body. The analysis uses the stream function-high precision triangular finite element representation for two-dimensional incompressible flow. The interactive forces and constraints between the fluid and body are developed from surface traction and compatibility conditions. This yields a set of coupled equations for the fluid-body system. Results are obtained for the linearized case of uni-directional motion of a spring-mass (of square shape) body system surrounded by still fluid and excited by a harmonic external force. Numerical results are obtained for the response and phase angle of the body as functions of frequency. At the same time, results for the associated flow fields are also obtained. As a by-product, results are obtained for the added mass and damping coefficients as a function of frequency.

1. INTRODUCTION

Much work has been done in recent years to develop numerical solutions for the two-dimensional flow around bodies using finite element and finite difference techniques. Results of acceptable accuracy have been obtained for flows and forces

---

Professor and MASc graduate, Dept. of Civil Engineering, U.B.C., Vancouver, Canada.

on fixed bodies for a finite range of Reynolds numbers. A logical extension of these methods would be to the study of the flow around a rigid body which is elastically supported, the movement of which would affect the flow field and the fluid forces themselves and raise the question of resonant oscillations of the body.

The purpose of this work is to extend the finite element method for viscous flow problems to cover interaction between viscous flow and a moving solid body. The problem configuration considered is that of an elastically supported rigid body surrounded by a two-dimensional incompressible viscous flow. High precision triangular elements are used to discretize the stream function form of the Navier-Stokes equations, and a system of nonlinear differential equations for the coupled system is obtained.

The first step towards a nonlinear solution of this system by the perturbation method is a zeroth order linear solution. In this paper, we restrict ourselves to the linear problem only. Thus the results are confined to highly viscous flow and small amplitude motion.

The example treated is a square shaped elastically supported mass which is excited by a harmonic force in a fluid otherwise at rest. The effect of the fluid viscosity and the mass to fluid density ratio on the response of the system is studied. The effective damping and added mass is also obtained as a function of a frequency parameter.

2. THEORY

In the following, the equations of motion for the coupled viscous fluid-solid body system are presented. The stream function form of the Navier Stokes equations is used for the fluid domain.

The relevant non-dimensional parameters that arise naturally from this equation are the Reynolds number $R = UL/\nu$ and the frequency parameter $\beta = \omega L^2/\nu$, where ω, U, L and ν are the characteristic frequency, velocity, length and kinematic viscosity, respectively. The form of the non-dimensional equation indicates that the nonlinear terms can be neglected when R is small and β is finite. It can be easily shown that these conditions are satisfied for any frequency provided the amplitude of the body oscillation is small compared to its length.

The equations of motion for the elasticity supported rigid body can be written as

$$[M_b]\{\ddot{r}\} + [K_s]\{r\} = \{F\} \qquad (1)$$

where $\{r\}^T = (r_x, r_y, r_\alpha)$ is the position vector of its mass centre. The time dependent load vector $\{F\}$ is representated as the sum of two components $\{F_1\}$ and $\{F_2\}$ where $\{F_1\}$ is the vector of fluid forces on the body and $\{F_2\}$ is the vector of other external forces on the body.

For each component of the fluid force, a consistent force vector is derived in terms of the finite element nodal variables by integrating the tangential and normal stresses on the boundary of each fluid element in contact with the body. Discretization of the stream function equation yields a set of equations for the fluid domain. Compatibility constraints are imposed between the fluid and the body by specifying the tangential and normal velocity components on the interface to be the same. This gives a set of equations for the coupled motion of the system.

2.1 Finite Element Representation

The stream function is represented by the 18 degree of freedom C^1 triangular elements following [1]. A typical element in local coordinates is shown in Figure 1. The basic element is formulated in terms of an incomplete quintic polynomial in the local ξ,η coordinates with coefficients a_i, and this is used in the following force calculations. Further it is assumed that only edge 1-2 of the element (Figure 1) will be in contact with the body.

The fluid force on the body is obtained by integrating the expressions for the tangential and normal stresses on the boundary of each fluid element in contact with the body. The force expressions are derived in terms of the polynomial coefficients, for each element, then transformed to the global variables. The force vectors obtained for each element are assembled in the usual manner into global force vectors representing the fluid forces on the body.

The shear stress in local coordinates is

$$\tau_{\xi\eta} = \mu(\psi_{\eta\eta} - \psi_{\xi\xi}) \qquad (2)$$

and hence the shearing force on the body will be (parallel to the 1-2 edge)

$$F_\tau = -\int_{-b}^{a} [\tau_{\xi\eta}|_{\eta=0}]d\xi = \mu h_i a_i \qquad (3)$$

$$i = 1,2,\ldots 20$$

where the coefficients h_i are easily found explicitly.

The normal stress σ_η can be expressed as the sum of the pressure p and a frictional component also known as the deviatoric stress component σ'_η where

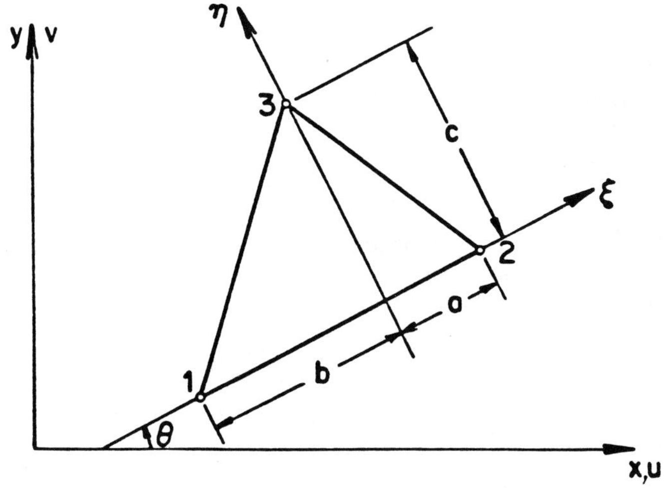

Figure 1. Triangular Finite Element and Local Coordinates.

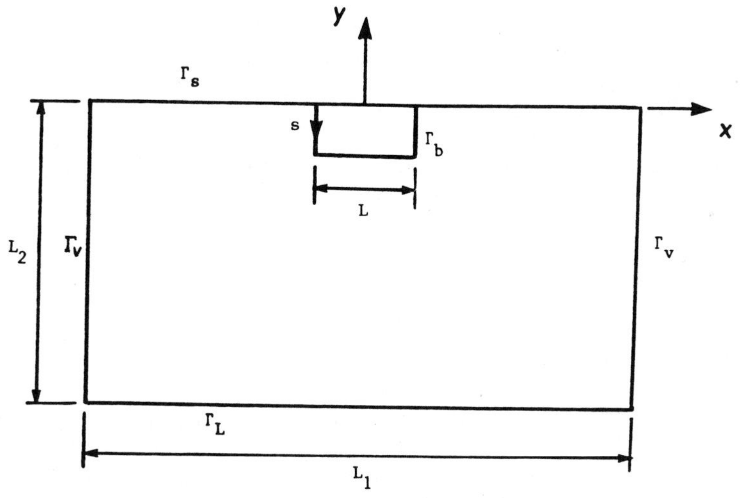

Figure 2. Problem Geometry.

$$\sigma_\eta = -p + \sigma'_\eta \qquad (4)$$

The deviatoric stress σ'_η is given by

$$\sigma'_\eta = -2\mu \ \psi_{\xi\eta} \qquad (5)$$

and following the same procedure as before, the force normal to edge 1-2 of the element is

$$F_\sigma = -\int_{-b}^{a} [\sigma'_\eta|_{\eta=0}]d\xi = \mu d_i a_i \qquad (6)$$

The expression for the pressure distribution along the element edge 1-2 is obtained by integrating the momentum equation in the ξ-direction from the Navier Stokes equations

$$\partial p/\partial\xi = \mu(\psi_{\eta\xi\xi} + \psi_{\eta\eta\eta}) - \rho\dot{\psi}_\eta - \rho(\psi_\eta\psi_{\xi\eta} - \psi_\xi\psi_{\eta\eta}) \qquad (7)$$

Integrating this expression gives the pressure distribution along edge 1-2 of the element for $\eta=0$

$$p = \int_0^\xi [\mu(\psi_{\eta\xi\xi}+ \psi_{\eta\eta\eta}) - \rho\dot{\psi}_\eta - \rho(\psi_\eta\psi_{\xi\eta}- \psi_\xi\psi_{\eta\eta})]\Big|_{\eta=0} d\xi + g(\eta) \qquad (8)$$

where $g(\eta)$ is a function of η which is a constant for $\eta=0$, equal to the pressure at $\xi=0$.

From equation (8) the pressure at node 2 of an element can be expressed in terms of the pressure at node 1 of that element. It is apparent that the presure, p_k, at any node k will depend on the datum value assigned to the pressure at the first node of the first element. The net force on the body will depend on the pressure difference rather than the absolute pressure. Thus the datum value can be arbitrarily set to zero.

Integrating the pressure expression gives the force for one element

$$F_p = (a+b)p_1 + \mu\ell_i a_i + \rho w_{ij} a_i a_j + \rho z_i \dot{a}_i \qquad (9)$$

where (a+b) is the length of the edge 1-2 of the element and p_1 is the pressure at node 1 of the element. Note that this pressure force naturally involves time dependent terms as well as nonlinear ones. The latter are neglected in the following.

After all the fluid force terms are transformed to stream function variables in the global system, they are assembled for the entire body in the form

$$\{F_1\} = \begin{Bmatrix} F_x \\ F_y \\ F_\alpha \end{Bmatrix} = \begin{Bmatrix} (F_x)_\tau \\ (F_y)_\tau \\ (F_\alpha)_\tau \end{Bmatrix} + \begin{Bmatrix} (F_x)_\sigma \\ (F_y)_\sigma \\ (F_\alpha)_\sigma \end{Bmatrix} + \begin{Bmatrix} (F_x)_p \\ (F_y)_p \\ (F_\alpha)_p \end{Bmatrix} \qquad (10)$$

where some of the terms are proportional to ψ_i and others to $\dot{\psi}_i$, where $\{\psi\}$ is the global vector of nodal variables for the entire problem.

3. EXAMPLE APPLICATION

The foregoing formulation is applied to the linear solution of a single degree of freedom spring-mass system immersed in viscous fluid and excited by a harmonic force. The system of equations for the fluid domain becomes

$$[M]\{\dot{\psi}\} + [K]\{\psi\} = \{F^1\}\dot{r} + \{F^2\}\ddot{r} + \{h_o\} \qquad (11)$$

where r is the displacement of the body, $\{F^1\}$ and $\{F^2\}$ represent the fluid forces acting on the body and $\{h_o\}$ is the force vector resulting from boundary conditions on the outer boundaries of the domain. Here the latter is zero.

In [2], the effect of the body pushing back on the fluid was investigated via a virtual power equation and it was found that the same force vectors $\{F^1\}$ and $\{F^2\}$ come into play. Hence for this case, the equation of motion for the body becomes

$$M_b\ddot{r} + K_s r = \{F^1\}^T\{\psi\} + \{F^2\}^T\{\dot{\psi}\} + f_o \qquad (12)$$

where M_b, K_s are the mass and spring constant for the body and f_o is any external force acting on it. Hence the coupled set of equations for the body-fluid interaction becomes

$$\begin{bmatrix} [M] & -\{F^1\} \\ -\{F^1\}^T & M_b \end{bmatrix} \begin{Bmatrix} \{\dot{\psi}\} \\ \ddot{r} \end{Bmatrix} + \begin{bmatrix} [K] & -\{F^2\} \\ -\{F^2\}^T & 0 \end{bmatrix} \begin{Bmatrix} \{\psi\} \\ \dot{r} \end{Bmatrix} + K_s \begin{Bmatrix} 0 \\ r \end{Bmatrix} = \begin{Bmatrix} \{h_o\} \\ f_o \end{Bmatrix} \qquad (13)$$

Note the symmetry of the mass and stiffness matrices.

3.1 Frequency Domain Analysis

Since the objective of this investigation is the solution of the problem for harmonic flows, where the forcing terms can be expressed as

$$\{h_o\} = \{\tilde{h}_o\}e^{i\omega t}, \qquad f_o = \tilde{f}_o\, e^{i\omega t}, \qquad (14)$$

the solution of the linear problem may also be represented in the harmonic form

$$\{\psi\} = \{\tilde{\psi}\}e^{i\omega t} , \qquad r = \tilde{r}\, e^{i\omega t} \tag{15}$$

and a complex system of equations is obtained

$$\begin{bmatrix} i\omega[M] + [K], & -i\omega\{F^1\} - \{F^2\} \\ -i\omega\{F^1\}^T - \{F^2\}^T, & i\omega M_b + \frac{K_s}{i\omega} \end{bmatrix} \begin{Bmatrix} \{\tilde{\psi}\} \\ i\omega\tilde{r} \end{Bmatrix} = \begin{Bmatrix} \{\tilde{h}_o\} \\ \tilde{f}_o \end{Bmatrix} \tag{16}$$

A simple dimensional analysis yields the appropriate non-dimensional parameters as

$$\text{response} = \{\frac{r_{max}}{r_{st}} , \phi\} = f\{\frac{\nu}{L^2\omega_n} , \frac{\omega}{\omega_n} , \frac{M_b}{\rho L^3}\} \tag{17}$$

where r_{max} is the amplitude of the mass, r_{st} is the equivalent static deflection of the spring due to a static force f_o, ϕ is the phase angle, ρ, the density of the fluid, and ω_n, the natural frequency of the spring mass system.

3.3 Problem Geometry-Square Cylinder

The example problem geometry is shown in Figure 2 where Γ_s = symmetry axis, etc. The domain sides L_1 and L_2 were varied to simulate an infinite fluid domain.

Zero flow or no slip conditions were assumed on Γ_v and Γ_L. The no slip condition on the body was enforced by equating the flow velocity $u = \psi_y$ to that of the body ($\dot{r}$), whereas $v = -\psi_x = 0$ everywhere on Γ_b. On the vertical edges of the body, the stream function varies linearly with y with coefficient ($\dot{r}$), such that on the horizontal part of Γ_b it becomes

$$\psi = \psi_{\Gamma_s} - (\frac{L}{2})\dot{r} \tag{18}$$

where ψ_{Γ_s} is the value of ψ on Γ_s.

At the corners of the body, there is an abrupt change in geometry. Therefore, triple noding is used at these corners in order to properly specify the boundary conditions along both the adjacent edges.

3.3 Numerical Results

Numerical results for the response of the mass and the flow patterns were obtained for the two grids shown in Figure 3. The computational domain was chosen with $L_1 = 11.0$, $L_2 = 5.0$ and $L = 1.0$.

Figure 4 shows the added mass coefficient $K_1 = M_a/\rho L^3$ and damping coefficient $K_2 = C/\omega\rho L^3$ versus frequency plots for the square along with those for the cylinder [3]. Grid 2(b) had L_1 and L_2 increased by 50%, whereas grid 2(c) and the no-slip

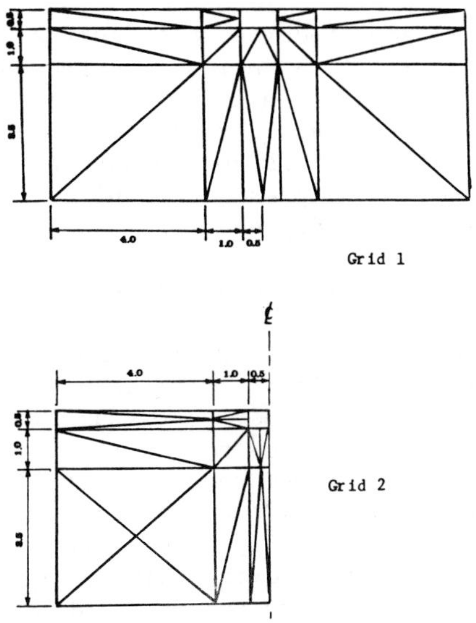

Figure 3. Finite Element Grids.

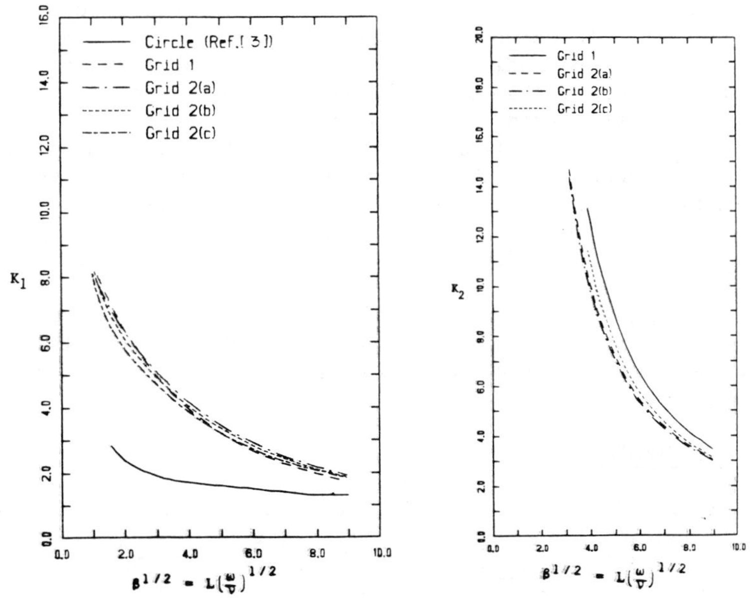

Figure 4. Added Mass and Damping Coefficients

velocity boundary conditions relaxed to free flow. It is clear that neither change had much effect on the results.

Figure 5 shows amplitude response curves for the specific mass ratio $M_b/\rho L^3 = 10$ and a series of different viscosities. They are typical of a damped single degree of freedom system affected by the effect of added mass, ie., the resonance peaks are shifted to the left. This became more pronounced as the mass ratio decreased. The phase angle plots (not shown) were also much like the single degree of freedom ones.

Figure 6 shows the flow velocity vectors obtained at different times $\tau = \omega t/\pi$ over about one half a cycle of the body motion. Figure 6(a) shows the developed flow when the body has its maximum velocity to the left, $\tau = 0.5$. Figure 6(b) is at $\tau = 1.0$ when the body is virtually stopped but there is still some residual flow due to its inertia. Figures 6(c),(d) and (e) show the flow reversal details for small time steps at $\tau = 1.05$, 1.1 and 1.2, respectively. Note the scale changes in each figure. Finally, Figure 6(f) shows the developed flow in the opposite direction when the body is nearly at its maximum velocity to the right at $\tau = 1.3$.

4. CONCLUDING REMARKS

A finite element method for the analysis of the interaction of a viscous fluid and an elastically supported rigid body was presented. As the first approximation for a solution by the perturbation method, the equations were linearized and the linear system was applied to the problem of an elastically supported, square shaped mass subjected to a harmonic force in otherwise still fluid. Response plots for the coupled system were obtained. The effect of the fluid, in the form of added mass and damping, was clearly evident in the shift of the resonance peaks of the amplitude response curves of the mass. The added mass and damping coefficients for the square shape were presented as a function of the frequency parameter.

Having proven the applicability of the theory developed for the linear case, further effort is required in solving the nonlinear problem by considering the effects of the higher order terms.

5. REFERENCES

1. OLSON, M.D., "Variational Finite Element Methods for Two Dimensional and Axisymmetric Navier Stokes Equations", Finite Elements in Flow Problems - Vol. 1, Eds. Gallagher, R.H., et al., John Wiley, New York, 1975.

2. IRANI, M.B., "Finite Element Analysis of Viscous Flow and Rigid Body Interaction", M.A.Sc. Thesis, Dept. of Civil Engineering, University of British Columbia, Vancouver, Canada, 1982.

3. STUART, J.T., "Periodic Boundary Layers", Fluid Motion Memoirs-Laminar Boundary Layers, Rosenhead, L., Ed., Oxford University Press, 1963, pp. 390.

6. ACKNOWLEDGEMENT

This work was supported by the Natural Sciences and Engineering Research Council of Canada.

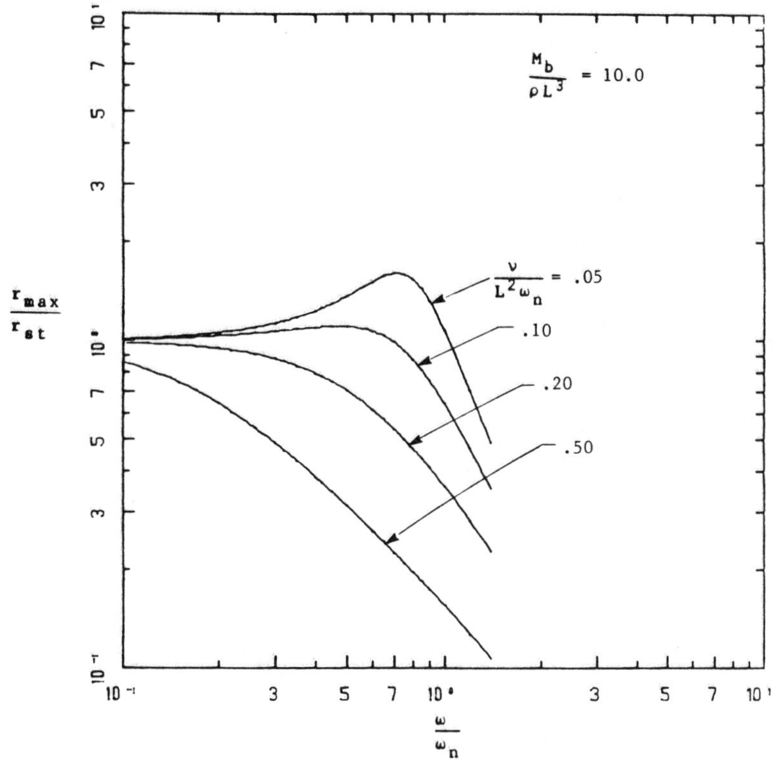

Figure 5. Amplitude Response Curves

(a) τ = 0.5
Scale = 1.0

(b) τ = 1.0
Scale = 1.0

(c) τ = 1.05
Scale = 3.5

(d) τ = 1.1
Scale = 3.0

(e) τ = 1.2
Scale = 1.8

(f) τ = 1.3
Scale = 1.3

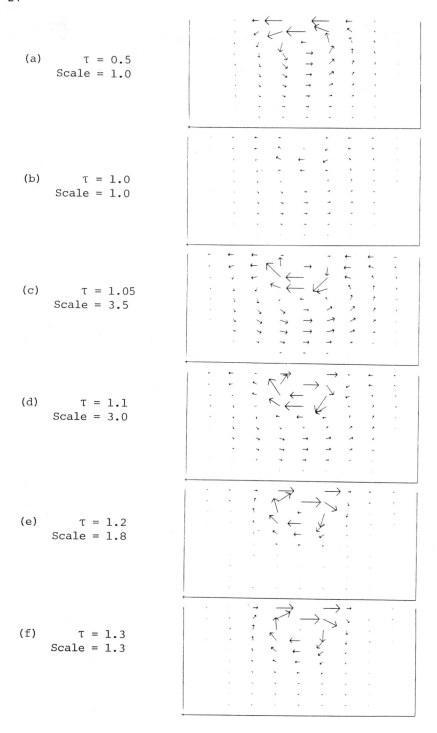

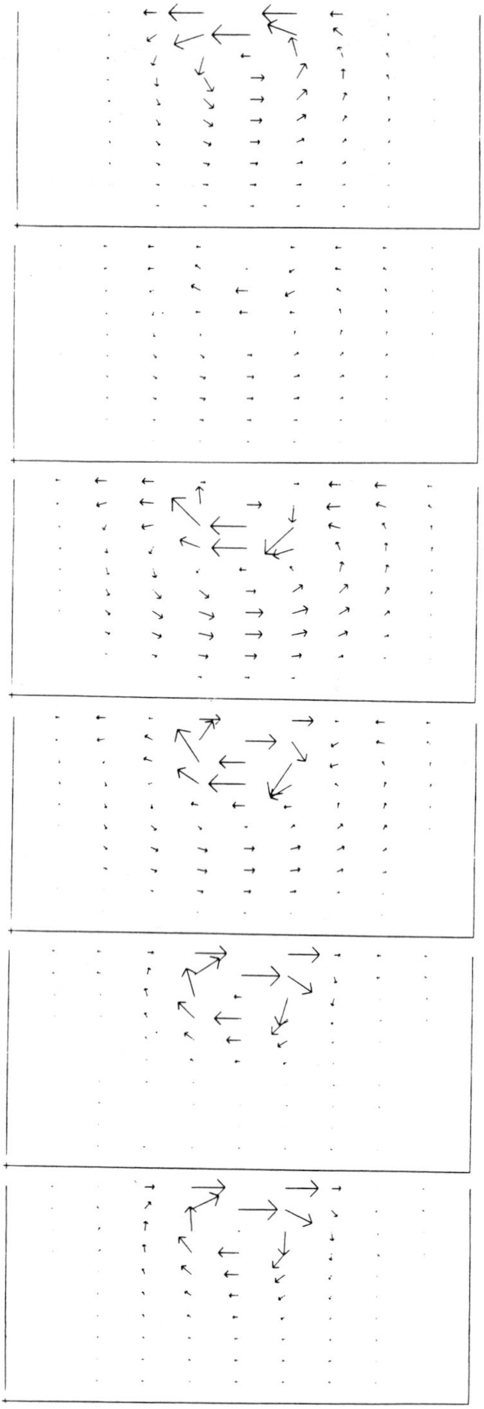

Figure 6. Flow Velocity Vectors over One-Half Cycle.

NUMERICAL ASPECTS OF FINITE-ELEMENT CALCULATIONS OF FLOW PAST A SPHERE

K.A. CLIFFE and D.A. LEVER

Theoretical Physics Division, AERE Harwell, England.

SUMMARY

The steady axisymmetric isothermal incompressible flow past a sphere in unbounded fluid is investigated for Reynolds numbers in the range 1 to 100. A stream function - vorticity formulation of the flow equations in spherical polar co-ordinates is used and solved by a Galerkin finite-element method. The drag coefficient is calculated and compared with the results of previous investigations. Plots of the perturbation stream function are given and are used to examine the effect of different boundary conditions far from the sphere.

1. INTRODUCTION

The combustion of liquid fuels is important in a number of areas of technology, such as furnaces and internal combustion engines. In developing physical and numerical models to describe these complex combustion processes, it is essential to understand the motion of the fuel drops. The models used to date either neglect the droplet drag or incorporate a very simple model [1], which is taken from steady isothermal flow past a non-evaporating sphere. A simplified model is also used for the evaporation rate. This is assumed to be the same as for an isolated drop in a stagnant ambient fuel vapour atmosphere. A study is underway to examine the validity of these models by looking at the effect of droplet evaporation on drag, and the effect of the motion of the droplet on evaporation, by solving the equations for flow around a sphere numerically.

The actual droplet motion is complex: the drop is decellerating and evaporating as it travels through a turbulent and varying gas flow. The motion is described by a number of parameters; the most important being the Reynolds number ($Re = d\,U\rho/\mu$, where d is the droplet diameter, U the velocity relative to the surrounding gas, ρ the density and μ the dynamic viscosity). Typical values of Re are less than 1000, for example a 100 μm drop moving at 8 ms^{-1} through kerosene at 1000°K gives Re = 160. If the droplet decelleration is sufficiently slow, and the droplet life-time sufficiently long, then the motion can be examined by a steady drag calculation. For steady isothermal flow around a sphere, the flow at low Reynolds number is stable, and it starts to become unstable at a Reynolds number of 130, when oscillations appear in the wake of the sphere. Therefore for Re $\lesssim$ 130 the drag can be examined by looking at the steady flow past a spherical drop.

The second important quantity is the strength of evaporation at the droplet surface; this leads to a radial component of the gas velocity there. This is determined by the temperature gradient at the drop surface and by the latent heat of vaporisation. It can be greater than the velocity of the drop, and as a consequence have a considerable effect on the flow and on the drag.

In the initial stage of the study we are looking at isothermal flow around a sphere, first around a non-blowing sphere and secondly around a sphere with a blowing boundary condition, modelling evaporation. This gives a qualitative measure of the effects of evaporation. Later we intend to do full thermal calculations to make this more quantitative for different temperatures and fuels.

There have been a number of calculations of isothermal flow around non-blowing cylinders and spheres, but very few with a blowing boundary condition [2-6]. So the first step is particularly important in validating the finite-element method used by comparing results with earlier calculations. In this paper we concentrate on this aspect: isothermal flow around the sphere with zero velocity on the boundary.

In the next section the finite-element formulation of the problem is described. Then in the last section the results are presented. Values of the stream function, vorticity and drag are given for Reynolds numbers in the range 1 to 100, and are compared with previous calculations [3-5].

2. FINITE-ELEMENT FORMULATION

In this section we describe the finite-element method used to solve the equations for flow around a sphere. Attention is given to the form of the Navier-Stokes equations used, the coordinate system and the boundary conditions. In particular we discuss the boundary condition to be applied at a large distance from the sphere. A brief summary of the computational details is also given.

There is a basic choice between a stream function-vorticity and velocity-pressure form of the Navier-Stokes equations. For the case of isothermal external flows there seem to be clear advantages to using the stream function-vorticity form. This is mainly to do with the nature of the flow far from the body, outside the wake region, where viscous forces are small and the flow is essentially irrotational. In this region the vorticity is very small and so the stream function-vorticity equations simplify to a single linear, elliptic equation for the stream function. With the velocity-pressure form, the equations are still nonlinear in the inviscid region, being essentially the Euler equations. A further point, which is related to the above point, is that it is somewhat easier to develop and apply a physically meaningful boundary condition for the stream function than for the velocities. The details of this boundary condition are discussed below.

The natural coordinate system to use is spherical polar coordinates (r,θ,ϕ), and, with axisymmetry, the equations for the stream function and vorticity are:

$$\frac{1}{\sin\theta}\frac{\partial^2\psi}{\partial r^2} + \frac{1}{r^2}\frac{\partial}{\partial\theta}\frac{1}{\sin\theta}\frac{\partial\psi}{\partial\theta} = -r\zeta \quad , \tag{1}$$

$$\frac{1}{r^2\sin\theta}\frac{\partial\psi}{\partial\theta}\frac{\partial\zeta}{\partial r} - \frac{1}{r^2\sin\theta}\frac{\partial\psi}{\partial r}\frac{\partial\zeta}{\partial\theta}$$
$$- \frac{\zeta}{r^3\sin\theta}\frac{\partial\psi}{\partial\theta} + \frac{\cot\theta}{r^2\sin\theta}\frac{\partial\psi}{\partial r}$$
$$- \frac{2}{Re}\left\{\frac{1}{r^2}\frac{\partial}{\partial r}r^2\frac{\partial}{\partial r}\zeta + \frac{1}{r^2\sin\theta}\frac{\partial}{\partial\theta}\sin\theta\frac{\partial\zeta}{\partial\theta}\right.$$
$$\left. - \frac{\zeta}{r^2\sin^2\theta}\right\} = 0 \quad , \tag{2}$$

where ψ is the stream function, ζ the vorticity and Re the Reynolds number. The radial coordinate has been made dimensionless by dividing by the radius of the sphere. These equations are solved in the region given by $1 \leq r \leq r_\infty$ and $0 \leq \theta \leq \pi$. For computational purposes we introduce the variable $\xi = \exp(r)$ and treat ξ and θ as the independent variables. This change of variables produces a natural compression, in real space, of the grid lines near the sphere.

The equations (1) and (2) must be supplemented by appropriate boundary conditions. Along the lines $\theta = 0$ (downstream symmetry axis) and $\theta = \pi$ (upstream symmetry axis), ψ and ζ are both zero. Along the surface of the sphere $(r = 1)$ the no-slip condition implies that ψ and $\frac{\partial\psi}{\partial r}$ are zero. One of the difficulties associated with external flow calculations is the specification of boundary conditions far from the body (sphere in this case). This question is much more acute in two-dimensional calculations where the disturbance to the free stream flow decays more slowly than in three dimensions. Nevertheless, insufficient care can lead to inaccurate results in three-dimensional calculations.

The boundary condition, far from the sphere, on vorticity is that $\zeta = 0$ for $\pi/2 \leq \theta \leq \pi$ and $\frac{\partial\zeta}{\partial r} = 0$ for $0 \leq \theta \leq \pi/2$. Vorticity is created at the surface of the sphere and then diffuses away from the surface and is convected downstream by the flow. The part of the boundary $\pi/2 \leq \theta \leq \pi$ at $r = r_\infty$ is the inflow boundary and the above boundary condition expresses the fact that there is no vorticity in the free stream. The condition on $0 \leq \theta \leq \pi/2$ is numerically convenient, and because of the form of the vorticity equation, any error introduced by this condition decays exponentially away from the boundary.

Many previous calculations [3-6] of flow past a sphere have employed the free stream boundary condition on the stream function at $r = r_\infty$. This implies that r_∞ is sufficiently large so that the stream function is unaffected by the presence of the sphere. The error induced by this boundary condition decreases as r_∞ increases, but for practical values of r_∞ may still be significant. If we write

$$\psi = \psi_{FS} + \psi_p \quad ,$$

where ψ_{FS} is the stream function in the absence of the sphere, and ψ_p is the perturbation due to the presence of the sphere then the free stream boundary condition implies that $\psi_p = 0$ at $r = r_\infty$. A better boundary condition may be developed by noting that at large distances from the sphere and for arbitrary Reynolds number, the perturbed component of the flow has two parts [7]. There is an inflow in the wake region which is associated with the momentum defect; the momentum removed from the free stream which produces the drag on the sphere. To compensate this inflow there is a uniform radial flow out from the sphere which, at large distances, looks like a point source of mass. This behaviour can be seen in the streamline pattern for ψ_p shown in figure 1(ii). At sufficiently large distances the perturbed flow is radial and this implies

$$\frac{\partial\psi_p}{\partial r} = 0 \quad . \qquad (3)$$

The condition (3) is much easier to apply in a stream function-vorticity formulation than in a velocity-pressure

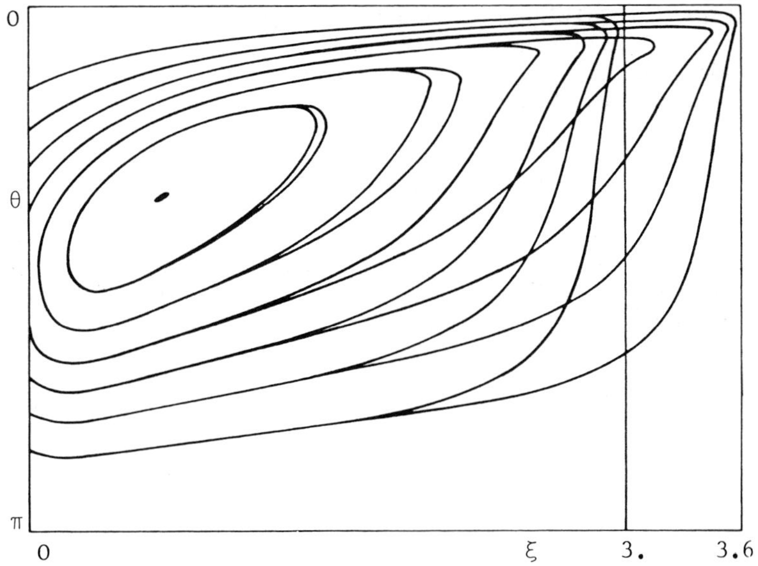

(i) free stream condition

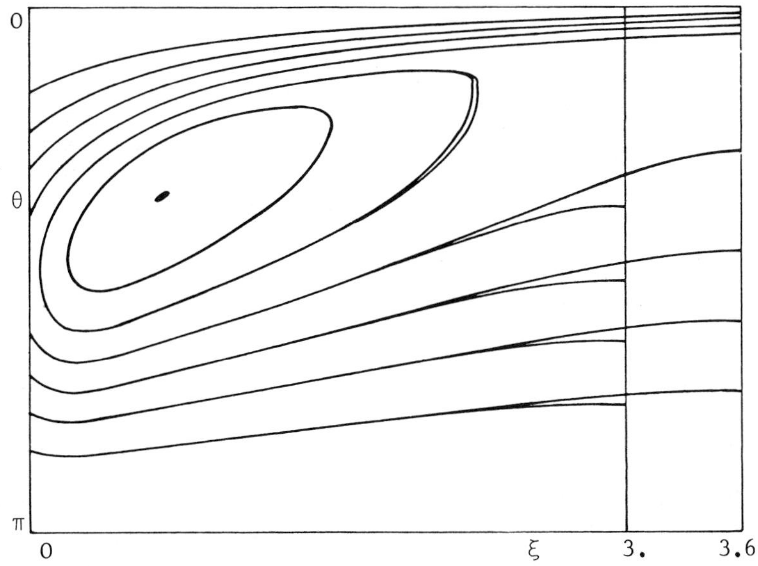

(ii) zero normal derivative condition

Figure 1 The effect of varying the position of the outer boundary on the perturbation stream function with two different boundary conditions for Re = 40.

formulation. This condition has been used previously by Fornberg [2] in a study of flow past a circular cylinder. It has a more secure physical basis than the free stream boundary condition and we thus expect to be able to apply it closer to the sphere.

A Galerkin finite-element method using nine-node biquadratic elements is used to discretise equations (1) and (2). The method is similar to that of Tong [8] and has been used previously by Winters and Cliffe [9] amongst others. Further details of the method can be found in Cliffe and Lever [10] where the implementation of the radial perturbed-flow boundary condition (3) is discussed in more detail.

The quantity that we are most interested in is the drag coefficient C_D. This has two components: a viscous component C_V and a pressure or 'form drag' component C_P. C_V and C_P are given by

$$C_V = -\frac{8}{Re}\int_0^{\pi} \zeta \sin^2\theta \quad d\theta \quad , \tag{4}$$

$$C_P = \frac{4}{Re}\int_0^{\pi} \left(\zeta + \frac{\partial\zeta}{\partial r}\right) \sin^2\theta \; d\theta \quad . \tag{5}$$

Since C_V contains only values of ζ on the surface of the sphere we expect $O(h^3)$ convergence, where h is the element size. For C_P, which involves derivatives of ζ, we expect an $O(h^2)$ component in the error. These error estimates for the drag have not, as far as we know, been rigorously established yet.

The nonlinear algebraic equations for the nodal values of stream function and vorticity are solved by Newton's method. The linear system at each iteration is solved using the frontal method [11]. The program was run on the Cray-1 computer at Harwell.

3. DISCUSSION OF RESULTS

In this section we describe some of the results obtained. We give values of the drag coefficient and present plots of the stream function and vorticity.

First we discuss calculations of the drag coefficient for Re = 1, 5, 10, 40 and 100 using the boundary condition (3) specifying ψ_p has zero normal derivative. The results are shown in Table 1, where they are also compared with previous calculations [3-5]. A variety of uniform grids were used in our calculations. In general, 20 elements were found to be adequate in the θ direction, except for Re = 100, where 30

Reynolds number	Le Clair et al.[4]	Dennis and Walker[3]*	Present calculation	Error estimate (% error)
1	27.315	27.44	27.317	0.005(0.02)
5	7.03	7.21	7.138	0.002(0.03)
10	4.29	4.42	4.308	0.001(0.03)
40	1.86[5]	1.808	1.789	0.001(0.06)
100	1.096	-	1.088	0.002(0.2)

Table 1. Comparison of drag coefficients with previous investigations (* these values are twice those quoted in [3], because a different definition of C_D is used).

were needed, and solutions were found to be independent of the outer boundary when ξ_∞ was set to 3.6 ($r_\infty = \exp(\xi_\infty) = 37$) except for Re = 1 where $\xi_\infty = 4.2$ ($r_\infty = 67$) was used. Independence of the number of elements in the radial direction was not so easy to achieve. As we commented in the last section, the viscous drag (4) depends only on the surface vorticity and so converges as $O(h^3)$, where h is the element size in the radial direction. However, the form drag (5) converges as $O(h^2)$, as it depends on derivatives of the surface vorticity. The results for different grids, for example 50, 80 and 100 radial elements, can be used to calculate an extrapolated value of the drag coefficient, fitting quadratic and cubic error terms. It is this extrapolated value which is given in Table 1, together with an estimate of its accuracy. This is not a strictly rigorous refinement procedure, it is impractical to do a proper refinement for all values of r_∞. The finest grids used were 100 x 30 and 120 x 20 elements, corresponding to 201 x 61 and 241 x 41 nodes. More details are given in Cliffe and Lever [10].

Finally, we examine some contour plots. The perturbation stream function is plotted in figure 1 as a function of ξ (= lnr) and θ for Re = 40, looking at the effects of the position of the boundary far from the sphere and the condition applied there (free stream or zero normal derivative (3)). The incorrect form of the perturbation stream function near the outer boundary, caused by the free stream boundary condition, is seen. It is also clear that, as r_∞ is increased, the solution away from the boundary changes less with the zero normal derivative condition. This is reflected in the calculation of the drag. Increasing ξ_∞ from 3 to 3.6 changes the drag for the zero normal derivative condition by only 0.02%, whereas for free stream condition it changes by 0.3%.

This is further illustrated in figure 2 for Re = 100.

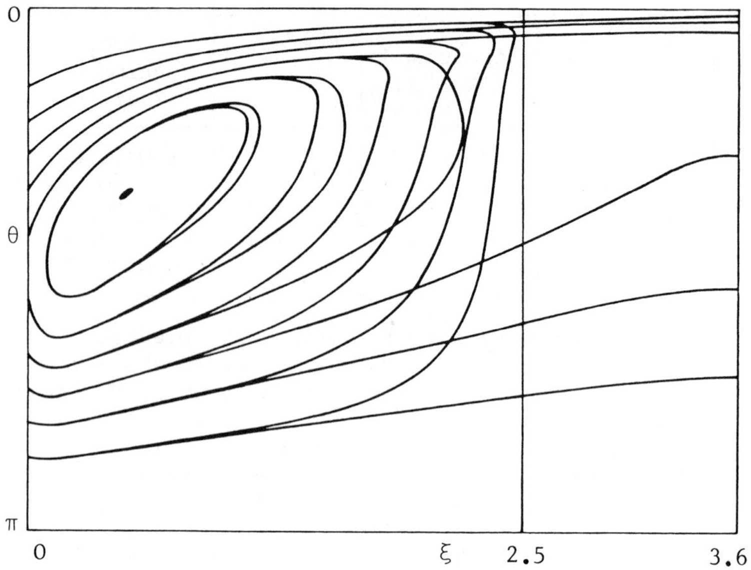

Figure 2 A comparison of ψ_p calculated using the zero normal derivative condition applied at ξ = 3.6 with a simulation of a calculation by Le Clair et al. [4] for Re = 100.

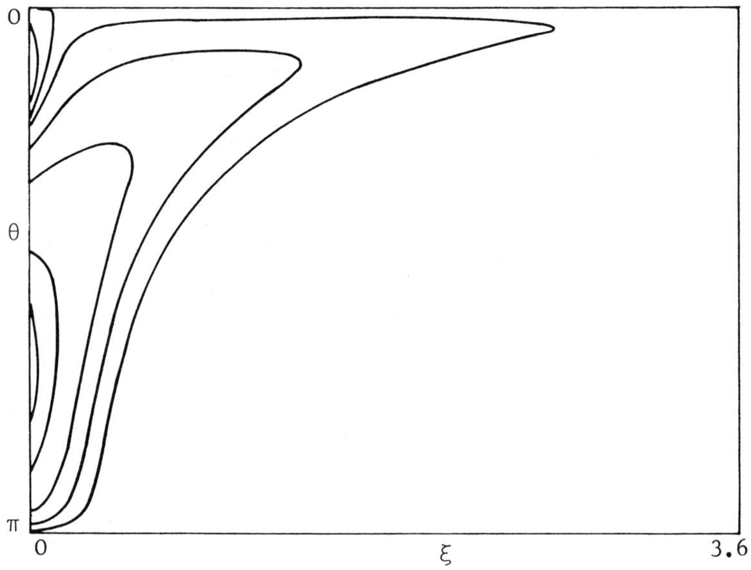

Figure 3 Vorticity contours for Re = 40, ζ = -5, -3.2, -1.1, -0.4, -0.1, 0., 0.1.

A solution with the zero normal derivative condition at ξ_∞ = 3.6 is compared to one with the free stream condition at ξ_∞ = 2.5. This latter solution was designed to simulate the finite-difference solution of Le Clair et al. [4], having the same boundary position at infinity, and comparable grid spacing near the sphere. It gives precisely the same value of C_D (1.096). It also shows the advantage of the exponentially stretched coordinate system. Le Clair et al. used grid points at equally spaced values of r, and had 480 points going out to r_∞ = 12 (ξ_∞ = 2.5), whereas our solution had only 100 elements (201 nodes), but with comparable spacing in the crucial region near the sphere.

In figure 4 the vorticity is plotted in ξ,θ coordinates for Re = 40. It can be seen diffusing away from the sphere and being convected downstream. In figures 5 and 6 the full stream function and the vorticity are plotted in real space for Re = 1 and 40. The values close to the sphere are shown. In the Re = 40 plots the small recirculation region behind the sphere can be seen.

ACKNOWLEDGEMENT

This work is supported by the IEA Combustion Project.

REFERENCES

1. GOSMAN, A.D., IOANNIDES, E., LEVER, D.A. and CLIFFE, K.A. - A Comparison of "Continuum" and "Discrete Droplet" Finite-Difference Models used in the Calculation of Spray Combustion in Swirling, Turbulent Flows. Harwell Report AERE-TP.865/HTFS-RS 308, AERE Harwell, 1980.

2. FORNBERG, B. - A Numerical Study of Steady Viscous Flow past a Circular Cylinder. J. Fluid Mech., 98, 819-855, 1980.

3. DENNIS, S.C.R. and WALKER, J.D.A. - Calculation of the Steady Flow past a Sphere at Low and Moderate Reynolds Numbers. J. Fluid Mech., 48, 771-789, 1971.

4. LE CLAIR, B.P., HAMIELEC, A.E. and PRUPPACHER, H.R. - A Numerical Study of the Drag on a Sphere at Low and Intermediate Reynolds Numbers. J. Atmos. Sci., 27, 308-315, 1970.

5. HAMIELEC, A.E., HOFFMAN, T.W. and ROSS, L.L. - Numerical Studies of the Navier-Stokes Equation for Flow past Spheres I. A.I.Ch.E.J., 13, 212-219, 1967.

6. RENKSIZBULUT, M. - Energetics and Dynamics of Droplet Evaporation in High Temperature Intermediate Reynolds

Number Flows, Ph.D. Thesis, Northwestern University, Evanston, Illinois, 1981.

7. BATCHELOR, G.K. - An Introduction to Fluid Dynamics, Cambridge University Press, 1970.

8. TONG, P. - On the Solution of the Navier-Stokes Equations in Two Dimensional and Axial Symmetric Problems. Proc. 1st. Int. Conf. on Finite Element Methods in Flow Problems, Swansea, 1974.

9. WINTERS, K.H. and CLIFFE, K.A. - A Finite Element Study of Driven Laminar Flow in a Square Cavity. Harwell Report AERE-R.9444, H.M.S.O., 1979.

10. CLIFFE, K.A. and LEVER, D.A. - A Finite Element Study of Isothermal, Laminar Flow past a Sphere at Low and Intermediate Reynolds Numbers. Harwell Report AERE-R.10868, In preparation, 1983.

11. DUFF, I.S. - MA32 - A Package for Solving Sparse Unsymmetric Systems using the Frontal Method. Harwell Report AERE-R.10079, H.M.S.O., 1981.

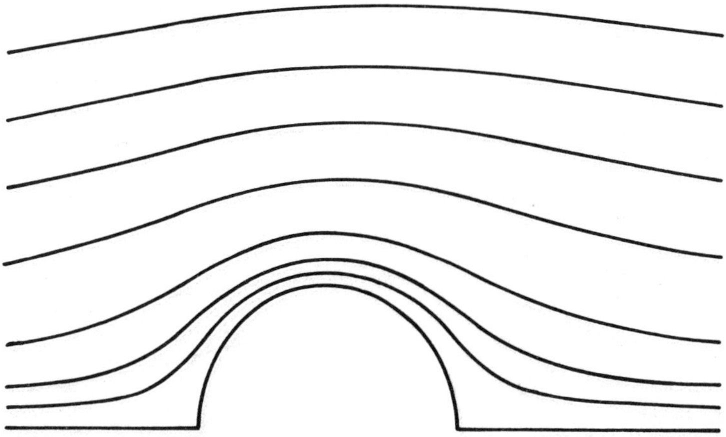

(i) Streamlines, ψ = 2.53, 1.62, 0.91, 0.41, 0.101 0.025, 0.0063, 0.

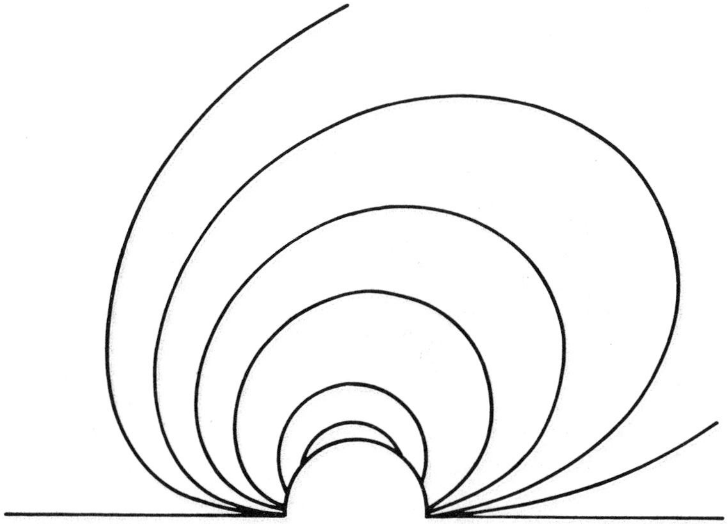

(ii) Vorticity contours, ζ = -1.2, -0.6, -0.2, -0.1, -0.05, -0.025.

Figure 4 Flow round a sphere for Re = 1.

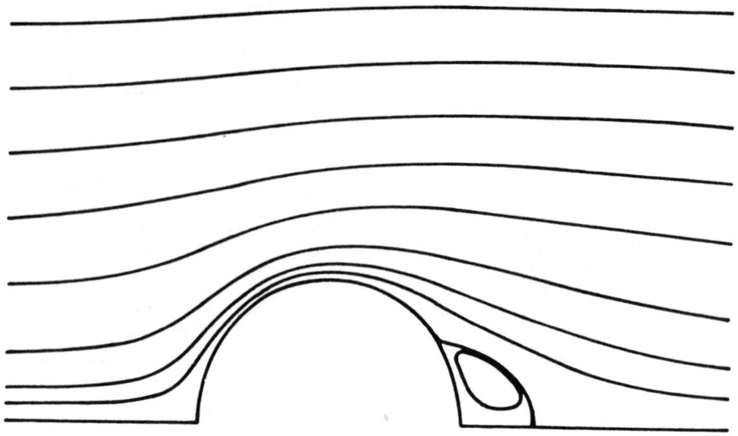

(i) Streamlines, ψ = 3.65, 2.53, 1.62, 0.91, 0.41, 0.101, 0.025, 0.0063, 0., -0.0001.

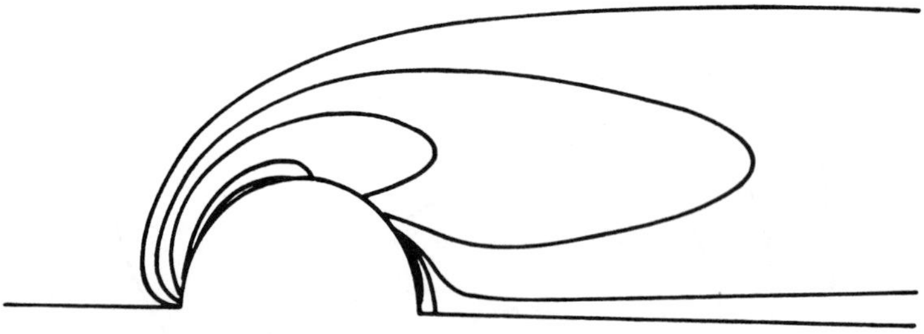

(ii) Vorticity contours, ζ = -5., -3.2 , -1.1, -0.4, -0.1, 0., 0.1.

Figure 5 Flow round a sphere for Re = 40.

A DIRECT METHOD FOR CALCULATING LAMINAR DUCT FLOW WITH INJECTION OR SUCTION

W. W. Baumann and F. Thiele*

Hermann-Föttinger-Institut, Technische Universität Berlin, Berlin, West-Germany

ABSTRACT

The laminar flow in straight, two-dimensional and axisymmetric ducts is calculated from the entrance region to the fully developed flow. The numerical method applied is based on the boundary-layer equations in both physical and transformed coordinates. At the boundary-layer edge the velocity which is related to the pressure gradient has to be chosen such that the overall mass balance is satisfied. In this way the unknown pressure gradient is directly incorporated in the calculation procedure. The stream function equation is solved numerically by using the finite-difference method of Hermitian type. The numerical results obtained demonstrate that the calculation procedure combines accuracy with efficiency and is capable of predicting duct flows with injection or suction.

1. INTRODUCTION

The laminar flow in the entrance region of straight, axisymmetric and plane ducts with porous walls can be described by the boundary-layer equations. For the solution of the parabolic differential equations the centerline velocity has to be chosen such that the overall mass balance is satisfied. The studies of Raithby [1] and Rhee and Edwards [2] contain a discussion about similar solutions for these flow problems. Without the similarity assumptions, the parabolic boundary-layer equations have in general to be solved by means of a numerical method. The investigations of Cebeci and Chang [4] and Blottner [5] have shown that it is advantageous to divide the flow field into an entrance region and a developed region. In addition, they introduced a coordinate transformation which removes the singularity at the inlet and stretches the radial coordinate in the entrance region.

---

*Presently at Department of Mechanical Engineering, Imperial College, London SW7, U.K.

Cebeci and Keller [6] considered the pressure drop as an eigenvalue. Their method leads to the solution of a 3 x 3 block-tridiagonal system of algebraic equations for the stream function and its first and second derivative. It requires a form of iteration based on Newton's method and the solution of a variational equation.

The Mechul approach of Cebeci and Chang [4] treats the pressure drop as an unknown, and hence, eliminates the need for an additional iteration cycle. However, it is somewhat more difficult to solve the resulting 4 x 4 block-tridiagonal system.

In this paper the pressure drop is related to the centerline velocity by applying the differential equation at the boundary-layer edge. In this way the centerline velocity is included as an unknown of the calculation procedure. The numerical solution of the differential equation is obtained by a modified finite-difference method of Hermitian type.

The developed calculation method will be used to predict the laminar flow in plane and axisymmetric ducts with suction or injection. The numerical results are compared with theoretical and experimental findings of other authors.

2. BASIC EQUATIONS AND COORDINATE TRANSFORMATION

We consider the entrance flow in a straight plane or axisymmetric duct of length ℓ and half-width r_0, as shown in Fig.1. The flow is assumed to be laminar, steady and incompressible with constant density ρ_0 and viscosity ν_0. The porous duct wall allows suction or injection. Starting with uniform axial velocity w_0 at the inlet, the boundary-layer thickness increases downstream until it reaches the middle of the duct. At the same time the velocity w_e at the outer edge of the boundary layer

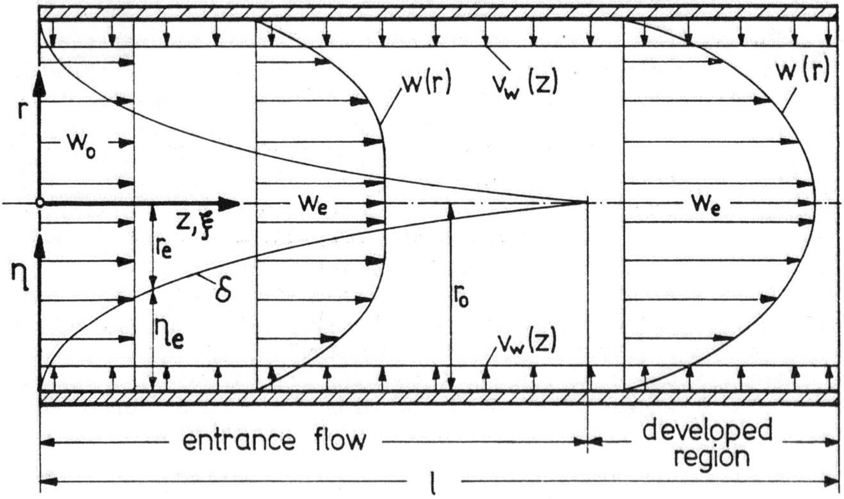

Figure 1: Schematic representation of the flow problem

changes with respect to the overall mass balance. Without mass transfer the velocity profile becomes fully developed further downstream.

For two-dimensional boundary-layer flows the equations can be written in the form

$$\frac{\partial}{\partial z}(r^\kappa w) + \frac{\partial}{\partial r}(r^\kappa v) = 0 \, , \tag{2.1}$$

$$w\frac{\partial w}{\partial z} + v\frac{\partial w}{\partial r} = -\frac{1}{\rho_0}\frac{dp}{dz} + \frac{\nu_0}{r^\kappa}\frac{\partial}{\partial r}\left(r^\kappa \frac{\partial w}{\partial r}\right) . \tag{2.2}$$

The governing equations have to be solved with respect to the following boundary conditions:

$$z = 0 : \qquad w = w_0(r) \, , \tag{2.3}$$

$$r = r_0 : \qquad w = 0 \, , \qquad v = -v_w(z) \, , \tag{2.4}$$

$$0 \leqq r \leqq r_e : \qquad w = w_e(z) \, . \tag{2.5}$$

Because of symmetry we require

$$r = 0 : \qquad \frac{\partial w}{\partial r} = 0 \tag{2.6}$$

for the flow in the developed region. Eqs. (2.1) and (2.2) describe plane ($\kappa = 0$) as well as axisymmetric ($\kappa = 1$) flows. Whilst the distribution of $v_w(z)$ is prescribed, the velocity $w_e(z)$ is determined by integration of the continuity equation (2.1) which results in

$$\int_0^{r_0} r^\kappa \, w \, dr - \frac{r_0^{\kappa+1}}{\kappa+1} w_0 - \int_0^{z} r_0^\kappa \, v_w \, dz = 0 \, . \tag{2.7}$$

Regarding the numerical solution of the partial differential equations (2.1) and (2.2) it is beneficial (see e.g. [4], [5]) to introduce a coordinate transformation. Here, we apply different transformations for the entrance region ($\lambda = 1/2$) and the developed region ($\lambda = 0$). As proposed by Cebeci and Chang [4] the coordinates are transformed through

$$d\xi = \left(\frac{r_0}{\ell}\right)^{2\kappa} \frac{1}{\ell} \, dz \, , \tag{2.8}$$

$$d\eta = -\left(\frac{r}{r_0}\right)^{\kappa} \left(\frac{w_0}{\nu_0 \ell}\right)^{1/2} \left(\frac{\ell}{z}\right)^{\lambda} dr \, , \tag{2.9}$$

which is a combination of the Mangler and Falkner-Skan transformation. In addition a stream function

$$\psi = - r_0^{\kappa} (\nu_0 \; w_0 \; \ell)^{1/2} \left(\frac{z}{\ell}\right)^{\lambda} f \tag{2.10}$$

is introduced which satisfies the continuity equation by

$$\frac{\partial \psi}{\partial r} = r^{\kappa} \; w \; , \qquad \frac{\partial \psi}{\partial z} = - r^{\kappa} \; v \; . \tag{2.11}$$

The application of the transformation to the differential equations (2.1) and (2.2) generates one parabolic equation of third order

$$\xi^{2\lambda} \left(f' \frac{\partial f'}{\partial \xi} - f'' \frac{\partial f}{\partial \xi} \right) - \lambda \, f \, f'' = - \xi^{2\lambda} \frac{d\bar{p}}{d\xi} + \kappa \alpha' \, f'' + \alpha^{\kappa} \, f''' \, . \tag{2.12}$$

Here, the prime denotes differentiation with respect to η and the abbreviations represent

$$\alpha = \left(\frac{r}{r_0}\right)^2 \left(\frac{\ell}{r_0}\right)^{2(1-2\lambda)} \; , \tag{2.13}$$

$$\bar{p} = \frac{p}{\rho_0 \; w_0^2} \; . \tag{2.14}$$

Applying eq. (2.12) at the outer edge of the boundary-layer or at the axis, depending on z, the pressure gradient yields

$$- \xi^{2\lambda} \frac{d\bar{p}}{d\xi} = \xi^{2\lambda} \, f'_e \frac{df'_e}{d\xi} - \kappa \alpha'_e \, f''_e - \alpha_e^{\kappa} \, f'''_e \; . \tag{2.15}$$

This relation is used to eliminate $d\bar{p}/d\xi$ from eq. (2.12)

$$\begin{aligned} \xi^{2\lambda} \left(f' \frac{\partial f'}{\partial \xi} - f'' \frac{\partial f}{\partial \xi} \right) - \lambda \, f \, f'' &= \xi^{2\lambda} \, f'_e \frac{df'_e}{d\xi} - \\ &- \kappa \alpha'_e \, f''_e - \alpha_e^{\kappa} \, f'''_e + \kappa \alpha' \, f'' + \alpha^{\kappa} \, f''' \; . \end{aligned} \tag{2.16}$$

The boundary conditions (2.3) and (2.4) can be written in terms of stream function

$$\xi = 0 : \qquad f' = 1 \; , \tag{2.17}$$

$$\eta = 0 : \qquad f' = 0 \; , \qquad f = f_w(\xi) \; . \tag{2.18}$$

The overall mass balance (2.7) transforms to

$$\begin{aligned} &\left[(\nu_0 w_0 \ell)^{1/2} \; r_0^{\kappa} \left(\frac{z}{\ell}\right)^{\lambda} \right] f_e + \left[w_0 \frac{r_e^{\kappa+1}}{\kappa + 1} \right] f'_e = \\ &= \ell^{2\kappa+1} \int_0^{\xi} v_w d\xi + \frac{r_0^{\kappa+1}}{\kappa + 1} \, w_0 + (\nu_0 w_0 \ell)^{1/2} \; r_0^{\kappa} \left(\frac{z}{\ell}\right)^{\lambda} f_w \; . \end{aligned} \tag{2.19}$$

This equation which contains the unknowns f_e and f'_e determines the velocity $w_e = w_0 f'_e$ and replaces boundary condition (2.5).

For the plane flow we obtain from eq. (2.6) in the developed region

$$\eta = \eta_e : \qquad f''_e = 0 . \tag{2.20}$$

A similar expression for the axisymmetric case leads to a singularity. Hence, it cannot be used as a boundary condition. Therefore, eq. (2.6) is directly applied together with the relation $w/w_0 = f'$

$$\eta = \eta_e : \qquad \frac{\partial f'}{\partial r} = 0 . \tag{2.21}$$

With the definition of the stream function (2.10) and eq. (2.11) the following relation can be derived for the boundary condition (2.18)

$$\lambda f_w + \left(\frac{r_0}{\ell}\right)^{2\kappa(1-2\lambda)} \xi^{2\lambda} \frac{\partial f_w}{\partial \xi} = - v_w \left(\frac{z}{\ell}\right)^{\lambda} \left(\frac{\ell}{\nu_0 w_0}\right)^{1/2} . \tag{2.22}$$

Here, the derivative $\partial f_w/\partial \xi$ needs to be approximated and this will be outlined in the next section. In order to predict the duct flow the stream function equation (2.16) with the boundary conditions (2.17), (2.18), (2.20), (2.21) and the continuity equation (2.19) have to be solved numerically.

3. NUMERICAL METHOD

Due to the parabolic nature of the boundary-layer equations the derivatives can be approximated separately in each direction. With respect to the discretization in the ξ-direction the first-order accurate scheme has greater streamwise damping than the second order accurate Crank-Nicolson scheme. This is important in suppressing streamwise oscillations in conjunction with boundary-layer calculations subject to discontinuous wall mass transfer [7]. For this reason backward finite-differences will be used throughout. This results in ordinary differential equations at each ξ-station which are solved by a finite-difference method.

The flow field is divided into the rectangular grid

$$\begin{aligned} &\xi_1 = 0 \qquad \xi_{i+1} = \xi_i + k_i \qquad i = 1,\dots,N-1 \\ &\eta_1 = 0 \qquad \eta_{j+1} = \eta_j + h_j \qquad j = 1,\dots,M-1 \quad . \end{aligned} \tag{3.1}$$

For calculating laminar flow we apply constant mesh spacing throughout.

To derive the finite-difference approximation to eqs. (2.16) and (2.22) one mesh rectangle is considered. The backward finite-difference expression for the function g

$$\frac{\partial g}{\partial \xi} \cong \frac{g_{i+1} - g_i}{k_i} \tag{3.2}$$

yields from eq. (2.16) the nonlinear ordinary differential equation for the station ξ_{i+1}.

$$\begin{aligned} &\alpha^{\kappa} f''' + (\lambda + \bar{\xi}) f f'' + (\kappa\alpha' - \bar{\xi} f_i) f'' - \bar{\xi}(f')^2 \\ &+ \bar{\xi} f' f_i' = - \bar{\xi}(f_e')^2 + \bar{\xi} f_e' f_{e,i}' + \kappa \alpha_e' f_e'' + \alpha_e^{\kappa} f_e'' \quad , \end{aligned} \tag{3.3}$$

where the abbreviation $\bar{\xi} = \xi_{i+1}/k_i$ has been used. Similarly, the boundary condition (2.22) gives a relation for the stream function at the wall

$$f_w = \frac{1}{\lambda + \bar{\xi}} \left[- v_w \left(\frac{\ell}{\nu_0 w_0} \right)^{1/2} \left(\frac{z}{\ell} \right)^{\lambda} \left(\frac{\ell}{r_0} \right)^{2\kappa(1-2\lambda)} + \bar{\xi} f_{w,i} \right] . \tag{3.4}$$

To linearize eq. (3.3) we apply the fast converging Newton-Raphson method to all nonlinear terms

$$\begin{aligned} &\alpha^{\kappa} f''' + \left[\kappa\alpha' - \bar{\xi} f_i + (\lambda + \bar{\xi})\hat{f}\right] f'' + \left[\bar{\xi} f_i' - 2\bar{\xi}\hat{f}'\right] f' \\ &+ \left[(\lambda + \bar{\xi})\hat{f}''\right] f + \left[2\bar{\xi}\hat{f}_e' - \bar{\xi} f_{e,i}'\right] f_e' = \\ &= (\lambda + \bar{\xi})\, \hat{f}\, \hat{f}'' - \bar{\xi}(\hat{f}')^2 + \alpha_e^{\kappa} f_e''' + \bar{\xi}(\hat{f}_e')^2 + \kappa\alpha_e' f_e'' . \end{aligned} \tag{3.5}$$

Here, "^" denotes the value of any variable in the previous iteration step. The iteration is continued until the convergence criterion

$$\left| f_w'' - \hat{f}_w'' \right| < \varepsilon\, f_w'' \tag{3.6}$$

is satisfied where ε is a small value. The direct solution of the linear equation (3.5) is performed with an accurate finite-difference method of Hermitian type [8]. As the grid point values of the function and the first derivative are the unknowns of the finite-difference approximation, complex boundary conditions such as eq. (2.19) can be directly included in the numerical scheme.

Following the presentation of Thiele and Wagner [9] the linearized stream function equation can be written in the general form

$$b^* Y''' + c^* Y'' + d^* Y' + e^* Y + g^* Y_M' = r^* \tag{3.7}$$

with the boundary conditions (2.18)

$$\eta = 0 : \qquad Y = f_w(\xi) \ , \qquad Y' = 0 \ . \tag{3.8}$$

In the region of developed flow eqs. (2.20) and (2.21) yield

$$\eta = \eta_e: \qquad Y'' = 0 \qquad \text{for} \quad \kappa = 0 \,, \tag{3.9}$$

$$B_{2M-3}\, Y'_{M-2} + D_{2M-3}\, Y'_{M-1} + F_{2M-3}\, Y'_M = 0 \quad \text{for} \quad \kappa = 1 \,. \tag{3.10}$$

Here, the boundary condition $\partial f'/\partial r = 0$ for axisymmetric flow is represented by a backward finite-difference expression. With $Y_M = f_e$ and $Y'_M = f'_e$ the relation (2.19) becomes

$$\eta = \eta_e: \quad E_{2M-1}\, Y_M + F_{2M-1}\, Y'_M = R_{2M-1} \,. \tag{3.11}$$

The derivation of the finite-difference equations and their solution are similar to that of [9], except at the outer edge where $\eta = \eta_e$. The system of algebraic equations

$$A_i Y_{j-1} + B_i Y'_{j-1} + C_i Y_j + D_i Y'_j + E_i Y_{j+1} + F_i Y'_{j+1} + g^* Y'_M = R_i \qquad j = 2,\dots,M-1 \tag{3.12}$$

where the indices are defined by

$$i = 2j + \ell \quad \text{with} \quad \begin{cases} \ell = -3,-2,-1 & \text{for} \quad j = 2 \\ \ell = -2,-1 & \text{for} \quad j = 3,\dots,M-2 \\ \ell = -2,-1,0,1 & \text{for} \quad j = M-1 \end{cases}$$

can be solved by means of a Gaussian elimination procedure. Essentially the system has a 2 x 2 block-tridiagonal structure as shown in Fig. 2. Further details of the numerical method and overall solution procedure are given in [10].

$Y'_1\ Y_2\ Y'_2\ Y_3\ Y'_3\ Y_4\ Y'_4 \dots Y_{M-3}\ Y'_{M-3}\ Y_{M-2}\ Y'_{M-2}\ Y_{M-1}\ Y'_{M-1}\ Y_M$	Y'_M	
$B_1\ C_1\ D_1\ E_1\ F_1$	g^*	$= R_1 - A_1Y_1$
$B_2\ C_2\ D_2\ E_2\ F_2$	g^*	$= R_2 - A_2Y_1$
$B_3\ C_3\ D_3\ E_3\ F_3$	g^*	$= R_3 - A_3Y_1$
$A_4\ B_4\ C_4\ D_4\ E_4\ F_4$	g^*	$= R_4$
$A_5\ B_5\ C_5\ D_5\ E_5\ F_5 \dots$	g^*	$= R_5$
$\dots$		$\vdots$
$A_{2M-6}\ B_{2M-6}\ C_{2M-6}\ D_{2M-6}\ E_{2M-6}\ F_{2M-6}$	g^*	$= R_{2M-6}$
$A_{2M-5}\ B_{2M-5}\ C_{2M-5}\ D_{2M-5}\ E_{2M-5}\ F_{2M-5}$	g^*	$= R_{2M-5}$
$A_{2M-4}\ B_{2M-4}\ C_{2M-4}\ D_{2M-4}\ E_{2M-4}$	$F_{2M-4}+g^*$	$= R_{2M-4}$
$A_{2M-3}\ B_{2M-3}\ C_{2M-3}\ D_{2M-3}\ E_{2M-3}$	0	$= 0$
$A_{2M-2}\ B_{2M-2}\ C_{2M-2}\ D_{2M-2}\ E_{2M-2}$	$F_{2M-2}+g^*$	$= R_{2M-2}$
$0\ \ 0\ \ 0\ \ 0\ \ E_{2M-1}$	F_{2M-1}	$= R_{2M-1}$

Figure 2: System of algebraic equations

4. RESULTS

The numerical procedure developed is used to calculate flows in plane and axisymmetric ducts for constant values of the injection number $v_w^* = 2\ v_w\ Re_0/w_0$. Here, the Reynolds-number is defined by $Re_0 = 2\ r_0\ w_0/\nu_0$. For reasons of comparison the values of v_w^* are chosen such that the plane duct flow is completely sucked off at the same location as the axisymmetric flow.

In order to demonstrate the features of the method the number of iterations is compared with those of the eigenvalue procedure in Table 1 [10]. Typical computing time for the

flow	v_w^*	eigenvalue method	present method
pipe	5	12.2	6.1
	0	12.3	6.3
	-5	13.0	6.3
duct	0	9.6	5.8

Table 1: Average number of iterations required to solve the linearized stream function eq. (3.5)

accurate calculation of a duct with $M = 49$ and $N = 200$ is about 20 sec on a CYBER 170. In comparison to the eigenvalue method the direct procedure requires half of the computing time.

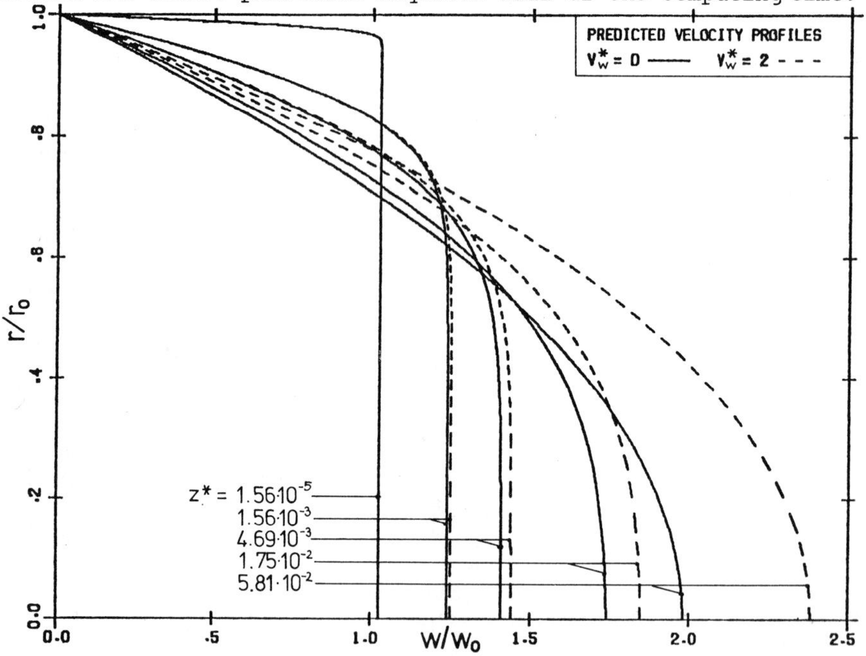

Figure 3: Development of the velocity profiles for axisymmetric duct flow

This is due to the fact that we have only to solve the stream function equation (3.5) instead of two differential equations for the eigenvalue method.

Fig. 3 shows the development of the velocity profile in a round tube. It clearly demonstrates the increase of the centerline velocity for the flow with injection (dashed line). In the case of no mass transfer the value $w_e/w_0 = 2$ is achieved for the fully developed flow. The centerline velocity along the axial coordinate $z^* = z/2r_0Re_0$ in Fig. 3 is in good agreement with the results of Reichle [3]. As expected, the wall mass transfer has a strong influence on the velocity distribution. For injection ($v_w^* > 0$) the centerline velocity continually increases whereas for suction it reaches a maximum value and decreases further downstream until the fluid is completely sucked off. The pressure drop $\Delta p^* = 2(\bar{p}_0 - \bar{p}(z))$ is presented in Fig. 4. Again, the agreement with the data of Reichle [3] is

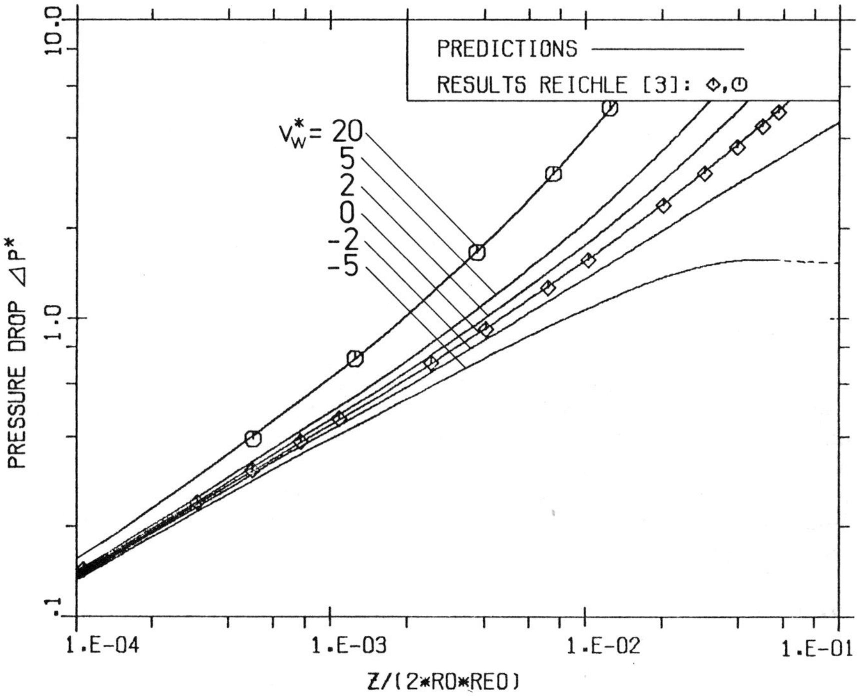

Figure 4: Pressure drop along the axis of a tube

good. The curve corresponding to $v_w^* = 0$ approaches the logarithmic law in the fully developed region which is in our notation $dp^*/dz^* = 16/Re_0$. From the plot it is obvious that injection increases the pressure drop whereas suction has the opposite influence. In Fig. 5 the friction factor $c_f = 2\tau_w/\rho_0 w_0^2$ is plotted along the plane duct. The friction factor decreases in the case of suction. But for injection the curves indicate a minimum. This behaviour can also be observed in the velocity profiles. Further details of the results are presented in [10].

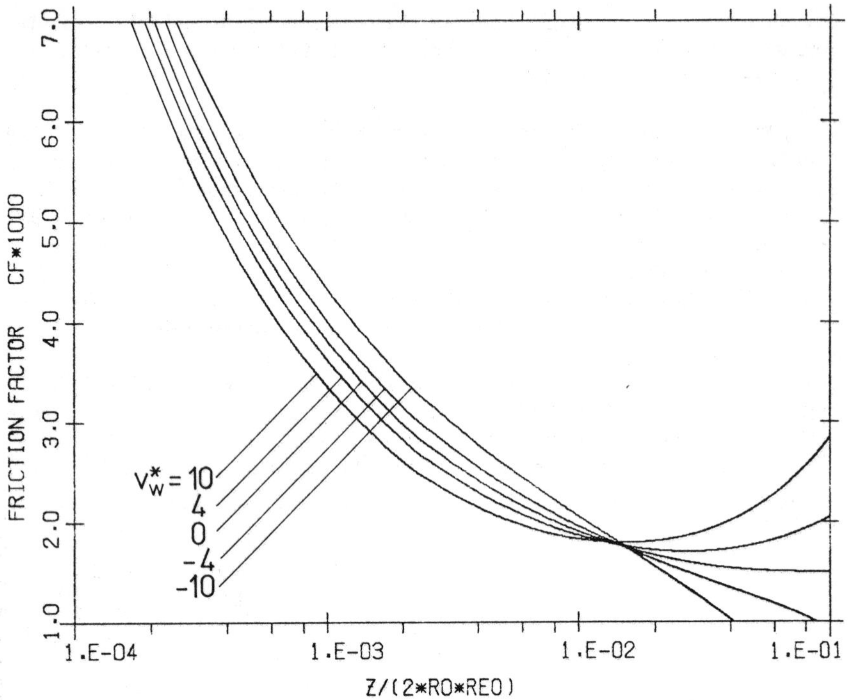

Figure 5: Friction factor c_f along the centerline of a plane channel

5. CONCLUSION

A numerical procedure which is based on the boundary-layer equations has been developed. It predicts the laminar duct flow with suction or injection. The main feature of the method is the direct incorporation of the pressure drop in the numerical scheme. From the present study it is evident that this direct method allows the efficient and accurate calculation of plane and axisymmetric flows.

6. REFERENCES

1. RAITHBY, G. - Laminar Heat Transfer in the Thermal Entrance Region of Circular Tubes and Two-dimensional Rectangular Ducts with Wall Suction and Injection. Int. J. Heat Mass Transfer, Vol. 14, pp. 223-243, 1971.

2. RHEE, S. J. and EDWARDS, D. K. - Laminar Entrance Flow in a Flat Plate Duct with Asymmetric Suction and Heating. Num. Heat Transfer, Vol. 4, pp. 85-100, 1981.

3. REICHLE, L. - Berechnung des gekoppelten Wärme- und Stoffübergangs bei Verdunstung eines Flüssigkeitsfilms in eine laminare Gasströmung im Einlaufgebiet eines porösen Kreisrohres. Dissertation, Universität Kaiserslautern, 1975.

4. CEBECI, T. and CHANG, K. C. - A General Method for Calculating Momentum and Heat Transfer in Laminar and Turbulent Duct Flows. Num. Heat Transfer, Vol. 1, pp. 39-68, 1978.

5. BLOTTNER, F. G. - Entry Flow in Straight and Curved Channels with Slender Channel Approximations. J. Fluids Engng., Vol. 98, pp. 666-674, 1977.

6. CEBECI, T. and KELLER, H. B. - Flow in Ducts by Boundary Layer Theory. Proc. 5th Australas. Conf. Hydraulics Fluid Mech. University of Canterbury, Christchurch, New Zealand, pp. 538-543, 1974.

7. CARTER, J. E. - STAYLAM: A Fortran Program for the Suction Transition Analysis of a Yawed Wing Laminar Boundary Layer. NASA TM X - 740 13, 1977.

8. THIELE, F. - Accurate Numerical Solutions of Boundary Layer Flows by the Finite-Difference Method of Hermitian Type. J. Comp. Phys., Vol. 27, pp. 138-159, 1978.

9. THIELE, F. und WAGNER, H. - Ein Berechnungsverfahren für den laminaren zweidimensionalen Wandstrahl entlang gekrümmter Oberflächen. Hermann-Föttinger-Institut, IB-01/81, TU Berlin, 1981.

10. BAUMANN, W. W. und THIELE, F. - Laminare Kanal- und Rohreinlaufströmungen mit Einblasen oder Absaugen. Hermann-Föttinger-Institut, IB-02/83, TU Berlin, 1983.

SOME NEW PENALTY ELEMENTS FOR INCOMPRESSIBLE FLOWS

Gouri Dhatt

Civil Engineering
Laval University
Québec, Canada

Guy Hubert

Département de Génie Mécanique
Université de Technologie
Compiègne, France

We present a penalty finite element formulation to study the two and three dimensional incompressible laminar flows. The efficiency of the penalty method is pretty well demonstrated for two dimensional flows, which is primarily due to elimination of pressure variables at element level and a better conditionning of matrices. Historically, these methods were first introduced in finite elements through reduced integration technique [1, 2]. However in recent works [3, 4], we find a generalised version of the penalty formulation, the reduced integration technique being a special case. Such a generalized approach leads to a better understanding of penalty methods and opens the doors for development of a complete series of new elements.

A major difficulty for incompressible problems has been a correct choice of finite element approximation which is compatible with the incompressibility constraints. Whether the formulation is based on the concepts of Lagrange multipliers or penalty parameter, finite element formulation should satisfy the discrete version of LBB conditions [5] in order to obtain numerically stable solutions. Though the mathematical expression of LBB conditions is pretty straight forward, the verification of these conditions for a given type of element approximations is another story. It is very interesting and amusing to scan through the works of various finite element researchers, starting with Taylor [6] to include Sani et al. [7], Oden et al. [8] Fortin [9], etc... on different direct and indirect techniques of verification and understanding of LBB conditions.

The actual volume of publications of this subject shows that the question of establishing standard techniques of these conditions is not yet fully answered or understood. Furthermore, the development of faster computers is shifting the interest of researchers from two dimensional to realistic three dimensional flows along with study of LBB conditions for 3D elements.

In this study, we develop 3 new penalty elements : 9 d.o.f. triangle with constant pressure (T6C), 11 d.o.f. triangle with linear pressure (T7L), 30 d.o.f. hexaedre with constant pressure (HFC). The development of these elements has been influenced by a recent work of Fortin [10] where he presented a simple technique for the verification of LBB conditions. Though this technique is not a general one, it is still sufficiently powerful and simple to be employed for developing efficient engineering elements. Finally we present a number of two and three dimensional numerical examples in order to demonstrate the efficiency and reliability of these elements.

1. FORMULATION

Equilibrium and continuity equations are :

$$\left.\begin{array}{l} U \text{ grad } U_i + 1/\rho \text{ grad } p - \nu \, \Delta \, U_i = f_i \\ \text{Div } U - p/\lambda = 0 \qquad 1/\lambda \to 0 \end{array}\right\} \text{ on } V \tag{1}$$

with relevant boundary conditions on ∂V

and U : $<u_1, u_2, u_3>$ or $<u, v, w>$ velocity components

p : pressure ; ρ : density (mass/volume)

f : $<f_1\ f_2\ f_3>$ volume forces

ν : cinematic viscosity

λ : penalty parameter ($10^6 < \lambda < 10^{10}$)

The variationnal expression is obtained by using standard method of weighted residus with Galerkin type weighting functions $\Psi = <\Psi_1, \Psi_2, \Psi_3>$, Ψ_p (that is, Ψ, Ψ_p and U,p belong to identical functionnal spaces). We perform as well necessary integration by parts in order to reduce order of derivatives on U and p (no pressure derivative appear in the integral).

$$W = \int_V \sum_{i=1}^{3} (\Psi_i \, U \text{ grad } U_i - 1/\rho \text{ div } \Psi_i \, p + \nu \text{ grad } \Psi_i \text{ grad } U_i$$

$$- \Psi_i \, f_i) \, dV + \int_V \Psi_p \, (\text{div } U - p/\lambda) \, dV + \int_{\partial V} (\Psi_n \cdot p$$

$$- \Psi_i \, \partial U_i/\partial n) \, ds \tag{2}$$

The solution of (2) approaches that of problem (1) as $\lambda \to \infty$, the variational formulation satisfying LBB conditions in the continuum space.

1.1. Finite element discretisation

The finite elements discretisation leads to $W = \Sigma W^e = 0$ where W^e is evaluated on each element V^e. For the mixed formulation (2), the velocity-pressure finite elements approximations should not only satisfy the necessary continuity requirement but the discretised equivalence of LBB conditions as well for numerically stable solutions. We choose C^0 approximation for velocity and C^{-1} for pressure, discontinuous field, which allows us to eliminate pressure terms at element level by static condensation.

The approximations are :

$$\Psi_1 = \langle N \rangle \{\Psi_1^n\}, \quad u_1 = \langle N \rangle \{u_1^n\}$$

and similar approximation for (Ψ_2, u_2) and (Ψ_3, u_3)

$$\Psi_p = \langle N_p \rangle \{\Psi_p^n\}, \quad p = \langle N_p \rangle \{p^n\}$$

The discretised representation of W^e may be written as :

$$W^e = \langle \Psi^n \ \Psi_p^n \rangle \begin{bmatrix} k^* & c \\ c^t & 1/\lambda \ m \end{bmatrix} \begin{Bmatrix} U^n \\ p^n \end{Bmatrix}$$

The discontinuous pressure approximation permits us to eliminate it by static condensation at element level.

$$\{p^n\} = \lambda \ [m]^{-1} \ [c]^T \ \{U^n\}$$

$$W^e = \langle \Psi^n \rangle \ [k] \ \{U^n\} \quad \text{with} \quad [k] = [k^*] - \lambda \ [c] \ [m^{-1}] \ [c^T]$$

One may remark that for constant pressure approximation, $[m]$ is simply a scalar (area or volume of element) and for linear pressure approximation, $[m]$ is diagonal if three pressure nodes are located at three integration points (T7L element).

In order to obtain correct solution, velocity-pressure approximation should lead to discretised formulation satisfying LBB conditions. These are :

$$\sup_{\forall U_h} \frac{\int_V p_h \operatorname{div} U_h \, dv}{\|U_h\|} > \alpha \ \|p_h\| \tag{3}$$

where α is a positive constant independent of element size h and U_h, p_h are finite element approximations of U, p.

(Eq. (3) is verified if the rank of [C] is equal to number of pressure terms). Fortin [10] has shown that relation (3) is verified if U_h field satisfies the following relation :

$$\int_V p_h \operatorname{div} U_h \, dV = \int_V p_h \operatorname{div} U \, dV \tag{4}$$

where U is a general velocity field.

The tricks is to transform verification of (3) to the verification of (4) which employs the fact that the continuum formulation (2) already satisfies LBB conditions. The relation (4) is generally difficult to verify, however for a special case of constant p_h, it becomes :

$$\int_V \operatorname{div} U_h \, dV = \int_{\partial V} U_h \cdot n \, dS = \int_{\partial V} U \cdot n \, dS = 0 \tag{5}$$

Furthermore if eq. (4) or (5) is satisfied at element level, it is automatically satisfied over the whole volume

$$\int_{V^e} U_h \cdot n \, dS = \int_{V^e} U \cdot n \, dS$$

or in a more restrictive sense, for each side S_i of an element :

$$\int_{S_i} U_h \cdot n \, dS = \int_{S_i} U \cdot n \, dS \tag{6}$$

1.2. Triangular elements

T6C Element

The condition (5) or (6) leads to an element T6C having 9 dof with constant pressure, and linear variation of tangentiel velocity and a quadratique variation of normal velocity along each side. For such a choice, Eq. (6) is verified for any U_h. This is shown by assuming velocities on corner nodes equal to U, and chosing freely the quadratic components of U_h, i.e. a_i such that (6) is satisfied which is always possible since a_i may be chosen in any manner.

The approximations for u, v are :

$$\begin{aligned} \langle N_u \rangle &= \langle N_1 \;\; 0 \;\; N_2 \;\; 0 \;\; N_3 \;\; 0 \;\; \ell_1 \cdot N_4 \;\; \ell_2 \cdot N_5 \;\; \ell_3 \cdot N_6 \rangle \\ \langle N_v \rangle &= \langle 0 \;\; N_1 \;\; 0 \;\; N_2 \;\; 0 \;\; N_3 \;\; m_1 \cdot N_4 \;\; m_2 \cdot N_5 \;\; m_3 \cdot N_6 \rangle \end{aligned} \tag{7}$$

avec $N_1 = 1 - \zeta - \eta$, $N_2 = \zeta$, $N_3 = \eta$,

$N_5 = \zeta\eta$, $N_6 = (1 - \zeta - \eta)\eta$, $N_4 = (1 - \zeta - \eta)\zeta$

$$\langle u^n \rangle = \langle u_1 \; v_1 \; u_2 \; v_2 \; u_3 \; v_3 \; a_1 \; a_2 \; a_3 \rangle$$

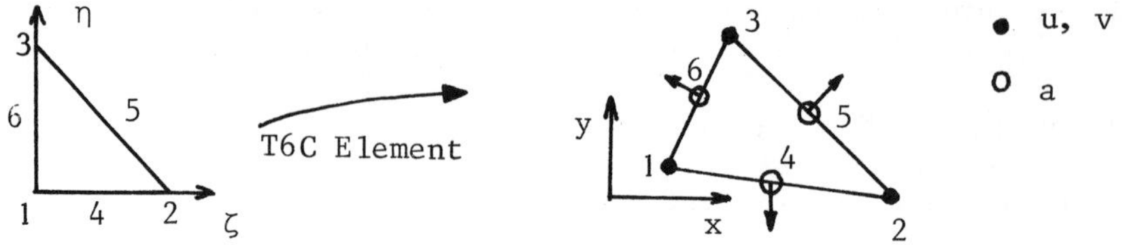

T7L Element

This is an extension of T6C with a linear pressure and an addition of an internal node at the centre of gravity (11 dof)

$N_7 = 27\,(1 - \zeta - \eta)\zeta\eta$ for u and v

$p = a + bx + cy$

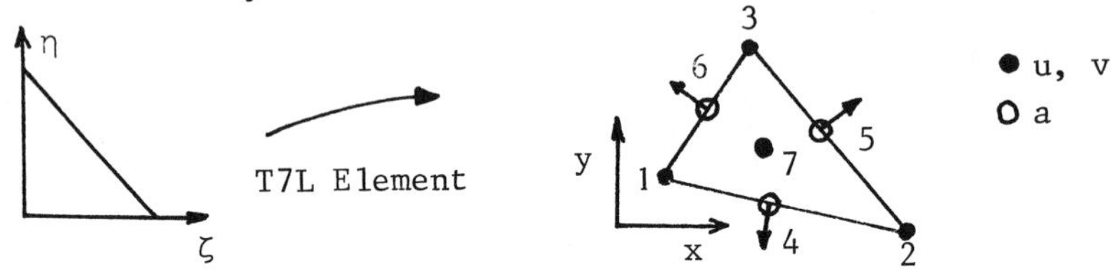

For this element, eq. (5) becomes :

$$\int_{V^e} \operatorname{div} U_h \, dV = \int_{V^e} \operatorname{div} U \, dV = 0 \qquad (8.a)$$

$$\int_{V^e} x \operatorname{div} U_h \, dV = \int_{V^e} x \operatorname{div} U \, dV = 0 \qquad (8.b)$$

$$\int_{V^e} y \operatorname{div} U_h \, dV = \int_{V^e} y \operatorname{div} U \, dV = 0 \qquad (8.c)$$

We can show (8.a) is verified by normal velocity components and (8.b), (8.c) are verified by internal velocity components. We employ as well the standard 7 nodes (14 dof) T7L* element having quadratic variation of u, v with a bubble function N_7 at the centroid for each velocity component.

1.3. <u>3-Dimensional element : HFC</u>

We develop a new hexaedral element with 8 corner nodes (u,v,w) and 6 mid face nodes having generalized variable similar to normal velocity as dof. This element has 30 degrees of freedom with constant pressure. The LBB conditions are fully satisfied for any hexaedre geometry. The velocity approximations are :

$$\langle N_u \rangle = \langle N_1, N_2, \ldots, N_8, \overline{N}_1\, \ell_1, \overline{N}_2\, \ell_2, \ldots, \overline{N}_6\, \ell_6 \rangle$$

where $N_1, \ldots, N_8$ are standard trilinear functions and $\overline{N}_1, \ldots, \overline{N}_6$, are bubble functions on each face, for example :

Face $\xi = -1$ $\overline{N}_1 = \frac{1}{2}(1 - \xi)(1 - \eta^2)(1 - \zeta^2)$

Face $\eta = 1$ $\overline{N}_4 = \frac{1}{2}(1 - \xi^2)(1 + \eta)(1 - \zeta^2)$

and in a similar way for other faces.

ℓ_i, m_i, n_i are direction cosines of the normal on face i

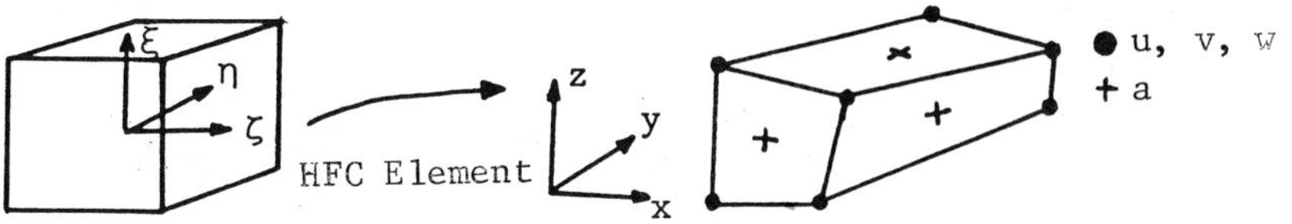

2. NUMERICAL APPLICATIONS

It may be remarked that for elements T6C, T7L and HFC, the normal directions should be chosen in a global manner so that for all elements having common boundaries, it is unique. For triangles, we employed the convention that normal is oriented in a trignometrical direction with respect to the side oriented from its lower node number to the higher node number.

Nonlinear equations are solved by an incremental technique with Newton-Raphson correctors [11] . Pressure is calculated from the penalisation of the divergence terms.

We present a number of two and three dimensional examples in order to assess the efficiency and reliability of these elements.

2.1. Two-dimensional flows

Stokes flow : we investigate the flow with volume forces such that analytical solution corresponds to :

$u = \sin y$; $v = \sin x$; $p = \sin(x + y)$ with

$f_x = \sin y + \cos(x + y)$; $f_y = \sin x + \cos(x + y)$

on $0 < x < 1$ and $0 < y < 1$

Velocity values are imposed along the four boundary lines. We evaluate the precision for velocity and pressure using different mesh sizes with elements T6C, T7L, T7L* and T6-3 (quadratic velocities with linear C° pressure approximation [12]).

Results obtained with T7L are very disappointing on coarse mesh sizes. This may be due to undesirable influence of bubble functions coupled with incomplete quadratic approximations for velocity components. The superior results obtained with T7L* seem to confirm this argument. In Fig. 1, we present variations of error pressure calculations for different elements.

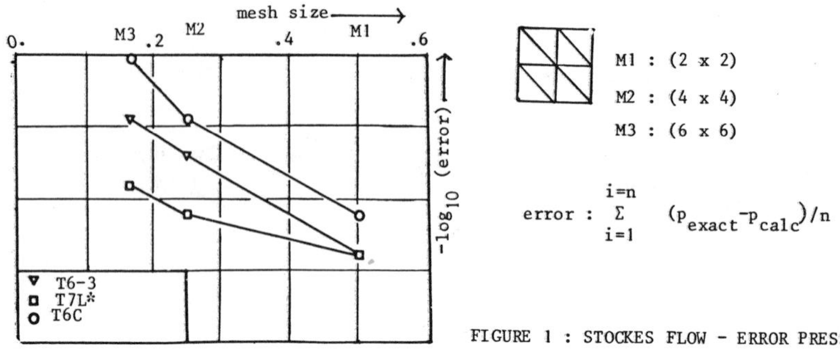

FIGURE 1 : STOCKES FLOW - ERROR PRESSURE

2.2. Recirculating cavity flow

We study the classical recirculating flow in a unit square cavity using T6C and T7L elements for R_e = 0, 100, 400.

The boundary conditions are : u, v = 0 on all sides except on the sliding plate where u = 1 (y = 1). The u-velocity profileat R_e = 400 is shown in Fig. 2, using 6 x 6 mesh size.

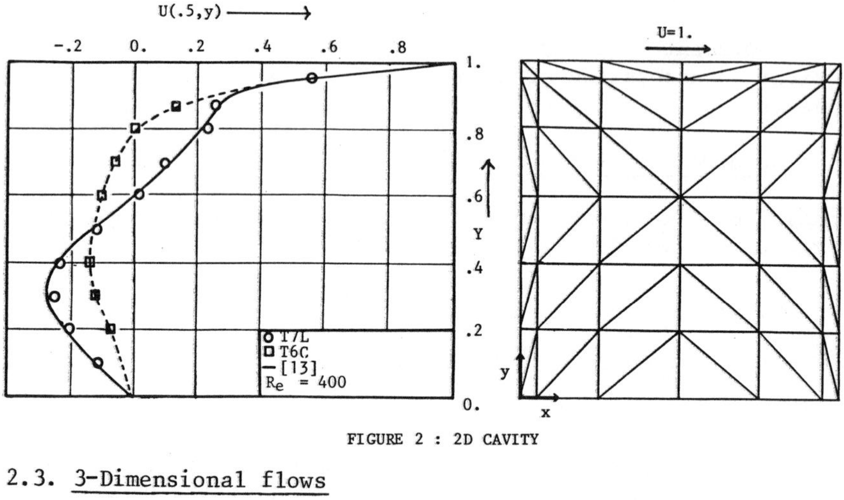

FIGURE 2 : 2D CAVITY

2.3. 3-Dimensional flows

Poiseuille flow at R_e = 0 : we use 8 nodes cubic elements HB8 and HFC for studying simple Poiseuille flow beetween two plates. Two meshes employed are :

M1 : 3 x 3 x 3 cubics elements
M2 : like M1, with distorted central element

	HB8	HFC
M1	U : $0(10^{-6})$; p : $0(10^{-4})$	U : $0(10^{-6})$; p : $0(10^{-4})$
M2	U : $0(10^{-3})$; p : meaningless	U : $0(10^{-5})$; p : $0(10^{-3})$

FIGURE 3 - VELOCITY AND PRESSURE ERROR

2.4. 3-Dimensional recirculating cavity flow

This problem is studied using 8-nodes cubic element (HB8) and the new HFC element. We employ 224 elements with 8 x 7 x 4 (symmetry) mesh along with boundary conditions as shown in the Fig. 3. Comparative velocity profiles for R_e = 100, 400 are given in the same figure.

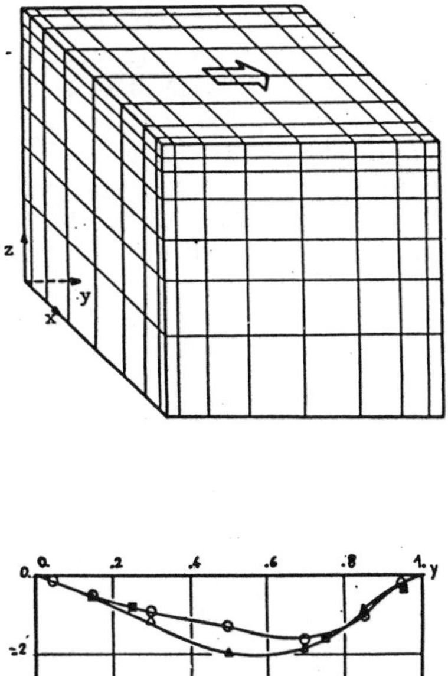

Symetry : x = .5

4 x 8 x 7 elements

Boundary conditions

u = 0, v = 1, w = 0 on z = 1.

U = 0, v = 0, w = 0 on z = 0, y = 0
y = 1, x = 0

U = 0 on x = .5 Symmetry

FIGURE 4 - CUBIC CAVITY

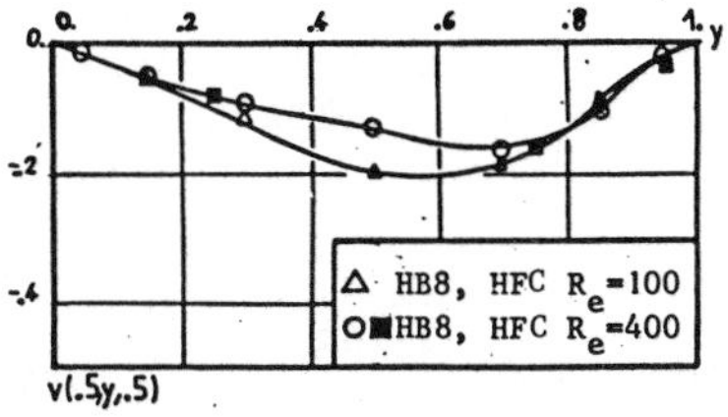

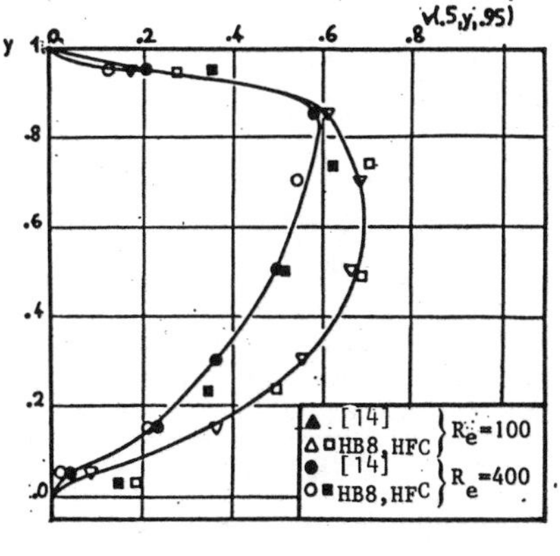

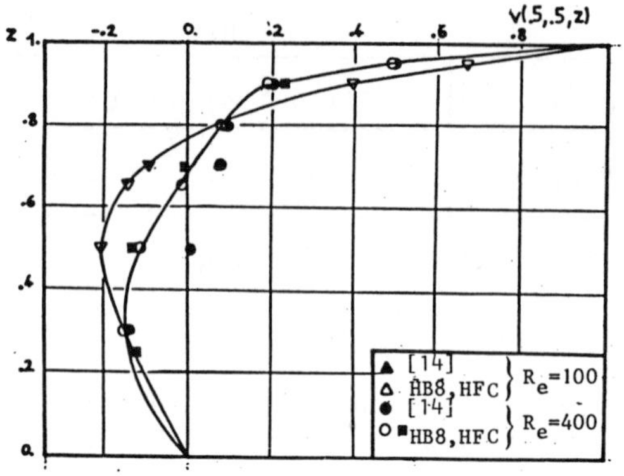

FIGURE 4 (end)

CONCLUSION

We observe that the element T6C is a good compromise for precision and problem size. However our limited study with T7L is very discouraging and we thus recommand T7L* for better precision. The interesting part of this study is the development of HFC for general hexaedral geometry. This is the simplest element which satisfies LBB conditions for any geometry. It is still to be verified how this element compares with other higher precision 3-D elements from point of view of computational costs.

REFERENCE

1. ZIENKIEWICZ, O.C. - Constrained variational principles and penalty function methods in finite element analysis. Lecture notes in Mathematics, 363 (Springer, 1974)

2. HUGHES, T.J.R., TAYLOR, R.L. & LEVY, J.F. - High Reynolds number steady incompressible Flows by a finite element method, Finite Element in Fluids, ed. R.H. Gallagher et al. chap. 3 (Wiley, 1977)

3. ENGELMAN M.S. et al. - Consistent vs - Reduced integration penalty methods for incompressible media using several old and new elements. I.J.N.M. in Fluids, Vol. 2, 25-42 (1982).

4. COCHET J.F., DHATT, G., HUBERT, G. & TOUZOT, G. - River and estuary flows by a new penalty element. 4th Int. Symposium on finite element in flow problems, july 26-29th, 1982, Tokyo, Japan.

5. LADYSZHENSKAYA, O.A. - The mathematical theory of viscous incompressible flows. Gordon and Breach, New York, 1969.

6. TAYLOR, C. & HOOD, P. - Numerical solution of the Navier Stockes equations using the finite element technique. Computers and Fluids, Vol. 1, 1973, pp 1-28.

7. SANI, R.L., GRESHO, P.M., LEE, R.L. & GRIFFITHS, D.F. - The cause and cure of the spurious pressures generated by certain FEM solutions of the incompressible Navier-Stockes equations : Part 1, Vol. 1, 17, Part 2, Vol. 2, 171, I.J.N.M. in Fluids (1981)

8. ODEN, J.T., KIKUCHI, N., SONG, Y.J. - Penalty-finite element methods for the analysis of Stokesian flows. Com. Meth. in Applied Mech. and Eng., Vol. 31, 297-329 (1982)

9. FORTIN, M. - An analysis of the convergence of mixed finite element methods. R.A.I.R.O. Numerical Analysis, Vol. 11, N° 4, 341-354, 1977.

10. FORTIN, M. - Old and new finite elements for incompressible flows. I.J.N.M. in Fluids, Vol. 1, 347-364 (1981).

11. DHATT, G., TOUZOT, G. - Une présentation de la méthode des éléments finis. Maloine, France 1981.

12. COCHET, J.F. - Modélisation d'écoulements stationnaires et non stationnaires par éléments finis. Thèse de Docteur-Ingénieur, 1979, Université de Compiègne, France

13. BURGRAFF, O.R. - Analytical and numerical studies of structure of steady separated flows. J. Fluids Mech. Vol. 24, pp 113-151, 1966.

14. REDDY, J.N. - Penalty finite element analysis of 3D Navier-Stokes equations. Comp. Meth. in Applied Mech. and Eng., Vol. 35, pp 87-106 (1982)

DEVELOPING FLOW IN A RECTANGULAR DUCT

Vijay K. Garg

Department of Mechanical Engineering
Naval Postgraduate School, Monterey, CA 93940

SUMMARY

The developing laminar flow of an incompressible, viscous fluid in a rectangular duct is analyzed numerically. Finite difference representation of the nonlinear equations of motion for the problem is solved using variable mesh technique. A uniform velocity distribution is assumed at the duct inlet. Results are presented for the velocity and pressure distribution in the entrance region, and comparison made with previous work.

1. INTRODUCTION

It is well known that a fluid entering a duct of constant cross section alters its velocity distribution from some initial profile to a fully developed profile which is thereafter invariant in the downstream direction. The analysis of this hydrodynamically developing region has been the subject of extensive study, especially for laminar flow. However, owing to the presence of nonlinear inertia terms in the equations of motion, it has not been possible to find exact solutions. Various types of approximations [1,2] have therefore been devised to get results.

This work is concerned with the laminar flow of an incompressible fluid in the hydrodynamic entrance region of a rectangular duct. The developing flow in a rectangular duct has been studied theoretically [1,3-6] as well as experimentally [7,8]. While Lundgren et al. [1] provide only pressure drop results, Han's results [3] deviate considerably from the experimental data [7,8]. References [4,5] use a method that linearizes the inertia terms, and it is known [9] that such a linearization yields results inferior to those given by the finite-difference method. While Curr et al. [6] do use a finite-difference method, it is much too involved for the relatively simple problem under discussion here.

The objective therefore is to formulate and apply a relatively simple but accurate finite-difference method to the entrance-region analysis of rectangular ducts, and to compare the results with available experimental and theoretical data. We first develop the finite-difference method, and then discuss the results obtained by this method.

2. ANALYSIS

We consider the steady laminar flow of an incompressible fluid with constant physical properties. Taking a representative length in the axial direction x to be of the order of development length L, and in the transverse direction y to be of the order of duct half-width a, we assume that $L >> a$. Then, if terms of $O(a/L)$ or lower compared to those of $O(1)$ are neglected, the Navier-Stokes equations reduce to

$$\frac{\partial u}{\partial y} + \frac{\partial v}{\partial y} + \frac{\partial w}{\partial z} = 0 \quad , \tag{1}$$

$$\rho\left(u\frac{\partial u}{\partial x} + v\frac{\partial u}{\partial y} + w\frac{\partial u}{\partial z}\right) = -\frac{dp}{dx} + \mu\left(\frac{\partial^2 u}{\partial y^2} + \frac{\partial^2 u}{\partial z^2}\right) \quad , \tag{2}$$

$$\partial p/\partial y = \partial p/\partial z = 0, \tag{3}$$

where (u,v,w) are the velocity components in the (x,y,z) directions respectively, p is the pressure, ρ is the density and μ the dynamic viscosity of the fluid.

This model corresponds to the usual boundary layer assumptions but is incomplete since (1) and (2) represent two equations in the three unknowns u, v, and w. Note that $p = p(x)$ is not a true unknown in the same sense as u, v, and w. Placing the origin of the coordinate system at the center of the duct inlet, and following Carlson and Hornbeck [10], the additional relation in order to complete the set is taken to be

$$wy = vz. \tag{4}$$

This relation implies that the net transverse velocity at any point is directed toward the duct centerline. Although this relation has been verified only for the square duct [10], it is felt that it may reasonably be extended to ducts of moderate aspect ratio, although certainly not to large aspect ratios. Good agreement between the present results and experimental data bears it out. Note that (4) holds on all lines of symmetry in the duct cross-section. Defining the dimensionless variables as

Developing Flow in a Rectangular Duct

$$X = \frac{\mu x}{\rho a^2 u_o},\ Y = \frac{y}{a},\ Z = \frac{z}{a},\ U = \frac{u}{u_o},\ V = \frac{\rho a v}{\mu},\ W = \frac{\rho a w}{\mu},\ P = \frac{p-p_o}{\rho {u_o}^2},$$

where u_o and p_o are the uniform axial velocity and pressure at duct inlet, eqs. (1), (2) and (4) can be easily written in dimensionless form. The boundary conditions are the no-slip conditions at the duct walls and symmetry conditions at the duct center-lines. The initial conditions are taken to be

$$U(0,Y,Z) = 1, \quad P(0) = 0. \tag{5}$$

Before writing the basic equations in finite-difference form, it is found advantageous to introduce the integral form of the continuity equation. In dimensionless form it is

$$\int_0^\sigma \int_0^1 U dY dZ = \sigma , \tag{6}$$

where $\sigma = b/a$ is the aspect ratio of the duct; b being the duct half-width in the z-direction.

3. FINITE DIFFERENCE METHOD

Equations (1), (2), (4) and (6) are now written in finite-difference form by superposing a three-dimensional rectangular mesh on the flow field. Let indices (i,j,k) indicate positions, and let ΔX, ΔY, ΔZ be the mesh spacing in the X,Y,Z directions respectively. Due to symmetry along duct center lines, only a quarter of the duct cross-section need be considered. Let the duct center lines at Y=0 and Z=0 be designated by j=1 and k=1, respectively, and the duct walls at Y=1 and Z=σ be designated by j=n+1 and k=m+1, respectively. Using the notation $\beta 2(j,k)$ to represent the variable β at a point (i+1,j,k) in the mesh, and $\beta 1(j,k)$ to represent β at (i ,j,k), the finite-difference forms of eqs. (2), (1), and (4) at (i+1,j,k) are written as

$$U1(j,k)\,\frac{U2(j,k)-U1(j,k)}{\Delta X} + V1(j,k)\,\frac{U2(j+1,k)-U2(j-1,k)}{2(\Delta Y)}$$

$$+ W1(j,k)\,\frac{U2(j,k+1)-U2(j,k-1)}{2(\Delta Z)} = -\frac{P2-P1}{\Delta X}$$

$$+ \frac{U2(j+1,k)-2U2(j,k)+U2(j-1,k)}{(\Delta Y)^2}$$

$$+ \frac{U2(j,k+1)-2U2(j,k)+U2(j,k-1)}{(\Delta Z)^2}\ , \tag{7}$$

$$\frac{U2(j,k)-U1(j,k)}{\Delta X} + \frac{V2(j+1,k)-V2(j,k)}{\Delta Y} + \frac{W2(j,k+1)-W2(j,k)}{\Delta Z} = 0, \quad (8)$$

and

$$(j-1)(\Delta Y)W2(j,k)=(k-1)(\Delta Z)V2(j,k). \quad (9)$$

Application of the trapezoidal rule to eq. (6) yields

$$\frac{4\sigma}{(\Delta Y)(\Delta Z)} = 4\sum_{j=2}^{n}\sum_{k=2}^{m} U2(j,k) + 2\sum_{j=2}^{n} U2(j,1) + 2\sum_{k=2}^{m} U2(1,k) + U2(1,1). \quad (10)$$

Eq. (7) written for j=1(1)n and k=1(1)m plus eq. (10) represent (mn+1) equations for the (mn+1) unknowns U2(j,k) and P2 in terms of known quantities at axial position i. The solution can therefore be obtained in a straight forward marching fashion, starting from the inlet of the duct. This marching technique does require specification of V and W at the inlet section while no such conditions were specified for the differential equation. Following Gupta and Garg [11], however, V and W can be set to zero at the inlet section without loss of accuracy or generality.

Once the values of U2(j,k) and P2 have been determined, the transverse velocities V2(j,k) and W2(j,k) must be found. This is achieved by solving the continuity equation (8) for V2(j,k). Substituting for W2(j,k) from (9) into (8) yields

$$V2(j,k) = \frac{V2(j+1,k)+(\Delta Y)W2(j,k+1)/(\Delta Z)+(\Delta Y)[U2(j,k)-U1(j,k)]/\Delta X}{1+(k-1)/(j-1)} \quad (11)$$

This equation is used in a stepwise manner, starting at the corner (Y=1, Z=σ) and moving in the -Z direction to Z=0, then moving down one step in the -Y direction and repeating the process. As each new V2(j,k) is computed, a W2(j,k) is found from eq. (9). This procedure is carried down to and including j=2. For j=1 (the Z-axis), V2(1,k) = 0, and continuity equation (8) may be solved for W2(1,k) in a stepwise manner starting at k=m and moving in the -Z direction. This completes the solution at the present value of X. An additional step downstream is now taken and the process repeated. Thus the solution is "marched" downstream. The difference equations are found to be consistent and stable for all mesh sizes as long as the axial velocity remains everywhere non-negative.

With the initial condition U1(j,k)=1 for all j=1(1)n and k=1(1)m at the inlet section, the discretization of the cross-section leads to a reduction in the volumetric flow rate because of the no slip boundary condition at the duct walls

Therefore, U1(j,k) at the inlet section was taken to be U_o, for all j=1(1)n and k=1(1)m, with U_o given by

$$U_o = 4mn(4mn - 2m - 2n + 1)^{-1}. \qquad (12)$$

This follows directly from (10) for a dimensionless discharge of σ.

The relations given so far in this section are based on uniform values of ΔY and ΔZ across the duct cross-section. For an accurate solution, both ΔY and ΔZ should be quite small but that implies the solution of a large number of simultaneous equations, which not only requires more computer time but also involves larger roundoff error. The only solution to this problem is the use of fine mesh size near the walls where the velocity varies rapidly, and a relatively coarse mesh size near the duct center line where the velocity varies slowly. Variable mesh technique was therefore used to cut down on the number of points in the cross-section and still maintain reasonable accuracy. Two different mesh sizes were used in each of the two transverse directions. The technique does become a little cumbersome since four different regions with different mesh sizes result, and the corresponding difficulties with bridging the mesh size changes arise. While eqs. (7) and (8) get modified only at nodes across which mesh size changes occur, eqns. (9)-(12) change considerably. Such modifications are, however, easy to derive. For example, see Hornbeck [12].

After some experimentation, $\Delta Y = \Delta Z/\sigma = 0.05$ for $0 \leqslant$ Y or $Z/\sigma \leqslant 0.8$ and $\Delta Y = \Delta Z/\sigma = 0.01$ for $0.8 <$ Y or $Z/\sigma \leqslant 1.0$ were selected while ΔX was slowly increased from 10^{-5} at duct inlet to 10^{-3} for $X > 0.03$. Points on the duct walls were excluded from consideration so as to reduce the matrix size and thereby computer time. A subroutine in the Harwell Library for solving a sparse system of linear equations was used; the matrix size being 1297 x 1297 but the number of non-zero elements being only 8928.

4. RESULTS AND DISCUSSION

Results were obtained for two values of the aspect ratio, σ=2 and σ=5, since most of the experimental data is available for these values. They are displayed in terms of a dimensionless axial distance $\overline{X}$ defined as

$$\overline{X} = \frac{\mu x}{\rho D^2 u_o}, \qquad (13)$$

where D is the hydraulic diameter of the duct, given by

$$D = 4a\sigma(\sigma+1)^{-1}, \qquad (14)$$

so that $\overline{X} = (1+1/\sigma)^2 X/16.$ (15)

4.1 Velocity Distribution

Figures 1-4 show the development of axial velocity U at the Y and Z axes (i.e., at the symmetry planes of the duct cross section) for the two ducts (σ = 5 and 2). While the solid curves in these figures represent the present theoretical results, the data points correspond to the experimental results of Sparrow, et al. [7]. The agreement between the two is remarkably good.

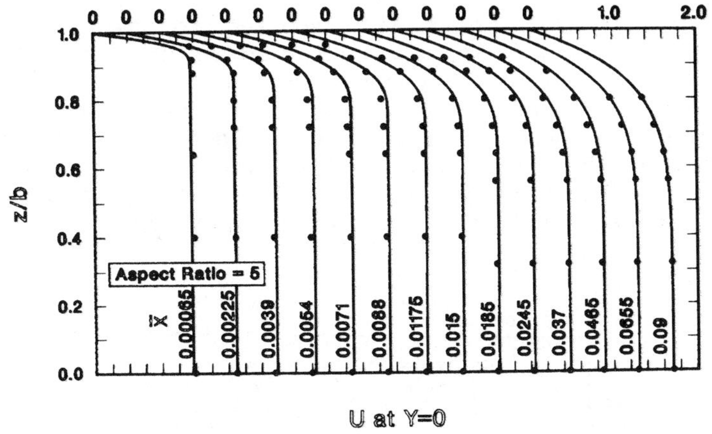

Fig. 1 Axial velocity development in a rectangular duct.

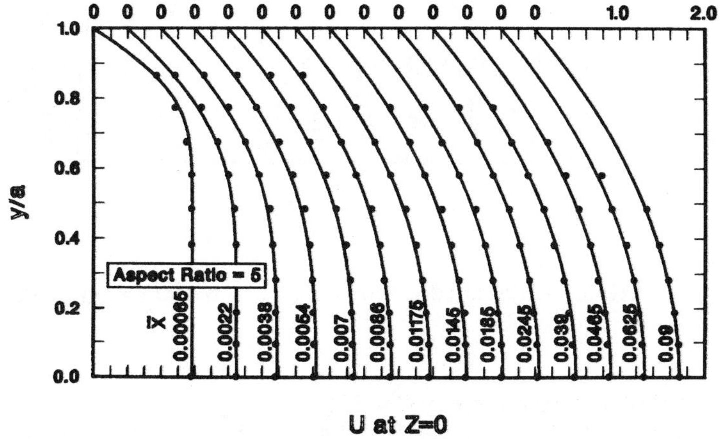

Fig. 2 Axial velocity development in a rectangular duct.

Developing Flow in a Rectangular Duct

The succession of velocity profiles graphically depicts the hydrodynamic development of the flow field over half the duct cross section; the profiles being similar over the other half. Considering Fig. 1, which displays profiles across the wide dimension of the duct of $\sigma = 5$, we find that the profile at the first reported station ($\bar{X}$=0.00065) is very nearly flat over the entire cross section. With increasing downstream distance, a boundary layer develops along the wall and spreads into the uniform core flow. Consequently, the velocity in the core increases in order to satisfy the continuity equation. Thus, the velocity profiles grow successively more rounded. However, since the aspect ratio is large, there still remains a relatively flat portion even under fully developed conditions ($\bar{X} \gtrsim 0.09$).

This general trend also applies to the sequence of velocity profiles in Fig. 2, which displays profiles across the narrow dimension of the duct of $\sigma = 5$. There are, however, differences in detail since, relative to the dimension a, the boundary layer appears to be five times thicker than it does relative to the dimension b. Thus with increasing downstream distance, the flat portion of the velocity profile is ultimately engulfed by the growing boundary layer.

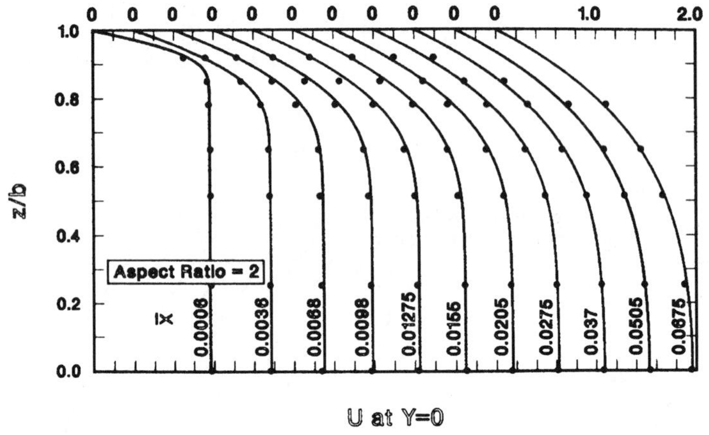

Fig. 3 Axial velocity development in a rectangular duct.

Similar remarks apply to Figs. 3 and 4 for the 2:1 duct but the rate of hydrodynamic development and the ratio of the maximum to the mean velocity are somewhat different for the two ducts. Due to a small aspect ratio, the differences between the profiles in Figs. 3 and 4 are also small compared to those in Figs. 1 and 2. Needless to say, the selection of stations

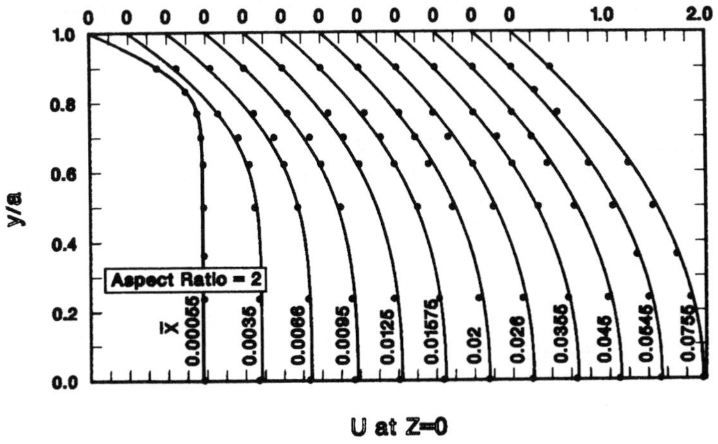

Fig. 4 Axial velocity development in a rectangular duct.

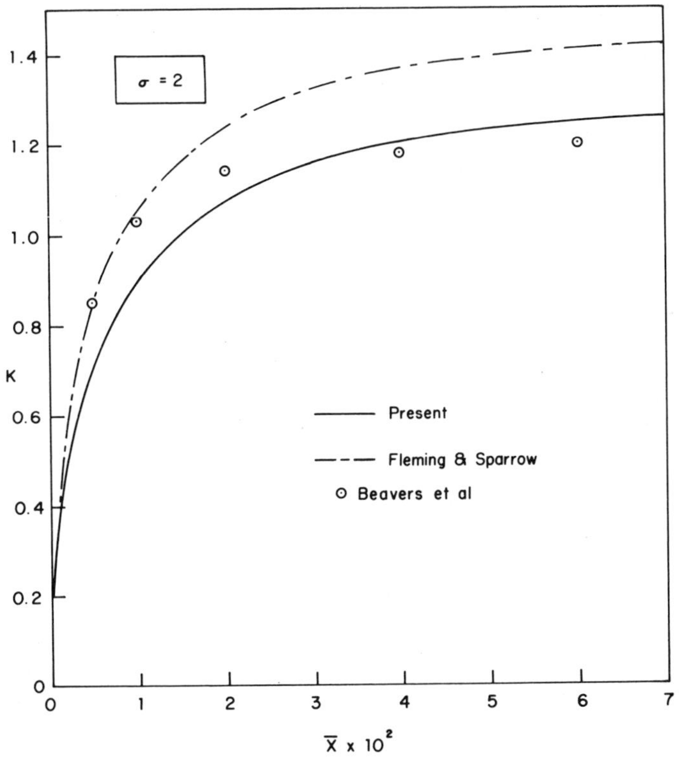

Fig. 5 Variation of pressure drop parameter with axial distance.

($\overline{X}$ values) at which the profiles are displayed in Figs. 1-4 was dictated by the experimental data in [7].

4.2 Pressure Distribution

Figures 5 and 6 show the variation of the pressure drop parameter $K(\overline{X})$ with $\overline{X}$ for the 2:1 and 5:1 ducts respectively. Besides the present theoretical results, these figures also display previous theoretical [4,5] and experimental data [8].

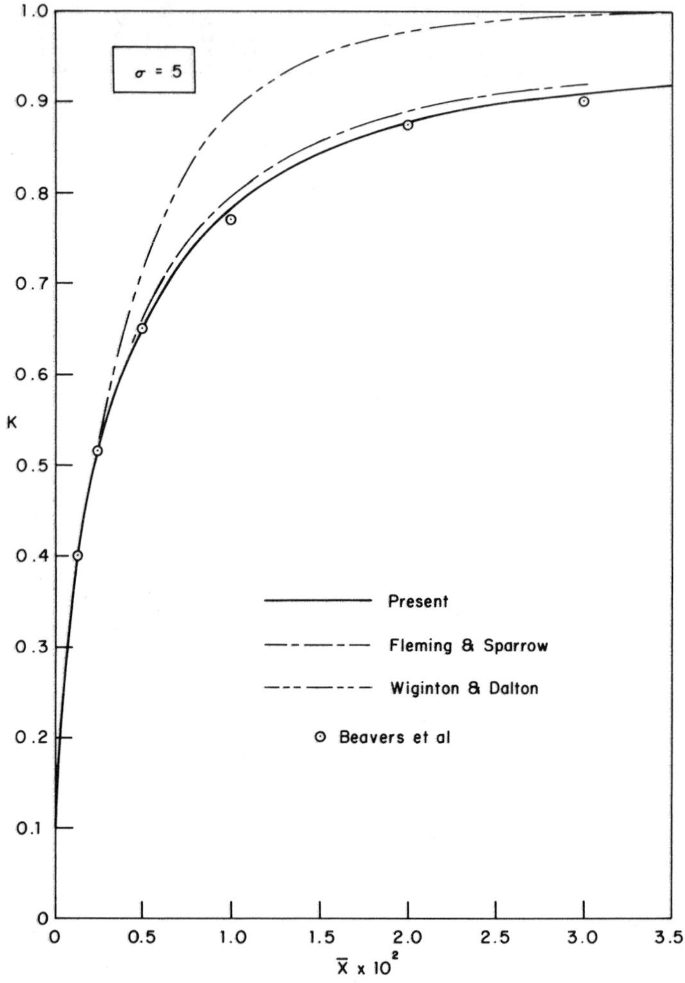

Fig. 6 Variation of pressure drop parameter with axial distance.

Here $K(\bar{X})$ is related to $P(\bar{X})$ by

$$K(\bar{X}) = -2P(\bar{X}) - (fRe)\,\bar{X}\ ,$$

where fRe is the product of Reynolds number and friction factor for fully developed flow. Specifically, fRe = 62.192 for $\sigma = 2$, and fRe = 76.282 for $\sigma = 5$. $K(\bar{X})$ represents the pressure drop due to flow development over and above that which would be there had the flow been bully developed right from the inlet section. $K(\bar{X})$ is seen from figs. 5 and 6 to increase rapidly in the near-entry region but as the flow becomes fully developed it gradually approaches a constant value, as it should. The present results match well with the experimental data [8]. It is relevant to note that differences between the theoretical and experimental values of K are accentuated by the subtractive process used to determine K. Thus, what appears to be a substantial deviation in K is of much lesser significance in P. The present results seem to compare better with experimental data than previous theoretical results.

REFERENCES

1. LUNDGREN, T.S., SPARROW, E.M. and STARR, J.B. - Pressure Drop Due to the Entrance Region in Ducts of Arbitrary Cross Section. J. Basic Eng., Vol. 86, pp. 620-626, 1964.

2. GUPTA, S.C. and GARG, V.K. - Effect of Velocity Distribution on the Stability of Developing Flow in a Channel. J. Phys. Soc. Japan, Vol. 50, pp. 673-680, 1981.

3. HAN, L.S. - Hydrodynamic Entrance Lengths for Incompressible Laminar Flow in Rectangular Ducts. J. Appl. Mech., Vol. 27, pp. 403-409, 1960.

4. FLEMING, D.P. and SPARROW, E.M. - Flow in the Hydrodynamic Entrance Region of Ducts of Arbitrary Cross Section. J. Heat Transfer, Vol. 91, pp. 345-354. 1969.

5. WIGINTON, C.L. and DALTON, C. - Incompressible Laminar Flow in the Entrance Region of a Rectangular Duct. J. Appl. Mech., Vol. 37, pp. 854-856, 1970.

6. CURR, R.M., SHARMA, D. and TATCHELL, D.G. - Numerical Predictions of Some Three-Dimensional Boundary Layers in Ducts. Comp. Meth. Appl. Mech. Eng., Vol. 1, pp. 143-158, 1972.

7. SPARROW, E.M., HIXON, C.W. and SHAVIT, G. - Experiments on Laminar Flow Development in Rectangular Ducts. J. Basic Eng., Vol 89, pp. 116-124, 1967.

Developing Flow in a Rectangular Duct

8. BEAVERS, G.S., SPARROW, E.M. and MAGNUSON, R.A. - Experiments on Hydrodynamically Developing Flow in Rectangular Ducts of Arbitrary Aspect Ratio. Int. J. Heat Mass Transfer, Vol. 13, pp. 689-701, 1970.

9. SHAH, R.K. - A Correlation for Laminar Hydrodynamic Entry Length Solutions for Circular and Noncircular Duct. J. Fluids Eng., Vol. 100, pp. 177-179, 1978.

10. CARLSON, G.A. and HORNBECK, R.W. - A Numerical Solution for Laminar Entrance Flow in a Square Duct. J. Appl Mech., Vol. 40, pp. 25-30, 1973.

11. GUPTA, S.C. and GARG, V.K.- Developing Flow in a Concentric Annulus. Comp. Meth. Appl. Mech. Eng., Vol 28, pp. 27-35, 1981.

12. HORNBECK, R.W. - Numerical Marching Techniques For Fluid Flows with Heat Transfer, NASA SP-297, 1973.

SURFACE ROUGHNESS EFFECT ON A VERY THIN CURVED SLIDE BEARING LUBRICATION

Chih Wu+ and L. K. Chi*

1. INTRODUCTION

When the lubricant film which separates the surfaces of a lubricated contact or bearing becomes very thin, the surface roughness that are present on all engineering surfaces begin to interfere. An approach to predicting behavior in the very thin lubrication region is to assume that the total load applied normal to the plane of the lubricated surface is carried partly by the hydrodynamic action of the lubrication film and partly by asperity contacts. In the same fashion, the total friction force between the lubricated surface is taken to be partly due to viscous friction and partly due to asperity contacts.

Christensen[1], Berthe and Godet[2], Tseng and Saibel[3], Weyler and Wu[4], and others have proposed very thin lubrication theories based on this model and the use of the Reynolds equation for incompressible fluid flow. Their work includes the ensemble average of the incompressible lubrication equation and the solution of lubricant pressure in bearings. The roughness is treated as a random quantity with a prescribed probability distribution function.

This paper aims to develop equations for determination of lubrication pressure in a very thin lubricating film and to develop a numerical method for calculating the lubricant pressures in a slide bearing with a convex pad surface. The side flow of the bearing is neglected. The geometric surface equation of the slide bearing surface is assumed to be parabola for small height of the pad crown curvature. The film-thickness

+Professor, Mechanical Engineering Department, U. S. Naval Academy, Annapolis, MD 21402
*Associate Professor, Computer Science Department, U. S. Naval Academy, Annapolis, MD 21402

equation for the inclined chord of the crown segment is determined. Hydrodynamic region and interference region of the bearing in the very thin lubrication film is obtained. A correlation between the film spacing and transverse pressure gradient is hypothesized. The classical Reynolds equation for hydrodynamic lubrication is modified in the interference region because the bearing surface is not perfectly smooth. Statistical expectation operators are needed in the calculation of the pressure gradient of the very thin lubricant film. A numerical finite difference method is employed to carry out the expectation operators, pressure gradient and pressure distribution of the fluid flow. It is shown that the pressure distribution of the lubricant film in the interference region is significantly different than the pressure distribution in the hydrodynamic region.

2. MATHEMATICAL MODEL FOR BEARING GEOMETRY

In order to predict the load carrying capacity and the friction to be generated in the mixed lubrication region, a mathematical model must be proposed for the random surface asperities found on most engineering surfaces.

For the present analysis, the geometry of a convex bearing will be used, as shown in Figure (1). The lubricant flow exists between an upper plate moving a constant speed U in the x-direction and a lower pad crown segment[5]. Flow in the y-direction will be assumed negligible and the length ℓ will be assumed to be very much larger than h_o or h_ℓ, the lubricant film thicknesses at the entrance and exit to the slider bearing, respectively. The nominal film thickness, h(x), turns out to be a function of x alone.

In the model being developed, it is further assumed that the bearing surfaces are covered with one-dimensional roughness in the form of ridges and valley running in the direction of the flow. The surface roughness varies in a direction transverse to the flow, i.e., the y-direction, but remains constant in the direction of flow, i.e., the x-direction. Figure (2) shows how the assumed roughness appears on a y-z plane perpendicular to the direction of flow. While h(x) in Figure (2) is constant for any value of y, the presence of the surface roughness results in different values of total film thickness for each value of y. Therefore, H is defined as the total film thickness, and is described by the equation.

$$H = h(x) + h_s , \qquad (1)$$

where h_s is the portion of the total film thickness due to the random surface asperities.

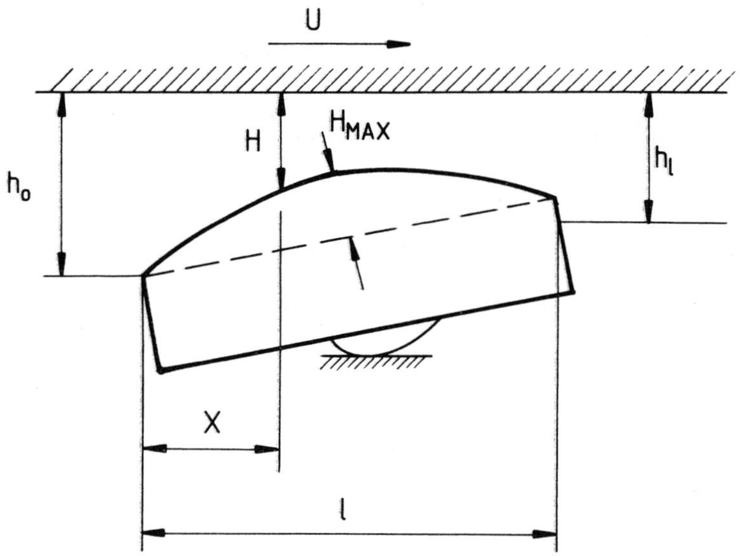

FIGURE 1 GEOMETRY OF CONVEX BEARING

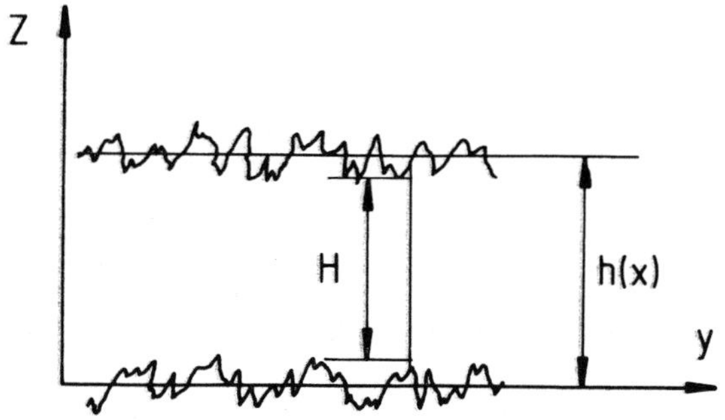

FIGURE 2 ASSUMED ROUGHNESS PROFILE

Figure (2), as shown, is for a value of x in the purely hydrodynamic region where no interference is present. Obviously, H is always positive in this region. As the value of x increases, the value of h(x), the nominal film thickness decreases, and at some point the surface asperities will start to interfere. The value of x where interference begins will be referred to as x_c, the critical length. For values of x greater than x_c, interference increases. In this interference region, a value of H = 0 will exist at points of asperity contact.

While the function h(x) may be determined from the geometry of the bearing, a function for the distribution of h_s must be proposed in order to complete the model. Experiments indicate that many engineering surfaces have roughness that may be described by an truncated normal probability density function given by

$$f(h_s) = \frac{K}{\sqrt{2\pi}\,\sigma} e^{-\frac{1}{2}\left(\frac{h_s-\mu}{\sigma}\right)^2} \quad , \text{ for } -c \leq h_s \leq c \quad , \tag{2}$$

and $f(h_s) = 0$ for all other values of h_s. K is a constant, μ is the mean value of the random variable h_s given by

$$\mu = \frac{1}{n} \sum_{i=1}^{n} h_{s_i} \quad , \tag{3}$$

and σ is the standard deviation of h_s given by the equation

$$\sigma^2 = \frac{1}{n-1} \sum_{i=1}^{n} (h_{s_i} - \mu)^2 \quad , \tag{4}$$

where $c = \max(h_{s_i})$, and h_{s_i} is the ith sample of the n samples.

From statistics, the probability density function of a random variable x having a normal distribution is expressible as

$$f(x) = \frac{1}{\sqrt{2\pi}\,\sigma} e^{-\frac{1}{2}\left(\frac{x-\mu}{\sigma}\right)^2} \quad , -\infty<x<\infty \quad , \tag{5}$$

and

$$\int_{-\infty}^{+\infty} f(x)\, dx = 1. \tag{6}$$

In order to have $F(h_s)$ represent a probability density function, then

$$\int_{-c}^{+c} f(h_s)\ dh_s = 1 \tag{7}$$

must also be true. This is accomplished by multiplying the probability density function of equation (2) by a constant K, such that

$$\int_{-c}^{+c} \frac{K}{\sqrt{2\pi}\ \sigma}\ e^{-\frac{1}{2}\left(\frac{h_s-\mu}{\sigma}\right)^2}\ dh_s = 1\ . \tag{8}$$

Therefore,

$$f(h_s) = \frac{K}{\sqrt{2\pi}\ \sigma}\ e^{-\frac{1}{2}\left(\frac{h_s-\mu}{\sigma}\right)^2}\ . \tag{9}$$

K is determined by first assuming a dummy variable z, such that

$$z = \frac{h_s-\mu}{u}\ , \tag{10}$$

and $h_s = \sigma z + \mu$, such that

$$dh_s = \sigma dz\ . \tag{11}$$

Thus, equation (9) becomes

$$f(z) = \frac{K}{\sqrt{2\pi}\ \sigma}\ e^{-\frac{1}{2}z^2}\ ,\quad \frac{-c-\mu}{\sigma} \leq z \leq \frac{c-\mu}{\sigma}\ . \tag{12}$$

Equations (11) and (12) then allow equation (8) to be rewritten as

$$\int_{\frac{-c-\mu}{\sigma}}^{\frac{c-u}{\sigma}} \frac{K}{\sqrt{2\pi}}\ e^{-\frac{1}{2}z^2}\ dz = 1\ . \tag{13}$$

Equation (13) can be restated as

$$K\int_{-\infty}^{\frac{c-\mu}{\sigma}} \frac{1}{\sqrt{2\pi}}\ e^{-\frac{1}{2}z^2}\ dz - K\int_{-\infty}^{\frac{-c-\mu}{\sigma}} \frac{1}{\sqrt{2\pi}}\ e^{-\frac{1}{2}z^2}\ dz = 1 \tag{14}$$

Again, from statistics, the relationship

$$\int_{-\infty}^{n} \frac{1}{\sqrt{2\pi}}\ e^{-\frac{1}{2}z^2}\ dz = Pr(n)\ , \tag{15}$$

where Pr(n) is the normal probability of n. Thus, equation (14) can be restated as

$$K[\Pr(\frac{c-\mu}{\sigma}) - \Pr(\frac{-c-\mu}{\sigma})] = 1 \quad , \tag{16}$$

or

$$K = \frac{1}{\Pr(\frac{c-\mu}{\sigma}) - \Pr(\frac{-c-\mu}{\sigma})} \tag{17}$$

For the symmetric case where μ, the mean value of h_s, is equal to zero, equation (17) is simpler still, since $\Pr(\frac{-c}{\sigma}) = 1 - \Pr(\frac{c}{\sigma})$. In this case, equation (17) would become

$$K = \frac{1}{2\Pr(\frac{c}{\sigma})-1} \tag{18}$$

Since c, μ, and σ can be calculated on the basis of data gathered from a microscopic examination of the h_s distribution on a given surface, a value of K is obtainable from equation (17) or equation (18).

3. FLUID DYNAMICS

The classical Reynolds' Equation for hydrodynamic lubrication is used to find the pressure distribution in a bearing with incompressible lubricant. For the geometry of Figure (1), and assuming smooth surfaces, Reynolds' Equation for a one-dimensional, steady flow reduces to

$$\frac{dp}{dx} = 6\eta U \frac{h(x) - h'}{[h(x)]^3} \quad , \tag{19}$$

where p is the lubricant pressure, η is the lubricant viscosity, U is the surface relative velocity, and h' is the value of h(x) where dp/dx = 0[6].

Since the surfaces to be used are not perfectly smooth, and since the value of total film thickness, H, is a random variable, Christensen[1] uses the mathematical expectation operator E, and has rewritten equation (19) to read

$$\frac{dp}{dx} = 6\eta U \frac{E(H) - E(H')}{E(H^3)} \quad , \tag{20}$$

where E(H), $E(H^3)$, and E(H') are the "expected values" of H, H^3, and H' at the point where dp/dx = 0, respectively. The expectation operator is defined by

$$E[u(x)] = \int_{-\infty}^{+\infty} u(x)\ f(x)\ dx \quad , \tag{21}$$

where x is a random variable having a probability density function of f(x), and u(x) is any function of x. For smooth plates, i.e., for plates where h_s reduces to zero, the "expected value" of H, that is E(H), is obviously going to be h(x). Likewise, $E(H^3)$ would reduce to $[h(x)]^3$. Thus, equation (20) would reduce to equation (19) as roughness was reduced.

The Reynolds' Equation, equation (20) is clearly applicable for the hydrodynamic region of the bearing being considered, since one-dimensional flow has been assumed. In the interference region, where the ridges contact, lubricant flow through the valleys can still be seen to be one-dimensional. Therefore, equation (2) will again be assumed to be applicable in the interference region, as well as in the hydrodynamic region.

4. PRESSURE EQUATIONS

While equation (20) may apply in both the hydrodynamic region and the interference region, the calculations of the expectations are significantly different in each region. Therefore, it becomes useful to separate the two regions for calculation purposes. For small values of H_{max}, the curved surface is assumed to be a portion of a parbola. The critical length, $x_c < \ell$ can be determined by[5].

$$H_{max}\left[4\left(\frac{x_c}{\ell}-\frac{1}{2}\right)^2-1\right]+h_o-\frac{h_o x_c}{\ell}+\frac{h_\ell x_c}{\ell}-c=0 \qquad (22)$$

x_c can be calculated on the basis of geometric data, and indicates the separation point between the two regions. Further, it should be noted that since $dp/dx = 0$ represents the single point where pressure is a maximum, $E(H')$ is a constant during the integration of equation (20).

In the hydrodynamic region, the pressure p at any point x can be found from

$$p-p_o=\int_0^x\left(\frac{dp}{dx}\right)dx=\int_0^x 6\eta U\frac{E(H)}{E(H^3)}\,dx-E(H^1)\int_0^x\frac{6\eta U}{E(H^3)}\,dx, \qquad (23)$$

$0<x<x_c$, where p_o is the pressure at the entrance to the bearing at $x = 0$. In like manner, in the interference region the pressure p at any point x can be found from

$$p_\ell-p=\int_x^\ell\left(\frac{dp}{dx}\right)dx=\int_x^\ell 6\eta U\frac{E(H)}{E(H^3)}\,dx-E(H^1)\int_x^\ell\frac{6\eta U}{E(H^3)}\,dx, \qquad (24)$$

$x_c<x<\ell$ where p_ℓ is the pressure at the exit of the bearing at $x = \ell$. $E(H^1)$ can be determined from the fact that the pressure at the critical length x_c can be calculated from both directions, that is, can be calculated from both equation (23) and equation (24). If p_o and p_ℓ, the pressures at the inlet and outlet to the slider bearing, are assumed to be equal, then the pressure at the critical length, p_c is given by

$$p_c = \int_o^{x_c} 6\eta U \frac{E(H)}{E(H^3)}\,dx - E(H^1)\int_o^{x_c} \frac{6\eta U}{E(H^3)}\,dx =$$

$$\int_{x_c}^{\ell} 6\eta U \frac{E(H)}{E(H^1)}\,dx - E(H^1)\int_{x_c}^{\ell} \frac{6\eta U}{E(H^3)}\,dx \quad , \tag{25}$$

or

$$E(H^1) = \frac{\displaystyle\int_o^{x_c} 6\eta U \frac{E(H)}{E(H^3)}\,dx + \int_{x_c}^{\ell} 6\eta U \frac{E(H)}{E(H^3)}\,dx}{\displaystyle\int_o^{x_c} \frac{6\eta U}{E(H^3)}\,dx + \int_o^{\ell} \frac{6\eta U}{E(H^3)}\,dx} \tag{26}$$

In summary then, the pressure distribution through the bearing will be found by the use of

$$p = \int_0^{x} 6\eta U \frac{E(H)}{E(H^3)}\,dx - E(H^1)\int_o^{x} \frac{6\eta U}{E(H^3)}\,dx \tag{27}$$

for $0 < x < x_c$ and $p_o = 0$, and by the use of

$$p = -\int_x^{\ell} 6\eta U \frac{E(H)}{E(H^3)}\,dx + E(H^1)\int_x^{\ell} \frac{6\eta U}{E(H^3)}\,dx \tag{28}$$

for $x_c < x < \ell$ and $p_\ell = 0$, where $E(H^1)$ is found from equation (26).

5. THE NUMERICAL METHOD

Pressures in the hydrodynamic region will be determined from equation (27). For each value of x, pressure is found by integrating from 0 to x. E(H) and $E(H^3)$ in equation (27) are functions of x, and thus vary during integration.

In the interference region, the pressures can be determined from equation (28). For each value of x, pressure is found by integrating from x to ℓ. The E(H) and $E(H^3)$ in equation (28) are functions of x, both of which involve integration over surface roughness.

Finally, both equation (27) and (28) include the term $E(H^1)$, which is solved by the four integrations indicated in equation (26). While all of these integrations would be tedious if done by hand, they can be reapidly accomplished by finite difference techniques in a computer program.

Basically, all of the integrals can be handled by considering each continuous function as broken up into a finite number of areas, such as the example in Figure (3) indicates. The integration of f(x) in Figure (3) would be approximated

$$\int_o^L f(x)\ dx = \sum_{i=1}^{n} f(x_i)\ \Delta x \quad ,$$

where $n = L/\Delta x$.

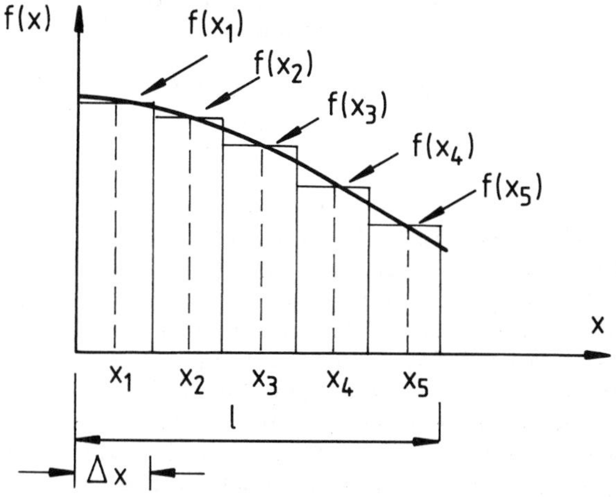

Figure (3). Finite Difference Method Applied to Continuous Function

6. THE NUMERICAL SOLUTION

A program was written in the programming language BASIC for use on the Naval Academy's computer. The program performed the numerical integration of the equations discussed above, and solved for pressure as a function of position in the bearing.

The data fed into the program were as follows:

1. The breadth of the pad surface; ℓ in Figure (]).
2. The plate spacing at the rear edge of the bearing; h_ℓ in Figure (1).
3. The plate spacing at the front edge of the bearing;

h_o in Figure (1).

4. The value of c from equation (2), i.e., half the total range of the random film thickness variable h_s. This is determined by measuring the roughness on both surfaces, finding the range of roughness on each, and adding the two ranges together to get the total range of roughness.
5. σ, the standard deviation of surface roughness, determined from observation of the previously gathered surface roughness data, and the use of equation (4).
6. The viscosity of the lubricant.
7. The speed of the upper surface in Figure (1).
8. The number of finite increments to be used in the numerical integration of the surface roughness probability density function.
9. The number of finite increments to be used in the numerical integration of functions integrated over the length of the bearing surface.

Figure (4) is a plot of pressure over the length of the bearing using the data listed in the figure.

7. APPLICATION

As discussed in the introduction, the present model for mixed lubrication assumes that part of the bearing load is carried by the hydrodynamic action and the remainder through the solid to solid contacts. With the pressure distribution obtained from the method developed in this paper, a bearing's hydrodynamic load carrying capability can be calculated by integrating the pressure over that part of the bearing area subject to hydrodynamic action. This includes all area in the hydrodynamic region, as well as the area in the interference region where solid to solid contact does not exist. Christensen[1] develops just such an approach, and shows how the roughness probability density function can be used to calculate the percentage of the area in the interference region subject to hydrodynamic action. Several works on Tribology[6], [7], [8], [9] also present methods for calculating the loads carried by the solid to solid contact.

In a fashion similar to the normal load, the shear load, or drag is assumed to be partly due to hydrodynamic drag and partly due to solid friction. For the bearing used in this paper, Streeter[10] shows that the shear stress exerted on the lower, moving plate by the lubricant is given by

$$\tau = -\tfrac{1}{2}h(x)\frac{dp}{dx} - \frac{\eta U}{h(x)} \quad ,$$

and that the force required to move the lower plate is obtained by integrating the shear stress over the area of the subject to hydrodynamic action. Since equation (20) expresses the pressure gradient dp/dx in terms of expectation operatios discussed earlier, the hydrodynamic drag is readily obtained. The standard works on Tribology[6], [7], [8], [9] also address calculation of solid friction, thus allowing determination of the total drag in a slider bearing under mixed lubrication conditions.

Ratio of inlet to outlet film thickness - - - - - 3.0

Dimensionless crown ratio - - - - - - - - - - - - - 0.6

Viscosity of lubricant - - - - - - - - - - - - - - 0.2 poise

Speed of upper place - - - - - - - - - - - - - - 100cm/sec

Standard deviation of surface roughness - - - - - $1.667X10^{-4}$cm

Transition to mixed lubrication at X = 8.33 cm

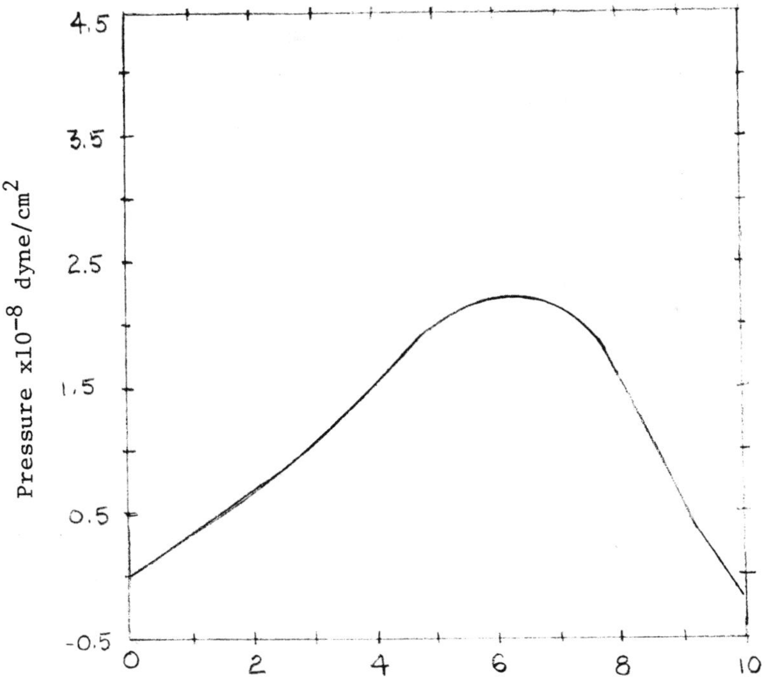

Figure (4). Mixed Lubrication with Convex Surface

8. REFERENCES

1. CHRISTENSEN, H., A Theory of Mixed Lubrication, Proceedings The Institution Mechanical Engineers, 1972, 186, 41.

2. BERTHE, D. and GODET, M., A More General Form of Reynolds Equation--Application to Rough Surfaces, Wear, 1974, 27, 345.

3. TSENG, S. T. and SAIBEL, E., Surface Roughness Effect on Slide Bearing Lubrication, ASLE Transaction, 1967, 10, 334.

4. WEYLER, M. E. and WU, C., A Numerical Method for the Calculation of Lubricant Pressures in Bearings with Mixed Lubrication, Tribology International, 1982, 15, 89.

5. ABRAMOVITZ, S., Theory for a Slider Bearing with a Convex Pad Surface; Side Flow Neglected, Journal of Franklin Institute, 1955

6. CZICHOS, H., Tribology, Elsevier Scientific Publishing Company, Amsterdam, 1978.

7. HALLING, J., Ed., Principles of Tribology, MacMillan Press, London, 1975.

8. FULLER, D., Theory and Practice of Lubrication for Engineers, Wiley, New York, 1956.

9. HERSEY, M., Theory and Research in Lubrication, Wiley, New York, 1966.

10. STREETER, V., Fluid Mechanics, McGraw-Hill, New York, 1958.

SECTION 2

TURBULENT FLOW

ASPECTS OF A PENALTY FINITE ELEMENT ALGORITHM FOR THE TURBULENT THREE-DIMENSIONAL PARABOLIC NAVIER-STOKES EQUATIONS

A. J. Baker†

University of Tennessee
Knoxville, TN USA

SUMMARY

The formal derivation of the three-dimensional parabolic Navier-Stokes equations for subsonic turbulent flow is reviewed. A penalty finite element algorithm is established for numerical solution of the sixteen dependent variable system. Key numerical results are summarized documenting applications in various problem definitions.

INTRODUCTION

The time-averaged, three-dimensional Navier-Stokes equations, for steady, turbulent, subsonic flow of a compressible, heat conducting fluid, can be simplified to admit an efficient space marching numerical solution procedure under certain restrictions. Baker, et al [1], documents the derivation of this simplified equation system, the so-called "parabolic" Navier-Stokes equations, using formal ordering arguments. References [2-7] document application of a penalty, finite numerical solution algorithm for the parabolic Navier-Stokes equations, for a variety of subsonic flow configurations, including an embedding within an interaction algorithm to impose axial pressure gradient feedback. These results have provided the basic assessments of accuracy, convergence and versatility aspects of the finite element penalty algorithm, as well as detailed comparison between experimental data and prediction for various turbulent flow geometries.

The basic requirements for any numerical solution algorithm construction for the three-dimensional parabolic Navier-Stokes (3DPNS) equation set are accuracy, geometric versatility and efficiency. Any theoretical basis, eg., finite difference, finite volume, finite element, etc, applied to constructing a 3DPNS algorithm ultimately yields the linear algebra statement $\{FI\} = \{0\}$, where

† Professor of Engineering Science and Mechanics

elements of { FI } are strongly non-linear functions of the dependent variable set $q_i(x_j)$, $1 \leq i \leq 16$, $1 \leq j \leq 3$. Dependent upon the algorithm designer's decisions, members of the set q_i can include density, mean velocity vector, stagnation enthalpy, turbulent kinetic energy, isotropic dissipation function, pressure and/or scalar potential fields, and six components of the Reynolds stress tensor.

For the subject finite element penalty algorithm statement, for the 3DPNS equation set, the functional form of the linear algebra statement is

$$\{FI(k, \lambda, \Theta, \Delta x_1, \{QI\})\} = \{0\} \tag{1}$$

In equation 1, k is the completeness of the (polynomial) basis selected to construct the semi-discrete approximation $q_i(x_j)$ to $q_i^h(x_j)$, λ is the penalty function scalar multiplier, Θ is the implicitness factor of the downstream integration step Δx_1, where $\Theta = ½$ is the trapezoidal rule, and $\{QI(x_1)\}$ is the array of expansion coefficients of $q_i^h(x_j)$, evaluated at the nodal coordinates of the (spatial) discretization $\cup R_e$ of the 3DPNS solution domain $\Omega = R^2 \times x_1$.

For all $\Theta > 0$, equation 1 is a non-linear algebraic equation system eligible for solution using any of a multitude of (approximate) procedures. Most of these can be interpreted within the framework of the basic Newton iteration algorithm matrix solution statement,

$$[J(\{QI\})]^p_{j+1} \{\delta QI\}^{p+1}_{j+1} = - \{FI\}^p_{j+1} \tag{2}$$

where p is the iteration index at step x_{j+1}, and

$$\{QI\}^{p+1}_{j+1} \equiv \{QI\}^p_{j+1} + \{\delta QI\}^{p+1}_{j+1} \tag{3}$$

The Jacobian [J], appearing in equation 2, is by definition the square matrix,

$$[J(\{QI\})] \equiv \frac{\partial\{FI\}}{\partial\{QJ\}} \tag{4}$$

where both I and J range $1 \to 16$. This paper establishes the functional forms for equations 1-4, for the penalty algorithm construction, and summarizes key aspects.

PARABOLIC NAVIER-STOKES EQUATIONS

The three-dimensional parabolic Navier-Stokes (3DPNS) equations are a simplification of the steady, three-dimensional time-averaged Navier-Stokes equations, which in Cartesian tensor conservation forms are

$$L(\rho) = \frac{\partial}{\partial x_j}\left[\rho u_j\right] = 0 \tag{5}$$

$$L(\rho u_i) = \frac{\partial}{\partial x_j}\left[\rho u_i u_j + p\delta_{ij} + \rho\overline{u_i u_j} - \sigma_{ij}\right] = 0 \qquad (6)$$

$$L(\rho H) = \frac{\partial}{\partial x_j}\left[\rho H u_j - u_i\sigma_{ij} + \rho\overline{H u_j} - \overline{u_i\sigma_{ij}} + q_j\right] = 0 \qquad (7)$$

$$L(\rho k) = \frac{\partial}{\partial x_j}\left[\rho u_j k + (C_k \frac{k}{\varepsilon}\rho\overline{u_i u_j} - \mu\delta_{ij})\frac{\partial k}{\partial x_i}\right]$$
$$+ \rho\overline{u_i u_j}\frac{\partial u_i}{\partial x_j} + \rho\varepsilon = 0 \qquad (8)$$

$$L(\rho\varepsilon) = \frac{\partial}{\partial x_j}\left[\rho u_j\varepsilon + C_\varepsilon \frac{k}{\varepsilon}\rho\overline{u_i u_j}\frac{\partial\varepsilon}{\partial x_i}\right] + C_\varepsilon^1 \rho\overline{u_i u_j}\frac{\varepsilon}{k}\frac{\partial u_i}{\partial x_j}$$
$$+ C_\varepsilon^2 \frac{\rho\varepsilon^2}{k} = 0 \qquad (9)$$

In equations 5-9, the usual superscript bar notation denoting time-averaged quantities [8] has been deleted for clarity. The time-averaged dependent variables are density (ρ), mean momentum vector (ρu_i), pressure (p) and stagnation enthalpy (H). Further, δ_{ij} denotes the Kronecker delta, and the Stoke's stress tensor (σ_{ij}) and heat flux vector (q_j) are defined as,

$$\sigma_{ij} \equiv \frac{\mu}{Re}\left[E_{ij} - \frac{2}{3}\delta_{ij}E_{kk}\right] \qquad (10) \qquad\qquad q_j \equiv -\kappa\frac{\partial H}{\partial x_j} \qquad (11)$$

where μ and κ are laminar viscosity and heat conductivity respectively. E_{ij} is the symmetric mean flow strain rate tensor.

$$E_{ij} \equiv \frac{\partial u_i}{\partial x_j} + \frac{\partial u_j}{\partial x_i} \qquad (12)$$

Finally, $-\overline{u_i u_j}$ is the symmetric Reynolds stress tensor with trace equal to 2k, where k is the turbulent kinetic energy. For present purposes, $\overline{u_i u_j}$ is assumed correlated in terms of k, u_j and ε, the isotropic dissipation function, in the form,

$$\overline{u_i u_j} \equiv C_{\underline{i}}\, k\delta_{ij} - C_4\frac{k^2}{\varepsilon}E_{ij} - C_2 C_4\frac{k^3}{\varepsilon^2}E_{ik}E_{kj} \qquad (13)$$

where the C_α, $1 \leq \alpha \leq 4$, are known constants [1].

The parabolic approximation to the steady flow Navier-Stokes set, equations 5-13, is generated by assuming a principal direction of the flow persists, say parallel to the (curvi-linear) coordinate x_1. Assuming the corresponding mean velocity component u_1 is of order unity, ie., O(1), and that the other two orthogonal components u_ℓ are smaller, say of $O(\delta)$, then for modest density variation the continuity equation 1 confirms that for $\partial/\partial x_1 \equiv O(1)$, then $\partial/\partial x_\ell = O(\delta^{-1})$. Proceeding through the analysis details [9] confirms that the O(1) 3DPNS equation set is

$$L(\rho) = \frac{\partial}{\partial x_j}[\rho u_j] = 0 \tag{14}$$

$$L(\rho u_1) = \frac{\partial}{\partial x_j}\left[\rho u_1 u_j + p\delta_{1j}\right] + \frac{\partial}{\partial x_\ell}\left[\overline{\rho u_1 u_\ell} - \sigma_{1\ell}\right] = 0 \tag{15}$$

$$L(\rho u_\ell) = \frac{\partial}{\partial x_k}\left[p\delta_{k\ell} + \rho\,\overline{u_k u_\ell}\right] = 0 \tag{16}$$

$$L(\rho H) = \frac{\partial}{\partial x_j}\left[\rho H u_j\right] + \frac{\partial}{\partial x_\ell}\left[\overline{\rho H u_\ell} - \kappa\frac{\partial H}{\partial x_\ell} - u_i\sigma_{i\ell} - \overline{u_i\sigma_{i\ell}}\right] = 0 \tag{17}$$

$$L(\rho k) = \frac{\partial}{\partial x_j}\left[\rho k u_j\right] + \frac{\partial}{\partial x_\ell}\left[\left(\rho C_k \frac{k}{\varepsilon}\overline{u_j u_\ell} - \mu\delta_{j\ell}\right)\frac{\partial k}{\partial x_j}\right] + \rho\overline{u_1 u_\ell}\frac{\partial u_1}{\partial x_\ell} + \rho\varepsilon = 0 \tag{18}$$

$$L(\rho\varepsilon) = \frac{\partial}{\partial x_j}\left[\rho\varepsilon u_j\right] + \frac{\partial}{\partial x_\ell}\left[\rho C_\varepsilon \frac{k}{\varepsilon}\overline{u_j u_\ell}\frac{\partial\varepsilon}{\partial x_j}\right] + C_\varepsilon^1\,\rho\overline{u_1 u_\ell}\frac{\varepsilon}{k}\frac{\partial u_1}{\partial x_\ell} + C_\varepsilon^2\,\rho\varepsilon^2/k = 0 \tag{19}$$

In equations 14-19, the tensor index summation convention is $1 \leq (i,j) \leq 3$ and $2 \leq (k, \ell) \leq 3$. The Reynolds stress tensor constitutive equation 13 also becomes considerably simplified under the ordering analysis. For example, in rectangular Cartesian coordinates, and retaining the first two orders of terms yields equation 20. An equation of state $\rho = \rho(p,H)$ completes the basic 3DPNS statement.

$$O(\delta) \qquad\qquad\qquad\qquad O(\delta^2)$$

$$\begin{aligned}
\overline{u_1u_1} &= C_1k - C_2C_4\frac{k^3}{\varepsilon^2}\left[\left(\frac{\partial u_1}{\partial x_2}\right)^2 + \left(\frac{\partial u_1}{\partial x_3}\right)^2\right] && - 2C_4\frac{k^2}{\varepsilon}\left[\frac{\partial u_1}{\partial x_1}\right]\\
\overline{u_2u_2} &= C_3k - C_2C_4\frac{k^3}{\varepsilon^2}\left[\frac{\partial u_1}{\partial x_2}\right]^2 && - 2C_4\frac{k^2}{\varepsilon}\left[\frac{\partial u_2}{\partial x_2}\right]\\
\overline{u_3u_3} &= C_3k - C_2C_4\frac{k^3}{\varepsilon^2}\left[\frac{\partial u_1}{\partial x_3}\right]^2 && - 2C_4\frac{k^2}{\varepsilon}\left[\frac{\partial u_3}{\partial x_3}\right]\\
\overline{u_1u_2} &= - C_4\frac{k^2}{\varepsilon}\left[\frac{\partial u_1}{\partial x_2}\right] && - C_2C_4\frac{k^3}{\varepsilon^2}\left[\frac{\partial u_1}{\partial x_3}\left(\frac{\partial u_2}{\partial x_3} + \frac{\partial u_3}{\partial x_2}\right)\right.\\
& && \left. + 2\frac{\partial u_1}{\partial x_2}\left(\frac{\partial u_1}{\partial x_1} + \frac{\partial u_2}{\partial x_2}\right)\right]\\
\overline{u_1u_3} &= - C_4\frac{k^2}{\varepsilon}\left[\frac{\partial u_1}{\partial x_3}\right] && - C_2C_4\frac{k^3}{\varepsilon^2}\left[\frac{\partial u_1}{\partial x_2}\left(\frac{\partial u_2}{\partial x_3} + \frac{\partial u_3}{\partial x_2}\right)\right.\\
& && \left. + 2\frac{\partial u_1}{\partial x_3}\left(\frac{\partial u_1}{\partial x_1} + \frac{\partial u_3}{\partial x_3}\right)\right]\\
\overline{u_2u_3} &= - C_2C_4\frac{k^3}{\varepsilon^2}\left[\frac{\partial u_1}{\partial x_2}\frac{\partial u_1}{\partial x_3}\right] && - C_4\frac{k^2}{\varepsilon}\left[\frac{\partial u_2}{\partial x_3} + \frac{\partial u_3}{\partial x_2}\right]
\end{aligned} \tag{20}$$

Equations 14-20 do not represent a well-posed, initial-boundary value problem statement for the dependent variable set. In particular, equation 14 is the sole definition for u_ℓ, yielding an underdetermined system. The finite element penalty algorithm construction of Baker [1] yields a well-posed problem statement by inclusion of both $O(\delta)$ transverse momentum equations, definition of an auxiliary harmonic function for a penalty constraint, and definition of complementary and particular solutions to a pressure Poisson equation formed from equation 16.

The derived pressure Poisson equation is,

$$L(p) \equiv \frac{\partial L(\rho u_\ell)}{\partial x_\ell} = \frac{\partial^2 p}{\partial x_\ell^2} + \frac{\partial^2}{\partial x_\ell \partial x_k}\left[\rho\overline{u_k u_\ell}\right] = 0 \tag{21}$$

the solution to which is defined as

$$p(x_j) \equiv p_c(x_\ell, x_1) + p_p(x_\ell . x_1) \tag{22}$$

The complementary pressure $p_c(x_\ell, x_1)$ is the solution to the homogeneous form of equation 21 with exterior flow boundary conditions [1]. The particular pressure is computed throughout the 3DPNS solution, from equation 21, and added in a delayed manner into equation 22, when used in the u_1 momentum equation solution, yielding a multi-pass interaction algorithm.

The retained $O(\delta)$ transverse momentum equation set is

$$L^{\delta}(\rho u_\ell) = \frac{\partial}{\partial x_j}\left[\rho u_\ell u_j\right] - \frac{\partial}{\partial x_k}\left[\sigma_{k\ell}\right] = 0 \qquad (23)$$

The auxiliary harmonic variable ϕ is defined as the solution to a Poisson equation, driven by the continuity equation 13, in the form

$$L(\phi) \equiv \frac{\partial^2 \phi}{\partial x_\ell^2} - \frac{\partial}{\partial x_j}\left[\rho u_j\right] = 0 \qquad (24)$$

FINITE ELEMENT PENALTY ALGORITHM

As a consequence of the 3DPNS equation set completion, equations 15, 16+23, 17, 18, 19, 21 and 24 define a well-posed, initial-boundary value statement for the dependent variable set $q_i(x_j) \equiv \{q(x_j)\} = \{u_1, u_\ell, H, k, \varepsilon, p_c, p_p, \phi\}$. The equation of state, $\rho = \rho(p,H)$, and equation 20 are algebraic definitions for the remaining seven members $\{\rho, \overline{u_i u_j}\}$. Therefore, the first nine members of $\{q(x_j)\}$ are eligible for constraint on the solution domain boundary, $\partial\Omega = \partial R \times x_1$, by a linear combination of Dirichlet and Neumann boundary conditions of the form

$$\ell(q_i) \equiv a_1^i q_i + a_2^i \frac{\partial q_i}{\partial x_\ell} \hat{n}_\ell + a_3^i = 0 \qquad (25)$$

In equation 25, the a_j^i are defined to enforce the appropriate constraint for each variable, cf., Baker [2,12]. Since the remaining seven members of $\{q(x_j)\}$ are defined by algebraic equations, no boundary conditions are appropriate. Finally, the first six members of $\{q(x_j)\}$ are required defined on the initial solution plane, $\Omega_0 = R^2 \times x_1(0)$, by an appropriate initial condition $\{q_i(x_\ell, x_1(0))\}$.

The complete derivation of the finite element penalty constraint numerical solution algorithm is given in [1,9]. Briefly, the semi-discrete approximation for each member of the set $q_i(x_j)$ is formed by the union of elemental approximations $q^e(x_j)$ as,

$$q_i(x_j) \approx q_i^h(x_j) \equiv \bigcup_e q_i^e(x_j) \qquad (26)$$

In turn, each elemental semi-discrete approximation, valid on the representative finite element domain $\Omega_e = R_e^2 \times x_1$, is formed as an expansion in the cardinal basis $\{N_k(x_\ell)\}$, the members of which are (typically) polynomials complete to degree k, in the form

$$q_i^e(x_j) \equiv \{N_k(x_\ell)\}^T\{QI(x_1)\}_e \tag{27}$$

In equation 27, $\{\cdot\}$ denotes a column matrix, superscript T its transpose, subscript e denotes pertaining to R_e^2, and the elements of $\{QI\}_e$ are the evaluation of the semi-discrete approximation at the nodal coordinates of R^2 at any coordinate station x_1.

A basic requirement in any algorithm construction is a formal statement regarding constraint on the error formed by employing the semi-discrete approximation for the differential equation set. The finite element algorithm requires this error ε_i^h to be orthogonal to the basis employed to construct $q_i(x_j)$. For all members of $\{q^h\}$ except u_ℓ^h, the resultant error constraint statement is,

$$\int_{R^2}\{N_k(x_\ell)\}L(q_i^h)d\vec{x} + \beta\int_{\partial R}\{N_k(x_\ell)\}\ell(q_i^h)dx \equiv \{0\} \tag{28}$$

where β is a scalar multiplier selected to achieve cancellation of the middle term in equation 25. The error extremization statement for the members u_ℓ^h of $\{q^h\}$ is,

$$\int_{R^2}\{N_k(x_\ell)\}\left[L(u_\ell^h) + L^\delta(u_\ell^h)\right]d\vec{x} + \beta\int_{\partial R}\{N_k(x_\ell)\}\ell(u_\ell^h)dx$$
$$+ \lambda\int_{R^2}\frac{\partial\{N_k\}}{\partial x_\ell}L(\phi^h)d\vec{x} \equiv \{0\} \tag{29}$$

where λ is an arbitrary parameter modifying the penalty term, which constrains the error extremization by the continuity equation (error).

Equations 28-29 define the finite element penalty algorithm semi-discrete error constraint statement for the 3DPNS equation set. For the non-initial-valued dependent variables, equation 28 yields the linear algebra statement $\{FI\} = \{0\}$, recall equation 1. For the initial-valued variables, equations 28-29 yield a coupled ordinary differential equation set,

$$[C]\frac{d\{QI\}}{dx_1} + \{B(QI)\} = \{0\} \tag{30}$$

which is transformed to a linear algebra statement using a Taylor series, for example,

$$\{FI\} \equiv \{QI\}_{j+1} - \{QI\}_j - \Delta x_1 \{QI\}'_{j+\Theta} + \ldots = \{0\} \tag{31}$$

where superscript prime denotes the ordinary derivative and $\Theta > 0$ implies an implicit statement, since equation 30 is quite non-linear.

Hence, the final fully-discrete approximation error constraint statement is the non-linear algebraic equation set,

$$\{FI(k, \lambda, \Theta, \Delta x_1, \{QI\})\} = \{0\} \tag{32}$$

where $1 \leq I \leq 16$, see equation 1. The Newton algorithm solution for equation 32 is given in equations 2-4.

COMPUTATIONAL IMPLEMENTATION

Equations 1-4, see also 32, define the linear algebra problem statement of the penalty finite element algorithm. The complexity of the computational implementation of the algorithm, into a computer code, is a strong function of the few remaining basic decisions on the independent parameters defined in equation 32, namely k, λ, and Θ. The earliest work on this algorithm, cf. [10-11], used an explicit ($\Theta = 0$) integration procedure, whereupon equation 32 becomes linear, see equation 31, upon definition and use of "mass lumping" in equation 30. This construction was soon determined to be of extremely limited capability, and discarded in favor of an implicit formulation using, for example, the trapezoidal integration rule ($\Theta = ½$). Thereupon, procedures are required defined to embody a suitable (accurate and efficient) approximation to the Newton algorithm statement, equations 2-4.

The accuracy and complexity of this implicit development is significantly influenced by the choice of the particular semi-discrete approximation basis $\{N_k(x_\ell)\}$, recall equation 27. There are basically two choices, 1) the natural coordinate basis spanning element domains R_e^2 which are triangular in cross-section, or 2) the tensor product basis $\{N_k^+\}$ spanning element domains R_e^2 which are four-sided. In either case, k=1 yields a linear approximation on elements with straight sides, while k > 1 can employ elements with curved sides, the so-called isoparametric family. For k=1, and $\Theta = ½$, a linearized truncation error analysis would show that the basic algorithm is second-order accurate in space (x_ℓ), and in the initial-value coordinate (x_1) direction. For k=2, the linearized analysis predicts spatial fourth-order accuracy for the linear differential operators, for example the diffusion terms. Conversely, in the energy norm, the linearized analysis semi-discrete approximation error is bounded by the extremum aspect ratio element measure to the power 2k, and by Δx_1, to the exponent 2, cf. [9].

The computer program evaluation of the algorithm statement, equation 32, with k, λ, Θ, and Δx_1 fixed by this decision, and $\{QI\}_{j+1}^{p}$ available, see equation 2, is a straight-forward coding operation. On the other hand, the construction of a suitable approximation to the Jacobian, equation 4, is considerably more detailed. The exact evaluation of equation 4, for direct solution of equation 2, is prohibitively demanding on current (or near term) computer resources for all but trivially coarse discretizations. The basic structure of the Newton Jacobian is block-sparse, of the form

$$[J] = \begin{bmatrix} [J1,1] & [J1,2] & [J1,3] & \cdots & [J1,16] \\ [J2,1] & [J2,2] & \cdots & & \\ \cdot & & & & \\ \cdot & & & & \\ & & & & [J16,16] \end{bmatrix} \qquad (33)$$

where $[J1,1] \equiv [JU1,U1] \equiv \partial\{F1\}/\partial\{Q1\}$, for example, see equation 4. Each of the block Jacobians $[JI,J]$ has a bandwidth equal to 2M + 2, where M is the minimum member of nodes spanning a coordinate direction x_ℓ on R^2. For example, for a 21 x 21 nodal mesh, the bandwidth equals 44, and the rank of $[JI,J]$ equals 441. The size of $[J]$ in a direct formulation is clearly prohibitive.

There are any number of ways to approximate solution of the Newton algorithm statement, equations 2-4. A code designer with a finite difference background would probably employ a variant of an alternating direction implicit (ADI) procedure, wherein approximations would also be made in the construction of components of $\{FI\}$. A second approximate procedure, which would not compromise $\{FI\}$, could employ the decomposition of $[J]$ into its matrix tensor products in the form $[J_2] \otimes [J_3]$. This is easy to accomplish when employing the finite element tensor product basis, $\{N_k^+(x_\ell)\}$ in equation 27, and each of the matrices $[J_\ell]$ is either block tri-diagonal (k=1) or block penta-diagonal (k=2). This procedure is not reported in the literature for the 3DPNS equations, but is documented [12] for the complete Navier-Stokes equations.

A third alternative, which has been reduced to practice in an operational computer program [13], is to approximate equation 33 using only the diagonal block matrices $[JI,I]$. This decouples the dependent variables in the $\{\delta QI\}$ solution, but retains the dominant cross-coupling of the non-linear convection terms in equations 15, 17-19 and 16+23. Further, since there are three linear Poisson equations in the algorithm, a single (constant) Jacobian can be employed for multiple right-side evaluations. This is also possible with the six Reynolds stress component solutions, equation 20. This option was the choice in construciton of CMC:3DPNS computer code [13]. As cited in the Introduction, numerous data are reported in the literature comparing prediction to theory, and/or experimental data, including turbulent three-dimensional flows.

ACKNOWLEDGMENTS

The author wishes to acknowledge the significant contributions of Messrs. Joe Orzechowski and Paul Manhardt in development of the 3DPNS theory and the computer program embodiment. This long term research project has been principally supported by NASA Langley Research Center, Hampton, VA, USA.

REFERENCES

1. Baker, A.J., and Orzechowski, J.A., "An Interaction Algorithm For Three-Dimensional Turbulent Subsonic Aerodynamic Juncture Region Flow," AIAA Journal, V. 21, 1983, to appear.
2. Baker, A.J., "Why (?) a Finite Element Algorithm For the Parabolic Navier-Stokes Equations," in T. Cebeci (ed.), Proceedings, 2nd Sym. Numerical and Physical Aspects of Aerodynamic Flows, Cal. St. Univ./Long Beach, 1983.
3. Baker, A.J., Orzechowski, J.A., and Stungis, G.E., "Prediction of Secondary Vortex Flowfields Induced by Multiple Free-Jets Issuing in Close Proximity," Paper AIAA-83-0289, 1983.
4. Baker, A.J., and Orzechowski, J.A., "A Penalty Finite Element Method For Parabolic Flow Prediction," ASME, App. Mech. Div. AMD-Vol. 51, 1982, pp. 137-142.
5. Baker, A.J., Yu, J.C., Orzechowski, J.A., and Gatski, T.B., "Prediction and Measurement of Incompressible Turbulent Aerodynamic Trailing Edge Flows," AIAA Journal, V. 20, 1982, pp. 51-59.
6. Baker, A.J., and Orzechowski, J.A., "A Continuity-Constraint Finite Element Algorithm For Three-Dimensional Parabolic Flow Prediction," in Ghia, K.N. et.al. (eds), Proceedings Joint ASME-AIAA Sym. on Computers in Flow Predictions and Experiments, ASME, 1981, pp. 103-117.
7. Baker, A.J., and Orzechowski, J.A., "An Assessment of Factors Affecting Prediction of Near-Field Development of A Subsonic VSTOL Jet In Cross-Flow," U.S. Navy Technical Report NADC-81177-60, 1982.
8. Cebeci, T., and Smith, A.M.O., Analysis of Turbulent Boundary Layers, Academic Press, New York, 1974.
9. Baker, A.J., Finite Element Computational Fluid Mechanics, McGraw-Hill/Hemisphere, New York, 1983.
10. Baker, A.J., "Finite Element Solution Theory for Three-Dimensional Boundary Flows," Comp. Mtd. Appl. Mech. & Engr., V. 4, 1974, pp. 367-386.
11. Baker, A.J., Rogers, R.C., and Zelazny, S.W., "Analytical Study of Mixing and Reacting Three-Dimensional Supersonic Combustor Flowfields," NASA Report SP-347, 1975, pp. 251-315.
12. Baker, A.J., and Soliman, M.O., "On A Finite Element Algorithm For Computational Fluid Dynamics," AIAA Journal, V. 21, 1983, to appear.
13. Manhardt, P.D., "The CMC:3DPNS Computer Program For Prediction of Three-Dimensional, Subsonic, Turbulent Aerodynamic Juncture Region Flow - Volume 1 - User's Guide," NASA Report CR-165997, 1982.

THE LIMITING CASE OF RODS TOUCHING IN TURBULENT FLOW THROUGH ROD BUNDLES

C. W. Rapley

Department of Mechanical Engineering
Sunderland Polytechnic, U.K.

SUMMARY

An orthogonal-curvilinear-mesh-based finite volume calculation method has been applied to the problem of turbulent flow in the tri-cusped cornered duct formed when parallel circular rods touch in triangular array. Algebraic stress relations combined with the k - ε turbulence model are used for calculation of the required stresses. A single circulation of turbulence driven cross-plane secondary flow from the core into the duct corner has been predicted in a one-sixth symmetry region of the duct and the convective transport effects of this flow are seen to have much influence on local mean flow distributions. The turbulence predicted by the k - ε model showed significant damping in the cusped corner region where turbulent viscosities approached the laminar value. Satisfactory agreement was obtained with the limited local and overall mean flow measurements available.

1 INTRODUCTION

The problem of fully developed turbulent flow in closely spaced triangular array rod bundles has received much attention from both experimenters and analysts with its important applications in nuclear reactor and other compact heat exchanger systems. When the clearance between the rods becomes very small, as occurs in the most compact arrangements, the peripheral variations in axial velocity and wall shear stress become large and thus feature markedly in design considerations.

Experimental evidence of the local mean flow is limited for rod pitch/diameter (P/D) ratios below 1.1 and theoretical analysis is hampered by the difficult geometry of the passages and the significant influence of turbulence driven cross-plane secondary flow which makes the prediction of a three-dimensional velocity field usually necessary. No theoretical calculations which include prediction of the cross-plane secondary flow appear to be available for P/D ratios below 1.1 and this situation has been the main spur to the present work. Of particular interest is the limiting case of P/D = 1.0 where rods touch and form ducts with cusped corners since this should represent one of the severest tests of a rod bundle calculation method. In the present work a general prediction method for fully developed non-circular passage turbulent flows has been applied to the compact triangular array rods touching arrangement which yields the passage shapes shown in figure 1.

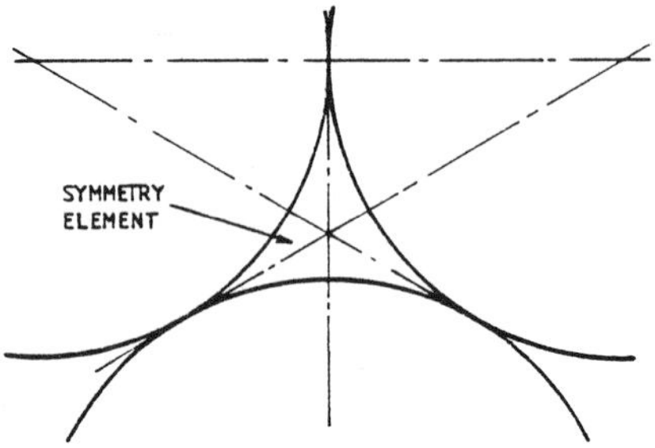

Figure 1. Passage shape

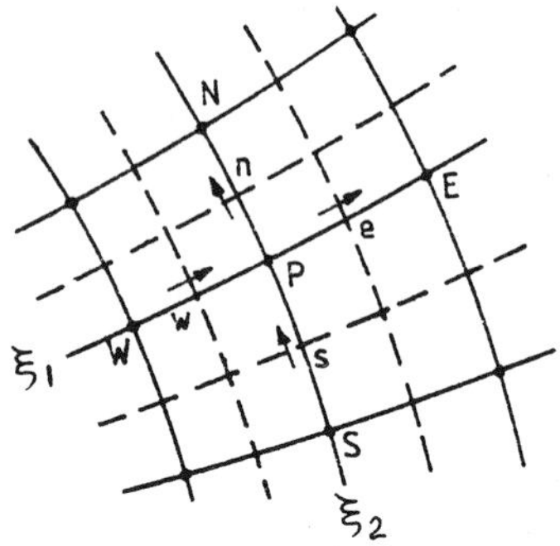

Figure 2. Orthogonal mesh

Experimental work on turbulent flow in this duct geometry is mainly confined to friction factor measurements [1,2,3,4] with conflicting results which range from 30% to 60% below the circular duct data, using equivalent diameter. The only local measurements that appear to be available are the axial velocity and wall shear stress distributions reported by Levchenko et al [1]. These measurements, which form the main basis of comparison with the present predictions, did not include secondary flow measurement but yielded axial velocity contours which bulge markedly into the duct corners, thus implying convective transport by secondary motion from the core towards the corners. These effects, including the implied secondary flow circulation, will be discussed in more detail later.

The present prediction method solves the Reynolds equations by finite volumes on an orthogonal curvilinear mesh which is generated numerically to fit the duct cross-section. The turbulent (Reynolds) stresses required are calculated with an algebraic stress transport model (ASTM) which was first derived for square duct calculations by Launder and Ying [5] and Gessner and Emery [6]. In Cartesian co-ordinates, with directions x_1 and x_2 in the duct cross-plane and x_3 in the axial direction, the ASTM yields the following kinetic stress relations :

$$
\begin{aligned}
\overline{u_1'^2} &= C_3k - C_2C_4(k^3/\varepsilon^2)(\partial u_3/\partial x_1) \\
\overline{u_2'^2} &= C_3k - C_2C_4(k^3/\varepsilon^2)(\partial u_3/\partial x_2) \\
\overline{u_1'u_2'} &= -C_2C_4(k^3/\varepsilon^2)(\partial u_3/\partial x_1)(\partial u_3/\partial x_2) \\
\overline{u_1'u_3'} &= -C_4(k^2/\varepsilon)\partial u_3/\partial x_1 \\
\overline{u_2'u_3'} &= -C_4(k^2/\varepsilon)\partial u_3/\partial x_2
\end{aligned}
\qquad (1)
$$

where u and u' are time-averaged mean and fluctuating component velocities respectively and C_2, C_3 and C_4 are related constants. The turbulence kinetic energy k and its dissipation rate ε were calculated here with the $k - \varepsilon$ two equation turbulence model. The cross-plane stresses $\rho\overline{u_1'^2}$, $\rho\overline{u_2'^2}$ and $\rho\overline{u_1'u_2'}$ are seen to depend on axial strain rates and the axial shear stress ($\rho\overline{u_1'u_3'}$ and $\rho\overline{u_2'u_3'}$) relations are seen to be of conventional turbulent viscosity form where

$$\text{turbulent viscosity } \mu_t = C_4k^2/\varepsilon \qquad (2)$$

Full details of the mathematical problem and the present solution procedure are set out in Rapley [7] so only the following brief outline will be given here.

2 THE GOVERNING EQUATIONS

The Reynolds equations for steady time-averaged incompressible turbulent flow can be written in Cartesian tensor form as

$$\partial(\rho u_i u_j)/\partial x_i = -\partial p/\partial x_j + \partial(T_{ij})/\partial x_i \qquad (3)$$

The continuity equation is

$$\partial(\rho u_i)/\partial x_j = 0 \qquad (4)$$

The stress tensor T_{ij} represents the sum of the viscous and turbulent (Reynolds) stresses i.e.

$$T_{ij} = \mu(\partial u_i/\partial x_j + \partial u_j/\partial x_i) - \rho\overline{u_i'u_j'} \qquad (5)$$

The Reynolds stresses $\rho\overline{u_i'u_j'}$ were calculated with the ASTM (equations (1)) which requires values of k and ε. These were obtained here from their modelled partial differential transport equations as conventionally used in the $k - \varepsilon$ turbulence model viz:

$$\partial(\rho u_i k)/\partial x_j = \partial((\mu_t/\sigma_k)\partial k/\partial x_i)/\partial x_i + P - \rho\varepsilon \qquad (6)$$

$$\partial(\rho u_i \varepsilon)/\partial x_i = \partial(\mu_t/\sigma_\varepsilon)\partial\varepsilon/\partial x_i)/\partial x_i + \varepsilon(C_{\varepsilon 1}P - C_{\varepsilon 2}\rho\varepsilon)/k \qquad (7)$$

where P is the production rate of k, given as

$$P = -\rho\overline{u_i'u_j'}(\partial u_i/\partial x_j) \qquad (8)$$

The transport equations (3), (6) and (7), transformed to general orthogonal co-ordinate form (e.g. [8]), expanded for each direction and specialised to fully developed flow can all be written in the following common form :

$$\partial(h_2\rho u_1\phi)/\partial\xi_1 + \partial(h_1\rho u_2\phi)/\partial\xi_2 = (h_2D_\phi\partial\phi/h_1\partial\xi_1)/\partial\xi_1 + (h_1D_\phi\partial\phi/h_2\partial\xi_2)/\partial\xi_2 + C_\phi \quad (9)$$

where ϕ stands for either u_1, u_2, u_3, k or ε. Co-ordinate directions ξ_1 and ξ_2 are in the duct cross-plane and have metric coefficients h_1 and h_2 respectively. Co-ordinate ξ_3 is in the straight axial direction. The exchange coefficients D_ϕ and source terms C_ϕ appropriate to each variable ϕ are summarised in table 5.3.1 in [7]. The turbulent stresses required were calculated with an orthogonal co-ordinate form of the ASTM given in equation (1).

The statement of the mathematical problem is completed with specification of the boundary conditions. The commonly used method of employing wall functions to bridge the gap between the interior solution and boundary surfaces was also adopted here with conventional functions derived from assumptions of one-dimensionality and local equilibrium in the wall region. This led to the familiar logarithmic 'velocity law of the wall' from which appropriate functions for the other variables were also derived [7,9]. Experimental measurements have confirmed the applicability of the logarithmic velocity law in near wall regions in a wide range of non-circular passages [10].

3 THE SOLUTION METHOD

The finite volume mesh consisted of orthogonally intersecting main grid lines in the passage cross-plane with straight axial lines passing through the points of intersection to complete the three dimensional mesh. A typical portion of cross-plane mesh is shown in figure 2 with the main grid lines as full lines. The intersections form the main nodes with a typical such node designated 'P' and surrounded by nearest neighbours 'N', 'E', 'S' and 'W'. The main nodes are surrounded by contiguous control volumes, the cross-plane boundaries of which are indicated with broken lines. These control volume boundaries are placed mid-way between the main grid lines. Variables p, u_3, k and ε were calculated at the main grid nodes whereas the cross-plane velocities u_1 and u_2 were calculated at intermediate locations designated by the intersections between the control volume boundaries and the main grid lines in the direction of the velocity component e.g. in figure 2, u_1 was calculated at w, e etc. whereas u_2 was calculated at n, s etc.. The cross-plane boundaries of the control volumes for u_1 and u_2 are formed by planes represented by the nearest grid lines and main grid control volume boundaries. This staggered grid arrangement is conventional practice in finite volumes for fluid flow since it enhances stability and convergence of the solution [11]. With the inherent non-uniform spacing of orthogonal meshes, much care was needed to ensure that the proper momentum was calculated at each required control volume face.

The finite volume form of the common transport equation (equation (9)) was obtained by linearising the source and integrating each term over the appropriate control volume across the interior calculation domain. This micro-integration technique ensured that the relevant conservation principles were satisfied and was carried out in such a way that all areas and volumes were expressed in terms of curvilinear arc lengths in the mesh. This led to conventional finite volume equations with coefficients that contained proportions of convection and diffusion transport according to the differencing scheme. In the present work a hybrid scheme was used which was essentially central differencing with the provision to switch to upwind when convection dominated [e.g. 12].

The finite volume equations were solved with a conventional line-by-line ADI method based on the tri-diagonal matrix algorithm [e.g. 13]. The cross-plane momentum equations were handled with the 'SIMPLE' method [12] in which the continuity equation was used to derive pressure corrections which were in turn used to correct the velocities. Due to the coupling and non-linearity of the equations to be solved, convergence of the solution was found to be uncertain. These difficulties were also experienced by previous workers solving the full 3D momentum equations in other passage geometries [e.g. 14,15,16], who were eventually forced to simplify or to prescribe the sign of sources in the equations containing cross-plane stresses, which were the main source of the instabilities. In the present method however, no simplifications were made or restrictions imposed as convergence of the solution was obtained through careful linearising of source terms together with extensive under-relaxation, programme control and the use of block adjustment [7,17].

The various empirical constants used were taken from previous work [5,7,9,15,17,18] and were: $C_2 = 0.013$, $C_3 = 0.562$, $C_4 = 0.085$, $\sigma_k = 1.0$, $\sigma_\varepsilon = 1.34$, $C_{\varepsilon 1} = 1.55$, $C_{\varepsilon 2} = 2.0$.

4 PREDICTIONS

Calculations were made in a one-sixth symmetry element of a triangular array rod bundle with rods touching, as indicated in figure 1. A finite difference method was used to generate the orthogonal curvilinear mesh fitted into the cross-plane. In this method Cartesian co-ordinates x_1 and x_2 are related to the general orthogonal co-ordinates ξ_1 and ξ_2 by the equations [19]:

$$\left.\begin{aligned}
&\alpha_2 \partial^2 x_1/\partial\xi_1{}^2 + \alpha_1 \partial^2 x_1/\partial\xi_2{}^2 = 0 \\
&\alpha_2 \partial^2 x_2/\partial\xi_1{}^2 + \alpha_1 \partial^2 x_2/\partial\xi_2{}^2 = 0 \\
&\text{where} \\
&\alpha_1{}^2 = (\partial x_1/\partial\xi_1)^2 + (\partial x_2/\partial\xi_1)^2 \\
&\alpha_2{}^2 = (\partial x_1/\partial\xi_2)^2 + (\partial x_2/\partial\xi_2)^2
\end{aligned}\right] \qquad (11)$$

These equations were solved by finite differences using a similar micro-integration technique and ADI algorithm to that used in the flow solution. Boundary conditions were calculated by imposing the Cauchy-Riemann conditions mid-way between the boundary and the nearest internal nodes and simultaneously with the boundary shape equation [7]. The nodes of a mesh generated in this way for the present case are located at the tails of the vectors shown in figure 3.

Numerical accuracy of the flow solution was tested with laminar flow calculations, grid refinement and comparisons of solutions obtained with meshes of different spacings. The laminar flow solutions obtained were found to be within ½% of previously published point matching solutions [20]. In turbulent flow calculation, careful differencing of the source terms on the orthogonal curvilinear mesh was found necessary, particularly the terms containing cross-plane stresses, to avoid local effects due to mesh geometry. Extensive symmetry tests with different passage geometries were needed to establish differencing methods that eliminated these effects [7]. With the present passage shape, tests with different meshes were used to obtain mean flow solutions that were substantially independent of the mesh. As may be expected, the secondary flow field was found to be more sensitive to mesh changes than mean flow particularly the regions where nodes became sparse. Of particular interest was the calculated mean flow and turbulence fields in the cusped corner region where the near wall node was close to the viscous sub-layer and thus could not be considered in the fully turbulent region as assumed in the modelling of the turbulence equations and in the logarithmic velocity law based wall functions. However, as will be shown, the $k - \varepsilon$ model yielded low values of turbulent viscosity in this region and whether the logarithmic law or a simple laminar flow relation was used in the near wall region, negligible changes occurred in the solution. Thus no special modifications of the turbulence model were found necessary to cope with this region. Check calculations were made using a range of meshes in the corner region and only negligible changes occurred in the main flow solution. It appears that this relatively stagnant region has only a minor influence on the main flow.

A single circulation of secondary flow was predicted in the symmetry element as seen in the plot of calculated secondary velocity vectors in figure 3. Flow is from the core into the corner, recirculating to the core via the wall with maximum secondary velocities of 1½% of the mean axial velocity, occurring along the corner bisector and near the wall. Although no measurements are available for this geometry, this pattern is similar to that measured by Tahir and Rogers [21] in a triangular array rod bundle with P/D = 1.06 and is consistent generally with observed secondary flow which is from the core into the corners in non-circular passages [17]. The predicted pattern of figure 3 can thus be considered plausible.

The effect of this circulation on the axial velocity contours can be seen in figure 4 which compares predictions with experiment and with calculations with secondary flow suppressed. The predictions are seen to be in good agreement with experiment and show marked bulging of the contours into the duct corners due to the convective transport of flow into the corner from the core by secondary motions. An opposite effect can be seen in the near wall region towards $\theta = 30^{o}$ where the contours bend away from the wall due to secondary flow convection of near wall fluid in that direction. Both of these effects appear to be well predicted.

More detail of the axial velocity field is given in figure 5 which compares profiles plotted along the main symmetry planes and confirms the satisfactory agreement between predictions and measurements. This agreement is continued in the wall shear stress profiles displayed in figure 6.

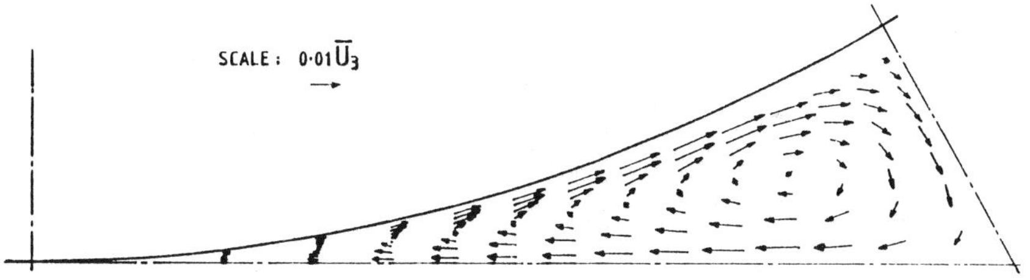

Figure 3. Secondary velocity vectors, R_e = 95000.

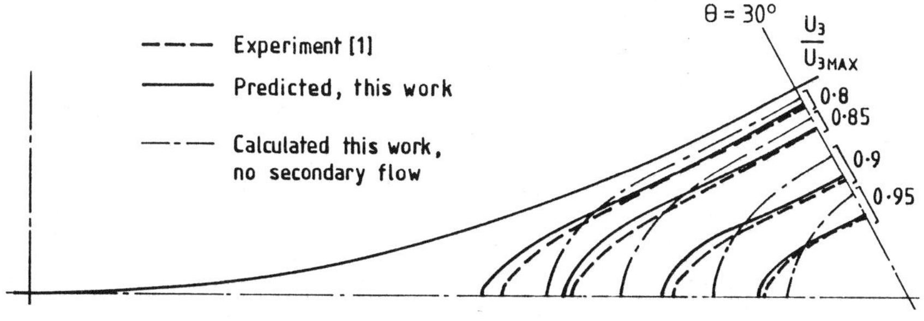

Figure 4. Axial velocity contours, R_e = 42600.

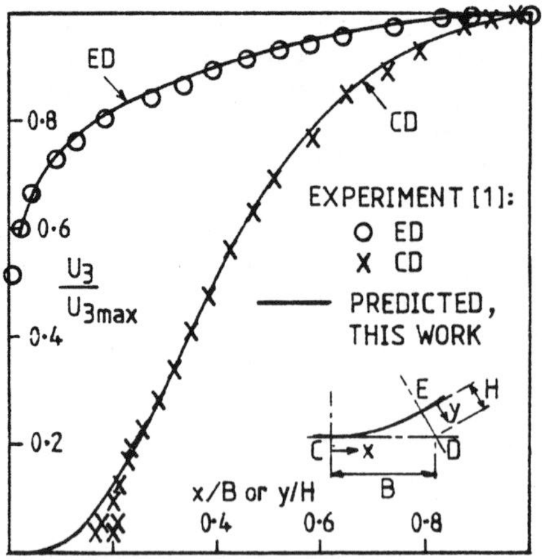

Figure 5. Axial velocity profiles, R_e = 42600.

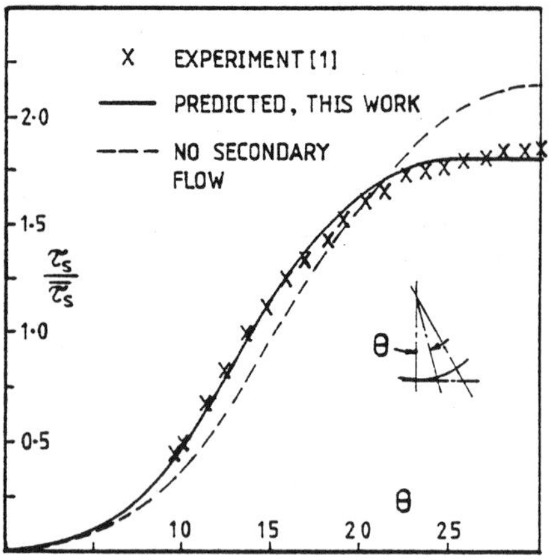

Figure 6. Wall shear stress, R_e = 95000.

The effect of secondary flow in reducing the peripheral variation is evident with a reduced shear stress as θ approaches 30°, due to the peripheral coupling of the flow field provided by the secondary flow and also to the reduced axial velocity gradients in wall region near θ = 30° due to convective transport by secondary motion away from the wall.

Figures 7 and 8 give some details of the turbulence field as predicted by the k - ε turbulence model. Unfortunately there are no experimental measurements available for comparison. The increased axial velocity gradients into the corner along the corner bisecting centre plane have increased turbulence kinetic energy generation and have thus shifted the point of maximum k nearer to the corner as seen in figure 7. The levels of turbulence kinetic energy and turbulent viscosity are seen to decay rapidly from this point into the corner to give viscosities approaching laminar in the final 20% of the centre plane. This is the expected behaviour, due to the damping

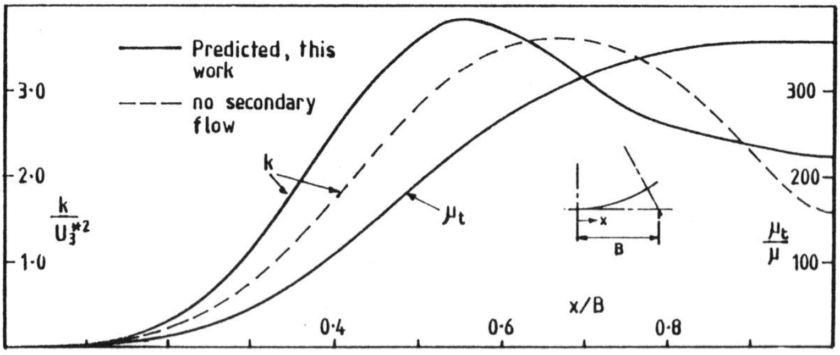

Figure 7. Turbulence profiles along centre-plane, Re = 95000.

effect of the walls in this cusped corner region and is consistent with the previously mentioned checks made with wall functions at the near wall node in this region, where replacing the logarithmic velocity law by a simple laminar relation had only a negligible effect on the solution.

Predicted turbulence kinetic energy and cross-plane normal stress profiles along the $\theta = 30^{o}$ radial plane are shown in figure 8. As may be expected from the features of mean flow already discussed in this region, secondary flow has significantly reduced the level of k in this wall region. The calculated cross-plane normal stresses $u_1{}^2$ tangential to and $u_2{}^2$ normal to the rod surface show some of the expected anisotropy due to the damping effect of the surface on $u_3{}^2$, but only close to the wall.

Finally, the predicted friction factor characteristic is compared with the available measurements in figure 9 where it is seen to be in reasonable agreement with the measurements of Eifler and Nijsing [3] and Levchenko [1]. The Sutherland and Kays [4] measurements appear to be significantly lower than these, a result for which there does not appear to be any explanation - particularly since the measured characteristics for P/D ratios of 1.15 and 1.25 presented by them on the same plot are in fair agreement with other published measurements for similar P/D ratios. The characteristics in figure 9 are seen to be all well below the circular duct data as represented

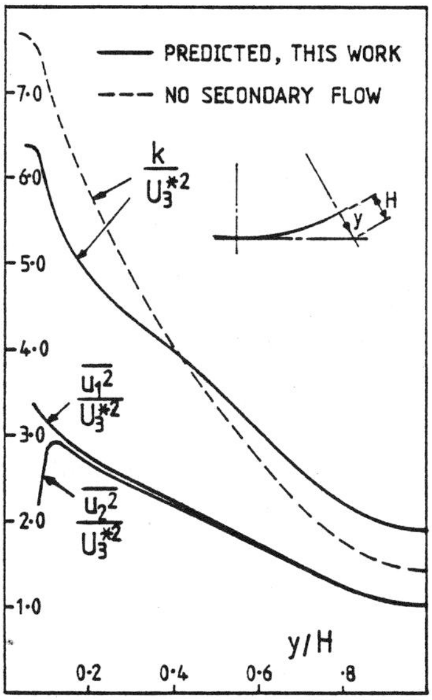

Figure 8. Turbulence profiles along radial plane, $R_e = 95000$.

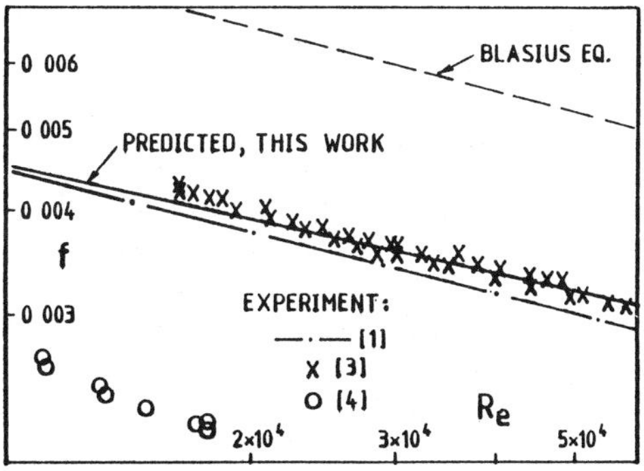

Figure 9. Friction factors

by the Blasius equation thus indicating that the equivalent diameter concept would over-estimate axial pressure drop by more than 70% in this case. This is not surprising since the cusped shape of the duct gives rise to significant variations in wall shear stress, as seen in figure 6, which makes it far removed from the circular duct case.

5. CONCLUSIONS

The prediction method developed for fully developed flow in arbitrary shaped passages has been successfully applied to the difficult case of turbulent flow in the cusped cornered duct formed when rods touch in triangular array. A single swirl of secondary flow was predicted in a symmetry one-sixth of the duct and good agreement between predicted and measured local and overall mean flow has been obtained with the convective transport effects of secondary flow clearly evident in the predictions. The $k - \varepsilon$ two equation turbulence model, which was coupled with the algebraic stress transport model for stress calculations, appears to have performed adequately. The difficulties expected in the cusped corner region which should contain transition flow did not materialise in the solutions obtained which yielded plausible turbulence fields. However, since the turbulence model used was valid only for fully turbulent flow it cannot be expected to calculate transition flow properly and the implication is that the relatively stagnant cusped corner region flow does not have a significant influence on flow in the remainder of the duct.

REFERENCES

1. LEVCHENKO, Y. D. et al - The Distribution of Coolant Velocities and Wall Stresses in Closely Packed Rods. Soviet Atomic Energy, Vol.22, p 262, 1968.
2. SUBBOTIN, V. I., USHAKOV, P. A. and GABRIANOVICH, B. N. - Hydraulic Resistance to the Flow of a Liquid Along a Bundle of Rods. Soviet J. Atomic Energy, Vol. 9, p 848, 1961.
3. EIFLER, W. and NIJSING, R. - Experimental Investigation of Velocity Distribution and Flow Resistance in a Triangular Array of Parallel Rods. Nuclear Eng. and Design, Vol.5, p 22, 1967.
4. SUTHERLAND, W. A. and KAYS, W. M. - Heat Transfer in Parallel Rod Arrays. Report GEAP-4637, 1965.
5. LAUNDER, B. E. and YING, W. M. - Prediction of Flow and Heat Transfer in Ducts of Square Section. Proc. I.Mech.E., Vol.187, p 455, 1973.
6. GESSNER, F. B. and EMERY, A. F. - A Reynolds Stress Model for Turbulent Corner Flows, Part 1. J. Fluids Eng., Trans. ASME 76-FE-C, 1976.
7. RAPLEY, C. W. - Fluid and Heat Flow in Tubes of Arbitrary Cross-Section, Ph.D. Thesis, University of London, 1980.
8. POPE, S. B. - The Calculation of Turbulent Recirculating Flows in General Orthogonal Co-ordinates. J. Comp. Phys., Vol.26, p 197, 1978.

9. LAUNDER, B. E. and SPALDING, D. B. - Mathematical Models of Turbulence, Academic Press, 1972.
10. RAPLEY, C. W. - A Summary of Experimental Turbulent Non-Circular Passage Flow and Heat Transfer. Mech. Eng. Report FS/80/41, Imperial College, London, 1980.
11. HARLOW, F. H. and WELCH, J. E. - Numerical Calculation of Time Dependent Viscous Incompressible Flow of Fluids with Free Surface. Phys. of Fluids, Vol.8, p 2182, 1965.
12. CARETTO, L. S. et al - Two Calculation Procedures for Steady Three - Dimensional Flows with Recirculation. Proc. 3rd Int. Conf. Numerical Meth. in Fluid Mech., Paris, p 60, 1972.
13. AMES, F. A. - Numerical Methods for Partial Differential Equations, Academic Press, 1977.
14. CARAJILESCOV, P. and TODREAS, N. E. - Experimental and Analytical Study of Axial Turbulent Flows in an Interior Sub-Channel of a Bare-Rod Bundle. J. Heat Trans., Trans. ASME, Vol.98, p 262, 1976.
15. ALY, A. M. M., TRUPP, A. C. and GERRARD, A. D. - Measurement and Prediction of Fully Developed Flow in an Equilateral Triangular Duct. J. Fluid Mech., Vol.85, p 57, 1978.
16. TRUPP, A. C. and ALY, A. M. M. - Predicted Turbulent Flow Characteristics in Triangular Rod Bundles. Mech. Eng. Report ER25.26, University of Manitoba, 1978.
17. RAPLEY, C. W. - The Simulation of Turbulent Flow and Heat Transfer in Bare Rod Bundles and Other Non-Circular Passages. vKI Lecture Series Comp. Fluid Dynamics, Belgium, 1982.
18. GOSMAN, A. D. and RAPLEY, C. W. - Fully Developed Flow in Passages of Arbitrary Cross-Section, Recent Advances in Numerical Methods in Fluids, Ed. Taylor, C. and Morgan, K., Pineridge Press, Swansea, 1980.
19. GOSMAN, A. D. and JOHNS, R. - A Simple Method for Generating Curvilinear-Orthogonal Grids for Numerical Fluid Mechanics Calculations. Mech. Eng. Report FS/79/23, Imperial College, 1979.
20. AXFORD, R. A. - Multi-Region Analysis of Temperature Fields in Reactor Tube Bundles. Nuc. Eng. and Design, Vol.6, p 25, 1967.
21. TAHIR, A. and ROGERS, J. T. - The Mechanism of Secondary Flows in Turbulent Interchange in Rod Bundles. Proc. 7th Canadian Cong. App. Mech., p 773, 1979.

METHOD OF INTEGRAL RELATIONS FOR CURVED COMPRESSIBLE TURBULENT MIXING LAYERS WITH LATERAL DIVERGENCE

by

P.N. Green(1) and P.W. Carpenter(2)

(1) Research Student, (2) Lecturer.

Department of Engineering Science, Exeter University, Exeter, U.K.

SUMMARY

A method is presented for predicting the development of mixing layers subjected to extra strain rates. The calculations are performed using an adaptation of the Method of Integral Relations, in which one strip is used and the velocity profile is approximated by an exponential function. A modified version of the Prandtl-Görtler eddy-viscosity model is used, which takes into account the effects of upstream conditions and the extra strain rates due to curvature, lateral divergence and mean flow dilation. Results are presented showing the development of mixing layer width and maximum shear stress.

1. INTRODUCTION

The present study of the mixing layer region near the exit slot of the Indair gas flare(I) has been undertaken as part of a research programme on the noise sources in such gas flares. The flare tip is a tulip-shaped body of revolution with gas emerging from an annular nozzle at the "tulip" base. The gas adheres to the surface by means of the Coanda effect, and near the exit slot a mixing layer forms as shown in Fig.1. The mixing layer depicted in Fig.1 is subjected to other rates of strain besides simple shear. These arise due to the effects of streamline curvature, radial expansion and compressibility. In a recent analysis [1] of turbulent mixing noise emitted by the flare, these extra strain rates were neglected. A comparison between theoretical predictions and data measured in the field on Indair flares showed that the geometric scaling of the noise is underestimated by the theory. The present results indicate that theory and experiment would compare more favourably when the effects of extra strain rates on the turbulence are taken into account.

(I) The Indair flare is designed and marketed by Kaldair Ltd., U.K.

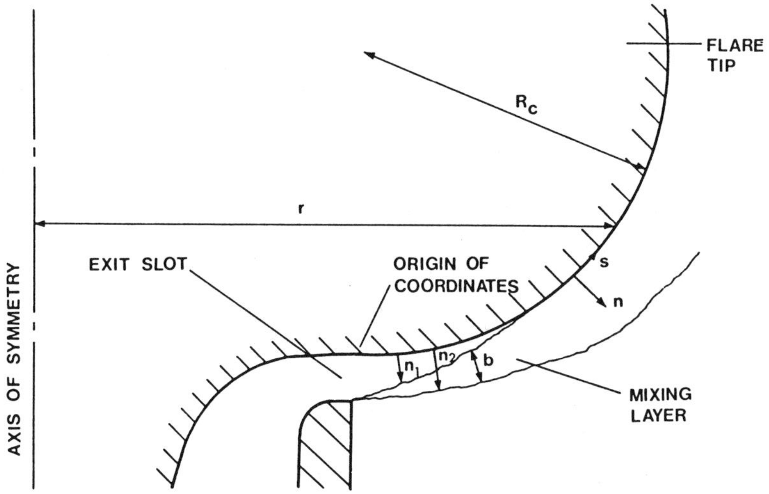

Fig.1 Flare Geometry and Coordinate System.

2. THE EFFECTS OF EXTRA STRAIN RATES ON TURBULENCE

It is well known that extra rates of strain have significant effects on turbulence [2]. In general, it seems that extra strain rates cause changes in Reynolds stress roughly ten times greater than would be expected by examining the additional terms that appear in the governing equations. The physics of these effects are unclear, and so Bradshaw [2] has proposed a semi-empirical method of accounting for them. He has used this technique, in conjunction with his own calculation procedure [3], to yield substantially improved agreement between prediction and experiment for a number of distorted shear flows [2].

Bradshaw has suggested that the effects of extra strain rates can be modelled by multiplying the turbulence length scale by a factor of the form

$$F = 1 + \alpha e/(\partial\bar{u}/\partial y) \tag{1}$$

where e is an additional strain rate and α is a constant of O(10). As this factor takes no account of upstream events, it was proposed that αe, written below as E, should be obtained from the following semi-empirical equation

$$2b(dE/dx) = \alpha_o e - E \tag{2}$$

where b is the mixing layer width and α_o is the value of α obtained after a prolonged application of e. Eqn (2) allows for a lag between the application of an extra strain rate, and a change in the turbulence level. Eqns (1) and (2) form the basis of the present method of accounting for extra strain rates.

3. THE GOVERNING EQUATIONS

The compressible turbulent boundary layer equations have been derived in (s,n) coordinates (Fig.1). Surface curvature is assumed to be small, so $\partial\bar{p}/\partial n$ is neglected, as is viscous shear stress. The present calculations are for comparison with model tests, so the gas in the jet is assumed to be air. The turbulent Prandtl number is taken as unity, and the jet's stagnation temperature is assumed to be equal to that of the surroundings. Using the eddy diffusivity concept, it can be shown, using the same assumptions as [4], that the density profile is given by

$$\rho_a/\rho = 1 - m(\bar{u}/u_e)^2 \; ; \qquad m = \tfrac{1}{2}(\gamma-1)(u_e^2/a_a^2) \tag{3}$$

where ρ_a and a_a are the ambient density and sound speed respectively. The subscript e denotes quantities at the mixing layer's inner edge. The non-dimensional governing equations are written in (ξ,η) coordinates, where $\xi = s/h_s$ and $\eta = (n-n_1)/b$, and take the form:

$$\text{Continuity:} \quad \frac{\partial}{\partial\xi}(\hat{r}RU) - \hat{r}\left(\frac{1}{\hat{b}}\frac{d\hat{n}_1}{d\xi} + \frac{\eta}{\hat{b}}\frac{d\hat{b}}{d\xi}\right)\frac{\partial}{\partial\eta}(RU) + \frac{\hat{r}}{\hat{b}}\frac{\partial}{\partial\eta}(RV) = 0 \tag{4}$$

$$\text{Momentum:} \quad RU\frac{\partial U}{\partial\xi} - \left(\frac{1}{\hat{b}}\frac{d\hat{n}_1}{d\xi} + \frac{\eta}{\hat{b}}\frac{d\hat{b}}{d\xi}\right)RU\frac{\partial U}{\partial\eta} + \frac{RV}{\hat{b}}\frac{\partial U}{\partial\eta} = \frac{1}{\hat{b}}\frac{\partial}{\partial\eta}\left(R\kappa\frac{\partial U}{\partial\eta}\right) \tag{5}$$

Hatted quantities have been non-dimensionalized with respect to slot width h_s, and $R = \rho/\rho_a$, $U = \bar{u}/u_e$, and $V = \{1 + \overline{\rho' v'}/(\bar{\rho}\bar{v})\}\bar{v}/u_e$. The boundary conditions are:

$$U = 0 \quad \text{for} \quad \eta > \hat{n}_2 \; ; \quad U = 1 \quad \text{for} \quad \eta < \hat{n}_1 \tag{6}$$

$$V = V_e = -\hat{n}_1(d\hat{r}/d\xi)/\hat{r} \quad \text{at} \quad \eta = \hat{n}_1 \tag{7}$$

(7) is obtained by applying a mass balance across the potential core, ignoring the thin surface boundary layer.

Eddy viscosity is given by

$$\varepsilon = \kappa(\xi)\hat{b}\, h_s u_e \tag{8}$$

which is a modification of the Prandtl-Görtler model. The introduction of a variable κ accounts for upstream effects and is derived from the turbulence kinetic energy equation. Using the approximations and assumptions of [3], the TKE equation can be converted into an equation for $\hat{\tau} = \tau/(\rho U_e^2)$. In (ξ,η) coordinates it becomes:

$$RU\frac{\partial\hat{\tau}}{\partial\xi}-\left(\frac{1}{\hat{b}}\frac{d\hat{n}_1}{d\xi}+\frac{\eta}{\hat{b}}\frac{d\hat{b}}{d\xi}\right)RU\frac{\partial\hat{\tau}}{\partial\eta}+\frac{RV}{\hat{b}}\frac{\partial\hat{\tau}}{\partial\eta}=\frac{a_1R\hat{\tau}}{\hat{b}}\frac{\partial U}{\partial\eta}-\frac{a_1}{\rho_a U_e^3\hat{b}}\frac{\partial\Lambda}{\partial\eta}-\frac{a_1R\hat{\tau}^{3/2}}{\hat{L}} \tag{9}$$

where $\Lambda = \bar{\rho}\,\overline{kv'} + \overline{\rho' kv'} + \overline{p'v'}$. a_1 is defined by the assumption [3] that

$$\tau/(2\rho k) = a_1 = \text{constant}.$$

$\hat{L} = 0.089\hat{b}$ and is non-dimensional dissipation length scale. It is $\hat{L}$ that must be multiplied by F (Eqn (1)) to account for the extra strain rates. The F factor used has the form

$$F = 1 + \left[E_{\ell d} + E_d + \left(1 + \frac{(\gamma-1)}{2}M_{av}^2\right)E_c\right] \Big/ \frac{\partial\bar{u}}{\partial n} \tag{10}$$

$E_{\ell d}$ is the value of αe for lateral divergence, E_d that for dilation and E_c the value for curvature; all are obtained from equations like (2). The factor premultiplying E_c accounts for the fact that compressibility slightly amplifies the curvature effect [2]. M_{av} is the average Mach number across the mixing layer. The extra strain rates are given by:

$$e_c = -0.5\, u_e/R_c \tag{11}$$

$$e_{\ell d} = 0.5\, u_e(dr/ds)/r \tag{12}$$

$$e_d = \text{div}\, \underline{V} = \frac{u_e}{b}\left(\frac{d\hat{n}_1}{d\xi}(1-R_e) + \frac{d\hat{b}}{d\xi}(0.5-I_5) + \frac{\hat{b}}{\hat{r}}(0.5-I_5)\frac{d\hat{r}}{d\xi} + \frac{\hat{n}_1}{\hat{r}}(1-R_e)\frac{d\hat{r}}{d\xi}\right) \tag{13}$$

$$I_5 = \int_0^1 RU d\eta \tag{14}$$

(11) and (12) can be derived by considering the distortion of a fluid element, while (13) is the result of using the velocity profile given in (16). All three expressions have been averaged across the mixing layer. The integral I_5 can be evaluated analytically by using the profiles (16) and (17).

The problem of the mixing layer in Fig.1 thus involves the solution of equations (4) and (5), and equation (9) with $\hat{L}$ multiplied by F. The method is described in the next section.

4. ANALYSIS

The calculations are based on the method of integral relations (MIR) as adapted by Carpenter [4]. For a detailed discussion of the MIR see Holt [5]. The basic approach is to reduce eqns (4), (5) and (9) to a set of ordinary differential equations which can be solved by some suitable technique.

The analysis proceeds by defining a set of linearly independent weighting functions $f_j(U)$ ($j=0,1,2.....$). Eqn (4) is multiplied by f_j, (5) by $\hat{r}f_j'$ (the dash indicates differentiation with respect to U) and the resulting equations added and integrated across the mixing layer:

$$\frac{d}{d\xi}\int_o^1 \hat{r}f_j RU d\eta + \frac{1}{\hat{b}}\frac{d\hat{b}}{d\xi}\int_o^1 \hat{r}f_j RU d\eta - \frac{\hat{r}}{\hat{b}}\frac{d\hat{n}_1}{d\xi}\Big[f_j RU\Big]_o^1$$

$$+ \frac{\hat{r}}{\hat{b}}\Big[f_j RV\Big]_o^1 = -\frac{\hat{r}}{\hat{b}}\int_o^1 R\kappa f_j'' \left(\frac{\partial U}{\partial \eta}\right)^2 d\eta \qquad (14)$$

The weighting functions are required to satisfy the condition

$$f_j(1) = 0 \;;\quad j = 0,1,2, \ldots\ldots$$

So, let $\quad f_j = U(U-1)^j \;;\quad j = 0,1,2 \ldots\ldots \qquad (15)$

At this point a functional form for U must be chosen. There is evidence ([6] and [7]) that the velocity profiles of distorted jets differ very little from those of plane jets. This suggests that the velocity profile of the mixing layer in Fig.1 should not be significantly different from that in a plane mixing layer, and so U is assumed to be given by:

$$U = \frac{1}{2}\big(1 - \tanh(k_1\eta + k_2)\big) \qquad (16)$$

where $k_1 = 2\ \mathrm{arctanh}(0.98)$ and $k_2 = -\ \mathrm{arctanh}(0.98)$. This expression has been successfully used by Carpenter [4] in calculating the properties of compressible plane and axisymmetric mixing layers with heat transfer. From (6) it is seen that

$$R = 1/(1-mU^2) \qquad (17)$$

Substituting (16) and (17) into (14), it is clear that the derivatives of only two unknowns remain. This suggests that only two integral relations are required, so j = 0 and 1 in eqn (15). The resulting integral relations are:

$$\frac{d\hat{b}}{d\xi} = -\frac{\hat{b}}{\hat{r}}\frac{d\hat{r}}{d\xi} - \frac{2\kappa I_2}{I_1 - I_o} \qquad (18)$$

$$\frac{d\hat{n}_1}{d\xi} = V_e + \frac{2\kappa I_o I_2}{\big(R_e(I_1 - I_o)\big)} \qquad (19)$$

$$I_o = \int_o^1 RU^2 d\eta \;;\quad I_1 = \int_o^1 RU^3 d\eta \;;\quad I_2 = \int_o^1 R\left(\frac{dU}{d\eta}\right)^2 d\eta \qquad (20)$$

These integrals can be evaluated analytically.

$\kappa(\xi)$ is obtained by deriving a set of integral relations using eqns (4) and (9). Another set of weighting functions $g_j(\hat{\tau})$ is defined, and a similar procedure to that resulting in (14) is followed. The functional form chosen for $\hat{\tau}$ is

$$\hat{\tau} = \kappa(\xi)\, dU/d\eta \qquad (21)$$

The equation contains only one unknown, so only one integral relation is required. This yields

$$\frac{d\kappa}{d\xi} = \frac{a_1 I_2 \kappa}{\hat{b}\ I_3} - \frac{a_1 I_4 \kappa^{3/2}}{F\hat{L}I_3} - \frac{\kappa}{\hat{r}}\frac{d\hat{r}}{d\xi} - \frac{\kappa}{\hat{r}}\frac{d\hat{b}}{d\xi} \tag{22}$$

$$I_3 = \int_o^1 RU\frac{dU}{d\eta}\,d\eta \;; \qquad I_4 = \int_o^1 R\left(\frac{dU}{d\eta}\right)^{3/2} d\eta \tag{23}$$

I_3 can be found analytically. I_4 must be computed numerically.

The boundary layer equations have thus been reduced to a set of ordinary differential equations, empirically corrected for extra strain rates. The equations have been solved using a fourth-order Runge-Kutta method. Gaussian quadrature was used to evaluate I_4.

5. RESULTS AND DISCUSSION

Results are presented for the growth rates and maximum shear stress in the flare mixing layer, and in plane, radial and curved two-dimensional mixing layers. All calculations have been performed using the same exit Mach number, namely 1.2, and a slot width, h_s, of 15 mm. It has been found that the exit Mach number varies little with h_s, and that the growth rates are virtually the same for different values of h_s.

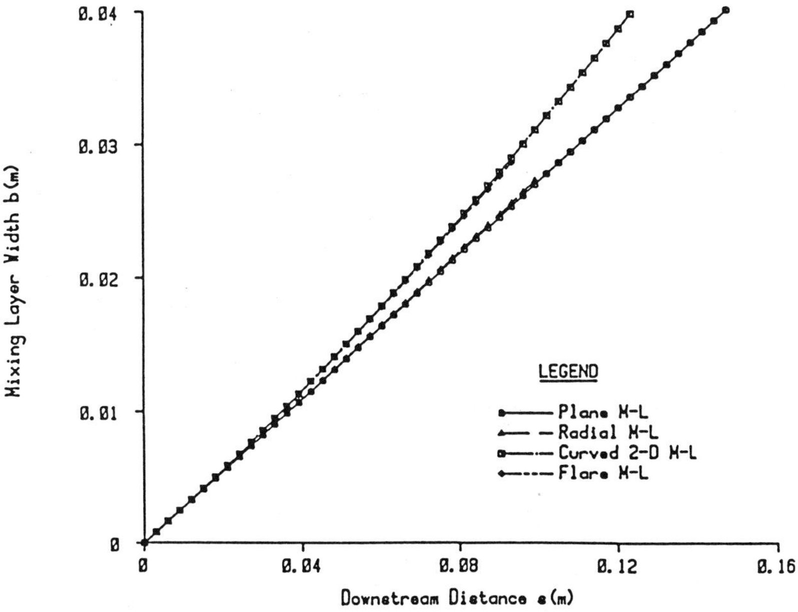

Fig. 2 Mixing Layer Width vs Downstream Distance.

The results for the plane mixing layer are virtually the same as those of Carpenter [4], which have been shown to be in reasonable agreement with experiment. From Fig.2 it can be seen that the growth rates of the plane and radial mixing layers are almost identical. This can be explained by referring to eqn (18). For plane flow the first term on the right-hand side is zero, in the radial case it is always negative, so that any increase of the second term due to lateral divergence is offset by the other term. It has been found [7] that the growth rates of radial and plane jets are similar.

It is also evident from Fig.2 that the potential core of a radial jet is shorter than that of a plane jet. This is because, to conserve mass, the boundaries of an inviscid radial jet must become closer with increasing downstream distance. As a mixing layer can be considered to grow around the boundary of an inviscid jet, the inward displacement of the latter leads to a corresponding displacement of the former, resulting in a shorter potential core. This effect has also been observed in [7].

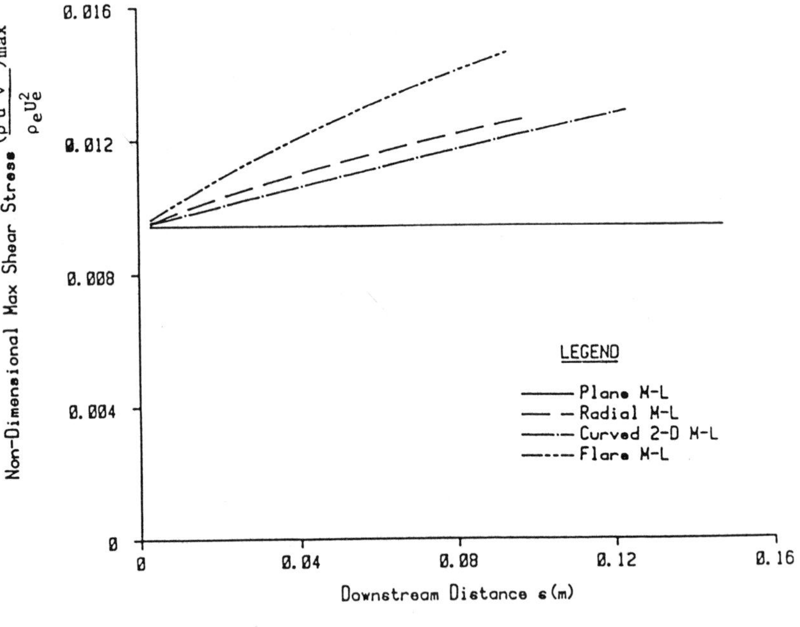

Fig.3 Maximum Shear Stress vs Downstream Distance.

Fig.2 also shows that the growth rates of the curved two-dimensional and flare mixing layers are almost the same, and that the potential core of the former is the longer, in accordance with the above explanation. At the end of the potential core the maximum shear stress in the curved two-dimensional flow is about 27 percent larger than that in the plane layer, as can be seen from Fig.3. Since there does not seem to be any experimental data concerning mixing layers with convex curvature, it is difficult to assess the accuracy of this result. Giles, Hays and Sawyer [8] found that shear stress in the outer part of a very highly curved incompressible wall jet reached about twice the value in a plane layer, after a considerable development region. It does seem as though the maximum shear stress in the calculated two-dimensional flow would continue to rise in the wall jet region, but whether it would ever become double the plane layer value is difficult to predict. The maximum shear stress at the end of the flare potential core is about 44 percent higher than that of the plane layer. This fact, coupled with the increased growth rate, could very well explain the discrepancy between theoretical noise calculations and experimental data [1].

6. CONCLUSIONS

The above method seems to produce results in broad agreement with the limited experimental data which is available. Unfortunately, owing to the lack of experimental data, it is not yet possible to make a conclusive evaluation of the present procedure.

ACKNOWLEDGEMENTS

The authors would like to thank Dr. J.C. Boden and others at the B.P. Research Centre and Kaldair Ltd. for their help. They would also like to thank the Science and Engineering Research Council (U.K.) and British Petroleum Co. plc. for supporting the research leading to this paper.

REFERENCES

1. CARPENTER, P.W. and GREEN, P.N. - Noise sources in external Coanda-type gas flares. AIAA Paper 83-0758, 1983.

2. BRADSHAW, P. - Effects of streamline curvature on turbulent flow. AGARDograph 169, 1973.

3. BRADSHAW, P. and FERRISS, D.H. - Calculation of boundary-layer development using the turbulent energy equation: compressible flow on adiabatic walls. J.Fluid Mech, Vol.46, part 1, pp.83-110, 1971.

4. CARPENTER, P.W. - A theoretical investigation of high-speed axisymmetric turbulent mixing layers with large temperature differences. Aero Research Council, C.P. No.1345, 1976.

5. HOLT, M. - Numerical Methods in Fluid Dynamics. Springer-Verlag, 1977.

6. BRADSHAW, P. and GEE, M.T. - Turbulent wall jets with and without an external stream. Aero Research Council, R and M No.3252, 1960.

7. TANAKA, T. and TANAKA, E. - Experimental study of a radial turbulent jet. JSME Bulletin, Vol.19, pp.792-799, 1976.

8. GILES, J.A., HAYS, A.P., and SAWYER, R.A. - Turbulent wall jets on logarithmic spiral surfaces. Aero. Quart. Vol.17, pp.201-215, 1966.

NUMERICAL SIMULATION OF TURBULENT FLOWS IN PLANE CHANNEL

M.Antonopoulos-Domis *

Nuclear Engng Dept., Queen Mary College, Univ. of London.

ABSTRACT

The three-dimensional Navier-Stokes equations are solved numerically in plane channel geometry. A method is presented with which the equations are decoupled and reduced to a set of one-dimensional ordinary differential equations. The method was implemented with (a) finite difference schemes and (b) spectral methods. The relevant computer codes were tested at both: low (laminar flows) and high Reynolds numbers.

1. INTRODUCTION

Following the success of numerical simulations, with (Large Eddy Simulation) or without (Direct Simulation) subgrid models, of homogeneous isotropic turbulence |1,2,3,4| interest is now being focused in wall bounded flows and in particular in plane channel flows |5,6,7,8|. The aim is to prove the feasibility of such simulations in realistic geometries of practical interest and to use such a numerical apparatus for numerical experiments which could assist and/or replace laboratory experiments. In this paper, a general method for studying channel flows will be presented. The 3-dimensional Navier - Stokes equations are reduced to a set of one-dimensional uncoupled ordinary differential equations. Non-slip conditions are imposed directly-exactly. We report here on the method itself and the codes

* Present Address: School of Engng, Aristotelio Univ. of Thessaloniki, Greece

developed to test it. Use of the codes for turbulence research is now in progress at QMC/Nuclear Engng. Dept. .

2. THE METHOD

Incopressible flow between two parallel plates at $x_3 = \pm 1$ is considered.

$$\frac{\partial u_i}{\partial t} = S_i - \frac{\partial p}{\partial x_i} + \frac{1}{R_e}\frac{\partial^2 u_i}{\partial x_i^2} \quad , \quad i = 1,2,3 \tag{1}$$

$$S_i = -\frac{\partial\, u_i u_j}{\partial x_j} - \Pi\, \delta_{i1} \tag{2}$$

$$\partial u_i / \partial x_i = 0 \tag{3}$$

$$u_i\ (x_1, x_2, x_3 = \pm 1) = 0 \tag{4}$$

Where Π is the mean pressure gradient and all variables have been normalized with channel half width and some velocity (say centre line or $u_\tau = \sqrt{\tau_w/\rho}$). Periodic boundary conditions are assumed in the streamwise x_1 and spanwise x_2 directions. In this case it is convinient and efficient to expand the variables in Fourier series. Doing so and uncoupling the equations the following equivalent system is finally obtained. ($u \equiv u_1$, $v \equiv u_2$, $w \equiv u_3$, $z \equiv x_3$). All variables are from here onwards Fourier transforms in the x_1, x_2 directions, functions of (k_1, k_2, z) where k_1, k_2 the Fourier variables in the x_1, x_2 directions respectively

$$\frac{\partial \varphi}{\partial t} = H - \frac{k^2}{R_e}\varphi + \frac{1}{R_e}\frac{\partial^2 \varphi}{\partial z^2} \tag{5}$$

with conditions on φ,

$$\int_{-1}^{+1} e^{\pm kz}\, \varphi(z)\, dz = 0, \qquad k \neq 0$$

$$\left.\begin{aligned} &\int_{-1}^{+1} z\, \varphi(z)\, dz = 0 \\ &\int_{-1}^{+1} \varphi(z)\, dz = 0 \end{aligned}\right\} \quad k = 0 \tag{6}$$

$$k^2 = k_1^2 + k_2^2$$

$$H = -(\partial S_{1,2}/\partial z + k^2 S_3) \tag{7}$$

$$S_{1,2} = i\,k_1 S_1 + i\,k_2 S_2 \quad , \quad i = \sqrt{-1}$$

$$\frac{\partial^2 w}{\partial z^2} - k^2 w = \varphi \qquad\qquad w(z=\pm 1) = 0 \tag{8}$$

and

$$k^2 p = -\frac{\partial}{\partial t}\frac{\partial w}{\partial z} - S_{1,2} + \frac{1}{R_e}\left(\frac{\partial^3 w}{\partial z^3} - k^2 \frac{\partial w}{\partial z}\right) \tag{9}$$

and

$$\frac{\partial u}{\partial t} = S_1 - i\,k_1\,p - \frac{k^2}{R_e}u + \frac{1}{R_e}\frac{\partial^2 u}{\partial z^2} \tag{10}$$

$$u(k_1, k_2, \pm 1) = 0$$

and

$$\frac{\partial \upsilon}{\partial t} = S_2 - i\,k_2\,p - \frac{k^2}{R_e}\upsilon + \frac{1}{R_e}\frac{\partial^2 \upsilon}{\partial z^2} \tag{11}$$

$$\upsilon(k_1, k_2, \pm 1) = 0$$

or

$$ik\upsilon = -ik_1 u - \frac{\partial w}{\partial z}$$

Advancing the fields in a discrete approximation to time step (n+1) the terms $S_{1,2}$ and H can be computed explicitly or semi-implicitly from fields at steps n, n-1. Having done this, φ at (n+1) is uniquely determined from (5) and (6). Then w is computed from (8). Then p is computed from (9) and finally u and υ from (10) and (11). The system is thus reduced to a set of one-variable, one-dimentional second order ordinary differential equations and on discretization all matrices are nearly tridiagonal. Note finally that for k=0 the pressure can not be obtained from (9). It can be shown that,

$$p(k=0, z) + w^2(k=0, z) = C$$

where the constant C is indeterminant and can be set to any value we wish. (note that all variables are Fourier transforms).

The formalism of this section was derived without reference to any particular numerical scheme. It can be, and it was, implemented with both: finite difference schemes and spectral methods. In all caces time advancing was done with: Adams-Bashforth for the non-linear terms and Crank-Nicolson for the pressure and viscous terms.

3. IMPLEMENTATION AND RESULTS

The method was implemented in two codes and the aim of the first set of runs was to verify the method by comparison with known results.

(a) A two dimensional code was developed to simulate the viscous wall region of a turbulent flow, following the model of Hatziavramidis and Hanratty |9|; z-derivatives were approximated by finite difference methods.

The results were in good agreement with those in |9|.

(b) Direct (without subgrid or any other model) simulation of the 3-dimensional channel flow was implemented with the developement of CHANEL which is a spectral code (Fourier expansions in x_1,x_2 and Chebyshev polynomial expansion in x_3). The advantages of Chebyshev expansion are: (i) high resolution near the walls, lower tesolution in the bulk of the flow (ii) high accuracy and (iii) computational efficiency.

The following tests were done (all Reynolds Nos quoted here are based on centre line velocity and channel half width).

Tests at low R_e (laminar flow). First it was verified that CHANEL does maintain exact Poiseuille flow for any number of time steps at R_e= 5/100/500.

Next, initial fields identical to those of Moin and Kim |10| were used. Those had a flattened mean-u profile ($\langle u\rangle = 1-z^8$) with small perturbations imposed on u,υ,w. Runs at R_e = 5/100 and 500 with 8 Fourier modes in x_1,x_2 and 9 Chebyshev polynomials in $x_3 \equiv z$, decayed to laminar Poiseuille flow ($\langle u\rangle = 1-z^2$) as they should (Fig.1).

The asymptotic decay was in very good agreement with the theoretical prediction ($\exp(-\pi^2 t/4R_e)$ as can be seen in Fig.2.

At large R_e CHANEL was tested with a ($1-z^2$) initial mean velocity profile on top of which a large two-dimensional disturbance, studied previously by George et al |11|, was imposed. George et al |11|, used Fourier series in x_1 and finite-difference methods in x_3 and the disturbance was generated by the stream function:

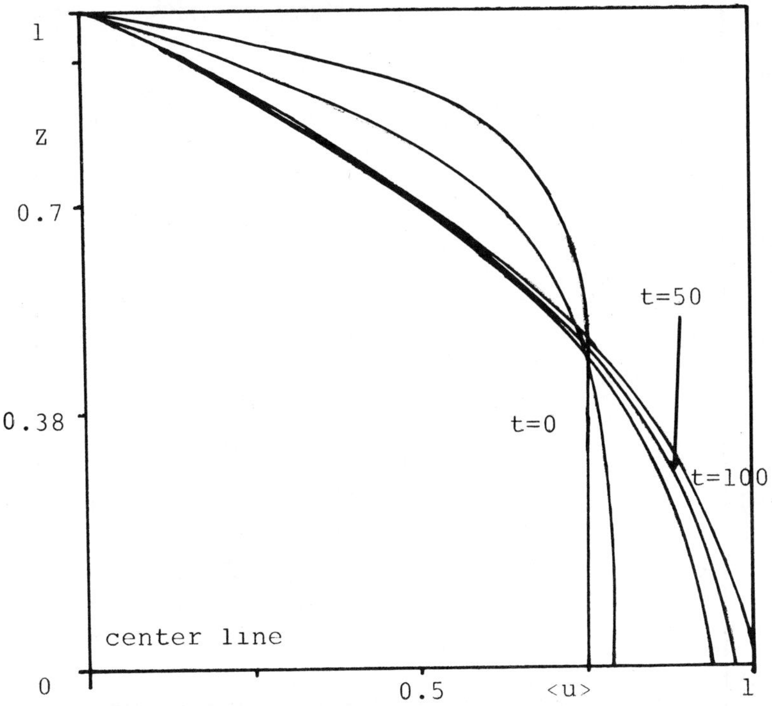

Fig.1. Decay of initial mean-u profile from $\langle u \rangle = 1-z^8$ to Poisseuille flow.

$$\Psi = k_A \left(\frac{\cosh az}{\cosh a} - \frac{\cos az}{\cos a} \right) \cos bx \qquad (13)$$

The box size in x_1 is set to accomodate one fundamental wavelength of the disturbance ($L_1 = 2\pi/b$). Depending on the values of the amplitude constant k_A and R_e , the flow can be unstable or neutrally stable. Our runs did reproduce the results of George et al |11| at $R_e = 4000$ (ex.Figure 3). These runs were essentially two-dimensional, as we used one only Fourier mode in x_2, 8 Fourier modes in x_1 and 33 Chebyshev polynomials in x_3.

In the next set of runs the initial field was laminar ($1-z^2$ mean-u profile) on top of which we imposed a 3-dimensional disturbance of the form (13) with cosbx replaced by (cosbx + cosby). A set of runs at $R_e = 2000$ and $R_e = 4000$ was done with 16×16 Fourier modes and 17 Chebyshev polynomials. Bearing in mind that no model at all is used, it is clear that this resolution is inadequate to resolve the

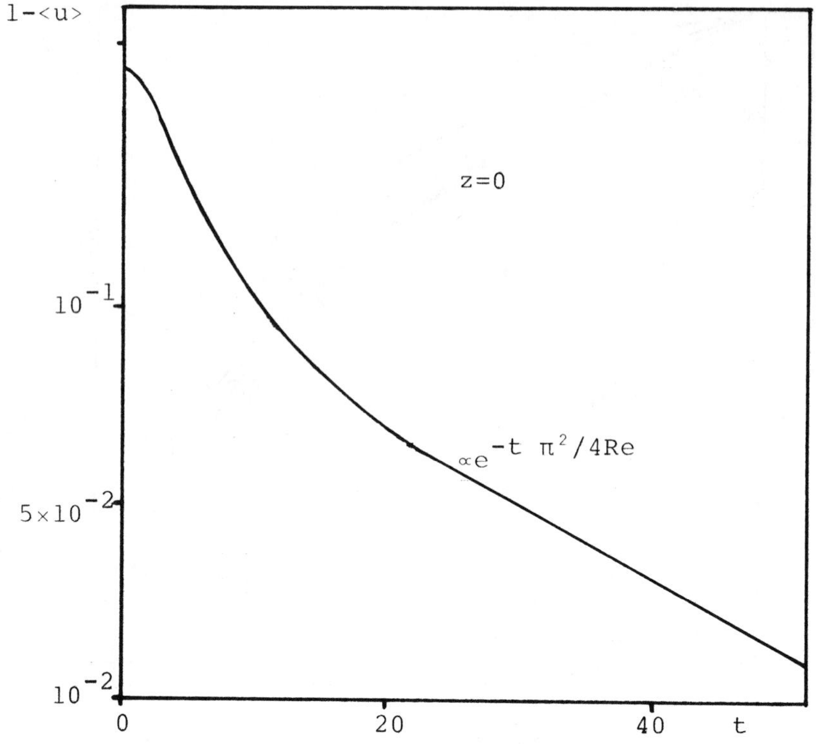

Fig.2. Asymptotic decay to Poisseille flow

viscous layer and the important small scales in x_1, x_2 directions. The aim was to study the feasibility of transitioning from the deterministic laminar initial fields to steady random turbulent fields which could then be used as initial fields in further numerical experiments. For that, runing time of a few diffusion time scales $t_D = h/u_\tau$ (h = channel half width, u_τ = shear velocity) is required and $t_D = O(R_e^{0.5})$ which is prohibitevely large. The process could be accelerated with the use of a large value of the amplitude constant k_A together with an artifitially large R_e (in relation to grid resolution). Too large values of k_A, R_e are expected to lead to blow-up of the computation, and they do. Too small values give slow transition. We are currently experimenting on this and other methods of acceleration and we will report on these experiments. Some of the results are shown in Figs. 4 to 6. Time units quoted are in terms of the convective time scale $t_c = h/u_c = 1$, u_c = centre line velocity. In Fig.4 the Reynolds stress is presented: it can be seen that although in the initial field <uw> is to-

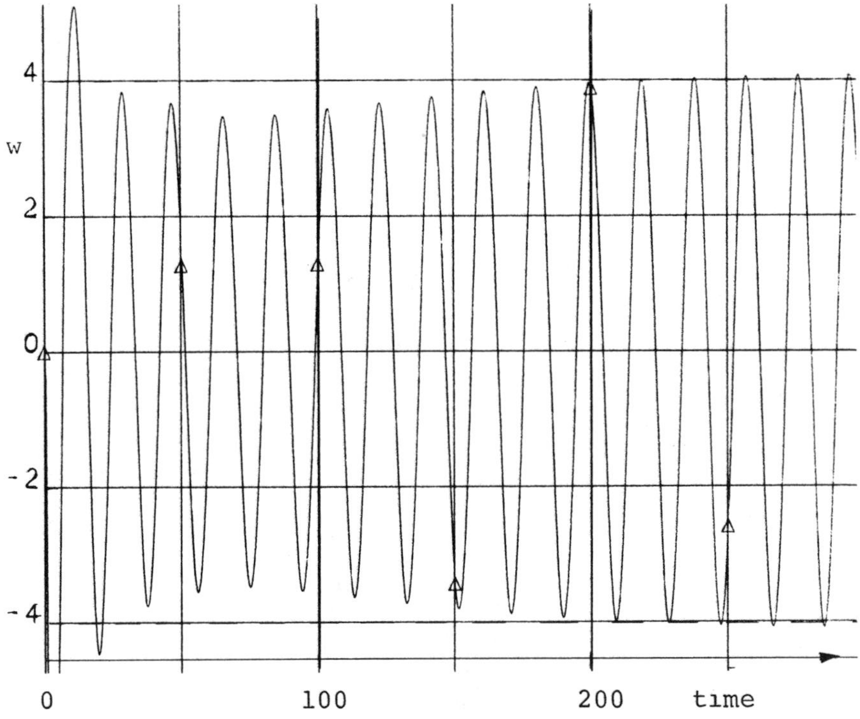

Fig.3. The flow with disturbance that of (17) is stable at Re = 4000 for long times.

tally unrealistic, it quickly developes to the right "turbulence-like" shape. Bearing in mind that we are still in transition, the value of the slope of <uw> is in good agreement with theoretical prediction. This fact coupled with the development of the mean velocity ptofile (Fig.6) and the u-fluctuation profile (Fig.5), showes that CHANEL does indeed transition from laminar to turbulent flow. With the poor resolution (16×16×17) to which we are at present restricted, these results are particularly encouraging.

AKNOWLEDGEMENTS

The finite difference code, as well as a significant part of the channel code was developed by Dr A.Splawski to whom I am indebted for his invaluable contribution. I am also grateful to Professor D.C.Leslie and Dr S.T.B.Young for many fruitful discussions. This work is supported by the SRC.

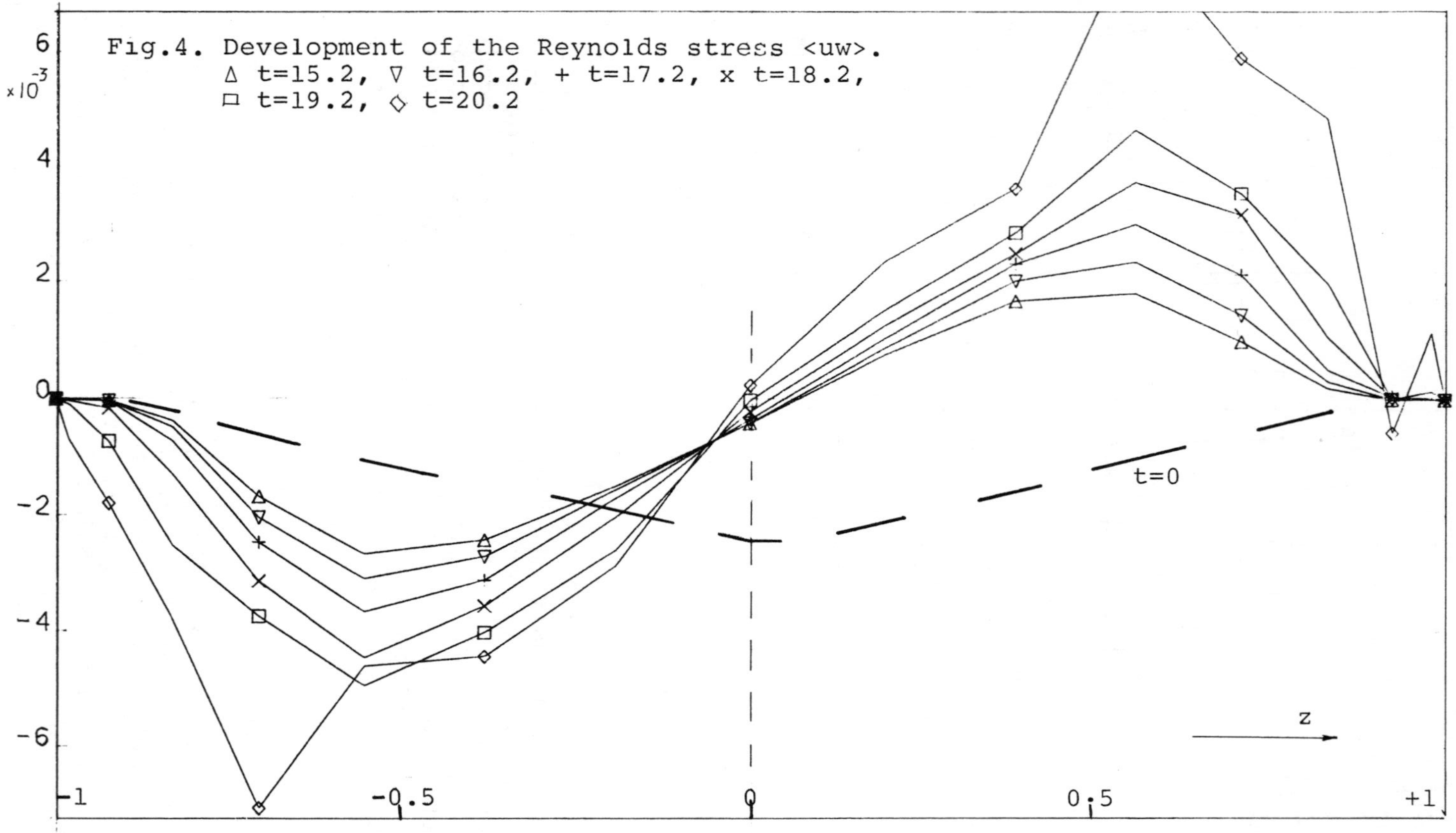

Fig.4. Development of the Reynolds stress <uw>.
Δ t=15.2, ∇ t=16.2, + t=17.2, x t=18.2,
□ t=19.2, ◇ t=20.2

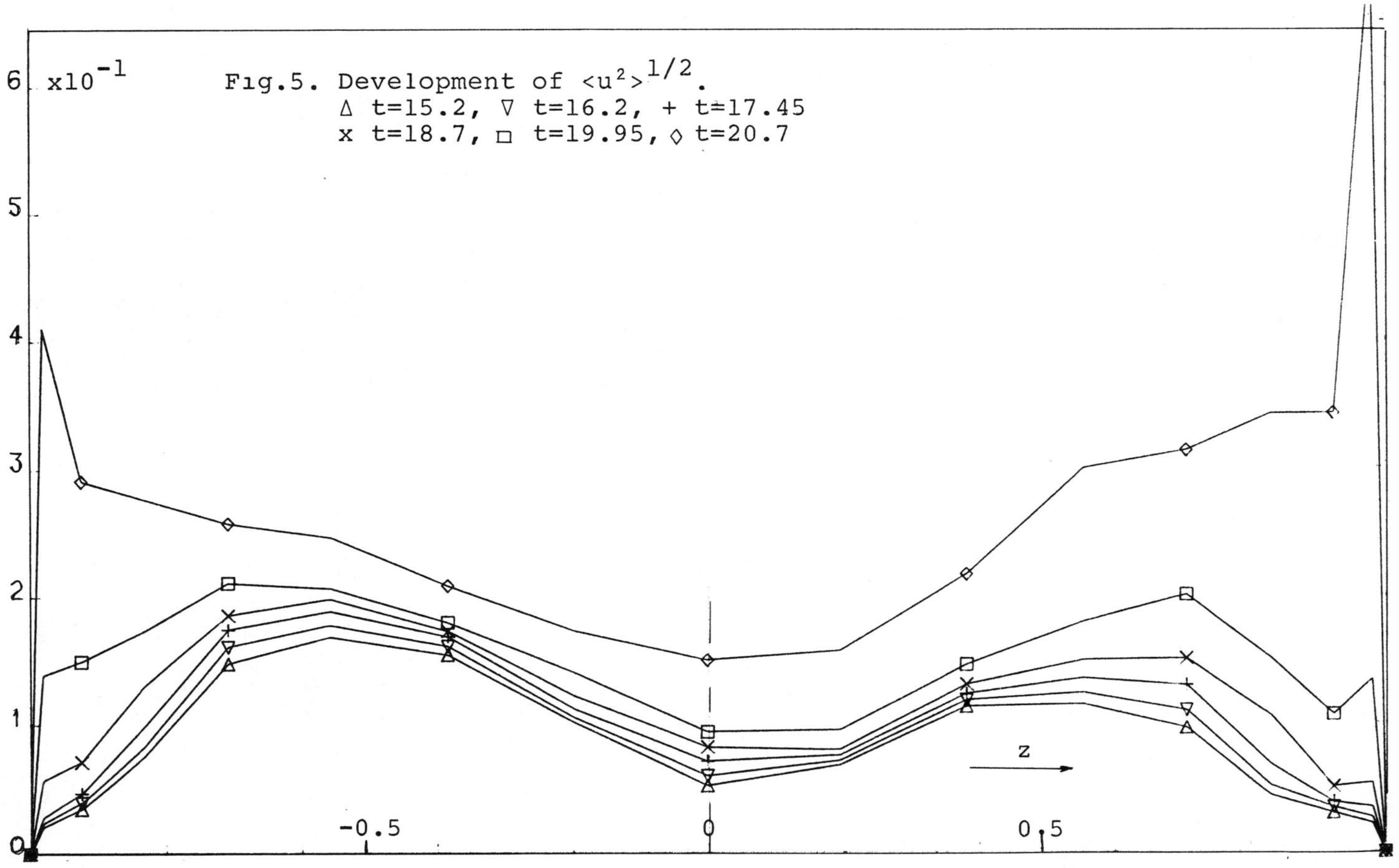

Fig.5. Development of $<u^2>^{1/2}$.
△ t=15.2, ▽ t=16.2, + t=17.45
x t=18.7, □ t=19.95, ◇ t=20.7

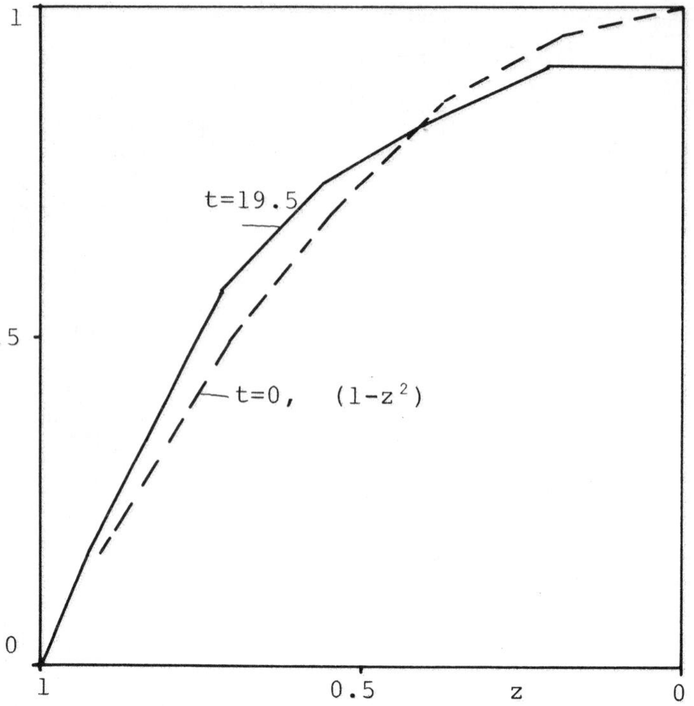

Fig.6. Development of the mean streamwise velocity profile <u>.

REFERENCES

1. SCHUMANN U. et al Prediction methods for turbulent flows. V.Karman Inst. Lecture Notes (1979)

2. KWAK D. et al Three dimensional, time dependent Computation of turbulent flows. Rep. No TF-5, Mech. Engng Dept., Stanford Univ., 1975.

3. ANTONOPOULOS-DOMIS M. - Large Eddy Simulation of a passive scalar in isotropic turbulence. J.Fluid Mech., vol.104, pp 55-79, 1981.

4. ANTONOPOULOS-DOMIS M. - Aspects of Large Eddy Simulation of homogeneous isotropic turbulence. Int. J. Numerical Methods in Fluids, vol.1 , pp 273-290, 1981.

5. SCHUMANN U. - Subgrid scale model for finite difference simulations of turbulent flows in plane channels and annuli. J.Comp. Phys., vol.18, pp 376-404, 1975.

6. MOIN P. et al - Large Eddy Simulation of a turbulent channel flow. Rep. TF-12, Dept. Mech. Engng, Stanford Univ., 1978

7. KIM J, MOIN D. Large Eddy Simulation of turbulent channel flow-ILLIAC IV calculation. AGAR Conf. Proc. No 271, 1979.

8. ORSZAG S.A., KELLS LC. - Transition to turbulence in plane Poiseuille and plane Couette flows. J.Fluid Mech. vol.96, part 1. pp 159-205, 1980.

9. HATZIAVRAMIDIS D.T., HANRATY T.J. - The representation of the viscous wall region by a regular eddy pattern. J.Fluid Mech., vol.95, part 4, pp 655-679, 1980.

10. MOIN P, KIM J. - On the numerical solution of time dependent viscous incompressible fluid flows involving solid boundaries. J.Comp. Phys. vol.35, No 3,pp 381-391 , 1980

11. GEORGE W.D., HELLUMS J.D. - Hydrodynamic stability in plane Poiseuille flow with finite amplitude disturbances. J.Fluid Mech., vol 51, part 4, PP 687-704, 1972.

SECTION 3

BOUNDARY LAYERS

ASYMPTOTIC FINITE ELEMENT METHOD FOR BOUNDARY LAYER PROBLEMS

P. Bar-Yoseph,(i) A. Ostronov(ii) and M. Israeli(iii)

Technion - Israel Institute of Technology
Haifa 32000 Israel

SUMMARY

A scheme that uses singular perturbation theory to improve the performance of existing finite element methods is presented. The proposed scheme improves the L_∞ error bound of the standard Galerkin finite element scheme by a factor of $O(\varepsilon^{n+1})$ (where ε is the "small" parameter and n is the order of the asymptotic approximation). Numerical results for linear second order O.D.E.'s are given and are compared with several other schemes.

1. FORMULATION

Let Ω be a bounded open set of R^N with boundary $\partial\Omega$. We consider the following stiff boundary value problem

$$Ly = \varepsilon\nabla^2 y + Dy = f \quad \text{in } \Omega \tag{1}$$

$$y = g(\varepsilon) \quad \text{on } \partial\Omega \tag{2}$$

where D denotes a given first order differential operator. We represent the region Ω by an assemblage of elements. The variation of y over an element is approximated by

$$y \sim y^h = N_i y_i^h \tag{3}$$

where y_i^h are the nodal values of the approximate solution and N_i are the basis functions (we use the summation convention, with summation over the nodes within the respective element). The Galerkin-Petrov approximation of equation (1) is given by

$$(W_j\ ,\ LN_i y_i^h)_h = (W_j\ ,\ f)_h \quad j=1,2,..,m \tag{4a}$$

or in the weak form

$$-\varepsilon(\nabla^T W_j\ ,\ \nabla N_i y_i^h)_h + (W_j\ ,\ DN_i y_i^h)_h = (W_j, f)_h \quad j=1,2,..,m \tag{4b}$$

(i) and (ii) Senior Lecturer and Graduate Instructor, respectively. Computational Mechanics Group, Faculty of Mechanical Engng.
(iii) Associate Professor, Computer Science Department.

where, W_j are the test functions, N_i are the corresponding basis functions, m is the number of inner nodel points and $(\cdot\ ,\ \cdot)$ denotes the usual scalar product in $L_2(\Omega)$. The index h in $(\cdot\ ,\ \cdot)_h$ denotes an approximation to $(\cdot\ ,\ \cdot)$ obtained by a quadratic rule.

Suppose that we have an asymptotic solution of eq.(1) which satisfies the following equations

$$Ly^\varepsilon = f^\varepsilon \qquad \text{in } \Omega \tag{5}$$

$$y^\varepsilon = g^\varepsilon \qquad \text{on } \partial\Omega \tag{6}$$

such that

$$y - y^\varepsilon = O(\varepsilon^{n+1}) \tag{7}$$

Our asymptotic Galerkin formulation for eq.(1) is the following

$$(W_j\ ,\ LN_i y_i^h)_h = (W_j\ ,\ f)_h +$$

$$\{(W_j\ ,\ LN_i y_i^\varepsilon)_h - (W_j\ ,\ f^\varepsilon)_h\} \qquad j=1,2,..,m \tag{8a}$$

or in the weak form

$$-\varepsilon(\nabla^T W_j\ ,\ \nabla N_i y_i^h)_h + (W_j\ ,\ DN_i y_i^h)_h = (W_j\ ,\ f)_h +$$

$$\{-\varepsilon(\nabla^T W_j\ ,\ \nabla N_i y_i^\varepsilon)_h + (W_j\ ,\ DN_i y_i^\varepsilon)_h - (W_j\ ,\ f^\varepsilon)_h\}$$

$$j=1,2,..,m \tag{8b}$$

Here, the terms included in the first line coincide with the standard finite element (S.F.E.) scheme, while the terms included in the wavy parenthesis represent the correction term which is the essence of the present asymptotic finite element (A.F.E.) scheme.

For a linear operator, L, the A.F.E. scheme, eq.(8a), is equivalent to the following

$$(W_j\ ,\ LN_i(y_i^h - y_i^\varepsilon))_h = (W_j\ ,\ f - f^\varepsilon)_h \qquad j=1,2,..,m \tag{9}$$

Thus, we apply the method of weighted residuals on the error of the asymptotic solution. Based on the above equations, we prove, for appropriate operators L and classes of functions f and f^ε the following

if the S.F.E. error satisfies a bound of the form

$$\|e_{S.F.E.}\|_\infty \leqq G(h,\varepsilon) \tag{10}$$

then, for the A.F.E. method we have the following error estimate

$$\|e_{A.F.E.}\|_\infty \leqq C\,\varepsilon^{n+1}\,G(h,\varepsilon) \tag{11}$$

where C is a constant independent of ε and h.
Such operators L and classes of functions f and f^ε, where used

by Israeli and Ungarish[1] to establish estimates of type (10) to (11) for a finite difference analogue of the present method (called the "booster method"). Similar estimates for the present A.F.E. scheme will be the subject of a forthcoming paper[2]. It follows that for properly constructed y^{ε}, every existing L_{∞} error estimate for the S.F.E. scheme, will give rise to a corresponding estimate for the present A.F.E. scheme.

In the next section we demonstrate numerically the validity of (10) and (11) for a number of 1-D problems.

2. NUMERICAL EXAMPLES

Here we shall restrict ourselves to simple O.D.E.'s where the main features of the A.F.E. method are already exhibited. The problem on which the present numerical experiments were conducted consists

$$Ly = \varepsilon y'' + p(x)y' + q(x)y = f(x,\varepsilon) \text{ , } x \text{ in } (0,1) \qquad (12)$$

together with the boundary conditions

$$y(0) = \alpha(\varepsilon) \text{ , } y(1) = \beta(\varepsilon) \qquad (13)$$

Several examples of type (12) were run with $\varepsilon = h^p$ for various values of p, using the S.F.E. and A.F.E. methods. In the following the Galerkin-Bubnov method ($\tilde{W} = \tilde{N}$) has been employed. The interval (0,1) has been discretized by a uniform mesh of linear elements. We consider a partition Π of (0,1), $0 = x_0 < x_1 < x_2 < \ldots < x_m < x_{m+1} = 1$, and introduce the maximum pointwise error norm denoted by $\| e \|_{\pi,\infty}$. For each value of p, the element length, was successively halved. Here, the rate of convergence, R_{∞}, is based on the maximum pointwise error norm.

Example 1 (Nayfeh[3]):

$$p(x) = q(x) = \beta = 1 \text{ ; } f = \alpha = 0 \qquad (14)$$

The performance of the A.F.E. scheme based on the one term (Table 1) and two terms matched asymptotic solutions (Table 2) shows that the improvement of the rate of convergence approaches (n+1)p.

Example 2 (Berger et al.[4]; Israeli and Ungarish[1]):

$$p(x) = (x+1)^3 \text{ ; } q(x) = 0$$

$$f(x,\varepsilon) = [-0.5(1+x)^3 + \varepsilon/4] \exp(-x/2) + 12\varepsilon(1+x)^{-5} \exp\{-[(x+1)^4-1]/4\varepsilon\} \qquad (15)$$

$$y(0) = 2; \; y(1) = \exp(-0.5) + 1/8 \exp(-15/4\varepsilon)$$

The numerical results are presented in Table 3.

The two examples cited here serve to illustrate that our

A.F.E. scheme has been effective for the whole range of the so-called Peclet number.

Comparisons with other schemes and numerical results for the present Galerkin-Petrov scheme using higher order elements are presented in Reference [2]. The implementation of the present approach to the F.E. solution of the non-linear similarity equations in rotating compressible flows is presented in Reference [5]. A.F.E. schemes for 2-D incompressible flow problems are presented in Reference [6].

No. of elements	p	$\|\|y-y^{\varepsilon}\|\|_{\pi,\infty}$	$\|\|y-y^h\|\|_{\pi,\infty}$		R_∞	
			A.F.E.	S.F.E.	A.F.E.	S.F.E.
32	0.5	2.22E-1	4.30E-4	2.38E-3	1.29	0.55
	0.75	1.38E-1	1.11E-3	1.49E-2	1.23	0.52
	1.0	7.05E-2	2.91E-3	9.36E-2	0.98	-0.002
	1.5	1.33E-2	7.36E-3	1.32	1.07	-0.43
	2.0	2.50E-3	2.74E-3	2.6	1.85	-0.16
64	0.5	1.84E-1	1.62E-4	1.32E-3	1.41	0.85
	0.75	9.40E-2	4.66E-4	1.06E-2	1.25	0.49
	1.0	3.80E-2	1.46E-3	9.37E-2	0.99	-0.002
	1.5	5.17E-3	3.22E-3	1.64	1.19	-0.31
	2.0	6.43E-4	6.99E-4	2.96	1.97	-0.19
128	0.5	1.54E-1	5.77E-5	6.6E-4	1.48	0.99
	0.75	6.07E-2	1.94E-4	7.37E-3	1.27	0.52
	1.0	1.39E-2	7.33E-4	9.38E-2	1.0	-0.002
	1.5	1.85E-3	1.32E-3	1.91	1.29	-0.22
	2.0	1.63E-4	1.83E-4	3.09	1.93	-0.06
256	0.5	1.23E-1	2.04E-5	3.28E-4	1.50	1.01
	0.75	3.81E-2	8.19E-5	5.24E-3	1.24	0.46
	1.0	1.02E-2	3.67E-4	9.38E-2	1.0	0.0
	1.5	6.59E-4	5.17E-4	2.12	1.35	-0.15
	2.0	4.11E-5	4.67E-5	3.08	1.97	0.005
512	0.5	9.40E-2	7.20E-6	1.63E-4	1.50	1.01
	0.75	2.35E-2	3.44E-5	3.70E-3	1.25	0.50
	1.0	5.20E-3	1.83E-4	9.39E-2	1.0	0.0
	1.5	2.34E-4	1.97E-4	2.28	1.39	-0.10
	2.0	1.03E-5	1.18E-5	3.10	1.98	-0.01

Table 1. Example 1 - Numerical results for the A.F.E. scheme based on the one term matched asymptotic solution.

No. of elements	p	$\|\|y-y^{\varepsilon}\|\|_{\pi,\infty}$	$\|\|y-y^{h}\|\|_{\pi,\infty}$		R_{∞}	
			A.F.E.	S.F.E.	A.F.E.	S.F.E.
32	0.5	1.28E-1	2.35E-4	2.38E-3	2.07	0.55
	0.75	2.52E-2	2.05E-4	1.50E-2	2.04	0.51
	1.0	5.45E-3	2.28E-4	9.36E-2	1.98	-0.002
	1.5	1.96E-4	1.02E-4	1.33	2.57	-0.43
	2.0	6.06E-6	6.69E-6	2.80	3.85	-0.16
64	0.5	5.95E-2	5.26E-5	1.32E-3	2.16	0.85
	0.75	1.03E-2	5.15E-5	1.06E-2	1.99	0.50
	1.0	1.47E-3	5.72E-5	9.37E-2	1.99	-0.002
	1.5	2.52E-5	1.57E-5	1.64	2.70	-0.31
	2.0	3.92E-7	4.41E-7	2.96	3.92	-0.08
128	0.5	3.35E-2	1.27E-5	6.60E-4	2.05	1.0
	0.75	3.95E-3	1.27E-5	7.37E-3	2.02	0.52
	1.0	3.88E-4	1.43E-5	9.38E-2	2.00	-0.002
	1.5	3.19E-6	2.29E-6	1.91	2.78	-0.22
	2.0	2.49E-8	2.83E-8	3.04	3.96	-0.038
256	0.5	1.89E-2	3.18E-6	3.28E-4	2.00	1.01
	0.75	1.48E-3	3.20E-6	5.24E-3	1.99	0.49
	1.0	9.97E-5	3.58E-6	9.39E-2	2.00	-0.002
	1.5	4.02E-7	3.16E-7	2.12	2.85	-0.15
	2.0	4.57E-9	1.79E-9	3.08	3.98	-0.02
512	0.5	1.03E-2	7.95E-7	1.63E-4	2.00	1.01
	0.75	5.43E-4	7.98E-7	3.70E-3	2.00	0.50
	1.0	2.53E-5	8.95E-7	9.39E-2	2.00	0.00
	1.5	5.04E-8	4.25E-8	2.28	2.89	-0.11
	2.0	9.85E-11	1.13E-10	3.10	3.99	-0.01

Table 2. Example 1 - Numerical results for the A.F.E. scheme based on two terms matched asymptotic solution.

No. of elements	p	$\|\|y-y^{\varepsilon}\|\|_{\pi,\infty}$	$\|\|y-y^h\|\|_{\pi,\infty}$		R_∞	
			A.F.E.	S.F.E.	A.F.E.	S.F.E.
64	0.5	1.57E-1	2.79E-4	7.61E-4	1.59	1.25
	0.75	6.68E-2	7.27E-4	4.52E-3	1.32	0.69
	1.0	2.49E-2	2.21E-3	3.66E-2	1.05	0.095
	1.5	7.05E-5	2.16E-3	5.9E-1	1.14	-0.32
	2.0	*	4.70E-5	1.025	2.88	-0.12
128	0.5	1.20E-1	9.50E-5	3.32E-4	1.55	1.20
	0.75	4.22E-2	2.95E-4	2.99E-3	1.30	0.60
	1.0	1.27E-2	1.09E-3	3.55E-2	1.02	0.04
	1.5	1.77E-6	6.04E-4	6.92E-1	1.84	-0.23
	2.0	*	6.12E-6	1.07	2.94	-0.06
256	0.5	9.08E-2	3.25E-5	1.50E-4	1.55	1.15
	0.75	2.6E-2	1.23E-4	2.03E-3	1.26	0.56
	1.0	6.41E-3	5.40E-4	3.5E-2	1.01	0.02
	1.5	1.13E-8	1.60E-4	7.73E-1	1.92	-0.16
	2.0	*	7.82E-7	1.09	2.97	-0.03
512	0.5	6.75E-2	1.12E-5	7.05E-5	1.54	1.09
	0.75	1.58E-2	5.14E-5	1.40E-3	1.26	0.53
	1.0	3.22E-3	2.69E-4	3.48E-2	1.00	0.008
	1.5	1.04E-11	4.17E-5	8.35E-1	1.94	-0.11
	2.0	*	*	1.11		-0.03
1024	0.5	4.96E-2	3.91E-6	3.34E-5	1.52	1.08
	0.75	9.53E-3	2.15E-5	9.76E-4	1.26	0.52
	1.0	1.61E-3	1.34E-4	3.47E-2	1.00	0.004
	1.5	*	1.07E-5	8.81E-1	1.96	-0.08
	2.0	*	*			

*Round off error

Table 3. Example 2 - Numerical results for the A.F.E. scheme based on the one term matched asymptotic solution.

3. REFERENCES

[1] ISRAELI, M. and UNGARISH, M. - Improvement of Numerical Solution of Boundary Layer Problems by Incorporation of Asymptotic Approximations. Numer.Math. (to appear).

[2] BAR-YOSEPH, P., ISRAELI, M. and OSTRONOV, A. - Improvement of Finite Element Solution of Boundary Layer Problems by the Asymptotic Finite Element Method (to appear).

[3] NAYFEH, A. H. - Perturbation Methods, John Wiley, 1973.

[4] BERGER, A. E., SOLOMON, J. M., CIMENT, M., LEVENTHAL, S. H. and WEINBERG, B. C. - Generalized O.C.I. Schemes for Boundary Layers Problems, Math.Comp., Vol.35, pp. 695-731, 1980.

[5] BAR-YOSEPH, P. and OLEK, S. - Analytical and Numerical Studies of Heat Transfer in Rotating Compressible Flow over an Infinite Disk, submitted to Computers and Fluids.

[6] BAR-YOSEPH, P. - Standard and Asymptotic Finite Element Methods for Incompressible Viscous Flows, ASME AMD, Vol. 51, pp. 143-155, 1982.

THE EFFECTS OF WALL DEFORMATION AND HEAT TRANSFER ON THE STABILITY OF LAMINAR BOUNDARY LAYERS

Leonidas Sakell[I] and Robert J. Hansen[II]
Naval Research Laboratory
Washington, DC

SUMMARY

The stability characteristics of laminar boundary layers over an isothermal elliptical cross section body with and without surface deformation and a smooth elliptical cross section body with spatially varying wall temperature are presented. Both compressible air and fresh water are used as working fluids. Results for the spatial growth of the disturbance amplification factor for Tollmien-Schlichting waves are presented using disturbance frequency as parameter.

BACKGROUND

In [1] a qualitative analogy was established between the effects of wall deformation and wall heating on the characteristics of laminar boundary layers. The boundary layer properties over an isothermal, two dimensional ellipse with and without a wavy wall section were numerically calculated using the Transition Analysis Program System (TAPS) [2] code. It was determined that with an appropriate choice of spatial

---

[I] Aerospace Engineer, Boundary Layer Hydrodynamics Section, Fluid Dynamics Branch, Naval Research Laboratory, Washington, DC

[II] Head, Boundary Layer Hydrodynamics Section, Naval Research Laboratory, Washington, DC variation of wall temperature the boundary layer displacement thickness distribution over a smooth 2D ellipse could be made to match that over the ellipse with sinusoidally varying wall waves.

The present work continues that described above by determining the stability response of the laminar boundary layer to the surface deformations and to the spatially varying wall temperature distribution for compressible air and fresh water as working fluids. Using the TAPS code the boundary layer disturbance amplification factor distributions were obtained.

RESULTS

As before a fineness ratio 10, two dimensional elliptical cross section body was used. It is shown in Figure 1.

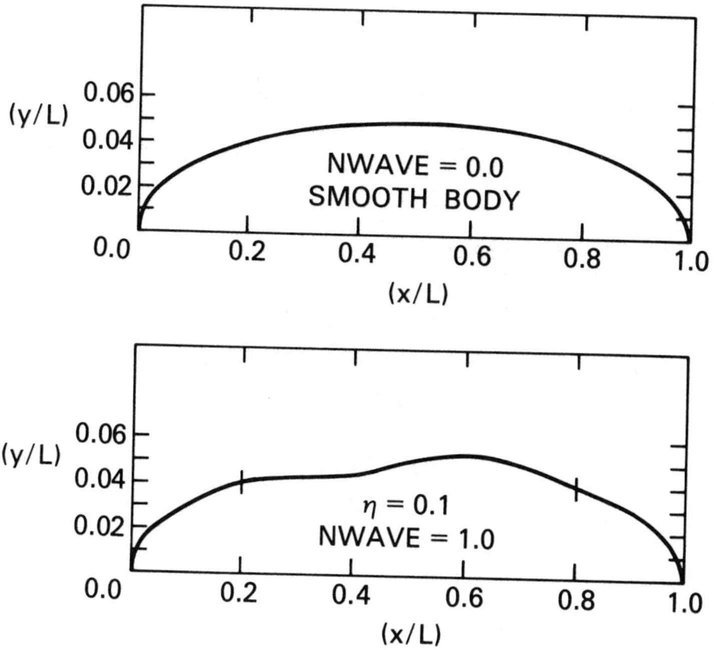

Figure 1: Body Geometries with and without surface deformation

The number of wall deformation waves (NWAVE) and their amplitude as a perturbation (η) of the smooth ellipse thickness, together define the wavy wall geometry.

For air, two subsonic free stream Mach numbers were used, namely 0.2 and 0.6. The corresponding unit freestream Reynolds numbers are 1.382 and 4.145 x 10^6 per foot. Results for the smooth isothermal ellipse (T_w = 532°R) are shown in Figures 2 and 3. Using a value of nine for the logarithm of the amplification factor where transition takes place, neither of these two cases have reached transition. As is expected, the higher Mach number case comes closer. The lower frequencies (non-dimensionalized by the free stream velocity squared, divided by the kinematic viscosity) experience the largest growth of amplification.

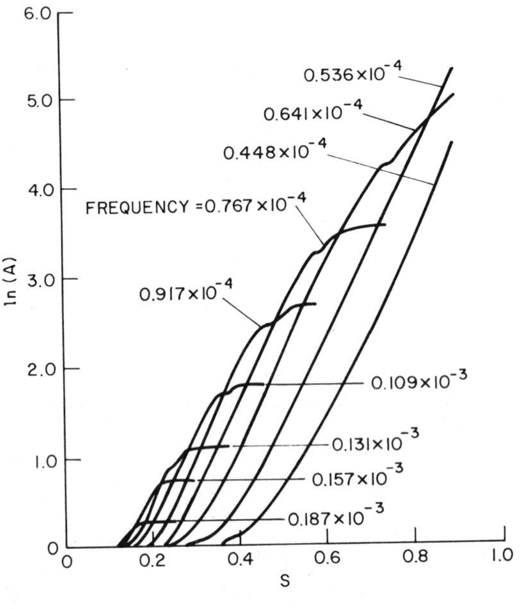

Figure 2. Stability results, M_∞ = 0.2, NWAVE = 0.0, ETA = 0.0, T_w = 532°R

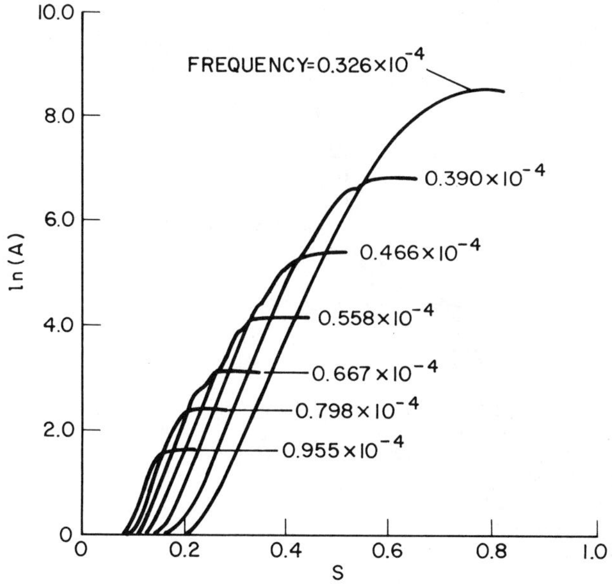

Figure 3. Stability results, M_∞ = 0.6, NWAVE = 0.0, ETA = 0.0, T_w = 532°R

The effect of wave amplitude (i.e. pressure gradient since for all cases the wavy wall extends over $0.2 \leq S \leq 0.8$) is shown in the following two figures.

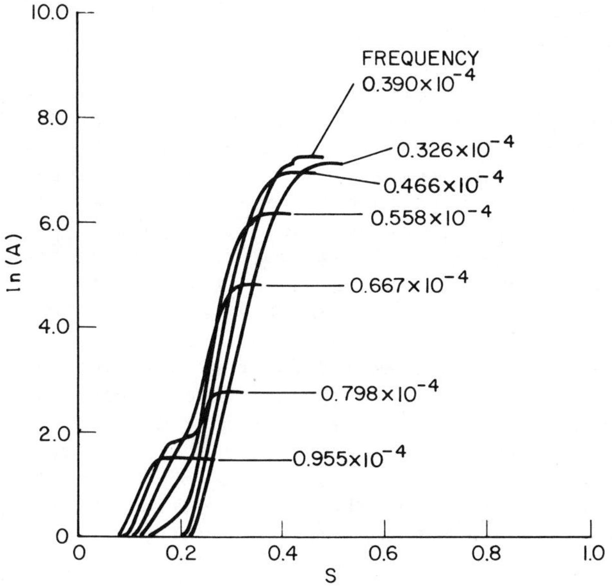

Figure 4. M_∞ = 0.6, NWAVE = 1.0, ETA = .03, T_w = 532°R

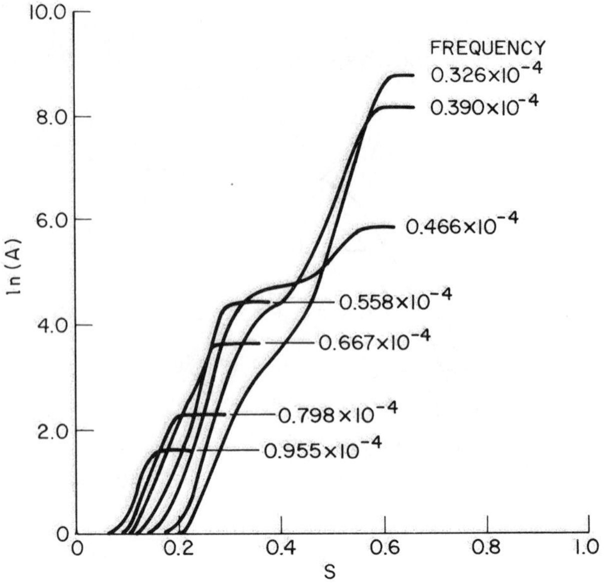

Figure 5. $M_\infty = 0.6$, NWAVE = 2.0, ETA = .005, $T_W = 532°R$

The value of ETA in Figure 5 is so small because separation results at larger values. With two surface waves and ETA = .005 an amplifiction factor of almost nine (8.7) is reached, while with only one surface wave of amplitude ETA = .030 the corresponding amplification factor approaches 7.3.

Figure 6 presents results for the smooth ellipse with a wall temperature distribution which matches the displacement thickness distribution of the wavy wall body (NWAVE = 1.0, ETA = .030) whose results were shown in Figure 4. Though the displacement thickness distributions are the same for Figures 4 and 6 the stability results are much different. The variable wall temperature case reaches amplification factors as high as 11.2 while the same disturbance frequency in the wavy wall reaches only 7.4, clearly a negative result. However, other frequencies are favorably affected. For example, the frequency .6679 x 10^{-4} reaches 4.8 for the isothermal wavy wall while being reduced to 3.6 for the smooth spatially varying wall temperature body. This is indeed a general result. The amplification experienced by some frequencies is increased while for others decreased. Therefore the utility of wall heating as a means of simulating wall deformation, while still hopeful, is not yet clearly resolved.

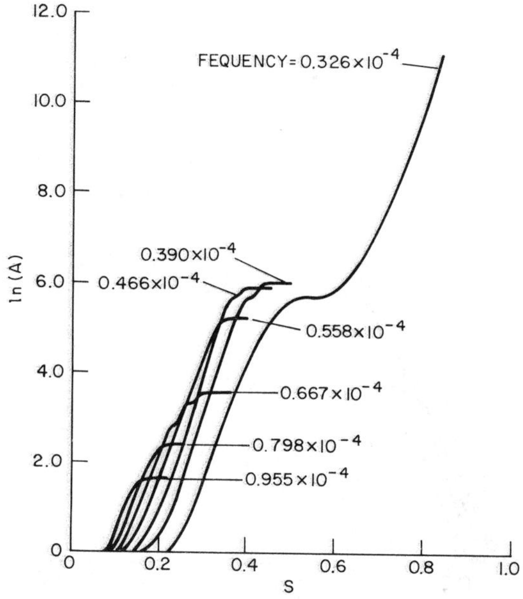

Figure 6. M_∞ = 0.6, NWAVE = 0.0, ETA = 0.0, T_w = variable, (δ^* matched with Figure 4)

Stability results for the smooth isothermal ellipse using fresh water as working fluid are shown in Figures 7 and 8 for free stream velocities of ten and twenty feet per second with corresponding unit Reynolds numbers of .9755 x 10^6 and 1.951 x 10^6 per foot.

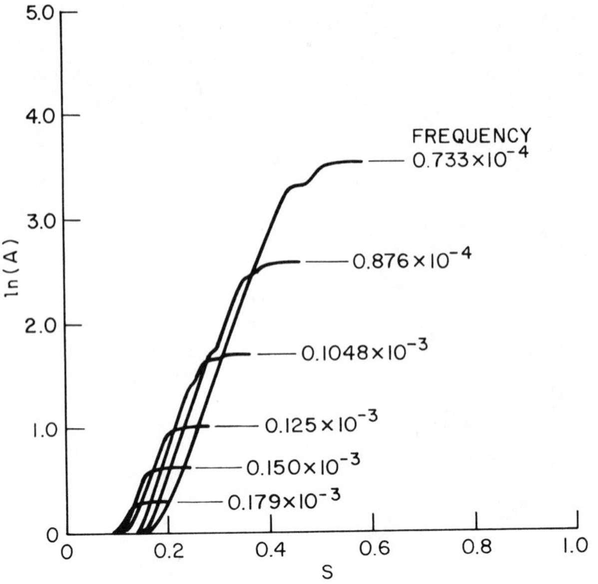

Figure 7. U_∞ = 10 FT/SEC, NWAVE = 0.0, ETA = 0.0. T_W = 532°.

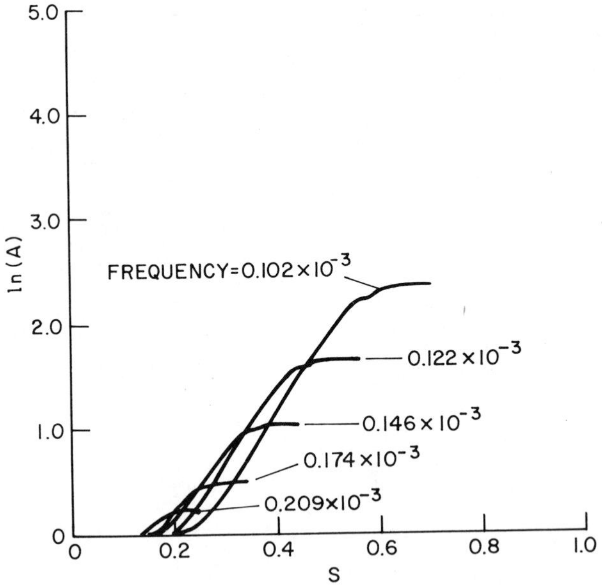

Figure 8. U_∞ = 20 FT/SEC, NWAVE = 0.0, ETA = 0.0, T_W = 532°

The frequencies selected are not the same due to problems in the eigenvalue iterations which arose with fresh water as a working fluid. However, the larger velocity case yields higher amplification factors, though not nearly high enough to infer that transition has occurred.

Figure 9 shows the effect of wall deformation on the stability results for fresh water at a free stream velocity of ten feet per second. By comparing Figure 7 and 9 one can see the effect of adding wall deformation on the stability results. (Again, due to eigenvalue convergence difficulties the frequencies are not quite the same.) The peak value of amplification is increased for each frequency by the wall deformation. Furthermore, the slopes are also increased.

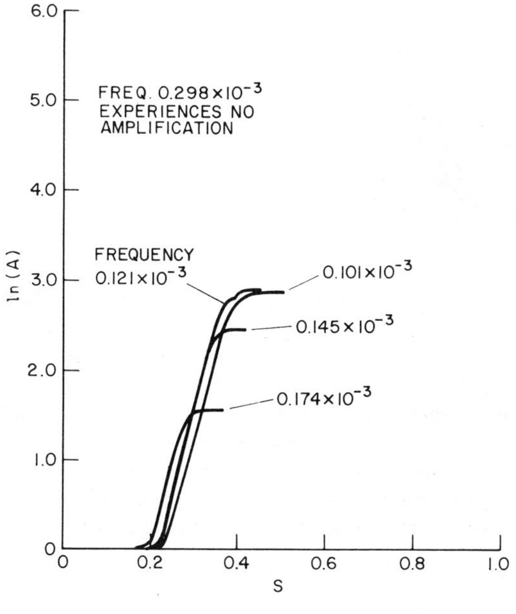

Figure 9. U_∞ = 10 ft/sec, NWAVE = 1.0, ETA = .030, T_W = 532°R

Figure 10 shows the results for water when a sinusoidal wall temperature distribution is applied. A two percent amplitude about 532°R is used resulting in a maximum wall temperature of 543°R and a minimum of 522°R. The minimum occurr first, followed by the overshoot further down the surface.

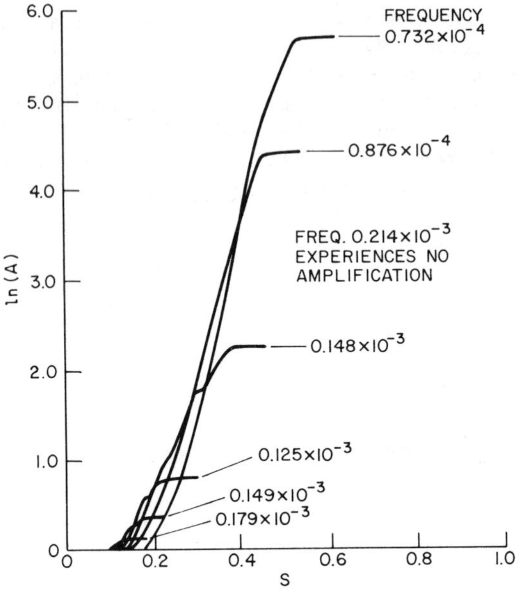

Figure 10. U_∞ = 20 ft/sec, NWAVE = 0.0, ETA = 0.0
T_W = variable

As can be seen by comparing Figures 8 and 10, this small wall temperature variation produces large changes in the stability characteristics. Frequencies less than or equal to .10485 x 10^{-3} experience increases in disturbance amplification while those greater experience decreases in disturbance amplification.

ACKNOWLEDGEMENT

This work was supported by ONR, Compliant Coating Drag Reduction Program, Dr. Michael Reischman, Program Manager.

REFERENCES

1. "A Numerical Study of the Analogy Between Wall Heating and Wall Deformation", L. SAKELL and R.J. HANSEN, paper presented at the 7th Biennial Symposium on Turbulence, University of Missouri, Rolla, Missouri, Sept. 21-23, 1981.

2. "The Transition Analysis Program System, Volume I, II", GENTRY, A.E., Douglas Aircraft Company, McDonnell Douglas, Report No. MDC J7255, June 1976

INCORPORATION OF BOUNDARY CONDITIONS IN BOUNDARY LAYER FOURTH ORDER SOLUTIONS WITH STRETCHING IN THE NORMAL DIRECTION

by

O. Pade*, A. Postan*, D. Anshelovitz* and
M. Wolfeshtein**

SUMMARY

Compact fourth order numerical schemes for boundary layer problems and coordinate stretching are studied. General Algorithms for an easy implementation of arbitrary stretching fun-functions and fourth order boundary conditions are discussed and presented. Convergence rates are shown. The results indicate a very considerable reduction in the number of requi-red mesh points by these methods

1. INTRODUCTION

1.1 Background

Fourth order finite difference approximations to the boundary layer equations appear to be highly desirable, in order to economize in computer space and time. Although the algorithms are more complex and the number of arrays may be bigger than in second order solution, a prescribed accuracy may be usually obtained with much fewer mesh points, with the result of much more efficient computer runs. Two general approaches are possible: Multi-point approximations are easy to implement, but they cause difficulties near the boundaries, and may produce fluctuations or instabilities. Compact methods use more equations, and require the specification of higher derivatives on the boundary, and at high order accuracy. This last requirement may pose severe demands on the programmer, especially when solving complex equations, e.g. turbulence models equations. The problem becomes even more acute when coordinate stretching is incorporated, in order to condense the mesh points in regions of high gradients. Yet in these methods

* Ministry of Defence, Scientific Dept. P.O.B. 2250, Haifa Israel
**Technion, Israel Institute of Technology, Haifa, Israel.

the computational molecule require only three mesh points in the direction of integration for a fourth order accurecy. Therefore they became very attractive and popular in recent years.

Compact methods usually require the solution of additional difference equations for the derivatives. Although these equations appear to formally require additional boundary conditions, such boundendary conditions are superfluous, and should be chosen so as to retain compatibility with the original differential equations and their boundary conditions. Therefore the formulation of these conditions is not necessarily a straightforward matter, and may require special techniques. In this paper we present a general method for the formulation of these boundary conditions, which facilitates easy implementation of such schemes.

1.2 Literature survey

Standard algorithms for the solution of boundary layer problems may be found in the literature. Patankar and Spalding [1] solve the equations using the Von Mises transformation, and a self-adjusting thickness of the boundary layer. Their method is first order, and may require a great number of points if a complex boundary layer with steep gradients is to be resolved. Cebeci and Smith [2] presented a method which is based on the Keller box scheme [3]. Their solution is second order in the normal direction and first order in the streamwise direction. These methods are not very efficient for boundary layers with large pressure gradients, where the velocity profile tends to become extremely steep on the wall, and they require a large number of mesh points, or the application of wall functions, or both.

Compact fourth order schemes have been described by various authors [4, 5, 6, 7, 8, 9]. These methods were applied to laminar boundary layer flow by Postan et. al. [10], who used a simple coordinate stretching to improve the computer time economy. These authors used fourth order accuracy in the normal direction and second order accuracy in the streamwise direction. The method was implicit, and produced good agreement with existing experimental and theoretical data for large pressure gradients. However, the treatment of the boundary conditions was cumbersome and gradient boundary conditions could not be accommodated. The stretching function used was a rational function which did not allow arbitrary stretching near the boundary.

1.3 Purpose of the paper

In this paper we present a general method for the formulation of the boundary conditions for compact fourth order numerical scheme for the boundary layer equations. The method accounts for arbitrary coordinate stretching, and allows easy numerical calculation of the derivatives on the boundary. This procedure is suitable to any boundary value problems which may be expressed in terms of second order differential equations, and may be applied for any boundary conditions (the derivative, the function, or mixed).

1.4 Outline of the paper

The objective of the paper is the generalized formulation of the boundary conditions in fourth order schemes. However, the mathematical formulation and the numerical method will be briefly persecuted as well here for completeness, in chapter 2 and 3. Some results and discussions will be presented in chapter 4 and 5.

2. MATHEMATICAL FORMULATION

2.1 The boundary layer euqations

We consider a two dimensional compressible boundary layer. The governing equations for the conservation of mass, momentum and energy are:

$$\frac{\partial}{\partial x}(\rho u) + \frac{\partial}{\partial y}(\rho v) = 0 \qquad (1)$$

$$\frac{\partial}{\partial x}(\rho uu) + \frac{\partial}{\partial y}(\rho vu) = -\frac{\partial p}{\partial x} + \frac{\partial}{\partial y}\left[\left(\mu \frac{\partial u}{\partial y}\right)\right] \qquad (2)$$

and the energy equation is

$$\rho u \frac{\partial H}{\partial x} + \rho v \frac{\partial H}{\partial y} = \frac{\partial}{\partial y}\left[\mu_e\left(1 - \frac{1}{P_R}\right) u\frac{\partial u}{\partial y} + \frac{\mu_e}{P_R}\frac{\partial H}{\partial y}\right] \qquad (3)$$

where u and v are the velocity components in the streamwise (x) and the normal (y) direction; H is the stagnation enthalpy; p and ρ are the pressure and density, and P_R is the Prandtl number. The viscosity μ_e stands for the effective viscosity, for laminar or turbulent flow. The boundary conditions are

$$\begin{array}{lll} \text{at } y = o & u = v = 0 & H = H_s \\ \text{at } y \to \infty & u = u_\infty & H = H_\infty \end{array} \qquad (4)$$

2.2 The Stewartson-Illingworth transformation

Equations (1) - (3) contain variable properties and are therefore difficult to solve. Conventionally we transform them by a modified Lees-Dorodnytzin transformation [10].

$$\left.\begin{aligned} \xi &= \xi_o + \int_o^x \rho_\infty \mu_\infty u_\infty dx \\ \eta &= \frac{1}{\delta(x)} \int_o^y \frac{\rho}{\rho_\infty} dy \quad . \\ \delta(x) &= \sqrt{2\xi}/\rho_\infty u_\infty \end{aligned}\right\} \tag{5}$$

where the constant ξ_o is defined by $\xi_o = 0.5\beta\, M_{\infty,o}/(dM_\infty/d\xi)_o$ and $M_{\infty,o}$ is the edge Mach number at ξ_o.

To complete the transformation we define the following dimensionless quantities:

$$\begin{aligned} f' &= u/u_\infty & g &= H/H_\infty \\ c &= \rho\mu/\rho_\infty\mu_\infty & \beta &= (2\xi/M_\infty)(dM_\infty/d\xi) \end{aligned} \tag{6}$$

Under this transformation the momentum and energy equation become

$$(c\hat{\varepsilon}_m f'')' + ff'' + \beta(g-f'^2) = 2\xi(f'\frac{\partial f'}{\partial \xi} - f''\frac{\partial f}{\partial \xi}) \tag{7}$$

$$(c\hat{\varepsilon}_h g')' + fg' = 2\xi(f'\frac{\partial g}{\partial \xi} - g'\frac{\partial f}{\partial \xi}) \tag{8}$$

$$\begin{aligned} &\text{at } \eta = 0 \quad f = f' = 0, \quad g = T_s/T_{o,\infty} \\ &\text{at } \eta \to \infty \quad f' = g = 1 \end{aligned} \tag{9}$$

Equations (7) and (8) form a set of equations which should be solved subject to boundary conditions (9). The solution will be obtained by marching in the streamwise direction. The normalised stream function f is calculated by quadrature of f'. The solution will be of fourth order in the normal direction, and therefore some additional boundary conditions will be required.

3. THE NUMERICAL METHOD

3.1 Discretization in the x direction

In order to obtain stable marching and second order accuracy in the x direction, we use an extension of the three level scheme proposed by Livne et. al.[11].The scheme applies to any differential equation of the form

$$\frac{\partial f}{\partial x} = Lf$$

where L is a nonlinear operator of order two. Proceeding from x_1 to x_2, and using the solution at the previous station x_o as well, we get

$$\left.\frac{\partial f}{\partial x}\right|_{x=x_2} = -\frac{\lambda}{1+\lambda}\frac{f_1-f_0}{h_1} + \frac{1+2\lambda}{1+\lambda}\frac{f_2-f_1}{h_2} = Lf_2 \tag{10}$$

where

$h_1 = x_1 - x_0$

$h_2 = x_2 - x_1$

$\lambda = h_1/h_2$

Equation (10) may be regarded as a second order ordinary differential equation for f_2, which will be solved using a fourth order scheme.

3.2 Stretching in the y - direction

As the velocity gradients become very steep near solid walls, it becomes necessary to use non-uniform meshes, with a fine grid near the wall, and a coarse grid away from it. If the order of the method is to be retained, it is required to replace the non uniform mesh by a suitable coordinate transformation. In general we choose a new coordinate ζ such that

$$\zeta = \zeta(\eta) \tag{11}$$

The governing equations (7) - (8) become, after the transformation:

$$c\hat{\varepsilon}_m \zeta'^2 \frac{\partial^2 v}{\partial \zeta^2} + [c\hat{\varepsilon}_m \zeta'' + \frac{\partial(c\hat{\varepsilon}_m)}{\partial \zeta}\zeta' + f]\,\zeta' \frac{\partial v}{\partial \zeta} + J\beta(g-v^2) =$$

$$2\xi(v\frac{\partial v}{\partial \zeta} - \zeta'\frac{\partial v}{\partial \zeta}\frac{\partial f}{\partial \zeta}) \tag{12}$$

where $J = 0$ when $v = g$, and $J = 1$ when $v = \partial f/\partial \eta$.
The numerical solution is performed on these equations, where the derivatives $d\zeta/d\eta$, $d^2\zeta/d\eta^2$ and $d^3\zeta/d\eta^3$ should be supplied to the computer program. We have used a number of transformations, which will be discussed below:

<u>For laminar flow</u>

$$\zeta = 1 - \exp(-A\eta) \qquad \zeta' = A\exp(-A\eta)$$
$$\zeta'' = -A^2\exp(-A\eta) \qquad \zeta''' = A^3\exp(-A\eta) \tag{13}$$

This transformation is linear near the wall, where $\zeta \simeq A\eta$, and varies very slowly near the outer boundary. Moreover, $\zeta = 1$ corresponds to $\eta \to \infty$. Experience shows that the quality of the solution is not very sensitive to the value of A. Typically $A \simeq 1$ for $\beta = 0$, and increases to $A = 5$ for $\beta \to \infty$. Typical ζ - distributions are shown in fig. 1. The dots show the computed uniform pressure velocity profile $f'(\eta)$, while the solid line is $\zeta(\eta)$ for A=1. The open circles show the computed velocity profile at a high pressure gradient, of $\beta = 100$, and the dotted line is the $\zeta(\eta)$ distribution for A=5. The $\zeta(\eta)$ distributions are shown to be very similar to the corresponding velocity profiles, and appear therefore to be good stretching functions.

For turbulent flows

In this case we define, a double-layer transformation, namely

$$\zeta = A\eta + B\eta^2 + C\eta^3 + D\eta^4 \quad \text{for } \eta \leq \eta_{in}$$

$$\zeta = (\eta/\eta_\infty)^\alpha \quad \text{for } \eta \geq \eta_{in} \qquad (14)$$

with $\eta_{in} \simeq 0.3$; $\alpha \simeq 1/10$

$\eta_\infty = 20$

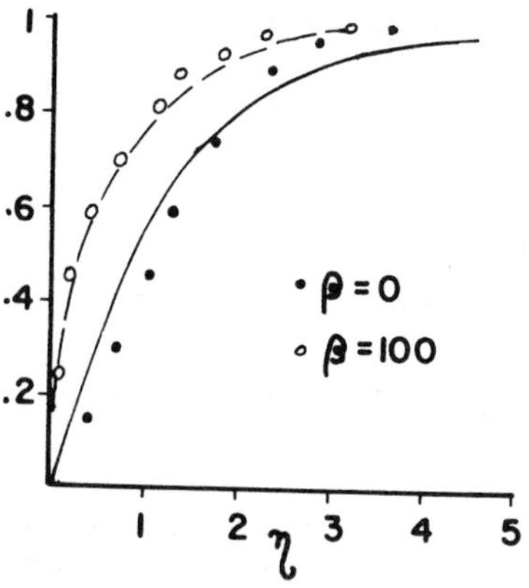

Fig. 1: Laminar Coordinate stretching

The constants A,B,C,D are obtained by equating ζ, ζ', ζ'' and ζ''' at η_{in} for both formulations of the stretching. This stretching accounts for the linear velocity profile in the sub-layer, and turns to a power law in the outer part of the flow from η=0.3 and out. As the turbulent boundary layer thickness is about η=20, the power law profile starts from about one hundredth of the boundary layer thickness which seems a reasonable estimate.

This stretching function does not require modification with the pressure gradient, because the turbulent velocity profile is not very sensitive to the pressure gradient parameter . Typical results are shown in fig. 2, where the dots and the open circles represent computed velocities at β=0, and β=100 respectively calculated with the mixing length model and the solid line is $\zeta(\eta)$ for the values of the parameters shown above. It is easy to see that the transformation is suitable for both value of β.

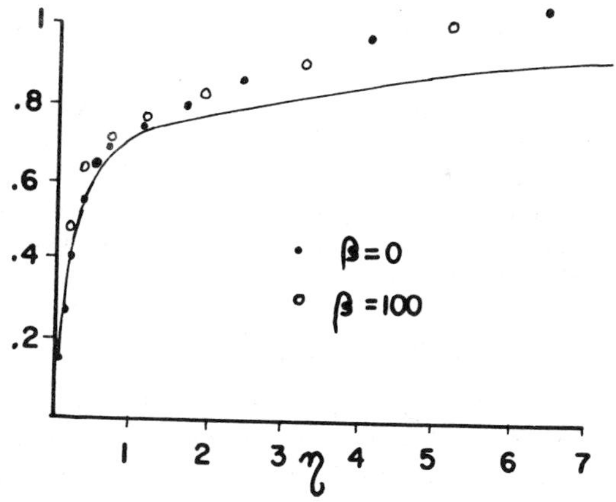

Fig. 2: Turbulent Coordinate stretching

3.3 The fourth order approximation to the ordinary differential equation

The fourth order solution of eq (12) is obtained by using the fourth order equations for the first and second derivatives

$$\frac{v_{i+1} - v_{i-1}}{2h} = \frac{v'_{i+1} + 4v'_i + v'_{i-1}}{6} \tag{15}$$

$$\frac{v_{i+1} - 2v_i + v_{i-1}}{h^2} = \frac{v''_{i+1} + 10v''_i + v''_{i-1}}{12} \tag{16}$$

The discretized form of eq (12) for the energy or momentum at the j-th station downstream is given by

$$(c\hat{\varepsilon}_m)_i \zeta_i'^2 \left(\frac{\partial^2 v}{\partial \zeta^2}\right)_i + \left[c_i \hat{\varepsilon}_{m_i} \zeta''_i + \frac{\partial}{\partial \zeta}(c\hat{\varepsilon}_m)_i + f_i + 2\xi_j \zeta'_i \frac{\partial f}{\partial \xi}_i \right] \zeta'_i \left(\frac{\partial v}{\partial \zeta}\right)_i$$

$$-(J\beta f'_i + 2\xi_j x_{2_i}) v_i + J\beta g_i + 2\xi_j f_i [x_1 v_i^{j-2} + (x_2 - x_1) v_i^{j-1}] = 0 \tag{17}$$

where $x_1 = -\lambda/(1+\lambda)\Delta\xi_j \qquad x_2 = (1+2\lambda)/(1+\lambda)\Delta\xi_j$

$\lambda = \Delta\xi_{j-1}/\Delta\xi_j$ and $\Delta\xi_j = \xi_j - \xi_{j-1}$ $\Delta\xi_{j-1} = \xi_{j-1} - \xi_{j-2}$

f'_i is the solution from the previous iteration step. v_i^{j-1} and v_i^{j-2} are the solutions at the j-1 and j-2 stations.

By substitution of (17) in (16) we get a system of two equations in v and v'. This is a block-tri-diagonal system which can be easily solved by standard algorithms. As equation (17) represents two non-linearly coupled equations, the tri-diagonal solver is applied iteratively until convergence is obtained.

3.4 The boundary conditions

The method presented in section 3.3 above is based on two steps. Firstly we discretize the parabolic partial differential equation in the marching direction β. Thus we obtain a second order ordinary differential equation for the variables in the downstream station. The fourth order solution of such equations is performed by numerical solution of two coupled equations, for the function and its derivative. Thus new boundary conditions are required for the first derivative, if the function is given, or for the function if the first derivative is given.

The method outlined below allows the specification of an additional relation between the function and its first derivative, by using the differential equation. Let us consider the original ordinary equation (12) which may be written in the following form:

$$v_{\eta\eta} + \hat{a}v_{\eta} + \hat{b}v + \hat{c} = 0 \tag{18}$$

One differentiation of (18) and substitution of $v_{\eta\eta}$ from (18) gives

$$v_{\eta\eta\eta}+(-\hat{a}^2+\hat{a}_{\eta}+\hat{b})v_{\eta}+(-\hat{a}\hat{b}+\hat{b}_{\eta})v+\hat{c}_{\eta}-\hat{a}\hat{c} = 0 \tag{19}$$

Thus we can write, for a point on the wall ("s")

$$\begin{aligned} v_{\eta\eta,s} &= Av_{\eta,s} + Bv_s + C \\ v_{\eta\eta\eta,s} &= Dv_{\eta,s} + Ev_s + F \end{aligned} \tag{20}$$

where

$$\left.\begin{aligned} A &= -\hat{a}_s & D &= \hat{a}_s^2 - \hat{a}_{\eta,s} - \hat{b}_s \\ B &= -\hat{b}_s & E &= \hat{a}_s\hat{b}_s - \hat{b}_{\eta,s} \\ C &= -\hat{c}_s & F &= \hat{a}_s\hat{c}_s - \hat{c}_{\eta,s} \end{aligned}\right\} \tag{21}$$

In order to trasnform eq (20) to the stretched coordinate we use the transformation derivatives:

$$\left.\begin{aligned} v_{\eta} &= v_{\zeta}\zeta' \\ v_{\eta\eta} &= v_{\zeta\zeta}\,\zeta'^2 + v_{\zeta}\,\zeta'' \\ v_{\eta\eta\eta} &= v_{\zeta\zeta\zeta}\,\zeta'^3 + 3v_{\zeta\zeta}\,\zeta'\,\zeta'' + v_{\zeta}\,\zeta''' \end{aligned}\right\} \tag{22}$$

and now by substitution of (20) in (22) we obtain

$$\begin{aligned} v_{\zeta\zeta,s} &= \bar{A}\,v_{\zeta,s} + \bar{B}v_s + \bar{C} \\ v_{\zeta\zeta\zeta,s} &= \bar{D}v_{\zeta,s} + \bar{E}v_s + \bar{F} \end{aligned} \tag{23}$$

where

$$\left.\begin{aligned} \bar{A} &= (A\zeta_s'-\zeta_s'')/\zeta_s'^2 & \bar{D} &= (D-3\zeta_s'\zeta_s''\bar{A}-\zeta_s''')/\zeta_s'^3 \\ \bar{B} &= B/\zeta_s'^2 & \bar{E} &= (E-3\zeta_s'\zeta_s''B)/\zeta_s'^3 \\ \bar{C} &= C/\zeta_s'^2 & \bar{F} &= (F-3\zeta_s'\zeta_s''C)/\zeta_s'^3 \end{aligned}\right\} \tag{24}$$

To obtain the new $v \sim v_{\zeta}$ relation on the boundary we expand v in ζ near the wall as follows

$$16v_1 - v_2 = 15v_s + 14h\, v_{\zeta,s} + 6h^2 v_{\zeta\zeta,s} + \frac{4h^3}{3} v_{\zeta\zeta\zeta,s} + \ldots \quad (25)$$

where "1" and "2" are the two points adjacent to the wall. By substitution of (23) in (25) we get

$$16v_1 - v_2 = R + Sv_s + Tv_{\zeta,s} \quad (26)$$

where

$$\left.\begin{aligned} R &= 6h^2\bar{C} + 4h^3\bar{F}/3 \\ S &= 15 + 6h^2\bar{B} + 4h^3\bar{E}/3 \\ T &= 14h + 6h^2\bar{A} + 4h^3\bar{D}/3 \end{aligned}\right\} \quad (27)$$

Eq. (26) is the additional relation between the three edge points, "s", "1" and "2".

As an illustration of the method we list below all the coefficients for the case of turbulent boundary layer:

we use (17) and see that 17 has the form of

$$v_\xi\, v_{\zeta\zeta} + \hat{a}v_\zeta + \hat{b}v_\zeta + \hat{c} = 0$$

where $\hat{a} = [\, c\hat{\epsilon}_m \zeta'' + (c\hat{\epsilon}_m)_\zeta + f + 2\xi\, \zeta' \frac{\partial f}{\partial \xi}]/(c\hat{\epsilon}_m \zeta')$

$$\hat{b} = -(J\beta f' + 2\xi x_2)/(c\hat{\epsilon}_m \zeta'^2)$$

$$\hat{c} = \{2\xi f[\, x_1 v^{j-2} + (x_2 - x_1)v^{j-1}] + J\beta g\}/(c\hat{\epsilon}_m \zeta'^2)$$

on the wall these expressions become much simpler

$$\hat{a}_s = (c_s \zeta''_s + c_{\zeta,s})/c_s \zeta'_s$$

$$\hat{b}_s = -2\xi x_2/c_s \zeta'^2_s$$

$$\hat{c}_s = J\beta g_s/c_s \zeta'^2_s$$

when we have these expressions we use them to calculate $A,B,C,D,E,F,\ \bar{A},\bar{B},\bar{C},\bar{D},\bar{E},\bar{F}$ according to (21),(24), and (27) to calculate R,S,T, the form of T for example is

$$T = 14h - 6h^2 \frac{c'_s}{c_s} + \frac{4}{3}\frac{h^3}{c_s}\left(2\frac{c'_s}{c_s} - c''_s\right)$$

where $c'_s = \left(\frac{\partial c}{\partial \eta}\right)_s$ and $c''_s = \left(\frac{\partial^2 c}{\mu \eta^2}\right)_s$

The method as described allows easy formulation of an "additional boundary condition" at any side at boundary layer, and is applicable also to other equations, as those governing the concentration or the turbulant quantities. It may also be noted that all the coefficients are calculated in the original (unstretched coordinate system, and may, therefore be calculated regardless of the transformation. Only when R, S and T are calculated is the transformation taken into account, and in a universal and easy way.

3.5 The numerical procedure

Description of the algorithm

Equation (17) represents a system of two coupled non-linear equations, which is solved iteratively. The two equations are first iterated to resolve the linear coupling between them. Then all the non-linear coefficients are updated, and the linear iterations are renewed. The inner iterations are stopped when the condition $||v_n - v_{n-1}||_\infty < \varepsilon_1$ is satisfied, where $\varepsilon_1 \simeq 10^{-7}$. The outer iteration is stopped when the residue of eq. (17) is smaller than $\varepsilon_2 \simeq 10^{-3}$.

Initial values for the iterative process

At low pressure gradient, when moving, downstream from x^k to x^{k+1} we calculate an initial guess of the coefficients using the value of the functions g and f at x^k. However, at the presence of large pressure gradient, it turns out that the coefficients and the values of $\frac{\partial f}{\partial \xi}$ and $\frac{\partial f'}{\partial \xi}$ are better approximated by the self-similar laminar solution for the local β, obtained from tabulated solutions.

Numerical integrations

All the integrals for evaluation of parameters as θ, δ^* are computed by the Simpson's rule. To compute it we use the following method: Since f' and f" are known, we apply (15) and (16) to compute f and f and then by (15) compute f for i>3 (i denotes here the points along the lateral direction, i=0 denotes the point at the wall).

4. RESULTS

As we are interested in a general treatment of stretching transformations and fourth order boundary conditions, we shall examine the convergence rates of the skin friction coefficient (C_f) with and without the stretching transformations. This will be done for the different cases, as follows:

4.1 Laminar flow

In this case C_f is strongly dependent on β. Therefore we calculated two cases of β=0 and β=100, using the stretching given by eq (13). The results are shown in fig. 3 in the form of a Richardson extrapolation for C_f at a given distance of 0.0564 m. from the initial station and starting from a self-similar velocity profile. The "exact" solution was obtained by extrapolating the computed Cg to $Cf_{h=0}$, at a zero mesh size, and the ordinate, ε, is given by

$$\varepsilon = c_{f_h} - c_{f_{h=0}}$$

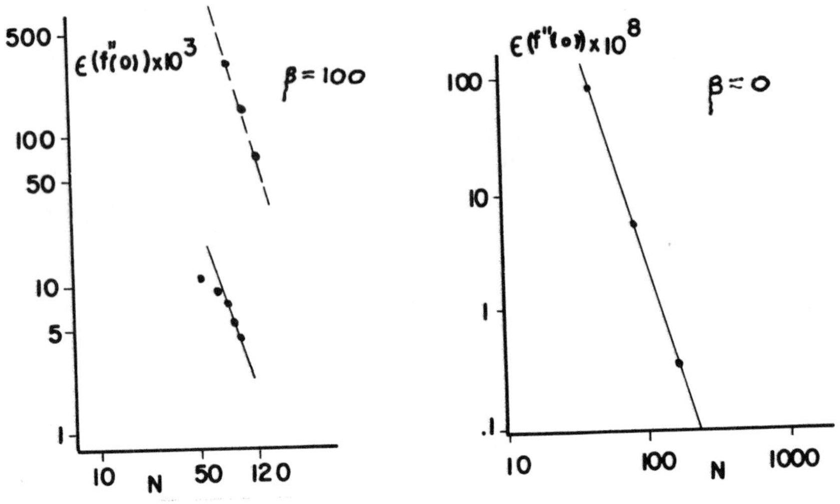

Fig. 3: Richardson extrapolation for Laminar flows

lines represent the convergence with stretching while the dashed lines are for the case without stretching. The fourth order convergence is easily identified. The influence of the fourth order method allows a great reduction in the number of mesh points. It is very interesting to note that about 6 mesh points are sufficient for highly accurate zero pressure-gradient solution, and about 10 are required for β=100. Much finer meshes are required for unstretched coordinates. The fourth order convergence of the boundary values is clearly demonstrated. We should note that the method is entirely independent of the type of boundary condition, or the stretching used.

4.2 Turbulent flow

Here the solution is not strongly influenced by β. Therefore a solution with β=0 is sufficient to demonstrate the method. Results for the mixing length model are shown in fig. 4. About 20 mesh poins are sufficient to obtain a good accuracy. Similar conclusions may be drawn from results with a two equation turbulence model shown in fig. 4 as well. We

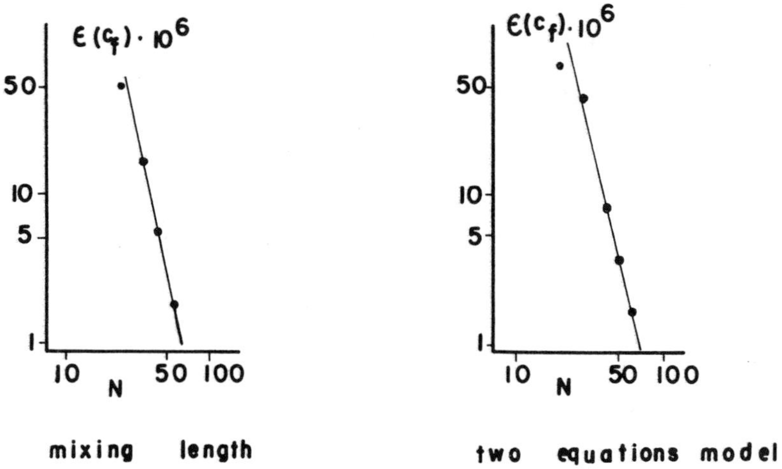

Fig. 4: Richardson extrapolation for turbulent flow, β=0

found that if we want to obtain the same accuracy with a second order scheme the number of mesh points should be increased by an order of magnitude, as reported by Arad et. el. [12].

5. CONCLUSIONS

The paper demonstrated a simple universal method for the incorporation of fourth order boundary conditions and coordinate stretching in boundary layer flows. The method has been demonstrated in Laminar and turbulant flow, for various pressure gradients, and it is suitable also for compressible flow. Some coordinate transformations were presented, and their usefullness was demonstrated.

Our main conclusion from this study is that coordinate stretching and compact high order schemes form an essential and inevitable tool for the numerical solution of boundary layer problems, and that carefull analysis and programming allow for an easy implementation of these techniques.

REFERENCES

1. PATANKAR, S.V. and SPALDING, D.B. - Heat and Mass Transfer in Boundary Layers, Intertext Books, London, 1971.

2. CEBECI, T. and SMITH, A.M.O. - Analysis of Turbulent Boundary Layers, Academic Press, New York, 1974.

3. KELLER, H.B. and CEBECI, T. - Accurate Numerical Methods for Boundary Layer Flows - II. Two Dimensional Turbulent Flows, AIAA J., Vol. 10, p. 1193, 1972.

4. KREISS, H.O. - Methods for the approximate Solution of Time Dependant Problems, GARP Report No. 13, Geneva, 1975.

5. KRAUSE, E., HIRSCHEL, E.H., and KORDULLA, W. - Fourth Order "Mehrstellen" - Integration for Three-Dimensional Turbulent Boundary Layers, Computers and Fluids, Vol. 4, pp. 77-92, 1976.

6. HIRSH, R.S. - Higher Order Accurate Difference Solutions of Fluid Mechanics Problems by a Compact Differencing Technique, J. of Computational Physics, Vol.19, pp.90-109, 1979.

7. ADAM, Y. - Highly Accurate Compact Implicit Methods and Boundary Conditions, J. of Computational Physics, Vol.24, pp.10-22, 1977.

8. CIMENT, M., LEVENTHAL, S.H., and WEINBERG, B.C. - The Operator Compact Implicit Method for Parabolic Equations, J. of Computational Physics, Vol.28, pp.135-166, 1978.

9. BERGER, A.E., SOLOMON, J.M., CIMENT, M., LEVENTHAL, S.H., and WEINBERG B.C., - Generalised OCI Schemes for Boundary Layer Problems, Math. of Comp., Vol.35, No. 151, pp. 695-731, 1980.

10. ROSTAN, A., PADE, O., ANSHELOVITZ, D., and WOLFSHTEIN, M., Accurate solutions for Laminar and turbulent boundary layers at very large pressure gradients, in Numerical Methods in Laminar and Turbulent Flows, Ed. C. Taylor and B.A. Schrefler, Pineridge Press, 1981.

11. LIVNE, A. and ISRAELI, M., The development of a super-stable scheme of second order in the streamwise direction and fourth order in the normal direction for boundary layer flows, Aerodynamic Laboratory Report No. 0-166, Technion (in Hebrew), 1977.

12. ARAD, E., BERGER, M., ISRAELI, M., and WOLFSHTEIN M., Numerical calculation of transitional boundary layers, Int. J. Num. Math. in Fluids, Vol. 2, pp. 1-23, 1982.

NUMERICAL CALCULATION OF THE EFFECT OF PROCESS VARIABLES ON RIBBON THICKNESS FORMATION DURING MELTSPINNING

L. Katgerman and W.E. Zalm*

ABSTRACT

To determine quantitatively the effect of meltspinning process variables (liquid metal superheat, substrate velocity) on ribbon thickness formation a mathematical model has been developed. The equations of liquid metal flow and heat flow - which are coupled by the moving solid - liquid interface - are solved numerically using a finite difference scheme. From the computed velocity and temperature distribution the momentum and thermal boundary-layers are calculated. To relate the momentum boundary-layer thickness to the ribbon thickness as a function of substrate velocity the usually applied 1% boundary-layer definition has been modified by application of the dynamic pressure concept. The prediction of the effects of melt superheat and wheel velocity on ribbon thickness show a good agreement with experimental data.

1. INTRODUCTION

Rapid quenching techniques to solidify molten metals and alloys at high cooling rates in the order of 10^5-10^7 K/s have shown a growing interest in the last years. Improvements in structure and properties by rapid solidification have been readily reported [1-4]. However many rapid solidification techniques only produce laboratory specimens and cannot be used for continuous large scale processing.

One of the continuous processes becoming widely used is the meltspinning technique. This process involves the impingement of a molten metal stream onto a cold rotating wheel of high thermal conductivity. A continuous ribbon is formed by solidification and extraction of liquid metal from the pool of

* Laboratory of Metallurgy, Rotterdamseweg 137
Delft University of Technology, 2628 AL Delft, The Netherlands

molten metal (melt puddle) at the stagnation point on the wheel. (Fig. 1). Meltspun ribbons with typical thicknesses of 50-100 μm usually exhibit dimensional variations which can give non-uniform properties [5,6]. These variations can often be associated with casting parameters and have been investigated experimentally [7-10].

In order to interprete the experimental data the development of a mathematical model is desirable to quantify the effects of process variables. In the analyses so far the interaction between fluid flow and heat flow has been neglected [10-13]. When this interaction is taken into account it was demonstrated [14,15] that the thermal boundary-layer plays a more significant role than may be concluded from non-interaction calculations.

In the present paper the main objective will be to determine the effects of meltsuperheat and wheel velocity on ribbon thickness by means of a mathematical description of fluid flow and heat flow phenomena during the process. The predictions based on this model will be compared with experimental results.

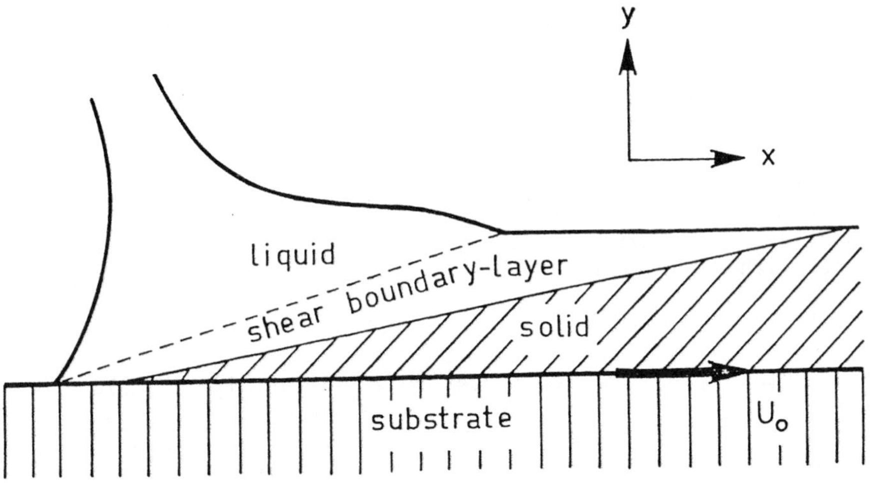

Fig. 1 Schematic representation of the melt puddle.

2. MATHEMATICAL FORMULATION

To describe the fluid flow and heat flow in the meltpuddle mathematically the following assumptions are made:
- the puddle is semi-infinite in the y-direction
- thermal and physical properties are independent of temperature
- the fluid flow is laminar
- the velocity components in y and z directions are zero and

consequently the x-component is a function of y and t only
- normal forces are neglible as compared with shear forces
- the solid-liquid interface is plane and the position is given by $y = \delta_T(t)$
- the heat flow is geometrically in one dimension (y-direction) and can be characterized by a convective heat-transfer coëfficiënt h

For a Cartesian co-ordinate system the governing partial differential equations relative to the moving solidification front take then the following form:

$$\frac{\partial u}{\partial t} = \nu \frac{\partial^2 u}{\partial y'^2} + R(t)\frac{\partial u}{\partial y'} \qquad y' \geqq 0 \tag{1}$$

$$\frac{\partial T}{\partial t} = \alpha_L \frac{\partial^2 T}{\partial y'^2} + R(t)\frac{\partial T}{\partial y'} \qquad y' \geqq 0 \tag{2}$$

$$\frac{\partial T}{\partial t} = \alpha_S \frac{\partial^2 T}{\partial y'^2} + R(t)\frac{\partial T}{\partial y'} \qquad -\delta_T \leqq y' \leqq 0 \tag{3}$$

in which $y' = y - \delta_T(t)$ is the distance from the solid-liquid interface; δ_T the solidified thickness; $R(t) = d\delta_T/dt$ the solidification rate; α_L and α_S thermal diffusivity of liquid and solid metal respectively and ν the kinematic viscosity.
The initial condition is (see appendix for nomenclature):

$$t = 0 \qquad \text{all } y' \qquad u = 0 \qquad T = T_p \tag{4}$$

and the boundary conditions for $t > 0$ are:

$$\text{chill surface} \qquad y' = -\delta_T \quad u = U_o \qquad K_S \frac{\partial T}{\partial y'} = h(T - T_w) \tag{5}$$

$$\text{solid-liquid interface} \qquad y' = 0 \quad u = U_o \qquad T = T_M \tag{6}$$

$$y' = \infty \quad u = 0 \qquad T = T_P \tag{7}$$

The conservation of heat at the solid-liquid interface in addition requires that

$$\rho L_f \frac{d\delta_T}{dt} = K_s \left.\frac{\partial T}{\partial y'}\right|_{y'=0^-} - K_L \left.\frac{\partial T}{\partial y'}\right|_{y'=0^+} \tag{8}$$

To obtain the velocity distribution from eq. (1) first the temperature distribution and the solidification rate have to be calculated. For zero-superheat ($T_P = T_M$, the liquid metal is at the melting point) approximate analytical solutions are available [16,17]. In the Megerlin approximation [17] the solidification rate can be expressed as:

$$\frac{d\delta_T}{dt} = \frac{K_s}{\rho L_f} \frac{\partial T}{\partial y'}\bigg|_{y'=0^-} = M \frac{h(T_M - T_W)}{1 + h\delta_T/K_s} \tag{9}$$

Substituting this expression into the heat-balance equation at the solid-liquid interface yields

$$\frac{d\delta_T}{dt} = \frac{1}{\rho L_f} \left(M \frac{h(T_M - T_W)}{1 + h\delta_T/K_s} - K_L \frac{\partial T}{\partial y'}\bigg|_{y'=0^+} \right) \tag{10}$$

The solidification rate and solidified thickness can be obtained from the combined solution of eqs. (2) and (10) and further the velocity distribution is determined from eq. (1). Consequently the total ribbon thickness δ is related to the thermal and shear boundary-layer thicknesses by $\delta = \delta_T + \delta_M$.

The momentum contribution δ_M to the total ribbon thickness is usually calculated from the boundary-layer thickness $\delta_{0.01}$ defined as that thickness for which $u(y',t)$ has dropped to a value of $0.01U_o$. This is an arbitrary definition and it is more appropriate as proposed by Vincent et al [13] to use the displacement thickness δ_1 defined by [18]

$$U_o\delta_1 = \int_o^\infty u(y',t)dy' \tag{11}$$

In both definitions the calculated momentum thickness is independent of the substrate velocity U_o and consequently the physical condition that the liquid metal has to be dragged out the meltpuddle is not satisfied. To include this condition the boundary-layer thickness is defined as that distance for which the average dynamic pressure exceeds the surface tension of the puddle. Mathematically this can be expressed as follows:

$$\tfrac{1}{2}\rho_L < u^2(\delta_M, t)> = 4\gamma_{LV}/\delta \tag{12}$$

where < > denotes the average in the y' direction.

3. SOLUTION

Eqs. (1), (2) and (10) with initial and boundary conditions (4)-(7) are solved simultaneously by numerical techniques. For eqs. (1) and (2) the finite difference method with a 4-point explicite difference scheme and forward time difference was used. With the same time step Δt different mesh sizes for velocity and temperature field were used in order to fulfill the stability requirements. The solidified thickness and solidification rate are calculated from eq. (10) by the Runga-Kutta method. The liquid thermal gradient in eq. (10) is obtained from the finite difference discrete temperature distribution by application of a two-dimensional Lagrange interpolation formula.

The boundary-layer thicknesses according to eqs. (11) and (12) were calculated from the velocity distribution using a 5-point Lagrange interpolation scheme in combination with a regula-falsi procedure.

The numerical procedure was first checked with the only available analytical solution in the case of ideal cooling ($h = \infty$). In that case the solidification rate is given by [19]:

$$\frac{d\xi_T}{dt} = \gamma(\alpha_S/t)^{\frac{1}{2}} \tag{13}$$

in which γ is determined from the heat-balance equation (8) by an implicite expression given in ref. [19].
The resulting velocity distribution is then given in ref. [14]. The agreement between analytical and numerical solutions of heat flow and fluid flow was highly satisfactory and the relative error in the calculated boundary-layers was less than 1%. The calculations were carried out using the AMDAHL 470/V7B computer system of the Delft University of Technology.

4. RESULTS AND DISCUSSION

The computations have been carried out in the case of aluminium. The data used in the calculations are listed below. Density ρ 2.7 10^3 (solid), 2.36 10^3 kg/m^3 (liquid), latent heat of fusion L_f = 3.9 10^5 J/kg, specific heat C_p 1.04 10^3 (solid), 1.08 10^3 J/kg/K (liquid), thermal conductivity K 209 (solid), 92 W/m/K (liquid), dynamic viscosity 1.3 10^{-3} Ns/m^2, T_M melting point 660°C, T_W substrate temperature 25°C, γ_{LV} surface tension 0.866 N/m.

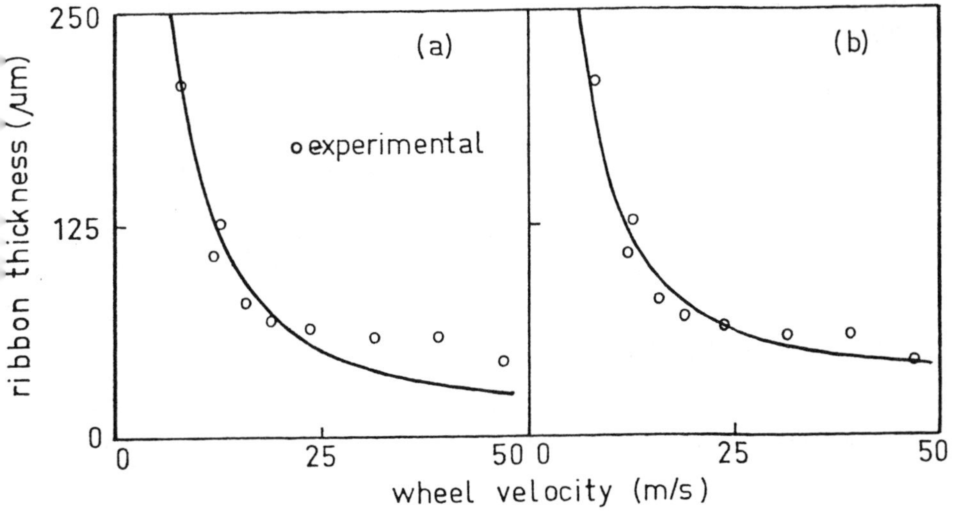

Fig. 2 Experimental and calculated ribbon thickness vs. substrate velocity. T_p = 780°C.
(a) displacement thickness; $h = 8.25\ 10^5$ W/m^2/K
(b) dynamic pressure ; $h = 7.23\ 10^5$ W/m^2/K

To compare the predicted ribbon thickness with the experimental values the calculations were carried out for different values of the heat-transfer coefficient h. In the calculations the empirical derived expression for the puddle length [20] $L_p = 8.61\ 10^{-3}\ U_o^{-0.42}$ (m) was used.
The results are shown in Fig. 2a and b. The ribbon thickness calculated with the displacement thickness (eq. (11)) largely underestimates the ribbon thickness at higher velocities. The calculated boundary-layers according to the dynamic pressure definition (eq. (12)) are in better agreement with the experimental data especially at higher wheel velocities, which demonstrates the applicability of the dynamic pressure concept. The obtained values of h of about 7.10^5 W/m^2/K are in agreement with independently determined heat-transfer coefficients in the range 10^5–10^6 from X-ray experiments [21].

In Figs. 3-5 the results for different degrees of melt superheat are given. For conveniency the following dimensionless variables are used:

$$\tau = \frac{h^2}{K_s^2}\alpha t \ ; \ \xi_M = \frac{h\delta_M}{K_s} \ ; \ \xi_T = \frac{h\delta_T}{K_s}$$

The total ribbon thickness is hardly influenced by the melt superheat. This is in agreement with our experimental experience [20] where no significant change in ribbon thickness is found as a function of meltsuperheat. However the different contributions to the total ribbon thickness are sensitive to the superheat. When the ratio of momentum thickness to thermal thickness ξ_M/ξ_T is plotted against time we can see that momentum transport becomes more dominant as the superheat increases. The momentum boundary-layer coming out the puddle solidifies under different conditions and as a result can give a different as-quenched microstructure. From Fig. 4 it is clear that the characterization by thickness (cooling rate) alone is incomplete and that the meltsuperheat should be included. Fig. 5 shows the development of the momentum boundary-layer. At the beginning of solidification the boundary-layer is equal to the non-interaction thickness. Due to the moving solid-liquid interface the momentum boundary-layers is substantially decreased and approaches the ideal cooling boundary-layer thickness for that specific melt superheat. This is analoque to previous calculations where the superheat was not included [15].

ADDITIONAL NOMENCLATURE

h	heat-transfer coefficient	(W/m^2/K)
L_f	latent heat of fusion	(J/kg)
M	constant in Megerlin-approximation	
R	solidification rate	(m/s)

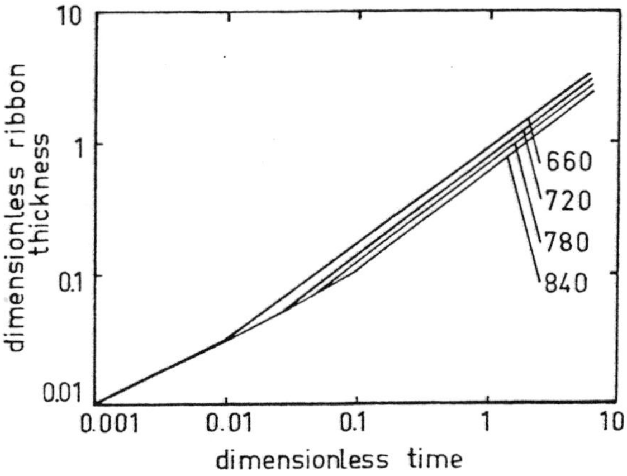

Fig. 3 Calculated dimensionless ribbon thickness vs. dimensionless time for different casting temperatures.

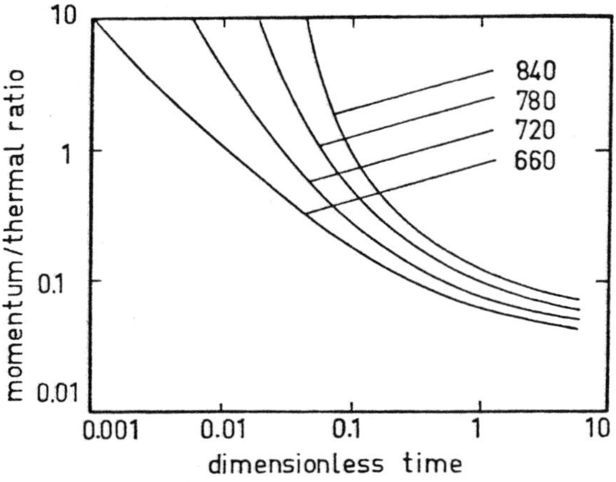

Fig. 4 Calculated momentum to thermal thickness ratio vs. dimensionless time for different casting temperatures.

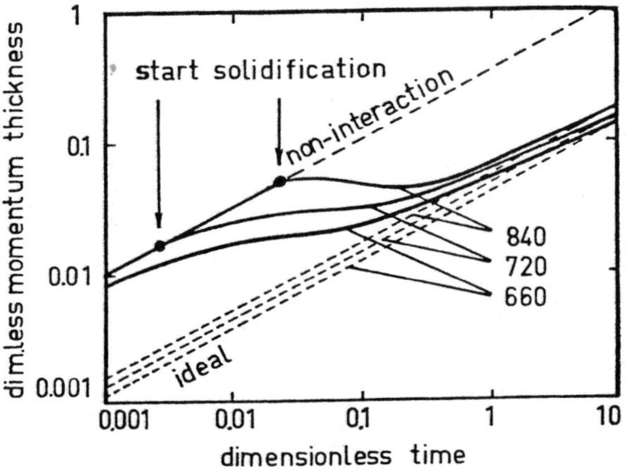

Fig. 5 Calculated dimensionless momentum boundary-layer thickness vs. dimensionless time for different casting.

T_M	melting point	(^oC)
T_p	casting temperature	(^oC)
T_W	substrate temperature	(^oC)
α	thermal diffusivity	(m^2/s)
γ_{LV}	surface tension	(N/m)
δ_M	momentum boundary-layer thickness (m)	
δ_T	solidified thickness	(m)
ρ	density	(kg/m^3)

ACKNOWLEDGEMENTS

The authors are very grateful to Miss Anneke van Veen for accurate typing of the manuscript and the Alcoa Foundation Pittsburgh, USA for partial financial support.

REFERENCES

1. Rapid Solidification Processing: Principles and Technologies, Ed. R. Mehrabian et al, Claitor's Publ. Div., Baton Rouge, La (1978).
2. Rapid Solidification Processing: Principles and Technologies II, Ed. R. Mehrabian et al, Claitor's Publ. Div., Baton Rouge, La (1980).
3. Rapidly Quenched Metals IV, Eds. T. Masumoto, K. Suzuki, The Japan Institute of Metals, Sendai, Japan (1982).
4. JONES, H. - Rapid Solidification of Metals and Alloys, Monograph no 8, The Institution of Metallurgist, London (1982).
5. BENDIJK, A., R. DELHEZ, L. KATGERMAN, Th. H. de KEIJSER, E.J. MITTEMEIJER, N.M. van der PERS - Characterization of Al-Si-alloys rapidly quenched from the melt, J. Mater. Sci. 15, 2803(1980).
6. DELHEZ, R., Th.H. de KEIJSER, E.J. MITTEMEIJER, P. van MOURIK, N.M. van der PERS, L. KATGERMAN, W.E. ZALM - Structural inhomogeneities of AlSi alloys rapidly quenched from the melt, J. Mater. Sci. 17, 2887(1982).
7. LIEBERMANN, H.H., C.D. GRAHAM - Production of Amorphous Alloy Ribbons and Effects of Apparatus Parameters on Ribbon Dimensions, IEEE. Trans. on Magnetics MAG-12, 921(1976).
8. CHARTER, S.J.B., D.R. MOONEY, R. CHEESE, B. CANTOR - Meltspinning of crystalline alloys, J. Mater. Sci. 15, 2658(1980).
9. LIEBERMANN, H.H. - The dependence of the geometry of glassy alloy ribbons on the chillblock meltspinning process parameters, Mat. Sci. Eng. 43, 203(1980).
10. HILLMAN, H., H.R. HILTZINGER - On the formation of amorphous ribbon by the meltspin technique, Rapidly Quenched Metals III Vol. 1, 22, Ed. B. Cantor, The Metals Society, London (1978).

11. KAVESH, S. - Principles of Fabrication, Metallic Glasses, ASM, Metals Park OH, USA (1978).
12. ANTHONY, T.R., H.E. CLINE - On the uniformity of amorphous metal ribbon formed by a cylindrical jet impinging on a flat moving substrate, J. Appl. Phys. 49, 829(1978).
13. VINCENT, J.H., H.A. DAVIES, J.G. HERBERTSON - A study of the meltspinning process, Continuous casting of small cross sections, 103, TMS-AIME, Warrendale, PA, USA (1981).
14. KATGERMAN, L. - Theoretical analysis of ribbon thickness formation during meltspinning, Scripta Met. 14, 861(1980).
15. KATGERMAN, L. - A numerical solution of ribbon thickness formation during meltspinning, Numerical Methods in Turbulent and Laminar Flow, 615, Eds. C. Taylor, B.A. Schrefler, Pineridge Press, Swansea, UK (1981).
16. JONES, H. - A comparison of approximate analytical solutions of freezing from a plane chill, J.I.M. 97, 38(1969).
17. MEGERLIN, F. - Geometrish Eindimensionable Warmeleitung beim Schmelzen und Erstarren, Diss., T.H. Aachen, Germany (1965).
18. SCHLICHTING, H. - Boundary Layer Theory, ch. 2, 7, 7th ed. McGraw-Hill, London (1979).
19. CARSLAW, H.S., J.C. JAEGER - Conduction of Heat in Solids, Oxford University Press, Oxford (1959).
20. KATGERMAN, L., P.J. van den BRINK - A mathematical model for meltspinning of crystalline alloys, Rapidly Quenched Metals IV, Vol. 1, 61, Eds. T. Masumoto, K. Suzuki, The Japan Institute of Metals, Sendai, Japan (1982).
21. KATGERMAN, L. - Microsegregation and extended solid solutions after rapid solidification of aluminium alloys, Scripta Met. accepted for publication.

A NUMERICAL INVESTIGATION INTO BOUNDARY-LAYER STABILITY ON COMPLIANT SURFACES

by
P.W. Carpenter(1), M. Gaster(2) and G.J.K. Willis(3)

(1) Lecturer in Engineering Science, University of Exeter, U.K.
(2) Senior Principal Scientific Officer (I.M.), NMI Ltd., Teddington, U.K.
(3) Scientific Officer, NMI Ltd.

SUMMARY

The effects, that a viscous fluid substrate in a Kramer-type compliant surface has on boundary-layer stability, are investigated numerically. The previously encountered difficulties are overcome by using the compound matrix method following Davey to integrate the Orr-Sommerfeld equation coupled with an appropriate eigenvalue search scheme. Contouring is used to confirm that two instability modes coalesce. Contrary to previous theoretical results it is found that viscous damping reduces the growth rate of the instabilities.

1. INTRODUCTION

The original compliant surfaces tested by Kramer [1] were constructed in the manner suggested schematically in Fig.1. Kramer found that with respect to drag reduction there was an optimum viscosity for the substrate fluid. In his opinion the function of the substrate fluid was to damp out "boundary layer" waves. In theoretical analyses Benjamin [2,3] and Landahl [4] showed that damping destabilises Tollmien-Schlichting instability (hereafter denoted by TSI) but there is another mode of instability which is stabilised by damping. This additional mode is similar to the travelling-wave flutter instability in aeroelasticity. It was shown in Refs. [5] and [6] that yet another mode of instability also appears to occur on Kramer-type surfaces. This takes the form of either a slow travelling wave or a divergence. In Ref. [6] it was shown that a viscous fluid substrate may have either a destabilising or stabilising effect on this mode. Despite, or perhaps because of, the theoretical work mentioned above the precise function of the viscous fluid in Kramer's surfaces still remains obscure.

Carpenter and Garrad [7] have recently studied the effect of a viscous fluid substrate on TSI on Kramer-type surfaces.

It appeared from their results that some sort of interaction occurs between the TSI and travelling-wave flutter modes. However, owing to the shortcomings of their numerical methods it was not possible to investigate this phenomenon properly. The problem is re-examined in the present paper by means of more suitable numerical techniques.

2. FORMULATION OF PROBLEM

A Kramer-type compliant surface is depicted schematically in Fig.1. It consists of a tensionless elastic plate of thickness, b, and density, ρ_p, with flexural rigidity, B. It is supported on springy stubs which are represented as a continuous spring foundation of stiffness, K_s. There is a fluid substrate below the plate having a kinematic viscosity, ν_s, and density, ρ_s, which, in general, are different from the main stream values, ν_e and ρ_e. The equation of motion for the plate is [6,7]:

$$\rho_p b \frac{\partial^2 Y_p}{\partial t^2} + B \frac{\partial^4 Y_p}{\partial x^4} + K_E Y_p = \delta p_s - \delta p_e \tag{1}$$

The coordinate system and other notation are defined in Fig.1, δp represents the perturbations in dynamic pressure acting on the plate, K_E is an equivalent spring stiffness defined by $K_E = K_s - g(\rho_e - \rho_s)$.

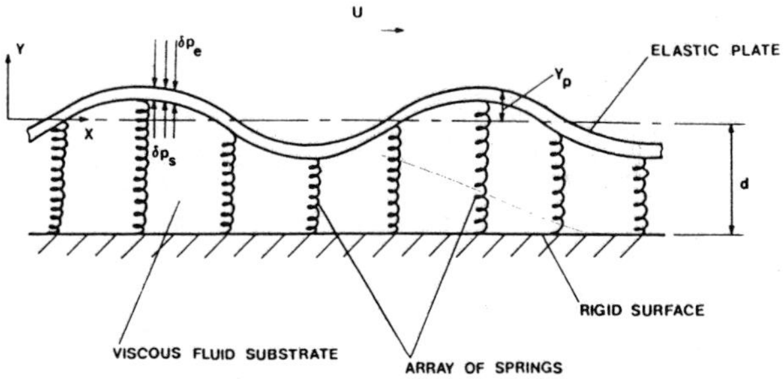

Fig.1 Schematic representation of a Kramer-type compliant surface.

Let the surface deflection take the form:

$$Y_p = Y_{po} \exp\{i\alpha(x-ct)\} \tag{2}$$

Substitution of (2) in (1) gives:

$$-\alpha^2 c^2 \rho_p b + \alpha^4 B + K_E = (\delta p_s - \delta p_e)/Y_p \tag{3}$$

It is assumed that the substrate fluid motion is unaffected by the presence of the springy stubs in which case the linearised Navier-Stokes equations can be solved for the substrate fluid motion. In this way it can be shown [6,7] that

$$\frac{\delta p_s}{Y_p} = \frac{\alpha\rho_s c^2\left[\beta(\beta+\alpha)\{e^{(\beta-\alpha)d}-e^{-(\beta-\alpha)d}\}+\beta(\beta-\alpha)\{e^{(\beta+\alpha)d}-e^{-(\beta+\alpha)d}\}\right]}{8\alpha\beta+(\beta-\alpha)^2\{e^{(\beta+\alpha)d}+e^{-(\beta+\alpha)d}\}-(\beta+\alpha)^2\{e^{(\beta-\alpha)d}+e^{-(\beta-\alpha)d}\}} \tag{4}$$

where d is the substrate depth and $\beta^2 = \alpha^2 - i\alpha c/\nu_s$.

The boundary-layer stability is investigated by assuming the small disturbances are characterised by a stream function of the form $f(y)\exp\{i\alpha(x-ct)\}$. f satisfies the well-known Orr-Sommerfeld equation. The usual outer boundary conditions can be used but the wall conditions for a flexible surface are quite different from the rigid case. Following Landahl [4] the no-slip condition takes the form:

$$f_w U_w' + c\, f_w' = 0 \tag{5}$$

where $(\ )_w$ denotes conditions at the wall, $(\ )'$ denotes differentiation with respect to y and U is the undisturbed velocity in the boundary layer. The other wall condition takes the form [7]:

$$\frac{v_w'}{\delta p_e} = \frac{\alpha^2 Re\, f_w}{f_w''' - \alpha^2 f_w'} = \frac{i\alpha c\, Y_p}{\delta p_e} \tag{6}$$

where v' is the velocity normal to the wall and $Re = U_\infty \delta^*/\nu_e$. The right-hand side of (6) can be evaluated in terms of the mechanical properties of the surface, etc., by using eqn. (3) with (4).

3. NUMERICAL METHODS

The Orr-Sommerfeld equation is a stiff differential equation and as such requires special procedures for its numerical integration. The rapid variation of f within certain parts of the range is the main cause of difficulty. One approach to the problem is to use an ortho-normalisation scheme. This approach was used successfully in Ref. [7] for many cases but problems were experienced in finding solutions for Kramer-type surfaces with viscous substrates. The problems appeared to be caused partly by the loss of analycity in the solutions due to repeated orthonormalisation which caused the eigenvalue search scheme to fail, and partly by modal interaction which is characterised by the existence of singular points amongst other things. So the present methods have been developed to avoid loss of analycity and to deal with the singular points.

The present results were obtained by using compound matrix method following Davey [8]. Six first-order differential equations are derived which involve cross-products of the derivatives of the two fundamental solutions. The problem can be cast in initial-value form starting at the outer boundary. A characteristic equation of the form:

$$F(\omega;\ \alpha,Re) = 0 \qquad (7)$$

(where $\omega = \alpha c$) can then be derived from the wall conditions (5) and (6). The compound matrix method avoids the necessity of repeated orthonormalisation so that $F(\omega)$, derived from the numerical integration, is analytic to within the limits of rounding errors.

The six equations are integrated by means of a fourth-order Runge-Kutta method. The undisturbed velocity profile and its derivatives are obtained by integrating numerically the Blasius equation.

Assuming that the function F is entire it can be expanded:

$$F(\omega) = F(\omega_o) + (\omega-\omega_o)F'(\omega_o) + \frac{1}{2}(\omega-\omega_o)^2F''(\omega_o) + \ldots.. \qquad (8)$$

where ω_o is an eigenvalue. A first-order Newton-Raphson scheme was used when $F'(\omega_o) \neq 0$ or a second-order scheme in special cases when $F'(\omega_o) \simeq 0$.

The first-order eigenvalue search scheme was used with success to construct the ω_i contours shown in Fig.2. At certain points in the α,Re-plane, however, it became extremely difficult to find eigenvalues indicating that $F'(\omega)$ was very small in the neighbourhood of a zero. This implied the coalescence of two eigenmodes. Accordingly contouring was used to investigate the solutions near these points.

For a point in the α,Re-plane where eigenvalues are hard to find a circle in the ω plane is chosen such that

$$\omega - \omega_o = re^{i\theta} \qquad (9)$$

where ω_o is an estimate of one of the eigenvalues. Using values of $F(\omega)$ calculated on this circle a Fourier series representation of $F(\omega)$ is made [9]. The number of zeroes, N_z, i.e. eigenvalues, (assuming no poles) inside the circle can be determined by evaluating numerically the contour integral:

$$\oint \frac{F'(\omega)}{F(\omega)}\, d\omega = N_z \qquad (10)$$

where $F(\omega)$ and $F'(\omega)$ are calculated from the series representation. Having established the existence of two close eigenvalues in these 'difficult' parts of the α,Re-plane the second-order search scheme was used to obtain the eigenvalues using values of $F(\omega)$ calculated from the series and checking the solution with values of $F(\omega)$ calculated from the compound matrix method.

4. RESULTS AND DISCUSSION

The present results all correspond to Kramer's [1] best coating, Type C. In Ref. [7] it was estimated that for this coating B = 0.217 MN m^{-2} and K_E = 115 MN m^{-3}, b = 2 mm, d =

1 mm, $\rho_p = 930$ kg m^{-3}, $\rho_e = 1025$ kg m^{-3} and $\rho_s = 970$ kg m^{-3}. Kramer's top operational speed of 18 m/s was assumed throughout.

It was found in Ref. [7] that for an inviscid substrate there are two independent modes of instability. One corresponds to TSI and the other to a type of travelling-wave flutter. It appeared as if some sort of interaction occurs between the modes but owing to the shortcomings of the numerical techniques it was not possible to verify this or to obtain a complete set of neutral curves for any viscosity ratio, ν_s/ν_e between 1 and 50.

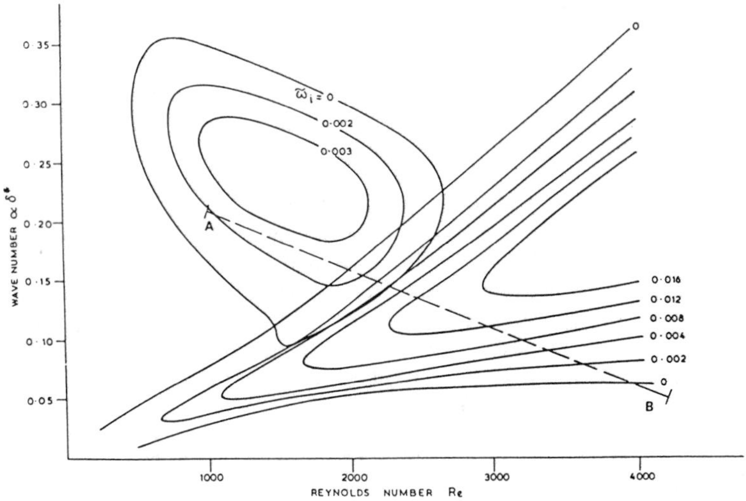

(a) $\nu_s/\nu_e = 10$

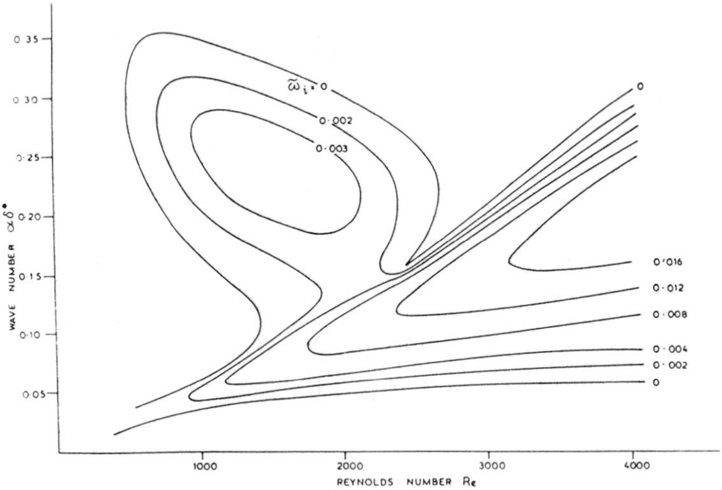

(b) $\nu_s/\nu_e = 50$

Fig.2 Contours of temporal amplification rate.

The present methods have been used to obtain the curves of constant $\tilde{\omega}_i$ ($\tilde{\omega} = \omega\delta^*/U_\infty$) shown in Fig.2. In Fig.2a the two modes still appear to be independent in that two unstable eigenvalue surfaces clearly exist over part of the α,Re-plane. However, the modes can be shown to interact since a double circuit round the branch point ($\alpha\delta^* = 0.09$, Re = 1500), following one eigenvalue surface, traverses both surfaces before closing on itself. This is illustrated in Fig.3.

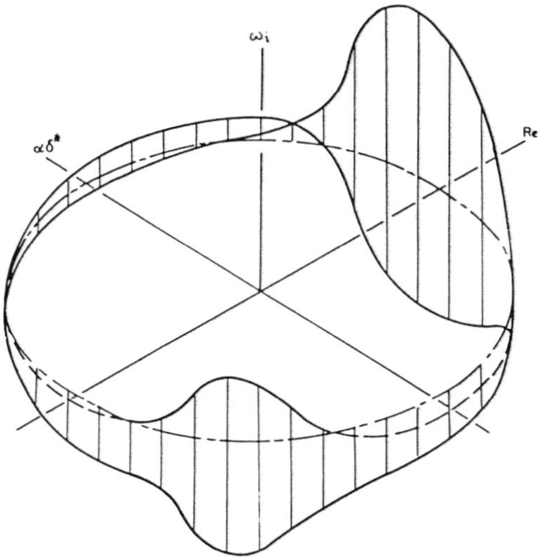

Fig.3 Circuit around the point $\alpha\delta^* = 0.09$, Re = 1500. $\nu_s/\nu_e = 10$.

In Ref. [7] a 'second island' of TSI was found to exist for $\nu_s/\nu_e = 0$ and 1 at fairly high Re and low $\alpha\delta^*$. The present results show that this 'island' shrinks rapidly with an increase in ν_s/ν_e and disappears completely for $\nu_s/\nu_e \simeq 2{\cdot}5$. This is illustrated in Fig.4 which shows the variation of $\tilde{\omega}_i$, corresponding to TSI and various values of ν_s/ν_e, along the cross-section AB defined in Fig.2a.

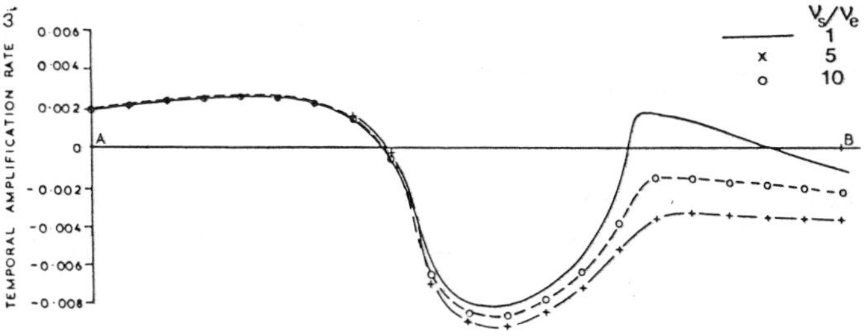

Fig.4 Variation of $\tilde{\omega}_i$ along cross-section AB of Fig.2a.

So, in contrast to previous theoretical results, it appears that viscous damping can have a stabilising effect on TSI. The present results also confirm the tentative conclusion made in Ref. [7] that viscous damping reduces the growth rates of the instabilities when the two modes coalesce. This may explain how a viscous substrate can have a beneficial effect on hydrodynamic stability.

REFERENCES

1. KRAMER, M.O. - Boundary layer stabilisation by distributed damping. J.Amer.Soc.Naval Engrs., Vol.72, pp.25-33. 1960.

2. BENJAMIN, T.B. - Effects of a flexible boundary on hydrodynamic stability. J.Fluid Mechs., Vol.9, pp.513-532. 1960.

3. BENJAMIN, T.B. - The threefold classification of unstable disturbances in flexible surfaces bounding inviscid flows. J.Fluid Mechs., Vol.16, pp.436-450. 1963.

4. LANDAHL, M.T. - On the stability of a laminar incompressible boundary layer over a flexible surface. J.Fluid Mechs., Vol.13, pp.609-632. 1962.

5. GARRAD, A.D. & CARPENTER, P.W. - A theoretical investigation of flow-induced instabilities in compliant coatings. J.Sound & Vib., Vol.85, pp.483-500. 1982.

6. CARPENTER, P.W. & GARRAD, A.D. - Effect of a viscous fluid substrate on the flow-induced vibrations of a compliant coating. Proc.Int.Conf. on Flow Induced Vibrations in Fluid Engineering, ed. H.S. Stephens and G.B Warren, BHRA Fluid Engineering, Cranfield, U.K., pp.369-482. 1982.

7. CARPENTER, P.W. & GARRAD, A.D. - Hydrodynamic stability of flow over Kramer-type surfaces. Part 1. Tollmien-Schlichting instabilities. Engineering Science Dept., University of Exeter, Technical Note 83/1. 1983.

8. DAVEY, A. - A difficult numerical calculation concerning the stability of the Blasius boundary layer. Proc.IUTAM Symp. on stability in mechanics of continua, ed. F.H. Schroeder, Nümbrecht, Germany, Springer Verlag, pp.365-372, 1981.

9. GASTER, M. & JORDINSON, R. - On the eigenvalues of the Orr-Sommerfeld equation. J.Fluid Mechs., Vol.72, pp.121-133, 1975.

COMPUTER SIMULATION OF THICK INCOMPRESSIBLE BOUNDARY LAYERS

C. Kleinstreuer(I), A. Eghlima(II) and J. E. Flaherty(III)
Rensselaer Polytechnic Institute, Troy, N.Y. 12181, U.S.A.

SUMMARY

The accurate analysis of laminar and turbulent boundary layer flow developments along strongly curved surfaces is of great importance for superior contour designs of submerged bodies. Although incompressible higher-order boundary layer theory is a mature subject, discrepancies exist in the literature between predicted results and measured data points.

In this paper, simulation results based on a new set of second-order equations are compard with Van Dyke's equations for laminar boundary layer flow and Huang's measurements of turbulent flow in regions where $v \ll u$. In addition to the excellent predictive capabilities, the turbulent flow version of the new equations reduce the dependence of correction factors commonly employed to include curvature effects in zero-equation models for the Reynolds stresses. The numerical scheme used to solve the system of nonlinear PDE's is Keller's box method together with an improved matrix equation solver which is discussed in some detail.

1. INTRODUCTION

For many practical boundary-layer flow developments the thin shear layer assumption breaks down since the boundary-layer becomes thick, i.e. $\delta/L_r = \mathcal{O}(1)$, where L_r is a (body) reference length such as the length of the submerged body, its longitudinal radius or transverse radius of curvature. External flows past marine crafts, airfoils or turbine blades and internal flows in diffusors, bent conduits or pressure-driven membrane units might serve as examples. At the high Reynolds numbers of interest, non-uniformities of (attached) boundary-layer flow such as curvature effects, normal stress and

(I) Assistant Professor, Dept. of Chem. & Env. Eng.
(II) Research Associate, Dept. of Mechanical Engineering
(III) Associate Professor, Dept. of Mathematical Sciences

pressure variations, viscous-inviscid flow interactions, additional rates of strain and the effects of thickening on (the) turbulence intensity have to be accounted for. Previous investigators [1-4] employed Van Dyke's higher-order boundary-layer equation [5] or its various spin-offs to compute flow parameters along curved surfaces but simulation results could not predict laboratory observations [6-9]. On the other hand, Kleinstreuer et al. [10] and Eghlima and Kleinstreuer [11] derived a new set of parabolic second-order equations with which thick laminar and turbulent boundary-layer flows were more accurately represented than before. These equations are valid for steady incompressible, two-dimensional or axisymmetric flow in regions where $v \ll u$, i.e. the mean flow streamlines remain nearly parallel to the wall regardless of the relative thickness of the boundary layer. Ongoing studies concentrate on regions in which the normal velocity reaches the order of magnitude of the tangential velocity [12] and where separation and near-wake effects dominate the flow fields [13].

In this paper, simulation and laboratory results of thick laminar and turbulent boundary layer flows are compared and the effect of various corrective functions in mixing-length models on integral flow parameters are investigated.

2. ANALYSIS OF THICK BOUNDARY LAYERS IN REGION I

2.1 Governing Equations and Submodels

The external flow fields of interest together with the coordinate system used, are schematically depicted in Fig. 1. For Region I, the portion of the attached thick boundary layer where $v \ll u$, Kleinstreuer et al. [10] derived a new set of second-order boundary layer equations from the general conservation equations using the relative order of magnitude analysis (ROMA). They read

$$\text{(continuity)} \quad \frac{\partial}{\partial s}(r^i u) + \frac{\partial}{\partial n}[v(r^i - r_o^i(1-h^j))] + O(Re^{-2m}) = 0 \qquad (1)$$

$$\text{(s-momentum)} \quad (2-h^j)\, u\frac{\partial u}{\partial s} + v\frac{\partial u}{\partial n} + Kuv = -\frac{1}{\rho}(2-h^j)\frac{\partial p}{\partial s} + \nu\frac{\partial^2 u}{\partial n^2} + \nu\frac{\partial u}{\partial n}\frac{\partial}{\partial n}\left[\left(\frac{r}{r_o}\right)^i + h^j\right] + O(Re^{-2m}) \qquad (2)$$

$$\text{(n-momentum)} \quad K\, u^2 = \frac{1}{\rho}\frac{\partial p}{\partial n} + O(Re^{-2m}) \qquad (3)$$

where n is the distance normal to the (curved) body surface and s is the distance measured along the wall starting with

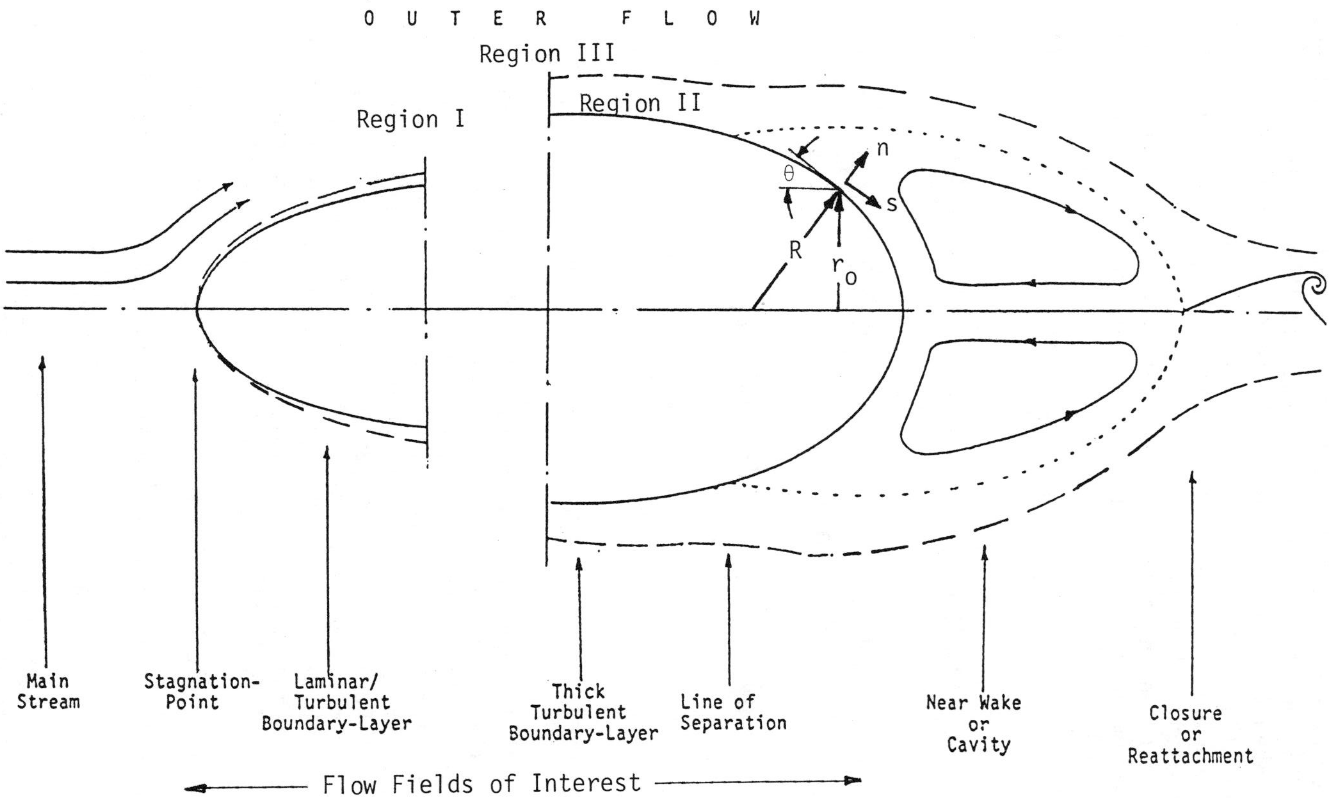

Figure 1 Schematics of Flow Domain.

the stagnation point, for example. The flow indices i and j indicate the cases i=j=0 for flow past a flat plate and i=1, j=0 for axisymmetric bodies with no longitudinal curvature and i=0, j=1 for flow without transverse curvature. Finally, i=j=1 when longitudinal and transverse curvature effects exist. The exponent m is equal to 1/2 for laminar and m = 4/5 for turbulent thin boundary-layer flow with zero pressure gradient along a flat plate.

Splitting the instantaneous variables into time-smoothed and random components the turbulent flow version of Eqns (1 to 3) read [10]:

$$\frac{\partial}{\partial s}(r^i\,\bar{u}) + \frac{\partial}{\partial n}(\bar{v}\,M) = 0 \tag{4}$$

$$(2-h^j)\,u\frac{\partial\bar{u}}{\partial s} + v\frac{\partial\bar{u}}{\partial n} + K\bar{u}\bar{v} = -\frac{1}{\rho}(2-h^j)\frac{\partial\bar{p}}{\partial s} +$$

$$+\frac{\partial^2 u}{\partial n^2} + \frac{\partial\bar{u}}{\partial n}\frac{\partial}{\partial n}\left[\left(\frac{r}{r_o}\right)^i + h^j\right] - \frac{1}{M}\left[\left(\frac{M(2-h^j)+r}{2}\right)\frac{\partial}{\partial s}\left(\overline{u'^2}\right)\right. \tag{5}$$

$$\left. + h^j\sin\theta\left(\overline{u'^2}\right) + \frac{\partial}{\partial n}\left(\overline{u'v'M}\right) + KM\left(\overline{u'v'}\right)\right]$$

$$K\,\bar{u}^2 = \frac{1}{\rho}\frac{\partial p}{\partial n} - K\left(\overline{u'^2}\right) \tag{6}$$

where $M = r^i - r^i{}_o\,(1-h^j)$, $K = \left(\frac{1}{R}\right)^{1/j}$ and $\frac{\partial}{\partial s}(u'^2) \approx 0$

since u'^2 is approximately constant for attached boundary layers [14]. This simplification which actually holds for flat plate flow, makes Eqn (5) parabolic. An alternative approach for obtaining (the) turbulent flow modeling equations would be to apply the relative order of magnitude analysis to the exact Reynolds equations as oultined by Patel [15] for flow past cylinders and cones. When this is done, an additional turbulence term, $\partial/\partial n\,(v'^2)$, appears in Eqn (6). When Eq. (6) is inserted into Eqn (5), $\partial/\partial n\,(v'^2)$ can be combined with $\partial/\partial n\,(u'v')$ following Bradshaw's hypothesis that

$$u'v' = a(u'^2 + v'^2 + w'^2)\ ;\ v'^2 = b\,u'^2 \text{ and } w'^2 \approx c u'^2.$$

At the wall, the "no-slip" condition is usually invoked, i.e.

$$n = 0 \quad \rightarrow \quad \overline{u} = \overline{v} = 0\ . \tag{7}$$

In case of suction/injection $\bar{v}(n=0) = \pm\,v_{wall}$. The condition for the tangential velocity at the boundary layer edge is

$$\overline{u/u_e} \rightarrow 1 \quad \text{for} \quad n \rightarrow \infty\ . \tag{8}$$

In order not to mask the fluid mechanics of thick boundary-layer flow developments in Region I, only zero-equation models representing turbulence were considered. In general, the shear stresses are a function of mean-time velocity and velocity gradient as well as mixing lengths which in turn are dependent upon the radii of curvature, viz:

$$\overline{u'v'} = \overline{u'v'}\,[\nu_t(R,r_o);\ R,r_o;\ \overline{u},\ \nabla\overline{u}\,].$$

Regarding the thick turbulent boundary-layer as a composition of an inner and outer layer, we write for the variation of the eddy-viscosity in the inner region [16]

$$(\nu_t)_i = (\ell_m^2)_i\ S^2\left(\frac{r}{r_o}\right)^i \left|\frac{\partial u}{\partial y}\right|_{\gamma_{tr}} ,\quad 0 \leqslant y \leqslant y_c \qquad (9)$$

where γ_{tr} is an intermittency factor that accounts for the transitional region. Streamline curvature is incorporated in the function S(R) as postulated by Eide and Johnston [17] or Bradshaw [18]. Bradshaw's correction to the eddy viscosity reads

$$S = \frac{1}{1 + \beta\, R_i} \quad \text{where} \quad R_i = \frac{2u}{R}\left(\frac{\partial u}{\partial y}\right)^{-1} . \qquad (10)$$

The coefficient β can range from 2 to 10 or higher. Its significance and best value will be investigated below with a sensitivity analysis. A suitable expression for the inner mixing length for axisymmetric flow is given by Cebeci and Smith [16] as:

$$(\ell_m)_i = k\ r_o\ \ell n\left(\frac{r}{r_o}\right) \left\{1 - \exp\left[-\frac{r_o}{A}\ \ell n\left(\frac{r}{r_o}\right)\right]\right\} \qquad (11)$$

where k = 0.4 and A is a damping length.

Huang et al. [3] proposed that the mixing length of an axisymmetric turbulent boundary-layer is proportional to the square root of the area of the turbulent annulus which is formed by the body surface and the edge of boundary-layer. Based on their results, the existing thin turbulent boundary-layer differential methods can be applied to the forward portion of the axisymmetric body up to the station where the boundary-layer thickness increases to about 20 percent of the body radius. Further downstream, the apparent mixing length of the thick axisymmetric stern boundary-layer (ℓ_m) can be approximated to the mixing length for a thin flat boundary-layer (ℓ_{mf}) by

$$\alpha = \frac{(\ell m)_0}{\ell_{mf}} = \frac{\sqrt{(r_0 + \delta)^2 - r_0^2}}{3.33\ \delta} < 1\ . \tag{12}$$

Now, α^2 is used as a correction factor to the flat plate eddy-viscosity formula

2.2 Numerical Analysis

There is a large number of numerical schemes available for solving (parabolic) boundary-layer flow problems. Eghlima [19] discussed the merits of Keller's box method [20] with an improved equation solver over other competitive algorithms such as the Crank-Nicolson finite difference method or the Galerkin finite element method.

Equations (4 to 6) are discretized for an arbitrary rectangular grid using centered difference quotients and averages at the midpoints of the net rectangles. The box scheme gives rise to a large block bi-diagonal linear algebraic system. The standard technique for solving this system is to use a variant of the block tridiagonal algorithm with row pivoting performed only within blocks. However, this procedure is not guaranteed to be numerically stable and any attempts to use stable row pivoting algorithms will introduce fill in outside of the blocks. We use a stable numerical technique that was proposed by Varah [21] and implemented by Flaherty and Mathon [22] to solve the block bi-diagonal system. This approach uses an alternating row and column pivoting strategy to guarantee stability without introducing fill in. Furthermore, the complexity of the algorithm is comparable to the unstable tridiagonal procedure. Some additional improvements have been obtained by Diaz, Fairweather, and Keast [23]. The main sources of computational errors are caused by round-off error, iteration error, the error induced due to linearization, and the truncation error of the numerical scheme. The first three errors are smaller than the fourth one because the program is in double precision and the iteration and linearization errors are of the order of 10^{-4}. The local accuracy of the numerical algorithm is $O(h^2)$ where h is equal to the grid spacing, due to the centered difference quotients employed. To check the overall accuracy of the numerical solution, we have divided the mesh spacing by two and four and the results have shown that the global error is $O(h^2)$ as well.

The procedure for simulating a general flow system is best illustrated with reference to Figs. 1 and 2. Starting from the stagnation point, a suitable computer code (e.g. Ref. 24) is employed until the boundary layer becomes thick, i.e.

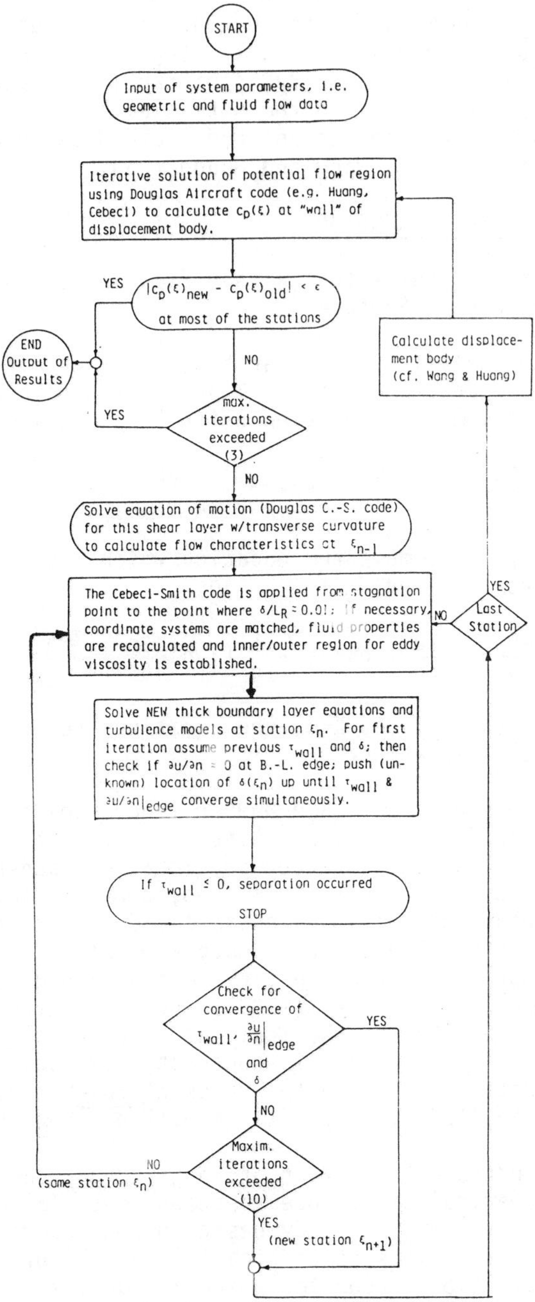

Figure 2 Computation Flow Sheet

$\delta/R << 1$ is not valid anymore. These intermediate results are then conveyed to our program which solves the new parabolic equations of Region 1. This model remains valid as long as $v<<u$, which may include the point of separation. It has to be noted that boundary layer thickening might occur for certain body geometries at quite an early stage as demonstrated later for a laminar flow case study. Interactions between the growing boundary layer and the potential flow (Ref. 24) is computed with the displacement body method using an iterative procedure outlined by Wang and Huang [25].

3. RESULTS AND DISCUSSIONS

3.1 Simulation Results for Laminar Flow in Region I past a Convex/Concave Surface

For the analysis of laminar boundary layer flow with longitudinal curvature effects Van Dyke [5] developed a higher-order momentum equation using perturbation theory as

$$u \frac{\partial u}{\partial s} + h^j \; v \frac{\partial u}{\partial n} + Kuv = - \frac{1}{\rho} \frac{\partial p}{\partial s} + \nu h^j \frac{\partial^2 u}{\partial n^2} \; \nu K \frac{\partial u}{\partial n} \qquad (13)$$

In contrast, our s-momentum equation which can be directly obtained from (2) for $r_0 \to \infty$, reads

$$(2-h) \; u \frac{\partial u}{\partial s} + v \frac{\partial u}{\partial n} + Kuv = - \frac{1}{\rho} (2-h) \frac{\partial p}{\partial s} + \nu \frac{\partial^2 u}{\partial n^2} + \nu K \frac{\partial u}{\partial n} \qquad (14)$$

For $h=1$, which includes $K=0$, the equations reduce to Prandtl's boundary layer equation.

Both equations have the same number (and type) of terms but different coefficients, i.e. emphasis or weight for four terms. This leads to a better physical representation of the fluid dynamics when Eqn (14) is used, provided that the following assumptions hold: Newtonian fluid, thick laminar shear layers, and streamlines approximately parallel to the curved surface. Both equations are mathematically equivalent and in error of $O(\varepsilon^2)$, where $\varepsilon = O(Re^{-m})$. However, with the derivation technique ROMA the physical integrity of all retained terms has been preserved. In contrast, asymptotic analysis requires that $\varepsilon \to 0$ but in the cases of thick boundary layer flow ε may be (locally) of $O(1)$ as will be shown below.

To demonstrate the effect of longitudinal curvature and to show the numerical differences between Eqns (13) and (14), we selected an axisymmetric body designed and documented in Ref. [9]. Figure 3 depicts schematically one-half of the symmetric, two-dimensional body for which in our case $r_0 \to \infty$. Of much higher resolution and physical interest is Fig. 4. It shows the local $\delta/R = K\delta$ vs. the chord length of the submerged body. The curve peaks early due to a sharp decrease of the radius of longitudinal curvature and then falls to zero (despite the growing boundary layer) when $R \to \infty$. When the flat middle section is over, δ/R increases again since R becomes finite and δ is quite thick ($\approx$ 0.4 ft) by then. Figure 5 demonstrates that the new coefficients of our equation do indeed make a significant difference, actually 10% at one point in this case study. It is evident that the maximal difference occurs after $\delta/R(x)$ has peaked.

3.2 Theoretical and Experimental Results for Turbulent Flow in Region I past an Axisymmetric Anti-Separation Body

Huang et al. [3] have conducted their measurements at a Reynolds number ($Re = UL/\nu$) of 6.8×10^6 with an axisymmetric convex afterbody without stern separation.

Figure 6 shows a comparison between measured and computed displacement plus boundary layer thickness, using Eqns (4) to (6). Representative predicted and measured velocity profiles for u(r) and v(r) at a particular station x where v is still much smaller than u are given in Figures (7) and (8).

In another study, the effect of a correction [see Eqns (9 and 10)] to the inner eddy-viscosity due to boundary layer thickening on flow parameters has been investigated. The effect of the correction factor β on the computation of velocity profiles is limited to a region close to the wall and diminishes downstream where $\delta/R \approx \mathcal{O}(1)$. Figures 9 and 10, which compare the computed and measured skin friction coefficients along the body, indicate that, by applying the outer layer correction, α of Eqn (12), we have computed a skin friction coefficient which is about six percent less than the measured value for $x/L > 0.90$. The reason is that Huang's correction factor to the outer eddy-viscosity is based on fitting experimental results with the theoretical computation of boundary-layers in which transverse curvature has been taken into account only. This correction basically compensates the extra rate of strain which is caused by longitudinal curvature. Hence, when we use a more complete set of equations which takes into account transverse as well as longitudinal curvature, applying this correction will introduce too much artificial loss of kinetic energy close to the wall. Therefore, we recommend that (this) correction to the outer eddy-viscosity is not profitable, when the new set of boundary-layer equations is applied.

As shown in this analysis, the MLH may actually be restricted to (attached) thick shear layers for which the

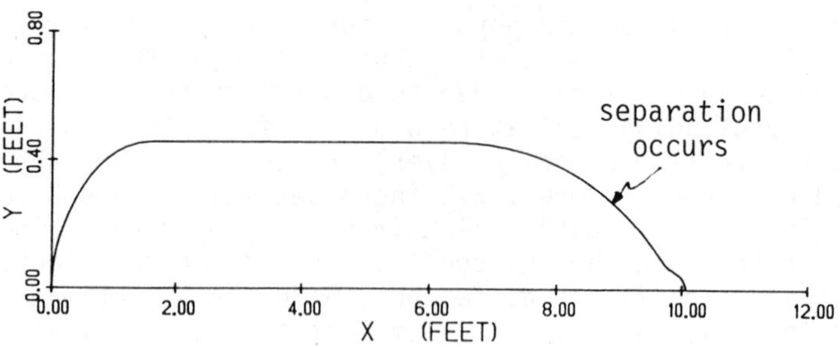

Fig. 3. Semi-infinite body with longitudinal curvature.

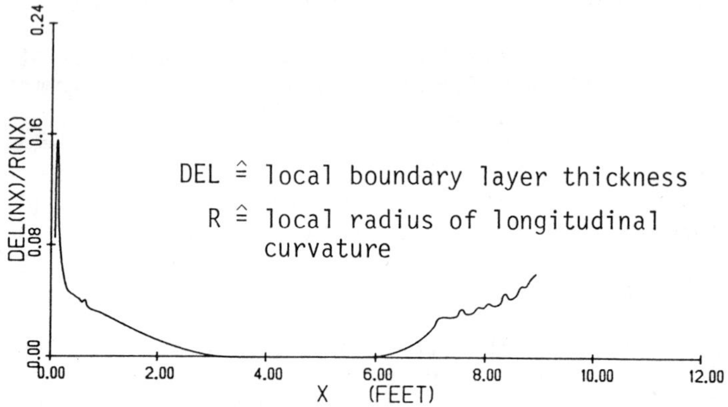

Fig. 4. Local $\frac{\delta}{R} = K\delta$ vs. chord length of submerged body.

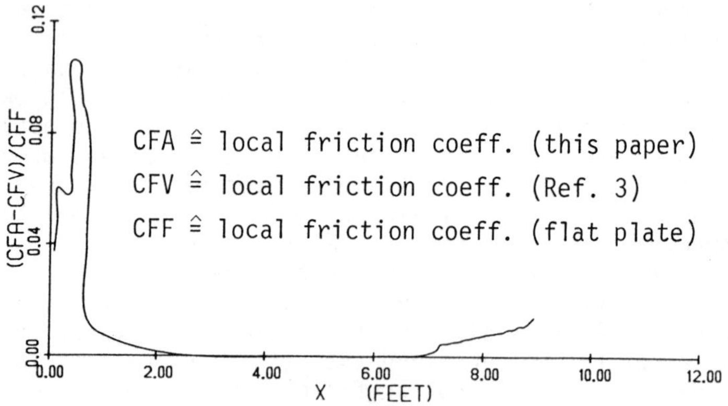

Fig. 5. Effects of thick laminar boundary layer development on the local friction coefficient: a comparison between equations (13) and (14) at $Re \approx O(10^3)$.

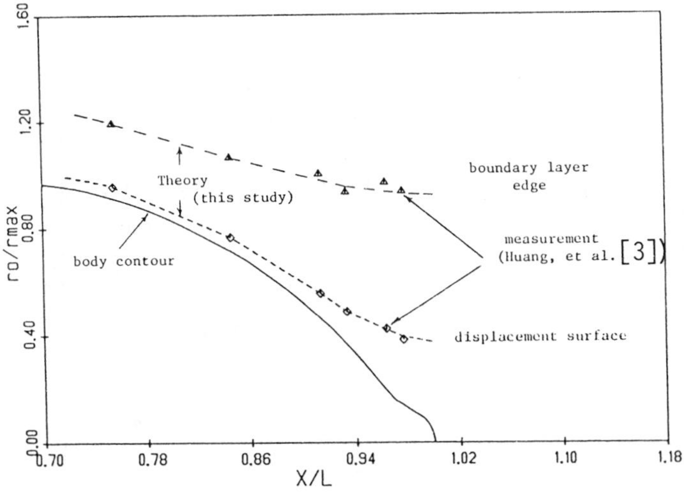

Fig. 6. Schematic of the tail of the body, computed and measured displacement thickness and boundary-layer thickness.

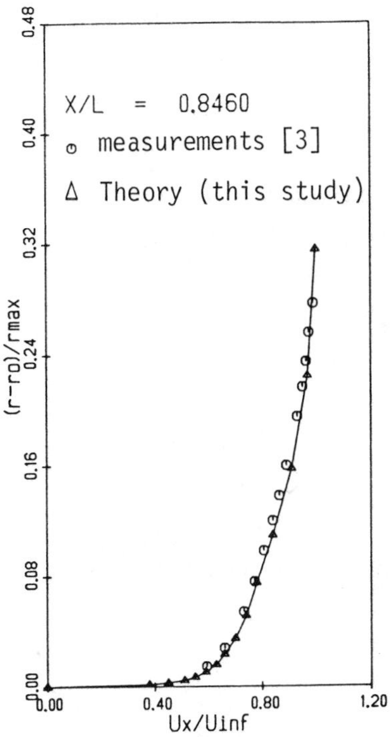

Fig. 7 Computed and measured mean axial velocity distribution across stern boundary-layer (this study).

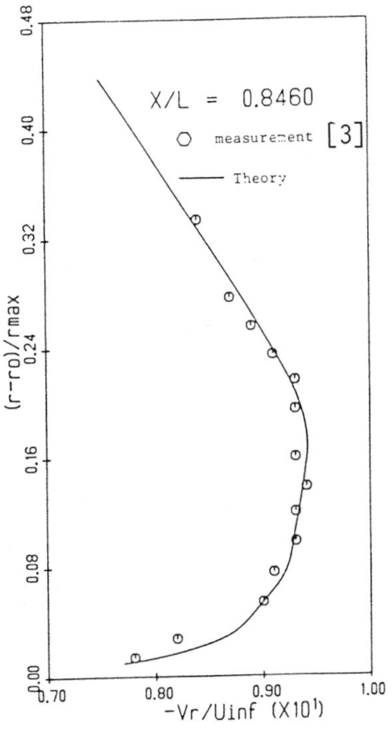

Fig. 8. Computed and measured mean radial velocity distribution across stern boundary-layer (this study).

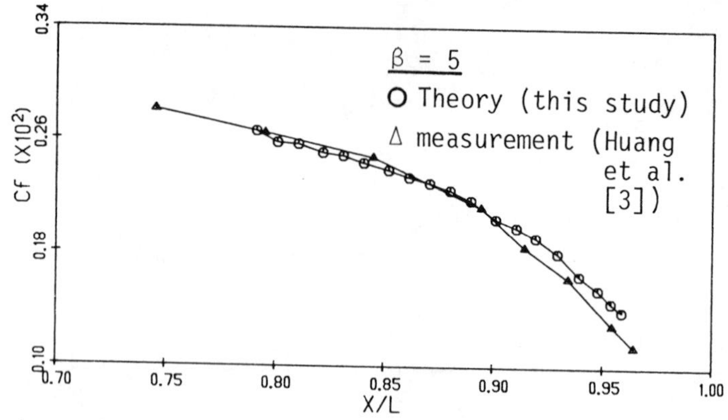

Fig. 9. Computed and measured skin friction coefficient along the body.

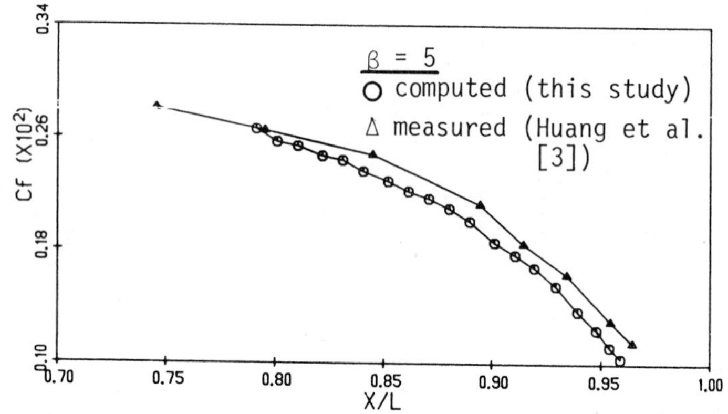

Fig. 10. Computed and measured skin friction coefficient along the body (Huang et al. correction has been applied to the outer eddy-viscosity).

assumption $v \ll u$ holds. In other words, strong variations of flow parameters within the boundary layer, i.e. Region II where $v \approx u$, would lead to turbulence phenomena which require more sophisticated submodels as discussed in Section 3.

REFERENCES

1. PATEL, V.C. and LEE, Y.T. - Thick Axisymmetric Boundary Layers and Wakes: Experiment and Theory. Paper 4, Int. Symp. on Ship Viscous Resistance, Göteborg, Sweden, 1978.

2. CEBECI, T., HIRSH, R.S. and WHITELAW, J.H. - On the Calculation of Laminar and Turbulent Boundary-Layers on Longitudinally Curved Surfaces. AIAA Journal, Vol. 17, 4, 1973.

3. HUANG, T.T., et al. - Stern Boundary-Layer Flow on Axisymmetric Bodies. 12th Symp. on Naval Hydrodynamics, Wash., D.C., June 1978. Avail. N.A.S., Wash., D.C., 1979, p.127.

4. GIBSON, M.M., JONES, W.P. and YOUNIS, B.A. - Calculation of Turbulent Boundary-Layers on Curved Surfaces. Phys. Fluid, Vol. 24, 3, 1981.

5. VAN DYKE, M. - Higher-order Boundary-layer Theory. Ann. Rev. Fluid Mech., p. 265, 1969.

6. SO, R.M.C. and MELLOR, G. - J. Fluid Mech., Vol. 60, 43, 1973.

7. MERONEY, R.N. and BRADSHAW, P., Turbulent Boundary-layer Growth over a Longitudinally Curved Surface. AIAA Journal, Vol. 13, 11, 1975.

8. GILLIS, J.C. and JOHNSTON, J.P. - In: Turbulent Shear Flows II, Springer-Verlag, 1980.

9. HUANG, T.T., et al. - Boundary-Layer Flow on an Axisymmetric Body with an Inflected Stern. DTNSRDC-80/064, 1980.

10. KLEINSTREUER, C., EGHLIMA, A. and FLAHERTY, J.E. - New Higher-Order Boundary-Layer Equations for Laminar and Turbulent Flow Past Axisymmetric Bodies. Phys. of Fluids (submitted 1983).

11. EGHLIMA, A. and KLEINSTREUER, C. - Numerical Analysis of Attached Turbulent Boundary Layers Near the Tail of Axisymmetric Bodies. AIAA (submitted 1982).

12 KLEINSTREUER, C. and EGHLIMA, A - Thick Turbulent Boundary Layer Flow Near the Stern of an Anti-Separation Body. (in preparation).

13. EGHLIMA, A. and KLEINSTREUER, C. - Thick Turbulent Boundary Layer Flow Developments with Separation. (in preparation).

14. BRADSHAW, P. - personal communication, 1981.

15. PATEL, V.C. - On the Equations of a Thick Axisymmetric Turbulent Boundary Layer. IIHR Rept. No. 143, The Univ. of Iowa, Jan. 1973.

16. CEBECI, T. and SMITH, A.M.O. - Analysis of Turbulent Boundary Layers, Academic Press, New York, 1974.

17. EIDE, S.A. and JOHNSTON, J.P. - Rept. PD-19, Thermoscience Division, Mech. Eng. Dept., Stanford Univ., Calif., 1974.

18. BRADSHAW, P. - The Analogy Between Streamline Curvature and Buoyancy in Turbulent Shear Flow. J. Fluid Mech., No. 33, 1969.

19. EGHLIMA, A. - Theoretical Analysis and Computer Simulation of Thick Incompressible Laminar/Turbulent Boundary-Layer Flows Along Curved Surfaces. Ph.D. thesis, RPI, Troy, New York, 1982.

20. KELLER, H.B. - A New Difference Scheme for Parabolic Problems, Numerical Solutions of Partial Differential Equations, Vol. II, Academic Press, New York, 1970.

21. VARAH, J.M. - Alternate Row and Column Elimination for Solving Certain Linear Systems. SIAM J. Numer. Anal., Vol. 13, pp. 71-75, 1976.

22. FLAHERTY, J.E. and MATHON, W. - Collocation with Polynomial and Tension Splines for Singularly-Perturbed Boundary Value Problems. SIAM J. Sci. Stat. Comput., Vol. 1, pp. 260-289, 1980.

23. DIAZ, J.C., FAIRWEATHER, G. and KEAST, P. - personal communication, 1982.

24. CEBECI, T., KAUPS, K., JAMES, R.M. and MACK, D.P. - Boundary-Layer and Inverse Potential-Flow Methods for Axisymmetric Bodies. Rept. No. MDC J7895, Douglas Aircraft Co.; prepared for ONR, Arlington, VA, 1978.

25. WANG, H.T. and HUANG, T.T. - Calculation of Potential Flow/Boundary-Layer Interaction on Axisymmetric Bodies. Report by the Ship Performance Dept. of the David W. Taylor Naval Ship R & D Center, Bethesda, Md., 1979.

CALCULATION OF UNSTEADY BOUNDARY LAYERS

A.N. Menendez(1) and B.R. Ramaprian(2)

ABSTRACT

A relatively simple, yet efficient and accurate finite difference method is developed for the solution of the unsteady boundary layer equations for both laminar and turbulent flows. Using this method calculations of periodic laminar boundary layers are performed for zero and adverse mean pressure gradients, for various oscillation frequencies, for small and large amplitudes, and even in the presence of significant flow reversal. The results are compared with exact analytical solutions, asymptotic solutions and experimental data. The numerical method is then applied to predict an experimental periodic turbulent boundary layer, using simple quasi-steady closure models. The present numerical results are shown to be in better agreement with the measurements than others obtained using identical turbulence models but a different numerical procedure.

1. INTRODUCTION

There has been increased interest in the study of unsteady flows in recent years. This is due to the relevance of this study to such varied applications as biofluid flows, missile aerodynamics, aircraft flutter, helicopter rotor blade flows, turbomachinery flows, etc. Various numerical methods for laminar/turbulent boundary layers subjected to a periodic freestream velocity have been developed over the last several years [1,2,3,4,5,6,7,8]. The turbulent boundary layer calculations have given inconclusive results when

---

(1)Graduate Research Assistant, Iowa Institute of Hydraulic Research, University of Iowa, Iowa City, Iowa

(2)Professor of Mechanical Engineering and Research Engineer, Iowa Institute of Hydraulic Research, The University of Iowa, Iowa City, Iowa compared with some of the experimental data that became recently available. Furthermore, different calculation methods, using identical turbulence models, have given substantially different results [8]. There is thus a need for the development of a calculation method whose accuracy is tested with rigor.

In this paper, a numerical procedure is developed and checked for accuracy in laminar flow. The method is shown to perform very well even in extreme situations. It is then applied to predict a periodic turbulent boundary layer. The results are compared with experimental data. Only a brief description of the method and some typical results are presented here. More details can be found in [10].

2. THE NUMERICAL PROCEDURE

The incompressible, two-dimensional, unsteady, ensemble-averaged turbulent boundary layer equations are

$$\frac{\partial\langle u\rangle}{\partial x} + \frac{\partial\langle v\rangle}{\partial y} = 0 \tag{1}$$

$$\frac{\partial\langle u\rangle}{\partial t} + \langle u\rangle\frac{\partial\langle u\rangle}{\partial x} + \langle v\rangle\frac{\partial\langle u\rangle}{\partial y} = -\frac{1}{\rho}\frac{\partial\langle p\rangle}{\partial x} + \nu\frac{\partial^2\langle u\rangle}{\partial y^2} - \frac{\partial}{\partial y}\langle u'v'\rangle \tag{2}$$

$$\frac{\partial\langle k\rangle}{\partial t} + \langle u\rangle\frac{\partial\langle k\rangle}{\partial x} + \langle v\rangle\frac{\partial\langle k\rangle}{\partial y} = \frac{\partial}{\partial y}[\langle(k + p'/\rho)v'\rangle] - \langle u'v'\rangle\frac{\partial\langle u\rangle}{\partial y} - \langle\varepsilon_1\rangle \tag{3}$$

with ε_1 = dissipation of turbulent kinetic energy. The ensemble averaged values, denoted by the notation $\langle\ \rangle$, can be regarded as the average obtained from a number of realizations. The equations are analogous to the time-averaged Reynolds equations for steady turbulent flows, with $\langle k\rangle$ and $\langle u'v'\rangle$ being interpreted as the ensemble averaged turbulent kinetic energy and Reynolds shear stress, respectively. Also, with $\langle k\rangle = \langle u'v'\rangle = 0$, they reduce to the laminar boundary layer equations. The initial and boundary conditions are

$$\langle u\rangle(x,o,t) = 0, \qquad \langle v\rangle(x,o,t) = 0 \qquad \langle k\rangle(x,o,t) = 0$$

$$\langle u\rangle(x,y,t) \to u_e(x,t) \quad \text{as } y \to \infty$$

$$\langle k\rangle\,(x,y,t) \rightarrow k_e = \text{freestream turbulence intensity as } y \rightarrow \infty \tag{4}$$

$$\langle u\rangle\,(x,y,o) = \phi_u(x,y), \qquad \langle k\rangle\,(x,y,o) = \phi_k(x,y)$$

$$\langle u\rangle\,(x_o,y,t) = \psi_u(y,t) \qquad \langle k\rangle\,(x_o,y,t) = \psi_k(y,t)$$

where ϕ and ψ are known functions, and the edge velocity $u_e(x,t)$ is related to the pressure gradient via the inviscid flow equation

$$-\frac{1}{\rho}\frac{\partial\langle p\rangle}{\partial x} = \frac{\partial\langle u_e\rangle}{\partial t} + \langle u_e\rangle\frac{\partial\langle u_e\rangle}{\partial x} \tag{5}$$

Equations (1) to (3) can now be solved provided a suitable turbulence closure model is introduced.

If the flow is periodic (oscillatory), the calculations can, in principle, be started from any set of initial conditions at t = 0, and be expected to yield eventually (after the transient effects die down) a truly periodic solution independent of those particular conditions. In the special case of periodic boundary layers, the above equations are nondimensionalized using the following dimensionless variables:

$$\hat{x} = \omega x/U_o, \qquad \eta = y/b(x) \qquad \hat{t} = \omega t$$

$$\hat{u} = \langle u\rangle/U_o, \qquad \hat{v} = \langle v\rangle/\omega b \qquad \hat{u}_e = \langle u_e\rangle/U_o \tag{6}$$

$$\hat{k} = \langle k\rangle/u_\tau^2$$

where U_o is the streamwise velocity scale, $\omega = 2\pi f$, f = frequency associated with the periodicity, b(x) is a prescribed length scale that encloses the region where significant changes in flow properties are expected to occur, and $u_\tau(x)$ is a prescribed turbulent velocity scale of the order of the shear velocity.

The transformed equations in the $\hat{x}$-η plane are:

$$\frac{\partial\hat{u}}{\partial\hat{x}} + \frac{\partial\hat{v}}{\partial\eta} - \frac{\eta}{b}\frac{db}{d\hat{x}}\frac{\partial\hat{u}}{\partial\eta} = 0 \tag{7}$$

$$\frac{\partial \hat{u}}{\partial \hat{t}} + \hat{u}\frac{\partial \hat{u}}{\partial \hat{x}} + \hat{v}\frac{\partial \hat{u}}{\partial \eta} - \frac{\hat{u}\eta}{b}\frac{db}{d\hat{x}}\frac{\partial \hat{u}}{\partial \eta} =$$

$$\frac{\nu}{\omega b^2}\frac{\partial}{\partial \eta}\left(\frac{\partial \hat{u}}{\partial \eta}\right) - \frac{1}{\omega b U_o}\frac{\partial}{\partial \eta}(\langle u'v'\rangle) + \hat{F} \tag{8}$$

$$\frac{\partial \hat{k}}{\partial \hat{t}} + \hat{u}\frac{\partial \hat{k}}{\partial \hat{x}} + \hat{v}\frac{\partial \hat{k}}{\partial \eta} - \frac{\hat{u}\eta}{b}\frac{db}{d\hat{x}}\frac{\partial \hat{k}}{\partial \eta} + \frac{2}{u_\tau}\frac{\partial u_\tau}{\partial \hat{x}}\hat{u}\,\hat{k} =$$

$$\frac{1}{\omega b u_\tau^2}\frac{\partial}{\partial \eta}[\langle p'/\rho + k)v'\rangle] - \frac{U_o}{\omega b u_\tau^2}\langle u'v'\rangle\frac{\partial \hat{u}}{\partial \eta} - \frac{\langle \varepsilon_1\rangle}{\omega u_\tau^2} \tag{9}$$

$$\hat{F} = \frac{\partial \hat{u}_e}{\partial \hat{t}} + \hat{u}_e\frac{\partial \hat{u}_e}{\partial \hat{x}} \tag{10}$$

These equations are discretized using an adaptation of the implicit finite difference scheme proposed by Oskolkov [9]. To illustrate the procedure, the discretized versions of the continuity and momentum equations for laminar flow, i.e. Eqs. (7) and (8) with $\langle u'v'\rangle = 0$, are shown below, with $b(x) = \sqrt{\nu x/U_o}$ (the Blasius boundary layer thickness),

$$\frac{\hat{u}_{ij}^{\ell+1} - \hat{u}_{i-ij}^{\ell+1}}{\Delta\hat{x}} - \frac{\eta_j}{2\hat{x}_i}\frac{(\hat{u}_{ij}^{\ell+1} - \hat{u}_{ij-1}^{\ell+1})}{\Delta\eta_{jj-1}} + \frac{\hat{v}_{ij}^{\ell+1} - \hat{v}_{ij-1}^{\ell+1}}{\Delta\eta_{jj-1}} = 0 \tag{11}$$

$$\frac{\hat{u}_{ij}^{\ell+1} - \hat{u}_{ij}^{\ell}}{\Delta\hat{t}} + \left[\hat{u}\frac{\partial \hat{u}}{\partial \hat{x}}\right]_{ij} - \frac{\hat{u}_{ij}^{\ell}\eta_j}{2\hat{x}_i}\frac{(\hat{u}_{ij}^{\ell+1} - \hat{u}_{ij-1}^{\ell+1})}{\Delta\eta_{jj-1}} + \left[\hat{v}\frac{\partial \hat{u}}{\partial \eta}\right]_{ij}$$

$$= \frac{2}{\hat{x}_i(\Delta\eta_{j+1j} + \Delta\eta_{jj-1})}\left(\frac{\hat{u}_{ij+1}^{\ell+1} - \hat{u}_{ij}^{\ell+1}}{\Delta\eta_{j+1j}} - \frac{\hat{u}_{ij}^{\ell+1} - \hat{u}_{ij-1}^{\ell+1}}{\Delta\eta_{jj-1}}\right) + \hat{F}_{ij}^{\ell+1} \tag{12}$$

with

$$\left[\hat{u}\frac{\partial \hat{u}}{\partial \hat{x}}\right]_{ij} = \begin{cases} \hat{u}_{ij}^{\ell}\dfrac{(\hat{u}_{ij}^{\ell+1} - \hat{u}_{i-ij}^{\ell+1})}{\Delta\hat{x}} & \text{for} \quad \hat{u}_{ij}^{\ell} > 0 \\ \hat{u}_{ij}^{\ell}\dfrac{(\hat{u}_{i+1j}^{\ell} - \hat{u}_{ij}^{\ell+1})}{\Delta\hat{x}} & \text{for} \quad \hat{u}_{ij}^{\ell} < 0 \end{cases} \tag{13}$$

$$\left[\hat{v}\frac{\partial \hat{u}}{\partial \eta}\right]_{ij} = \begin{cases} \hat{v}_{ij}^{\ell}\dfrac{(\hat{u}_{ij}^{\ell+1} - \hat{u}_{ij-1}^{\ell+1})}{\Delta\eta_{jj-1}} & \text{for} \quad \hat{v}_{ij}^{\ell} > 0 \\ \hat{v}_{ij}^{\ell}\dfrac{(\hat{u}_{ij+1}^{\ell+1} - \hat{u}_{ij}^{\ell+1})}{\Delta\eta_{j+1j}} & \text{for} \quad \hat{v}_{ij}^{\ell} < 0 \end{cases} \tag{14}$$

where i, j, and ℓ are the indices in the $\hat{x}$, η, and $\hat{t}$ directions, respectively, $\Delta\hat{t}$, $\Delta\hat{x}$ and $\Delta\eta_{jj-1} = \eta_j - \eta_{j-1}$ are the corresponding discretization steps and $i = 1,2,...I$; $j = 1,2,...J$; $\ell = 1,2,....$ The discretized equations (12) with $j = 1,2,...J$, are solved simultaneously for each column i. The solution is obtained by means of the efficient tridiagonal algorithm. Once $\tilde{u}_{ij}^{\ell+1}$ is known from j=1 to J, $\tilde{v}_{ij}^{\ell+1}$ is calculated from to Eq. (11).

Note that the convection terms in the momentum equation have been linearized in their discrete version, Eq. (12). It was found that for oscillatory flow this is accurate enough. Note also that the scheme can handle negative velocity components. This is essential to deal with mean adverse pressure gradient flows, large amplitudes of oscillation or high frequencies for which flow reversal (but not separation) is likely to occur during part of the cycle.

The numerical method was first tested by applying it to various simple laminar shear flow problems for which exact solutions are known. These included a steady flow (the flat plate boundary layer), an unsteady nonperiodic flow (the evolution of Couette flow) and an unsteady but periodic flow (fully developed periodic flow in a two-dimensional channel). In all cases the results were quite satisfactory. These are described in detail in [10].

3. RESULTS FOR PERIODIC LAMINAR BOUNDARY LAYERS

Some typical results are presented here for periodic laminar boundary layers in a freestream whose velocity is given by

$$u_e(x,t) = \overline{u}_e(1 - b_1\, x/\, U_o + \varepsilon \sin \omega t) \qquad (15)$$

The case $b_1 = 0$, with $\overline{u}_e = U_o$ = const., corresponds to zero-mean pressure gradient and will be referred to as the Blasius-Mean-Flow. If $\varepsilon \ll 1$ (small amplitude of oscillation), the velocity profile can be approximated, to first order in ε, by

$$u\,(x,y,t) = u_B(x,y) + \varepsilon\Delta u(x,y) \sin\,[\omega t + \phi(x,y)] \qquad (16)$$

where u_B is the Blasius velocity distribution, and the second term represents the periodic perturbation. Calculations were performed for this condition from low ($\hat{x} < 1$) to very high ($\hat{x} \gg 1$) frequencies.

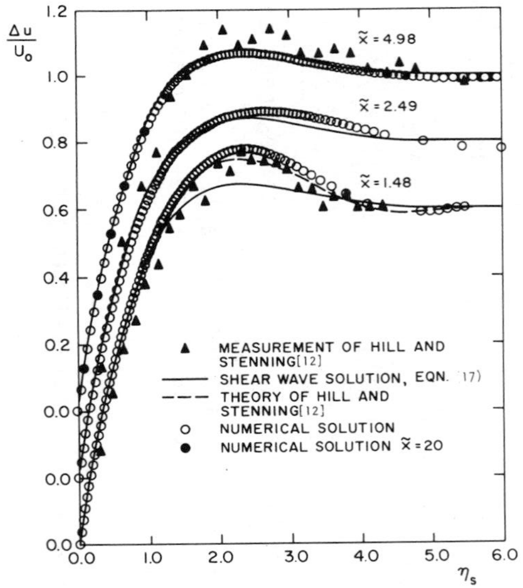

Fig. 1a. Amplitude results for Blasius-Mean-Flow

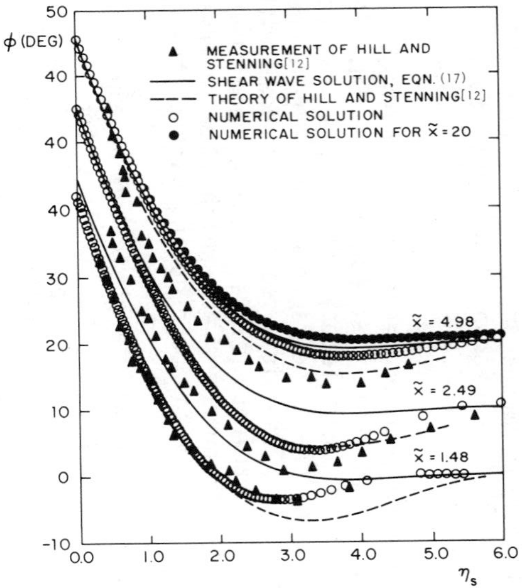

Fig. 1b. Phase results for Blasius-Mean-Flow

Figs. 1a and 1b show some results for intermediate and high oscillation frequencies ($\tilde{x} > 1$) . They are presented in terms of the Stokes coordinate $\eta_s = \sqrt{\omega/2\nu}\ y$, and compared with the Lighthill-high frequency solution [11], which is essentially the shear wave-solution, namely

$$\tilde{u}_{osc} = (u - u_B)/\varepsilon U_o = \cos \tilde{t} - \exp(-\eta_s) \cos(\tilde{t}-\eta_s) \qquad (17)$$

Figures 1a and 1b also show the experimental data of Hill and Stenning [12] as well as the results of their theoretical analysis which is an extension of the Lighthill analysis to moderately large frequencies. The present numerical calculations and the calculations of Hill and Stenning agree reasonably well with each other and with the experimental results, and all these approach the shear wave solution for very large values of $\tilde{x}$. Similar results were obtained for the low frequency range and these are reported in [10].

When the amplitude of oscillation is large, the velocity profile depends on the two parameters $\tilde{x}$ and ε. Then, the asymptotic solutions referred to previously, will not be applicable except at very high frequencies. As far as the authors are aware, the only solution available for such flows is that obtained by Pedley [13] using both regular (for $\omega \to o$) and singular (for $\omega \to \infty$) asymptotic expansions. Even these are essentially numerical solutions and are available for the skin friction and surface heat transfer only. No experimental data are available, at present, for such flows.

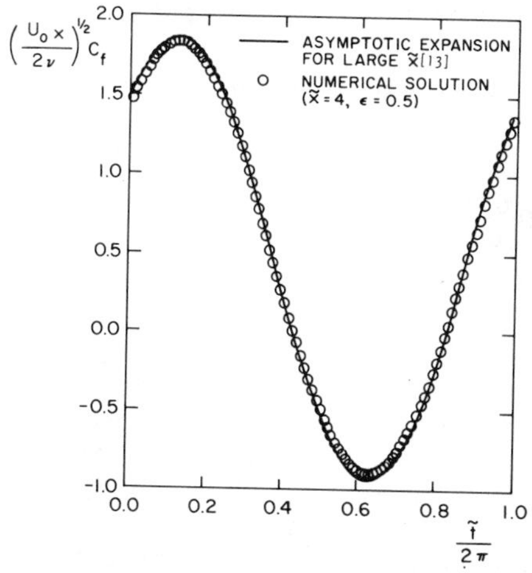

Fig. 2. Skin friction for large amplitudes

Calculations for $\varepsilon = 0.5$ were performed for various values of $\tilde{x}$ [10]. As a typical example, Fig. 2 shows the calculated evolution of the wall shear stress with time during the oscillation cycle for $\tilde{x} = 4$. Here, C_f is the skin friction coefficient. The theoretical solution of Pedley [13] for the same amplitude, $\varepsilon = 0.5$, is also presented. Good agreement is observed. Negative values of the wall shear stress indicate the existence of a backflow region. It is seen that the numerical scheme works satisfactorily even with strong flow reversals (large negative wall shear stress).

Calculations were also made for the case $b_1 x/U_o = 0.1$ (Howarth-Mean-Flow problem). Some typical results and comparisons for the amplitude and phase of the velocity for high oscillation frequencies are shown in Figs. 3a and 3b, respectively. The correspondence between the different theoretical results is as expected. The agreement with experimental results varies from moderate to poor.

4. RESULTS FOR PERIODIC TURBULENT BOUNDARY LAYERS

The method was extended to calculate periodic turbulent boundary layers, using relatively simple turbulence closure models. These are the Prandtl-mixing length model [5] and the Prandtl-energy (k-L) model [14]. Both turbulence models were used in a quasi-steady form, by relating the ensemble averaged shear stress $\langle u'v' \rangle$ to the ensemble averaged flow properties at the same phase position. Recent experiments by Parikh, Reynolds and Jayaraman [15] have been used as the test cases. In these experiments there existed a region with steady zero-pressure gradient followed by a zone with a periodic adverse pressure gradient. Four oscillation frequencies were studied in these experiments.

Equations (7) to (10) were used with the scaling length $b(x)$ and velocity $u_\tau(x)$ being chosen as

$$b(x) = 0.14 \frac{\nu}{U_o} Re_x^{6/7} \qquad (18)$$

$$u_\tau(x) = \sqrt{0.013}\, U_o\, Re_x^{-1/14} \qquad (19)$$

The above expressions are well-know relations (see White [16]) for the boundary layer thickness and the shear velocity, respectively, along a flat plate, with $Re_x = U_o x/\nu$. A typical result for the amplitude of oscillation is shown and compared with experiments in Fig. 4 (there, δ is the boundary layer thickness). Comparisons are also shown with similar calculations made by Orlandi [6] using a nearly identical k-L model but a different numerical scheme.

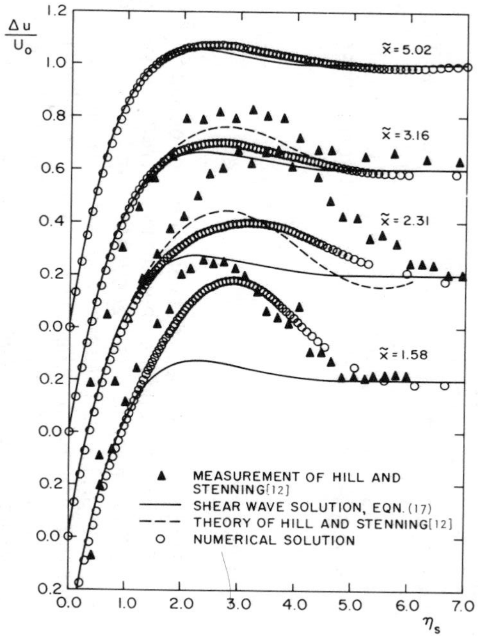

Fig. 3a. Amplitude results for Howarth-Mean-Flow

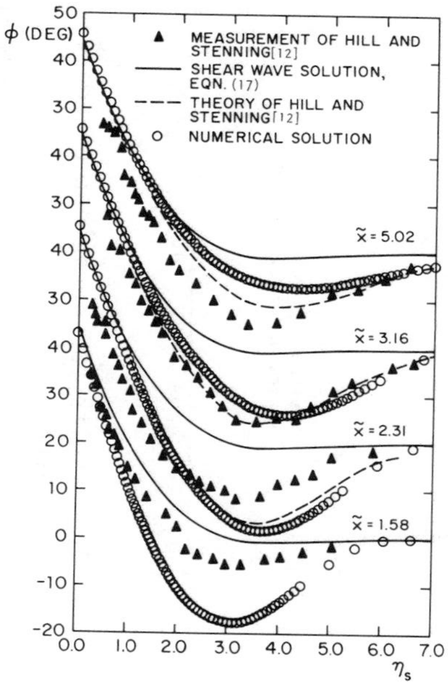

Fig. 3b. Phase results for Howarth-Mean-Flow

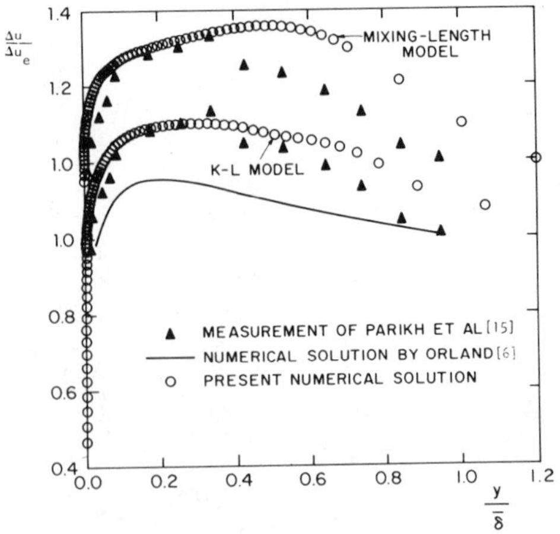

Fig. 4. Results for turbulent boundary layers

The prediction of the amplitude is satisfactory in an overall sense though the two-models deviate from the experimental results in detail. Of particular significance is the difference between the present calculations and those of Orlandi. The latter calculations are seen to underpredict grossly the amplitude, even though essentially the same k-L model has been used. In fact, the same was found to be true for the quasi-steady case [10]. However, this discrepancy between the present calculations and those of Orlandi appeared to decrease with increasing frequency. The reason for the underprediction of the amplitude at low frequencies by Orlandi's calculations cannot be explained, except as possibly being due to numerical errors in his handling of convection terms. More calcuations and comparisons can be found in [10].

5. CONCLUSIONS

An implicit, finite difference procedure for the prediction of time-dependent boundary layer flows has been developed. The method is accurate at small and large frequencies, at all amplitudes and even when there is flow reversal during a part of the cycle in a periodic flow. When applied to turbulent flow, its predictions seem to be in better agreement with experimental results than previous calculations using a different numerical scheme.

ACKNOWLEDGEMENTS

The support for this study by the U.S. Army Research Office through Grants No. DAAG29-79-G-0017 and DAAG29-83-K-0004 is gratefully acknowledged.

REFERENCES

1. McCROSKEY, W.J., PHILIPPE, J.J., "Unsteady Viscous Flow on Oscillating Airfoils", AIAA Journal, 13, 71-79 (1975).
2. CEBECI, T., "Calculation of Unsteady Two-Dimensional Laminar and Turbulent Boundary Layers with Fluctuations in External Velocity", Proc. Roy. Soc., 355A, 225-238 (1977).
3. TSAHALIS, D. TH., TELIONIS, D.P., "Oscillating Boundary Layers with Large Amplitude", in Unsteady Flows in Jet Engines, F.O. Carta (Ed.), 1974.
4. NASH, J.F., PATEL, V.C., "Three-Dimensional Turbulent Boundary Layers", Scientific and Business Consultants, Inc., Atlanta (1972).
5. COUSTEIX, J., HOUDEVILLE, R., DESOPPER, A., "Resultats experimentaux et methodes de calcul relatifs aux couches limites turbulentes en ecoulement instationnaire", ONERA T.P. No. 1977, 134 (1977).
6. ORLANDI, P., "Unsteady Adverse Pressure Gradient Turbulent Boundary Layers", Unsteady Turbulent Shear Flows, Ed. R. Michel, J. Cousteix, R. Houdeville, Springer-Verlag, Berlin, 159-170 (1981).
7. ORLANDI, P., FERZIGER, J.H., "Implicit Noniterative Schemes for Unsteady Boundary Layers", AIAA Journal, 19, 1408-1414 (1981).
8. MURPHY, J.D., PRENTER, P.M., "A Hybrid Computing Scheme for Unsteady Turbulent Boundary Layers", Proceedings of the Third Symposium on Turbulent Shear Flows, Davis, 8.26-8.34 (1981).
9. OSKOLKOV, A.P., "Certain Finite-Difference Schemes for Equations of the Nonstationary Laminar Boundary Layer", Foreign Technology Division, FTD-ID (RS)T-0880-77 (1977).
10. MENENDEZ, A.N., RAMAPRIAN, B.R., "Calculation of Unsteady Boundary Layers", IIHR Report No. 248, 1982.
11. LIGHTHILL, M.J., "The response of laminar skin friction and heat transfer to fluctuations in the stream velocity", Proc. Roy. Soc., 224A, 1-23 (1954).
12. HILL, P.G., STENNING, A.H., "Laminar Boundary Layers in Oscillatory Flow", Jr. of Basic Engrg., 82, 593-608 (1960).
13. PEDLEY, T.J., "Two-dimensional Boundary Layer in a Freestream which Oscillates without Reversing", J. Fluid Mech., 55, 359-383 (1972).
14. ACHARYA, M., REYNOLDS, W.C., "Measurements and Predictions of a Fully Developed Turbulent Channel Flow with Imposed Controlled Oscillations", Stanford University Technical Report TF-8 (1975).
15. PARIKH, P.G., REYNOLDS, W.C., JAYARAMAN, R., "Behavior of an Unsteady Turbulent Boundary Layer", AIAA Journal, 20, 769-775 (1982).
16. WHITE, F.M., Viscous Fluid Flow, McGraw-Hill, New York, 1974.

STUDY OF TURBULENCE MODELING IN TRANSONIC SHOCK WAVE-BOUNDARY LAYER INTERACTION***

by L. Cambier* and J. Délery**

Office National d'Etudes et de Recherches Aérospatiales (ONERA)
29 Avenue de la Division Leclerc,
BP 72 - 92322 CHATILLON Cédex (France)

SUMMARY

First, an experimental study shows some turbulence properties. Then, several turbulence models are tested in the boundary-layer approximation. Finally, a method solving the full Navier-Stokes equations in a multi-domain approach is presented.

1. INTRODUCTION

Phenomena associated with Shock-Wave/Turbulent Boundary Layer Interaction (SW-TBLI) play an important part in many domains of practical interest : wings and airfoils, turbomachinery, helicopter blades... This kind of interaction is a complex process still very difficult to compute correctly either by coupling methods or by global approach. Even in the last case, where the full time-averaged Navier-Stokes equations are still frequently disappointing. The poor agreement with experiment is mainly due :

- to the inability of the presently used turbulence models to correctly depict the very strongly out of equilibrium situations met in a SW-TBLI ;

- to the possible inadequacy of the mesh, in particular too large a mesh size in the flow direction compared to the characteristic interaction length.

---

*Research Engineer
**Division Deputy Head
***We would like to thank Pr. B.E. Launder for his contribution about turbulence modeling. The solving method of the Navier-Stokes equations has been developed in collaboration with W. Ghazzi, J.P. Veuillot and H. Viviand.

Improvement of turbulence models depends largely on reliable experimental information. In this field, decisive progress have been made in the past few years with the advent of Laser Doppler Velocimetry.

At ONERA, fundamental researches have been undertaken to improve the prediction of SW-TBLIs either in the framework of viscid-inviscid interactive method [1] or by the solution of the full Navier-Stokes equations [2-3]

This paper presents results of studies made to remedy the above cited limitations, i.e :

- improvement of turbulence modeling,
- development of accurate numerical techniques.

2. EXPERIMENTAL EVIDENCES

The experimental part of the study has been conducted in a 2-D transonic channel, the test section of which is shown in Fig. 1.

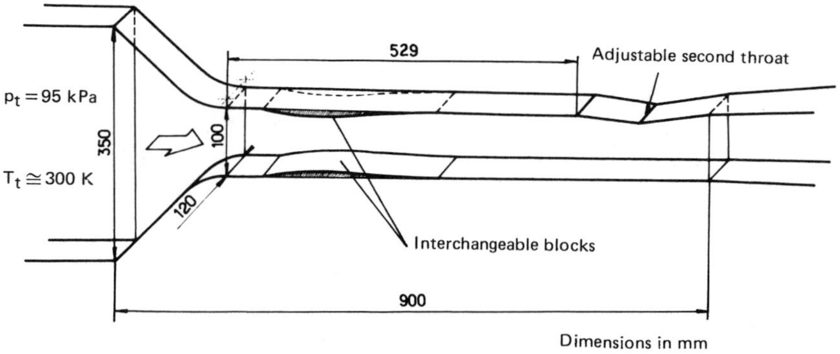

Fig. 1 — Experimental setup.

Interchangeable nozzle blocks or bumps can be mounted in the working section in order to accelerate the flow up to slightly supersonic velocities. A second throat, of adjustable cross-section, is placed at the test section outlet to produce, by choking effect, a quasi-normal shock-wave whose position, and hence intensity, can be adjusted in a precise and continuous manner. The SW-TBLI phenomena under investigation take place on the lower wall of the channel. Detailed measurements of mean and turbulent quantities in the interaction region have been made by using a two-color laser velocimeter [4].

Three SW-TBLIs have been investigated. For the first one (Flow ⓐ), the Mach number Me_0, at the start of the interaction is equal to 1.30. This situation corresponds to incipient shock-induced separation. The second case (Flow ⓑ) is a stronger interaction (Me_0 = 1.45) leading to separation. The last interaction (Flow ⓒ) takes place on a bump the shape of which is such that a large separated bubble develops. In this case, Me_0 = 1.37. More complete presentation of the results can be found in Ref. [5]. Here, we will focus on some turbulence properties. Fig. 2 shows the streamwise variations of the maximum levels of turbulent kinetic energy - (k) max - and shear stress - $(-\overline{u'v'})$max - reached at each station. The interaction entails a large growth of k and $(-\overline{u'v'})$ which, thereafter, slowly decrease towards a new equilibrium state. Also, as shown by Fig. 3, the flow initially exhibits a strong anisotropy. This fact, in conjonction with the strong retardation taking place in the first part of the interaction, makes the contribution of the normal stresses essential in the turbulence production mechanism, as shown in Fig. 4. The experiment shows also that normal stresses play a major part in the momentum equation. These two facts have to be taken into consideration in developing turbulence models.

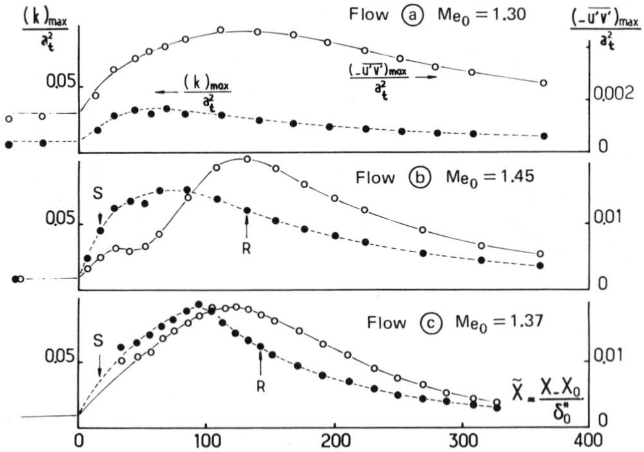

Fig. 2 – Maximum kinetic energy and shear stress variation (δ_0 : displacement thickness at the start of interaction, a_t : sound velocity for stagnation conditions).*

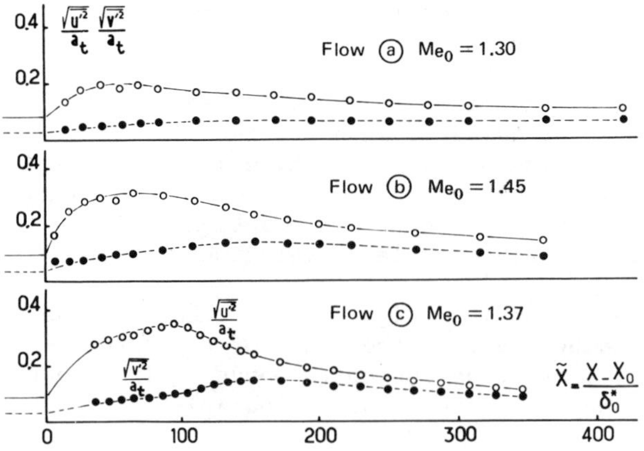

Fig. 3 – Maximum turbulence intensity variation.

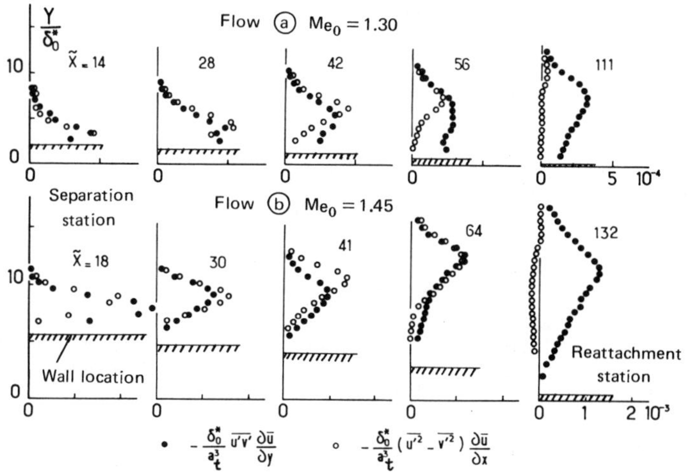

Fig. 4 – Turbulence production terms.

3. THEORETICAL INVESTIGATIONS

3.1. The boundary-layer approach

A thorough study of turbulence modeling in SW-TBLI entails a large number of calculations to adjust the different components and constants of the model. In this point of view, the boundary-layer approximation provides an economical numerical "tool" when compared to the solution of the full Navier-Stokes equations. Of course, the boundary-layer approach does not solve the problem entirely since it necessitates some external input which must be provided by experiment. It is only a way to rapidly detect the most promising model which will be then more accurately tested with the help of the full Navier-Stokes equations.

In the present applications, where separation is likely to occur, the boundary-layer equations (along with the turbulence transport equations) are solved by an inverse procedure in which the displacement thickness δ is prescribed [6]. The output of the calculation consists of the wall pressure (or external Mach number) distribution and of the boundary-layer properties. This kind of approach is justified, provided that the (first order) boundary-layer level of approximation, in which the normal pressure gradient is neglected, does not lead to an inaccurate description of the flow. In order to make this point clear, an inverse boundary-layer calculation has been executed by prescribing the δ^* (x) distribution given by the Navier-Stokes calculation* (the same algebraic turbulence model being used in the two codes). Fig. 5 shows that there is a very good agreement between the "wall" Mach number distributions (M_W is computed from the pressure at the wall by assuming an isentropic relation).

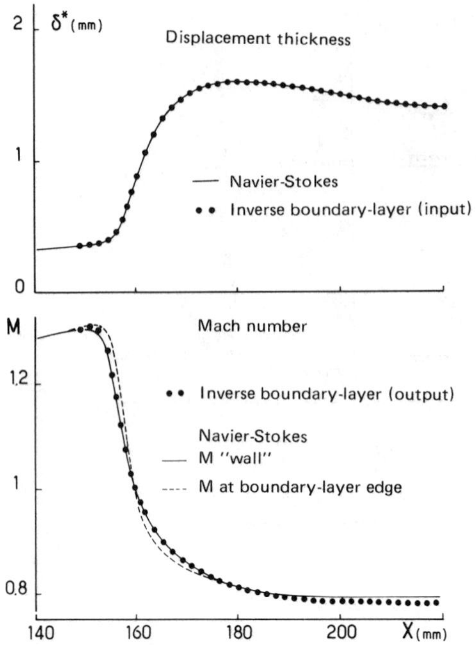

Fig. 5 – Comparison between Navier-Stokes and inverse boundary-layer calculations.

Several turbulence models of increasing level of sophistication have been tested within the framework of the boundary-layer approach. The first one is a classical algebraic model making use of the eddy viscosity concept along with a mixing length type of formulation [7]. The eddy viscosity concept being highly questionable in the flows under consideration, the second investigated model uses three transport equations for k, the dissipation rate ε, and $-\overline{u'v'}$. The adopted formulation is the usual thin layer formulation in which the transport equations are written :

$$\rho\bar{u}\frac{\partial k}{\partial x} + \rho\bar{v}\frac{\partial k}{\partial y} = -\rho\overline{u'v'}\frac{\partial \bar{u}}{\partial y} - \rho\varepsilon + \frac{\partial}{\partial y}\left[(\mu + \mu_T)\frac{\partial k}{\partial y}\right]$$

$$\rho\bar{u}\frac{\partial \varepsilon}{\partial x} + \rho\bar{v}\frac{\partial \varepsilon}{\partial y} = -1.57\,\rho\overline{u'v'}\frac{\varepsilon}{k}\frac{\partial \bar{u}}{\partial y} - 2\rho\frac{\varepsilon^2}{k} + \frac{\partial}{\partial y}\left[(\mu + \frac{\mu_T}{1.3})\frac{\partial \varepsilon}{\partial y}\right]$$

$$\rho\bar{u}\frac{\partial \overline{u'v'}}{\partial x} + \rho\bar{v}\frac{\partial \overline{u'v'}}{\partial y} = -0.135\,\rho k\frac{\partial \bar{u}}{\partial y} - 1.5\,\rho\frac{\varepsilon}{k}\overline{u'v'} + \frac{\partial}{\partial y}\left[(\mu + \frac{\mu_T}{0.9})\frac{\partial \overline{u'v'}}{\partial y}\right]$$

In these equations : $\mu_T = 0.09\,\rho\,(k^2/\varepsilon)$.

The above equations are valid only if the turbulence Reynolds number $R_T = \rho k^2/\mu\varepsilon$ is high enough (say. $R_T \gg 200$). Consequently, a special treatment must be adopted in the near wall region. In the present study, a patching approach has been used : in the wall region, a mixing length expression is

* described in section 3.2.

employed up to an ordinate y_R where R_T has reached a sufficiently high value (R_T = 200). Above y_R, the transport equations are used, the boundary conditions on y_R being provided by the mixing length model.

As demonstrated by experiment, a more realistic representation of the flow physics necessitates the consideration of normal stresses. The Algebraic Stress Model (ASM) provides a rather simple way to evaluate the complete Reynolds tensor [8]. In this approach, one solves two transport equations for k and ε which contain the normal stress terms. Then, the Reynolds tensor components are derived from algebraic relations involving k, ε and the derivatives of the mean velocity field.

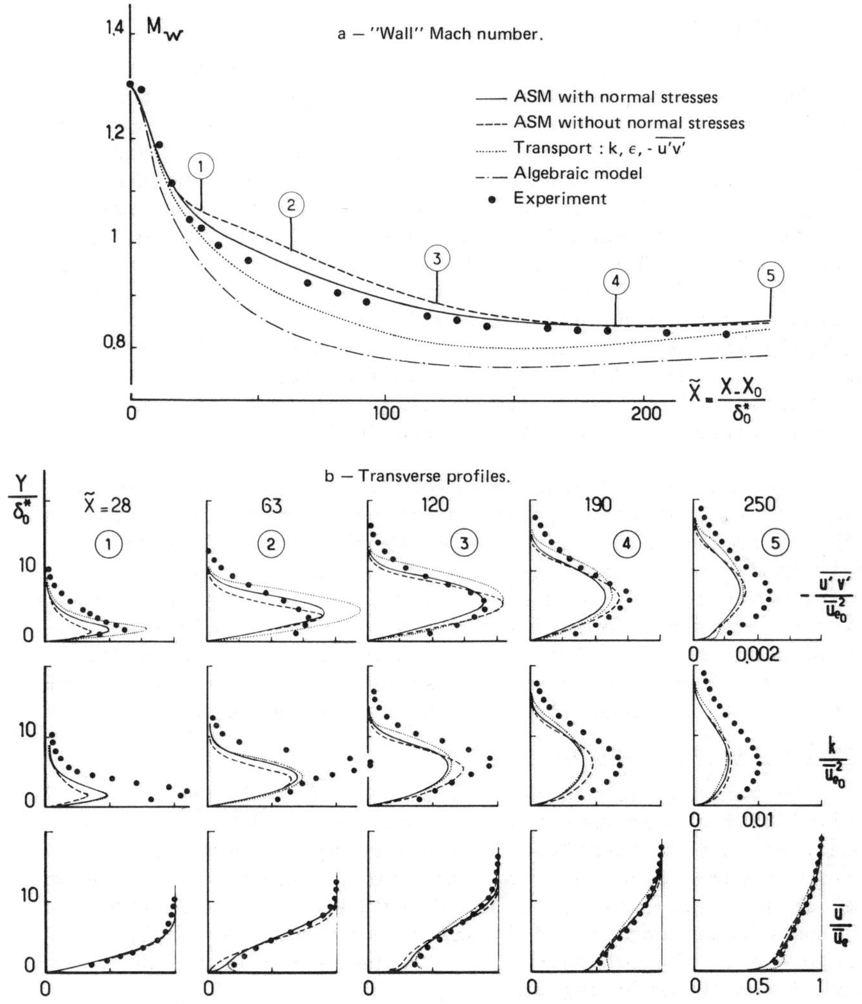

Fig. 6 — Test of turbulence models. Boundary-layer approach.

The above model has been tested for conditions corresponding to flow a . As shown in Fig. 6a, generally speaking, the algebraic and the 3-equation models give a compression steeper than in reality. On the other hand, the ASM model without the normal stresses leads to an unsufficient compression. The best agreement with experiment is obtained with the full ASM model including normal stresses. However, even in this case, the agreement is far from being entirely satisfactory. As shown by Fig. 6b, the level of k is largely under-predicted at the beginning of interaction. Furthermore, the slow decrease of $-\overline{u'v'}$ during the relaxation process is not correctly represented.

3.2. Solution of the Navier-Stokes equations

Generalities

A numerical simulation of a SW-TBLI has been carried out, solving the time averaged Navier-Stokes equations. A multi-domain approach has been developed. The computational domain (Fig. 7), limited by the wall and the symmetry axis, and also by an upstream section and a downstream section (located far enough from the interaction region), is divided into three sub-domains. A sub-domain $\mathcal{D}_V$ where viscous effects are important is defined below a longitudinal cut line Σ_1. The remaining part $\mathcal{D}_p$ of the computational domain, where perfect fluid approximation is justified, is divided into two sub-domains $\mathcal{D}_p(1)$ and $\mathcal{D}_p(2)$, separated by the shock-wave Σ_2. The position and the shape of Σ_2 are unknowns of the problem ; they are calculated using a shock-fitting technique. So, the shock-wave is fitted in $\mathcal{D}_p$ and captured in $\mathcal{D}_V$.

Such a multi-domain approach has some important advantages. First, we can separately define the mesh in $\mathcal{D}_p$ and in $\mathcal{D}_V$. Secondly, the full Navier-Stokes equations are solved only in $\mathcal{D}_V$ and Euler equations are solved in $\mathcal{D}_p$. Hence, computation cost is reduced. At last, the fitting of the major part of the shock wave enables to save up mesh points near Σ_2 in $\mathcal{D}_p$ and moreover improves the accuracy of the results. (The shock-wave thickening in a shock-capturing approach probably tends to lessen the interaction phenomenon).

Coupling of sub-domains

A general technique for coupling two sub-domains $\mathcal{D}(1)$ and $\mathcal{D}(2)$ has been developed. We only recall here its main features. A more detailed description may be found in [3]. This technique requires the systems of equations which are solved in both sub-domains to be t-hyperbolic, which is the case of Euler equations solved by unsteady methods. To apply this technique to the cut line Σ_1, it is necessary to assume that Σ_1 lies in an inviscid region.

At a point P of the boundary Σ between $\mathcal{D}(1)$ and $\mathcal{D}(2)$, the two sets of values U_1 and U_2 of the variables ($U = (\rho, \rho u,$

ρv, ρE) in 2-D Euler equations) and possibly the normal speed w of Σ are derived from a system consisting of suitable compatibility relations and coupling relations. The compatibility relations used are those associated with the normal to Σ at P, and we only retain in $\mathcal{D}(1)$ (resp. in $\mathcal{D}(2)$) the ones which transport information from $\mathcal{D}(1)$ into $\mathcal{D}(2)$ (resp. from $\mathcal{D}(2)$ into $\mathcal{D}(1)$).

In the case of the fixed boundary Σ_1(w = 0), four compatibility relations are to be retained and the coupling relations simply express the continuity of $\overline{U}$. These four independent compatibility relations enable us to determine $\overline{U} = \overline{U}_1 = \overline{U}_2$. In the case of the moving boundary Σ_2 (w unknown), the four compatibility relations must be considered on $\mathcal{D}_p(1)$ side. These four relations are equivalent to Euler equations and enable us to determine $\overline{U}_1$ irrespective of $\overline{U}_2$ and w. One compatibility relation is to be retained on $\mathcal{D}_p(2)$ side. The coupling relations are the classical Rankine-Hugoniot shock-equations. Thus, the system consisting of the compatibility relation on $\mathcal{D}_p(2)$ side and of these four coupling relations is a non-linear system for the unknowns U_2 and w.

Boundary conditions

On the wall, we impose a no-slip condition and a zero heat flux. Wall pressure is derived from the normal momentum equation at the wall.

On the symmetry axis (located in $\mathcal{D}_p$), we impose symmetry conditions which are equivalent in perfect fluid to a slip condition.

On the upstream boundary, where the flow is supersonic except for a very small region near the wall, all flow properties are prescribed.

At last, on the downstream boundary, where the flow is subsonic, we set the pressure value. The shock position essentially depends on this value. The other flow properties are derived from the three compatibility relations to be used.

Numerical method

The Navier-Stokes equations in $\mathcal{D}_v$ and the Euler equations in $\mathcal{D}_p(1)$ and $\mathcal{D}_p(2)$ are solved by a time-marching method with an explicit predictor-corrector type scheme. The equations are discretized using the MacCormack's finite difference scheme [9] directly in the physical plane on an arbitrary curvilinear mesh [10 - 11].

The computation is carried out on a moving mesh. The transverse mesh lines move in order to follow the shock-wave displacement in $\mathcal{D}_p$ like in $\mathcal{D}_v$. The longitudinal mesh lines are invariable. In $\mathcal{D}_p$, the shock-wave is fitted ; it is a moving mesh line, of fixed index, which intersects the cut line Σ_1 at the moving point A. In $\mathcal{D}_v$, the transverse mesh lines are vertical, the vertical from A being a mesh line of fixed index $i = i_A$. So the mesh contraction introduced near $i = i_A$ in order to correctly represent the interaction zone, automatically follows the shock-wave motion. Practically, we compute $\overline{U}^{n+1}$ and w on a fixed mesh defined at time $t^n = n\Delta t$ and we project this solution $\overline{U}^{n+1}$ upon a new mesh defined according to the shock wave displacement, at time $t^{n+1} = (n + 1)\ \Delta t$.

Near the wall, a very important mesh refinement is required to represent the rapid variations of the flow properties in the viscous sublayer. This refinement is obtained by combining two techniques : a) a regular contraction defined by a geometrical progression ; b) a technique of subgrid with several zones [12], consisting of successive halvings of the mesh size from one zone to the following one. This technique of dichotomy provides a rapid transition from a relatively coarse mesh in the outer quasi-inviscid flow to a very fine mesh in the viscous sublayer with a smoothly varying mesh size in each zone. Coupling of adjacent zones is insured by overlapping. A systematic procedure has been set up to sweep the computational domain and insure the numerical matching in such a way that the maximum allowable time step can be used in each zone.

Results

The code has been tested for conditions corresponding to flow ⓐ. The turbulence model employed is the algebraic model already cited [7]. Two computations have been carried out : a first one (C1) with a constant maximum mixing length l_{max} ; a second one (C2) with a l_{max} (x) distribution deduced from experimental values of $-\overline{u'v'}$ and of mean velocity.

Figure 7 represents the meshes (at convergence). They are made up of (22 x 13) points in $\mathcal{D}_p(1)$, (30 x 13) points in $\mathcal{D}_p(2)$ and (61 x 33) points in $\mathcal{D}_v$. An enlargement near the wall (Fig. 8) shows the important mesh refinement. The subgrid is made up of 5 zones.

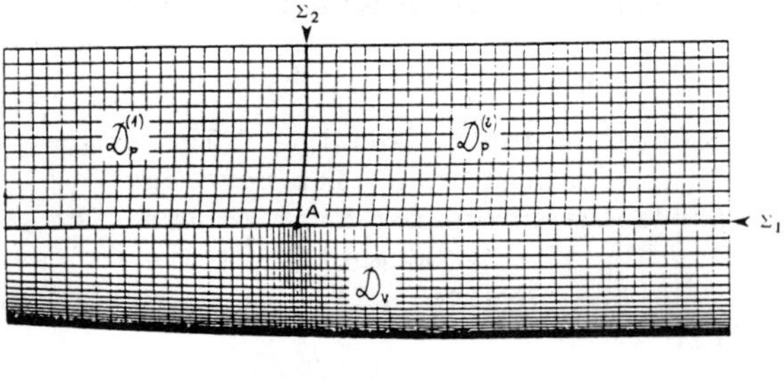

Navier-Stokes calculations — Computational domain and meshes.

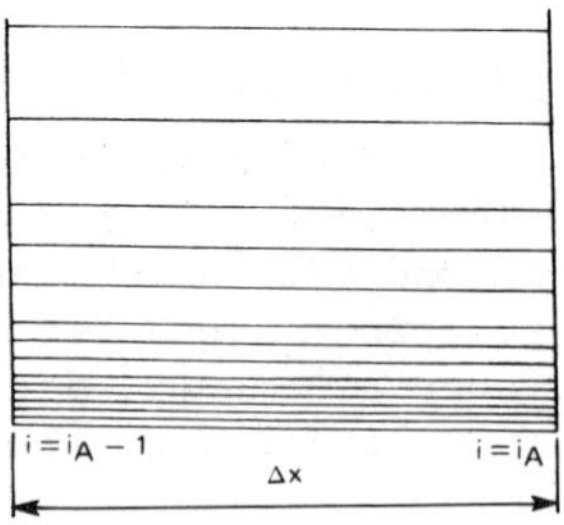

Fig. 8 — Mesh refinement near the wall.

On the upstream boundary, flow conditions have been provided by a preliminary computation, carried out in the upstream part of the nozzle, in a shockless choked configuration. On the downstream boundary, the pressure p, is imposed such that : $p = 0.6625\ p_t$. This value is different from the experimental one ($p = 0.6357\ p_t$) because of side effects resulting from the presence of lateral boundary layers in the experiment.

A qualitative comparison between experiment and computation C2 is shown on Fig. 9a and 9b, where the experimental (interferogram) and computed iso-density lines are plotted. Fig. 9c and 9d represent the computed isomach lines and isobaric lines. A rapid thicknening of the boundary layer, due to the interaction with the shock wave may be noticed on Fig. 9c. (Only a part of the computational domain is shown on these figures).

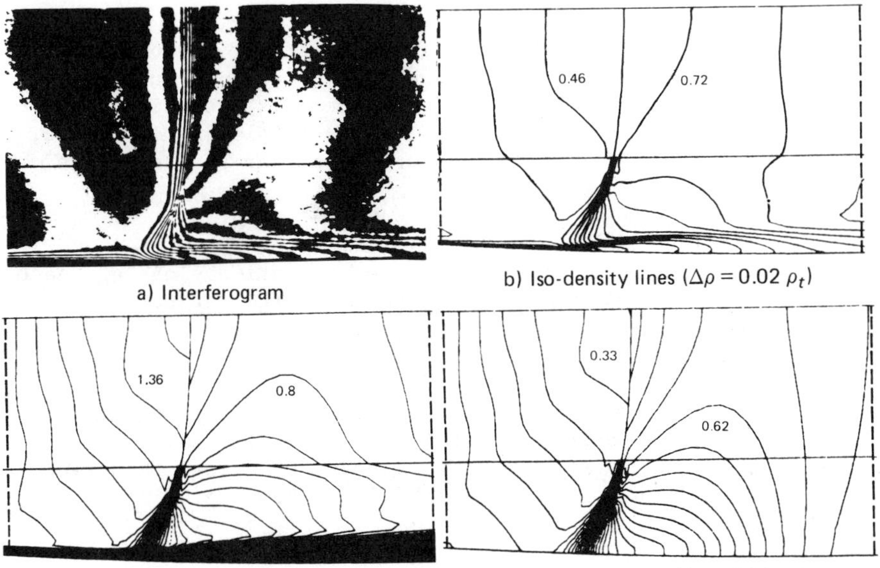

a) Interferogram

b) Iso-density lines ($\Delta\rho = 0.02\ \rho_t$)

c) Iso-Mach lines ($\Delta M = 0.02$)

d) Isobaric lines ($\Delta p = 0.01\ p_t$)

Fig. 9 – Navier-Stokes calculations – Flow field in the interaction region
p_t and ρ_t : pressure and density for stagnation conditions.

Fig. 10 shows a quantitative comparison between experiment and the calculations C1 and C2. The distributions of the wall pressure are plotted in Fig. 10a and the streamwise velocity profiles in Fig. 10b. It should be noticed that the downstream pressure value imposed in the calculations has not proved accurate enough to provide computed shock-positions identical with the experimental one. Taking into account this shift in the shock positions, the C2 pressure distribution is seen to be closer to the experimental one than the C1 pressure distribu-- tion. However, the discrepancy between the velocity profiles should clearly not be imputed only to the variations in shock-positions, but also to the deficiencies of the employed turbulence model.

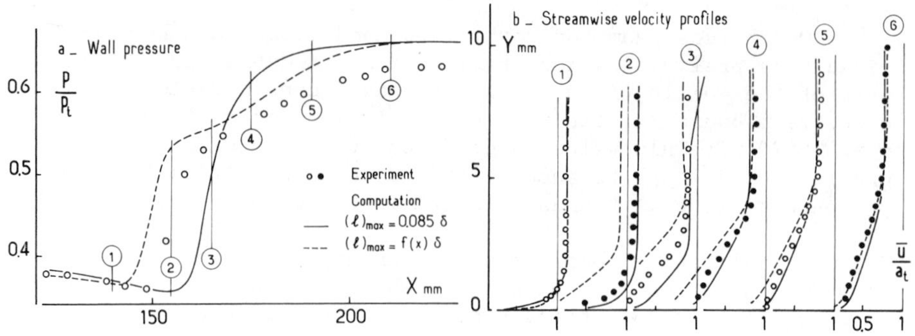

Fig. 10 — Navier Stokes calculations — Comparison with experiment — Algebraic turbulence model.

4. CONCLUSION

Experiment shows that the initial part of a shock/turbulent boundary-layer interaction entails a large turbulence production which mainly affects the streamwise velocity component. In this region, the normal stresses play a major part. The non-equilibrium effects are very important in the entire domain.

A critical study of turbulence modeling is carried out in the framework of boundary-layer approximation. This approach constitutes an economical tool to test a large variety of turbulence models. The calculations presently done confirm the influence of the normal stresses.

In order to solve the problem more completely and more exactly, a method solving the full time-averaged Navier-Stokes equations has been developed. The adopted numerical technique uses a multi-domain approach which allows computer time saving and better accuracy.

The Navier-Stokes calculation has succeeded in providing a fine description of the flowfield. However, agreement with experiment is still unsatisfactory because of the rusticity of the turbulence model used in this first calculation. Study is presently carried on to incorporate more sophisticated models selected from tests made within the framework of the boundary-layer approach.

5. REFERENCES

1. LE BALLEUR, J.C., Numerical viscid-inviscid interaction in steady and unsteady flows. Proc. of the "2nd Symposium on Numerical and Physical Aspects of Aerodynamics Flows". 17-20 jan. 83, California State University, Long-Beach, to be published by Springer-Verlag.

2. VIVIAND, H., Traitement des problèmes d'interaction fluide parfait - fluide visqueux en écoulement bidimensionnel compressible à partir des équations de Navier-Stokes. Conférences AGARD/VKI 20-24 février 1978. ONERA-TP n° 1978-4.

3. CAMBIER, L., GHAZZI, W., VEUILLOT, J.P. and VIVIAND, H. Une approche par domaines pour le calcul d'écoulements compressibles. INRIA, 1981. ONERA-TP n° 1981-143.

4. BOUTIER, A., Two-dimensional laser velocimeter 20 H61. ONERA Data Sheet 1981-4, 1981.

5. DELERY, J., Investigation of strong shock/turbulent boundary-layer interaction in 2-D transonic flows with emphasis on turbulence phenomena. AIAA Paper 81-1245, 1981.

6. DELERY, J. and LE BALLEUR, J.C., Interaction et couplage entre écoulement de fluide parfait et écoulement visqueux. ONERA-RSF 4/3073 AY, 1980.

7. MICHEL, R., QUEMARD, C. and DURANT, R., Application d'un schéma de longueur de mélange amélioré à l'étude des couches limites d'équilibre. ONERA-NT 154, 1969.

8. LAUNDER, B.E., An improved algebraic stress model in turbulence. Imperial College, Mech. Eng. Dept. Report n° TM/TN/A/9.

9. MacCORMACK, R.W., The effect of viscosity in hyper velocity impact cratering. AIAA Paper n° 69-354, 1969.

10. VIVIAND, H. and VEUILLOT, J.P., Méthodes pseudo-instationnaires pour le calcul d'écoulements transsoniques. ONERA Publication n° 1978-4 (English Translation, ESA-TT 561), 1978.

11. HOLLANDERS, H. and VIVIAND H., The numerical treatment of compressible high Reynolds number flows. Kollman, W. (ed.), Computational Fluid Dynamics, 2, Hemisphere Publ. Corp. 1980.

12. VIVIAND, H. and GHAZZI, W., Numerical solution of the Navier-Stokes equations at high Reynolds numbers with applications to the blunt body problem. Lecture Notes in Physics, 59, Springer-Verlag, 1976.

PSEUDOSPECTRAL CALCULATION OF SHOCK TURBULENCE INTERACTIONS

T. A. Zang(I) D. A. Kopriva(II) M. Y. Hussaini(II)

College of William and Mary ICASE ICASE

SUMMARY

A Chebyshev-Fourier discretization with shock fitting is used to solve the unsteady Euler equations. The method is applied to shock interactions with plane waves and a simple model of homogeneous isotropic turbulence. The plane wave solutions are compared to linear theory.

1. INTRODUCTION

The effects of a shock wave upon a turbulent boundary layer have been studied extensively by both experimental and computational means. A persistent experimental result is that the turbulence levels downstream of the shock are higher, by at least a factor of 2, than the upstream levels [1]. The computational results, however, have consistently failed to exhibit this dramatic turbulence amplification [2]. This failure is almost surely due to inadequacies in the turbulence models used in the computations. Current turbulence models have achieved their greatest success in low-speed flows and attached high-speed flows which agree well with experiment even up to Mach 20 [3]. Evidently, some essential physics of separated compressible flow is missing from the models. Their

---

(I)Research Associate at the College of William and Mary, Williamsburg, VA 23185, NASA Grant No. NAG1-109.

(II)Staff Scientist and Senior Staff Scientist, respectively, at the Institute for Computer Applications in Science and Engineering at NASA Langley Research Center, Hampton, VA 23665, NASA Contract Nos. NAS1-16394, NAS1-17070 and NAS1-17130 .

improvement must await a better understanding of the interaction between shock waves and turbulence.

Two theoretical tools are now available for the study of the inviscid aspects of this interaction. The basically linear effects may be examined analytically in a way developed by Ribner [4] and others, and used by Anyiwo and Bushnell [5] in their assessment of turbulence amplification in the shock-boundary layer interaction. Numerical computations based on the nonlinear, two-dimensional Euler equations are more recent. Pao and Salas [6] used this approach to study the shock-vortex interaction. So did Zang, Hussaini and Bushnell [7] in their examination of the interaction of plane waves with shocks. Both sets of numerical computations used second-order finite differences for the spatial discretization. In this paper we will describe a pseudospectral spatial discretization.

2. NUMERICAL METHOD

The model problem which is used to study the turbulence amplification and generation mechanisms is illustrated in Figure 1. At time $t = 0$ an infinite, normal shock at $x = 0$ separates a rapidly moving, uniform fluid on the left from the fluid on the right which is in a quiescent state except for some specified fluctuation. The initial conditions are chosen so that in the absence of any fluctuation the shock moves uniformly in the positive x-direction with a Mach number (relative to the fluid on the right) denoted by M_s. In the presence of fluctuations the shock front will develop ripples. The structure of the shock is described by the

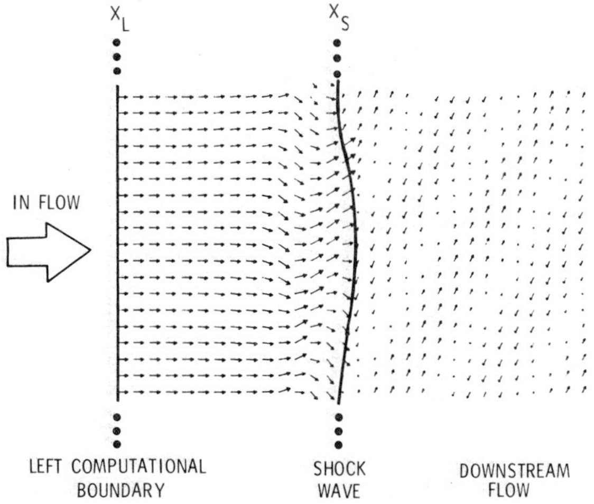

Figure 1: Model problem in the physical domain.

function $x_s(y,t)$. The numerical calculations are used to determine the state of the fluid in the region between the shock front and some suitable left boundary $x_L(t)$ and also to determine the motion and shape of the shock front itself.

In the applications given below the physical problem is periodic in y with period y_L. The change of variables

$$\begin{aligned} X &= \frac{x - x_L(t)}{x_s(y,t) - x_L(t)} \\ Y &= y/y_L \\ T &= t \end{aligned} \tag{1}$$

produces the computational domain

$$\begin{aligned} &0 \leqslant X \leqslant 1 \\ &0 \leqslant Y \leqslant 1 \\ &T \geqslant 0. \end{aligned} \tag{2}$$

In terms of the computational coordinates the two-dimensional Euler equations are

$$Q_T + BQ_X + CQ_Y = 0, \tag{3}$$

where $Q = (P,u,v,S)$,

$$B = \begin{bmatrix} U & \gamma X_x & \gamma X_y & 0 \\ \frac{c^2}{\gamma} X_x & U & 0 & 0 \\ \frac{c^2}{\gamma} X_y & 0 & U & 0 \\ 0 & 0 & 0 & U \end{bmatrix} \tag{4}$$

and

$$C = \begin{bmatrix} V & \gamma Y_x & \gamma Y_y & 0 \\ \frac{c^2}{\gamma} Y_x & V & 0 & 0 \\ \frac{c^2}{\gamma} Y_y & 0 & V & 0 \\ 0 & 0 & 0 & V \end{bmatrix}. \tag{5}$$

The contravariant velocity components are given by

$$U = X_t + uX_x + vX_y$$

and (6)

$$V = Y_t + uY_x + vY_y.$$

A subscript denotes partial differentiation with respect to

the indicated variable. P, c and S are the natural logarithm of pressure, the sound speed and the entropy (divided by the specific heat at constant volume), respectively, all normalized by reference conditions at upstream infinity; u and v are velocity components in the x and y directions, both scaled by the characteristic velocity defined by the square root of the pressure-density ratio at downstream infinity. The ratio of specific heats is denoted by γ; a value $\gamma = 1.4$ has been used for all the calculations in this paper.

Let n denote the time level and Δt the time increment. The time discretization of Eq. (3) is

$$\tilde{Q} = \left[1-\Delta t L^n\right]Q^n \tag{7}$$

$$Q^{n+1} = \frac{1}{2}\left[Q^n+\left(1-\Delta t\tilde{L}\right)\tilde{Q}\right], \tag{8}$$

where L denotes the spatial discretization of $A\partial_X + B\partial_Y$. The solution Q has the Chebyshev - Fourier series expansion

$$Q(X,Y,T) = \sum_{p=0}^{M} \sum_{q=-N/2}^{N/2-1} Q_{pq}(T)\tau_p(\xi)e^{2\pi iqY}, \tag{9}$$

where $\xi = 2X-1$ and τ_m is the Chebyshev polynomial of degree m. The derivatives Q_X and Q_Y are approximated by

$$Q_X = 2\sum_{p=0}^{M} \sum_{q=-N/2}^{N/2-1} Q_{pq}^{(1,0)}(T)\tau_p(\xi)e^{2\pi iqY} \tag{10}$$

$$Q_Y = 2\pi\sum_{p=0}^{M} \sum_{q=-N/2}^{N/2-1} Q_{pq}^{(0,1)}(T)\tau_p(\xi)e^{2\pi iqY} \tag{11}$$

where

$$Q_{pq}^{(1,0)} = \frac{2}{c_p} \sum_{\substack{m=p+1 \\ p+m \text{ odd}}}^{M} m\, Q_{mq} \tag{12}$$

with

$$c_p = \begin{cases} 2 & p = 0 \\ 1 & p > 1 \end{cases} \tag{13}$$

and

$$Q_{pq}^{(0,1)} = iq\, Q_{pq}. \tag{14}$$

The most critical part of the calculation is the treatment of the shock front. The shock-fitting approach used here is desirable because it avoids the severe post-shock oscillations that plague shock-capturing methods. The time derivative of the Rankine-Hugoniot relations provides an equation for the shock acceleration. This equation is integrated to update the shock position (see [8] for details). The right boundary at $x_s(y,t)$ is a supersonic inflow boundary. Hence it is appropriate to prescribe all variables there. For $M_s > 2.08$, the left boundary at $X_L(t)$ is also a supersonic inflow boundary. Lower Mach numbers require more sophisticated boundary conditions as discussed in [8]. Spectral methods for compressible flows typically require some sort of filtering. The type of filtering employed in the present calculations is described in [9].

3. INTERACTION OF PLANE WAVES WITH A SHOCK

The non-linear interaction of plane waves with shocks was examined at length in [7]. The numerical method used there was similar to the one described above but employed second-order finite differences in place of the present pseudospectral discretization. Detailed comparisons were made in [7] with the predictions of linear theory [4]. The linear results turned out to be surprisingly robust, remaining valid at very low (but still supersonic) Mach numbers and at very high incident wave amplitudes. The only substantial disagreement occurred for incident waves whose wave fronts were nearly perpendicular to the shock front. This type of shock-turbulence interaction is a useful application of the pseudospectral technique. The method can be calibrated in the regions for which linear theory has been shown to be valid. The questionable regions can then be gainfully investigated by means of this alternative method.

We concentrate on Mach 3 normal shocks interacting with incident vorticity waves. The perturbation velocities ahead of the shock are taken to be

$$\left(u',v'\right) = A'\left(-k_{1,y},k_{1,x}\right)\left(c_1/p_1 k\right)\cos\left(\underline{k}_1 \cdot \underline{x}\right),$$

where the wavevector $\underline{k}_1 = \left(k_{1,x},k_{1,y}\right)$ and A' is the amplitude. In terms of the incidence angle θ_1, $\underline{k}_1 = \left(k_1\cos\theta_1, k_1\sin\theta_1\right)$. The transmitted vorticity behind the shock can be expressed in the same manner with all subscripts 1 changed to 2. The vorticity transmission coefficient is defined to be the ratio A_2'/A_1'. At Mach 3 the ratio of the rotational part of the perturbation velocities is 0.158 times the transmission coefficient; if these perturbation velocities are scaled by the respective mean velocities (in the frame in which the shock is at rest), then this ratio is 0.611 times the transmission coefficient.

A typical numerical simulation is shown in Figure 2. At time $t = 0$ all the fluid in the computational domain, which then extended from $x_L = -0.5$ to $x_s = 0.0$, was in uniform

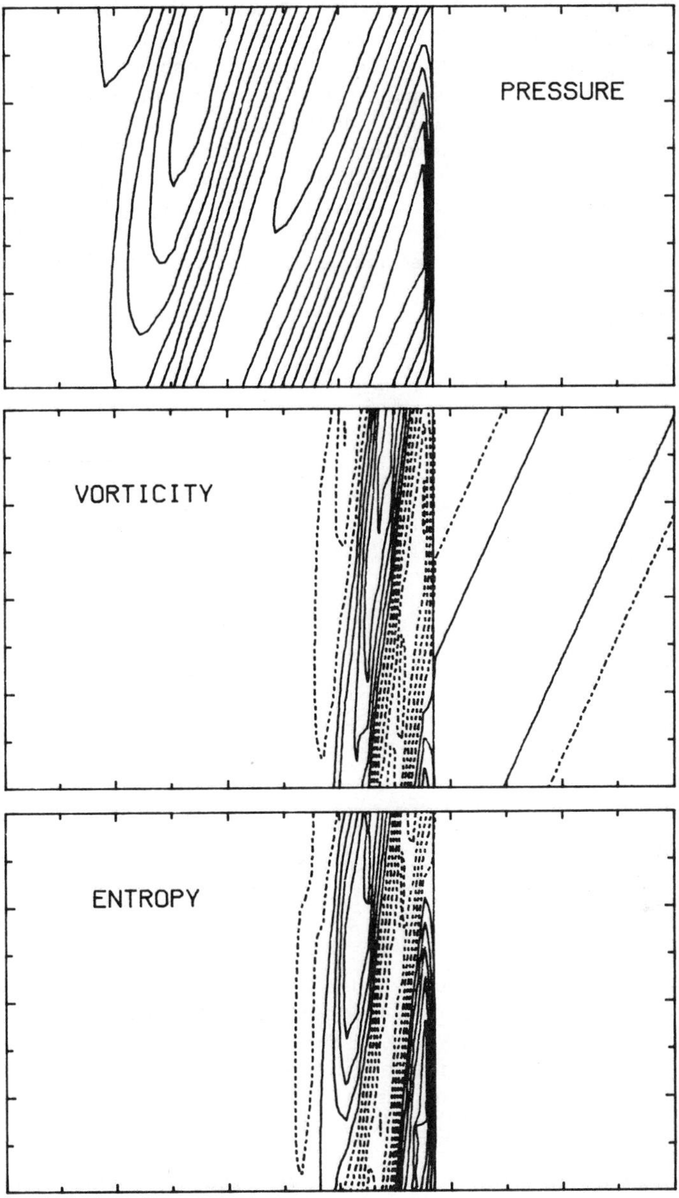

Figure 2: Responses at time $t = 0.80$ to a 0.1% vorticity wave striking a Mach 3 shock at a 30^0 angle of incidence. The shock front is denoted by the solid line connecting the top and bottom boundaries. Dashed lines indicate negative contour levels.

flow. The flow ahead of the shock was given by Eq. (15) with an amplitude $A' = 0.001$. In order to reduce the transients that would result if the fluid were to encounter this wave in a sudden fashion, the amplitude was turned on smoothly from $A' = 0.000$ to $A' = 0.001$ during the passage of the shock over the first half-wavelength of the incident wave. The interaction of the shock with the incident vorticity wave has produced a transmitted vorticity wave as well as generated acoustic and entropy waves. The acoustic wave clearly has a larger inclination angle and a faster propagation speed than the other two waves.

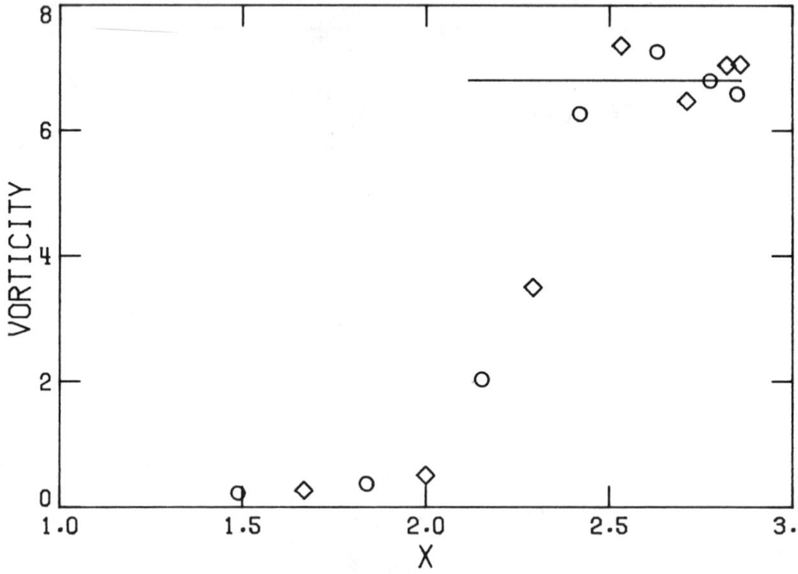

Figure 3: Post-shock dependence at time $t = 0.80$ of the vorticity response to a vorticity wave incident at 30^0 to a Mach 3 shock. The solid line is the linear theory prediction. Computed responses are given for 0.1% (circles) and 50% (diamonds) waves.

Figure 3 indicates how we measure the transmitted vorticity. The computed velocities are differentiated pseudospectrally and combined to produce the response vorticity at each point on the computational grid. At each fixed value of X we perform a Fourier analysis in Y of the vorticity. The wavenumber of interest is $q = 1$ (see Eq. (9)). This amplitude is then scaled as indicated above to yield A_2' and thence the ratio A_2'/A_1' which is plotted in Figure 3 as a function of a typical value of the physical coordinate x corresponding to X. In the case shown in Figure 3 the incident wave does not reach full strength until $t = 0.56$. A single value representing the transmission coefficient is obtained by averaging the x-dependent responses. The standard deviation of these values serves as an error estimate, indicated in the following two figures.

The angular dependence of the vorticity transmission coefficient for Mach 3 is displayed in Figure 4. Corresponding results from the finite difference discretization used in [7] are shown for comparison. The amplitude dependence for 30^0 incident waves is shown in Figure 5.

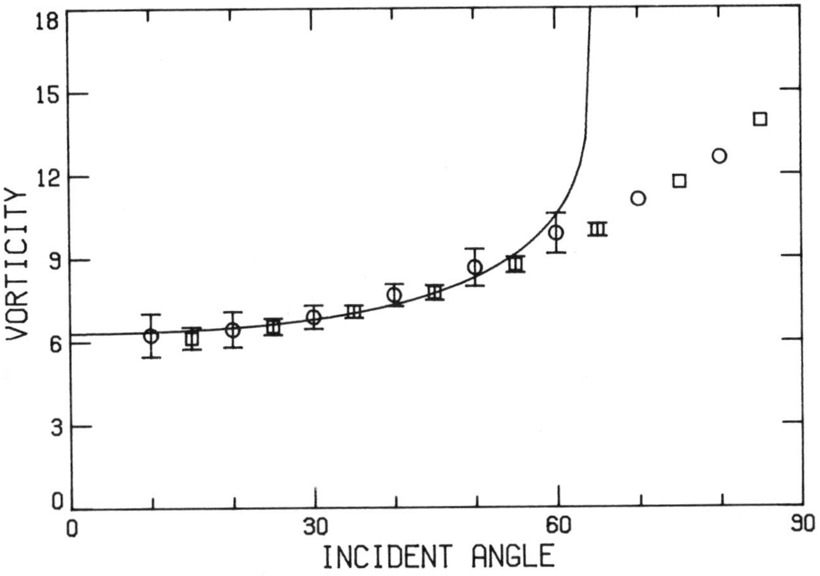

Figure 4: Incident angle dependence of the vorticity response to 0.1% vorticity waves incident upon a Mach 3 shock. Pseudospectral (circles) and finite difference (squares) results are given. The solid line is the linear theory prediction for sub-critical angles.

Linear theory predicts that the vorticity responses are peaked in the vicinity of $\theta_1 = 62^0$, the so-called critical angle. At larger angles the acoustic responses are evanescent. These predictions do not appear in the non-linear calculations. Tentative explanations for this failure of linear theory are provided in [7].

Below the critical angle, however, the linear theory is extremely robust. Figure 5 indicates that it remains valid for incident relative velocity fluctuations approaching 50%. (In the frame in which the shock is at rest, the ratio of the fluctuating velocity to the mean velocity is one-third of the amplitude A_1'.)

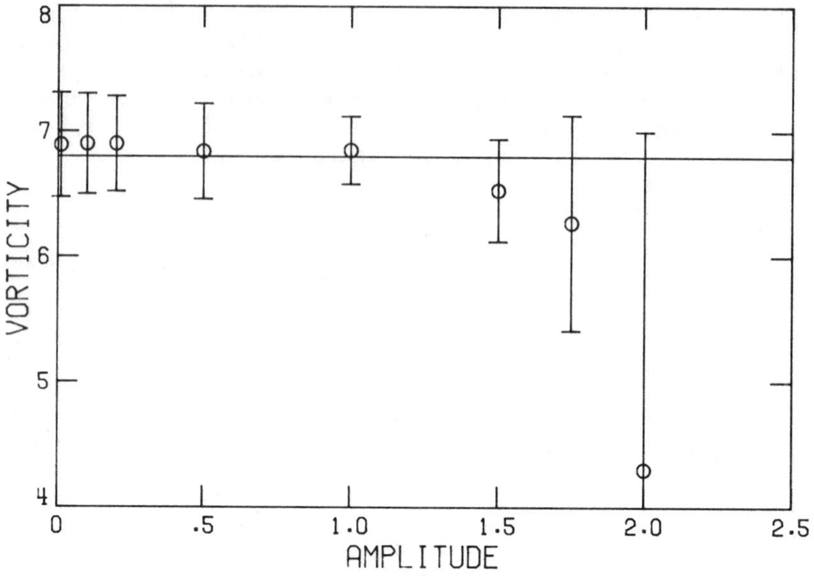

Figure 5 Amplitude dependence of the vorticity response to vorticity waves incident at 30^0 to a Mach 3 shock. The solid line is the linear theory prediction.

4. INTERACTION OF TURBULENCE WITH A SHOCK

Low intensity turbulence can be viewed as the superposition of plane shear waves just discussed. Figure 6 shows velocity vector plots of the interaction of a shock with a simple homogenous isotropic turbulence model in which the velocities are defined as sums of sinusoidal waves with random phases. The frame at $t = 0.38$ shows the undisturbed velocity field downstream of the shock and $t = 1.6$ shows the result after the shock has traveled approximately one wavelength of the longest wave component of the undisturbed turbulence. The development of the post-shock vorticies seen in the figure coincides with the generation of rows of pressure and entropy spots behind the shock.

5. CONCLUSION

We have solved the two-dimensional unsteady non-linear Euler equations with a Chebyshev-Fourier pseudospectral method. The linear theory of the interaction of a shock and plane shear waves has provided a test of the method. Details of the smoothing methods and questions of accuracy have been omitted due to lack of space. But accuracy comparable to the finite difference calculations can be obtained with about half the number of Chebyshev modes in X and with only four modes in Y. The results indicate that it should be possible to compute more complicated shock-turbulence interactions. Quantitative results will be discussed in a future paper.

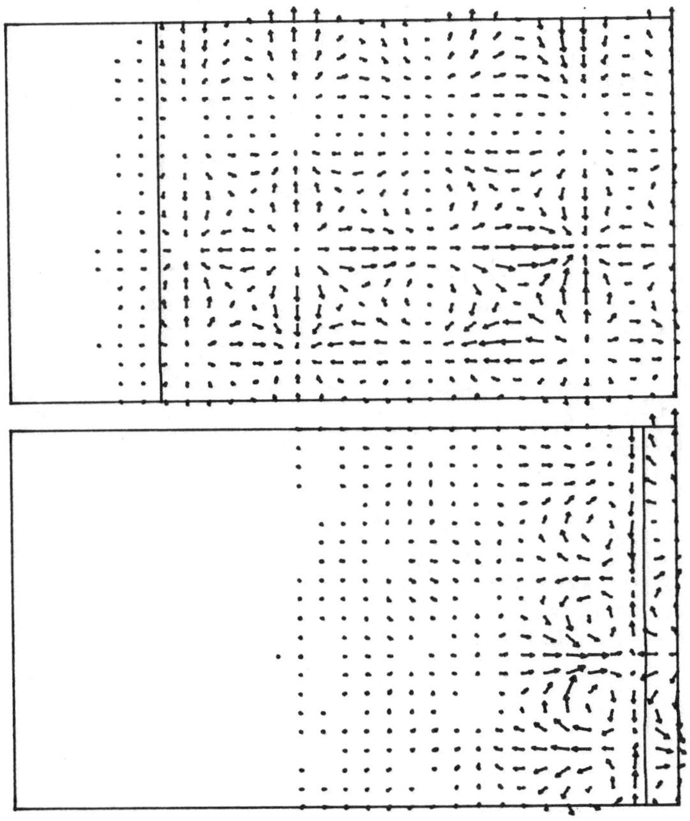

Figure 6: Velocity vector plots at t = 0.38 (top) and t = 1.6 (bottom) of the interaction of a Mach 3 shock with low intensity homogeneous isotropic turbulence.

REFERENCES

1. ROSE, W. C. - The Behavior of a Compressible Turbulent Boundary Layer in Shock-Wave-Induced Adverse Pressure Gradient. NASA TN D-7092, 1973.

2. HORSTMAN, C. C., HUNG, C. M., SETTLES, G. S., VAS, I. E. and BOGDONOFF, S. M. - Reynolds Number Effects on Shock-Wave/Turbulent Boundary-Layer Interactions - A Comparison of Numerical and Experimental Results. AIAA Paper, 77-42, 1977.

3. BUSHNELL, D. M. CARY, A. M., JR., and HARRIS, J. E. - calculation Methods for Compressible Turbulent Boundary Layers. NASA SP-422, 1976.

4. RIBNER, H. S. - Shock-Turbulence Interaction and the Generation of Noise. NACA Report 1233, 1955.

5. ANYIWO, J. C. and BUSHNELL, D. M. - Turbulence Amplification in Shock Wave Boundary Layer Interactions. AIAA Journal, 20, pp. 893-899, 1982.

6. PAO, S. P. and SALAS, M. D. - A Numerical Study of Two-Dimensional Shock Vortex Interaction. AIAA Paper, 81-1205, 1981.

7. ZANG, T. A., HUSSAINI, M. Y. and BUSHNELL, D. M. - Numerical Computations of Turbulence Amplification in Shock Wave Interactions, AIAA Journal, to appear.

8. HUSSAINI, M. Y., SALAS, M. D. and ZANG, T. A. - Spectral Methods for Inviscid Compressible Flows in Advances in Computational Transonics. Ed. Habashi, W. G., Pineridge Press, Swansea, United Kingdom, 1983.

9. KOPRIVA, D. A., ZANG, T. A. and HUSSAINI, M. Y. - A Practical Assessment of Spectral Accuracy for Hyperbolic Problems with Discontinuous Solutions. Submitted to 6th AIAA Computational Fluid Dynamics Conference, Danvers, MA., July, 1983.

COMPUTER SOLUTIONS OF NAVIER-STOKES EQUATIONS FOR SHOCK WAVE TURBULENT BOUNDARY LAYER INTERACTIONS FAR AWAY FROM THE LEADING EDGE

V. Swaminathan(1) and N.S. Madhavan(2)
Vikram Sarabhai Space Centre, Trivandrum, India.

SUMMARY

An accurate and efficient numerical algorithm for the solution of two-dimensional viscous compressible unsteady Navier-Stokes equations for obtaining flow parameters of oblique shock wave turbulent boundary layer interaction far away from the leading edge is presented. The conditions at the upstream entrance boundary were kept fixed at the boundary layer values obtained either through detailed computations or by making use of separate empirical relationships for the laminar sublayer and turbulent layer. MacCormack's explicit time-splitting second-order accurate algorithm was made use of for integrating the governing equations in the required time and space intervals. The results thus obtained for a typical case through the relevant computer program exhibit good comparison with those got by a calculation starting from the leading edge, testifying to the utility of the scheme described in the paper.

1. INTRODUCTION

A fluid flow problem of considerable interest and significance in the context of satellite launch vehicles, particularly those with strap-on motors, is the interaction of the shock wave generated by the strap-on booster with the core vehicle boundary layer, which is generally turbulent in character (the flow being at high speed and consequently at high Reynolds number), and which invariably occurs

---

(1) Head, Applied Mathematics Section
(2) Research Scientist, Applied Mathematics Section

far away from the leading edge. In cases of near-leading edge interactions, the unsteady Navier-Stokes equations are solved using an explicit[1] or implicit [2] time-marching algorithm with initial condition of uniform flow parameters everywhere, except at the aft-shock points on the top mesh boundary and the boundary condition of uniform flow along the upstream entrance boundary which is placed a few mesh points ahead of the leading edge (eg. [1]). This practice becomes too costly in terms of computer time, if the shock impingement occurs, as in many practical situations, at a considerably large distance away from the leading edge. An attempt has been made in the present paper to obtain numerical solutions of the Navier-Stokes equations for the flow parameters corresponding to this problem for the flat plate case by making use of the boundary layer values evaluated either by detailed calculations or by separate empirical formulae for the laminar sublayer and turbulent layer, as initial and upstream entrance boundary conditions. Although the technique has been presented only for the two-dimensional case, its extension to axi-symmetric and three-dimensional situations are straightforward.

2. DESCRIPTION OF THE PROBLEM

The flow field under investigation is depicted in Fig. 1. A shock wave, incident on a turbulent boundary layer formed by air-flow over a flat plate, produces (if the shock is strong enough) a separation bubble within which there is reversed flow. Streamlines deflected by the plate aft of reattachment give rise to reflected shockwaves. The aim of the computational procedure in this paper is to obtain steady flow parameters in a small region enclosing the interaction, using time-dependent Navier- Stokes equations and appropriate initial and boundary conditions.

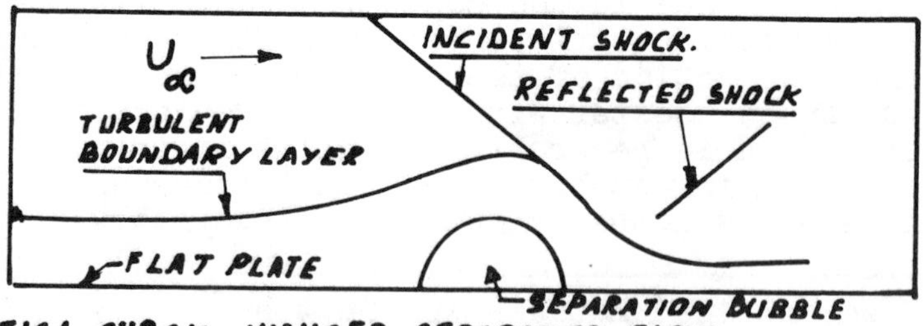

FIG.1. SHOCK-INDUCED SEPARATED FLOW

3. MEAN FLOW EQUATIONS

The time-averaged viscous compressible Navier-Stokes equations, cast in terms of mass-averaged variables modified for turbulence using Cebeci-Smith model [3] and neglecting body forces and heat sources, may be written in conservation form as (cf. [4])

$$\frac{\partial U}{\partial t} + \frac{\partial F}{\partial x} + \frac{\partial G}{\partial y} = 0, \tag{1}$$

where

$$U = \begin{bmatrix} \rho \\ \rho u \\ \rho v \\ e \end{bmatrix}, \quad F = \begin{bmatrix} \rho u \\ \rho u^2 + \sigma_x \\ \rho u v + \tau_{xy} \\ (e+\sigma_x)u + \tau_{xy} v + q_x \end{bmatrix},$$

$$G = \begin{bmatrix} \rho v \\ \rho u v + \tau_{yx} \\ \rho v^2 + \sigma_y \\ (e+\sigma_y)v + \tau_{yx} u + q_y \end{bmatrix}, \tag{2}$$

ρ is a mean value and u, v are mass-averaged values.

The bulk viscosity in the various terms are taken to be zero. Thus the fluxes σ_x etc. are given by

$$\sigma_x = p + \tfrac{2}{3}(\mu+\epsilon)\left(\frac{\partial u}{\partial x}+\frac{\partial v}{\partial y}\right) - 2(\mu+\epsilon)\frac{\partial u}{\partial x} \tag{3}$$

$$\sigma_y = p + \tfrac{2}{3}(\mu+\epsilon)\left(\frac{\partial u}{\partial x}+\frac{\partial v}{\partial y}\right) - 2(\mu+\epsilon)\frac{\partial v}{\partial y} \tag{4}$$

$$\tau_{xy} = \tau_{yx} = -(\mu+\epsilon)\left(\frac{\partial u}{\partial y}+\frac{\partial v}{\partial x}\right) \tag{5}$$

$$q_x = -\left(\mu/P_R + \epsilon/P_{RT}\right)\gamma\,\frac{\partial e_i}{\partial x} \tag{6}$$

$$q_y = -\left(\mu/P_R + \epsilon/P_{RT}\right)\gamma\,\frac{\partial e_i}{\partial y} \tag{7}$$

The mass-averaged specific internal energy e_i is related to the mean total energy per unit volume e by

$$e_i = e/\rho - (u^2+v^2)/2 \,. \tag{8}$$

The perfect gas relations

$$p = (\gamma - 1)\,\rho\, e_i \,, \tag{9}$$

$$T = e_i / c_v \,, \tag{10}$$

and Sutherland's visoosity relation

$$\mu = \frac{2.27 \times 10^{-8}\, T^{3/2}}{198.6 + T} \quad \text{slug/ft sec} \tag{11}$$

are also made use of for our analysis.

The eddy viscosity ϵ is evaluated using Cebeci-Smith's zero-equation equilibrium model of turbulence by applying

$$\epsilon = \min(\epsilon_i, \epsilon_o)\,, \tag{12}$$

where

$$\epsilon_i = 0.16\, y^2 \left[1 - \exp\left(-y\sqrt{\tau_w \rho}/26\mu\right)\right]^2 \left|\frac{\partial u}{\partial y}\right| \rho\,, \tag{13}$$

$$\epsilon_o = 0.0168\, u_{max}\, \rho \int_0^{y_\delta} (1 - u/u_{max})\, dy\,, \tag{14}$$

$$\tau_w = \mu_w \left(\frac{\partial u}{\partial y}\right)_{y=0}\,, \tag{15}$$

and u_{max} the maximum velocity in the profile extending from $y = 0$ to $y = y_\delta$.

4. INITIAL AND BOUNDARY CONDITIONS

Initial flow parameters were fixed by making use of boundary layer evaluation. Detailed boundary layer calculations may, if desired, be carried out as elucidated in [3] to obtain these initial estimates or, as an approximation, empirical relations may also be made use of. One such approximation (cf. [5]) is as follows:

The boundary layer thickness $\delta(x)$ is given by

$$\delta(x) = 0.37\, x\,(u_\infty x/\nu)^{-1/5}. \tag{16}$$

The velocity distribution in boundary layer is calculated using

$$u/u_\infty = (y/\delta)^{1/7} \tag{17}$$

everywhere except in the inner laminar sublayer, wherein we define

$$y^+ = \frac{y}{\mu_w/\sqrt{\rho_w \tau_w}} = \frac{u}{u_\infty}\sqrt{\frac{2}{c_{fw}}}, \tag{18}$$

from which u is evaluated. The laminar sublayer extends upto $y^+ = 11.5$.

Also the temperature distribution in the boundary layer is obtained by using

$$T = T_0 + (u_\infty^2 - u^2)/2c_p, \tag{19}$$

from which parameters like enthalpy, viscosity and density are also evaluated.

The boundary layer parameters, thus obtained, are held as the upstream entrance boundary condition. At the top mesh boundary, free stream conditions are assumed until the shock and shock parameters obtained using Rankine-Hugoniot relations afterwards. Zero-order extrapolation was applied at the downstream boundary and no-slip condition, numerically employed as mirror symmetry, was made use of on the flat plate.

5. COMPUTATIONAL MESH

The computational mesh, depicted in Fig. 2, consisted of 32 mesh points each in x- and y-directions. Along the x-direction, the mesh points were at equal intervals, whereas along the y-direction, a two-mesh system, consisting of

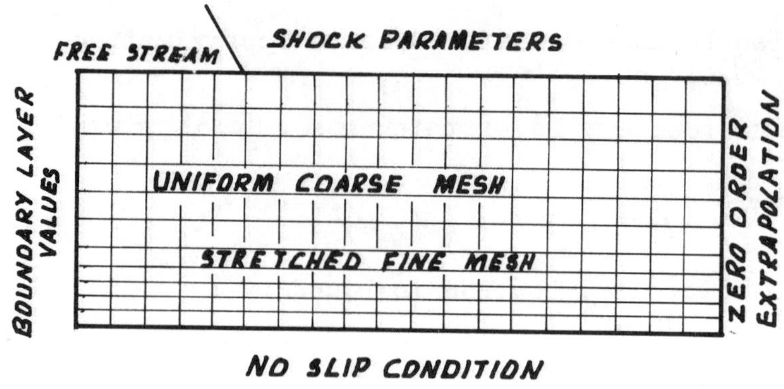

FIG. 2. COMPUTATIONAL MESH

stretched fine mesh near the plate (to have adequate resolution of the viscous effects) and uniform coarse mesh away from the plate, was made use of. Since the shock impingement point is taken to be much distant from the leading edge, the boundary layer will be considerably thick and the ordinates of the mesh system will have to be adjusted to cover the boundary layer and its subsequent growth by the interaction; in such cases, the number of points in the y-direction may have to be appropriately increased.

6. COMPUTATIONAL ALGORITHM

MacCormack's explicit time-splitting second-order accurate predictor-corrector algorithm [1] was made use of for solving the problem described above. The scheme involves the splitting of the time increment into parts during each of which only one space direction is considered for integration and combining the integrals in each part to obtain a more accurate total integral. The broad features of this scheme are spelt out below:

<u>The operator $L_x(\Delta t_x)$ along x-direction</u>

Predictor:

$$\overline{U_{ij}^{n+1/2}} = U_{ij}^{n} - \frac{\Delta t_x}{\Delta x}\left(F_{ij}^{n} - F_{i-1,j}^{n}\right)$$

Corrector:

$$U_{ij}^{n+1/2} = \frac{1}{2}\left[U_{ij}^{n} + U_{ij}^{\overline{n+1/2}} - \frac{\Delta t_x}{\Delta x}\left(F_{i+1,j}^{\overline{n+1/2}} - F_{ij}^{\overline{n+1/2}}\right)\right] \tag{20}$$

The operator $L_y(\Delta t_y)$ along y-direction

Predictor:

$$U_{ij}^{\overline{n+1/2}} = U_{ij}^{n} - \frac{\Delta t_y}{\Delta y}\left(G_{ij}^{n} - G_{i,j-1}^{n}\right)$$

Corrector:

$$U_{ij}^{n+1/2} = \frac{1}{2}\left[U_{ij}^{n} + U_{ij}^{\overline{n+1/2}} - \frac{\Delta t_y}{\Delta y}\left(G_{i,j+1}^{\overline{n+1/2}} - G_{ij}^{\overline{n+1/2}}\right)\right], \tag{21}$$

where Δx, Δy, Δt denote, respectively, the increments in space and time, the subscripts x and y represent the space direction of time split and the superscript indicates the advance in time. In practice, only one Δt is used instead of Δt_x and Δt_y and that is selected for stability reasons as

$$\Delta t \leqslant \min\left[\frac{\Delta y}{(|v|+c)_{max}}, \frac{\Delta x}{(|u|+c)_{max}}\right]. \tag{22}$$

In order to obtain second order accuracy, L_x and L_y are combined to form the total integral as

$$U_{ij}^{n+1} = \left[L_x\left(\frac{\Delta t}{M}\right)L_y\left(\frac{2\Delta t}{M}\right)L_x\left(\frac{\Delta t}{M}\right)\right]^M U_{ij}^{n}, \tag{23}$$

where M is an integer. In the fine mesh, the value of M is of the order 10 depending on the stability conditions and in the coarse mesh, the value is unity; this difference allows more computational time to be spent on the fine mesh, thus enhancing the efficiency of the algorithm.

7. COMPUTER PROGRAM

A computer program was developed in FORTRAN language for carrying out the above calculations and was run on the CDC CYBER 170/730 Computer of Vikram Sarabhai Space Centre, Trivandrum, for various shock angles, Mach numbers, Reynolds numbers and distances from the leading edge. The input parameters of the program consist of the Mach number, Reynolds number, distance from the leading edge, shock angle, extents of the mesh in x- and y-directions, and freestream and wall temperatures; properties of air as assumed in the program are, specific heat at constant volume C_v = 4290 ft. lb/slug^{0}R, specific heat ratio $\gamma = 1.4$, $P_R = 0.72$ and turbulent Prandtl number $P_{RT} = 0.9$. The print out of the program consists of the complete flow field characteristics such as ρ, u, v, p, e_i at every mesh point, skin friction coefficient ($c_f = \tau_w / 0.5 \rho_\infty u_\infty^2$) and pressure ratio values at every x-mesh point on the wall.

8. RESULTS

A comparison of the results obtained using the above program for a near-leading edge shock with those got from the usual method [4] is shown in Fig. 3. As can be seen from the figure, the pressure distributions obtained through the two numerical schemes are almost identical but there appears to be a slight variation in the skin friction values, attributable essentially to the coarser resolution in the x-direction employed in the computer program using the conventional method. The results from typical calculations corresponding to shocks at distances 10.6 feet and 1.5 feet away, respectively, from the leading edge are illustrated in Figs. 4 and 5.

9. CONCLUSION

This paper is concerned with an efficient scheme for obtaining numerical solutions of two-dimensional viscous compressible Navier Stokes equations as applicable to the shock wave-turbulent boundary layer interaction problem, the impingement of the shock on the flat plate being quite far from the leading edge. Typical numerical results for pressure distribution and skin friction obtained through a computer program specially developed for the purpose are also included. A comparison of the results for the flow parameters for a specific case with those got by the conventional method is presented, which suggests the feasibility of the technique for handling such problems. Besides, the method also appears to be of value owing to the substantial reduction in computation time it is likely to entail.

10. ACKNOWLEDGEMENT

The authors are grateful to Dr. Vasant R. Gowariker, Director, Vikram Sarabhai Space Centre, Trivandrum, for his kind encouragement in the preparation of this paper.

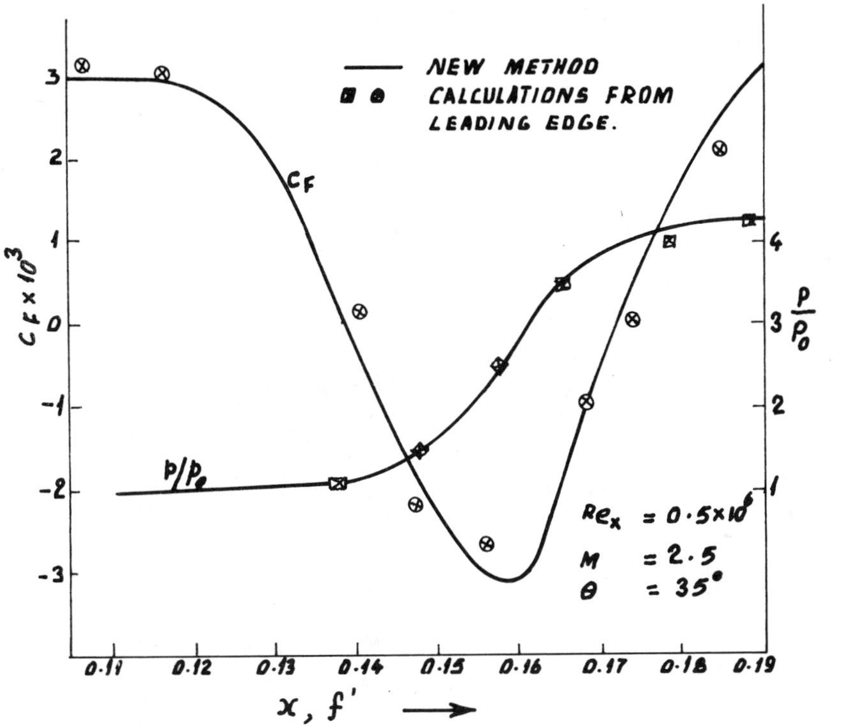

FIG. 3. COMPARISON OF SKIN FRICTION AND PRESSURE RATIO VALUES OBTAINED FROM NEW METHOD AND CALCULATIONS FROM LEADING EDGE, (DISTANCE OF SHOCK FROM LEADING EDGE = 0.1625')

11. REFERENCES

(1) MacCORMACK, R.W. - Numerical Solution of the Interaction of a Shock wave with a Laminar Boundary Layer. Lecture Notes in Physics, Springer Verlag, Vol. 8, pp. 151-163, 1971.

(2) MacCORMACK, R.W. - A Numerical Method for Solving Equations of Compressible Viscous Flow. AIAA Journal, Vol. 20, pp. 1275-1281, 1982.

(3) CEBECI, T., SMITH, A. M.O. and MOSINSKIS, G. - Calculation of Compressible Adiabatic Turbulent Boundary Layers. AIAA Journal, Vol. 8, pp. 1974- 1982, 1970.

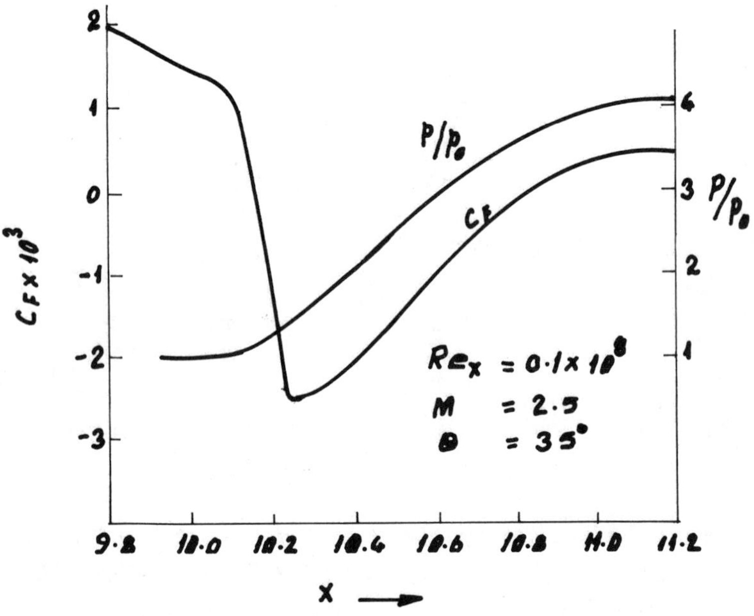

FIG. 4. SKIN FRICTION AND PRESSURE RATIO WHEN IMPACT POINT IS AT A DISTANCE 10.6' AWAY FROM LEADING EDGE.

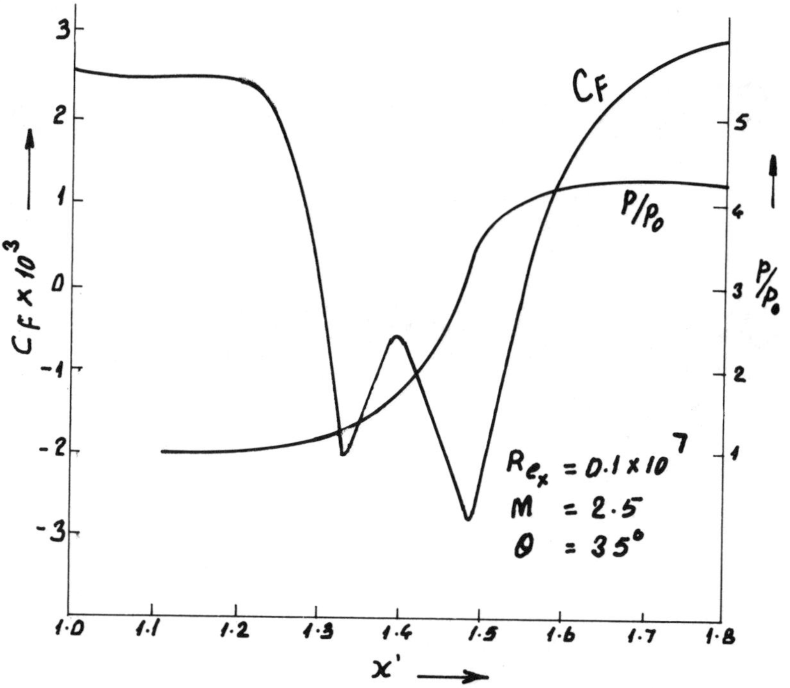

FIG. 5 SKIN FRICTION AND PRESSURE RATIO WHEN IMPACT POINT IS AT A DISTANCE OF 1.5' AWAY FROM THE LEADING EDGE.

(4) BALDWIN, B.S. and MacCORMACK, R.W. - Numerical Solution of a Strong Shock wave with a Hypersonic Turbulent Boundary Layer. AIAA Paper No.74-558, AIAA 7th Fluid and Plasma Dynamics Conference, 1974.

(5) SHAPIRO , A .H. - <u>The Dynamics and Thermodynamics of Compressible Fluid Flow</u>. The Ronald Press Company, New York, 1954.

SECTION 4

FLOW WITH SEPARATION

A STUDY OF TWO-EQUATION TURBULENCE MODELS FOR AXI-SYMMETRIC RECIRCULATING FLOW

A. G. Hutton and R. M. Smith[(I)]

SUMMARY

The performance of two-equation turbulence models in simulating turbulent flow through a sudden pipe expansion with a complex inlet geometry is systematically examined. Two such models are considered, namely the well known k-ε model and a new q-f model (where q is $k^{\frac{1}{2}}$ and f is a characteristic frequency of the large-scale motions) which has been devised as a suitable basis for Galerkin approximations. The models are evaluated by performing finite difference and finite element calculations respectively which are then compared with experiment. It is found that, using the standard k-ε model, the mean velocity field is not accurately predicted. The cause of this discrepancy is traced to the ε-equation. By adjusting the constant controlling the source terms in both the ε and f equations, both model predictions can be made to agree closely with experiment.

1. INTRODUCTION

The importance to a wide range of engineering interests of being able to analyse turbulent recirculating flows is evidenced by the ever increasing number of computer simulations which are reported in the literature [1, 2,3]. Almost without exception these calculations are based upon the k-ε description of turbulence [2], probably because it is one of the simplest levels of turbulence closure which in principle can handle such flows. It is therefore highly desirable to establish the performance of the model in the simplest of recirculating flow geometries (e.g. downstream-facing steps, pipe expansions,etc.) and isolate those model features which are the cause of any observed discrepancies. Gosman et al.[3] have compared predictions against available data in a range of

(I) Central Electricity Generating Board, Berkeley Nuclear Laboratories, Berkeley, Gloucestershire, U.K.

such geometries and, more recently, Rodi [2] has discussed the performance of the model in the light of reported experience. It appears that, although the broad features of recirculating flows can be reproduced, the model predictions are deficient in detail (e.g. recirculation lengths in unconfined flows are underpredicted). Pope et al [4] have proposed the ε-equation as a principal source of error. However there is difficulty in isolating the cause of such discrepancies with any certainty due to a scarcity of highly detailed experimental data. Furthermore, as pointed out by Rodi [2], it is often difficult to disentangle the effects of numerical error, since most prediction methods use upwind differencing and thus introduce the complication of numerical diffusion.

The aim of the present paper is two-fold. It is firstly to examine and resolve some of these uncertainties in a systematic way by performing finite difference (FD) calculations in a sudden pipe expansion with a complex inlet geometry. A wealth of detailed experimental data has recently become available for this particular flow configuration [5] thus rendering such an exercise possible. The second aim is to validate in the same geometry a novel finite element (FE) scheme incorporating a q-f two-equation turbulence model (q is the square-root of turbulence energy and f is a frequency of the large scale motions). This was recently introduced by Smith [6] and Hutton and Smith [7] in order to remedy the stability problems reported by various authors in applying Galerkin-FE discretisations of the k-ε equations to recirculating flows [8, 9 10].

2. THE EXPERIMENTAL CONFIGURATION

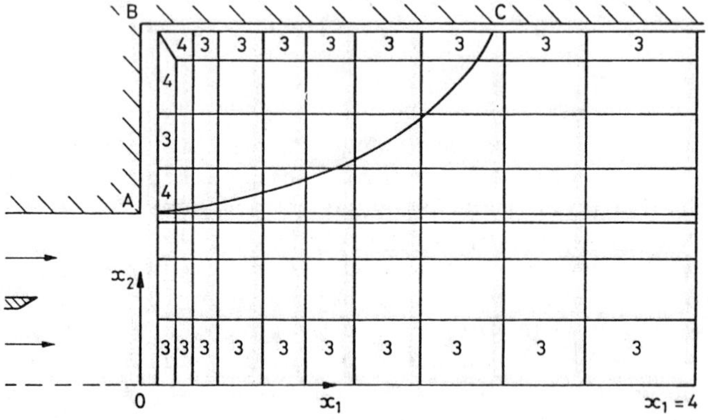

FIGURE 1 Schematic of the flow configuration and finite-element grid.

The geometry and basic features of the experiment are portrayed in Figure 1. A turbulent jet emerges from a nozzle which is placed within and is coaxial with a pipe. This jet mixes with a turbulent co-flowing annular stream before passing through a sudden expansion at the step AB. A shear layer spreads downstream from A and reattaches at C producing a region of recirculation ABC. The expansion ratio (downstream to upstream pipe diameter) is 2.1 and the Reynolds number (Re) based on diameter, D, and bulk velocity, U, downstream of the step is 62000. Further details of this arrangement are available in reference [5] which also reports measurements (using a two-channel laser-Doppler anemometer) of the mean velocity field and Reynolds stresses throughout the region downstream of the step.

3. MATHEMATIC MODELS

Let u_1, u_2 represent the mean velocity components in the axial and radial co-ordinate directions x_1, x_2 respectively and let p be the pressure. Then, if all variables are rendered dimensionless with respect to U and D, the models used to calculate the above described flow can be formulated as follows.

3.1 Dynamic Equations.

$$u_m \frac{\partial u_n}{\partial x_m} + \frac{\partial p}{\partial x_n} - \frac{1}{x_2}\frac{\partial}{\partial x_m}(x_2 \tau_{mn}) + 2\frac{(1+\mu_T)}{Re}\frac{u_2}{x_2^2}\delta_{n2} = 0 \tag{1}$$

$$\frac{1}{x_2}\frac{\partial}{x_m}(x_2 u_m) = 0 \tag{2}$$

$$\tau_{mn} = \frac{(1+\mu_T)}{Re}\left(\frac{\partial u_m}{\partial x_n} + \frac{\partial u_n}{\partial x_m}\right) \tag{3}$$

3.2 Turbulence Model Equations.

The effective viscosity, μ_e is given by the relation,

$$\mu_e = 1+\mu_T = 1+Re k^{\frac{1}{2}} \ell = 1+Re\,|\,q\,|\,\ell \tag{4}$$

where q is the square root of the turbulence kinetic energy, k, and ℓ is a length scale associated with the energy-bearing eddies. Two separate equation systems for determining these parameters will be considered. The first constitutes the well known k-ε model (ε denoting the rate of dissipation of k) which can be written,

$$\ell = C_\mu k^{3/2}/\varepsilon \tag{5}$$

$$u_m \frac{\partial k}{\partial x_m} - \frac{1}{Re}\frac{1}{x_2}\frac{\partial}{\partial x_m}\left(\frac{x_2 \mu_T}{\sigma_k}\frac{\partial k}{\partial x_m}\right) = \frac{\mu_T}{Re} S_u - \varepsilon \tag{6}$$

$$u_m \frac{\partial \varepsilon}{\partial x_m} - \frac{1}{Re}\frac{1}{x_2}\frac{\partial}{\partial x_m}\left(\frac{x_2 \mu_T}{\sigma_\varepsilon}\frac{\partial \varepsilon}{\partial x_m}\right) = C_\mu C_{1\varepsilon} k S_u - C_{2\varepsilon}\frac{\varepsilon^2}{k} \tag{7}$$

where $C_\mu, \sigma_k, \sigma_\varepsilon, C_{1\varepsilon}$, $C_{2\varepsilon}$ are model constants and the source term S_u is given by:

$$S_u = \left(\frac{\partial u_n}{\partial x_m} + \frac{\partial u_m}{\partial x_n}\right)\frac{\partial u_n}{\partial x_m} + 2\left(\frac{u_2}{x_2}\right)^2 \tag{8}$$

The second system of equations considered constitutes the q-f model and is written,

$$\ell = C_\mu |q|/f \tag{9}$$

$$2u_m \frac{\partial q}{\partial x_m} - \frac{1}{x_2}\frac{\partial}{\partial x_m}\left(x_2 \pi_q \frac{\partial q^2}{\partial x_m}\right) = P_q - D_q \tag{10}$$

$$u_m \frac{\partial f}{\partial x_m} - \frac{1}{x_2}\frac{\partial}{\partial x_m}\left(x_2 \pi_f \frac{\partial(\ell n f^2)}{\partial x_m}\right) = P_f - D_f + R_f \tag{11}$$

$$\pi_q = \ell/\sigma_q; \quad \pi_f = C_\mu q^2/2\sigma_f \tag{12}$$

$$P_q = \ell\left(S_u + \frac{2}{\sigma_q}\left(\frac{\partial q}{\partial x_m}\right)^2\right); \quad P_f = C_\mu C_{1f} S_u \tag{13}$$

$$D_q = C_\mu q^2/\ell; \quad D_f = C_{2f} f^2 \tag{14}$$

where, once again, C_μ, σ_q, σ_f, C_{1f}, C_{2f} are model constants to be determined by experiment and R_f is set to zero. As already mentioned, this model was recently introduced by Smith [6] and Hutton and Smith [7] as a more suitable basis than k-ε for Galerkin-FE turbulent flow calculations (a feature which is further discussed later in the paper). It is very similar to k-ε. Indeed, with suitable choice of constants and provided,

$$R_f = \frac{C_\mu}{\sigma_\varepsilon}\frac{1}{f^2}\frac{\partial f^2}{\partial x_m}\frac{\partial q^2}{\partial x_m} - \frac{f}{x_2 q^2}\frac{\partial}{\partial x_m}\left(x_2\mu_T\left(\frac{1}{\sigma_k} - \frac{1}{\sigma_\varepsilon}\right)\frac{\partial q^2}{\partial_{xm}}\right) \tag{15}$$

it is formally equivalent to k-ε[6].

3.3 Boundary Conditions.

The boundary conditions applied at the walls (or to be more exact, at the wall edge of the computational mesh) conform with the usual wall-function treatment in the case of FD [11] and the slip-conditions described by Smith [6] in the case of FE. The remaining boundaries are treated as follows

$$u_m = \hat{u}_m,\ q = \hat{q},\ f = \hat{f}, \varepsilon = \hat{\varepsilon} \quad x_1 = X, 0 \le x_2 \le 0.238 \ [\text{INLET}] \tag{16}$$

where $\hat{u}_m$, $\hat{q}$, $\hat{f}$ and $\hat{\varepsilon}$ are experimentally determined functions at the first measuring station downstream of the step (i.e. X = 0.07).

$$u_2 = 0;\ \frac{\partial u_1}{\partial x_2} = \frac{\partial q}{\partial x_2} = \frac{\partial f}{\partial x_2} = \frac{\partial \varepsilon}{\partial x_2} = 0 \quad x_2=0,\ 0 \le x_1 \le 4 \ [\text{AXIS}] \tag{17}$$

$$-p+\tau_{11} = \tau_{12} = \frac{\partial q}{\partial x_1} = \frac{\partial f}{\partial x_1} = 0 \quad x_1=4,\ 0 \le x_2 \le \tfrac{1}{2} \ [\text{OUTLET, FE}] \tag{18}$$

$$\frac{\partial u_1}{\partial x_1} = \frac{\partial u_2}{\partial x_1} = \frac{\partial k}{\partial x_1} = \frac{\partial \varepsilon}{\partial x_1} = 0 \quad x_1=4 \quad 0 \le x_2 \le \tfrac{1}{2} \ [\text{OUTLET, FD}] \tag{19}$$

4. NUMERICAL MODELS

As already explained, the experiment (see Figure 1) was simulated using both a FD calculation procedure (incorporating the k-ε model) and a FE calculation procedure (incorporating the q-f model). These will now be briefly described.

4.1 Finite Difference Model.

The finite difference calculation procedure employed is very similar to that described in [3]and [13]. The conservative forms of equations (1) to (8) are discretised on a staggered FD mesh using hybrid upwind differencing for the convective terms and central differencing for the diffusive terms. The continuity equation is incorporated and the whole system solved by means of the SIMPLE algorithm [13]. Wall boundaries are treated by employing wall functions in the usual manner [11]. Thus if y is distance measured normal to the wall in question and y^+ denotes $yReC_\mu^{\frac{1}{4}}k^{\frac{1}{2}}$, then the momentum flux at the wall is obtained from the "law of the wall", invoking the laminar asymptote if $y^+<11$ at the wall adjacent node and the turbulent asymptote ("log-law") otherwise. The wall flux of turbulent energy is set to zero and the dissipation rate, ε, at the wall adjacent node is determined from eq.(5) with ℓ given by:

$$\ell = C_\mu^{\frac{1}{4}}\kappa y \qquad (20)$$

where κ represents the von-Karman constant, taken to be 0.419.

The computations were performed on a thirty-by-thirty grid which was uniform in the x_1-direction ($\Delta x_1 = 0.143$) and uniform in the x_2-direction on either side of the step ($\Delta x_2 = 0.018$ for $x_2<x_2^A$, $\Delta x_2=0.017$ for $x_2>x_2^A$, see Figure 1).

4.2 Finite Element Model.

The mesh used for the finite element calculations is shown in Figure 1. Un-numbered elements are of the eight-noded quadratic velocity/linear pressure type (type-2 elements). Those numbered 3 are of a special type (type-3) introduced in ref.[14]. Their main distinguishing feature is the introduction of the normal derivatives of velocity as additional nodal variables at the boundary (so that the $\underline{u}$ variation is cubic in the normal direction, pressure remaining bilinear). Elements numbered 4 (type-4) are essentially type-3 elements with the derivative parameters suppressed at one of the nodes (either because they are not required as at B(Figure 1) or to merge a type-3 with a type-2 element e.g. point A)[14]. Each of these elements is regarded as type-2 as far as the μ_e-variation is concerned. The momentum equations (1) - (3) are discretised on this mesh by invoking the standard Galerkin procedure, continuity being handled by the PALM method[12] with a penalty parameter of unity.

Wall boundaries are treated by invoking the slip conditions described by Smith [6] based on the treatment introduced in [12]. Accordingly the mesh boundary is displaced a distance $\Delta h = 0.07$ from the step (inlet is therefore at $x_1 = 0.07$ in order to coincide with the first measurement station) and $\Delta h = 0.02$ from the pipe wall (this value ensured fully turbulent conditions everywhere at the mesh edge). The numerical solution is then "matched" to the law of the wall by employing the normal derivative parameters which are available at each edge node, with Δh^+ defined as $C_\mu^{\frac{1}{4}}k^{\frac{1}{2}}Re\Delta h$. This avoids Δh^+ becoming small near reattachment and the "log law" is then appropriate everywhere [6,7]. The point B(Figure 1) is dealt with by simply setting u_1 and u_2 to zero.

The q-f equation system (eqs (9) - (14)) is discretised in the manner described by Smith [6]. The essence of the approach can be conveyed by considering eq.(10). On each element, q in the convective and q^2 in the diffusion terms are interpolated in the same way as $\underline{u}$. The source terms P_q, D_q and the diffusion coefficient π_q are all discretised with type-2 interpolation. An algebraic system of equations is then developed by weighting the equation residual with the "velocity" basis functions in the usual Galerkin fashion. The required nodal values of P_q, D_q and π_q are generated by simply evaluating the right hand sides of eqs.(12) to (14) at each node (where the values of S_u and $(\partial q/\partial x_m)^2$ at a node are taken as the area averages of these quantities over the elements sharing the node). The f equation (eq.(11)) is treated in a similar fashion, with f in the convective and $(\ell n f^2)$ in the diffusive terms being interpolated on the same level as $\underline{u}$. At the wall edge of the mesh q is left free (approximating zero flux of k) and f is determined from eqs. (9) and (20).

This discretisation strategy leads to a large non-linear algebraic system of equations for the nodal values of $\underline{u}$, p, q and f. As explained in [6] and [7], a solution can be effected by the following iterative procedure:

Suppose approximations to $\underline{u}$, μ_T, ℓ (and hence q, f from eqs.(4) and (9)) are known. Then,

A. With μ_T fixed, solve(II) the dynamical equations to update $\underline{u}$, p

B. With $\underline{u}$ fixed, update μ_T by solving the turbulence model equations using the following sequence (repeated until μ_T converged).

B1. With q fixed, solve(II) the f-equation to update ℓ via eq.(9).

B2. With ℓ-fixed, solve(II) the q-equation to update q.

Steps A and B are operated in sequence until overall convergence is achieved.

This algorithm has proved robust and reliable on several problems where other authors, using Galerkin-FE methods have reported great difficulties [8, 9, 10]. The principal reason for this success is the adoption of q-f rather than k-ε as a basis for the turbulence model. Smith [8] has shown that whilst solutions to Galerkin-type approximations of the k-ε system of equations do exist in pipe expansion geometries

(II) The Newton-Raphson iterative method in conjunction with a frontal solver is used.

(provided the source terms are discretised appropriately), these are extremely difficult to obtain using a Newton-Raphson iterative scheme. The mathematical properties of the q-f formulation renders it considerably more attractive in this respect so that, when coupled with the above described discretisation strategy, it gives rise to algebraic systems which individually display comparatively large Newton Raphson radii of convergence [6]. As a consequence the solution algorithm outlined here displays considerable stability even with relatively poor initial guesses.

5. EVALUATION OF THE TURBULENCE MODELS

The comprehensiveness of the experimental measurements reported in [5] has enabled the ratio $-\overline{u_1'u_2'}/e_{12}$ to be deduced from the data everywhere downstream of the expansion (here the usual notation $\overline{u_i'u_j'}$ has been adopted for the Reynolds stresses and e_{ij} represents $(\partial u_i/\partial x_j + \partial u_j/\partial x_i)$). Now, implicit in both the turbulence models described above is the assumption that the Reynolds stress tensor can be related to the mean rate of strain tensor via the mechanism of a scalar turbulent viscosity, μ_T (eqs.(3) and (4)). Thus if we (tentatively) identify this experimentally derived ratio with μ_T we can test the validity of this assumption, at least as a basis for predicting mean flow distributions. Figure 2 compares the measured u_1-velocity field with that calculated using these experimentally provided values of μ_T (the calculations were performed on a highly refined FE mesh). As can be seen, the agreement is very satisfactory except perhaps in the immediate lee of the step face, which probably signifies

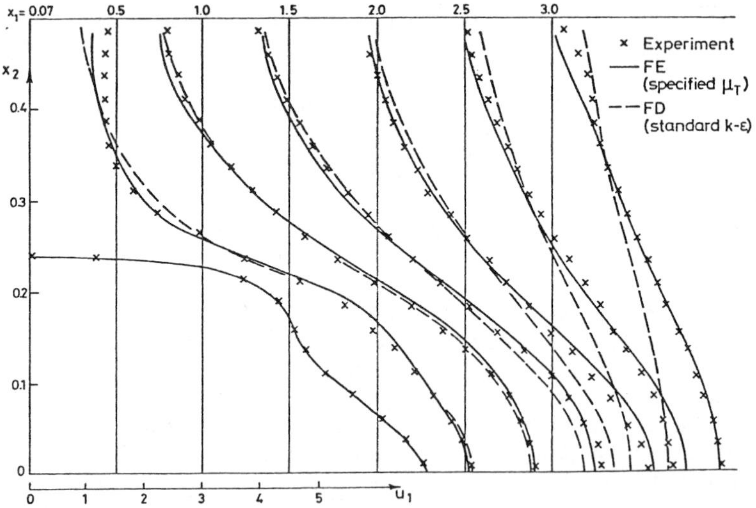

FIGURE 2 Specified μ_T and standard k-ε model predictions of u_1-profiles compared with experiment.

the inadequacy of the wall treatment rather than μ_T in this region. Although this agreement does not in itself validate the scalar turbulent viscosity as a mechanism for interpreting the Reynolds stresses, it nonetheless confirms its adequacy for the prediction of mean velocity data in the geometry under discussion, which is usually sufficient for engineering purposes.

5.1 The k-ε Model.

Having addressed the uncertainty attaching to the adoption of a scalar turbulent viscosity, its interpretation in terms of k and ε as governed by eqs.(5) to (7) can be investigated. Accordingly the recommended values 0.09, 1.44, 1.92, 1.0 and 1.3 for the model constants C_μ, $C_{1\varepsilon}$, $C_{2\varepsilon}$, σ_k and σ_ε respectively[2] were adopted and finite difference calculations performed using the procedure described in 4.1. The resulting u_1-velocity field is compared with experiment in Figure 2 and, as can be seen, for $x_1>1.5$ (the point where the shear layer reaches the centre line) the agreement is not good. The predicted k-field (or to be more exact, $q = k^{\frac{1}{2}}$) is compared with experiment in Figure 3 and, somewhat surprisingly in view of Figure 2, the agreement is quite reasonable. If it is taken for granted that μ_T can be represented by way of eq.(4) then it is a simple matter to derive an experimental distribution for ℓ (both μ_T and k data are available). This distribution is compared against that derived from eq.(5) in Figure 4, from which it is clear that the length scale is increasingly overpredicted as x_1 increases beyond 1.5. As a consequence, the turbulent viscosity is also overpredicted

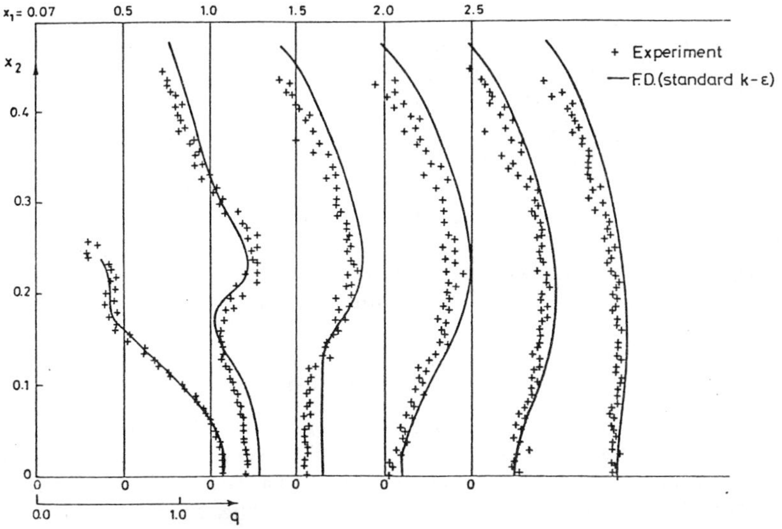

FIGURE 3 Standard k-ε model prediction of q-profiles compared with experiment.

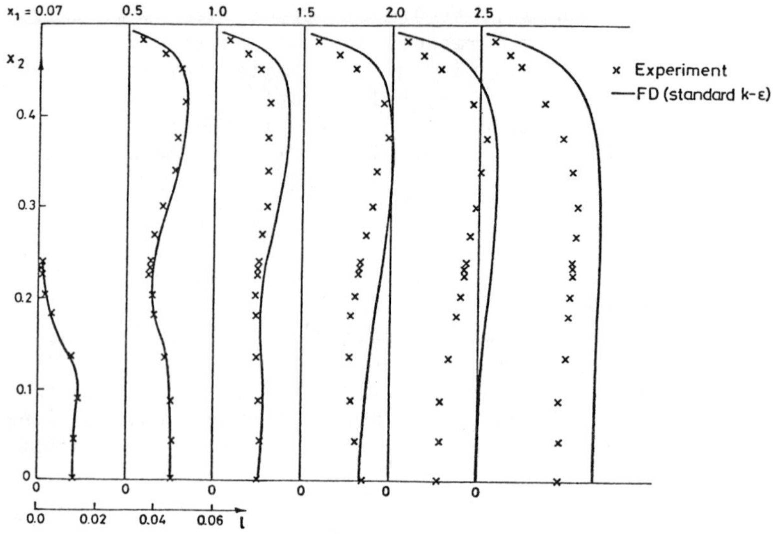

FIGURE 4 Standard k-ε model prediction of ℓ-profiles compared with experiment.

(q is in reasonable agreement) resulting in too rapid a relaxation of the u_1-velocity field beyond x_1= 1.5 as already noted from Figure 2.

These results suggest that the principal source of error in the k-ε determination of μ_T is attributable to the ε variable. This was confirmed by repeating the calculations, invoking the experimental length scale to dispense with ε altogether (i.e. replacing ε in eq.(6) by $C_\mu k^{3/2}/\ell$). The resulting distributions of u_1 and q are presented in Figures 5 and 6 respectively. As can be seen, the agreement with experiment is good, validating the physics of the k-ℓ model (it is later argued that the calculations are free from significant errors due to numerical diffusion).

It remains to improve the k-ε predictions by adjusting the model for ε. The model constant $C_{2\varepsilon}$ can be determined directly from the measured rate of decay of k behind a grid [2] whereas $C_{1\varepsilon}$ is related to σ_ε and $C_{2\varepsilon}$ by

$$C_{1\varepsilon} = C_{2\varepsilon} - \kappa^2/(\sigma_\varepsilon C_\mu^{\frac{1}{2}}) \tag{21}$$

and its value can perhaps be considered less certain [2]. It was observed that the decay of the centre line value of u_1 was extremely sensitive to the value of $C_{1\varepsilon}$ (a feature which has been noted previously in connection with the spreading rates of jets [2]). Furthermore it was found that this decay

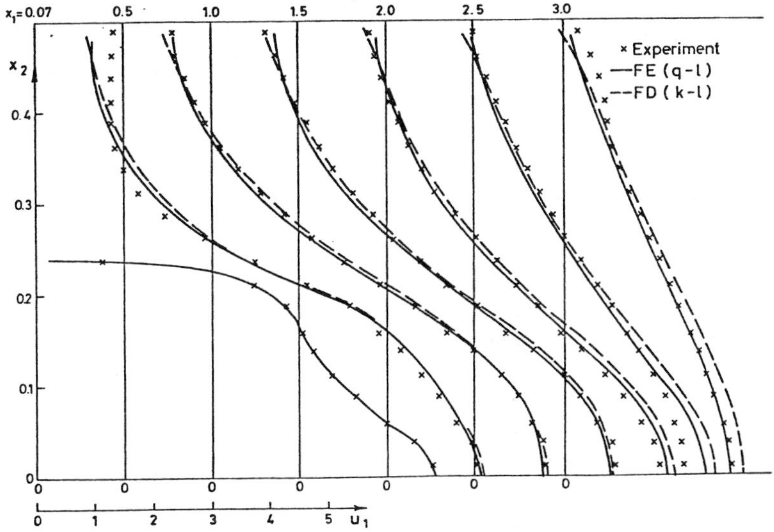

FIGURE 5 One-equation model predictions of u_1-profiles compared with experiment.

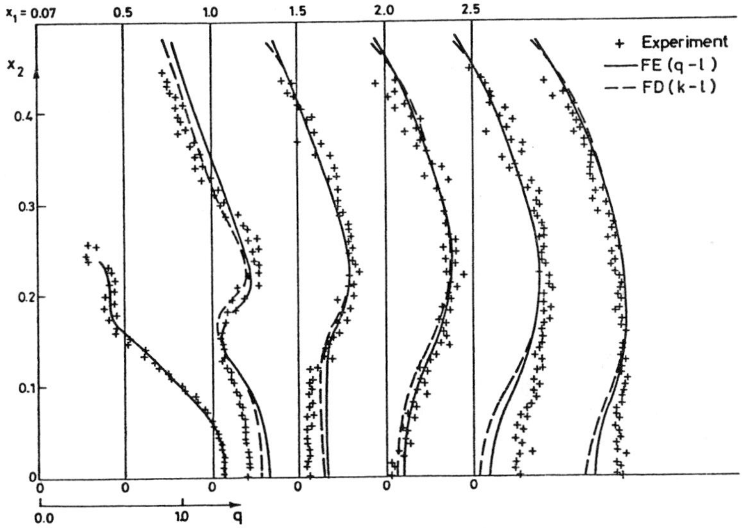

FIGURE 6 One-equation model predictions of q-profiles compared with experiment.

could be predicted reasonably well on setting $C_{1\varepsilon}$ to 1.5. The radial profiles of u_1 and ℓ (derived from eq.(5)) which are predicted when this value is adopted (the other constants remaining unaltered except for σ_ε which was adjusted to satis-

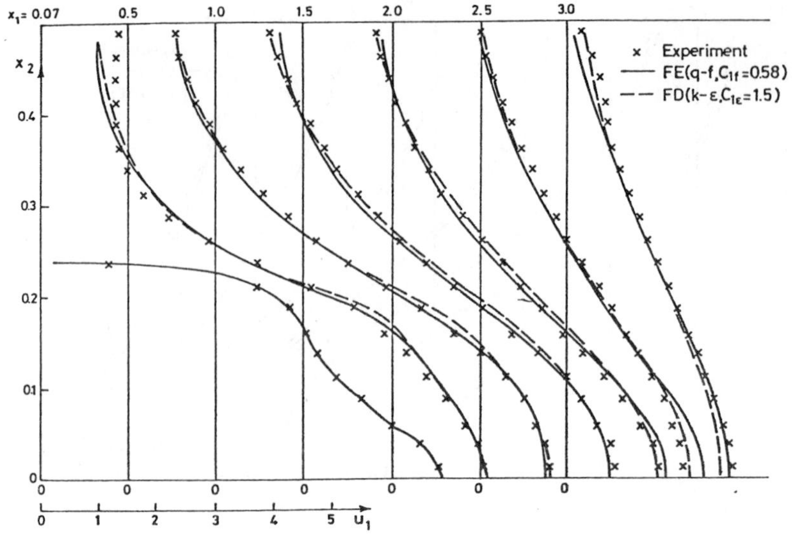

FIGURE 7 Tuned two-equation model predictions of u_1-profiles compared with experiment.

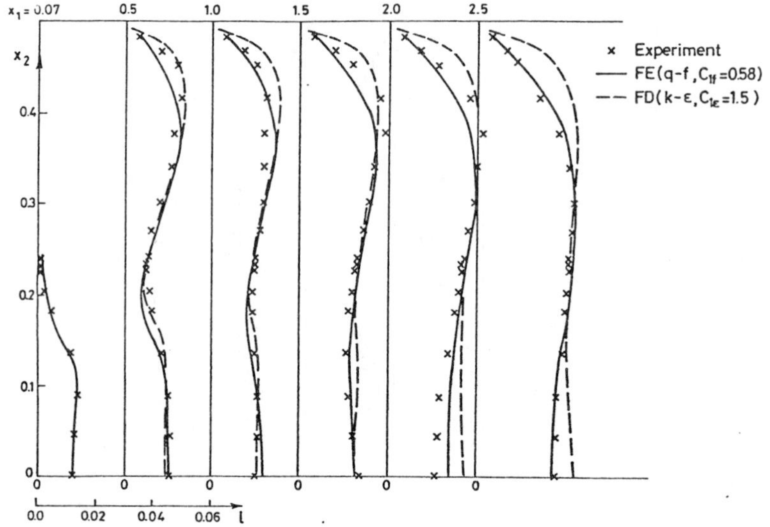

FIGURE 8 Tuned two-equation model predictions of ℓ-profiles compared with experiment.

fy eq.(21)) are plotted in Figures 7 and 8 respectively. It is evident that the agreement with experiment is now significantly better than the results plotted in Figures 2 and 4.

5.2 The q-f Model.

Since the length scale appearing in eqs.(12) - (14) is available from the experimental data, the q-equation (eq.(10)) and its FE discretisation can be directly evaluated. Accordingly the constants C_μ, σ_q were given values 0.09 and 1.0 respectively and, dispensing with the f-equation, FE predictions were made of the velocity and q-fields. These are compared with experiment in figures 5 and 6 and are seen to be in close agreement. Further refinement of the mesh resulted in insignificant change. Since the FE discretisation is free from numerical diffusion errors, this validates not only the physics of the q-ℓ model but also the physics of the k-ℓ model (the latter is formally equivalent to the former). Consequently it can be deduced that, on the chosen grid, the above described FD calculations are also free from significant numerical diffusion (since the FD predictions based on k-ℓ also agreed closely with experiment).

It remains to evaluate the f-equation. Smith [6] has shown that the full q-f turbulence model is formally equivalent to the k-ε model provided $\sigma_f=\sigma_\varepsilon, C_{1f}=C_{1\varepsilon}-1$ and $C_{2f}=C_{2\varepsilon}-1$, and provided R_f in eq.(11) is given by eq.(15). In view of the discussion in 5.1, this would suggest 0.09, 0.5, 0.92, 1.0 and 1.4 as suitable values for the model constants C_μ, C_{1f}, C_{2f}, σ_q and σ_f respectively. However, since R_f is set to zero in the present analysis, further adjustment of the f-constants might well be anticipated. This was indeed confirmed by numerical experiment and optimal agreement between full q-f predictions and experiment was achieved on setting C_{1f} to 0.58. The resulting velocity field and length scale distribution are plotted in Figures 7 and 8 respectively.

6. CONCLUSIONS

A comprehensive set of experimental data for turbulent flow through an axisymmetric pipe expansion with a complex inlet geometry has been used to evaluate the performance of two-equation turbulence models. It has been established that the use of a scalar turbulent viscosity (relating the Reynolds stress to the mean rate of strain tensor) is valid for the purpose of predicting the mean velocity field. However, using the standard values for the model constants, the (k-ε)-turbulence model gives rise to too high a turbulent viscosity value downstream of the point where the detached shear layer reaches the centre-line, thus failing to predict the mean flow development accurately. The principal source of error can be traced to the ε-equation since, using experimental data to provide a length scale closure, predictions for both velocity and turbulence energy were found to agree closely with experiment. Further investigation has revealed a remarkable sensitivity of the k-ε predictions to the value of the source term constant ($C_{1\varepsilon}$) in the ε-equations. Accurate solutions were generated when this was set to 1.5 rather than the recommended value of 1.44.

Finally, the q-f turbulence model, recently introduced as a basis for stable Galerkin-finite element turbulent flow calculations, was found to perform in a very similar manner. Close agreement between experiment and prediction was achieved on suitably adjusting the value of the f-equation source term constant, the remaining constants being fixed at values suggested by equivalence with k-ε.

7. ACKNOWLEDGEMENT

This paper is published by permission of the Central Electricity Generating Board.

8. REFERENCES

1. TAYLOR,C. and SCHREFLER,B.A.(eds) - Numerical Methods in Laminar and Turbulent Flow, Pineridge Press, 1981.
2. RODI,W. - Turbulence Models and their Application in Hydraulics, IAHR State-of-the-Art Paper, Delft, Netherlands, 1980.
3. GOSMAN,A.D., KHALIL,E.E.and WHITELAW,J.H. - The Calculation of Two-dimensional Turbulent Recirculating Flows, Turbulent Shear Flows 1, Springer Verlag, Heidelberg, 1979.
4. POPE,S.B.and WHITELAW,J.H. - The Calculation of Near-Wake Flows, J.Fluid Mech.,73,9-32,1976.
5. FREEMAN,A.R.and SZCZEPURA,R.T. - Mean and Turbulent Velocity Measurements in an Abrupt Axi-symmetric Pipe Expansion with Complex Inlet Geometry Proc.Int.Symp.on Applications of Laser-Doppler Anemometry to Fluid Mechanics, Lisbon, 1982.
6. SMITH,R.M. - A Practical Method of Two-Equation Turbulence Modelling Using Finite Elements, CEGB Report TPRD/B/0182/N82.
7. HUTTON,A.G.and SMITH,R.M. - Turbulent Flow in a Pipe Expansion, The Finite Element Handbook, McGraw-Hill (to appear).
8. SMITH,R.M. - On the Finite-Element Calculation of Turbulent Flow using the k-ε Model, CEGB Report TPRD/B/0161/N82.
9. LAROCK,B.E.and SCHAMBER,D.R. - Approaches to the Finite Element Solution of Two-dimensional Turbulent Flows, Computational Techniques in Transient and Turbulent Flow,Eds.Taylor, C.and Morgan,K., Pineridge Press, 1981.
10. TONG,G.D. - Computation of Turbulent Recirculating Flow Ph.D.Thesis, Dept.Civil Engineering, University College of Swansea, U.K., 1982.
11. LAUNDER,B.E.and SPALDING,D.B. - The Numerical Computation of Turbulent Flows Comp.Meth.Appl.Mechs.Engng.,3,269-289,1974.

12. HUTTON, A.G. and SMITH, R.M. - On the Finite Element Simulation of Incompressible Turbulent Flow in General Two-dimensional Geometries, Numerical Methods in Laminar and Turbulent Flow, Eds Taylor, C. and Schrefler, B.A., Pineridge Press, 1981.
13. PATANKAR, S.V. - Numerical Heat Transfer and Fluid Flow, Hemisphere Publishing Corp., 1980.
14. HUTTON, A.G. - Finite Element Boundary Techniques for Improved Performance in Computing Navier-Stokes and Related Heat Transfer Problems, Finite Elements in Fluids Vol.IV, Ed. R.Gallagher, Wiley, 1982.

An Evaluation of Higher-Order Upwind Differencing for Elliptic Flow Problems

N. S. Wilkes C. P. Thompson

Engineering Sciences Division Computer Science and Systems Division

AERE Harwell, Oxfordshire. OX11 0RA

SUMMARY

The development of higher-order discretisation methods is often a balance between accuracy and ancillary requirements such as resistance to spurious numerical oscillations and ease of solution. In turbulent flow computations these two effects become increasingly important. In this paper we present the higher-order upwind difference scheme which is $O(h^2)$, robust, and which can easily be solved using a "pseudo-deferred correction" method. Applications of this method to laminar and turbulent flows over a backward facing step show significant reductions in errors per unit cost in comparison with hybrid differencing. It is important to note that it was not necessary to use excessive under-relaxation nor an especially good initial approximation for the new scheme.

1. INTRODUCTION

The development of discretisations for computational fluid dynamics problems requires a compromise between accuracy, robustness, and efficiency of solution. Here "robust" is used to describe a scheme which will not exhibit spurious, numerical oscillations in any circumstances. These wiggles usually arise when the ratio of convection to diffusion in a cell is large; this ratio is usually characterised by mesh-Peclet (mesh-Reynolds) number, defined in Section 2. Simple analysis of non-robust discretisations shows that the usual source of difficulty is the treatment of the advection terms. The higher-order upwind finite-difference scheme (HUW) approximates the first derivative terms by fitting a parabola to the point and its two upwind neighbours. The aim of this paper is the assessment of this scheme from a practical, as well as a theoretical, viewpoint.

For the class of problems which most interest the authors, namely k-ε turbulence modelling, robustness is very important since one frequently has to deal with very large mesh-Peclet numbers (of the order of 100) which can arise during the iterative solution of equations. We also think (though there are merits on both sides of this argument) that a code should be able to produce some form of approximation to a fluid flow problem even if this estimate is not very accurate; low accuracy calculations have an important role in design processes. Moreover, a "qualitatively right" solution allows intelligent adaptation of meshes, which in many cases is an integral part of the solution process.

The majority of calculations reported here have used a TEACH-like computer code [1]. Versions of this program using hybrid (upwind/central) differencing are widespread. This form of discretisation is extremely stable: it will not exhibit wiggles for any mesh-Peclet number and it displays a (3-D) maximum principle. Furthermore, it is compact, involving only a five-point molecule, and does not introduce any ancillary unknowns. It is well known that the major drawback of this approach is that it can be extremely diffusive and frequently one operates in a regime where the errors are O (h) (i.e. when the mesh-Peclet number is greater than 2).

Recently there have been a large number of proposals for improved methods of discretisation; particularly worthy of note (among finite difference applications) are: Compact (Hermitian) methods [2,3], Operator compact implicit (OCI) schemes [4,5] and the QUICK scheme [6]. The first two of these allow fourth order accuracy to be obtained whilst remaining compact; however, these methods are susceptible to "wiggles" (the scheme outlined by Leventhal et al does not have this drawback but is not easily extended to problems with more than one space dimension). Leonard's QUICK scheme achieves increased accuracy by enlarging the stencil size; because of partial upwinding this method exhibits only bounded wiggles and this represents a significant improvement. The bigger molecule creates some difficulties for the solution algorithm and some authors [7] have reported difficulties.

The higher-order upwind scheme represents a compromise between these methods. It is conventional in the sense that only function values are used; it is wiggle-proof, and has only a slight propensity for overshoots; and it is formally (consistently) O (h^2). This has been achieved at the expense of compactness. However, experiments show that equivalent runs cost only 60% more than hybrid in CPU time and an approximate analysis based on an equal error basis shows the costs are significantly less for HUW.

2. NUMERICAL DESCRIPTION

Firstly, we consider the one dimensional convection-diffusion operator as an indicator of the likely behaviour of discretisations in more realistic problems.

Thus we have:

$$\mathcal{L}(\phi) = u\frac{\partial\phi}{\partial x} - K\frac{\partial^2\phi}{\partial x^2} \qquad \text{on the domain} \quad 0 \leqslant x \leqslant L \; .$$

The ratio of diffusion to convection is measured by the Peclet number Pe = (u L/K).

The mesh-Peclet number (P_g) is defined analogously by P_g = u h/K where h is now the mesh size.

Difficulties arise in the case of large mesh-Peclet number in similar ways as are posed by the singular perturbation problem. Basically there are regions of rapid change which cannot be well represented by polynomial interpolation.

More careful analysis shows that the usual second-order approximation to the diffusion term is stable and the source of the oscillations is due to the treatment of the convective term. Our approach is to mimic the uni-directional nature of convection (following the heuristic of upwinding for stability) and fit a quadratic to the point and its two upwind neighbours. This gives:

$$u\frac{\partial\phi}{\partial x} = \frac{u}{h}\left\{\frac{3\phi_i}{2} - 2\phi_{i-1} + \frac{\phi_{i-2}}{2}\right\}$$

$$+ u\frac{h^2}{3}\frac{\partial^3\phi}{\partial x^3} + O(h^3) \qquad u \geqslant 0 \qquad (1a)$$

$$= \frac{u}{h}\left\{-\frac{3\phi_i}{2} + 2\phi_{i+1} - \frac{\phi_{i+2}}{2}\right\}$$

$$+ u\frac{h^2}{3}\frac{\partial^3\phi}{\partial x^3} + O(h^3) \qquad u < 0 \qquad (1b)$$

It can be shown that this scheme will not give rise to wiggles (in fact it is monotone in 1-D) and is O (h^2), matching the order of the diffusion operator. Overshoots are possible but have not given difficulties in practice.

The drawback, of course, is the increased size of the molecule. However, if we write:

$$u \frac{\partial \phi}{\partial x} \simeq \frac{3u}{2h}(\phi_i - \phi_{i-1}) - \frac{u}{2h}(\phi_{i-1} - \phi_{i-2}) \qquad u \geqslant 0$$

(similarly for $u < 0$)

and put the second term on the right hand side of our equations, then we have several more nice properties:

(1) The method can now be reduced to systems of tri-diagonal equations.

(2) These are diagonally dominant and usually converge rapidly. (In one dimension we can prove that the maximum eigen-value of the iteration matrix is 2/3).

(3) The formulation is extremely easy to implement since we are not introducing any extra unknowns; and it is simple to modify the expressions on a flux correction basis if maximum principles are vital.

A range of discretisations (including most of those mentioned in Section 1) have been tested for accuracy and robustness. This work, described in [8], included 1-D problems and the passive convection-diffusion of a scalar. This preliminary investigation led us to the conclusion that HUW represented a reasonable compromise and we implemented this on a range of fluid-flow test problems.

3. LAMINAR FLOW OVER A BACKWARD-FACING STEP

This well known test problem was studied at a recent GAMM meeting in Paris.

The geometry for the problems is shown in Diagram 1. Two pairs of values for the quantities (H, h) were specified: (1.5, 1.0) and (1.0, 0.5), and a parabolic profile was

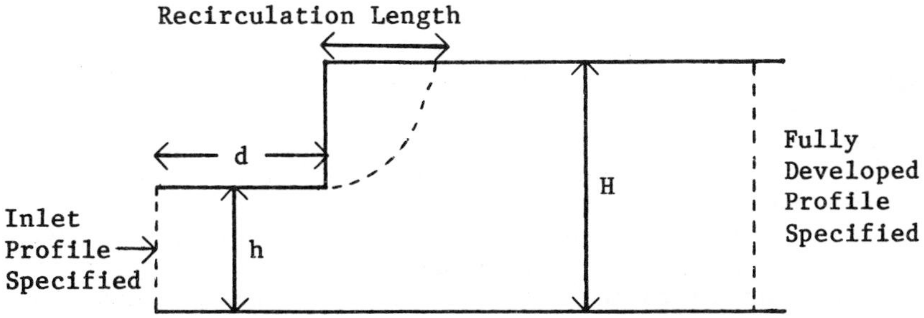

Diagram 1: Geometry for flow over a backward-facing step

prescribed upstream (at d = 3). Two Reynolds numbers were defined, 50 and 150, which gave a set of four benchmarks.

These problems were attempted at AERE Harwell using a variety of methods including some extremely detailed finite element calculations (with 160 x 80 nodes) which, for the purposes of this comparison, we take as exact; and finite difference calculations using hybrid, central, and higher-order upwind schemes, see [9].

The implementation of the HUW scheme involves altering the usual boundary treatments to avoid the introduction of fictitious points. Derivation of the modified difference stencil requires the application of the continuity equation at the wall. Discrete conservation laws can also be incorporated. As usual one integrates over control volumes and uses Green's theorem. Diffusion is estimated in the normal way and convective transport is carried out using linear upwind interpolation to obtain values at cell faces. This reduces to the expressions in equation (1) in the case of a constant velocity field and a uniform mesh. However, this gives rise to a reasonably simple system of equations, which is $O(h^2)$ everywhere and employs a stable discretisation.

The robustness of this new operator is further demonstrated by the fact that it was not necessary to use a good initial approximation for either the hybrid or HUW runs, nor were the under-relaxation parameters altered. This was definitely not the case with central differencing where lack of diagonal dominance made severe under-relaxation essential and instability in the difference scheme required good starting guesses to be employed. On the whole, the use of central differences was not economic for this problem.

On the other hand, the performance of the HUW discretisation was encouraging. Figures 1 and 2 show the wall shear stresses for 4 different calculations at a Reynolds number 150: hybrid and HUW on a 43 x 20 grid; HUW on a 25 x 14 mesh; the finite element calculations on the 160 x 80 grid. When one compares the first two calculations, the extra accuracy of HUW over hybrid differencing is obvious; furthermore, the increase in cost of the better results was less than 70%. Roughly, we require 60% more iterations using the "pseudo-deferred correction" approach outlined earlier, and each iteration requires only 6% more work (on a scalar processor). The coarse grid HUW results have errors which are similar in size to those of the low-order calculations. However, the cost of the novel scheme is now only 40% of the hybrid results. It is of interest to note that if one wishes to reduce the errors further then the cost advantage in favour of the higher-order upwind discretisation would increase, at least until the mesh-Peclet number fell below 2.

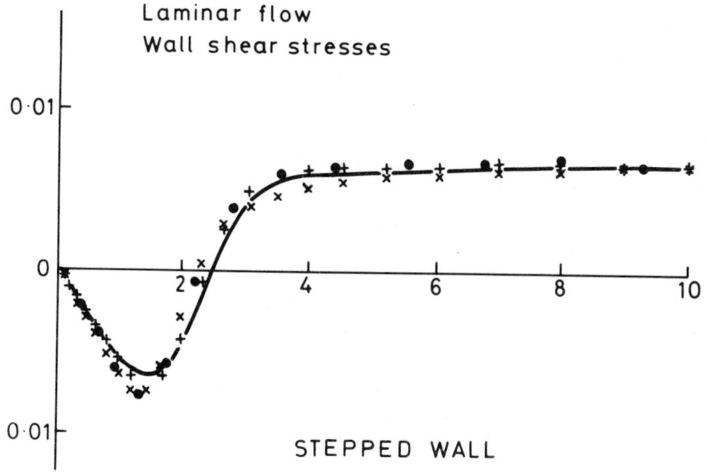
Laminar flow
Wall shear stresses
0·01
0
0·01
2
4
6
8
10
STEPPED WALL

Figure 1

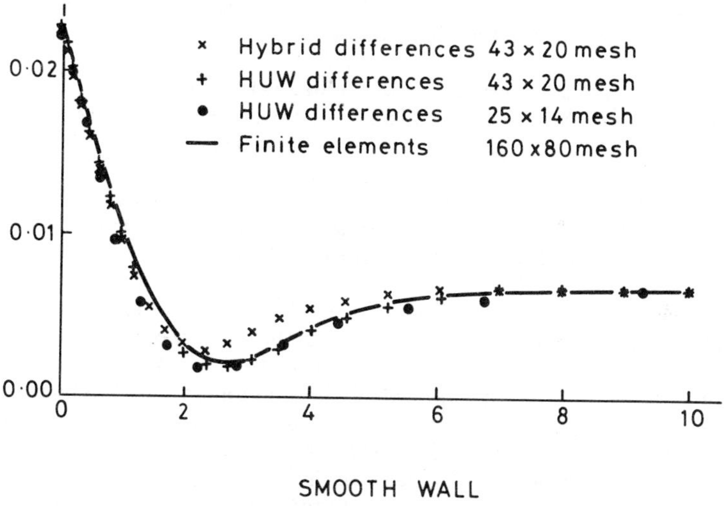
× Hybrid differences 43 × 20 mesh
+ HUW differences 43 × 20 mesh
● HUW differences 25 × 14 mesh
— Finite elements 160 x 80 mesh
0·02
0·01
0·00
0
2
4
6
8
10
SMOOTH WALL

Figure 2

4. TURBULENT FLOW OVER A BACKWARD-FACING STEP

The regime of turbulent flow over a backward-facing step is considerably more difficult than the laminar case. Physically one has far shorter length scales and rapidly fluctuating components of velocity. These have been simulated using the k-ε turbulence model. One of the fundamental assumptions of this model is that the turbulent kinetic energy (k) and the dissipation (ε) are positive everywhere. The hybrid difference scheme guarantees this; however, it is possible with the new method to obtain overshoots on coarse meshes in regions of rapid variation [8]. This did not happen in these flow calculations, possibly because of the source terms. From a numerical viewpoint the extra equations make convergence slower and the effective viscosity may vary significantly over the course of the iteration. Non-robust schemes can have severe difficulties in these circumstances where the mesh-Reynolds number can be large because of rapid changes in the initial approximations.

In this section we compare computations with hybrid and HUW differencing performed on the same grid with experimental results obtained by Bates and Yeoman [10]. The geometry is as shown in Diagram 1 with the following dimensions: d = 10 mm, h = 15 mm and H = 25 mm.

The Reynolds number, for the data presented here, is 3×10^4 although very similar observations and predictions are available for Reynolds number 5×10^4.

LDA velocity measurements of stream-wise velocity-components (both mean and fluctuating) are available and are compared with predictions in Figures 3 and 4.

The predictions use a standard k-ε model and inlet boundary conditions which are deduced from experimental observations 10 mm upstream of the step. (Values of k are derived from $(u')^2$ by assuming homogeneous turbulence; any 'intelligent' guess for ε will suffice.) The standard logarithmic wall treatments were used for both the hybrid and HUW calculations and a border of hybrid differences was used adjacent to the walls. It is possible to adopt more subtle strategies in this region; however there is very little point in using quadratic interpolation in areas of logarithmic variation!

It is clear, from the two sets of figures provided, that the higher-order upwind differences represent a significant improvement in the calculation of this type of flow. The mean velocity profiles are slightly improved, and the recirculation length is 4.9 step-heights with hybrid and 5.4

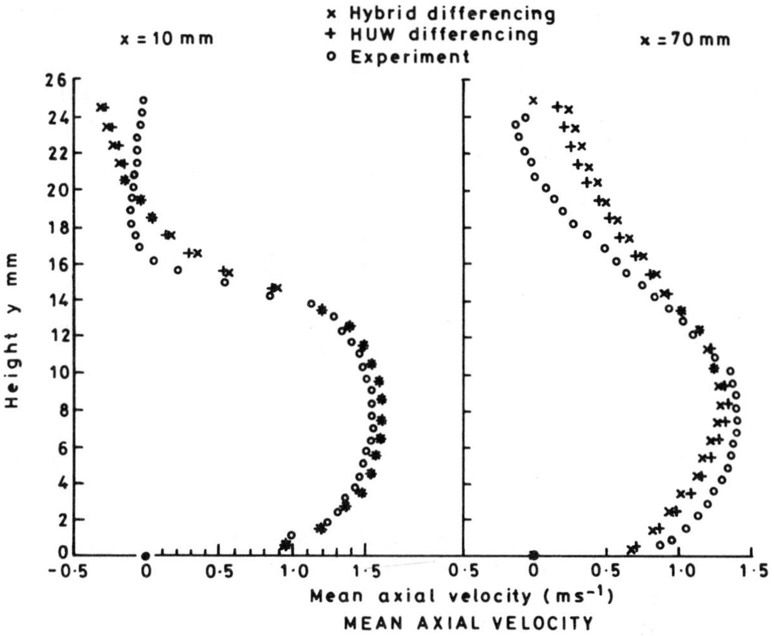

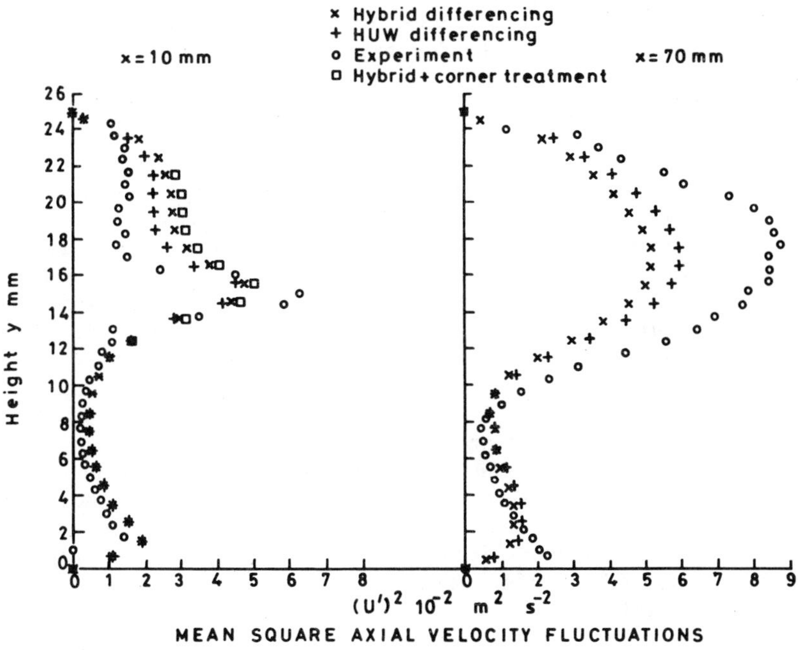

Figure 3

Figure 4

with HUW (the experimental value is about 7). Fluctuating components are estimated using the Boussinesq formula: $(u')^2 = 2k/3 - \nu_T \partial u/\partial x$. Higher levels of turbulence are predicted behind the step, since the shear layer is less diffuse with the more accurate method; and even in the recirculation region, where the mean flows are poorly predicted, the values of u' are in better agreement.

These runs were performed on a CRAY 1-S computer and the increase in costs of the HUW over the hybrid calculations is again about 60%. As before it was not necessary to modify the initial approximation or the under-relaxation factors to achieve convergence with the more accurate scheme.

Although the new predictions are more accurate it is clear that significant discrepancies exist between theory and experiment. The nature of the errors committed with both schemes allows some criticism of the turbulence model. The behaviour of the logarithmic boundary conditions is good in regions of fully developed flow, but is poor in the recirculation region where the velocities are over-predicted. Attention to this aspect, and to the model equations, is likely to provide significant improvements. (For example: changing the coefficient of shear generation in the dissipation equation to 1.5 - a 4% alteration - improves the agreement dramatically for this flow.)

5. CONCLUSIONS

The preliminary results presented here indicate that the higher-order upwind difference scheme represents a reasonable compromise between accuracy and robustness. Moreover, the solution method outlined in Section 2 seems to be reasonably efficient. The cost/error comparisons with results using hybrid differencing are encouraging, as is the fact that (as yet) there have been no unusual convergence difficulties associated with the new scheme.

Of course, it is not as accurate as the QUICK scheme although the importance of this depends on the relative sizes of the convective and diffusive errors. The Leonard scheme may present more difficulties to some solution algorithms.

The HUW scheme does seem to be a significant improvement over hybrid differencing, and merits serious, further investigation.

6. REFERENCES

1. GOSMAN, A.D. and PUN, W.M. - Calculation of Recirculating Flows, Lecture Notes, Rep. No. HTS/74/2, Dept. Mech. Eng., Imperial College, London, 1974.

also:

CLIFFE, K.A., RAE, J. and SYKES, J. - Private communication.

2. HIRSCH, R.S. - Higher Order Accurate Difference Solution of Fluid Mechanics Problems by a Compact Differencing Technique, J. Comp. Phys., Vol.9, No.1, pp.90-109, 1975.

3. ROUX, B., BONTOUX, P., GILLY, B. and GRONDIN, J.C. - Natural Convection in an Enclosed Cavity, Numerical Solutions for a Comparison Problem on Natural Convection in an Enclosed Cavity, AERE-R 9955, Ed. Jones, I.P. and Thompson, C.P., HMSO, 1981.

4. CIMENT, M., LEVENTHAL, S.H. and WEINBERG, B.C. - The Operator Compact Implicit Method for Parabolic Equations, J. Comp. Phys., Vol.28, No.2, pp.135-166, 1978.

5. BERGER, A.E., SOLOMON, J.M., CIMENT, M., LEVENTHAL, S.H. and WEINBERG, B.C. - Generalized OCI Schemes for Boundary Layer Problems, Math. Comp., Vol.35, No.151, pp.695-731, 1980.

6. LEONARD, B.P. - A Stable and Accurate Convective Modelling Procedure Based on Quadratic Upstream Interpolation, Computer Methods in Appl. Mech. and Eng., Vol.19, pp.59-98, 1979.

7. HAN, T., HUMPHREY, J.A.C. and LAUNDER, B.E. - A Comparison of Hybrid and Quadratic-Upstream Differencing in High Reynolds Number Elliptic Flows, Computer Methods in Appl. Mech. and Eng., Vol.29, pp.81-95, 1981.

8. THOMPSON, C.P. and WILKES, N.S. - Experiments with Higher-Order Finite Difference Formulae, AERE-R 10493, HMSO, 1982.

9. CLIFFE, K.A., JONES, I.P., PORTER, J., THOMPSON, C.P. and WILKES, N.S. - GAMM Workshop, Bievres 1983, Solution to Test Problem. (To appear)

10. BATES, C.J. and YEOMAN, M.L. - Measurements in the Harwell Water Channel Rig of Turbulent Flow over a Backward Facing Step, (HTFS RS 436), HMSO, 1982.

CALCULATION OF FLOW REATTACHMENT WITH VORTEX INTEGRAL EQUATIONS

M. Ribaut

Brown Boveri Ltd., Switzerland

SUMMARY

A vortex integral equation method is applied to the calculation of the separated and reattaching flow past an airfoil. The method solves a mixed boundary condition problem in which the vortex strength or the distance of the separated vortex sheet to the profile surface is considered as explicit unknown. This provides a system of Fredholm equations of the first and second kind which can be solved by means of iteration and yields the shape and extension of the region underlying the separated flow and the vorticity distribution. Results of calculations for the flat and curved plate at different incidences are presented. The influence of vorticity diffusion on the location of the reattachment point is analysed. Calculated velocity distributions are compared with the experiment.

1. INTRODUCTION

The need of performance predictions at off-design conditions has considerably increased the interest for flow separation and reattachment. In particular, for profiles at large incidences and having a sharp leading edge, the introduction of a separation bubble is certainly the only realistic way to avoid the nearly infinite velocities obtained by calculating the inviscid attached flow. As there are excellent review papers in the field (e.g. [1], [2]) only the basic ideas justifying the present approach will be mentionned in the following.

The attached flow of an inviscid or viscous fluid can be calculated by considering the vorticity associated to each point of the boundary surface as the explicit variable allowing to satisfy the boundary conditions. If the flow and the rotational region next to the wall are detached and have no more "contact" to this wall, the amount of vorticity carried away in this region will remain constant downstream of the separation point. Consequently, the position of this vorticity becomes the variable of a mixed boundary condition problem. All this is very clear for the inviscid fluid flow, which can be calculated either with conformal mapping methods (e.g. [3], [4]) or by means of vortex integral equations [5], provided that some pressure distribution is given along the jet lines. But for the viscous fluid flow, it is more difficult to define the state at which the wall ceases to influence the strength of the vorticity of the boundary layer breaking away from the wall. Also, there are two points of view in the boundary layer approach of flow separation and reattachment. The first recognizes honestly the departure of the vortex layer by considering a non-rotational zone close to the wall [6]. This leads to a two-parameter family of velocity profiles and to the necessity of an additional equation defining the thickness of this zone. The second avoids this situation by assuming implicitly that the vorticity is still in "contact" to the wall downstream of the separation point and considers a purely rotational backflow produced by an increased wake component [7]. Nevertheless, by making use of zonal models, displacement surfaces or some other interaction process, all methods recognize the necessity of some information provided by the kinematics of the flow.

The experiment supports the idea of a non-rotational zone in the backflow region of a bubble [8] and this zone must exist if the amount of shed vorticity is larger than the velocity induced at the outer edge of the separated shear layer [9]. It will be assumed here that, for a given rate of vorticity diffusion, the position of the separated and "lost" vorticity is essentially determined by the kinematics of the flow and that the location of the reattachment point can therefore be calculated by formulating inverse boundary conditions at the centre-line of the separated vortex sheet.

2. FLOW MODEL

The time-averaged velocity field of a separated two-dimensional flow may be divided into four regions. From the stagnation point to the point where the limiting streamline leaves the surface, a region of attached flow for which the boundary layer theory seems to be valid everywhere [7]. Then a region of separated flow, limited downstream by the point where the inner edge of the vortex sheet rejoins the surface (Fig. 1), followed by a reattachment zone in which the vorticity deposits on the wall. Lastly, a region of attached flow characterized by the fact that the boundary layer theory is again applicable. This unusual definition of the reattachment point allows to formulate the following assumptions about the vorticity distribution:

$$\zeta_t = Z\zeta_t + (1-Z)\zeta_t \frac{\pi}{2\delta_0} \int_0^{\delta_0} \sin(\frac{n}{\delta_0}\pi)dn \qquad (1)$$

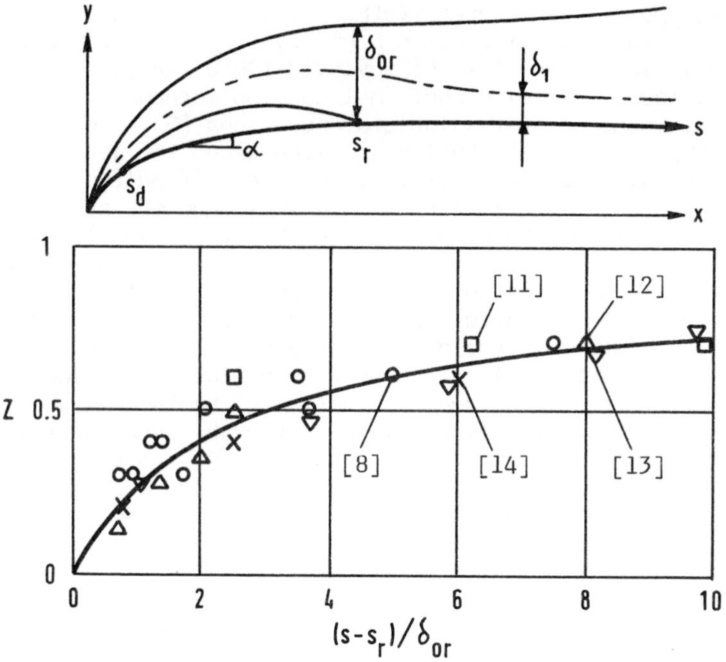

Fig. 1 Vortex sheet model and wake recovery in the reattachment zone.

where ζ_t denotes the local strength of the vortex sheet of thickness δ_o and Z is a profile parameter determining the size of the wall and wake components [10]. In the region of separated flow, the value of Z equals zero and the thickness of the shear layer may be approximated by:

$$\delta_o = \delta_{od} + \lambda_f(s_r - s_d)\left(\frac{s - s_d}{s_r - s_d}\right)^q \tag{2}$$

where λ_f is the vorticity diffusion coefficient, q takes into account the laminar or turbulent character of the shear layer ($0.5 < q < 1$) and s_d , s_r correspond to the detachment resp. reattachment point. In the reattachment zone, the vortex sheet spreads on the wall, thus producing a strong reduction of the wake vorticity and a corresponding increase of the profile parameter Z. If one considers the measurements in [8],[11],[12] [13] and [14] illustrated in the Fig. 1, it appears that this phenomenon has a local and relatively universal character, which can be represented by the following expression:

$$Z = \frac{s - s_r}{1.1(s - s_r) + 2.7\delta_{or}} \tag{3}$$

δ_{or} being the thickness of the vortex sheet at the reattachment point. It will be assumed here that the vorticity involved in the wall component is concentrated on the profile surface, whereas the vorticity associated to the wake component is still distributed in accordance to the equation (1).

The kinematical model proposed for the separated and reattaching flow can then be summarized as follows:

- In the region of separated flow, the position of the vortex sheet centre-line is considered as explicit unknown. The associated vorticity, the distribution of which is given by the equations (1) and (2), has a constant strength and moves like a stiff body with the centre-line.
- In the reattachment zone, the vortex strength is considered as explicit unknown, whereas the distance of the centre-line to the profile surface is set equal to the displacement thickness of the vortex sheet:

$$\delta_1 = 0.5\delta_{or} - (\delta_{1r} - \delta_1) = 0.5(1 - Z)\delta_o \tag{4}$$

This last assumption accounts for the reduction of the wake component downstream of the reattachment point and reduces to about one half the diffusion of the vorticity in this region. It connects also the position of the centre-line to the last unknown position just up-stream of the reattachment point, thus taking into account that the separated fluid moves as a whole by seeking its position of equilibrium against the wall.

3. BASIC EQUATIONS

The mixed boundary condition problem described above can be solved by formulating one or both of the following Fredholm integral equations:

$$\int_s K_1 \zeta_t(x_s, y_s) ds = -w_{x\infty}\sin\alpha + w_{y\infty}\cos\alpha \tag{5}$$

$$\int_s K_2 \zeta_t(x_s, y_s) ds - \frac{1}{2} Z\, \zeta_t(x, y) = -w_{x\infty}\cos\alpha - w_{y\infty}\sin\alpha \tag{6}$$

with

$$K_1 = \frac{1}{2\pi} \frac{(y - y_s)\sin\alpha + (x - x_s)\cos\alpha}{(x - x_s)^2 + (y - y_s)^2} \tag{7}$$

$$K_2 = \frac{1}{2\pi} \frac{(y - y_s)\cos\alpha - (x - x_s)\sin\alpha}{(x - x_s)^2 + (y - y_s)^2} \tag{8}$$

at discrete points x,y of the profile surface and the equation (5) on the centre-line of the vortex sheet [9]. The integrals, which involve the vorticity distribution defined by the equations (1) and (2), extend on the vortex sheets, from the stagnation point to their last junction point at the end of the wake. The right hand side represents the unperturbed flow. In order to obtain a linear representation of the variable y defining the unknown position of the vortex sheet, a Taylor development is applied to the geometrical kernel of the integrals, which allows to rewrite the equations (5) and (6) in the following form:

$$\int_s [K_1\zeta_t + (\frac{\partial K_1}{\partial y}\Delta y + \frac{\partial K_1}{\partial \alpha}\frac{\partial \alpha}{\partial y}\Delta y + \frac{\partial K_1}{\partial y_s}\Delta y_s)\,\zeta_t^*]ds$$

$$= -w_{x\infty}\sin\alpha + w_{y\infty}\cos\alpha \tag{9}$$

$$\int_s [K_2\zeta_t + (\frac{\partial K_2}{\partial y}\Delta y + \frac{\partial K_2}{\partial \alpha}\frac{\partial \alpha}{\partial y}\Delta y + \frac{\partial K_2}{\partial y_s}\Delta y_s)\,\zeta_t^*]ds$$

$$-\frac{1}{2}Z\,\zeta_t(x,y) = -w_{x\infty}\cos\alpha - w_{y\infty}\sin\alpha \tag{10}$$

where ζ_t and $\Delta y = y - y^*$ are the explicit unknowns of the problem and the asterisk refers to initial conditions or a precedent iteration. The derivative of the angle α results from simple geometrical considerations [5]:

$$\frac{\partial \alpha}{\partial y}\Delta y = \frac{\Delta y_{x+\Delta x} - \Delta y_{x-\Delta x}}{2\Delta x}(\cos\alpha^*)^2 \tag{11}$$

If the equations (9) or (10) are formulated at the solid boundary surface, the second and third terms on the left hand side drop out. Similarly, the fourth term must be omitted if the current point of the integral is situated on this surface. It was also supposed that in the region of reattached flow, only the vorticity involved in the wake part contributes to the last term on the left hand side of these equations.

4. NUMERICAL RESULTS

The present method was applied to the flow past an infinitely thin plate at angles of attack comprised between zero and 10 degrees. The kinematical condition (9) was formulated in the mid-point of 20 equidistant intervals on the plate and corresponding points (with same abscissa x) on the centre-line of the separated vortex sheet. In the region of reattached flow, the equation (10) was formulated at the end points of the plate intervals. Let us note that the formulation of a Fredholm equation of second kind in the region of separated flow

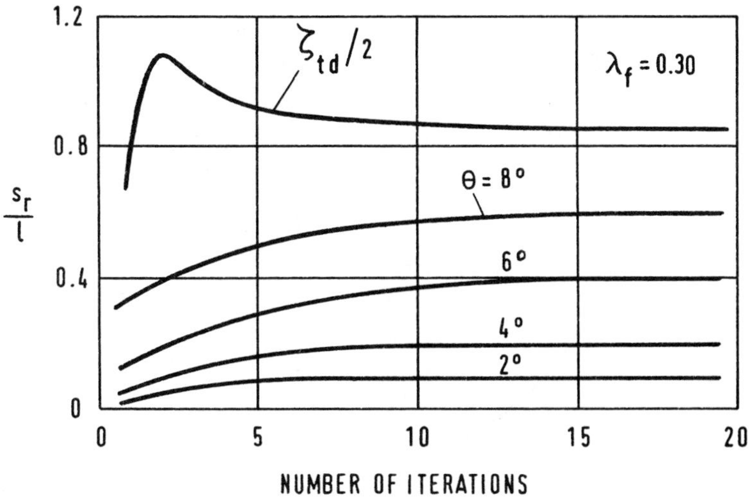

Fig. 2 Convergence of the solution

would have no sense in the case of an infinitely thin airfoil, without having a precise knowledge of the amount of reversal flow in this region. These equations were complemented by the usual steadiness condition of zero vorticity discharge at the trailing edge. The shape of the vortex sheet centre-line was represented by a cubic spline extending from the leading edge to the wake end and passing through the end points of the intervals at which the ordinate y was considered as unknown of the equation system. Lastly, it was supposed, by calculating the integrals of the equations (9) and (10), that the reattachment point coincides with one of the end points of the plate intervals. This procedure has introduced some scatter of the results, say up to five percent.

A number of calculations were performed for different plate inclinations, with q=1 in the equation (2) and neglecting the boundary layer on the windward side of the plate. The computation time was about 8 minutes for twenty iterations, using an IBM 370-168. It will be seen, in the Fig. 2, that for small plate inclinations, convergence of the reattachment length was achieved within 10 iterations, whereas for larger inclinations up to 20 iterations were necessary for having less than one percent variation of the solution. The vorticity ζ_{td} shed at the leading edge has shown to converge to the same value for all plate inclinations. This corroborates the results obtained in [9] for the entirely separated flow at plate inclinations comprised between 15 and 90 degrees. However, for the reattaching flow and the same vorticity diffusion, the calculated amount of shed vorticity is about 10 percent superior. This may be explained by the reattachment process which increases the camber of the free vortex sheet, thus inducing

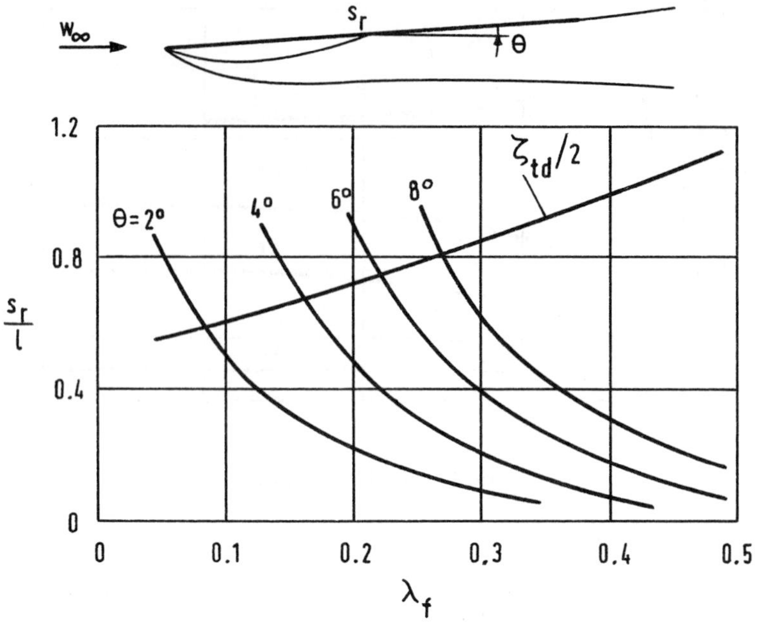

Fig. 3 Influence of vorticity diffusion on the reattachment length and the vorticity shed at the leading edge.

more excess velocity at the leading edge.

The influence of the vorticity diffusion on the location of the reattachment point and the amount of vorticity leaving the leading edge is shown in the Fig. 3, for different plate inclinations. As for the flow without reattachment [9], the curvature of the centre-line of the free vortex sheet and therefore the reattachment length are essentially determined by the amount of vorticity diffusion. In order to compare these results with the experiment, the velocity field in the regions of separated and reattaching flow was calculated for two plate inclinations, still using the value of 0.3 for the vorticity diffusion coefficient. The exponent q in the equation (2) was set equal 0.65 for the free part of the vortex sheet and equal 1 downstream of the reattachment point. It will be seen, in the Fig. 4, that these values yield good agreement with the experiment in [8], if one considers that relatively strong non-rotational gradients reduce the apparent vortex sheet thickness in the region of separated flow. Let us note, that using a value of t=1 everywhere, one obtains practically the same reattachment length. The agreement between the calculated and measured backflow confirms the existence of a non-rotational and purely kinematical flow reversal, arising from the difference between the amount of vorticity produced at the leading edge and the velocity induced outside of the vortex

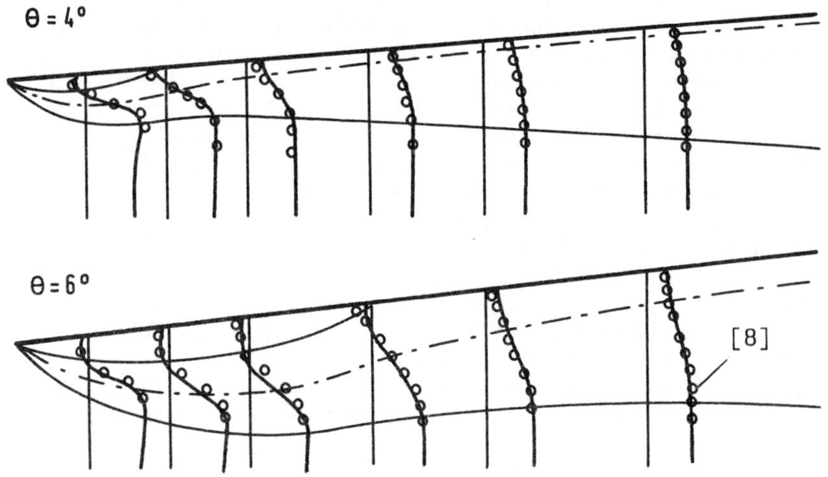

Fig. 4 Calculated velocity profiles in the shear layer, compared to the experiment in [8].

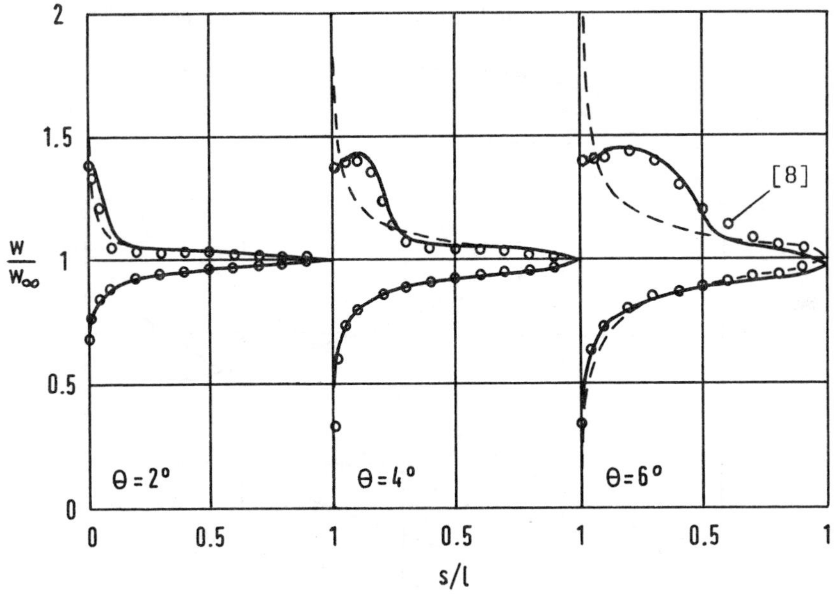

Fig.5 Calculated mainstream velocity at the plate surface, compared to the experiment in [8].

sheet. This latter is illustrated in the Fig. 5 together with the velocity distribution of the attached inviscid fluid flow (dotted line). It may be seen that for angles of attack higher than 3 degrees, a boundary layer calculation using the computed inviscid solution would have little sense. On the other hand, it may be observed that the calculated reattachment points are situated too much up-stream, which can be seen also from the velocity profiles of the reattachment zone, illustrated in the Fig. 4. A possible explanation of this situation could lie in the fact that the vorticity diffusion coefficient used in the equation (2) is depending not only on the turbulent vorticity diffusion, but also on large scale fluctuations of the vortex sheet position, the size of which reduces probably for the reattaching flow. Therefore, it would be justified to use a slightly reduced value of the vorticity diffusion coefficient, thus improving immediately the agreement with the experiment.

Finally, a number of calculations were made for the curved plate, again considering different plate inclinations and varying the amount of camber between -10 and +10 percents of the chord length. It will be seen, in the Fig. 6, that the amount of vorticity shed at the leading edge is still independent of the plate inclination, but varies with the camber of the plate. This latter seems to add to the curvature of the vortex sheet induced by the vorticity diffusion, thus producing more or less excess velocity at the leading edge. It must

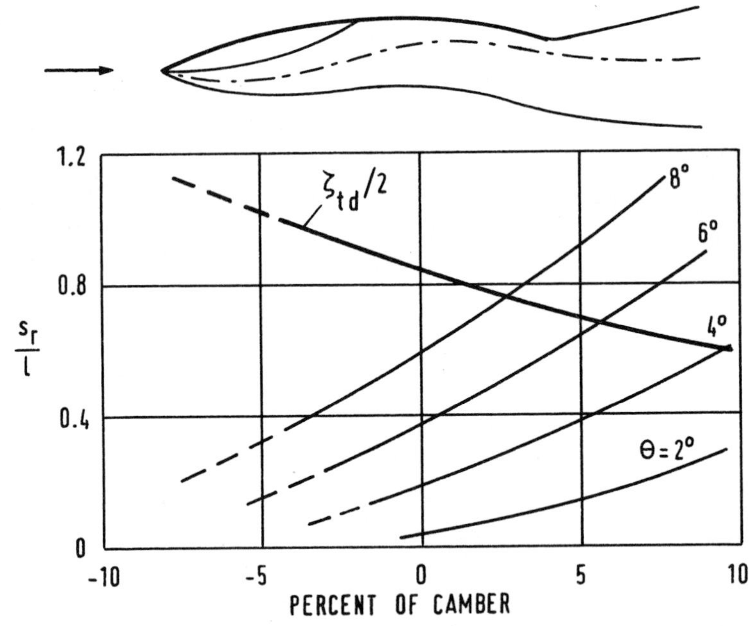

Fig. 6 Influence of plate curvature on the reattachment length and the vorticity shed at the leading edge.

be remarked that for large negative cambers, the convergence of the solution was no more possible, the calculated vortex sheet centre-line having the tendency to jump on the "false" side of the plate.

5. CONCLUSIONS

The results obtained in the present study or with earlier computations [5],[9] for the separated flow past a plate demonstrate that accurate solutions to the kinematic problem can be obtained for all plate inclinations by using the same kinetic model, with a constant amount of vorticity diffusion. This indicates that the behaviour of a separated flow at high Reynolds numbers and especially the flow reattachment are essentially determined by kinematical conditions. In the present method , these conditions are not only formulated at the boundary surface but also in the field, according to the mixed boundary condition problem which defines the solution at infinitely large Reynolds numbers [3],[4]. The encouraging results of this approach, which is just the opposite of classical methods solving the direct boundary condition problem (e.g. [15]), permits to suggest that the difficulties encountered by calculating separated flows are probably more related to an ineffective formulation of the kinematical conditions as to the lack of suitable turbulence models.

7. REFERENCES

1. HORTON, H.P. - Fundamental Aspects of Flow Separation Under High-Lift Conditions. AGARD LS. 43, No. 4, Von Karman Inst. Brussels, 1971.
2. SIMPSON, R.L. - A Review of Some Phenomena in Turbulent Flow Separation. J. of Fluids Engineering, Vol. 103, No. 12, pp. 520-533, 1981.
3. WOODS, L.C. - The Theory of Subsonic Plane Flow. Cambridge University Press, pp. 357-361, 1961.
4. TOKEL, H., SISTO, F. - Dynamic Stall of An Airfoil with Leading Edge Bubble Separation Involving Time Dependent Reattachment. ASME Paper 78-GT-194, 1978.
5. RIBAUT, M., - A Kinematical Study of Kirchhoff-Rayleigh Flow. ZAMP, Vol. 31, No. 1, pp. 83-93, 1980.
6. GREEN, J.E. - Two-Dimensional Turbulent Reattachment as A Boundary Layer Problem. AGARD C.P., No. 4, Von Karman Inst. Brussels, 1966.
7. KLINE, S.J., BARDINA, J., STRAWN, R., - Correlation and Computation of Detachment and Reattachment of Turbulent Boundary Layers on Two-Dimensional Faired Surfaces. AIAA Paper 81-1220, 1981.

8. GAULT, D.E., - An Investigation at Low Speed of The Flow over A Simulated Flat Plate at Small Angles of Attack Using Pitot-Static and Hot-Wire Probes. NACA TN3876, 1957.

9. RIBAUT, M., - A Vortex Sheet Method for Calculating Separated Two-Dimensional Flows. AIAA Paper 82-1030, 1982.

10. ESCUDIER, M.P., NICOLL, W.B. - The Entrainment Function in Turbulent Boundary Layer and Wall-Jet Calculations. J. of Fluid Mechanics, Vol. 25, No. 2, pp. 337-366, 1966.

11. BRADSHAW,P., Wong, F.Y.F., - The Reattachment and Relaxation of a Turbulent Shear Layer. J. of Fluid Mechanics, Vol. 52, No. 1, pp. 113-135, 1972.

12. MASBERNAT, L., - Contribution à l'Etude du Décollement dans les Ecoulements de Fluides Incompressibles. Thèse Univ. de Toulouse, 1968.

13. SETTLES, G.S., WILLIAMS, D.R.,BACA, B.K., BOGDONOFF, S.M. , - Reattachment of a Compressible Turbulent Free Shear Layer. AIAA Journal, Vol. 20, No. 1, 1982.

14. KIM, J., KLINE, S.J., JOHNSTON, J.P., - Investigation of Separation and Reattachment of a Turbulent Shear Layer: Flow over a Backward-Facing Step. REP. MD-37, Thermosciences Division, Stanford Univ. California, 1978.

15. SUGAVANAM, A., WU, J.C., - Numerical Study of Separated Turbulent Flow over Airfoils. AIAA Journal, Vol. 20, No. 4, pp. 464-470, 1982.

A COMPARISON OF NUMERICAL SIMULATION AND EXPERIMENTAL VISUALIZATION OF THE EARLY STAGE OF THE FLOW GENERATED BY AN IMPULSIVELY STARTED ELLIPTIC CYLINDER

TA PHUOC LOC[1], O. DAUBE[1], P. MONNET[2], M. COUTANCEAU[2]
1 LIMSI BP 30 91406 ORSAY CEDEX FRANCE
2 LMF-Université de POITIERS 86022 POITIERS CEDEX FRANCE

ABSTRACT

The early stage of incompressible viscous flow around an elliptic cylinder with an angle of attack is studied in this paper by means of experimental and numerical techniques. The experimental technique is a visualization one which may also give some quantitative informations such as measurements of velocity. The numerical scheme is a high order compact finite difference one. Qualitative and quantitative comparisons between the results that were obtained by both methods are presented and are found to be quite satisfactory. These comparisons enable us to give an accurate description of the separation phenomenon which occur in the first instants of the flow.

INTRODUCTION

Unsteady measurements in fluid mechanics represent one of the greater difficulties encountered in experimental works. An alternate way is to use numerical simulation in conjunction with experimentation in order to complete the analysis of the physical phenomenon. Some phenomenon as the very first instants of a flow cannot be reached by experimental investigation and in such a case the numerical simulation may be very useful. Numerical simulation may also be a way to check the validity of a model derived from experimental analysis. On the other hand, the comparison between numerical results and experimental data is, when it is possible, one of the best way to check the validity of a numerical method.

The paper which is presented here has got these two aspects. The same problem – flow around an elliptic cylinder – was treated by the first two authors in a numerical way and by the two others in an experimental way using a visualization technique. The goal of this study was to get an accurate description of the separation phenomenon in the early stages of the flow. We were more particularly interested in the "starting vortex" at the trailing edge and in the appearance and the development of the secondary vortices at the leading edge. These secondary vortices were first found by experimental techniques in the case of a circular cylinder [1] and were recovered for the first time in a numerical work by the proposed numerical method. [2] This fact proves the ability of the proposed numerical method to deal with physical problems.

A description of the numerical technique and of the experimental technique is given. Then some results concerning the time evolution of the flow structure for two Reynolds numbers (Re =1000 & 3000) and two angles of attack (α=30°&50° are reported. The agreement between experimental and numerical results is found to be quite satisfactory. In addition,some purely numerical results about the time evolution of the aerodynamic coefficients are shown .

I. EXPERIMENTAL TECHNIQUE

The flow is produced by causing the elliptic cylinder having a horizontal axis to rise vertically along the median plane of a tank with a rectangular cross-section. To obtain the desired range of Reynolds number,the tank is filled with a liquid of suitable viscosity. The visualization is carried out by illuminating along a thin right plane,solid tracers uniformly put in suspension in the fluid.

In the present work,the geometric characteristics of the ellipse are b=7cm and a/b=0.1 ("b" is the great axis and "a" is the small axis of the ellipse). The tank was filled with water and powder of rilsan was used as solid tracers. In order to minimize the effects of the walls of the tank and of the free surface,the experiments were carried out in a sufficiently large tank(56x46x100 cm) and the photographs were made whenthe cylinder was about half-way up of the tank. The motions of the cylinder and of the camera are coupled by a rigid support and are generated by a system of pulleys and balance weights. Start up is almost instantaneous and the pictures are taken in regular time intervals (only one picture per experiment) equal to 0.5. The unit dimensionless time is,in the experiments,the time needed by the cylinder to move along a distance equal to the great axis "b". Because of the size of the tank the dimensionless time for experiments is limited to about 5.

In contrast with most of the experimental works,this visualization technique is used here not only to get qualitative results but also to get some quantitative information. In particular,the velocity modulus is determined from the measurements of the white dashes which appear on the pictures. These dashes constitute parts of the tracers trajectories and of the fluid particles if the visualization is faithful,during the time of exposure. These length measurements are made manually with the help of a binocular lens. Some examples of such measurements can be found in reference 3 .

Furthermore, the streamline pattern and the main geometric characteristics of the flow structure can be determined. Particularly, the bifurcation points of the streamlines and the eddies centers can be located and their time evolution can be analyzed.

II. PHYSICAL PROBLEM AND GOVERNING EQUATIONS

II.1 Description of the problem

Let us consider an elliptic cylinder moving with a constant forward speed $V_o\vec{i}$ and with an angle of attack α with respect to the x-axis (see figure 1). Let b be the great axis of the ellipse. All variables are made dimensionless with respect to the forward speed V_o and to the half great axis b/2. The Reynolds number Re is defined as

$$Re=V_o b/\nu$$

where ν is the kinematic viscosity of the fluid. In vorticity-function ζ and stream function Ψ formulation, the Navier-Stokes equations in divergence form are:

$$\frac{\partial\zeta}{\partial t} + \nabla.(\zeta\vec{V}) = \frac{2}{Re}\nabla^2\zeta \qquad (1)$$

$$\nabla^2\Psi = \zeta \qquad (2)$$

II.2 Transformed equations

In order to get a rectangular domain of computation, the outside of the ellipse is mapped onto a semi-infinite strip in a Λ-computation plane. this is achieved with the help of two conformal mapping defined by

$$z = x+iy = Z+c^2/Z$$

$$Z = Re^{i\theta} = e^{\pi\Lambda} = e^{\pi(\xi+i\eta)} \quad ; \ \xi \geq 0 \quad 0 \leq \eta \leq 2$$

Let $h^2(\xi,\eta) = |dz/d\Lambda|^2$, then equations (1) and (2) become

$$h^2\frac{\partial\zeta}{\partial t} + \frac{\partial}{\partial\eta}(\zeta\frac{\partial\Psi}{\partial\xi}) - \frac{\partial}{\partial\xi}(\zeta\frac{\partial\Psi}{\partial\eta}) = \frac{2}{Re}(\frac{\partial^2\zeta}{\partial\xi^2} + \frac{\partial^2\zeta}{\partial\eta^2}) \qquad (3)$$

$$\frac{\partial^2\Psi}{\partial\xi^2} + \frac{\partial^2\Psi}{\partial\eta^2} = h^2\zeta \qquad (4)$$

with boundary conditions :

$$\frac{\partial\Psi}{\partial\eta} = \frac{\partial\Psi}{\partial\xi} = 0 \quad \text{on} \quad \xi = 0 \qquad (5)$$

$$\vec{\nabla}\Psi \longrightarrow -\vec{j} \quad \text{as} \quad \xi \longrightarrow \infty \qquad (6)$$

The initial condition is an abrupt start of the cylinder.

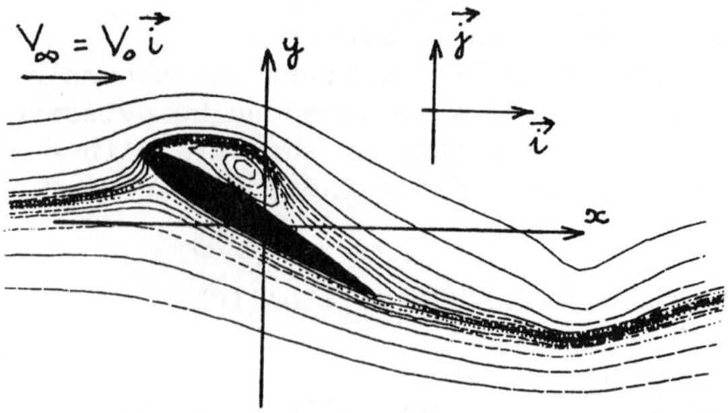

FIGURE 1
The reference frame

III. NUMERICAL METHOD

The numerical method used is a combination of two finite-difference schemes. The vorticity transport equation is solved by a second order accurate A.D.I. scheme and the Poisson equation of the stream function is solved by a compact fourth order accurate one. This so-called combined method was proposed by the first two authors [4] and has been used to study various problems [5] [6]. Further details can be found in references above and only the main features of the method will be outlined in the following section.

III.1 Grid and notations

The domain of calculation in the Λ-plane is $(0 \leq \xi \leq \xi_\infty$; $0 \leq \eta \leq 2$) . ξ_∞ is chosen so that the points $M(\xi_\infty, \eta)$ are far enough from the body. More precisely,this means that the computations will be valid only for $t < t_\infty$. The time t_∞ is the time needed by the cylinder to move along a distance equal to the radii of the computation domain. The mesh increments in ξ and η directions are constant and equal to $\Delta\xi = \xi_\infty/NX$,$\Delta\eta = 2/NY$ where NX and NY are the number of nodes in each direction. The well-known advantage of the (ξ,η) computation is that a regular grid in the Λ-plane gives a refined mesh near the body in the physical plane.

III.2 Resolution of the vorticity transport equation

In the following f^n_{ij} will denote the value of any function f at time $t_n = n\Delta t$ and at node $((i-1)\Delta\xi, (j-1)\Delta\eta)$.

The time marching procedure is an A.D.I. one. In order to minimize the "false viscosity" effect,the divergence form (3) of the transport equation is discretized by a second order upwind predictor-corrector scheme. Assuming all the flow variables are known at time $n\Delta t$, the way to compute ζ^{n+1} can be described as follows:

Let $(\delta_x^+ f)_i = \frac{f_{i+1}-f_i}{\Delta x}$; $(\delta_x^- f)_i = \frac{f_i - f_{i-1}}{\Delta x}$; $(\delta_x^2 f)_i = \frac{f_{i+1}-2f_i+f_{i-1}}{\Delta x^2}$

for any function f and independant variable x.

Let $v_{ij}^n = \left(\frac{\partial\Psi}{\partial\xi}\right)_{ij}^n$ $\quad \alpha_{ij}=1$ if $v_{ij}<0$ and $=-1$ if $v_{ij}\geq 0$

$u_{ij}^n = -\left(\frac{\partial\Psi}{\partial\eta}\right)_{ij}^n$ $\quad \beta_{ij}=1$ if $u_{ij}<0$ and $=-1$ if $u_{ij}\geq 0$

Each time step is divided in two half-steps

1st Half-step : on each line ξ=constant we solve the tridiagonal linear system:

$$\frac{Re}{2}\left[h_{ij}^2 \frac{\zeta_{ij}^{n*}-\zeta_{ij}^n}{\Delta t/2} + 0.5((1+\alpha_{ij})\delta_\eta^- + (1-\alpha_{ij})\delta_\eta^+)(v^n\zeta^{n*})_{ij}\right] - \delta_\eta^2\zeta_{ij}^{n*} =$$

$$-\frac{Re}{2}0.5((1-\beta_{ij})\delta_\xi^- + (1+\beta_{ij})\delta_\xi^+)(u^n\zeta^n)_{ij} + \delta_\xi^2\zeta_{ij}^n \qquad (7)$$

2nd Half-step : on each line η=constant we solve the tridiagonal linear system

$$\frac{Re}{2}\left[h_{ij}^2 \frac{\zeta_{ij}^{n+1}-\zeta_{ij}^{n*}}{\Delta t/2} + 0.5((1+\beta_{ij})\delta_\xi^- + (1-\beta_{ij})\delta_\xi^+)(u^n\zeta^{n+1})_{ij}\right] - \delta_\xi^2\zeta_{ij}^{n+1} =$$

$$-\frac{Re}{2}\left[0.5((1-\alpha_{ij})\delta_\eta^- + (1+\alpha_{ij})\delta_\eta^+)(v^n\zeta^{n*})_{ij}\right] + \delta_\eta^2\zeta_{ij}^{n*} \qquad (8)$$

III.3 Resolution of the stream function equation

Once the values ζ^{n+1} of the vorticity function at time $(n+1)\Delta t$ are known, the values Ψ^{n+1} of the stream function at this time are calculated through the use of a fourth order accurate scheme. This scheme is based on the use of tridiagonal relations between the values of a function and of its first and second derivatives at three adjacent nodes [7]. Let us note Δx the constant spatial step of discretization, f_i, f'_i, f''_i the values of the function f and of its first and second derivative f' and f" at node i. Then the following tridiagonal relations hold:

$$f'_{i+1}+4f'_i+f'_{i-1} = \frac{3}{\Delta x}(f_{i+1}-f_{i-1}) + O(\Delta x^4) \qquad (9)$$

$$f''_{i+1}+10f''_i+f''_{i-1} = \frac{12}{\Delta x^2}(f_{i+1}-2f_i+f_{i-1}) + O(\Delta x^4) \qquad (10)$$

Relation (10) used in conjunction with the Poisson equation yields a system with three unknown functions : the stream function Ψ and its second derivatives $\partial^2\Psi/\partial\xi^2$ and $\partial^2\Psi/\partial\eta^2$. The A.D.I. technique is used with a set of optimum cyclic relaxation parameters. At each half step of the iterative A.D.I. procedure,a set of independant tridiagonal sytems is solved. These tridiagonal sytems are obtained by eliminating the second derivative to be computed in this half-step between the Poisson equation and the relation (10). Further details about this resolution can be found in [4]. Once the values Ψ^{n+1} of the stream function are calculated,the values of its first derivatives $\partial\Psi/\partial\xi$ and $\partial\Psi/\partial\eta$ are computed with the help of the relation (9)

IV. RESULTS

The results which are presented in this paper were obtained for an ellipse with a 10% thickness. Two Reynolds numbers Re — 1000 and 3000 — and two angles of attack α — 30° and 50°— are considered. The outer boundary of the computation domain was chosen in order to be approximatively at a distance of 10 half-great axis from the body. A grid of 61x61 was used for both Re and in each case the time step was taken equal to 0.01. As already noticed,we were primarly interested in the "starting vortex" and in the appearance and the development of secondary vortices. These phenomenon are described below.

IV.1 Starting vortex

In the very first instants of the flow after the abrupt start ($t<0.2$),a "starting vortex" is created at the trailing edge and is immediatly shed. This phenomena can be seen on figure 2. It can be noticed that this "starting vortex" is more easily seen when the incidence α increases.

IV.2 Secondary vortices

On figures 3 to 7 different steps of the separation phenomena are shown. First (step 1),a vortex (a) is created on the upper part of the leading edge. Then (step 2) a pair of secondary vortices (b) and (b') appear close to the leading edge while the main vortex (a) continues to grow,its reattachment point moving towards the trailing edge. Then (step 3) the main vortex (a) begins to detach from the body while a small vortex (c) appears at the trailing edge. The secondary vortex (b') seems to open itself in its upper part,while the trailing edge vortex (c) continues to grow,causing the absorption of the secondary vortex (b') (step 4). the other secondary vortex (b) grows too and forces the vortex (c) to be shed in the downstream flow,starting a Karman's vortex street (step 5).

All these phenomenon were found by both ways.Some examples of this agreement between numerical and experimental results are plotted on figures 8 and 9 for times t=2 and 7 .

In the last sections we present another comparison between numerical and experimental results concerning the velocity on the x-axis behind an ellipse perpendicular to the free stream. At last some results which were obtained by numerical means are reported.

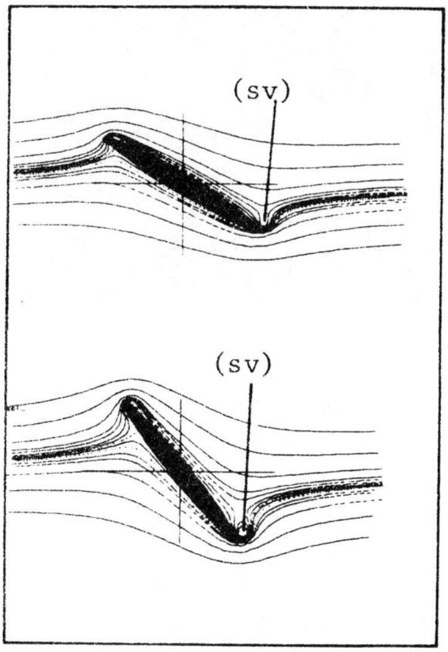

Figure 2
Re = 3000
α = 30°
α = 50°

(sv) : starting vortex

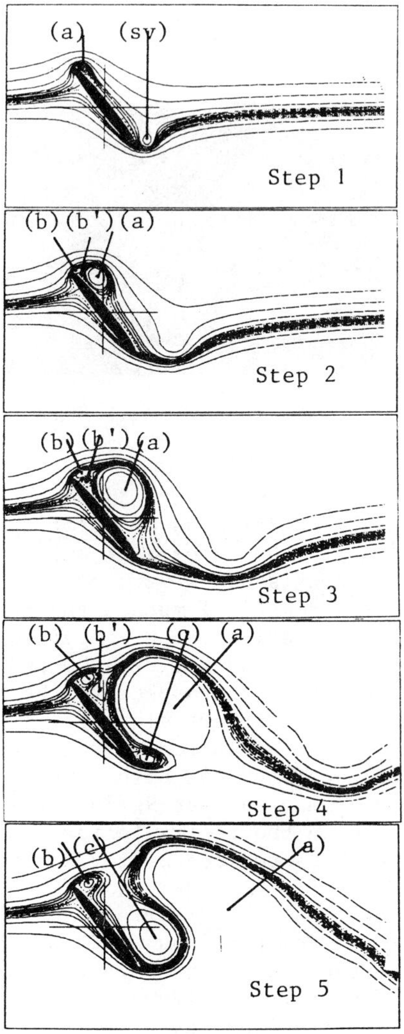

Figures 3 – 7
Re = 10^3 α = 50°

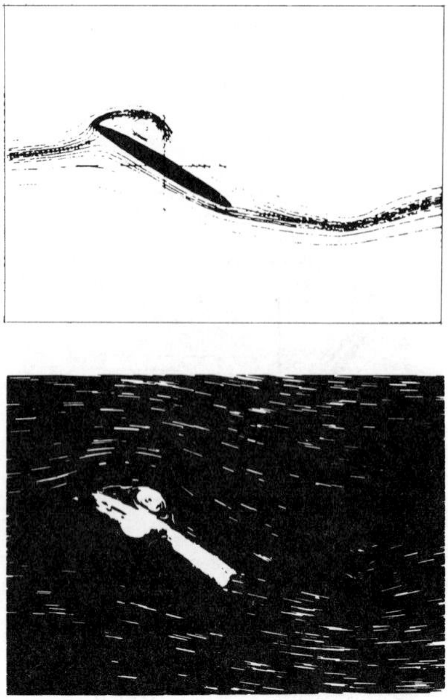

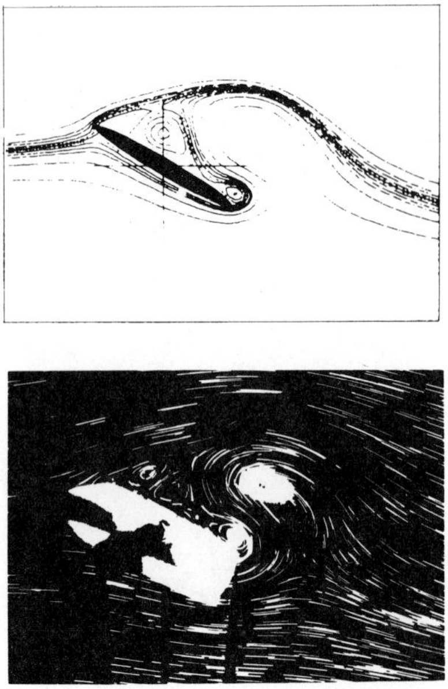

Re = 1000 $\alpha = 30°$

Figure 8; t = 2

Figure 9 ; t = 7

IV.3 Comparisons on velocity profiles

In the case of an angle of attack $\alpha=90°$, the velocities on the x-axis downstream of the ellipse were computed by the proposed numerical method and determined from the experimental visualizations. The comparisons are achieved at different time t=1,2,3,4 and for Re=1000 and 3000 (All quantities are dimensionless with respect to the half-chord of the ellipse and to the free stream velocity). (see figures 10 and 11)

The evolution of wake length is well recovered by the computations for both Reynolds numbers. The discrepancy is more important at Re=3000 for the maximum modulus of the velocity in the recirculating zone. This is likely due to the grid which is too coarse for Re=3000.

IV.4 NACA 0012 Airfoil

Computations were carried out for a Joukowski airfoil very close to the NACA 0012 airfoil (our method cannot deal with sharp trailing edge, so we considered an airfoil whose trailing edge has a very small radius of curvature).

Some examples of the streamline patterns which were obtained are plotted on figure 12. These patterns are similar to those which were obtained in the case of an elliptic airfoil. This is true at least for the first instants of the flow : at the time of this paper, the computations were carried out only for values of the time $t < 4$.

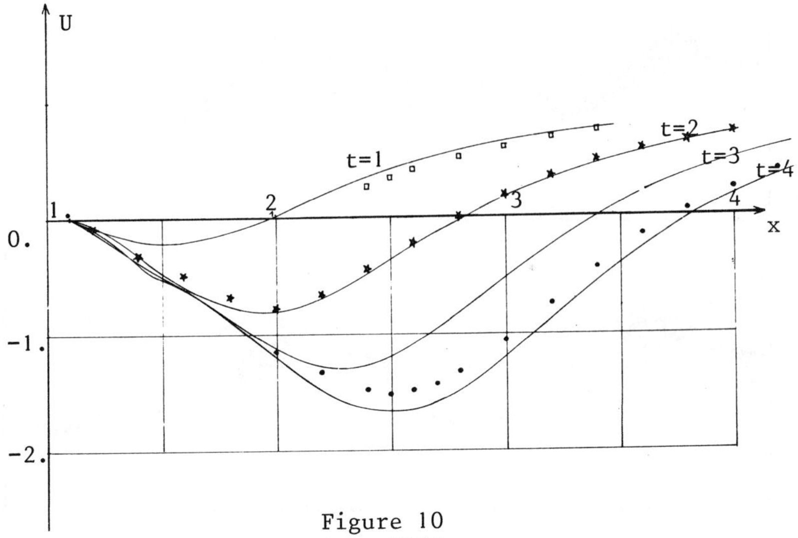

Figure 10
Re = 3000

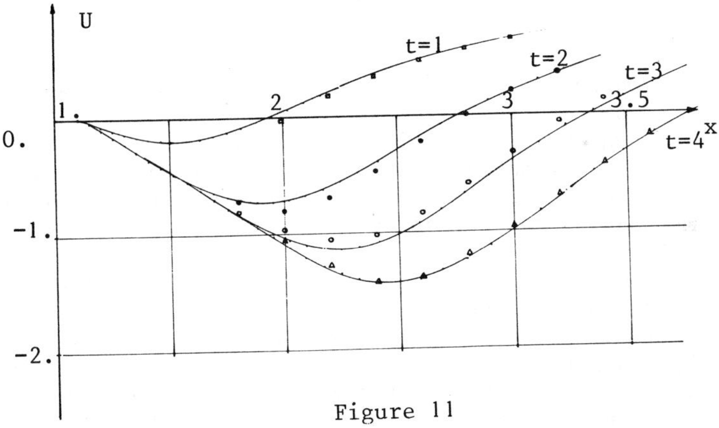

Figure 11
Re = 1000

Figures 10 and 11: Velocity U on the x-axis behind an elliptic cylinder . The angleof attack α is 90°

In both figures the full lines represent numerical values.

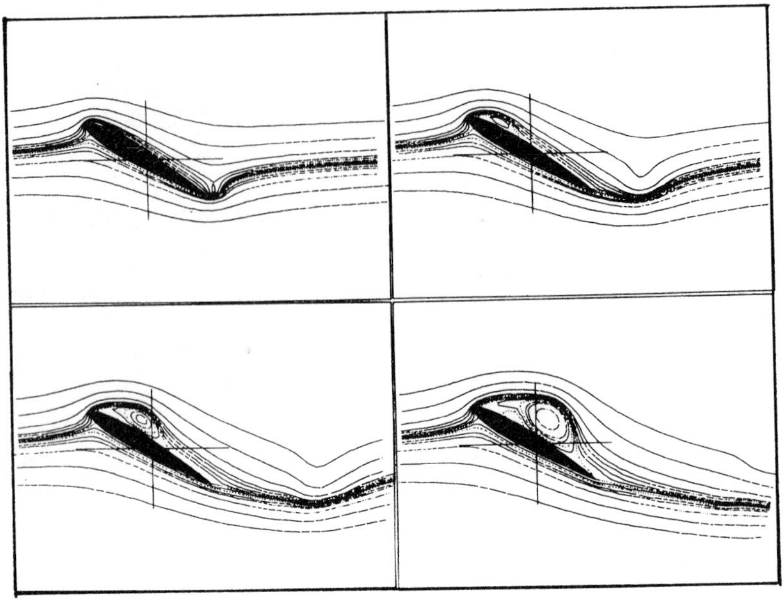

Figure 12
NACA 0012 Re = 1000 $\alpha = 30°$

CONCLUSION

The combination of both numerical and experimental techniques has proven to be efficient for the understanding of basic phenomena as separation. Different parameters as the Reynolds number,the angle of attack,the shape of the body have been tested. In the studied range of these parameters the numerical method was able to describe in an accurate way the physical phenomena.

Up to now,the experimental study can be achieved only for the first instants of the flow because of the size of the tank. From a numerical point of view,it would be easier to go further in time by pushing away the artificial outer boundary of the computationnal domain. Doing this requires,if no loss of accuracy is accepted,to work on grids with more than 61x61 nodes. this would be possible with vectorized computers (It must be noticed that all the computations which are reported here,were carried out on an UNIVAC 1110 with 64K words)

ACKNOWLEDGEMENTS

This work was supported by the Direction des Recherches et Etudes Techniques, French minister of Defense,under contract n° 80/586

REFERENCES

[1] BOUARD,R. and COUTANCEAU,M. -*The early stage of development of the wake behind an impulsively started cylinder for* $40<Re<10^4$ - J. Fluid Mech. ,Vol 101,n°3,pp 583-607,1980

[2] TA PHUOC,L. - *Numerical analysis of unsteady secondary vortices generated by an impulsively started circular cylinder* - J. FLUID MECH. ,Vol 100,n° 1,pp 111-128,1980

[3] COUTANCEAU,M. and BOUARD,R. - *Experimental determination of the main features of the viscous flow in the wake of a circular cylinder in uniform translation,part 2,unsteady flow*- J. Fluid Mech.,Vol 79,n°2,pp 257-272,1977

[4] DAUBE,O. and TA PHUOC,L. -*Etude numérique d'écoulements instationnaires de fluide visqueux incompressible autour de corps profilés par une méthode combinée* $O(h^2)$ *et* $O(h^4)$- J. de Mécanique,Vol 17,n°5,pp 651-678,1978

[5] TA PHUOC,L. and DAUBE,O. -*Higher order accurate numerical solution of unsteady viscous flow generated by a transversely oscillating elliptic cylinder*- Vortex flows,WAM of ASME, Ed. Swift,W.L.,Barna,P.S. and Dalton,C. ,ASME,1980

[6] DAUBE,O.,TA THUOC,L.,COUTANCEAU,M. and MONNET,P. -*Numerical and experimental study of the viscous flow generated by an impulsively started elliptic cylinder* - Computational methods and experimental measurements,Ed. Keramidas,G.A. and Brebbia,C.A. ,Springer-Verlag,1982

[7] HIRSH,R.S. - *Higher order accurate difference solutions of fluid mechanic problems by a compact differencing technique* J. Comp. Phys. ,Vol 19,n°1,pp 90-100,1975

SELECTION OF RELAXATION FACTORS FOR COMPUTING STEADY STATE TURBULENT FLOWS

A.W. Neuberger, A.U. Chatwani, H. Eickhoff, J. Koopman

Institut für Antriebstechnik der DFVLR, Köln, W.-Germany

SUMMARY

Modified versions of computer codes TEACH and CHAMPION, developed at Imperial College, are used to calculate 2-dimensional steady-state turbulent flows. These codes employ an underrelaxation procedure and it is known, that their convergence properties are very sensitive to the choice of the relaxation factors.

In the present paper a method, based on the long known analogy between the solution of the unsteady state equations and the iterative solution of the steady state equations, is reported which can be used for the solution of both unsteady parabolic equations and steady state elliptic equations. In the later case, the time step is a formal parameter for calculating the relaxation factors which may change from point to point and from iteration to iteration. The selection of the time step is based on a criterion which ensures an optimum convergence behavior. Initial test results show that the present procedure may lead to considerable saving of the computational time (up to a factor of 9).

1. INTRODUCTION

The analogy between the iterative solution of the elliptic equations and time dependent method for solving the parabolic equations is known for a long time [1]. It was discussed as early as 1950 by Frankel [2]. In two previous reports [3,4], a method was developed which can be used for solution of both unsteady state parabolic equations as well as steady state elliptic equations. Similar results have been reported by many other authors, e.g. Chorin [7] and McDonald and Briley [6]. The utility of the method reported here is in the fact that a class of readily available computer codes, e.g. [5,8,9,11], can be so modified by incorporation of small changes that on one hand an optimum convergence behavior is ensured and on the other hand the same code can be easily used for calculating the unsteady flow. Convergence and accuracy of the method depend strongly on the optimum selection of the time step. For its appropriate selection a criterion, which is also applicable to three dimensional transport processes, is derived. The results reported in the present paper for simple two dimensional flows show a saving of computational time up to one order in magnitude compared to the method employing constant relaxation factors.

2. METHOD

2.1. Discretisation of Equations

The method can be best illustrated by means of a one-dimensional transport equation:

$$\frac{\partial \phi}{\partial t} = \frac{\Gamma}{\rho} \frac{\partial^2 \phi}{\partial x^2} - u(t) \frac{\partial \phi}{\partial x} + \frac{s'}{\rho} \phi + \frac{s_o}{\rho} \quad (2.1.1)$$

This equation is valid under the assumption of constant density and diffusion-coefficient. The source density s is linearised in the form $s = s(\phi) = s'\phi + s_o$ where s' and s_o are also constants and $s' \leqq 0$.

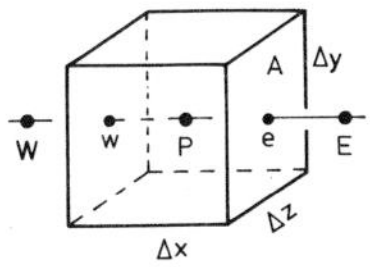

Fig. 1 Control Volume V_P

Using the standard nomenclature, which has also been used in the discription of many computer codes [5, 8, 9, 11], the integration of Eq. (2.1.1) over the control volume V_P, shown in Fig. 1, leads to

$$\rho V_P \left(\frac{\partial \phi}{\partial t}\right)_P = a_E \phi_E + a_W \phi_W + S_o - (a_P - S') \phi_P \quad (2.1.2)$$

$S_o = s_o \cdot V_P$ and $S' = s'V_P$ are the integrated source terms. The coefficients a_E, a_W and a_P consist of convection terms $C(t) = \rho u(t) A$ and diffusion terms $D = \Gamma A/\Delta x$

$$a_E = D \cdot \tilde{q}(Pe) \; ; \; a_W = C(t) + D \cdot \tilde{q}(Pe) \; ; \; a_P = a_E + a_W = C(t) + 2D \cdot \tilde{q}(Pe) \quad (2.1.3)$$

Here $\tilde{q}(Pe)$ is a function of the Péclet number

$$P_e = P_e(t) = \frac{C(t)}{D} = \frac{\rho u(t) \Delta x}{\Gamma} \quad (2.1.4)$$

Spalding [10] has shown on the basis of the one dimensional heat conduction equation that $\tilde{q}(Pe)$ can be approximated as follows:

$$\tilde{q}(Pe) = \begin{cases} 0 & \text{for } Pe > 2 \\ 1 - Pe/2 & \text{for } |Pe| \leqq 2 \\ - Pe & \text{for } Pe < -2 \end{cases} \quad (2.1.5)$$

Using the following notations

$$Q_P = \frac{a_P - S'}{\rho V_P} \; ; \quad \phi_{P,0} = \frac{a_E \phi_E - a_W \phi_W + S_o}{a_P - S'} \quad (2.1.6)$$

the discretised Eq. (2.1.2) can be expressed as

$$(2.1.7) \qquad \left(\frac{\partial\phi}{\partial t}\right)_P = Q_P\,(\phi_{P,0} - \phi_P)$$

2.2 Relaxation Method

If the steady state solution is required, then Eq. (2.1.7) is to be solved for every point P in the domain of interest by considering the steady state boundary conditions such that $(\partial\phi/\partial t)_P$ vanishes for every point. If n is the number of grid points, then n-2 equations of the form

$$(2.2.1) \qquad \phi_P - \phi_{P,0} = 0$$

with $\phi_{P,0}$ given by (2.1.6) are to be solved. Starting from an initial guess, which satisfies the boundary conditions but in general does not satisfy Eq. (2.2.1) over the whole field, one tries to calculate iteratively a sequence of solutions which converge to the final state. At the level of the ith iteration it is

$$(2.2.2) \qquad \rho V_P\, Q_P^i\,(\phi_P^i - \phi_{P,0}^i) = (Res)_P^i$$

$(Res)_P^i$ is called as the residual source [5]. A comparison of (2.2.2) and (2.1.7) shows, that the residual source of the relaxation method and $\partial\phi/\partial t$ for the time dependent method are related by

$$(2.2.3) \qquad (Res)_P^i = -\rho V_P \left(\frac{\partial\phi}{\partial t}\right)_P^i$$

The solution of Eq. (2.2.1) is obtained by a relaxation method. If ϕ_P^{i+1} and ϕ_P^i are the solutions at two successive iteration levels, the relaxation equation is given by

$$(2.2.4) \qquad \phi_P^{i+1} = \theta\phi_{P,0}^{i+1} + (1-\theta)\,\phi_P^i$$

Here θ is a positive real number. Values of $\theta < 1$ and $\theta > 1$ correspond to underrelaxation and overrelaxation prozesses. Selection of appropriate relaxation factors is important for the convergence and the rate of convergence. The magnitude of θ is determined through the initial iterations where the residual sources are very high and exhibit large changes. Lilley and Rhode [11], for example, attempt to increase the relaxation factors with the iteration number and modify them according to the residuals at each level. Such a process is purely empirical due to lack of a valid criterion for modification of the relaxation factors.

2.3 Time dependent method

A relatively simple, but general, time dependent method

can be formulated by discretising Eq. (2.1.7) at two successive time levels differing by time step Δt^i. Both equations are multiplied by weighting factors ψ and $1-\psi$, respectively, and added. The weighting factor ψ is a real number between 0 and 1. After linearisation of coefficient Q_P and development of the left hand side by a Taylor series one obtains

$$(2.3.1)\quad \frac{\phi_P^{i+1} - \phi_P^i}{\Delta t^i} = \psi Q_P^i (\phi_{P,0}^{i+1} - \phi_P^{i+1}) + (1-\psi) Q_P^i (\phi_{P,0}^i - \phi_P^i) + R_P^i$$

where the remaining term is

$$(2.3.2)\quad R_P^i = -\sum_{k=2}^{\infty} \frac{1}{k!} (k\psi-1) \cdot \overset{(k)}{\phi}{}_P^i \cdot (\Delta t^i)^{k-1}$$

$\overset{(k)}{\phi}{}_P^i$ is the kth derivative of function $\phi_P(t)$ at time t^i. Multiplication of Eq. (2.3.1) with Δt^i, introduction of quantities

$$(2.3.3)\quad \tau_P^i = Q_P^i \cdot \Delta t^i \qquad \text{and} \qquad \theta_P^i = \frac{\tau_P^i}{1 + \psi \tau_P^i}$$

and neglecting the remaining term R_P^i leads to

$$(2.3.4)\quad \phi_P^{i+1} = \theta_P^i \{\psi \phi_{P,0}^{i+1} + (1-\psi)\, \phi_{P,0}^i\} + (1-\theta_P^i)\, \phi_P^i$$

For $\psi = 1$, one obtains the purely implicit relaxation method

$$(2.3.5)\quad \phi_P^{i+1} = \theta_P^i\, \phi_{P,0}^{i+1} + (1-\theta_P^i)\, \phi_P^i$$

Comparison of Eq. (2.3.5) and (2.2.4) shows that the time dependent method and the relaxation method are formally related to each other. Instead of constant relaxation factor in Eq. (2.2.4), one obtains variable relaxation factors, which are related to time step Δt^i through (2.3.3), variing from point to point and iteration to iteration.

2.4 Ambivalence of the method

Both relaxation and time dependent methods for calculating transport phenomenon can be expressed in the form of the Eq. (2.3.4) with θ_P^i given by Eqs.(2.2.3). The method expressed by Eq. (2.3.4) is ambivalent. A relaxation method with a constant relaxation factor θ corresponds to a time dependent method with local time steps

$$\Delta t_P^i = \frac{\theta}{Q_P^i (1-\psi\theta)}$$

and a time dependent method with constant Δt^i corresponds to a relaxation method with local relaxation factors

$$\theta_P^i = \frac{Q_P^i \, \Delta t^i}{1 + \psi Q_P^i \, \Delta t^i}$$

This ambivalence is important for accelerating the convergence of existing computer codes, e.g. [5,8,11], used for the calculation of 2-D turbulent steady flows. It may also be used for extention to 3-D problems as well as for computing unsteady flows. The constant relaxation factor θ in Eq. (2.2.4) should only be replaced by the local value given by Eqs.(2.3.3). Introduction of an additional free parameter ψ and the extra term $(1-\psi) \cdot \phi_{P,0}^i$ gives the possibility of freely variing the difference method from a purely explicit to a purely implicit scheme. It is only necessary to specify a criterion for the selection of the appropriate time step.

2.5 Allowable Time Step

The discretised transport Eq. (2.1.7) describes the change in the variable ϕ at grid point P with time. In this respect, this equation can also been considered as an ordinary differential equation of first order

$$\frac{d\phi_P}{dt} = f(t, \phi_P) \tag{2.5.1}$$

where f is given by

$$f(t, \phi_P) = - Q_P(t) \cdot \phi_P + Q_P(t) \cdot \phi_{P,0}(t) \tag{2.5.2}$$

Development of derivative $d\phi_P/dt$, in a fashion described in Sec. 2.3, leads to

$$\phi_P^{i+1} = \phi_P^i + \{ \psi f_P^{i+1} + (1 - \psi) f_P^i \} \Delta t + R_P^i \cdot \Delta t \tag{2.5.3}$$

with the remaining term given by Eq. (2.3.2). The function $\phi_{P,0}(t)$ contains the influence of the neighbouring points E and W. If the values of ϕ at these points are incorrect, they lead to an error in the calculation of $f(t, \phi_P)$. One can look at the conditions to be fulfilled so that this error does not lead to an unlimited deviation of the approximate solution of Eq. (2.5.3), respectively Eq. (2.3.4), and that in the limit of $\Delta t \to 0$ the approximate solution approaches the exact solution of Eq.(2.5.3). Convergence conditions for the approximate solutions of this type are well known. They are based on the requirement that function $f(t, \phi_P)$ should be limited by Lipschitz condition. Detailed description is given for example in Collatz [12] and Zurmühl [13]. A particular discussion of the method, described by Eq. (2.3.4), is carried out in [14]. The obtained results can also be extended to 2- and 3-dimensional nonlinear transport equations, as these equations can also be brought into the form of Eqs. (2.5.1) and (2.5.3) or Eq. (2.3.4).
If $\kappa = K \cdot \Delta t$ defines the characteristic time step, where K is

the upper limit of the Lipschitz condition, then it is necessary to fulfill the following condition:

$$(2.5.4)\qquad \left|\frac{\partial f}{\partial \phi_P}\right|_P^i \Delta t < \kappa$$

Here κ depends not only on the parameter ψ, which defines the difference scheme, but also on the special problem. Experience has shown that κ is in the range $0.1 \leqq \kappa \leqq 1.0$.

The derivatives of function $f(t,\phi_P)$ are connected by

$$\frac{df}{dt} = \frac{\partial f}{\partial t} + f\frac{\partial f}{\partial \phi_P} \qquad \Longrightarrow \qquad \frac{\partial f}{\partial \phi_P} = \frac{1}{f}\left(\frac{df}{dt} - \frac{\partial f}{\partial t}\right)$$

Under the assumption that the total derivative of f is given by

$$\frac{df}{dt} = \frac{\partial f}{\partial t} + \vec{v}\nabla \cdot f \quad , \qquad \text{where } \vec{v} = (u\, v\, w)$$

the partial derivative $\partial f/\partial \phi_P$ is given by

$$(2.5.5)\qquad \frac{\partial f}{\partial \phi_P} = \frac{1}{f}\vec{v}\nabla \cdot f$$

The relationship between function $f(t,\phi_P)$ and the residual source $Res(t,\phi_P)$ can be expressed from Eq. (2.2.3) as

$$f(t,\phi_P) = -\frac{1}{\rho V_P} Res(t,\phi_P)$$

and leads to

$$(2.5.6)\qquad \frac{\partial f}{\partial \phi_P} = \frac{\vec{v}\nabla \cdot Res}{Res}$$

Equation (2.5.4) provides the allowed time step at point P as

$$(2.5.7)\qquad (\Delta t)_P^i = \frac{|Res|_P^i}{|\vec{v}\nabla \cdot Res|_P^i}$$

In case of axisymmetric flow, when using cylindrical coordinates

$$|\vec{v}\nabla \cdot Res|_P^i = \left|u\frac{\partial}{\partial x}(Res) + \frac{v}{r}\frac{\partial}{\partial r}(Res)\right|_P^i$$

and defining the following difference formulas

$$\Delta_x(Res)_P^i = (Res)_E^i - (Res)_W^i \quad ; \quad \Delta_r(r\,Res)_P^i = r_N(Res)_N^i - r_S(Res)_S^i$$

one obtains the necessary criterion for the allowable time step as

$$(2.5.8)\qquad (\Delta t)_P^i < 2\kappa\frac{|Res|_P^i}{|u_P^i|\,|\Delta_x(Res)|_P^i} \cdot \left\{1 + \frac{1}{r_P}\frac{|v_P^i|}{|u_P^i|}\frac{|\Delta_r(r\,Res)|_P^i}{|\Delta_x(Res)|_P^i}\frac{\Delta x}{\Delta r}\right\}^{-1}$$

Substitution of this time step in (2.3.3) gives the value of the local relaxation factor which ensures the optimum convergence behavior of the method.

3. RESULTS

3.1 Turbulent Flow Through a Sudden Enlargement of a Circular Pipe

The type of computed flow is represented in the upper right corner of Fig. 2. The cross section of the pipe is expanded in the ratio 1:2. The inlet Reynolds number is $Re = 10^5$. Downstreams of the enlargement a recirculating flow is generated.
The calculation is executed using the CHAMPION code of Pun and Spalding [5]. An aequidistant grid of 34 x 27 points is used. The turbulent shearstresses are introduced by the 2 Eqs. k,ε-model of Harlow and Nakayama [15]. Four coupled partial differential equations have to be solved simultaniously: the equations of the velocity components u,v, and the equations of the kinetic turbulence energy k, and its dissipation rate ε. The calculation of pressure is performed by means of the SIMPLE-procedure, described by Patankar and Spalding [16].
In this case, the ε-equation shows the slowest convergence rate. For this reason the logarithm of the residualsum

$$(RES)^i = \sum_P (Res)_P^i \,/(RES)^0 \qquad (3.1.1)$$

of the ε-equation has been introduced as a measure of the con-

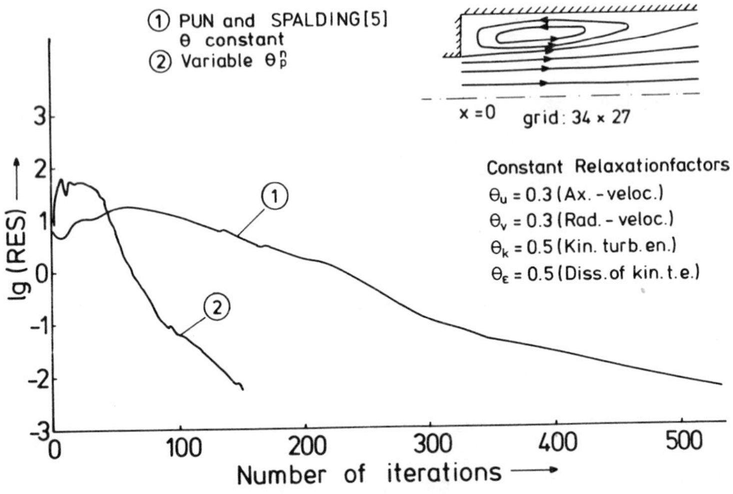

Fig. 2 Comparison of the convergence rate of the CHAMPION-code of Pun and Spalding [5], using constant relaxation factors ①, and the same code but using variable relaxation factors ②

vergence rate. It contains the sum of the residual sorces $(Res)_P^i$ according to Eq.(2.2.2) and is related on a reference value $(RES)^0$ which is constant for all iterations. The calculation is stopped if $(RES)^i < 0.5 \cdot 10^{-2}$ has been reached.
Curve 1 shows the decrease of the residualsum with the number of executed iterations, using the constant relaxation factors tabulated in Fig. 2. An increase of these factors would lead to instability. Curve 2 shows the result of the calculation using the variable relaxation factors given by Eqs.(2.3.3), and calculated with the local time step according to relation (2.5.8).

In the first case 532 iterations are needed to reach the required measure of convergence. On the other hand in the second case only 150 iterations are needed. Nearly in the same ratio the computational time is reduced, because all quantities which are needed to calculate the time step $(\Delta t)_P^i$ must also be calculated in case 1.

3.2 Swirling Turbulent Flow Through a Sudden Enlargement of a Circular Pipe

The original CHAMPION code has been extended with regard to the azimuthal velocity component w. This permits the computation of the swirling flow, represented in the upper right corner of Fig. 3. The ratio of the expansion and the inlet Reynolds number are the same as in the case described in section 3.1. The swirl number is D = 0.631. Apart from the recirculating flow in the corner, a second zone of recirculation is generated near the center line.

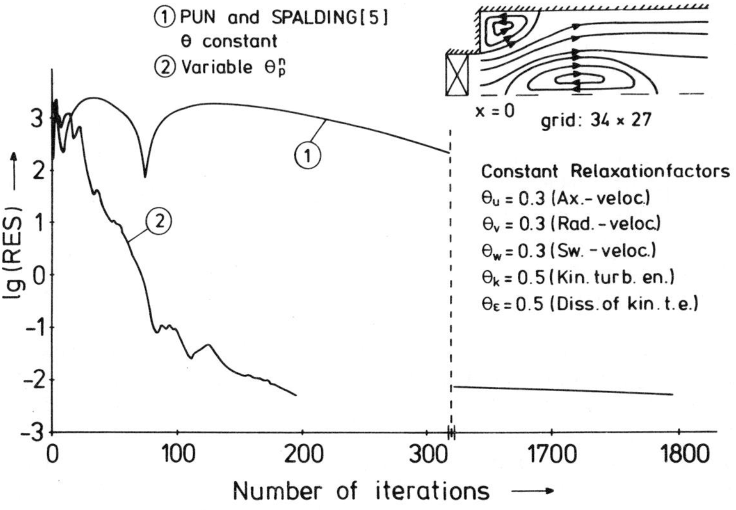

Fig. 3 Swirling Flow. ① Convergence rate using constant relaxation factors. ② Convergence rate using variable relaxation factors

Curve 1 is representing the decrease of the residualsum using the constant relaxation factors tabulated in Fig. 3. When 320 iterations have been executed, the residualsum is still larger than 10^2. It needs 1475 iterations in addition, to reach a value lower than $0.5 \cdot 10^{-2}$. All in all 1795 iterations are required. An increase of the constant relaxation factors would lead to instabilities.
Curve 2 shows the rate of convergence using the variable relaxation factors, defined by Eqs.(2.3.3). In this case it needs only 195 iterations to reach the required measure of convergence. The ratio of iterations needed in case 1 to the iterations needed in case 2 is 9.2:1 and the ratio of computational time is 8.1:1.

4. REFERENCES

1. ROACHE,P.J. - Computational Fluid Dynamics, Hermosa Publishers Albuquerque, 1972.

2. FRANKEL,S.P. - Convergence Rates of Iterative Treatments of Partial Differential Equations. Math. Tables and Other Aids to Computation, Vol. 4, pp. 65-75, 1950.

3. NEUBERGER,A.W. - Optimierung eines Differenzenverfahrens zur Berechnung elliptischer Strömungen. DFVLR - Inst. für Antriebstechnik, Rep. IB 325/3/81, Köln, 1981.

4. NEUBERGER,A.W. - Zur Entwicklung von Berechnungsverfahren für elliptische Drallströmungen. GAMM-Tagung, Ed. János Bolai Math. Gesellschaft, Budapest, 1982.

5. PUN,W.M. and SPALDING,D.B. - A General Computer Program for Two Dimensional Elliptic Flows. Imperial College, Mech. Eng. Deptm., Rep. No. HTS/76/2, London, 1976.

6. McDONALD,H. and BRILEY,W.R. - Computational Fluid Dynamic Aspects of Internal Flows. AIAA Computational Fluid Dynamics Conference, Pap. No. 79-1445, 1979.

7. CHORIN,A.J. - Numerical Solution of the Navier-Stokes Equations. Math. of Computation, Vol. 22, pp. 745-762, 1968.

8. GOSMAN,A.D. and IDERIAH,F.J.K. - TEACH-2E: A General Computer Program for Two Dimensional,Turbulent, Recirculating Flows. Imperial College, Mech. Eng. Dptm., London, 1976.

9. LILLEY,D.G. - Primitive Pressure-Velocity Code for the Computation of Strongly Swirling Flows. AIAA-Journal, Vol. 14, pp. 748-756, 1976.

10. SPALDING,D.B. - A Novel Difference Formulation of Differential Expressions Involving Both First and Second Derivatives. Int. J. Num. Meth. in Eng., Vol. 4, pp. 551-559, 1972.

11. LILLEY,D.G. and RHODE,D.L. - A Computer Code for Swirling Turbulent Axisymmetric Recirculating Flows in Practical Isothermal Combustor Geometries, NASA Rep. CR-3442, 1982.

12. COLLATZ,L. - Numerische Behandlung von Differentialgleichungen, Springer, Berlin, Göttingen, Heidelberg, 1955.

13. ZURMÜHL,R. - Praktische Mathematik für Ingenieure und Physiker, Springer, Berlin, Heidelberg, New York, 1965.

14. NEUBERGER,A.W. - Zum Konvergenzverhalten eines Differenzenverfahrens zur Lösung der nichtlinearen Transportgleichung, DFVLR - Inst. für Antriebstechnik, Rep. IB 325/1/83, Köln, 1983.

15. HARLOW,F.H. and NAKAYAMA,P.I. - Transport of turbulent energy deray rate. Los Alamos Sci. Lab., University of California, LA-3854, 1968.

16. PATANKAR,S.V. and SPALDING,D.B. - A Calculation Procedure for Heat, Mass and Momentum Transfer in Three-Dimensional Parabolic Flows. Int. J. Heat and Mass Transfer, Vol. 15, pp. 1787-1806, 1972.

NUMERICAL PREDICTIONS OF TURBULENT FLOW SEPARATION

E.E.KHALIL*

Faculty of Engineering, Cairo University,
CAIRO - EGYPT

SUMMARY

This paper is intended as a demonstration that the designer can use an alternative method,namely a numerical one, to experimental program in performing parametric study of practical problems such as separated flows in confined coaxial jet configuration with sudden enlargement and in straight walled diffusers. The present numerical method solves the governing equations,in finite difference form. Turbulence simulation is by way of a two-equation turbulence model. Computational results show the interesting effects of various design parameters on the subsequent flowfield development,separation and performance.

1. INTRODUCTION

1.1. Preamble

Although there are many forms of recirculating flows, nevertheless,a useful distinction may be made betweem those which stem from the presence of bluff bodies and rapid(sudden) changes in shape and the ones which stem from adverse pressure gradients and swirl action. The flow in straight walled diffusers,burner quarls with diverging angles fall within the category where flow separation occurs along the wall due to the effects of area changes and the consequent adverse pressure gradients. Flows of that type were investigated experimentally, references [1] to [3], where flow separation was observed.

The flows considered here can be represented,

* Associate Professor, Member ASME,ISCME & IMechE.

for practical situations,by the steady state form of the continuity and the Navier-Stokes equations.Boundary layer assumptions are appropriate upstream of separation zones,but they become increasingly inappropriate as separation is approached;thence the cross-stream pressure gradient and longitudinal diffusion are of increasing importance[4].

The solution of the time-averaged Navier-Stokes equations for axisymmetric and two-dimensional flows is becoming more feasible through mathematical modelling which is increasing steadily in realism and refinement. Models that represent the turbulent nature of the flow are essentially used to yield the governing equations soluable[4]. The present work has made use of a two-equation turbulence model of Launder et al[5]. The model computes the Reynolds and shear stresses from the local values of the kinetic energy of turbulence "k" and its dissipation rate "e" ; these are in turn obtained from the solution of their corresponding transport equations. Inlet and boundary conditions are necessarily specified with a modified wall function [4] to represent the wall shear stresses, particularly at inclined walls.

1.2. The Way Forward

In order to approach the goals of achieving the improved design and performance of diffusers and burner quarls,assessment of the effects of the flow and geometrical parameters on the flow properties is essentially needed. Among the important design features are the divergence angle,divergence length,inlet flow characteristics and flow exit conditions. The present work employs a numerical algorithm that solves the finite difference governing equations at a mesh of orthogonal gridlines superimposed on the flow domain. The level of agreement that can be obtained with the present procedure can be easily observed in the results obtained for the flows of references[1]and[3]. A parametric investigation of the effects of some design factors on the flow pattern is also included in this present work.

The second section of this paper describes the phenomenon of flow separation,while section 3 describes the governing equations and physical modelling, finite difference equations,boundary conditions and solution procedure. Section 4 describes the various parametric investigations of the factors affecting the diffusers and burner quarl performances. The paper ends with a brief account of the more important conclusions.

2. FLOW SEPARATION PHENOMENON

Flow separation occurs in many practical flow situations such as flow in boiler furnaces,flow behind bluff bodies and wake flows. The presence of recirculation in the flow can be due to the various geometrical configurations and can be also due to the momentum balance of the various streams in the flow. Swirl generators of appropriate strength result in flow reversal due to the presence of an adverse pressure gradient. The flow may separate from a surface at one location and reattaches at another downstream location,along the same surface;the location of such a separation zone,its size and strength are dependent on the momentum and mass ratios of the two coflowing streams, [4].

Flow separation downstream sudden enlargement geometries occurs at the location where the duct size (diameter) changes;the reattachment point occurs at a further location downstream the larger duct(tube). The location at which such reattachment occurs depends ,among many factors,on the enlargement ratio,Re, and the type of the flow. Separation of the flow may occur remote from the surface , such a separation occurs in the wake of strongly swirling coaxial jets. That kind of flow separation occurs due to the strong adverse pressure gradient along the centreline of the flow due to swirl action. The swirl action imparts a tangential momentum to the flow,hence the flow tends to rotate causing momentum and mass diffusion towards the confining walls away from the centreline.

Other types of flow separation occur,for example, in straight walled diffusers and stalled blading. In such flows, separation appears at the diffuser half angles greater than 7° degrees. The separation in two-dimensional diffusers is not summetrical and is quite frequently of the non-steady flow type with the interaction between the recirculation zones at large divergence angles.

The examples given here for separated flows are of direct relevence to practical flow situations where the flow recirculation plays an important role in the flow,mixing,heat transfer and combustion characteristics. The existance of separated flow in some diffusers has a harmful effect on the overall performance of the diffuser. This is due to the reduction in the pressure coefficient along the diffuser and the increase of the eddy currents in the diffuser. In other flow situations,the creation of recirculation zones is an asset,such as in furnace flows where central recirculation zones help to stabilize the flames and result in steady flow combustion.

Hence in such two widely different applications, the study of the flow separation is viewed from different angles,nevertheless the differential equations that govern the transport of mass,momentum and energy are the same. These equations are described in the following section.

3. GOVERNING EQUATIONS AND PHYSICAL MODELLING

3.1. General Equations

The flow in confined conduits whose configuration is two-dimensional or axisymmetric,is governed by the following conservation equations represented, [4], by the general tensor form, j=1,2,3 ;

$$\frac{\partial}{\partial x_j}(\bar{\rho}\,\bar{u}_j\,\bar{\emptyset} + \bar{\rho}\,\overline{u'_j\varphi}) = \frac{\partial}{\partial x_j}\,\Gamma_\emptyset\,\frac{\partial\bar{\emptyset}}{\partial x_j} + \bar{S}_\emptyset \quad \ldots\ldots\ldots\ldots(1)$$

This form represents the continuity and the three momentum equations as well as the equations governing any scalar entities. Putting $\bar{\emptyset} = 1$ and $\bar{S}_\emptyset$, the source term of any general dependent variable $\bar{\emptyset}^\emptyset$,to zero yields the continuity equation. Also replacing $\bar{\emptyset} = \bar{u}_i$ and i = 1,2 and 3 in the general form(1) results in the three momentum equations in the axial,radial and tangential directions. The term $\Gamma_\emptyset$ represents the laminar diffusion coefficient and the second term on the right hand side represents the laminar diffusion of the entity $\bar{\emptyset}$ in question. Due to the time averaging of the conservation equations,the second term on the left hand side of equation(1) emerges,it represents the turbulent diffusion of $\bar{\emptyset}$. The terms $\overline{u'_i\varphi}$ represent the shear and Reynolds stresses $\overline{u'_i u'_j}$ in the three momentum equations.

3.2. Physical Modelling

The analytical solution of the above equation is ruled out due to the complexity of the flow,ellipticity and non-linearity of the equations used. The present work utilized the finite difference technique with a finite difference grid superimposed on the flow field under consideration. The grid comprises two sets of intersecting orthogonal grid lines,at each intersection point ,nodes are formed where the conservation equations are numerically solved and the properties obtained. The solution of the equations is complemented by a set of inlet and boundary conditions.

To solve the conservation equations,in the finite difference form,the terms $\overline{u'_i\varphi}$ should be determined and specified to render the equations soluable. Alternative approaches of various degrees of complexity are available; the use of any one of these is governed by the physical flow situation and economical running of computers. These approaches are characterized by formulating a turbulence model to yield a closure to the momentum equations.

3.2.1 Turbulence Model

The turbulence model is a formulation of the physical phenomena of turbulence into a mathematical equation that represents the shear and Reynolds stresses in the flow. Such representation can be obtained through simple rigorous algebraic formulae that yield the stresses in terms of the mean velocity gradient and a viscosity[5]. Alternatively, a more complicated model[6] where the eddy viscosity is calculated as a function of the local turbulence properties can be used. A third approach[7], is to solve transport equations for the stresses themselves rather than to use the eddy viscosity concept. The validity of the Reynolds stresses model in recirculating flows is not well documented on top of which is added the complications involved in the solution of these equations and the assumptions embodied.

To describe the turbulent transport, the present work employed a two-equation k-e turbulence model, whereby the turbulent viscosity was calculated from:

$$\mu_t = C_\mu \bar{\rho} k^2/e \quad \ldots\ldots\ldots\ldots\ldots\ldots\ldots\ldots\ldots\ldots (2)$$

Where C_μ is a constant = 0.09 .
The local values of k and e were obtained from the solution of their corresponding transport equations expressed in the form (1) with $\emptyset$ replaced by k and e respectively. The constants of the model were given the values recommended in reference[5] and were kept unchanged throughout the computations. The Reynolds and shear stresses were obtained from the product of the turbulent viscosity and the local velocity gradients , [4]. The two-equation turbulence model has been widely applied to axisymmetric and two-dimensional flows with and without separation and its validity was assessed via comparison test cases against well documented experiments, [3], [4], [5]. The level of agreement varied according to the complexity of the flow situation and the adequacy of the assumptions imposed. Simple shear and boundary layer flows were very well predicted[4]&[5] while wake and swirling flows predictions exhibited some discrepancies, particularly so in the vicinity of central recirculation zones.

The proper treatment of the presence of a wall

results in better representation of the flow properties in the wall vicinity where the wall functions are used to link the wall values of shear stresses to the momentum flux at the neighbouring nodes. A modified log law of the wall that accounts for the effect of the pressure gradient[8]in the wall vicinity was used.

3.3 Finite Difference Equations and Boundary Conditions

The finite difference equations were formed by integrating the general equation(1) over a cell control volume for each Ø at each grid intersection point "p" in the flow field,and expressing the resulting terms through grid-point values. In axisymmetric and two-dimensional flows the grid node "p" is surrounded by four neighbouring nodes(N,S,E &W) in the four compas directions. The general finite difference equation is given as;

$$A_p^{\emptyset}\,\emptyset_p = \sum_n A_n^{\emptyset}\cdot\emptyset_n + S_o^{\emptyset} \quad \& \quad A_p^{\emptyset} = \sum_n A_n^{\emptyset} - S_p^{\emptyset} \quad \ldots\ldots\ldots(3)$$

where $\sum_n$ is the summation over N,S,E and W neighbours. and the term $A_n^{\emptyset}$ is a coefficient representing the net convection and diffusion at node n for a variable Ø. The terms $S_o^{\emptyset}$ and $S_p^{\emptyset}$ are the two components of the linearized volume integral of the source/sink term;

$$\int_V S_{\emptyset}\, dV = S_p^{\emptyset}\cdot\emptyset_p + S_o^{\emptyset} \quad \ldots\ldots\ldots\ldots\ldots\ldots\ldots\ldots\ldots\ldots\ldots(4)$$

The finite difference counterpart of equation(1) was then solved using the SIMPLE algorithm[4,5]described by Launder et al[5]which has been embodied in the TEACH-T computer code of Gosman et al[9],and was modified and tested,[10].

Insertion of correct boundary conditions requires certain amendments to the finite difference formulation for the near boundary nodes. The usual link (coupling coefficient) with the value of an external point was broken at a wall,for example in the case of a western wall boundary the normal W-P link was broken by setting $A_W^{\emptyset} = 0$. Also the source terms were modified at the boundaries and the values of the variables were set to the prescribed values. At the centrelines in symmetric flows, normally zero gradient assumptions were made in the form $\partial\emptyset / \partial r = 0$. The inlet flow conditions were prescribed to match the measurements or were assumed in accordance with the recommendations of reference[4]. For inclined walls, the usual link was broken for the resultant velocity component at the flow angles dictated by the velocity components in question.

3.4 Solution Procedure

The solution procedure was based on the SIMPLE algorithm. Due to the ellipticity of the flow, the specification of the inlet and boundary conditions was essential and a procedure of guessing and correction was needed. The above equations and the boundary conditions constituted a system of strongly coupled simultaneous algebraic non-linear equations. An initial guess of the flow field and pressures was made; then solving the momentum equations using the prevailing pressure field, yielded new velocity field. The continuity was then enforced, thus yielding the required adjustment of the velocities and pressures and so new values of the velocities and pressures were obtained. This iterative procedure was repeated until the governing equations were satisfied to within 0.01%. To speed up the convergence and to prevent instability and divergence, under-relaxation factors were employed. A typical grid comprised 25x25 grid nodes, required core size of 40 kilowords, and on the ICL 2956, the run time was nearly 25 minutes.

4. PREDICTIONS AND DISCUSSION

In this work, greater emphasis is placed on the application of this procedure to diffuser flows, where separation occurs, in order to assist in their design and development. General predictions are shown to illustrate the effects of the following parameters; Divergence angle, inlet velocity profile, divergence length and the exit boundary conditions.

However since the predictive capability of the present code has been demonstrated in a wide range of flow situations, references [3], [4], [5], [8] and [10], a couple of simple test cases are shown here to demonstrate the level of performance of the code. Figure 1a shows the centreline velocity variation in the coaxial confined jet flow of Baker et al [11]. The present predictions are in close agreement with the measured velocity distribution. Figure 1b shows the predicted and measured axial velocity profiles in a straight walled diffuser reported by Mobarak et al [12]. The shown agreement is fair and suggests that the numerical procedure can be employed in a parametric investigation to evaluate the relative influence of inlet flow and geometrical characteristics on the flow pattern.

The basic flow configuration considered here is that of a straight walled two-dimensional diffuser whose inlet width is W_1 and the divergence half angle is α. The divergent part of the diffuser is of a

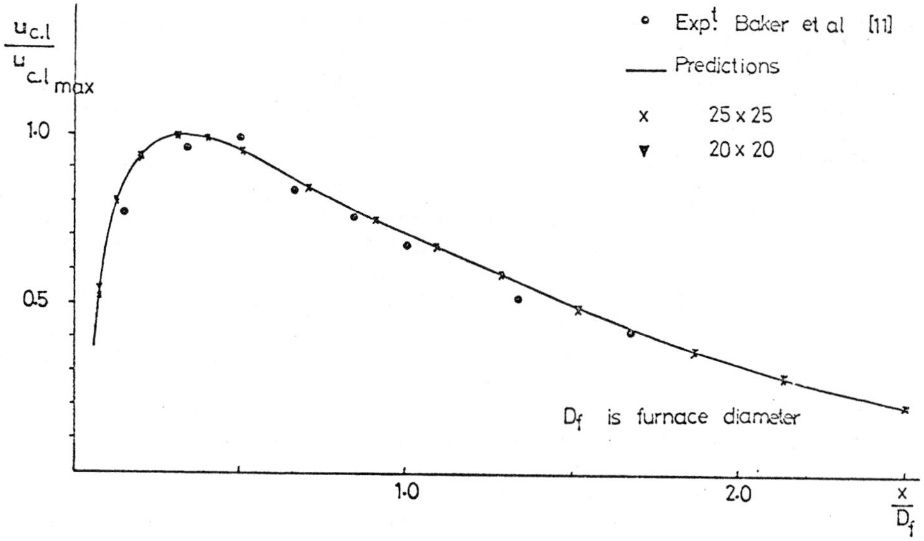

Fig.1a:Centreline Velocity Distribution in a Confined Jet

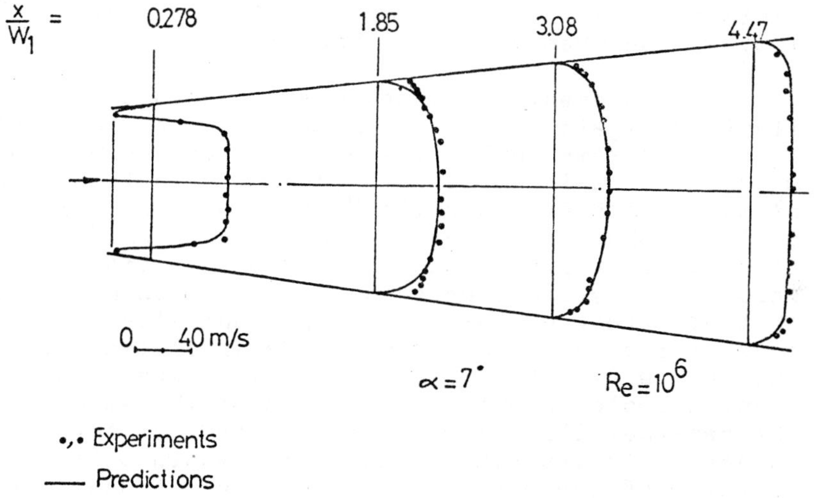

Fig.1b:Axial Velocity Profiles in a Straight Walled Diffuser

length l_d and is followed by a parallel wall duct. Figure 2 shows the normalized mean axial velocity profiles at different axial locations along the diffuser and for different values of α. A wall recirculation zone appeared at α greater than about 7°, and for α=12.93° the zone extended over a distance of $0.33W_1$. At larger values of α, the length of this zone increased to 1.45 W_1 when α= 22.88°. Such flow separation resulted in a decreased pressure coefficient and lower diffuser performance characteristics. Experimental data[1] indicated that at larger values of α,the two recirculation zones at the divergent walls become unsymmetrical and interact with each other,thus producing an unsteady and asymmetric phenomenon. The treatment of such problem requires a three-dimensional time dependent numerical procedure.

The effect of inlet velocity profile(inlet boundary layer thickness assumptions) on the flow pattern and separation was numerically investigated. Two extreme cases were considered,and corresponded to two profiles, generally,

$u = u_o(1 - 2y/W_1)^n$; n was put to zero and to 1/7

When n = 1/7 the wall separation occured at an earlier distance of x/W_1 = 0.17 for α= 22.88° rather than at x/W_1 = 0.34 when n= 0 as shown in figure 2.

The present procedure was arranged to numerically investigate the effect of the divergence length l_d for the same angle α. The consequent changes of W_d due to the increase in l_d,resulted in larger separation zones at the walls. The normalized velocity contours shown in figure 3 exhibited an unaffected core region represented by the u/u_o=0.5 lines; the zero velocity lines were changed. It should be noted that at larger values of l_d, the separation zones cover larger areas,and hence a deteriorated performance can be expected. Similar results were found at other Reynolds numbers and show that the strength of the recirculation zone was independent of the Reynolds number while its size slightly increased with Re. The location of the exit boundary of the diffuser may raise some doubts about the accuracy of the calculations near the exit and particularly in situations where recirculation zones were extended further downstream. The results of a parametric study conducted to investigate the influence of exit boundary conditions indicated that,for the flow situation of figure 2 (α =22.88°, Re =1.6x10^6),flow pattern was unaffected when exit was assumed at x/W_1 = 2.

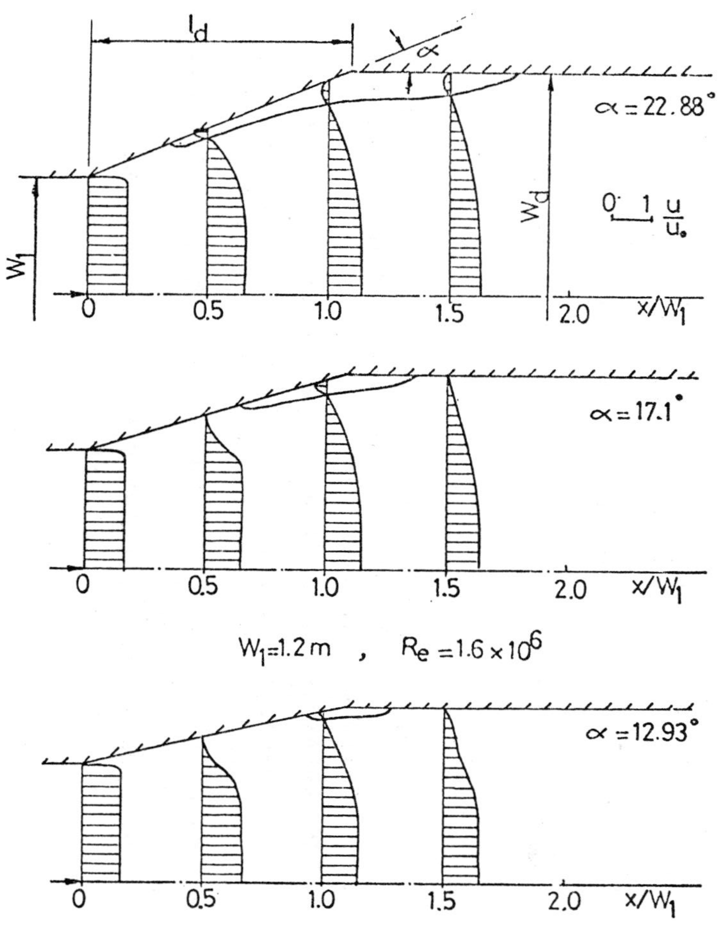

Fig.2 : Effect of Divergence Angle on Separation in Diffusers

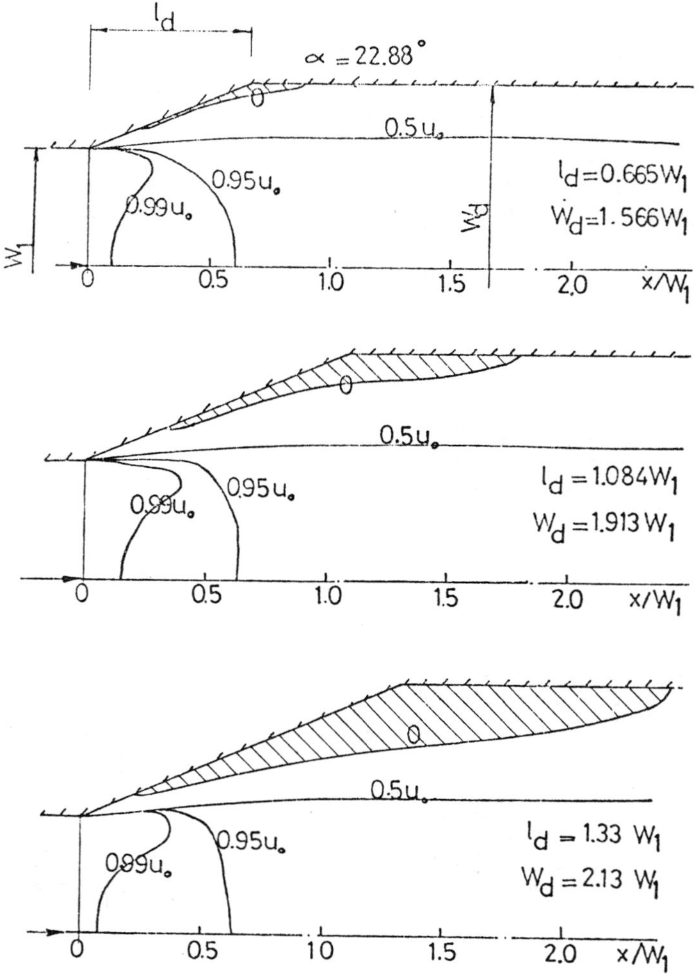

Fig.3 : Effect of Divergence Length on Separation in Diffusers

All the results were obtained via the use of a non-uniform grid system which was determined by way of systematic studies illustrating the effect of the grid system on the convergence rate and the accuracy of the final solution. It should be noted that a dense grid was used in regions of great activity ,i.e. high velocity gradients,high shear stresses,near the boundaries,etc. There was also a gradual grid expansion in the downstream direction;the solution grid independency was tested with grids of 20x20,25x20 and 30x30. The difference between the predictions obtained with the 30x30 grid and the 25x25 were less than 2% in velocities and 3% in k and e . However,considerable differences were shown for the coarser grid of 20x20, details of grid refinement investigations can be found in references[4]and[8].

5. CONCLUDING REMARKS

The present work discussed the development of a primitive-variable,finite difference technique together with its application to separated flows of the types found in confined coaxial jets,and separated flow in straight walled diffusers. The analysis and computer code involve a staggered grid system for the axial and radial velocities,underrelaxation,a two-equation turbulence model and a modified wall function. Finite difference predictions,which are now possible, show the interesting effects of design parameters on subsequent performance. Predictions of this type,which are trendwise indicative,allow some design concepts to be investigated more economically and faster than by the exclusive use of experimental means. Further development and application of the flow field calculations will provide a valuable supplementary technique for designers of diffusers,burner quarl and flame tubes.

REFERENCES

1. ASEM,M. Experimental Study of Flow in Straight Walled Diffuser with Different Angles, MSc.Thesis, Cairo University, 1980.
2. HASANJAK, J. , Private Communication , 1977.
3. KHALIL,E.E.,Flow and Combustion in Axisymmetric Furnaces, PhD Thesis, London University, 1977.
4. KHALIL,E.E.,Modelling of Furnaces and Combustors, Abacus Press,England,1982.
5. LAUNDER,B.E. and SPALDING,D.B.,Comput.Methods in Appl.Mech.and Engng,Vol.3,pp. 269-289,1974.

6. LAUNDER,B.E. and SPALDING,D.B., Mathematical Models of Turbulence, Academic Press, 1973.

7. LAUNDER,B.E.,REECE,G.J. and RODI,W., J.Fluid Mechanics, Vol. 68 ,pp. 537-566 ,1975.

8. KHALIL,E.E. Numerical Computations of Turbulent Flow Structure in a Cyclone Chamber, Proc.18th National ASME Heat Transfer Conference,ASME-79-HT -31 , 1979

9. GOSMAN,A.D.and PUN,W.M., Calculation of Recirculating Flows, Imperial College report,HTS/74/2,1974.

10. GOSMAN,A.D.,KHALIL,E.E. and WHITELAW,J.H., In Turbulent Shear Flows,Edt Durst et al, 1979,

11. BAKER,R.J.,HUTCHINSON,P.,KHALIL,E.E.and WHITELAW, J.H., Proc.15th Symposium(Int.) on Combustion, pp. 553-559 ,1974.

12. MOBARAK,A.,KHALIL,E.E.and ASEM,M.,Proc.3rd Mech. Power Engng.Conf.,Cairo,1980.

AN ANALYSIS OF FLOW OVER A BACKWARD-FACING STEP BY AN ASSUMED STRESS MIXED FINITE ELEMENT METHOD

C-T. Yang[1] and S. N. Atluri[2]

1. SUMMARY

In this paper, a detailed analysis of flow over a backward facing step is presented for two types of outlet boundary conditions, (i) traction free and (ii) zero vertical velocity and zero horizontal traction, and for Re numbers up to 229. The computational scheme that is used is a mixed finite element method based on "assumed-deviatoric-stress, -pressure, -velocity". In this method, the final finite element equations involve the nodal velocities and the constant term in the pressure field in each element as unknowns. The convective acceleration term is treated via a standard Galerkin scheme.

2. INTRODUCTION

Experimental results on this problem have been reported by Denham and Patrick [1], while finite difference calculations using "upwind" difference schemes were reported by Atkins, Maskell, and Patrick [2]. In [2] it was found that the upwind difference scheme underestimates both the length as well as intensity of the recirculation zone as compared to the conventionally stable central difference method. A finite element analysis of this problem was recently presented by Thomas, Morgan, and Taylor[3], who used a "velocity-pressure" type primitive variable mixed method. The convection term was treated in [3] using both the conventional Galerkin technique as well as the "upwinding" technique. The specific elements used in [3] were eight-noded elements with quadratic velocity fields and linear pressure fields.

---

[1] Doctoral Candidate
Georgia Institute of Technology

[2] Regents' Professor of Mechanics
Georgia Institute of Technology

Recently the authors have presented the formulation of a new mixed method for the analysis of incompressible viscous flows with significant convection at high Reynold's numbers [4, 5]. This mixed method is based on assumed deviatoric-stress, hydrostatic pressure, and velocity fields in each element, and thus is radically different from the velocity-pressure mixed formulation used, for instance, in [3]. The deviatoric stress and hydrostatic pressure are subject to the constraint of homogeneous momentum balance (sans inertial terms) in each element, a priori, and the convective acceleration term is treated by a standard Galerkin technique. In this method, which does not involve selective-reduced-integration, the final system of equations invokes only the constant term of the pressure field (which can otherwise be an arbitrary polynomial) in each element, in addition to nodal velocities. Thus for a given spatial finite element discretization, the present method, in general, results in fewer equations than in the "velocity-pressure" formulations. However, as compared to the reduced-integration penalty methods, the present method leads, for a given mesh, to a system of equations which is larger by the same number as that of the elements in the mesh.

The object of the present paper is to compare the performance of (i) the present method of satisfying incompressibility constraint through the "deviatoric stress-pressure-velocity" formulation and (ii) the present method of treating convective acceleration (with no "upwinding") with the previously presented approaches [2,3] for the same problem.

The rest of the contents of the present paper, in the order of their appearance, are: (a) a synopsis of the present approach, (b) the detailed results for the title problem, and (c) their comparison with previously published [1,2,3] results.

3. A SYNOPSIS OF THE FORMULATION

We consider the Navier-Stokes flow in a fluid domain V with spatial coordinates x_i. We consider the domain to be descritized into finite elements V_m, m=1,..M, each with a boundary ∂V_m. In general, $\partial V_m = \rho_m + S_{tm} + S_{vm}$ where ρ_m is the interelement boundary, and S_{tm} and S_{vm} are those segments of ∂V_m where tractions, and velocities, are respectively prescribed. We use the notation: ρ the fluid density, $\overline{F}_i$ are body forces excluding inertia, σ_{ij} the fluid stress, σ'_{ij} the deviatoric stress, p the hydrostatic pressure, v_i the velocity of a fluid particle, $\overline{T}_i$ are prescribed tractions, $\overline{v}_i$ are prescribed velocities, and $( ),_j$ denotes partial differentiation w·r·t cartesian x_i. Then, in a finite element assembly, the following equations must be satisfied.

$$\left(\begin{matrix}\text{interelement velocity}\\ \text{compatibility}\end{matrix}\right) \qquad v_i^+ = v_i^- \quad \text{at } \rho_m \tag{1}$$

(interelement traction reciprocity) $(\sigma_{ij}n_i)^+ + (\sigma_{ij}n_j)^- = 0$ at ρ_m

or $[(\sigma'_{ij} - p\delta_{ij})n_j]^+ + [(\sigma'_{ij} - p\delta_{ij})n_j]^- = 0$ at ρ_m (2)

(incompressibility) $v_{i,i} = 0$ in V_m (3)

(momentum balance) $\sigma_{ij,j} + \rho\overline{F}_i = \rho\left(\frac{\partial v_i}{\partial t} + \rho v_{i,j}v_j\right)$ in V_m (4a)

$$\sigma_{ij} = \sigma_{ji} \quad \text{in } V_m \tag{4b}$$

(compatibility) $V_{ij} = V_{ji} = v_{(i,j)} \equiv \frac{1}{2}(v_{i,j} + v_{j,i})$ in V_m (5)

(constitutive law) $\sigma_{ij} = \partial A/\partial V_{ij}$ in V_m (6a)

where $A[V_{ij},p] = - pV_{RR} + \mu V_{ij}V_{ij}$ (6b)

such that $\sigma_{ij} = - p\delta_{ij} + 2\mu V_{ij}$ (6c)

(traction b·c) $\sigma_{ij}n_j = T_i = \overline{T}_i$ at S_{tm} (6d)

(velocity b·c) $v_i = \overline{v}_i$ at S_{vm} (6e)

In (1) the superscripts (+) and (-) refer, arbitrarily, to the right and left-hand side vicinities of the interelement boundary ρ_m, and n_j are components of a unit outward normal to S. We define, from (6), the deviatoric (or viscous) fluid stress, σ'_{ij} to be:

$$\sigma'_{ij} = 2\mu V_{ij} \tag{7}$$

such that $\sigma_{ij} = \sigma'_{ij} - p\delta_{ij}$ (8)

In Eq. (6a) the potential A may be referred to as the stress-working density. We may consider the contact transformation,

$$-B(\sigma_{km}) = A(p,V_{km}) - \sigma_{km}V_{km} \tag{9}$$

such that $V_{ij} = \frac{\partial B}{\partial \sigma_{ij}}$ (10)

Using (6c), (7), and (8) in (9), we find that

$$B(\sigma_{km}) = \frac{1}{4\mu}\sigma'_{km}\sigma'_{km} \tag{11}$$

or $$V_{ij} = \frac{1}{2\mu}\sigma'_{ij} \tag{12}$$

If σ'_{ij}, p, and v_i are treated as the basic unknowns, the field equations can now be expressed in terms of these, as:

(interelement compatibility) $$v_i^+ = v_i^- \quad \text{at } \rho_m \tag{13}$$

(interelement traction reciprocity) $$[(\sigma'_{ij} - p\delta_{ij})n_j]^+ + [(\sigma'_{ij} - p\delta_{ij})n_j]^- = 0 \quad \text{at } \rho_m \tag{14}$$

(momentum balance) $$\sigma'_{ij,j} - p_{,i} + \rho\overline{F}_i = \rho\left(\frac{\partial v_i}{\partial t} + v_{i,j}v_j\right) \tag{15}$$

(incompressibility) $$v_{i,i} = 0 \tag{16}$$

(compatibility) $$\partial\frac{1}{2\mu}\sigma'_{ij} = \tfrac{1}{2}(v_{i,j} + v_{j,i}) \equiv v_{(i,j)} \tag{17}$$

(traction b·c) $$\sigma_{ij}n_j = \overline{T}_i \quad \text{at } S_{tm} \tag{18}$$

(velocity b·c) $$v_i = \overline{v}_i \quad \text{at } S_{vm} \tag{19}$$

In the present development, we assume a C^0 continuous velocity field v_i at ∂V_m such that Eqs. (13) and (19) are satisfied, a priori. We will also assume fields σ'_{ij} and p, to start with, such that they are arbitrary within each element (i.e. they are not subject to either interelement constraint as in Eq. (14) or any intra-element constraint. We shall generate a weak or Galekin form of weighted residual equation governing Eqs. (14-18).

To this end, we choose the weighting functions: v_i^* at ρ_m in Eq. (14); v_i^* in V_m for Eq. (15); p^* for Eq. (16); $\sigma^{*\prime}_{ij}$ for Eq. (17); and v_i^* for Eq. (18). Thus the weighted residual equations are:

$$\int_{\rho_m}\left\{[(\sigma'_{ij} - p\delta_{ij})n_j]^+ + [(\sigma'_{ij} - p\delta_{ij})n_j]^-\right\} v_i^* d\partial V = 0 \tag{20}$$

$$\int_{V_m}\left[\sigma'_{ij,j} - p_{,i} + \rho\overline{F}_i - \rho\left(\frac{\partial v_i}{\partial t} + v_{i,j}v_j\right)\right]v_i^* dV = 0 \tag{21}$$

$$\int_{V_m}(v_{i,i})p^* dV = 0 \tag{22}$$

$$\int_{V_m}\left[\frac{1}{2\mu}\sigma'_{ij} - v_{(i,j)}\right]\sigma^{*\prime}_{ij}\, dV = 0 \tag{23}$$

$$\int_{S_{tm}}(\sigma_{ij}n_j - \bar{T}_i)v_i^* = 0 \tag{24}$$

To arrive at global weighted residual equations, we sum Eqs. (20-24) over the elements in the system. In doing so, we now introduce the following restrictions on the test (weight) functions v_i^*.

$$v_i^* = 0 \qquad \text{at } S_{vm} \tag{25}$$

$$(v_i^*)^+ = (v_i^*)^- \qquad \text{at } \rho_m \tag{26}$$

Under restrictions (25) and (26), the following identity is seen to be valid:

$$\sum_m \left\langle \int_{\rho_m}\left\{[(\sigma'_{ij} - p\delta_{ij})n_j]^+ + [(\sigma'_{ij} - p\delta_{ij})n_j]^-\right\}v_i^*\, d\partial V \right.$$
$$\left. + \int_{S_{tm}}\sigma_{ij}n_jv_i^*d\partial V\right\rangle \equiv \sum_m\int_{\partial V_m}[(\sigma'_{ij} - p\delta_{ij})n_j]v_i^*d\partial V \tag{27}$$

The above is true because, when the summation is carried over each element boundary ∂V_m as on the right-hand side of Eq.(27), the interelement boundary ρ_m occurs twice, by definition, since ρ_m is common to two adjoining elements. Recognizing (27) we write the globally summed weighted residual (or weak) form as:

$$\sum_{elem,m=1}^{M}\left\{\int_{\partial V_m}[(\sigma'_{ij} - p\delta_{ij})n_j]v_i^*d\partial V\right.$$
$$- \int_{V_m}\left[\sigma'_{ij,j} - p_{,i} + \rho\bar{F}_i - \rho\left(\frac{\partial v_i}{\partial t} + v_{i,j}v_j\right)\right]v_i^*dV$$
$$+ \int_{V_m}(v_{i,i})p^*dV + \int_{V_m}\left[\frac{1}{2\mu}\sigma'_{ij} - v_{(i,j)}\right]\sigma^{*\prime}_{ij}dV$$
$$\left. - \int_{S_{tm}}\bar{T}_iv_i^*\right\} = 0 \tag{28}$$

The negative sign on the second integral in Eq. (28) is motivated by the consistentcy of the form of virtual work balance relation, when v_i^* is also C^0 continuous in V_m [see Ref. 4 for an entirely different, but equivalent, explanation of the method].

To bring further simplicity, and a certain novelty to be immediately evident, we subject the trial functions (σ'_{ij}, p) and the test functions $(\sigma^{*'}_{ij}, p^*)$ in each element to the constraints.

$$\sigma'_{ij,j} - p_{,i} = 0 \quad \text{in } V_m \tag{29}$$

and $$\sigma^{*'}_{ij,j} - p^*_{,i} = 0 \quad \text{in } V_m \tag{30}$$

However, both the sets (σ'_{ij}, p) and $(\sigma^{*'}_{ij}, p^*)$ are not subject to any interelement constraints, such as of the traction reciprocity type.

A consequence of the assumptions in (29) and (30) is that the fields p and p^* in each V_m are determined, to within an arbitrary constant in each case, in terms of the fields σ'_{ij} and $\sigma^{*'}_{ij}$ respectively. Thus only the constant term in the pressure field in each element remains an independent unknown. The higher order, and arbitrary, variation in the pressure field is dependent on the deviatoric stress field.

Now we introduce the assumptions for trial and test functions. First the trial function v_i is assumed (so as to be C^o continuous) in each element as,

$$v_i = B_{ir} q_r \tag{31}$$

where q_r are nodal velocities, and B_{ir} are appropriate interpolants.

To keep the formulation within the framework of conventional Galerkin approach, we assume the test function v_i^* to be similar to that in Eq. (31) i.e.,

$$v_i^* = B_{ir} q^*_r \tag{32}$$

Again, within the conventional Galerkin approach, we assume trial functions (σ'_{ij}, p) and test functions $(\sigma^{*'}_{ij}, p^*)$, to obey the constraints (29) and (30), respectively, as:

$$\sigma'_{ij} = \sigma'_{ijn} \beta_n \quad ; \quad p = \alpha_1 + A_m \alpha_m \tag{33}$$

$$\sigma^{*'}_{ij} = \sigma'_{ijn} \beta_n^* \quad ; \quad p^* = \alpha_1^* + A_m \alpha_m^* \tag{34}$$

Note that the interpolants σ'_{ijn} and A_m are identical in Eqs. (33) and (34) respectively. Constraints (29) and (30) imply that

$$A_m \alpha_m = D_n \beta_n \quad ; \tag{35}$$

$$A_m \alpha_m^* = D_n \beta_n^* \tag{36}$$

where A_m and D_n are functions of spatial coordinates. For explicit representations of relations of type (35) and (36) for some two-dimensional elements, see, for instance, Ref. [4]. Use of Eqs. (31-34) in (28) results in the equation:

$$\sum_{elem,m=1}^{M}\left\{\int_{\partial V_m}[\sigma'_{ijn}\beta_n - (\alpha_1 + D_n\beta_n)\delta_{ij}]n_j B_{ir} q_r^* \, d\partial V\right.$$
$$- \int_{V_m}\left[\rho\bar{F}_i - \rho\left(B_{ik}\frac{\partial q_k}{\partial t} + B_{ir,j}B_{js}q_r q_s\right)\right]B_{i\ell}q_\ell^* \, dV$$
$$+ \int_{V_m} B_{ir,i}q_r(\alpha_1{}^* + D_n\beta_n{}^*)\, dV$$
$$+ \int_{V_m}\left[\frac{1}{2\mu}\sigma'_{ijn}\beta_n - \tfrac{1}{2}(B_{ir,j} + B_{jr,i})q_r\right]\sigma'_{ijk}\beta_k^* \, dV$$
$$\left.- \int_{S_{tm}} \bar{T}_i B_{ir} q_r^*\right\} = 0 \tag{37}$$

Now, since β_k^* are not subject to any nodal connectivity, we obtain from Eq. (37) that

$$-\int_{V_m}(B_{ir,i}D_k)q_r \, dV + \int_{V_m}\tfrac{1}{2}(B_{ir,j} + B_{jr,i})\sigma'_{ijk}q_r \, dV$$
$$= \int_{V_m}\frac{1}{2\mu}\sigma'_{ijn}\sigma'_{ijk}\beta_n \, dV \quad \text{in each } V_m \tag{38}$$

$$\text{or } G_{kr}q_r = H_{kn}\beta_n \qquad \text{in each } V_m \tag{39}$$

with apparent definitions for matrices G_{kr} and H_{kn} (which is symmetric). Further, it can be immediately seen that H_{kn} is positive definite. Thus,

$$\beta_n = H_{kn}^{-1}G_{kr}q_r \tag{40}$$

Note also, that because of constraints (33) and (34), we have:

$$\int_{V_m} v_{i,i}^* p - \int_{V_m} v_{(i,j)}^* \sigma'_{ij} \, dV$$
$$= -\int_{V_m} v_{(i,j)}^*[\sigma'_{ij} - p\delta_{ij}]\, dV$$
$$= -\int_{\partial V_m}(\sigma'_{ij} - p\delta_{ij})n_j v_i^* \, d\partial V \tag{41}$$

Thus,

$$- \int_{\partial V_m} (\sigma'_{ij} - p\delta_{ij}) n_j v_i^* = - \int_{\partial V_m} [\sigma'_{ijn}\beta_n - (\alpha_1 + D_n\beta_n)\delta_{ij}] n_j B_{ir} q_r^*$$

$$= - \int_{V_m} \tfrac{1}{2}(B_{ir,j} + B_{jr,i}) [\sigma'_{ijk}\beta_k - \delta_{ij}(\alpha_1 + D_k B_k)] q_r^*$$

$$= - \beta_k G_{kr} q_r^* + \alpha_1 S_r q_r^* = - q_S G_{Sm} H_{mk}^{-1} G_{kr} q_r^* + \alpha_1 S_r q_r^* \tag{42}$$

$$\equiv - q_S R_{Sr} q_r^* + \alpha_1 S_r q_r^* \tag{43}$$

Using (38) and (43) in (37), we obtain:

$$\sum_{\text{elem},m=1}^{M} \left[R_{S\ell} q_S q_\ell^* - \alpha_1 S_\ell q_\ell^* - Q_\ell q_\ell^* + m_{k\ell} \frac{\partial q_\ell}{\partial t} q_\ell^* \right.$$

$$\left. + c_{rS\ell} q_r q_S q_\ell^* \right] = 0 \tag{44}$$

where $\quad M_{r\ell} = \int_{V_m} B_{ik} B_{i\ell} \, dV$

$$C_{rS\ell} = \int_{V_m} B_{ir,j} B_{i\ell} B_{jS} \, dV \tag{45a,b}$$

Noting the nodal connectivity of q_ℓ^*, we finally obtain the global system of equations:

$$\underset{\sim}{M} \frac{\partial \underline{q}}{\partial t} + [\underset{\sim}{K}_{\text{linear}} + \underset{\sim}{K}_{\text{conv}}(\underline{q})]\underline{q} - \underset{\sim}{S}\underline{\alpha} = \underline{Q}^* \tag{46a}$$

$$- \underset{\sim}{S}^T \underline{q}^* = 0 \tag{46b}$$

where the usual "assembly" is understood.

Several remarks concerning the distinguishing features of the present approach are in order.

Remark 1: For a given finite element mesh, the system of equations (46a,b) is larger than that in the reduced-integration-penalty methods by only the number of elements in the mesh, N.

Remark 2: Here, the pressure field in each element can be an arbitrary polynomial. However, it is only the constant term of this polynomial that remains as an unknown to be solved for in the global finite element equations (See Eq. 46a). In contrast in the standard "velocity-pressure" mixed formulation, all the parameters in the polynomial pressure field remain as unknowns in the global system. Thus, the present method, in general, results in a smaller system of equations than the standard "velocity-pressure" formulation.

Remark 3: As can be seen from the development presented above, the present method, even though labelled as a mixed method, is radically different from the standard "velocity-pressure" mixed method, as in [3].

Remark 4: No reduced or selective-reduced integrations are used here; all integrations are performed with necessary order quadrature rules.

Remark 5: No "upwinding" technique is used here; the convective acceleration is still treated by the standard Galerkin technique.

We now present and discuss the presently obtained numerical results for the title problem.

4. FLOW OVER A BACKWARD FACING STEP

The problem definition is given in Fig. 1, which shows a backward facing step of expansion ratio (2:3). At the top as well as bottom walls, the boundary conditions are $v_x = v_y = 0$. At the inlet, the boundary conditions are $v_y = 0$ and $v_x = 4_y(1-y)$ with the origin of y as shown in Fig. 1. Two different types of outlet conditions are used: (i) traction-free conditions $t_x = t_y = 0$; (ii) $v_y = 0$ and $t_x = 0$.

Experimental work on this problem has been reported in [1]; "upwind" finite difference calculations were reported in [2]; and "velocity-pressure" mixed finite element results with and without upwinding, using eight-noded elements with quadratic velocity and linear pressure, were reported in [3].

The results for velocities and stream-functions presented in [3] correspond to a finite element mesh (Fig. 8 of [3]) with: (i) the inlet length (prior to expansion) of 3 units, (ii) the outlet length (after expansion) of 22 units, and (iii) total equations (velocity and pressure) of about 520. The authors of [3] state that for this mesh: (i) they could not obtain convergence for Re≥100 when the outlet conditions were $t_x = 0$ and $v_y = 0$; (ii) convergence for Re≥100 was obtained when the outlet condition was changed to $t_x = 0$ and $t_y = 0$, and, in addition, a condition that $v_y = 0$ was imposed at the first node down stream of the step and at the same level (See Fig. 2 of [3]); (iii) upwinding was necessary to obtain reasonably converged results at Re=125 even with outlet conditions as in (ii); and (iv) even "upwinding" resulted in a failure of convergence at Re=125 when the outlet conditions were $t_x = 0$, $v_y = 0$. However, even though the results in [3] did not converge for Re=125, $t_x = 0$, $v_y = 0$, it appears that "upwinding" "stabilized and smoothed" out the results. It should also be remarked that the results in [3] failed to converge for Re≥73 when (i) the inlet length was 6.5 units and (ii) the outlet length was 12 units.

In contrast, in the present set of results, the following apply: (i) the inlet length has 2 units, (ii) the outlet length is only 8 units, (iii) the inlet conditions are $v_x =$

$4y(1-y)$, and $v_y = 0$, (iv) the results are presented for both sets of outlet conditions ($t_x = 0$, $v_y = 0$) as well as ($t_x = t_y = 0$, and Re values of 73, 125, 191, and 229, respectively.

Based on the studies of Leone and Gresho [6] (See Fig. 5 therein), it may be surmised that the inlet length has no noticeable effect on the obtained results. However, the results of [6] do indicate that the longer the outlet length, the smoother and more stable the solutions are. Further, the numerical experiments of [6] tend to suggest that traction-free conditions at the outlet result in better, smoother, and more stable solutions. Even though the studies in [6] are for a flow in a channel past a rectangular obstacle, the effect of outlet conditions can be seen to be similar to the present problem of flow past a backward facing step. Thus, the above comments should be kept in mind while comparing the present results with those of [3].

The finite element mesh used in the present computations is shown in Fig. 2. It consists of 172 four-noded elements, 201 nodes, and total number of (2 x 201 + 172) = 574 equations prior to the imposition of boundary conditions. Fig. 3 shows the velocity vector plots as well as the stream lines for R=73 and the outlet condition of $t_x = t_y = 0$. Likewise, Figs. 4, 5, 6, and 7 show the velocity vector plots as well as stream lines for Re=125, 155, 191, and 229, respectively, and for b·c of $t_x = t_y = 0$. Note that in all these computations no "upwinding" was used.

More interestingly, Figs. 8, 9, and 10 show, respectively, the velocity vector plots and stream line countours for Re=125, 155, and 191 and for outlet conditions $t_x = v_y = 0$. Note that these results are rather smooth, even for the present outlet of 8 units.

Finally, in Fig. 11, the presently computed length of the recirculation zones, as well as those from [2,3], are compared with the experimental results of [1]. It should be borne in mind, however, that the inlet condition is $v_x = 4y(1-y)$ in the present calculation as well as in [2,3], but this may be somewhat different from the experimental condition. Also, [2,3] use upwinding schemes, while the present method uses a conventional Galerkin technique. From Fig. 11 it is evident that the present results are in best agreement with the experimental measurements.

5. CLOSURE

A new assumed "deviatoric stress-velocity-pressure" mixed method is presented and rationalized in terms of a Galerkin weighted residual formalism. Numerical results, based on the present method, are presented for flow over a backward facing step. The present results, which use no upwinding, have been demonstrated to be much more superior compared to those based on a "velocity-pressure" mixed formulation with upwinding.

6. ACKNOWLEDGEMENTS

The results herein were obtained during the course of investigations supported by the Air Force Office of Scientific Research under a grant to Georgia Tech. The authors gratefully acknowledge this support, as well as the encouragement of Dr. A. Amos. It is a pleasure to acknowledge with thanks the skillful assistance of Ms. J. Webb in the preparation of this manuscript.

7. REFERENCES

1. DENHAM, M.K. and PATRICK, M.A. - Laminar Flow Over a Downstream-Facing Step in a Two-Dimensional Flow Channel. Trans. Inst. Chem. Engineers (Britain) Vol. 52, pp. 361-367, 1974.

2. ATKINS, D.J., MASKELL, S.J., and PATRICK, M.A. - Numerical Prediction of Separated Flows. International Journal for Numerical Methods in Engineering, Vol. 15, pp. 129-144, 1980.

3. THOMAS, C.E., MORGAN, K., and TAYLOR, C. - A Finite Element Analysis of Flow Over a Backward Facing Step. Computers & Fluids, Vol. 9, pp. 265-278, 1981.

4. YANG, C-T. and ATLURI, S.N. - An "Assumed Deviatoric Stress-Pressure-Velocity" Mixed Finite Element Method for Unsteady, Convective, Incompressible Flow: Part I - Theoretical Development. International Journal for Numerical Methods in Fluids, 1983 (in press).

5. YANG, C-T. and ATLURI, S.N. - An "Assumed Deviatoric Stress-Pressure-Velocity Mixed Finite Element Method for Unsteady, Convective, Incompressible Flow: Part II - Computational Studies. International Journal for Numerical Methods in Fluids, 1983 (in press).

6. LEONE, J.M., JR. and GRESHO, P.M. - Finite Element Simulations of Steady, Two-Dimensional, Viscous Incompressible Flow Over a Step. Journal of Computational Physics, Vol. 32, pp. 167-191, 1981.

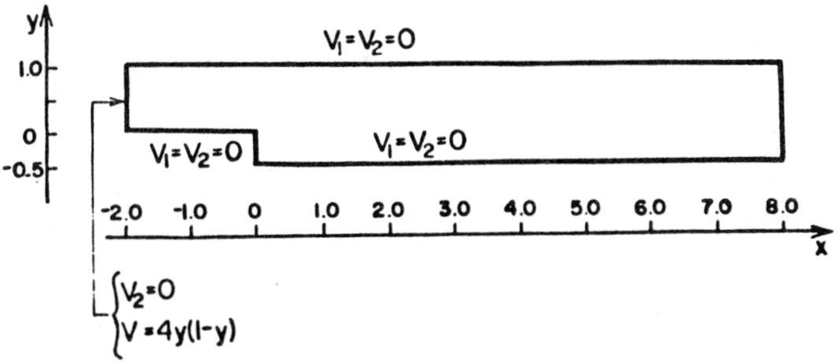

B.C.

(i) Parabolic inlet profile at x = -2.0
i.e. $V_1 = 4y(1-y)$, $V_2 = 0$

(ii) Two different outlet boundary conditions at x = 8.0
(1) traction free (2) $V_2 = 0$

Figure 1 . Problem Definition for Flow over a Backward Facing Step

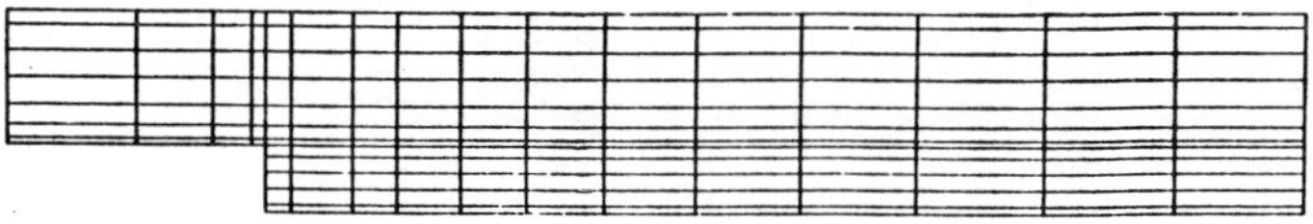

No. of Elements = 172 No. of Points = 201

Figure 2 . Finite Element Mesh for Flow over a Backward Facing Step

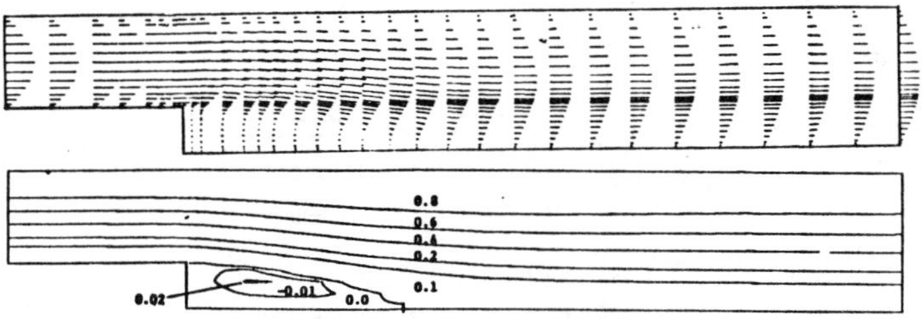

Figure 3 . Flow Over a Backward Facing Step with Traction Free Outlet B.C. at Re = 73: (a) Velocity Vectors; (b) Streamlines, (Numerical values are of dimensionless stream function ψ/ψ_{max})

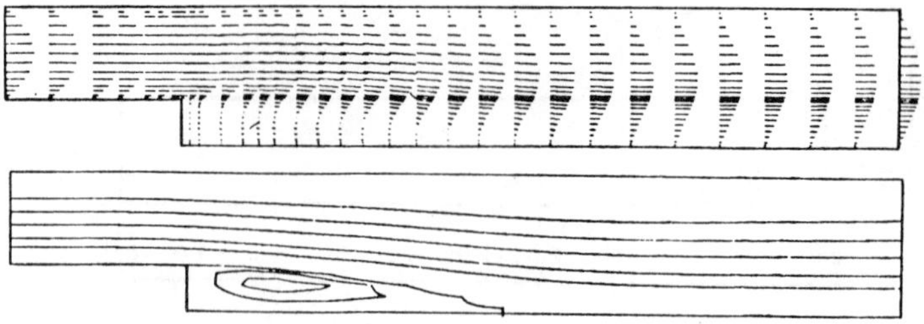

Figure 4 . Flow Over a Backward Facing Step with Traction Free Outlet B.C. at Re = 125: (a) Velocity Vectors; (b) Streamlines

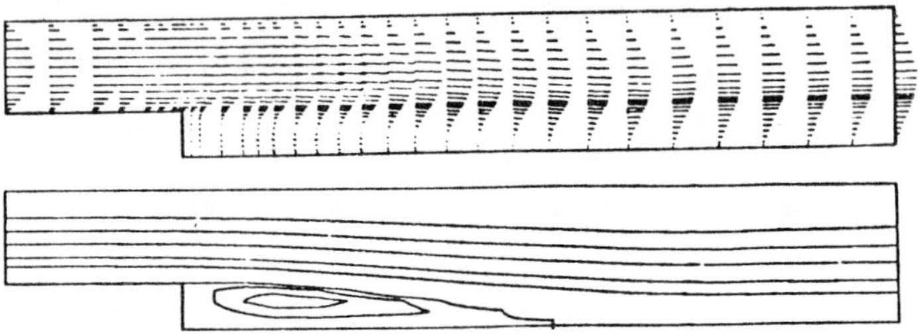

Figure 5. Flow over a Backward Facing Step with Traction Free Outlet B.C. at Re = 155: (a) Velocity Vectors; (b) Streamlines

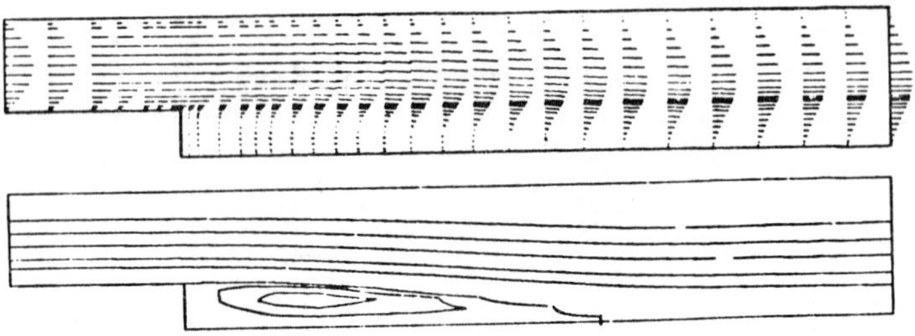

Figure 6. Flow over a Backward Facing Step with Traction Free Outlet B.C. at Re = 191: (a) Velocity Vectors; (b) Streamlines

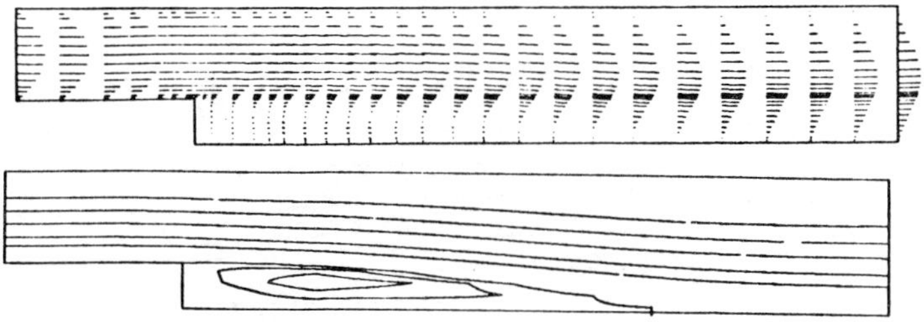

Figure 7. Flow over a Backward Facing Step with Traction Free Outlet B.C. at Re = 229: (a) Velocity Vectors; (b) Streamlines

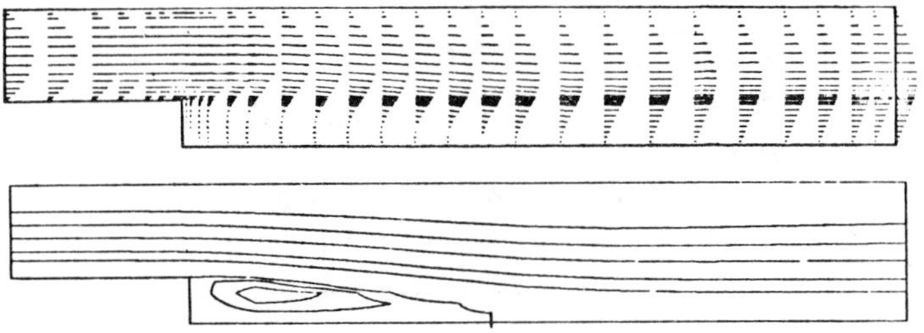

Figure 8. Flow over a Backward Facing Step with Outlet B.C. of $V_2 = 0$; Re = 125: (a) Velocity Vectors; (b) Streamlines

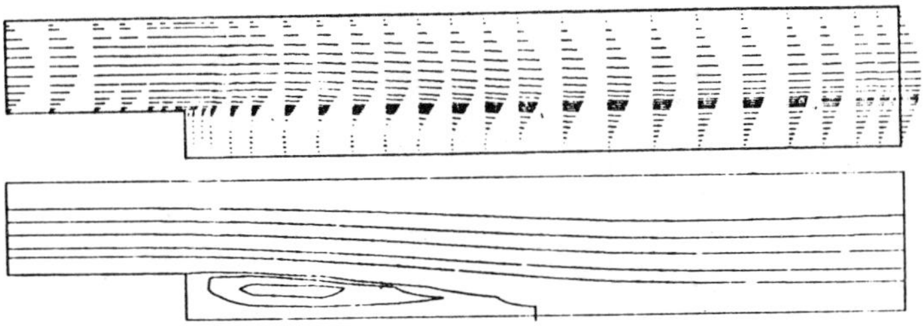

Figure 9. Flow over a Backward Facing Step with Outlet B.C. of $V_2 = 0$; Re = 155: (a) Velocity Vectors; (b) Streamlines

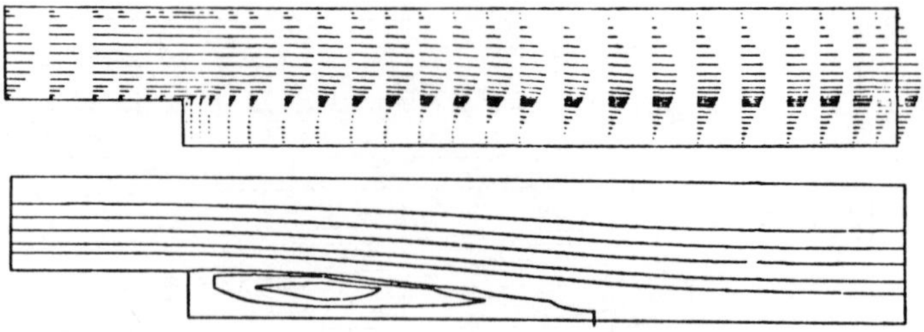

Figure 10. Flow over a Backward Facing Step with Outlet B.C. of $V_2 = 0$; Re = 191:

(a) Velocity Vectors; (b) Streamlines

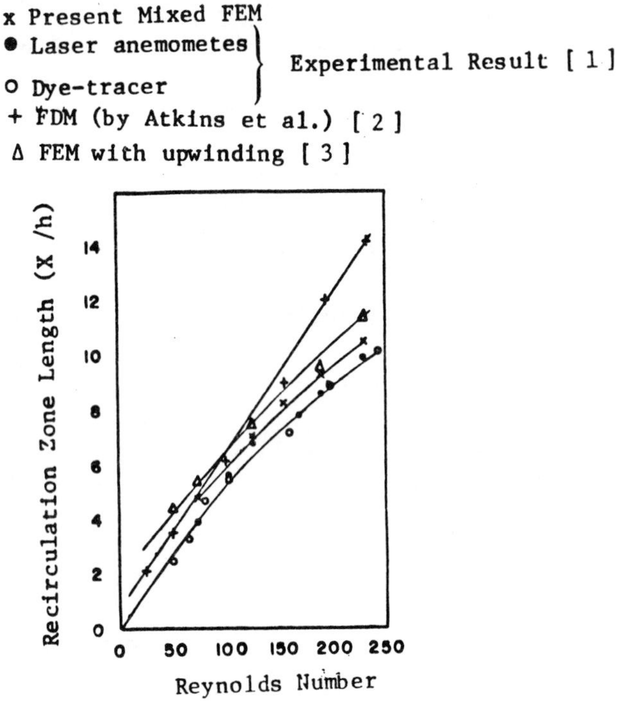

Figure 11. Length of Recirculation zone vs Re Number

ON THE ACCURACY AND STABILITY OF THE QUADRATIC UPSTREAM DIFFERENCING IN LAMINAR ELLIPTIC FLOWS

S. Elghobashi*, M. Prud'homme**
Mechanical Engineering Department
University of California, Irvine, CA 92717

ABSTRACT

Predictions by both the QUICK and hybrid schemes are presented for two laminar elliptic flows, the flow downstream of a sudden expansion in a plane duct, and natural convection in a rectangular enclosure with two adiabatic-and two isothermal-walls. Comparison with detailed experimental data covering a range of Reynolds numbers of 100 to 1020 is made in the first case. This provides a more challenging test for the ability of the QUICK scheme to reduce numerical diffusion than reported previously. Some details are given about the treatment of boundary conditions in the QUICK scheme. Comments on the stability of this scheme are provided in the appendix.

I. INTRODUCTION

The QUICK (Quadratic Upstream Interpolation for Convective Kinematics) scheme of Leonard [1] has emerged in recent years as a promising alternative to the well established hybrid (upwind/central difference) scheme because the former reduces numerical (false) diffusion at large Peclet numbers. False diffusion occurs, in the presence of finite gradients of the dependent variable, when the streamlines are oblique to the finite-difference grid lines. Numerical solutions made in this case can contain serious errors. The need to eliminate these errors becomes pressing especially when critical evaluation of turbulence models is considered. This is because discrepancies due to false diffusion may, in some cases, be orders of magnitude higher than those originating from a particular turbulence model.

* Associate Professor
** Research Assistant (Ph.D. Student)

Several studies of the QUICK and hybrid schemes (referred to hereinafter as Q and H schemes) have been reported, a large number of which were mainly concerned with the transport of a scalar quantity in a given flow field [2,3,4]. Comparisons were made with exact solutions whenever possible or with predictions of other schemes and QUICK always performed well. A test in turbulent unconfined recirculating flow is described in [5] where the momentum equations were solved with QUICK for annular and plane jets and a comparison with experimental data clearly favors Q over H. So far no comparison of the predictions of QUICK with detailed experimental data has been reported for a laminar recirculating flow where exact solutions cannot be obtained. In recent papers [6,7] predictions of the Q and H schemes were compared with each other but seldom with experiment [7] in wall driven cavity flows or sudden pipe expansions. This paper presents the predictions made by the Q and H schemes for two laminar flows. The first is the flow downstream of a sudden expansion in a plane duct where detailed measurements are reported by Haas [8] for Re = 100,715,1020. The second is the natural convection flow in a rectangular cavity reported by Yin et al. [9]. In the first flow the length of the recirculation region is shorter than that found in the sudden pipe expansion flow of Han et al. [6], referred to hereinafter as H.H.L. The larger streamline curvature encountered in the plane duct expansion, presented here, provides a more critical test of the performances of both the Q and H schemes. Some discussions of the treatment of boundary conditions in the Q scheme will also be presented.

2. FINITE-DIFFERENCE EQUATIONS

The conservation equation of a dependent varible ϕ can be written in the form:

$$\frac{\partial}{\partial x}\left[\rho u\phi-\Gamma\frac{\partial\phi}{\partial x}\right]+\frac{1}{y^j}\frac{\partial}{\partial y}y^j\left[\rho v\phi-\Gamma\frac{\partial\phi}{\partial y}\right]=S(x,y) \tag{1}$$

where $j = 1$ for cylindrical coordinates
and $j = 0$ for Cartesian coordinates.
Integration over a given control volume (Fig. 1) yields:

$$\overline{\left[\rho u\phi-\Gamma\frac{\partial\phi}{\partial x}\right]}_e A_e-\overline{\left[\rho u\phi-\Gamma\frac{\partial\phi}{\partial x}\right]}_w A_w+\overline{\left[\rho v\phi-\Gamma\frac{\partial\phi}{\partial y}\right]}_n A_n$$

$$-\overline{\left[\rho v\phi-\Gamma\frac{\partial\phi}{\partial y}\right]}_s A_s=\bar{S}\cdot V \tag{2}$$

The overbar denotes an average value over a face except for $\bar{S}$ which is an average over the control volume. We will now consider the first term on the left of (2) and discretize it according to both schemes. Let us write (c.f. fig 2):

$$\overline{\left[\rho u\phi-\Gamma\frac{\partial\phi}{\partial x}\right]}_e A_e=A_e[F_e\phi_e-\Gamma_e(\phi_E-\phi_P)/\Delta x_{i+1}] \tag{3}$$

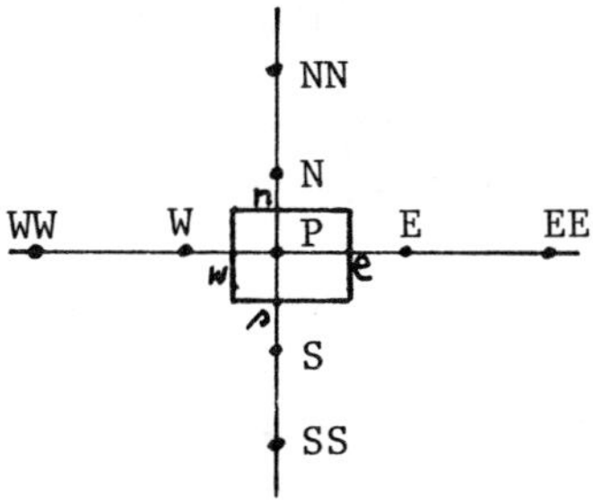

Fig. 1 Control Volume Cell

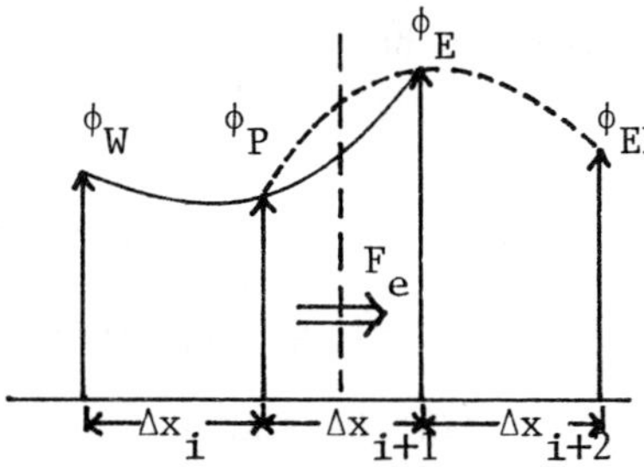

Fig. 2 Interpolation for ϕ_e

The H-scheme evaluates ϕ_e as:

$$\phi_e = (\phi_E+\phi_P)/2 \text{ for } |F_e|\Delta x_{i+1}/\Gamma_e \leq 2$$

$$\phi_e = \phi_P \text{ for } |F_e|\Delta x_{i+1}/\Gamma_e > 2,\ F_e > 0$$

$$\phi_e = \phi_E \text{ for } |F_e|\Delta x_{i+1}/\Gamma_e > 2,\ F_e < 0 \quad (4)$$

where the control volume face e is located midway between P and E. The Q scheme evaluates ϕ_e as:

$$\phi_e = (\phi_E+\phi_P)/2 - C_i$$

where

$$C_i = a_i[(\phi_E/2 + a_i\phi_W)/(1+2a_i) - \phi_P/2] \text{ for } F_e \geq 0$$

$$C_i = b_i[(\phi_P/2 + b_i\phi_{EE})/(1+2b_i) - \phi_E/2] \text{ for } F_e < 0 \quad (5)$$

and

$$a_i = \Delta x_{i+1}/2\Delta x_i,\ b_i = \Delta x_{i+1}/2\Delta x_{i+2}$$

The main difference between the Q and H schemes is that the former evaluates ϕ_e by parabolic interpolation using two nodes upstream of the e-location and one downstream. If the control volume faces are not located midway between the nodes, as it is the case for staggered nonuniform grid, then slightly more complicated expressions of ϕ_e and C_i result. Substitution of the expressions (4) or (5) will yield the discretized equation for ϕ_P:

$$A_P\phi_P = A_E\phi_E + A_W\phi_W + A_N\phi_N + A_S\phi_S + S_u \cdot V + S'_u \cdot V \quad (6)$$

where the source term $\bar{S}$ of (2) has been split as

$$\bar{S} = S_u + S_P\phi_P, \quad (7)$$

and for the Q scheme: $A_E=(-.5F_e+D_e)A_e$, $A_W=(.5F_w+D_w)A_w$,

$$S'_u \cdot V = F_n A_n C_n - F_s A_s C_s + F_e A_e C_e - F_w A_w C_w,^*$$

for the H scheme:
$$A_E = [0, -.5F_e + D_e, -.5F_e] A_e \qquad (8)$$
$$A_W = [0, .5F_w + D_w, .5F_w] A_w$$
$$S'_u V = 0$$

for both: $A_P = \Sigma_j A_j - S_p + F_e A_e - F_w A_w + F_n A_n - F_s A_s$

Expressions for A_N, A_S are obtained from symmetry.

3. TREATMENT OF THE BOUNDARY CONDITIONS

Solution methods for elliptic flows require the specification of either the value of ϕ or its gradient along all the solution domain boundaries. Prescription of the boundary value of ϕ requires no special attention in the Q or H schemes.

On the other hand when the ϕ-gradient is prescribed at the boundary due care is needed when using the Q scheme as discussed below.

An upwind differencing method doesn't need a value at L+1 (Figure 3) while central differencing would require a value. The conventional paractice assumes a linear ϕ-distribution between L and L+1. The Q scheme uses parabolic interpolation where three nodes are involved and two conditions will be needed in general.

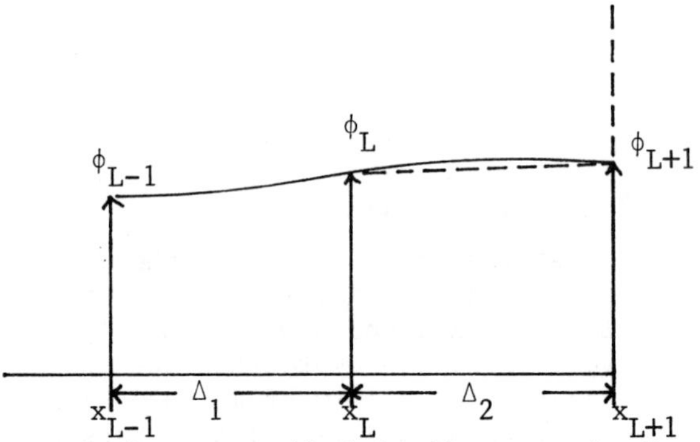

Figure 3. Prescribed Gradient at the Boundary

Let G be the prescribed value of the gradient at L+1.

$$G = \frac{\partial\phi}{\partial x}\Big|_{x=x_{L+1}} = [\phi_{L-1}/r(\Delta_1+\Delta_2)] - [\phi_L(1+r)/r\Delta_2] + \left\{\phi_{L+1}[1/\Delta_2 + 1/(\Delta_1+\Delta_2)]\right\} \qquad (9)$$

where $r = \Delta_1/\Delta_2$

*SU' is an explicit source term, evaluated with the latest available values during a computation.

Now assume that ϕ varies linearly between L and L+1, (i.e. zero-curvature at boundary) yet retaining the parabolic variation of ϕ between L-1, L, L+1. Thus:

$$G = (\phi_{L+1} - \phi_L)/\Delta_2 \tag{10}$$

Combination of (9) and (10) yields

$$\phi_L = (\phi_{L-1} + r\,\phi_{L+1})/(1+r) \tag{11}$$

The following sequence of steps is then followed to obtain the value of ϕ_{L+1} at the current iteration (m+1) based on its value at the previous iteration (m):

i) Calculate the field values ϕ_i^{m+1} using ϕ_{L+1}^{m} as a boundary value

ii) Override ϕ_L^{m+1} from $\phi_L^{m+1} = (\phi_{L-1}^{m+1} + r\phi_{L+1}^{m})/(1+r)$

iii) Set $\phi_{L+1}^{m+1} = G\cdot\Delta_2 + \phi_L^{m+1}$

While step iii) alone is sufficient for the hybrid scheme it will not be enough for QUICK (with one important exception: when the boundary is located midway between L and L+1 then (9) and (10) are identical and step (iii) is sufficient.)

The difference between Q and H schemes lies in how they treat the convective fluxes of ϕ. So when convection across a boundary vanishes, as in the case of a symmetry axis, say along, x, it is immaterial whether Q or H is used. Therefore it is enough to set $\phi_{j+1} = \phi_j$ to satisfy $\frac{\partial\phi}{\partial y} = 0$ on the axis.

4. RESULTS AND DISCUSSION

Figure (4) shows the flow configuration of the first test case. The expansion ratio (S/H) is 1:2. Computations were carried out on a CDC 7600 computer for Re values of 100, 715 and 1020 (based on average velocity and duct width). Figure (5) compares the measured axial velocity (for Re=100) with the predictions of the Q and H schemes for a fully developed inlet profile at the step section. A staggered 17x17 non-uniform grid mesh was used. The finite difference equations (6) were solved for the axial and lateral velocities in an iterative manner. The pressure field is obtained from the SIMPLER algorithm of Patankar [10,11]. The predictions of the velocity field and the reattachment point location x_r are in good agreement with experimental data for both schemes. Figures (6), (7) show the predicted velocity for Re = 715 and 1020, using 21x25 and 23x25 non-uniform grids respectively.

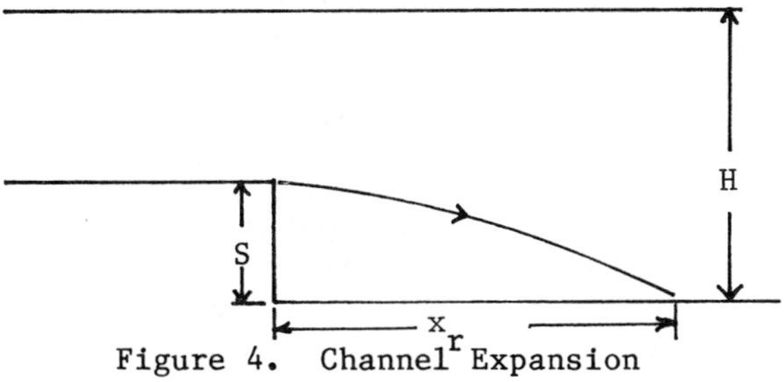

Figure 4. Channel Expansion

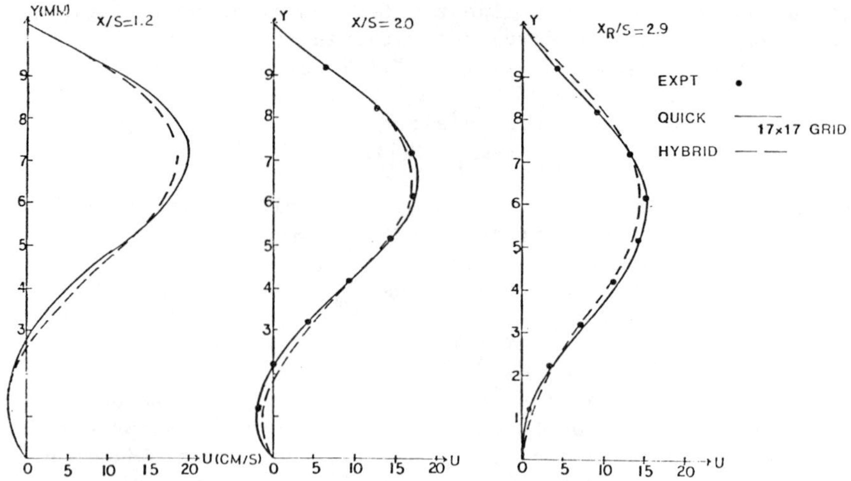

Figure 5. Channel Step Flow, Re=100

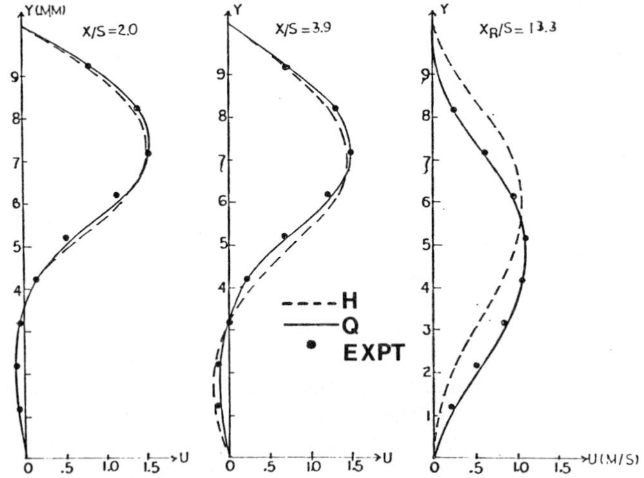

Figure 6. Channel Step Flow, Re=715

In order to obtain an accurate velocity profile at the step section, an additional duct of width S and length 3.2S was included in the flow domain upstream of the step. The velocity profile at the inlet of this additional duct was fully developed. The predicted velocity profile at the step section was thus allowed to "feel" the adverse pressure effects in the expanded section rather than being artificially fixed as a fully-developed profile. The superior accuracy of Q is very clear near the reattachment point (Fig. 7-c). The hybrid scheme overpredicts x_r by 8-10% while QUICK underpredicts x_r by 4-5% at Re=1020. Both the Q and H schemes required the same number of iterations but Q required about 10% more time per iteration than H. No stability problems were encountered during this test.

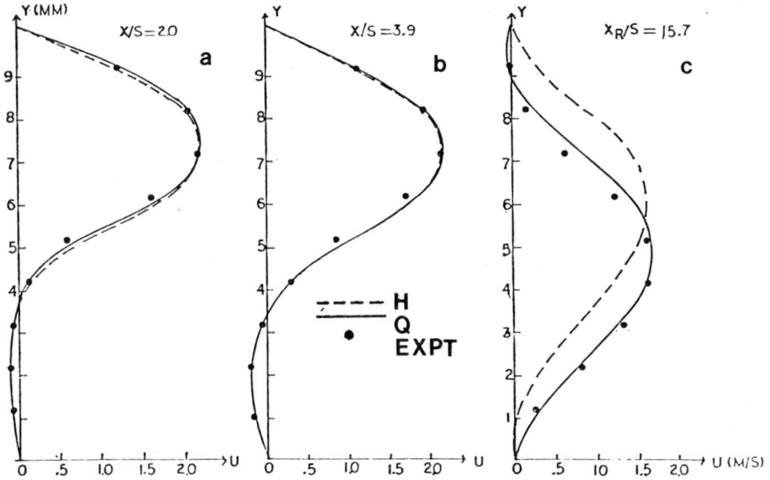

Figure 7. Channel Step Flow, Re=1020

It is interesting to compare briefly the Q and H schemes in a sudden pipe expansion (1:2) (see also ref. [6]). Figure (8) shows the axial velocity prediction of both schemes for a 20x12 non-uniform grid at Re of 1500 and 2000 (based on small diameter and uniform velocity profile at the step). The streamline curvature in this case was less than that in the channel step flow. This results in only minor false diffusion effects and little discrepancy between the solutions obtained by the Q and the H schemes.

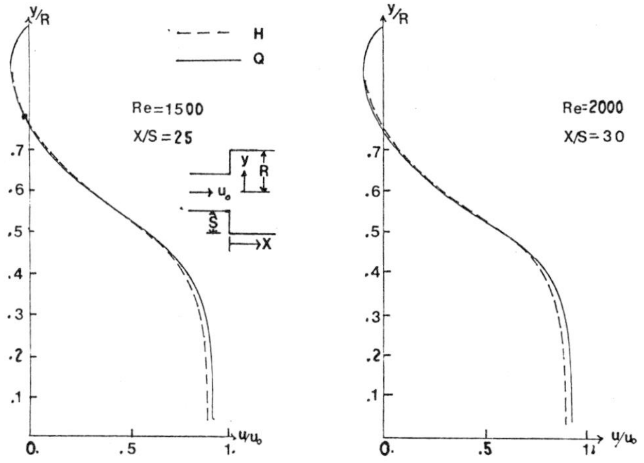

Figure 8. Pipe Expansion, Re 1500 and 8000

Computation with QUICK for Re=5000 yielded unrealistic results and divergence occurred for Re=8000. Thus while the hybrid scheme is unconditionally stable at high Re numbers QUICK is not. H.H.L. defined the coefficients Aj's and the curvature terms C_i for Q in a form, different from that used here (eq. (5)), which renders QUICK more stable at high Re although slower to converge. The reader is referred to the appendix for a comparison of the stability of the H.H.L. form with the present one.

Natural convection flow in a rectangular enclosure was the second test problem considered. Figure (9) compares the predictions of normalized temperature profiles with the data of Yin et al. [9] for $Gr=3.5x10^5$, Pr=.73, $T=20^oC$, L/D=19.7. The u and v momentum equations were solved by either the Q or H schemes while the enthalpy equation, which is not very sensitive to the choice of scheme used, was always solved by the H scheme. A non-uniform 22x22 grid was used. The boundary conditions were prescribed to simulate two adiabatic- and two isothermal-walls for the enthalpy equation and no-slip for the momentum equations. In this case both velocities are of the same order through most of the solution domain. Both the Q and H schemes predictions match the data fairly well although the Q scheme performs better, especially near the corners of the enclosure where the curvature of the streamlines is significant.

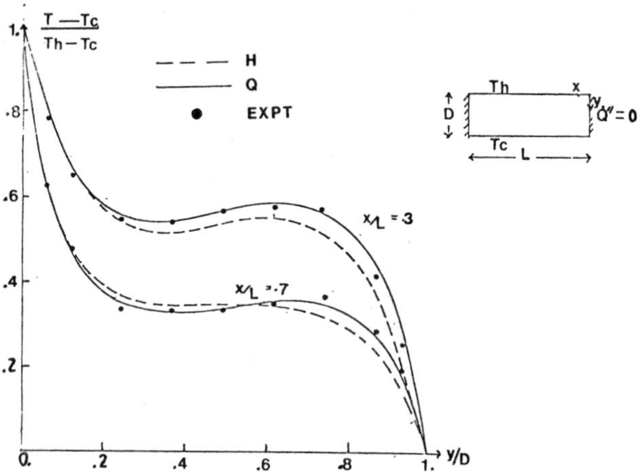

Figure 9. Temperature Profiles

5. CONCLUDING REMARKS

Comparison of the predictions of both the QUICK and hybrid schemes with detailed experimental data has been made for two laminar elliptic flows. The following remarks can be drawn from this study:

1) The results obtained so far show clearly the ability of the QUICK scheme to minimize false diffusion effects as compared to the hybrid scheme.

2) It is possible to use the QUICK scheme in a straight-forward manner without encountering stability problems for grid-cell Peclet numbers of up to 15000 while the number of iterations required remains the same as hybrid's.

3) The QUICK scheme will yield unrealistic results when the grid-cell Peclet number locally exceeds 25000 and will eventually diverge for higher Peclet numbers.

ACKNOWLEDGEMENTS:

This work has been supported by a National Science Foundation grant number CME-8018407. The authors would like to thank Verna Bruce for typing the manuscript. Authors' names appear alphabetically

NOMENCLATURE

A_e, A_w, A_m, A_s	areas of control volume faces
A_P, A_E, A_W, A_N, A_S	coefficients in finite difference equations
D_e, D_w, D_m, D_s	diffusion coefficients
F_e, F_w, F_m, F_s,	convection coefficients
Gr	Grashoff number
D	cavity height
L	cavity length
Pe	grid-cell Peclet number
Re	Reynolds number
V	volume of a grid-cell
S, S_u, S_P, S'_u	source term and components
u,v	velocities
x,y	coordinate directions
ρ	density
Γ	diffusivity
ϕ	any variable
(m)	m^{th} iteration value
Pr	Prandtl number

REFERENCES

1. LEONARD, B.P. A Stable and Accurate Convective Modelling Procedure Based on Quadratic Upstream Interpolation. Comf. Meth. Appl. Mech. Eng., No. 19, pp. 59-98, 1979.

2. SMITH, M.R., HUTTON, A.G. The Numerical Treatment of Advection: A Performance Comparison of Current Methods. Num. Heat Transfer, Vol. 5, pp. 439-461, 1982.

3. LEONARD, B.P., LESCHZINER, M.A., MC GUIRK, J. Third Order Finite Difference Method for Steady Two-Dimensional Convection. Proceedings of First International Conference on Numerical Methods in Laminar and Turbulent Flow, Swansea, G.B., pp. 807-819, 1978.

4. LESCHZINER, M.A. Practical Evaluation of Three Finite Difference Schemes for the Computation of Steady State Recirculating Flows. Comp. Meth. Appl. Mech. Eng., No. 23, pp. 293-312, 1980.

5. LESCHZINER, M.A., RODI, W. Calculation of Annular and Twin Parallel Jets Using Various Discretization Schemes and Turbulence Model Variations. Jour. Fluids Eng., Trans. ASME, Vol. 103, pp. 325-360, 1981.

6. HAN, T., HUMPHREY, J.A.C., LAUNDER, B.E. A Comparison of Hybrid and Quadratic-Upstream Differencing in High Reynolds Number elliptic Flows. Comp. Meth. Appl. Mech. Eng., No. 29, pp. 81-95, 1981.

7. POLLARD, A., SIU, A.L.-W. The Calculation of Some Laminar Flows Using Various Discretization Schemes. Comp. Meth. Appl. Mech. Eng., No. 35, pp. 293-313, 1982.

8. HAAS, W. Wärmeübertragung in Einer Kanalströmung mit Plötzlicher Querschmittserweiterung. Institut für Hydromechanik, Sonderforschungsbereich 80, University of Karlsruhe Report, 1981.

9. YIN, S.H., WUNG, T.Y., CHEN, K. Natural Convection in an Air Layer Enclosed Within Rectangular Cavities. Int. J. Heat Mass Trans., Vol. 21, pp. 307-315, 1978.

10. PATANKAR, S.V. A Calculation Procedure for Two-Dimensional Elliptic Situations. Num. Heat Transfer, Vol. 4, pp. 409-425, 1981.

11. PATANKAR, S.V. Numerical Heat Transfer and Fluid Flow. Hemisphere Publishing, 1980.

APPENDIX

The discretization form proposed by H.H.L.[6]

H.H.L. used the QUICK scheme but claimed they could not achieve convergence in any flow if they treated the curvature terms (eq. 5) as an explicit source term involving the latest available ϕ's. They proposed and used the following discretization form, given here for a uniform grid distribution.

$$\phi_e = 1/8(6\phi_p + 4\phi_E) - 1/8(\phi_W+\phi_E), \quad F_e \geq 0$$

include in Su

$$\phi_e = 1/8(3\phi_p + 4\phi_E) - 1/8(\phi_{EE}-2\phi_E), \quad F_e < 0$$

include in Su (A.1)

Pollard and Siu [7] used a similar approach. We can consider a very simple problem in one dimension and compare the stability of H.H.L.'s grouping and the present one using the ratio $\sum_i |A_i|/A_p$, which is usually a good indicator of stability, as a criterion for comparison. Let us start with:

$$F\frac{d\phi}{dx} = \frac{d}{dx}\left(\Gamma\frac{d\phi}{dx}\right), \quad F = \rho u, \text{ a constant.} \qquad (A.2)$$

According to the present practice, the finite-difference equivalent of (A.2) is:

$$(F_e/2-F_w/2+2\Gamma/\Delta x)\ \phi_p = (-F_e/2+\Gamma/\Delta x)\ \phi_E +$$

$$(F_w/2+\Gamma/\Delta x)\ \phi_W + S_u\cdot\Delta x \qquad (A.3)$$

where Su contains the curvature terms. Since $|F_e|=|F_w| = F$,

$$2\phi_p/P_e = (-1/2+1/P_e)\ \phi_E + (1/2+1/P_e)\phi_W + S_u\Delta x/F, \quad P_e=F\Delta x/\Gamma \qquad (A.4)$$

Let us assume that Pe > 2, we get at once:

$$\sum_i |A_i|/A_p = \frac{Pe}{2} \quad \text{(same as central difference)} \qquad (A.5)$$

The H.H.L. grouping would give the following:

$$(1/4+2/P_e)\ \phi_p = (-1/2+1/P_e)\ \phi_E + (3/4+1/P_e)\ \phi_W + S_u\Delta x/F \qquad (A.6)$$

Assume that Pe > 2, this yields:

$$\sum_i |A_i|/A_p = \frac{5}{1+8/P_e} \qquad (A.7)$$

Comparing (A.4) and (A.7) the H.H.L. grouping is more stable than the present one as Pe increases, however it is significantly slower. H.H.L. needed 35% to 50% more iterations than that of the hybrid scheme while we needed the same number as hybrid's.

FINITE ELEMENT SOLUTION OF STEADY NAVIER-STOKES EQUATIONS FOR LAMINAR RECIRCULATING FLOW WITH AN ACCELERATED PSEUDO-TRANSIENT METHOD

E. DICK*, D. DESPLANQUES**

STATE UNIVERSITY OF GHENT, DEPARTMENT OF MACHINERY,
SINT PIETERSNIEUWSTRAAT 41, 9000 GENT, BELGIUM

SUMMARY

The steady Navier-Stokes equations in primitive variables are solved by the integration in time of the artificial compressibility equations of Chorin. The Petrov-Galerkin method is used, employing bilinear shape functions and weighting functions with parabolic modifications in upwind sense. Identical shape functions are used for velocity components and pressure. Due to this equal order interpolation, spurious pressure modes are to be expected. These pressure modes are surpressed by the use of a smoothing technique in the artificial continuity equation. The resulting scheme is slow due to the severe time step limitation caused by the artificial compressibility. To accelerate the technique, a relaxation equation for pressure is added. For optimum relaxation parameters, the convergence rate loss due to artificial compressibility can completely be compensated.

PSEUDO-TRANSIENT EQUATIONS

The steady Navier-Stokes equations, in primitive variables are solved by the integration in time of the well-known artificial compressibility equations of Chorin [1] :

$$\frac{\partial p}{\partial t} + \delta\left(\frac{\partial u}{\partial x} + \frac{\partial v}{\partial y}\right) = 0 \qquad (\delta>0) \tag{1}$$

$$\frac{\partial u}{\partial t} + u\,\frac{\partial u}{\partial x} + v\,\frac{\partial u}{\partial y} + \frac{1}{\rho}\,\frac{\partial p}{\partial x} - \nu\left(\frac{\partial^2 u}{\partial x^2} + \frac{\partial^2 u}{\partial y^2}\right) = 0 \tag{2}$$

---

*Senior Research Assistant, **Research Assistant

$$\frac{\partial v}{\partial t} + u\,\frac{\partial v}{\partial x} + v\,\frac{\partial v}{\partial y} + \frac{1}{\rho}\,\frac{\partial p}{\partial y} - \nu\left(\frac{\partial^2 v}{\partial x^2} + \frac{\partial^2 v}{\partial y^2}\right) = 0 \qquad (3)$$

The set of equations (1)(2)(3) differs from the unsteady Navier-Stokes equations for incompressible flow by the non-physical equation of continuity (1). However, the steady solutions of the Chorin equations and the Navier-Stokes equations are identical.

The mathematical character of the Chorin equations is identical to that of compressible Navier-Stokes equations. The reduced set of equations ($Re \to \infty$) is hyperbolic with respect to time, the full set is parabolic.

The reduced set of equations can be written as :

$$\frac{\partial}{\partial t}\begin{pmatrix} p \\ u \\ v \end{pmatrix} + \begin{pmatrix} 0 & \delta & u \\ \frac{1}{\rho} & u & 0 \\ 0 & 0 & u \end{pmatrix} \frac{\partial}{\partial x}\begin{pmatrix} p \\ u \\ v \end{pmatrix} + \begin{pmatrix} 0 & 0 & \delta \\ 0 & v & 0 \\ \frac{1}{\rho} & 0 & v \end{pmatrix} \frac{\partial}{\partial y}\begin{pmatrix} p \\ u \\ v \end{pmatrix} = 0 \qquad (4)$$

This set has the same form as the Euler equations :

$$\frac{\partial \xi}{\partial t} + A_1\,\frac{\partial \xi}{\partial x} + A_2\,\frac{\partial \xi}{\partial y} = 0 \ ; \quad \xi^T = \{p,u,v\} \qquad (5)$$

The characteristic matrix is :

$$A = k_1 A_1 + k_2 A_2 \ ; \quad k_1^2 + k_2^2 = 1$$

A has the eigenvalues :

$$\lambda_1 = k_1 u + k_2 v = \vec{k}.\vec{V}$$

$$\lambda_{2,3} = \vec{k}.\vec{V}/2 \pm \left((\vec{k}.\vec{V}/2)^2 + c^2\right)^{.5} \qquad (6)$$

c is called the artificial velocity of sound :

$$c^2 = \delta/\rho$$

Figure 1 shows the loci of the endpoints of the vectors $\vec{\Lambda}_1 = \lambda_1\vec{k}$, $\vec{\Lambda}_2 = \lambda_2\vec{k}$, $\vec{\Lambda}_3 = \lambda_3\vec{k}$ for variable $\vec{k} = (k_1,k_2)$.

The locus of the endpoint of $\vec{\Lambda}_1$ is a circle with centre $(u/2,\ v/2)$ and radius $\left((u^2+v^2)/4\right)^{.5}$. The locus of the endpoints of $\vec{\Lambda}_2$ and $\vec{\Lambda}_3$ is a circle with the same centre but with radius $\left((u^2+v^2)/4 + c^2\right)^{.5}$.

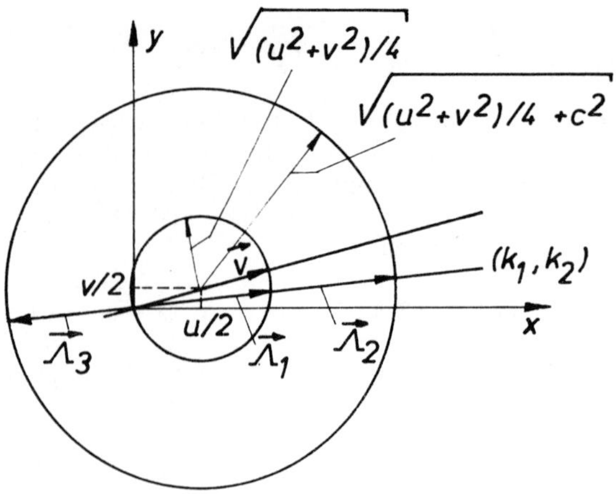

Fig. 1. Eigenvalues of the characteristic matrix for the artificial compressibility equations

For the Euler equations, the eigenvalues of the characteristic matrix are :

$$\lambda_1 = \vec{k}.\vec{V}$$

$$\lambda_{2,3} = \vec{k}.\vec{V} \pm c$$

Figure 2 shows the loci of the endpoints of the vectors $\vec{\Lambda}_1$, $\vec{\Lambda}_2$ and $\vec{\Lambda}_3$ for subsonic Euler equations ($u^2+v^2 < c^2$).

Fig. 2

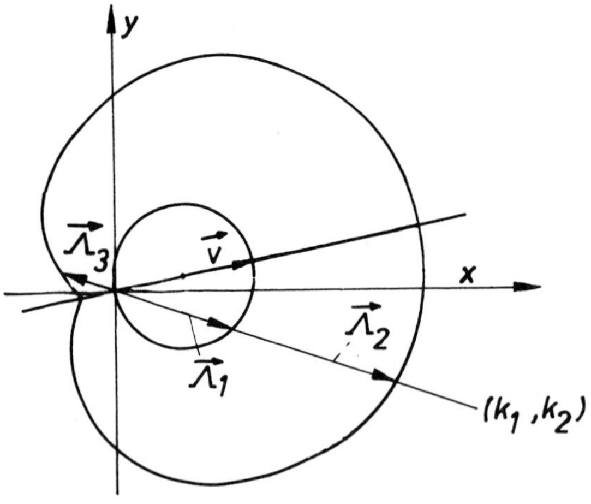

Comparison of figures 1 and 2 reveals that the mathematical character of the reduced artificial compressibility equations and the subsonic Euler equations is identical.

As a consequence, any method, which is suitable for subsonic compressible Navier-Stokes equations, can be used for the Chorin equations, taking into account the difference in magnitude of the eigenvalues of the characteristic matrices. For example, the maximum transfer velocities in x- and y-directions for the Chorin equations are :

$$u^* = u/2 + \left((u^2+v^2)/4 + c^2\right)^{.5}$$
$$v^* = v/2 + \left((u^2+v^2)/4 + c^2\right)^{.5} \qquad (7)$$

instead of $u^* = u+c$ and $v^* = v+c$ for Euler equations.

Using a time-marching scheme, the time step limitation has to be based on the expressions (7).

PROBLEM GEOMETRY

Figure 3 shows a backward facing step, used as test geometry, discretized by fixed Δx and Δy. The step height h is half the inlet width d. $\Delta x = d/3$, $\Delta y = d/6$. Calculations were also done on a coarser grid ($\Delta x = 2d/3$, $\Delta y = d/6$) and on a finer grid ($\Delta x = d/6$, $\Delta y = d/12$).

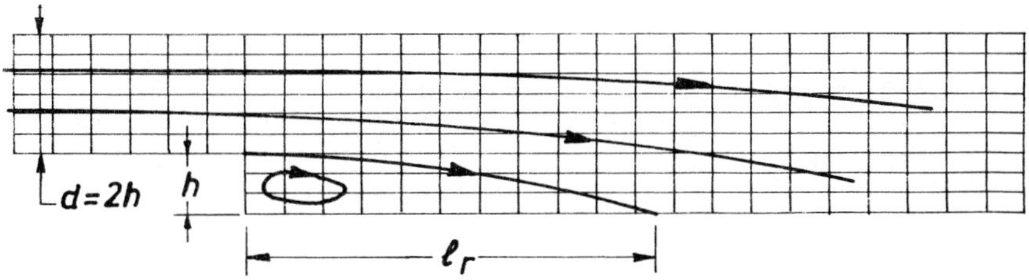

Fig. 3. Problem geometry

FINITE ELEMENT DISCRETIZATION

In order to circumvent the cell-Reynoldsrestriction in regions of large velocity, the equations (2) and (3) are discretized by the Petrov-Galerkin method, using bilinear shape functions and weighting functions with the parabolic modifications in upwind sense, as suggested by Christie et al. [2] . Identical shape functions are used for velocity components and pressure. The upwind parameter associated to the direction i, vanishes when the cell-Reynoldsnumber in this direction $R_{c,i}$

is smaller than 2 and is $1-2/R_{c,i}$ for $R_{c,i}$ larger than 2. In the equation (1), there is no need for upwinding. Therefore the weighting functions are chosen to be the shape functions in (1).

BOUNDARY CONDITIONS

The classical boundary conditions for subsonic compressible flow calculations are to be used :

- inflow boundary : $u = u_o(y)$, $v = 0$
- outflow boundary : $\frac{\partial u}{\partial x} = 0$, $\frac{\partial v}{\partial x} = 0$, $p = 0$ (8)
- solid boundaries : $u = 0$, $v = 0$.

TIME INTEGRATION TECHNIQUE

Due to the equal order interpolation of velocity components and pressure, spurious pressure modes are to be expected [3]. In the finite difference method, these modes are eliminated by the use of a staggered grid, by filtering with one sided differences or by smoothing.

A typical method using the filtering technique by alternating downwind and upwind differencing is the well known MacCormack method. This filtering approach is undoubtedly the most popular technique for regular grids, but its extension to the finite element method is not obvious.

The staggered grid technique corresponds to the mixed interpolation approach in the finite element method. This technique is currently used. The objective of this paper is to test the smoothing technique.

In a smoothing technique, the wiggles which are to be expected are surpressed by the explicit use of a viscosity type damper. Because such schemes essentially have a residual artificial viscosity, in principle, they cannot be used on Navier-Stokes equations. These techniques however have some popularity for inviscid calculations. An example is the corrected viscosity technique or damping surface technique introduced by Couston et al. [4].

On the model equation :

$$\frac{\partial f}{\partial t} + \frac{\partial g}{\partial x} = 0$$

the scheme is :

$$f(t+\Delta t) = f(t) - \frac{\Delta t}{2\Delta x}\left(g(x+\Delta x,t) - g(x-\Delta x,t)\right)$$
$$+ \frac{1}{2}\left(f(x+\Delta x,t) - 2f(x,t) + f(x-\Delta x,t)\right)$$
$$- \frac{\alpha}{2}\left(f(x+\Delta x,t^*) - 2f(x,t^*) + f(x-\Delta x,t^*)\right)$$

t^* is a time level fixed during a cycle of time steps, equal to the time level in the beginning of the cycle. α is a factor as close as possible to 1. Here it is chosen to be .99. This technique can be used on the artificial continuity equations (1), introducing a pressure smoother. It is unallowable in the momentum equations due to its residual dissipation. Therefore, equations (2) and (3) are advanced in time by the explicit Euler method. To facilitate the explicit time integration used in all equations, the mass matrices are lumped with the row-sum technique.

The resulting scheme is stable under the usual time step restriction for explicit schemes, based on the maximum transfer velocities (7) :

$$\Delta t < 1/(2\nu/\Delta x^2 + 2\nu/\Delta y^2 + u^*/\Delta x + v^*/\Delta y) \qquad (9)$$

In order to obtain maximum convergence rate, the number of steps in the smoothing cycle and the value of δ (or c) is to be chosen. The optimum number of steps in the cycle is approximately equal to half the number of elements in the longitudinal direction of the flow field (7, 14 and 28) [4]. When δ is chosen to be constant in the flow field, the optimum is such that the pseudo velocity of sound c is approximately equal to the mean inlet velocity $\bar{u}$. That c has to be of the order of magnitude of a typical velocity can be seen on the one-dimensional model problem :

$$\frac{\partial p}{\partial t} + \delta \frac{\partial u}{\partial x} = 0$$
$$\frac{\partial u}{\partial t} + u \frac{\partial u}{\partial x} + \frac{1}{\rho}\frac{\partial p}{\partial x} = \nu \frac{\partial^2 u}{\partial x^2} \qquad (10)$$

Assuming that the steady state is given by p = 0 and u = 0, a Fourier component for the error can be written as :

$$\begin{pmatrix} p \\ u \end{pmatrix} = \begin{pmatrix} p_o \\ u_o \end{pmatrix} e^{\sigma t} e^{j\omega x} \qquad (11)$$

By substitution of (11) into (10), σ is found to be the solution of :

$$\begin{vmatrix} \sigma & j\omega\sigma \\ j\omega/\rho & \sigma + j\omega u + \nu\omega^2 \end{vmatrix} = 0$$

or $$\sigma^2 + \sigma(j\omega u + \nu\omega^2) + \omega^2 c^2 = 0$$

The σ with the smallest real part corresponds to the smallest wave number ω.
For small ω :

$$2\sigma = -j\omega u - \nu\omega^2 \pm (j\omega\sqrt{u^2+4c^2} + \nu\omega^2 \frac{u}{\sqrt{u^2+4c^2}})$$

$$\mathrm{Re}(\sigma_{1,2}) = -\frac{\nu\omega^2}{2}(1 \pm \frac{u}{\sqrt{u^2+4c^2}}) \tag{12}$$

In one time step, the damping of the error is proportional to :

$$e^{-\mathrm{Re}(\sigma).\Delta t}$$

Δt is limited by a condition which is approximately :

$$\frac{\Delta t}{\Delta x}(\frac{u}{2} + \sqrt{u^2/4 + c^2}) < 1$$

Hence, the convergence rate is maximized for maximum :

$$(1 - \frac{u}{\sqrt{u^2+4c^2}})/(u + \sqrt{u^2+4c^2})$$

This leads to :

$$\frac{c}{u} = (\frac{1+\sqrt{2}}{2})^{.5} \simeq 1.1$$

From (12) it is clear that for optimum c/u, the convergence rate is only dependent on the physical viscosity. The convergence rate decreases with increasing Reynoldsnumber.

INITIAL CONDITIONS

All calculations are started with p = 0 and v = 0 in all gridpoints. On all transversal gridlines in front of the step, the u-profile is equal to the parabolic inlet profile. On all transversal gridlines behind the step the u-profile is equal to a parabolic profile with the same flow rate as the inlet profile.

RESULTS

Figure 4 shows the velocity vectors and streamlines for $Re_d = \bar{u}d/\nu = 300$, $\delta/\rho = \bar{u}^2$ and $\bar{u}\Delta t/\Delta x = .2$ after 5240 time steps for the grid $\Delta x = d/3$, $\Delta y = d/6$.

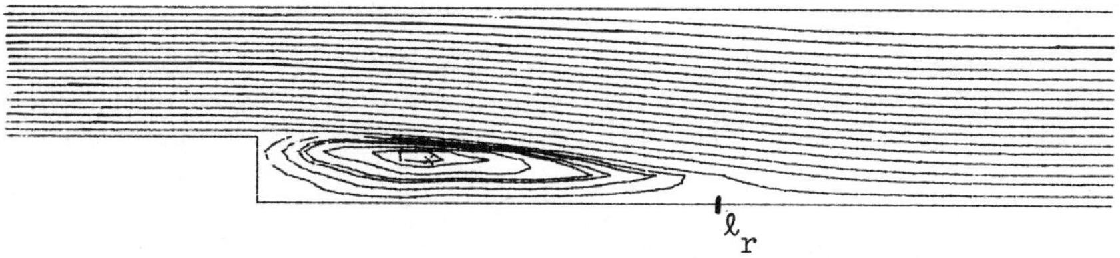

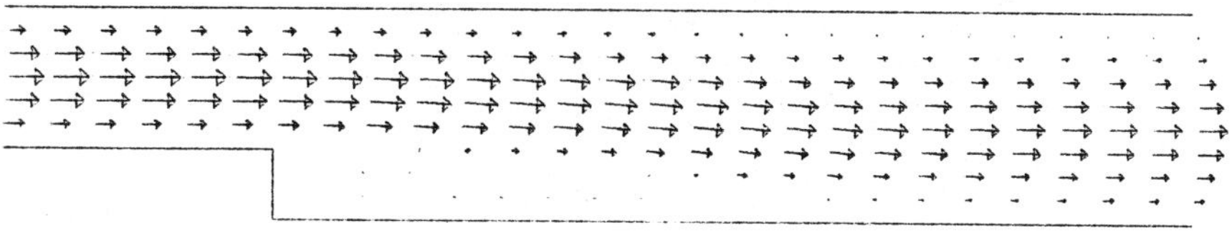

Fig. 4. Results for the scheme without relaxation equation.

The point of reattachment is determined as the point for which $\partial u/\partial y = 0$. Similar calculations were done on the coarser and the finer grid. Table 1 shows the results.

				Experiments [5]	Calculations [6]
Mesh	$\Delta x = 2d/3$ $\Delta y = d/6$	$\Delta x = d/3$ $\Delta y = d/6$	$\Delta x = d/6$ $\Delta y = d/12$		
Number of iterations	5040	5240	5351	-	-
L_r/h	7.08	7.26	7.74	$\sim$ 7.20	$\sim$ 9.10

Table 1. Results for the scheme without relaxation equation. The number of iterations indicated, is necessary for convergence of the velocity components and the pressure up to .001.

The increase of the computed reattachment length with mesh refinement reveals the effect of the artificial viscosity associated with the use of the upwind Petrov-Galerkin method and the pressure smoother. It is remarkable that the finest grids give reattachment lengths which are larger than the experimental values from Denham et al. [5] . The results of Atkins et al. [6] on the same testproblem show a dependence of the result upon the inlet velocity profile. The results shown in table 1 are all obtained with a fully developed (i.e. parabolic) inlet profile. Comparison with the result from [6] shows that even the finest grid is not free of artificial viscosity.

ACCELERATION OF THE CONVERGENCE

The algorithm used in the previous examples is rather slow. To accelerate the technique, a relaxation equation of the type suggested by Wirz [7] is added. The equations (1)(2)(3) are replaced by :

$$\frac{\partial p}{\partial t} + \delta\left(\frac{\partial u}{\partial x} + \frac{\partial v}{\partial y}\right) = 0 \qquad (13)$$

$$\frac{\partial u}{\partial t} + u\,\frac{\partial u}{\partial x} + v\,\frac{\partial u}{\partial y} + \frac{1}{\rho}\,\frac{\partial q}{\partial x} - \nu\left(\frac{\partial^2 u}{\partial x^2} + \frac{\partial^2 u}{\partial y^2}\right) = 0 \qquad (14)$$

$$\frac{\partial v}{\partial t} + u\,\frac{\partial v}{\partial x} + v\,\frac{\partial v}{\partial y} + \frac{1}{\rho}\,\frac{\partial q}{\partial y} - \nu\left(\frac{\partial^2 v}{\partial x^2} + \frac{\partial^2 v}{\partial y^2}\right) = 0 \qquad (15)$$

$$\frac{\partial q}{\partial t} + \lambda\delta\left(\frac{\partial u}{\partial x} + \frac{\partial v}{\partial y}\right) + \frac{1}{\tau}(q-p) = 0 \qquad (16)$$

λ is a positive constant, τ is the positive relaxation time and q is the relaxation pressure. The parameters δ, λ and τ are to be chosen to maximize the convergence rate. The relaxation equation (16) is to be discretized in the same way as the artificial continuity equation so that p and q are identical in steady state. The role of the relaxation equation can be analysed on the one-dimensional analogon of (13-16). Assuming that the steady state is given by p=q=u=0, a Fourier component for the error can be written as :

$$(p\ q\ u)^T = (p_o\ q_o\ u_o)^T\, e^{\sigma t}\, e^{j\omega x} \qquad (17)$$

By substitution of (17) into (13-16), σ is given by :

$$\sigma(\sigma + 1/\tau)(\sigma + j\omega u + \nu\omega^2) + \omega^2 c^2(\lambda\sigma + 1/\tau) = 0 \qquad (18)$$

with $u = c$, $s = \tau/\Delta t$, $\phi = \omega u \Delta t$, $\beta = (\nu/u\Delta x)(\Delta x/u\Delta t)$, (18) can be written as :

$$(\sigma\Delta t)^3 + (\sigma\Delta t)^2(1+js\phi+s\beta^2\phi^2) + (\sigma\Delta t)(j\phi+\beta\phi^2+\lambda s\phi^2) + \phi^2 = 0 \qquad (19)$$

For small wave number ω, the roots of (19) can be expanded as :

$$\sigma_1\Delta t = -\,1/s - (1-\lambda)s\phi^2 + \dots$$

$$\sigma_{2,3}\Delta t = \frac{\pm\sqrt{5}-1}{2}\,j\phi - \frac{\sqrt{5}\pm 1}{2\sqrt{5}}\,\left(s(\lambda-1) + \beta\right)\phi^2 + \dots$$

Due to the factor $(\lambda-1)s$, the damping of the error can be much larger than the damping due to viscous terms. Maximum convergence rate is obtained by maximizing $(\lambda-1)s$. In practice λ and s are bounded by stability limits which are dependent on the problem geometry and the discretization scheme, but only weakly dependent on Reynoldsnumber. The relaxation equation (16) can be combined with the artificial continuity equation (13), giving :

$$\frac{\partial q}{\partial t} - \lambda\,\frac{\partial p}{\partial t} + \frac{1}{\tau}\,(q-p) = 0 \qquad (20)$$

(21) can, for instance, be discretized by :

$$q^{(n+1)}-q^{(n)}-\lambda\left(p^{(n+1)}-p^{(n)}\right)+\left(q^{(n+1)}+q^{(n)}-p^{(n+1)}-p^{(n)}\right)/2s=0 \quad (21)$$

Using (21) for the geometry of figure 4, with $\delta=\rho\bar{u}^2$, the optimum values are approximately $\lambda=7$, $s=4$. Figure 5 shows q, p, ψ and $\vec{V}$ after 2200 time steps on the grid ($\Delta x=2d/3$), $\Delta y=d/6$). The wiggles in the relaxation pressure indicate that convergence is not yet obtained. Figure 6 shows q, p, ψ and $\vec{V}$ after 2400 time steps, corresponding to a convergence up to .001. Without the relaxation equation the necessary number of time steps is 5040 (table 1). Since equation (21) only adds a very limited amount of work, the saving in computer time is about 50 %.

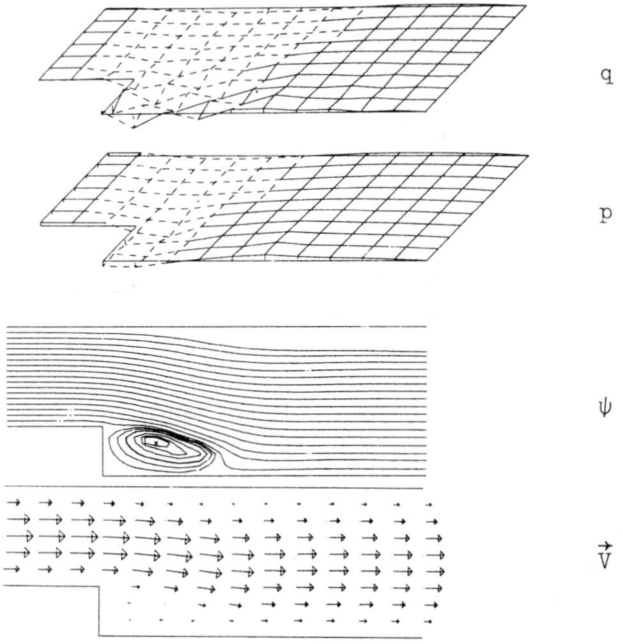

Fig. 5. Results after 2200 time steps for the scheme with relaxation equation, $\lambda = 7$, $s = 4$.

CONCLUSIONS

By the pseudo-transient corrected smoothing technique, a solution of the steady incompressible flow Navier-Stokes equations can be obtained which is free of spurious pressure modes, even when equal order interpolation functions are used for velocity and pressure. Due to the acceleration by relaxation, the time consumption is reduced to an acceptable level.

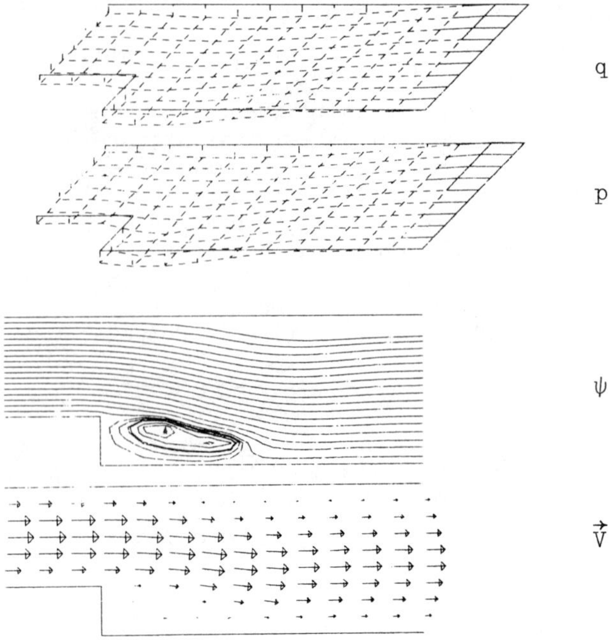

Fig. 6. Results after 2400 time steps for the scheme with relaxation equation, $\lambda = 7$, $s = 4$.

REFERENCES

1. CHORIN, A.J. - A numerical method for solving incompressible viscous flow problems. J. Comp. Phys. vol 2, pp.12-26, 1967.
2. CHRISTIE, I., GRIFFITHS, D.F., MITCHELL, A.R., ZIENKIEWICZ, O.C. - Finite element methods for second order differential equations with significant first derivatives. Int. J. Num. Meth. Engng. Vol. 10, pp. 1389-1396, 1976.
3. SANI, R.L., GRESHO, P.M., LEE, R.L., GRIFFITHS, D.F. - The cause and cure of the spurious pressures generated by certain FEM solutions of incompressible Navier-Stokes equations. Int. J. Num. Meth. in Fluids, Vol. 1, pp. 17-43 and 171-204, 1981.
4. COUSTON, M., McDONALD, P.N., SMOLDEREN, J.J. - The damping surface technique for time dependent solutions to fluid dynamic problems. Von Karman Institute, TN 109, 1975.
5. DENHAM, M.K., PATRICK, M.A. - Laminar flow over a downstream facing step in a two-dimensional flow channel. Trans. Inst. Chem. Engrs., Vol. 52, pp. 361-367, 1974.
6. ATKINS, D.J., MASKELL, S.J., PATRICK, M.A. - Numerical prediction of separated flows. Int. J. Num. Meth. Engng., Vol. 15, pp. 129-144, 1980.
7. WIRZ, H.J. - Relaxation methods for time dependent conservation equations in fluid mechanics. Von Karman Institute, LS 97, 1977.

SECTION 5

ESTUARY

AND

COASTLINE HYDRODYNAMICS

Euler-Lagrangian Computations in Estuarine Hydrodynamics

Ralph T. Cheng[I]

ABSTRACT

The transport of conservative and suspended matter in fluid flows is a phenomena of Lagrangian nature because the process is usually convection dominant. Nearly all numerical investigations of such problems use an Eulerian formulation for the convenience that the computational grids are fixed in space and because the vast majority of field data are collected in an Eulerian reference frame. Several examples are given in this paper to illustrate a modeling approach which combines the advantages of both the Eulerian and Lagrangian computational techniques. These examples include the calculations of tracer trajectories, the study of Lagrangian residual circulation, and the study of tidal mixing. Because the Lagrangian method more closely represents the actual physical processes, the computed results provide a clear picture of these phenomena. Results of simulations give further insights to aid formulation of the next level simulation models.

1. INTRODUCTION:

Generally, transport phenomena in fluid flows are processes of Lagrangian nature. The fundamental governing equations are the conservation laws of mass, momentum, and energy. The conservation equations have been commonly derived starting from a Lagrangian point of view. These equations are then transformed onto an Eulerian reference frame to give rise to a set of working governing partial differential equations [1]. This popular method of deriving the governing equations points out the fact that when dealing with transport phenomena in fluid flows, the Lagrangian viewpoint offers a clearer physical explanation of the conservation principles. On the other hand, in practice, the Lagrangian reference frame is not a convenient coordinate system in which solutions for fluid flow problems can be

[I] U.S. Geological Survey, Menlo Park, CA 94025

easily obtained. Of course, when the governing equations are transformed and written in Eulerian coordinates, all the necessary balancing terms which constitute the conservation laws are included. However, the physical meaning for each term in the equations may not be clear. For example, the process of convection cannot be easily visualized as written in the Eulerian form. Yet, the advantages of working with Eulerian coordinates are obvious in that the independent variables are fixed in space.

In research of estuarine processes, the vast majority of field data are collected using an Eulerian reference. Measurements of current using anchored recording current meters, and observations of tidal water elevation at tide stations are but a few examples. In ecological studies, biological samples and water quality samples must be collected throughout the seasons. For logistic reasons, when the field program is scheduled, these samples are often collected from the same fixed stations.

Lagrangian sampling techniques such as deployment of drifters, drogues, and release of dye are also used in estuarine research. Lagrangian sampling provides very useful information, but the field programs are usually rather labor intensive, and the Lagrangian field sampling techniques are not as well developed and advanced as in the case of Eulerian measurements.

Indeed, there is a direct analogy on methodology between field studies and theoretical investigations. In modeling research, Eulerian methods and techniques are much more popular and well developed than methods formulated based on Lagrangian viewpoint. Recent advances in mathematical modeling research have given model users confidence that the mathematical models can be used to describe tidal circulation in estuaries that is representative of the actual flow. Based on the known tidal flow field which is obtained by conventional Eulerian methods, additional Lagrangian numerical experiments can be carried out to simulate directly the transport phenomena in tidal basins.

The concept of combined Euler-Lagrangian computations is not exactly new. The method of characteristics using a fixed grid combines the Lagrangian technique (i.e., the method of characteristics) in an Eulerian data base (fixed computational grid), [2]. In this paper, several new applications which use combined Euler-Lagrangian methods are given. Some results given herein are preliminary; further research is presently in progress.

2. BASE-LINE FLOW MODEL:

Applications of Euler-Lagrangian method in estuarine hydrodynamics will be demonstrated by examples from the author's modeling studies of San Francisco Bay. The solutions of the shallow water equations have been used satisfactorily to represent tidal circulation in well mixed estuaries [3,4,5,6]. Numerical solutions of the shallow water equations have been the subject of numerous studies; both the finite difference [4] and finite element methods [7] can be used to obtain temporal and spatial distribution of tidal circulation. Which method is better suited for solving the shallow water equations is not the issue in this discussion.

The most commonly used and well documented estuarine circulation model is probably the model developed by Leendertse et al. [4,5,6]. A spatially staggered grid and a two time-level finite difference method is used. The method is known as the alternating-direction-implicit scheme (ADI) which has been shown to be computationally stable and efficient [4,6,8]. Basically, a two dimensional problem has been decomposed into a set of one-dimensional problems which is solved implicitly. Applications of such a model for computations of tidal circulation in South San Francisco Bay, (South Bay) California has been given by Cheng [9] in which a coarse grid (Δx = 1 km) was used. Additional results, which will be shown in this paper, are based on a fine grid system with Δx = 500 m. to give a more detailed spatial resolution.

South San Francisco Bay (South Bay) is a shallow, semi-enclosed embayment which, for the most part of the year, is isohaline. Therefore the inherent assumptions used in the shallow water equations are justifiable. South Bay is characterized by a deep relict channel (>10 m.) which is connected to a broad shoal east of the channel. The bathymetric contours shown in Fig. 1 through 4 are 2 and 5 m. at mean lower low water. Using a finite-difference grid of 500 m., a South Bay model of 43 x 86 points is set up and driven by tides specified at the open boundary near the north end of the model. Two types of tides have been simulated: (1) a semi-diurnal tide of 12-hour period with an amplitude of 1 m., and (2) a mixed semi-diurnal and diurnal tide with 2/3 m. and 1/3 m. amplitudes, respectively. Typical tidal circulation patterns near maximum flood and maximum ebb are shown in Fig. 1. The computed results are in good general agreement with field observations that the tidal circulation, both in magnitude and direction, is strongly affected by the basin bathymetry [10].

3. EXAMPLES OF LAGRANGIAN COMPUTATIONS:

(a) Tracer Studies

Conservative tracers such as dye, drifters, and drogues

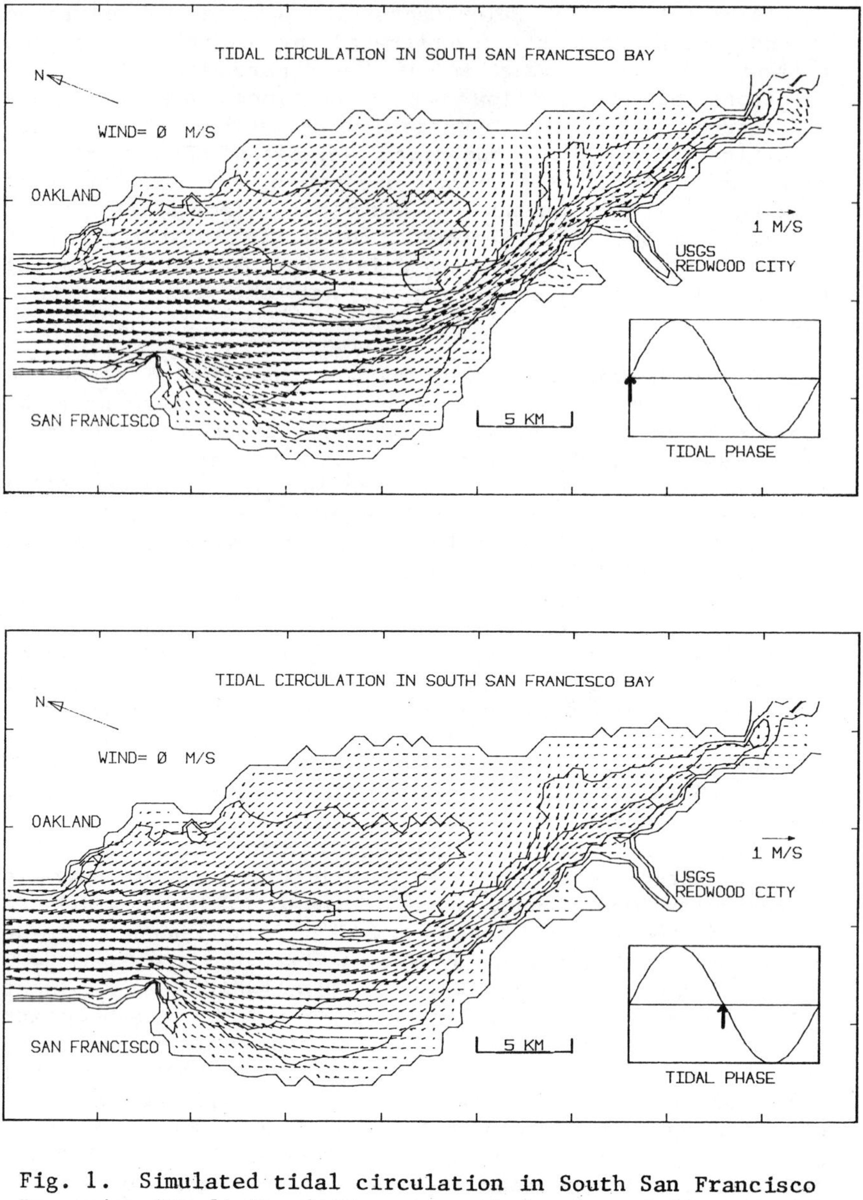

Fig. 1. Simulated tidal circulation in South San Francisco Bay using ADI finite difference method. A semi-diurnal tide, 1 m. amplitude was specified at the open boundary. (a) Tidal circulation near maximum flood, (b) near maximum ebb.

are often released in estuaries, and the subsequent movements of these tracers are followed for the purpose of getting insights into the complicated mixing process in a tidal basin. An equivalent tracer study is carried out numerically by releasing a set of labeled tracer particles. These particles are assumed to be massless, thus their movements are strictly functions of LOCAL (Lagrangian) water velocity. The movements of the tracer particles are tracked numerically at each time step, and the local velocity is determined by means of interpolation from known velocity values at neighboring grid points (Eulerian). Shown in Fig. 2a are the particle trajectories tracked over a 24-hour period while the circulation is driven by a mixed semi-diurnal tide of 2/3 m. amplitude and a diurnal tide of 1/3 m.

The progressive vector diagram is a plot of a series of displacement vectors with their tails joining the heads of the displacement vector generated from the previous time step. The displacement vector is computed from the Eulerian velocity calculated (or measured) at a fixed point. The progressive vector diagram is used in oceanography as an approximation to represent the movement of water mass in the Lagrangian sense using Eulerian data. This approximation may be satisfactory when the spatial gradient of the velocity field is small, as in the case of open ocean. Shown in Fig. 2b are the progressive vector diagrams derived from numerical simulations under the same conditions as Fig. 2a. It can be seen that there are significant differences between the "true" particle trajectories and the progressive vector diagrams. This difference is most pronounced in areas where there is a high gradient of velocity or high gradient of basin bathymetry. Thus, in a tidal estuary where the characteristic length of the basin is on the same order of magnitude as tidal excursion, the progressive vector diagram is a poor approximation as an indicator for the actual water mass movements. As recommended by Zimmerman [11] and Cheng and Casulli [8], one should avoid using Eulerian data (or calculations) for Lagrangian interpretations in such a situation.

(b) Lagrangian Residual Circulation

An immediate extension of the tracer experiment is the computation of Lagrangian residual circulation. Some authors have introduced Lagrangian residual circulation as the sum of Eulerian residual circulation and the Stokes' drift [12,13]. As pointed out by Zimmerman [11] and others [8,14], the above definition for Lagrangian residual circulation is conceptually incorrect, and it is a poor approximation in an estuarine situation. The only correct definition of Lagrangian residual circulation is

(a)

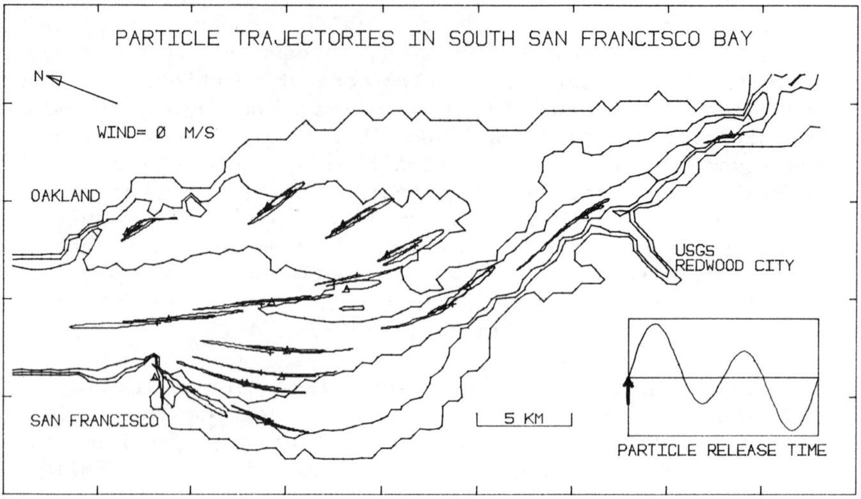

(b)

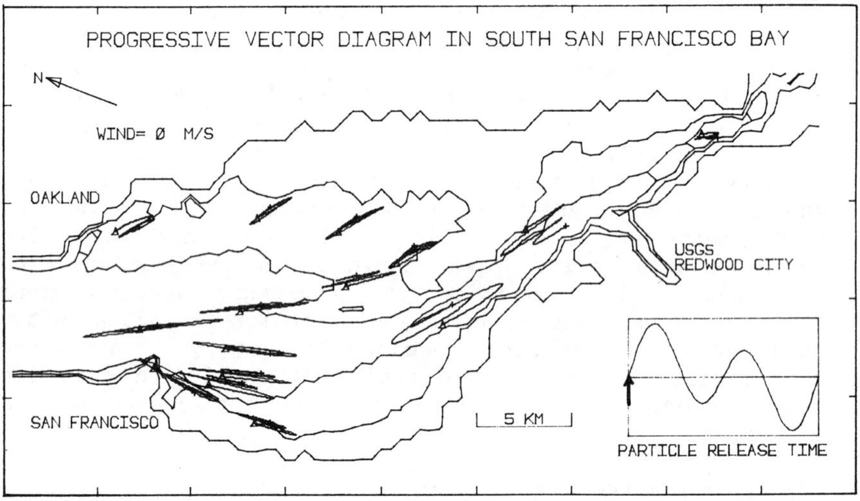

Fig. 2. (a) Particle trajectories over a diurnal tidal period at a few selected locations. A mixed semi-diurnal and diurnal tide was specified at the open boundary. (b) Progressive vector diagrams at the same locations which were generated by the same tide. The tracers were released from positions marked by + and were found at the end of 24 hours at positions marked by Δ.

$$\vec{u}_{1r}(\vec{X}_o,t_o) = \frac{\vec{X}(\vec{X}_o,t_o+T) - \vec{X}(\vec{X}_o,t_o)}{T} \qquad (1)$$

where
$\vec{X}(\vec{X}_o,t_o + t)$ = the position of a labeled particle at time t_o + t which was initially labeled at position X_o when $t = t_o$, and T = tidal period.

The position of the labeled particle is calculated by integrating with time following the particle trajectory [8]. As might be expected, the Lagrangian residual circulation, which is defined as the net displacement of the water mass over a complete tidal cycle divided by tidal period, is an order of magnitude smaller than tidal velocity (Fig. 2). Furthermore, the net displacement value depends on the phase of the tide when the particles are released. This property is strictly of Lagrangian nature and can only be revealed by Lagrangian computation. Figure 3 (a) and (b) depict some calculated results of Lagrangian residual circulation when the tracer particles are released at high and low waters at the open boundary, respectively. In these two extreme situations the tracers released from the same position actually covered two entirely different trajectories; thus it is not surprising that the final results differ as augmented by the local basin bathymetry.

(c) Mixing Induced by Tidal Circulation

Generally, the time scale for ecological processes is much longer than the tidal period, so it is not practical to formulate ecological models on the same time scale as tidal circulation models. For the same reason, the dispersion coefficient measured in the field over a short period of time cannot be used to represent adequately the mixing process on a time scale of several days. To study the long-term effects of mixing induced by tidal current, Awaji [15] conducted numerical experiments by following tracer particles over several tidal cycles. From the final distribution of the particles, the effective dispersion coefficient has been estimated to be an order of magnitude higher than the dispersion coefficient due to turbulent mixing. Although Awaji's study is indeed interesting, Cheng [16] has pointed out that further investigation is needed to include consideration of the variation due to the change of the phases of tide when the tracer particles are released.

To complement Awaji's investigation, the Lagrangian residual circulation is calculated with a complete set of particles released at hourly intervals over a period of 12 hours. A semi-diurnal tide of 12-hour period and 1 m. amplitude is used to generate the base-line tidal circulation. This simulation gives 12 Lagrangian residual vectors with respect to the specific release time. When all 12 residual vectors are plotted (Fig. 4), the distribution of

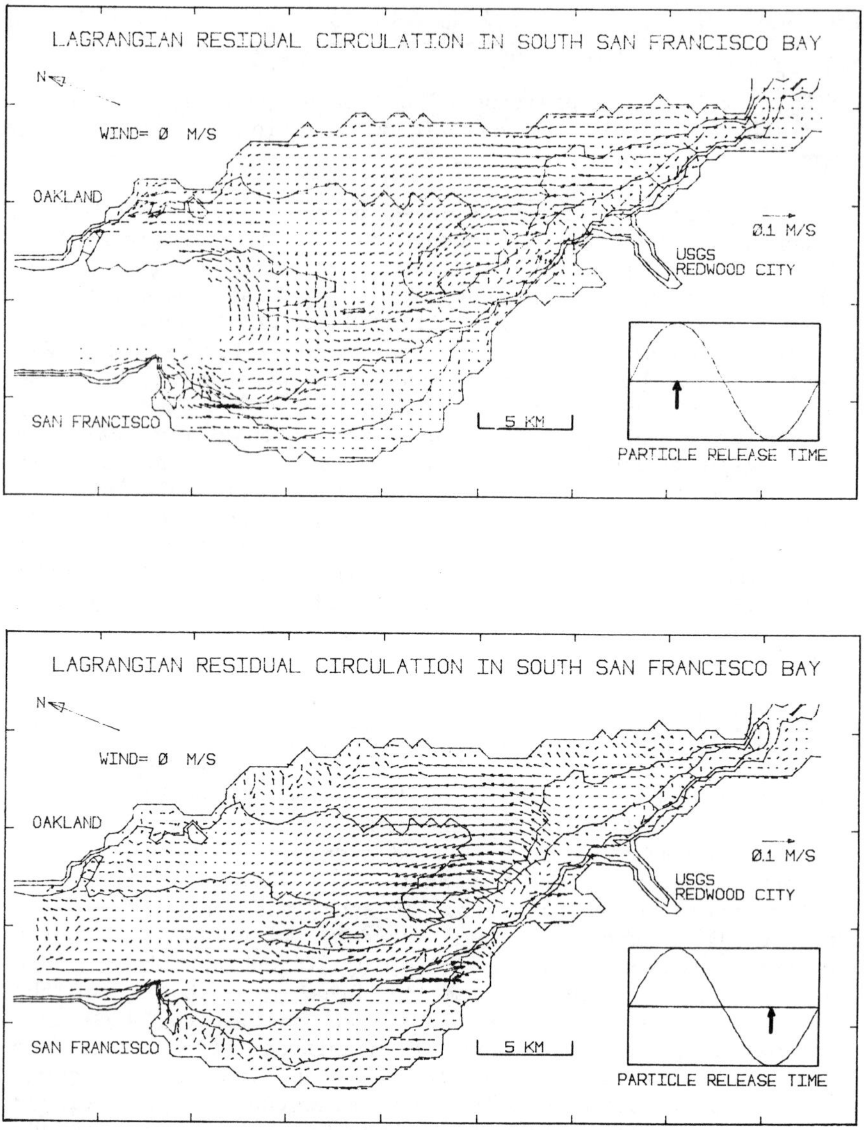

Fig. 3. Lagrangian residual circulation in South San Francisco Bay due to a semi-diurnal tide. The massless tracers were released when the high water (a), and low water (b) were at the open boundary. When there is no velocity vector plotted, the tracer particles have moved outside of the computation. The Lagrangian residual circulation there is unknown, not zero.

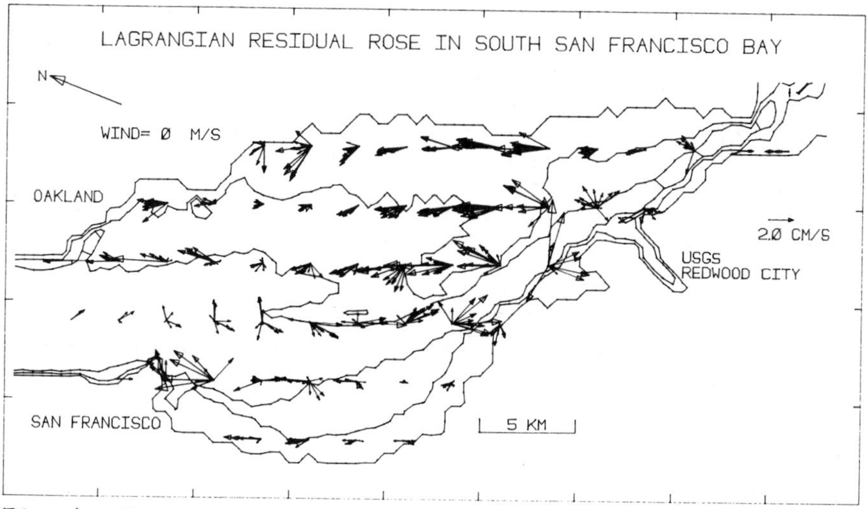

Fig. 4. Lagrangian residual circulation is calculated by releasing tracers at one-hour intervals over a 12-hour period. The base-line tidal flow was driven by a semi-diurnal tide of 12-hour period, and 1 m. amplitude. Distribution of the resultant vectors is a function of tidal phase. The spread of velocity vectors suggests a mechanism of tidal current induced mixing.

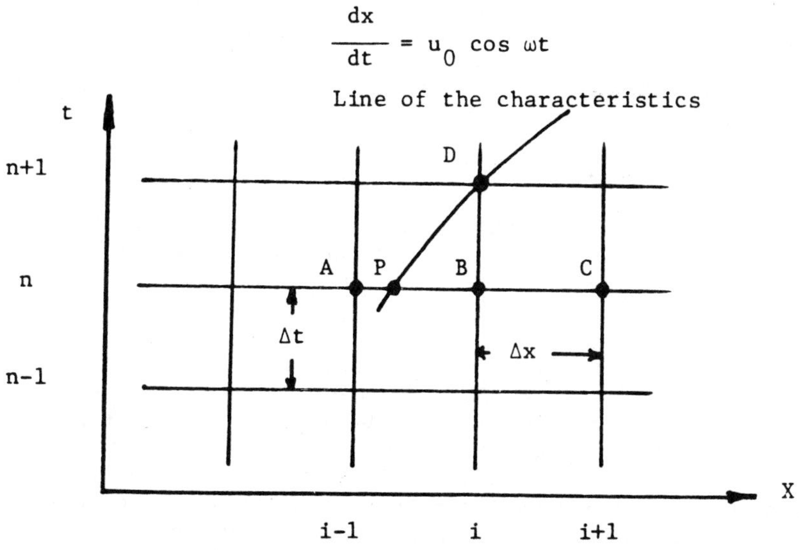

Fig. 5. Schematic diagram of the finite difference approximation. C_i^{n+1} is defined at $x = i\,\Delta x$, $t = (n+t)\,\Delta t$. C(P) is the value of C at point P which is defined by the line of characteristics.

the end points of the residual vectors gives the probable final position of tracers as if the tracers were released continuously. The net displacement of mass is, of course, the combined effects due to residual convection and tidal induced dispersion. The present author suggests that the tidal induced dispersion is characterized by the spreading of Lagrangian residual vectors. These results are preliminary; further quantification of the mixing coefficient is currently being considered.

(d) Convective Transport Equation

Consider, for simplicity, the one-dimensional convection dispersion equation as the last example of Euler-Lagrangian computations.

$$\frac{\partial C}{\partial t} + u_o \cos \omega t \frac{\partial C}{\partial x} = \kappa \frac{\partial^2 C}{\partial x^2} \qquad (2)$$

at $t \leqq 0$ $\quad C = 1$ for $\quad |x| < a$

$C = 0$ elsewhere

and for $t > 0$, $c \to 0$ as $\quad |x| \to \infty$

where u_o, ω, and κ are constants. The exact solution of the above initial-boundary value problem is

$$C(x,t) = \qquad (3)$$

$$1/2\{\mathrm{erfc}[-\frac{(x+a)-u_o/\omega \sin \omega t}{2\sqrt{\kappa t}}]-\mathrm{erfc}[-\frac{(x-a)-u_o/\omega \sin \omega t}{2\sqrt{\kappa t}}]$$

where erfc is the complementary error function.

When Eq. (2) is solved by a fixed grid method of characteristics, Eq. (2) is first rewritten in Lagrangian form as

$$\frac{DC}{Dt} = \kappa \frac{\partial^2 C}{\partial x^2} \qquad (4)$$

The substantial derivative can be treated by a central-time difference along the line of characteristics, giving

$$\frac{C_i^{n+1} - C(P)}{\Delta t} = \frac{C_{i-1}^n + C_{i+1}^n - 2C_i^n}{(\Delta x)^2} \qquad (5)$$

The finite difference approximations used are shown schematically in Fig. 5. Because a fixed grid system is used, the value of C(p) must be obtained by interpolation from neighboring points. When a linear interpolation is used, the method is equivalent to the up-wind differencing scheme. In this case the method is stable but the numerical dispersion induced by this method renders the scheme practically useless [17,18]. However, when a three-point interpolation (Fig. 5, between i-1, i, i+1) based on Lagrangian polynomials is used, the maximum relative error of the numerical solution is less than 4% after the integration is carried over 960 time steps using a grid Courant number $(U_o \Delta t/\Delta x) = 0.25$. Further extension of the Lagrangian method to solve higher dimensional convection dispersion equation and the analysis of the numerical properties of this scheme are being investigated.

4. CONCLUDING REMARKS:

Although Lagrangian methods are more descriptive for transport processes in fluid flows, both the field methods and computational techniques based on the Lagrangian viewpoint are not so well developed and advanced as Eulerian methods. The major difficulty in a pure Lagrangian computation is that the computational grids vary with the movements of fluid flows. The computational grids can be so distorted that the numerical errors stemming from discrete approximation of the governing equations erode the validity of the numerical solution.

Because modeling techniques based on Eulerian methods are well tested, Eulerian modeling results can be used as the basis for performing additional Lagrangian computations. The examples given above suggest that this combined Euler-Lagrangian method is very effective in revealing accurately properties of fluid flows which are truly of Lagrangian nature. It may be advantageous to use this technique for investigations of other aspects of transport processes in fluids.

5. REFERENCES:

1. BATCHELOR, G.K. - An Introduction to Fluid Dynamics, Cambridge Press, 1967.

2. MAHMOOD, K. and YEVJEVICH, V. - Unsteady Flow in Open Channel. Water Res. Pub., Fort Collins, Colorado, Vol. 1, 1975.

3. DRONKERS, J.J. - Tidal Computations in Rivers and Coastal Waters, John Wiley, New York, 1964.

4. LEENDERTSE, J.J. - Aspects of a Computational Model for Long Period Water Wave Propagation, Rep. RM-5294-PR, Rand Corp., Santa Monica, Calif., 1967.

5. LEENDERTSE, J.J. - A Water-quality Simulation Model for Well-mixed Estuaries and Coastal Seas. Vol. I, Principles of Computation, Rep. RM-6230-RC, Rand Corp., Santa Monica, Calif., 1970.

6. LEENDERTSE, J.J. and GRITTON, E.C. - A Water-quality Simulation Model for Well-mixed Estuaries and Coastal Seas, Vol. II., Computation Procedure, Rep. R-708-NYC, Rand Corp., Santa Monica, Calif., 1971.

7. CHENG, R.T., POWELL, T.M., and DILLON, T.M. - Numerical Models of Wind-driven Circulation in Lakes. Appl. Math. Modelling, Vol. 1, No. 4, pp. 141-159, 1976.

8. CHENG, R.T. and CASULLI, V. - On Lagrangian Residual Currents with Applications in South San Francisco Bay, California. Water Resources Research, Vol. 18, No. 6, pp. 1652-1662, 1982.

9. CHENG, R.T. - Modeling of Tidal and Residual Circulation in San Francisco Bay, California, Proceedings of a Seminar on Two-dimensional Flow Modeling, Hydrologic Eng. Center, U.S. Army Corps of Eng, Davis, Calif., pp. 172-185, 1982.

10. CHENG, R.T. and GARTNER, J.W. - Tides, Tidal and Residual Circulation in South San Francisco Bay, California. AGU, EOS, Vol. 63, No. 45, p. 946, 1982.

11. ZIMMERMAN, J.T.F. - On the Euler-Lagrangian Transformation and the Stokes Drift in the Presence of Ocillatory and Residual Currents. Deep Sea Res., Vol. 26A, pp. 505-520, 1979.

12. TEE, T.K. - Tide-induced Residual Current: A2-DNonlinear Numerical Model. J. Mar. Res., Vol. 31, pp. 603-628, 1976.

13. LONGUET-HIGGINS, M.S. - On the Transport of Mass by Time varying Ocean Currents. Deep Sea Res., Vol. 16, pp. 431-447, 1969.

14. ALFRINK, B.J. and VREUGDENHIL, C.B. - Residual Currents. Rep R1469-11, Delft Hydraul. Lab, the Netherlands, 1981.

15. AWAJI, T. - Mixing in a Tidal Current and the Effect of Turbulence on Tidal Exchange through a Strait. J. Phys. Ocean, Vol. 12, pp. 501-514, 1982.

16. CHENG, R.T. - Comments on "Water Mixing in a Tidal Current and the Effect of Turbulence on Tidal Exchange through a Straight," to appear in J. Phys. Ocean, 1983.

17. ROACHE, P.J. - Computational Fluid Dynamics, Hermosa Publ, Albuquerque, NM., 1972.

18. HUFFENUS, J.P. AND KHALETZKY, D. - The Lagrangian Approach of Advective Term Treatment and Its Appplication to the Solution of the Navier-Stokes Equations, Inter. J. or Num. Meth. in Fluids, Vol 1, pp. 365-387, 1981.

APPLICATIONS IN HYDRAULICS OF A CURVILINEAR FINITE-DIFFERENCE METHOD FOR THE NAVIER-STOKES EQUATIONS

B.J. Alfrink*, M.J. Officier, C.B. Vreugdenhil, H.G. Wind

Delft Hydraulics Laboratory, Delft, The Netherlands

SUMMARY

A short description is given of the program system ODYSSEE for the numerical solution of the Navier-Stokes equations in two-dimensional regions. Both truly 2-d and depth-averaged flows can be handled. Turbulence can be simulated using various formulations for the eddy viscosity. For the depth-averaged case, a rigid-lid assumption is made. A finite-difference scheme in fractional time steps on boundary-fitted curvilinear grids is used. Some examples are given, particularly for depth-averaged flows, in which the flexibility of the curvilinear grids is shown. Some numerical aspects are briefly discussed.

1. INTRODUCTION

The potential of numerical solutions of the Navier-Stokes equations has been recognized in many fields of application. Following earlier research activities, the Laboratoire National d'Hydraulique LNH (Chatou, France) and the Delft Hydraulics Laboratory DHL (Delft, The Netherlands) joined their efforts in the development of a powerful and reliable program system ODYSSEE for application in hydraulics. As far as the type of flows is concerned, ODYSSEE was designed to cover the following aspects:

- two-dimensional flows in complex geometries, either in a plane or in cases of cylindrical symmetry;
- two-dimensional depth-averaged flows in shallow water;
- either laminar or turbulent flow, in the latter case using standard turbulence models;
- flows with buoyancy influence by heat or salinity.

To reach these goals, use was made of curvilinear boundary-fitted coordinate systems for easy representation of complex

---

* presently at ENR Computing Center, Petten, The Netherlands

geometries, and of finite-difference methods in fractional time steps [1,7,8,9].

In this paper, a short description will be given, together with some applications that illustrate the potential and flexibility of the system.

2. FORMULATION

1.2 Equations

The basic equations are the Reynolds averaged equations for unsteady flow in two dimensions, possibly with an additional average over depth:

$$\frac{\partial u}{\partial t} + u\frac{\partial u}{\partial x} + v\frac{\partial u}{\partial y} - fv + \frac{1}{\rho}\frac{\partial p}{\partial x} + \beta\,(T-T_o)g_x + c_f\,Vu/h + - \frac{1}{h}\frac{\partial}{\partial x}(h\,T_{xx}/\rho) - \frac{1}{h}\frac{\partial}{\partial y}(hT_{xy}/\rho) = 0 \tag{1}$$

$$\frac{\partial v}{\partial t} + u\frac{\partial v}{\partial x} + v\frac{\partial v}{\partial y} + fu + \frac{1}{\rho}\frac{\partial p}{\partial y} + \beta\,(T-T_o)g_y + c_f\,Vv/h + - \frac{1}{h}\frac{\partial}{\partial x}(hT_{xy}/\rho) - \frac{1}{h}\frac{\partial}{\partial y}(hT_{yy}/\rho) = 0 \tag{2}$$

$$\frac{\partial}{\partial x}(hu) + \frac{\partial}{\partial y}(hv) = 0 \tag{3}$$

The notation is explained at the end of the paper. In case of pure 2-d flow, the depth h is constant and the bottom friction terms are omitted. For buoyant flows, the temperature T is computed from

$$\frac{\partial T}{\partial t} + u\frac{\partial T}{\partial x} + v\frac{\partial T}{\partial y} - \frac{1}{h}\frac{\partial}{\partial x}(hF_x) - \frac{1}{h}\frac{\partial}{\partial y}(hF_y) = 0 \tag{4}$$

2.2 Turbulence

For all practical purposes, the stresses T_{xx}, etc., and the heat fluxes F_x, F_y are evaluated by means of an eddy viscosity ν_t and an eddy diffusion coefficient κ_t:

$$\frac{1}{\rho}T_{xx} = 2\,\nu_t\frac{\partial u}{\partial x} - \frac{2}{3}k \qquad \frac{1}{\rho}T_{yy} = 2\,\nu_t\frac{\partial v}{\partial y} - \frac{2}{3}k$$

$$\frac{1}{\rho}T_{xy} = \nu_t\left(\frac{\partial u}{\partial y} + \frac{\partial v}{\partial x}\right) \tag{5}$$

$$F_x = \kappa_t\frac{\partial T}{\partial x} \qquad F_y = \kappa_t\frac{\partial T}{\partial y}$$

The eddy coefficients ν_t and κ_t may be formulated as constants for "quasi-laminar" flow, or expressed by mixing-length theory; finally, a k-ε model for turbulence may be used, if necessary with correction terms for depth averaging:

$$\nu_t = c_\mu \, k^2/\varepsilon \qquad \kappa_t = \nu_t/\sigma_t \tag{7}$$

The equations describing the evolution of k and ε are standard and they are not reproduced here (cf. [11]).

2.3 Rigid-lid approximation

For depth-averaged flows, the term $\partial h/\partial t$ has been omitted from the continuity-equation (3), and the depth h is assumed to be a known function of x and y. This amounts to the rigid-lid approximation, in which the free surface is replaced by a frictionless lid. The pressure head $p/\rho g$, exerted on the lid, will be an approximation of the actual free-surface elevation. The rigid-lid approach limits the application to relatively small regions in which important fluctuations of the water-level do not occur. Wavelike flows are excluded and the character of the equations changes from the hyperbolic long-wave equations to the elliptic Navier-Stokes equations.

2.4 Boundary conditions

The usual types of boundary conditions are used, with velocity components specified at inflow, no-slip, free-slip or law-of-the-wall conditions at fixed boundaries, and "free" conditions at outflow.

3. NUMERICAL METHODS

The previous experience with finite-difference methods has led to the choice of such methods on curvilinear grids, rather than finite-element methods. The great potential of curvilinear grids is shown by the proceedings of a recent conference on the subject [12]. In ODYSSEE, the region to be modelled is mapped onto a region consisting of rectangles in the transformed (ξ,η) plane (Fig. 1). A variety of techniques can be used, varying

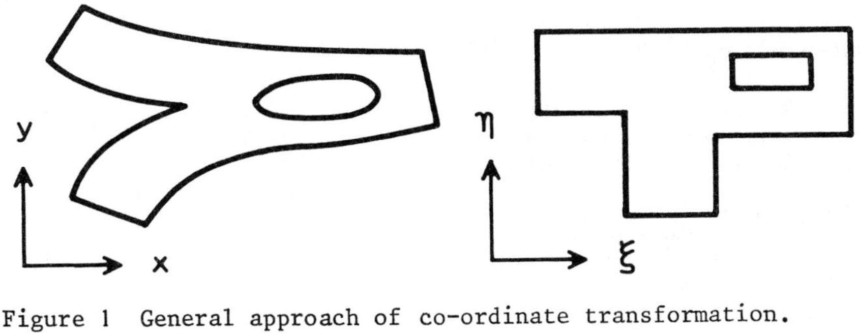

Figure 1 General approach of co-ordinate transformation.

from simple co-ordinate stretching to the solution of elliptic equations for the co-ordinates, as proposed by Thompson. The resulting co-ordinate systems need not be orthogonal. All computations are done on a rectangular grid in the (ξ,η) plane. The numerical grid is staggered in space, with (u,v) components defined in cell corners and pressure p in the cell center of gravity. For details on the co-ordinate transformations cf. [8]. Some examples are given below.

3.2 Finite-difference methods

The finite-difference methods used are outlined for the original Cartesian coordinates (x,y) without expanding the full expressions. For details, cf. [9]. The equations are split in functional parts and solved in fractional time steps.

Convection is treated by a method of characteristics in two dimensions [7]. Intermediate velocity components (u^*,v^*) are determined from

$$\frac{\partial u^*}{\partial t} + u^n \frac{\partial u^*}{\partial x} + v^n \frac{\partial u^*}{\partial y} = 0$$
$$\frac{\partial v^*}{\partial t} + u^n \frac{\partial v^*}{\partial x} + v^n \frac{\partial v^*}{\partial y} = 0 \qquad (8)$$

Using the known flow field (u^n, v^n), characteristics (flow paths) are traced backwards from the new grid points. At the origin of the characteristics, the velocities are interpolated cubically between the known grid values. As the characteristics may cross cell boundaries, the Courant number need not be restricted below unity.

Diffusion is treated together with bottom friction (if any) by an implicit method, using SOR iteration so far. More efficient methods are being studied. Due to the formulation of Eq. (5), the u and v equations are coupled, and solved together.

$$\frac{u^{**}-u^*}{\Delta t} + ru^{**} - \frac{1}{h}\left[\frac{\partial}{\partial x}\left(2h\,\nu_t \frac{\partial u}{\partial x}\right) + \frac{\partial}{\partial y}\left\{h\,\nu_t\left(\frac{\partial u}{\partial y} + \frac{\partial v}{\partial x}\right)\right\}\right]^{**} = 0$$
$$\frac{v^{**}-v^*}{\Delta t} + rv^{**} - \frac{1}{h}\left[\frac{\partial}{\partial x}\left(h\,\nu_t\left(\frac{\partial u}{\partial y} + \frac{\partial v}{\partial x}\right)\right) + \frac{\partial}{\partial y}\left(2h\,\nu_t \frac{\partial v}{\partial y}\right)\right]^{**} = 0 \qquad (9)$$

where $r = c_f\, V^*/h$.

To satisfy the continuity equation, two alternative methods can be used. One is the solution of a Poisson equation for the pressure [6], following from Eq. (3) together with

$$\frac{u^{n+1}-u^{**}}{\Delta t} + \frac{1}{\rho}\frac{\partial p}{\partial x} = 0$$
$$\frac{v^{n+1}-v^{**}}{\Delta t} + \frac{1}{\rho}\frac{\partial p}{\partial y} = 0 \qquad (10)$$

The resulting equation is solved by a biconjugate-gradients method; again, more efficient methods are being studied.

The second method to satisfy continuity is to introduce a stream-function ψ, based on the fact that the vorticity of the flow field has already been updated in the previous fractional steps, and the divergence remains to be adjusted.

$$hu^{n+1} = -\partial\psi/\partial y \qquad hv^{n+1} = \partial\psi/\partial x \tag{11}$$

$$\nabla^2\psi = \frac{\partial v^{**}}{\partial x} - \frac{\partial u^{**}}{\partial y} \tag{12}$$

The stream-function equation (12) is solved by SOR iteration in anticipation of more efficient methods.

4. STABILITY AND ACCURACY

Due to the set-up of the method of characteristics for the convection step, there are no stability limits on the Courant number

$$\sigma = V\,\Delta t/\Delta x \tag{13}$$

From an accuracy point of view, σ should not be too large. Note, however, that for depth-averaged flows this is much more favourable than the usual Courant condition based on the propagation speed of surface waves.

The diffusion and continuity steps, by their implicit character, do not impose stability limits either. It has been found, however, that the source terms in the k-ε equations do cause stability problems, which are solved by using a Runge-Kutta method with self-adjusting time steps.

As a consequence, the choice of the time step is mainly motivated by accuracy considerations. Some numerical experiments with respect to accuracy have been reported in [3]. The following sources of error may be discerned:

(a) Even for constant ("frozen") coefficients, it appears that the fractional step method introduces a truncation error of first order in Δt, which persists in the steady state.
(b) Coefficients are actually taken at "old" time levels, which introduces another first order error. This is no longer important in the steady state.
(c) Spatial accuracy is at least of second order in $\Delta\xi$, $\Delta\eta$. However, it has been shown that inaccuracies can be introduced by the coordinate transformation [10].

5. APPLICATIONS

5.1 Flow pattern in river junctions

In a study of hydropower in the Dutch river system, a location in the River Maas has been studied, which is given schematically in Fig. 2. The projected hydropower station bypasses a ship lock and flow patterns would be critical in two junctions indicated in the figure. Flow patterns as the upstream side were studied previously by different means. Detailed models of the two river junctions were set up using the ODYSSEE system.

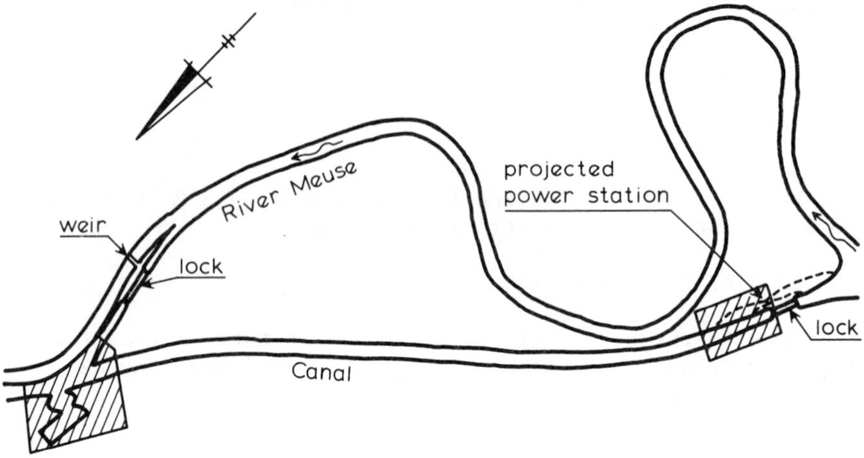

Figure 2 General situation of hydropower station in the River Maas

The co-ordinate systems generated for the two situations are shown in Fig. 3. Care was taken to have a grid as smooth as possible, with increased resolution at critical points. In both cases, the number of grid points was roughly 900. For the grid generation, the elliptical technique was used, with control by the specification of boundary grid points. No attraction functions were used for internal grid modification.

Note that these are rather small scale applications, which supports the rigid-lid approximation as discussed above. The average grid size is about 10-20 m in both cases.

Steady-state solutions were obtained for constant eddy viscosity $\nu_t = 1\ m^2/s$, starting from an initial situation at rest and using time steps ranging from 15 to 360 s. With velocities of 2 m/s near the outlet, this results in Courant numbers up to 12. In the greater part of the models, the velocities are much lower and so are the Courant numbers. Some time histories of the approach of steady state are given in Fig. 4. It is seen that not only the transient but also the steady state is influenced by the time step. This was also found in [3]. From the figure, a time step of 30 s appears to give satisfactory accuracy. The initial adjustment of the flow pattern is extremely quick due to the rigid-lid assumption. Final adjustment due to bottom friction and viscosity is much slower. Some of the resulting flow patterns are shown in Figure 5. Similar runs using the depth-averaged k-ε model were planned but not yet available at the completion of this paper. The required computer time for 900 grid points and 25 time steps was about

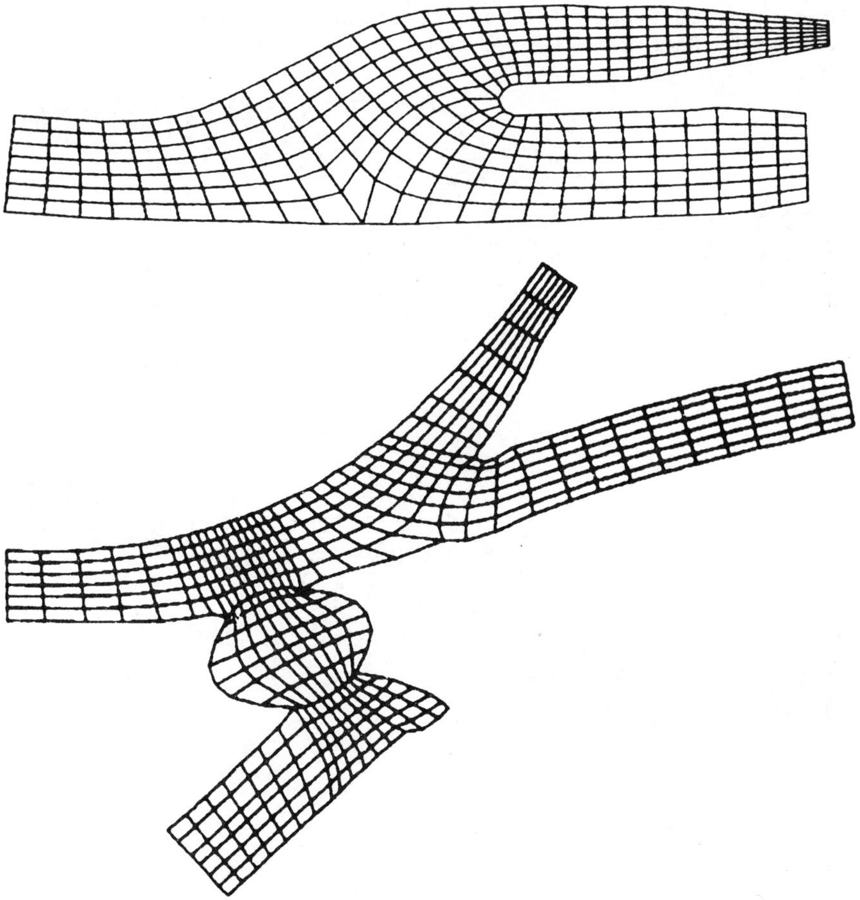

Figuur 3 Grids for the two detail models

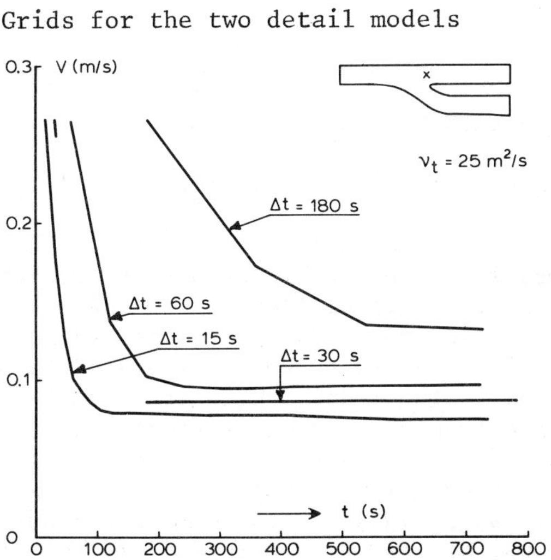

Figure 4 Time histories for selected points in case A

40 s on a CDC Cyber 175. No comparison with field data has yet been made. Some material in this respect has been presented in [9].

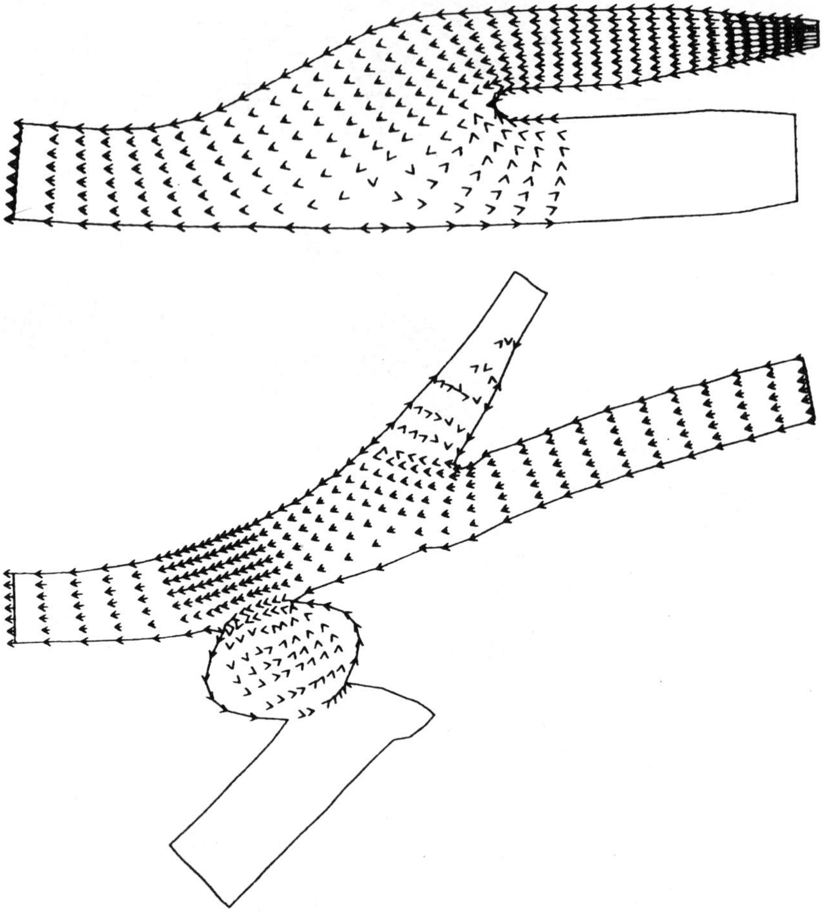

Figure 5 Computed flow patterns using Δt = 30 s.

5.2 Flow pattern near a harbour entrance

In order to assess the applicability of ODYSSEE to coastal currents, a model was set up of the Amsterdam harbour entrance near IJmuiden. The local flow pattern near the entrance is an essential link in the study of erosion and sedimentation in the area. The situation is shown in Figure 6. An area of 13x7 km was covered in the model, which is so small compared with the tidal wave length, that the rigid-lid approximation in ODYSSEE is considered to be acceptable. It is, however, to be expected that the velocity boundary conditions will have to be time dependent. A grid was generated by co-ordinate stretching in subregions, as illustrated in Fig. 7. This results in a less smooth grid than in the previous case and therefore

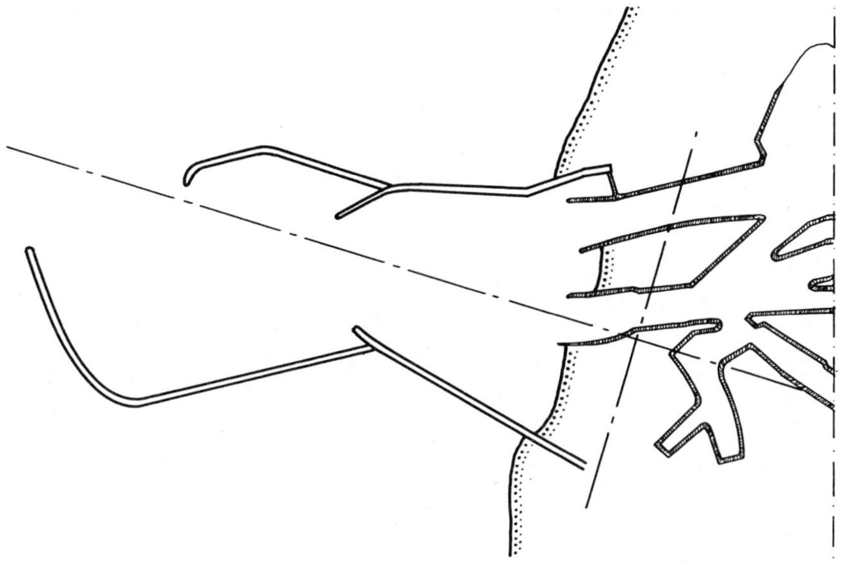

Figure 6 IJmuiden harbour entrance

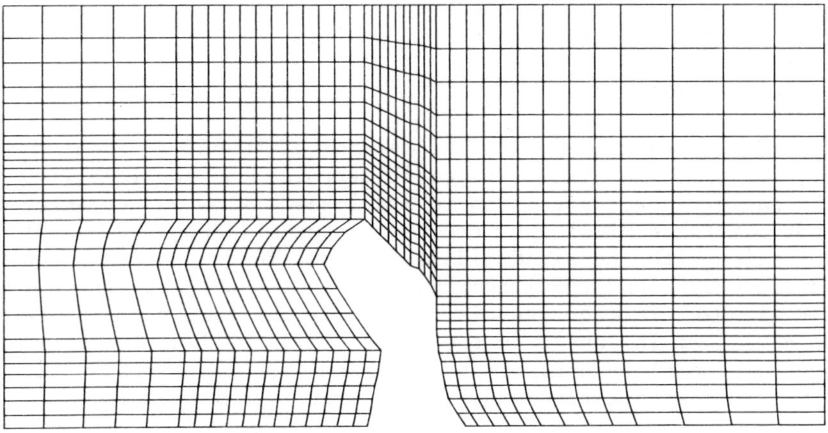

Figure 7 Grid of IJmuiden area

another experiment is planned in a later stage using a smooth grid. This will give experimental evidence of the importance of the grid for the final accuracy.

To start with, a steady flow of about 0.6 m/s from the North was applied, to get an impression of the time scale of the flow adjustment. A time step of 150 s was used, resulting in Courant numbers ranging from an average of 0.25 to 1 near the tips of the harbour moles. Some of the resulting flow patterns are shown in Fig. 8. It is found that the major features of the flow field adjust quite quickly. However, it takes considerably more time for the eddy downstream of the breakwaters to develop. This takes several hours. It is, therefore, to be

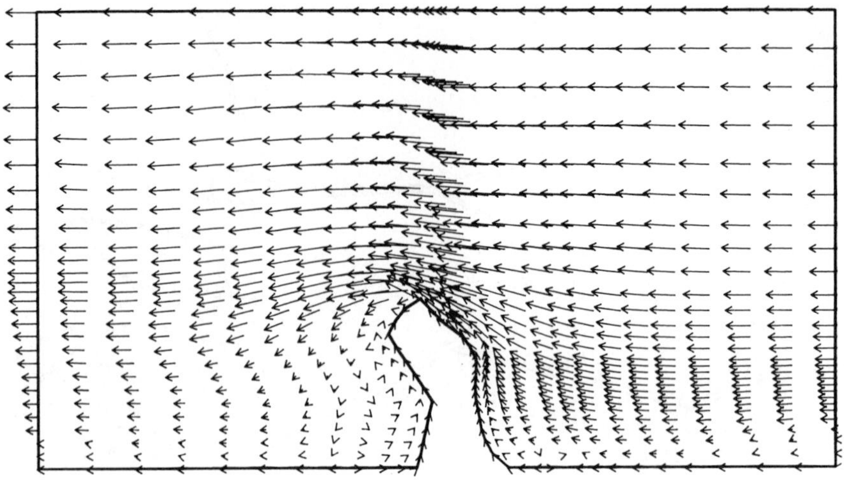

Figure 8 Computed flow patterns for IJmuiden area at time 3 hours

expected that the eddy does not fully develop in a tidal flow situation. This will be checked in a separate numerical experiment with tidal flow boundary conditions.

6. CONCLUSION

Following applications to several of two-dimensional flows [1,4,5,8,9], the present paper illustrates the applicability of the Navier-Stokes equations to small-scale flow patterns for depth-averaged flows. The program system ODYSSEE handles this situation with the rigid-lid assumption. The potential of using curvilinear boundary-fitted coordinates is once more shown. As far as numercial accuracy is concerned, the time step is found to have some influence in the transient as well as the steady state. A complete analysis of accuracy is not yet available. A comparison with field data is not given in this paper.

ACKNOWLEDGEMENT

The authors express their gratitude to NV. PLEM (Electricity Corporation of Limburg Province) for permission to present some results of the River Maas study.

7. REFERENCES

1. ALFRINK, B.J. The computation of turbulent recirculating flow using curvilinear finite differences, Lab. Nat. d'Hydraulique, Chatou, France, Report HE/41/81.22, 1981

2. ALFRINK, B.J. On the Neumann problem for the pressure in a Navier-Stokes model, Proc. Int. Conf. Numerical Methods in Laminar and Turbulent Flow, C. Taylor and B.A. Schrefler, eds. Pineridge Press 1981, 389-399, Also: Delft Hydraulics Lab., Publ. 257
3. ALFRINK, B.J. Flows in complex geometries, IAHR Working Group on Refined Modelling of Flows, Rome, 1982
4. ALFRINK, B.J. Importance of refined turbulence modelling for the flow over a trench, Proc. Int. Symp. Refined Modelling of Flows, J.P. Benqué et al, eds., Paris, 1982. Also Delft Hydraulics Lab., Publ. 268
5. ALFRINK, B.J. and VAN RIJN, L.C. A k-ε model for turbulent flow in dredged trenches, J. Hydr. Division ASCE, submitted for publication
6. CHORIN, A.J. Numerical solution of the Navier-Stokes equations, Mathematics of Computation, 22, 1968, 745-762
7. ESPOSITO, P. Resolution bidimensionelle des équations de transport par la méthode des charactéristiques, Lab. Nat. d'Hydraulique Chatou, France. Report HE/41/81.16, 1981
8. GOUSSEBAILLE, J. Modelisation d'écoulements et de transferts de chaleur par une méthode de differences finis en mailles curvilignes non orthogonales, Lab. Nat. d'Hydraulique, Chatou, France, Report HE/41/81.27, 1981
9. MARY, D. Modelisation numérique bidimensionelle des écoulements en rivière, application à l'étude des rejets thermiques, Thèse Ecole Nat. des Ponts et Chaussées, Paris, 1982
10. MASTIN, C.W. Error induced by co-ordinate systems. Ref [12], pp. 31-40
11. RODI, W. Turbulence models and their application in hydraulics, IAHR, Delft, The Netherlands, 1980
12. THOMPSON, J.F. (ed.) Symposium on Numerical Grid Generation, Nashville, 1982, North-Holland Publ. 1982

NOTATION

C_f	bottom resistance coefficient
f	Coriolin parameter
F_x, F_y	heat fluxes
g_x, g_y	components of gravitational acceleration
h	water depth
k	turbulent kinetic energy
p	pressure
t	time
T	temperature
T_{xx}, T_{xy}, T_{yy}	effective stresses
u, v	velocity components
V	magnitude of velocity vector
x, y	co-ordinates
β	coefficient of thermal expansion
ε	dissipation rate of turbulence energy
κ_t	eddy diffusion coefficient for heat
ν_t	eddy viscosity
ρ	fluid density
σ_t	turbulent Schmidt number
ψ	stream function

NUMERICAL PREDICTION OF SEA FLOW AND TEMPERATURE RANGE BETWEEN POWER STATION INTAKE AND DISCHARGE PORTS IN A BAY

N C Markatos, Concentration Heat and Momentum Limited
40 High Street, Wimbledon, London SW19 5AU, UK.

and R Simitovic, Electrowatt Engineering Services Limited
Bellerivestrasse 36, CH-8022 Zurich, Switzerland.

ABSTRACT

A power station uses sea water as primary coolant, drawing cool water through intakes at one part of the coast and discharging it at another part.

A numerical procedure is used to predict the sea temperature at the intakes, under prescribed ocean-current conditions. Of particular importance is the case where the ocean-current flow is likely to convect the hot water from the outlet back to the intakes. A two-dimensional, depth-averaged model was used; and the topography of the bay, and the profile of the ocean bed, were specified by means of non-unity porosities.

The results show deep penetration of the jet into the sea, a vortex motion between intake and discharge channels, and strong bending of the jet due to the momentum forces of ocean current. The same general method can be used for performing two-dimensional two-layer, and fully three-dimensional analyses.

1. INTRODUCTION

Increasing demand for efficient use of energy requires the optimisation of layout of the cooling-water structures of power plants, located at coastal zones or off-shore. The problem considered in this paper is the determination of the effect of thermal discharge from a coastal power plant, on cooling-water intake temperature, for various tidal conditions. The long-term objective of this work is to examine whether and by what mechanisms the wasted heat, emitted by discharging water, could be dispersed away from the coastal region; and also to study the layout of intake and discharge channels with regard to ensuring the optimum water temperature at the inlet of the intake channel.

The specific problem considered was of significant practical importance because the heat to be removed was very large and, at the same time, the distances between intake and discharge channels were short. Furthermore, the ocean current prevailing at the site of the plant was such that undesirable flow from discharge to intake could easily occur. For design purposes, it was attempted to examine the problem under the most extreme oceanographic, meteorological and operational conditions. A parametric study of the influence of important physical factors such as sea current and its temperature, and layout and temperature of the discharge jet was carried out, and sample results are presented and discussed.

2. THE MATHEMATICAL FORMULATION

The starting point of the analysis is the set of elliptic partial differential source-balance equations, that express the transport of mass, momentum, energy and other fluid variables in recirculating flows.

2.1 The Dependent and Independent Variables

In many environmental flows (eg rivers, lakes, oceans etc) the width to depth ratio is large, and thus the horizontal distribution of flow quantities is of more interest than the vertical distribution, which is often nearly uniform due to strong vertical mixing induced by the bottom shear. It is therefore possible to reduce the expense of the computations required for the solution of the actual three-dimensional problem, by integrating the equations over the water depth, without losing too much physical realism. Therefore, the independent variables considered are the two components of a cartesian coordinate system (y, z). The dependent variables are the horizontal velocities v, w, the pressure, p, and the sea temperature, T. The above variables are depth-averaged values, eg

$$\overline{\phi} = \frac{1}{h} \int_{o}^{h} \phi dx \qquad (1)$$

where h is the local depth, and x is the depth direction.

2.2 Differential Equations

The equations for all variables above, with the exception of pressure, take the following general form:

$$\text{div } \{(\rho\vec{v}\phi - \Gamma_\phi \text{ grad } \phi)\} = S_\phi \qquad (2)$$

where ρ, $\vec{v}$ Γ_ϕ and S_ϕ are density, velocity vector, "effective exchange coefficient of ϕ", and source rate per unit volume, respectively.

The sources and exchange coefficients for the variables considered are well-documented in the literature [1, 2, 3] and are not repeated here.

The pressure is associated with the continuity equation:

$$\text{div}\,(\rho\vec{v}) = 0 \qquad (3)$$

in anticipation of the so-called pressure-correction equation [4] which is deduced from the finite-domain form of the continuity equation.

2.3 Physical Properties

The density of the sea water is set to 1000 kg/m^3. The flow is turbulent (Re $\simeq 10^6$), arising as a consequence of natural sea motion (currents, tidal movements), morphology of the sea bed, coastal line geometry and disturbances caused by the discharging jets. A uniform "effective viscosity", fixed at 1000 times the laminar value, was used.

2.4 The Finite-Domain Equations

A finite-domain technique is used which combines features of the methods of [4, 5], and a whole-field pressure-correction solver. The space dimensions are discretised into finite intervals; and the variables are computed at only a finite number of locations, at the so-called "grid-points". These variables are connected with each other by algebraic equations, derived from their differential counterparts by integration over the control volumes or cells defined by the above intervals. This leads to equations of the form:

$$\sum_{\eta}(A_{\eta}^{\phi} + C)\phi_p = \sum_{\eta} A_{\eta}^{\phi}\phi_{\eta} + CV \qquad (4)$$

where the summation η is over the cells adjacent to a defined point p. The coefficients A_{η}^{ϕ}, which account for convective and diffusive fluxes across the elemental cell, are formulated using upwind differencing. The source terms are written in the linear form $S_{\phi} = C(V - \phi)$ where C, V stand for a coefficient and a value of the variable ϕ.

The "SIMPLEST" practice [5] is followed for the w-equation and the v-momentum equation is solved point-by-point.

The above solution procedure is incorporated into a general computer program for the solution of multi-dimensional, multi-phase problems, which has been described elsewhere [6 to 9].

2.5 Irregular Geometries

The irregular geometrical features (the land, the sea-bed variation and the intake channels) are modelled by use of "porosities" [6, 8]. In this approach, each cell in the domain is characterised by a set of fractions, normally in the range from 0 to 1. These fractions determine the proportion of the cell volume which is available for occupancy by the fluid, and the proportion of each cell-face area available for flow, by convection or diffusion, from the cell to its neighbour in a given direction.

It is worth mentioning that the variations in depth of the sea-bed were accounted by modifying the cell-face areas and volumes according to the local depth. Therefore, porosity settings greater than unity were used, since the depth is greater than 1 metre in most places.

Finally, as the size of the single cell of the finite-domain grid was sometimes larger than the width of the discharging jet, a special geometrical jet-inflow treatment was applied to preserve its momentum.

2.6 Boundary and Initial Conditions

Boundary conditions are specified as follows (see also Fig. 1): Depending on the ocean-current direction, one of the South, East or West boundaries is an inflow boundary. At this boundary a prescribed mass-flow rate of the sea current is used for the continuity equation (with proper account being taken of the variation in depth of the sea-bed); and given values for v, w and T, carried into the domain by this flow rate, are prescribed for the other equations. The other two of the above boundaries are then treated as fixed-pressure boundaries, condition that allows the computation of outflow/inflow through them. The coast is treated as an irregular solid boundary (see Section 2.5) and a constant friction factor of 3.10^{-3} is used to calculate the shear-stress there.

A fixed mass-flow rate is used to specify the discharge from the power station; and given values are used for w, v, T. Fixed flow rates are also used at the two intakes. Since the intakes represent outflows from the integration domain, there is no need to set values for the associated fluid properties. Surface heat loss appears as a heat sink and is calculated, throughout the calculation domain, based on a given heat-transfer coefficient.

Finally, the friction due to the sea-bed is applied throughout the calculation domain as a momentum sink (this frictional force acts on areas normal to the x-direction). The frictional force, per unit bulk sea speed, per unit sea-bed area, is taken as 0.04 Ns/m^3,value derived from the Darcy-

Weisbach expression for the representative sea-bed conditions considered.

The initial fields for p, v, w and T are set to the uniform ambient values throughout the domain.

3. APPLICATION OF THE PREDICTION PROCEDURE

3.1 Test Cases Considered

Several flow cases were modelled, differing from one another in one or more parameters, in such a way that the influence of each of them could be easily deduced and applied later for future similar work. Major variations were made to the sea-current direction and magnitude. The most detailed investigations were performed for the "worst-case" conditions, in which the ocean current moves parallel to the coastline, directed from the discharge port towards the two intake ports.

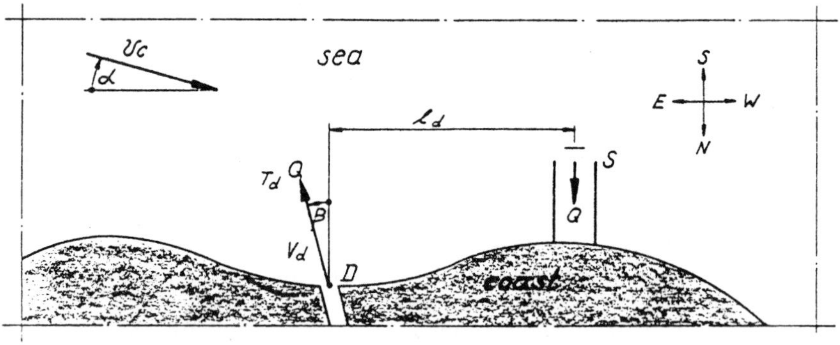

Reference values:

$Q = 100\ (m^3s^{-1})$, $\beta = 30\ (^o)$, $v_d = 5\ (ms^{-1})$,

$T_d = 8\ (^oC)$, $l_d = 1250\ (m)$, $v_c = 0.15\ (ms^{-1})$.

Figure 1: Layout of cooling-water structures

Figure 1 shows a schematic representation of the offshore region investigated. The coastline has an irregular contour, and the sea-bed depth varies unevenly, as shown in Figure 2. A submerged hillock is present directly ahead of the discharge. All of these factors are important in the simulation. Reference values of the important parameters are also given in Figure 1. Table 1 presents a summary of the pertinent characteristics of the flow cases presented.

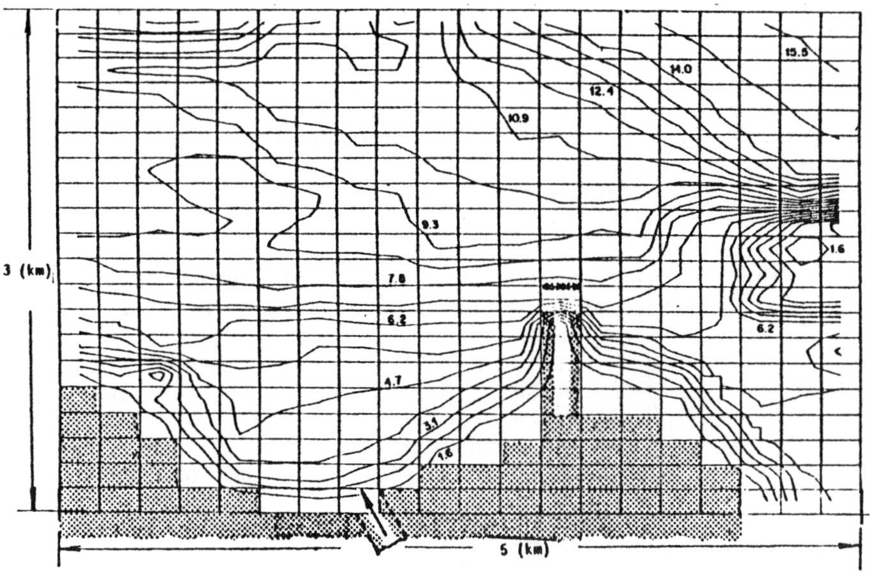

Figure 2: Finite-domain grid with sea-depth contours (in m) and coast shape

		Flow Cases					
		1	2	3	4	5	6
Sea Current	$\frac{v}{v_c}$	0	1				1.5
	α (deg)	-	0			90	135
Discharge jet	$\frac{Q_i}{Q}$	1		0.5	1		
Layout	$\frac{l}{l_d}$	1			1.4	1	

Table 1: Flow Cases studied

The flow case 1 refers to a zero-current sea; case 6 to the tide-in conditions of sea.

3.2 Grid-Dependence and Computer Storage

The reported results have been obtained using a non-uniform grid consisting of 31 cells in the y-direction and 22 in the z-direction. A typical grid is shown in Figure 2. The solutions reported are believed to be not fully independent of the grid. The extent of this dependence has not been ascertained.

The program required 290 kilobytes of storage on a Perkin-Elmer 3220 mini-computer, of which 77 kilobytes was for storage of variables, with the remainder taken by the program object code. It should be mentioned that the full program (eg three-dimensional, two-phase, see [9]) was loaded, and the above storage requirement can be significantly reduced in this particular case, by the omission of inactive subroutines. However, this possibility was not exploited in the work reported.

3.3 Convergence and Computer Time

After 200 sweeps performed, the sum of absolute cell-continuity volumetric errors is typically less than 0.5% of the total volumetric inflow of sea at the inlet boundary. The values of the dependent variables at that stage did not change significantly from one sweep to another and the procedure had converged.

This 200-sweep run took 16.4 minutes of CPU time to execute on a Perkin-Elmer 3220 mini-computer.

3.4 Results and Discussion

Some of the results of the study are presented here in the form of velocity vectors and of isotherms. Also given are the plots of temperature profiles along the line connecting the discharge and intake positions. The velocity vectors and the temperature contours have been obtained using GRAFFIC [10].

The zero current-velocity Case 1 (Fig 3) shows strong influence of the outgoing jet over the whole region eastwards from the structures. The jet momentum forces the entire discharge mass away from the shore, creating a large region of low pressure, located between intake and discharge channels. This causes a predominant local sea motion from west to east, leading to a large amount of "fresh" (cool) sea water entering the inflow channel.

Flow Case 2 refers to sea current flowing from east to west at a = 0°, and the discharge and intake channels are close to each other. As expected, this flow leads to strong convection of the hot water from the discharge channel back to the intakes. As a consequence, the temperature increase above the ambient at the intake channel is the largest observed, at

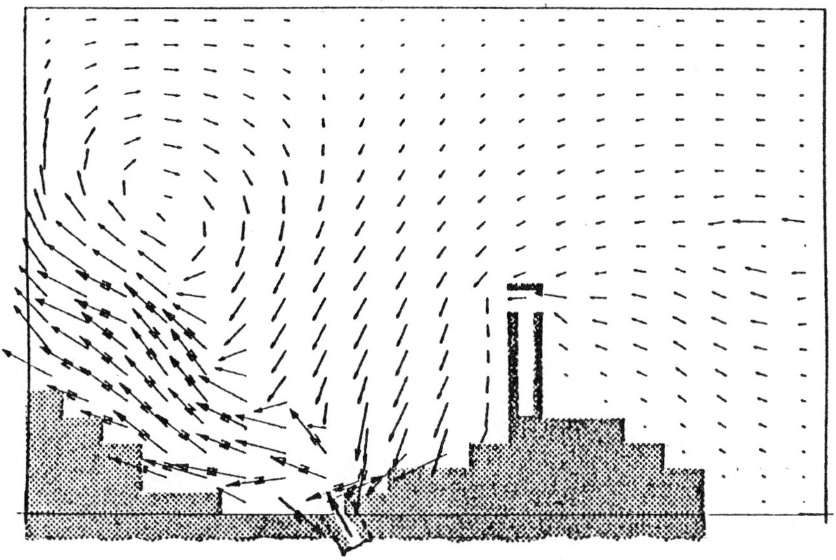

(a) Velocity vectors

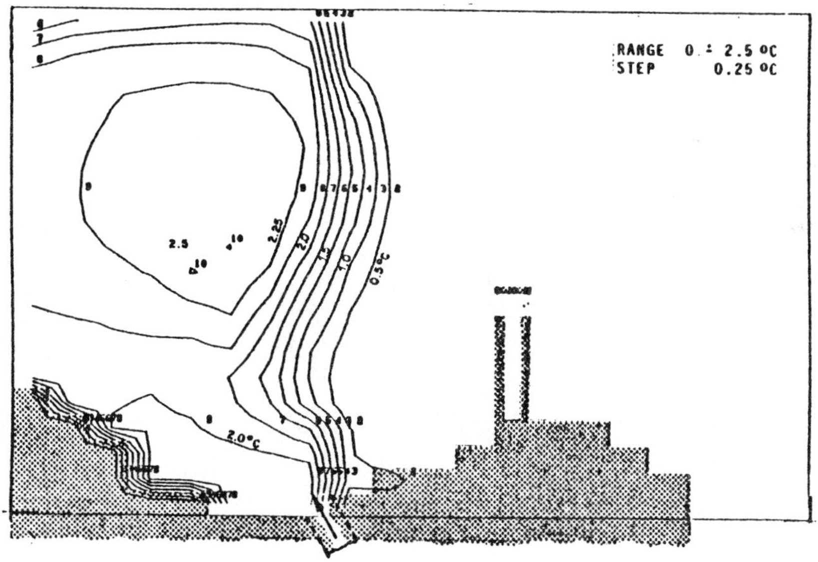

(b) Temperature contours

Figure 3: Flow patterns and Isotherms for Case 1

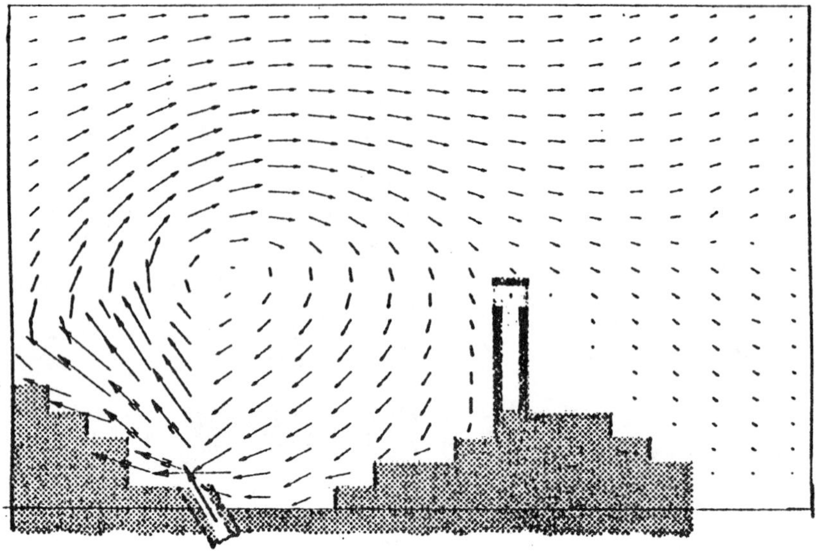

(a) Velocity vectors

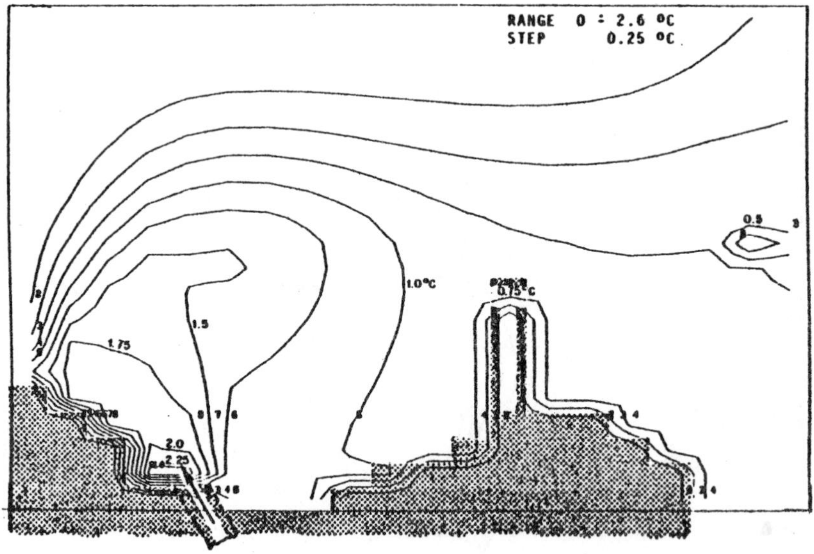

(b) Temperature cont urs

Figure 4: Flow patterns and Isotherms for Case 4

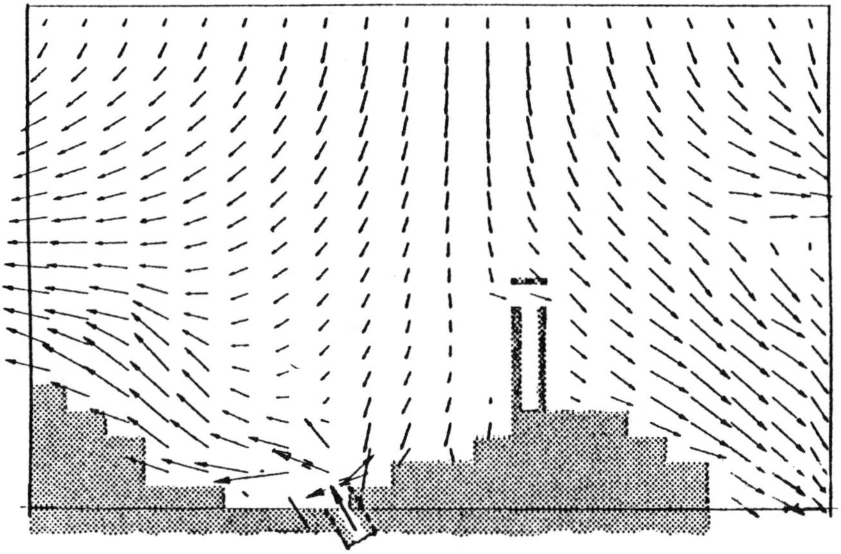

(a) Velocity vectors

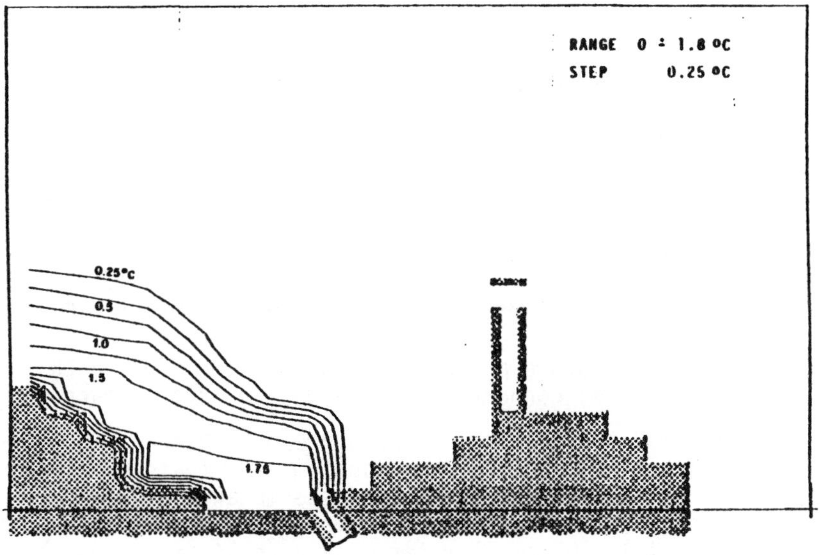

(b) Temperature contours

Figure 5: Flow patterns and Isotherms for Case 5

1.8°C. Case 3 is identical to Case 2 but the discharge flow rate is half of that before. The temperature increase at the intakes is now only 1.4°C above the ambient.

The spread of the jet into the open sea in Case 4 (Fig 4) is much better than before, due to the larger distance between the two cooling-water structures. As a consequence, the temperature increase at the intake is now only 0.9°C (indicated in Fig 4 by the temperature at the intake channels falling between the two temperature contours of 1°C and 0.75°C). A strong vortex is observed between discharge and intake channels. When the current is directed towards the coast, from south to north (Case 5), a strong partition of the current flow occurs, as is clearly shown in Fig. 5. The boundary of that partition is located between the discharge and intake structures. The discharged warm water leaves (together with "fresh" sea water) the region of cooling-water structures towards the east, away from the intakes, thus not causing any significant influence on the intake temperature.

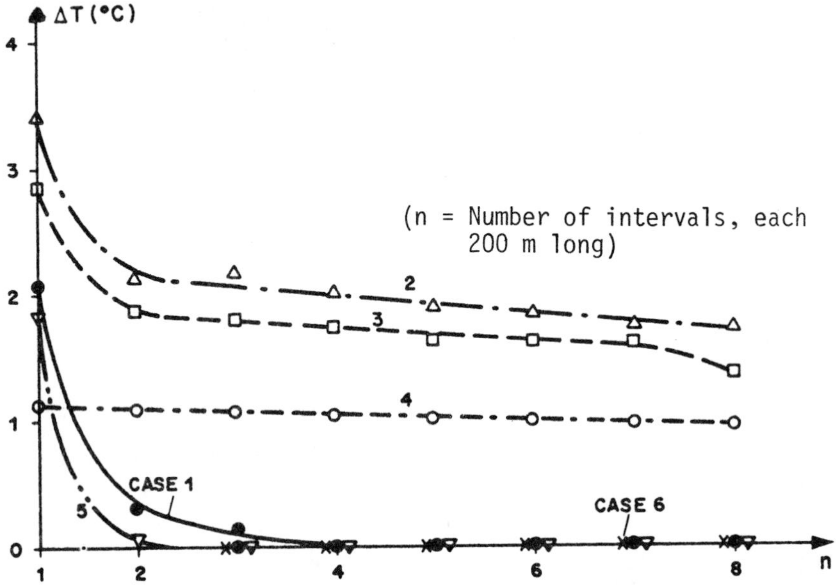

Figure 6: Temperature profiles along the line (distance 1.6 km) connecting discharge and intake channels

Figure 6 presents the temperature drop between the discharge and intake channels, and confirms the features already mentioned. The temperature rise at the intakes above the ambient is 1.8°C for Case 2 (full discharge flow rate) but it is approximately 20% smaller for Case 3 (half discharge flow rate) at 1.4°C. A larger distance between the cooling-water structures (case 4) causes a further decrease to the temperature rise (0.9°C). Finally Cases 1 (no sea-current), 5 (current directed towards the coast from south to north) and 6 (current directed towards the coast from SW to NE) do not show any appreciable temperature increase at the intake channels.

The study of all results obtained leads to the following general conclusions:

(a) For all strong-current cases the largest rate of mixing of warm and cold water occurs in a region of approximately 1 km by 1 km around the discharge position.

(b) Mechanisms for heat removal were the mixing of the hot jet into the cool open sea, sideways entrainment of "fresh" water towards the discharge channel and heat transfer to the atmosphere. The main mixing is caused by the side entrainment of "fresh" water, due to the low pressure generated in the region westwards of the entering jet.

(c) All flow cases considered, except the one of sea current moving towards the shore (Case 4), show a strong clockwise vortex formation. The centres of vortex move offshore with increasing ratio of jet to sea-current velocity and with increasing distance between discharge and intake structures [11,12]. On average, these locations are 1 to 2 km offshore.

(d) The maximum penalty that results from the given configuration is a 1.8^{o}C temperature rise at the intake when the discharge is placed upstream of the intakes. A limited number of calculations done on finer grids showed, as expected, changes in the values of the computed parameters. The trend observed was always towards lower intake temperatures for the finer grids. From this point of view, the predictions are therefore believed to be conservative.

4. CONCLUSIONS

Predictions were obtained of the flow and temperature fields in the sea surrounding the cooling-water structures of a coastal power plant.

Calculations were made using a general computer program in a two-dimensional, steady-state mode. The topography of the bay, including the irregular contour of coastline and the profile of the ocean bed, were specified by means of non-unity porosities. A fixed "effective viscosity" treatment was chosen (although the program provides for either the $k \sim \varepsilon$ or the $k \sim W$ two-equation turbulence models), in view of the coarseness of the finite-domain grid with respect to the dimensions of the computational domain. For such grids, velocity gradients cannot be resolved accurately and therefore solving for k becomes meaningless. The calculated variables are depth-averaged ones, and the most important consequence of this fact is that influential effects, such as fluid stratification, are neglected. Activation of the "two-phase" capability of the general program used can remedy this neglect, before proceeding to fully three-dimensional computations, also provided for by the same program.

Notwithstanding the above limitations, it was found that the simple models used provide a correct qualitative description of the problems under study, as judged by physical intuition and study of the performance of existing power plants. Furthermore the predictions have been obtained at a very low cost, each complete run requiring less than 20 mins on a minicomputer.

5 ACKNOWLEDGEMENTS

The authors wish to express their thanks to Mr. C. Aldham who performed many of the calculations reported in this paper.

6. REFERENCES

1. SERAG-ELDIN, M.A. and SPALDING, D.B. - A computational Procedure for Three-Dimensional Recirculating Flows Inside Gas Combustors. Numerical Methods in Heat Transfer, John Wiley and Sons Ltd, pp. 445-466, 1981.

2. MARKATOS, N.C. and MOULT, A. - The Computation of Steady and Unsteady, Turbulent, Chemically-Reacting Flows in Axisymmetrical Domains. Trans. Instn. Chem. Engrs, Vol. 57, No. 3, pp. 156-162, 1979.

3. MARKATOS, N.C. - Transient Flow and Heat Transfer of Liquid Sodium Coolant in the Outlet Plenum of a Fast Nuclear Reactor. Int. J. Heat Mass Transfer, No. 21, pp. 1565-1579, 1978.

4. PATANKAR, S.V. and SPALDING, D.B. - A Calculation Procedure for Heat, Mass and Momentum Transfer in Parabolic Flows. Int. J. Heat Mass Transfer, No. 15, pp. 1787-1806, 1972.

5. SPALDING, D.B. - Mathematical Modelling of Fluid Mechanics, Heat Transfer and Chemical-Reaction Processes, Imperial College, A Lecture Course, CFDU, Report No. HTS/80/1, 1980.

6. SPALDING, D.B. - A General Purpose Computer Program for Multi-Dimensional One- and Two-phase Flow. Mathematics and Computers in Simulation, North Holland Press, Vol. XXIII, pp. 267-276, 1981.

7. MARKATOS, N.C. and MUKERJEE, T. - Three-Dimensional Computer Analysis of Flow and Combustion in Automotive Internal Combustion Engines, Mathematics and Computers in Simulation, North Holland Press, Vol. XXIII, No. 4, pp. 354-366, 1981.

8. MARKATOS, N.C. - Computer Simulation of Turbulent Fluid Flow in Chemical Reactors. Adv. Eng. Software, Vol. 5, No. 1, pp. 32-38, 1983.

9. MARKATOS, N.C., RHODES. N. and TATCHELL, D.G. - A General Purpose Program for the Analysis of Fluid Flow Problems, Numerical Methods for Fluid Dynamics, Academic Press, pp. 463-480, 1982.

10. MARKATOS, N.C. and PERICLEOUS, C.A. - GRAFFIC - A Computer Package for the Interactive Graphical Representation of Fluid-Flow Phenomena, Proc. of 3rd Int. Conference and Exhibition on Engineering Software, London, April 1983.

11. SIMITOVIC, R., - Turbulent Flow in Rectangular Duct with an Obliquely Introduced Secondary Jet, Ph.D. Thesis, Univ. of London, 1977.

12. GOSMAN, A.D. and SIMITOVIC, R. - An Experimental Study of Confined Jet Mixing, Submitted for publication, Chem. Eng. Science, 1983.

RESIDUAL FLOW AND SALINITY STRATIFICATION IN THE ESTUARY OF SANTOS

Sérgio R. Signorini(I)

1. SUMMARY

This study is an application of a three-dimensional model developed by the author (1) in which the dimension has been reduced by considering only along-channel and vertical variations in the dynamics. This stems from the fact that, in the modeled domain, cross-channel variations in both velocity and salinity are small when compared with vertical and along-channel variations of the same variables. Supporting field data were used to provide initial and boundary conditions for the model. A 25-hour long time series measurement of the velocity and salinity vertical profiles at mid-channel, proved to be extremely valuable for the calibration of the combined river flow, eddy exchange coefficients and bottom stress in the model simulation. Very good agreement between the data and model results was reached.

2. MODEL FORMULATION

The governing equations consist of continuity, momentum and salt-balance equations. The fluid is assumed to be incompressible and hydrostatic, the Coriolis force is neglected in face of the small scale geometry, and the Boussinesq approximation is taken into account. These equations can be written as:

$$u_x + w_z = 0 \tag{1}$$

$$u_t + uu_x + wu_z = -1/\rho\ p_x + (A_v u_z)_z \tag{2}$$

$$S_t + uS_x + wS_z - (K_h S_x)_x - (K_v S_z)_z = 0 \tag{3}$$

Subscripts denote differentiation, except for A_v, K_h and K_v which denote the vertical momentum eddy exchange coefficient and along-channel and vertical diffusion coefficients,

(I). Assistant Professor, Department of Physical Oceanography, Instituto Oceanográfico, Universidade de S. Paulo, Brazil.

respectively. In this model, x is the along-channel Cartesian coordinate with the origin at the ocean boundary, and z is the vertical coordinate oriented upward.

Since the governing equations (1), (2) and (3) will be layer-by-layer vertically averaged, the vertical dimension in the model is taken into account by dividing the total water depth into multiple vertically averaged layers. Therefore, $z = -kh/b$, where k is the layer number, b the total number of layers and h is the water depth. If the water depth h is not constant, corrections must be applied to the horizontal derivatives to account for the "stretching" of the vertical coordinate. However, since in this model application the depth is kept constant, such corrections are not necessary. The layer-by-layer vertically averaged form of (1), (2) and (3) for each layer k, is given by:

$$\eta_t = -1/b(\,(h+\eta)\bar{u}(1))_x - h/b \sum_{k=2}^{b} (\bar{u}_x(k)) \tag{4}$$

$$\bar{u}_t(k) = - \bar{u}(k)\bar{u}_x(k) - b/h(\, w(k-\tfrac{1}{2})\bar{u}(k-\tfrac{1}{2}) - w(k+\tfrac{1}{2})\bar{u}(k+\tfrac{1}{2})) - \bar{p}_x(k)/\bar{\rho}(k) + A_v b/h(\bar{u}_z(k-\tfrac{1}{2}) - \bar{u}_z(k+\tfrac{1}{2})) \tag{5}$$

$$\bar{S}_t(k) = - \bar{u}(k)\bar{S}_x(k) - b/h(w(k-\tfrac{1}{2})\bar{S}(k-\tfrac{1}{2}) - w(k+\tfrac{1}{2})\bar{S}(k+\tfrac{1}{2})) + (K_h\bar{S}_x(k))_x + K_v\, b/h(\bar{S}_z(k-\tfrac{1}{2}) - \bar{S}_z(k+\tfrac{1}{2})) \tag{6}$$

where $\bar{u}(k)$, $\bar{S}(k)$, $\bar{p}(k)$ and $\bar{\rho}(k)$ are the vertically averaged values of the variables for each layer k, and η is the free-surface elevation. The layered vertical integration leads to the finite differencing of the vertical derivatives. Equations (4), (5) and (6) are numerically solved in a semi-implicit coupled form. The numerical scheme uses finite element techniques for the along-channel derivatives (with respect to x), and finite differences for the vertical and time derivatives (with respect to z and t, respectively).

The vertical velocity for the top layer w(1) is identical to η_t in (4) ($\eta_t = 0$ when the steady state is reached), whereas the vertical velocity w(k) for the remaining layers is given by:

$$w(k) = - h/b \sum_{i=k}^{b} (\bar{u}_x(i)) \tag{7}$$

From the vertical integration of the hydrostatic equation ($p = \rho g z$), we derive expressions for the pressure gradient $\bar{p}_x(k)$. For the top layer we have:

$$\bar{p}_x(1) = \bar{\rho}(1)g\, \eta_x + g(\eta + h/2b)\bar{\rho}_x(1) \tag{8}$$

and, for the remaining layers, we have the following recursion relation:

$$\bar{p}_x(k) = \bar{p}_x(k-1) + gh(\bar{\rho}_x(k-1) + \bar{\rho}_x(k))/2b \qquad (9)$$

The salinity and temperature of the sea water can be related to the density by the following linearized equation of state:

$$\bar{\rho}(k) = \rho_o - \alpha\bar{T}(k) + \beta\bar{S}(k) \qquad (10)$$

where ρ_o is the freshwater density, $\bar{T}(k)$ is the layer averaged temperature, assumed to be constant and homogeneous ($\bar{T}$ = 21° C derived from the field data), $\alpha = 1.7 \times 10^{-4}$ gr cm^{-3} °C^{-1} is the coefficient of thermal expansion, and $\beta = 8.0 \times 10^{-4}$ gr cm^{-3} ppt^{-1} is the coefficient of salt contraction. Therefore, (5) and (6) are coupled by (8), (9) and (10) through the pressure gradient term in (5).

3. BOUNDARY AND INITIAL CONDITIONS

At the free-surface and at the bottom there is no vertical diffusive or advective transport of salt, i.e.,

$$w\bar{S}_z = K_v\bar{S}_z = 0, \text{ at } z = \eta \text{ and } z = -h \qquad (11)$$

At the solid boundaries (channel margins), it is specified that the water flux and the advective and diffusive horizontal fluxes of salt are also zero in the direction perpendicular to the coastal boundaries. These conditions are automatically satisfied due to the assumed dimensionality of the problem, i. e., the cross-channel variations were neglected in the dynamics. Frictional effects are prescribed by a bottom stress $\tau_b = A_v\bar{u}_z(b) = g|\bar{u}(b)|\bar{u}(b)/C^2$, where $C = 4.6\ h^{1/6}/n$ (cm$^{1/2}$ s^{-1}) is the Chezy coefficient and n is the Manning factor.

At the ocean boundary (x = 0) and at the river mouth (x = ℓ), the salinity is specified at each layer according to the time (at least over a tidal cycle) and layer averaged salinities obtained from the field work ($\bar{S}(0,k) = \bar{S}_O(k)$ and $\bar{S}(\ell,k) = \bar{S}_R(k)$, respectively). The initial salinity stratification was obtained by linear interpolation between the boundary values for each layer at the internal grid nodes.

The initial velocity field is set equal to zero at all layers and the free-surface elevation adopted initially is a linear function of the distance to the ocean boundary, sloping towards the river mouth in order to provide the proper pressure gradient, which is in turn calibrated through the slope coefficient (δ), in order to obtain the proper river flow ($\eta(x) = \delta x$). At the ocean open boundary $\eta(0) = 0$, and at the river mouth $\eta(\ell) = \delta\ell$.

4. FINITE ELEMENT FORMULATION

The finite element form of equations (4), (5) and (6) is obtained by substituting the variables η, $\bar{u}$ and $\bar{S}$ by their approximated form, through the use of linear interpolation functions applied to three-node triangular elements, and integrating the equations over the whole solution domain, weighting the residuals at each element and imposing that they should vanish over the complete assemblage of elements. When the weighting functions, used as residual weights at each element, are replaced by the interpolation functions themselves, which have coefficients depending on the global grid coordinates alone, we arrive to the well known Galerkin form of the finite element version of the differential equations. This was the method used in this study, and the details of its formal derivation can be found in a large variety of references (for example: (2) and (3)), and therefore will not be reproduced in this paper for the sake of abbreviation.

The Galerkin procedure reduces the system of partial differential equations into a system of ordinary differential equations with respect to time. The spatial solution scheme seeks to determine the time derivatives η_t, $\bar{u}_t$ and $\bar{S}_t$, achieved by solving a system of algebrical equations assembled in a global matrix system, obtained by the summation of individual element contributions for each layer.

After the time derivatives are evaluated through matrix inversion for each nodal point and layer, a finite difference, semi-implicit time integration scheme (taking half time steps) is employed to couple the solution of the continuity, momentum and salt-balance equations, finally leading to the evaluation of η, $\bar{u}$ and $\bar{S}$ at each time step. In order to achieve stable numerical results, the time step Δt was upper bounded by the modified form of the Courant-Friedricks-Lewy condition, i.e., $\Delta t < \Delta x/(2gh)^{1/2}$, where Δx is the smallest side of the smallest triangular element located in the deepest water. Stable results were achieved with $\Delta t = 50$ s. The steady-state solution was defined at a time level when the values of $\bar{u}(k)$ and $\bar{S}(k)$ differed by only 0.5% of the values for the previous time step. The steady-state thus defined was reached after 17 minutes of CPU time in the Burroughs B-6700 Computer System.

5. FIELD DATA AND MODEL RESULTS

The data necessary to provide initial and boundary conditions were obtained through the collection of temperature and salinity data at 2 meter depth intervals along 13 hydrographic stations distributed along the Estuary. This was done during the flood and ebb tides. Fig. 1 shows the geometry of the Estuary, grid system adopted (50 triangular elements with

52 nodal points) and station locations.

The along-channel salinity stratification showed in Fig. 2 represents the average between the ebb and flood stratification conditions. In the lack of a salinity time series along a tidal cycle for all stations along the Estuary, the average obtained between the two extreme conditions gives a first approach to the salinity distribution free of advective tidal circulation effects.

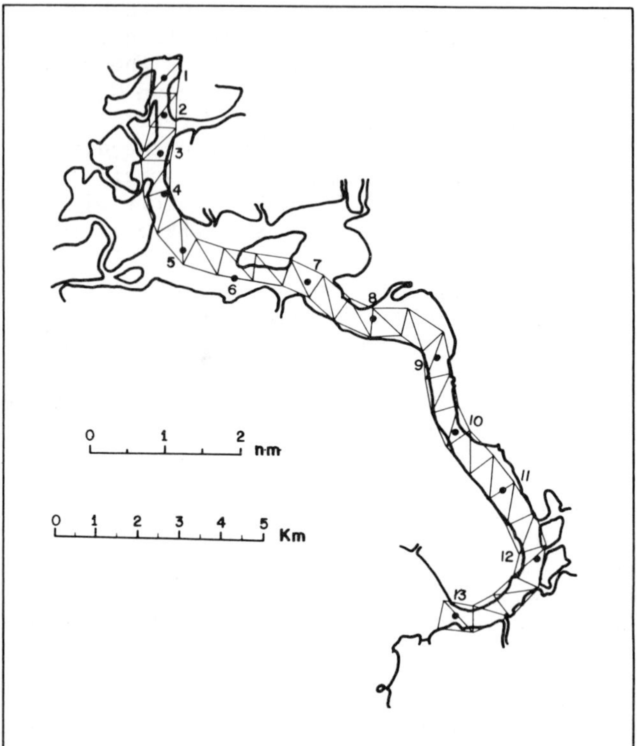

Figure 1. The geometry of the Estuary of Santos and the grid system used in the model. Station locations are numbered 1 through 13.

The salinity near the surface changes from $16^0/_{00}$ at the river mouth to $32^0/_{00}$ at the ocean boundary, 20 Km downstream. Near the bottom, the change in salinity along the Estuary is only $4^0/_{00}$, about a quarter of the surface variation.

The vertical salinity stratification is much more intense near the river mouth, about $14^0/_{00}$ in 12 meters, than in the ocean side, where the stratification is only $2^0/_{00}$ in 12 meters. This may be due to the fact that tidal currents are probably more intense near the ocean boundary, promoting a higher degree of vertical mixing, and decrease in intensity towards the river mouth, promoting much less vertical mixing. This is a characteristic of a standing tidal wave, which is a possible feature of the tidal circulation in the Estuary of Santos.

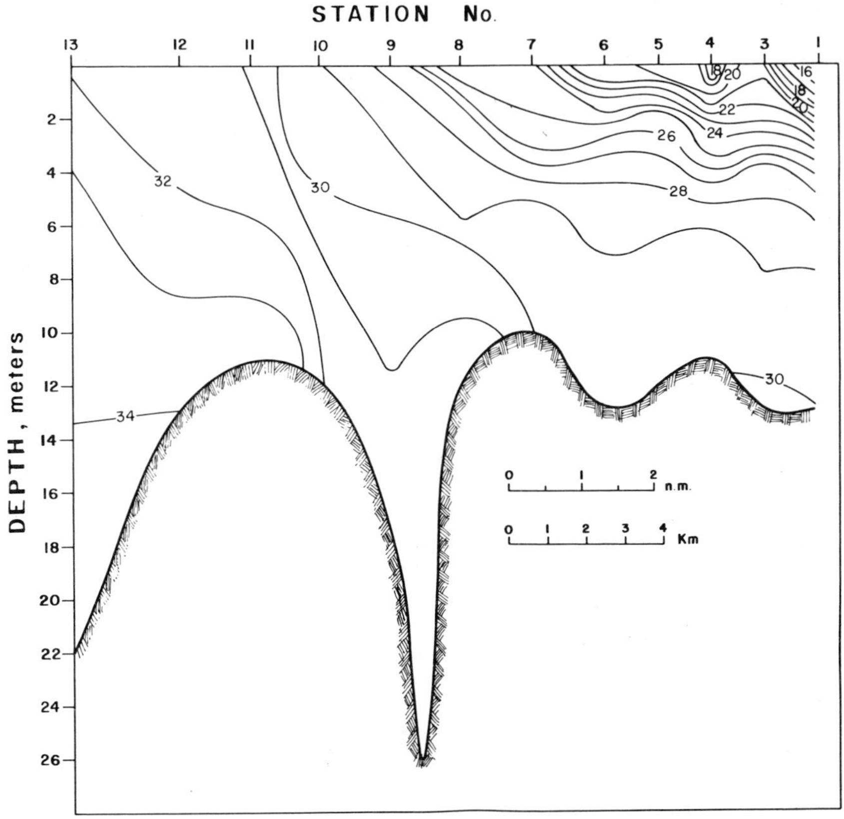

Figure 2. Along-channel salinity stratification (in parts per thousand, $^0/_{00}$) based on the field data. The numbers on top denote the station locations.

Current measurements were also made at station 9. They consisted of a 25-hour time series of hourly vertical profiles of velocity, with measurements taken at 2 meter depth intervals from surface to bottom. The velocity profiles were time-averaged and layer-averaged in order to make them suitable for comparison with the model results.

As it can be seen from Fig. 1, upstream from station 9,

there are a variety of rivers contributing to the freshwater input to the Estuary. In order to simplify the model geometry, all river contributions were combined into one single hypothetical river flowing into the Estuary through the upstream open boundary at station 1. The river flow was adjusted in order to obtain the integrated transport, $R_f = 820\ m^3\ s^{-1}$, measured at station 9. The 26 meter deep trench, located between stations 8 and 9 (see Fig. 2), was not included in the model's bottom topography. Instead, a 12 meter deep flat bottom topography was adopted without altering significantly the results.

Fig. 3 shows the along-channel salinity stratification and residual velocity fields predicted by the model. By comparing Figs. 2 and 3, we see that there is a good agreement between the field data and model salinity stratifications. The flow is directed towards the ocean from surface to bottom, except near the ocean boundary where bottom velocities are directed upstream. This result implies that, 2 Km upstream of the ocean boundary and up to the river mouth, the bottom salt transport is achieved by horizontal diffusion rather than horizontal advection.

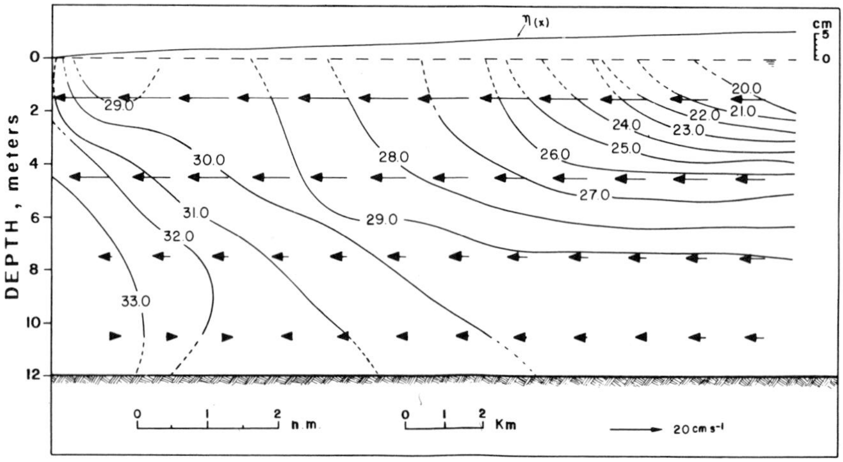

Figure 3. Along-channel salinity stratification and residual velocity field predicted by the model. Velocity vectors are scaled in cm s^{-1}.

The variation of the free-surface elevation $\eta(x)$ along the Estuary is shown on the top part of Fig. 3. Residual velocities have a strong vertical shear, varying from 20 cm s^{-1} in the top layer, to about 2 to 5 cm s^{-1} in the bottom layer.

The comparison between the salinity and velocity vertical profiles derived from the measurements made at station 9 with those predicted by the model, are illustrated in Fig. 4

for three different sets of computational parameters. Even though more than three model simulations were performed, the results showed in Fig. 4 are the closest match to the data and therefore were selected for comparison and sensitivity parameter analysis. The best agreement with the data is shown in Fig. 4 C, where $K_v = 5\ cm^2\ s^{-1}$, $K_h = 10^5\ cm^2\ s^{-1}$, $n = 0.02$, $\delta = 2.9 \times 10^{-6}$ and $R_f = 820\ m^3\ s^{-1}$.

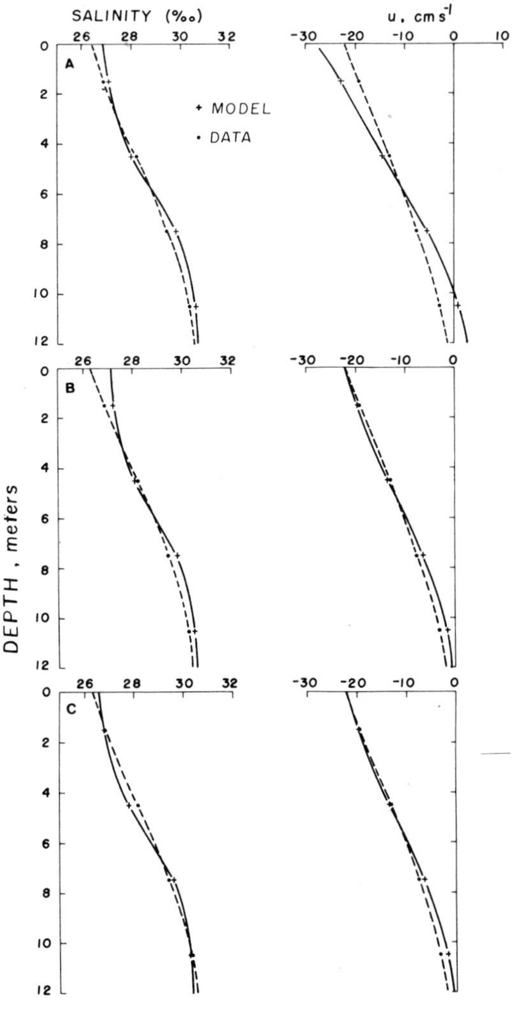

Figure 4. Salinity and velocity vertical profiles at mid-channel (station 9). Model profiles are compared with data profiles. A - $K_v = 5\ cm^2s^{-1}$, $A_v = 15\ cm^2s^{-1}$, $K_h = 10^5 cm^2s^{-1}$, $n = 0.02$, $\delta = 2.9 \times 10^{-6}$ and $R_f = 820\ m^3\ s^{-1}$. B - As in A, except for $A_v = 20\ cm^2s^{-1}$ and $K_h = 5 \times 10^5\ cm^2s^{-1}$. C - As in B, except for $K_h = 10^5 cm^2s^{-1}$.

5. CONCLUSIONS

The model results proved to be in good agreement with the data, also showing that, time averaged vertical profiles of salinity and velocity obtained through time series measurement along at least one tidal cycle (preferably more), are crucial to achieve ideal model calibration and performance in order to simulate and predict the residual flow in an estuary.

Therefore, the field data necessary to supply the model with proper boundary and initial conditions, as well as data to achieve model calibration, must be originated from at least three time series measurements: salinity profiles both at the ocean and river open boundaries, and a combination of velocity and salinity profiles at a site along the estuary where all river contributions are accounted for. With the above information available, all model parameters can be determined, allowing the prediction of the residual velocity field and salinity stratification for the whole estuary domain.

On the other hand, the residual circulation and salinity stratification observed during the field work used in this study may not be a permanent regime in the Estuary of Santos. These conditions may change with the strength of the tidal circulation and river runoff, and therefore further investigations must be performed in order to include these changes.

6. REFERENCES

1. SIGNORINI, S.R. - A Three-Dimensional Numerical Model of Circulation and Diffusion-Advection Processes for Estuarine and Coastal Application by the Finite Element Method, The Proceedings of the Fourth International Symposium on Finite Element Methods in Flow Problems, T. Kawai, Elsevier North-Holland, July 26-29th, 1982.

2. HUEBNER, K.H. - The Finite Element Method for Engineers, John Wiley & Sons, New York, 1975.

3. CONNOR, J.J. and BREBBIA, C.A. - Finite Element Techniques for Fluid Flow, Newnes-Butterworths, London, 1976.

SECTION 6

FREE SURFACE FLOW

FINITE ELEMENT SIMULATION OF INCOMPRESSIBLE FLUID FLOWS WITH A FREE/MOVING SURFACE

M. S. Engelman and R. L. Sani
CIRES and Dept. Chem. Engng.
University of Colorado
Boulder, Colorado 80309

1. INTRODUCTION

There are many important technological and engineering science applications in which a free and/or moving fluid interface plays a dominant role. Such systems encompass general static and dynamic applications encountered, for example, in capillarity, crystal growth, electrochemical plating and corrosion, coating and polymer technology, separation processes, metal and glass forming processes and many other areas of engineering and science. Specifically, the equilibrium shape and stability of menisci between pairs of immiscible fluids in containers; coating flows in which a viscous fluid is deposited on a rigid, or flexible, substrate as commonly encountered in the manufacturing of photographic films and plate glass or the coating of paper; solidification processes which are commonly encountered in the material science of crystal growth, e.g., in open boat configuration, Czochralski or floating zone methods; extrusion of liquid from nozzles as encountered in the continuous production of fibres or curtain coating operations; free surface flows in porous media associated with seepage and enhanced oil recovery processes. However, a *quantitative* description and solution of such problems has been elusive even in the simplest cases because of firstly, the often present static and dynamic contact lines whose continuum representation is a current area of activity and secondly, the inherent nonlinearities in the continuum equations. For example, the presence of the curved free surface makes the problem inherently nonlinear even in the Stokes flow limit of vanishing acceleration effects. To further complicate the situation many of the systems of interest possess not only flow but concomitant heat and/or mass transport which combined with the presence of a free surface can lead to a significant "surface tension driven" flow contribution.

In this paper we shall present a versatile, robust Galerkin-finite element algorithm appropriate for the simulation of the steady and time dependent behaviour of such systems. The basic approach adopted here is that of Saito and Scriven [1], Ettouney and Brown [2] and Ruschak [3]; the algorithm is generalised and implemented in a robust and versatile finite element code that can be employed with relative ease for the simulation of two or three dimensional free surface problems in complex geometries. This generalisation includes the elimination of the costly dependence on the Newton-Raphson algorithm by replacing it with a quasi-Newton iterative method, while nearly retaining the superior convergence properties of the Newton-Raphson method.

2. FORMULATION OF CONTINUUM PROBLEM

The equations of incompressible fluid flow are derived from the basic physical principles of conservation of mass and linear momentum. An Eulerian formulation will be employed and the equations expressed in cartesian tensor notation with implied summation over indices and "," representing differentiation with respect to position. We shall first consider the stationary two-dimensional form of these equations:

Momentum $$\rho u_j u_{i,j} = -p_{,i} + b_i + \mu(u_{i,j} + u_{j,i})_{,j} \tag{1}$$

Mass $$u_{i,i} = 0 \tag{2}$$

where u_i is the velocity component in the direction of x_i, b_i the body force component (typically gravity g_i), p is the pressure and μ the absolute viscosity of the fluid.

In order to form a complete system, boundary conditions must be adjoined to equations (1) and (4). At fixed boundaries, these can take the form of specified velocities and/or applied tractions (stresses), i.e.

$$u_i = \bar{u}_i(\mathbf{x},t)\ , \quad \mathbf{x}\varepsilon\Gamma_u \tag{3}$$

$$\sigma_{ij} n_j = \bar{\sigma}_i(\mathbf{x},t)\ , \quad \mathbf{x}\varepsilon\Gamma_t \tag{4}$$

where $\mathbf{n} = \{n_j\}$ is the *outward pointing* normal at the boundary, $\bar{u}_i$ and $\bar{\sigma}_i$ are prescribed functions, and Γ_u are Γ_t are portions of the boundary.

At the free surface, continuity of stress and velocity is required which leads to the conditions

$$u_n = 0 \quad , \quad \mathbf{x}\varepsilon\Gamma_f \tag{5}$$

$$f_n - p_0 = 2\sigma H \quad , \quad \mathbf{x}\varepsilon\Gamma_f \tag{6}$$

$$f_t = 0 \quad , \quad \mathbf{x}\varepsilon\Gamma_f \tag{7b}$$

where $u_n = u_i n_i$ is the normal component of velocity and $f_n = \sigma_{ij} n_j n_i$ the normal and $f_t = \sigma_{ij} n_j t_i$ the tangential component of the stress vector at the boundary; σ is the surface tension, p_0 the pressure in the adjacent vapour phase and H the mean Gaussian curvature of the surface. We have assumed that the viscosity of the exterior phase is negligible. From differential geometry $H = -\frac{1}{2}\nabla_s \cdot \mathbf{n}$ where ∇_s is the surface divergence operator, equation (7a) becomes

$$f_n = p_0 - \sigma\nabla_s \cdot \mathbf{n} \tag{8}$$

Equations (2) - (8) form a complete set for the determination of the velocity, pressure and the position of the free surface of the fluid in some region Ω, where the boundary of Ω is $\Gamma = \Gamma_u \oplus \Gamma_t \oplus \Gamma_f$. These equations may be written in dimensionless form as:

$$u_j u_{i,j} = -p_{,i} + \frac{1}{\mathrm{Re}}(u_{i,j} + u_{j,i})_{,j} + \frac{S_t}{\mathrm{Re}} b_i \qquad \mathbf{x} \varepsilon \Omega \tag{9a}$$

$$u_{i,i} = 0 \tag{9b}$$

$$u_i = \bar{u}_i(\mathbf{x},t) \ , \quad \mathbf{x} \varepsilon \Gamma_u \tag{9c}$$

$$t_i = \bar{\sigma}_i(\mathbf{x},t) \ , \quad \mathbf{x} \varepsilon \Gamma_t \tag{9d}$$

$$u_n = 0 \ , \quad \mathbf{x} \varepsilon \Gamma_f \tag{9e}$$

$$f_n = p_0 + \frac{1}{\mathrm{Ca}} \nabla_s \cdot \mathbf{n} \ , \quad \mathbf{x} \varepsilon \Gamma_f \tag{9f}$$

$$f_t = 0 \ , \quad \mathbf{x} \varepsilon \Gamma_f \tag{9g}$$

where u_i, p, p_0, b_i are now dimensionless quantities and $\mathrm{Re} = \dfrac{\rho L U}{\mu}$ is the Reynolds number and $\mathrm{Ca} = \dfrac{\mu U}{\sigma}$ is the capillary number. U is the characteristic velocity and L the characteristic length used to scale the velocity and length, respectively; $S_t = \dfrac{FL^2}{\mu U}$, where F is a characteristic force for the problem, is often called the Stokes number.

In order to make equation (9f) more tractable when developing the finite element equations of the system represented by (9) in the case of a 2D system we make use of a geometric property of the mean curvature of a surface (see [4]):

$$(\nabla_s \cdot \mathbf{n})\mathbf{n} = -\frac{\partial \mathbf{t}}{\partial l} \tag{10}$$

where $\mathbf{t}$ is the tangent vector associated with $\mathbf{n}$ and l is arc length (see figure 1); that is, the curvature of the line is proportional to the rate of change of the direction of the tangent vector. Thus equation (9f) becomes:

$$f_n n_i = p_0 n_i - \frac{1}{\mathrm{Ca}} \frac{\partial t_i}{\partial l} \tag{11}$$

3. FINITE ELEMENT FORMULATION

The procedure for generating the finite element equations for the Navier-Stokes equations is well described in many other references (e.g. [5]) so here we shall only concentrate on those aspects peculiar to treatment of the free surface. In each element, the velocity and pressure fields are approximated by:

$$u_i(\mathbf{x},t) = \boldsymbol{\varphi}^T \mathbf{U}_i(t) \tag{12a}$$

$$p(\mathbf{x},t) = \boldsymbol{\psi}^T \mathbf{P} \tag{12b}$$

where $\mathbf{U}_i$ and $\mathbf{P}$ are column vectors of the element velocity and pressure degrees of freedom and $\boldsymbol{\varphi} = \{\varphi_i\}$ and $\boldsymbol{\psi} = \{\psi_i\}$ are column vectors of the polynomial basis functions.

For a free surface problem, we need to introduce a new degree of freedom at each node of the free surface boundary. The value of this degree of freedom at such a node will enable the determination of the position of the node within the region and is an integral part of the representation of the free surface. The method employed is a generalisation of the technique developed by Saito and Scriven [1]. Each node (for example A in figure 1) on the free surface is taken to be the endpoint of a line segment which begins at some node (B in figure 1) within the mesh; this node, and all nodes below it, are considered to be fixed in space. Node A is constrained to lie on the line defined by AB throughout the simulation and and all the nodes on line AB are free to move. If the line AB is represented parametrically as:

$$x = \alpha_x r + \beta_x \tag{13a}$$

$$y = \alpha_y r + \beta_y \tag{13b}$$

such that $r = 0$ at point B and $r = 1$ at point A, then the natural candidate for the free surface degree of freedom at node A is the value of the parameter r. When a new value for r is determined equations (13a,b) immediately give the coordinates of the new position of node A. This parametric representation also allows the initial aspect ratio of the nodes in the interval AB to be easily maintained. This aspect ratio is simply the starting value of parameter r at each of the nodes in AB. Thus once a new value for r at A has been computed, r_A, the new value of r at each of the other nodes is this value multiplied by the initial r value for the node. It is also important to note that each of the boundary generator lines is independent one of the other. In practical terms, this method of representation of the free surface requires a minimal amount of additional input; all that is required is the the input of the nodes on the free surface, plus the input of the nodes defining each of the boundary generator lines.

Now consider the finite element representation of a line segment on the free boundary; each segment is the side of some element and consists of N nodes. For discussion purposes assume that the line segment is the side of a linear element, i.e. N=2. Let $C(x_1,y_1)$ and $D(x_2,y_2)$ be the coordinates of the endpoints of a line segment and $(\alpha_{x_1},\beta_{x_1})$, $(\alpha_{y_1},\beta_{y_1})$ and $(\alpha_{x_2},\beta_{x_2})$, $(\alpha_{y_2},\beta_{y_2})$ be the coefficients of parametric representation of the generator lines on which the points C and D lie. Then the free surface is approximated by:

$$x = \sum_{1=1}^{2} x_i \chi_i(s) = \sum_{1=1}^{2} (\alpha_{x_i} r_i + \beta_{x_i}) \chi_i(s) \tag{14a}$$

$$y = \sum_{1=1}^{2} y_i \chi_i(s) = \sum_{1=1}^{2} (\alpha_{y_i} r_i + \beta_{y_i}) \chi_i(s) \tag{14b}$$

where $\chi_i(s)$ are the basis functions. This representation corresponds to the velocity and pressure approximations of equations (12a) and (12b).

We are now in a position to form the finite element equations of the system (9). In the following sections, the basis functions φ_i, ψ_i, χ_i will denote global basis functions when appearing in integrals over the entire domain and element basis functions (i.e. global basis functions restricted to an element) when in integrals over an element domain. Applying the Galerkin finite element method to the equations of momentum and mass balances results in the equations:

$$\int_\Omega u_j u_{l,j} \varphi_i d\Omega - \frac{1}{\text{Re}}\int_\Omega (u_{l,j} + u_{j,l})\varphi_{i,j} d\Omega + \int_\Omega p\varphi_{i,l} d\Omega + \frac{S_t}{\text{Re}}\int_\Omega b_i\varphi_i d\Omega$$

$$+ \int_{\Gamma_f} (p_0 n_l - \frac{1}{\text{Ca}}\frac{dt_l}{dl})\varphi_i d\Gamma = \int_{\Gamma_t} \bar{\sigma}_l \varphi_i d\Gamma \quad (15)$$

All the integrals in these equations are the standard integrals obtained from the application of the GFEM to the Navier-Stokes equations with the exception of the last integral on the left hand side. If we apply Greens theorem to this term we obtain:

$$\int_{\Gamma_f} \frac{dt_l}{dl}\varphi_i d\Gamma = -\int_{\Gamma_f} t_l \frac{\partial \varphi_i}{\partial l} d\Gamma + \left[t_l\varphi_i\right]_{l_1}^{l_2} \quad (16)$$

This procedure has the advantage of eliminating the first derivative of the tangent vector, which eliminates the need for second derivatives of the basis function χ_i.

If the last term in equation (16) is written as:

$$\left[t_l\varphi_i\right]_{l_1}^{l_2} = \hat{t}_l\varphi_i\,|_{l_2} + \hat{t}_l\varphi_i\,|_{l_1} \quad (17)$$

then this term represents the boundary condition at the two ends of the free surface; l_1 and l_2 are the endpoints of the boundary Γ_f and the $\hat{t}_l$ on the right hand side of equation (17) are the components of the outward pointing tangent vector to Γ_f at either end. These vectors are closely related to the 'contact angle' of the fluid with the fixed boundary (see figure 2). In fact, this boundary condition is more general than that of a contact angle. For example, the free surface may be at a free outflow boundary; then the specification of the tangent vector as $\hat{\mathbf{t}} = (1,0)$ would correspond to a condition of vanishing slope at the boundary.

The requirement of no flow across the free surface (equation (9e)) is now left to complete the system of finite element equations. Its residual is to be made orthogonal to the basic functions defining the free surface, i.e.

$$\int_{\Gamma_f} u_n\chi_i d\Gamma = \int_{\Gamma_f} (n_x u_x + n_y u_y)\chi_i d\Gamma = 0 \quad (18)$$

The substitution of equations (12) and (14) into equations (15) and (18) results in a matrix system of nonlinear algebraic equations of the form

$$\mathbf{K(U)U} - \mathbf{CP} + \mathbf{BX} = \mathbf{F} \quad (19a)$$

$$\mathbf{C}^T\mathbf{U} = \mathbf{0} \quad (19b)$$

$$\mathbf{K}_n\mathbf{U} = \mathbf{0} \quad (19c)$$

where $\mathbf{X}$ is the global vector of the free surface unknowns, $\mathbf{K}(\mathbf{U})$ is a matrix which includes the effects from the convective and diffusive terms, $\mathbf{C}$ is the divergence matrix, $\mathbf{B}$ is the matrix representing the contribution of the normal stress balance boundary condition in the momentum equation, $\mathbf{K}_n$ contains the normal velocity boundary condition effects and $\mathbf{F}$ is a vector including the effects of the body force, applied tractions and contact angle boundary conditions.

4. SOLUTION OF THE FINITE ELEMENT EQUATIONS

It now remains to solve the nonlinear system of equations (19). A fixed point iterative procedure could be used, however, this would necessitate the interpolation of the nodal velocity degrees of freedom to the new nodal positions at each iteration; such a method would also probably exhibit a linear rate of convergence. On the other hand, if a Newton-Raphson procedure is utilised, and *complete* account of the variation with respect to the free surface degrees of freedom is incorporated into the system Jacobian, then at the end of each iteration the velocity solution will represent the velocities at the *new* nodal locations. The great advantage of such an approach is that the free surface location and the velocities at the new nodal locations are the direct result of the iterative procedure with no interpolation or updating being required. A major disadvantage of this approach is that it requires the computation, assembly and LU decomposition of the Jacobian matrix at each iteration.

Both the Newton-Raphson and quasi-Newton methods [6] require the computation of the Jacobian matrix and residual vector of the system of equations (19). This Jacobian can be written as

$$\mathbf{J} = \begin{pmatrix} \mathbf{K}+\mathbf{N} & -\mathbf{C} & \mathbf{B}+\mathbf{J}_K-\mathbf{J}_P-\mathbf{J}_F+\mathbf{J}_B \\ -\mathbf{C}^T & \mathbf{0} & -\mathbf{J}_C \\ \mathbf{K}_n & \mathbf{0} & \mathbf{J}_{K_n} \end{pmatrix} \tag{20}$$

where

$$\mathbf{N}=\frac{\partial \mathbf{K}}{\partial \mathbf{U}}\mathbf{U},\ \mathbf{J}_K=\frac{\partial(\mathbf{KU})}{\partial \mathbf{X}},\ \mathbf{J}_F=\frac{\partial \mathbf{F}}{\partial \mathbf{X}},\ \mathbf{J}_P=\frac{\partial(\mathbf{CP})}{\partial \mathbf{X}},\ \mathbf{J}_{K_n}=\frac{\partial(\mathbf{K}_n\mathbf{U})}{\partial \mathbf{X}},\ \mathbf{J}_B=\frac{\partial \mathbf{B}}{\partial \mathbf{X}}\mathbf{X},\ \mathbf{J}_C=\frac{\partial(\mathbf{C}^T\mathbf{U})}{\partial \mathbf{X}}$$

The partitioned upper left hand corner matrix is the regular Jacobian matrix obtained from the application of the GFEM to the Navier-Stokes equations on a fixed mesh. The sub-matrices $\mathbf{J}_K, \mathbf{J}_N, \mathbf{J}_P, \mathbf{J}_{K_n}, \mathbf{J}_B, \mathbf{J}_C$ and $\mathbf{J}_F$ are the additional contributions to the Jacobian matrix due to the fact that the nodal positions are no longer constant but may vary from iteration to iteration. For those elements defined by nodes fixed in space, these sub-matrices will be zero; only those elements whose nodes move will contribute to that part of the Jacobian which represents the variations with respect to the free surface degrees of freedoms.

The evaluation of the sub-matrices $\mathbf{J}_K, \mathbf{J}_P, \mathbf{J}_{K_n}, \mathbf{J}_C$ and $\mathbf{J}_F$ is particularly tedious as the matrices $\mathbf{K}, \mathbf{C}, \mathbf{K}_n$ and the vector $\mathbf{F}$ depend on $\mathbf{X}$ (or rather x and y by equations (13)), not only through the integrands but also through the limits of integration, i.e., the element domain. However, just as in a finite element code the integrals over different element domains are handled by mapping an element onto a 'parent' or reference element and performing the integration over its simpler, fixed geometry, the same technique is used to compute the variation of various matrices with respect to $\mathbf{X}$.

5. TIME DEPENDENT PROBLEMS

The technique for handling the free surface movement for a transient free surface problem is identical to that described in the previous sections for steady state problems. Care, however, must be taken in the treatment of the time derivatives which now appear in the momentum equations and the free surface boundary condition.

The momentum equation for a transient flow is

$$\rho(\frac{\partial u_i}{\partial t} + u_j u_{i,j}) = \sigma_{ij,j} + \rho b_i \tag{21}$$

Also the condition of no normal flow across the free surface, equation (9e), must be modified to

$$\frac{\partial S}{\partial t} + \mathbf{u} \cdot \nabla S = 0 \tag{22}$$

where $S(x,y,t)=0$ is the function defining the free surface. If this equation is normalised by dividing by $|\nabla S|$, it becomes

$$\frac{\partial \bar{S}}{\partial t} + \mathbf{u} \cdot \mathbf{n} = 0 \quad ; \quad \bar{S} = \frac{S}{|\nabla S|} \tag{23}$$

The time derivatives appearing in equations (21) and (23) are Eulerian time derivatives, i.e. the nodal velocity field must be for nodes fixed in space. However, our technique of parametrisation of the free surface is such that the nodes are not fixed in some frame of reference (Eulerian formulation), nor are they fixed in a frame of reference carried along by the fluid (Lagrangian formulation); rather each node is constrained to move along a particular line in space - a mixed Eulerian-Lagrangian formulation. Thus the time derivatives in equations (21) and (23) must be transformed to time derivatives which follow the moving nodes along these lines. Since any given node will always transform to the same *fixed* point in the (ξ,η) space, this derivative can be thought of as a time derivative at a fixed point in the (ξ,η) coordinate frame.

Denoting by $\frac{\delta}{\delta t}$ the time derivative following a moving node, the relationship between $\frac{\delta}{\delta t}$ and $\frac{\partial}{\partial t}$, the Eulerian time derivative, is given by

$$\begin{aligned} \frac{\delta}{\delta t} &= \frac{\partial}{\partial t} + \frac{\partial \mathbf{x}}{\partial t} \cdot \nabla \\ &= \frac{\partial}{\partial t} + \frac{\delta \bar{S}}{\delta t} \frac{\partial \mathbf{x}}{\partial \bar{S}} \cdot \nabla \end{aligned} \tag{24}$$

where $\mathbf{x} = \mathbf{x}(\overline{S},\xi,\eta)$ are the coordinates of a moving nodal point.

The finite element discretisation proceeds as outlined in the prvious sections and leads to:

$$\begin{bmatrix} \overline{\mathbf{M}} & \mathbf{O} & \mathbf{M}_U \\ \mathbf{O} & \mathbf{O} & \mathbf{O} \\ \mathbf{O} & \mathbf{O} & \mathbf{M}_B \end{bmatrix} \begin{Bmatrix} \dfrac{\delta \mathbf{U}}{\delta t} \\ \dfrac{\delta \mathbf{P}}{\delta t} \\ \dfrac{\delta \mathbf{X}}{\delta t} \end{Bmatrix} + \begin{bmatrix} \mathbf{K} & -\mathbf{C} & \mathbf{B} \\ -\mathbf{C}^T & \mathbf{O} & \mathbf{O} \\ \mathbf{K}_n & \mathbf{O} & \mathbf{O} \end{bmatrix} \begin{Bmatrix} \mathbf{U} \\ \mathbf{P} \\ \mathbf{X} \end{Bmatrix} = \begin{Bmatrix} \mathbf{F} \\ \mathbf{O} \\ \mathbf{O} \end{Bmatrix} \tag{25}$$

where

$$\overline{\mathbf{M}} = \begin{bmatrix} \mathbf{M} & \mathbf{O} \\ \mathbf{O} & \mathbf{M} \end{bmatrix} \quad \text{and} \quad \mathbf{M}_U = \begin{bmatrix} \mathbf{M}_{U_1} \\ \mathbf{M}_{U_2} \end{bmatrix}$$

This matrix equation represents a system of simultaneous non-linear ordinary differential equations which must be solved. The time derivatives will be discretised in the standard manner using a finite difference scheme. It is almost mandatory to use an implicit difference scheme as the mass matrix is now time dependent due to the changing nodal positions and the linkage of the velocity and free surface time derivatives makes the possibility of mass lumping questionable.

The difference scheme used to approximate the time derivatives is a predictor-corrector scheme developed by Gresho et al.[7]. The predictor step is the second order Adams-Bashforth rule and the corrector step is the second order unconditionally stable trapezoid rule. The scheme is implemented with a variable time increment, where the time increment at each step is determined by control of the local time truncation error to within some user specified tolerance. Since at each time step the trapezoid rule requires the acceleration vector from a previous time step, the time integration is initiated with a specified number of backward Euler steps, typically 3 or 4, before switching to the trapezoid rule. In the event that a penalty formulation is being employed, this procedure also helps to bypass any initial non-physical portion of the transient due to the penalty technique [8].

At each time step, a nonlinear system of algebraic equations must be solved. Once again the Newton-Raphson or quasi-Newton algorithms are available for this purpose. Note that the matrices $\mathbf{M}$ and $\mathbf{M}_B$ will also result in contributions to the system jacobian. Gresho et al. have shown that with this predictor-corrector scheme, if the user specified local time truncation error tolerance is set to .001-.005, i.e. .01-.05%, then the predictor is sufficiently accurate that only one Newton-Raphson iteration is required at each time step to achieve convergence. This, however, can be very expensive, since most transient calculations will require of the order of 100 time steps. The quasi-Newton method again provides a cost-effective alternative, as the reformation of the Jacobian matrix need only be performed every N time steps. Of course, a balance must be found between the number of steps N and the number of quasi-Newton iterations required at each time step to achieve convergence. Typically, if N=2 or 3, a savings in computer time of the order of 50% over the one-step Newton-Raphson method can be achieved.

6. COMPUTER IMPLEMENTATION

In this section we offer some comments on the actual computer implementation of the technique presented in the previous sections. The algorithm - both mixed and penalty formulations - has been implemented into the finite element code FIDAP (Fluid Dynamics Analysis Program)[9] which is a general purpose finite element code for the simulation of the steady state or transient flow of an incompressible fluid in two-dimensional, axi-symmetric or three-dimensional geometries. The implementation has been done in a completely general manner so that all the capabilities already existing in FIDAP would be available for free-surface simulations.

There is one additional feature of the implementation that needs to be detailed. Each node which lies on the free surface boundary has a free surface degree of freedom, r_A say, associated with it. Those nodes which lie on the same boundary generator line (AB in figure 1) as the node corresponding to r_A will also move but do *not* have individual free surface degrees of freedom associated with them. Once the movement of a free surface node is computed, then the movement of those nodes on the associated boundary generator line is completely determined by the initial aspect ratio of nodes on this line. In other words, the free surface degree of freedom for internal moving nodes lying on line AB, for example, is r_A times the initial value of the line parameter r for each of these nodes. Although the motion of such nodes is completely dependent on the movement of the free surface node, elements which contain them also contribute to the system Jacobian via the surface degrees of freedom which are associated with the nodes of the element through the boundary generator lines.

7. NUMERICAL RESULTS

As a benchmark case, the classical steady state two-dimensional dieswell problem was chosen to compare our technique quantitatively with existing results. A complete description of the die-swell problem can be found in Omedei [10]. In the die swell problem, attention is focused on the case of a Newtonian jet exiting from a capillary tube and the problem of determining the free-surface for the two-dimensional steady case. The boundary conditions employed and the computational mesh are shown in figures 3 and 4 respectively. A series of simulations for different Reynolds number and Capillary number combinations were performed and the results compared with those of Omedei [10] and Dupret [11]; the parameter of interest being the swelling ratio, r_s. A mixed mode formulation using a nine node quadrilateral with a linear pressure approximation was employed for the simulations. As can be seen in Table 1, our results are in close agreement with their results. A few interesting points about the computation need to be noted. The oscillation present in the results of Omedei [10] were absent in our results. Each simulation required only 2 or 3 quasi-Newton iterations to converge with a convergence criterion of .1% rms relative error between successive iterations and in the residual vector. The initial starting solution for each simulation was the solution from a previous run. This performance compares favourable with Dupret's results which required 6 or 7 iterations to converge in which he also used the solution from a previous run as the initial solution vector.

Additional numerical results, including the transient simulation of a transient capillary-gravity wave in a container, will be presented at the conference.

ACKNOWLEDGEMENTS

The authors would like to acknowledge Professor Scriven's research group at the University of Minnesota and, in particular, Haroon Kheshgi, for stimulating discussions and providing numerical experiments and data for comparison. R.L.S. would also like to acknowledge support from the U.S. Army Research Office (Grant DAAG29-82-C-0010).

REFERENCES

1. H. Saito and L. E. Scriven, Study of coating flow by finite element method, JCP, **42** ,53 (1981)

2. H. Ettouney and R.A. Brown, Finite element methods for steady solidification problems, JCP (1982).

3. K. Ruschak, A method for incorporating free boundaries with surface tension in finite element fluid-flow simulators, IJNME, **15** ,639-648 (1980).

4. E. Kreyszig, **Differential Geometry**, University of Toronto Press, Toronto, 1959.

5. O.C. ZIENKIEWECZ, **Finite Element Method in Engineering Science**, 3rd Edition, 19--.

6. M.S. Engelman, G. Strang and K.J. Bathe, The application of quasi-Newton methods in fluid mechanics, IJNME, **17**, 707-718 (1981).

7. P.M. Gresho, R.L. Lee and R.L. Sani, On the time-dependent solution of the incompressible Navier-Stokes equations in two and three dimensions, in **Recent Advances in Numerical Methods in Fluids**, Pineridge Press, Swansea, U.K., 1980.

8. R.L. Sani, B.E. Eaton, P.M. Gresho, R.L. Lee and S.T. Chan, On the solution of the time-dependent incompressible Navier-Stokes equations via a penalty method, Lawrence Livermore Laboratory Report, UCRL-85354.

9. M.S. Engelman, FIDAP - A fluid dynamics analysis package, Adv. Eng. Soft. **4** ,163-166 (1982) Soft, Oct. 82.

10. B. Omedei, Computer solutions of a plane Newtonian jet with surface tension, Comp. Fluids, **7**,79-96 (1979).

11. F. Dupret, A method for the computation of viscous flow by finite elements with free boundaries and surface tension, in. **Finite Element Flow Analysis**, Ed: T. Kawai, University of Tokyo Press, Tokyo 1982.

Re	Ca^{-1}	$r_s(\%)$	Dupret	Omedei
1.	0.	18.97	19.56	19.0
1.	0.4	16.67	16.92	16.6
1.	1.6	11.58	11.4	11.65
1.	3.6	7.3	7.41	6.4 (±osc)
4.	.4	15.47	15.69	15.5
4.	1.6	12.34	12.41	12.2
4.	2.4	10.57	10.62	10.4 (±osc)
18.	2.2222	1.31	1.35	1.3 (±osc)
18.	4.4444	2.00	2.02	2.2 (±osc)
75.	2.		-10.92	-10.48 (±osc)
300.	0.	-15.24		-15.52

Table 1: Dieswell Percentages

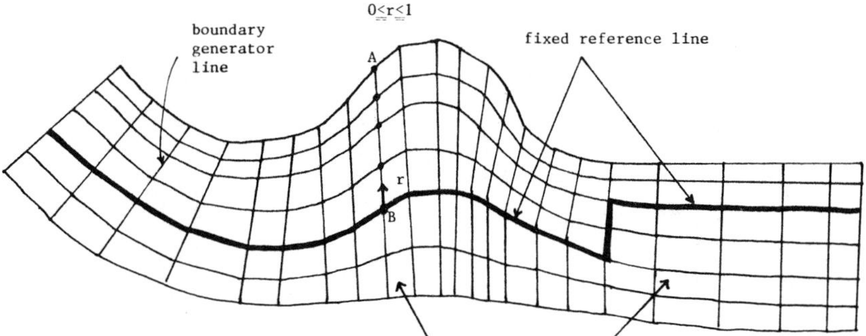

Figure 1

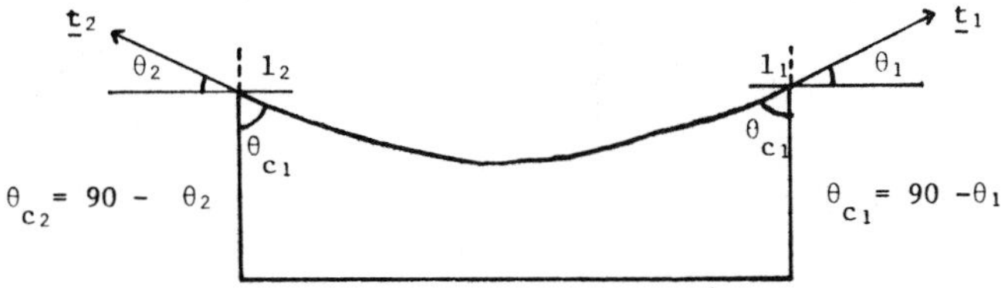

Figure 2: Contact angles

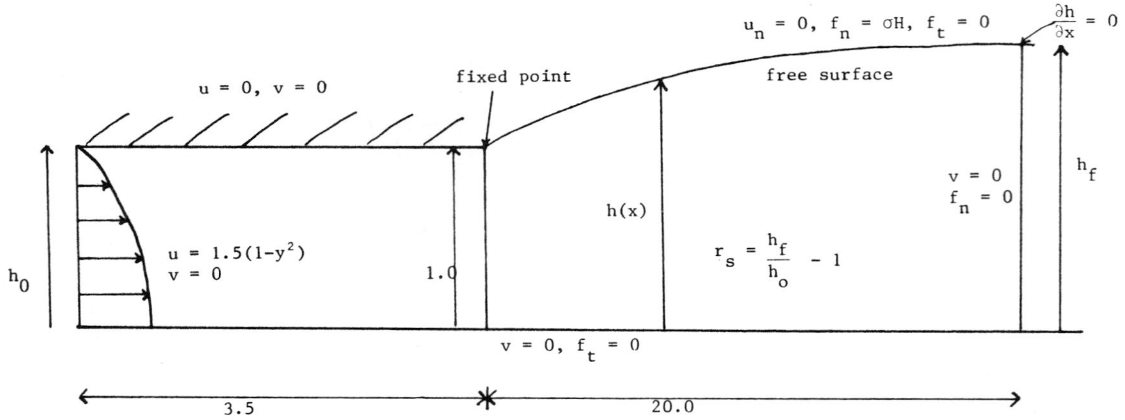

Figure 3: Dieswell problem

Figure 4: Computational mesh - dieswell problem

CONTINUATION IN A PARAMETER: EXPERIENCE WITH VISCOUS AND FREE SURFACE FLOWS

H. S. Kheshgi, O. A. Basaran, R. E. Benner, S. F. Kistler and L. E. Scriven

Department of Chemical Engineering & Materials Science, University of Minnesota, Minneapolis, MN 55455

1. SUMMARY

Continuation in a parameter allows prediction of accurate initial estimates for iterative solution of the nonlinear finite-difference or finite-element approximated equations of motion. Information from solutions at previous parameter steps is used to extrapolate to solutions at new parameter values. Step size can be automatically controlled to insure accurate estimates. Solution by continuation predictor/iterative corrector traces branches in parameter space along which steady flow states lie. Turning points, i.e. points at which a branch turns back upon itself, are passed by switching continuation parameters. Bifurcation points, i.e. points where one branch splits into more, are detected and multiple branches traced by special methods. Experiences with continuation applied to solution of the Navier-Stokes system for incompressible steady flow are reported for cases of flow through a rotating channel, formation of a falling liquid curtain, and gyrostatic equilibria of rotating cylindrical drops.

2. INTRODUCTION

Flow problems made nonlinear by momentum convection or a free surface must be solved iteratively when they are reduced to a set of nonlinear algebraic equations by finite difference discretization or finite element approximation. Solutions are generally sought over ranges of parameter values, e.g. Reynolds number, capillary number, boundary value, boundary position, or even velocity or pressure at some interior point. An iterative scheme to succeed has to generate at each step an initial estimate that falls in the scheme's domain of convergence for the parameter values at that step. The closer the estimate is to the solution sought, the fewer the interations needed to approximate that solution within the prescribed tolerance. Newton iteration is the scheme of choice, because it converges *quadratically* if the initial estimate is accurate enough and because it brings other benefits [1].

Continuation methods extrapolate from one or more acceptable approximations—'solutions'—at nearby parameter values to an initial estimate of the solution at a new set of parameter values (§3.1), much as the predictor step performs in predictor-corrector schemes for solving initial-value problems. Simplest is zeroth-order continuation: an available solution becomes the initial estimate at a not-too-different set of parameters [2]. First-order continuation extrapolates linearly from an available solution by means of its derivative with respect to the parameters that differ [3-10]; this usually provides a better initial estimate. By marching in the parameters, an entire branch of solutions in parameter space can be traced out. Indeed, continuation can track solution branches so highly nonlinear as to defy other methods [2-5]. Ways are known to select continuation parameters in order to maneuver around turning points, to detect bifurcation points, and to get on the new branches that join the old one at such points (§3.3)[5,6]. All of the branches of a solution tree can be mapped out, and this can be done efficiently by automatic selection of parameter step size so that initial estimates remain appropriately within the respective domains of convergence (3.2), although the optimal strategy is elusive. Moreover, stability of solutions to infinitesimal perturbations can often be deduced from the structure of the solution tree [7].

All of the foregoing is familiar in the computational mathematics of conservative systems, notably elastic structures [8], but has scarcely been exploited in computation of laminar and turbulent flow. Here we summarize our experience with standard continuation methods and improvements we innovated while solving Navier-Stokes systems for incompressible, steady, two-dimensional flows. We employed Galerkin weighted residuals, finite element basis functions, and Newton iterations. For examples (§4) we draw on our analyses of flow through a square duct rotating about an axis perpendicular to one wall [3], flow of a liquid film off the edge of an inclined plate [2], and shapes of gyrostatically rotating drops [4].

3. CONTINUATION METHODS FOR FLOW PROBLEMS

3.1 Parametric Continuation: Prediction of Solutions. The finite element solution of a problem is, in the first instance, a column matrix $\underline{u}$ of coefficients u_i of basis functions in which pressure, velocity, free surface (if present), etc. are expanded. The finite difference solution of a problem is a column matrix $\underline{u}$ of values u_i of pressure, velocity, etc. at meshpoints. In either case $\underline{u}$ is a function of the set of parameters of the problem $\{\alpha_1, \alpha_2, .., \alpha_n\}$; sometimes it is useful to include among the parameters one or more of the u_i

themselves, as in §3.3. The parameters can also be ordered into a column matrix $\underline{\alpha}$. Thus $u = u(\underline{\alpha})$ is the solution of the set of nonlinear algebraic equations $\underline{R}(\underline{u}\,;\underline{\alpha}) = \underline{0}$ where the members of $\underline{R}$ are the Galerkin weighted residuals or finite differenced residuals of the equations of motion. This section is restricted to continuation in a single parameter α.

Continuation is the prediction $\underline{u}^o(\alpha^{n+1})$ of a solution at a new parameter value α^{n+1}, the value reached by the n + 1-st step change in the parameter, from the preceding solution $\underline{u}(\alpha^n)$ and perhaps additional information such as $\underline{u}(\alpha^{n-1})$ or $d\underline{u}/d\alpha$ at $\alpha = \alpha^n$. In practice the prediction is from an acceptable approximation to $\underline{u}(\alpha^n)$ and any additional information being used. For the chosen iteration scheme there is a domain of convergence of $\underline{u}(\alpha^{n+1})$, the solution sought. If the prediction falls in this domain, the iteration will ultimately converge. Moreover, the 'closer' the prediction $\underline{u}^o(\alpha^{n+1})$ is to the solution $\underline{u}(\alpha^{n+1})$, the fewer the iterations $\underline{u}^i(\alpha^{n+1})$ required to reach an acceptable approximation of $\underline{u}(\alpha^{n+1})$, as estimated by some norm of the difference $\underline{u}^i(\alpha^{n+1}) - \underline{u}^{i-1}(\alpha^{n+1})$. From here on the ordinarily small discrepancy between $\underline{u}(\alpha^n)$ and $\underline{u}^i(\alpha^n)$ is ignored.

So-called zeroth-order continuation is prediction by

$$\underline{u}^o(\alpha^{n+1}) = \underline{u}(\alpha^n) = \underline{u}(\alpha^{n+1}) + O\,(\alpha^{n+1} - \alpha^n). \tag{1}$$

First-order continuation, like a forward-difference predictor, improves the prediction by making use of the set of derivatives that measure sensitivity to the parameter α:

$$\begin{aligned}\underline{u}^o(\alpha^{n+1}) &= \underline{u}(\alpha^n) + (\alpha^{n+1} - \alpha^n)\left[\frac{d\underline{u}}{d\alpha}\right]_{\alpha=\alpha^n} \\ &= \underline{u}(\alpha^{n+1}) - \frac{(\alpha^{n+1} - \alpha^n)^2}{2}\left[\frac{d^2\underline{u}}{d\alpha^2}\right]_{\alpha=\alpha^n} \\ &\quad + O[(\alpha^{n+1} - \alpha^n)^3].\end{aligned} \tag{2}$$

The sensitivity derivative $d\underline{u}/d\alpha$, which is also called the continuation vector, can be approximated by differencing solutions at successive parameter values, e.g.

$$\left[\frac{d\underline{u}}{d\alpha}\right]_{\alpha=\alpha^n} = \frac{\underline{u}(\alpha^n) - \underline{u}(\alpha^{n-1})}{\alpha^n - \alpha^{n-1}} - \frac{(\alpha^n - \alpha^{n-1})}{2}\left[\frac{d^2\underline{u}}{d\alpha^2}\right]_{\alpha=\alpha^n} + O[(\alpha^n - \alpha^{n-1})^2]\,. \tag{3}$$

Alternatively, when the iteration scheme is Newton's, $d\underline{u}/d\alpha$

can be evaluated exactly at little cost by using the Jacobian matrix $\underline{\underline{J}} \equiv (\partial\underline{R}/\partial\underline{u})$ already available with the converged solution $\underline{u}^i(\alpha^n)$ (cf. [9]):

$$\left[\frac{d\underline{u}}{d\alpha}\right]_{\alpha=\alpha^n} = \left[\underline{\underline{J}}^{-1}\frac{d\underline{R}}{d\alpha}\right]_{\alpha=\alpha^n} . \tag{4}$$

When the sensitivity becomes unduly large or multivalued, special tactics (§3.3) have to be used.

Methods of higher-order continuation, like Adams-Bashforth predictors, have been designed [9]; however, it appears that first-order continuation ordinarily suffices.

3.2 Automatic Choice of Parameter Step Size. Once the continuation parameter is selected and the iteration scheme chosen, the sizes of successive parameter steps can be adjusted so that successive predictions $\underline{u}^o(\alpha^{n-1})$, $\underline{u}^o(\alpha^n)$, $\underline{u}^o(\alpha^{n+1})$, . . . do not wander out of the zone of convergence. The tactic is to control parameter step-size to ensure that the continuation-prediction $\underline{u}^o(\alpha^n)$ always lies within a specified small 'distance' ε of the iteration-corrected solution $\underline{u}(\alpha^n)$.

With zeroth-order continuation, the experience in the n-th parameter step is used to estimate the root-mean-square ($|\ |$) of the sensitivity derivative in terms of the 'distance'

$$\begin{aligned}\delta^n &\equiv |\underline{u}^o(\alpha^n) - \underline{u}(\alpha^n)| \\ &= \left|(\alpha^n - \alpha^{n-1})\left[\frac{d\underline{u}}{d\alpha}\right]_{\alpha=\alpha^n}\right| + O[(\alpha^n - \alpha^{n-1})^2] .\end{aligned} \tag{5}$$

If $d\underline{u}/d\alpha$ does not change appreciably, δ^{n+1} can be estimated and the next parameter value α^{n+1} chosen so that

$$\delta^{n+1} \simeq \delta^n \, |\alpha^{n+1} - \alpha^n| \Big/ |\alpha^n - \alpha^{n-1}| = \varepsilon . \tag{6}$$

With first-order continuation, the 'distance' can be estimated from

$$\delta^n = \frac{(\alpha^n - \alpha^{n-1})^2}{2}\left|\left[\frac{d^2\underline{u}}{d\alpha^2}\right]_{\alpha=\alpha^n}\right| + O[(\alpha^n - \alpha^{n-1})^3] \tag{7}$$

when the sensitivity derivative is evaluated by (4), or by

$$\delta^n = \frac{|\alpha^n - \alpha^{n-1}||\alpha^n - \alpha^{n-2}|}{2}\left|\left.\frac{d^2\underline{u}}{d\alpha^2}\right|_{\alpha=\alpha^n}\right| + O[(\alpha^n - \alpha^{n-2})^3] \tag{8}$$

when the derivative is approximated by (3). In either case α^{n+1} is estimated as in (6) and the corresponding results are, respectively,

$$\alpha^{n+1} = \alpha^n + (\alpha^n - \alpha^{n-1})\sqrt{\varepsilon/\delta^n} \tag{9}$$

$$\alpha^{n+1} = \alpha^n - \frac{\alpha^n - \alpha^{n-1}}{2} + \sqrt{\frac{(\alpha^n - \alpha^{n-1})}{4} + \frac{(\alpha^n - \alpha^{n-1})(\alpha^n - \alpha^{n-2})\varepsilon}{\delta^n}} \; . \tag{10}$$

The same approach applies to higher-order continuation.

If at some value α^{n+1} of the parameter the iteration fails to converge within the allotted number of iterates, a smaller parameter step can be automatically generated. When the goal is to reach a particular parameter value from a given starting value with a minimum amount of computational work, the best policy may be to set coarser convergence tolerances along the way so as to remain well within the zone of convergence without investing in unnecessarily accurate intermediate solutions. Or the best policy may be the extreme of making just one iteration at each new value of α; then the 'distance' δ^{n+1} can be checked and if it is found to exceed ε by unacceptably much, e.g. a factor of 1.1, α^{n+1} can be recalculated from (6), (9), or (10) by replacing δ^n and the old step size with the unsuccessfully large δ^{n+1} and step size.

This last procedure, as well as those based on (6), (9) or (10), can lead to such small parameter steps that continuation fails to make progress. This is particularly likely to happen near turning points, at which a branch of solutions passes through a local maximum or minimum in a parameter, and near bifurcation points, at which a new branch (or branches) of steady solutions departs from the one being traced. Turning points and bifurcation points are singular points in the sense that at them one or more of the eigenvalues of $\underline{\underline{J}}$ becomes zero. Special procedures are needed to locate singular points accurately and to make efficient use of continuation near them.

3.3 <u>Continuation Near Singular Points</u>. As a turning point is approached, it is revealed by the magnitude of $d\underline{u}/d\alpha$ which grows unbounded at it. Bifurcation points at which an odd number of eigenvalues and turning points at which one eigenvalue of $\underline{\underline{J}}$ pass through zero can be isolated by noting the pair of values of $(\underline{u},\alpha)$ between which det $\underline{\underline{J}}$ changes sign [6]. As any singular point at $\alpha = \alpha_s$ is approached, it is signaled by one or more Gauss elimination pivots of $\underline{\underline{J}}$ approaching zero. In particular, bifurcation points at which an even number of eigenvalues pass through zero can be identified this way [7]. An alternative approach is to find the dominant eigenvalues of $\underline{\underline{J}}$, a difficult problem. Moreover, Hopf bifurcations to unsteady solution branches where only the real parts of complex pairs of eigenvalues pass through zero can be detected by solving the eigenvalue problem, and is often called for to determine stability [1].

The detectable singular points can be accurately located by augmenting the equation set $\underline{R}(\underline{u},\alpha) = \underline{0}$ with either $\det \underline{\underline{J}} = 0$ or $||d\underline{u}/d\alpha||_\infty^{-1} = 0$. The augmented set can be solved for α_s by Abbott's [6,7] technique, which ultimately converges quadratically to turning points and linearly to bifurcation points.

The singularity of the Jacobian matrix at a turning point in α can be removed by augmenting the unknown coefficients $\underline{u}$ with α, i.e. $\underline{\hat{u}} \equiv (\underline{u}^T,\alpha)^T$, and choosing automatically as the 'adaptive' parameter that element $\hat{u}_j$ of $\underline{\hat{u}}$ which changed most over the last parameter step. Then the equation set is

$$\begin{aligned} R_k(\underline{\hat{u}}^{n+1}) &= 0, \quad k = 1, \ldots, N+1 \text{ except } k \neq j \\ \hat{u}_j^{\,n+1} &= \hat{u}_j^{\,n} + (\hat{u}_j^{\,n+1} - \hat{u}_j^{\,n}) \end{aligned} \tag{11}$$

where the size of the parameter step $(\hat{u}_j^{\,n+1} - \hat{u}_j^{\,n})$ can be chosen as in §3. 2. Newton's iteration process is then

$$\hat{\underline{\underline{J}}} \begin{bmatrix} \Delta \underline{u} \\ \hline \Delta\alpha \end{bmatrix} = \left[\begin{array}{c|c} \dfrac{\partial \underline{R}}{\partial \underline{u}} & \dfrac{\partial \underline{R}}{\partial \alpha} \\ \hline \multicolumn{2}{c}{0 \ldots 1 \ldots 0} \end{array}\right] \begin{bmatrix} \Delta \underline{u} \\ \hline \Delta\alpha \end{bmatrix} = - \begin{bmatrix} \underline{R} \\ \hline 0 \end{bmatrix} \tag{12}$$

The nonzero entry in the last row of $\hat{\underline{\underline{J}}}$ falls in the column that corresponds to $\hat{u}_j$. This formulation works perfectly well for continuation through regular points, not just turning points.

The singularity of the Jacobian matrix at a bifurcation point $(\underline{u}_b^T, \alpha_b)$ cannot be removed in the foregoing way because there are multiple solutions in the neighborhood. Where just a single real eigenvalue of $\underline{\underline{J}}$ crosses zero the equation set $\underline{\underline{J}}\Delta \underline{u} = -R$ has a homogeneous solution, the null vector or critical eigenvector $\underline{\eta}$, which in the parameter space points in the direction of the tangent to the intersecting branch of solutions at the bifurcation point. There $\underline{\eta}$ is readily calculated by inverse iteration [6] and can then be used to jump the parameter α or $\hat{u}_j$ into the domain of convergence of a nearby solution $\underline{u}$ on the intersecting branch. The Newton corrector can be forced onto a solution $\underline{u}(\alpha)$ in the intersecting or bifurcating family by augmenting the equation set $\underline{R}(\underline{u},\alpha) = \underline{0}$ with the constraint [7]

$$[\underline{u}(\alpha) - \underline{u}_b(\alpha_b)]^T \underline{\eta} = \varepsilon , \quad \varepsilon \neq 0 \ . \tag{13}$$

The following examples are drawn from our experience with continuation methods for flow problems.

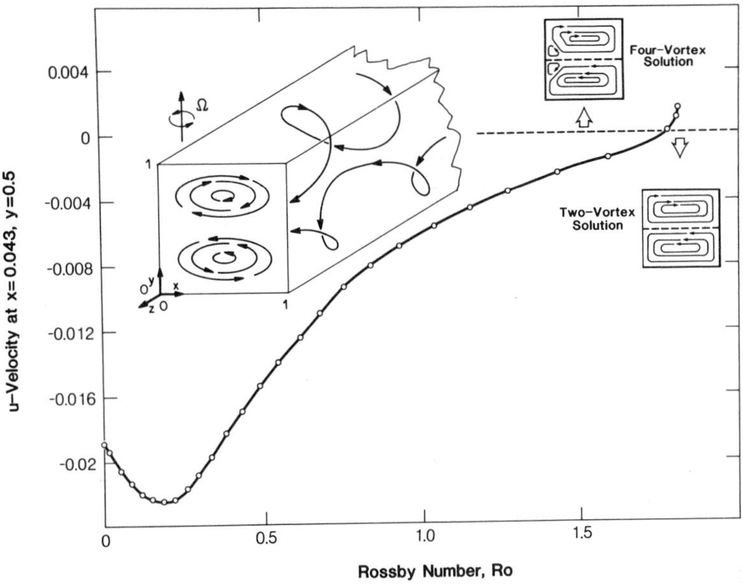

Figure 1. Flow through a rotating square channel.

4.1 Flow Through a Rotating Channel. Incompressible liquid is driven by a pressure gradient down a long, square channel [3]. Because the channel rotates about an axis perpendicular to its top, the streamlines are not rectilinear, a situation found in many channels and passages in rotating machinery (see Figure 1). Cast in the appropriate rotating frame the dimensionless Navier-Stokes equation is

$$\mathrm{Ro}\ \underset{\sim}{v} \cdot \nabla \underset{\sim}{v} + 2\ \underset{\sim}{j} \times \underset{\sim}{v} = -\ \nabla p + \mathrm{Ek}\ \nabla^2 \underset{\sim}{v} \qquad (14)$$

where the Rossby number $\mathrm{Ro} \equiv (\partial p/\partial z)/\rho L^2\Omega^2$ is a ratio of convective acceleration to Coriolis acceleration, and the Ekman number $\mathrm{Ek} \equiv \nu/L^2\Omega$ is a ratio of viscous force to Coriolis force (L is channel height; Ω the magnitude of angular velocity, which is in the $\underset{\sim}{j}$-direction; and the gyrostatic pressure gradient in the $\underset{\sim}{k}$-direction is taken as unity).

The unknown pressure and three components of velocity over the two-dimensional, square cross-section we found with the penalty/Galerkin finite element method and Newton iteration. At Ek = 0.01, Coriolis's effect is important. First-order continuation in Ro, Eqs. (2) and (3), with automatic step-size control, Eq. (10), followed the obvious solution branch from the unique solution at Ro = 0. ε was chosen to be $0.002\ |\underline{u}|$, where $\underline{u}$ is the set of velocity basis function coefficients. One Newton iteration was taken per parameter step. The Coriolis effect deflects the fast-moving liquid in the channel core to the lagging left wall (Figure 1, upper left), where it recirculates symmetrically to establish a double-vortex flow state. The results in Figure 1 show that as Ro rises, the x-velocity component near that wall at first falls, then climbs, and then abruptly rises above zero at an apparent turning point. Where the plotted velocity component is greater than zero, there is not a double-vortex flow but a four-vortex one. Such steady states have indeed been approached by solving unsteady flow problems at higher Ro [3]. The tactics described in §3.3 are called for.

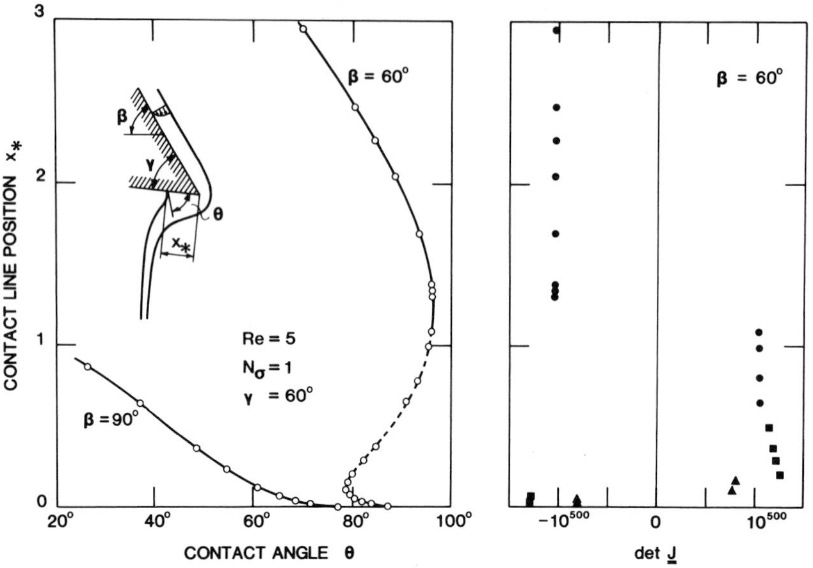

Figure 2. Formation of falling liquid curtains.

4.2 Formation of Falling Liquid Curtains. When liquid flows down an inclined plate and falls off as an unsupported sheet (cf. Figure 2), the sheet often deflects toward the underside of the plate [2]. Moreover, if wetting phenomena intrude, the contact line where the lower free surface separates from the solid is free to migrate along the solid underside instead of remaining pinned at the sharp edge. To solve the Navier-Stokes system for steady flow we used a spine parametrization of the free surface and the Galerkin finite element method with Newton iteration. Stress at the free surface is balanced by capillary force. For a given capillary parameter $N_\sigma \equiv \sigma[\rho/g(4\mu)^4]^{1/3}$ (σ is surface tension) and Reynolds number $Re \equiv q/\nu$ (q is volumetric flux), solution branches for two different plate inclinations β are shown (see Figure 2). The contact line position x_* (measured in units of upstream film thickness on the inclined plate) depends on the continuation parameter, the quasi-static contact angle θ which measures the wettability of the solid underside by the liquid.

Zeroth-order continuation in θ alone could not follow solution branches near turning points in θ ($\beta = 60°$ case) or near $x_* = 0$ (both cases). Changing the continuation parameter from θ to x_* allowed calculation of the solution branches shown. When the underside is horizontal ($\beta = 60°$) two turning points in θ exist. At each turning point the sign of det $\underline{\underline{J}}$ ($\underline{\underline{J}}$ constructed for fixed θ, free x_*) changes corresponding to an odd number of eigenvalues passing through zero. Flow stability also changes. Experiments establish that the solid portions of the curve correspond to stable states, and so the broken portion between must represent unstable states [2].

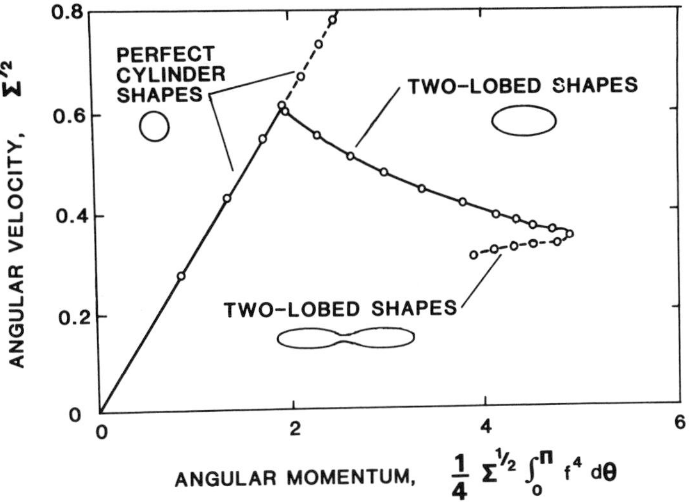

Figure 3. The gyrostatic shapes of rotating cylindrical drops.

4.3 Rotating Liquid Drops. Indefinitely long, rotating cylindrical drops are unstable to the axisymmetric Rayleigh mode of break-up, but if this mode is suppressed, such drops turning slowly are stable to translationally-symmetric disturbances but are unstable at higher angular velocity or angular momentum. The Young-Laplace equation describes gyrostatic equilibrium shapes ($f = f(\theta)$ is radial position of the free surface):

$$(f\,f_{\theta\theta} - 2f_\theta - f\,)(f_\theta{}^2 + f^2)^{-3/2} + 4\,\Sigma\,f^2 \;+ 2K = 0 \qquad (15)$$

The parameters are rotational Bond number $\Sigma \equiv \rho\,\Omega^2\,R^3/8\sigma$, a ratio of kinetic energy of rotation to surface energy, and dimensionless reference pressure K. Successive instabilities of the perfect cylinder are due to harmonic shape disturbances proportional to $\cos n\,\theta$. To study the bifurcating families of increasingly distorted n-lobed shapes we used adaptive spine parametrization of the free surface, Galerkin's method and finite element basis functions along with Newton iteration, adaptive continuation, Abbott's method, and jumping (§3.3)[4]. Third-order perturbation analysis confirmed the computed primary bifurcation. Figure 3 shows a portion of the solution tree.

5. ACKNOWLEDGEMENT

This research was supported by grants from NASA, Kodak Research Laboratories and the University of Minnesota Computer Center.

6. REFERENCES

1. BIXLER, N. E. and SCRIVEN, L. E.- Coating Flow Stability and Bifurcation by Finite Element and Asymptotic Analysis of the Navier-Stokes System, Proc. 3rd Int. Conf. Num. Meth. Laminar and Turbulent Flow, Seattle, WA, 1983.
2. KISTLER, S. F. and SCRIVEN, L. E.- The Teapot Effect and its Role in Buoyant Surface Jets. Bull. Amer. Phys. Soc. 27, 1173. 1982.
3. KHESHGI, H. S. and SCRIVEN, L. E. - Viscous Flow Through a Rotating Square Passage. Bull. Amer. Phys. Soc. 25, 1099. 1980.
4. BENNER, R. E., BASARAN, O. A. and SCRIVEN, L. E. - Stability and Bifurcation of Translationally Symmetric, Rotating Liquid Drops. Bull. Amer. Phys. Soc. 25, 1073. 1980.
5. BROWN, R. A., YAMAGUCHI, Y. and CHANG, C. J. - Finite Element Calculation of the Steady Axisymmetric Flows in a Vertical Cylinder Heated from Below, A.I.Ch.E. Ann. Mtg. Paper No. 21b, Los Angeles, 1982.
6. ABBOTT, J. P. - An Efficient Algorithm for the Determination of Certain Bifurcation Points. J. Comp. Appl. Math. 4, 19. 1978.
7. UNGAR, L. H. and BROWN, R. A. - The Dependence of the Shape and Stability of Captive Rotating Drops on Multiple Parameters. Phil. Trans. Roy. Soc. Lond. A 306, 347. 1982.
8. THOMPSON, J. M. T. - Instabilities and Catastrophes in Science and Engineering, John Wiley & Sons, New York, 1982.
9. KUBICEK, M. - Algorithm 502 — Dependence of Solution of Nonlinear Systems on a Parameter [C5]. ACM Trans. Math. Software 2, 98. 1976.

Momentum Control Volumes for Finite Difference Codes

L. G. Margolin and B. D. Nichols
Earth and Space Sciences Division
Los Alamos National Laboratory
Los Alamos, NM 87545

SUMMARY

We consider the integration of the Navier-Stokes equations in finite difference codes. We first describe the surface integral technique using momentum control volumes. Then we focus on the problem of the spherical expansion of a gas. We analyze the difference equations to see why the numerical calculation does not preserve spherical symmetry. Then we propose general, yet simple, techniques for reducing the errors that break the symmetry. We proceed to show numerical results to demonstrate the utility of our techniques. Finally, we consider questions of accuracy and stability of the integration for nonuniform meshes.

1. INTRODUCTION

We consider the integration of the momentum (Navier-Stokes) equations in finite difference computer codes. The spatial discretization leads to inaccuracies in this integration. Some of these errors are associated directly with the resolution of the computational mesh. Other errors arise, however, in the numerical techniques that are employed. These latter errors can be corrected to produce more accurate results with no increase in spatial resolution.

To illustrate this type of error, we have chosen to analyze the spherical expansion of a gas with point symmetry. Thus, the velocity magnitude, pressure and other physical variables at a point are a function only of the distance of the point from the center of symmetry. Also, the direction of the velocity vector must be everywhere radial. Rather than compare our numerical simulation with analytic results, we compare our calculations with the expected analytic symmetries. The SHALE code was used for these numerical calculations.

The SHALE code is a two-dimensional finite difference program for calculating transient fluid flow and wave propagation in solids [1] that may be run in cylindrical or plane coordinates. SHALE is based on the ALE technique [2]. In SHALE, the mesh may move with arbitrary velocity with respect to the fluid. However, all calculations in this paper are purely Lagrangian and utilize cylindrical coordinates.

In the following sections, we first describe the calculation of acceleration at nodes using momentum control volumes. Then we analyze the difference equations, with respect to the spherical expansion of a gas, for several choices of control volumes. We identify several causes for lack of point symmetry in the difference approximation, and describe general methods to regain this symmetry. We then present several sets of numerical calculations to illustrate the validity of our analysis. Finally, we discuss the difference forms with regard to stability.

2. MOMENTUM CONTROL VOLUMES

SHALE uses a staggered finite difference mesh of quadrilateral cells. Each cell consists of four nodes, or vertices, connected by straight lines. Coordinates and velocities are stored at the nodes. Density and internal energy are stored at cell centers. Thus, accelerations must be calculated at nodes, and pressure at cell centers.

The acceleration is calculated from the Navier-Stokes equations

$$1) \qquad \frac{du_k}{dt} = -\frac{1}{\rho}\frac{\partial p}{\partial x_k} \quad ,$$

where u_k is the velocity, p is the pressure including viscous (real or numerical) contributions and

$$2) \qquad \frac{du_k}{dt} = \frac{\partial u_k}{\partial t} + u_\ell \frac{\partial u_k}{\partial x_\ell}$$

is the Lagrangian time derivative.

Many finite difference codes calculate the acceleration by approximating the pressure gradient at the node point [3]. However, SHALE uses a different technique [1]. In SHALE, one chooses a volume containing the node point. This volume is called the momentum control volume (MCV). The average acceleration of the volume is then equated to the volume integral of the pressure gradient divided by the mass (M) of the MCV. The volume integral can be converted to a surface integral of pressure over the MCV. In this form

$$\frac{du_k}{dt} = -\frac{1}{M} \int_S p n_k \, dS \qquad (3)$$

where n_k is the outward pointing unit surface vector.

The surface integral technique described here has the advantage that it is easy to rigorously conserve momentum. All that is required is that the MCVs are constructed so that each area element, i.e., each side of the MCV, is common to two nodes. Then the momentum flux to one node is exactly balanced by the loss from the node across the common area. SHALE accomplishes this by constructing a secondary MCV mesh whose vertices are the cell centers of the computational mesh (figure 1). However, other versions of ALE codes [4,5] connect the four logically nearest nodes to form diamond shaped MCVs (figure 2). The diamond MCVs cover the mesh twice, i.e., each point belongs to two different MCVs and the average mass of an MCV is twice the average mass of a cell. We shall show that use of the diamond MCVs is unnecessarily diffusive.

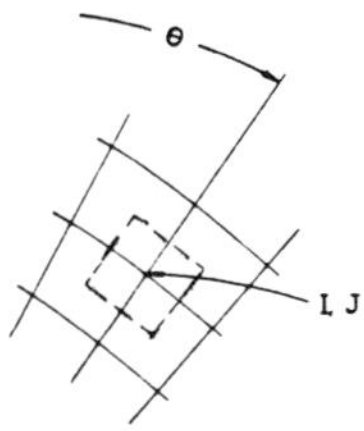

SHALE MOMENTUM CONTROL VOLUME

Figure 1

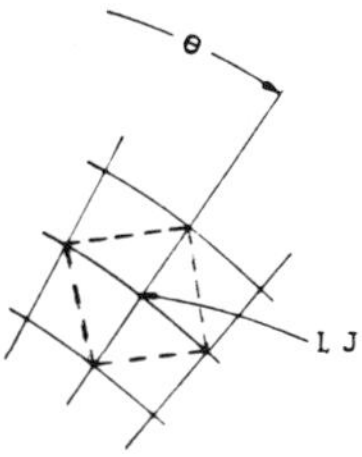

DIAMOND MOMENTUM CONTROL VOLUME

Figure 2

We see that the Navier-Stokes equations in the form of equation 3 are equivalent to Newton's law for the average acceleration of a body. However, we note that the average acceleration, by Newton's law, is correctly applied only to the center of mass of the body. The acceleration of other points in the body may differ from that of the center of mass by a term representing angular acceleration about the center of mass. Similarly, the velocity, calculated as the integral in time of the acceleration of any point - and in particular of the node point - will in general differ from the velocity of the center of mass. It is for this reason that, generally, finite difference codes do not conserve angular momentum.

In fact, the node point will differ from the center of mass of the MCV under almost all circumstances. For example, the difference may be the consequence of using a spatially nonuniform mesh, of having spatial gradients of density, or even from using uniform zoning with cylindrical coordinates. The main point of this paper is that the center of mass effect described above exists, and is important, and can be corrected by a simple, yet general, technique.

3. PROBLEM DESCRIPTION

We consider the spherical expansion of an ideal gas that is composed of two regions. The inner region is a sphere of radius 5 cm and is initially at a pressure of 3 x 10" dynes cm^{-2} (300 kbar). The outer region is a shell extending in radius from 5 cm to 15 cm. Gas in the outer region obeys the equation of state

$$p = (\gamma-1)\ \rho I \quad , \tag{4}$$

where we have chosen $\gamma = 5/3$. Initially, the specific internal energy is 1×10^9 dynes gm^{-1} and the density is 1.0 gm cm^{-3}. The gas in the inner region is governed by an adiabatic equation. In addition, the inner region is treated as a single cell to ensure sphericity during its expansion.

The numerical simulation is zoned in cylindrical coordinates (r,z). The nodes are numbered by logical coordinates (indices). Logical J lines are spherical surfaces spaced 1 cm apart. Logical I lines are cones spaced 6° apart. The I=1 line is the cylindrical axis of symmetry and the I=16 line is the equatorial plane of symmetry. Because of the symmetry of the problem, only one hemisphere is represented. All the I lines converge at the spherical center of the gas, so that all points with logical coordinate J=1 are coincident. The mesh is shown in figure 3.

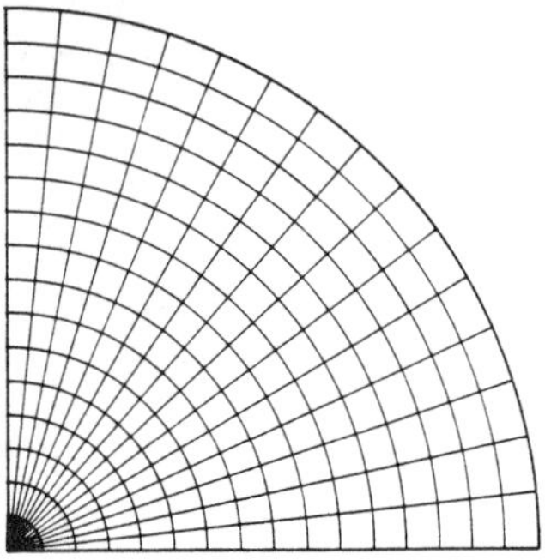

INITIAL ZONING FOR SPHERICAL EXPANSION

Figure 3

4. ANALYTIC RESULTS

In this section, we compute analytically the acceleration of a node based on the difference equations for several choices of MCVs. Recalling that each node lies on a spherical shell, we assume a uniform pressure inside the shell and zero pressure outside the shell. The cells are assumed to have uniform angular width. Thus, we are computing the equivalent of the acceleration of the first computational cycle in the computer codes.

The results for the acceleration are summarized in table 1. The two basic choices of MCVs are the diamond shaped MCVs and the MCV described above for the SHALE code. In each case, the areas and the masses are exact calculations for the MCVs in SHALE. However, SALE [5] uses the diamond MCVs, but approximates the masses and areas.

If spherical symmetry were exactly preserved, then the ratio of horizontal to vertical velocity would be tan θ, where θ is the polar angle measured from the cylindrical axis of symmetry. As noted above, SHALE uses Newton's law for the average acceleration of a body. It also uses exact formulae for the area and the mass of the body. Yet, a loss of symmetry is observed, which is manifested by the acceleration vector not moving radially outward. We identify only two reasons for this loss of symmetry. One reason is the distinction between the node and the center of mass of the MCV. The other reason is that our MCV has conical, not spherical, boundaries perpendicular to the radial boundaries. Thus, instead of the expansion of a shell, we are calculating the expansion of a more complicated body, which only approximates spherical symmetry.

To separate these effects, we also consider in table 1 the acceleration of a set of MCVs formed by connecting the centers of cells of equal radius with spherical surfaces instead of with conical surfaces as SHALE does. These results are listed in table 1 as SHALE/arcs. It is easy to show that the acceleration calculated for these MCVs is exactly radial for the appropriate center of mass of such a body, thus supporting our contention that the two effects listed above are the only reasons for lack of point symmetry in the standard SHALE code.

Table 1

The ratio of horizontal to vertical acceleration for several choices of control volume.   is the angular width of a cell, and θ is the polar angle. The exact answer is tan θ.

Momentum Control Volumes	Ratio of Horizontal to Vertical Acceleration
Diamond	$\dfrac{2\cos\theta\sin\theta\cos\delta}{2\cos^2\theta+\sin^2\delta\,(\cos^2\theta-\sin^2\theta)}$
SHALE	$\dfrac{2\cos\theta\sin\theta\cos\frac{\delta}{2}}{2\cos^2\theta+\sin^2\frac{\delta}{2}\,(\cos^2\theta-\sin^2\theta)}$
SALE	$\dfrac{\sin\theta}{\cos\theta\cos\delta}$
SHALE/arcs	$\dfrac{2\cos\theta\sin\theta\cos\frac{\delta}{2}\sin\frac{\delta}{2}}{\frac{\delta}{2}+\sin\frac{\delta}{2}\cos\frac{\delta}{2}\,(\cos^2\theta-\sin^2\theta)}$

From table 1, we can draw several important conclusions. For all the choices of MCVs, the errors, i.e., the deviation of the direction from tan θ, are of order $(\;)^2$ is the angular width of a cell. The error, then, is greatest for the cylindrically radial component of acceleration and for the vertices nearest the cylindrical axis of symmetry. That is, for vertices with logical coordinate I=2 where sin θ is itself of order (δ). For example, for cells of 6° angular width, the r component of acceleration has an error of 12.5% in SHALE.

Comparing the exact diamond shaped control volumes with SHALE, we see that the errors are of order $(2\delta)^2$ in the former, as opposed to $(\delta)^2$ in SHALE. This is not surprising, for in using the diamond control volumes, we couple vertices that are farther apart than is the case in SHALE. It appears that the smaller the MCVs, the smaller the error. On the other hand, to conserve momentum, it is necessary to have the MCVs exactly cover the mesh, i.e., the MCVs must form a dual mesh with the calculational mesh.

We cannot conclude that SHALE's MCVs are optimal, however, for it is possible that there is a better construction than connecting cell centers. In fact, if the z coordinates of the nodes of a cell are (z_1, z_2, z_3, z_4), then it is advantageous to choose

$$5) \qquad \bar{z} = \frac{z_1 + z_2 + z_3 + z_4}{\cos \frac{\delta}{2}} \quad .$$

This choice leads to accelerations nearly equal to those of SHALE/arcs in table 1. We do not recommend this choice for general meshes. However, for meshes with a center of convergence, i.e., meshes like ours where all nodes for J = 1 are coincident, this is a useful change. Further confirmation of the advantages of this construction is given in a later section when we consider meshes with cells of unequal angular width.

Finally, we note that SALE appears to have the best behavior for our spherical problem. Our calculations, of course, only represent the first cycle of the problem. One might expect that codes that begin with the most symmetric accelerations will preserve symmetry best. However, we will see that SALE actually does worst in preserving symmetry. We suspect this is because SALE does not use exact formulae for the area and mass of a MCV. We will discuss this point in more detail also when we consider meshes with cells of unequal angular width.

5. SUBROUTINE RAINIER

In this section we describe a subroutine (Rainier) we have added to SHALE to correct for the spatial difference between the vertex and the center of mass of an MCV. In the subroutine, we first exactly calculate the position of the center of mass for each MCV. The volume averaged acceleration, which we calculate in the Lagrangian phase of the code, is correctly applied to this point. Cycle by cycle, we integrate in time this acceleration to derive the velocity of the center of mass. In SHALE, we would then use this velocity to move the node. In Rainier, however, we take the additional step of interpolating the velocity from the center of mass to the node. This corrected velocity is then used to move the node.

The interpolation we have constructed is based on a Taylor series expansion of the velocity about the center of mass. We estimate the spatial gradients of velocity $\frac{\partial u_i}{\partial x_k}$ from the four nearest node points. The velocity at the vertex x_k is calculated from the velocity at the center of mass $\bar{x}_k$

$$u_\ell (x_k) = u_\ell (\bar{x}_k) + (x_k - \bar{x}_k) \frac{\partial u_\ell}{\partial x_k} \quad . \tag{6}$$

The velocities we use to construct the spatial gradients are themselves slightly in error, since they also are associated with the centers of mass of the neighboring MCVs rather than with the neighboring nodes. This error is, in some sense, of second order.

One might ask why we do not use Newton's law relating torque on the MCV to the angular acceleration. At first glance, it appears that one could use such an approach to derive an angular velocity of the vertex about the center of mass. The problem with this approach is that our MCVs are not rigid bodies. In fact, the MCV must move in a fashion consistent with its neighbors, which is not possible if all the MCVs were rigid bodies. Thus, it is not even approximately true that one could treat the MCVs in this manner to correct for the spatial difference between the vertex and the center of mass.

6. NUMERICAL RESULTS

In this section we present several sets of numerical simulations. These will serve to verify the conclusions we drew in section 4 based on analysis of the difference equations. These will also demonstrate the usefulness of the velocity interpolation scheme described in the previous section.

We have run three sets of problems, each based on the point symmetric expansion of a gas. In the first problem set, we choose a uniform mesh such that each cell has the same angular width (6°), as shown in figure 3. In the second problem set, we use the same uniform mesh, but introduce a density discontinuity into the problem. We set the density in the outer low pressure region to be five times larger than the density in the inner high pressure region. The density discontinuity should accentuate the effect as a result of the displacement of the center of mass of a MCV from the vertex point. In the third problem set, we rerun the uniform density problem, but with a nonuniform mesh. For this problem set, the angular width of a cell varies from 8° at the cylindrical axis to 4° at the equator. Users of Lagrangian codes know that it is extremely difficult to maintain spherical symmetry in a mesh with nonuniform angular widths.

In each problem set, we compare five separate runs. The runs are

1) Basic SALE, as described in Ref. 5.
2) Basic SHALE, as described in Ref. 1.
3) SHALE with the velocity interpolation scheme described in the previous section. These runs are designates SHALE/R.
4) SHALE with the slightly optimized control volumes described in section 4. These runs are designated SHALE/M.
5) SHALE with both the optimized control volumes and the interpolation. These runs are designated SHALE/R/M.

For each set of runs, we compare three figures of merit. First, we compare our numerical results with the analytic formulae of section 4. There we found that it was the first node off the cylindrical axis of symmetry which had the largest error. At cycle 1, then, the node on the interface between the high and low pressure regions that is next to the axis of symmetry should have an acceleration which differs most from being radial. Thus, for each run, we list the ratio of the horizontal to vertical velocity for the vertex (I=2,J=6) at the end of cycle 1. This will be compared with the correct result, which is tan θ (θ is the polar angle of line I=2), and with the analytic results of table 1.

Next we wish to compare the degree to which each code preserves spherical symmetry. We look at the variation of pressure in the shock front as a function of polar angle θ at a time of 2×10^{-5} sec. This time is arbitrarily chosen. However, it is a reasonable choice as the shock has moved out to about twice the radius of the initial high pressure region, i.e., about 6 cells.

Table 2

Summary of results for problem set 1 with uniform angular zoning (see text). The ratio of accelerations is for vertex I=2,J=6 after the first cycle.

Code	$\frac{a_x}{a_z}$	$\frac{\Delta p}{p}$ (in shock)	$\frac{\Delta p}{p}$ (behind shock)
SALE	.1057	.113	.027
SHALE	.1183	.073	.011
SHALE/M	.1117	.016	.006
SHALE/R	.1092	.053	.011
SHALE/R/M	.1027	.012	.008
Analytic	.1051	0	0

Table 3

Summary of results for problem set 2 with 5 to 1 density discontinuity (see text).

Code	$\frac{a_x}{a_z}$	$\frac{\Delta p}{p}$ (in shock)	$\frac{\Delta p}{p}$ (behind shock)
SALE	.1057	.240	.021
SHALE	.1183	.110	.015
SHALE/M	.1117	.046	.007
SHALE/R	.1090	.045	.008
SHALE/R/M	.1025	.005	.003
Analytic	.1051	0	0

Table 4

Summary of results for problem set 3 with variable angular zoning (see text).

Code	$\frac{a_x}{a_z}$	$\frac{\Delta p}{p}$ (in shock)	$\frac{\Delta p}{p}$ (behind shock)
SALE	.1418	.381	.028
SHALE	.1553	.050	.008
SHALE/M	.1479	.048	.003
SHALE/R	.1448	.012	.015
SHALE/R/M	.1375	.029	.006
Analytic	.1405	0	0

We finally look at the pressure variation in a shell near the original interface between low and high pressure. The point here is to see to what degree spherical symmetry has been achieved in the final equilibrium (long after the shock has passed). The data described above are tabulated in tables 2, 3, and 4. We point out several conclusions from the data in tables 2, 3 and 4.

1) The direction of the velocity vector in cycle 1 is exactly consistent with the analytic formulae of table 1.
2) SHALE with the velocity interpolation (SHALE/R) does consistently much better in maintaining spherical symmetry than either SHALE or SALE.

3) SHALE with the velocity interpolation and the optimized control volumes (SHALE/R/M) does slightly better than SHALE/R.
4) SALE appears to start each problem with the most nearly correct radial velocities. However, it does worst in preserving spherical symmetry. This is especially true in table 4 for the problem with variable angular zoning. This is true both in the shock front and far behind the shock.

The last point above seems somewhat hard to understand. One might expect that the code which starts out with the best radial velocities will do best at preserving sperical symmetry.

In figure 4, we have plotted $\Delta P/P$ for all cells in the shock. Comparing basic SHALE with SALE, we see an important difference. The error in SHALE is concentrated near the axis of symmetry. This is consistent with the results of table 1 which show the error is significant only when the polar angle is comparable to the angular width of a cell. SALE on the other hand has significant error for almost all polar angles. This suggests that the problem may not be one of accuracy.

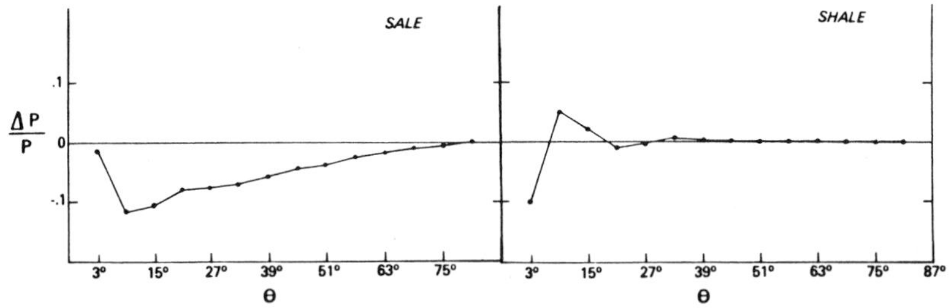

Figure 4
Normalized pressure deviation from equatorial pressure, plotted against polar angle.

Pursuing this line of thought, we generalized the results of table 1 for the case of variable angular widths. The results are that the errors in all versions of the SHALE code are still second order (i.e. of order $\Delta\theta^2$). This is due to the use of exact formulae for the area and mass of the MCVs. SALE, however, is only first order in the difference of angular widths. Again, this is because SALE approximates the area and mass of the MCVs.

Another important consideration is that the SALE calculation is unstable in the following way. The analysis shows that if one begins with equal angular widths, and if the node points wanders sightly from the center, then the SALE equations accentuate this deviation on the next cycle. This instability is bounded, for we have assumed the driving pressures are spherically symmetric. As the center point moves, a presure gradient arises which bounds the point's drift away from the center.

To further test this idea, we reran the first problem set (uniform zoning) with larger artificial viscosity. Because the viscous pressure is proportional to the velocity gradient (instead of change in volume which drives the real pressure), the viscous pressures arise more quickly to damp the instability. Indeed, the calculations further support our analysis. The SHALE problems all showed only insignificant change when rerun with large viscosity. The SALE code, on the other hand, showed significant improvement. Unfortunately, the amount of artificial viscosity required to allow SALE to maintain spherical symmetry to the same degree as SHALE is too diffusive. Furthermore, viscosity is not sufficient to correct the errors in the SALE run for the problem with variable angular zoning. This suggests that the approximation used in SALE for the area and mass of the MCV are not accurate enough.

REFERENCES

1. ADAMS, T. F., DEMUTH, R. B., MARGOLIN, L. G., and NICHOLS, B. D. - Simulation of Rock Blasting with the SHALE Code, to be published in Proceedings of First Intern. Symposium on Rock Fragmentation by Blasting, Lulea, Sweden, 1983.

2. HIRT, C. W., AMSDEN, A. A., and COOK, J. L. - An Arbitrary Lagrangian-Eulerian Computing Method for All Flow Speeds, J. Comp. Phys., Vol. 14, pp. 227-253, 1974.

3. MAENCHEN, G., and SACK, S. - The Tensor Code, Methods in Computational Physics, Ed. Alder, B., Academic Press, 1964.

4. AMSDEN, A. A., and HIRT, C. W. - YAQUI: An Arbitrary Lagranian-Eulerian Computer Program for Fluid Flow at All Speeds, Report LA-5100, Los Alamos Scientific Laboratory, Los Alamos, NM, 1973.

5. AMSDEN, A. A., RUPPEL, H. M., and HIRT, C. W. - SALE: A Simplified ALE Computer Program for Fluid Flow at All Speeds, Report LA-8095, Los Alamos Scientific Laboratory, Los Alamos, NM, 1980.

COMPUTATION OF TWO-DIMENSIONAL FRONTS IN SHALLOW-WATER FLOW

Nikolaos D. Katopodes[1]

A Petrov-Galerkin method is used for the simulation of steep fronts in shallow-water flow. The model is based on the complete, nonlinear laws for conservation of mass and momentum. The computation is second-order accurate with respect to the time step of integration. Space integrals are evaluated by exact quadrature and conservation of mass is continuously monitored for self-consistency. The amplification and phase portraits of the scheme are compared to a well known difference scheme. Applications include spontaneous surge formation in various test rigs and natural channels.

1. INTRODUCTION

Natural channels are characterized by highly irregular cross-sections accompanied by abrupt contractions and expansions in width. In the presence of such geometric complexities the assumptions of one-dimensional flow methods are invalidated and the use of a two-dimensional model, which is capable of capturing spontaneous bore and jump formation is very desirable. The majority of conventional finite-element models for shallow water flow suffer from excessive numerical damping, however, and their applicability to problems containing discontinuities is questionable.

The Petrov-Galerkin formulation originally suggested by Dendy [2] has been shown to result in optimal solutions for linearized, one-dimensional shallow-water waves. Its application to two-dimensional flow in the presence of complex bed topography and resistance, however, must be preceded by numerical experimentation in order to determine the behavior of the scheme under these conditions.

[1]Asst. Professor, Dept. of Civil Engineering, University of Michigan, Ann Arbor, MI.

2. CONSERVATION LAWS FOR SHALLOW WATER FLOW

Shallow-water or nearly horizontal flow represents a physical situation described extensively in the literature. The fundamental hypothesis concerns the distribution of pressure in the vertical, which is assumed to be hydrostatic. Physically this limits the flow to situations without appreciable free surface curvature, uniformity of the velocity vector in the vertical direction and relatively mild changes in bed elevation. Conservation of mass and momentum under these conditions leads to the following system of equations.

$$\frac{\partial h}{\partial t} + \frac{\partial p}{\partial x} + \frac{\partial q}{\partial y} = 0 \tag{1}$$

$$\frac{\partial p}{\partial t} + \frac{\partial}{\partial x}\left(\frac{p^2}{h} + g\frac{h^2}{2}\right) + \frac{\partial}{\partial y}\left(\frac{pq}{h}\right) + gh\frac{\partial Z_0}{\partial x} + gn^2\frac{p(p^2+q^2)^{1/2}}{h^{7/3}} = 0 \tag{2}$$

$$\frac{\partial q}{\partial t} + \frac{\partial}{\partial y}\left(\frac{q^2}{h} + g\frac{h^2}{2}\right) + \frac{\partial}{\partial x}\left(\frac{pq}{h}\right) + gh\frac{\partial Z_0}{\partial y} + gn^2\frac{q(p^2+q^2)^{1/2}}{h^{7/3}} = 0 \tag{3}$$

In Eqs. 1-3 p, q = the volumetric flow rates per unit width in the x and y directions respectively; h = depth of flow; t = time; g = the ratio of weight to mass; Z_0 = bed elevation from arbitrary datum; and n = the Manning coefficient of roughness.

Equations 1-3 may be written in matrix form as follows

$$\frac{\partial U}{\partial t} + A\frac{\partial U}{\partial x} + B\frac{\partial U}{\partial y} + F = 0 \tag{4}$$

in which $U = \begin{bmatrix} h \\ p \\ q \end{bmatrix}$ (5)

$$A = \begin{bmatrix} 0 & 1 & 0 \\ c^2-u^2 & 2u & 0 \\ -uv & v & u \end{bmatrix} \tag{6}$$

$$B = \begin{bmatrix} 0 & 0 & 1 \\ -uv & v & u \\ c^2-v^2 & 0 & 2v \end{bmatrix} \tag{7}$$

and

$$F = \begin{bmatrix} 0 \\ gh \dfrac{\partial Z_0}{\partial x} + gn^2 \dfrac{p(p^2 + q^2)^{1/2}}{h^{7/3}} \\ gh \dfrac{\partial Z_0}{\partial y} + gn^2 \dfrac{q(p^2 + q^2)^{1/2}}{h^{7/3}} \end{bmatrix} \tag{8}$$

In Eqs. 6-8 u, v = the velocity components in the x and y directions respectively; and $c = (gh)^{1/2}$ is the celerity of elementary gravity waves.

3. PETROV-GALERKIN FORMULATION

The finite element approximation to the solution of Eq. 4 can be expressed as

$$U = \langle N \rangle \{U\} \tag{9}$$

where $\langle N \rangle$ is a 3 x 12 matrix, whose elements are the bilinear shape functions N interpolating the nodal values $\{U\}$ of the solution vector,i.e.

$$N_i = \frac{1}{4}(1 + \xi\xi_i)(1 + \eta\eta_i) \quad ; \qquad i=1,2,3,4 \tag{10}$$

in which ξ and η are the local space coordinates. The Petrov-Galerkin approximation is based on the following discontinuous weighting function, which is a direct extension of the test function proposed by Dendy[2] for one-dimensional convection

$$\overset{*}{N_i} = N_i + \alpha_\xi \frac{\partial N_i}{\partial \xi} + \alpha_\eta \frac{\partial N_i}{\partial \eta}$$

$$= \frac{1}{4}[(1 + \xi\xi_i)(1 + \eta\eta_i) + \alpha_\xi\xi_i(1 + \eta\eta_i) + \alpha_\eta\eta_i(1 + \xi\xi_i) \tag{11}$$

in which α_ξ and α_η are general dissipation parameters to be identified in the following. If, for instance, the dissipation parameters are set equal to unity, the weighting function suggested by Eq. 11 takes the form shown in Fig. 1. The discontinuous nature of this function and the upstream weighting are responsible for the remarkable dissipation and phase characteristics of the Petrov-Galerkin method that are presented in following sections. The identification of optimum dissipation functions and levels adopted in here is similar to the procedures of Raymond and Garder[6] and Baker and Soliman[1] for one-dimesional convection. The extension to a system of equa-

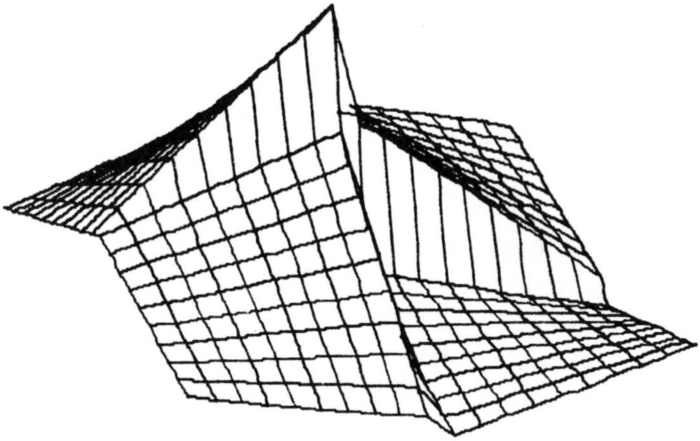

Figure 1. Dissipative Galerkin test function

tions is straightforward provided that the flow equations are converted in characteristic form, as shown by Katopodes[4]. The back transformation to the system of conservation laws is quite illuminating and has been shown by Morton and Parrott[5] and Hughes and Brooks to lead to generalizations of the well-known Lax-Wendroff finite difference scheme. Accordingly the weighting function used in this work takes the following form

$$\langle N_* \rangle = \langle N \rangle + \varepsilon_x A^T \frac{\partial \langle N \rangle}{\partial x} + \varepsilon_y B^T \frac{\partial \langle N \rangle}{\partial y} \tag{12}$$

and the corresponding variational statement reads

$$\int_\Omega \langle N_* \rangle^T \left(\frac{\partial U}{\partial t} + A \frac{\partial U}{\partial x} + B \frac{\partial U}{\partial y} + F \right) d\Omega = 0 \tag{13}$$

A better physical understanding of the implications of Eq. 13 can be obtained by completion of the matrix multiplications and expression of the result in scalar form. The continuity equation has taken the form

$$\int_\Omega \left(N^T E_1 + \varepsilon_x \frac{\partial N^T}{\partial x} E_2 + \varepsilon_y \frac{\partial N^T}{\partial y} E_3 \right) d\Omega = 0 \tag{14}$$

while the x and y-momentum equations respectively read

$$\int_\Omega \{N^T E_2 + \varepsilon_x \frac{\partial N^T}{\partial x}[(c^2-u^2)E_1 + 2uE_2] + \varepsilon_y \frac{\partial N^T}{\partial y}(-uvE_1 + vE_2 + uE_3)\}d\Omega = 0 \qquad (15)$$

$$\int_\Omega \{N^T E_3 + \varepsilon_x \frac{\partial N^T}{\partial x}(-uvE_1 + vE_2 + uE_3) + \varepsilon_y \frac{\partial N^T}{\partial y}[(c^2-v^2)E_1 + 2vE_3]\}d\Omega = 0 \qquad (16)$$

in which E_1, E_2 and E_3 are the approximation errors resulting from subtitution of interpolated values in the continuity, x-momentum and y-momentum equations respectively. It is obvious from Eq. 14-16 that the original flow equations have undergone a linear combination quite similar to the combination leading to compatibility relations of the method of characteristics. In addition, higher order derivatives are now present in the approximate equations as it is easily seen by rewriting Eq. 13 in the form

$$\frac{\partial U}{\partial t} + A\frac{\partial U}{\partial x} + B\frac{\partial U}{\partial y} + F - \varepsilon_x \left(A\frac{\partial^2 U}{\partial x \partial t} + A^2 \frac{\partial^2 U}{\partial x^2} + AB \frac{\partial^2 U}{\partial x \partial y} + A \frac{\partial F}{\partial x}\right) - \varepsilon_y \left(B \frac{\partial^2 U}{\partial y \partial t} + BA \frac{\partial^2 U}{\partial x \partial y} + B^2 \frac{\partial^2 U}{\partial y^2} + B \frac{\partial F}{\partial y}\right) = 0 \qquad (17)$$

Of course the addition of higher order dissipative terms is not desirable unless the dissipative mechanism is selective. Although there is no analytical support for the generalization of the one-dimensional optimum dissipation levels suggested by Raymond and Garder[6], their results are extended in this work to space dimensions as follows

$$\varepsilon_x = \frac{\Delta x}{[u+c]\sqrt{15}} \qquad (18)$$

$$\varepsilon_y = \frac{\Delta y}{[v+c]\sqrt{15}} \qquad (19)$$

where Δx and Δy are measures of the element dimensions in the x and y directions, respectively. Hughes and Brooks[3] have shown that the following expressions are compatible with the adopted isoparametric coordinates

$$\Delta x = 2\left\{\left(\frac{\partial x}{\partial \xi}\right)^2 + \left(\frac{\partial x}{\partial \eta}\right)^2\right\}^{1/2} \tag{20}$$

$$\Delta y = 2\left\{\left(\frac{\partial y}{\partial \xi}\right)^2 + \left(\frac{\partial y}{\partial \eta}\right)^2\right\}^{1/2} \tag{21}$$

It is obvious from Eqs. 18 and 19 that the dissipation mechanism is dependent on the characteristic speed and the mesh con figuration. In fact, the relation of the dissipation parameter to the local Courant number is quite remarkable.

4. ALGORITHMIC DAMPING AND PHASE PROPERTIES

The accuracy and stability of the numerical scheme can be predicted by performing a Fourier analysis of the one-dimensional linearized equations for shallow-water flow. The time derivatives are approximated by a central difference approximation of the form

$$\frac{U_{n+1} - U_n}{\Delta t} = \frac{1}{2}\left\{\left(\frac{\partial U}{\partial t}\right)_{n+1} + \left(\frac{\partial U}{\partial t}\right)_n\right\} \tag{22}$$

and the resulting difference equations are subjected to a harmonic analysis. The resulting amplification matrix is rather complicated and its eigenvalues are computed numerically. Adding and squaring the real and imaginary parts of the spectral radius of the amplification matrix results in an estimate of the algorithmic damping, while dividing the imaginary by the real part leads to the phase error to be expected. Figure 2 shows the algorithmic damping and celerity ratio of the one-dimesional flow problem compared to the Lax finite difference scheme. Both the Lax and Galerkin methods are not dissipative at Courant number one, but although the Lax scheme is predicting the exact celerity for unit Courant number, the Galerkin method is dispersive. Selective dissipation is of course a necessity for nonlinear problems. The Lax scheme achieves this at smaller Courant numbers at the expense of its celerity accuracy. The Petrov-Galerkin model, however, exhibits two remar kable characteristics. The algorithmic damping is strongly dependent on the wave frequency and, furthermore the dissipation results in improved celerity estimates. Both of these features are very desirable for an algorithm designed to model sharp fronts in shallow-water flow.

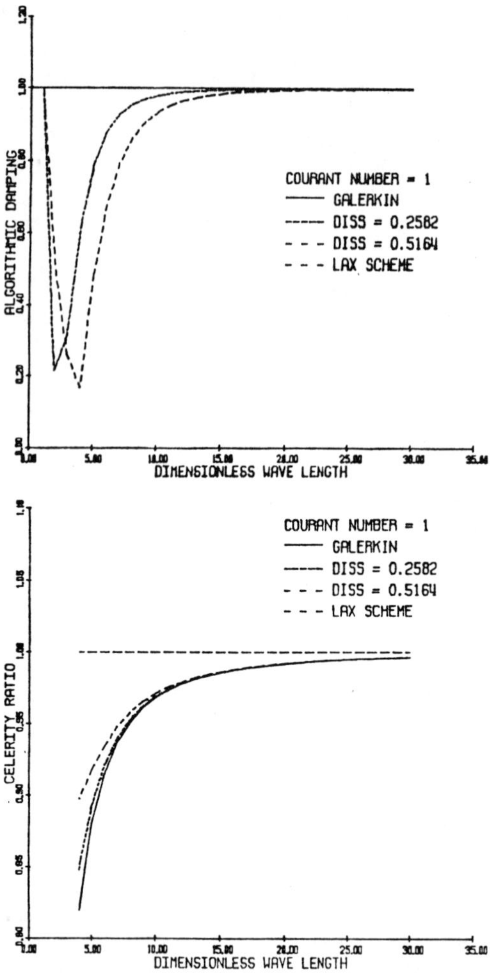

Figure 2. Dissipation and phase errors of the Petrov-Galerkin finite element scheme

5. SOLUTION OF THE NONLINEAR ALGEBRAIC EQUATIONS

The algebraic equations resulting form application of the Petrov-Galerkin method to the Shallow-Water Equations are non-linear and their solution may require massive computations. Traditionally the Newton-Raphson method is used to provide new estimates of the solution vector in the form

$$U^{i+1} = U^{i} - \left(\frac{\partial\{f\}}{\partial\{U\}}\right)^{i} \{f\}^{i} \tag{23}$$

The computation of the Jacobian matrix is rather complicated and results in a very expensive to execute algorithm. It was found more efficient to update the coefficients of the error terms in Eqs. 14-16 linearly, which results in a rather simple algorithm while maintaining overall second-order convergence. Therefore, the elements of the Jacobian matrix are computed by the following equations

$$\frac{\partial\{f_1\}}{\partial\{U_i\}} = \int_\Omega \left(N^T E_{1,i} + \varepsilon_x \frac{\partial N^T}{\partial x} E_{2,i} + \varepsilon_y \frac{\partial N^T}{\partial y} E_{3,i}\right) d\Omega \tag{23}$$

$$\frac{\partial\{f_2\}}{\partial\{U_i\}} = \int_\Omega \Big\{ \left[(c^2-u^2)\, \varepsilon_x \frac{\partial N}{\partial x} - uv\varepsilon_y \frac{\partial N}{\partial y}\right]^T E_{1,i} +$$

$$+ \left(N + 2u\varepsilon_x \frac{\partial N}{\partial x} + v\varepsilon_y \frac{\partial N}{\partial y}\right)^T E_{2,i} + u\varepsilon_y \frac{\partial N^T}{\partial y} E_{3,i} \Big\} d\Omega \tag{24}$$

$$\frac{\partial\{f_3\}}{\partial\{U_i\}} = \int_\Omega \Big\{ \left[(c^2-v^2)\, \varepsilon_y \frac{\partial N}{\partial y} - uv\varepsilon_x \frac{\partial N}{\partial x}\right]^T E_{1,i} + v\varepsilon_x \frac{\partial N^T}{\partial x} E_{2,i} +$$

$$+\left(N+2v\varepsilon_y \frac{\partial N}{\partial y} + u\varepsilon_x \frac{\partial N}{\partial x}\right)^T E_{3,i} \Big\}\, d\Omega \tag{25}$$

in which $E_{i,j}$ is the partial derivative of equation i with respect to the jth element of the solution vector.

6. COMPUTATIONAL EXPERIMENTS

The predictive ability and reliability of any numerical method for shallow-water flow is best evaluated by application to certain test problems, which combine some special flow and boundary conditions. The conservation characteristics, for instance, are easier described by modeling flows in frictionless horizontal channels. The propagation properties are amplified by use of perfectly reflecting boundaries, so that no "fresh" information is allowed to enter the solution domain following an initial impulse. All of the following tests represent surge formation in originally stationary water of uniform depth. The time step is controlled so that the Courant number behind the main surge is approximately unity. Bilinear elements and exact quadrature are used exclusively in this work although other possible formulations lead to successful results too. Figure 3 shows the spontaneous formation of a surge as computed by the Petrov-Galerkin model. The shock-capturing ability of the algorithm is very satisfactory and mass conservation for this example is represented almost exactly.

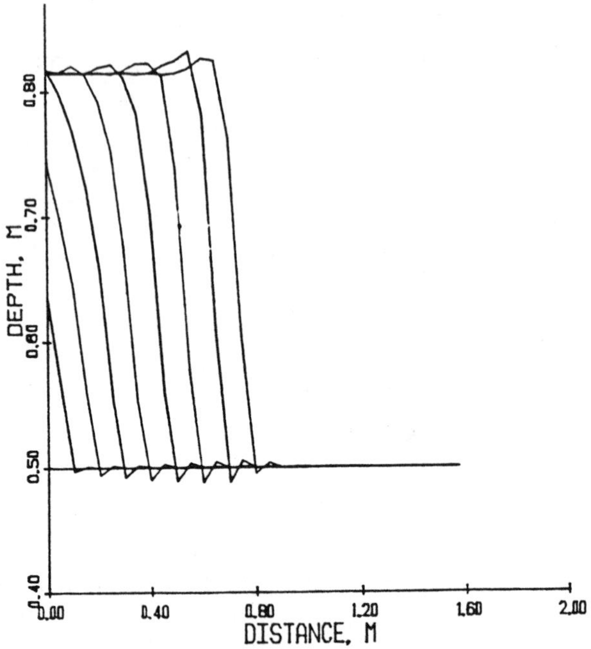

Figure 3. Spontaneous surge formation

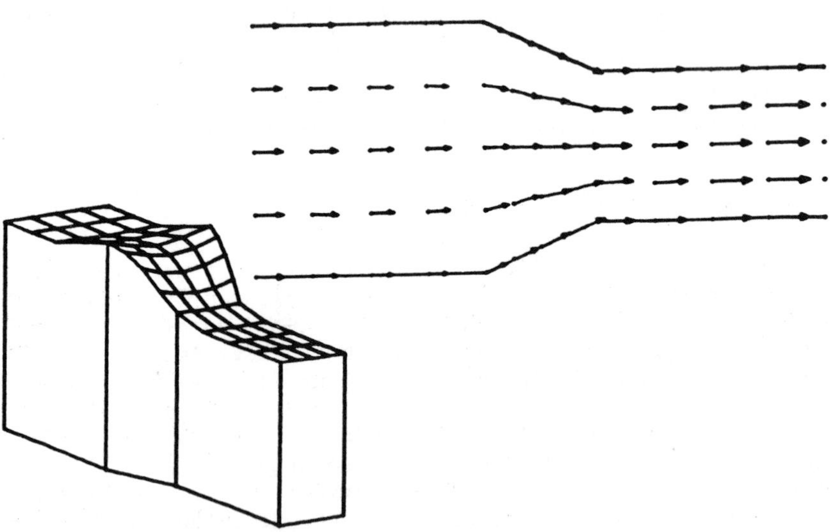

Figure 4. Surge through a gradual channel constriction

Figure 5 shows a surge moving through an abrubt constriction and the formation of spontaneous oblique shocks.

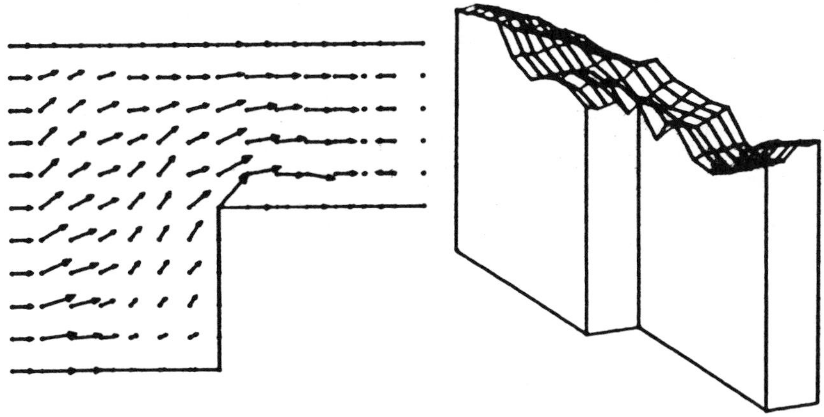

Figure 5. Surge through abrupt constriction

Finally, Figs. 6 and 7 show the formation and propagation of a surge in an abrupt expansion. In all tests the mass conservation is monitored by numerical integration and used as a the basis for evaluating the conservation characteristics of the scheme. In all problems tested the computed volume error is consistently less than 1%, which is judged very satisfactory for the rather coarse grids used in the test problems.

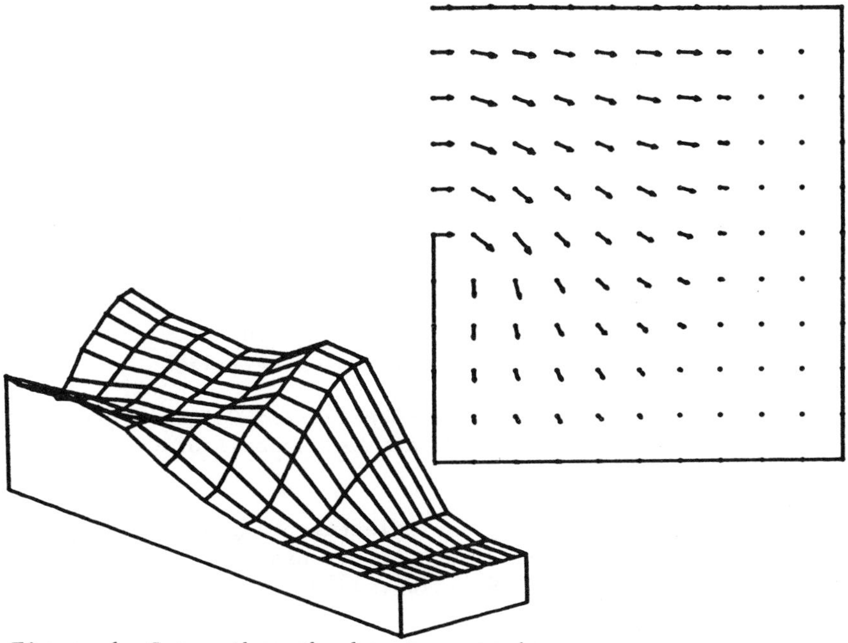

Figure 6. Surge through abrupt expansion

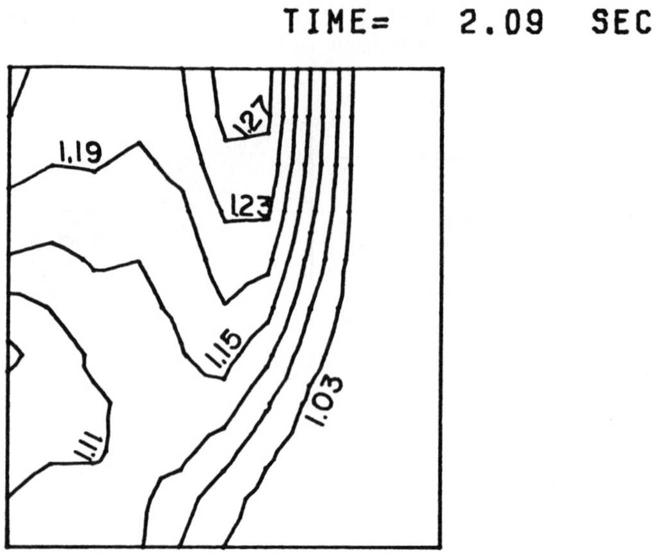

Figure 7. Depth contours; surge in expanding channel

7. REFERENCES

1. BAKER, A.J. and SOLIMAN, M.O-On the Accuracy and Efficiency of a Finite Element Tensor Product Algorithm for Fluid Dynamics Applications, Computer Methods in Applied Mechan. and Engineering, 27, 1981

2. DENDY, J.E.-Two Methods of Galerkin-Type Achieving Optimum L^2 Rates of Convergence for First Order Hyperbolics, SIAM Journal of Numerical Analysis, Vol.11, 1974

3. HUGHES, T.J.R. and BROOKS, A.N. -A Theoretical Framework for Petrov-Galerkin Methods with Discontinuous Weighting functions: Application to the Streamline-Upwind Procedure, Finite Elements in Fluids, Ed. Gallagher, R.H. et als., Vol 4, J. Wiley & Sons, 1982

4. KATOPODES, N.D. - Finite Element Computation of Inertia-Dominated Free-Surface Flow, Finite Element Flow Analysis, ed. Kawai, T., North Holland Publishing Co., 1982

5. MORTON, K.W. and PARROTT, A.K.-Generalized Galerkin Methods for First-Order Hyperbolic Equations, Journal of Computational Physics, Vol.36, 1980

6. RAYMOND, W.H. and GARDER, A.- Selective Damping in a Galerkin Method for Solving Wave Problems with Variable Grids, Monthly Weather Review, Vol. 104, 1976

FLUX CORRECTED TRANSPORT FOR ONE-DIMENSIONAL SIMULATION MODELS IN RIVERS AND ESTUARIES INCORPORATING THREE-DIMENSIONAL EFFECTS

J. Häuser, M. Lobmeyr, D. Eppel, A. Müller
Institute of Physics, GKSS-Research Center
D-2054 Geesthacht, Germany

F. Tanzer
1. Institute of Physcis, Giessen University
D-6300 Giessen, Germany

1. INTRODUCTION

There are many cases of application where one-dimensional mathematical models calculating the water level and the main flow component which is in the direction of the stream axis, provide sufficient information for the determination of the subsequent transport processes for various pollutants. Moreover, the one-dimensional model can be used to produce the time-dependent boundary conditions for two-dimensional models. These models then can be applied to regions with plume structures where one-dimensional models would yield unacceptable results. In many cases of practical interest the horizontal flow component possesses a vertical profile. This structure in combination with a varying width of the river gives rise to additional transport processes. This paper, following an idea of A.H. Eraslan et al. [1], uses the Blasius power law [2] to calculate the vertical velocity profile. Although this law was first developed for turbulent pipe flow, it is also applicable to open channel flow. The fact that the maximum velocity in the vertical plane occurs some distance from the free water surface can be neglected. The Blasius power law, however, is no longer valid when severe opposing pressure forces, such as salt intrusion, occur. Then the phenomenon of reversing flow occurs and a breakaway point exists. The flow then changes its direction along the breakaway line (velocity is zero) and a rotatory motion is generated. This phenomenon is well-known at the mouth of an estuary where ebb flow on the surface and flood at the bottom is possible. However, for such a situation the Blasius power law is not valid.

If any concentration of a constituent is modeled, the respective vertical distribution cannot be obtained by a simple formula as the Blasius power law; rather a formulation which fits the data (hopefully most of the time) has to be used. Normally a polynomial expression will be employed. In all cases similarity type profiles are used. The cross-section averaged concentration and velocity values are calculated from physical conservation laws. A dimensionless vertical coordinate $\xi = z/\overline{H}(s,y,t)$ is introduced (see below: $\overline{H}$ is the total height of the water column) which, in conjuntion with the mean values of the variables, determines the vertical profile of concentration and velocity. By means of this profile the non-convective transport rates can be modeled which consist of transport processes introduced by spatial and temporal averages [4]. Turbulent diffusion is modeled by the Boussignesq ansatz, including the effects of vertical variations of both velocity and constituent concentration. In calculating the bottom stress, the friction velocity has to be known which is obtained from Blasius' resistance formula [2, p. 600 ff.]. The value of the needed empirical constant is known from measurements but depends on the Reynolds number. The Reynolds number is calculated locally in the model so that the "right" empirical constant for every type of flow is used. Although numerous two-dimensional hydrodynamic models exist [e.q. 5 - 8] their application may not be advantageous for certain flow regimes where a one-dimensional model can be used with confidence. The model, presented in this paper, has the special feature that the detailed geometrical characteristics of the cross-sections and the effects of the vertical distributions are explicitly incorporated in the formulation of the resulting equations. The underlying assumptions based on the experience of laboratory and field measurements are that (i) the Blasius power law is a valid formulation for the vertical flow variation in open channel flow (a channel with a width more than ten times larger than its depth has the same velocity profile as an infinitely wide channel [3, p. 154], (ii) the vertical profile of a constituent is consistent with the measured data, (iii) the resistance law which gives the relationship between friction velocity and horizontal velocity components holds true for the range of Reynolds numbers encountered in the river flow, and (iv) a relation between the bottom friction coefficient and the bottom roughness height exist

The first part of this paper is concerned with the implementation of a numerical scheme which has sufficient accuracy for both the non-linear shallow water equations and the scalar transport where the modeling of the latter gives rise to numerical diffusion. If a source works continuously, numerical diffusion is not a severe problem for most numerical schemes, and, e.g., second upwind differencing can be used, resulting in an efficient algorithm. For the transport of peaked distributions this scheme, as many others, is not suited. Hence, the flux corrected transport (FCT) formulation [9 - 13] is employed and the discrete element method (DE) [14, 15] is reformulated using this approach. It is seen that this scheme handles the problem of numerical diffusion effectively and the clipping phenomenon (see below) does not severly affect the calculated distributions even for very large computation times.

2. BASIC EQUATIONS

In the following we present the equations for the water level, one-dimensional momentum transport and constituent mass conservation. The physical phenomena for the transport are convection and diffusion (turbulent) as well as bottom friction force and wind shear stresses for the momentum equation. Any source or sink terms have been omitted but can be readily included in the formulation. For the momentum equation we assume that the vertical and transverse velocity components can be neglected, and that diffusion terms are negligible in comparison with bottom stress. Pressure p is eliminated from the momentum equation using the hydrostatic pressure formulation. All quantities then are time-averaged (fluctuation time) and cross-section averaged. However, the additional transport terms resulting from these averaging processes are incorporated in the model equations. We start with the 3D Navier-Stokes equations

$$\frac{\partial}{\partial t}\int_V \rho\, v_i\, dV + \oint_{A(V)} \rho\, M_{ik}\, dA_k = 0 \tag{1}$$

where $\rho\, M_{ik} = \rho\, v_i\, v_k - p\, S_{ik} + 1/2\, \mu\, (\frac{\partial v_i}{\partial x_k} + \frac{\partial v_k}{\partial x_i})$ denotes the momentum flux (momentum per area and per time). The equation of continuity reads

$$\frac{\partial}{\partial t}\int_V \rho\, dV + \oint_{A(V)} \rho\, v_k\, dA_k = 0 \tag{2}$$

and the equation for any constituent taken the form

$$\frac{\partial}{\partial t}\int_V C\, dV + \oint_{A(V)} C\, v_k\, dA_k = \int_V q\, dV \tag{3}$$

where q describes the various sink or source terms (not of interest for the following). All quantities depend upon variables x_k, t', t where the x_k denote spatial variables, t' is fluctuation time and t (t >> t') is the time in which the fluctuation time averaged quantities change; i.e. Eqs. (1 - 3) are averaged over t' to eliminate this dependence. Since a 1D model is described, all quantities are averaged over the cross-section area A. Hence,we can separate the contribution of a quantity as follows

$$C(s,y,z,t,t') = \overline{C}(s,t) + \overline{C}_D(s,y,z,t) + C'(s,y,z,t,t') \tag{4}$$

where s denotes the spatial coordinate along the stream axis, y is the lateral coordinate and z is the vertical axis with respect to an arbitrarily specified datum plane, such that all z-values are positive (see Fig. 1). The coordinate system used is a right-hand system with orthogonal axes. The overbar specifies two averaging processes, namely, averaging over t' and over cross-section $\overline{A}(s,t)$. $\overline{C}(s,t)$ is the mean value over the cross-section and the fluctuation time, $\overline{D}_D(s,y,t)$ is the deviation from that mean and $C'(s,y,z,t,t')$ gives the turbulent fluctuation. The same procedure applies to all other quantities. Eqs. (1 - 3) are written in integral form in order to get the flux (conservation) form for the numerical formulation. To this end the equations are time and cross-section averaged using the following assumptions:

(i) average over fluctuation time t',

(ii) average over cross-section $\overline{A}(s,t)$,

(iii) pressure p is eliminated using the hydrostatic pressure ansatz

$$p = \rho\, g\, \overline{H}(s,y,t),$$

(iv) the velocity only has a component in direction of the stream axis

$$\vec{v} = (u,0,0),$$

(v) the flow is incompressible, i.e., ρ = constant and hence

$$\frac{\partial v_\ell}{\partial x_\ell} = 0\,,$$

(vi) Eq. (4) is used for each quantity to be modeled.

Eqs. (1 - 2) are used in the form

$$\frac{\partial}{\partial t}\int_V u\, dV + \oint_{A(V)} M_{11}\, S_{1k}\, dA_k = 0 \tag{5}$$

$$\frac{\partial}{\partial t}\int_V dV + \oint_{A(V)} u\, S_{1k}\, dA_K = 0\,. \tag{6}$$

Integrating over an elemental volume $\Delta V(s,t) = \overline{A}(s,t)\,\Delta s =$ $= \overline{B}(s)\,\overline{H}(s,t)\,\Delta s$ where $\overline{H}(s,t) := \frac{1}{B(s)}\int_0^{\overline{B}(s)} \overline{H}(s,y,t)dy$ and $\overline{B}(s)$ describes the width of the river. We define a quantity, called transport, J = u A and obtain from Eqs. (5):

$$\frac{\partial}{\partial t}\overline{J}(s,t) + \Delta s^{-1}\,[F_j(s - \Delta s/2,\ t) - F_j(s - \Delta s/2,t] = 0 \tag{7}$$

where index j indicates the flux for the transport J. The flux F_j is of the form

$$F_j(s,t) := \int_0^{\overline{B}(s)}\int_0^{\overline{H}(s,y,t)} \Big\{[u(s,y,z,t,t')]^2 - g\,\overline{H}(s,y,t) + \frac{\partial\, u(s,y,z,t,t')}{\partial z}\Big\}\, dy\, dz \tag{8}$$

Using Eq. (4) and replacing $\overline{H}(s,y,t')$ by $\overline{H}(s,t)$ in Eq. (8), one finds

$$F_j(s,t) = \overline{u}(s,t)\,\overline{J}(s,t) + F_j^D(s,t) + F_j'(s,t) - g\,\overline{H}(s,t)\,\overline{A}(s,t) + \\ + (\tau_S - \tau_B)\,\overline{B}(s) \qquad (9)$$

τ_S and τ_B are the surface wind stress and bottom shear stress, respectively. $\overline{H}_S$ and $\overline{H}_B$ denote the suface resp. bottom high.

$$F_j^D(s,t) := \int_0^{\overline{B}(s)} \int_0^{\overline{H}(y)} \overline{u}_D^{\,2}\,(s,y,z,t)\;dz\;dy;$$

$$F_j'(s,t) = \int_0^{\overline{B}(s)} \int_0^{\overline{H}(y)} \overline{u'^2}\;dy\;dz \qquad (10)$$

$$\rho\tau_S = \mu\left(\frac{\partial\overline{u}}{\partial z}\right)_{\overline{H}_S} = \alpha\rho_A\,v_W^2;\quad \rho\tau_B = \mu\left(\frac{\partial\overline{u}}{\partial z}\right)_{\overline{H}_B} = \rho\,u_*^2(z_o)$$

$\alpha \approx 2.6 \cdot 10^{-3}$ is a measured constant, ρ_A is air density above the water surface, v_W denotes wind speed at anemometer height and $u_*(z_o)$ is the friction velocity at bottom roughness height z_o (see below), $\mu = \rho\,\nu$ is dynamic molecular viscosity.

For the continuity equation we obtain

$$\frac{\partial\,\overline{H}(s,t)}{\partial t} + \Delta s^{-1}\,[F_H(s+\Delta s/2,t) - F_H(s-\Delta s/2,t)] = 0 \qquad (11)$$

$$F_H(s \pm \Delta s/2,t) = \overline{B}(s)^{-1}\,\overline{u}(s \pm \Delta s/2,t)\,\overline{A}(s \pm \Delta s/2,t). \qquad (12)$$

It is noted that the contributions of $\overline{u}_D$ und u' vanish, since the continuity equation is averaged over the fluctuation time and the cross-section.

For the constituent equation using Eq. (12) we get

$$\frac{\partial\,\overline{C}(s,t)}{\partial t} + \Delta s^{-1}\,[F_C(s+\Delta s/2,t) - F_C(s-\Delta s/2,t) - \overline{q}(s,t)] = 0 \qquad (13)$$

$$F_C(s \pm \Delta s/2,t) = \overline{A}(s,t)^{-1}\,[\overline{u}(s \pm \Delta s/2,t)\,\overline{A}(s \pm \Delta s/2,t)\,(\pm 1) \\ (\overline{C}(s \pm \Delta s/2,t) - \overline{C}(s,t)) + F_C^D(s \pm \Delta s/2,t) + F_C'(s \pm \Delta s/2,t)] \qquad (14)$$

The coupled system of Eqs. (7), (11) and (13) now has to be solved numerically specifying appropriate boundary and initial conditions.

3. NUMERICAL SOLUTION PROCEDURE

It is long known that the numerical solution of a scalar transport or continuity equation suffers from numerical diffusion, caused by the approximation of the convective terms. In particular, if a peaked distribution is to be transported, numerical diffusion can smoothe the profile of the distribution and so lead to gross errors after several time steps. Higher order schemes are plagued by the problem of over- and undershoots because of the non-linear interpolation which may generate new extrema. Recently, Boris and Book [9, 10, 13] invented the method of flux corrected transport (FCT) which was extended to multi-dimensions by Zalesak [11, 12]. This method gives excellent results for continuity type equations such that steep gradients can be resolved within a few grid spacings while their shape is not distorted during the transport process. The basic idea of FCT, which is a conservative scheme, is to first calculate a numerical solution from a low order (positive) scheme which possesses a high amount of numerical diffusion (e.g., second upwind). In the next step, the numerical solution is calculated by a high order scheme having only little numerical diffusion but being plagued by wiggles. From these two numerical solutions the so-called anti-diffusive fluxes are constructed such that the new numerical solution, obtained from these fluxes, does not exhibit new extrema or accentuate existing ones and remains positive as well. To this end, a so-called flux limiter is necessary. Since the construction of such a limiter is not unique, numerous schemes can be constructed. The important task of such a limiter is to preserve the shape of a distribution; that is, clipping must be avoided. In the following the numerical solution procedure for Eqs. (7, 11 and 13) is described, following the algorithm of Zalesak [11].

In the numerical formulation only the convective part of the fluxes is considered, since for diffusion like processes the FCT formulation is not needed. The procedure is outlined for the scalar transport equation and applies equally well to two-dimensions.

1. Calculate concentration values using a low-order (upwind differencing [15, 16]) scheme:

$$\frac{\partial \overline{C}^{L}}{\partial t} + \Delta s^{-1}\left[F_{C}^{L}(s+\Delta s/2,t) - F_{C}^{L}(s-\Delta s/2,t)\right] = 0$$

$$F_{C}^{L}(s\pm\Delta s/2,t) := \overline{A}(s,t)^{-1}\left[u(s\pm\Delta s/2,t)\,\overline{A}(s\pm\Delta s/2,t)\,(\pm 1)\right.$$
$$\left.(\overline{C}(s\pm\Delta s/2,t) - \overline{C}(s,t))\right] \qquad (1\text{:}$$

$$\overline{u}_{j\pm1/2} := \frac{\overline{u}_j\,\Delta s_{j\pm1} + \overline{u}_{j\pm1}\,\Delta s_j}{\Delta s_j + \Delta s_{j\pm1}}\;;\quad \overline{A}_{j\pm1/2} := \frac{\overline{A}_j\,\Delta s_{j\pm1} + \overline{A}_{j\pm1}\,\Delta s_j}{\Delta s_j + \Delta s_{j\pm1}}$$

$$C_{j+1/2}: = \begin{cases} \overline{C}_j & \text{for } \overline{u}_{j+1/2} > 0 \\ \overline{C}_{j+1} & \text{for } \overline{u}_{j+1/2} \leq 0 \end{cases} ; \quad C_{j-1/2}: = \begin{cases} \overline{C}_j & \text{for } u_{j-1/2} < 0 \\ \overline{C}_{j-1} & \text{for } u_{j-1/2} \geq 0 \end{cases}$$

where $s = s_j$; $s_{j+1/2} = s_j + \Delta s/2$ (Δs may depend on j).

2. Calculate the solution using a high order scheme (second order ZIP [12, 17]):

$$\frac{\partial \overline{C}^H}{\partial t} + \Delta s^{-1} [F_c^H(s \pm \Delta s/2,t) - F_c^H(s - \Delta s/2,t)] = 0 \qquad (16)$$

$$F_c^H(s \pm \Delta s/2,t): = \overline{A}(s,t)^{-1} [\overline{u}(s \pm \Delta s/2,t)\, \overline{A}(s \pm \Delta s/2,t)\ (\pm 1)\ (\overline{C}(s \pm \Delta s/2,t) - \overline{C}(s,t))]$$

$$\overline{u}_{j\pm 1/2}\, \overline{A}_{j\pm 1/2}\, \overline{C}_{j\pm 1/2} = 1/2\, \left(\overline{u}_{j\pm 1}\, \overline{A}_j\, \overline{C}_j + u_j\, \overline{A}_{j\pm 1}\, \overline{C}_j + \overline{u}_j\, \overline{A}_j\, \overline{C}_{j\pm 1}\right) \qquad (17)$$

$$\overline{u}_{j\pm 1/2}\, \overline{A}_{j\pm 1/2}\, \overline{C}_j = 1/2\, \left(\overline{u}_{j\pm 1}\, \overline{A}_j\, \overline{C}_j + \overline{u}_j\, \overline{A}_{j\pm 1}\, \overline{C}_j\right) \qquad (18)$$

Inserting Eqs. (17), (18) into Eq. (16) and expanding the difference formulas [17, 18] in a Taylor series about x_j, it can be easily seen that all spatial derivates of even powers cancel out, that is, no derivatives which produce numerical diffusion remain.

3. Construct the final solution $\overline{C}(s,t+\Delta t)$ from the low and high order solutions using constant $0 \leq C(s \pm \Delta s/2) \leq 1$ such that no new extrema are constructed or existing extrema are accentuated and positivity of the solution is retained:

$$\overline{C}(s,t+\Delta t) = \overline{C}^L(s,t) - \Delta s^{-1}\, \Delta t\, \left[F^A_{j+1/2} - F^A_{j-1/2}\right];$$

$$F^A_{j\pm 1/2}: = C_{j\pm 1/2}\, \left(F^H_{c;j\pm 1/2} - F^L_{c;j\pm 1/2}\right) \qquad (19)$$

$$\overline{C}(s,t+\Delta t) = \overline{C}(s,t) - \Big[\left(1 - C_{j+1/2}\right) F^L_{c;j+1/2} - \left(1 - C_{j-1/2}\right) F^L_{c;j-1/2} + C_{j+1/2}\, F^H_{c;j+1/2} - C_{j-1/2}\, F^H_{c;j-1/2}\Big] \Delta s^{-1}\, \Delta t \qquad (20)$$

where $C(s \pm \Delta s/2) = C_{j\pm 1/2}$ is used. The constants $C_{j\pm 1/2}$ have to constructed (locally) from the numerical solution itself. Eq. (20) is a weighted average of the low- and high-order solution.

If the constants $C_{j\pm1/2}$ equal 1 only the high-order solution remains; for $C_{j\pm1/2}$ equal 0 no influence of the high-order solution remains and the highly diffusive low-order solution is not corrected. The $F^A_{j\pm1/2}$ are the so-called anti-diffusive fluxes that normally are negative since the low-order solution produces a larger numerical diffusion than the high order scheme.

4. Construct the flux-limiter; i.e., determine constants $C_{j\pm1/2}$:

(i) The magnitude of all anti-diffusive fluxes into a discrete element centered at grid point j [11]:
(a DE in one dimension is formed by its two sides at j+1/2 and j-1/2 having mid point j)

$$F^+_j = \max(F^A_{j-1/2}, 0) - \min(F^A_{j+1/2}, 0)$$

where $F^A_{j-1/2} > 0$ and $F^A_{j+1/2} < 0$ are directed into element j; i.e., are counted positive.

(ii) The maximal allowable flux into a DE is given by:

$$F^+_{j,max} = (C^{max}_j - C^L_j)\ \bar{u}^{n+1}_j$$

where C^{max}_j is determined later.

(iii) Construct admissible correction (anti-diffusive) fluxes:

$$r^+_j = \begin{cases} 0 & \text{for } F^+_j \le 0 \\ \min(1, F^+_j / F^+_{j;max}) & \text{for } P^+_j > 0 \end{cases}$$

that is, the solution is not corrected if no anti-diffusive fluxes are produced (no undershot allowed) and on the other hand, the formulation prevents a correction larger than the maximal allowable flux. The respective quantities $\overline{F}_j$, $\overline{F}_{j,max}$, C^{min}_j and r^-_j are constructed for all anti-diffusive fluxes out of the discrete element j in the same manner. The constant $C_{j+1/2}$ is then determined from the fact that an anti-diffusive flux $F^A_{j+1/2}$ less than zero increases the concentration in element j, and decreases the concentration in element j+1, according to Eq. (19). For an anti-diffusive flux $F^A_{j+1/2}$ greater than zero, concentration values are decreased in element j and increased in element j+1. Since neither new extrema in these two elements should be created nor existing extrema accentuated, $C_{j+1/2}$ must be of the form

$$C_{j+1/2} = \begin{cases} \min(r^+_{j+1}, r^-_j) & \text{for } F^A_{j+1/2} > 0 \\ \min(r^-_{j+1}, r^+_j) & \text{for } F^A_{j+1/2} \le 0 \end{cases}$$

In the same way $C_{j-1/2}$ is found.

For the construction of the maximal and minimal allowable fluxes, it must be borne in mind that for Courant numbers less than one, the peak of a distribution can land between two successive grid points. If only grid point values (of the concentration) are used for determining the maximal anti-diffusive flux, its value can be too small. Hence, the peak of the distribution will be cut-off, leading to the so-called clipping phenomenon. In order to avoid this, interpolation between grid points is necessary. As Zalesak [11] pointed out, there are many possibilities in constructing a limiter. We present the following one:

$$\bar{C}_j^{a}: \; = \max(\bar{C}_j^{m}, \bar{C}_j^{L}); \quad \bar{C}_j^{b}: \; = \min(\bar{C}_j^{n}, \bar{C}_j^{L})$$

$$\bar{C}_j^{max}: \; = \max(\bar{C}_{j-1}^{a}, \bar{C}_j^{a}, \bar{C}_{j+1}^{a}, \bar{C}_{j-1/2}^{ext}, \bar{C}_{j+1/2}^{ext})$$

$$\bar{C}_j^{min}: \; = \min(\bar{C}_{j-1}^{b}, \bar{C}_j^{b}, \bar{C}_{j+1}^{b}, \bar{C}_{j-1/2}^{ext}, \bar{C}_{j+1/2}^{ext})$$

$$\bar{C}_{j+1/2}^{ext}: = \begin{cases} \bar{C}_j^{L} + 1/2\,(\bar{C}_j^{L} - \bar{C}_{j-1}^{L})\,\dfrac{\Delta s_j}{\Delta s_{j-1}} & \text{for } \bar{u}_{j+1/2}^{n} \geq 0 \\ \bar{C}_j^{L} + 1/2\,(\bar{C}_{j+2}^{L} - \bar{C}_{j+1}^{L})\,\dfrac{\Delta s_j}{\Delta s_{j+1}} & \text{for } \bar{u}_{j+1/2}^{n} < 0 \end{cases} \tag{21}$$

where Eq. (21) simply accounts for the selection of interpolation points depending upon the direction of the flow. The superscript n denotes (known) values time step n, values $\bar{C}_j^L$ are found from the low-order solution Eq. (15). Eqs. (15), (16) represent a coupled system of non-linear ordinary differential equations which is solved using a fourth-order Runge-Kutta method. If two space variables are used, Eq. (21) is replaced by a five-point interpolation surface, where the base-points for interpolation are dependent upon the flow direction [19, 20].

4. CALCULATION OF NON-CONVECTIVE TRANSPORT RATES AND BOTTOM STRESS

In the numerical solution process of the momentum and the scalar transport equation the following unknown quantities were encountered:

(1) momentum equation: $F_j^D(s,t)$, $F_j'(s,t)$, τ_B; Eqs. (9, 10)

(2) concentration equat.: $F_c^D(s,t)$, $F_c'(s,t)$; Eq. (14)

In this section a mathematical formulation of these integrals is given using the vertical profiles for velocity and concentration along with the Boussinesq ansatz in order to correlate the two-point correlations with the gradient of the mean concentration. The expressions derived for turbulent diffusion and for bottom stress only contain the Mannings number as unknown quantity. The value of the Mannings number was determined experimentally for various bottom types. Hence, the formulation is quasi-independent of the experiments thus reducing substantially the verification efforts for the one-dimensional model. The lamiar velocity is defined as:

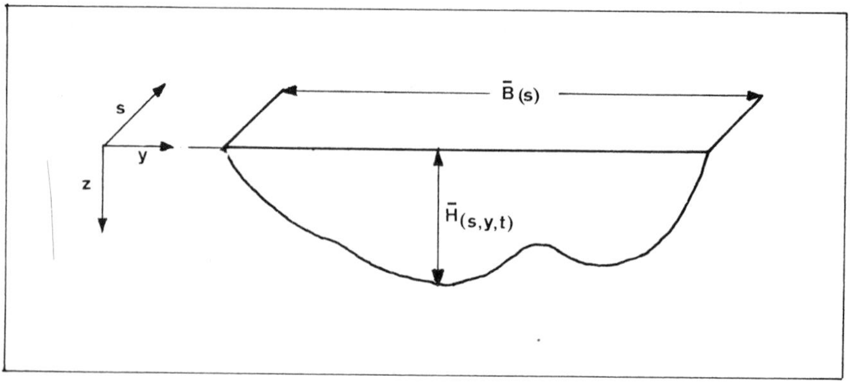

Fig. 1: Coordinate system

$$\bar{u}_L(s,z,t) := \bar{u}(s,t) + \bar{u}_D(s,y,z,t) \tag{22}$$

The vertical velocity profile is specified by the Blasius power law:

$$\bar{u}_L(s,y,z,t) = \left(1 - \frac{z}{\bar{H}(s,y,t)}\right)^{1/n} u_{max}(\bar{H}) \tag{23}$$

where $n \in \mathbb{N}$ depends upon the Reynolds number and lies in the range $5 \leq n \leq 11$. Averaging over the cross-section yields

$$0 = \int_0^{\bar{B}(s)} \int_0^{\bar{H}(y)} \bar{u}_D(s,y,z,t)\,dy\,dz = \int_0^{\bar{B}} \int_0^{\bar{H}(y)} u_{max}(\bar{H}) \left(1 - \frac{z}{\bar{H}(y)}\right)^{1/n} dz\,dy - \bar{u}(s,t)\,\bar{A}(s,t) \tag{24}$$

where the abbreviation $\bar{H}(y) = \bar{H}(s,y,t)$ was used.

Rewriting Eq. (24), one obtains

$$\bar{u}(s,t)\,\bar{A}(s,t) = \frac{n}{n+1} \int_0^{\bar{B}(s)} \bar{H}(s,y,t)\, u_{max}(\bar{H})\, dy\,. \tag{25}$$

The solution of Eq. (25) is not unique, since the ansatz $u_{max}(H) = \alpha_m \overline{H}^m$ $m \in \mathbb{N}$ is a solution. According to von Karman's similarity rule the flow should differ for various water depths only by length scale factors, that is $u_{max}(\overline{H}) = \alpha \overline{H}(s,y,t)$. From Eq. (25) one finds for α

$$\alpha = \frac{n+1}{n} \frac{\overline{u}(s,t)\ \overline{A}(s,t)}{<\overline{H}^2>\ \overline{B}(s)} \quad ; \quad <\overline{H}^2>: = \frac{1}{\overline{B}(s)} \int_0^{\overline{B}(s)} (\overline{H}(s,y,t))^2\ dy$$

The velocity profile then takes the form

$$\overline{u}_L(s,y,z,t) = \overline{u}(s,t) + \overline{u}(s,t) \left\{ \frac{n+1}{n} \frac{\overline{A}(s,t)\ \overline{H}(s,y,t)}{<\overline{H}^2>\ \overline{B}(s)} \left(1 - \frac{z}{\overline{H}(s,y,t)}\right)^{1/n} - 1 \right\} \tag{26}$$

The microscale turbulences are formulated in the form

$$F'_C = \int_{\overline{A}} \overline{u'C'}\ d\overline{A} = \int_0^{\overline{B}(s)} \int_0^{\overline{H}(y)} - \nu_T \left(\frac{\partial \overline{C}}{\partial s} + \frac{\partial \overline{C}_D}{\partial s}\right) dy\ dz \tag{27}$$

where ν_T is the turbulent diffusion coefficient (m^2/s) which has to be determined. With the use of Newton's law

$$\nu_T = \tau \left(\rho \frac{\partial}{\partial z} \mid \overline{u}_L \mid\right)^{-1} = \tau_B\ \xi\ \overline{H}(y) \left(\rho \frac{\partial}{\partial \xi} \mid \overline{u}_D(\xi) \mid\right)^{-1} \tag{28}$$

where a linear variation of the shear stress from the bottom to the free surface is assumed and $\xi = z/H(s,yt)$. For the sake of simplicity variables s and t will be omitted from hereon. From Eq. (26) we find that $\overline{u}_D(\xi)$ takes the form

$$\overline{u}_D(\xi) = \overline{u} \left(\frac{n+1}{n} \frac{\overline{A}\ \ \overline{H}(y)}{<\overline{H}^2>\ \overline{B}} (1-\xi)^{1/n} - 1\right) . \tag{29}$$

For the deviation $\overline{C}_D(s,y,z,t)$ of the concentration from its mean value a formulation similar to Eq. (23) is used

$$\overline{C}_D(s,y,z,t) - \Delta C_B(s,t) = \Delta C_{max}(s,t) \frac{\overline{H}(s,y,t)}{H_{max}(s,t)} g(\xi) \tag{30}$$

where the left-hand side is approximately zero at the bottom, that is

$$\overline{C}_D(s,y,\overline{H}(y),t) - \overline{\Delta C}_B(s,t) \approx 0 \tag{31}$$

and $\Delta C_B(s,t)$ denotes the averaged deviation (in y-direction) from the mean value $\overline{C}(s,t)$ at the bottom $(z = \overline{H}(y))$. Hence, Eq. (31) is only approximately zero. The right-hand side of Eq. (30) is derived from the same ansatz as the one used for u_{max} in Eq. (23), and therefore we have $g(1) = 0$ and $g(0) = 1$. Since the cross-section averaged value of $\overline{C}_D$ vanishes, $\Delta\overline{C}_B$ can be determined leading to the final form of $\overline{D}_D$

$$\overline{C}_D = \Delta C_{max}\, \overline{g}\, \frac{<\overline{H}^2> \overline{B}}{\overline{A}\, H_{max}} \left(1 + \frac{\overline{A}\;\overline{H}(y)}{\overline{g} < \overline{H}^2 > \overline{B}}\, g(\xi)\right) \tag{32}$$

where quantities $\Delta C_{max}(s,t)$, $\overline{g} = \int_0^1 g(\xi)d\xi$, $g(\zeta)$ and $H_{max}(s,t)$ have to be determined from experiment. $\Delta C_{max}(s,t)$ is the averaged maximal concentration deviation for a given cross-section. For the calculation of the turbulent kinematic viscosity we insert (29) into (28)

$$\nu_T = \tau_B \frac{<\overline{H}^2> \overline{B}}{\rho\, \overline{A}\, \overline{u}} \frac{n^2}{n+1} (1 - \xi)^{\frac{2n-1}{n}}. \tag{33}$$

For the bottom shear stress the following ansatz is used

$$\tau_B = :\rho\, (u_*^B)^2, \quad \text{where} \quad u_*^B = u_*(z_o) \tag{34}$$

where u_*^B denotes the bottom friction velocity. From [2, p. 600] one gets a relation between friction velocity and $\overline{u}_L(\xi_o)$ $(\xi_o: = = z_o/\overline{H}(y)$, where z_o denotes the bottom roughness length, de- determined below)

$$u_*^B = \left(\alpha\, \overline{u}_L(\xi_o)\right)^{\frac{n}{n+1}} (\xi_o\, \overline{H}(y)/\nu)^{-\frac{n}{n+1}} \tag{35}$$

Inserting Eq. (26) into (35)

$$u_*^B = \left(\frac{n+1}{n} \frac{\overline{H}(y)}{<\overline{H}^2> \overline{B}} (1 - \xi_o)^{1/n}\right)^{\frac{n}{n+1}} (\alpha\, \xi_o\, \overline{H}(y)/\nu)^{-\frac{n}{n+1}}\, \overline{J}^{\frac{n}{n+1}} \tag{36}$$

where u_*^B is a function of s,y and t. Under the assumption that $\xi_o << 1$

$$u_*^B = \left(\alpha\, \frac{n+1}{n} \frac{1}{<\overline{H}^2> \overline{B}}\right)^{\frac{n}{n+1}} \overline{H}(y)^{\frac{n-1}{n+1}} (\xi_o/\nu)^{-\frac{n}{n+1}}\, \overline{J}^{\frac{n}{n+1}} \tag{37}$$

The roughness height ξ_o is determined by the Colebrook-White equation, an empirical formula which relates de Chézy's coefficient and ξ_o [22]. The hydraulic radius R is approximated by $\overline{H}(y)$ thus leading to

$$\xi_o = 12 \times 10^{-1/_{18}C} \tag{38}$$

where C denotes de Chézy's coefficient. The values for this constant are well-known for different bottom types [23]. For a water depth of 10 m, which is a reasonable value for tidal rivers, we obtain C = 41.06 (mud), 48.04 (shell), 50.34 (sand), 60.25 (grass) The value α in Eq. (38) depends upon n.For n=10, which corresponds to a Reynolds number of some 3 x 10^6, we find α = 11.5 [2, p.601]. By means of Eq. (38) the roughness height ξ_o is determined by a quantity that is known from measurements and no further experiments are necessary in order to calibrate the model. F'_C is now calculated using the value of the bottom friction velocity u_*^B, Eq. (39), and the expression for the eddy viscosity, Eq. (38), where the bottom shear stress is replaced via Eq. (35).

$$\nu_T = (n+1)^{\frac{n-1}{n+1}} \, n^{\frac{2}{n+1}} \, \alpha^{\frac{2n}{n+1}} \, (\overline{H}(y))^{\frac{2(n-1)}{n+1}} \, (<\overline{H}^2> \overline{B})^{\frac{1-n}{n+1}} \, (\xi_o/\nu)^{-\frac{2}{n+1}} \, (1-\xi)^{\frac{2n-1}{n}} \, \overline{J}^{\frac{n-1}{n+1}} \tag{39}$$

$$F'_C = -\int_0^{\overline{B}} \int_0^1 \nu_T \left(\frac{\partial \overline{C}}{\partial s} + \frac{\partial \overline{C}_D}{\partial s}\right) \overline{H}(y) \, dy \, d\xi = - {}^1\nu_T \frac{\partial}{\partial s} \left(\overline{C} + \Delta C_{max} \, \overline{g} \, \frac{<\overline{H}^2> \overline{B}}{\overline{A} \; H_{max}}\right) + {}^2\nu_T \frac{\partial}{\partial s} \left(\frac{\Delta C_{max}}{H_{max}}\right) \tag{40}$$

where

$${}^1\nu_T = \left(\frac{n+1}{n}\right)^{\frac{n-1}{n+1}} \frac{n^3}{3n-1} \, \alpha^{\frac{2n}{n+1}} \, \xi_o^{-\frac{2}{n+1}} \, \nu^{\frac{2}{n+1}} <\overline{H}^{\frac{3n-1}{n+1}}> <\overline{H}^2>^{\frac{1-n}{n+1}} \, \overline{B}^{\frac{2}{n+1}} \, \overline{J}^{\frac{n-1}{n+1}} \tag{41}$$

$${}^2\nu_T = -\left(\frac{n+1}{n}\right)^{\frac{n-1}{n+1}} \beta(n) \, \alpha^{\frac{2n}{n+1}} \, \xi_o^{-\frac{2}{n+1}} \, \nu^{\frac{2}{n+1}} <\overline{H}^{\frac{4n}{n+1}}> <\overline{H}^2>^{\frac{1-n}{n+1}} \, \overline{B}^{\frac{2}{n+1}} \, \overline{J}^{\frac{n-1}{n+1}} \tag{42}$$

and $\beta(n) := \int_0^1 g(\xi) \, (1-\xi)^{\frac{2n-1}{n}} \, d\xi.$

For the bottom shear stress per unit length we find

$$f_B = \int_0^{\overline{B}} \rho (u_*^B)^2 \, dy = \rho \left(\frac{n+1}{n}\right)^{\frac{2n}{n+1}} < \overline{H}^2 >^{-\frac{2n}{n+1}} \xi_o^{-\frac{2}{n+1}} \nu^{\frac{2}{n+1}} \alpha^{\frac{2n}{n+1}} < \overline{H}^{\frac{2n-1}{n+1}} > \overline{B}^{-\frac{n-1}{n+1}} \tag{43}$$

The integrals $F_J^D(s,t)$ and $F_C^D(s,t)$ can be calculated in a straightforward manner, following their definitions

$$F_J^D(s,t) = \int_0^{\overline{B}} \int_0^1 \overline{H}(y) \overline{u}_D \, \overline{J}_D \, dy \, d\xi = \overline{A} \int_0^{\overline{B}} \int_0^1 \overline{H}(y) \overline{u}_D^{\,2} \, dy \, d\xi \tag{44}$$

$$F_C^D(s,t) = \int_0^{\overline{B}} \int_0^1 \overline{H}(y) \overline{u}_D \, \overline{C}_D \, dy \, d\xi \tag{45}$$

while $F_J'(s,t)$ is neglected in comparison with f_B.

5. CONCLUSIONS AND OUTLOOK

The first part of the paper describes a numerical solution technique for the one-dimensional momentum equations as well as the respective transport processes, including cases of peaked distributions. The technique used is a modified version of Zalesak's flux correction algorithm. By means of the numerical procedure, outlined in Sect. 3, numerical diffusion is substantially reduced and concentrations having high gradients can be simulated for long periods of time without loss of accuracy. The numerical technique employed can be directly used for two-dimensional problems.

In Sect. 4 the influence of the vertical profiles of velocity and concentration on the horizontal transport is modeled. The velocity profile is assumed to follow the Blasius power law [1]. This assumption is valid for most of the time as is shown by comparisons with field measured data for the Elbe river. The turbulent transport processes are modeled using by Boussinesq ansatz but account for the vertical variation. It turns out that the eddy diffusion coefficient depends upon the Reynolds number, the geometry and the flow conditions. De Chêzy's number is the only experimental constant which enters the formulation and is well-known for different bottom types [23]. The bottom shear stress is also calculated analytically and depends on the Reynolds number and the detailed geometry. The above formulation, however, does not assume time-varying eddy viscosity. Neglecting time-variations in viscosity in the case of oscillatory flow can result in underestimates of maximum bottom stress and distortion of the flow profile near times of flow reversal as was pointed out by Lavelle [24]. The application of the generalized bottom friction law using a phase lead Θ between bottom stress τ_B and mean velocity $\overline{u}$, as given in [24], could lead to an improved formulation.

ACKNOWLEDGEMENTS

The authors are grateful to W.C. Thacker, who presently is with the Institute of Physics, for his many valuable suggestions and discussions and for his numerous hints concerning the relevant literature.

REFERENCES

1. Hetrick, D.M.; Eraslan, A.H., and M.R. Patterson, 1979: SEDONE: A Computer Code for Simulating Tidal-Transient, One-Dimensional Hydrodynamic Conditions and Three Layer, Variable-Size Sediment Concentrations in Controlled Rivers and Estuaries, ORNL/NUREG/TM-256, 296 pp.

2 Schlichting, H., 1979: Boundary Layer Theory, McGraw Hill, 817 p.

3 Dronkers, J.J., 1964: Tidal Computations, North-Holland, 517 pp.

4 Schroeder, H., 1976: Lectures in Environmental Hydraulics, Danish Hydraulic Institute.

5 Thacker, W.C., 1978: Irregular-Grid Finite-Difference Techniques for Storm Surge Calculations for Curving Coastlines, Marine Forecasting (J.C.J. Nihone Ed.), Elsevier, pp. 261-283

6 Runchal, A.K., 1978: Mathematical Modeling Study of a Large Water Body, Proceedings of the Symposium on Technical, Environmental, Socioeconomic and Regulatory Aspects of Coastal Zone Management ASCE/San Francisco,Calif. March 14 - 16, pp. 1897-1916.

7 Leendertse, J.J., Alexander, R.C., and S.K. Lin, 1973: A Three-Dimensional Model for Estuaries and Coastal Seas: Vol. I, Principles of Computation, Rand Corporation, R-1417-OWRR, 57 pp.

8 Rodenhuis, G.S., Kjaer, O.B., and J.A. Bertelsen, 1977: A North-Sea Model that can Provide Detailed Hydrographic Information, Offshore Technology Conference, pp. 325-330.

9 Boris, J.P., Book, D.L., 1976: Flux Corrected Transport III. Minimal Error FCT Algorithms, J. of Comp. Physics, Vol. 20, No. 4, pp. 397-431.

10 Boris, J.P., Book, D.L., 1976: Solution of Continuity Equations by the Method of Flux Corrected Transport, Methods in Computational Physics, Vol.16, Academic Press, pp. 85-128.

11. Zalesak, S.T., 1979: Fully Multidimensional Flux Corrected Transport Algorithm for Fluids, J. of Comp. Physics, 31, pp. 335-362.

12. Zalesak, S.T., 1981: High Order "Zip" Differencing of Convective Terms, J. of Comp. Physics, 40, pp. 497-508.

13. Book, D.L. (Editor), 1981: Finite Difference Techniques for Vectorized Fluid Dynamics Calculations, Springer, 226 pp.

14. Eraslan, A.H., Kim, K.H., Dec. 4-6, 1978: Cost effective mathematical modeling for the assessment of hydrodynamic and thermal impact of power plant operations on controlled-flow reservoirs, Second Conference on Waste Heat Management and Utilization, Miami Beach.

15. Häuser, J., Eppel, D., and F. Tanzer, 1980: Analysis of Thermal Impact in Tidal Rivers and Estuaries. Water Research, Vol. 14, Pergamon Press, pp. 1409-1419.

16. Raithby, G.D., 1976: A Critical Evaluation of Upstream Differencing Applied to Problems Involving Fluid Flow. Computer Methods in Applied Mech. and Eng., 9, pp. 75-103.

17. Eppel, D., Häuser, J., and F. Tanzer, 1982: Numerische Lösung Partieller Differentialgleichungen der Physik, Teil 2, pp. 55-117, GKSS 82/E/4, GKSS-Research-Center.

18. Hirt, W.C., 1968: Heuristic Stability Theory for Finite-Difference Equations. J. of Comp. Physics, 2, pp. 339-355.

19. Roache, P.J., 1976: Computational Fluid Dynamics. Hermosa Publishers, 446 pp.

20. Leonhard, B.P., 1981: A Stable Accurate, Economical, and Comprehendible Algorithm for the Navier-Stokes and Scalar Transport Equations. Numerical Methods in Laminar and Turbulent Flow, Pineridge Press, pp. 543-553.

21. Fiedler, H., Müller, A., and D. Nolte, 1981: FLUSS - ein eindimensionales Modell des Wärme- und Stoffstransportes in Flüssen, GKSS 81/E/12, GKSS-Research-Center.

22. Querner, E.P., 1981: The Finite Element Method Applied to Turbulent Open Channel Flow in: Numerical Methods in Laminar and Turbulent Flow, Pineridge Press, pp. 693-704.

23. Palmer, S.L., 1978: A Calibrated and Verified Thermal Plume Model for Shallow Coastal Seas and Embayments, Second Conference on Waste Heat Management and Utilization, Miami.

24. Lavelle, J.W., Mofjeld, H.O., 1982: The Effect of Time-Varying Viscosity on Oscillatory Turbulent Channel Flow, Pacific Marine Environmental Laboratory, Seattle, Washington 98 105, No. 570, NOVA/ERL, 31 pp.

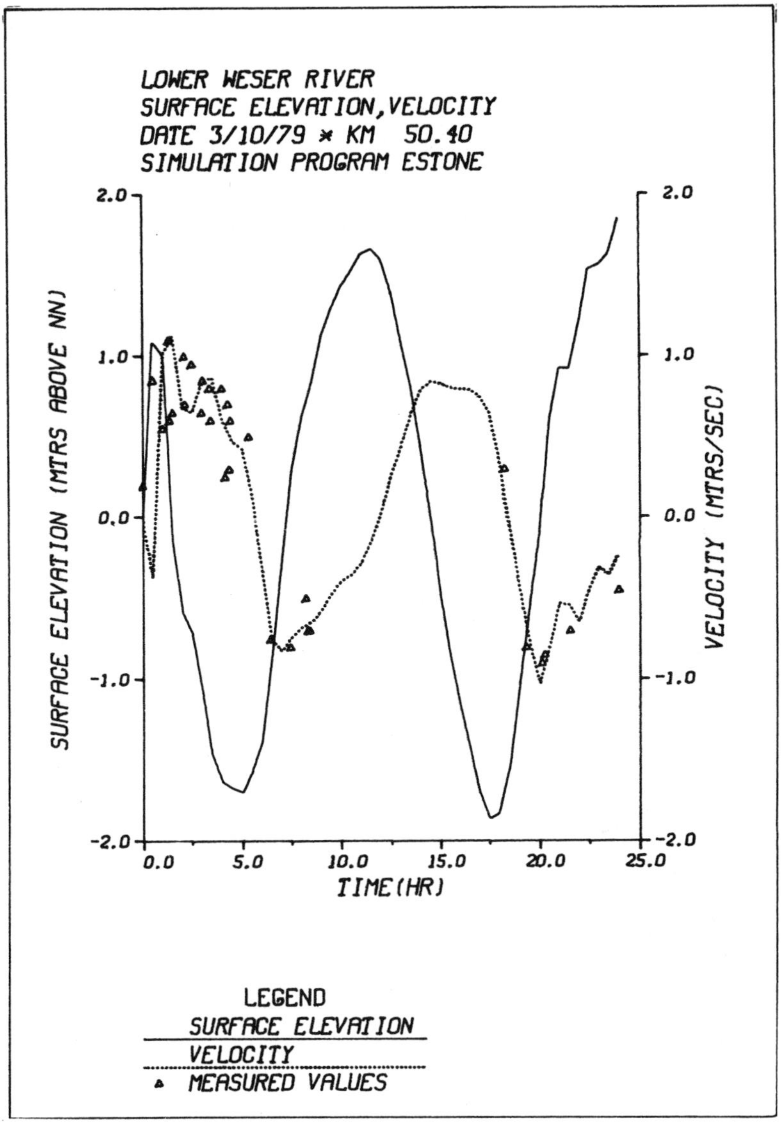

Fig. 2: Calculation of tidal elevation and mean velocity at river km 50.4 (Lower Weser), showing good agreement between measured and computed values.

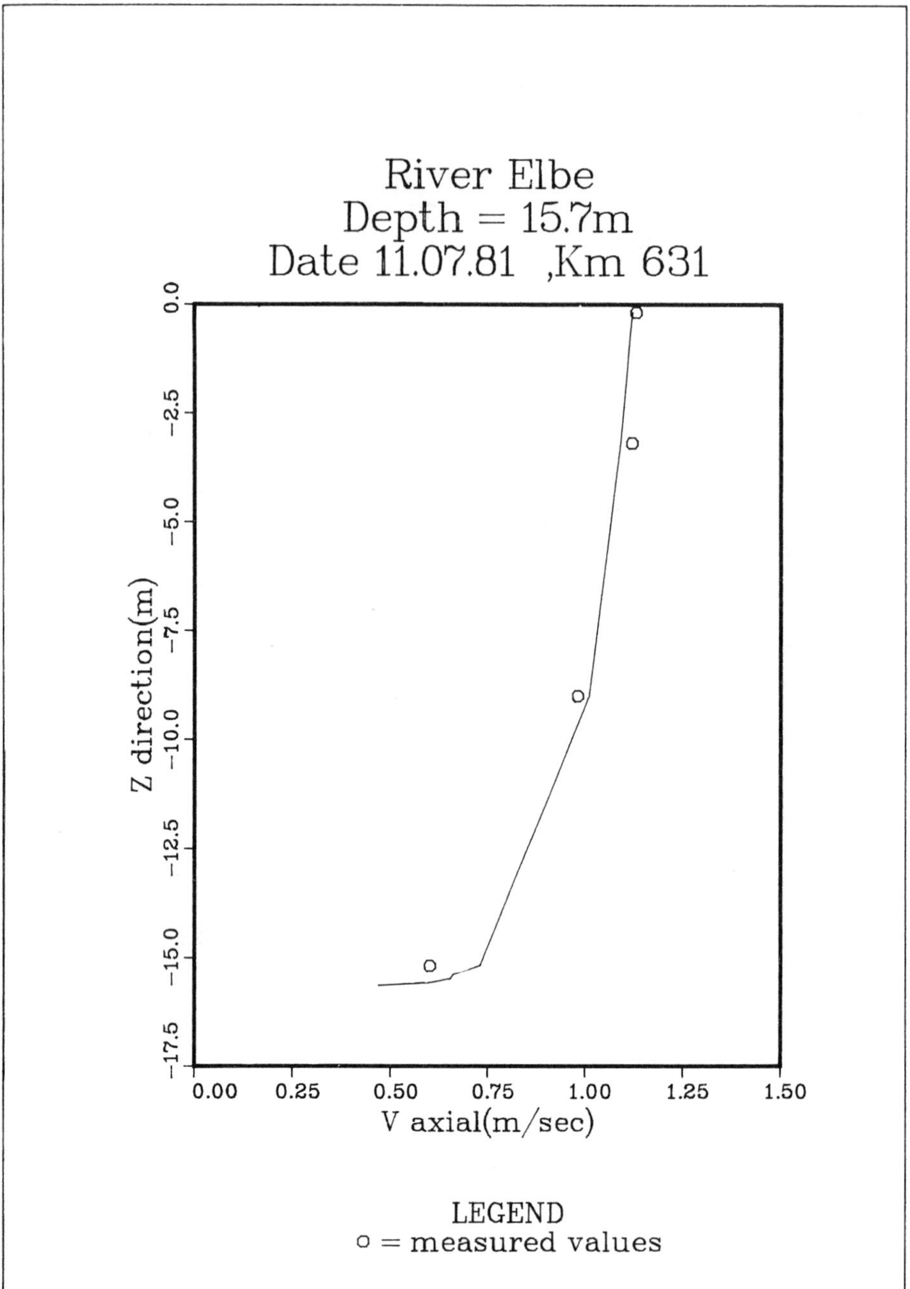

Fig. 3: Comparison of measured and calculated vertical velocity profiles at river km 631 (Lower Elbe). The computed profile was obtained by the Blasius power law using n=10 (Eq. (29))

PARAMETER IDENTIFICATION IN OPEN CHANNEL FLOW

J.M. DE KOK *

RIJKSWATERSTAAT DATAPROCESSING DIVISION

SUMMARY

A procedure is proposed to estimate values for bottom-friction and effective storage width in a 1-D-model for open channel flow when stage hydrographs are available beyond the boundary conditions.

The effect of infiltration in riverbanks is estimated by means of a coupled open channel - groundwater flow model.

A waterlevel-frictionfactor relationship is computed by way of direct substitution of observed waterlevels into the difference equations.

An application is exposed to an existing model of the river Maas for which stage hydrographs from 30 gauges, distributed over a range of 200 km were available.

1. INTRODUCTION

Mathematical models for surface water and groundwater flow always depend on values of parameters which can not be measured in a more or less direct way. Often those parameters have to annihilate the shortcomings of the formulation of the model or the lack of accurate boundary conditions.

When we have at our disposal, beyond the boundary conditions, observed values at certain points in space and time for which solutions can be computed, we can try to match them by changing the modelparameters. Commonly this is done by minimizing some error function which can be either the sum, or the maximum, of either squares or absolute values of differences of observed and computed values. When there are solutions for which the error vanishes we can speak of an inverse model. Existence and uniqueness of the inverse solution can be proven in some particular parabolic cases. |1|

* Project engineer

The minimum of the error is commonly determined by means of a method of steepest descent, and the gradient of the error- (or cost-)function is sometimes obtained by the solution of an adjoint system |2|,|3|, or approximated by varying the parameters and determining the change of the error. |5|,|6|,|7| Convergence is sometimes bad and Rasmussen and Badr |4| recommend the variable metric method which has better convergence but requires the evaluation of the second derivative of the cost function.

The cost function can also be minimized using a Linear Programming algorithm.|7|

In this study we do not minimize a cost function, but we substitute the observed waterlevel values directly into the system of difference equations. This gives us the opportunity to regard some Chezy values for bottom friction as unknown variables and to determine them as part of the solution of a nonlinear system. This requires only relatively simple changes in existing software for open channel flow. Interdependency of channel reaches between gauges is allowed, and therefore a whole river can in principle be calibrated in one single run. The Newton algorithm we used for the solution showed good convergence (4 iterations).

The disadvantage is that a great discrepancy between observed data and model can cause divergence.

Observed data are needed for all timelevels of the computation and this requires some interpolation technique.

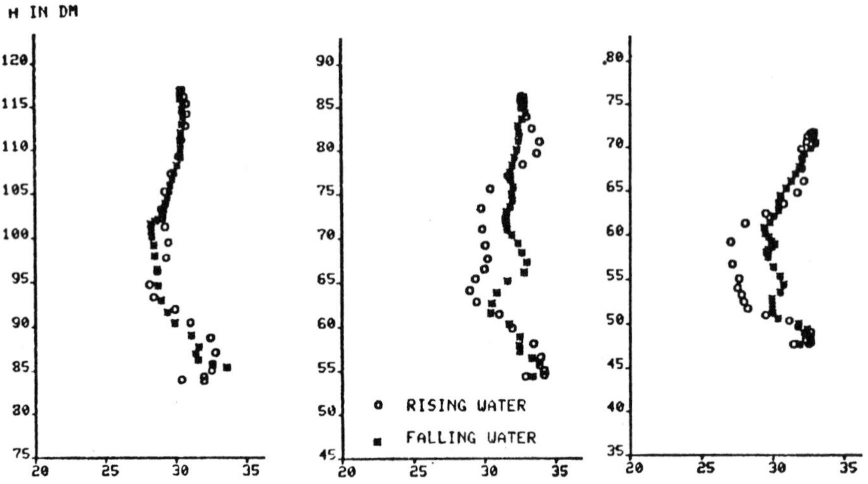

CHEZY (C) - WATERLEVEL (H) RELATION
COMPUTED WITH THE INVERSE MODEL
AT KM 160, 180, 200 FROM 20-2-'70 UNTIL 9-3-'70 .

The inverse model provides Chezy values as a function of time, and we try to relate them to fixed waterlevels. We suppose that during the simulated period no changes in the flow regime or bottom geometry have occurred and therefore expect a unique relationship between waterlevels and Chezy-values. If a hysteresis occurs the model gives rise to a phaseshift and values for storagewidth have to be adjusted. An alternative is to model the infiltration into riverbanks by a groundwatermodel for saturated flow.

Verma and Brutsaert |9| showed that if the horizontal scale is > 4 times the depth of the aquifer one can assume horizontal flow (Dupuit assumption) and use the one-dimensional Boussinesq equation to compute phreatic lines perpendicular to the channel axis, and from there the rate of infiltration. Values for permeability and porosity can than be varied in order to adjust the phases.

2. GOVERNING EQUATIONS

For open channel flow the following equations are valid :

Momentum :
$$\frac{\partial Q}{\partial t} + \frac{\partial Q^2/A}{\partial x} + gA\frac{\partial H}{\partial x} + \frac{gQ}{C^2AR} = 0 \qquad (1)$$

Continuity :
$$\frac{\partial Q}{\partial x} + B\frac{\partial H}{\partial t} = 0 \qquad (2)$$

$H(x,t)$ - piezometric waterlevel
$Q(x,t)$ - discharge
$R(x,H)$ - hydraulic radius
$A(x,H)$ - stream area of the cross section
$C(x,H)$ - Chezy value
$B(x,H)$ - storage width

We use a four-point implicit difference scheme :

$$\frac{H_{i+1}^{j+\theta}-H_i^{j+\theta}}{\Delta x_i} + \frac{Q_{i+\frac{1}{2}}^{j+1}-Q_{i+\frac{1}{2}}^{j}}{gA_{i+\frac{1}{2}}^{j}\Delta t} + \frac{2Q_{i+\frac{1}{2}}^{j}(Q_{i+1}^{j+\theta}-Q_i^{j+\theta})}{g(A_{i+\frac{1}{2}}^{j})^2\Delta x_i} - \frac{(Q_{i+\frac{1}{2}}^{j})^2(A_{i+1}^{j+\theta}-A_i^{j+\theta})}{g(A_{i+\frac{1}{2}}^{j})^3\Delta x_i}$$
$$+ \frac{Q_{i+\frac{1}{2}}^{j+1}|Q_{i+\frac{1}{2}}^{j}|}{(C^2A^2R)_{i+\frac{1}{2}}^{j}} = 0 \qquad (3)$$

$$\frac{Q_{i+1}^{j+\theta} - Q_i^{j+\theta}}{\Delta x_i} + B_{i+\frac{1}{2}}^{j}\frac{(H_{i+\frac{1}{2}}^{j+1} - H_{i+\frac{1}{2}}^{j})}{\Delta t} = 0 \qquad (4)$$

The scheme is of second order accuracy in space and of first (almost second) order in time. It is based upon the well known Preissman scheme. For $\theta=1$ it is equivalent to the third scheme of Dronkers. |11|,|12|
Boundary conditions can be a given discharge upstream and waterlevels downstream. Splittings and junctions can be modellec at gridpoints by imposing $\Sigma Q_i = 0$, Q_i positive when directed tc the node point.

For diffusive waves we can simplify (1) by omitting acceleration and convection terms and linearise (1) and (2) by substituting

$$H = H_0 + h \qquad u_0 = Q_0/A_0$$
$$Q = Q_0 + q \qquad Bs = \frac{\partial A}{\partial H}$$
$$A = A_0 + hBs$$
$$B = B_0 \qquad \frac{\partial H_0}{\partial x} + \frac{u_0^2}{C^2 R} = 0$$
$$\kappa = \frac{2Q_0}{C^2 A_0^2 R}$$

$H >> h$, $Q >> q$, neglecting h^2 and q^2 ,

resulting in
$$\frac{\partial h}{\partial x} + \kappa q - \kappa Bsu_0 h = 0 \qquad (5)$$

and
$$B_0 \frac{\partial h}{\partial t} + \frac{\partial q}{\partial x} = 0 \qquad (6)$$

This leads to the well known diffusion analogy : |8|

$$\frac{\partial h}{\partial t} + Ce \frac{\partial h}{\partial x} - D \frac{\partial h}{\partial x^2} = 0 \qquad (7)$$

with $Ce = \frac{Bs}{B_0} u_0$, $D = \frac{1}{\kappa B_0}$

3. INFILTRATION MODEL

Infiltration in river banks can cause some attenuation and phase shift of flood waves which can not be neglected. |11| We compute the infiltration rate by calculating the horizontal flow perpendicular to the channel axis with a 1-D-model for saturated groundwater flow. |9|

For every section between gridpoints of the open channel flow model we determine phreatic lines for all timelevels of the computation and compute the rate of infiltration in one bank

$$Qs(t+\Delta t) = \frac{\nu}{\Delta t} \int_{Yo(t+\Delta t)}^{Ym} \phi(y,t+\Delta t)-\phi(y,t)dy +$$
$$\frac{1}{2}\frac{\nu}{\Delta t} \{Yo(t+\Delta t)-Yo(t)\}\{\phi(Yo(t+\Delta t),t+\Delta t) - \phi(Yo(t+\Delta t),t)\}$$

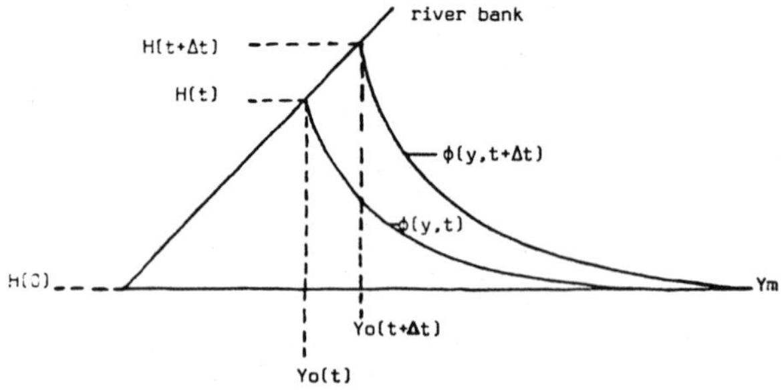

$\phi(y,t)$ - piezometric groundwaterlevel
ν - porosity of soil
$Yo(t)$ - horizontal position of river border at water surface
Ym - landinward position in the riverbank not influenced by infiltration

The river is considered to be symmetric.

Phreatic lines are determined by solution of a linearised Boussinesq equation |14| :

$$\frac{\partial\phi}{\partial t} = \frac{kd}{\nu}\frac{\partial^2\phi}{\partial y^2} \qquad (8)$$

k = saturated hydraulic conductivity
d = depth of the aquifer

$\phi(y,0) = H(0) \qquad y > Yo(0)$
$\phi(Yo(t),t) = H(t) \qquad t > 0$
$\phi(Ym,t) = 0 \qquad t > 0$
$Yo(t) = \frac{1}{2}B(H(t)) - \frac{1}{2}B(H(0))$ so with rising waterlevels there are increasing storage widths and moving boundaries to the groundwatermodel.

For the solution of (8) we used the explicit difference scheme :

$$\frac{\phi_i^{j+1} - \phi_i^j}{\Delta t} = \frac{kd}{\nu}\frac{(\phi_{i+1}^j - 2\phi_i^j + \phi_{i-1}^j)}{(\Delta y)^2} \qquad (9)$$

The condition for stability is $\Delta y^2 > 2\frac{kd}{\nu}\Delta t$ |10|

Because the moving boundary is located between gridpoints the size of the first computational cell has to be reduced or increased.

The infiltration discharge is added to the continuity equation :

$$\frac{\partial Q}{\partial x} + B\frac{\partial H}{\partial t} + 2Qs = 0 \qquad (10)$$

Since there is always an underestimation of reflow in the river as a result of the neglection of anisotropy of the soil and tributary inflows |11|, one can approximate :

$$Qs(t) = \tfrac{1}{2}\nu'(B(t)-B(0))\frac{\partial H(t)}{\partial t}$$

and simply increase the B-values.

For rising waterlevels one can take $\nu' = \nu(1+\frac{tg\alpha}{tg\beta})$ computing the amount of water between the parallel straight lines approximating the phreatic lines at subsequent timelevels.

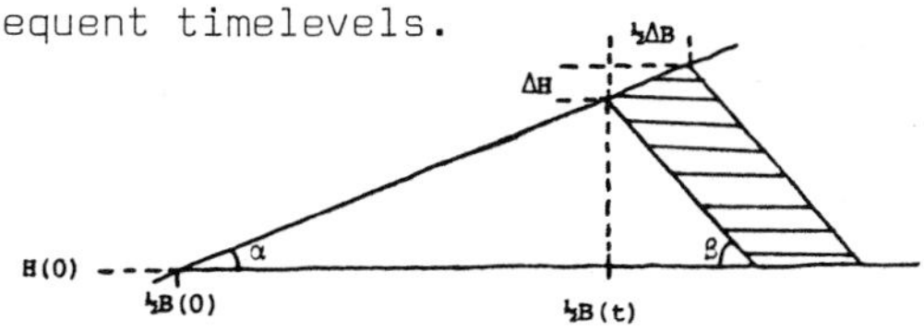

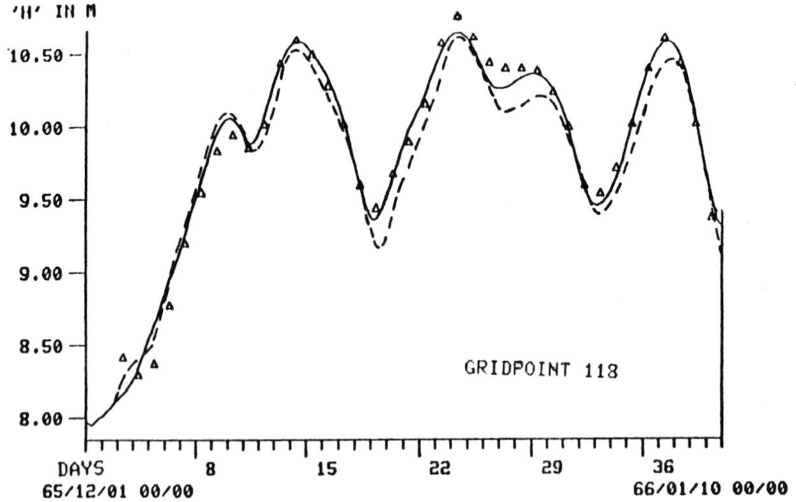

WATERLEVELS COMPUTED WITH COUPLED MODEL ———
WITH INCREASED STORAGE WIDTHS - - -
OBSERVED △△△

We see that a model with increased B-values does not reproduce the later maxima as well as the coupled model does, due to the saturation of the banks. In the coupled model this causes a decay of the infiltration which is not the case in the single model without groundwater computation.

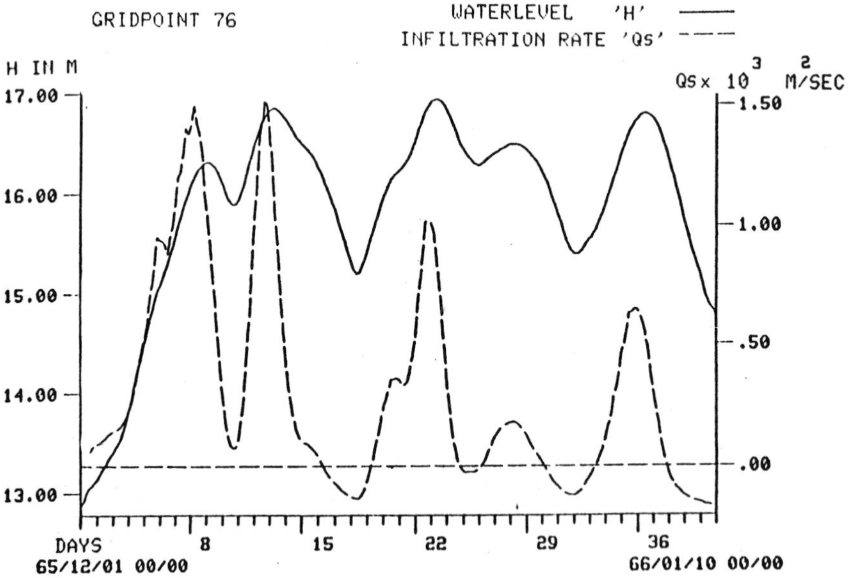

WATERLEVELS AND INFILTRATION RATE COMPUTED WITH COUPLED MODEL

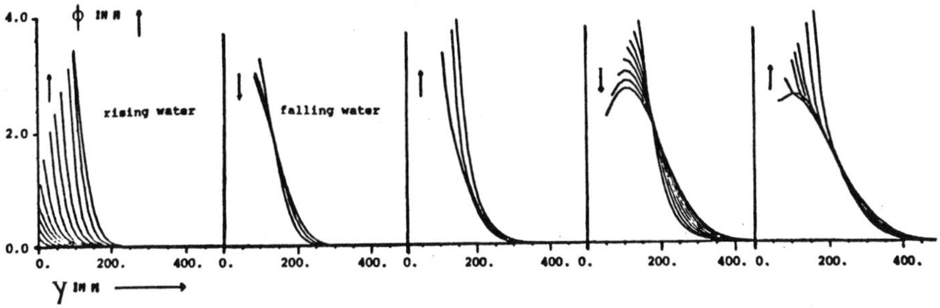

PHREATIC LINES PLOTTED EVERY 16 HOURS FROM 1-12-'65 UNTIL 24-12-'65

4. VARIATION OF PARAMETERS

One can estimate the global values of ν and ν' by adjusting the phases of the computed wave, when there is no ambiguity about the length of the path that wave follows. We see that increasing ν' has the same effect as increasing the B-values, that is a smaller wave speed, which is evident in the diffusion analogy. The effect on the waterlevel maxima is little, because we are dealing with very long waves. Waterlevels can locally be adjusted by locally changing the C-values.

Once we have estimated B-values, and C-values are globally adjusted, so there is no phaseshift to expect as a consequence of more detailed calibration we can apply the inverse model.

5. INVERSE MODEL

When the schematization consists of N sections (not necessarily of equal length, splittings, junctions and parallel sections are allowed) we have for each timestep j+1 a set of 2N coupled equations , linear in H_i^{j+1} , Q_i^{j+1} . By eliminating the Q_i^{j+1} we come out at a set of n equations, where n is the number of gridpoints.

At m gridpoints through the network there are gauges and observed waterlevels for all timesteps of the computation. We consider now C-values to be variable in time and constant for a whole channel reach between gauges. The inverse problem is to find for all timesteps n H-values and m C-values satisfying the n equations for open channel flow and

$H_{i_k}^{j+1} = G_k^{j+1}$, k=1,m , G the observed waterlevel. After elimination of the $H_{i_k}^{j+1}$ we have a system of n nonlinear equations F(x) = 0 that can be solved by Newton iteration :

$x^{(k+1)} = x^{(k)} - A^{-1}(x^{(k)})\ F(x^{(k)})$ with A(x) the G-derivative of F at x. For k=0 we can take the values of the preceding timelevel. |13|

Discrepancies between model and reality will be annihilated by a change in C-values and this can cause an error in the computed discharges that not will vanish with increasing x in the following example :

Suppose at timelevels j and j+1 an almost stationnary situation in a prismatic channel :

$$\frac{\partial H}{\partial x} + \frac{Q^2}{C^2A^2R} = 0 \quad , \quad Q > 0$$

The equations solved in the inverse system are in first approximation : (11)

$$\frac{h_{i+1}^{j+\theta}-h_i^{j+\theta}}{\Delta x} + \frac{q_{i+\frac{1}{2}}^{j+1}-q_{i+\frac{1}{2}}^{j}}{gA\Delta t} + 2Q\,\frac{q_{i+1}^{j+\theta}-q_i^{j+\theta}}{gA^2\Delta x} + Q\,\frac{q_{i+\frac{1}{2}}^{j+1}+q_{i+\frac{1}{2}}^{j}}{C^2A^2R} = \frac{2Q^2\Delta C}{C^3A^2R}$$

and $$\frac{q_{i+1}^{j+\theta}-q_i^{j+\theta}}{\Delta x} + B\,\frac{h_{i+\frac{1}{2}}^{j+1}-h_{i+\frac{1}{2}}^{j}}{\Delta t} = 0 \quad , \quad i = 0,1 \qquad (12)$$

A, B, C, R, Q, ΔC are constant for a channel reach consisting of two sections between the gridpoints 0 and 2 where waterlevels are imposed conform the stationnary situation. At gridpoint 0, timelevel j+1 a discharge is imposed with a small error $\hat{q}$.

So h_0^{j+1} , h_2^{j+1}, h_k^j , $q_k^j = 0$, k=1,2,3 , $q_0^{j+1} = \hat{q}$.

We eliminate ΔC and q_1^{j+1} and write $q_2^{j+1} = \lambda q$, $h_1^{j+1} = \tilde{h}$.

(11) gives $$4\,\frac{\theta\tilde{h}}{\Delta x} - \hat{q}\,\frac{\lambda-1}{gA\Delta t} - \frac{Q\hat{q}(\lambda-1)}{C^2A^2R} = 0 \qquad (13)$$

(12) gives $$\frac{B\tilde{h}}{2\Delta t} + \theta\,\hat{q}\,\frac{\lambda-1}{2\Delta x} = 0 \qquad (14)$$

The only solution is $\lambda=1$, $\tilde{h}=0$, so in this case an error propagates without damping.

For this reason it is desirable to have a few sections in the network for which no C-values are computed, in order to give errors (in general consisting of short oscillations) the opportunity to damp out in a 'natural' way.

6. APPLICATION

The model has been applied to the Dutch part of the river Maas over a length of 200 km. Stage hydrographs were available for 30 points in the netwotk consisting of 124 gridpoints and 147 sections. Flood plains were modelled as parallel sections. The hydraulic radius R was computed according to :

$$A^2R = \{\sum_i \frac{A_i^{5/3}}{O_i^{2/3}}\}^2$$

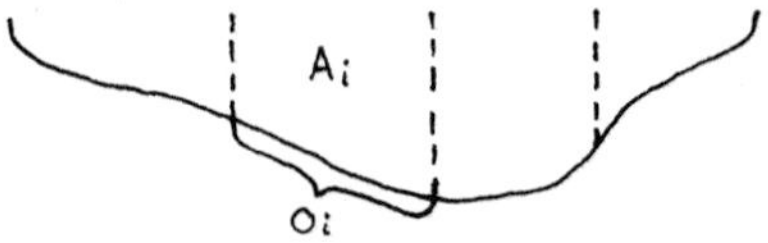

with A_i the stream area and O_i the wetted perimeter of part i of the cross section. |11|,|15|

Stepsize in time was 1800 sec and stepsizes in space varied between 200 and 4000 m with an average of 2000 m. Flow was always subcritical. The wave celerity Ce varied between .5 and 1.5 m/sec , the Courantnumber Cf = Ce $\Delta t/\Delta x$ was averaged 1. The parameter $\psi = D\Delta t/\Delta x^2$ was averaged 1. We took $\theta=1$. From appendix I we can see that the scheme is stable for $\theta=1$ and from appendix II we can see that for waves with periods in the order of 10 days the numerical accuracy is satisfactory.

First the coupled model was roughly calibrated to observed waterlevels of december '65. ν = .2 and kd = .004 m^2/sec gave a satisfactory fit. For the groundwater model we used Δy= 17 m Ym = $\frac{1}{2}$Bm + 1000 m with Bm the maximum storage width. Then the single model with increased storage widths was used with ν'= .25 and calibrated with the inverse model with observed values of feb/march '70. Then the calculated timedependant C-values were averaged around fixed waterlevels with intervals of .5 m and the model was verified a.o. with dec '65. The model showed a reasonable agreement except for reaches that not were calibrated. The effect of local adjustment of C-values is indeed only local.

CPU-times for the single open channel flow model were 10 sec per simulated day on a UNIVAC 1180 computer. The coupled model takes twice as much time and the inverse model (without groundwater) takes 5 times as much.

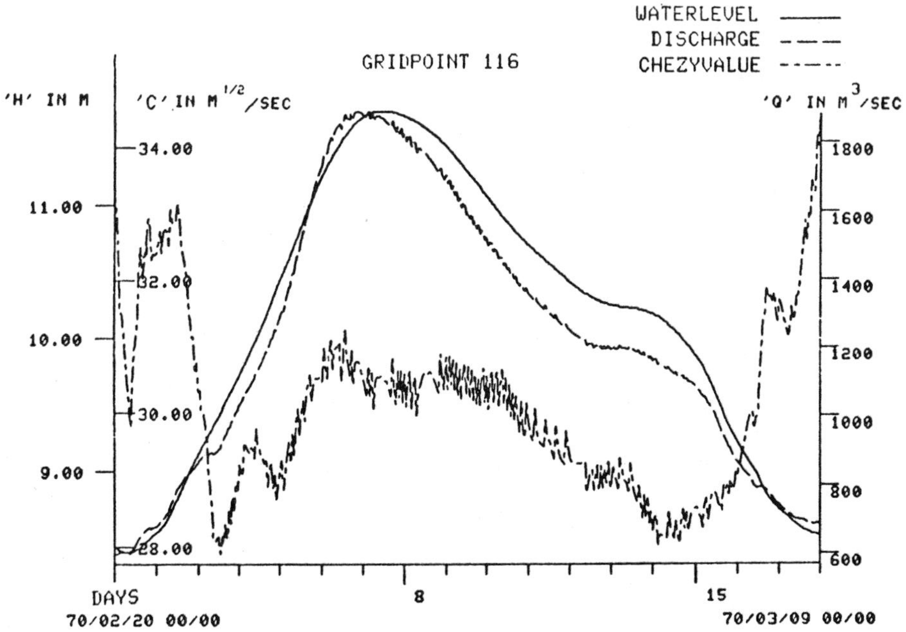

WATERLEVELS, DISCHARGES AND CHEZYVALUES COMPUTED BY THE INVERSE MODEL

ACKNOWLEDGMENT

The author wishes to express his thanks to the Department 'Maas" of the Direction of 'Waterhuishouding en Waterbeweging' of Rijkswaterstaat for providing the basic schematization and the observed data.

7. REFERENCES

1. CANNON,J.R. and DUCHATEAU,P - An inverse problem for a nonlinear diffusion equation. SIAM J.Appl.Math.Vol.39 No.2 October 1980 p.272
2. CHAVENT,G. , DUPUY,M. and LEMONNIER,P. - History Matching by use of Optimal Control Theory. Society of Petroleum Engineers Journal. feb.'75, p.74
3. VOLPI,G. and SQUAZZERO,P. - The linearization of the quadratic resistance term in the equations of motion for a pure harmonic tide in a canal and the identification of the Chezy parameter C.
   Proceedings of the 8th IFIP Conference on Optimization Techniques. Wurzburg, Germany 5-9 Sept. 1977. Springer Verlag, Berlin, 1978.
4. RASMUSSEN,H. and BADR,H.M. - Validation of numerical models of the unsteady flow in lakes. Appl.Math.Modelling. Vol.3 dec'79 p.416
5. FREAD,D.L. and SMITH,G.F. - Calibration Technique for 1-D Unsteady Flow Models. Journal of the Hydraulics Division. July '78 p. 1027
6. BECKER,L. and YEH.W.W.G. - Identification of Multiple Reach Channel Parameters. Water Resources Research.Vol.9 No.2 april '73
7. YEH,W.W.G. and BECKER,L. - Linear Programming and Channel Flow Identification. Journal of the Hydraulics Division.nov.'73,p.2013
8. VREUGDENHIL,C.B. and GRIJSEN J.G. - Numerical Representation of Flood Waves in Rivers. Publication No. 165 Dutch Hydraulics Laboratory. 1976
9. VERMA,R.D. and BRUTSAERT,W. - Unsteady Free Surface Groundwater Seepage. Journal of the Hydraulics Division. august'71,p.1213
10. RICHTMEYER,R.D. and MORTON,K.W. - Difference Methods for Initial Value Problems. Interscience Publishers. 1967
11. CUNGE,J.A.,HOLLY,F.M. and VERWEY,A. - Practical Aspects of Computational River Hydraulics.Pitman 1980
12. DRONKERS,J.J. - Tidal Computations for Rivers,Coastal Areas and Seas. Journal of the Hydraulics Division. jan.'69
13. ORTEGA,J.M. and RHEINBOLDT,W.C. - Iterative Solution of Nonlinear Equations in Several Variables. Academic Press 1970
14. BEAR,J. - Hydraulics of Groundwater. MacGrawHill 1979
15. RIJKSWATERSTAAT,Directie Wat. en Wat.Afdeling Maas. Beschrijving van de schematisatie van de Maas. Notitie 1978-H, 1978

8. APPENDIX I : STABILITY

Having (3) and (4) simplified following the diffusion analogy we obtain :

$$\frac{h_{k+1}^{j+\theta}-h_{k}^{j+\theta}}{\Delta x} + \kappa q_{k+\frac{1}{2}}^{j+\frac{1}{2}} - \kappa u_0 B s h_{k+\frac{1}{2}}^{j} = 0 \qquad (15)$$

$$\frac{q_{k+1}^{j+\theta} - q_{k}^{j+\theta}}{\Delta x} + B_0 \frac{h_{k+\frac{1}{2}}^{j+1} - h_{k+\frac{1}{2}}^{j}}{\Delta t} = 0 \qquad (16)$$

Applying the Fourier method to examine Neumann stability |10|, we substitute in (15) and (16) :

$h_n^m = \hat{h}\lambda^n\mu^m$, $q_n^m = \hat{q}\lambda^n\mu^m$, $\lambda = e^{-i\sigma\Delta x}$, $\mu = e^{i\omega\Delta t}$, σ real ,

to obtain : $\begin{pmatrix} \alpha & \beta \\ \gamma & \delta \end{pmatrix} \begin{pmatrix} q_m^n \\ h_m^n \end{pmatrix} = 0$, $\begin{matrix} \alpha = \kappa(\mu+1)\Delta x \\ \beta = 4(\theta\mu+1-\theta)\zeta - 2\Delta x u_0 \kappa B s \\ \gamma = 2\Delta t(\theta\mu+1-\theta)\zeta \\ \delta = \Delta x B(\mu-1) \end{matrix}$

$$\zeta = \frac{\lambda-1}{\lambda+1} \qquad (17)$$

For $\theta = 1$ this results in $|\mu| \leqslant (1-\psi')^{-\frac{1}{2}}$, $\psi' = 8\frac{\Delta t}{\Delta x^2}\zeta^2 D$

ψ' is real and negative because $\zeta^2 = -tg^2\frac{1}{2}\Delta x$ so for $\theta=1$ the scheme is unconditionnally stable.

9. APPENDIX II : NUMERICAL ACCURACY

We compare now harmonic solutions of eq. (5) and (6) and of eq. (15) and (16) with boundary conditions :

$h(0,t) = \hat{h}e^{i\omega t}$, $q(0,t) = \hat{q}e^{i\omega t}$, ω real

By substitution of $h(x,t) = h(0,t)e^{-i\sigma x}$ in (7) we get

$$\sigma\Delta x = i\frac{Cf}{2\psi} - i\sqrt{\left(\frac{Cf}{2\psi}\right)^2 + i\,\frac{2\pi}{\psi s}} \ ,$$

$s = T/\Delta t$, $T = 2\pi/\omega$, $Cf = Ce\Delta t/\Delta x$, $\psi = D\Delta t/\Delta x^2$.

By substitution of $h_n^m = \hat{h}\lambda^n\mu^m$, $q_n^m = \hat{q}\lambda^n\mu^m$, $\mu = e^{i\omega\Delta t}$

$\lambda = e^{-i\tilde{\sigma}\Delta x}$ in (15) and (16) we obtain again (17) and λ can be determined as a function of θ, Cf, ψ and s.

In the figure we see values of the 'amplitude factor' $\left(\frac{|\lambda|}{e^{im\sigma\Delta x}}\right)^n$.

$n = L/\Delta x$, $L = CeT$.

$|\lambda|^n$ is the remaining part of the amplitude of the numerical wave at 1 wavelength distance from the boundary.

$e^{im\sigma L}$ is the remaining part of the analytical wave after 1 wavelength L. s is the number of points (timelevels) per waveperiod T.

Next to it are plotted values of the 'phasefactor' $\frac{re\sigma\Delta x}{-\arg\lambda}$ against s. The 'phasefactor can be seen as the ratio of the celerities of the numerical and the analytical wave.

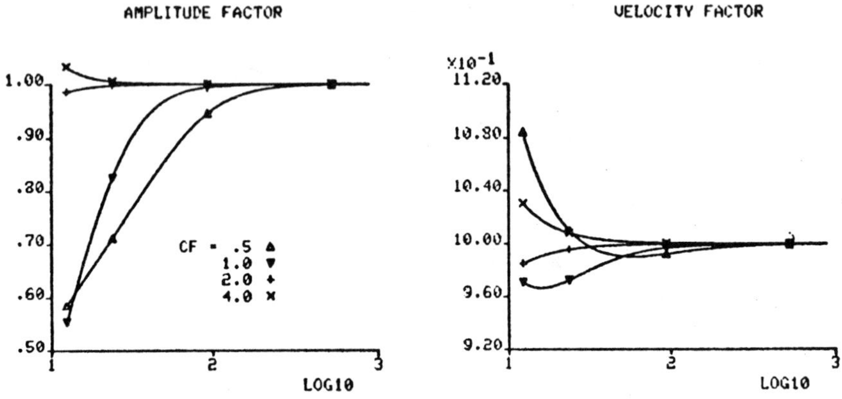

⟶ TIMESTEPS PER WAVEPERIOD. THETA = 1.00 PSI = 1.00

A FINITE DIFFERENCE, TIME MARCHING SOLUTION OF THE SLAMMING PROBLEM (WATER ENTRY PROBLEM)

P. Gallagher B.Sc. (Eng.)

Research Assistant
Department of Naval Architecture and Ocean Engineering,
University of Glasgow.

ABSTRACT

This paper describes the development of a time-marching finite difference model for the free-surface flow associated with the water entry problem. A short history of the authors early attempts at a solution are given. The lessons learned lead to modelling the flow using the compressible form of the mass and momentum conservation equations. A variational technique applied to energy conservation (due to Sasaki), is used to control the change in density and fluid velocity.

The fluid pressure is considered as three separate components; a 'modelled' accoustic pressure, hydrodynamic, and hydrostatic values are superimposed. The hydrodynamic pressure is found via a Poisson equation whose solution is obtained using the Multi-Level Adaptive techniques of Brandt et al.

General working efficiency guidelines are discussed and a sample problem presented.

1 INTRODUCTION

In the past few years, Naval Architects and Structural designers of offshore vehicles have become more interested in transient dynamic loads. One such dynamic load is that of the 'slam'. It is usually caused either as a result of heavy pitching and heaving of a vehicle, or by the impact of a wave. Such loadings can be responsible for high levels of stress in a vessel's structure, and in some cases, heavy damage is sustained.

The purpose of this author's research has been to produce a pressure distribution and time history of the slamming load. The results of such an algorithm would be used in a structural simulation program to derive stress/time histories for various designs of ship bottom structure. Indeed, the main aim of this research is to produce a fluid/structure interaction program for slamming.

This paper reports on work done to provide a model of the fluid dynamics of the problem.

2 PRESENT AUTHOR'S OWN EARLY INVESTIGATIONS

A model of the fluid behaviour based on the combination of continuity and the incompressible Navier Stokes equations was the starting point of this research.

The first attempt used a Lagrangian system of moving particles to represent the fluid and it's free surface. This however, proved far from ideal. Problems arose in the book-keeping effort required to avoid ill conditioning in the representation of spacial derivatives caused by 'cross-over' effects. Boundary conditions, particularly Neumann conditions at the free-surface, proved difficult to code satisfactorily.

The second attempt (SPLASH), used a fixed mesh, extending above the free-surface. The same formulation of the continuity equation and momentum transfer equations were used as before. This consisted of using Navier-stokes equations to time march in (U,V) explicitly and solving the Poisson Equation for pressure:

$$\nabla^2 p = S \quad , \qquad S \text{ DERIVED FROM NAVIER STOKES EQNS VIA CONTINUITY}$$

at each time-step. Book-keeping consisted of an algorithm to follow the free-surface. A number of methods such as;

(a) the kinematic free-surface condition,
(b) free-surface marker particle technique.
(c) the Marker and Cell technique, (10)
(d) the Volume of Fluid technique, (11)

were all analysed, the latter proving the most efficient.

SPLASH proved to be insufficient however. Problems were encountered as the initial conditions were physically incompatible with the discrete formulation of continuity. Any attempt to cure this via the introduction of a non-divergent velocity field violated energy conservation. As a result, pressure calculations proved unreliable. Boundary conditions as applied to the moving body, especially in the case of pressures, were difficult to evaluate and were a source of instability.

3 A MORE FLEXIBLE IMPACT MODEL

3.1 Consider an arbitrarily shaped section falling towards a quiescent fluid free surface. The fluid is considered compressible in the sense that, at zero March number, and depending upon the volumetric strain, an 'acoustic' pressure, given by the homentropic law:

$$\Delta p = C^2 \Delta \rho \tag{1}$$

can exist. This pressure is superimposed upon the hydrodynamic pressure resulting from any flow properties within the fluid.

3.2 Forms of Equation Used

The equations used in the time-marching are those for compressible mass and momentum conservation.

Mass

$$\frac{\delta \rho}{\delta t} = \rho(\nabla . \vec{U}) + \vec{U}.(\nabla \rho) \tag{2}$$

ie compression + convection of density. In this case, convection is usually small compared with compression.

Momentum in 2 dimensions

$$\frac{\delta \rho U}{\delta t} + \frac{\delta \rho U^2}{\delta X} + \frac{\delta \rho UV}{\delta Y} = - \frac{\delta p}{\delta X} + (\text{VISCOUS SHEAR})$$

$$\frac{\delta \rho V}{\delta t} + \frac{\delta \rho V^2}{\delta Y} + \frac{\delta \rho UV}{\delta X} = - \frac{\delta p}{\delta Y} + (\text{VISCOUS SHEAR}) \tag{3}$$

3.3 The mass conservation equation is used to explicitly march in density. In fact, the change in density, $\Delta \rho$, is used as a main variable. The Div (u) compression term is evaluated in the control volume sense, and includes a source term to represent the encroachment on the cell volume by the body.

3.4 The principle of superposition is applied for the overall pressure in the fluid, ie total pressure = ('acoustic' + hydrodynamic + hydrostatic). This is effectively the same as the solution of the following set of Poisson Equations:

1 $$\nabla^2 p = C^2 \frac{\partial^2 p}{\partial t^2}, \quad p = 0 \text{ ON } \Omega, \text{ (EXCEPT BODY)} \tag{4}$$

2 $$\nabla^2 p = S, \quad p = 0 \text{ ON } \Omega, \text{ (EXCEPT BODY)} \tag{5}$$

3 $$\nabla^2 p = 0, \quad p = \rho g h \text{ ON } \Omega . \tag{6}$$

where body (1) or (2) are superimposed in order to calculate the pressure on the body. In fact, since the homentropic pressure law is used to calculate the acoustic pressure, equation (4) is not required. Equation (6) is not required either, since the value:

$$p = \rho g h \tag{7}$$

may be easily calculated.

3.6 A staggered mesh is use, pressures and densities placed at cell centres, u and v components placed at cell sides in the conventional manner. This greatly aids the control volume approach used in the time marching of density. The mesh extends above the free surface.

3.7 Forward time differencing is used throughout. Spatial derivatives use a centred difference scheme. Where boundaries cross the mesh between modes, an "irregular star" method is adopted to evaluate derivatives.

3.8 The volume of fluid technique (11), slightly adjusted to deal with mass transfer in a compressible fluid, is used to deal with the free surface book-keeping.

3.9 The moving body boundary is represented by a series of sources. Each cell is considered to contain a source variable. These sources are "switched on" as the body moves through each cell, as indicated by a book-keeping process.

3.10 The free surface boundary conditions are modelled by: For velocities - Assuming constant density in free surface cells and consequently, continuity in such cells. For hydrodynamic pressure - Assuming zero pressure, or, if flow is viscous, zero normal stress (12).

3.11 The far right and left hand boundaries were modelled;
(a) as rigid, free slip, walls,
(b) using a discrete evaluation of the Sommerfield radiation condition due to Orlanski (13) to remove the problem of 'reflections' during any iterative processes over the domain.

3.12 The bed boundary was considered to be a rigid wall, mainly since experiments in the form of 'drop tests' were to be undertaken, and depth was one parameter that was fixed by the geometry of our tank at Glasgow University.

3.13 Variational Energy Algorithm

A simple time marching in (ρ,u,v) with a running calculation of hydrodynamic pressure, would not guarantee the accuracy of the model. A technique was required to restrict the build up of errors in the system caused by the finite difference truncations.

Various researchers have been examining the problems associated with the conservation of quadratic quantities, such as kinetic energy, in finite difference models (14).

In particuar, the work of Sasaki (14) and Arakawa was examined and adapted to provide a technique for energy conservation. The basic concepts can be summed up as:

(a) Restriction of the total energy of the system over the whole domain either;

- (i) to be constant or in water entry problem.
- (ii) to be calculable in terms of the total work done upon the system.

(b) The use of a variational model to share out any imbalance in energy at a particular time step between the variables, such that the discrete system obeys the same energy conservation laws as the continious system.

3.14 Two basic versions of the method exist for this problem.
A The functional J is given by

$$J = \sum[\alpha(U-\tilde{U})^2 + \alpha(V-\tilde{V})^2 + \beta(\Delta\rho-\tilde{\Delta\rho})^2] + \lambda_e\left\{\sum(\tfrac{1}{2}\rho(U^2+V^2) + \tfrac{1}{2}B\frac{\Delta\rho^2}{\rho^2}) - Q\right\} \quad (8)$$

where:

α, β are weights to be determined.
λ_e is the Lagrange multiplier.
$U, V, \Delta\rho$ represent the variables corrected to conserve energy.
$\tilde{U}, \tilde{V}, \tilde{\Delta\rho}$, represent the approximate values found from time marching.
B bulk modulus for the fluid.
Q total energy of fluid. It includes allowances for a change in total fluid potential energy and work done on the fluid by the falling body.

The first variation of (8) is taken and yields the following Euler-Lagrange equations:

$$U = \frac{2\alpha\tilde{U}}{2\alpha + \lambda_e\rho} \qquad V = \frac{2\alpha\tilde{V}}{2\alpha + \lambda_e\rho} \qquad \Delta\rho = \frac{2\beta\tilde{\Delta\rho}}{2\beta + \lambda_e B} \quad (9)$$

and:

$$\sum(\tfrac{1}{2}\rho(U^2+V^2) + \tfrac{1}{2}B\frac{\Delta\rho^2}{\rho^2}) - Q = 0 \quad (10)$$

By making assumptions about the relative sizes of (U,V) and ρ, and assuming the existence of a fractional adjustment (χ) rate for the variable, $\tilde{U}$, $\tilde{V}$, $\tilde{\Delta\rho}$ we may derive the value of χ by substitution of equations (9) into equation (10). Thus the new values $U, V, \Delta\rho$ may be found.

In general, for the linear case, the above process leads to the evaluation of the fractional adjustment as the square root of Q divided by the discrete summation of fluid kinetic energy plus strain energy.

A 'non-linear' formulation is possible which includes the effect of the change in density upon the total kinetic energy and potential enrgy of the fluid. Thus leads to a third order polynomial equation in fractional adjustment rate with one real root and two imaginary roots which are discarded.

B The functional J is given by

$$J = \sum[\alpha(U-\tilde{U})^2 + \alpha(V-\tilde{V})^2 + \beta(\Delta\rho-\Delta\tilde{\rho})^2] + \lambda_e\left\{\sum(\tfrac{1}{2}\rho(U^2+V^2) + \tfrac{1}{2}B\frac{\Delta\rho^2}{\rho^2}) - \tfrac{1}{2}M_bV_b^2 - Q\right\} \quad (11)$$

The analysis is as before with:

M_b = mass of body
V_b = velocity of body
Q = change in potential energy of fluid with allowance for change in potential energy of body.

This leads to an extra Euler-Lagrange equation and more assumptions to be made concerning the relative weights. However, this technique has proved more amenable to computer use since the computation of the work done on the body by the fluid via the integration of pressures upon the body is not required to find V_B.

4 IMPLEMENTATION OF THE CODE

4.1 The change in density per time step is implemented via the control volume approach for each cell, ie.

$$\Delta\rho = [((U_{ij}.h_y.\Delta t - U_{ij+1}.h_y.\Delta t - V_{ij}.h_x.\Delta t + V_{i+1,j}.h_x.\Delta t) + (\text{SOURCE})_b) / (\text{Volume of cell})] + \text{density convection terms.} \quad (12)$$

It is explicit, using old time values exclusively.

4.2 The change in velocity per time step was calculated explicitly from equations (3). Pressure gradients were found from the total pressure field as given in section 3.4.

4.3 As given in section 3.14, there are three levels of computation of the fractional adjustment rate X for the variational energy technique:

Linear Case A gives an equation of the form,

$X^2 = Q/(\text{fluid K.E.} + \text{Strain Energy})$ ......(13)

Non-linear Case A gives an equation of the form,

$PX^3 + CX^2 + DX - Q = 0$ ......(14)

where:

P = Perturbed kinetic energy of fluid caused by change in density.
C = fluid KE + Strain Energy
D = Perturbed potential energy of fluid caused by change in density.
Q = as before.

The roots of thus equation are found by Muller's Method, with a first guess being found from the linear formulation.

Linear Case B

Since the effect of the fractional change on V_B is considered to be the inverse of that upon the fluid variables, (ie an increase in total fluid energy causes a decrease in body kinetic energy), a fourth order polynomial is produced.

$$AX^4 + BX^2 - C = 0 \qquad \text{......(15)}$$

where:

A = total fluid (KE + Strain Energy)
B = change in fluid potential energy
C = body kinetic energy

4.4 Hydrodynamic Pressures

The time history of the growth of hydrodynamic pressures is an important aspect of the flow model. The rapid rise of 'acoustic' pressure in the first few time steps acts as a 'buffer' for the fluid. The transfer of energy from the comression of the fluid to its' kinetic energy provides a method of doing work on the fluid and the rise in hydrodynamic pressures reflects this aspect of the flow model.

4.5 The hydrodynamic pressures are solved for as a boundary value problem derived from the Poisson equation given in section (3.4) equation (5). A fast and efficient algorithm is required to solve this equation. Successive-Over-Relaxation was used at first. The relaxation factor was a function of convergence rate, a continious checking process being enacted using the relationship given by Carre (15). The technique proved slow however, so the multigrid techniques of Brandt et al have now been adopted (16,17,18).

4.6 Implementation of Multigrid

In essence, the idea is to solve not for the pressure field itself, but for the error in the field (defect eqn), as given by the difference between the true solution and the starting approximation. Relaxation sweeps are used to smooth away the various frequency components of the 'error' field on a number of grids of varying size. Convergence of any relaxation scheme is based upon the relative size of the wave-number of each error component to the grid size. The multigrid method attempts to decrease overall computational effort by the use of a series of grids, each one of which is in some way optional in its removal of a particular wave number ocmponent of error.

4.7 The presence of the free-surface complicates the implementation of Multigrid. In this case the problem of designing the various grids to be used was solved as follows.

A series of Integer array variables, representing the I and J indices of the cells involved in each level of computation, were formed. The resulting arrays vary in size as 2^n (n = 1,2,...P). The coarsest grid level simply consists of one point plus four boundary values (n = 1). The finest grid has n = 7 or 8 ie a 128 x 128 grid or 256 x 256.

4.8 For ease of coding, Dirichelet boundary conditions for hydronamic pressure were applied.

4.9 The subject of the multigrid method took up a large portion of the time applied to the investigation. Not only did Multigrid prove simple to code, but in overall C.P.U. time for the calculation was consistently lower than that required for the equivalent S.O.R. algorithm.

4.10 A flow chart for the code is shown in fig. 1. It shows a final check on dynamic equilibrium made after the calculation of the hydrodynamic pressure field. This simply consists of summing forces action on the body and equating them to its mass multiplied by its acceleration.

5 RESULTS OF AND LESSONS LEARNED FROM SIMULATION PROGRAM (SPLASH 2)

5.1 The program SPLASH 2 was programmed using FORTRAN 77 and run on an I.C.L. 2988 computer.

5.2 The example used in this paper is that of the impact of a circular cylinder of 1.0 meters radius moving with a range of impact velocities from 0.50 to 5.0 meters per second. Mesh size was 0.1 meters down to 0.01 meters.

5.3 A number of lessons were learned concerning the orders of magnitude of the various elements of the calculation which gave some insight into the problem.

(a) The DIV(U) term in the mass conservation equation (2) was predominant in the calculation of change in density.

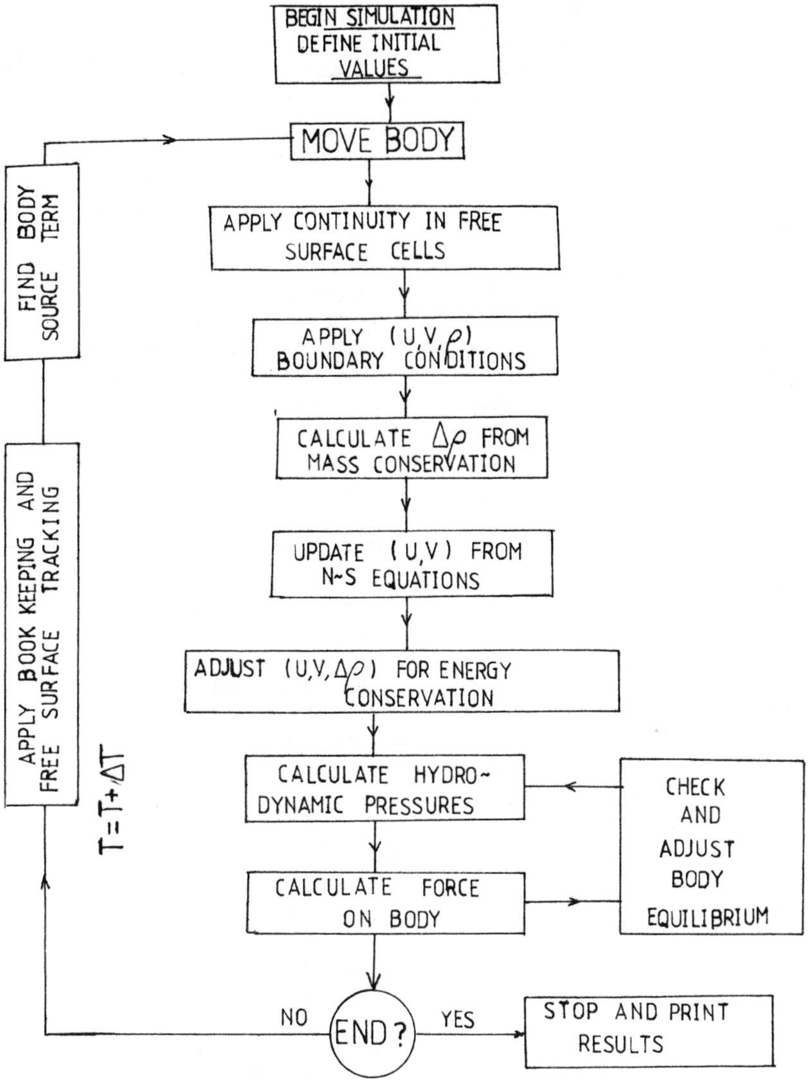

FIG 1 FLOW CHART

(b) Since the change in density, $\Delta\rho$, was calculated prior to the marching of (U,V), later values of 'acoustic' pressure component could be used in this calculation. Doing so led to increased accuracy, in the form of a lower energy defect in the variational technique.

(c) THE ENERGY METHOD

(1) Non-linear terms mentioned in technique A were always small. Since these caused problems in the numerical

evaluation of Mueller's method for the roots of the third order polynomial, the terms were dropped from the simulation until they attained a value equal a five or ten per cent of the fluid Kinetic energy (itself a small quantity).

(2) Method B, using a 'linear' approach proved more accurate as it did not rely on the calculation of body pressures to find the body velocity. If fitted easily within the framework given by the flow chart (fig. 1).

(3) When the correction factor is greater than unity, a general increase in fluid energy results. As the flow resumes its compressible behaviour, it is desirable that this factor be less than unity so as to reduce the value of DIV(U), which should tend to zero. This placed a constraint upon the time step such that the total fluid Kinetic energy plus compression energy should be greater than the change in the total excess energy Q.

(d) The Volume of Fluid or in this case mass of Fluid fluxing technique proved adequate, though the free-surface shapes provided by this simulation are still sqaiting verification by experiment.

(e) Following on from d, the total fluid potential energy was by far the largest single component ($O(10^5)$ KJ), and since only its change in value per time step was of importance, a normalisation process was developed such that small errors would not accumulate during the integration process used in the energy technique.

(f) Two Overall integration methods were used for the energy technique:

(1) Algebraic summation of cell values of Energy
(2) Simpson's first and second rules.

There was no significant difference in the results of using either (1) or (2). Whilst actual values of Energy components differed very slightly, their ratios removed nearly the same. Method (1) was adopted for reasons of computational speed.

5.4 The results of the pressure time histories for the cylinder impact problem agree well with experimental results obtained at the Wolfson Marine Craft unit.

When the simulation was left free to run, the maximum predicted pressure given by expt as

$$P_{max} = \frac{1}{2} K \rho V_b^2 , \qquad K \simeq 5.15$$

was reached with a sound speed slightly less than that for water.

The time history for the 'acoustic' phase showed good agreement. Hence, some more work needs to be done to properly evaluate the benefit of artificially controlling the compressibility of the flow.

The method of using the dynamic equilibrium of the body to check the hydrodynamic pressure contribution proved accurate in this case, giving the same type of pressure/time histories for both the simulation and the experiment.

6 CONCLUSIONS

As presented, the SPLASH2 code provides a flexible method of programming the water entry problem. However, more experimental data is required for the program to be tested against. The model needs to be given problems which extend its capability in order for all possible modes of fluid behaviour may be examined and checked for credibility.

One remaining task is to examine the development of the trapped air layer by simulating the compressible flow of the air between body and free surface.

7 ADKNOWLEDGMENTS

This work is funded by the Science and Engineering Research Council of Great Britain, under the auspices of a grant to the University of Glasgow, Department of Naval Architecture and Ocean Engineering.

8 REFERENCES

1 LEWISON, GRANT and MACLEAN W.M. (1966). On the cushioning of water impact by entrapped air. Jounal of Ship Research 1966.

2 VON KARMAN, T. (1929) The Impact of Seaplane Floats during Landing.

3 WAGNER, H (1932) Uber Stross nud Gleitvorgange und der Oberflache van Flussigkeiten, 2 angaw. Math Mech, Band 12 Helft 4, pp 192-215.

4 FABULA, A.G. (1957) Ellipse fitting approximation of two dimensional normal symmetric impact of rigid bodies in water. Fifth Midwestern conf. on Fluid Mechnanics, Univ of Mich. Press, Ann Arbor, Mich, pp. 299-315.

5 BISPLOINGHOFF, R.L. and DOHERTY, C.S. (1952). Some studies of the impact of Vee wedyes on a water surface, J. Franklin Inst, Vol 253 , pp 547-561.

6 OGILVIE,T.V. (1963) Compressibility effects in ship slamming, Shiffstechnik, Vol 10, no. 53, pp 147-154.

7 KOEHLER, B.R. KETTLEBOROUGH, C.F. Hydrodynamic Impact of a falling body upon a viscous Incompressible Fluid. Journal of Ship Research, Vol 21, No. 3, 1977, pp 165-181.

8 JOHNSON, R.S. The Effect of Air Compressibility in the Impact of a Flat Body on a Free Surface. Report No. NA-66-8 University of California, Berkeley 1966

9 NICHOLS, B.D., HIRT, C.W. NON-LINEAR HYDRODYNAMIC FORCES ON FLOATING BODIES. Second International Conference on Numerical Ship hydrodynamics . Sept. 1977, pp 382-394.

10 HARLOW, F.H. The Marker and Cell Method. V.K.I Lecture series no. 44, 1972.

11 HIRT, C.W. NICHOLS B.D. The Volume of Fluid Method for dynamic free boundaries. Journal of Computational Physics 39, pp 201-225, 1981.

12 NICHOLS, B.D. HIRT, C.W. IMPROVED FREE SURFACE BOUNDARY CONDITIONS FOR NUMERICAL INCOMPRESSIBLE FLOW CALCULATIONS Journal of Computational physics, 8, pp 434-448, 1971.

13 ORLANSKI, I. A Simple Boundary Condition for Unbounded Hyperbolic Flows. Journal of Computational Physics, 21, pp 251-269, 1976.

14 SASAKI, Y.K. Variational Design of Finite-Difference Schemes for Initial Value Problems with an Integral Invarient. Jour. Comp. Phys, 21, pp 270-278, 1976.

15 ALAN JENNINGS, Matrix Computation for Engineers and Scientists. John Wiley, 1980.

16 BRANDT, A. Multi-Lebel Adaptive Solutions to Boundary Value Problems. Math. Comp.31, pp 333-390, 1977.

17 BRANDT, A. DENDY J.E. RUPPEL, H. The Multigrid Method for Semi-Implicit Hydrodynamic Codes. Journal of Comp Phys. 34, pp 348-370, 1980.

18 HACKBUSCH, W. Trottenberg, Multigrid Methods, Proceedings, Koln-Porz, 1981.

19 CHORIN, A.J. A Numerical Method for Solving Incompressible Viscous Flow Problems. Journal of Comp Phys 2, pp 12-26, 1967.

SECTION 7

TURBO MACHINERY

FINITE ELEMENT ANALYSIS OF A SPLIT-FLOW PARTICLE SEPARATOR

D.S. Breitman (i), E.G. Dueck (ii), W.G. Habashi (iii)

ABSTRACT

This paper presents an analytical design method for inertial particle separators required for aviation gas turbines applications to helicopters. The solution of the flow field inside the separator is based on a Finite Element code for the radial equilibrium equation. The aerodynamic forces calculated are then used to predict the trajectories of solid particles of various sizes. Several separator designs are demonstrated and a very efficient final configuration is determined.

1.0 INTRODUCTION

Helicopters, especially military ones, make numerous take-off and landings from unprepared areas where they are exposed to ingestion of large amounts of sand and dust. This causes erosion of engine components, appreciably reducing the engine's life expectancy. Early efforts to protect the engine involved the use of filters located in the engine inlet. These filters provide excellent separation, are light-weight, and have a low initial cost. However, periodic cleaning and replacement is necessary to maintain tolerable pressure losses.

These considerations encouraged development of inertial

(i) Senior Aerodynamicist
Pratt and Whitney Canada, Mississauga, Ontario, Canada.

(ii) Chief, General Aerodynamics A
Pratt and Whitney Canada, Mississauga, Ontario, Canada.

(iii) Associate Professor, Concordia University, Montreal, Quebec, Canada; and Aerodynamics Consultant, Pratt and Whitney Canada.

particle separators which can be wholly integrated with the engine, are self-cleaning, and have a fixed pressure drop. To create a large inertia field for exceptional separation, swirl vanes have to be employed in the inlet. The inclusion of swirl induces high performance penalties, and the vanes require de-icing.

Thus, simplicity, performance and ruggedness are compelling reasons to develop a swirl-free, low-loss inertial type particle separator which still provides effective engine protection.

2.0 ANALYSIS

The analysis objective for a specific design concept is to predict the separation efficiency for a given sand distribution. This requires trajectory prediction for various particle sizes as they are acted upon by the separator aerodynamic forces. In turn, this requires an aerodynamic flow field solution inside the separator.

The analytical model presented here consists of two main codes; flow field computation by a finite element through-flow program and a trajectory analysis by a particle trajectory program.

2.1 Flow Field Numerical Computation

2.1.1 Governing equations and boundary conditions

The flow is governed by the radial equilibrium equation: [1,2].

$$C_z \left(\frac{\partial C_r}{\partial z} - \frac{\partial C_z}{\partial r}\right) = -\left(\frac{V_\theta}{r}\ \frac{\partial (rC_\theta)}{\partial r} + \frac{T\partial s}{\partial r} - \frac{\partial I}{\partial r}\right) \tag{1}$$

where C,V = absolute and relative velocity, respectively

T,I,S = temperature, rothalpy and entropy, respectively

By defining:

$$C_z = \frac{\dot{m}}{2\pi\rho rb}\ \frac{\partial\Psi}{\partial r} \tag{2}$$

$$C_r = -\frac{\dot{m}}{2\pi\rho rb}\ \frac{\partial\Psi}{\partial z} \tag{3}$$

equation (1) can be rewritten as:

$$\frac{\partial}{\partial r}\left(\frac{1}{\rho r b}\,\frac{\partial\Psi}{\partial r}\right) + \frac{\partial}{\partial z}\left(\frac{1}{\rho r b}\,\frac{\partial\Psi}{\partial z}\right)$$

$$= -\frac{2\pi}{\dot{m}C_z}\left(\frac{V_\theta}{r}\,\frac{\partial(rC_\theta)}{\partial r} + \frac{T\partial s}{\partial r} - \frac{\partial I}{\partial r}\right) \qquad (4)$$

where b = blockage factor and boundary conditions (Fig 1) are:

on hub,	BG	: $\Psi = 0$	
on shroud,	AC	: $\Psi = 1$	
on separator,	DEF	: $\Psi = \Psi_0$	; a constant
at inlet,	AB	: $\partial\Psi/\partial n = 0$	
at exit,	CD,FG	: $\partial\Psi/\partial n = 0$	(5)

n = outward normal to surface.

FINITE ELEMENT DISCRETIZATION

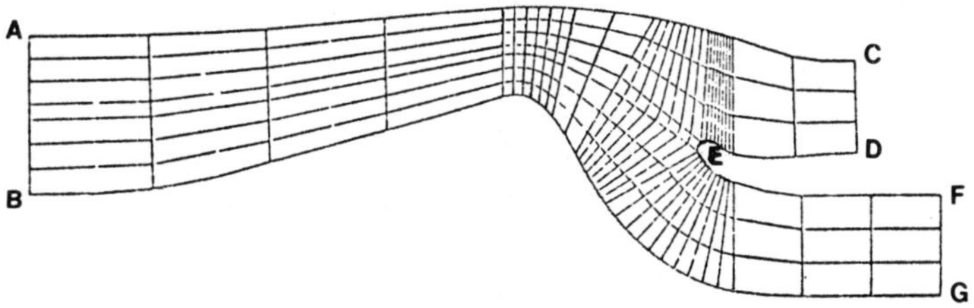

Fig. 1

2.1.2 Finite element discretization and solution procedure

Curvilinear eight-node isoparametric elements are used, where both the function, Ψ, and geometry have parabolic representation. In each element

$$\Psi = \sum_{i=1}^{8} N_i\ (\xi,\eta)\Psi_i\ ;\quad z = \sum_{i=1}^{8} N_i\ (\xi,\eta)z_i\ ; r = \sum_{i=1}^{8} N_i\ (\xi,\eta)\ r_i \qquad (6)$$

N_i being the parabolic shape functions written in terms of the local undistorted element coordinates ξ and η .

Applying the Galerkin weighted residual method to (4) results in the matrix equation

$$[K]\{\Psi\} = \{F\} \tag{7}$$

where $$k_{ij} = \iint \frac{1}{\rho r b} \left(\frac{\partial N_i}{\partial r} \frac{\partial N_j}{\partial r} + \frac{\partial N_i}{\partial z} \frac{\partial N_j}{\partial z} \right) drdz \tag{8}$$

$$F_i = \iint \frac{2\pi N_i}{\dot{m} C_z} \left(\frac{V_\theta}{r} \frac{\partial (rC_\theta)}{\partial r} + \frac{T \partial s}{\partial r} - \frac{\partial I}{\partial r} \right) drdz \tag{9}$$

The numerical integration of equations (8) and (9) is carried out using a 3x3 Gaussian integration procedure.

The mesh generation of the current split-flow is totally automated and an illustrative grid is shown in Figure 1.

The nonlinear algebraic system of equations represented by equation (7) is recast in the form:

$$[A]\,[\delta\Psi] = -\alpha R \tag{10}$$

where α = an under/over relaxation parameter
$\delta\Psi$ = change in the solution vector from one iteration to the next
A = symmetric, banded, matrix operator approximating the physics of the flow. Normally taken as the Laplacian operator scaled by the "nodal" densities
R = residual, defined as $[K][\Psi] - F$

Matrix [A] is decomposed at the first iteration by the Cholesky decomposition and only the backward substitution step is needed, at each iteration, for the solution of (10). All results shown in the paper are carried out to L-2 residual for R of less than 10^{-5}, where

$$R_{L-2} = \sum_{\text{nodes}} R_i^2 \leqslant 10^{-5} \tag{11}$$

2.2 Particle Trajectories

2.2.1 Governing equations

Particle motion within a flow field is determined by a resolution of forces. These include the aerodynamic drag, buoyancy, gravity, and kinematic forces. The buoyancy and gravitational forces are normally assumed to be negligible.

Written in cylindrical polar coordinates, the equations of motion are:

$$m\ddot{z} + F_{Dz} - (mg - F_B)\sin\phi = 0 \tag{12}$$

$$m\ddot{r} + F_{Dr} - mr\dot{\theta}^2 + (mg\cos\phi)\sin\theta - (F_B\cos\phi)\sin\theta = 0 \tag{13}$$

$$2m\dot{r}\dot{\theta} + mr\ddot{\theta} + F_{D\theta} + (mg\cos\phi)\cos\theta - (F_B\cos\phi)\cos\theta = 0 \tag{14}$$

where r, θ and z define the particle location. The corresponding drag forces, F_{Dr}, $F_{D\theta}$, and F_{Dz} are assumed positive if the particle velocity in these directions is greater than the air velocity. The angle ϕ gives the inclination of engine axis z from the horizontal. F_B is the buoyancy force. The complete force diagram is shown in Figure 2.

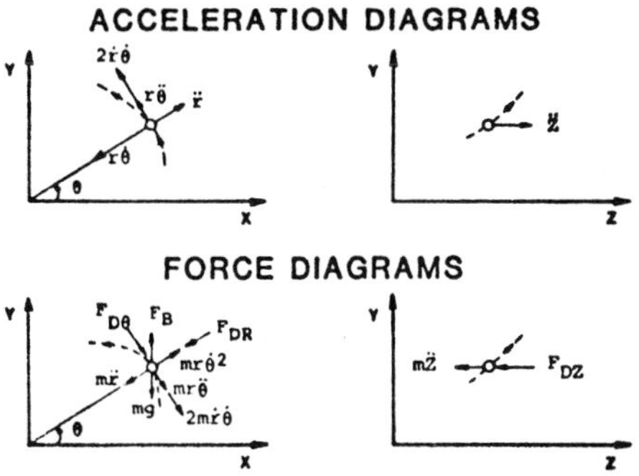

Fig. 2

2.2.2 Aerodynamic drag

The particle drag forces, in terms of a drag coefficient, C_D, are:

$$F_D = C_D \cdot \tfrac{1}{2} \cdot \frac{\rho_a}{g} V \; A_p \tag{15}$$

where V = the particle relative velocity
ρ_a = density of air
A_p = particle cross-sectional area
g = gravitational acceleration

Many empirical formulations for the drag coefficient of a spherical particle have been derived from experimental data. The formulations, in the form of a correction to Stokes' drag law, are valid for Reynolds' numbers corresponding to the entire flow region. The following [3] is used in the present analysis:

$$C_D = \frac{24}{Re} \; \frac{(1 + 0.15\,Re^{0.687})\,(1 + e^{-(0.427/M^{4.63}) - (3.0/Re^{0.88})})}{1 + (M/Re)\,(3.82 + 1.28\,e^{-1.25Re/M})} \tag{16}$$

2.2.3 Shape factor

Equation (16) has been derived for spherically shaped particles. Irregular shaped, non-spherical particles will, for an equivalent cross-sectional area, as in a flat plate, result in a drag equivalent to that of the sphere but with a lower particle mass. The larger mass of the spherical model gives the particle excess momentum resulting in over-prediction of the separation efficiency. General Electric [4] recognized this and empirically derived a "shape factor" based on test results from their R&D program. This shape factor is:

$$S = \text{true volume of particle}/d^3 \tag{17}$$

d is the diameter of the maximum cross-sectional area.

For an exact sphere, $S = \pi/6$. A shape factor of .26 to .28 was recommended for random particles [4]. A shape factor of .2618 ($\pi/12$) is used in this analysis. This number was derived from a trial and error matching of the code prediction with experimental results.

2.2.4 Bounce characteristics

The collision/reflection of a particle from a hard duct wall is, to some extent, a random process because of the irregular shape of the particle. Considerable effort [5-7] has been expended in deriving empirical expressions to model the collision/reflection of typical sand particles. The following expressions [8,9] predict reflected normal and tangential velocities as functions of incidence angle and incident normal and tangential velocities.

$$E_T = \frac{V_{T2}}{V_{T1}} = 1.0 - 2.12\beta_1 + 3.0775\beta_1{}^2 - 1.1\beta_1{}^3 \tag{18}$$

$$E_N = \frac{V_{N2}}{V_{N1}} = 1.0 - 0.4159\beta_1 - 0.4994\beta_1{}^2 + 0.292\beta_1{}^3 \tag{19}$$

where β_1 = the impingement angle in radians
V_T = component of particle velocity tangential to wall at point of impact
V_N = component of particle velocity normal to wall at point of impact
subscripts 1,2 represent conditions before and after impact, respectively

Recently acquired test data revealed a discrepancy with analysis for large particle sizes, where bounce is the predominant separating mechanism. The much simpler empirical expressions given below provide better agreement with test results and were incorporated into the analysis.

$$E_T = \frac{V_{T2}}{V_{T1}} = 1.0 \tag{20}$$

$$E_N = \frac{V_{N2}}{V_{N1}} = .85 \tag{21}$$

Two sample trajectory plots are presented. Figure 3 shows the trajectory plots for 5μ particles. Of particular interest is that the code shows some of the particles actually being carried back around the splitter into the core; a result of the stagnation point falling on top of the splitter.

Figure 4 shows the trajectory plots for 100μ particles. Here, where the particle mass, and therefore momentum, is much larger, a separation efficiency of 100% is predicted.

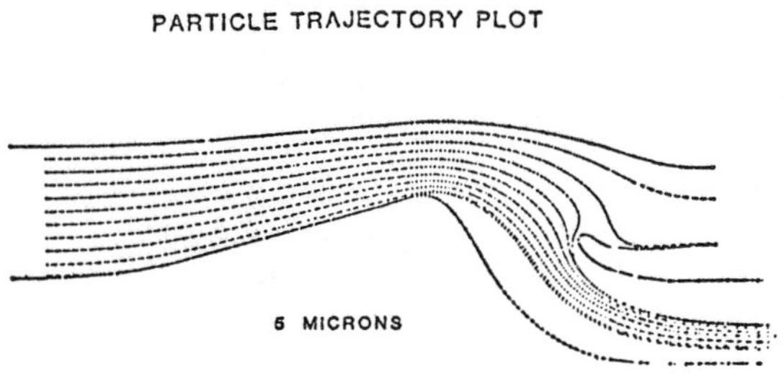

Fig. 3

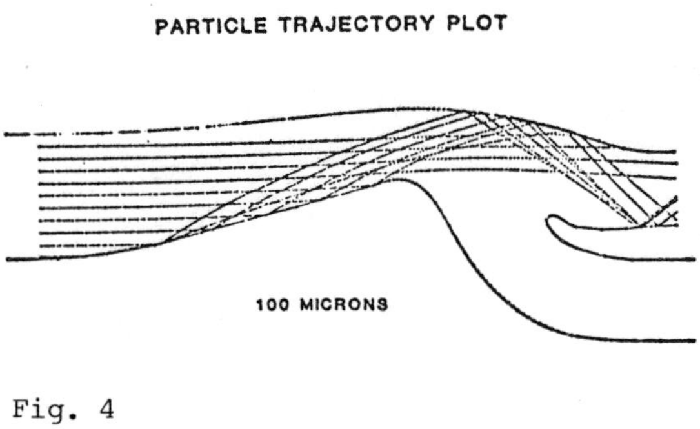

Fig. 4

3.0 GENERAL DESCRIPTION OF DESIGN

An axisymmetric inertial separator is designed such that the entering airflow is turned sharply inward to the engine centreline. The larger inertia of sand and dust propels them to the outer periphery where they are collected and dumped overboard. Large particles, which do not follow the airflow streamlines, hit the inner hub wall, which is angled to deflect the particles towards the outer periphery and hence out the bypass.

3.1 Use of Finite Element Through-Flow Analysis Code

This rather unique geometry results in an unusual intake to the engine core. Because of the splitter proximity and low-bypass ratio, there is a high positive splitter incidence inducing the stagnation point to fall on top of the lip. As the flow reverses and follows the sharp turn of the splitter lip, it undergoes rapid acceleration followed by abrupt deceleration. This may cause the flow to stall along the underside of the splitter lip as it enters the core duct. With the aid of the finite element through flow analysis program, several splitter lip designs were investigated with the objective being to design a splitter with correct leading edge incidence. Streamline curvature routines or programs requiring separate treatment of (or iterative treatment between) core and bypass are not able to correctly model this region [9,10]. The unique capability of this FEM program to model this region correctly and consistently in one pass lends immeasurably to the predictive capability of the total scheme.

Three splitter lip designs were selected for testing. Figure 5 shows a typical FEM streamline prediction. Figure 6 shows the predicted splitter lip Mach number along the lower surface of each design.

FINITE ELEMENT PROGRAM STREAMLINE PREDICTION

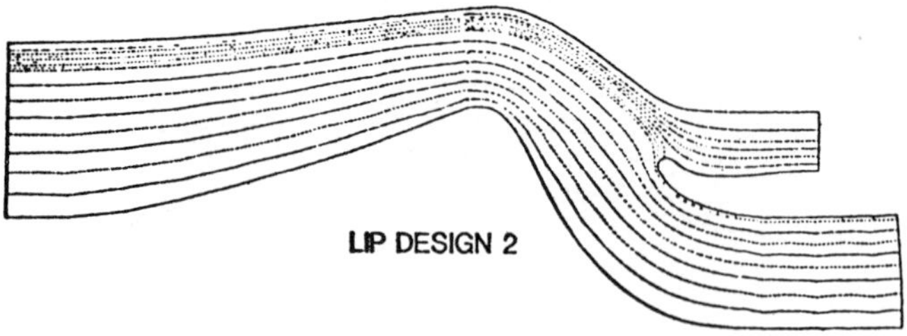

Fig. 5

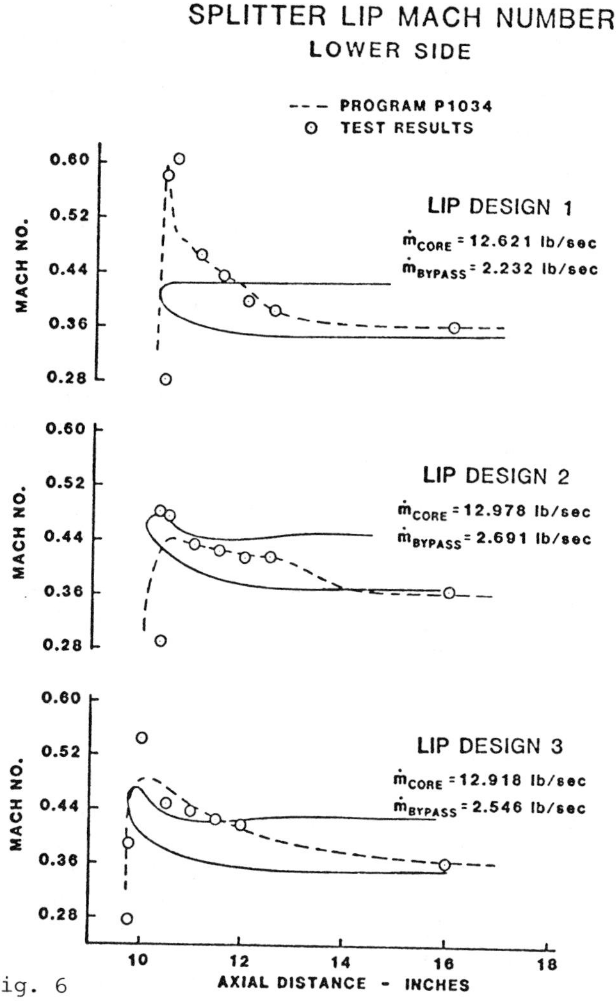

Fig. 6

Lip 1 is simply a $\sqrt{2}$:1 flat ellipse. If the stagnation point were to fall on the leading edge of the ellipse, the flow would experience constantly decreasing radii of curvature as it negotiated the turn into the core duct. However, since the stagnation point is off-centre, the flow will experience a large acceleration back as it passes over the peak curvature combined with area convergence as flow is drawn into the engine core, resulting in very high velocities throughout the turn. This may cause the flow to separate on the underside of the splitter lip, as the flow diffuses from this Mach number peak.

Lip 2 is a 2:1 ellipse, inclined at an angle of 47°, so as to have the stagnation point fall on the leading edge of the ellipse. Thus diffusion along the lower side is greatly reduced in comparison to design 1.

Lip 3 is a 2:1 ellipse inclined at an angle of 80° and extended .30 inches further upstream than lip 2. With this increased inclination, as the lip is moved upstream, the stagnation point can still be induced to fall on the lip leading edge. The higher Mach numbers are due in part to the smaller flow cross-section area into the core duct. The purpose of this option is to further improve the separator "step/gap" ratio, hence improve particle separation efficiency whilst exacting a small compromise on inlet pressure recovery.

4.0 TEST PROGRAM

Testing of the particle separator was conducted at the Pratt & Whitney Research and Development Centre located in West Palm Beach, Florida. The separator hardware was extensively instrumented to evaluate the aerodynamic performance and to study the flow behaviour of critical areas.

Sand ingestion tests were conducted using a variety of sand types and mixtures. In particular, two sand types were extensively used; military C-spec sand comprised of 0-1,000μ particles, and AC coarse test dust, comprised of 0-200μ particles.

5.0 RESULTS AND COMMENTS

Figure 6 also shows the splitter lip Mach numbers of the experimental results. Prediction discrepancy in the peak Mach numbers for lips 2 and 3 can be attributed to viscous flow effects, since stall is observed in the area of large diffusion along the shroud just past the hub apex. Since the finite element solution is inviscid, the large separation zone located on the shroud must be simulated by using displaced geometry. Any inaccuracy in the modelling of this stalled region would of course lead to inaccurate predictions. This displaced geometry was defined by an inverse integral boundary layer routine and correlated with experimental test measurement. This, and follow on work, will be reported in a future paper. In spite of this shortcoming, the analysis was successful in that both lips2 and 3 demonstrated better than a 10% improvement in aerodynamic performance (i.e. engine inlet total pressure drop) over lip 1.

To predict separation efficiency for a sand mixture, the analysis must be performed for the complete range of particle sizes in the sand mixture. The individual efficiencies are then integrated for the size distribution of a specified mixture. Figure 7 shows the predicted separation efficiencies correlated with particle size for one of the configuration tests.

Fine dust or other particles follow the flow field until they negotiate the sharp downward turn towards the engine core. When exposed to the inertial field created by the turn, the particle's momentum causes its path to diverge from the flow and hence on towards the bypass duct. For extremely fine particles (<10μ dia.) the inertial forces are small relative to drag forces, resulting in the particle following the flow field into the engine core.

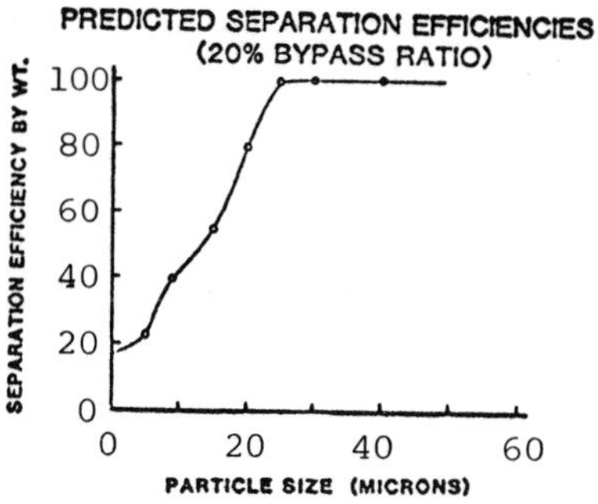

Fig. 7

For larger particles the aerodynamic drag force is small compared to the inertia forces, resulting in the particle trajectory being almost independent of the flow field. These particles collide with the duct walls which are designed to direct the particles into the bypass duct. As can be seen in Figure 7 for particles greater than 25 microns, the separation efficiency is predicted to be 100%.

Figure 8 compares predicted efficiencies with test results from a recent test program. Good agreement exists between the theory and test results. In the past, modelling of a particle separator would serve as a starting point, with hardware testing providing actual sand separation efficiencies. Although bench testing will always be a necessary part of any development program, with the new advances in numerical methods, theoretical predictions are becoming increasingly more accurate and reliable.

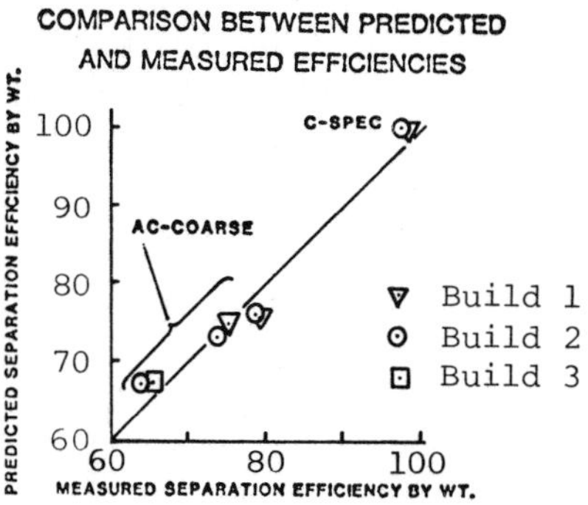

Fig. 8

ACKNOWLEDGEMENT

The authors wish to acknowledge the technical assistance of Messrs. R.A. Thompson, S. Zuquim of Pratt and Whitney, Canada and Mr. W. Haynie of Pratt & Whitney, Florida.

REFERENCES

1. HABASHI, W.G. - Numerical Methods of Turbomachinery, Chapter 8 in Recent Advances in Numerical Methods in Fluids, Vol. 1, C. Taylor, K. Morgan (Eds.), Pineridge Press (U.K), 1980.

2. HABASHI, W.G. and YOUNGSON, G.G. - A Transonic Quasi-3D Analysis for Gas Turbine Engines Including Split Flow Capability for Turbofans. Int. Journal for Numerical Method in Fluids, Vol. 3, No. 1, pp. 1-22, 1983.

3. CARLSON, D.J. and HOGLUND, R.F. - Particle Drag and Heat Transfer in Rocket Nozzles, AIAA Journal, Vol. 2, No. 11, 1964.

4. DUFFY, R.J., et al - Integral Engine Inlet Particle Separator, Volume II Design Guide, General Electric Company, USAAMRDL-TR-75-31B, Aug. 75.

5. WAKEMAN, T. and TABAKOFF, W. - Measured Particle Rebound Characteristics Useful for Erosion Prediction, ASME Paper, 82-GT-170.

6. GRANT, G. and TABAKOFF, W. - Erosion Prediction in Turbomachinery Resulting from Environmental Solid Particles, Journal of Aircraft, Vol. 12, No. 5, May 1975.

7. BEACHER, B., TABAKOFF, W. and HAMED, A. - Improved Particle Trajectory Calculations Through Turbomachinery Affected by Coal Ash Particles, Journal of Engineering for Power, Vol. 104, No. 1, Jan. 82.

8. TABAKOFF, W. and HAMED, A. - Aerodynamic Effects on Erosion in Turbomachinery, JSME and ASME Paper 70, Proceedings of the 1977 Joint Gas Turbine Congress, Tokyo, Japan, May 1977, pp. 574-581.

9. HAMED, A. - Particle Dynamics of Inlet Flow Fields with Swirling Vanes, Journal of Aircraft, Vol. 19, No. 9, Sept. 1982.

10. FABIAN, J.M. and OATES, G.G. - Analysis of Flows Within Particle Separators, ASME Paper 77-WA/FE-21.

THE EULER-QUICK CODE

B.P. Leonard

City University of New York

1. SUMMARY

For transonic flow about airfoils and turbine cascades, the compressible Euler equations are appropriate. Quasi-one-dimensional Laval nozzle flow forms an excellent test problem because of the known exact solutions. The EULER-QUICK code consists of writing channel-flow transport equations for density, velocity, and energy, with algebraic relations for pressure and temperature. The convection terms are nominally discretized using the QUICK algorithm; this produces nearly exact results in the isentropic regions. However, quadratic interpolation is not appropriate for the sudden jumps occurring at the imbedded shock wave. Near this region, the algorithm automatically changes over to an Exponential-Upwinding or Linear-Extrapolation Refinement of the basic QUICK scheme. The shock is captured at the correct location with the proper strength, and numerical shock structure is sharp (about four grid points) and essentially monotonic. By contrast, first-order upwinding produces very diffuse shocks and significant errors in the isentropic regions, and second-order central methods require ad hoc stabilization techniques resulting in impaired accuracy.

2. COMPRESSIBLE EULER EQUATIONS FOR QUASI-ONE-DIMENSIONAL FLOW

For quasi-one-dimensional flow of an ideal gas in a channel of varying cross-section, the frictionless (Euler) equations take the following form in nondimensional variables. The cross-sectional area A(x) is assumed given; e.g., as a converging-diverging Laval nozzle.

Density:

$$\frac{\partial \rho}{\partial t} = -\frac{1}{A}\frac{\partial(\rho u A)}{\partial x} \qquad (1)$$

Pressure:

$$p = \rho T \qquad (2)$$

Velocity:

$$\frac{\partial(\rho u)}{\partial t} = -\frac{1}{A}\frac{\partial(\rho u^2 A)}{\partial x} - \frac{1}{\gamma}\frac{\partial p}{\partial x} \qquad (3)$$

Energy:

$$\frac{\partial(\rho \mathcal{E})}{\partial t} = -\frac{1}{A}\frac{\partial\{\rho u[\mathcal{E} + (\gamma-1)T]A\}}{\partial x} \qquad (4)$$

Temperature:

$$T = \mathcal{E} - \frac{\gamma(\gamma-1)}{2}u^2 \qquad (5)$$

The following nondimensionalization has been used, based on upstream (reservoir) conditions and a reference channel length:

$$\left\{\begin{array}{llll} \rho = \bar{\rho}/\bar{\rho}_0; & u = \bar{u}/c_0 & = \bar{u}/\sqrt{\gamma R\bar{T}_0}; & \\ T = \bar{T}/\bar{T}_0; & \mathcal{E} = \bar{\mathcal{E}}/c_V\bar{T}_0 & = \bar{\mathcal{E}}(\gamma-1)/R\bar{T}_0; & (6) \\ p = \bar{p}/(\bar{\rho}_0 R\bar{T}_0); & x = \bar{x}/L; & t = \bar{t}/(L/c_0). & \end{array}\right.$$

Equations (1)-(5) are the governing coupled equations for the primary transport variables, ρ, u, and $\mathcal{E}$, and the thermodynamic variables, p and T. Auxiliary equations are also used for the Mach number, $M = u/\sqrt{T}$ and the mass flux, $\dot{m} = \rho uA$.

Numerical boundary conditions correspond to the physical conditions of reservoir values upstream and a specified downstream pressure. No other downstream variables are specified. The governing equations are written in explicit discretized form and stepped in time until a steady state has been reached. Although transient behaviour is correctly modelled, it is the steady state which is of primary interest here; the time-varying form of the equations is used simply as a convenient explicit iteration technique.

As is well known, for most pressure ratios of interest Laval nozzle flow involves an isentropic subsonic-supersonic transition (with sonic conditions at the throat), then a normal shock wave in the downstream section followed by a further subsonic isentropic expansion (at a higher value of entropy or lower stagnation pressure). Given the area variation A(x) and the overall pressure ratio, it is a staight-forward matter to compute the steady behaviour of all the variables of interest using theoretical algebraic formulas for isentropic flow and shock jump conditions. Thus a theoretically 'exact' solution is available for any pressure ratio. For this reason the Laval nozzle test problem is an excellent tool for measuring the performance of compressible Euler equation codes. In particular, the numerical method should adequately model the large smoothly varying isentropic regions as well as 'capturing' the shock at the correct location with the proper strength and a reasonable numerical shock structure--ideally, sharp and monotonic.

The QUICK discretization is quite accurate in modelling the smooth isentropic regions. However, quadratic interpolation for control-volume cell face values of the convected variable is not the most appropriate form in regions where sudden changes in value occur, as across a shock wave. In such regions, alternate interpolation schemes can be devised which nominally use the same three node values as the QUICK method. If the three adjacent node values are monotonic, an exponential upwind interpolation generates an appropriate face value. If the values are not monotonic, a form of linear extrapolation either from the two upwind nodes (second-order upwinding) or from the adjacent upwind node (first-order upwinding) represents a rational extension of exponential upwinding into the non-monotonic regime. This Exponential-Upwinding or Linear-Extrapolation Refinement of the basic QUICK scheme is the one used to simulate the compressible Euler equations as will now be described.

EXPONENTIAL-UPWINDING OR LINEAR-EXTRAPOLATION REFINEMENT OF QUICK

To understand the basic ideas of the EULER-QUICK method, refer to Figure 1, which shows quadratic upstream interpolation through three monotonic node vales to generate the right face value and gradient of the control-volume cell. It is constructive to define a normalized ϕ value given by

$$\tilde{\phi} = \frac{\phi - \phi_L}{\phi_R - \phi_L} \tag{7}$$

provided of course that $\phi_R \neq \phi_L$. Note that $\tilde{\phi}_R = 1$ and $\tilde{\phi}_L = 0$. In particular,

$$\tilde{\phi}_C = \frac{\phi_C - \phi_L}{\phi_R - \phi_L} \tag{8}$$

Then (for $u_r > 0$) the interpolated right-face value for the QUICK algorithm becomes, in terms of normalized variables

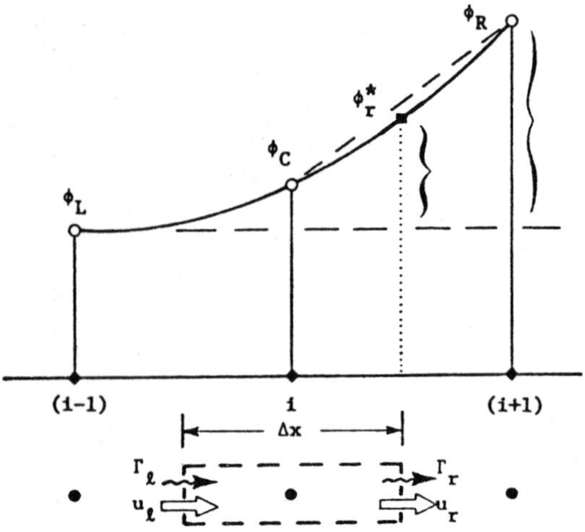

Figure 1. Quadratic upstream interpolation.

$$\text{QUICK:} \quad \tilde{\phi}_R = \frac{\phi_r^* - \phi_L}{\phi_R - \phi_L} = \frac{3}{8} + \frac{3}{4}\tilde{\phi}_C \quad (u_r > 0) \tag{9}$$

For comparison, second-order central differencing (given by linear interpolation between ϕ_C and ϕ_R) becomes

$$\text{SECOND-ORDER CENTRAL:} \quad \tilde{\phi}_r = \frac{1}{2} + \frac{1}{2}\tilde{\phi}_C \tag{10}$$

and first-order upwinding (given by the linear extrapolation, $\phi_r^* = \phi_C$, for $u_r > 0$) becomes

$$\text{FIRST-ORDER UPWINDING:} \quad \tilde{\phi}_r = \tilde{\phi}_C \quad (u_r > 0) \tag{11}$$

Finally, second-order upwinding (linear extrapolation from ϕ_L through ϕ_C to give ϕ_r^*, for $u_r > 0$) becomes

$$\text{SECOND-ORDER UPWINDING:} \quad \tilde{\phi}_r = \frac{3}{2}\tilde{\phi}_C \quad (u_r > 0) \tag{12}$$

Each of the above forms of $\tilde{\phi}_r$ is shown as a function of $\tilde{\phi}_C$ in Figure 2.

Consider now an entirely different type of interpolation for ϕ_r^* in the monotonic range, $0 \leq \tilde{\phi}_C \leq 1$. Figure 3 shows the same three node values as in Figure 1, but using a three-parameter exponential interpolation of the form

$$\phi = A + B \cdot \exp(C\xi) \tag{13}$$

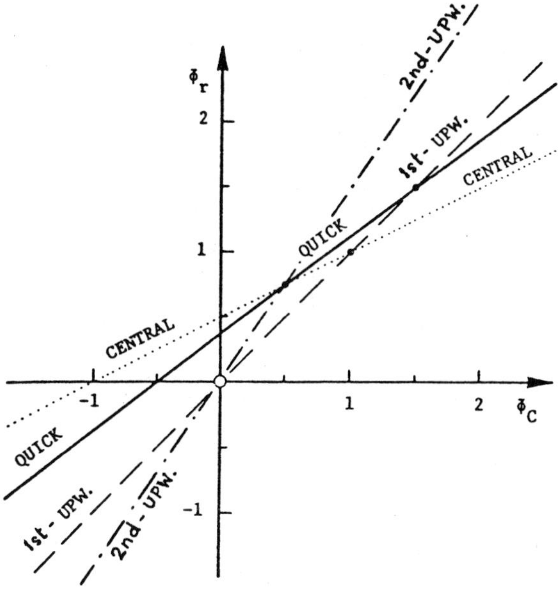

Figure 2. Normalized right-face value for various schemes.

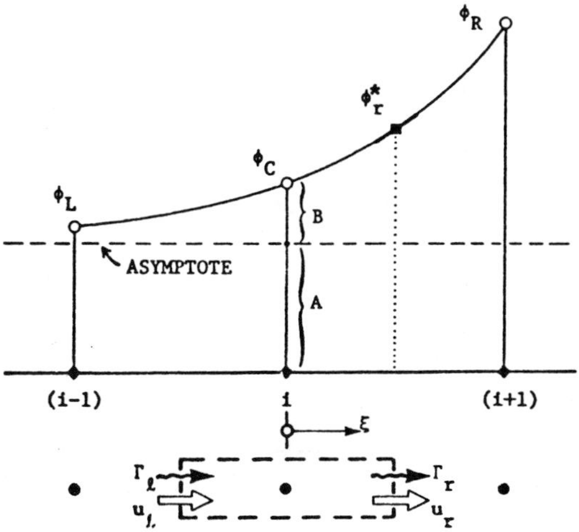

Figure 3. Exponential upwinding for right-face value.

to pass through the three values ϕ_L, ϕ_C, and ϕ_R, again assuming $u_r > 0$. It is not difficult to show that for constant grid spacing, the asymptote is given by

$$A = \frac{(\phi_R \phi_L - \phi_C^2)}{(\phi_R - 2\phi_C + \phi_L)} \tag{14}$$

and that the interpolated right-face value is

$$\phi_r^* = A \pm \sqrt{(\phi_R - A)(\phi_C - A)} \tag{15}$$

where the plus sign is used if $\tilde{\phi}_C < 0.5$ $(A < \phi_L)$ and the minus sign otherwise $(A > \phi_R)$. In terms of normalized variables, this can be written (for $u_r > 0$)

EXPONENTIAL UPWINDING:
$$\tilde{\phi}_r = \frac{\sqrt{\tilde{\phi}_C(1 - \tilde{\phi}_C)^3} - \tilde{\phi}_C^2}{(1 - 2\tilde{\phi}_C)} \tag{16}$$

which is shown in Figure 4 in relation to the QUICK line, to which the exponential curve is tangent at the point (0.5, 0.75)

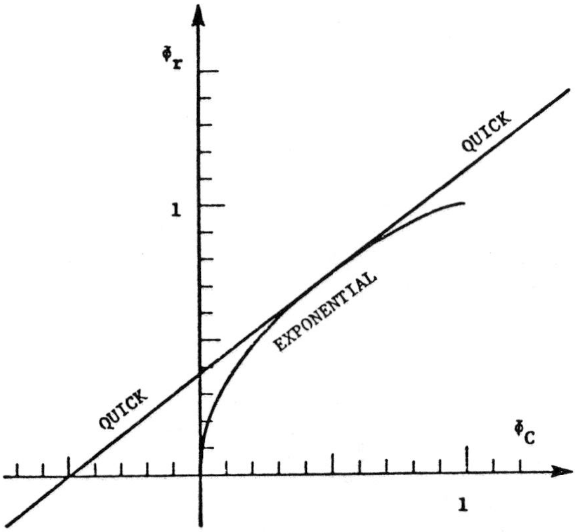

Figure 4. Normalized right-face value for exponential upwinding.

Exponential upwinding is of course not available outside the monotonic range. As seen from Figure 2, for $\tilde{\phi}_C > 1$, either second-order central or first-order upwinding would form a con tinuous extension. The EULER scheme uses first-order upwindin in the range $1 \leq \tilde{\phi}_C \leq 1.5$ and reverts to QUICK for larger values. For $\tilde{\phi}_C < 0$, either first or second-order upwinding would form a continuous extension of exponential. The EULER scheme uses second-order linear extrapolation in this regime. These choices maximize the damping of potential overshoots in regions adjacent to sudden jumps. It should be emphasized that the EULER version is used only in regions requiring sudden jumps in value; i.e., near the shock wave in the Laval nozzle problem, whereas in the smooth isentropic regions the unmodified QUICK algorithm is used. In the transition region, a weighting strategy is used to cross over between the EULER and QUICK schemes, based on the local curvature of the convected variable.

RESULTS FOR THE LAVAL NOZZLE SIMULATION

A prescribed area variation for a typical converging-diverging Laval nozzle is used as input data. The total lenght is divided into 100 computational cells with an interleaved grid so that velocity is computed at one set of nodes and density and energy (and pressure and temperature) at points midway between. For any specified pressure ratio, the theoretical solution for $p(x)$, $\rho(x)$, $T(x)$, $M(x)$, and other related variables can be calculated. In particular, for the pressure ratios of interest, this includes fixing the location and strength of a normal shock wave in the downstream section.

The solid curves in Figure 5 show the theoretical $p(x)$ for pressure ratios of 0.8, 0.6, 0.4, and 0.2. The figure also shows EULER-QUICK computed values at a number of points surrounding the shock waves. The computed values are not shown in the other (isentropic) regions because they are graphically indistinguishable from the exact results! Figures 6 and 7 show the corresponding behaviour of $u(x)$ and $T(x)$. Finally, Figures 8 and 9 show $M(x)$ and $\rho(x)$.

From these results, the following generalizations can be made regarding the EULER-QUICK scheme:

(i) In the isentropic regions the computed results are 'exact' for all practical purposes.

(ii) The code automatically 'captures' the shock wave at the correct location with the proper strength.

(iii) The numerical shock structure is relatively sharp (3 or 4 grid points) and without wiggles; and there is almost no overshoot as compared with the theoretical behaviour.

These are all highly desirable properties for a compressible Euler code. In addition, the algorithm is both conceptually and computationally straight-forward, with corresponding efficiencies in terms of computer usage. The development of a general purpose unsteady multidimensional version seems quite promising.

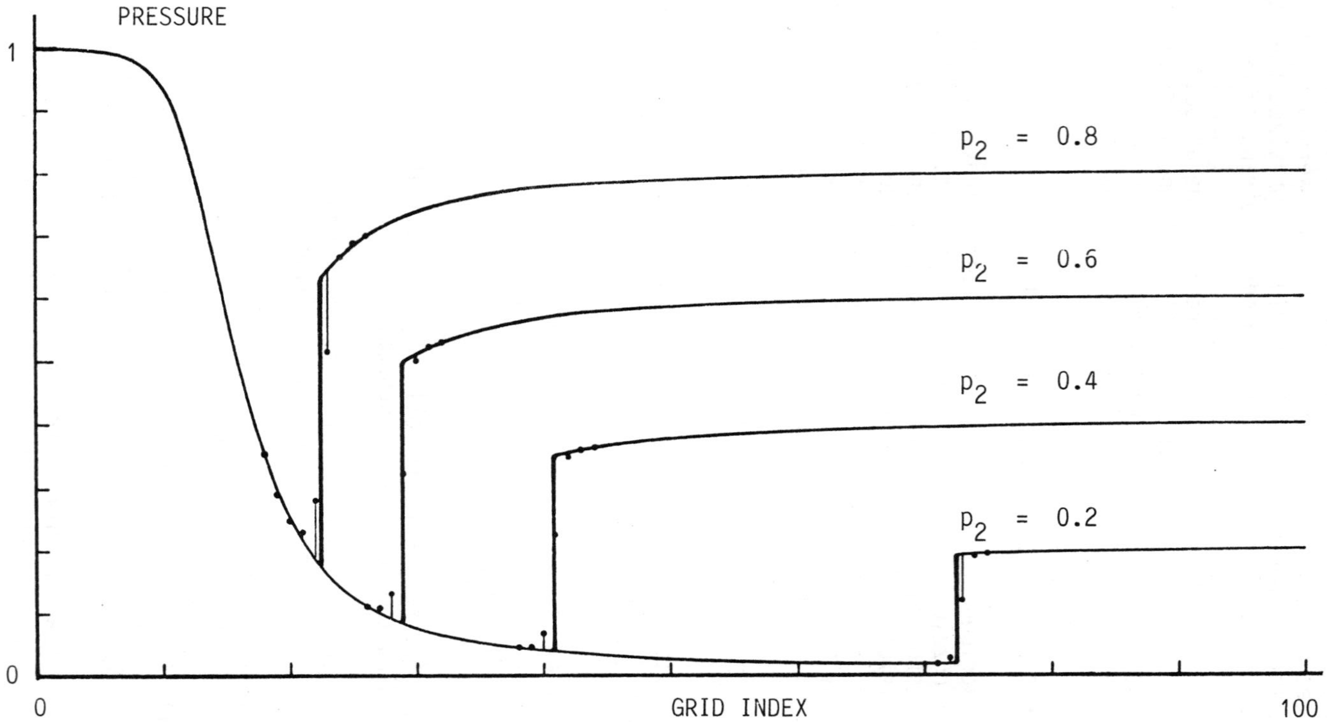

Figure 5. Pressure distribution.

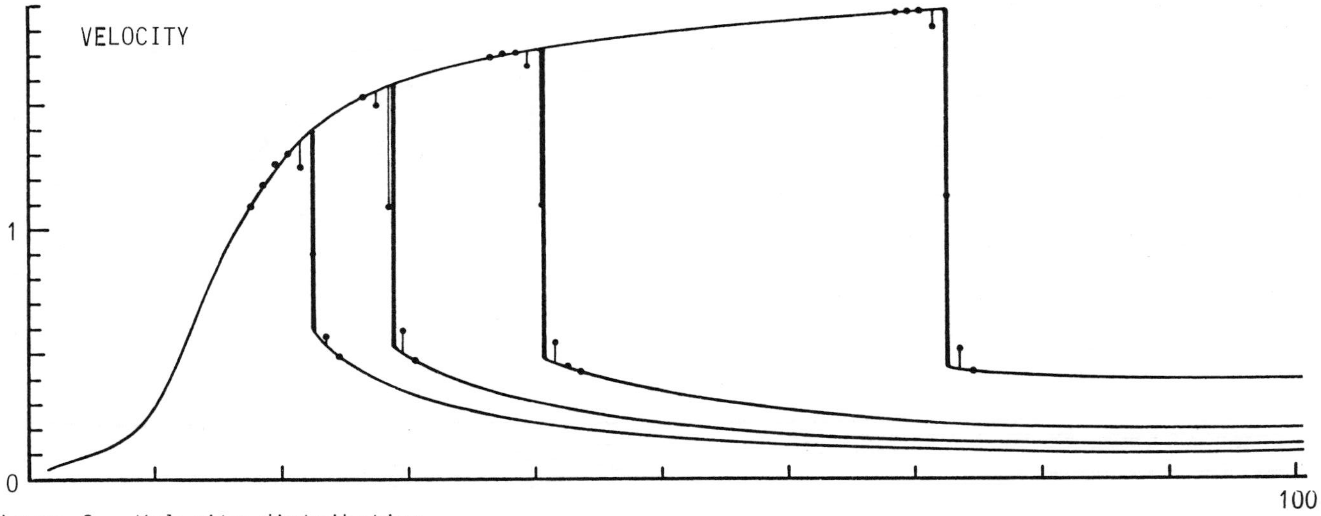

Figure 6. Velocity distribution.

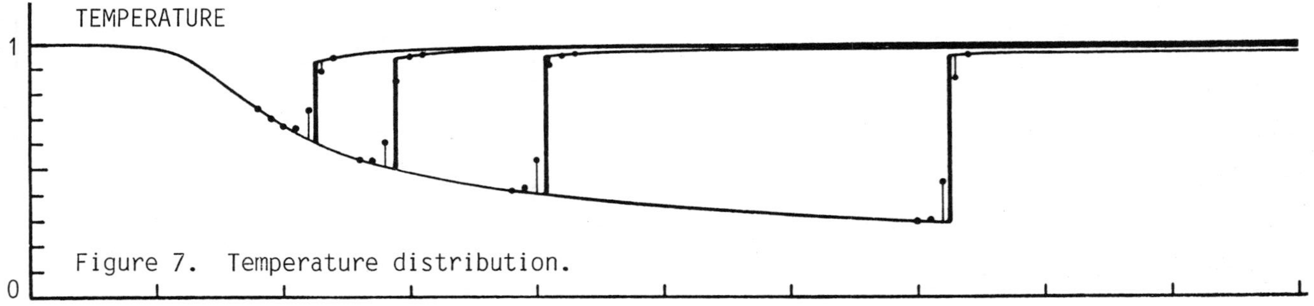

Figure 7. Temperature distribution.

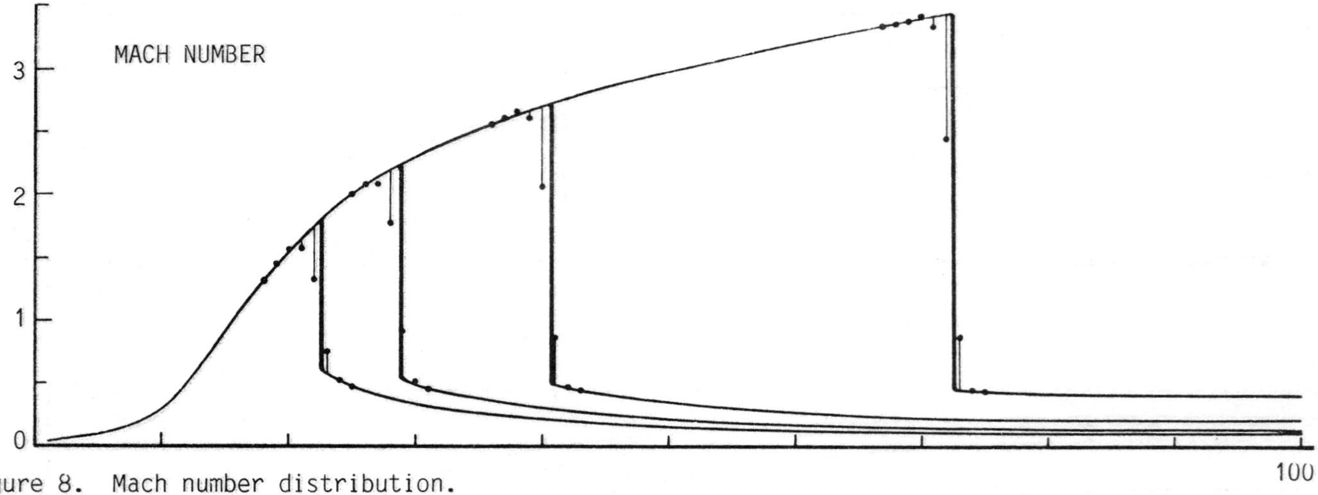

Figure 8. Mach number distribution.

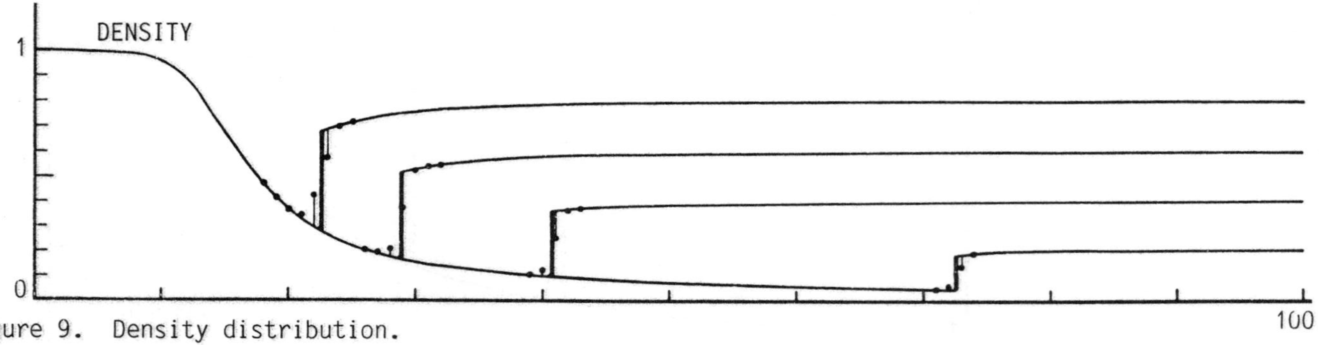

Figure 9. Density distribution.

Space does not permit the inclusion of graphical results using other convective discretization schemes. However, the following descriptive generalizations can be made.

First-order upwinding results in very diffuse numerical shock structure, as perhaps expected (because of inherent artificial diffusion). In addition, however, the isentropic regions are seriously in error--again apparently due to the pervading artificial dissipation wherever there is curvature in the convected variable.

Second-order upwinding produces better isentropic behaviour, but the shock structure is still fairly wide and involves a significant oscillation in the post-shock region. This is particularly pronounced in the pressure prediction.

The Lax-Wendroff type of second-order central differencing requires artificial stabilization in the form of unphysically large values of viscosity and thermal diffusivity in order to suppress otherwise wild oscillations in the pre-shock region. Optimal stabilization results in rather diffuse shock structure and a slight deterioration of isentropic behaviour. Similar remarks apply to the MacCormack scheme, although its behaviour is generally better than the Lax-Wendroff type.

Finally, it should be emphasized that the EULER-QUICK algorithm is a true compressible Euler code in that no (artificial) viscosity or thermal diffusion terms (involving control-volume cell gradients) are involved. It is a straight-forward matter to add (physical) terms of this type to generate a compressible Navier-Stokes code. For the Laval nozzle model problem studied here, this has essentially no effect on the results because the physical shock stucture is narrower than the numerical structure. However, it will clearly be important in modelling more general viscous compressible flows, particularly in regions involving shock-wave/boundary-layer interaction.

THEORETICAL INVESTIGATION OF THE OPERATING BEHAVIOUR OF ROTATING RADIAL BLADINGS DUE TO THE VARIATION OF THE REYNOLDS NUMBER

E. STECK/K.O. FELSCH (I)

Dr.-Ing./Prof. Dr.-Ing.

SUMMARY

The plane laminar flow through rotating radial cascades with backwards curved vanes is calculated. The density and viscosity of the fluid are supposed to be constant. Introducing the stream function and the vorticity, the equations of motion and continuity are transformed to non-orthogonal coordinates aligned with the blade contours. For solving the boundary value problem an implicit difference method is used. The operating behaviour of the impeller is investigated considering the total head and hydraulic efficiency as functions of the flow coefficient and the Reynolds number. In order to compare the numerical results with experimental data the vanes are described by logarithmic spirals. Taking into account three-dimensional effects the tendencies of theory and experiments are qualitatively well comparable.

1. INTRODUCTION

The flow through turbomachinery bladings is in general three-dimensional, turbulent and time-dependent. In many cases the compressibility of the fluid cannot be neglected. For these reasons the exact calculation of such processes will not be possible in the near future. Current computers, however, allow the theoretical investigation of complicated problems more and more, if appropriate models are used, which satisfy the character and the physics of the problem despite considerable simplifications.
Transporting highly viscous fluids, e.g. mineral oils, laminar flow can be expected according to low

(I) Institut fuer Stroemungslehre und Stroemungsmaschinen, Universitaet Karlsruhe, W. Germany

Reynolds numbers caused by viscosity. Thereby both the operating point of maximum efficiency and the total head are affected in a high degree due to the variation of the Reynolds number.

2. BASIC THEORY

2.1 Formulation of the Problem

Consider a purely radial impeller of constant breadth b and backwards curved blades, Fig. 1.

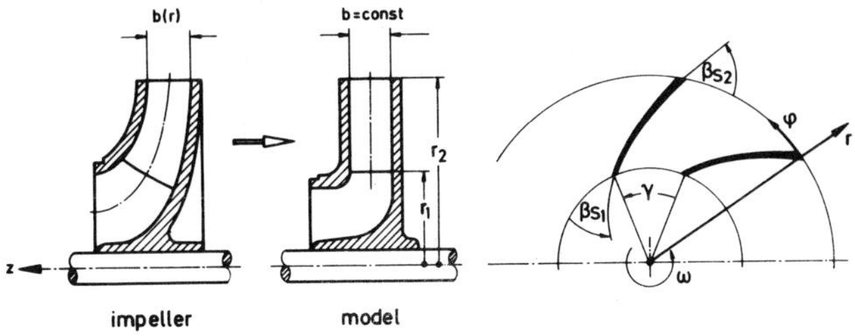

Fig. 1 Model of the Impeller

For a simple treatment of the blade shapes, the relative thickness D, which is the ratio of the blade thickness d(r) to the spacing $t = r\gamma$, is assumed to be constant and in this paper specially D = 0. Furthermore, due to a constant angular velocity ω , the following assumptions can be made:

- the relative flow be steady and plane,
- the density ρ and the dynamic viscosity μ be constant.

Moreover,

- the absolute flow far upstream and far behind the cascade be irrotational and uniform.

For numerical considerations it is convenient to express the equations of motion and continuity in a coordinate sytem aligned with the spiral vanes, Fig. 2. The equations of the transformation are

$$\tilde{r}(r,\varphi) = r$$

$$\tilde{\varphi}(r,\varphi) = F\left(\frac{r}{r_2}\right) + \varphi \quad . \qquad (1)$$

Herein the concentric circles r = const remain coordinate lines whereas the lines φ = const become spirals. The angle between the coordinate lines is

$$\cot\beta_s = f(\frac{r}{r_2}) = \frac{r}{r_2}\,\frac{d}{d(r/r_2)}\,F(\frac{r}{r_2}) \,. \tag{2}$$

Additionally an implicit stretching of the coordinates $\hat{r}(\tilde{r})$ and $\hat{\varphi}(\tilde{\varphi})$ is applied. With the outer radius r_2 and the circumferential velocity $r_2\omega$ the following dimensionless variables are used:

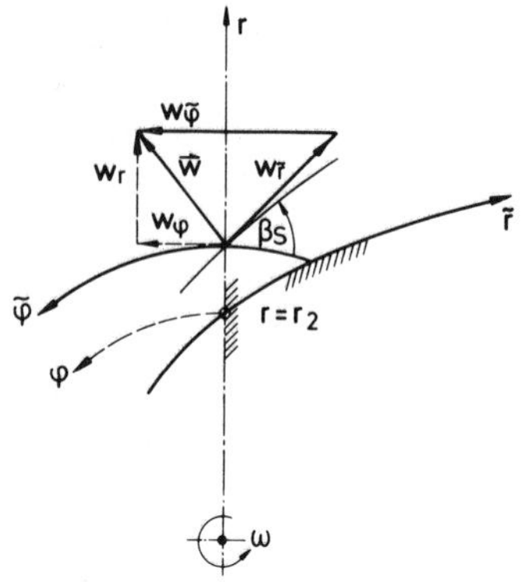

Fig. 2 Coordinates

$$\tilde{R} = \frac{\tilde{r}}{r_2}, \quad \tilde{\Phi} = \frac{\tilde{\varphi}}{\gamma}, \quad \gamma = \frac{2\pi}{z}, \quad \hat{R} = \frac{\hat{r}}{r_2}, \quad \hat{\Phi} = \frac{\hat{\varphi}}{\gamma}$$

$$W_{\tilde{R}} = \frac{w_{\tilde{r}}}{r_2\,\omega}, \quad W_{\tilde{\Phi}} = \frac{w_{\tilde{\varphi}}}{r_2\,\omega}, \quad \Psi = \frac{\psi}{r_2^2\,\omega\,\gamma}, \quad \Omega = \frac{\zeta}{\omega} \tag{3}$$

$$Re = \frac{r_2^2\,\omega}{\mu/\rho}, \quad \varphi_r = \frac{w_{r_2.m}}{r_2\,\omega}$$

wherein z is the number of blades, Re the Reynolds number and φ_r the flow coefficient. The essential output variables are the total pressure coefficient ψ_t and the hydraulic efficiency η_h, which is the ratio of the fluid power to the power input of the blades,

$$\psi_t = \frac{\Delta p_t}{(\rho/2)\,(r_2\,\omega)^2} \tag{4}$$

$$\eta_h = \frac{\Delta p_t\,\dot{V}}{m_s\,\omega} \tag{5}$$

The system of equations for the relative flow field consists of a vorticity transport equation and a

Poisson equation for the stream function, [1],

$$\begin{aligned} \underset{\hat{R}\hat{\Phi}}{\mathrm{div}}\,(\vec{W}\,\Omega) &= \frac{1}{\mathrm{Re}} \underset{\hat{R}\hat{\Phi}}{\Delta}\,\Omega \\ -\Omega &= \gamma \underset{\hat{R}\hat{\Phi}}{\Delta}\,\Psi \\ W_{\tilde{R}} &= \frac{\sqrt{1+f^2}}{\tilde{R}} \frac{d\hat{\Phi}}{d\tilde{\Phi}} \frac{\partial\Psi}{\partial\hat{\Phi}} \\ W_{\tilde{\Phi}} &= -\gamma \frac{d\hat{R}}{d\tilde{R}} \frac{\partial\Psi}{\partial\hat{R}} \,. \end{aligned} \tag{6}$$

2.2 Boundary Conditions

For complete mathematical description of the elliptic fluid flow problem appropriate boundary conditions are needed, see Fig. 3. At the entrance radius sufficiently far upstream, the components of the relative velocity $W_{\tilde{R}}$, $W_{\tilde{\Phi}}$ are determined by the operating point φ_r. The stream function Ψ is linear, the constant of integration being determined by the azimuthal component $W_{\tilde{\Phi}}$. The vorticity Ω is constant.

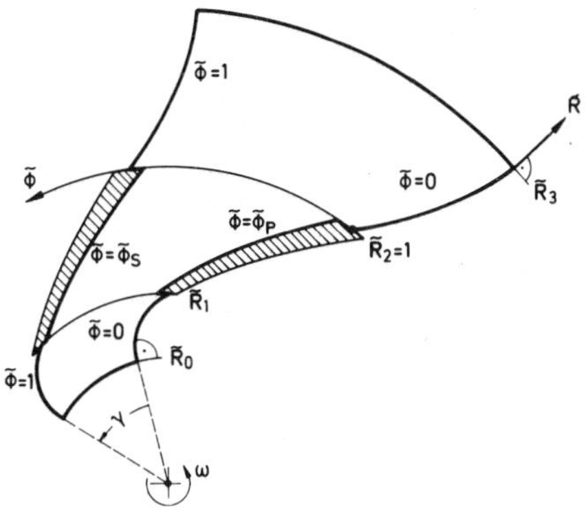

Fig. 3 Control Area

$\tilde{R} = \tilde{R}_0$:

$$W_{\tilde{R}} = \frac{\varphi_r}{\tilde{R}_0},\quad W_{\tilde{\Phi}} = W_{\tilde{R}} \cot a_0 - \tilde{R}_0\,,\quad \Psi = \varphi_r\,\tilde{\Phi} + \Psi_0\,,\quad \Omega = -2 \tag{7}$$

Far behind the cascade analogue prescriptions have to be made, whereas the flow angle α_3 is unknown and must be controlled by the law of angular momentum.

$\tilde{R} = \tilde{R}_3$:

$$W_{\tilde{R}} = \frac{\varphi_r}{\tilde{R}_3},\quad W_{\tilde{\Phi}} = W_{\tilde{R}} \cot a_3 - \tilde{R}_3\,,\quad \Psi = \varphi_r\,\tilde{\Phi} + \Psi_3\,,\quad \Omega = -2 \tag{8}$$

On the blade contours at the pressure and suction side the relative velocity vanishes. The Poisson equation yields the second derivative of the stream function with respect to the circumferential direction for the vorticity.

$\tilde{\Phi} = \tilde{\Phi}_P, \tilde{\Phi}_S$:

$$W_{\tilde{R}} = W_{\tilde{\Phi}} = 0, \qquad \Psi = \begin{cases} 0 : & \tilde{\Phi} = \tilde{\Phi}_P \\ \varphi_r : & \tilde{\Phi} = \tilde{\Phi}_S \end{cases}$$

$$\Omega = -\frac{1}{\gamma}\frac{1+f^2}{\tilde{R}^2}\frac{d\hat{\Phi}}{d\tilde{\Phi}}\frac{\partial}{\partial\hat{\Phi}}\left(\frac{d\hat{\Phi}}{d\tilde{\Phi}}\frac{\partial\Psi}{\partial\hat{\Phi}}\right) \tag{9}$$

At the lateral upstream and downstream boundaries defining the control area between an element of two blades,periodic boundary conditions have to be established.

$\tilde{R}_0 < \tilde{R} < \tilde{R}_1, \tilde{R}_2 < \tilde{R} < \tilde{R}_3$:

$$\begin{aligned} W_{\tilde{R}}, W_{\tilde{\Phi}}, \Omega(\tilde{\Phi}=1) &= W_{\tilde{R}}, W_{\tilde{\Phi}}, \Omega(\tilde{\Phi}=0) \\ \Psi(\tilde{\Phi}=1) &= \Psi(\tilde{\Phi}=0) + \varphi_r \end{aligned} \tag{10}$$

2.3 Dimensionless Parameters of the Problem

Solving the system of equations and taking into account the corresponding boundary conditions, the laminar flow can be calculated expressed by the integral quantities of the pressure coefficient and hydraulic efficiency as functions of the flow coefficient, the number of blades, the Reynolds number, the vane angle, the angle of the absolute velocity far upstream and the ratio of radii R_1:

$$\psi_t,\ \eta_h\ (\varphi_r,\ z,\ Re,\ \beta_S,\ \alpha_0,\ R_1) \tag{11}$$

Assuming $\alpha_0 = 90^\circ$, the influence of the flow coefficient φ_r and the Reynolds number Re is investigated in this paper. The vane angle $\beta_S = 30^\circ$ is constant (infinitely thin log-spiral blade), the number of blades is $z = 6$, and the ratio of radii is $R_1 = 0.6$.

2.4 Numerical Solution

The boundary value problem is numerically

solved by a second order finite difference method using one sided difference schemes to approximate the convective terms of the vorticity transport equation (upwind differencing of the second kind, see [2]). The grid of coordinate lines is shown in Fig. 4. The algebraic difference equations are either solved by a pointwise successive relaxation method (low Reynolds numbers) or by a simultaneous solver for the vorticity transport equation and the Poisson equation along lines $\tilde{R}$ = const (moderate Reynolds numbers).

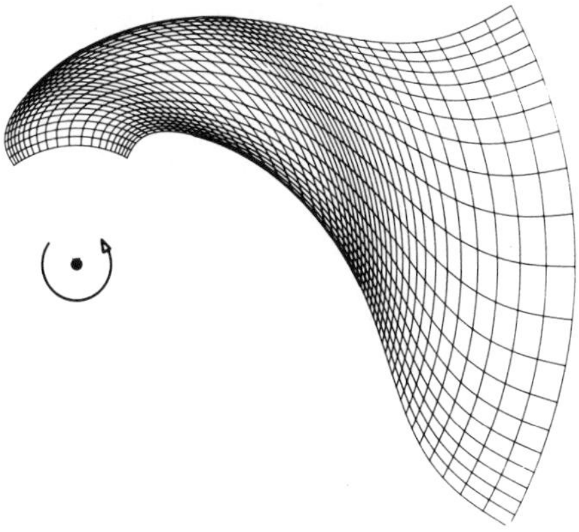

Fig. 4 Mesh of Coordinate Lines

3. RESULTS

3.1 Pressure Coefficient and Hydraulic Efficiency

Fig. 5 shows the characteristics of the pressure coefficient and the hydraulic efficiency for Re = 200, 600, 2400. For large values of the flow coefficient, the characteristics of the pressure coefficient move up due to increasing Reynolds number and thus decreasing losses. As for small flow coefficients this phenomenon turns around. Increasing the Reynolds number, the operating points of maximum efficiency increase as well and move to higher values of the flow coefficient. The behaviour of both systems of characteristics agrees well with the experimental experience [3, 4].

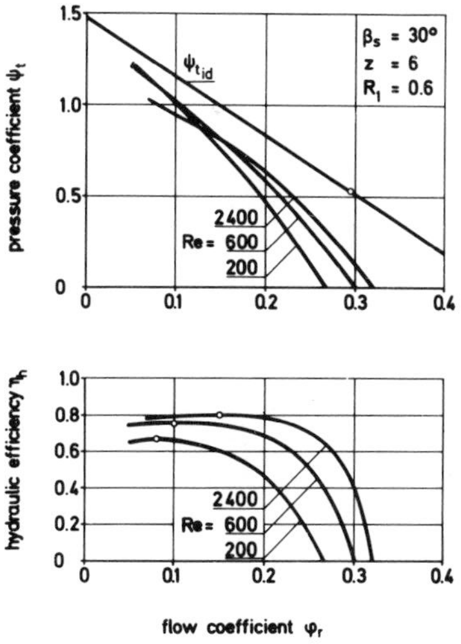

Fig. 5 Characteristics of the Pressure Coefficient and the Hydraulic Efficiency

3.2 Relative Streamlines in the Operating Points of Maximum Efficiency

The inherent streamlines of the operating points of maximum efficiency are plotted in Fig. 6. The developing boundary layer of the cascade flow along the pressure and suction sides of the blades can be studied respectively in dependence of the Reynolds number. In each case the stagnation point of the incident flow is situated at the pressure side moving to the entrance tip of the blade as the Reynolds number decreases.

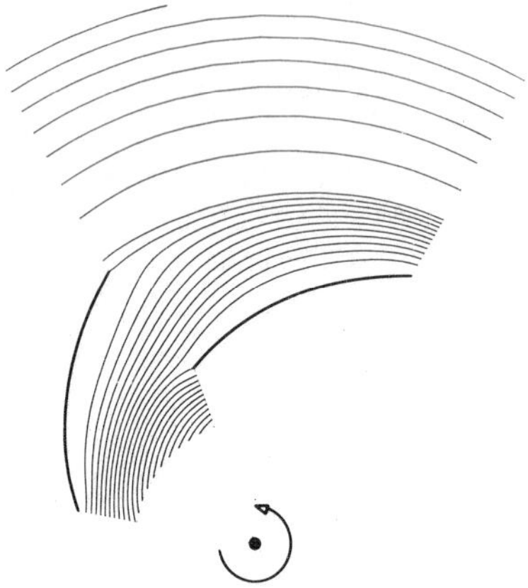

Re = 200

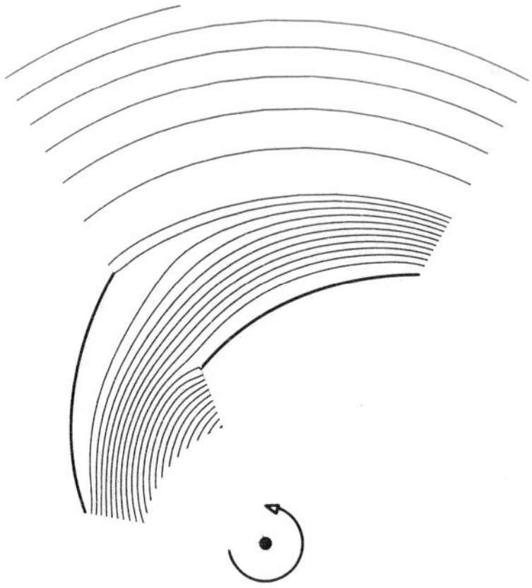

Re = 600

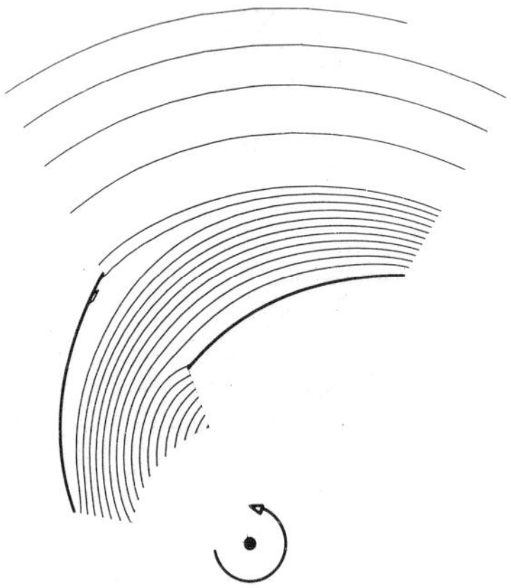

Re = 2400

Fig. 6 Relative Streamlines in the Operating Points of Maximum Efficiency

3.3 Comparison of Theoretical with Experimental Data

Fig. 7 shows the value of the flow coefficient $\varphi_{r.opt}$ concerning the operating points of maximum efficiency in dependence of the Reynolds number Re. The operating point of shockless entry (stagnation point at the tip of the blade) predicted by the two-dimensional theory for ideal fluids of Busemann [5] is indicated additionally. Both curves increase as the Reynolds number increases. The range of the theoretically and experimentally found curves do not overlap unfortunately. As it is expected, however, corresponding values $\varphi_{r.opt}$ given by the two-dimensional theory are higher than those measured. This fact can be seen for $Re \approx 7 \cdot 10^3$.

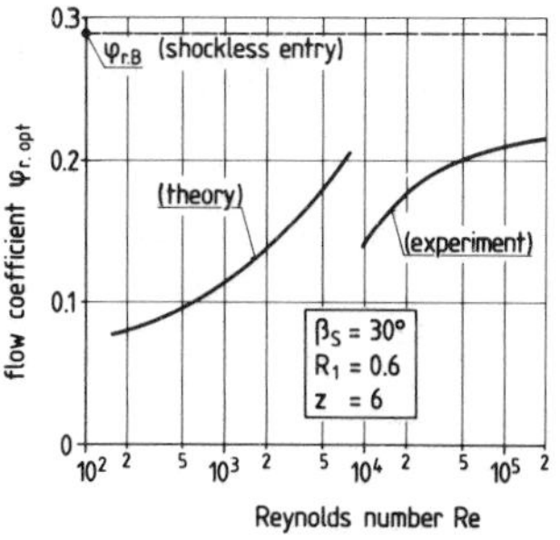

Fig. 7 Comparison of Calculated Values of the Flow Coefficient $\varphi_{r.opt}$ with Experiments after Kamimoto/Matsuoka/Shirai [4]

In Fig. 8 the maximum hydraulic efficiency $\eta_{h.max}$ is plotted versus the Reynolds number Re. Both the theoretical and the experimental curve increase monotonically for higher Reynolds numbers, which can be explained by decreasing losses. The causes for the difference between corresponding values in the vicinity of $Re \approx 7 \cdot 10^3$ are based firstly on the three-dimensional effects in [4] and secondly on the tremendous influence of the boundary layers for smaller Reynolds numbers at the side walls of the impeller.

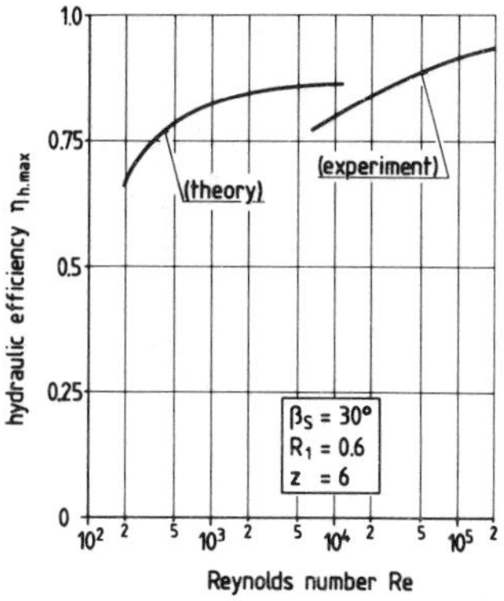

Fig. 8 Comparison of Calculated Values of the Hydraulic Efficiency $\eta_{h.max}$ with Experiments after Kamimoto/Matsuoka/Shirai [4]

4. CONCLUDING REMARK

The theoretical investigations are going to be extended to stream surfaces of revolution with arbitrary slope dr/dz.

5. ACKNOWLEDGEMENTS

The author is thankful to his friend and collegue Dr.-Ing. M. Abboud for many helpful discussions.

6. REFERENCES

1. STECK, E. - Berechnung des Betriebsverhaltens rotierender Radialgitter. Stroemungsmechanik und Stroemungsmaschinen, 30 (1981)
2. ROACHE, P.J. - Computational Fluid Dynamics, Hermosa Publishers, Albuquerque, 1972
3. CAGLAR, S. - Messungen zur Ermittlung des Reynoldszahl-Einflusses auf die Druckerhoehung eines Radialgitters, Internal Report, Institut fuer Stroemungslehre und Stroemungsmaschinen, Univ. Karlsruhe, 1979
4. KAMIMOTO, G./MATSUOKA, Y./SHIRAI, H. - Effects of Fluid Viscosity on the Flow in the Impeller of Centrifugal Type Hydraulic Machinery. Bulletin of ASME, Vol. 2, 8 (1959)
5. BUSEMANN, A. - Das Foerderhoehenverhaeltnis radialer Kreiselpumpen mit logarithmisch-spiraligen Schaufeln. Z. f. angew. Math. Mech. 8, 5 (1928)

TIME-SPLIT FINITE ELEMENT METHOD FOR COMPRESSIBLE AEROFOIL TRAILING-EDGE FLOWS

K. Srinivas* and C.A.J. Fletcher**

SUMMARY

A time-split finite element method for solving the compressible, Navier-Stokes equations in generalised coordinates is introduced. The method makes use of the group formulation which has been shown to be more economical and accurate than the conventional finite-element method for a related problem. The group formulation is combined with Lagrange elements which permits the directional mass and difference operators to be separated and an efficient three-level time-split computational algorithm to be constructed. The present procedure is made more economical by applying the finite element method in the (ξ,η) plane and using the isoparametric formulation to evaluate the transformation parameters, only. The method is illustrated by computing the flow at the trailing-edge of an aerofoil and the results are discussed.

1. INTRODUCTION

Applications of the time-split finite-element method introduced in [1] bring out the potential of the method for handling fluid flows. These applications include the computation of compressible, laminar and turbulent flows over a forward-facing step in [1] and over a flat plate and a backward-facing step in [2]. In these applications emphasis was given to flow separation and turbulence. In the present study the trailing-edge flow past an aerofoil is considered.

The previous version of the time-split finite element method made use of the group formulation [3] to avoid the computationally inefficient treatment of the convective terms

*Research Associate, Dept. of Mechanical Engineering, University of Sydney, N.S.W. 2006, Australia.
**Senior Lecturer, Dept. of Mechanical Engineering, University of Sydney, N.S.W. 2006, Australia.

by the conventional finite element method. The present problem features cubic nonlinearities and estimates [4,5] indicate that the group finite element formulation is approximately seventeen times more economical than the conventional finite element method. Experiments with Burgers' equation [4] suggest that there will also be a gain in accuracy.

In the previous applications [1,2] of the time-split method time was used merely as a convenient iteration device for reaching the steady state solution. To achieve this objective the terms associated with the mass matrix were lumped which allowed the stability to be enhanced but at the expense of the accuracy of the transient solution.

Combination of the group formulation with Lagrange elements permits directional mass and difference operations to be separated [5] and a consistent time-split formulation to be developed. Such a consistent time-split formulation is equally suitable for the transient solution while giving an increase in efficiency compared with the previous time-split formulation. In addition, here, we replace the two-level (in time) time-split scheme with a three-level time-split scheme which is more robust and converges faster than the two-level scheme. The consistent three-level time-split scheme has also been applied to incompressible separated flows [6].

The present problem introduces the further complication of a distorted physical domain. It is well known that the conceptually convenient isoparametric formulation is inefficient when applied, in the physical plane, to flow problems and is inaccurate if elements become too distorted. Here we pursue the more efficient route of expressing the equations of motion in generalised coordinates (ξ,η). The Galerkin group finite-element formulation with linear Lagrange elements is applied in the (ξ,η) plane. The resulting equations are converted into an efficient computational algorithm by application of the consistent time-split formulation in the (ξ,η) plane. Information about the physical domain is contained in nodal transformation parameters. It turns out that these can be evaluated particularly economically by using the isoparametric formulation in the (ξ,η) plane.

Thus the present finite element formulation will solve the Navier-Stokes equations governing compressible flow (laminar or turbulent) in an irregular domain without the need for the expensive factorisation of global Jacobian matrices. The computational efficiency of the present formulation follows from three important features:

(i) the group finite element formulation [3]

(ii) the consistent three-level time-split algorithm [6]

(iii) the implementation of the Galerkin finite element method in generalised coordinates. (ξ,η)

The third feature is described here for the first time.

The problem of the flow adjacent to aerofoil trailing-edges is of practical interest and has been studied previously by various investigators [7,8]. Here we apply the time-split finite element method to compute laminar flow about symmetric and asymmetric trailing edges to demonstrate its effectiveness.

2. GOVERNING EQUATIONS

The flow is governed by the two-dimensional, compressible Navier-Stokes equations. In the Cartesian coordinates these are

$$\frac{\partial\bar{q}}{\partial t} + \frac{\partial\bar{F}}{\partial x} + \frac{\partial\bar{G}}{\partial y} = \frac{\partial^2\bar{R}}{\partial x^2} + \frac{\partial^2\bar{S}}{\partial x\partial y} + \frac{\partial^2\bar{T}}{\partial y^2} \tag{2.1}$$

where

$$\begin{aligned}
\bar{q} &= \{\rho,\ \rho u,\ \rho v\} \\
\bar{F} &= \{\rho u,\ p + \rho u^2,\ \rho uv\} \\
\bar{G} &= \{\rho v,\ \rho uv,\ p + \rho v^2\} \\
\bar{R} &= \{\theta_\rho^d\ \rho,\ \tfrac{4}{3}\ \mu.u,\ \mu v\} \\
\bar{S} &= \left\{0,\ \tfrac{\mu}{3}\ v,\ \tfrac{\mu}{3}\ u\right\} \\
\bar{T} &= \left\{\theta_\rho^d\ \rho,\ \mu u,\ \tfrac{4}{3}\ \mu v\right\}
\end{aligned} \tag{2.2}$$

ρ = density, u,v = velocity components in x and y directions, p = pressure, μ = molecular viscosity.

It is expected that temperature changes in the solution domain are small; an assumption which is justified for the flows of interest which are transonic. Hence the molecular viscosity, μ is assumed to be constant everywhere. Further, the following relationship between pressure and velocity is assumed.

$$p = \rho\left\{RT_o - \frac{\gamma-1}{2\gamma}\ (u^2 + v^2)\right\} \tag{2.3}$$

which is valid for ideal gases.

As a consequence of Eqn. (2.3) the energy equation (not shown above) need not be solved.

The equations are nondimensionalised with respect to a characteristic length L, free stream velocity U_∞ and free stream density ρ_∞. After that μ is replaced by 1/Re where Re is the Reynolds number $(\rho_\infty U_\infty L/\mu)$. The nondimensional form of the pressure equation is

$$1 + \gamma M_\infty^2 p = \rho\left\{1 + \frac{\gamma-1}{2} M_\infty^2 (1 - u^2 - v^2)\right\} \tag{2.4}$$

3. TRANSFORMATION OF GOVERNING EQUATIONS

Through the isoparametric formulation the finite element method effectively introduces distorted elements in the physical plane to compute in irregular domains. But by solving the equations of motion in the (ξ,η) plane we retain the high accuracy inherent in uniform-grid finite element applications. We also achieve a significant economy through reduced convectivity [5] when the group finite element formulation [3] is exploited. Accordingly, a general transformation

$$\xi = \xi(x,y) \quad , \quad \eta = \eta(x,y) \tag{3.1}$$

is introduced. Under this transformation,

$$u_x = \xi_x u_\xi + \eta_x u_\eta \quad \text{and etc.}$$

Upon transforming eqns. (2.1) to the (ξ,η) plane, one gets

$$\bar{q}^*_t + \bar{F}^*_\xi + \bar{G}^*_\eta - \bar{R}^*_{\xi\xi} - \bar{S}^*_{\xi\eta} - \bar{T}^*_{\eta\eta} = 0 \tag{3.2}$$

where

$$\bar{q}^* = \frac{q}{J} \quad ,$$

$$\bar{F}^* = \frac{1}{J}\begin{Bmatrix} \rho U_c \\ \xi_x p + \rho u U_c + \left(\frac{4\mu}{3}\xi_{xx} + \mu\xi_{yy}\right)u + \left(\frac{\mu}{3}\xi_{xy}\right)v \\ \xi_y p + \rho v U_c + \left(\mu\xi_{xx} + \frac{4\mu}{3}\xi_{yy}\right)v + \left(\frac{\mu}{3}\xi_{xy}\right)v \end{Bmatrix} \tag{3.3}$$

$$\bar{G}^* = \frac{1}{J}\begin{Bmatrix} \rho V_c \\ \eta_x p + \rho u V_c + \left(\frac{4\mu}{3}\eta_{xx} + \mu\eta_{yy}\right)u + \left(\frac{\mu}{3}\eta_{xy}\right)v \\ \eta_y p + \rho v V_c + \left(\mu\eta_{xx} + \frac{4\mu}{3}\eta_{yy}\right)v + \left(\frac{\mu}{3}\eta_{xy}\right)u \end{Bmatrix} \tag{3.4}$$

$$\bar{R}^* = \frac{1}{J}\begin{Bmatrix} \rho \\ \left(\frac{4\mu}{3}\xi_x^2 + \mu\xi_y^2\right)u + \frac{\mu}{3}\xi_x\xi_y v \\ \left(\mu\xi_x^2 + \frac{4}{3}\mu\xi_y^2\right)v + \frac{\mu}{3}\xi_x\xi_y u \end{Bmatrix} \tag{3.5}$$

$$\bar{S}^* = \frac{1}{J}\begin{Bmatrix} 0 \\ 2\left(\frac{4\mu}{3}\xi_x\eta_x + \mu\xi_y\eta_y\right)u + \frac{\mu}{3}\left(\xi_x\eta_y + \xi_y\eta_x\right)v \\ 2\left(\mu\xi_x\eta_x + \frac{4\mu}{3}\xi_y\eta_y\right)v + \frac{\mu}{3}\left(\xi_y\eta_x + \eta_y\xi_x\right)u \end{Bmatrix} \tag{3.6}$$

$$\bar{T}^* = \frac{1}{J}\begin{Bmatrix} \rho \\ \left(\frac{4\mu}{3}\eta^2_x + \mu\eta^2_y\right)u + \left(\frac{\mu}{3}\eta_x\eta_y\right)v \\ \left(\mu\eta^2_x + \frac{4}{3}\mu\eta^2_y\right)v + \left(\frac{\mu}{3}\eta_x\eta_y\right)u \end{Bmatrix} \tag{3.7}$$

In the above equations J is the Jacobian of the transformation and U_c and V_c are the contravariant velocities along ξ and η directions [7].

4. TIME-SPLIT FINITE ELEMENT METHOD

An application of the group Galerkin finite element method [3] with linear Lagrange elements gives the following set of ordinary differential equations -

$$M_\xi \otimes M_\eta \, \bar{q}^*_t + M_\eta \otimes L_\xi \bar{F}^* + M_\xi \otimes L_\eta \, \bar{G}^* - M_\eta \otimes L_{\xi\xi} \, \bar{R}^*$$

$$- L_\zeta \otimes L_\eta \, \bar{S}^* - M_\xi \otimes L_{\eta\eta} \, \bar{T}^* = 0 \tag{4.1}$$

where $\otimes$ denotes the tensor product. (See Fig. 1)

In eqn. (4.1)

$$M_\xi = \left\{\frac{1}{6}, \frac{2}{3}, \frac{1}{6}\right\}, \quad M^t_\eta = \left\{\frac{1}{6}, \frac{2}{3}, \frac{1}{6}\right\} \tag{4.2}$$

and

$$L_\xi = \frac{1}{2\Delta\xi}\{-1, 0, 1\}, \quad L^t_\eta = \frac{1}{2\Delta\eta}\{1, 0, -1\},$$

$$L_{\xi\xi} = \frac{1}{\Delta\xi^2}\{1, -2, 1\},$$

$$L^t_{\eta\eta} = \frac{1}{\Delta\eta^2}\{1, -2, 1\} \tag{4.3}$$

These operators M,L are for uniform, rectangular grid in ξ and η directions. Corresponding operators for a non-uniform grid can also be written.

Now, we develop a time-split algorithm to solve the eqn. (4.1). A general three-level evaluation of the time derivative allows eqn. (4.1) to be written

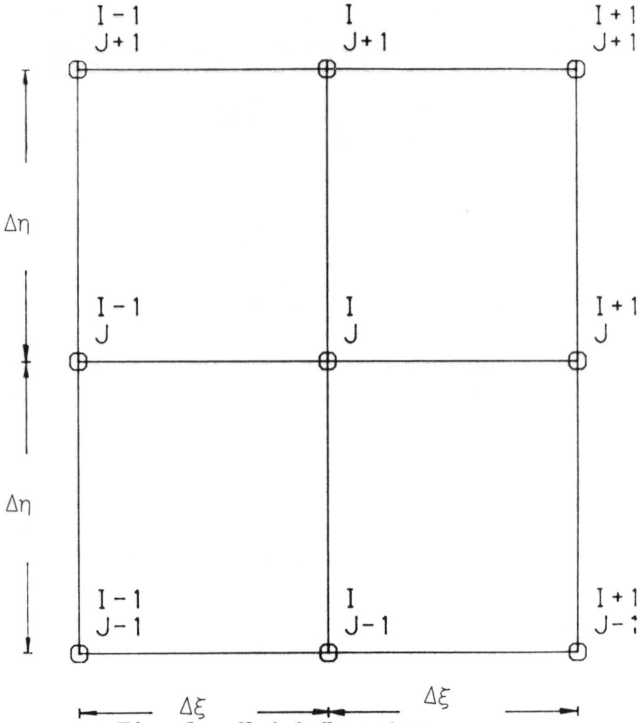

Fig. 1. Nodal Geometry

$$M_\xi \otimes M_\eta \left[\alpha \frac{\Delta \bar{q}^{*n+1}}{\Delta t} + (1-\alpha) \frac{\Delta \bar{q}^{*n}}{\Delta t} \right] = \beta \; RHS^{n+1} + (1-\beta) \; RHS^n$$

where

$$RHS = M_\eta \otimes L_{\xi\xi} \; \bar{R}^* + L_\xi \otimes L_\eta \; \bar{S}^* + M_\xi \otimes L_{\eta\eta} \; \bar{T}^* - M_\eta \otimes L_\xi \; \bar{F}^* - M_\xi \otimes L_\eta \; \bar{G}^* \tag{4.4}$$

α, β are free parameters. The choice $\alpha = 1$ and $\beta = 0.5$ gives the Crank-Nicolson scheme and $\alpha = 1.5$ and $\beta = 1.0$ gives a three-level implicit scheme which is used in the present studies. A comparison of the two algorithms for a closely related formulation is provided by Fletcher and Srinivas [6].

To obtain a linear system of equations for $\Delta \bar{q}^*$, i.e. $\left\{ \frac{\Delta \rho}{J}, \frac{\Delta u}{J}, \frac{\Delta v}{J} \right\}$ a Taylor expansion of the equation about nth level is introduced.

i.e.

$$(RHS)^{n+1} = (RHS)^n + \frac{\partial}{\partial \bar{q}^*} (RHS) \; \Delta \bar{q}^{*n+1} \tag{4.5}$$

where

$$\Delta\bar{q}^{*n+1} = \bar{q}^{*n+1} - \bar{q}^{*n} \tag{4.6}$$

∴ after some manipulation the eqn. (4.4) becomes

$$M_\xi \otimes M_\eta \left[\alpha\ \Delta\bar{q}^{*n+1}\right] - \beta\Delta t \left[M_\eta \otimes L_{\xi\xi} \frac{\partial\bar{R}^*}{\partial\bar{q}^*} + M_\xi \otimes L_{\eta\eta} \frac{\partial\bar{T}^*}{\partial\bar{q}^*} - M_\eta \otimes L_\xi \frac{\partial\bar{F}^*}{\partial q^*} - M_\xi \otimes L_\eta \frac{\partial\bar{G}^*}{\partial q^*}\right] \Delta\bar{q}^{n+1}$$

$$= \Delta t.(RHS)^n + \beta\ \Delta t\ L_\xi \otimes L_\eta \frac{\partial\bar{S}^*}{\partial\bar{q}^*} . \Delta\bar{q}^{*n} - (1-\alpha) M_\xi \otimes M_\eta \Delta\bar{q}^{*n}$$

$$= \Delta t.(RHS)^A - (1-\alpha)\ M_\xi \otimes M_\eta \Delta\bar{q}^{*n} \tag{4.7}$$

The contributions to $\bar{R}^*$ etc. arising from the transformation parameters ξ_x, η_x etc. are evaluated using the isoparametric formulation in the (ξ,η) plane.

In practice, eqn. (4.7) is implemented as

$$\left[M_\xi - \frac{\beta}{\alpha}\Delta t \left(L_{\xi\xi} \frac{\partial\bar{R}^*}{\partial\bar{q}^*} - L_\xi \frac{\partial\bar{F}^*}{\partial\bar{q}^*}\right)\right] \Delta\bar{q}^{*i} = \frac{\Delta t}{\alpha} (RHS)^A - \left(\frac{1-\alpha}{\alpha}\right) M_\xi^t \otimes M_\eta \Delta\bar{q}^{*n} \tag{4.8}$$

$$\left[M_\eta - \frac{\beta}{\alpha}\Delta t \left(L_{\eta\eta} \frac{\partial\bar{T}^*}{\partial\bar{q}^*} - L_\eta \frac{\partial\bar{G}^*}{\partial\bar{q}^*}\right)\right] \Delta\bar{q}^{*n+1} = \Delta\bar{q}^{*i} \tag{4.9}$$

where additional terms of $O(\Delta t^2)$ have been introduced to permit the splitting algorithm to be constructed. Eqns.(4.8) and (4.9) constitute a decoupled implicit local system of equations associated with each grid line in the ξ and η directions. This system is solved in O(N) operations without the need for global factorisation at each time-step.

The solution to the steady state is reached when $(RHS)^A = 0$ and so the evaluation of RHS provides a measure of 'closeness' of the solution to the steady state. Further when the steady state is reached $\Delta\bar{q}^* = 0$. This indicates that choices other than the present one for the left hand side are possible without altering the steady state solution [1].

5. RESULTS AND DISCUSSION

The physical domain of computation AEFG is shown in Fig. 2. BCD is the trailing edge over and downstream of which the flow is computed. The thickness BD is equal to 11 per cent of the chord. The flow Reynolds number is 100 based on unit chord length, and the Mach number being 0.4.

Fully developed boundary layer velocity profiles were prescribed as initial conditions along AB and DE. The boundary conditions prescribed on various boundaries are identical to the ones described in [2] and are not detailed here. A detailed discussion of one of the boundary conditions applied, i.e. the non-reflecting conditions along GF, has been given in [9].

In the physical plane a uniform mesh is adopted in the x-direction, the mesh width being 30 per cent of the boundary layer thickness at D. The mesh is also uniform in the y-direction and the mesh width is 10 per cent of the boundary layer thickness. The region AEFG was divided into 34 × 42 linear elements. DE was set equal to twice the boundary layer thickness at D and GF was five chords away from C.

The region AEFG was transformed into another rectangle in the computational plane and BCD was mapped into a straight line. To start with the mesh was uniform in the computational plane.

Both symmetric and asymmetric trailing edges have been considered (see Fig. 2). Computations were carried out using a time step of 0.1 for both the cases, till $(RHS)^A$ in eqn. (4.8) reduced to 1×10^{-6}. The computed u-velocities for different x-stations downstream of the trailing edge are plotted in Figs. 3 and 4. The profiles exhibit qualitative agreement with those obtained by Viswanath and Brown [10]. In the near wake, considerable changes take place close to the centreline and the boundary layer type flow slowly develops into a wake flow. Far downstream of the trailing edge changes also occur away from the centreline. In the asymmetric case, it is noticed that the wake is considerably thicker above the centreline than below. Also, for the assymmetric case, the minimum velocity at any x-station occurs above the centreline.

6. CONCLUDING REMARKS

The present study demonstrates the capability of the time-split finite element method in generalized coordinates to handle the trailing-edge flow. As the next stage, it is proposed to compute turbulent trailing-edge flows with an appropriate model for turbulence and check the accuracy of the method by comparing the results with those obtained by other investigators.

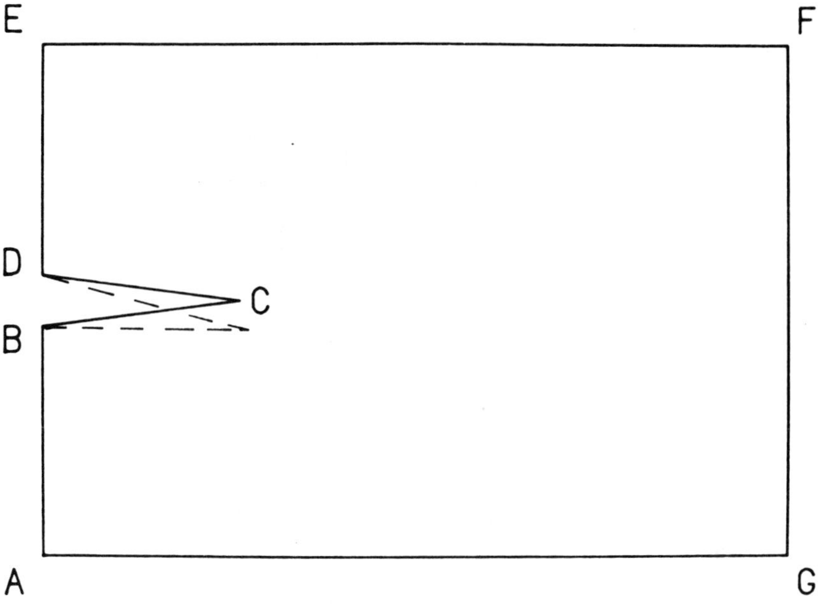

Fig. 2. Physical domain of computations.

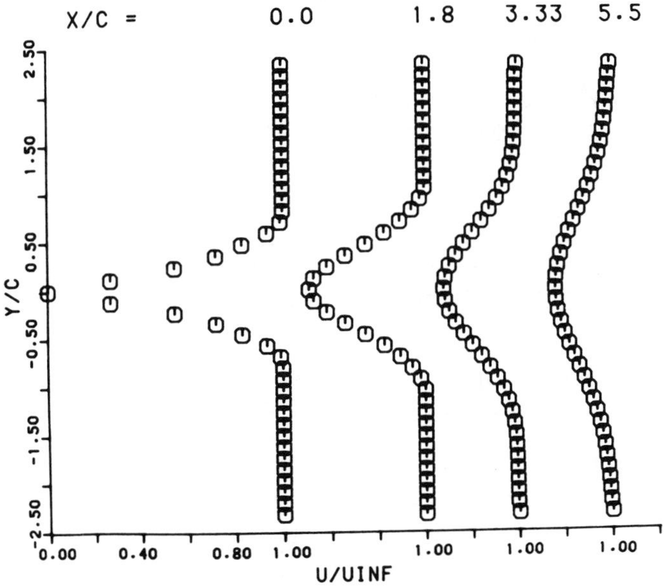

Fig. 3. u-velocity profiles downstream of the symmetric trailing-edge.

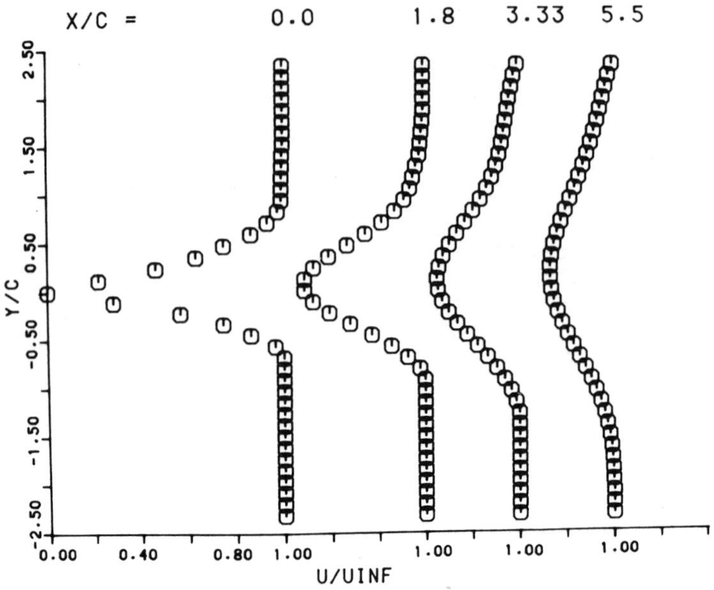

Fig. 4. u-velocity profiles downstream of the asymmetric trailing edge.

7. ACKNOWLEDGEMENTS

The authors are grateful to the Australian Research Grants Committee for their continued financial support.

REFERENCES

1. FLETCHER, C.A.J. - On an Alternating Direction Implicit Finite Element Method for Flow Problems. Comp. Meth. Appl. Mech. Eng., 30, pp.307-322, 1982.

2. SRINIVAS, K. and FLETCHER, C.A.J. - Finite Element Solutions for Laminar and Turbulent Compressible Flow, Int. J. Num. Meth. Fluids, 1983 (to appear).

3. FLETCHER, C.A.J. - The Group Finite Element Formulation. Comp. Meth. Appl. Mech. Engn., 1983 (to appear).

4. FLETCHER, C.A.J. - A Comparison of Finite Element and Finite Difference Solutions of the One and Two-Dimensional Burgers' Equations. J. Comp. Physics, 1983 (to appear).

5. FLETCHER, C.A.J. - Computational Galerkin Methods, Springer-Verlag, Heidelberg, 1983.

6. FLETCHER, C.A.J. and SRINIVAS, K. - Stream Function Vorticity Revisited. Comp. Meth. Appl. Mech. Engn., (1983, submitted).

7. STEGER, J.L. - Implicit Finite-Difference Simulation of Flow about Arbitrary Two-Dimensional Geometries, AIAA J, Vol. 16, No. 7, pp. 679-686, 1978.

8. BAKER, A.J., YU, J.C., ORZECHOWSKI, J.A., and GATSKI, T.B. - Prediction and Measurement of Incompressible Turbulent Aerodynamic Trailing Edge Flows, AIAA J, 20, pp. 51-59, 1982.

9. RUDY, D.H. and STRIKWERDA, J.C. - Boundary Conditions for Subsonic Compressible Navier-Stokes Calculations, Computers and Fluids, 9, pp. 327-338, 1981.

10. VISWANATH, P.R. and BROWN, J.L. - Separated Trailing-Edge Flow at a Transonic Mach Number - AIAA paper, 82-0348, 1982.

Numerical simulation of viscous flows in turbomachines

A. Fortin*, M. Fortin*, P. Tanguy**

* Département de mathématiques, Université Laval, Québec, Canada.
** Département de génie chimique, Université Laval, Québec, Canada.

SUMMARY: This paper presents the first results of a long term project aiming to the simulation of viscous incompressible flows in turbomachines. We used two finite element approaches for the discretization of the problem, one of them chosen for its possible extension to 3-D problems at a still reasonable cost. The numerical solution were computed by Newton's method and by a less standard conjugate gradient method which was temporarily rejected as too much dependent on the Reynolds number.

The final aim is to compute flows in an industrial context. We discuss how the programs developed will be integrated in a complex set of methods that should help the designer of turbomachines in an interactive way.

1. INTRODUCTION

Modern turbomachinery design is involved with computer simulation of highly turbulent flows in various complex geometries. Such a simulation requires ideally the solution of the full Navier-Stokes equations with complex boundary conditions which is clearly an impossible task in the present state-of-the-art. Simplifying assumptions must be introduced in the modelling with the hope that they will not change too much the validity of the predictions.

In this paper, we attempt to study the flow pattern inside a Kaplan turbine, and at the outlet. In an industrial context, these flows are 3-D problems at high Reynolds number and it is out of reach to try a complete solution without preliminary successful tests in simplified situations. The usual approach is to compute potential flows [1] even though it is not a realistic model. Our goal is to use the full Navier-Stokes equations. Then, in a first step, we describe the simulation of a classical 2-D cascade flow and the simulation of the flow in the deversoir, at the outlet of the turbine runner. In a second part, we will present how these numerical procedures will be integrated in a complex multi-level set of programs. The next section presents these codes and their results.

2. FINITE ELEMENT APPROXIMATION OF NAVIER-STOKES EQUATIONS

Computing 2-D incompressible viscous flows by finite element methods has now become a fairly standard procedure even if the choice of the elements must be done cleverly.

We work directly with a steady state velocity pressure formulation

(1) $-\nu\Delta\underline{u} + \underline{u}\cdot\nabla\underline{u} + \frac{1}{\rho}\nabla p = \underline{f}$

(2) $\nabla\cdot\underline{u} = 0$

where $\underline{u} = (u,v)$ is the velocity field and p the pressure. We present in figures 1 and 2 the geometries that we have considered along with the boundary conditions.

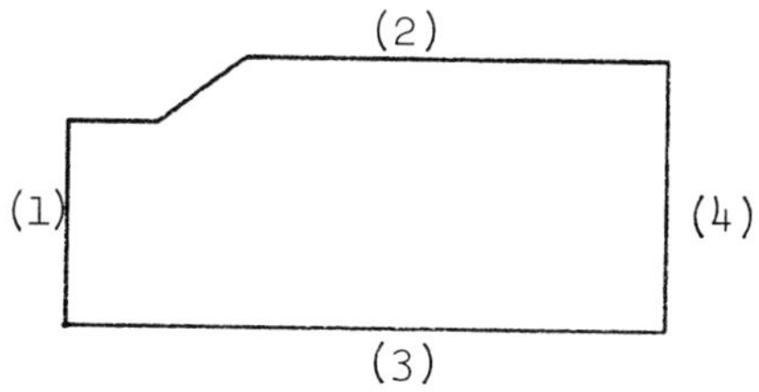

Figure 1. Deversoir

(1) $u = \bar{u}$, $v = 0$

(2) $u = v = 0$

(3) $v = 0 \quad \frac{\partial u}{\partial n} = 0$

(4) $\tau_{nn} = 0$

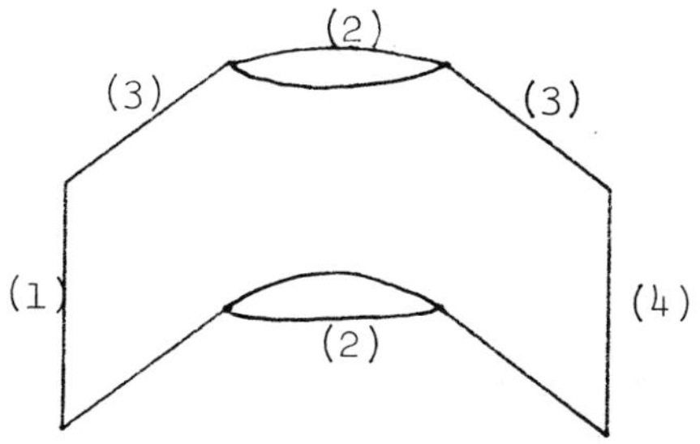

Figure 2. Cascade flow

(1) $u = \bar{u}$, $v = \bar{v}$

(2) $u = v = 0$

(3) periodicity conditions

(4) $\tau_{nn} = 0$

Let us consider an approximation of (1), (2) by a Galerkin type method. The associated variational formulation in the case of a Dirichlet problem is:

(3) $a(\underline{u},\delta\underline{u}) + b(\underline{u},\underline{u},\delta\underline{u}) + c(p,\delta\underline{u}) = (\underline{f},\delta\underline{u}) \quad \forall\delta u\in V$

(4) $c(\delta p,\underline{u}) = 0 \quad \forall\delta p\in Q$

where

(5) $a(\underline{u},\delta\underline{u}) = 2\nu\int_\Omega \dot{\underline{\underline{\gamma}}}(\underline{u}):\dot{\underline{\underline{\gamma}}}(\delta\underline{u})dx$

(6) $b(\underline{u},\underline{u},\delta\underline{u}) = \int_\Omega (\underline{u}\cdot\nabla\underline{u}).\delta\underline{u}\ dx$,

(7) $c(p,\delta\underline{u}) = -\int_\Omega \frac{p}{\rho}(\nabla.\delta\underline{u})dx$

(8) $(\underline{f},\delta\underline{u}) = \int_\Omega \underline{f}\cdot\delta\underline{u}\ dx$

and V and Q are the spaces of test velocities and test pressures respectively. To solve (3) (4), we use two types of elements.

The first is now a standard one, velocity being approximated using a 9-node biquadratic element and pressure with discontinuous piecewise linear function (figure 3) [2] [3] [4] .

Figure 3.

Q_2, P_1

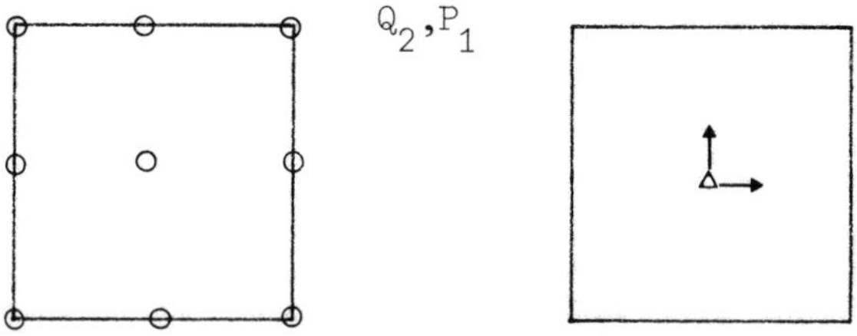

o (u,v)

△ piecewise linear pressure

The second is less standard and is described in Fortin [4]. It uses piecewise constant pressure and an incomplete biquadratic velocity, only the normal component of velocity being used as a degree of freedom at mid-side points (figure 4).

Figure 4.

R_2, P_0

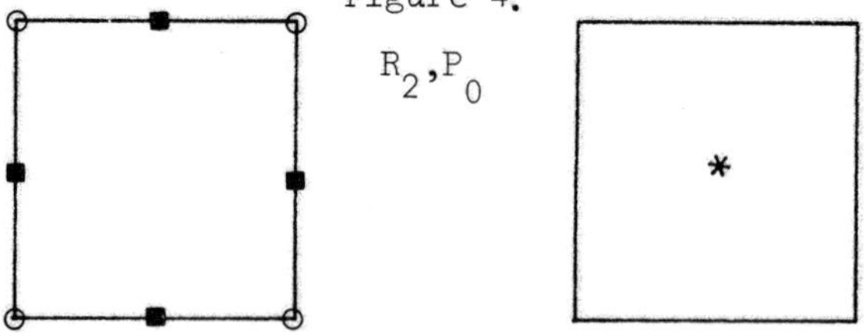

o (u,v)

■ $\underline{u}.\underline{n}$

* piecewise constant pressure

These elements have 3-D analogues described in Fortin [4] and can be used for axisymmetric problems. The first one can be shown to be $O(h^2)$, where h is the mesh size, while the second one is only $O(h)$.

Discretising the Navier-Stokes equations with these elements yields a set of non-linear equations. Denoting $\underline{U}$ the vector of nodal values of $\underline{u}$ and P the vector of nodal pressures, this set can be written

(9) $A\underline{U} + B(\underline{U})\underline{U} + C^tP = \underline{F}$

(10) $C\underline{U} = 0$.

We consider two different solution methods.

1.1. Newton's method.

We use an iterative algorithm already described in [3], which uses an exact penalty method associated with Uzawa's algorithm and a classical Newton-Raphson scheme. This gives

(11) $A\underline{U}^{n+1} + B(\underline{U}^n)\underline{U}^{n+1} + B(\underline{U}^{n+1})\underline{U}^n + \frac{1}{\lambda} C^tS^{-1}C\underline{U}^{n+1} = \underline{F} - C^tP^n - B(\underline{U}^n)\underline{U}^n$

(12) $P^{n+1} = P^n + \frac{1}{\lambda} S^{-1}C\underline{U}^{n+1}$,

where λ is the penalization parameter and S any non-singular matrix.

1.2. Least-squares-conjugate gradient method [5].

Let $W = \{\underline{v} | \nabla \cdot \underline{v} = 0$, $\underline{v}$ satisfying the boundary conditions$\}$. For $\underline{v} \in W$, we solve the associated problem:

Find $\underline{y} = \underline{y}(\underline{v})$ such that

(13) $-\nu\Delta\underline{y} + \frac{1}{\rho}\nabla p = \nu\Delta\underline{v} - \underline{v}\cdot\nabla\underline{v} + \underline{f}$

(14) $\nabla\cdot\underline{y} = 0$

(15) $\underline{y} = 0$ on the boundary.

It is now clear that $\underline{v}$ is a solution of (1), (2) if and only if $\underline{y}(\underline{v}) = 0$. It is thus natural to introduce the following least squares formulation:

Find $\underline{u} \in W$ such that

(16)

$J(\underline{u}) \le J(\underline{v}) \quad \forall \underline{v} \in W$

where $J(\underline{v}) = \frac{\nu}{2} \int_\Omega |\nabla\underline{y}(\underline{v})|^2 d\underline{x}$.

If $\underline{u}$ is a solution of (1),(2) then $J(\underline{u}) = 0$ and $\underline{u}$ is a solution of (16). Conversely, if $\underline{u}$ is such that $J(\underline{u}) = 0$, then $\underline{u}$ is a solution of (1),(2).

To this minimization problem, we apply a conjugate gradient method. The complete algorithm is described in [6].

Newton's method is classical and is very efficient. It requires the assembly and factorization of a new matrix at each step which is a costly procedure. Moreover it may fail to converge near a bifurcation point where the matrix becomes singular. The conjugate gradient method uses only a fixed matrix that can be factorized once for all.

3. NUMERICAL RESULTS

We present in figures 5,6 the computational results of the cascade flow obtained by means of Newton's scheme at Re = 200 using the restricted biquadratic velocity element (R_2,P_0) and at Re = 400 using the 9-node biquadratic element (Q_2,P_1).

It can be observed that turbulence is generated at the exit of the blade and downsteam. Though this turbulence has an obvious physical origin, it is also partly due to unrealistic boundary conditions (free outlet), particularly at high Reynolds number. In all cases, we observed that Newton's scheme is very efficient. On the other hand, the conjugate-gradient method was found to be strongly dependent on the Reynolds number and on the initial velocity fields and needs further improvement. The main interest of the (R_2-P_0) element is its low cost generalization to 3-D problems.

The flow in the deversoir was tested for Re between 0 and 2000 without any stability problem when Dirichlet boundary conditions are imposed. On the other hand, when Neumann conditions are prescribed at the inlet, instability occurs at Re = 140, probably due to a solution bifurcation. We present in figure 7, a typical result (Dirichlet conditions) at Re=400.

4. FUTURE TASKS

The question now arises of what remains to be done.

- Moving to 3-D problems: the elements are known and can be programmed readily. The open question is to find an efficient solving procedure.
- Simulation of turbulence: a k-ε turbulence model is now-being introduced in the 2-D simulations and other models are under study.
- Integration into industrial codes

Using a 3-D Navier-Stokes code to simulate a full cascade flow is not thinkable in the present state-of-the-art. We are considering the use of a set of programs.

Level 1: Potential flow computations.

They have already proved their usefulness and can be used as detectors of problems.

Level 2: Boundary layers computations.

They can be used to sharpen the results of Level 1.

Level 3: Navier-Stokes computations.

We think of restricting them to sub-regions where the preceding levels have detected possible troubles such as cavitation (low pressure zones). A window would be computed using as boundary conditions the results of Level 1 or Level 2.

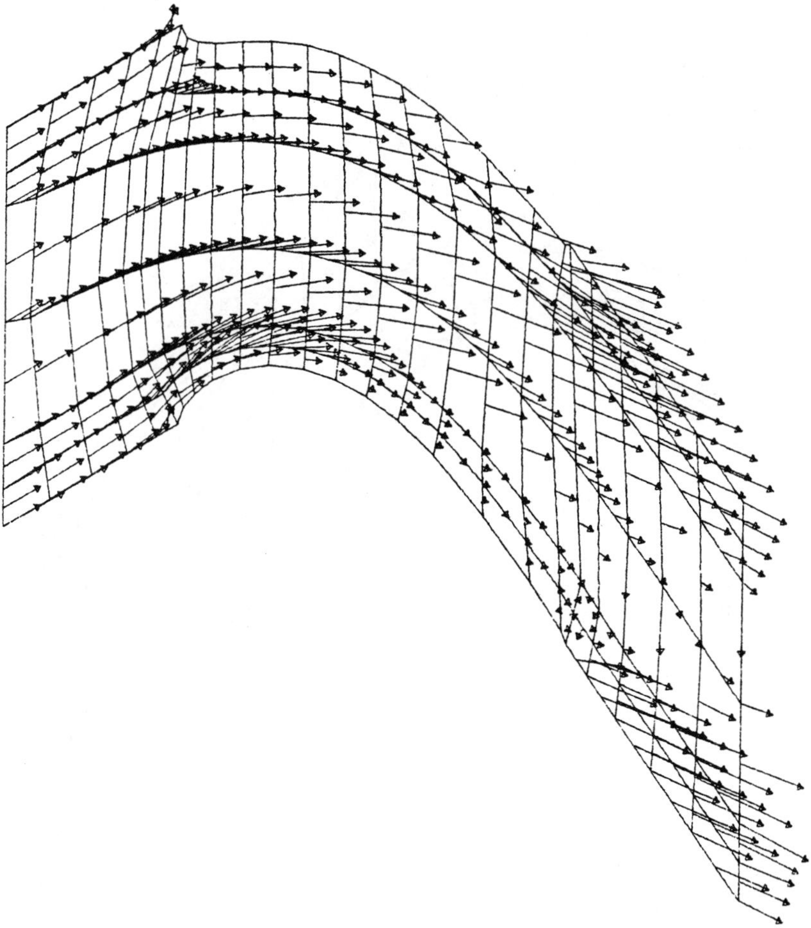

Figure 5.
Cascade flow at Re = 200
(computed with R_2,P_0)

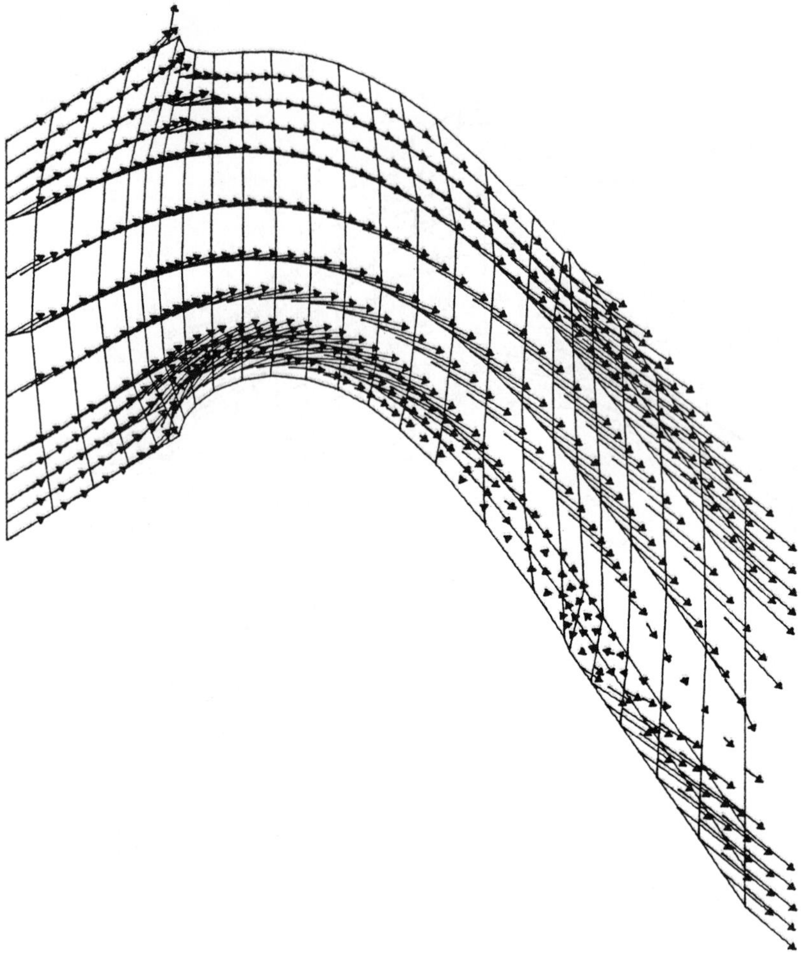

Figure 6.
Cascade flow at Re = 400
(computed with Q_2, P_1)

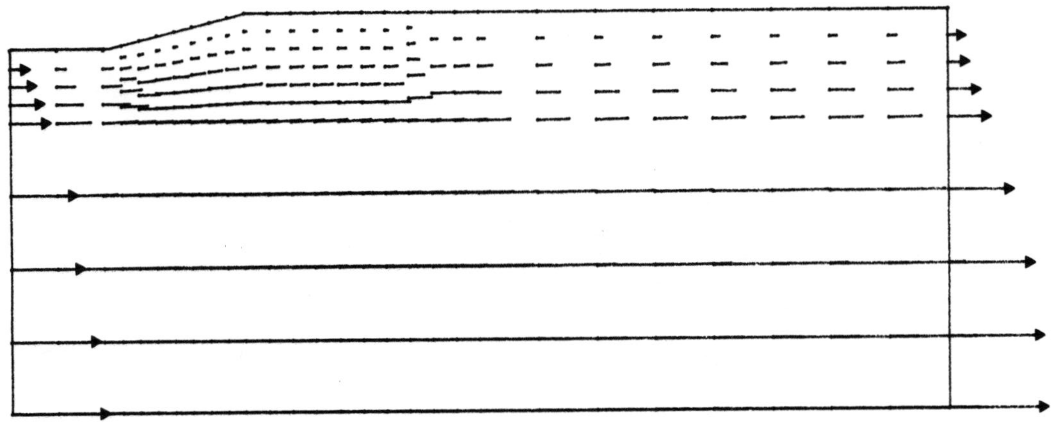

Figure 7. Deversoir at Re = 400
(computed with Q_2, P_1)

1. McNAB J.Y., HOLMES G. Application of three-dimensional finite element potential flow analysis to hydraulic turbines, Proc. Int. Symp. Refined Modeling of flows, Paris, Sept. 1982.

2. ENGELMAN M.S. et al. Consistent vs. reduced integration penalty method for incompressible media using several old and new elements, Int. J. Num. Methods Fluids, 2: 25 (1982).

3. TANGUY P. et al. Finite Element Simulation of dip coating. I-Newtonian fluids, Int. J. Num. Methods Fluids, in press.

4- FORTIN M. Old and new finite elements for incompressible flows, Int. J. Num. Meth. Fluids, 1: 347 (1981).

5. GLOWINSKI M. et al. Numerical solution of the time dependant Navier-Stokes equations for incompressible viscous fluids by finite element and alternating direction methods, Proc. Conf. Num. Meth. Aeronautical Fluid Dynamic, Reading, U.K., March-April 1981.

6- FORTIN A, Ph. D. Thesis, Laval University, Québec, to be submitted.

INTERNAL SWIRLING FLOWS BY THE DORODNITSYN FINITE ELEMENT FORMULATION

C.A.J. Fletcher[(i)]

Senior Lecturer

SUMMARY

The Dorodnitsyn boundary layer formulation is combined with a modified Galerkin finite element method to obtain the solution for swirling turbulent boundary layer flow inside a conical diffuser. The Dorodnitsyn formulation introduces a nondimensional normal velocity gradient as the primary independent variable, consequently skin friction is computed particularly accurately. The group finite element approach is used to generate a very efficient algorithm when combined with an implicit second-order, three-level, non-iterative marching algorithm. The systems of implicit equations, at each downstream location, are solved sequentially for the normal velocity gradient, circumferential velocity and pressure. An anisotropic eddy viscosity formulation is used to represent the Reynolds stresses. Results are presented for skin friction and displacement thickness for both a constant-area duct and a 10° included-angle conical diffuser with various initial swirl values. It is established that the addition of swirl helps to delay boundary layer separation.

1. INTRODUCTION

Diffuser-augmented wind-turbines are well-suited [1] to the problem of wind-generated electricity, both on the ground and in the jet-stream [2,3]. The flow through the diffuser (downstream of the turbine station) is characterised by an adverse pressure gradient and some swirl in the flow induced by the turbine.

The satisfactory performance of the diffuser-augmented wind turbine is dependent on the ability of the flow to remain

(i) Department of Mechanical Engineering, University of Sydney, N.S.W. 2006, Australia attached in the diffuser in the presence of the adverse pressure gradient. It is known that the swirl assists in the suppression of the boundary layer separation, but that the energy introduced into the flow as swirl is unavailable to generate electricity.

Therefore a computational investigation is being undertaken of the effect of adverse pressure gradient and swirl on the development of the internal boundary layer. In particular for given values of swirl one would like to know what the most adverse pressure gradient is that can be tolerated without separation occurring.

The Dorodnitsyn finite element formulation [4] is used to analyse the swirling boundary layer flow. For two-dimensional, turbulent incompressible flow it has already been demonstrated [5] that this formulation is about ten times more efficient than a representative finite difference package, STAN5.

In the Dorodnitsyn boundary layer formulation x and u are the independent variables and the nondimensional normal velocity gradient, $\partial u/\partial \eta$, becomes the dependent variable. Consequently a uniform grid in u gives high resolution close to the wall and automatically captures boundary layer growth. Since the normal velocity gradient, $\partial u/\partial \eta$, is obtained directly, the skin friction can be computed very accurately.

For internal swirling flows, the augmented axial, radial and circumferential momentum equations, are given a modified Galerkin interpretation [6] and the group finite element technique]7] is used extensively to minimise the algebraic complexity, and hence execution time. At each station in the flow direction the axial momentum equation is solved first for $\partial u/\partial \eta$ using a non-iterative implicit algorithm. Subsequently the circumferential momentum equation is solved for the swirl velocity, w. In the Dorodnitsyn formulation the major influences of the swirl velocity, w, on the axial momentum equation occur through the pressure gradient term and the eddy viscosity. The pressure variation across the boundary layer is obtained from the radial momentum equation. An anisotropic eddy viscosity formulation [8] is used to represent the Reynolds stresses.

In this paper we present the governing equations, in Dorodnitsyn form, in Section 2. The Dorodnitsyn finite element formulation is introduced in Section 3 and, subsequently, an efficient implicit marching algorithm is constructed. Results for representative swirling flows in a constant-area axisymmetric duct and a conical diffuser are discussed in Section 4.

2. EQUATIONS OF MOTION

In polar coordinates $(x, r, \theta; u, v, w)$ the equations of motion governing the flow in swirling boundary layers can be written, in the following nondimensional form,

$$\frac{\partial}{\partial x}(ru) + \frac{\partial}{\partial r}(rv) = 0 \tag{1}$$

$$ru\frac{\partial u}{\partial x} + rv\frac{\partial u}{\partial r} = - r\frac{\partial p}{\partial x} + \frac{1}{Re}\frac{\partial}{\partial r}\left\{r\left(1 + \frac{\nu_x}{\nu}\right)\frac{\partial u}{\partial r}\right\} \tag{2}$$

$$\frac{\partial p}{\partial r} = \frac{w^2}{r} \tag{3}$$

$$ru\frac{\partial w}{\partial x} + rv\frac{\partial w}{\partial r} + vw = \frac{1}{Re}\frac{\partial}{\partial r}\left\{r\left(1 + \frac{\nu_\theta}{\nu}\right)\frac{\partial w}{\partial r}\right\} . \tag{4}$$

In equations (1) to (4) lengths x,r have been nondimensionalised with the boundary layer development length, L; u,v and w have been nondimensionalised with the upstream velocity, U_∞. The Reynolds number, $Re = U_\infty L/\nu$ and ν_x, ν_θ are the axial and circumferential eddy viscosities introduced to represent the Reynolds stresses. It may be noted that in eqs (1) to (4) the dependent variables, u,v,w and p, are functions of x and r only; there is no circumferential dependence.

The equation set, (1) to (4), is parabolic in character, with the downstream direction, x, having a time-like role. Consequently initial conditions, at some value of x, are required in the form u(r), w(r) given. Boundary conditions at the wall, u=v=w = 0, are enforced. At the edge of the boundary layer the flow is inviscid and the Bernoulli equation holds i.e.

$$\frac{P_\infty}{\rho} + \frac{1}{2}u_\infty^2 = \frac{P_e}{\rho} + \frac{1}{2}u_e^2 + \frac{1}{2}w_e^2 . \tag{5}$$

In the present paper it is assumed that the angular momentum of the initial inviscid swirl distribution does not vary with radius, i.e. $w_e = \frac{k}{r}$, and that angular momentum is conserved (in the inviscid flow) at all downstream stations, i.e.

$$\frac{\partial w_e}{\partial r} = - \frac{w_e}{r} .$$

Consequently the inviscid circumferential momentum equation indicates that $w_e = w_e(r)$ only i.e. w_e is not a function of x. Combination of the radial momentum equation and the Bernoulli equation then indicates that $u_e = u_e(x)$ only, i.e. u_e is not a function of r. Thus at the outer edge of the boundary layer we have the following relationships:

$$w_e = k/r_e \;\; ; \;\; \partial P_e/\partial r = - w_e\, \partial w_e/\partial r = w_e^2/r, \qquad \text{and } \partial P_e/\partial x = - u_e\, \partial u_e/\partial x . \tag{6}$$

The Dorodnitsyn formulation is introduced in two stages. In the first stage the independent variables (x,r) are replaced by (ζ,η) where

$$\zeta = x \quad \text{and} \quad \eta = Re^{\frac{1}{2}}\, u_e(x)\,(r_w - r) \quad .$$

In addition a general test function, $f_k(u)$, is introduced and the following composite equations formed

$$f_k \times \text{eq.}(1) + \frac{df_k}{du} \times \text{eq.}(2) = 0 \tag{7}$$

$$\text{and} \quad wf_k \times \text{eq.}(1) + \frac{wdf_k}{du} \times \text{eq.}(2) + f_k \times \text{eq.}(4) = 0 \quad . \tag{8}$$

In the second-stage, the auxiliary variable, $q = P_e - P$, is introduced. Eqs. (7) and (8) are integrated across the boundary layer and the variable of integration changed from η to u. The result is

$$\begin{aligned}\frac{\partial}{\partial\zeta}\int_0^1 f_k u\, \Theta\, du = {} & \frac{u_{e_\zeta}}{u_e}\int_0^1 \frac{df_k}{du}\,\Theta\,(1-u^2)\,du + \frac{1}{u_e^{\,2}}\int_0^1 \frac{df_k}{du}\, Q\, \Theta\, du \\ & + u_e \int_0^1 \frac{df_k}{du}\frac{d}{du}\left(\left(1 + \frac{\nu_x}{\nu}\right)T\right)du \\ & - \frac{1}{Re^{\frac{1}{2}}}\left[\int_0^1 \frac{df_k}{du}\,\frac{(1+\nu_x/\nu)}{r}\,du + \int_0^1 f_k\left(\frac{v}{rT}\right)du\right]\end{aligned} \tag{9}$$

and

$$\begin{aligned}\frac{\partial}{\partial\zeta}\int_0^1 f_k uw\,\Theta\, du = {} & \frac{u_{e\zeta}}{u_e}\int_0^1 \frac{df_k}{du}(1-u^2)\; w\,\Theta\, du + \frac{1}{u_e^2}\int_0^1 \frac{df_k}{du}\, w\, Q\, \Theta\, du \\ & + u_e \int \frac{df_k}{du}\, w\, \frac{\partial}{\partial u}\left\{\left(1 + \frac{\nu_x}{\nu}\right)T\right\} du \\ & - \frac{u_{e\zeta}}{u_e}\int_0^1 f_k\; uw\,\Theta\, du + u_e \int_0^1 f_k \frac{\partial}{\partial u}\left\{\frac{\partial w}{\partial u}\left(1 + \frac{\nu_\theta}{\nu}\right)T\right\} du \\ & - \frac{1}{Re^{\frac{1}{2}}}\left[\int_0^1 \frac{df_k}{du}\, w\, \frac{(1+\nu_x/\nu)}{r}\,du + \int_0^1 f_k \frac{\nu_\theta/\nu}{r}\,\frac{\partial w}{\partial u}\,du + 2\int_0^1 f_k\,\frac{vw}{r}\,\Theta\, du\right] .\end{aligned} \tag{10}$$

In equations (9) and (10) $T = 1/\Theta = \partial u/\partial\eta$ and $u_{e\zeta} \equiv du_e/d\zeta$. In equations (9), (10) Θ , T and w are dependent variables and x,u are the independent variables. The variable Q is defined as follows:

$$Q = \partial q/\partial\zeta + \left[(r_w - r)u_{e\zeta}/u_e + r_{wx}\right]\left[u_e^2\, w^2/r - w_e^2/r_e\right] .$$

To obtain the variation of the pressure across the boundary layer eq. (3) is manipulated to give

$$\left[f_k q\right]_o^1 - \int_o^1 \frac{df_k}{du}\, q\, du = \frac{u_e}{Re^{\frac{1}{2}}} \int_o^1 f_k\, \Theta \left\{ \frac{w^2}{r} - \frac{w_e^2/u_e^2}{r_e} \right\} du \quad . \tag{11}$$

In eqs (9) and (10) the eddy viscosities, ν_x and ν_θ, are given by (after Lilley [8]),

$$\frac{\nu_x}{\nu} = \{\ell^+ \ (1 + \gamma_1 w)\}^2 \frac{r}{r_w} \left[\left(\frac{T}{T_1}\right)^2 + \left(\frac{\partial w}{\partial u}\frac{T}{T_1} - \frac{w/r}{Re^{\frac{1}{2}} u_e T_1}\right)^2 \right]^{\frac{1}{2}} ,$$

where ℓ^+ is the nondimensional mixing length which is subjected to the van Driest damping factor close to the wall. In the outer region of the boundary layer a modified Clauser formulation is used i.e.

$$\nu_x/\nu = 0.0168\ (1 + \gamma_1 w_{av})^2\ u_e Re\,\delta^*/L \quad ,$$

where γ_1 is an empirical constant and δ^* is the displacement thickness. Following Lilley [8] the eddy viscosity, ν_θ, is given by

$$\nu_\theta/\nu = \nu_x/\nu/(1 + \gamma_2\, w^{1/3})$$

where γ_2 is an empirical constant.

3. DORODNITSYN FINITE ELEMENT FORMULATION

The formulation used here is similar to that applied to two-dimensional laminar [4] and turbulent [5] boundary layer flow. Trial solutions are introduced for various groups [7] of terms, e.g.

$$(1 + \nu_x/\nu)T = \sum_{j=1}^{M} (1 - u) N_j(u)(1 + \nu_x/\nu)_j \tau_j(\zeta) \quad . \tag{12}$$

Also the test function $f_k(u)$ is represented by

$$f_k(u) = (1 - u)\ N_k(u) \ . \tag{13}$$

In eqs (12) and (13), N_j, N_k are one-dimensional linear shape functions. Substitution into eqs (9), (10) and (11) indicates that a modified Galerkin formulation [6] is produced. Equations (9), (10) and (11) become

$$\sum_j CC_{kj}\frac{d\theta_j}{d\zeta} = \frac{u_{e\zeta}}{u_e}\sum_j EF_{kj}\theta_j + \frac{1}{u_e^2}\sum_j GG_{kj}\, Q_j\theta_j + u_e\sum_j AA_{kj}\left\{\left(1+\frac{\nu_x}{\nu}\right)\tau\right\}_j$$

$$- \frac{1}{Re^{\frac{1}{2}}}\left\{\sum_j EE_{kj}\left\{\frac{v\theta}{r}\right\}_j + \sum_j GG_{kj}\left(\frac{1+\nu_x/\nu}{r}\right)_j\right\} \qquad (14)$$

$$\sum_j CC_{kj}\frac{d}{d\zeta}(w\theta)_j = u_{e\zeta}/u_e \sum_j (EF_{kj}-CC_{kj})(w\theta)_j + (\sum_j GG_{kj}(wQ\theta)_j)/u_e^2$$

$$+ u_e\sum_j GG_{kj}\left[w\left\{(1+\nu_x/\nu)\tau\right\}_u\right]_j$$

$$+ u_e\sum_j FF_{kj}\left((1+\nu_\theta/\nu)w_u\tau\right)_j - \left[\sum_j GG_{kj}\left((1+\nu_x/\nu)w/r\right)_j\right. \qquad (15)$$

$$\left. + \sum_j BB_{kj}\left((1+\nu_\theta/\nu)w_u/r\right)_j + 2\sum_j EE_{kj}(vw\theta/r)_j\right] \Big/ Re^{\frac{1}{2}}$$

and

$$- \sum_j GG^m_{kj} q_j = u_e \sum_j EE_{kj}\theta_j \left\{w^2/r - w_e^2/u_e^2/r_e\right\} \Big/ Re^{\frac{1}{2}} \qquad (16)$$

where the various algebraic coefficients, like CC_{kj}, arise from the evaluation of the Galerkin integrals. For example

$$CC_{kj} = \int_o^1 N_j N_k\, u\, du \ .$$

The equations, (14) and (15) are marched downstream using a three-level implicit algorithm. Equations (14) and (15) can be written in the following way (eq. 14 shown)

$$\Sigma CC_{kj}\left\{\alpha\, \Delta\theta_j^{n+1} + (1-\alpha)\Delta\theta_j^n\right\} = \Delta\zeta(\beta\ RHS^{n+1} + (1-\beta)\ RHS^n)\ , \qquad (17)$$

where $\Delta\theta_j^{n+1} = \theta_j^{n+1} - \theta_j^n$ and α and β are chosen to weight the evaluations at the time levels n and n+1. Equation (17) is to be manipulated to give an implicit system of equations for $\Delta\theta_j^{n+1}$. To achieve this RHS^{n+1} is linearised about time-level n. The final algorithm can be written

$$\sum_j \left\{CC_{kj} - \beta\Delta\zeta\partial\ RHS/\partial\theta\right\}\Delta\theta_j^{n+1} = \Delta\zeta\ RHS^{n,\beta} - (1-\alpha)\ \sum_j CC_{kj}\Delta\theta_j^n\ . \quad (18)$$

$RHS^{n,\beta}$ is evaluated at $\zeta^n + \beta\Delta\zeta$ for terms other than θ and τ. At each downstream station eq. (18) is solved for $\Delta\theta^{n+1}$. Subsequently a similar equation, based on eq. (15), is solved to obtain w_j. Then eq. (16) is solved for q_j. In the present study the values $\alpha = 1.5$, $\beta = 1.0$ have been used which produce a stable second-order marching algorithm.

4. RESULTS AND DISCUSSION

Two flow geometries have been considered. In Fig. 1 profile ABD represents the flow at the entrance of a constant-area axisymmetric duct. Profile ABC represents a conical diffuser of 10° included-angle. The computation in each case starts at B where the Reynolds number, based on the upstream boundary layer length, is Re = 0.386 x 10^6. This matches the value for the experimental data of Fraser [9] for the flow in 10° included-angle conical diffuser.

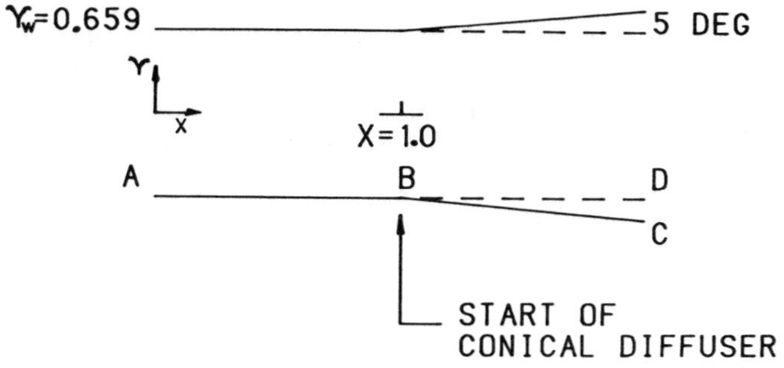

Fig. 1 Flowfield geometry

The skin friction variation with downstream position for the constant-area duct case is shown in Fig. 2 $(w_e/u_e)_i$ refers

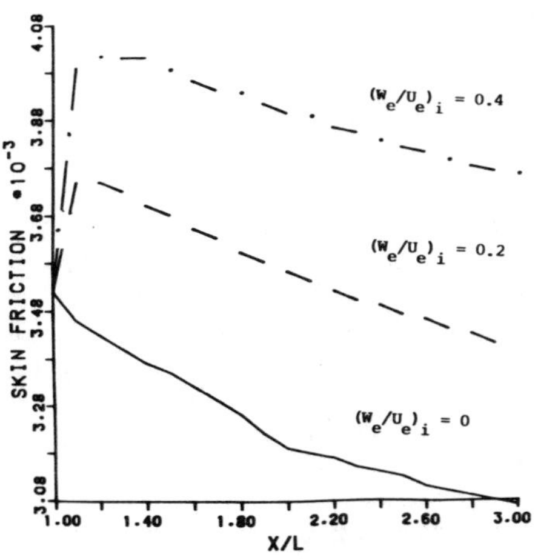

Fig. 2 Skin friction variation for constant-area duct.

to the initial (atB) ratio of circumferential velocity to axial velocity at the outer edge of the boundary layer. The initial velocity profile (at B) is taken from Fraser's experimental data [9] plus a linear (with u) variation of w across the boundary layer.

It is clear that the introduction of swirl increases the skin friction without changing the expected gradual reduction in skin friction in the axial direction. For steady flow separation corresponds to zero skin friction. Thus the introduction of swirl delays the occurrence of separation. Of course, for the constant-area duct, separation will not occur in the zero swirl case. The rapid initial increase in skin friction is presumably due to the response of the zero swirl initial axial velocity distribution to the imposition of swirl.

The corresponding variation of displacement thickness is shown in Fig. 3. A gradual increase in displacement thickness with downstream position is evident. The numerical results indicate a very small reduction in displacement thickness with increasing swirl.

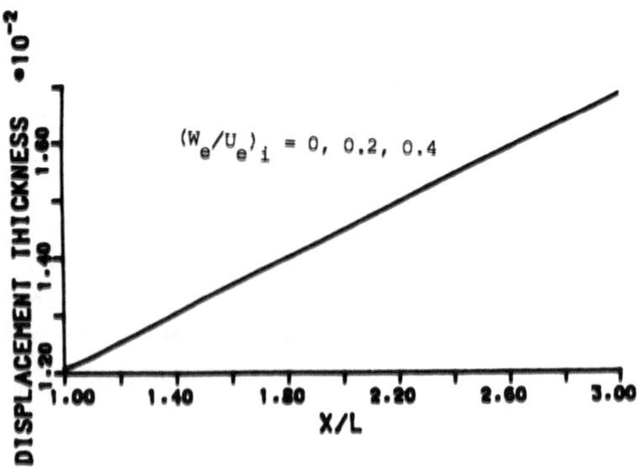

Fig. 3 Displacement thickness variation for a constant-area duct.

For the boundary layer development inside the 10° conical diffuser the variation of skin friction with downstream position is shown in Fig. 4. Included in Fig. 4 are the experimental results of Fraser [9] at zero swirl. The present method is seen to predict the experimental results satisfactorily. This flow is operating in a severe adverse pressure gradient which would cause the boundary layer to separate at some location further downstream.

The starting w profile for the swirl cases is taken from the downstream solutions (x = 3.00) of the constant-area duct results. As with the constant-area duct case the introduction

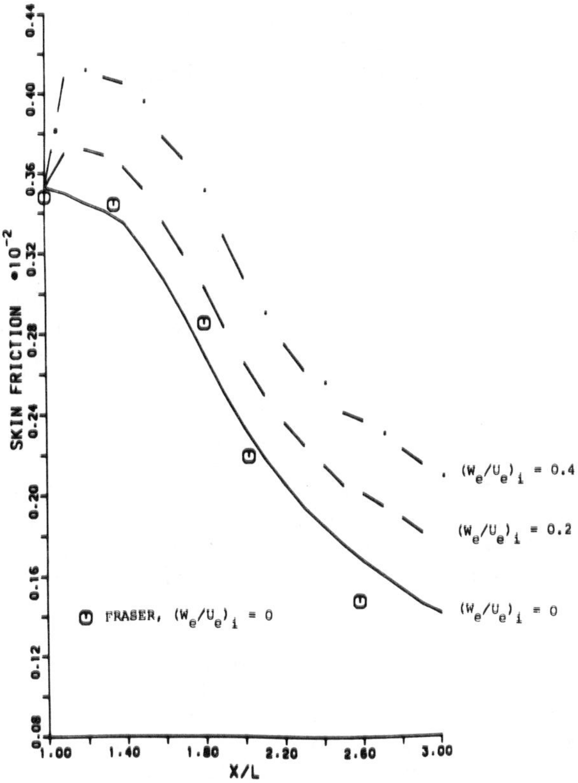

Fig. 4 Skin friction variation for 10° conical diffuser

of swirl increases the magnitude of the skin friction at a particular axial location. By inference separation of the boundary layer is delayed.

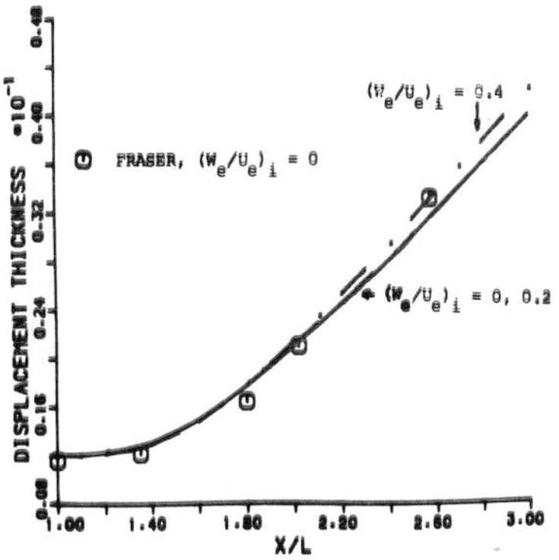

Fig. 5 Displacement thickness variation for a 10° conical diffuser

The variation of the displacement thickness for the 10° conical diffuser case is shown in Fig. 5. Due to the adverse pressure gradient the displacement thickness grows rapidly with axial location. For the zero swirl case there is good agreement with the experimental results of Fraser [9]. The effect of swirl is to cause a slight increase in the boundary layer (and displacement) thickness.

The good agreement with the experimental data [9] for the zero swirl case has been achieved with only 11 nodal points across the boundary layer. The high accuracy per number of nodal points for the Dorodnitsyn finite element formulation has been observed elsewhere [5].

However it does not follow that the swirl results are as accurate. The accuracy of the swirl results also depends on the choice of the empirical parameters, γ_1 and γ_2, introduced to represent the dependence of the eddy viscosities on the swirl. Here we have used correlations obtained by Lilley [8] from swirling jets. Therefore the results with swirl are considered to demonstrate a qualitative trend, only, at this stage.

It is concluded that the Dorodnitsyn finite element formulation is well-suited to the calculation of swirling boundary layer flows and that the introduction of swirl will delay boundary layer separation. However quantitative prediction of swirl effects requires validation of the empirical constants used in the eddy viscosity turbulence model.

ACKNOWLEDGEMENT

The author is grateful to Dr. K. Srinivas for his assistance.

REFERENCES

1. FLETCHER, C.A.J. - Computational Analysis of Diffuser-Augmented Wind Turbines, Energy Conversion and Management, Vol. 21, pp. 175-184, 1981.

2. FLETCHER, C.A.J. and ROBERTS, R.W. - Electricity Generation from Jet-stream Winds, J. of Energy, Vol. 3, pp. 241-249, 1979.

3. FLETCHER, C.A.J., HONAN, A., and SAPUPPO, J. - Aerodynamic Platform Comparison for Jet-stream Electricity Generation, J. of Energy, Vol. 7, 1983.

4. FLETCHER, C.A.J. and FLEET, R.W. - The Dorodnitsyn Boundary Layer Formulation for Laminar Boundary Layers, Int. J. Num. Meth. Fluids, to appear, 1983.

5. FLETCHER, C.A.J. and FLEET, R.W.- The Dorodnitsyn Boundary Layer Formulation for Turbulent Boundary Layers, Comp. and Fluids, submitted 1982.

6. FLETCHER, C.A.J.- Computational Galerkin Methods, Springer-Verlag, Heidelberg, 1983.

7. FLETCHER, C.A.J. - The Group Finite Element Formulation, Comp. Meth. Appl. Mech. Eng., to appear, 1983.

8. LILLEY, D.G. - Prediction of Inert Turbulent Swirl Flows, AIAA J., Vol. 11, pp. 955-960, 1973.

9. COLES, P. and HIRST, E. (eds.), Computation of Turbulent Boundary Layers - 1968, AFOSR-ISP Stanford Conference, Vol. 2, pp. 451- 465, 1968.

NUMERICAL SOLUTION OF TURBULENT FLOW THROUGH VANELESS DIFFUSERS OF CENTRIFUGAL COMPRESSORS.

A. ABIR + and A. WHITFIELD *

+ Research Products, Rehovot, Israel
* University of Bath, U.K.

SUMMARY

The operating range of turbocharger centrifugal compressors is partially dependent upon the stability of the vaneless diffuser. The performance of the diffuser is in turn dependent upon the nature of the flow at diffuser inlet, i.e. that issuing from the compressor impeller. In order to study the effect of inlet flow profiles and diffuser design upon the internal flow structure a finite element based prediction procedure has been developed. The technique solves the Navier Stokes equations using a stream function vorticity formulation, with the $k - \varepsilon$ model used to represent the turbulence.

A finite element Galerkin and variational approach has been used with the equations solved separately. A dynamic mesh system is used which solves for the centre node of each mesh. The mesh is then regenerated so that the computed values are no longer centre nodes and are used instead to compute the values at the new centre nodes. The mesh is modified in steps so that all nodes, except the boundary values, become centre nodes.

The numerical investigation has been carried out to study the effect of different inlet flow conditions upon the internal flow structure and stability of the diffuser. Good agreement has been demonstrated with available published experimental and theoretical results.

1. INTRODUCTION

The turbocharged automotive Diesel engine has to operate through a wide range of conditions; consequently the turbocharger compressor must also have a broad operating range. The compressor is usually a centrifugal design and the operating range is limited by choke conditions at high flow rates and surge conditions at low flow rates. Surge is a highly unstable flow condition, with the air flow periodically passing through the compressor in the reverse direction. The precise cause of surge is not fully understood, although it is well known that flow instabilities occur in either the compressor impeller or vaneless diffuser or both prior to surge. Good vaneless diffuser performance is, therefore, essential not only to provide good overall efficiencies but also to maintain a broad operating range.

At the University of Bath a significant research effort has been applied to the development of turbocharger compressor design with a specific view to high efficiency and broad operating range. This has led to the design of mixed flow as well as the more conventional radial flow impellers. Whilst the radial impellers lead quite naturally to the use of conventional radial vaneless diffusers, for the mixed flow designs a choice exists between a straight conical diffuser and a curved diffuser, with varying degrees of curvature possible, see fig. 1.

The nature of the flow through a radial compressor is extremely complex being three dimensional, turbulent and unsteady. In order to develop theoretical solutions simplifying assumptions are essential. Early solutions were based upon the model given by Wu [1] and employed streamline curvature techniques [2]. This method of solution does not, however, readily lend itself to the incorporation of turbulence models nor can it be extended to the discharge volute downstream of the vaneless diffuser. More recent theoretical approaches have been applied to the solution of the Navier-Stokes equations using either finite difference or finite element techniques [4], [5].

Whilst the duct geometry encountered with vaneless diffusers is not complex, it was decided to employ the finite element method with a view to its later development to include the compressor volute. A stream function-vorticity formulation has been employed, with the $k - \varepsilon$ model used to represent the turbulence. The equations have been solved separately with the Galerkin method used to solve the stream function equation and the variational principle applied to the other equations.

The experimental investigation associated with this theoretical study has been carried out on six times full size models of turbocharger diffusers, thereby making it possible to measure the detailed internal flow structure using five hole probes and hot wire anemometers.

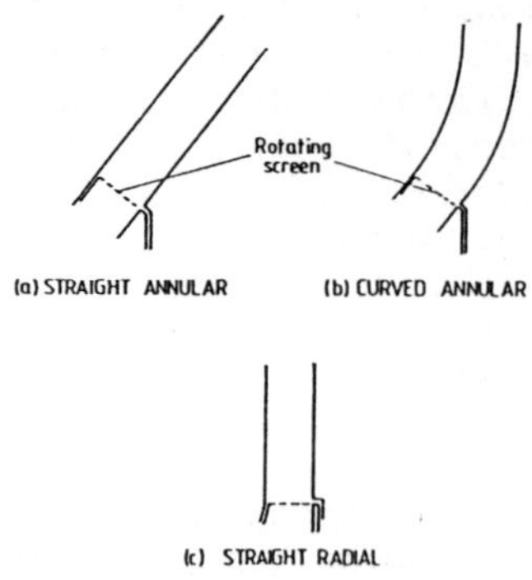

Fig.1 - Types of Diffuser

2. COMPUTATIONAL MODEL OF THE TWO DIMENSIONAL FLOW FIELD.

The flow through the three types of annular diffusers under investigation was considered to be steady, incompressible and axisymmetric. The assumption of axisymmetric flow leads to a two dimensional flow problem with no property variations in the θ direction, i.e. $\frac{\partial}{\partial \theta} = 0$.

2.1 The basic governing equations

The equations governing laminar and turbulent flow of a fluid are the conservation of momentum equations, the Navier-Stokes equations, and the continuity of mass equation. These equations are well established, Hinze [6]; by introducing the eddy viscosity concept and ignoring the second order derivates of viscosity, the equations become

$$U_r \frac{\partial U_r}{\partial r} + U_z \frac{\partial U_r}{\partial Z} - \frac{U_\theta^2}{r} =$$

$$-\frac{1}{\rho}\frac{\partial P}{\partial r} + \nu_e \left(\frac{1}{r}\frac{\partial U_r}{\partial r} + \frac{\partial^2 U_r}{\partial r^2} + \frac{\partial^2 U_z}{\partial Z^2} - \frac{U_r}{r^2} \right) \tag{1}$$

$$U_r \frac{\partial U_Z}{\partial r} + U_z \frac{\partial U_Z}{\partial Z} = -\frac{1}{\rho}\frac{\partial P}{\partial Z} + \nu_e \left(\frac{1}{r}\frac{\partial U_Z}{\partial r} + \frac{\partial^2 U_Z}{\partial r^2} + \frac{\partial^2 U_Z}{\partial Z^2} \right) \tag{2}$$

$$U_r \frac{\partial U_\theta}{\partial r} + U_Z \frac{\partial U_\theta}{\partial Z} + \frac{U_r U_\theta}{r} =$$

$$\nu_e \left(\frac{\partial^2 U_\theta}{\partial r^2} + \frac{\partial^2 U_\theta}{\partial Z^2} + \frac{1}{r}\frac{\partial U_\theta}{\partial r} - \frac{U_\theta}{r^2} \right) \tag{3}$$

and for continuity $\frac{\partial}{\partial r}(r\, U_r) + \frac{\partial}{\partial Z}(r\, U_Z) = 0$ (4)

where U_r, U_Z and U_θ are the time average velocities in the r, Z and θ directions respectively. To solve these equations the pressure terms in equations (1) and (2) were eliminated by differentiating with respect to Z and r respectively and subtracting. The dependent variables U_r and U_Z were replaced by the stream function ψ and vorticity ξ defined as :-

$$\frac{1}{r}\frac{\partial \psi}{\partial r} = \rho\, U_Z \qquad \frac{1}{r}\frac{\partial \psi}{\partial Z} = -\rho\, U_r \tag{5}$$

and for vorticity $\omega = \zeta/r = \frac{1}{r}\left(\frac{\partial U_r}{\partial Z} - \frac{\partial U_Z}{\partial r}\right)$ (6)

The definition of stream function ensures that the continuity equation is satisfied, leaving equation (3), the combined form of equations (1) and (2), and equation (6) to be solved for ψ, ω, and rU_θ. Following Gosman et al [7] the resulting equations can be expressed in the general form as

$$a\left[\frac{\partial}{\partial Z}\left(\phi\frac{\partial\psi}{\partial r}\right) - \frac{\partial}{\partial r}\left(\phi\frac{\partial\psi}{\partial Z}\right)\right] - \frac{\partial}{\partial Z}\left[br\,\frac{\partial(c\phi)}{\partial Z}\right]$$
$$- \frac{\partial}{\partial r}\left[br\,\frac{\partial(c\phi)}{\partial r}\right] + rd = 0 \qquad (7)$$

where ϕ represents any of the variables ψ, ω, rU_θ and the corresponding parameters a,b,c and d are given in table 1.

ϕ	a	b	c	d
ψ	0	$\frac{1}{\rho r^2}$	1	$-\omega$
ω	r^2	r^2	μ_e	$-\frac{\rho}{r^2}\frac{\partial(rU_\theta)^2}{\partial Z} - r^2 S_\omega$
rU_θ	1	$\mu_e r^2$	$1/r^2$	0
k	1	$\mu + \frac{\mu_t}{\sigma_k}$	1	$\rho\varepsilon - G.\mu_t$
ε	1	$\mu + \frac{\mu_t}{\sigma_\varepsilon}$	1	$\frac{C_2\varepsilon^2\rho}{k} - \frac{C_1\varepsilon\,\mu_t G}{k}$

$$G = 2\left[\left(\frac{\partial U_Z}{\partial Z}\right)+\left(\frac{\partial U_r}{\partial r}\right)^2+\left(\frac{U_r}{r}\right)^2\right]+\left(\frac{\partial U_Z}{\partial r}+\frac{\partial U_r}{\partial Z}\right)^2+\left(\frac{\partial U_\theta}{\partial Z}\right)^2+\left(r\,\frac{\partial}{\partial r}\left(\frac{U_\theta}{r}\right)\right)^2$$

and S_ω is assumed to be zero.

TABLE 1 Coefficients of the basic equations for turbulent flow.

The term S_ω in the vorticity equation has been assumed to be zero. This procedure was adopted by Gosman [7] and Oliver [10] although it is not clear whether this is acceptable for the highly swirling flows encountered in this study.

2.2 The Turbulence Model

In order to solve equation (7) the effective turbulent viscosity ν_e must be computed through a turbulence model. A broad review of turbulence models were presented by Launder and Spalding [9] and for this investigation the two equation, k, ε model has been adopted [10]. This model consists of two dynamic equations for the turbulent kinetic energy, k, and the turbulent energy dissipation, ε. The transport equations for k and ε in cylindrical co-ordinates following [8] and [11] can be represented in the general form by equation (7) with the coefficients a, b, c and d as given in table 1. Knowing local values of k and ε, the turbulent viscosity μ_t may be determined from the Prandtl-Kolmogorov relationship

$$\mu_t = C_\mu \rho\, k^2/\varepsilon \qquad (8)$$

For the empirical coefficients used in the above equations the following magnitudes have been applied

$$C_\mu = 0.09; \; C_1 = 1.44; \; C_2 = 1.92; \; \sigma_\varepsilon = 1.3; \; \sigma_k = 1.0$$

2.3 Boundary Conditions

In order to solve the basic equations it is necessary to specify the magnitude of all the unknown parameters on the inlet and discharge boundaries and on the solid wall boundaries.

2.3.1 Inlet and discharge boundary conditions

For the inlet boundary most of the variables were specified directly or readily calculated. The stream function was calculated from the measured or specified inlet velocity profile.

The vorticity was calculated through equation (6) with the assumption that

$$\frac{\partial U_Z}{\partial r} << \frac{\partial U_r}{\partial Z} \quad \text{then } \omega = \frac{1}{r}\frac{\partial U_r}{\partial Z} \qquad (9)$$

In order to specify the turbulence parameters k and ε detailed experimental data would be needed for each flow condition. Instead of this k was determined from the inlet velocity profile using a correlation proposed by Bobkov and given by Yamamoto [12] as

$$k = \bar{U}^2 \left(1 - \frac{U_{av}}{U_{max}}\right)^2 A\, e^{-B\bar{n}} \qquad (10)$$

The constants A and B were quoted by Yamamoto as 0.7225 and 2.54 respectively. $\bar{n}$ is the distance normal to the wall non-dimensionalized by the distance to the maximum velocity U_{max}, and

$$\bar{U}^2 = U^2_r + U^2_Z + U^2_\theta$$

The dissipation rate ε was calculated through equation (8) from a specified viscosity μ_t and the turbulence kinetic energy k, as given by equation 10.

At the downstream boundary Neumann boundary conditions have been applied i.e. the gradient normal to the boundary was specified.

2.3.2 Wall boundary conditions

The stream function and swirl velocity boundary conditions

can be readily specified. A constant value of stream function was specified for each wall and due to the no slip condition the swirl velocity is zero.

The wall vorticity cannot be specified directly but must be computed from known stream function values. Gosman et al [7] derived the wall boundary vorticity through a Taylor series expansion of the stream function, with the no slip condition applied at the wall, and by assuming a linear variation of vofticity. This led to the expression

$$\omega_A = - \frac{3}{r_A^2 \rho n^2} (\psi_B - \psi_A) + \frac{\psi_B}{2} \tag{11}$$

where subscripts A and B represent nodal points on the wall and distance n from the wall respectively.

In order to specify the turbulence parameters in the wall regions the technique known as 'the wall function' has been applied. In this approach the first nodal point is placed at a distance y from the wall in the logarithmic region of the boundary layer. The velocity profile is then assumed to be given by the logarithmic law of the wall

$$U^+ = 2.5 \ \ell n y^+ + 5.5 \tag{12}$$

where $U^+ = U/\sqrt{\tau_s/\rho}$ and $y^+ = \frac{Z}{\nu_e} \cdot \sqrt{\frac{\tau_s}{\rho}}$

Following Oliver [18] the turbulence parameters k and ε were derived from:

$$k = \tau_s/\rho \ C_\mu^{\frac{1}{2}} \tag{13}$$

and

$$\varepsilon = \left(\frac{\tau_s}{\rho}\right)^{3/2} \frac{1}{y\left[\sigma_\varepsilon C_\mu^{\frac{1}{2}} (C_2 - C_1)\right]^{\frac{1}{2}}} \tag{14}$$

The use of the wall function also leads to an alternative to equation (11) for the vorticity as :

$$\omega = \frac{1}{r} \tau_s/\mu$$

3. SOLUTION PROCEDURE

For low Reynolds number constant viscosity flows the equations were solved by applying the Galerkin Method; however, at high Reynolds numbers using the full turbulence model it was not possible to obtain a converged solution. To overcome this a variational approach was developed for the solution of all equations except the stream function equation. The numerical solution employed an iterative process, each equation being solved separately, the computed magnitudes from one equation being used in the next. For the Galerkin method, applied for the solution of the stream function, a nine noded Lagrangian element was used. For the variational principle, applied to the other equations, the same nine noded element was used but instead of describing the variables by second order interpolation functions the element was divided into a number of triangular elements with a common node at the centre. Each triangle was then considered as a three noded element with linear interpolation functions. During the development of the procedure two other elements, as shown in fig. 2, were used. The numerical solution procedure adopted was based upon that reported by Moult [13]. The procedure yielded a solution for the required variable at the centre nodes only. Consequently the mesh had to be systemically regenerated so that each node, other than boundary nodes, became centre nodes in turn, the procedure is illustrated in fig. 3. Full details of the theoretical and numerical procedure is given in reference 14.

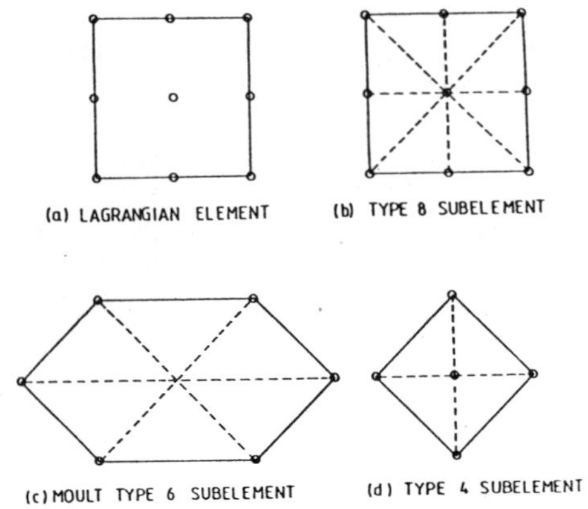

Fig. 2 Types of elements used.

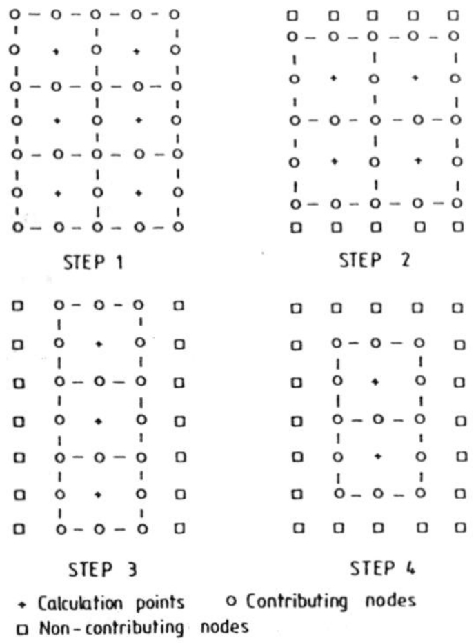

Fig. 3 Mesh regeneration.

With this numerical procedure stability and convergence problems were overcome by adjusting the mesh size, adjusting relaxation factors and/or by applying the 'upwinding' technique. Of the equations, it was found to be most difficult to obtain convergence of the vorticity field; however, as the vorticity was considered important computation was continued until convergence was within 0.5%. For the turbulence model k and ε were converged to 1% and the stream function and swirl parameters were always better than 0.5%.

3.1 Assessment of the numerical procedure

In order to assess the full numerical procedure a typical problem of a uniform flow through a simple radial diffuser was studied. The objectives were to determine the effects of mesh size and relaxation factors, and to assess the variational sub element method used at high Reynolds numbers. The predicted stream function and vorticity distribution across the diffuser width, 0.25 diffuser widths from inlet, is shown in fig. 4 for a number of mesh sizes. In addition a solution using the Galerkin method for all equations is shown. The vorticity is affected by the mesh size most significantly closed to the wall; this is to be expected as the boundary values are dependent upon the near wall magnitudes of vorticity. The predicted distributions obtained with the Galerkin and Variational sub element method show satisfactory agreement when a mesh size of 15 x 19 was used.

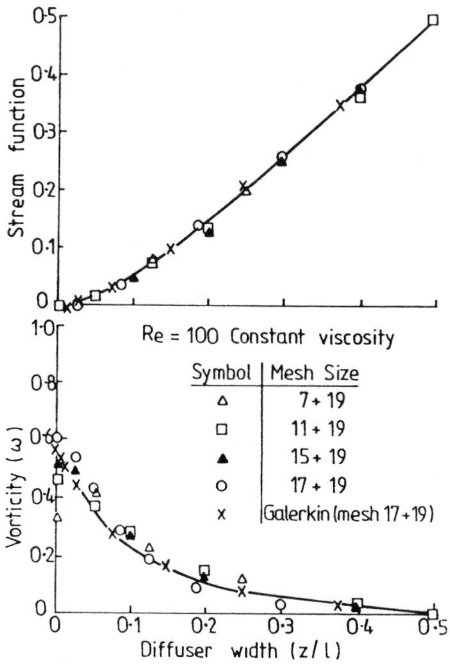

Fig. 4 Effect of mesh size on numerical solution.

To assess the Variational sub element method computations were carried out over a range of Reynolds number using each of the element types shown in fig. 2 and with and without the upwinding technique. Results are tabulated in table 2 showing vorticity magnitudes at two nodal points, one near the inlet and the other near the discharge. At the low Reynolds Number, R_e=10, good agreement is shown between both numerical techniques no matter which type of sub-element is employed. However, the introduction of upwinding leads to a worsening of the comparison. This was most significant when using 8 sub-elements and least when using 4 sub-elements. At a Reynolds number of 100 predicted magnitudes are similar in all cases. As the number of sub-elements was increased from 4 to 8 the comparison with the Galerkin method deteriorated slightly. In this case, however, the application of upwinding improved the comparisons. With further increase of the Reynolds number to 1000 the two methods did not agree satisfactorily unless the upwinding technique was applied to the variational sub-element method. The results shown for the Galerkin method did not employ the upwinding technique; however, it was necessary to increase the mesh size from 17 x 19 to 31 x 31. At the high Reynolds number of 10,000 no comparative solutions were obtained with the Galerkin method.

POSITION	METHOD	Re=10	Re=100	Re=1000	RE=10000
2,2	G.M	0.250117	0.5423	0.593	no sol.
	VE4(c)	0.2501	0.5424	no sol	no sol.
	VE6(c)	0.2501	0.5481	no sol	no sol.
	VE8(c)	0.2501	0.5513	no sol	no sol.
	VE4(w)	0.2512	0.5418	0.5931	0.6762
	VE6(w)	0.2734	0.5421	0.5822	0.6781
	VE8(w)	0.3169	0.5321	0.6772	no sol.
19,2	G.M	0.069	0.0687	0.0695	no sol.
	VE4(c)	0.069	0.0687	0.0685	no sol.
	VE6(c)	0.069	0.0694	no sol	no sol.
	VE8(c)	0.069	0.0694	no sol	no sol.
	VE4(w)	0.0688	0.06878	0.0592	0.05868
	VE6(w)	0.0668	0.06876	0.0593	0.05868
	VE8(w)	0.0761	0.06851	0.0754	no sol.

G.M- Central Galerkin method
VE4- Variational method with type 4 sub-element
VE6- Variational method with type 6 sub-element
VE8- Variational method with type 8 sub-element
(c)- Central method (w)- Upwind method

TABLE 2 Comparison of the vorticity values.

When using 8 sub-elements the numerical method proved to be unstable. Whilst the stability was improved by reducing the width of the element it was not always possible to obtain a converged solution. The cause of this instability has not been investigated. The use of 6 sub-elements proved to be the most stable. However, a large number of iterations were required when compared to the use of 4 sub-elements (48 iterations compared to 35). Generally, therefore, 4 sub-elements have been used in the theoretical investigations.

4. APPLICATION OF THE THEORETICAL ANALYSIS

The theoretical analysis was applied widely to the study of the three types of diffusers shown in fig. 1. In particular the effect of inlet velocity profiles upon diffuser performance and stability was studied in some detail [14]. To assess the capability of the prediction procedure a number of comparisons were made with experimental results. The experimentally determined inlet conditions were used to specify the theoretical inlet boundary whilst the discharge boundary was set sufficiently far from the actual discharge so that it did not significantly affect the computed results within the diffuser. The results for one test with an inlet swirl angle of approximately 62 deg., are shown in fig. 5. The differences shown in fig. 5 between the measured and theoretical inlet meridional velocity profiles reflects the process of converting measured velocities to stream function and back to velocity. The comparison between theory and experiment is satisfactory except for the tangential velocity profile at the second radius ratio. This difference could be due to errors and inaccuracies in either the experimental measurements or the numerical analysis. In order to assess the accuracy of the experimental measurements the conservation of mass and angular momentum at the three measurement stations was checked. The mass flow rate obtained from an integration of the velocity profiles proved to be reasonably satisfactory. The angular momentum is given by

$$2\pi\int \rho\, r^2\, U_\theta\, U_r\, dn$$

where the integration is carried out across the passage width.

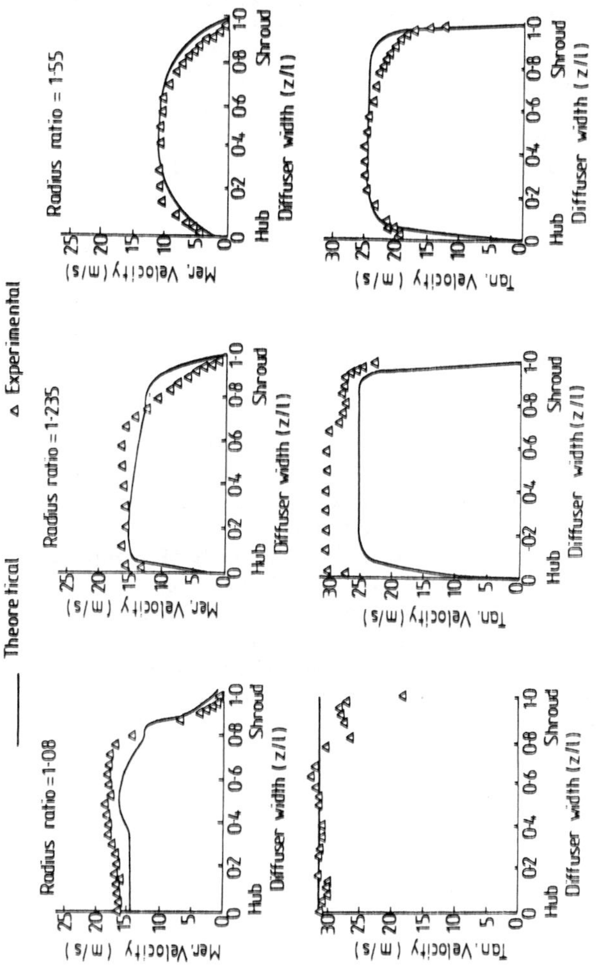

Fig. 5 Radial diffuser results test No. 2

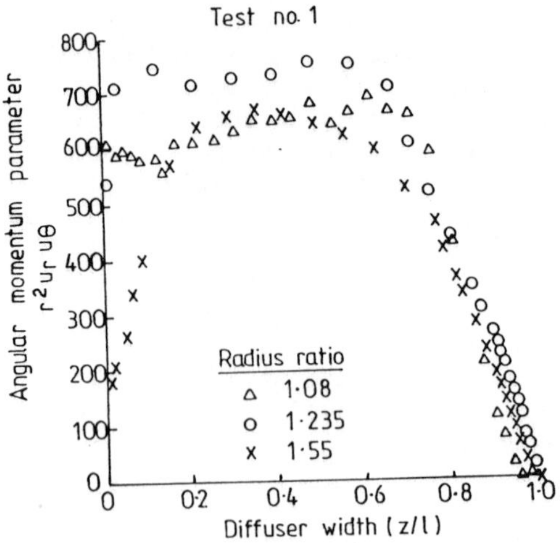

Fig. 6 Angular momentum distribution.

The parameter $r^2\ U_\theta\ U_r$ is therefore, plotted in fig. 6. If the angular momentum is conserved through the diffuser then the area under each curve should be equal. The relatively high values at the second measurement station in fig. 6 are clearly shown, and the discrepancies shown in fig. 5 can be attributed to experimental errors. In the theoretical analysis no attempt has been made to distinguish between the effective radial and tangential viscosities. Yamamoto [12] showed that the predicted decay in tangential velocity exceeded that obtained experimentally due to the use of a high effective viscosity. Discrepancies shown in other tests between the discharge theoretical and experimental tangential velocities were attributed to this.

5. CONCLUSIONS

A theoretical procedure for the study of laminar and turbulent flow through annular diffusers associated with radial and mixed flow impellers is presented. The stream function - vorticity model employed along with the k - ε turbulence model proved to yield satisfactory velocity profiles. Whilst a number of uncertainties exist with respect to suitable turbulence parameters and wall functions for flows with a high degree of swirl the predicted results compared well with those obtained experimentally. Problems associated with the convergence of the iterative process have been overcome through the application of a variational method with a dynamic mesh system, and by the use of relaxation factors and the 'upwinding' technique. The detailed cause of the convergence problems was not investigated in detail and this, along with a more detailed study of the applicability of the procedure in general, remains to be investigated further.

REFERENCES

1. WU, C.H. A General Theory of Three Dimensional Flow in Subsonic and Supersonic Turbomachines of Axial, Radial and Mixed Flow Types. NACA TN 2604 1952.

2. KATSANIS, T. Use of Arbitrary Quasi-Orthogonals for Calculating Flow Distributions in the Meridional Plane of a Turbomachine. NACA TN D2546, 1964.

3. WALLACE, F.J., ATKEY, R.C. and WHITFIELD, A. A Pseudo Three Dimensional Analysis of Flows in Vaneless Diffusers for Mixed Flow Compressors. J.Mech.Eng.Sci. I.Mech.E. 1975.

4. SWANSON, B.W. Solutions of the Three Dimensional Navier-Stokes Equations in Non-Orthogonal Co-Ordinates to Calculate the Flow in a Log Spiral Impeller. A.S.M.E. Paper No. 82-GT-268.

5. HIRSCH, C.H. and WARZEE, G. A Finite Element Method for through Flow Calculations in Turbomachines. Trans. A.S.M.E. J. of Fluids Eng. Sept. 1976, p.403.

6. HINZE, J.O. Turbulence. McGraw Hill Inc.

7. GOSMAN, A.D. et al. Heat and Mass Transfer in Recirculating Flows. Academic Press 1969.

8. OLIVER, A.J. The Prediction of Heat Transfer and Fluid Flow in the Entrance Region of an Annulus with the Inner Cylinder Rotating. Ph.D. Thesis CNAA, Leatherhead CERL 1975.

9. LAUNDER, B.E. and SPALDING, D.B. Mathematical Models of Turbulence. Academic Press, London 1972.

10. LAUNDER, B.E. and SPALDING, D.B. The Numerical Computation of Turbulence Flows. Comp.Meth.App.Math. Engng.3, 1974, pp.269-289.

11. LIN, A. and WEINSTEIN, H. Numerical Analysis of Confined Turbulent Flow. Computers and Fluids, Vol.10, No. 1 1982, pp.27-50.

12. YAMAMOTO, A., and MILLAR, D.J. A Calculation of Laminar and Turbulent Swirling Flows in Cylindrical Annuli. ASME Winter Annual Meeting N.Y. 1979.

13. MOULT, A., BURLEY, D., RAWSON, H. The Numerical Solution of a Two Dimensional Steady Flow Problem by the Finite Element Method. I.J. for Numerical Method in Eng. V.14 1979.

14. ABIR, A. Investigation of Steady and Unsteady Flow in Annular Vaneless Diffusers. Ph.D. Thesis, University of Bath 1983.

SECTION 8

DRIVEN CAVITY FLOW

NEW DECOMPOSITION FINITE ELEMENT METHODS FOR THE STOKES PROBLEM AND THE NAVIER-STOKES EQUATIONS

by J.P. BENQUE, P. ESPOSITO, G. LABADIE

Laboratoire National d'Hydraulique
Electricité de France
CHATOU - FRANCE

SUMMARY

Two iterative decomposition algorithms for the STOKES problem are introduced in the paper. They are interesting both for the computer storage savings and for the efficiency in computing time that they provide. The first one is a generalisation of UZAWA's method and the second one applies the principles of decomposition with coordination.
Both algorithms rely on the very simple idea that solving a POISSON equation with homogeneous NEUMAN boundary condition yields a pressure field very close to the consistent solution of the STOKES problem. Numerical experiments are presented for the NAVIER-STOKES equations, solved by using those algorithms coupled with a method of characteristics.

1. INTRODUCTION

In recent research works on the resolution of time-evolution Navier-Stokes equations, one trend seems to exploit the idea of decomposition, as well domain decomposition (see for example [5], [8], [11]) as equation decomposition (see for example the fractional step methods in [2], [9], [12], [14]). For the latter, decomposing the global difficulty of the whole set of equations so as to isolate more simple and specific problems [7], enables one to tackle each elementary problem with appropriate "natural" tools.

We introduce in this paper two families of algorithms providing a decomposition of the STOKES problem into a problem on pressure and a problem on velocity. The STOKES problem itself arises in resolution methods for NAVIER-STOKES equation as the result of a decomposition into a convection problem and the STOKES one. In that case, a method of characteristics has been proposed [2][15] for the convection equation in order to provide a good treatment of the non linear part of NAVIER-STOKES equations. We focus here on the STOKES problem resulting of such a decomposition which has the general following form :

$$(1)\quad \begin{cases} \alpha u - \nu\Delta u + \nabla p = S \\ \nabla . u = 0 \\ u|_{\Gamma} = u_d \end{cases}$$

where u is the velocity vector, p the pressure, ν the viscosity coefficient, and α some positive coefficient (either the inverse of the time step if we consider the time-discretised unsteady Stokes problem or zero for the steady state case). The computational domain Ω is bounded and Γ denotes its boundary.

In section 2, we describe two algorithms for the STOKES problem and numerical results of NAVIER-STOKES computations using these algorithms are given in section 3.

2. ALGORITHMS FOR THE STOKES PROBLEM

2.1. A preconditioned UZAWA algorithm

A classical finite element discretisation of problem (1) leads to the matrix system.

$$(2)\quad \begin{cases} Au - {}^{T}Bp = b \\ Bu = 0 \end{cases}$$

where A is a symmetric positive definite matrix and B is the rectangular matrix resulting from the discretisation of the divergence operator. The unknown (u, p) now belong to discretised finite dimensional spaces and T denotes the transposition.
The system (2) is known to be equivalent to the saddle point problem :

$$(3)\quad \underset{v}{\mathrm{Min}}\ \underset{q}{\mathrm{Max}}\ \mathcal{L}_r(v,q) = \frac{1}{2}\,{}^{T}vAv - {}^{T}qBv - {}^{T}bv + \frac{r}{2}\,|Bv|^2$$

where q is a lagrangian multiplier of the constraint Bu = 0 and r is a positive coefficient (or zero) used to increase the influence of the constraint in the functional (like in penalty method, but r has not to be infinite to provide the right solution). $\mathcal{L}_r(v,q)$ is called the Augmented Lagrangian [10].
UZAWA's method is a steepest descent algorithm applied to the minimization of the functionnal $J_r(q)$ [1] :

$$(4)\quad J_r(q) = -\underset{v}{\mathrm{Min}}\ \mathcal{L}_r(v,q)$$

The problem on $J_r(q)$ is equivalent to the resolution of the linear system.

(5) $BA_r^{-1T}Bp = - B A_r^{-1}b$

with $Ar = A + \frac{r}{2}{}^{T}B\,B$

In order to accelerate the convergence rate of the method, we introduce a symmetric positive definite matrix C used as a conditioner.

Then the algorithm is (p_m given)

| (i) find u_m solution of $A_r u_m = {}^{T}Bp_m + b$
| (ii) find p_{m+1} solution of $Cp_{m+1} = Cp_m - \rho_m Bu_m \qquad (\rho_m > 0)$

The conditioner C must be close to the matrix $BA_r^{-1T}B$ of the linear system (5) in order to be efficient, and the closer, the more efficient. We propose to choose the matrix resulting from the standard discretisation of the POISSON equation with a NEUMAN boundary condition. The main reason of such a choice is that, in finite difference analysis, we have been using for years a splitting up algorithm due to CHORIN [6] in which the pressure field is obtained from a POISSON equation with a zero normal derivative boundary condition. CHORIN's algorithm giving good results, its matrix must be close to the actual matrix of the pressure equation.

Although UZAWA method gives a good convergence rate only for r > 0 (strictly augmented lagrangian), we found that the above preconditionned algorithm allows the use of a simple lagrangian formulation (r = 0) with a very fast convergence rate (2 or 3 iterations at most for the examples given in section 3). In the case of prescribed velocity or prescribed stress boundary conditions for system (1), the choice r = 0 leads to the decoupling of the two components of the velocity vectors and then, the STOKES problem is split up into 3 N smaller size problems on scalar unknown functions, where N is the number of iterations of the method.

It can be noticed that several well known methods can be interpreted in terms of the above general scheme. For example the finite element version corresponding to CHORIN's algorithm is obtained with 1 iteration of the above procedure and, when applied to a STOKES problem, the "pressure correction" algorithm in TEACH [13] is obtained if one chooses a new conditioner :

$C = B\,D^{-1}\,{}^{T}B$ where D is the diagonal part of the matrix A.

2.2. Decomposition with coordination

a) Method

Let us consider now the saddle point problem equivalent to the STOKES problem without discretisation.

$$(6)\quad \begin{cases} \underset{u}{\text{Min}}\ \underset{p}{\text{Max}}\ \mathcal{L}(u,p) = \int_\Omega \left[\frac{1}{2}\alpha u^2 + \frac{\nu}{2}(\nabla u)^2 - p\ \text{div}\ u - Su \right] dw \\ \text{with } u|_\Gamma = u_d \end{cases}$$

Using two different functions u_1 and u_2 for the velocity we can introduce the new saddle point problem :

$$(7)\quad \begin{cases} \underset{u_1,u_2}{\text{Min}}\ \underset{p,q}{\text{Max}}\ J_r(u_1,u_2,p,q) = \int_\Omega \Big[\frac{1}{2}\alpha u_1^2 + \frac{\nu}{2}(\nabla u_1)^2 - Su_1 \\ \qquad - p\text{div}\ u_2 + q(u_1-u_2) \\ \qquad + \frac{r}{2}(u_1-u_2)^2 \Big] dw \\ \text{with } u_1|_\Gamma = u_d \quad \text{and } u_2 \cdot n|_\Gamma = u_d \cdot n|_\Gamma \quad \text{(normal component).} \end{cases}$$

where q is the Lagrange multiplier associated to the constraint $u_1 = u_2$. The constraint is reinforced by the term $\frac{r}{2}(u_1-u_2)^2$ in the functional according to the augmented Lagrangian technique [10].

The problem (7) is equivalent to the set of equations :

$$\begin{cases} \alpha u_1 - \nu\Delta u_1 = -q - r(u_1 - u_2) + S & (8.1) \\ u_1|_\Gamma = u_d & (8.2) \\ \nabla p = q + r(u_1 - u_2) & (8.3) \\ \text{div}\ u_2 = 0 & (8.4) \\ u_2 \cdot n|_\Gamma = u_d \cdot n|_\Gamma & (8.5) \\ u_1 = u_2 & (8.6) \end{cases}$$

and it is easy to see that this set of equations is equivalent to the STOKES problem (1).

A simple iterative method to solve the problem (7) can be :

q^m given at iteration m

(i) Solve $\begin{cases} \underset{u_1,u_2}{\text{Min}}\ \underset{p}{\text{Max}}\ J_r(u_1,u_2,p,q^m) \\ \text{with } u_1|_\Gamma = u_d \text{ and } u_2 \cdot n|_\Gamma = u_d \cdot n|_\Gamma \end{cases}$

(ii) $q^{m+1} = q^m + \rho_m (u_1^{m+1} - u_2^{m+1})$

Then the step (i) leads to solve

$$
\begin{cases}
\alpha u_1^{m+1} - \nu\Delta u_1^{m+1} = -q^m - r\,(u_1^{m+1} - u_2^{m+1}) + S & (9.1)\\
u_1^{m+1}|_\Gamma = u_d & (9.2)\\
\nabla p^{m+1} = q^m + r\,(u_1^{m+1} - u_2^{m+1}) & (9.3)\\
u_1^{m+1} \cdot n|_\Gamma = u_d \cdot n|_\Gamma & (9.4)\\
\operatorname{div} u_2^{m+1} = 0 & (9.5)
\end{cases}
$$

As this set of equations couples u_1^{m+1} u_2^{m+1} and p^{m+1}, we have not actually derived a simpler method ! But, indeed, there are several ways to get rid of this coupling by means of a little modification of the above algorithm. For example one can replace u_1^{m+1} in equation (9.1) by u_1^m. Then, taking the divergence of (9.3) and its normal component on the boundary one obtains a POISSON equation on the pressure and can write the following algorithm :

q^m, p^m, u_1^m, u_2^m given at iteration m,

(i) compute u_1^{m+1} solution of

$$
\begin{cases}
(r+\alpha)\, u_1^{m+1} - \nu\Delta u_1^{m+1} = -\, q^m + r\, u_2^m + S\\
u_1^{m+1}|_\Gamma = u_d
\end{cases}
$$

(ii) compute p^{m+1} solution of

$$
\begin{cases}
\Delta p^{m+1} = \operatorname{div} q^m + r \operatorname{div} u_1^{m+1}\\
\dfrac{\partial p^{m+1}}{\partial n}\Big|_\Gamma = q^m \cdot n
\end{cases}
$$

(iii) $u_2^{m+1} = \frac{1}{r}\,(q^m - \nabla p^{m+1}) + u_1^{m+1}$

(iiii) $q^{m+1} = q^m + \rho_m\,(u_1^{m+1} - u_2^{m+1})$

b) <u>Finite element discretisation</u>

Using a classical discretisation of u_1 u_2 and p (for example TAYLOR-HOOD Triangles) and choosing for q the same discretisation as for u_1 or u_2 we obtain the matrix algorithm :

u_1^m, u_2^m, p^m, q^m given at iteration m

$$(10.i)\quad A_r\, u_1^{m+1} = -Mq^m + rMu_2^m + S$$

$$(10.ii)\quad C\, p^{m+1} = Bq^m + rB\, u_1^{m+1}$$

$$(10.iii)\quad M\, u_2^{m+1} = M\, u_1^{m+1} + \frac{1}{r} M\, q^m - \frac{1}{r}\, {}^TB\, p^{m+1}$$

$$(10.iiii)\quad q^{m+1} = q^m + \rho_m (u_1^{m+1} - u_2^{m+1})$$

where $A_r = A + (\alpha + r)M$; C is the matrix of the discretised POISSON equation on the pressure ; M results from the finite element discretisation of identity and A and B are matrices encountered in system (2).

Let us now consider the particular case where $\rho_m = r$, which is a good choice for the convergence of the method [10].

Then $q^{m+1} = M^{-1}\, {}^TB\, p^m$

$$u2^m = u1^m + \frac{1}{r} M^{-1}\, {}^TB\, (p^{m-1} - p^m)$$

and the algorithm becomes

$$(11.i)\quad (A + (\alpha + r)M)\, u_1^{m+1} = -{}^TBp^m + rMu_1^m + {}^TB\, (p^{m-1} - p^m) + S$$

$$(11.ii)\quad C\, p^{m+1} = B\, M^{-1}\, {}^TB\, p^m + r\, B\, u_1^{m+1}$$

If we notice that the terms $r\, M\, u_1^{m+1}$ and $r\, M\, u_1^m$ in equation (11.i) denote a certain relaxation between iterations m and m+1 we can find a certain similarity between algorithm (11) and the preconditioned UZAWA algorithm of section 2.1.

2.3. Remarks :

1 - Both algorithms involve the solution of a Poisson equation on pressure and a problem of the form :

$$\alpha u - \nu \Delta u = \text{r.h.s.}$$

for velocity. For prescribed velocity on the boundary, contrary to the penalty formulation method, the latter problem can be written as uncoupled systems on each component of velocity, with the same matrix for each system : this provides appreciable storage savings.

2 - More elaborate versions of both algorithms can be given, using conjugate gradient descent with optimal parameter.

3. APPLICATION TO THE NAVIER-STOKES EQUATIONS

A fractional step algorithm is used with :

- a "convection" step with a processing of convective terms only, by a method of characteristics ([2], [3], [14]) which avoids forming a non-symmetric matrix varying at each time step.

- a Stokes step using the algorithms described above.

3.1. Square wall driven cavity - Re = 400

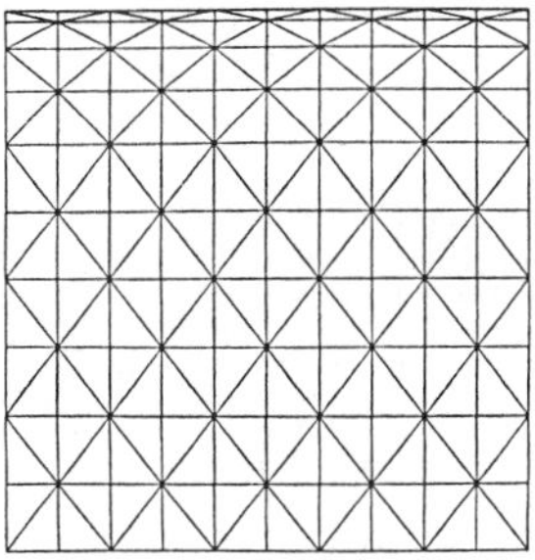

figure 1

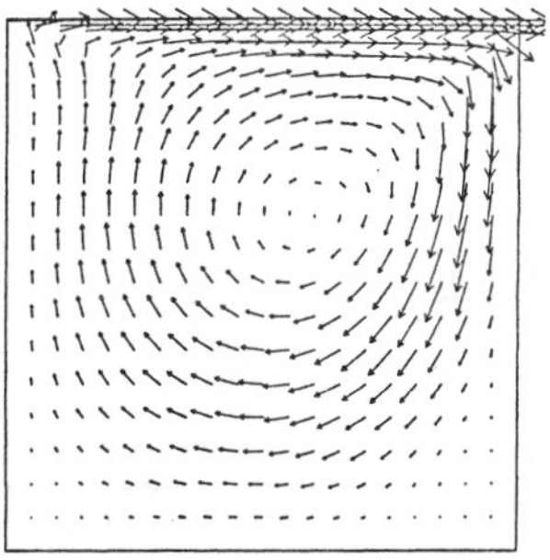

figure 2

The finite element grid of figure 1 contains 200 triangles, 121 pressure nodes and 441 velocity nodes. Figure 2 presents the steady state obtained at Reynolds = 400 with the preconditioned UZAWA algorithm for the STOKES problem.
Figure 3 shows a comparison between the velocity profiles along the vertical axis obtained by means of the two schemes presented in this paper and a reference from BURGGRAF [16].

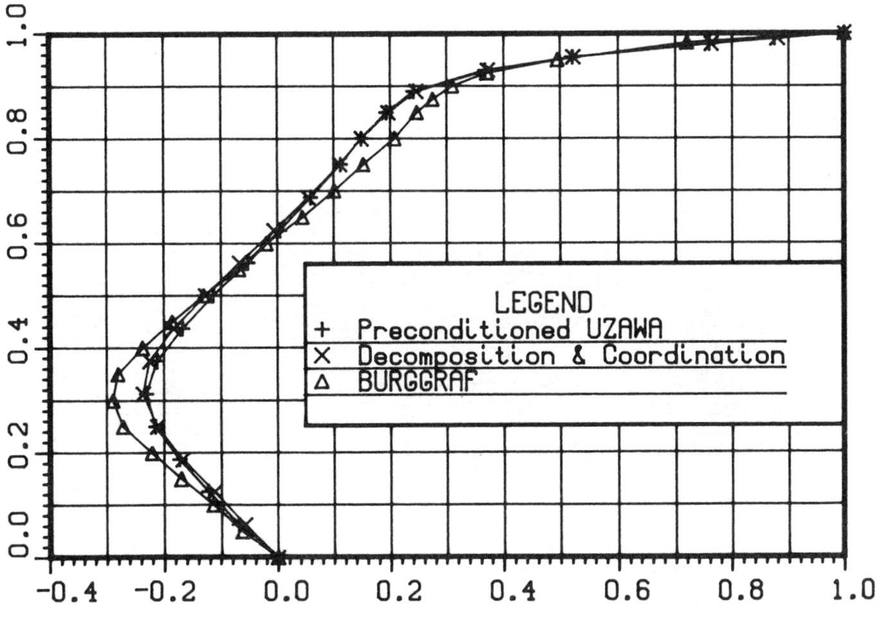

figure 3

3.2. Flow across a circular cylinder

a) Computational conditions

DOMAIN

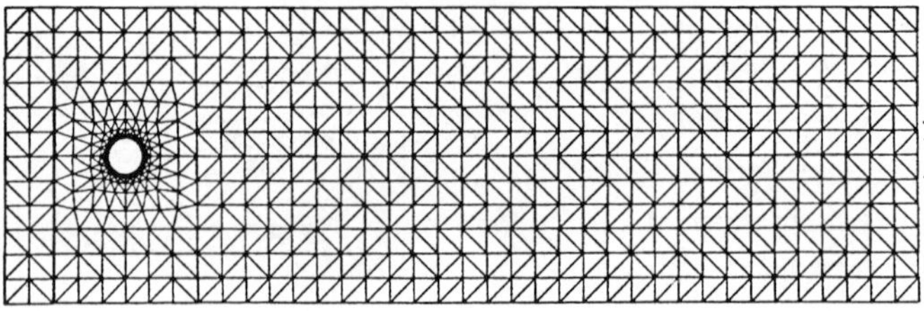

u_c and v_c are given by the advection step.

1128 triangles
2380 velocity nodes
626 pressure nodes
Reynolds number : 400

b) Results

Algorithm 1 (Preconditioned UZAWA) is used for the STOKES problem with optimal parameter and two iterations per time step. A vortex street is obtained without any numerical excitation.

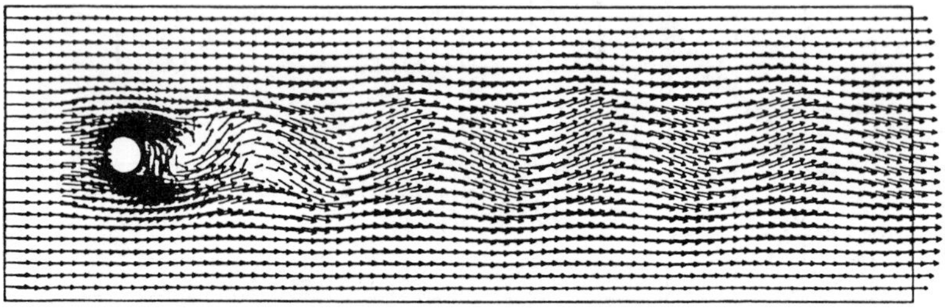

Figure 4 : Regime of fluid flow obtained

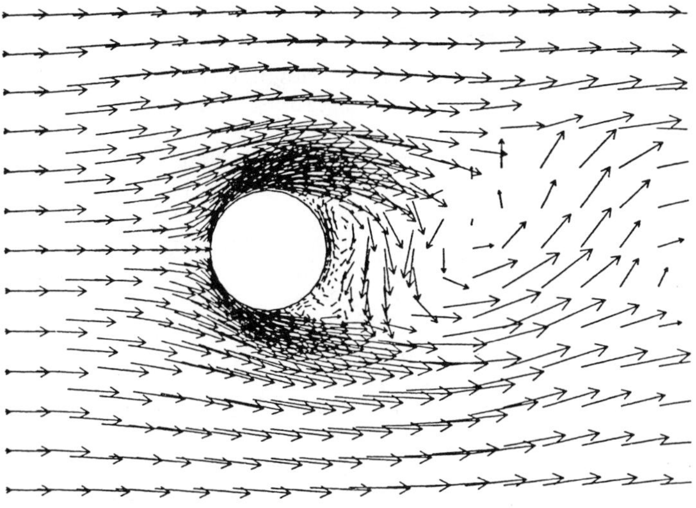

Figure 5 : Detail of the area of vortex generation

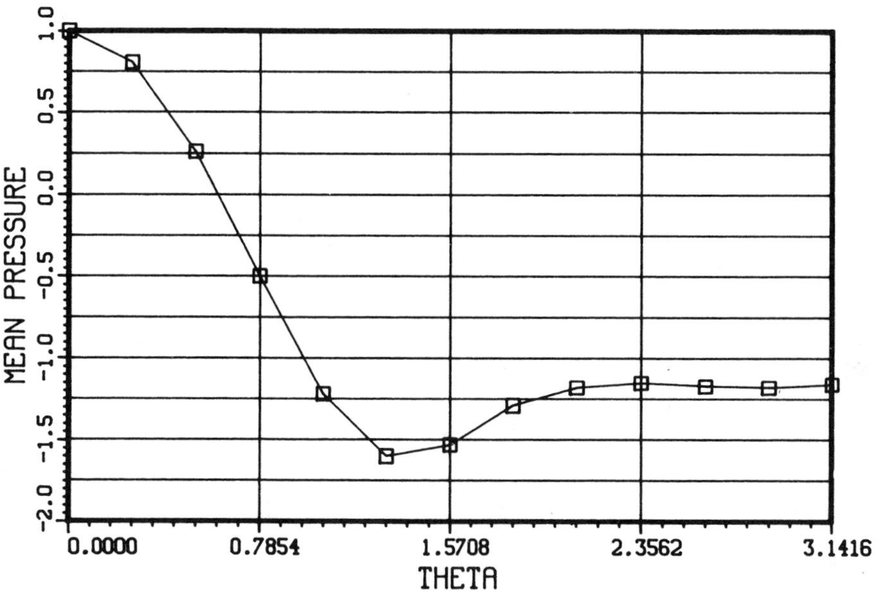

Figure 6 : Mean pressure field around the cylinder

The computed STROUHAL number is S = 0.195 .It lays in the range given in [17] [18] taking into account the confinement of the flow, and the drag coefficient obtained from figure 6 is about $\overline{C}_D$ = 1.2 which is a value close to different experimental results recorded in [18].

3.3. Computational aspects

In the two numerical examples presented, we used an optimal parameter ρm for the preconditioned UZAWA method. With 2 iterations per time step, it leads to solve 3 problems on velocity (6 problems on 1 component) and two problems on pressure. But the observed drastic reduction of $|B\ u|$ at the end of the first optimal gradient step suggests that we could avoid the computation of the optimal parameter (and then skip 1 velocity problem) taking for example $\rho m = \alpha$.
All the linear problems solved have constant matrices and are solved with an Incomplete Cholesky Preconditioned Conjugate Gradient method. The use of such an iterative method for the linear problems turned out to be very interesting because we start the resolution at step m with the solution of step m-1 and it provides a very fast overall convergence.
It we try to sum up the main characteristics of algorithms developped in this paper, it can be said that they seem to be very convenient for big problems (complex refined geometries with many nodes) thanks to :

- a limited computer storage,
- a fast convergence rate,
- an accuracy of the solution which can be tuned easily without destroying the performances of algorithms.

Of course, other numerical trials are needed to improve the methods, and among them, we can mention :

- a better (?) preconditioner for UZAWA method : $C = BA_L^{-1}\ {}^T B$ where A_L is the diagonal mass lumped matrix obtained from matrix A.
- a better way to solve problems where the mass matrix M is involved (decomposition and coordination : (11)) since Incomplete Cholesky factorisation seems to give a bad conditioner and destroy the convergence rate of the Conjugate Gradient method.

4. CONCLUSION

Algorithms presented here can be, according to the first results, considered as fast STOKES solvers. Moreover, they require a limited enough computer storage to allow computations with a large number of nodes. Then, they seem to be helpfull for refined analysis of flows in complex geometries (and 3.D. geometries). Finally one can mention another general advantage of decomposition : it provides modularity and makes easier to change one part of the algorithm. For example, we recently introduced in the computing code a new method for the convection equation called the "weak convection" [15] and we quickly obtained improved results.

REFERENCES

[1] ARROW K.J., HURWICZ L., UZAWA H. : Studies in non linear programming, Standford University Press, 1958.

[2] BENQUE J.P., IBLER B., KERAMSI A., LABADIE G. : A Finite Element Method for the Navier-Stokes equations. 3rd Int^{l} Conference on F.E.M. in flow problems. Banff. Alberta (CANADA) - June 1980.

[3] BENQUE J.P., RONAT J. : Quelques difficultés des modèles numériques en Hydraulique. 5th Int^{l} Symposium on Computing Methods in Applied Sciences and Engineering - Versailles (FRANCE). December 1981.

[4] BENSOUSSAN A., LIONS J.L., TEMAM R. and LEMONNIER P. : Sur les méthodes de décomposition et coordination in : LIONS J.L., MARCHOUK G.I. : Sur les méthodes numériques en Sciences Physiques et Economiques. Dunod, 1974.

[5] CAMBIER L., GHAZZI W., VEUILLOT J.P., VIVIAND H. : Une approche par domaines pour le calcul d'écoulements compressibles. 5th Int^{l} Symposium on Computing Methods in applied Sciences and Engineering. Versailles (FRANCE). December 1981.

[6] CHORIN A.J. : Numerical solution of the Navier-Stokes equations. Math. Comput. 22 (1968) 745-762.

[7] DESCARTES R. : Le discours de la méthode. Leyde 1637.

[8] DINH R.V., GLOWINSKI R., PERIAUX J. : Applications of domain decomposition techniques to the numerical solution of the Navier-Stokes equations. 2nd Int^{l} Congress on Numerical Methods for Engineering (G.A.M.N.I. 2). Paris (FRANCE), december 1980.

[9] DONEA J., GIULIANI S., LAVAL H., QUARTAPELLE L. : Finite Element Solution of the Unsteady Navier-Stokes equations by a fractional step method. Computer Methods in applied Mechanisms and Engineering 30 (1982) 53-73.

[10] FORTIN M., GLOWINSKI R. : Méthodes de Lagrangien augmenté. Applications à la résolution numérique de problèmes aux limites. Dunod, (1982).

[11] HARUMI K., KANO T., OKADA H., OOTSUKI A. : New method combining F.E.M. and F.D.M. for tidal flow computation 4th Int^{l} Symposium on F.E.M. in flow problems. Tokyo (JAPAN), July 1982.

[12] GLOWINSKI R., MANTEL B., PERIAUX J. : Numerical solution of the time dependent Navier-Stokes equations for incompressible viscous fluids by finite elements and alternating direction methods. Conference on Numerical Methods in Aeronautical Fluid Dynamics. Reading (U.K.), March 1981.

[13] GOSMAN A.D. : Computational Fluid Dynamics and Heat and Mass Transfer. Cours de l'Ecole d'été d'Analyse Numérique. EDF - INRIA - CEA, July 1979.

[14] LABADIE G., BENQUE J.P., LATTEUX B. : A Finite Element method for the Shallow Water equations. 2nd Int^{l} Conference on Numerical Methods in laminar and turbulent Flow. Venice (ITALY), July 1981.

[15] BENQUE J.P., LABADIE G. : A new Finite Element method for the NAVIER-STOKES equations with a temperature equation - Submitted to the International Journal of Numerical Methods in Fluids.

[16] BURGGRAF O.R. : Analytical and numerical studies of the structure of steady separate flows. J. Fluid Mech. (1966). Vol. 24.

[17] SHAW T.L. : Wake dynamics of two dimensional structures in confined flows (I.A.H.R.).

[18] Vibrations de structures dans un écoulement sous l'effet du détachement tourbillonnaire de sillage. Document de la Société Hydrotechnique de France.

A THREE-DIMENSIONAL LID-DRIVEN CAVITY FLOW: EXPERIMENT AND SIMULATION

by

J. R. Koseff (i), R. L. Street (i), P. M. Gresho (ii), C. D. Upson (ii), J. A. C. Humphrey (iii), and W.-M. To (iii)

SUMMARY

A facility has been constructed to study shear-driven, recirculating flow, namely, a three-dimensional lid-driven cavity flow. In the extant case, the cavity depth-to-width aspect ratio is 1:1, while the span-to-width aspect ratio is 3:1. A description of the circulation cell structure obtained by flow visualization is given for two Reynolds numbers with and without temperature-induced density stratification. The experimental results are compared to simulations by two different numerical codes, one employing finite differences and one employing finite elements. The numerical codes reproduce the overall behavior observed in the experiments, but fail to simulate observed longitudinal vortices. There are some significant differences also between the numerical results, highlighting the effects of using the HYBRID (upwind/central) differencing scheme in one code.

1. INTRODUCTION

Complex recirculating flows are common in engineering. Such flows are difficult to simulate because of their three-

---

(i) Department of Civil Engineering, Stanford University, Stanford, CA, U. S. A.

(ii) Lawrence Livermore National Laboratory, University of California, Livermore, CA, U. S. A.

(iii) Department of Mechanical Engineering, University of California, Berkeley, CA, U. S. A.

dimensionality and the various small scale flow structures which often occur. A unique facility has been constructed to study a shear-driven, recirculating flow. The resulting flow is a three-dimensional cavity flow driven by a moving lid. This configuration provides an ideal test case for numerical simulation because there are no irregular boundaries and the lid-wall intersection singularity is comparatively minor. Accordingly, the above three groups collaborated to compare the results of flow visualization experiments and calculations from two state-of-the-art numerical simulation codes for three-dimensional, unsteady flows.

In the following sections, a brief introduction to the experimental facility is given, and then the two numerical simulation codes are briefly described. Next, the key experimental results and the simulations of the appropriate cases are presented. The flows represented include both (a) isothermal and (b) thermally-stratified cases. Finally, a summary of the results and conclusions is offered.

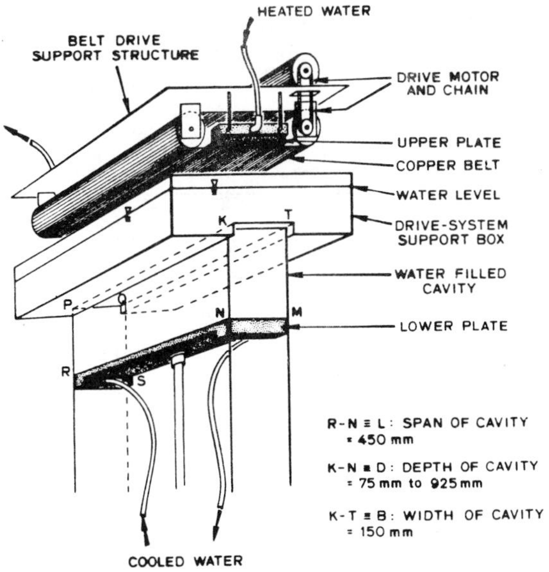

Figure 1: Schematic of Experimental Facility

2. THE EXPERIMENTAL FACILITY

A detailed description of the experimental facility is given in Koseff and Street [9]. This lid-driven cavity flow facility consists of two rectangular boxes (see Figure 1), the lower one being the cavity where the flow occurs and the upper one housing the belt drive-system that provides the shear at the top of the flow. The 0.08 mm thick copper belt is driven at a selected speed between 7 mm/s and 230 mm/s (corresponding to a Re range of 1000 to 35,000) and the speed can be hel constant to within ± 2 percent. The upper and lower plates are heat exchangers which provide the temperature difference to induce and maintain the stratification which was used in two of the three reported experiments. The cavity in which the flow takes place has a width B of 150 mm, a depth D of 15 mm, and a lateral span L of 450 mm. This leads to a depth-to width aspect ratio D/B of 1:1 and a span-to-width aspect rati L/B of 3:1 (see Figure 2). The Reynolds number Re is defined as $V B/\nu$, where V is the lid speed and ν is the average kinematic viscosity of the fluid.

The facility is constructed of 12.5 mm thick plexiglas; this facilitates flow visualization. The primary visualization method used for the flows studied herein is the thymol blue pH-indicator technique as described by Baker [2]. Passage of a DC current through the properly prepared solutio turns the solution from yellow to deep blue in the region of the cathode of an electric circuit with electrodes in the fluid. Because the thymol-blue molecule always remains an io in solution there is no density change accompanying the color change. Thus, one obtains a neutrally buoyant tracer.

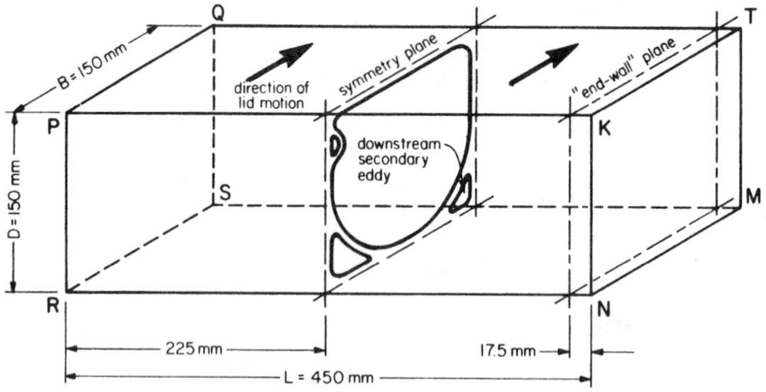

BOUNDARY DEFINITIONS

END-WALL : PQSR   SIDE-WALL UPSTREAM : PKNR
: KTMN   DOWNSTREAM : QTMS

LID : PQTK   LOWER HORIZ. : SMNR

Figure 2: Perspective Sketch of the Cavity

The "start up" or initial transient period was visualize by using the moving belt as the cathode and a graphite rod placed near it as the anode. Then blue "dye" is generated along the surface of the belt. When the belt motion begins, the colored ions are stripped off at the downstream, upper li of the cavity (line QT in Figure 2) and move downward with th starting vortex. The belt was used in a similar fashion at later stages in each experiment to visualize flow throughout the cavity by leaving the DC current on for a period of time to provide a continuous tracer.

Both still photographs (taken with a Pentax 35 mm camera) and movies (taken with a 16 mm Paillard-Bolex camera) were used to record the thymol-blue dyed flows. General impressions were made by direct observation of the flow by eye; more quantitative data was obtained by scaling from the photographs and movies. For special cases, additional visualization was performed by use of a rheoscopic liquid technique as described by Katsaros, et al. [8].

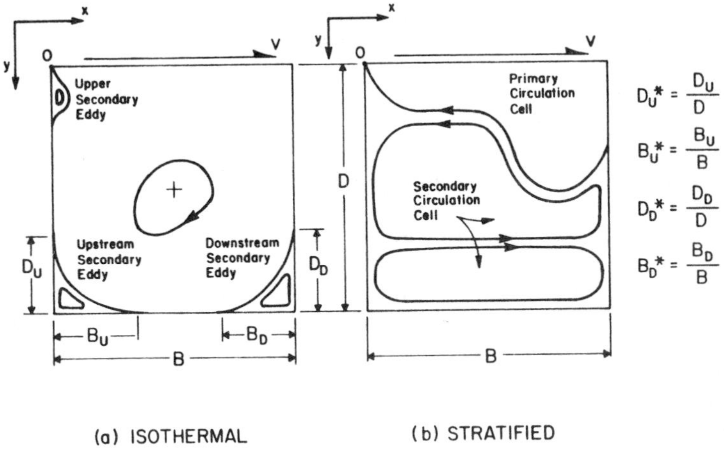

Figure 3: Flow Definition Schematic

3. THE FINITE DIFFERENCE METHOD

The finite-difference calculations were performed using REBUFFS (recirculating buoyant and forced flows solver) [12], which was extended to three dimensions for this comparative study. The code solves a set of elliptic in space and parabolic in time partial differential equations (the Navier-Stokes equations) for the primitive variables governing a single-phase, variable property flow. The code originated from the TEACH-2E code developed at Imperial College, London [5]. The difference equations are derived according to the TEACH principles by volume integration of the differential equations about cells surrounding nodes of the calculational mesh. Although REBUFFS works according to the principles of its parent, many changes have been made and are detailed in Le Quere, et al. [12]. They include changes due to the time-dependent nature of the equations to be solved and the strong link existing between the energy and momentum equations through the density dependence on the fluid temperature. The

HYBRID scheme, which employs upwind or central differencing for convection terms depending on the grid Peclet number, was used for all the three-dimensional simulations. Both the HYBRID scheme and the higher-order quadratic upstream interpolation (QUICK) scheme of Leonard [11] were used for two-dimensional simulations [7]. The latter scheme gives better spatial resolution for a given distribution of grid nodes. Finally, a nonuniform grid spacing was used in the near-wall regions in order to attempt to resolve adequately the sharp velocity gradients expected there. In the two-dimensional calculations and particularly with the QUICK scheme, the resolution was sufficient. Limitations on computer time and memory space prevented full satisfaction of our criteria in the three-dimensional cases.

4. THE FINITE-ELEMENT METHOD

In this scheme, the time-dependent incompressible Navier-Stokes, Boussinesq-approximated, equations are solved in primitive variable form via a modified finite-element method and mixed interpolation on a non-uniform mesh [6]. The method was originally derived from the Galerkin finite element method, but has been significantly modified in the interest of cost-effectiveness so that it now is a blend of finite element and finite difference techniques. The principal modifications from the Galerkin method include mass lumping and one-point quadrature. In mass lumping the "mass" matrix is diagonalized via lumping. All Galerkin integrals are then evaluated approximately by use of one-point quadrature on the 8-node (trilinear) isoparametric brick element employed here to represent the flow field. The mass lumping approximation affects only the transient solutions and does not affect steady-state results.

The resulting spatially-discretized equations are integrated in time by use of a compensated explicit Euler scheme in which the pressure is first obtained by solving the appropriate and consistent discretized Poisson equation. The conventional Euler scheme introduces a negative diffusion coefficient in the streamline direction owing to truncation error; this deleterious effect is cancelled (in principle) by the addition of a velocity-dependent tensor diffusivity to the physical scalar diffusivity, a procedure which increases both accuracy and stability. Further computational gains result from a technique called subcycling, in which the pressure updates are computed less frequently than would otherwise be the case. The variable major time step (which includes the pressure update) is automatically selected based on specified accuracy via local time truncation error estimates, whereas advection and diffusion are computed using a fixed and smaller (minor) time step based on stability (typically a Courant condition). In spite of these cost saving measures, the high cost of computer I/O (related to manipulation of the pressure matrix for the Poisson equation) has restricted the three-dimensional grid resolution and the simulations which could be accomplished using a reasonable amount of computer time [6].

5. OBSERVATIONS

5.1 General Comments

In this section we present the results of the experiments and numerical simulations for three different flow cases: (1) Reynolds number (Re) = 2000, isothermal ($\Delta T = 0^oC$); (2) Re = 1000, stratified ($\Delta T = 1^oC$), and (3) Re = 2100, stratified ($\Delta T = 1.15^oC$). Our original intention was to compare the experiments exclusively with three-dimensional simulatons. However, the prohibitive cost of performing the three-dimensional simulations with sufficient grid resolution led us to include some two-dimensional solutions in lieu of more extensive 3-D calculations.

At the University of California, Berkeley (UCB) the calculations were performed on a Control Data Corporation (CDC) 7600 system. At the Lawrence Livermore National Laboratory (LLL) a Cray-1A was used. The 2-D simulations were done for symmetry plane conditions and each 3-D simulation was done for half of the span (between the symmetry plane and the end wall) on the assumption that the flow is symmetric; the experiments confirm this assumption. Calculations performed at UCB on a (23x23) 2-D grid typically required 120 k_8 words of storage and 100 CPU seconds for 400 seconds of simulation for the Re = 2100, $\Delta T = 1.15^oC$ case. Corresponding requirements on a (23x23x23) 3-D grid were 1130 k_8 words of storage and 1000 CPU seconds. Calculations performed at LLL for the Re = 1000, $\Delta T = 1^oC$ case required 210 CPU seconds and 90 I/O seconds per hour of simulation in 2-D and 60 CPU minutes and 150 I/O minutes per hour of simulation in 3-D.

The results presented in this paper were obtained by interpretation of the various streamline, velocity vector and isotherm plots from simulations and from pictures of thymol blue dye streaks. The uncertainties indicated on the figures reflect both the grid sizes used in the simulations and our ability to read from the plots and photographs. In order to present the results efficiently we have adopted the following symbol system. The two- and three-dimensional simulations are referred to as 2D and 3D, respectively, while the HYBRID and QUICK differencing schemes are simply designated by H and Q. A three-dimensional HYBRID calculation performed at Berkeley, for example, is therefore referred to as UCB-3DH. The Stanford experiments are called EXP.

The basic time-averaged flow structure on the symmetry plane (see Figure 2) is sketched in Figure 3 for both the isothermal (3a) and stratified (3b) cases. In the case of the stratified flow the number of secondary circulation cells is a function of the degree of stratification (which is proportional to ΔT). Using these definitions as a basis, we present some of the most interesting results of this comparative study in the following sub-section. These results are then discussed further in Section 6 and the conclusions are drawn in Section 7.

TABLE 1: Size of secondary eddies at symmetry plane; Re = 2000, isothemal (ΔT = 0°C)

Re = 2000	Downstream Secondary Eddy D_D^*	B_D^*	Upstream Secondary Eddy D_U^*	B_U^*
Stanford-Exper.	0.33±0.03	0.21±0.03	0.24±0.03	0.30±0.03
LLL - 2D (20x 20)*	0.42±0.03	0.43±0.03	0.22±0.03	0.22±0.03
- 3D (17x16x10)*	0.41±0.03	0.34±0.03	0.22±0.03	0.23±0.03
UCB - 2DQ (23 x 23)+	0.31±0.05	0.25±0.05	0.34±0.05	0.33±0.05
-3DH (23x23x23)+	0.35±0.05	0.20±0.05	0.24±0.05	0.34±0.05

* Uncertainty is half the grid spacing
\+ Uncertainty is half the grid spacing plus that due to interpretation

5.2 Results

The results for the Re = 2000, isothermal (ΔT = 0°C) case are given in Table 1 and Figure 4. Table 1 contains a summary of the normalized dimensions of the upstream and the downstream secondary eddies (see Figure 3a) as measured on the symmetry plane. The results display a fair amount of scatter, with the respective numerical simulations disagreeing somewhat with each other. This measure of the secondary eddy size puts a premium on the representation of the corner and boundary-layer zones. From the table, it is seen that different grids

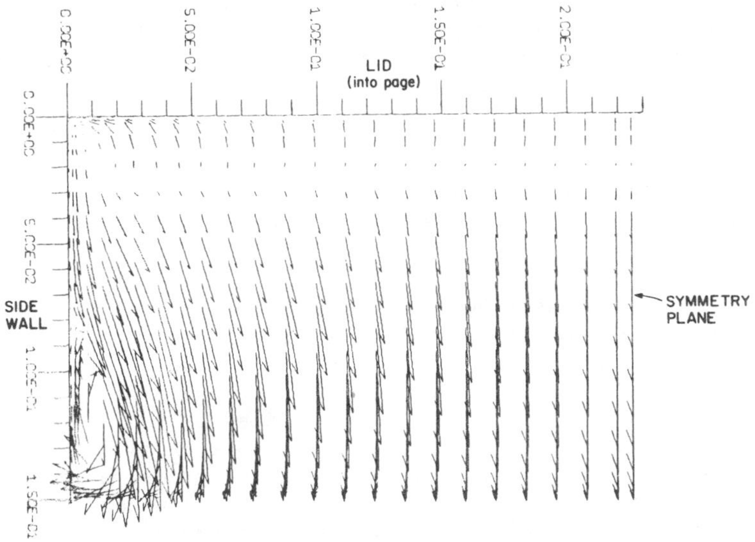

(a)

(b)

Figure 4: Velocity Vector Plot (UCB-3DH) and Streakline Photograph (EXP) Showing the Flow in a Plane Parallel to QTMS and 0.25B from it: Re = 2000, $\Delta T = 0°C$

were used in each simulation. It is likely that the discrepancy can in large part be attributed to the grid resolution near the walls, even though variable grids, refined near boundaries, were used. The UCB-3DH has the greatest number of grid points and does the best job of representing the eddy sizes. [The FEM of LLL uses linear elements so the accuracy of its spatial representation on a given grid is roughly equivalent to that of the UCB finite difference scheme (except when the low order HYBRID scheme comes into play in the latter).]

Figure 4a shows a UCB-3DH velocity vector plot on a plane parallel to the downstream wall QTMS (perpendicular to the direction of the belt motion; see Figure 2) and 25 % of the cavity width B from the downstream wall. Figure 4b is the same plane in the experiment, as visualized by use of the rheoscopic liquid technique. The corner vortex is believed to be caused by an adjustment of the shear and pressure forces acting on the recirculating fluid to the no-slip condition imposed by the end wall. This vortex is present in both the experiments and the 3-D simulations. Indeed, it was found first in the simulations and then sought and shot (photographically-like big game!) in the experiments.

Unfortunately, the longitudinal vortices (Taylor-Gortler-like vortices), first reported and described by Koseff and Street [9], are not evident in either 3-D numerical simulation. These vortices, which originate in the zone of separation between the primary eddy and the downstream secondary eddy, are shown clearly in the photograph and in fact persist as they are convected around the cavity.

Figure 5 shows the shape of the primary circulation cell for the Re = 1000, stratified ($\Delta T = 1^{o}C$) case 30 minutes after flow start up. [In both experiment and simulation the lid achieves its selected speed essentially instantaneously.] The agreement is good, except for the region of the internal wave ($x/B = 0.65$) where the UCB-3DH solution does not follow the experimental curve. The smoothing or diffusion of the wave appears to result from the numerical diffusion introduced by the HYBRID differencing scheme in regions of high velocity or sharp gradients. Figure 6 shows the flow (on the symmetry plane) after 2 hours as depicted by a LLL-2D velocity vector-isotherm plot (Figure 6a) and the experiment using the thymol-blue technique (Figure 6b). [The diagonal dye line in Figure 6b should be ignored for the purposes of this discussion; it indicates motion in the secondary circulation cell over a period of two hours.] The qualitative agreement between the experiment and the simulation is very good.

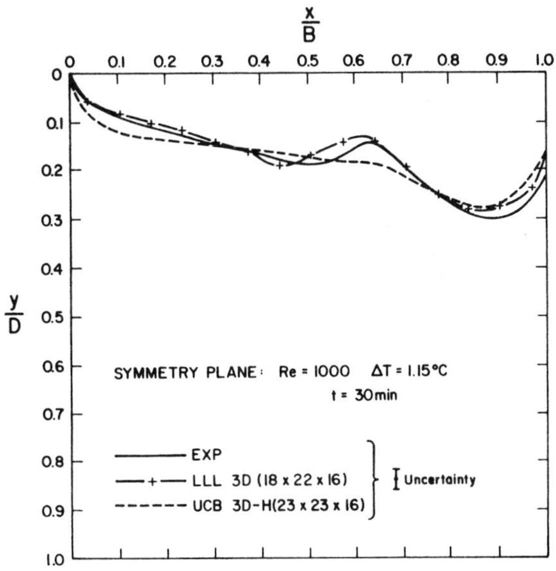

Figure 5: Approximate Shape of Primary Circulation Cell

In the higher speed, Re = 2100, stratified ($\Delta T = 1.15^{\circ}C$) case, however, the agreement is not as good, particularly for the 3D calculation (which uses the HYBRID scheme). The results for this case are shown in Figures 7 through 10. While the 2-D calculations are reasonable for both the UCB and LLL computations, the UCB-HYBRID simulation is erroneous, reflecting strongly a weakness of the HYBRID scheme. Two additional points are evident. First, at a time of 360 s (t * N = 41.9, where N is the Brunt-Vaisala frequency defined in Figure 7) after start up, the experiment shows a deeper penetration than the 2-D results. This is shown in Figures 7 and 9. Second, the shape of the primary circulation cell as predicted by the LLL-2D simulation at t = 150 min. is noticeably different from that of the experiment (see, e. g., the location of the maximum penetration point in Figure 10). However, the flow, as predicted by LLL-2D, is approaching a periodic steady state at t = 150 mins. and there are indications in the calculations of the existence of repetative Kelvin-Helmholtz type instabilities at the thermocline. These points, as well those raised previously, are discussed further in the following section.

6. DISCUSSION

Generally, there is good qualitative agreement between the numerical simulations and the experiments, especially for the more "moderate" cases, viz., Re = 2000, isothermal (ΔT =

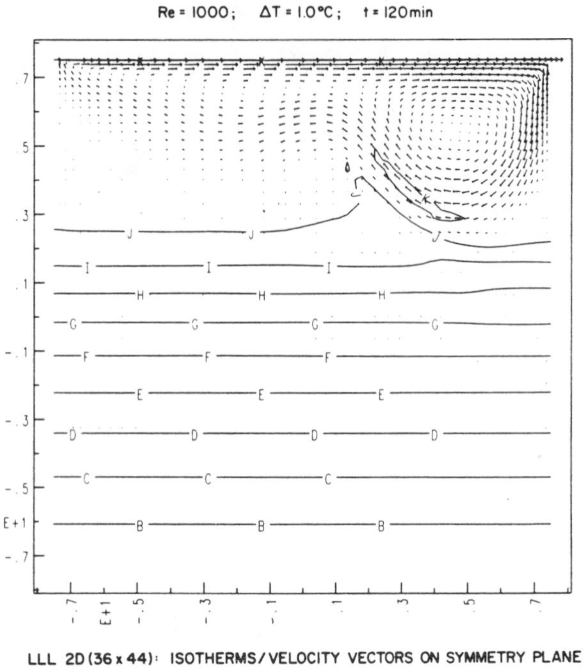

(a)

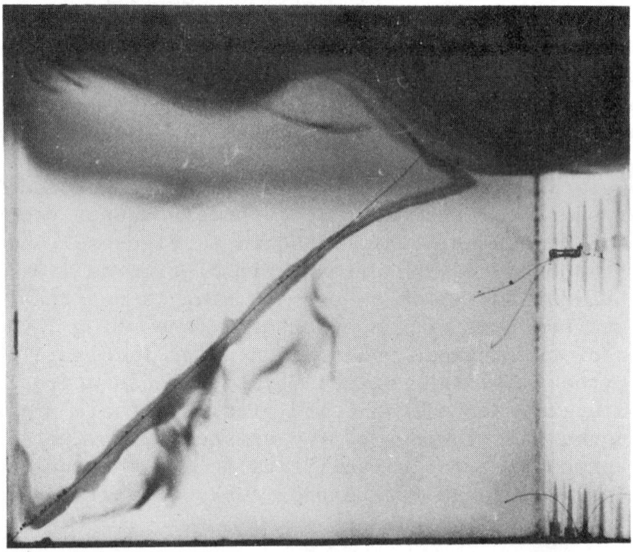

(b)

Figure 6: Flow on the Symmetry Shown by (LLL-2D) Plot and Thymol Blue Photograph: Re = 1000, ΔT = 1°C, t = 120 mins.

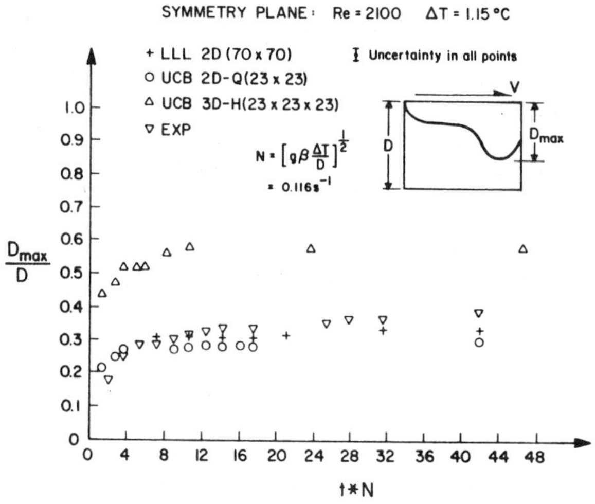

Figure 7: Penetration of Primary Circulation Cell as a Function of Time.

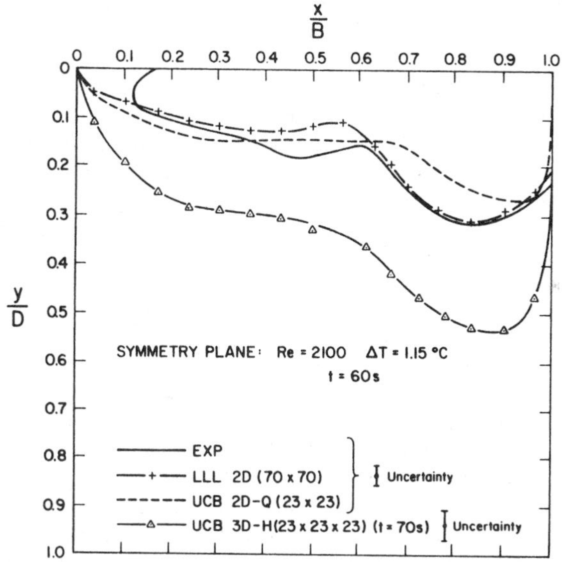

Figure 8. Approximate Shape of Primary Circulation Cell

$0^{o}C$) and Re = 1000, stratified (ΔT = $1.0^{o}C$). Both codes show the presence of the secondary eddies and the corner vortices (see Figure 4) for the isothermal case, as well as the variation of the size of the corner vortex with position along the end wall in the direction of the belt motion. The penetration trends (predicted by the 3-D simulations) for the

Re =1000, stratified ($\Delta T = 1^oC$) case are very close to those observed in the experiments. The 2-D simulations do a reasonable job of predicting penetration in the Re = 2100, stratified ($\Delta T = 1.15^oC$) case. Because the Prandtl number Pr = 7.1 in this flow at Re = 2100, the Peclet number is approximately 15,000, which most modelers acknowledge leads to a very diffucult simulation problem. The internal wave present at the thermocline at the bottom of the primary circulation cell for the Re =1000, stratified ($\Delta T = 1.0^oC$) case is also captured in both simulations, being especically well done by the LLL solution which does not suffer from the apparent diffusion effects of the HYBRID differencing scheme.

However, there are some important differences to be considered. The first is the differences in the measured and simulated dimensions of the secondary eddies (see Table 1) as predicted by the codes for the Re = 2000, isothermal case. For D_D^* (see Figure 3a) the LLL 2-D and 3-D results are essentially the same. Neither the experiments nor the UCB simulations give this result. Indeed, Koseff, et al. [10] show that the flow as measured on the symmetry plane (see Figure 2) is markedly weaker than that predicted by a 2-D solution. Therefore, the dimensions predicted by 2-D and 3-D numerical simulations should be different. The UCB-3DH solution uses the same vertical plane resolution as the UCB-2DQ solution, does show this expected difference, and is in good agreement with the experiments. The LLL-3D solution, which appears to be dominantly two dimensional when examined carefully, uses considerably less resolution than its 2-D counterpart and than the UCB solutions. One is led to suspect then that the problem is not in the model physics, but is a result of using too coarse a grid.

Further insight can be gained by examination of the 2-D solutions for the Re =2000, isothermal case in comparison to available, high resolution numerical solutions. Winters and Cliffe [14] solved the Re =2000 case, while Ghia, et al. [4], Agarwal [1], Schreiber and Keller [13], and Benjamin and Denny [3] solved the Re = 1000 and 3200 cases. Interpolating linearly we find an average D_D^* of 0.37 with a variance of 0.04 for the latter group; Winters and Cliffe give D_D^* = 0.38. The LLL-2D value for D_D^* is 0.42. However, the LLL-2D values for B_U^* and D_U^* are in good agreement with the other 2-D solutions, whereas those of the UCB-2DQ are not. Thus, it appears that both of our 2-D solutions are somewhat inaccurate (when compared to solutions using many more node points), but are still reasonable representations of the flow.

Returning to the 3-D simulations we observe that it is quite possible to conclude now that the difference between the experimental data and the LLL-3D solution is due (at least

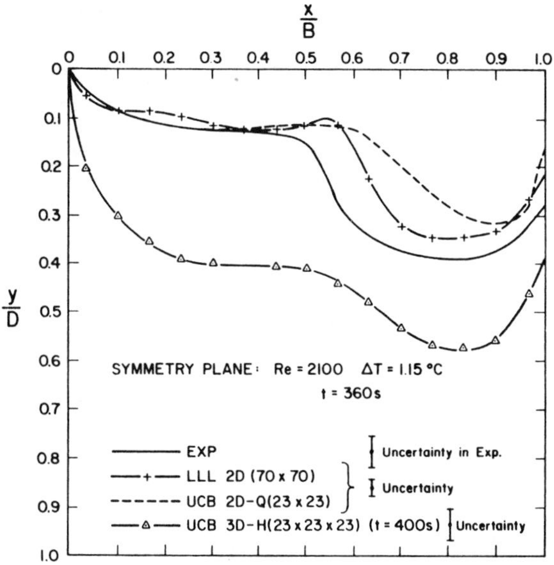

Figure 9. Approximate Shape of Primary Circulation Cell

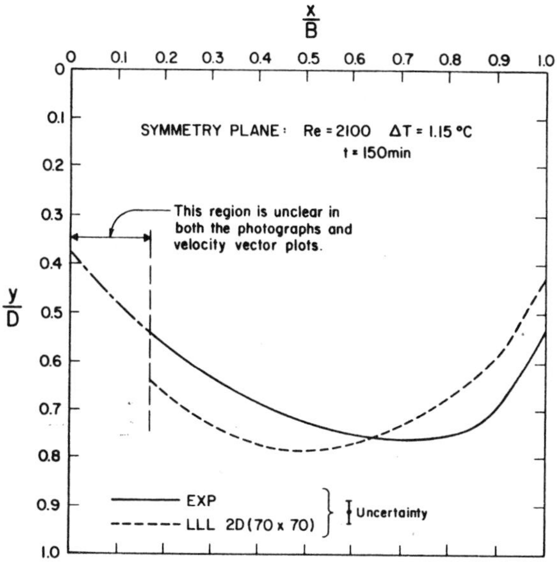

Figure 10. Approximate Shape of Primary Circulation Cell

partially) to the grid used. In comparison to the LLL-3D grid, the grid used by the UCB-3DH simulations has twice as many points in any comparable vertical plane parallel to the symmetry plane and twice as many spanwise points.

The problem of grid resolution becomes very evident when one considers that both the UCB-3D and LLL-3D simulations resolve the corner vortex in the region of the end wall, but do not resolve the Taylor-Gortler-like (TGL) vortices (see Figure 6) in the core of the flow. Both codes use a variable grid system in which the density of grid points near the boundaries is much higher than in the core of the flow. Thus, it is distinctly possible that use of a refined grid (2 or 3 times more dense than at present) would resolve the TGL vortices because they are of the same order of size as the corner vortex.

Also of significance is the apparent failure of the HYBRID differencing scheme when used in calculations involving the temperature-induced density gradients at these Reynolds numbers. The shortcomings of this scheme are illustrated by three examples. First, the wave on the thermocline at the bottom of the primary circulation cell in the Re = 1000, stratified ($\Delta T = 1^{o}C$) case is diffused away in the HYBRID calculation. Second, the size of the primary circulation cell given by the UCB-3DH for the Re = 2100, stratified ($\Delta T = 1.15^{o}C$) case is much larger than that observed in the experiments or predicted by the 2-D simulations. It is, however, of the right shape and proportions. Third, the isotherm pattern predicted by the UCB-3DH solution for this case is hardly perturbed from its original position despite the strong penetration. This last observation offers a clue as to the effects of the HYBRID scheme in regions of strong temperature gradients such as those present in this physical situation. There is an apparent strong numerical diffusion which occurs whenever the HYBRID scheme employs its upwind difference mode and this effectively reduces the influence of the temperature gradients on the convecting fluid to the point where the expected inhibition of the motion by the stratification is virtually eliminated, allowing the extra deep penetration shown in Figure 9. As expected, at a lower Reynolds number (Figure 5) fewer nodes exceed the criteria for invoking the upwind differencing mode of the HYBRID scheme and the numerical diffusion is much reduced. Han, et al. [7] recognize the weakness of the HYBRID scheme and show the improvements which can be obtained in the REBUFFS code in 2-D by implementation of the QUICK scheme; the 3-D REBUFFS version is now being modified to incorporate quadratic upstream differencing.

Finally, we turn again briefly to the question of using two-dimensional calculations to predict this three-dimensional recirculating flow. As seen from Table 1, 2-D solutions agree qualitatively, but not quantitatively, with the isothermal experiments where the spanwise-aspect-ratio is 3:1. Figures 8 through 10 show that 2-D solutions for the higher Re, stratified flow case are accurate soon after flow start up (Figure 8), are less so after some period of time (Figure 9), and seem to lose some essence of the physics in the long time frame (Figure 10). There appears to be a plausible explanation of this trend which is elucidated below.

The curved (concave when viewed from the lid) shape of the separation between the primary and secondary circulation cells promotes the formation of longitudinal vortices (TGL vortices) above the separation surface (Koseff and Street [9]). These vortices, which were observed in the experiments, but not is the numerical simulations, increase the vertical mixing at the interface between the circulation cells and, thereby, increase the rate of penetration. Since it takes time for the TGL vortices to develop, the 2-D simulation and experiment agreement at t = 60 s (Figure 8) was good. At t = 360 s (Figure 9) the experimental penetration is greater because the TGL vortices are established and active in mixing.

Figure 10 shows the point of maximum penetration predicted by LLL-2D (at t = 150 mins [(iv)]) to be similar to the experimental value, but their location along the direction of the belt motion are different. At the maximum penetration point (which is variant with time), the buoyancy force exerted by the lower and denser fluid balances the downward momentum generated by the belt motion. At this point the experiment shows that the flow turns upward more sharply than is predicted by the calculation. From the flow geometry, it is expected that the region of maximum vertical momentum flux will be close to the downstream side-wall QTMS (Figure 2). Thus, as in the experiments, the maximum penetration point should be close to that side wall as well. The spanwise motion induced by the end walls and observed by Koseff and Street [9] may play a role in preserving this asymmetry. In the 2-D case such a motion is not possible and the 2-D flow is then fundamentally different. The symmetric position of the maximum penetration reflects this difference.

7. CONCLUSIONS

The three-dimensional numerical simulations are in good agreement with the experiments for this complex recirculating, lid-driven cavity flow for the isothermal case with Re = 2000 and for the stratified case at the lower Reynolds number of 1000. For the Re = 2100 stratified case, the HYBRID

(iv) Corresponding calculations at UCB were terminated at 400 s.

differencing scheme appears to introduce numerical diffusion which ruins the solution. Two-dimensional solutions of the three-dimensional flow are in good qualitative agreement with the experiments. However, they cannot reproduce all of the physics of the flow and, therefore, there are key differences.

Greater numbers of grid or node points, especially in the spanwise direction, will be required to capture all the structures in the real flow. This appears to be important because the three-dimensional structures appear to play an important role in the stratified flow case. The probabiltiy of success with a refined grid is high because the refined grid near the end walls does resolve a corner vortex which is of the same size as the longitudinal vortices not resolved in the main flow body.

8. ACKNOWLEDGMENTS

This work was facilitated by a number of sponsors. The work of JRK and RLS was supported by the NSF through grant No. CEE-7921324 and by the DOE through contract No. DOE-DE-AT03-81-ER-10867. The work of PMG and CDU was supported by the DOE through contract No. W-7505-ENG-48. The work of JACH and WMT was supported by Sandia National Laboratories through contract No. 20-1012 and by the DOE through contract No. DE-AC03-76SF-0098.

9. REFERENCES

1. AGARWAL, R.K. - A Third-Order-Accurate Upwind Scheme for Navier-Stokes Solutions at High Reynolds Numbers, AIAA 19th Aerospace Sciences Meeting, St. Louis, MO, Jan. 12-15, 1981.

2. BAKER, D.J. - A Technique for the Precise Measurement of Small Fluid Velocities. J. of Fluid Mechanics, Vol 26, pp. 573-575, 1966.

3. BENJAMIN, A.S., and DENNY, V.E. - On the Convergence of Numerical Solutions for 2-D Flows in a Cavity at Large Re, J. Comp. Physics, Vol. 12, pp. 348-363, 1973.

4. GHIA, U, GHIA, K.N., and SHIN, C.T. - Solution of Incompressible Navier Stokes Equations by Coupled Strongly Implicit Multi-Grid Method, Symp. on Multigrid Methods, NASA-AMES Research Center, Calif., Oct. 21-22, 1981.

5. GOSMAN, A.D., and PUN, W.M. - Lecture Notes for Course Entitled Calculation of Recirculating Flows No. HT5/74/2, Imperial College, London, 1974.

6. GRESHO, P.M., and UPSON, C.D. - Current Progress in Solving the Time-Dependent, Incompressible Navier-Stokes Equations in Three-Dimensions by (almost) the FEM, Proc. Fourth Int'l Conf. on FEM in Water Resources, Hannover, Germany, 1982 (see also 1983 Seattle Conference Proceedings).

7. HAN, T., HUMPHREY, J.A.C., and LAUNDER, B.E. - A Comparison of Hybrid and Quadratic-Upstream Differencing in High Renolds Number Elliptic Flows. Computer Methods in Applied Mechanics and Engineering, Vol. 29, pp. 81-95, 1981.

8. KATSAROS, K.B., LIU, W.T., BUSINGER, J.A., and TILLMAN, J.E. - Heat Transport and Thermal Structure in the Interfacial Boundary Layer Measured in an Open Tank of Water in Turbulent Free Convection, J. Fluid Mech., Vol. 83, No. 2, pp. 311-335, 1977.

9. KOSEFF, J.R., and STREET, R.L. - Visualization Studies of a Shear Driven Three-Dimensional Recirculating Flow, Three Dimensional Turbulent Shear Flows, ASME, pp. 23-31, 1982.

10. KOSEFF, J.R., STREET, R.L., and RHEE, H. - Velocity Measurements in a Lid-Driven Cavity Flow, to be presented at ASME Conf. on Numerical and Experimental Investigations of Confined Recirculating Flows, Houston, June 20-22, 1983.

11. LEONARD, B.P. - A Stable and Accurate Convective Modeling Procedure Based on Quadratic Upstream Interpolation Computer Methods in Applied Mechanics and Engineering, Vol. 19, pp. 59 ff., 1979.

12. LE QUERE, P., HUMPHREY, J.A.C., and SHERMAN, F.S. - Numerical Calculation of Thermally Driven Two-Dimensional Unsteady Laminar Flow in Cavities of Rectangular Cross Section. Numerical Heat Transfer, Vol. 4, pp. 249-283, 1981.

13. SCHREIBER, R., and KELLER, H.B. - Efficient Numerical Techniques for Two-Dimensional Steady Flow, to appear J. Comp. Physics, 1983.

14. WINTERS, K.H., and CLIFFE, K.A. - A Finite Element Study of Laminar Flow in a Square Cavity, UKAERE Harwell Report R-9444, 1979.

NODAL INTEGRAL METHOD SOLUTIONS AND SINGULAR POINTS FOR DRIVEN CAVITY PROBLEMS

Y. Y. Azmy(i) and J. J. Dorning(ii)
Nuclear Engineering Program
University of Illinois
Urbana, IL 61801

SUMMARY

The recently developed nodal integral method (NIM) for the numerical solution of the steady-state, isothermal, incompressible Navier-Stokes equations in two-dimensional cartesian geometry is briefly reviewed, extended and applied to several driven cavity problems. The results for various Re are compared first with results previously obtained using other numerical methods to demonstrate the superior computational efficiency of the NIM, and then with experimentally observed flow fields to establish the veracity of the computed flow fields. Singular points and multiple solutions have been detected for relatively high Reynolds numbers and are discussed briefly.

1. INTRODUCTION

A nodal integral approach to the solution of linear partial differential equations has been developed elsewhere [1]. When extended to the nonlinear Navier-Stokes equations in the primitive variables $(\underline{u},p)$ for steady-state, isothermal, incompressible flow in two-dimensional cartesian geometry, the Nodal Integral Method (NIM) is obtained [1,2]. A brief description of this recently developed method [1,2] is given in Sec. 2. Test problems solved and reported here and elsewhere [1-3] are described, and the modification of the method to accomodate the closed driven cavity problem is described. In Sec. 3 the results for several modified driven cavity problems are presented and discussed. First, we review the comparison [1] of the computational efficiency of the NIM with those of the Nodal Green's Tensor Method (NGTM) [3] and a finite difference method in the form of the SOLA

(i) Research Assistant
(ii) Professor of Nuclear Engineering

code [4] for Re = 100 and various meshes. Then flow maps for Re = 3,000 are presented to show the method is capable of converging to reasonable solutions at high Reynolds numbers. Also presented are flow maps for Re = .01 which are in very good agreement with experimentally observed flow fields [5]. The results for the closed driven cavity are presented in Sec. 4 and include two significantly different solutions to that problem for the same Reynolds number. Sec. 5 is devoted to the discussion of these multiple solutions and the two singular points that are observed. Finally, some remarks are made on the interpretation of divergence and non-physical solutions in the literature.

2. FORMALISM OF THE NODAL INTEGRAL METHOD

The detailed development of the NIM equations for the numerical solution of the non-linear Navier-Stokes equations, along with one for general linear PDE's, already has been presented [1]. However, for the sake of completeness, the development will be described here very briefly. First, the domain of physical interest is divided into rectangular computational elements or nodes. The two-dimensional, steady-state, isothermal, incompressible continuity and momentum equations are transverse averaged [3] locally over each node, separately in each direction. This results in six ODE's in the six transverse-averaged variables defined by:

$$\overline{g}^q(r) = \frac{1}{2s}\int_{-s}^{+s} g(q,r)dq, \tag{1}$$

where $g \equiv u_x, u_y, T_x,$ or T_y; $q \equiv x, y$; $r \equiv y, x$; and $s = a,b$ is the dimension of the node in the q-direction. Here T_x and T_y are the normal stress components in the x- and y- directions respectively (defined by $T_q = \nu(\partial/\partial q)u_q - p$, where $q \equiv x,y$) and are introduced to formally decouple the transverse-averaged ODE's to obtain

$$\frac{d}{dr}\overline{u}_r^q(r) = \frac{1}{2s}\left[u_q(r,+s) - u_q(r,-s)\right], \tag{2}$$

$$\frac{d}{dr}\overline{T}_r^q(r) = \frac{1}{2s}\left[\int_{-s}^{+s}\left(u_r\frac{\partial u_r}{\partial r} + u_q\frac{\partial u_r}{\partial q} - f_r\right)dq - \nu\left.\frac{\partial u_r}{\partial q}\right|_{-s}^{+s}\right], \tag{3}$$

$$\nu\frac{d^2}{dr^2}\overline{u}_q^q(r) = \frac{1}{2s}\left[\int_{-s}^{+s}\left(u_r\frac{\partial u_q}{\partial r} + u_q\frac{\partial u_q}{\partial q} - f_q\right)dq - \nu\left.\frac{\partial u_q}{\partial q}\right|_{-s}^{+s} + \left.p(r,q)\right|_{-s}^{+s}\right], \tag{4}$$

where r, q, and s have the same meanings as above, and $\underline{f}$ is the external body force per unit mass.

The linear parts of these equations are inverted exactly to obtain node-interior expressions for the transverse- averaged velocity components and normal stresses in terms of their node-surface values, and integrals of the non-linear advection terms. Discrete-variable equations are obtained by approximating the latter in terms of the transverse-averaged, surface-evaluated velocity components, and taking the limits of the node-interior expressions to the appropriate node surfaces. After elimination of the transverse-averaged shear stresses via their continuity across node surfaces, a set of nonlinear algebraic equations in the transverse-averaged, surface evaluated variables is obtained. These equations, augmented by the appropriate set of boundary conditions for a specific problem are then solved using a Newton-Raphson procedure; the two-dimensional node averages of the primitive variables are constructed from the converged solution via the equations obtained for these node-averaged quantities by integrating (averaging) inverted forms of Eqs. (2-4).

The NIM has been applied in the form described above to three test problems. The exact (to within an error of 10^{-12}) analytical solution to the problem of fully developed flow between parallel plates [2] was computed via the method in two dimensions for a wide variety of Reynolds numbers as a rudimentary test of the method. The problem of inlet flow between parallel plates (developing flow) also was solved [1,2]. The results were identical to those previously obtained by the NGTM [3] and were obtained in shorter computing times [1,2]; for comparable accuracies they were obtained in computing times that were dramatically shorter than those required [3] by the finite difference code SOLA [4]. A variety of modified driven cavity problems for Re = 1, 10, 100 and 1,000 also were solved [1,2] and the solutions compared to those obtained previously by the NGTM [3]. Additional modified driven cavity problems have been solved now for higher Re and very low Re and are reported in Sec. 3 where the latter are compared with experimentally observed flows [5].

A fourth and final test problem to which the new method has been applied is the closed driven cavity problem. (See Sec. 4). However, in order to apply the method to that problem the equations had to be modified. To motivate the modification necessary to solve the closed driven cavity problem we note that the set of boundary conditions (zero velocity on all walls, except for $u_0 \neq 0$ at the plate at the top of the cavity) together with the Navier-Stokes equations leaves the pressure (and hence the normal stresses) indeterminate to within an additive constant. The absolute pressure is not relevant to the flow field, and only pressure drops are meaningful. Hence, we added an equation specifying the normal stress (either T_x or T_y) at one surface of one node (can be chosen arbitrarily) which is equilvalent to specifying the arbitrary constant in the pressure. [Equivalently one may choose to specify the average pressure over the whole cavity, however, the above-mentioned choice provides the simplest additional equation, which is actually treated as a boundary condition]. Finally, to restore the well-posedness of the algebraic problem, we must remove the equation (in the augmented set) that makes the problem singular. Otherwise the algebraic problem will be over determined. Summing up the discrete-variable form of the mass continuity equations for all the nodes, and making use of the boundary conditions results in a trivial equation. Hence we excluded the continuity equation for one of the nodes (equivalently the boundary condition on the surface of one node) from the set of algebraic equations. The resulting set is well posed, and the excluded continuity equation is automatically satisfied by virtue of the linear dependance of that equation on the remaining continuity equations and the set of boundary conditions employed. This modified form of the NIM equations has been used to solve the closed driven cavity problems reported in Sec. 4.

3. THE MODIFIED DRIVEN CAVITY PROBLEM

The square modified driven cavity problem (a driven cavity with the moving plate above but not adjacent to the top of the cavity) was solved previously using the NIM for Re = 1, 10, 100, and 1,000, on 2x2, 4x4, 6x6 and 8x8 meshes [1,2]. The resulting surface-averaged variables were numerically identical to those that had been obtained previously by the NGTM [3], but were obtained in shorter computing times [1,2] (since the NIM has fewer unknowns per node than the NGTM and is mathematically equivalent [1,2]). For Re = 100, the ratios of NGTM computing times (CPU) to NIM computing times were 4.3, 3.5, 3.2, and 2.5 for 2x2, 4x4, 6x6 and 8x8 meshes [6,1]. Since the computational efficiency of the NGTM, based on CPU computing time required for comparable accuracies of cavity, quadrant-averaged velocities, had previously been compared [3] with that of a finite difference method in the form of the SOLA code [4] and shown to be vastly superior, the NIM computational efficiency is even more superior. The superior accuracy, and resulting superior computational efficiency, of the NIM and the NGTM over primitive-variable, finite-difference methods as represented by the SOLA code is manifested in two ways. For Re = 1, 10, 100, the NIM and NGTM converged to (identical) solutions that were far more accurate in cavity, quadrant-averaged velocities, with respect to infinitessimal mesh extrapolated solutions, than those obtained by the SOLA code on the same and finer meshes [3]. For a higher Reynolds number, Re = 1,000, the NIM [1,2] and the NGTM [3] resolved the upstream, lower-corner, secondary vortex on meshes as coarse as 6x6, whereas the SOLA did not resolve it even on a 16x16 mesh [6].

In Fig. 1 we present the flow in a square modified driven cavity at Re = 3,000. The arrows shown have lengths (tip to tail) proportional to the magnitude of the transverse-averaged, surface-evaluated velocities, except when this procedure results in arrows that are too small to be useful visually, in which case a "minimum length" arrow is used. As expected the secondary vortex is larger than it was for Re = 1,000 [3,1,2].

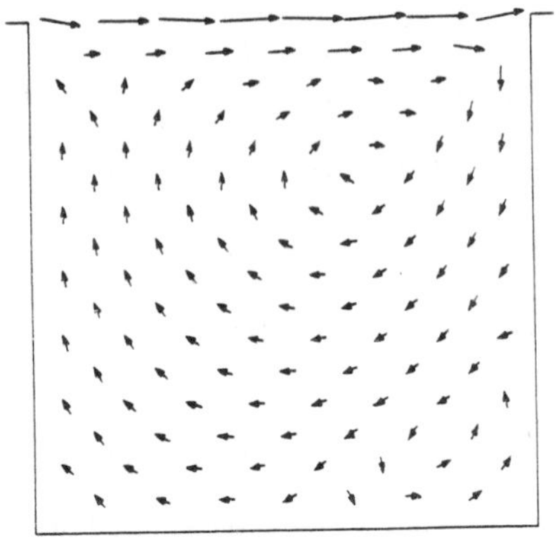

Fig. 1 Flow map (transverse-averaged, surface-evaluated velocities) for the modified driven cavity problem as computed on an 8x8 mesh by the nodal integral method for $Re=u_pL/\nu$ = 3,000, where u_p is the plate velocity, ν the kinematic viscosity, and L the length and depth of the square cavity. Not shown in the figure is the inlet-outlet region between the moving plate and the top of the cavity, comprised of ten horizontally adjacent elements: eight middle elements directly above the cavity elements each (L/8)x(L/2); and inlet and outlet elements each (L/2)x(L/2). The counter flow pattern (secondary vortex) in the lower right corner of the cavity is larger than the one which also was resolved by the nodal integral method even on a rather coarse mesh (6x6) for Re = 1,000 [3,2].

The definition of the Reynolds number for the modified driven cavity ($Re = u_pL/\nu$) used here and elsewhere in the literature [3,5] is deceptive since it does not take into account the distance between the moving plate and the top of the cavity. In order for the Reynolds number to characterize the cavity flow it should be based on a velocity that is characteristic of the cavity, or, at the very least, one that is in, or on the boundary of, the cavity. Although this is not practical in general since the velocities in the cavity are not known until after the problem is solved, it might still be useful to calculate an "effective cavity Reynolds number", after the problem is solved, based on the maximum velocity in the cavity, the near-horizontal velocity at the top. For the flow shown in Fig. 1, this effective Re ≅ 300, an order of magnitude lower than that based on the velocity of the plate located a distance L/2 above the cavity top. Hence the rotational flow depicted in Fig. 1 is more characteristic of a cavity with an effective "cavity" Re ≅ 300 than

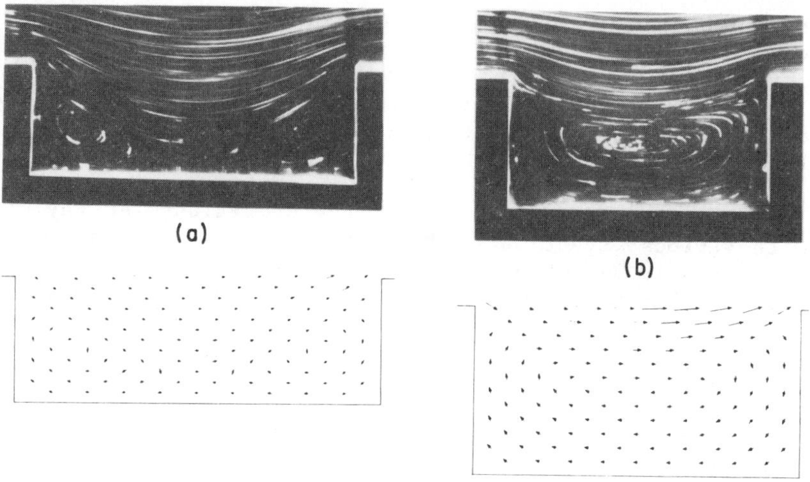

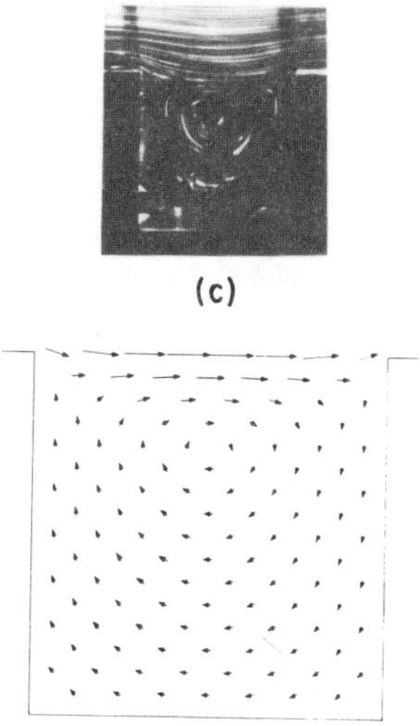

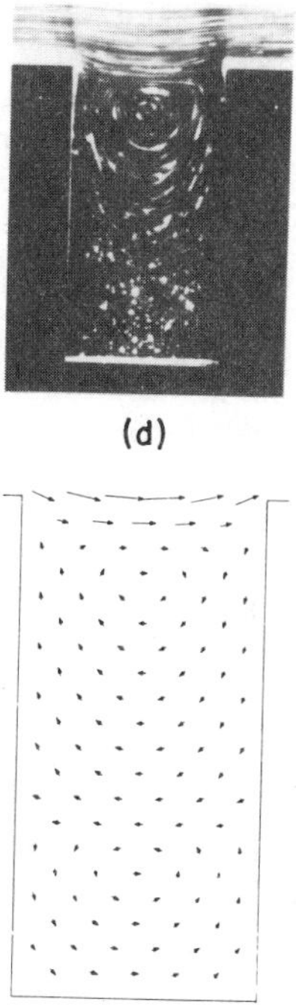

Fig. 2 A sequence of NIM computed flow maps compared to experimentally observed flow fields [5] for four different aspect ratios: A=3, 2, 1, and .5 respectively, for the modified driven cavity. Non-uniform meshes analogous to that described in Fig. 1, were used with cavity grids 10x6, 10x6, 8x8, and 6x10 respectively. Re, based on plate speed and cavity depth, was held constant at .01. Note the development of the two corner vortices (a) into one main vortex (b,c) then into two vertically adjacent vortices (d), in both the experimental and computed flows. Note also that the positions of the centers of the vortices in the NIM computed flow maps are in good agreement, to within the resolution of the coarse computational meshes used, with those in the photographs of the experimental flows.

one with a Re = 3,000. In fact, as will be seen in Sec. 4 the closed driven cavity flow map for Re = 300 shown in Fig. 3 is very similar to that shown in Fig. 1, especially in the lower half of the cavities where the effects of the details of the velocity distribution along the top of the cavity is least important.

The flow fields in a sequence of rectangular cavities with various aspect ratios and Re = 0.01, were computed and are compared to experimentally observed flow fields [5] in Fig. 2. The development of the two corner vortices (A=3) into one vortex (A=2,1) then into two vertically adjacent vortices (A=0.5) in the NIM computed flow maps is consistent with the actual photographs of the experimental flows. Moreover, the overall agreement between the computed and experimental flows is very good. The lower vortex in the case A=0.5 does not show in the picture because its magnitude is very small, however it is known to exist [5].

4. THE CLOSED DRIVEN CAVITY PROBLEM

The closed driven cavity problem is one of the simplest yet most important problems in the numerical study of fluid flow. Over the years it has become a common test problem for newly developed numerical methods. Hence it was important to modify the NIM equations (as described in Sec. 2) in order to solve this problem and compare the results to those in the existing vast literature (see Ref. 7). The results obtained are in good qualitative agreement with results obtained earlier [8,9] for a wide range of Re (400-1,000), even though a very coarse mesh was used in the present work (8x8 maximum). One example of these NIM calculations for Re = 300 is presented in Fig. 3.

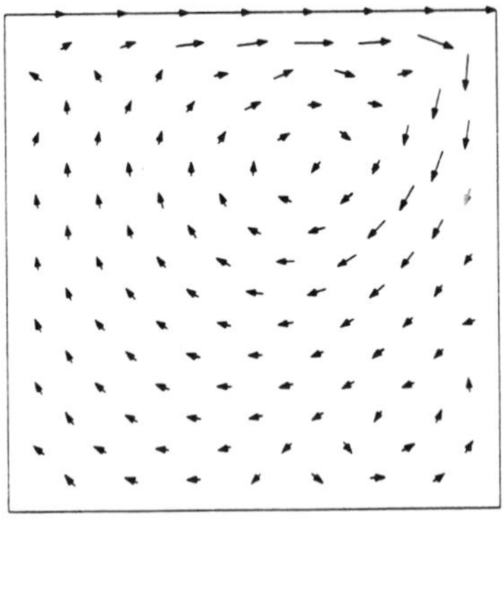

Fig. 3 Flow map (transverse-averaged, surface evaluated velocities) for the closed driven cavity problem as computed on an 8x8 uniform mesh by the NIM for $Re=u_pL/\nu=300$. Notice the qualitative similarity to the flow map in Fig. 1, especially in the lower half of the cavity.

Figure 4 is the predicted flow map at Re=900 on a 6x6 mesh. It was found that the Newton-Raphson iterations would not converge to a solution starting from a uniform, zero initial guess for the velocity field (as was used for all calculations described so far) for Re > 925. However, the same uniform, zero initial guess was then rather arbitrarily used for a calculation at Re=1,500, and the iterative procedure did converge. The solution thus obtained, was then tracked from Re=1,500 to lower Re in successive steps in which the converged solution for one Re was then used as an initial guess for the calculation at the next lower Re. This continued solution, to which we shall refer as Solution C, converged easily for all Re $\gtrsim$ 789 at which value it no longer could be converged. The first solution, shown in Fig. 4, to which we shall refer as Solution A, was tracked in the same way up to Re $\lesssim$ 1,469 at which value it no longer could be converged. The flow map corresponding to Solution C at Re=900 is shown in Fig. 5. It clearly differs significantly from that for Solution A at Re=900 shown in Fig. 4. Solution A and Solution C, and the Reynolds numbers at which they no longer could be converged will be discussed further in Sec. 5.

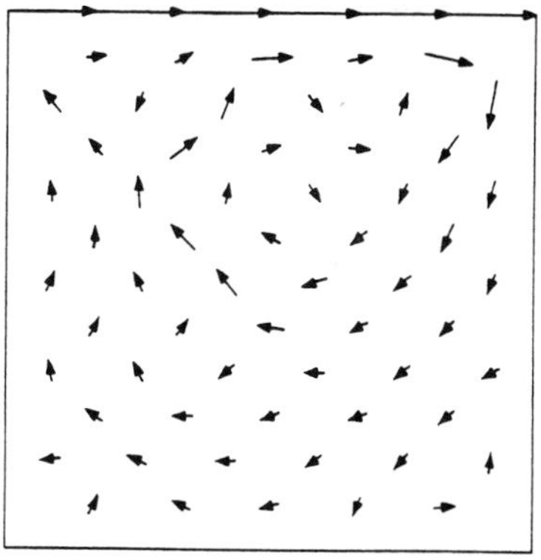

Fig. 4 Flow map (transverse-averaged, surface-evaluated velocities) for the closed driven cavity problem as computed on a 6x6 mesh by the NIM for Re=900. Note the main vortex and the two lower corner vortices, and the direction of rotation of each.

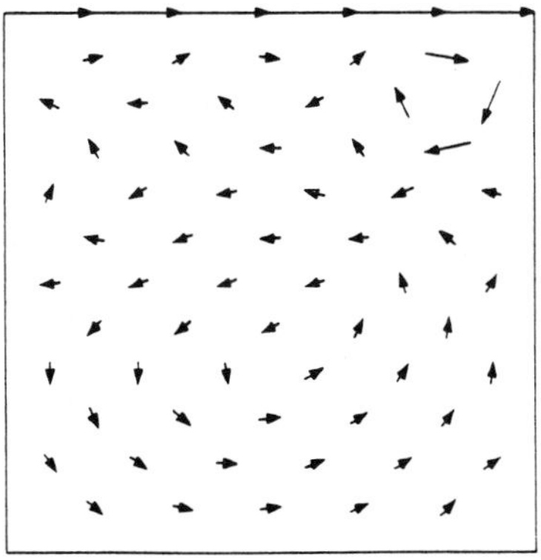

Fig. 5 Flow map (transverse-averaged, surface-evaluated velocities) for Solution C to the closed driven cavity problem as computed by the NIM on a 6x6 mesh for Re=900 (same Re as Fig. 4 for Solution A). Note the smaller size of the strong main vortex in the upper right corner, and the aglomeration of the two secondary vortices at the lower corners into a single weak, large vortex occupying the lower two-thirds of the cavity.

5. SINGULAR POINTS AND MULTIPLE NIM SOLUTIONS TO THE CLOSED DRIVEN CAVITY PROBLEM

The set of algebraic nonlinear equations obtained in Sec. 2 can be written as

$$\underline{F}(\underline{U}(Re),Re) = 0, \tag{5}$$

where $\underline{F}$ and $\underline{U} \in R^n$. The Newton-Raphson iterative procedure used to solve Eq. (5) is summarized by the equation

$$\underline{U}^{n+1} = \underline{U}^n - \underline{\underline{J}}^{-1}(\underline{U}^n, Re)\ \underline{F}(\underline{U}^n, Re), \tag{6}$$

where n is the index of iteration, and $\underline{\underline{J}}(\underline{U}^n, Re)$ is the Jacobian matrix ($J_{ij} \equiv \partial F_i/\partial U_j$), evaluated at $\underline{U}^n$.

A singular point in UxRe space is one at which $\det(\underline{\underline{J}}(\underline{U}^*, Re^*)) = 0$, where $\underline{F}(\underline{U}^*(Re^*),Re^*) = 0$. Hence $\underline{\underline{J}}^{-1}(\underline{U}^*,Re^*)$ does not exist at such a point, and Eq. (6) cannot be used to solve Eq. (5) for $\underline{U}^*$. The numerical method diverges for Re=Re*. Figure 6 is a plot of $\det(\underline{\underline{J}})$ vs Re, for the two solutions, Solution A and Solution C, shown for Re=900 in Figs. 4 and 5, respectively. This plot indicates singular points for Solution A and Solution B at Re*≅1,468.643 and 789 respectively. Although it is difficult to see in the figure, the determinant of J for Solution A does not go to zero as Re is decreased from 1,000; rather it remains positive even for very small Re. The singular point in Solution A at Re* ≅ 1,468.643 indicates that this is possibly a bifurcation point or a turning point (limit point) for this solution. Similarly, the singular point in Solution C at Re* ≅ 789 indicates that it is possibly a bifurcation point or a turning point for that solution. Various methods [10-12] have been developed recently to continue solutions through singular points. Keller [10] has presented a method that seems to be less complicated than most and particularly suitable to a problem as large as the present one. This method [10] guarantees continuation of a solution branch across regular points and also singular (bifurcation and turning) points. Furthermore it supplies a procedure to switch solution branches at bifurcation points (if they exist). This procedure, however, has not been carried out yet for the NIM solution to the closed driven cavity problem to determine the nature of the singular points reported here. However, it is possible to risk a conjecture based on other recent results obtained for the Navier-Stokes equations for the closed driven cavity problem. In a very recent study of this problem, using a stream function formulation, Schreiber and Keller [13] also detected two singular points (actually two consecutive turning points) on what appears to be the physical solution branch. As they refined their mesh, the two singular points moved towards higher Re [13]. Armed with a knowledge of their results we can conjecture that the two

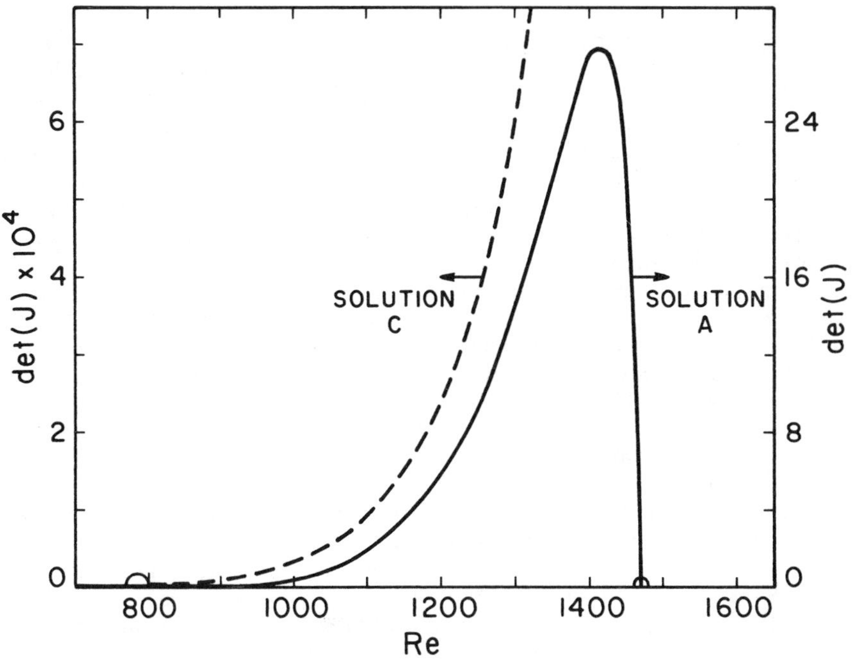

Fig. 6 A plot of det(J) vs Re to predict singular points (det(J)=0). As a singular point is approached the Newton-Raphson iterative procedure becomes difficult to converge; hence semicircles are included to indicate that there is some uncertainty in the exact value of the singular point.

NIM singular points discovered here are also turning points, that Solution A and Solution C are just different portions of the same solution branch and that Solution B, which we have not obtained yet, is the third portion which exists only between the two turning points and connects the other two portions of this solution branch. If this conjecture is correct then there are three solutions (A, B and C), all on the same branch, between the two turning points at $Re^* \cong 789$ and 1468.643, which is not inconsistant with the fact that our Newton-Raphson iteration procedure converged easily from a uniform, zero velocity field initial guess for $Re<900$ and also for Re=1,500 at which point there would be one solution, Solution C, on this branch. Future studies should determine whether this conjecture is correct. If so, "other solutions" such as the one obtained by Bozeman and Dalton [14] for the closed driven cavity at Re=1,000 should not be attributed to an inadequacy of the numerical method being used, but should be recognized as other legitimate solutions, perhaps even on the same branch as the physical solution.

6. CONCLUSIONS

Clearly the nonlinearities in the Navier-Stokes equations lead to many computational difficulties even for such a simple "test" problem as the driven cavity problem. The nodal integral method [1,2] which is capable of very high accuracies on very coarse meshes still has two singular points for the 6x6 mesh solution to the closed cavity problem (and is likely also to have them at comparable "effective" cavity Reynolds numbers for modified driven cavity problems). Hence convergence to a nonphysical solution should not be used as a criterion for discarding a numerical method as inadequate [14]. After all, any nonlinear problem must have multiple solutions. Unfortunately, there does not appear to be any simple way of insuring that a numerical method using a specific mesh will converge to the physical solution for a reasonably large Reynolds number. Thus to obtain reliable solutions for large Re, it appears that the only safe approach is to continue [10-12] the solutions, using a fixed mesh, from small Re (where it might be possible to prove uniqueness) to locate singular points and trace the curve of the solution, then refine the mesh and repeat the process until the mesh necessary to obtain the physical solution for a specific large Re is deduced. Starting from an arbitrary initial guess using a specific mesh for a high Re risks the possibility of converging to an "other" solution that might be difficult to distinguish on an intuitive basis from the solution to the continuous variable problem [15].

ACKNOWLEDGEMENT

We wish to thank Dr. W. C. Horak for communicating to us, prior to publicaiton, his NGTM results included here and Dr. R. Schreiber and Prof. H. B. Keller for communicating the results in their paper to us prior to publication. We also gratefully acknowledge useful discussions with Prof. H. B. Keller, and thank Prof. S. Taneda and the Physical Society of Japan for their permission to use the photographs shown in Fig. 2, and Prof. M. Van Dyke for making them available for us.

REFERENCES

1. AZMY, Y. Y. and DORNING, J. J. - A Nodal Integral Approach to the Numerical Solution of Partial Differential Equations, to be published in Advances in Reactor Computations, American Nuclear Society, LaGrange, IL, March 1983.

2. AZMY, Y. Y. - A Nodal Integral Method for the Numerical Solution of Incompressible Fluid Flow Problems. M.S. Thesis, Univ. of Illinois, 1982.

3. HORAK, W. C. and DORNING, J. J. - A Nodal Green's Tensor Method for the Efficient Numerical Solution of Laminar Flow Problems, Numerical Methods in Laminar and Turbulent Flow, Eds. Taylor, C. and Schrefler, B. A., Pineridge Press, Swansea, U.K., 1981. See also: HORAK, W. C. and DORNING, J. J. - A Nodal Coarse-Mesh Method for the Efficient Numerical Solution of Laminar Flow Problems. Submitted to J. Comp. Phys.

4. HIRT, C. W., NICHOLS, B. D. and ROMERO, N. C. - SOLA-A Numerical Solution Algorithm for Transient Fluid Flows, LA-5852, 1975.

5. TANEDA, S. - Visualization of Separating Stokes Flows. J. Phys. Soc. Jpn., Vol. 46, No. 6, pp. 1935-1942, June 1979. See also: VAN DYKE, M. - An Album of Fluid Motion. The Parabolic Press, Stanford, CA, 1982.

6. HORAK, W. C. - Personal Communication, June 1982.

7. BURGGRAFF, O. R. - Analytical and Numerical Studies of the Structure of Steady Separated Flows. J. Fluid Mech. Vol. 24, part 1, pp. 113-151, Jan. 1966.

8. PAN, F. and ACRIVOS, A. - Steady Flows in Rectangular Cavities. J. Fluid Mech., Vol. 28, part 4, pp. 643-655, 1967.

9. GHIA, U., GHIA, K. N. and SHIN, C. T. - Solution of Incompressible Navier-Stokes Equations by Coupled Strongly Implicit Multi-Grid Method. Sym. on Multigrid Methods, NASA-AMES Research Center, CA, Oct. 21-22, 1981.

10. KELLER, H. B. - Numerical Solution of Bifurcation and Nonlinear Eigenvalue Problems. Applications of Bifurcation Theory, Rabinowitz, P. H., ed., Academic Press, New York, 1977.

11. STEPHENS, A. B. and SHUBIN, G. R., - Multiple Solutions and Bifurcation of Finite Difference Approximations to Some Steady Problems of Fluid Dynamics. SIAM J. Sci. Stat. Comput., 2, pp. 404-415, 1981.

12. MOORE, G. and SPENCE, A. - The Calculation of Turning Points of Nonlinear Equations. SIAM J. Numer. Anal., 17, pp. 567-576, 1980.

13. SCHREIBER, R. and KELLER, H. B. - Spurious Solutions in Driven Cavity Problems. J. Comp. Phys., in press.

14. BOZEMAN, J. D. and DALTON, C. - Numerical Study of Viscous Flow in a Cavity. J. of Comp. Phys., Vol. 12, pp. 348-363, 1973.

15. KELLOGG, R. B., SHUBIN, G. R. and STEPHENS, A. B. - Uniqueness and the Cell Reynolds Number. SIAM J. Numer. Anal., 17, pp. 733-739, 1980.

PRESSURE METHODS FOR THE APPROXIMATE SOLUTION OF THE NAVIER-STOKES EQUATIONS

U. Bulgarelli[(i)], V. Casulli[(i)] and D. Greenspan[(ii)]

(i) Istituto per le Applicazioni del Calcolo del C.N.R. Viale del Policlinico, 137, 00161 Roma, Italy

(ii) Department of Mathematics, The University of Texas at Arlington, Arlington, Texas 76019

SUMMARY. In this paper we will develop and apply a family of powerful finite difference methods for the numerical solution of the Navier-Stokes equations for incompressible fluids. The power is derived by combining advantageous aspects of the Marker-and-Cell (MAC) method with several special techniques for hyperbolic and parabolic equations. The stability of the methods, in the discrete L_2 norm, is demonstrated to be independent of the pressure and is shown to depend only on the discretizations chosen for the convective and viscous terms. Appropriate choices of such discretizations are then suggested. A numerical example is described and discussed.

1. INTRODUCTION.

The partial differential equations governing the laminar flow of an incompressible fluid are obtained from the physical principles of conservation of linear momentum and volume. Such equations, in their primitive variables, and in two space dimensions, can be written as

$$(1)\qquad \begin{cases} \dfrac{\partial u}{\partial t} + u\,\dfrac{\partial u}{\partial x} + v\,\dfrac{\partial u}{\partial y} = -\,\dfrac{\partial p}{\partial x} + \nu\left(\dfrac{\partial^2 u}{\partial x^2} + \dfrac{\partial^2 u}{\partial y^2}\right) \\[2ex] \dfrac{\partial v}{\partial t} + u\,\dfrac{\partial v}{\partial x} + v\,\dfrac{\partial v}{\partial y} = -\,\dfrac{\partial p}{\partial y} + \nu\left(\dfrac{\partial^2 v}{\partial x^2} + \dfrac{\partial^2 v}{\partial y^2}\right) \\[2ex] \dfrac{\partial u}{\partial x} + \dfrac{\partial v}{\partial y} = 0, \end{cases}$$

where $u(x,y,t)$ and $v(x,y,t)$ are the velocity components in the x and y directions, respectively. The normalized pressure $p(x,y,t)$ is defined as the ratio of pressure to density, where the density is assumed to be constant. The kinematic viscosity coefficient ν is assumed to be constant and nonnegative.

If $\Omega \subset \mathbb{R}^2$ denotes the spatial domain for equations (1), and Γ is the boundary of Ω, the following initial and boundary conditions are associated with equations (1):

$$(2)\qquad \begin{cases} u(x,y,0) = u_0(x,y) \\ v(x,y,0) = v_0(x,y) \end{cases}, \quad (x,y) \in \Omega$$

$$(3)\qquad \begin{cases} u(x,y,t) = u_b(x,y,t) \\ v(x,y,t) = v_b(x,y,t) \end{cases}, \quad (x,y) \in \Gamma,\ 0 < t < T$$

where $u_0(x,y)$, $v_0(x,y)$, $u_b(x,y,t)$ and $v_b(x,y,t)$ are given.

An analytical solution of equations (1) would yield values for the unknowns u, v and p at each instant in time and at every point in the flow field. However, except for very few particular cases, an analytical solution can rarely be constructed and a numerical, computer oriented technique must be introduced.

2. SEMIDISCRETE FORMULATION.

In order to solve equations (1) numerically, we introduce, as in the MAC method [3], a finite difference mesh which consists of rectangular cells of width Δx and height Δy which cover Ω. The discrete pressure and fluid velocities are located at the cell positions shown in Fig. 1: the horizontal velocity u in the x-direction is defined at the center of each vertical side of a cell, the vertical velocity v in the y-direction is defined at the center of each horizontal side, and the pressure p is defined at each cell center. Each cell is numbered at its center with indicies i and j, where i is the cell's column number in the x-direction from left to right, and j is the cell's row number in the z-direction from bottom to top.

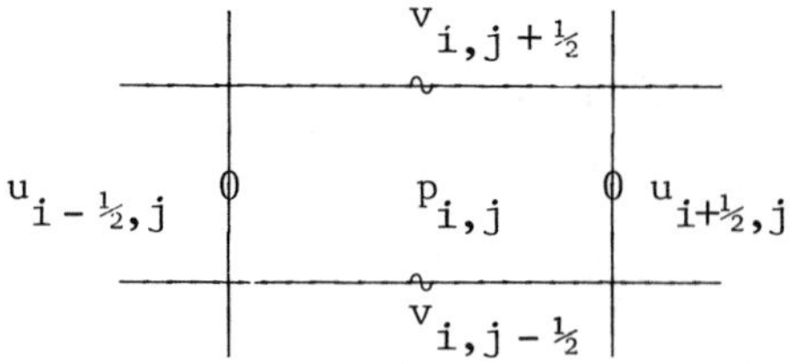

Fig. 1: Position of field variables.

If equations (1) are spatially discretized via the

conventional finite difference method by using centered differences for the grad p terms and for the continuity equation, there results the following matrix system of coupled first order differential equations

$$(4)\quad \begin{cases} \dfrac{du_{i+\frac{1}{2},j}}{dt} = -F_{i+\frac{1}{2},j} - \dfrac{p_{i+1,j}-p_{i,j}}{\Delta x} \\ \dfrac{dv_{i,j+\frac{1}{2}}}{dt} = -G_{i,j+\frac{1}{2}} - \dfrac{p_{i,j+1}-p_{i,j}}{\Delta y} \\ \dfrac{u_{i+\frac{1}{2},j}-u_{i-\frac{1}{2},j}}{\Delta x} + \dfrac{v_{i,j+\frac{1}{2}}-v_{i,j-\frac{1}{2}}}{\Delta y} = 0, \end{cases}$$

where $F_{i+\frac{1}{2},i}$ and $G_{i,j+\frac{1}{2}}$ contain the finite differences corresponding to the spatial discretization of the convective and viscous terms.

Once a discretization for the convective and for the viscous terms has been chosen, $F_{i+\frac{1}{2},j}$ and $G_{i,j+\frac{1}{2}}$ are determined and the differential equations in (4) are readily solvable numerically with any one of the several finite difference methods for ordinary differential equations which are abundantly available. The selected method can be one step or multistep and explicit or implicit with respect to $u^{n+1}_{i+\frac{1}{2},j}$ and $v^{n+1}_{i,j+\frac{1}{2}}$.

3. EXPLICIT METHODS.

A very simple way to solve system (4) is to discretize the ordinary differential equations of (4) with the explicit Euler scheme as follows:

$$(5)\quad \begin{cases} \dfrac{u^{n+1}_{i+\frac{1}{2},j}-u^{n}_{i+\frac{1}{2},j}}{\Delta t} = -F^{n}_{i+\frac{1}{2},j} - \dfrac{p^{n+1}_{i+1,j}-p^{n+1}_{i,j}}{\Delta x} \\ \dfrac{v^{n+1}_{i,j+\frac{1}{2}}-v^{n}_{i,j+\frac{1}{2}}}{\Delta t} = -G^{n}_{i,j+\frac{1}{2}} - \dfrac{p^{n+1}_{i,j+1}-p^{n+1}_{i,j}}{\Delta y}, \end{cases}$$

and to discretize the continuity equation, which we require to be satisfied at each time step, as follows:

$$(6)\quad \frac{u^{n+1}_{i+\frac{1}{2},j}-u^{n+1}_{i-\frac{1}{2},j}}{\Delta x} + \frac{v^{n+1}_{i,j+\frac{1}{2}}-v^{n+1}_{i,j-\frac{1}{2}}}{\Delta y} = 0.$$

In (5) and (6), the superscript n or $n+1$ indicates the time at which the variable is to be evaluated, and Δt denotes the time step. Thus, for example, $u^n_{i+\frac{1}{2},j}$ is to be evaluated at time $n\,\Delta t$.

Equations (5)-(6), at each time step, constitute a linear system of equations with unknowns $u^{n+1}_{i+\frac{1}{2},j}$, $v^{n+1}_{i,j+\frac{1}{2}}$ and $p^{n+1}_{i,j}$. However, the order of this system can be reduced by substituting expressions for u^{n+1} and v^{n+1} from (5) into (6) to yield

$$\frac{p^{n+1}_{i+1,j}-2\,p^{n+1}_{i,j}+p^{n+1}_{i-1,j}}{(\Delta x)^2}+\frac{p^{n+1}_{i,j+1}-2\,p^{n+1}_{i,j}+p^{n+1}_{i,j-1}}{(\Delta y)^2} \tag{7}$$

$$= -\left(\frac{F^n_{i+\frac{1}{2},j}-F^n_{i-\frac{1}{2},j}}{\Delta x}+\frac{G^n_{i,j+\frac{1}{2}}-G^n_{i,j-\frac{1}{2}}}{\Delta y}\right)$$

$$+\frac{1}{\Delta t}\left(\frac{u^n_{i+\frac{1}{2},j}-u^n_{i-\frac{1}{2},j}}{\Delta x}+\frac{v^n_{i,j+\frac{1}{2}}-v^n_{i,j-\frac{1}{2}}}{\Delta y}\right).$$

On the cells, the set of equations (7) constitutes a linear system with unknowns $p^{n+1}_{i,j}$. Since the velocity, not the pressure, is given as a boundary condition on Γ, the matrix of coefficients for the linear system of equations (7) is singular and hence the pressure field $p^{n+1}_{i,j}$ is determined up to an arbitrary additive constant which cancels in (5). Thus, the velocity field $u^{n+1}_{i+\frac{1}{2},j}$ and $v^{n+1}_{i,j+\frac{1}{2}}$ is determined uniquely. In practice, linear system (7) can be solved iteratively by using the successive overrelaxation iterative method in a very efficient way to yield, simultaneously, u^{n+1}, v^{n+1} and p^{n+1} (see, e.g., [2]).

In all previous considerations the structure of the terms $F^n_{i+\frac{1}{2},j}$ and $G^n_{i,j+\frac{1}{2}}$ has been left out. However, the structure of such terms plays a very important role for the consistency, the accuracy and for the stability of the numerical method (5), (6). Note that, since $F^n_{i+\frac{1}{2},j}$ and $G^n_{i,j+\frac{1}{2}}$ contain the finite differences corresponding to the spatial discretization of the convective and viscous terms of the Navier-Stokes equations (1), an appropriate choice of $F^n_{i+\frac{1}{2},j}$ and $G^n_{i,j+\frac{1}{2}}$ is implied from considerations of the parabolic (hyperbolic if $\nu = 0$) equations that one obtains from the momentum equations in (1) by neglect of the pressure terms.

Let us give, as examples, two different expressions to $F^n_{i+\frac{1}{2},j}$ and $G^n_{i,j+\frac{1}{2}}$ and, as a result, derive two different methods with different properties for the numerical solution of (1).

First assume that the viscosity ν is strictly positive. In this case, by use of centered finite differences to discretize both the convective and the viscous terms, second order accurate formulas for $F^n_{i+\frac12,j}$ and $G^n_{i,j+\frac12}$ are

(8)

$$\begin{cases} F^n_{i+\frac12,j} = u^n_{i+\frac12,j}\dfrac{u^n_{i+3/2,j}-u^n_{i-\frac12,j}}{2\Delta x} + v^n_{i+\frac12,j}\dfrac{u^n_{i+\frac12,j+1}-u^n_{i+\frac12,j-1}}{2\Delta y} \\ \quad -\nu\left[\dfrac{u^n_{i+3/2,j}-2u^n_{i+\frac12,j}+u^n_{i-\frac12,j}}{(\Delta x)^2}+\dfrac{u^n_{i+\frac12,j+1}-2u^n_{i+\frac12,j}+u^n_{i+\frac12,j-1}}{(\Delta y)^2}\right] \\ G^n_{i,j+\frac12} = u^n_{i,j+\frac12}\dfrac{v^n_{i+1,j+\frac12}-v^n_{i-1,j+\frac12}}{2\Delta x} + v^n_{i,j+\frac12}\dfrac{v^n_{i,j+3/2}-v^n_{i,j-\frac12}}{2\Delta y} \\ \quad -\nu\left[\dfrac{v^n_{i+1,j+\frac12}-2v^n_{i,j+\frac12}+v^n_{i-1,j+\frac12}}{(\Delta x)^2}+\dfrac{v^n_{i,j+3/2}-2v^n_{i,j+\frac12}+v^n_{i,j-\frac12}}{(\Delta y)^2}\right]. \end{cases}$$

It will be shown later that, if the viscosity is zero and $F^n_{i+\frac12,j}$, $G^n_{i,j+\frac12}$ are given by (8), the numerical method (5), (6) is always unstable.

Finite difference formulas for $F^n_{i+\frac12,j}$ and $G^n_{i,j+\frac12}$, which are only first order accurate but applicable for any nonnegative viscosity ν, are given as follows:

(9)

$$\begin{cases} F^n_{i+\frac12,j} = \dfrac{u^n_{i+\frac12,j}}{\Delta x}\begin{Bmatrix} u^n_{i+\frac12,j}-u^n_{i-\frac12,j} \\ u^n_{i+3/2,j}-u^n_{i+\frac12,j}\end{Bmatrix} + \dfrac{v^n_{i+\frac12,j}}{\Delta y}\begin{Bmatrix} u^n_{i+\frac12,j}-u^n_{i+\frac12,j-1} \\ u^n_{i+\frac12,j+1}-u^n_{i+\frac12,j}\end{Bmatrix} \\ \quad -\nu\left[\dfrac{u^n_{i+3/2,j}-2u^n_{i+\frac12,j}+u^n_{i-\frac12,j}}{(\Delta x)^2}+\dfrac{u^n_{i+\frac12,j+1}-2u^n_{i+\frac12,j}+u^n_{i+\frac12,j-1}}{(\Delta y)^2}\right] \\ G^n_{i,j+\frac12} = \dfrac{u^n_{i,j+\frac12}}{\Delta x}\begin{Bmatrix} v^n_{i,j+\frac12}-v^n_{i-1,j+\frac12} \\ v^n_{i+1,j+\frac12}-v^n_{i,j+\frac12}\end{Bmatrix} + \dfrac{v^n_{i,j+\frac12}}{\Delta y}\begin{Bmatrix} v^n_{i,j+\frac12}-v^n_{i,j-\frac12} \\ v^n_{i,j+\frac32}-v^n_{i,j+\frac12}\end{Bmatrix} \\ \quad -\nu\left[\dfrac{v^n_{i+1,j+\frac12}-2v^n_{i,j+\frac12}+v^n_{i-1,j+\frac12}}{(\Delta x)^2}+\dfrac{v^n_{i,j+3/2}-2v^n_{i,j+\frac12}+v^n_{i,j-\frac12}}{(\Delta y)^2}\right]. \end{cases}$$

In (9) the upper or lower line of each brace is to apply as the corresponding coefficient is nonnegative or not.

The discrete variables $u^n_{i,j+\frac12}$ and $v^n_{i+\frac12,j}$ in (8) and

(9) are not defined (see Fig. 1), thus a simple average, using the two closest scalar grid points, is to be used.

Formulas for $F^n_{i+\frac{1}{2},j}$ and $G^n_{i,j+\frac{1}{2}}$ other than (8) and (9) can be given (see e.g. [2]).

4. STABILITY CONSIDERATIONS

Once a space mesh has been chosen, the choice of the time increment and, at times, even the space steps have to satisfy special conditions for stability.

Consider the explicit difference scheme (5),(6). In matrix notation, (5) can be written in the following way:

$$(10)\qquad \frac{W^{n+1}-W^n}{\Delta t} = -H(W^n)W^n - AP^{n+1},$$

where W^n denotes a vector whose elements are the velocities $u^n_{i+\frac{1}{2},j}$ and $v^n_{i,j+\frac{1}{2}}$ in each cell. $H(W^n)$ is a square matrix whose structure depends on the formulas used in $F^n_{i+\frac{1}{2},j}$ and $G^n_{i,j+\frac{1}{2}}$. P^{n+1} denotes a vector whose elements are the pressure values $p^{n+1}_{i,j}$ in each cell, and rectangular matrix A is the matrix analogue of the finite difference gradient.

The discrete continuity equation (6) in matrix notation is

$$(11)\qquad A^TW^{n+1} = 0.$$

The system of finite difference equations (7) for the pressure in matrix notation is derived by combining (10) with (11):

$$(12)\qquad A^TAP^{n+1} = \frac{1}{\Delta t} A^TW^n - A^TH(W^n)W^n.$$

Assuming the pressure to be given at, at least, one boundary point, the matrix A^TA is nonsingular and a solution for P^{n+1} can be expressed as follows:

$$(13)\qquad P^{n+1} = (A^TA)^{-1}A^T[\frac{1}{\Delta t} I - H(W^n)]W^n$$

Substitution of (13) into (10) then yields

$$(14)\qquad W^{n+1} = [I - A(A^TA)^{-1}A^T][I - (\Delta t)H(W^n)]W^n.$$

To determine the condition under which (14) is stable, consider the discrete L_2 norm of W^{n+1}:

$$(15)\qquad \|W^{n+1}\|_2 \leq \|I - A(A^TA)^{-1}A^T\|_2\|I - (\Delta t)H(W^n)\|_2\|W^n\|_2$$

Note now that since the matrix $[I - A(A^TA)^{-1}A^T]$ is a projector, its L_2 norm is unity and hence (15) implies that (14) is stable provided the following inequality is satisfied at each time step:

$$\text{(16)} \qquad \| I - (\Delta t)H(W^n) \|_2 \leq 1.$$

The stability condition (16) is independent of the pressure. It depends only on the structure of the matrix $H(W^n)$, that is, on the expressions chosen for $F^n_{i+\frac{1}{2},j}$ and $G^n_{i,j+\frac{1}{2}}$. In particular, if $F^n_{i+\frac{1}{2},j}$ and $G^n_{i,j+\frac{1}{2}}$ are given by (8) then (16) is satisfied if

$$\text{(17)} \qquad \begin{cases} (\Delta x) \max\limits_{i,j} |u^n_{i,j}| \leq 2\nu \\ (\Delta y) \max\limits_{i,j} |v^n_{i,j}| \leq 2\nu \\ 2\nu(\Delta t)[\frac{1}{(\Delta x)^2} + \frac{1}{(\Delta y)^2}] \leq 1. \end{cases}$$

If $F^n_{i+\frac{1}{2},j}$ and $G^n_{i,j+\frac{1}{2}}$ are given by (9), then (16) is satisfied if

$$\text{(18)} \qquad (\Delta t) \max_{i,j} \left\{ \frac{|u^n_{i,j}|}{\Delta x} + \frac{|v^n_{i,j}|}{\Delta y} + 2\nu \left[\frac{1}{(\Delta x)^2} + \frac{1}{(\Delta y)^2} \right] \right\} \leq 1.$$

In general, if one uses different formulas for $F^n_{i+\frac{1}{2},j}$ and $G^n_{i,j+\frac{1}{2}}$, the matrix $H(W^n)$ is different and the stability restrictions on the finite difference steps are obtained by requiring that (16) be satisfied.

5. IMPLICIT METHODS.

The use of implicit numerical methods for systems of evolution equations is usually recommended when large time steps are essential. The price one pays for this capability is that one has to solve, at each time step, a system of algebraic or transcendental equations involving all the dependent variables.

A simple implicit method for the Navier-Stokes equations (1) is obtained by discretizing the differential equations in (4) with a modified implicit Euler scheme. Such a method, in matrix notation, can be written as

$$\text{(19)} \qquad \frac{W^{n+1} - W^n}{\Delta t} = -H(W^n)W^{n+1} - AP^{n+1}$$

$$\text{(20)} \qquad A^T W^{n+1} = 0,$$

where, again, the structure of the matrix $H(W^n)$ depends upon the discretization chosen for the convective and viscous terms. Note that system (19)-(20), though implicit, is linear with respect to W^{n+1} and P^{n+1}.

With regard to the stability of the implicit formulas (19)-(20), assume that $H(W^n)$ is chosen in such a fashion that $[I + (\Delta t)H(W^n)]$ is nonsingular. Thus (19) implies

$$W^{n+1} = [I + (\Delta t)H(W^n)]^{-1}[W^n - (\Delta t)AP^{n+1}]. \tag{21}$$

Substitution of (21) into (20) yields a finite difference system of equations for the pressure:

$$(\Delta t)A^T[I + (\Delta t)H(W^n)]^{-1}AP^{n+1} = A^T[I + (\Delta t)H(W^n)]^{-1}W^n. \tag{22}$$

Assuming the pressure to be known at, at least, one boundary point, system (22) has a unique solution P^{n+1} provided Δt is sufficiently small. In practice, this sufficient condition appears to be unnecessarily restrictive [1]. Thus, the solution P^{n+1} of (22) is given by

$$P^{n+1} = \frac{1}{\Delta t}(A^TR^nA)^{-1}A^T R^nW^n, \tag{23}$$

where R^n denotes the matrix $[I + (\Delta t)H(W^n]^{-1}$. Substitution of (23) into (21) yields

$$W^{n+1} = R^n[I - A(A^TR^nA)^{-1}A^TR^n]W^n. \tag{24}$$

Note now that since $[I - A(A^TR^nA)^{-1}A^TR^n]$ is a projector, (24) implies

$$\|W^{n+1}\|_2 \leq \|R^n\|_2\|W^n\|_2, \tag{25}$$

so that (24) is stable if the following inequality is satisfied

$$\|[I + (\Delta t)H(W^n)]^{-1}\|_2 \leq 1. \tag{26}$$

Thus, one can observe that the stability condition (26) is independent of the pressure. It depends only on the structure of the matrix $H(W^n)$, that is, on the discretization chosen for the convective and viscous terms. Moreover, since (26) is also the stability condition for the implicit scheme that one obtains from (19) by neglecting the pressure term, we can state that any stable implicit discretization for the parabolic (hyperbolic if $\nu = 0$) equations that one obtains from the momentum equations in (1) by neglecting the pressure gradient terms can be adapted to obtain a stable method of the form (19)-(20) for the complete Navier-Stokes equations (1).

The previous considerations also apply to the Navier-Stokes equations in three space dimensions (see, e.g., [2]).

6. A DRIVEN CAVITY CALCULATION.

As an application of the methods discussed above, consider the familiar driven cavity problem. Let us follow the flow which develops in a square cavity whose sides have length 2. Assume that the fluid has viscosity $\nu = 0.4$ and that at the initial time $t = 0$ the velocities are

$$u_0(x,y) = v_0(x,y) = 0.$$

In addition, the bottom and side boundaries are fixed, while the top boundary is assumed to be moving with constant velocity $u_T = 0.5$. The discretization parameters are taken to be $\Delta x = \Delta y = 0.2$ and $\Delta t = 0.01$.

Figures 2,3,4 and 5 show the computed solution obtained at times $t = 0.1$, $t = 0.2$, $t = 0.5$ and $t = 1$, respectively, by using the explicit method (5), (6) with the choice (9) for the terms $F^n_{i+\frac{1}{2},j}$ and $G^n_{i,j+\frac{1}{2}}$. The numerical solution shown in Fig. 5 represents the steady solution for this problem.

Because the maximum velocity in this cavity flow problem is at the top, moving boundary, our choice of space and time steps is consistent with stability condition (18).

REFERENCES

[1] AMIT, R., HALL, C. A., PORSCHING, T. A. - An Application of Network Theory to the Solution of Implicit Navier-Stokes Difference Equations. Journal of Computational Physics, Vol. 40, No. 1, pp. 183-201, 1981.

[2] BULGARELLI, U., CASULLI, V., GREENSPAN, D. - Pressure Methods for the Numerical Solution of Free Surface Fluid Flows, book in preparation.

[3] HARLOW, F. H., WELCH, F. E. - Numerical Calculation of Time-Dependent Viscous Incompressible Flow, Physics of Fluids, Vol. 8, No. 12, pp. 2182-2189, 1965.

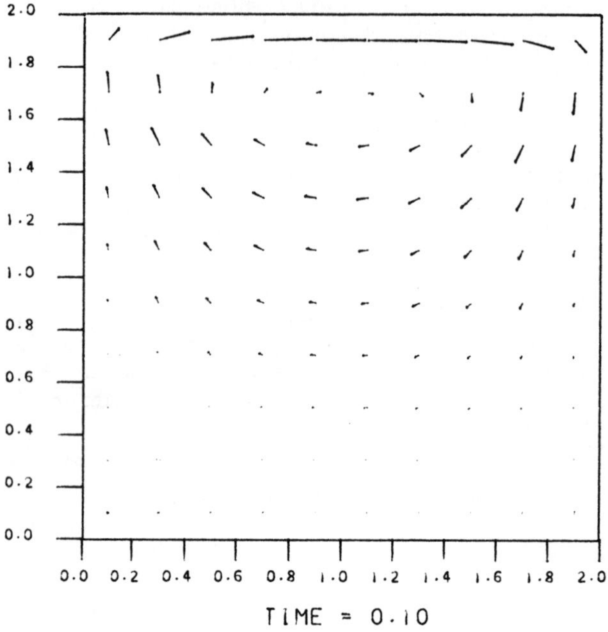

Fig. 2

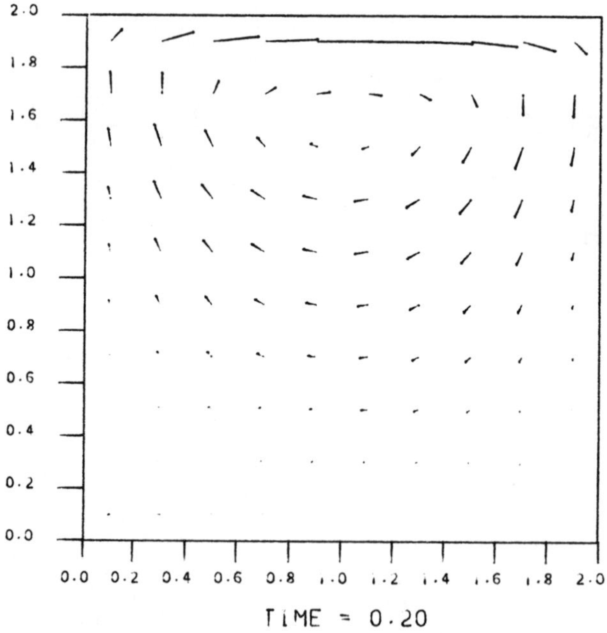

Fig. 3

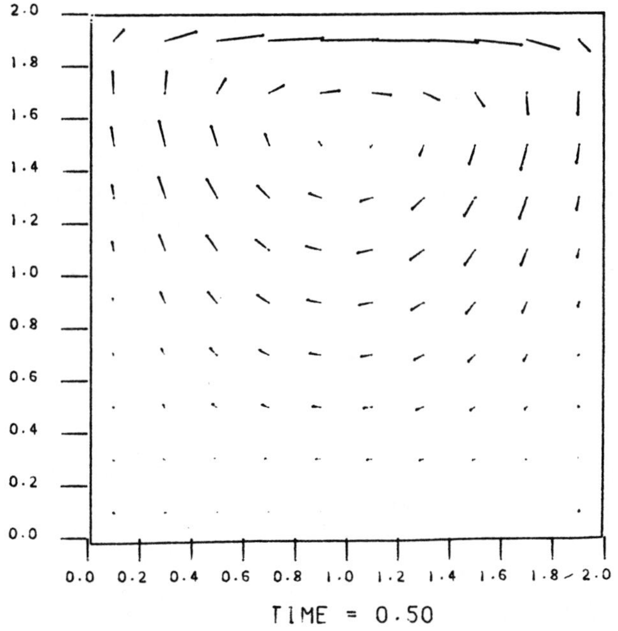
2.0
1.8
1.6
1.4
1.2
1.0
0.8
0.6
0.4
0.2
0.0
0.0 0.2 0.4 0.6 0.8 1.0 1.2 1.4 1.6 1.8 2.0
TIME = 0.50

Fig. 4

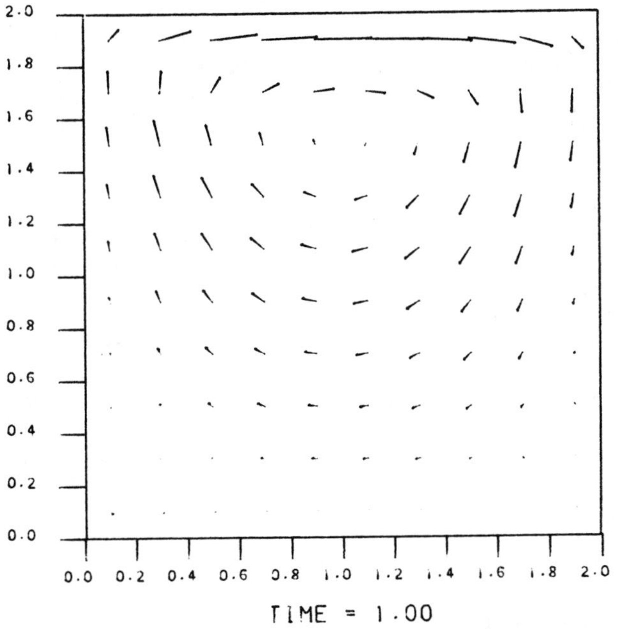
2.0
1.8
1.6
1.4
1.2
1.0
0.8
0.6
0.4
0.2
0.0
0.0 0.2 0.4 0.6 0.8 1.0 1.2 1.4 1.6 1.8 2.0
TIME = 1.00

Fig. 5

FAST AND ACCURATE SOLUTION OF TIME-DEPENDENT INCOMPRESSIBLE FLOW

Arne Karlsson[1) and Laszlo Fuchs[2)

SUMMARY

The stream function and vorticity formulation of the governing equations for non-steady incompressible viscous flows is considered. The time accurate solution to the differential problem requires that the transients shall be resolved. It is shown that the time scales of the transients are related to the spatial scales of the different components of the transients. By taking advantage of this relation we have developed a multi-grid method, in which both the time steps as well as the mesh spacing vary simultaneously. The resulting method is not only accurate, but very fast and machine round-off accuracy (reduce of the residuals by 16 orders of magnitude) can be achieved by a computational effort which is equivalent to about 100 successive point relaxation sweeps on the difference equations. Some results for the non-steady driven cavity flow are also given.

1. INTRODUCTION

The simulation of non-steady flow problems is more difficult than that of steady-state flows. In the steady-state case the length scales of the problem must be resolved to get a meaningful solution. In the non-steady state case, the transients may or may not be resolved, but still the steady-state, or the periodic solution can be accurate. The physical transients can be decomposed into Fourier components in space. It is shown here that for the heat equation, the (decay) time-constant of different components of the transient is related to the spatial fre-

----

1. Research student, Dep. of Gasdynamics
The Royal Institute of Technology, Stockholm, Sweden

2. Assistant professor, Dep. of Gasdynamics
The Royal Institute of Technology, Stockholm, Sweden

quency of that component. Rapid transients have short spatial wave-lengths while slowly varying transients have long spatial wave-lengths.

Such relations are not utilized by standard methods. Usually, constant or variable time steps are used to advance the solution in time but all the computations are done on one, fine grid. Thus, even the slowly varying components are computed on fine meshes, even when these grids are unnecessary from accuracy point of view. Furthermore, the convergence of standard iterative methods is slower on finer grids than on coarser ones. These facts make accurate time dependent solutions of non-steady flow problems, very expensive.

An approach which is suitable for solving slowly varying periodical problems, is to use semi-spectral methods. The time variations are Fourier transformed, and the series is truncated keeping only a limited number of terms. The resulting, coupled system of partial differential equations has to be solved by a finite difference method. Such a method has been used by Duck [1], where at most 5 terms of the Fourier expansion are used. The method in [1] has been applied to the computation of the oscillating flow in a driven cavity.

In this work we solve the same problem by using a new multi-grid (MG) method which utilizes the relation between the time and the spatial scales of the transients. In the following we describe the details of the MG method. (As a matter of fact, since we use an implicit scheme two MG solvers are used, but only the time-MG solver is described.)

The dependence of the rate of convergence of the method, on the number of levels, is described. Our computed results for the flow in a cavity with a periodically oscillating driving wall are also given. It has been found, that these results differ qualitatively from those which are obtained by Duck [1]. Our computed results are checked for discretization errors and the solution procedure has been terminated at machine round-off accuracy.

2. GOVERNING EQUATIONS AND BOUNDARY CONDITIONS

We consider plane and axisymmetric, incompressible viscous flows. The Navier-Stokes equations are written as a coupled system of a vorticity transport- and a Poisson-equation. In a coordinate system (ξ,η), where $(\xi,\eta) = (x,y)$ for plane flows and $(\xi,\eta) = (z,r)$ for axisymmetric flows, the governing equations are

$$S\omega_t + u\omega_\xi + v\omega_\eta - i\frac{v\omega}{\eta} = \frac{1}{Re}\left[\omega_{\xi\xi} + \left(\frac{1}{\eta^i}(\eta^i\omega)_\eta\right)_\eta\right] \qquad (1)$$

$$\psi_{\xi\xi} + \eta^i\left(\frac{1}{\eta^i}\psi_\eta\right)_\eta = \eta^i\omega \qquad (2)$$

where Re $(=UL/\nu)$ is the Reynolds number and S $(=L/UT)$ is the Strouhal number, with U, L and T as the characteristic velocity, length and time, respectively. Equations (1) and (2) may be used for both plane (i=0) and axisymmetric cases (i=1). For both cases the stream function and the vorticity are defined by

$$\omega = v_\xi - u_\eta \quad ; \quad u = -\psi_\eta/\eta^i \quad ; \quad v = \psi_\xi/\eta^i \tag{3}$$

The following boundary conditions are used

i) no-slip at solid boundaries, and

ii) the velocity components are given on in- and out-flow boundaries.

From the boundary conditions on the velocity vector, the stream function can be computed on all the boundaries. The vorticity on the boundaries satisfy equation (2) and is computed explicitly during the solution process.

In this paper we consider flows which vary in time periodically, with a period T. That is, any dependent variable f, is assumed to satisfy

$$f(\xi,\eta,t) = f(\xi,\eta,t+T) \tag{4}$$

for all times t. In the method described here the flow is computed as an initial-value problem, and condition (4) may be used as a convergence test.

3. FINITE-DIFFERENCE APPROXIMATIONS

Equations (1) and (2) are discretized using second-order accurate central differences except for the convective terms of equation (1) where first-order up-wind differences are used. The time derivative is approximated by a Crank-Nicholson scheme, which is also second-order accurate.

The boundary values of the vorticity are computed by using equation (2), and the boundary conditions on the velocity. At the boundaries the central five-point formula of the Laplace operator, cannot be used. Instead we use the following relations.

A. For a boundary parallel to the ξ-axis:

$$\omega_w = \frac{2(\psi_{w+j} - \psi_w)}{\eta_w^{\,i}(\Delta\eta)^2} + \left(\frac{i}{\eta_w} + \frac{2j}{\Delta\eta}\right)u_w + (v_\xi)_w + O(\Delta\eta) \tag{5a}$$

where subscript w stands for the boundary, subscript w+j stands for the nodal point inside the flow field adjecent to the boundary, j=1 when the computational domain is above the boundary (positive η-direction) and j=-1 when the computational domain is below the boundary. If the lower boundary is a symmetry axis the definition of the vorticity directly gives $\omega_w = 0$.

B. Similarly, for a boundary parallel to the η-axis:

$$\omega_w = \frac{2(\psi_{w+j} - \psi_w)}{\eta_w{}^i (\Delta\xi)^2} - \frac{2j}{\Delta\xi} v_w - (u_\eta)_w + O(\Delta\xi) \qquad (5b)$$

where j=1 or j=-1 when the computational domain is to the right or to the left of the boundary, respectively.

4. NUMERICAL PROCEDURE

First we consider the solution of linear equations of diffusion-convection types:

$$\varphi_t = \alpha(\varphi_{xx} + \varphi_{yy}) - \beta(u\varphi_x + v\varphi_y) \qquad (6)$$

The boundary conditions may be constants or depend on time. Assume that φ^* is also a solution to (6) with $\varphi^* = \varphi$ on the boundaries for all $t > 0$, and $\varphi^*(x,y,0) = \varphi(x,y,0) + g(x,y)$. Under these conditions, $\varepsilon = \varphi^* - \varphi$ may be considered as the evolution error. For the heat equation ($\beta = 0$, with $\varepsilon(x,y,0) = g(x,y)$ and vanishing ε on the boundaries) ε has an analytical solution:

$$\varepsilon(x,y,t) = \sum_{m,n} a_{mn} \exp[-(m^2+n^2)\pi^2\alpha t]\sin(m\pi x)\sin(n\pi y) \qquad (7)$$

Equation (7) is the Fourier decomposition of the error of evolution. It is evident that the high frequency components of the error (large m and n) decay faster than error components with low frequency (small m and n). If one wishes to resolve accurately the time variations, one has to use small time and space steps such that all the significant components are retained. The use of such small time steps throughout the evolution time requires much computational effort. A constant short time step is completely unnecessary since the high frequency components (which describe the fast transients) decay very fast and after a while only relatively slowly varying components remain. This fact may be utilized by choosing non-constant time steps. We also observe that the slowly varying transients are associated with longer wavelengths also in space, and therefore they can be represented accurately on coarse grids. Such a behaviour is very suitable for MG solution procedures where both the time and space steps are increased, simultaneously.

Some different finite-difference approximations to the time-derivative may be used. When the time transients have high frequency fluctuations it is natural to use small time steps and then explicit schemes are the most cost-effective. The DuFort-Frankel scheme may be used with small time steps or for problems with steady state solutions. Implicit methods, on the other hand, are stable for large time steps, but then a large system of finite-difference equations has to be solved at each time step. We note that independently of the finite-difference approximation to the vorticity transport equation (1), the Poisson equation (2), for the stream function has to be solved in each time step. Therefore, in this case, even by using an explicit scheme, a Poisson equation has to be solved many times. By using a MG method we solve both equations, for the stream function and the vorticity, simultaneously, in each time step. The MG method which is used is similar to the one which is described by Thunell and Fuchs [2]. To distinguish between the MG method which is used in each time step and the method which is used to solve the system for all times t, we introduce the following notation: The solution of the inner problem is the solution of the equations (1) and (2) at a given time step (that is, the solution of the implicit or the explicit time steping of equation (1) and the corresponding solution to equation (2)). In the outer MG solver we use a sequence of increasing time steps with increasing mesh spacing. Both MG solvers use a sequence of grids in space and time. The mesh spacing ratio between a coarse and a fine grid is 2. In such a way the additional number of node points on coarse grids (i.e. the required memory storage) increase very slowly as additional coarse grids are introduced.

As stated above, the Crank-Nicholson scheme is used to approximate the vorticity transport equation. The outer MG procedure has the following basic steps:

- i) In each time step the spatial problem is solved by an inner, MG solver (see [2]).
- ii) When the fast transients are smoothed out one transfers the problem to a coarser grid in space and larger time steps.
- iii) Steps i) and ii) are repeated until a coarsest grid is reached.
- iv) The corrections to the approximations are interpolated to the finer grid where step i) is repeated.
- v) Steps iv) and i) are repeated until the finest grid is reached.

Steps i)-v) define a MG cycle. Such a MG cycle is enough to solve the problem to the accuracy of the truncation errors. If the computed results are to be a better approximation to the solution of the finite-difference equations, several MG-cycles may be carried out sequentially.

In the computer program we use an a-priori fixed sequence of grid relaxations. That is, the number of time steps (periods) on each grid is predetermined. The fixed MG method is simpler and no switching criteria should be computed. On the other hand, the fixed MG method can be less efficient, especially in those cases when different error components has largely different amplitudes.

The sequence of outer MG cycling which is described above can be used to compute time accurate variations (including fast transients) in an efficient manner. If, on the other hand, the transients are not of interest, an even faster method may be used. The basic principle is to compute a solution on a coarse grid (and that can be done with the same MG solver), and then interpolate the solution to the fine grid. The interpolated solution contains mainly high frequency error components, which are eliminated by some relaxation sweeps (or some time steps) on that fine grid. The resulting approximations has an accuracy within that of the truncation errors.

The computation of the solution on a coarse grid can be repeated by starting on even coarser grids. We have applied this nesting solution procedure in the inner MG solver, which can be solved when the transients are not too large, by a computational effort which is equivalent to 3-4 relaxation sweeps on both equations on the finest grid. The total effects of the number of grids on MG relaxations is discussed in Section 5.

5. COMPUTATIONAL RESULTS

As a model problem, the flow in a square cavity, is computed (see Figure 1). The upper wall which drives the flow in the cavity, has an oscillatory motion with a period T. The boundary conditions for this problem are:

$$u(x,a) = U\cos(2\pi t/T)$$
$$u(0,y) = u(a,y) = u(x,0) = 0$$
$$v(0,y) = v(a,y) = 0$$
$$v(x,0) = v(x,a) = 0$$

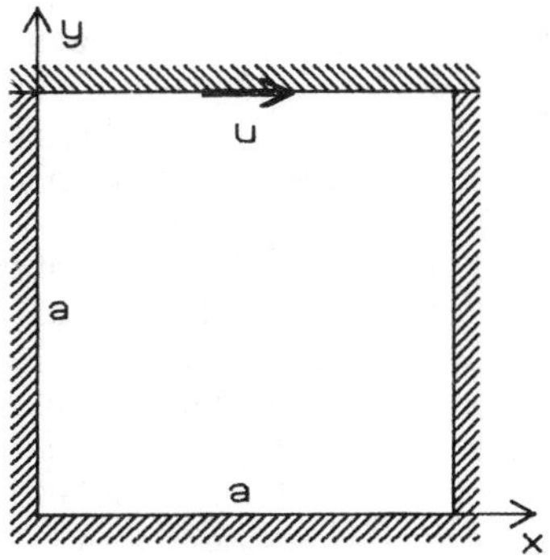

Figure 1

From these conditions on the velocity vector, it is possible to define a (constant) streamline value on all the walls of the cavity. The wall vorticity is computed by equations (5) and the boundary velocity. The Reynolds number is based upon the dimension of the cavity (a) and the maximal speed of the driving wall (U). The Strouhal number (S=a/UT) uses the period T. We assume in the following (as in [1]), that the convective terms and the time derivative term in equation (1) have the same importance, and hence we choose S = 1 for all the cases which are presented here.

First, we study the effects of using the MG solvers on the convergence rate. We define a work unit to be equal to the computational effort which is needed to make one relaxation sweep, on both equations, on the finest grid. Figures 2-4 shows the convergence history of the method, for different number of levels in the inner and the outer MG solvers (Re = 20 or Re = 200).

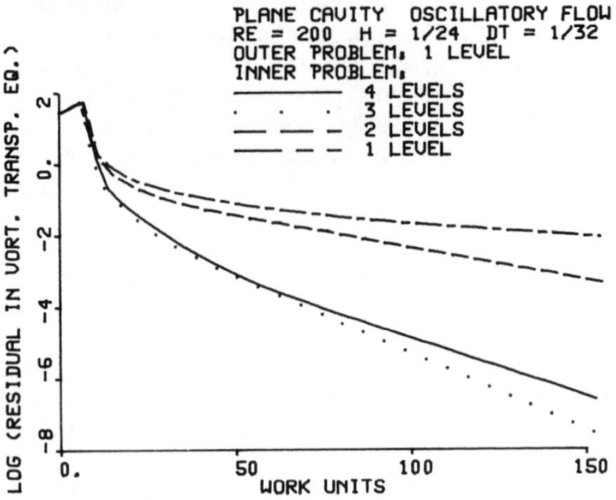

Figure 2

The finest grid which has been used for the cases in Figures 2-4, has 24×24 intervals and 32 time steps per period.

Figure 2 shows the convergence history when a single grid is used for the outer problem, while the inner problem (the implicit problem in each time step) is solved approximately. This approximate solution is obtained by a MG solver with varying number of levels. One level relaxation is equivalent to a standard Gauss-Seidel scheme. In all these cases we use 3.2 to 4.5 work units for the inner problem. It is noted that initially the convergence is fast, independently of the number of levels. However, as the computations goes on, standard (single grid) relaxation methods have very slow convergence (1 level). As the number of levels increase the convergence becomes faster. This phenomena can be explained on the basis of local mode analysis (see e.g. [3]). It can be shown that high frequency error components are eliminated efficiently by the relaxation scheme (and this gives rise to a fast decrease in the residual, initially). Low frequency error components, which are left after the initial stage, are eliminated slowly by a single relaxation operator. By the MG method, these low frequency errors are eliminated by little computational effort on coarse grids.

For the time dependent problem, as it is discussed in Section 3, there is a relationship between the spatial and the temporal error components. Figure 3 shows that even if the inner problem is solved efficiently by a MG method, the convergence can be accelerated by using larger time steps (MG in space and time). Once the initial transients are eliminated, larger time steps on coarser grids may be used. In the exemple which is shown in Figure 3, the four level (outer) MG solver converges to machine round-off error in 100 work units. For comparison, the (outer) single grid solver needs, when an inner MG solver is

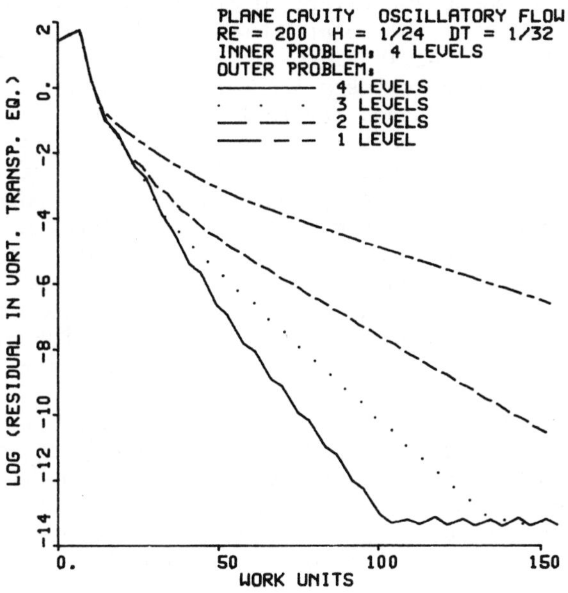

Figure 3

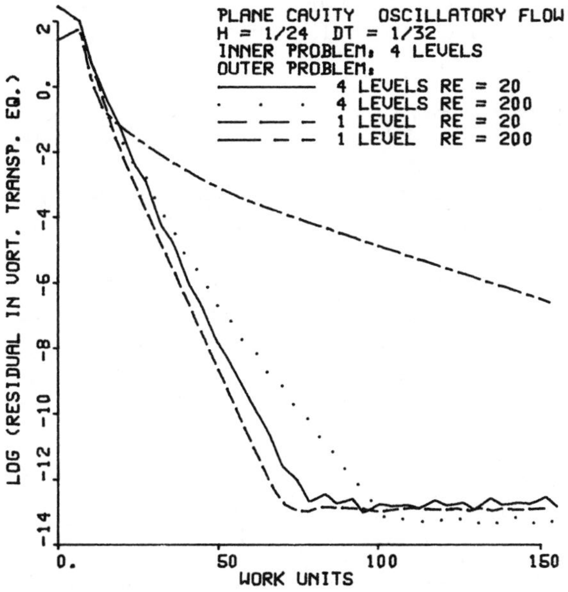

Figure 4

used, more than 300 work units to achieve a similar accuracy.

Figure 4 shows the effect of the Reynolds number and the number of levels of the outer MG solver. For Re = 200, the plots are the same as in Figure 3. For Re = 20, there is only a minimal difference between the single grid and the MG convergence history. The reason for this is that the real transient decay, which needs a certain number of periods, is shorter for smaller Re. The transients (for Re=20) decay so fast that they cannot be followed by the method (unless more accurate solutions are computed by the inner MG solver). Therefore, marching with the outer MG solver on a coarser grid (and hence, computing more periods with a given computational effort) does not accelerate the convergence, and a one-level MG- is equivalent to a four-level MG-solver. It should also be noted that for smaller time steps, the outer MG solver should be used even for low Re.

To check accuracy, we solved the driven cavity problem when the wall has a constant speed. The solution of the time dependent program has been compared to an accurate solution (on a 128×128 mesh) obtained by a primitive variable steady state MG-Navier-Stokes solver [4]. The agreement between these solutions was as good as expected due to truncation errors.

The oscillating motion in the cavity was also computed by Duck [1], using a semi-spectral method. The Fourier transformed solutions were truncated, keeping only at most 5 terms. The resulting coupled system of steady PDE's were solved by a relaxation method iteratively. Our method in contrast to the method of Duck, can be used to follow transients, and it is not neces-

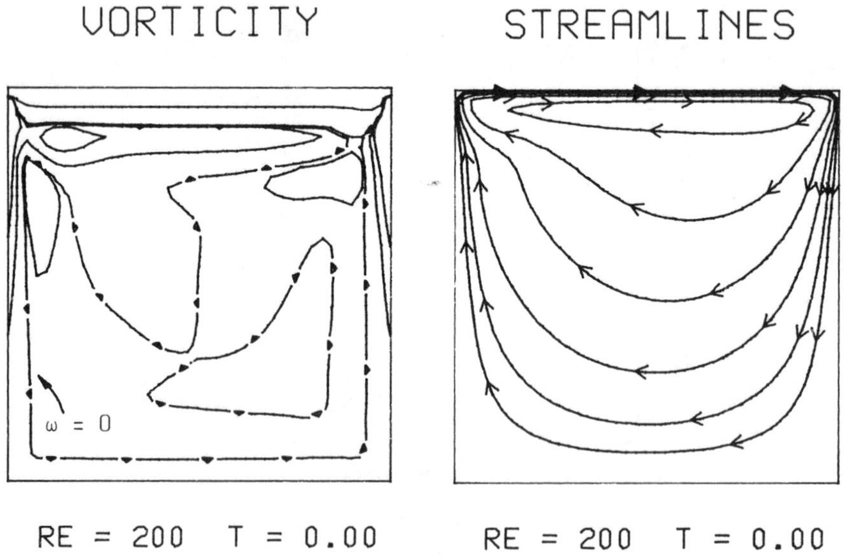

Figure 5a

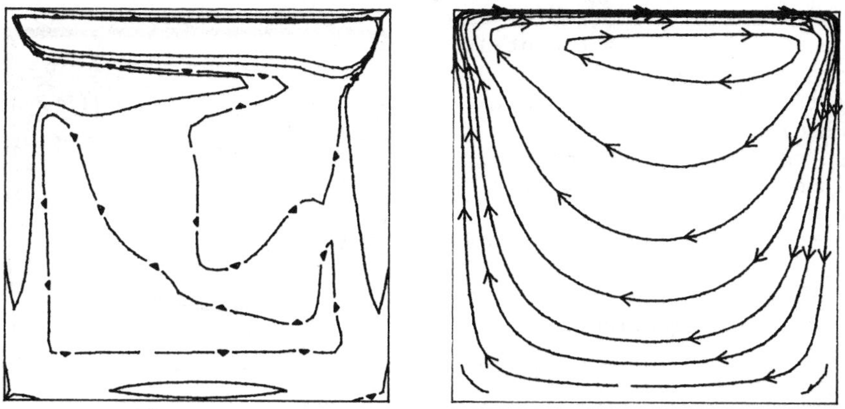

RE = 200   T = 0.25   RE = 200   T = 0.25

Figure 5b

sary at all to have a periodical motion to compute a time dependent solution. If only the periodical motion is to be computed then as already mentioned, an even faster MG method may be used.

Figures 5 show the periodical solution of the cavity problem at different phases, for $Re = 200$, $\Delta x = \Delta y\ (= h) = 1/40$ and $\Delta t = 1/16$. In Figures (5a) and (5b) we display the streamlines and the iso-vorticity lines, for $t = 0$ and $t = 0.25$, respectively. A qualitative comparison of our results with those in [1], show some discrepancies, mainly for $t = 0.25$. We have checked the periodicity of the solution by finding that $f(x,y,t) = -f(1-x,y,t+0.5)$ to round-off (where f is ω or ψ).

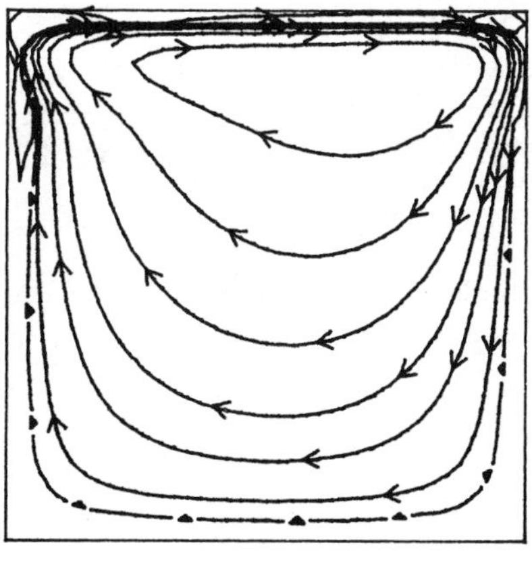

RE = 200   T = 0.31

Figure 5c

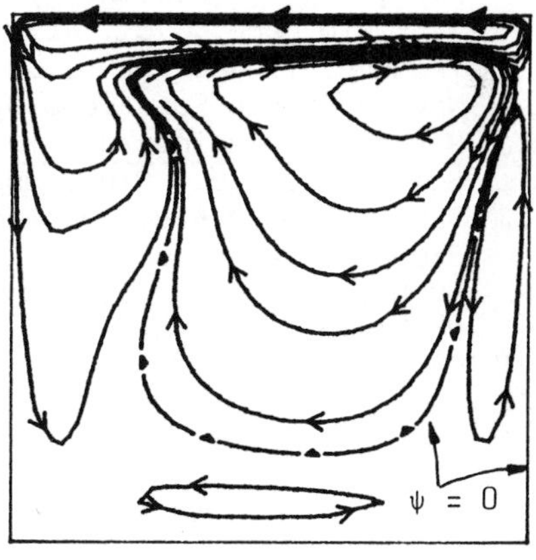

RE = 200   T = 0.38

Figure 5d

We have also computed the same problem, on the same spatial mesh, and varying time steps ($\Delta t = 1/4$ to $\Delta t = 1/16$). Results have also been computed with a given Δt and some h. The qualitative discrepancies between our solution and the one in [1] cannot be attributed to truncation errors, of the finite-difference equations.

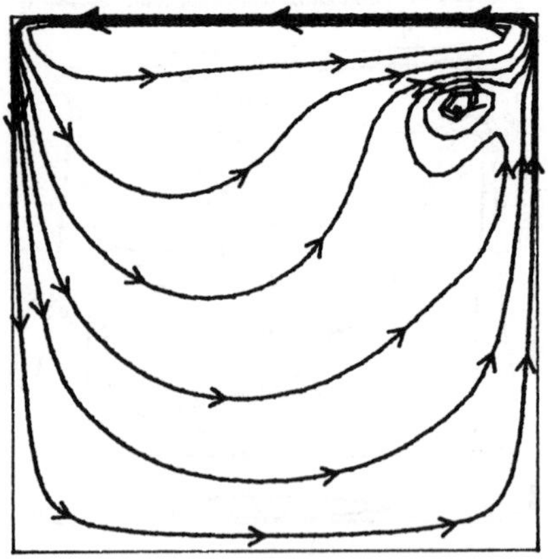

RE = 200   T = 0.44

Figure 5e

Figures (5b) - (5e) show the changes in the flow field when and immediately after the driving wall has changed its direction of motion. The large vortex, in the cavity and near the moving wall, at $t = 0.25$ decreases in size and moves toward the right wall as time goes on. At $t = 0.4375$ only a little and well confined vortex is left. At $t = 0.5$ the picture is the same as the mirror image of the picture at $t = 0$.

In summary, our results show that the MG method which has been described here, is fast, so that time (and space) accurate results can be computed by a limited computational effort.

REFERENCES

1. DUCK, P.W. - Oscillatory Flow Inside a Square Cavity, J. Fluid Mech., Vol. 122, pp. 215-234, 1982.
2. THUNELL, T. and FUCHS, L. - Numerical Solution of the Navier-Stokes Equations by Multi-Grid Technics, Numerical Methods in Laminar and Turbulent Flow, Eds. Taylor, C. and Schrefler, A.B., Pineridge Press, pp. 141-152, 1981.
3. FUCHS, L. - A Fast Numerical Method for the Solution of Boundary Value Problems, TRITA-GAD-4, 1980.
4. FUCHS, L. - New Relaxation Methods for Incompressible Flow Problems, To appear in this proceedings, 1983.

ON THE DIVERGENCE-FREE (i.e., MASS CONSERVATION, SOLENOIDAL) CONDITION IN COMPUTATIONAL FLUID DYNAMICS: HOW IMPORTANT IS IT?(1)

Karl Gustafson and Kadosa Halasi(2)

SUMMARY

In computational methods for the solution of the viscous incompressible Navier-Stokes equations, an underlying problem has been the effect of the modeling of the discretized divergence-free condition. Sometimes evidence of this at first subtle problem may be found in discussions of numerical instabilities, of ambiguities in assigning boundary conditions for the pressure equations, or in the absence of such discussions altogether. In capacitance matrix and panel methods for fast elliptic solvers on arbitrary regions, the problem shows up in how to acceptably model a divergence-free condition on partial boundary boxes, and in a related question of compatibility of discretizations of the divergence, gradient, and Laplacian.

We examine some aspects of these questions, including the interplay of incompressibility with vorticity and boundary conditions.

1. INTRODUCTION.

In numerical simulation of such phenomena as hydrodynamics care should be taken to ensure that algebraic properties governing the continuous model be preserved in the discrete model. More specifically, the discrete versions of operators such as divergence, gradient, curl, Laplacian, etc., should obey the same algebraic properties as do their continuous counterparts. Thus, for example, the discrete divergence operator when applied to the discrete gradient operator should yield the discrete Laplacian. Perhaps even more

(1) Partially supported by NSF Grant MCS 80-12220A2 and an NCAR Computing Resources Grant.

(2) Department of Mathematics, University of Colorado, Boulder, Colorado 80309.

important is that a discrete compatability condition hold whenever such a continuous condition exists. The necessity of preserving algebraic structure will become more evident when we discuss the numerical integration of the Navier-Stokes equations.

Further difficulties arise when treating boundary value problems of Neumann type, where in particular, the Neumann condition itself contains derivatives, as is the case for the pressure equation in Navier-Stokes flows. Here the boundary conditions for the pressure field are given by the momentum equations evaluated at the boundary. The existence of derivatives in the pressure gradient at the boundary requires the introduction of fictitious velocities outside the region of flow. The assignment of these fictitious velocity values should be consistent with given boundary conditions and mass conservation requirements.

A stated goal of this paper is to investigate how the incompressibility condition (divergence-freeness) of a flow is met in numerical schemes purporting to model such a flow. Furthermore, we hope to show by way of Helmholtz decomposition the interplay between incompressibility and vorticity--or the lack of them--with boundary conditions. In addition, we shall see how these influence the numerical stability of schemes used to model them.

2. THE FINITE DIFFERENCE EQUATIONS.

We consider the incompressible Navier-Stokes equations in primitive variable, conservative form for flow in a two-dimensional domain Ω

$$V_t = (1/Re)\,\Delta V - (V\cdot\nabla)V - \nabla P \tag{1}$$

$$\nabla \cdot V = 0 \tag{2}$$

where V is the velocity field, Re the Reynolds number, and P the pressure field. Boundary conditions, usually specified for the flow, include the no-slip and impermeable constraints. Thus for a nonmoving boundary we use $V \cdot n \equiv 0$ and $V \cdot T \equiv 0$, where n and T denote the normal and tangential vectors, respectively.

In discretizing equations (1) and (2) our first task is to establish a grid system on the domain, and then to define discrete versions of the continuous operators $\nabla\cdot$ and Δ in such a way that the algebraic structure of the continuous model is reflected in the discrete equations. We shall denote the discretized versions of the operators $\nabla\cdot$, ∇ , and Δ by the symbols D, G and L , respectively. These discrete operators will employ centered differences on a staggered MAC (Marker and Cell) mesh. The discretized versions of Eq. (1) and (2) are then given by

(1') $$V^{n+1} = V^n + \Delta t\left[F(V^n) - G(P^{n+1})\right]$$

(2') $$0 = D(V^{n+1}) \quad .$$

Here, for sake of brevity, we write $F(V^n)$ to denote $-(V^n \cdot G)V^n + (1/Re)\, L(V^n)$.

The MAC method of numerical integration first advances in time the advection-diffusion component of the flow

(3) $$\tilde{V}^{n+1} = V^n + \Delta t F(V^n)$$

and, second readjusts the pressure field so that the intermediate value V^{n+1} is made imcompressible. From equation (1') we get

$$D(V^{n+1}) = D(\tilde{V}^{n+1}) - \Delta t DG(P^{n+1}) \quad .$$

The requirement that $D(V^{n+1}) \equiv 0$ gives the desired pressure equation

(4) $$-\,DG(P^{n+1}) = (1/\Delta t)D(\tilde{V}^{n+1}) \quad .$$

The fact that on the MAC grid $DG = L$ allows us the use of fast direct Poisson solvers, developed in the last decade, to calculate efficiently, at each time step, the adjusting pressure field. Since, in general, a great many time steps are required, the use of iterative methods to invert Eq. (4) would be prohibitive.

3. BOUNDARY CONDITIONS FOR THE PRESSURE EQUATION.

Boundary conditions for the pressure equation are obtained by applying the momentum equations, and the incompressibility condition, at the boundary walls. This gives the Neumann condition

(5) $$-\frac{\partial P}{\partial n} = n \cdot (V \cdot \nabla)V$$

for the continuous model, and

(5') $$-\,n \cdot G(P) = n \cdot (V \cdot G)V$$

for the discrete model.

The evaluation of Eq. (5') requires that we assign fictitious velocity values at points outside the domain. When the domain is made up of rectangular subregions this assignment is made in accordance with the boundary wall conditions on the velocity, and with the requirement that these fictitious velocities conserve mass. The resulting pressure gradients at the boundary are of a simple form amenable to use in available, fast, direct Poisson solvers. However, when a boundary segment is no longer parallel to one of the grid lines, an ambiguity in the meaning of incompressibility results in an uncertainty of boundary conditions for the pressure field.

Suppose, for example, we use the MAC scheme, in which velocities are calculated at mid-edges of a uniform rectangular grid and pressures at the center of the grid box. As duplicated in Figure 1, consider the instance in which $\partial\Omega$ is linear but inclined through two grid boxes. Let P_i denote the desired pressure values, given the velocity values (U_i, V_i) . Let us assume that the matter of how to calculate the right side of Eq. (5') has been agreed upon, and that P_2 , P_3 and P_5 will be available to us from the interior box calculation on which the divergence-free condition has already been imposed in accordance with the above discussion. We seek P_1 such that the divergence-freeness is maintained in the next time step computation of V in some sense on the partial box, given the Neumann boundary condition Eq. (5') .

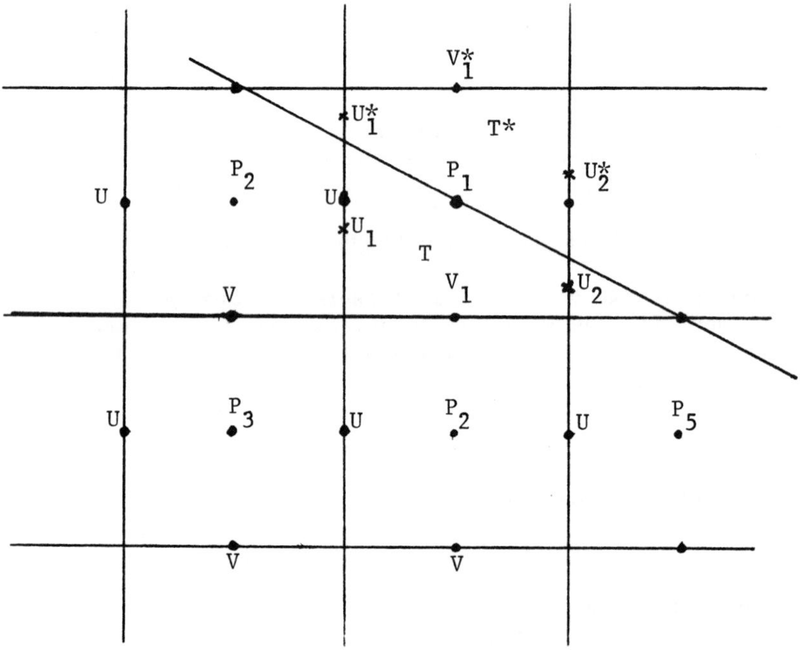

FIGURE 1. An instance of divergence-freeness on partial boundary boxes defined on a MAC grid system.

One approach is to assign imaginary velocity values U_1 and U_2 at the mid-edges of the sides of the trapezoidal partial box T and ask that

(6) $$U_2\delta_2 h - U_1\delta_1 h - V_1 h = 0$$

where h is the mesh size and where $\delta_1 h$ and $\delta_2 h$ are fractional heights of the two vertical partial sides. Corresponding to U_1, U_2 and V_1 we assign fictitious values U_1^*, U_2^* and V_1^* at complementary points outside the partial box. In order to guard against leakage we ask that the complement T^* of the partial box also be divergence-free. Thus

(6') $$(1-\delta_2 h)U_2^* - (1-\delta_1 h)U_1^* + V_1^* h = 0 \quad .$$

To preserve the no-slip and impermeable boundary condition along the inclined boundary we require that

$$\delta_1 h U_1 + (1-\delta_1 h)U_1^* = 0$$

and

$$\delta_2 h U_2 + (1-\delta_2 h)U_2^* = 0 \quad .$$

This gives us a value for V_1^* :

$$V_1^* h = V_1 h + U_1\delta_1 h - U_2\delta_2 h - (1-\delta_2 h)U_2^* + (1-\delta_1 h)U_1^*$$

$$= V_1 h \quad .$$

These fictitious values allow us to compute an expression similar to the right side of Eq. (5') , namely $n \cdot (V{\cdot}G')V$ where G' is a discrete gradient operator no longer using centered differences.

A more complete analysis of boundary divergence-freeness consistent with the factorization $L = DG$ is in progress.

4. HELMHOLTZ DECOMPOSITION OF THE VELOCITY FIELD.

Writing the intermediate velocity field V^{n+1} of the MAC algorithm as the sum of its irrotational and solenoidal components we get

$$\tilde{V}^{n+1} = \tilde{V}_1^{n+1} + \tilde{V}_2^{n+1} + \tilde{V}_3^{n+1}$$

where $\tilde{V}_1^{n+1}$ is curl-free, $\tilde{V}_2^{n+1}$ is divergence-free, and $\tilde{V}_3^{n+1}$ is both curl-free and divergence-free.

The desired incompressible field V^{n+1} is then obtained by readjusting $\tilde{V}^{n+1}$ in the presence of an appropriate pressure field P^{n+1}

$$V^{n+1} = \tilde{V}^{n+1} - G(P^{n+1})\Delta t \quad .$$

Since the curl of a gradient is the nul vector, we see that the intermediate solution V^{n+1} has the same vorticity as does the final, incompressible solution V^{n+l}. Thus the three velocity fields defined by $W_1^{n+l} = \tilde{V}^{n+l} - \tilde{V}_1^{n+l}$, $W_2^{n+l} = \tilde{V}^{n+l} - \tilde{V}_1^{n+l} - \tilde{V}_3^{n+l}$ and $V^{n+l} = \tilde{V}^{n+l} - GP^{n+l}$ have the same vorticity, and are all divergence-free.

To resolve this possible dilemma of uniqueness we begin by investigating the irrotational components of the flow. Since $\tilde{V}_1^{n+l}$ is curl-free, we may write it as the gradient of a scalar field

$$\tilde{V}_1^{n+l} = \nabla S \quad ,$$

from which we obtain

$$\nabla^2 S = \nabla \cdot (\tilde{V}_i^{n+l}) = \Delta t \nabla^2 P \quad .$$

For boundary conditions for S we have

$$\partial S/\partial n\Big|_{\partial\Omega} = n \cdot \tilde{V}_1^{n+l}\Big|_{\partial\Omega} \quad .$$

It is not clear what the values $n \cdot \tilde{V}_i^{n+l}$, $i = 1, 2, 3$, should be at the boundary. Howover, if we assume only that the normal component of divergence-free terms $\tilde{V}_2^{n+l}$ and $\tilde{V}_3^{n+l}$ vanish at the boundary, we get

$$\begin{aligned} \partial S/\partial n\Big|_{\partial\Omega} = n \cdot \tilde{V}_1^{n+l}\Big|_{\partial\Omega} &= n \cdot \tilde{V}^{n+l}\Big|_{\partial\Omega} \\ &= \Delta t n \cdot F(V^n)\Big|_{\partial\Omega} \\ &= \Delta t n \cdot \partial P/\partial n\Big|_{\partial\Omega} \quad . \end{aligned}$$

Thus $S/\Delta t \equiv P$, and so $W_1 = V^{n+l}$. Similarly, under the present assumption, we can show that

$$\tilde{V}_3^{n+l} \equiv 0 \quad ,$$

from which we have $W_2 = W_1 = V^{n+l}$.

This analysis demonstrates that the impermeable boundary condition need apply only to the divergence free component of a flow.

5. COMPRESSIBILITY AND STABILIZING FORCE FIELDS.

The final corrector step of the MAC scheme discussed here may be interpreted as introducing a force field GP to ensure that mass is conserved throughout the domain. Being that this force field is conservative, its presence does not introduce any changes in the vorticity of the existing flow. This force field is the result of superposing sinks and sources of density $DG(P) = D(\tilde{V}^{n+1}/\Delta t)$, and can be said to be governed by the compressibility of the irrotational components of the flow. One might thus expect to find, as in the case of static electric fields, field lines emanating from points of nonzero source/sink densities.

Figure 2(a) shows the normalized velocities of a driven cavity flow circulating about the primary vortex. Figure 2 (b) shows the normalized pressure gradient field lines originating from vortices. The behavior of these force lines near the boundary is not unlike force lines arising from the superposition of sources near the boundary.

Furthermore, from

$$\nabla^2 P = \nabla\cdot(V\cdot\nabla V) = \nabla\cdot(\tilde{V}_1^{n+1}/\Delta t)$$

we see that these sources and sinks are due entirely to the non-linear terms of the Navier-Stokes equation. In addition, it may be observed that the simple evolution of the advection-diffusion terms, i.e.,

$$\tilde{V}^{n+1} = \tilde{V}^n + \Delta t F(\tilde{V}^n)$$

is numerically, highly unstable. Whether this instability is due to the absence of a conservative force field such as ∇P, or to the absence of incompressibility, or to both, remains speculative.

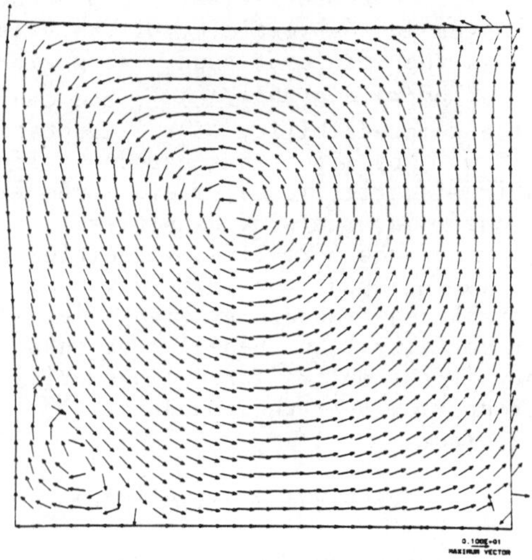

(a)

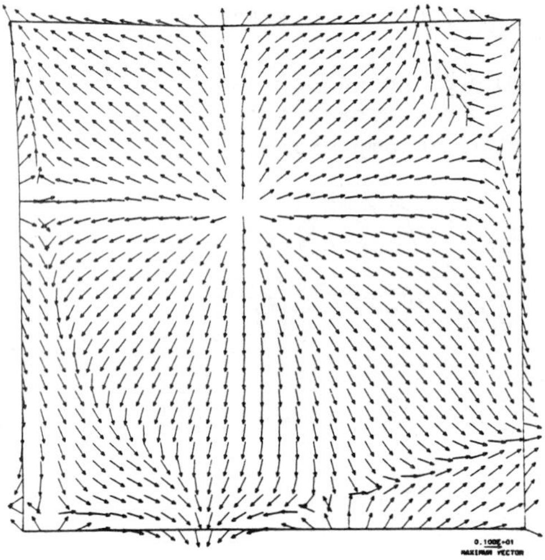

(b)

FIGURE 2. (a) Normalized velocities in cavity flow induced by moving top lid, at Reynolds number 400 , on 32 x 32 grid at 600 time steps with $\Delta t = 0.04$. (b) Corresponding normalized pressure gradient.

6. CONCLUDING REMARKS.

(a) The MAC schemes were introduced in Harlow and Welch [1] and have seen much development since then. From [1] we quote: " A more careful utilization of the correct [pressure] boundary condition awaits further development of methodology." As concerns the importance of the divergence-free conditions: "Since the computing technique is meant to be accurate for a variety of problems lacking crucial comparisons, it has been useful to develop an internal consistency accuracy check. The rational choice is verification of overall incompressibility."

(b) Others have brought attention to the importance of understanding the effects of divergence-freeness in fluid flow. Although we cannot give any comprehensive list here, important relevant papers include Orszag and Israeli [2] in 1974, Sani, Gresho, Lee, Griffiths, and Engelman [3] in 1981, and Fix [4] in 1982.

(c) For a previous recent discussion of methods that have been used to cope with the divergence-free condition see Gustafson and Young [5] . For its connections and implications in a hybrid scheme employing panel methods currently under development for flows in arbitrary irregular regions see Gustafson [6] . For a recent graph-theoretic treatment of the French FEM and further references to the treatment of the divergence-free condition by those methods, see Gustafson and Hartman [7] and Gustafson [8] .

(d) Substantially more detail and other aspects of the analysis presented here will be found in the paper Gustafson and Halasi [9] currently under preparation.

REFERENCES

1. HARLOW, F., and WELCH, J., Numerical Calculation of Time-Dependent Viscous Incompressible Flows of Fluid with Free Surface, The Physics of Fluids 8, pp. 2182-2189, 1965.

2. ORSZAG, S., and ISRAELI, M., Numerical Simulation of Viscous Incompressible Flows, Annual Rev. Fluid Mech. 6, pp. 281-318, 1974.

3. SANI, R., GRESHO, P., LEE, R., GRIFFITHS, D., and ENGELMAN, M., The Cause and Cure (?) of the Spurious Pressures Generated be Certain FEM Solutions of the Incompressible Navier-Stokes Equations: Parts I and II, Int. J. Numer. Methods Fluids 1, pp. 17-43 and 171-204, 1981.

4. FIX, G., Finite Element Methods for Flow Problems, Proc. 10th IMACS World Congress on System Simulation and Scientific Computation, Vol. 1, pp. 28-29, 1982.

5. GUSTAFSON, K., and YOUNG, D., Computation of Solenoidal (Divergence-free) Vector Fields, Numerical Integration of Differential Equations and Large Linear Systems, Ed. Juergen Hinze, Springer Lec. Notes in Math. No. 968, pp. 95-113, 1982.

6. GUSTAFSON, K., Hybrid Fast Poisson Solvers for Fluid Dynamics, Proc. 10th IMACS World Congress on System Simulation and Scientific Computation, Vol. 1, pp. 233-235, 1982.

7. GUSTAFSON, K., and HARTMAN, R., Divergence-free Bases for Finite Element Schemes in Hydrodynamics. SIAM J. Numer. Anal., to appear, 1983.

8. GUSTAFSON, K., Graph Theory and FLuid Hydrodynamics, with an Appendix on the Work of G. Kron, Proc. Ulam Chair Special Semester in Graph Theory, to appear, 1983.

9. GUSTAFSON, K., and HALASI, K., to appear.

NEW RELAXATION METHODS FOR INCOMPRESSIBLE FLOW PROBLEMS

Laszlo Fuchs(1)

SUMMARY

The primitive variable form of the steady incompressible Navier-Stokes equations are solved by three different variants of a multi-grid method. These variants differ in the way the pressure field is updated during the relaxation process. Two of the methods use the pressure as dependent variable, while the third one uses the pressure gradients as dependent variables. The efficiency of the methods is tested on the driven cavity problem. The speed of the multi-grid method is demonstrated for different numbers of levels which are utilized during the relaxations. The relative efficiency of the three methods is compared for some meshes and Reynolds numbers.

1. INTRODUCTION

The steady state Navier-Stokes (NS) equations are solved numerically. The primitive variable form of the equations has been solved for two- [1] and three-dimensional problems [2] by discretizing the differential equations on a staggered grid and solving the discrete problem by a Multi-Grid (MG) method which has been developed from the method described by Brandt and Dinar [3]. The MG method, which is also described here, includes a relaxation step where the continuity equation is relaxed and the pressure field is updated according to some distribution formula. This formula is obtained by adopting certain assumptions which are not accurate for large Reynolds numbers (Re) and when the flow field contains large gradients (see [2]). In this work we present two alternative approaches to the pressure field updating. One method

1. Assistant professor, Dep. of Gasdynamics
The Royal Institute of Technology, Stockholm, Sweden

uses the Poisson equation for the pressure to compute a correction pressure field. The Poisson equation has not to be solved accurately (only relaxed) so that the errors become smooth. Thus the computational effort is limited and is equivalent to some relaxation sweeps on a Poisson equation. In another method which has been tested we use the pressure gradients as dependent variables. These pressure gradients are corrected so that their cross derivatives shall be equal.

These methods have been incorporated into a MG solver. The efficiency of the MG method has been studied on the driven cavity problem. The MG method has been tested, utilizing different number of levels. The use of MG relaxations is shown to be much faster than relaxations on one (or even two) levels. All three variants are comparable for the test problem, with the fastest method being only twice faster than the slowest one. The relative merits of the methods become more clear when the flow field has a more complicated geometry with large velocity gradients.

2. THE DIFFERENTIAL EQUATIONS AND THEIR APPROXIMATION

The steady state incompressible NS equations in two pace dimensions are given by:

$$\nabla^2 u - Re(u\,u_x + v\,u_y) - p_x = 0 \qquad (1.a)$$

$$\nabla^2 v - Re(u\,v_x + v\,v_y) - p_y = 0 \qquad (1.b)$$

$$u_x + v_y = 0 \qquad (1.c)$$

We use no-slip boundary conditions on all solid boundaries.

Equations (1) are discretized on a staggered grid (see [1] and [2]). All the derivatives are approximated by central differences, except the first derivatives of the convective terms which are approximated by upwind differences. The finite-difference equations are written as

$$Qu - \partial_x p = 0 \qquad (2.a)$$

$$Qv - \partial_y p = 0 \qquad (2.b)$$

$$\partial_x u + \partial_y v = 0 \qquad (2.c)$$

where $Q = \partial_x^2 + \partial_y^2 - Re(u\overleftarrow{\partial}_x + v\overleftarrow{\partial}_y)$,

and ∂_x, ∂_y and ∂_x^2, ∂_y^2 are the central difference approximations to the first and the second derivatives, respectively. $\overleftarrow{\partial}_x$ and $\overleftarrow{\partial}_y$ denote the upwind difference approximations to the first derivatives.

3. THE MULTI-GRID SOLUTION PROCEDURE

A system of M grids, with mesh spacings $h_1, h_2, \ldots, h_M$ $h_{k-1}/h_k = 2, 1 < k \leqslant M$ is constructed. On each grid one applies a relaxation operator which works as a (low frequency) band-pass filter (for the error). The relaxation operator is applied on the sequence of grids, thus covering the whole frequency spectrum. Only few relaxation sweeps are needed because of the efficient filtering of high frequency error components. Usually, the high frequency errors decay by a factor greater than two while low frequency components decay as 1-e, where e is related to the mesh size.

We denote equations (2) on the grid M as:

$$L_M \varphi_M = R_M \tag{3}$$

$$L_M = \begin{bmatrix} Q & 0 & -\partial_x \\ 0 & Q & -\partial_y \\ \partial_x & \partial_y & 0 \end{bmatrix}$$

$$\varphi = (u,v,p)^T$$

and

$$R_m = (0,0,0)^T$$

The MG procedure consists of the following steps:

I. On each grid k $(1 \leqslant k \leqslant M)$ the problem

$$L_k \varphi_k = R_k$$

is smoothed out by an appropriate relaxation procedure.

II. The problem is transfered to a coarse grid, k-1, by defining

$$\varphi_{k-1} = I_k^{k-1} \varphi_k \tag{4}$$

$$R_{k-1} = L_{k-1}\varphi_{k-1} + \bar{I}(R_k - L_k\varphi_k) \tag{5}$$

where I_k^{k-1}, $\bar{I}_k^{k-1}$ are the transfer operators from the fine grid k to the coarse grid k-1. The problem is relaxed on the current grid according to step I.

The smoothing procedure preceeds on the different grids until the coarsest one is reached.

III. The corrections on a grid k-1 are interpolated and added to the current approximations on the grid k. The new approximation is smoothed out according to step I. In all the following cases we transfer the corrections to a fine grid by linear interpolations.

Steps III and I are repeated until the finest grid is reached. A MG-cycle is a sequence of steps I to III starting on the finest grid, going to coarser grids until the coarsest is reached and then visiting all the grids until the finest one is reached again.

The MG cycles are repeated until convergence to a prescribed accuracy is reached. It can be noted that improved initial approximations can be obtained (especially for smooth problems) if the solution is known on a coarse grid. The interpolated values on the fine grid give, usually rise to an initial approximation with relatively high frequency errors. One MG cycle on such initial approximation is enough to reduce the error under that of the truncation errors. Here, we have solved the problems much beyond the truncation errors and no coarse grid approximations are used. In this way the true asymptotic behaviour of the methods can be studied.

Consider a cell of a grid with spacings h_{k-1}. This cell is subvided into four cells in the finer grid with spacing h_k (see Sketch a).

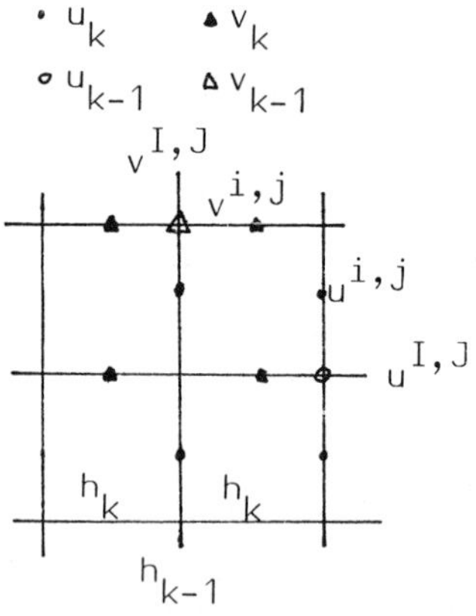

Sketch a

We define $u_{k-1}^{I,J} = (u_k^{i,j} + u_k^{i,j-1})\, h_k/h_{k-1}$

$$v_{k-1}^{I,J} = (v_k^{i,j} + v_k^{i-1,j})\, h_k/h_{k-1} \qquad (6)$$

Relations (6) ensure a mesh independent mass flux through any line segment. These relation define I_k^{k-1} in equation (4). The transfer of the residuals of the momentum equations is similar to the transfer of the corresponding velocity components. That is, for the momentum equations we take $\bar{I}_k^{k-1} = I_k^{k-1}$. The transfer relations (6) has a filtering property for some high frequencies, by which aliasing is reduced.

For the transfer of the mass conservation equation we choose another $\bar{I}_k^{k-1}$:

$$A^{I,J}_{k-1} = \bar{I}^{k-1}_{k} A^{i,j}_{k} = (A^{i,j}_{k} + A^{i,j-1}_{k} + A^{i-1,j}_{k} + A^{i-1,j-1}_{k}) h^2_k / h^2_{k-1} \qquad (7)$$

By equations (2.c), (4), (6) and (7) we get that for the mass conservation equation

$$R_{k-1} = \bar{I}^{k-1}_{k} R_k \qquad (8)$$

Since $R_M = 0$, the right hand side of the mass conservation equation vanishes identically (up to round-off errors) on all the grids.

The MG solver with these transfers needs only marginally larger memory compared to the number of unknowns. For two-dimensional cases we have three dependent variables and the required memory for these is denoted by 3T. The dependent variables and the right hand sides of the two momentum equations has to be stored also on coarse grids. Since each grid contains roughly only 1/4 of the number of nodes of the finer grid, the total number of nodes in all the coarse grids is bounded by

$$T/4 + T/4^2 + T/4^3 + \ldots\ldots + T/4^{M-1} < T/3$$

Thus the total required memory is less than (3 5/3)T. That is, this MG method requires 55% more memory than the memory which is required to store the solution itself.

4. THE SMOOTHING PROCEDURE

Efficient smoothing (relaxation) of the high frequency error components is fundamental and crucial for MG methods. We define high frequencies as those which can be resolved on a given grid but not on a coarser one. It is also important that the smoothing operator is such that low frequency errors decrease in their amplitude, but we accept relaxation operators with small low-frequency amplification (provided that the amplification on fine grids is over-compensated by the attenuation on coarse grids, such that totally all the error components decrease in each MG cycle).

Each momentum equation can be relaxed for the respective velocity component: relaxation (and MG solution) of equations of advection-diffusion type has been considered by Fuchs [4,5]. In many cases, (mostly uniform grids and relatively small Reynolds numbers) the simplest and the most cost-effective relaxation procedure is that of Succesive Point Relaxation (SPR) type. For large Re, the direction of the relaxation should be alligned (as much as possible) with the flow direction. For non-uniform meshes and non-simple flows, efficient smoothing may be obtained by other schemes (such as 'convective line relaxation' or some other scheme which is described in [4] and [5]).

Here, for the driven cavity problem we use SPR starting at the corner near the moving wall and marching parallel and away from the wall. The efficient smoothing of the choosen operator has been checked on a test problem for the convective-diffusion equation (with a solution simulating a real cavity flow) in a MG mode and found to be acceptable for all the test Re ($\leq$ 500).

The mass balance equation does not have a natural dependent variable which can be updated during the relaxations. One approach which has been developed by Harlow and Welch [6], is to replace the mass equation by a Poisson equation for the pressure. This equation is found by deriving the i-th momentum equation with respect to the i-th independent variable and adding the resulting equations. The new system of equations is of order 6 for which 3 conditions must be specified on the whole boundary. The original system (1) is only of order 4 and 2 conditions on the boundaries are enough.

Here, we adopt the method of Brandt and Dinar [3] that the correction velocity field can be approximated by an irrotational correction. Under this assumption there exists a correction potential χ, such that the correction velocity components are:

$$\begin{aligned} \Delta u &= \partial_x \chi \\ \Delta v &= \partial_y \chi \end{aligned} \qquad (9)$$

By the mass balance equation:

$$(\partial_x^2 + \partial_y^2)\chi = -\partial_x u - \partial_y v \qquad (10)$$

On the boundaries both Δu and Δv should vanish, but only one of these may be used if a solution to (10) is to be found.

Equation (10) is never solved. The correction function is applied only locally during the relaxation sweep (using a SPR scheme). In this way the high frequency errors are eliminated from equation (2.c).

In the following we describe three alternatives for updating the pressure field, or its gradients, during the relaxations.

5. DISTRIBUTIVE RELAXATION

Brandt and Dinar [3] have suggested that when the mass balance equation is relaxed pointwise, one may update both the velocity and the pressure fields simultaneously. At each node point (i,j) a special correction $\chi = a\delta_{ij}$ is choosen, where δ_{ij} is the Kronecker delta. Using the notation of equation (2) the corrections Δu, Δv and Δp are defined as follows:

$$\Delta u = \partial_x \chi \qquad (11.a)$$

$$\Delta v = \partial_y \chi \qquad (11.b)$$

$$\Delta p = Q \chi \qquad (11.c)$$

If relations (11) are satisfied everywhere then the change in the residuals of the momentum equations, after the relaxation of equation (2.c) vanish. (Note that this is only an approximation, since Q is a linearized operator, and that relations (11.a) and (11.b) cannot be used on the boundaries.)

This scheme ensures that the relaxation of the mass balance equation has little effect on the momentum equations and therefore the different relaxation steps can be treated and evaluated independently of each other.

By equations (11) one gets that the velocity components and the pressure are corrected at several neighbouring points simultaneously. This way of updating the dependent variables has been called by Brandt and Dinar [3] as Distributive Gauss-Seidel (DGS) relaxations. In this way the computational effort at each node point, in relaxing the mass equation, is about three times of SPR computations (of a Poisson equation).

The method has found to be efficient for many test problems. For some other problems, especially for large Reynolds numbers (and very much dependent on the flow field itself) the DGS seems to be less efficient, and in some cases no convergent solution could be obtained (see Sec. 8). The reason for this behaviour is the result of the underlying assumptions which have been used implicitly: The crude linearization of the momentum equations which leads to equations (11) is valid for low Re or if the flow gradients are not large. The irrotational assumption is not correct especially near the boundaries.

6. PRESSURE CORRECTION RELAXATIONS

By this variant of the relaxation procedure we relax the momentum equations and the mass balance equation (using the irrotational correction field assumption) as in the DGS scheme. Instead of distributing the pressure by relation (11.c) we use the following equation for the pressure correction field:

$$\nabla^2 \Delta p = \mathrm{Re}\ \Delta[(x\text{-momentum})_x + (y\text{-momentum})_y] \qquad (12)$$

where the right hand side of equation (12) is the change in the sum of the x- and the y-derivatives of equation (2.a) and (2.b), respectively, due to the relaxation of (2.c).

Equation (12), as equation (10) is not solved exactly but only relaxed by a standard succesive point relaxation (SPR) method. Usually, only few SPR sweeps are done in each MG relaxation step. In this way the errors in the pressure field are smoothed out. We denote this method variant as PCR (Pressure Correction Relaxation) method.

The total computational effort of the PCR method is larger than that of the DGS per MG-relaxation step. With PCR one has to compute the right hand side of equation (12), and do some (about 2-3) SPR sweeps on the pressure correction field. In all, the PCR variant requires totally less than 50% more computational effort than the DGS method, for each MG cycle.

7. PRESSURE GRADIENTS AVERAGING

In both DGS and PCR variants we use the pressure as a dependent variable. Another possibility is to use the pressure gradients, and the velocity vector, as dependent variables. Thus, the number of dependent variables is increased to four and hence the number of equations must be increased by one. The fourth governing equation (in addition to (1)) is

$$(p_x)_y = (p_y)_x \tag{13}$$

On the staggered grid (p_x) and (p_y) are defined at the same position as u and v, respectively. The finite-difference approximation to equation (13) can be written, using the notations of equation (2), as:

$$\partial_y(p_x) - \partial_x(p_y) = 0 \tag{14}$$

Thus, the finite-difference equation (14) is satisfied at the center of each compuational cell (Sketch a).

The relaxation procedure, on each grid, is sequential: The momentum equations are relaxed by SPR. The mass balance equation is relaxed pointwise, using equation (10). The change in the momentum equations, after relaxing (2.c), is atributed to changes in the pressure gradients. These changes are smoothed out in such a way that the corrections will satisfy equation (14). It is assumed that at the center of each cell, the corrections of the pressure gradients can be updated by a (local) 'stream-function' ψ. That is, we define

$$\Delta(p_x) = \partial_y\psi \tag{15.a}$$
$$\Delta(p_y) = -\partial_x\psi \tag{15.b}$$

By these definitions at each cell a Poisson equation is relaxed, and the corrections are distributed to the appropriate pressure gradients according to (15). At each MG relaxation step, only one sweep is done on equation (14). The method of relaxing the pressure gradients may be regarded as an averaging procedure and therefore we call this method as PGA (Pressure Gradient Averaging).

The transfer of the corrections from coarse to fine grid is done, as for DGS and PCR, by linear interpolations. The PGA variant needs some additional memory compared to the other methods. The additional computational effort, per a MG relaxation step, for the PGA is only marginally larger than that for the DGS. The performances of the three variants are discussed in the next Section.

8. NUMERICAL EXPERIMENTS

The three methods has been tested on the driven (square) cavity problem. First, we study the effects of MG relaxations (using the DGS variant as the smoothing scheme). Computations are done on a grid with 64×64 intervals, utilizing from 1 to 5 levels (M).

In Figures 1 we show the effects of the number of levels (keeping the finest mesh the same) and the Reynolds number on the convergence. The convergence rate (Θ) is defined to be equal to the factor by which the residuals decrease per a work unit (WU). The work unit is defined to be equivalent to the computational effort of one relaxation sweep on the finest grid. The rates of convergence which are given in all the figures are computed from:

$$\Theta = \left[\frac{\text{final residual}}{\text{initial residual}}\right]^{1/WU}$$

This averaged convergence factor is bounded by its asymptotic

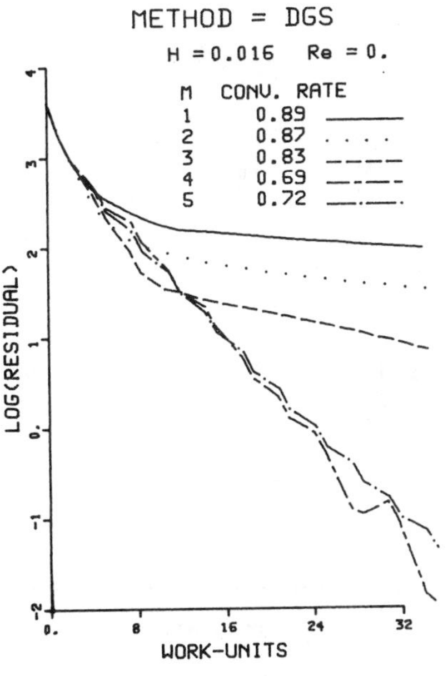

Figure 1.a

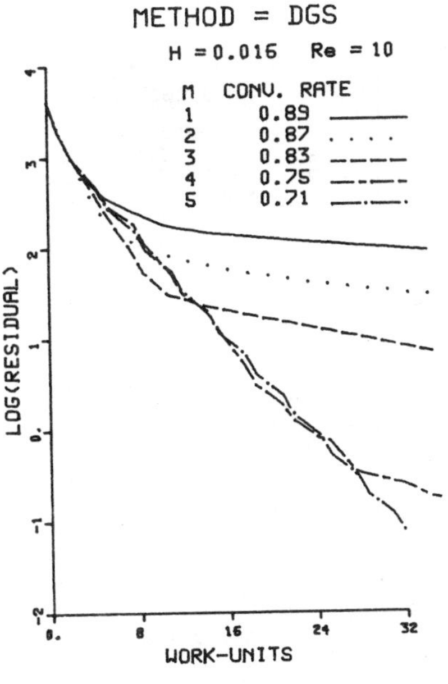

Figure 1.b

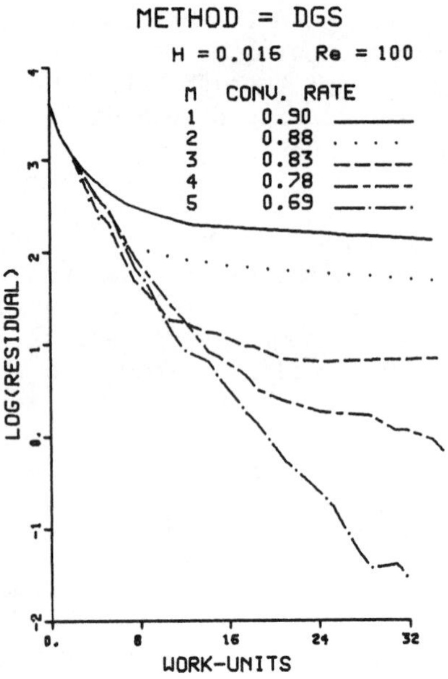

Figure 1.c

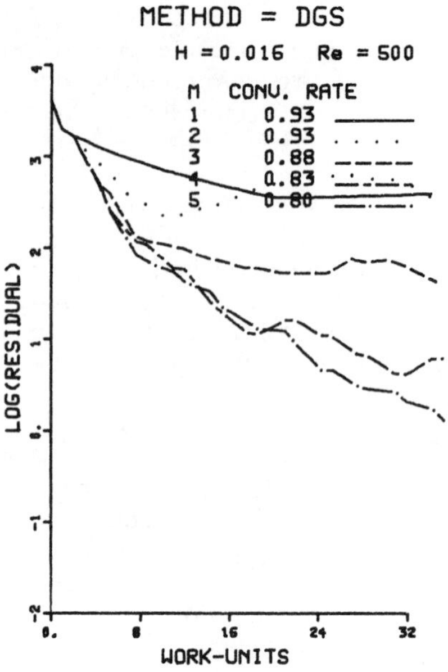

Figure 1.d

Figure 1: The convergence history of the MG-DGS method for different number of levels (M) and some Re.

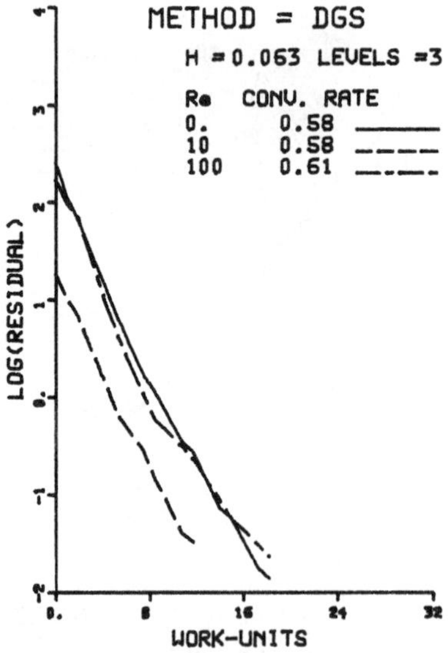

Figure 2.a

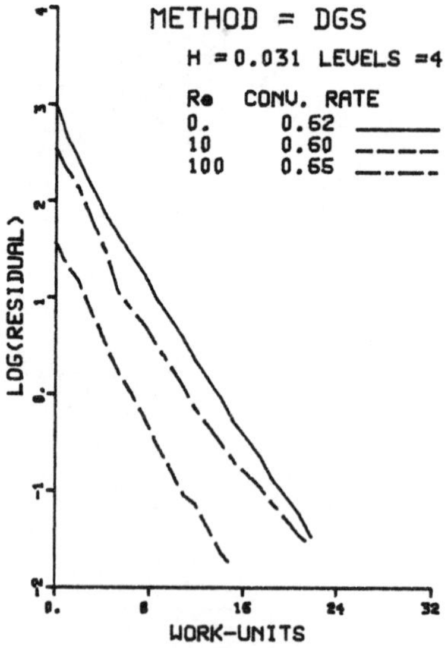

Figure 2.b

Figure 2: The convergence history of the MG-DGS method.

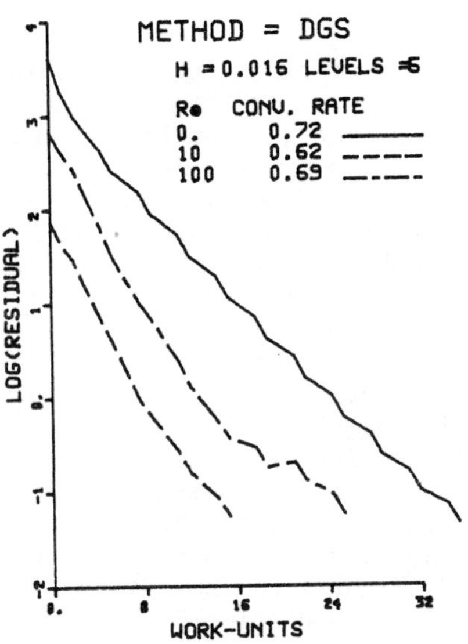

Figure 2.c

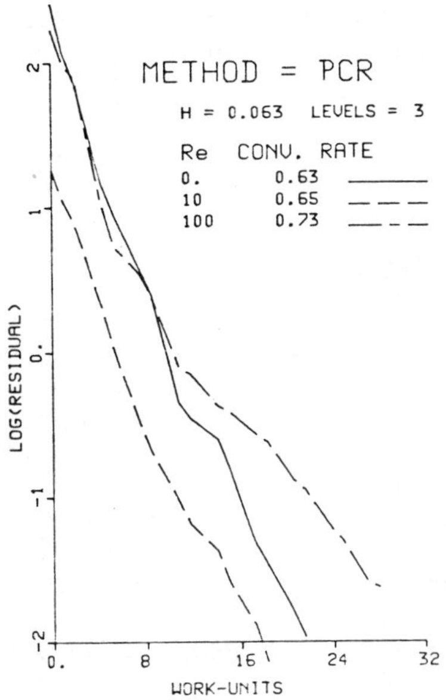

Figure 3.a

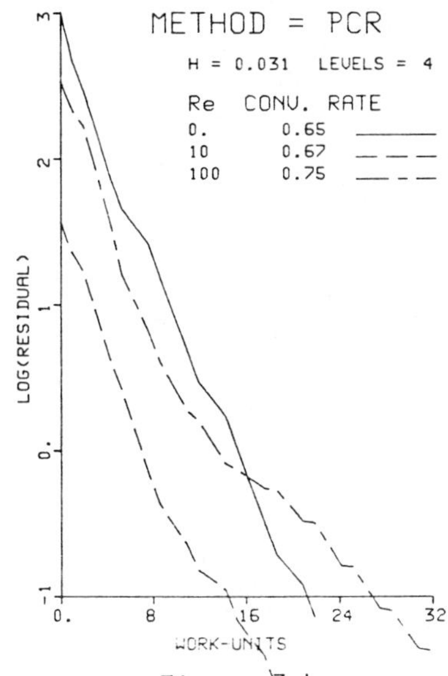

Figure 3.b

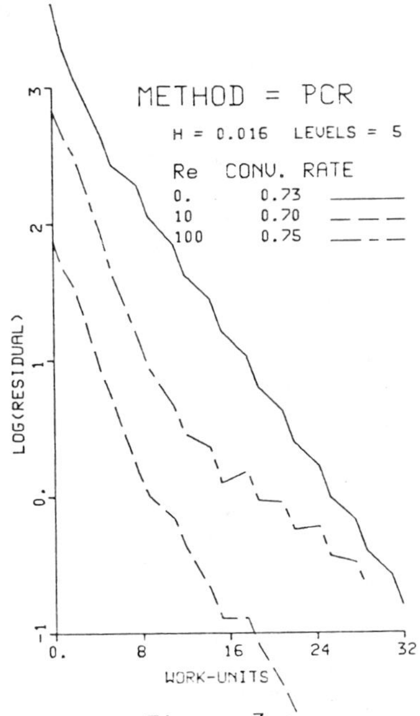

Figure 3.c

Figure 3: The convergence history of the MG-PCR method.

value (the slope of the curvs when the residuals have been decreased by some orders of magnitudes).

The superiority of the MG method over single grid relaxations becomes clear from Fig. 1. It is also clear that the increase in the number of levels (M) improves the convergence rate in general. For large M (M>5) the improvements are less significant. For increasing Re the need to larger M becomes more clear. Even if the MG-DGS method looses some of its efficiency for large Re (=500) it is the only reasonable way to get truely convergent solutions down to machine round-off.

In Figures 2-4 we compare the three variants of the MG methods which we have described in this paper. Each figure displays the rate of convergence for three Reynolds numbers (0, 10 and 100). In these figures we consider the effects of the mesh size (H) (and the number of levels) on the rate of convergence. The coarsest grid has 16×16 intervals (and 3 levels), the medium mesh is 32×32 (4 levels) and the fine mesh is of 64×64 (5 levels) intervals. A general trend with DGS (Figures 2) is that the convergence is slower for finer meshes and increasing Re. The sensitivity of this method to changes in H and Re are explained by the crude linearization which is used for the distribution formula. It is also clear that Δp behaves as $\Delta u/H$ and $Re\ \Delta u \cdot \partial_x \cdot u$ (eq. (11.c)) and therefore it becomes more sensitive to smaller H and large Re.

In Figures 3 the convergence of the PCR method is shown. The efficiency (in terms of method specific work units) is less **than** for the DGS but it is less sensitive to Re and almost

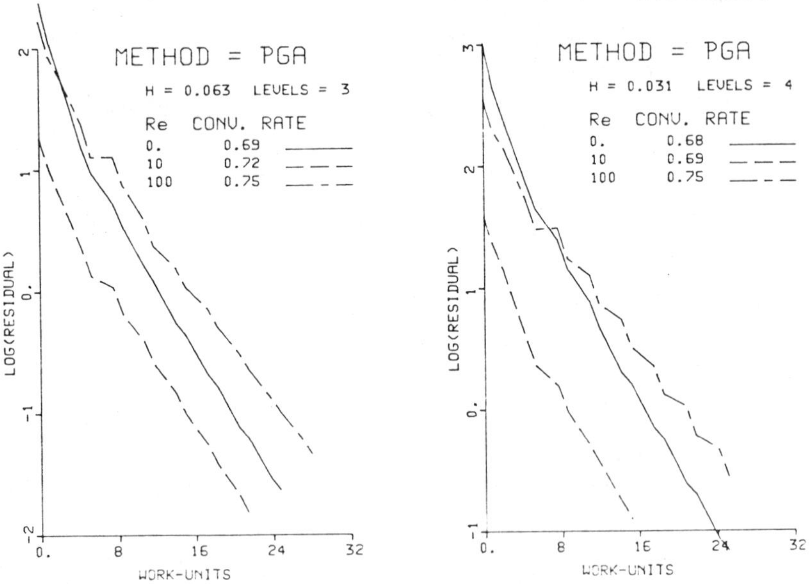

Figure 4.a Figure 4.b

Figure 4: The convergence history of the MG-PGA method.

independent of H. The method, because of the derivation of the momentum equations, reaches to the point where single precision is not enough (Fig. 3.c, Re=100 and H=0.016). In this case the convergence stops at a certain stage when the corrections are of the order of the round-off errors.

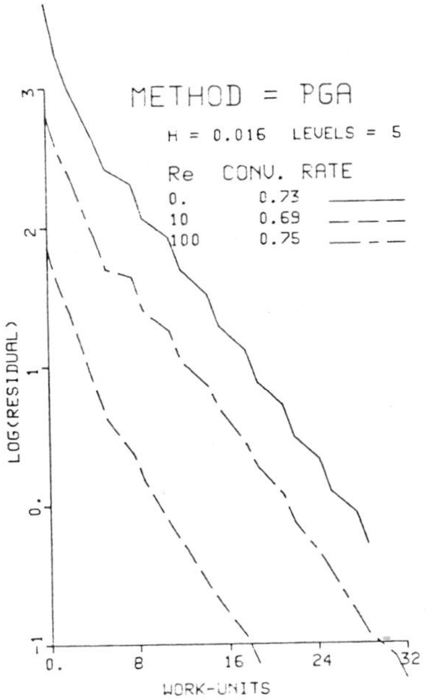

Figure 4.c

The PGA method (Fig. 4) is insensitive to changes in both Re and H. The total experience with the PGA variant is more limited compared to the other methods. The possible superiority of this method, for more complex problems has to be studied.

A difficult test problem for any numerical method is the computation of flows with large velocity gradients. Such a case is the computation of the flow in a separator [7]. For this problem the DGS method has a slow rate of convergence (θ=0.96) and for larger Re (>50) no converged solution could be obtained. The PCR method has for these problems a rate of convergence of about 0.86 and converged solutions have been obtained for Re$\leq$200.

In all our extensive tests we have found that the MG method results in a fast solution, whenever a proper relaxation method is used. The specific smoothing technique may be problem dependent if optimal results are to be achieved. For flows without too large velocity gradients the DGS method has proven itself to be reliable and fast even though some loss of efficiency may occur for large Re. The PCR variant works well and it is a good alternative to DGS for cases when distribution is too complicated or inaccurate.

REFERENCES

1. THUNELL, T. and FUCHS, L. - Numerical Solution of the Navier-Stokes Equations by Multi-Grid Techniques, Numerical Methods in Laminar and Turbulent Flow, Eds. Taylor, C. and Schrefler, A.B., Pineridge Press, pp. 141-152, 1981.
2. FUCHS, L. and ZHAO, H-S. - Solution of Three-Dimensional Viscous Incompressible Flows by a Multi-Grid Method, 1983.
3. BRANDT, A. and DINAR, N. - Multigrid Solution to Elliptic Flow Problems, Numerical Methods in PDE, Ed. Parter, S.V., Academic Press, pp. 53-147, 1977.
4. FUCHS, L. - A Fast Numerical Method for the Solution of Boundary Value Problems, TRITA-GAD-4, 1980.

5. FUCHS, L. - Multi-Grid Solution of the Navier-Stokes Equations on Non-Uniform Grids, NASA CP-2202, pp.83-100, 1981.
6. HARLOW, F.H. and WELCH, J.E. - Numerical Calculation of Time-Dependent Incompressible Flow, Physics of Fluids, Vol. 8, pp. 2182-2187, 1965.
7. THUNELL, T. and FUCHS, L. - Numerical Computation of Incompressible Flow in Separators, to appear, 1983.

TURBULENT FLOW OVER A RECTANGULAR CAVITY

G. D. Catalano[I]
C. Shih[II]
Louisiana State University

SUMMARY

A numerical and experimental investigation of turbulent flow over a rectangular cavity is reported. The numerical portion uses a Schwarz-Christoffel transformation for the centerline flow with the free shear layer represented by periodic insertion of point vortices near the separation point. The evolution of the point vortices is traced from a Lagrangian point of view. A vortex decaying function based upon the turbulent energy spectrum equation is introduced to simulate the viscous dissipation effect on a point vortex. The experimental portion of the research consists of examining the rectangular cavity flow field with flow visualization and hot wire anemometry techniques.

1. INTRODUCTION

Cavity-type flow fields have received considerable attention during recent years because of the large number of occurrences of these configurations in real world engineering problems. Slotted wall wind and water tunnels and flows in and around aircraft components are typical cases. The pressure oscillations associated with cavity type flows are believed to be responsible for the generation of noise, large

---

[I]Assistant Professor, Mechanical Engineering Department, Louisiana State University, Baton Rouge, Louisiana, U.S.A. 70803.

[II]Graduate Research Assistant, Aerospace Engineering Department, University of Southern California, Los Angeles, California U.S.A.

drag increases, and having a significant impact on the fatigue life of nearby structures.

A review article by Rockwell and Naudascher [1] categorizes the self sustaining oscillations of cavity flows into the following three types: (1) fluid-dynamic, where oscillations arise from inherent instability of the shear layer; (2) fluid-resonant, which are governed by the resonant effects associated with compressibility or free surface wave phenomena; (3) fluid-elastic, where oscillations are coupled with the elastic motion of a solid boundary. In practice, all three types of oscillations can occur simultaneously and interact with one another. For simplicity, only the pure fluid dynamic type oscillation will be considered. Fluid-dynamic type oscillations are attributable to the velocity and pressure flucturations generated by an initial disturbance which is amplified during convection downstream in the cavity. An effective feedback mechanism is provided by the trailing edge in that pressure fluctuations are reflected back upstream triggering a new disturbance near the leading edge. Thus, the process is self-sustaining.

Various attempts to attenuate the amplitude of the cavity oscillations have been made. Geometrical variations as well as mass injection schemes have been examined. Ethembabaoglu [2] found the gradual ramp with recess to be the most effective in reducing the pressure oscillations.

2. MATHEMATICAL DEVELOPMENT

Complex variables and conformal mapping technique are used to simplify the problem. The Schwarz-Christoffel transformation is introduced to transform the rectangular cavity geometry into the upper half plane. One of the important properties of the conformal transformation is that vortices map into vortices of the same strength. The complex velocities are not, in general, invariant but are proportional to each other, and the proportionality factor depends on the transformation function. Nevertheless, the continuity equation is satisfied.

In real viscous flow, the oncoming fluid will not turn around the leading corner but instead will separate from the edge creating a highly vortical shear layer. This shear layer appears to be dominated by a two-dimensional deterministic structure which arises initially from the Kelvin-Helmholtz instability of a vortex sheet. Analytically, the model uses a collection of point vortices to represent the roll-up of the vortex sheet, even though the use of finite point vortices to simulate the vortex sheet is still questionable [3]. However, several satisfactory results [4] still encourage the utilization of the so-called discrete vortex model.

Consider a two-dimensional rectangular cavity geometry of length, L, and depth, D (Figure 1) where d = 2D/L is the non-dimensionalized cavity depth. Introducing the following transformation function,

$$Z = \frac{E[\sin^{-1}(\lambda)/\sin^{-1}(1/a)]}{E(1/a^2)} \qquad (1)$$

the cavity geometry in Z-plane can be transformed into the real axis of the λ-plane where E[/] is the incomplete elliptical integral of the second kind with complex amplitude and E( ) is the complete elliptical integral of the second kind [5]. The four corners $Z = \pm 1$ and $Z = \pm 1 + id$ are mapped into the points $\lambda = \pm 1$ and $\lambda = \pm a$, respectively. The relation between a and d is given by the following equation [6].

$$d = \frac{K(1 - 1/a^2) - E(1 - 1/a^2)}{E(1/a^2)} \qquad (2)$$

where K( ) is the complete elliptical integral of the first kind. The determination of the parameter a will also determine the cavity depth to length ratio.

The ideal inviscid cavity flow field can be represented by the complex potential,

$$F(\lambda) = A\lambda \qquad (3)$$

where A is an undetermined constant. Thus, in the Z-plane

$$F(Z) = Af(Z) \qquad (4)$$

where $\lambda = f(Z)$ is the inversion of the Schwarz-Christoffel transformation.

In determining the constant A, the boundary condition at infinity must be used. The complex velocities are

$$U - iV = \frac{dF}{dZ} = \frac{dF}{d\lambda}\frac{d\lambda}{dZ} = Af'(Z) \qquad (5)$$

where

$$f'(Z) = \frac{d\lambda}{dZ} = aE(1/a^2)\left[\frac{(\lambda^2 - 1)}{(\lambda^2 - a^2)}\right]^{\frac{1}{2}} \tag{6}$$

At inifinity, as $Z \to \infty$, $u \to U_o$ and $v \to 0$ which implies that $A = U_o/[aE(1/a^2)]$.

In the real flow, the shear layer beginning at the leading edge will be approximate by a system of point vortices as mentioned in the preceding discussion. Each point vortex can be represented by a logarithmic-type singularity in the complex plane. For any closed contour which does not include the singularity, the circulation will be zero and the flow will be irrotational [7].

Suppose that at a given time, there are N point vortices locating at the positions Z_j with strengths Γ_j (j = 1, 2, ...N). If λ_j is the transform of Z_j, the complex potential function can then be written as

$$F(\lambda) = A\lambda - \frac{i}{2\pi}\sum_{j=1}^{N}\Gamma_j \log(\lambda - \lambda_j)$$

$$+ \frac{i}{2\pi}\sum_{j=1}^{N}\Gamma_j \log(\lambda - \lambda_j^*) \tag{7}$$

where an asterisk indicates the complex conjugate and image vortices have been introduced to satisfy the boundary condition. The complex velocities are given as the derivative of F( ).

$$U - iV = [A - \frac{i}{2\pi}\sum_{j=1}^{N}\frac{\Gamma_j}{\lambda - \lambda_j}$$

$$+ \frac{i}{2\pi}\sum_{j=1}^{N}\frac{\Gamma_j}{\lambda - \lambda_j^*}]f'(Z) \tag{8}$$

If the point whose velocity is required is also a vortex location then the return to the Z-plane is calculated by Routh's rule [8]. Suppose $Z = Z_k$ is a singular point, then

$$(U - iV)_{Z=Z_k} = \frac{dF}{dZ}\Big|_{Z=Z_k} = \frac{dF}{d\lambda}\Big|_{\lambda=\lambda_k} \frac{d\lambda}{dZ}\Big|_{Z=Z_k} - \frac{i\Gamma_k}{4\pi} \frac{f''(Z_k)}{f'(Z_k)} \tag{9}$$

where $\lambda_k = f(Z_k)$, and Γ_k is the strength of the point vortex at that point.

With such an arrangement, the vortex velocities at any point can be calculated in the physical plane. The motion of a point vortex marked at x_o, y_o at time t_o is governed by the equations,

$$\frac{dx}{dt} = U(x,y,t), \quad \frac{dy}{dt} = V(x,y,t) \tag{10}$$

where U and Vare the velocity components in the x and y directions, respectively, and their locations at time t are easily calculated.

The are two ways to determine the strength of the vortex shedding from the leading edge. First one makes use of the Kutta condition, which requires that the transformed leading edge be a stagnation point in the transform plane. Suppose a initial vortex is introduced at the point Z_k, and there are K - 1 vortices already in the cavity, then the circulation of the Kth vortex should be

$$\Gamma_k = \frac{-\pi A - \sum_{j=1}^{N-1} \frac{{}_j\mathrm{Im}(\lambda_j)}{[a^z + 2a\mathrm{Re}(\lambda_j) + \lambda_j^{\,z}]}}{\frac{\mathrm{Im}(\lambda_k)}{a^2 + 2a\mathrm{Re}(\lambda_k) + \lambda_k^{\,z}}} \tag{11}$$

where $\lambda_k = f(Z_k)$, and Re( ) and Im( ) indicate the real and imaginary parts of a complex number, respectively.

A second approach used is by Clements [8]. The rate of vorticity shedding into the shear layer is determined by the relationship

$$\frac{d\Gamma}{dt} = \frac{1}{2} U_+ z \tag{12}$$

where U_+ is the velocity at the outer edge of the boundary layer. The strength of a newly created vortex is then equal to $\Sigma \frac{1}{2} U_+ z\Delta t$, where Δt is the time step used in the calculation and the sum is over all the time steps since the previous vortex was introduced. For identification purposes it will be called the vortex shedding rate method. However, the flow is considered inviscid and the boundary layer thickness is thus unknown. Therefore, an artificial thickness is proposed, and the effect of varying various parameters is studied. In fact, the bulk flow motions are insensitive to variation of the thickness, and there is little difference between the two vortex shedding models.

Instead of assigning a physical unrealistic position as the centroid of a point vortex shedding from the leading edge, which has been used by some other authors [6, 8], here the use of 5 to 10 point vortices to represent a small length of vortex sheet is considered. The sheet is allowed to roll up by itself. The rolling-up vortex sheet motion can be seen clearly from Figure 2. The seemingly chaotic motion can be explained by the finite number of point vortices used in the approximation, or simply due to the fundamental difference between vortex sheets and an assembly of point vortices as remarked by Saffman and Baker [9].

With nonzero viscosity, the exact vorticity equation is recovered. Chorin [10] has suggested that it is equal to a diffusion equation in a moving medium. However the problem remaining is the increase in the CPU time. A different physical consideration of the viscous effecton the vorticity is therefore adopted. First, the turbulent energy spectrum equation is

$$\frac{\partial E(\kappa)}{\partial t} = T(\kappa) - Z\nu\kappa^2 E(\kappa) \tag{13}$$

where $E_S(\kappa)$ is the turbulent energy associated with eddies of wave number κ, $T(\kappa)$ is the transfer of the energy between different wave numbers, and κ is the wave number. Neglecting the transfer term in the equation, the following heat conduction type equation is obtained:

$$\frac{\partial E_S(\kappa)}{\partial t} = -2\nu\kappa^2 E_S(\kappa) \tag{14}$$

where the right hand side is the spectrum of the dissipation [7].

In order to solve the above equation, the functional relationship between $E_S(\kappa)$ and κ must be known. Lin [11] has been able to show that

$$E_S(\kappa) = C\kappa^4 \tag{15}$$

for $\kappa \to 0$, with C a constant. It can be assumed that the dissipation of turbulent energy is mostly due to the very large persistent eddies. It seems to be a plausible assumption in the vortex model because the coherent structures dominate. Using this relation and substituting in equation (15), the solution is obtained as [12],

$$E_S(\kappa,t) = E_S(\kappa,t_o)\exp[-2\nu\kappa^2(t - t_o)] \tag{16}$$

Where $E_S(\kappa,t_o)$, and thus $E_S(\kappa,t)$ would be of the form $C\kappa^4$.

The next task is to relate the turbulent energy $E_S(\kappa)$ and the vorticity strength Γ. It can be seen that the turbulent fluctuations in the model used are caused by the insertion of vortices. Consequently, a direct relationship between $E_S(\kappa)$ and κ can be expected. By definition

$$\Gamma = \oint \vec{u} \cdot d\vec{r} \tag{17}$$

where an increase in Γ will produce proportional increment of u, therefore, the assumption $\Gamma \propto u$, and $\Gamma^2 \propto E_S(\kappa)$ can be made. A simplified vortex decaying function will then be written as

$$\Gamma(t) = \Gamma(t_o)\exp[-\nu\kappa^2(t - t_o)] \tag{18}$$

In order to evaluate the wave number, the characteristics length scale must be found. From the required conservation of angular momentum consideration, the vorticity (circulation) is expected to be inversely proportional to its length scale if the viscous effects are absent. Thus the final relationship becomes,

$$\Gamma(t + \Delta t) = \Gamma(t)\exp[-\beta\Gamma^2(t)\Delta t] \tag{19}$$

where β is a properly chosen constant.

3. EXPERIMENT DESCRIPTION

The present experiment is performed in the newly constructed low turbulence subsonic wind tunnel at Louisiana State University. The tunnel is open-return type and driven by a centrifugal fan. Tunnel speeds can be varied by adjusting the angle of attack of fan blades. Speeds range from 35.0 to 190.0 ft/sec (10.7-57.9 m/sec) at the test section can be obtained. The air enters a 5.25 x 7.0 ft (1.6 x 2.13 m) metal inlet chamber, passes through two fine mesh screen and goes through a 12.25:1 contraction. The working section of the tunnel is 18.0 in. (0.46 m) high, 24.0 in. (0.61 m) wide, and 8.0 ft (2.44) long. Extra care taken in constructing the test section, and a low turbulence level (less than 0.3%) is found.

A shallow cavity model was designed and constructed with a fixed depth of 2.0 in. (5.08 cm), a fixed width W of 2.0 in. (5.08 cm), and a variable streamwise length L. The length to depth (L/D) ratio can be varied from L/D = 1 to L/D = 7. The cavity model was set in a 0.0625-in. (0.16 cm)-thick aluminum plate, and was mounted off side wall of the test section in the wind tunnel.

4. COMPARISON OF THEORY AND EXPERIMENT

A simple flow visualization technique is used to identify the bulk flow motions inside the cavity. Wool tufts were attached to the surfaces of the cavity at different positions. Sketches of the typical flow patterns are shown in Figure 3. A sketch of the estimated top view flow pattern is shown in Figure 4. A clear picture of the reverse motion near the bottom can be observed from these figures. In contrast to the two-dimensional assumption made in the paper, three dimensionality can be important.

Figure 5 gives an instantaneous picture of distribution of the point vortices at the final stage. The symbols in the figure are centers of the point vortices. The velocity field at the same time instant is given in Figure 6, where the lengths are proportional to the vector's magnitudes. A single, stable vortex flow pattern is clearly illustrated. Mean velocity and turbulent intensity profiles calculated from the

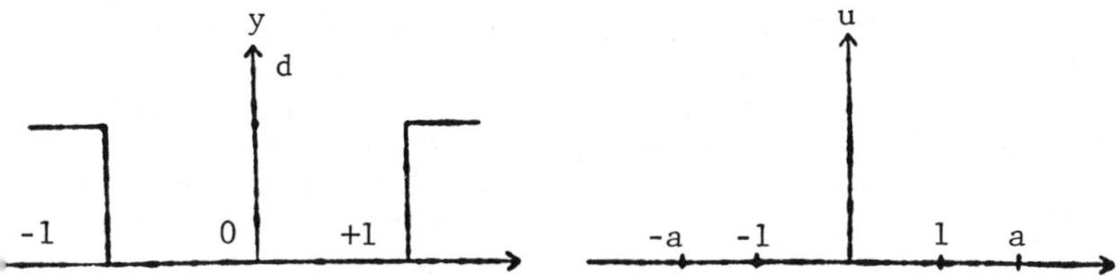

Figure 1. Cavity Geometry in Real & Transformed Plane

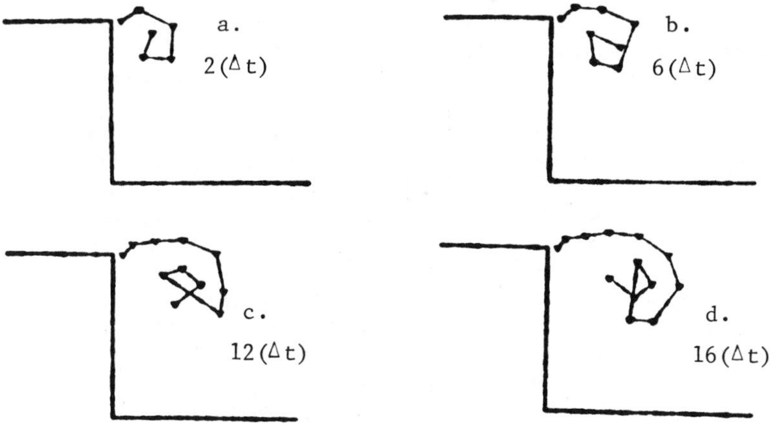

Figure 2. The Rolling-Up Motion of a Vortex Sheet

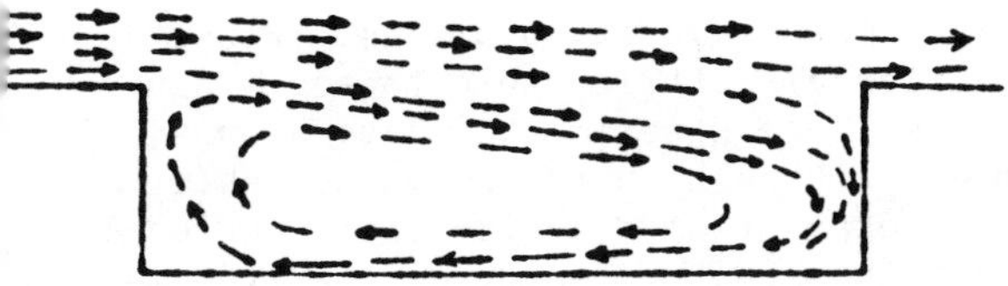

Figure 3. Side View Perspective of Cavity Flow

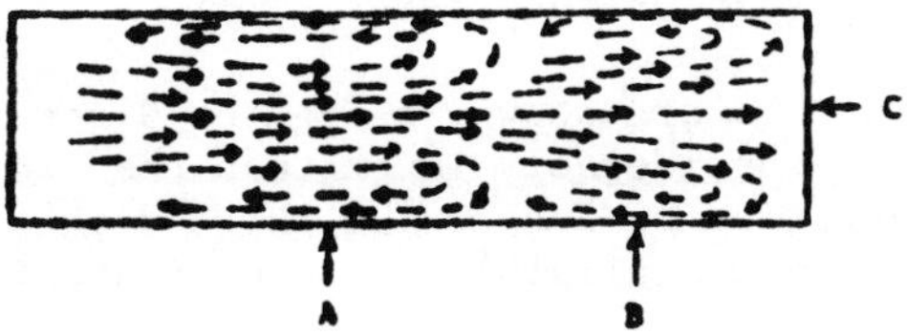

Figure 4. Top View Perspective of Cavity Flow

model are then compared with the experimental results at the corresponding positions. The comparison is presented in Figures 7 through 10. Here, the Kutta condition at the leading edge is used to calculate the initial vortex strengths. Viscosity decaying constant β is chosen as 0.01; it seems to be the most appropriate choice. In these figures, UMAX represents free stream velocity, and Y/D is the non-dimensionalized normal distance measured from the cavity bottom. Figure 7 presents the mean velocity profiles in the streamwise direction. The existence of reserse flow along the bottom is observed readily from both theoretical and experimental data. Maximum deviations are found near the transition region between the inside cavity area and the uniform flow. The abrupt jump in the velocity in the vortex model can be explained by a special feature of the model. A vortex cloud (a cluster of point vortices) behaves just like a large discrete vortex which implies that the velocity induced by vortices is going to be cancelled by others inside the cloud. However, outside the cloud the superposition effect is important. The transition point (where reverse motion starts to occur) is located for Section 1 at about Y/D = 0.8, indicating that the recirculation flow occupies a large portion of the space there. The transition point moves to approximately Y/D = 0.6 at Section 2. This indicates that the shear layer which separated at the leading edge is continually diffusing into the cavity and the reverse flow area is decreasing downstream accordingly. Also shown is that the discontinuity of the profile at section 1 is diminished The cavity shear layer is gradually developing downstream. The same profile at the downstream section (section 3), again showing that the transition point moves further into the cavity (approximately Y/D = 0.5). As shown by Figure 7, the mean reverse flow velocity increases monotonically downstream. This can be explained in two different ways. First, the decrease in size of the recirculation area requires the flow to accelerate in order to satisfy the continuity criterion. Secondly, the friction effects also tend to retard the reverse flow along the bottom. Figure 8 illustrates the cross-stream mean velocity profiles. Peaks appearing among these profiles are attributable to the singular nature of point vortices. The experimental data show a similar pattern as the model although there is considerable scatter. The increase in cross-stream velocity from section 1 to section 3 indicates that the kinetic energy of the flow is transfering from the streamwise direction to the cross-stream direction downstream. This energy transfer process is also supported by the streamwise velocity profiles, which are more uniform downstream and serve as an energy source.

Turbulent intensity profiles are shown in Figures 9 and 10. The predicted values are considerably lower than the measured data. This is thought to be due to the third dimensional fluctuations which are not negligible and contribute to

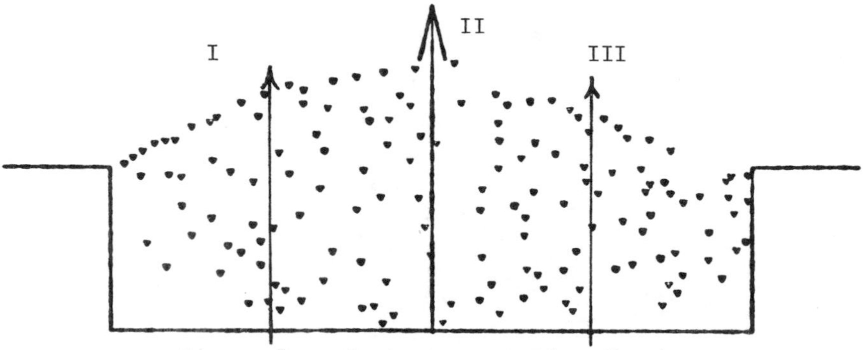

Figure 5. Instantaneous Distribution of Point Vortices

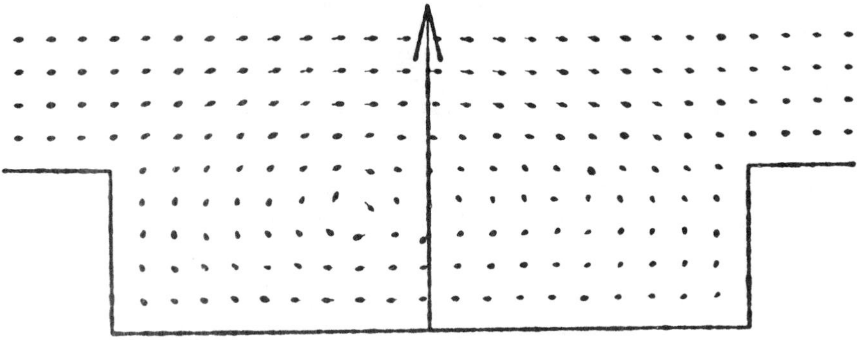

Figure 6. Instantaneous Velocity Field at Same Time

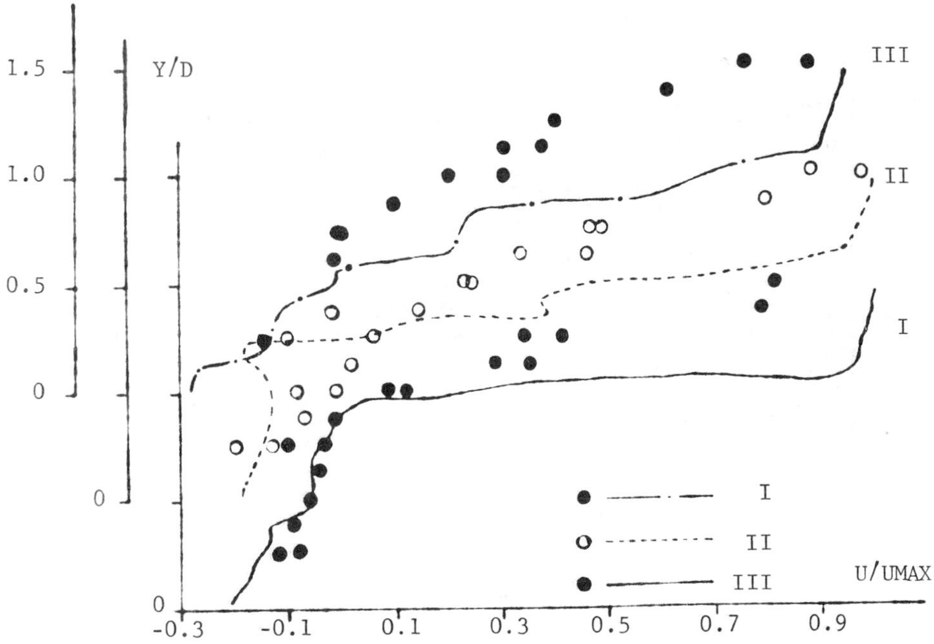

Figure 7. Longitudinal Mean Velocity Profiles at Different Downstream Locations

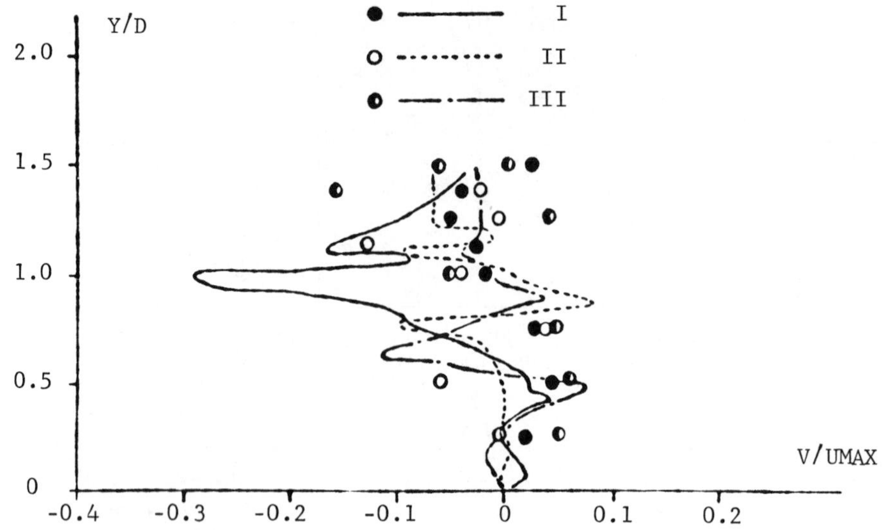

Figure 8. Cross Stream Velocity Field at Different Downstream Locations

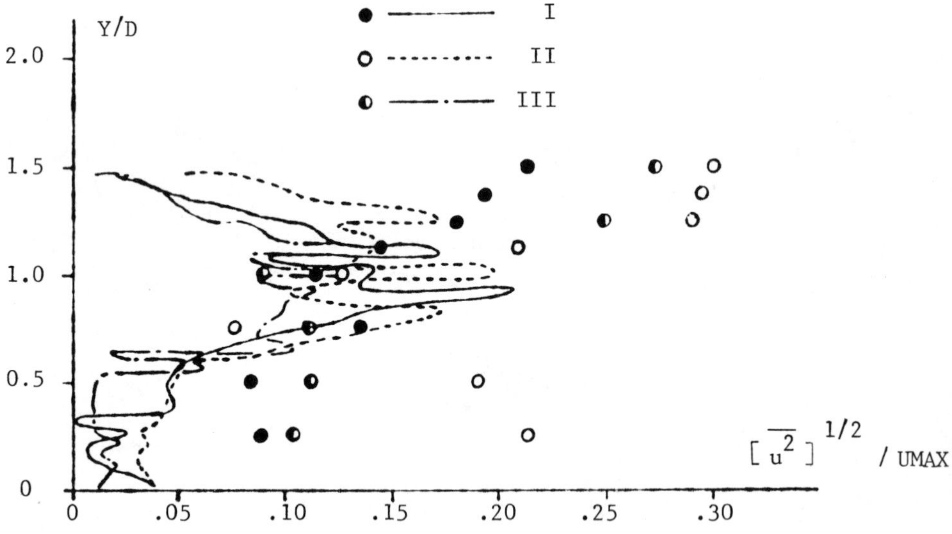

Figure 9. Longitudinal Turbulent Intensities

the measured fluctuations. Another possible origin is the self-sustaining oscillations inside the cavity which the vortex model cannot simulate. The turbulent intensity level should decrease gradually when approaching the free stream flow. However, the maximum turbulent intensity values are found at Y/D = 1.5, which is located above the cavity. The fluctuations convecting downstream and disturbances transmitting back from the downstream edge may be responsible for the high turbulent fluctuations above the cavity. Relatively low turbulent fluctuations are found along the cavity bottom. One reasonable explanation is the existence of two stagnation pressure regions near the two edges serve as a stable mechanism which tends to suppress the turbulent fluctuations inside the cavity. The cavity flow is neither homogeneous nor isotropic as shown by these figures. The highest turbulent RMS values appear near the downstream edge. It is suggested that the shear layer which separates at the leading edge is periodically pulled into and pushed out of the cavity. This periodic change in the flow direction may result in the peak occurring near the downstream edge.

The effect of the superimposed viscous dissipation mechanism is shown in Figure 11, where streamwise velocity profiles of zero viscosity $\beta = 0$, $\beta = 0.01$, and $\beta = 0.02$ are compared. The discontinuities which exist in the profile without viscosity are significantly reduced by the addition of the viscous mechanism.

5. SUMMARY

A discrete vortex model has been developed to predict the behavior of turbulent flow over a rectangular cavity. A simple experimental investigation was also made in support of the theoretical estimation. The following conclusions are obtained:

(1) There is a strong tendency to form a single, stable vortex within a cavity. This is confirmed by both theoretical and experimental results.

(2) The vortex model seems to be able to simulate the large scale structure of the flow, but is not as efficient in some complex areas such a the near wall region.

(3) The third dimension is believed to be important in the cavity flow and may contribute approximately 15% of measured experimental errors.

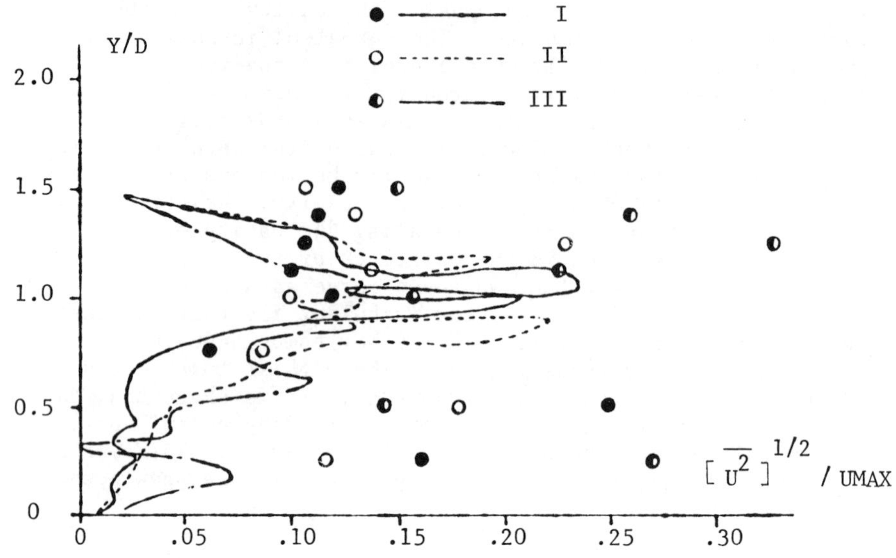

Figure 10. Cross Stream Turbulent Intensities

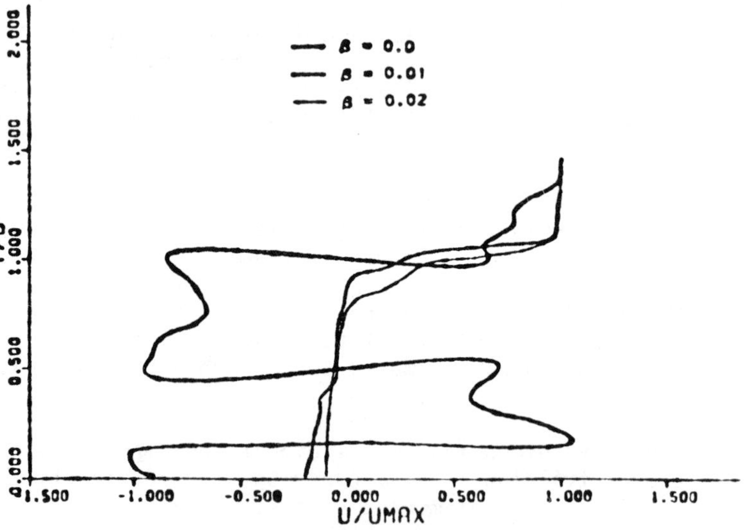

Figure 11. Effect of Varying Vortex Viscosity Parameter on Velocity Field

REFERENCES

1. ROCKWELL, D. and NAUDASCHER, E. - Review of Self Sustaining Oscillations of Flow Past Cavities. J. Fluids Eng., Trans ASME, 100, pp. 152-165, June 1978.

2. ETHEMBABAOGLU, S. - On the Fluctuating Flow Characteristics in the Vicinity of Gate Slots. Division of Hydraulic Engineering, University of Trondheim, Nowergian Institute of Technology, June 1973.

3. CLEMENTS, R. R., and MAULL, D. J. - The Rolling Up of a Trailing Vortex Sheet. Aero. J., 77, pp. 46-51, 1973.

4. MOOR, D. W. - A Numerical Study of the Roll-Up of a Finite Vortex Sheet. J. Fluid Mechanics, 63, pp. 225-235, August 1973.

5. ABRAMOWITZ, M. and STEGUN, I. A. - Handbook of Mathematical Functions, National Bureau of Standards, June 1964.

6. HARDIN, J. C. and MASON, J. P. - Broadband Noise Generation by a Vortex Model of Cavity Flow. AIAA Journal, 15, pp. 632-637, 1976.

7. TENNEKES, H. and LUMLEY, J. L. - A First Course in Turbulence, MIT Press, Cambridge, Mass., pp. 262-264, 1970.

8. CLEMENTS, R. R. - An Inviscid Model of Two Dimensional Vortex Shedding. J. Fluid Mechanics, 57, pp. 321-336, February 1973.

9. SAFFMAN, P. G. and BAKER, G. R. - Vortex Interactions. Annual Review of Fluid Mechanics, 11, pp. 95-122, 1979.

10. CHORW, A. J. and BERNARD, P. S. - Discretization of a Vortex Sheet with an Example of Roll-Up. J. Comput. Physics, 13, pp. 423-429, 1973.

11. LIN, C. C. - Proc. First Symp. on Applied Mathematics, American Mathematics Society, pp. 81, Providence, R.I.

12. BRODKEY, R. S. - The Phenomena of Fluid Motions, pp. 297-298.

CALCULATING THREE-DIMENSIONAL FLUID FLOWS USING NONORTHOGONAL GRIDS

C. R. MALISKA [1] G. D. RAITHBY [2]

ABSTRACT

This paper describes a new solution technique for the prediction of two-dimensional elliptic and three-dimensional parabolic flows. The method solves the conservation equations in a general curvilinear coordinate system maintaining the Cartesian velocity components as dependent variables. The disadvantage in using the Cartesian velocities as dependent variables in nonorthogonal grids is by-passed by forming finite difference equations for the contravariant components in the new system. The PRIME technique [1,2] (update PRessure Implicitly and Momentum Explicitly) is used to handle the pressure velocity coupling problem in conjunction with a new strategy which employs nonorthogonal grids. The method requires slightly higher computer storage but shows virtually the same computational performance when compared with schemes specially designed for Cartesian meshes [3]. The method is tested by solving two-dimensional elliptic and three-dimensional parabolic flows.

1. INTRODUCTION

The prediction of "incompressible" [4] fluid flows in arbitrary regions poses two major problems to the numerical analyst. They are; the strong coupling between pressure and velocity, for flows in which compressibility effects do not dominate, and the domain discretization. The domain discretization has important consequences related to the complexity and generality of the computer code to be designed. If the grid is obtained using an orthogonal system the advantage is that the conservation equations are written in a simple form. This apparent advantage, however, is lost when one faces the complexity in dealing with irregular elements at the boundaries. To overcome this difficulty it seems to require the governing equations be written in a coordinate system which matches the boundary of the domain, defined here a geometrically natural, or boundary fitted coordinate system [5]. These systems may be either orthogonal or nonorthogonal, thus necessitating a decision to be made as which type to use. Orthogonal discretization is numerically attractive since the solution techniques already developed for Cartesian grids can be applied to any orthogonal system with little extra effort.

1 Mechanical Engineering Department, Federal University of Santa Catarina - Brazil
2 Mechanical Engineering Department, University of Waterloo - Canada

However, the generation of orthogonal meshes is not a simple matter due to the necessity in satisfying the orthogonality constraint. In addition, it may be impossible to find an orthogonal mesh for a complex region if some concentration of the coordinate lines is required. In the other hand, the use of nonorthogonal systems has the disadvantage that the equations of motion are somewhat more complex because the presence of the cross derivative terms. These terms will lead to 9-point finite difference equations (for 2-D problems), opposed to a 5-point equations for orthogonal systems. Also, special care is required in choosing the storage location for the dependent variables in the computational grid. A non suitable choice may cause the solution of the 9-point equation for pressure to require excessive computer effort, or it may not converge at all [6].

In spite of that, if fast automatic methods for generating nonorthogonal grids are available, in conjunction with well designed and simple finite difference models, the use of nonorthogonal meshes may be about an optimum alternative for handling fluid flow problems in arbitrary complex geometries. A method which attempts to fullfil the above requirements is now addressed.

2. COORDINATE SYSTEM GENERATION

Consider a three-dimensional irregular domain defined in the Cartesian system, Figure 1a. One wants to map this domain onto a parallelepiped in the (ξ,η,Γ) transformed system, Figure 1b. To realize this objective the following transformation is used

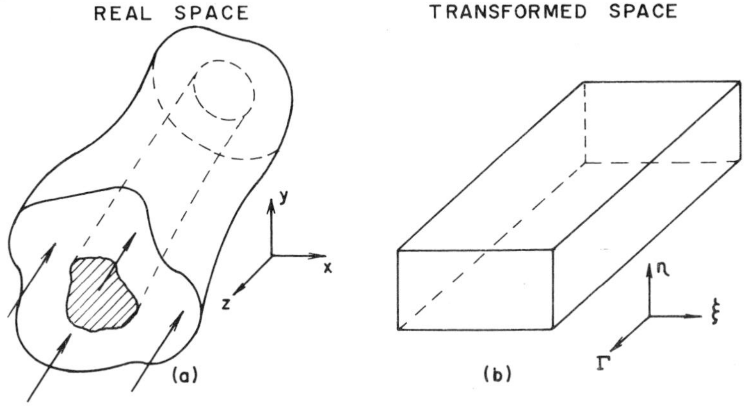

Figure 1. Physical and transformed domains

$$\xi = \xi(x,y,z) \quad ; \quad \eta = \eta(x,y,z) \quad ; \quad \Gamma = z \qquad (1)$$

Following Thompson et al [5], the nonorthogonal coordinate system was generated solving the following linear system of equations

$$\xi_{xx} + \xi_{yy} = P(\xi,\eta) \tag{2}$$

$$\eta_{xx} + \eta_{yy} = Q(\xi,\eta) \tag{3}$$

The above system was transformed and numerically solved in the new coordinate system. For details see [5,7].

3. TRANSFORMATION OF THE CONSERVATION EQUATIONS

At this point a decision which has important influences on the overall performance of the model must be made. The question is wheter to have the Cartesian or the contravariant velocity components as dependent variables in the new system. The natural choice would be to use the contravariant components since they are normal to the control volume surfaces and so they carry the mass flow across the element boundaries. This route gives rise to a complex set of transformed equations difficulting the finite differencing process.

If the Cartesian velocities are employed all velocity components need to be calculated at the same point on the grid, so that mass flow across element surfaces can be calculated. The great advantage, however, is that the transformed equations are very simple, making the finite differencing process and easy task. The latter approach is adopted here.

3.1. Transformed Set of Equations

A conservative form of the conservation equations can be written as [8]

$$\frac{\partial}{\partial t}(\rho\phi) + \frac{\partial}{\partial x}(\rho u\phi) + \frac{\partial}{\partial y}(\rho v\phi) + \frac{\partial}{\partial z}(\rho w\phi) + P^{\phi} = \frac{\partial}{\partial x}(\Gamma^{\phi}\frac{\partial\phi}{\partial x}) + \frac{\partial}{\partial y}(\Gamma^{\phi}\frac{\partial\phi}{\partial y}) + \frac{\partial}{\partial z}(\Gamma^{\phi}\frac{\partial\phi}{\partial z}) + S^{\phi} \tag{4}$$

where ϕ is any scalar field and $P\phi$ and $S\phi$ are the pressure and source terms when appropriate. For many duct flow problems encountered in engineering practice the downstream flow conditions have little effect on the upstream flow parameters. The parabolic approximation appears then as a good alternative. In this case, the pressure is split into two terms as follow

$$P(x,y,z) = \tilde{P}(x,y;z) + \bar{P}(z) \tag{5}$$

Equation (6) is transformed to the new system following the procedure described in [8]. The resulting equation is in the following conservative form

$$\frac{\partial}{\partial t}(\rho\phi) + \frac{\partial}{\partial \xi}(\rho U\phi) + \frac{\partial}{\partial \eta}(\rho V\phi) + \frac{\partial}{\partial \Gamma}(\rho W\phi) + \hat{P}^{\phi} = \frac{\partial}{\partial \xi}(C_1\frac{\partial\phi}{\partial\xi} + C_2\frac{\partial\phi}{\partial\eta}) + \frac{\partial}{\partial \eta}(C_3\frac{\partial\phi}{\partial\eta} + C_4\frac{\partial\phi}{\partial\xi}) + \hat{S}^{\phi} \tag{6}$$

where U, V and W are the contravariant velocity components written without metric normalization. They are related to the Cartesian velocities by

$$U = y_\eta u - x_\eta v + (y_\Gamma x_\eta - x_\Gamma y_\eta)w \tag{7}$$

$$V = x_\xi v - y_\xi u + (x_\Gamma y_\xi - y_\Gamma x_\xi)w \tag{8}$$

$$W = \frac{w}{J} \tag{9}$$

where $J = (x_\xi y_\eta - x_\eta y_\xi)^{-1}$ is the Jacobian of the transformation. The $\hat{P}^\phi$ and $\hat{S}^\phi$ terms can be found in [7,9]. The coefficients in Equation (6) are given by

$$C_1 = \Gamma^\phi J\alpha \quad ; \quad C_2 = C_4 = -\Gamma^\phi J\beta \quad ; \quad C_3 = \Gamma^\phi J\gamma \qquad (10,11,12)$$

4. FINITE DIFFERENCE PROCEDURE

4.1. Storage Location of the Variables on the Grid

The storage layout where all variables (velocity, pressure, etc.) are stored at the same point, trend followed by most of the recent works dealing with nonorthogonal grids is not adopted in this study. The reason is the danger this approach can pose in having unrealistic pressure and velocity fields [10]. The recommended remedy is the use of a staggered layout [10]. However, the proper staggered layout is not easily recognized when the Cartesian velocities are used as dependent variables in nonorthogonal grids. Then, a storage layout which promotes good convergence characteristics for the Poisson-like equation for pressure should be the main target in design models to predict incompressible fluid flows. Another important feature to be pursued is to have the model reverting to a 5-point scheme when the grid employed is orthogonal. The latter characteristic is numerically attractive since many non-orthogonal grids are, in many situations, quasi-orthogonal or

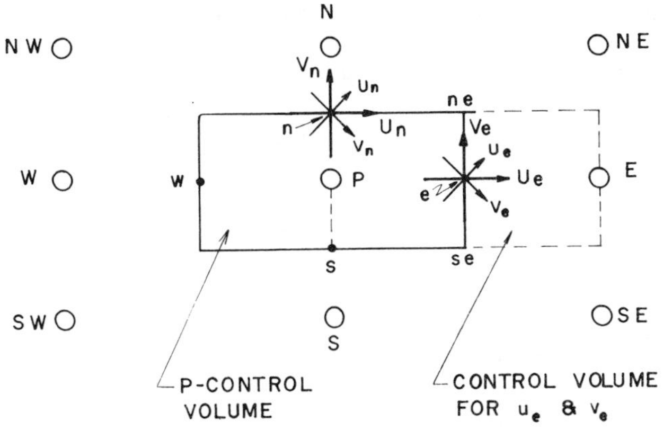

Figure 2. Grid layout

nonorthogonal only in certain regions. The desired flexibility and generality is thus achieved since the scheme is not restricted to the requirements of an orthogonal grid but shares many of its advantages. Figure 2 shows the storage layout adopted in this study. A detailed discussion on this subject can be found in [7,9].

4.2. Discrete Equations

To obtain the discrete equations, Equation (4) is integrated over the elemental control volume shown in Figure 3b. Approximations are introduced [4,11] to reduce the integral equation to the algebraic equation

$$A_p\Phi_P^{n+1} = A_e\Phi_E^{n+1} + A_w\Phi_W^{n+1} + A_n\Phi_N^{n+1} + A_s\Phi_S^{n+1} + A_u\Phi_{P,U}^{n+1} + \frac{A_p}{1+E}\Phi_P^n - L\,[\hat{P}^\phi]\Delta V + L\,[\hat{ST}^\phi]\Delta V \quad (13)$$

where

$$A_e = -(\rho U)_e\,\Delta\eta\Delta\Gamma\,(\tfrac{1}{2} - \bar{\alpha}_e) + \bar{\beta}_e C_{1e}\,\Delta\eta\,\frac{\Delta\Gamma}{\Delta\xi} \quad (14a)$$

$$A_w = +(\rho U)_w\,\Delta\eta\Delta\Gamma\,(\tfrac{1}{2} + \bar{\alpha}_w) + \bar{\beta}_w C_{1w}\,\Delta\eta\,\frac{\Delta\Gamma}{\Delta\xi} \quad (14b)$$

$$A_p^* = A_e + A_w + A_n + A_s + A_u \quad (14c)$$

and

$$A_p = A_p^*\,\frac{(1+E)}{E} \quad (14d)$$

$$\hat{ST}^\phi = \hat{S}^\phi + \frac{\partial}{\partial\xi}(C_2\,\frac{\partial\phi}{\partial\eta}) + \frac{\partial}{\partial\eta}(C_4\,\frac{\partial\phi}{\partial\xi}) \quad (14e)$$

In the above equations L[ ] denotes the finite difference approximation of the quantity in brackets and E is a constant that can not exceed unity for explicit formulation.

The discrete equations for the contravariant velocity components are obtained using Equation (13) with ϕ equal to u and v and Equations (7) and (8). For the control volume depicted in Figure 2 the U_e and V_n equations are

$$U_e = \hat{U}_e - (\frac{\Delta\Gamma}{A_p^u}\alpha)_e(P_E - P_P) + (\frac{\Delta\Gamma}{4A_p^u}\beta)_e(P_{NE} + P_N - P_{SE} - P_S) \quad (15)$$

$$V_n = \hat{V}_n - (\frac{\Delta\Gamma}{A_p^v}\gamma)_n(P_N - P_P) + (\frac{\Delta\Gamma}{4A_p^v}\beta)_n(P_{NE} + P_E - P_{NW} - P_W) \quad (16)$$

Similar equations can be written for U_w and V_s, completing then the four velocity components needed for the mass conservation balance applied to the elemental volume shown in Figure 3b. The equation is

$$\left((\rho U)_e - (\rho U)_w\right)\Delta\eta\Delta\Gamma + \left((\rho V)_n - (\rho V)_s\right)\Delta\xi\Delta\Gamma + \left((\rho W)_p - (\rho W)_U\right)\Delta\eta\Delta\xi \quad (17)$$

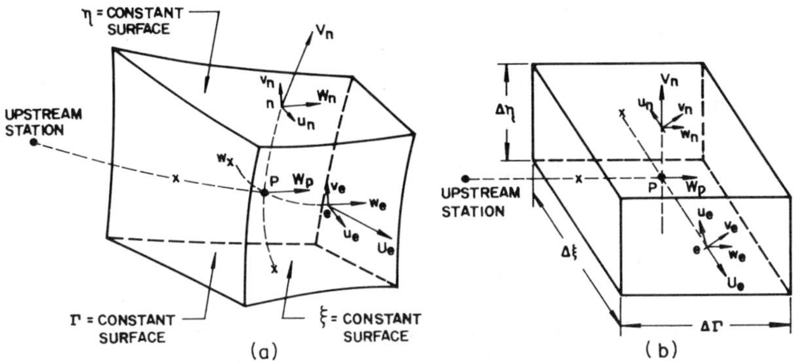

Figure 3. Physical and transformed elemental volumes

4.3. Solution Procedure

In the present work only 3-D flows with a predominant flow direction are considered. Computer storage for the variables are needed only in the calculation and upstream plane. The pressure velocity coupling in the axial direction is solved using the method developed by Raithby and Schneider [4] which removes the necessity of iteration by taking advantage of the linear relationship between the axial velocity and the pressure gradient. The elliptic coupling problem is solved using the PRIME method [1,2]. This method simplifies considerably the iteration cycle since it eliminates the standard two-step solution procedure in which an estimate of the velocity field is obtained by solving the momentum equation with a guessed pressure field, followed by the solution of two Poisson equations for velocity correction and pressure updating [4,10,9].

In the PRIME procedure a Poisson equation is formed by inserting Equations (15) and (16) into continuity equation. The solution of the resulting equation will give a pressure field which is used to correct the hat-velocity vector and at the same time is recognized as the update pressure field. In this technique the momentum equations are solved in a Jacobi iteration fashion. The great attractiveness of this approach is the simplicity it introduces in the computer code. The determination of the hat-velocities uses the best estimates of the velocity field.

The following steps constitutes one iteration cycle performed in the equation set when solving 3-D parabolic fluid flow problems.

(a) Guess the cross pressure field in the inlet section and the pressure gradient $d\bar{P}/dz$. Specify a 3-D velocity profile at the duct inlet.
(b) Compare the coefficients for the axial momentum equation. Solve for w and compute the mass flow. Applying the method described in [4] determine w and $d\bar{P}/dz$. Compute W.

(c) Compute the coefficients for the equations for U and V and compute the hat-velocities. See [7] for details.
(d) Form and solve the Poisson-like equation for pressure. The equation is

$$A_p P_P = A_e P_E + A_n P_N + A_s P_S + A_w P_W + A_{ne} P_{NE} + A_{se} P_{SE} + A_{sw} P_{SW} + A_{nw} P_{NW} + B \quad (18)$$

and

$$A_p = A_e + A_w + A_n + A_s \quad (19)$$

It is seen that the pressure at a point P is strongly linked with the four parallel pressure points and not with the diagonal ones. This is the type of structure exhibited by the Poisson equation derived for orthogonal grids using the standard staggered layout. This similarity is probably the reason why this equation shows similar convergence rate as the equations for orthogonal meshes. In addition, all the diagonal coefficients vanish when the grid employed is orthogonal. The equation was successfully solved using the S.O.R. point iteration method.
(e) Correct the contravariant velocities using Equations (15) and (16). These velocities now satisfy mass conservation.
(f) Determine the contravariant velocities that do not enter the mass conservation balance using an interpolation process.
(g) Determine the Cartesian velocities using Equations (7), (8) and (9).

Iteration, by cycling back to step (b) is required to account for nonlinearities and inter-equation coupling.

5. APPLICATIONS

The assessment of a 3-D parabolic model must begin by testing its secondary flow model. The reason is because the main flow is very little affected by changes in the cross flow and, consequently, the checking of the flow parameters in the predominant direction only does not validate the full model. Furthermore, the experience gained in dealing with the 2-D elliptic flows in nonorthogonal grids is, in principle, extendable to 3-D elliptic problems. These facts require that the model be tested by first solving 2-D elliptic flows such that the cross flow velocities can be compared quantitatively.

The driven flow in a square cavity with a moving lid is a good test problem since flow conditions ranging from dominant diffusion to dominant convection can be analysed. Figure 4a shows the geometric parameters and the boundary conditions for the problem. A nonorthogonal grid with 28×28 grid points shown in Figure 4b was used. The problem was also solved in a 28×28 Cartesian grid and the solutions compared. Figure 5 shows the u-velocity along the lines A-A and B-B in the nonorthogonal grid for $R_e = 400$. In the Cartesian grid the lines A-A and B-B

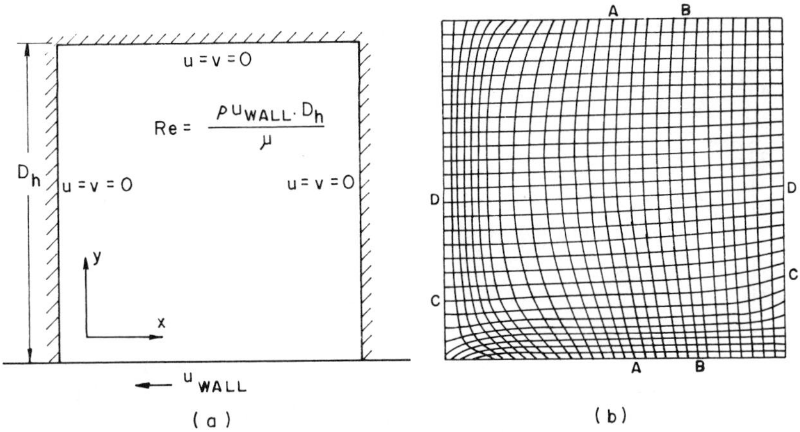

Figure 4. Square cavity problem

are constant x lines with the values 14Δx and 20Δx. The results obtained using orthogonal and nonorthogonal grids are in very good agreement. A slightly disagreement is observed, for a small region close to the maximum reversal velocity, for the results obtained in this study compared to those of Burggraf [12]. This disagreement is noted for both coordinate systems used, what suggests that the problem is not due to the presence of nonorthogonalities. The most probable cause is that the grid was not refined to the point where numerical diffusion no longer influences the solution. Figure 6 shows the v-velocity along the C-C and D-D lines. In the Cartesian grid these are lines of y = 6Δy and y = 14Δy. Again the results using both grids compares very well. Figure 7 brings to the reader the comparisons of the convergence rate obtained using both coordinate systems. It can be seen that the number of iterations needed to obtain a solution of the equation set, to the same level of accuracy, is practically the same for both grids. This finding encourages further developments in this area. As a second test problem the parabolic flow in a converging-diverging duct with changing cross section was solved. The duct geometry

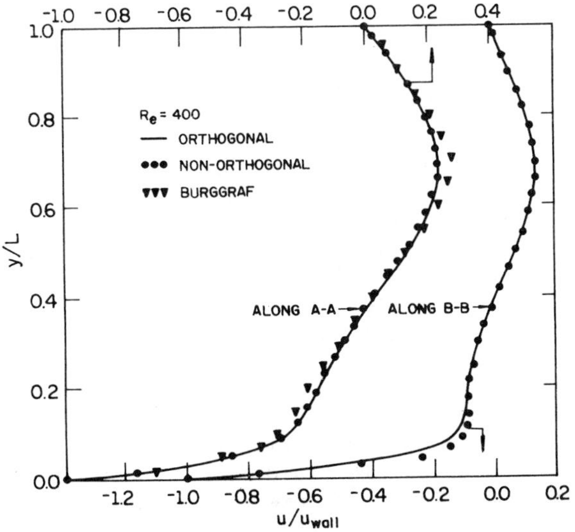

Figure 5. Square cavity, u - velocity

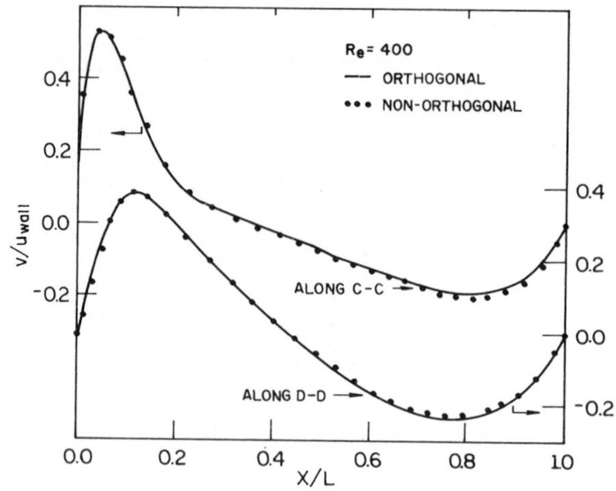

Figure 6. Square cavity, v - velocity

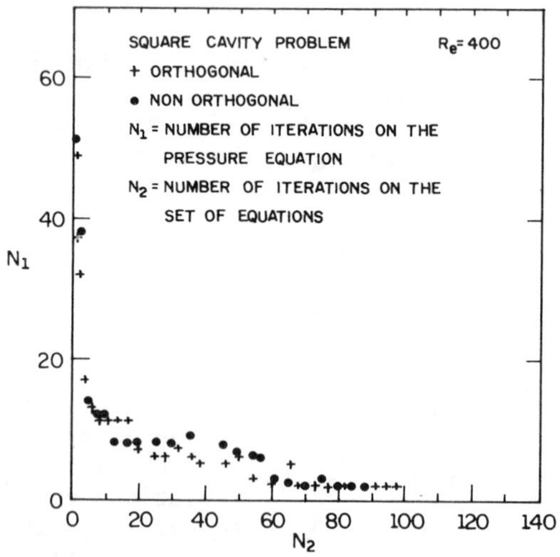

Figure 7. Convergence rate comparison

is shown in Figure 8 and the geometric characteristics of the intermediate calculation planes can be found in [7]. The objective in solving this problem was two-fold. Firstly, due to the duct convergence the flow exhibits strong cross velocities, situation not encountered in the entrance region of constant cross section duct flows. Secondly, the existence of a different nonorthogonal coordinate system for each calculation plane constitutes a good test for the generality of the technique. The axial step was 1/6th of the inlet hydraulic diameter and nine solution planes were used. The velocity profile of the fully developed flow in a rectangular duct was prescribed at the inlet. Figure 9 shows the axial velocity profiles along the symmetry plane x = 0. Comparisons are not made but it can be seen that the axial velocity shows the expected trends. The cross flow velocities, not reported here, also shows the expected trends.

6. CONCLUSIONS

A solution technique for the prediction of three-dimensional parabolic incompressible fluid flow in arbitrary geometry has been presented. The procedure employs non-orthogonal grids and transforms the conservation equations to the new curvilinear system maintaining the Cartesian velocity components as dependent variables.

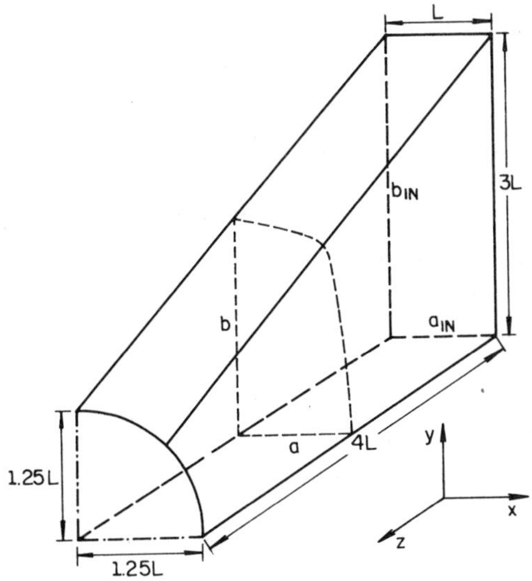

Figure 8. Converging-diverging duct

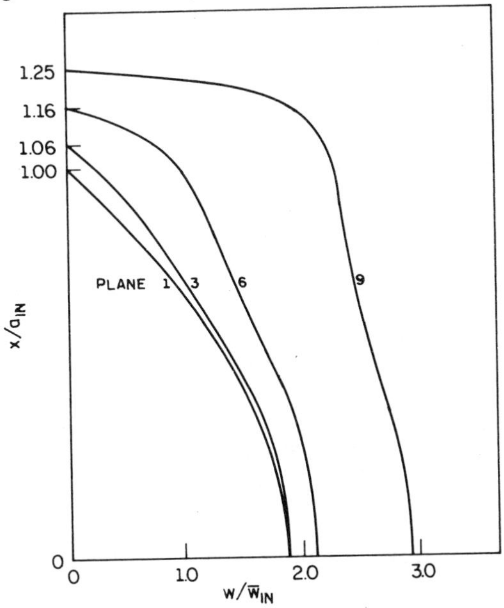

Figure 9. Axial velocity profiles

In the new system finite difference equations for the momentum equations are written in terms of the contravariant velocity components. These equations, togheter with the mass conservation equation, are solved according to the PRIME procedure. The incorporation of the PRIME procedure into a method which employs non-orthogonal grids was of great benefit since it introduced the desirable simplicity to the solution procedure as a whole.

By the tests performed if is seen that the numerical technique presented here is very promising. Currently, tests are being performed where inflow-outflow and thermal problems are being solved.

ACKNOWLEDGEMENTS

The first author wish to thank the National Council of Research and Development of Brazil (CNPq) for supporting in part this work.

7. REFERENCES

1. HACKMAN, L. - Personal Communication. 1981.

2. VAN DOORMAAL, J.P. and RAITHBY, G.D. - The Application of the Quasi-Continuum Approach to Three-Dimensional Elliptic Flows in Complex Geometries. Waterloo Research Institute, Report to Atomic Energy of Canada Ltd., September 1981.

3. NOGUEIRA, A.C.R. - Personal Communication. 1983.

4. RAITHBY, G.D. and SCHNEIDER, G.E. - Numerical Solution of Problems in Incompressible Fluid Flow: Treatment of the Velocity-Pressure Coupling, Numerical Heat Transfer, vol. 2, pp. 417-440, 1979.

5. THOMPSON, J.F., THAMES, F.C. and MASTIN, C.W. - Automatic Numerical Generation of Body-Fitted Curvilinear Coordinate System for Field Containing Any Number of Arbitrary Two-Dimensional Bodies. J. Comp. Phys., vol. 15, pp. 299-319, 1974.

6. VANKA, S.P., CHEN, C.J. and SHA, W.T. - A Semi-Implicit Calculation Procedure for Flows Described in Boundary Fitted Coordinate Systems, Numerical Heat Transfer, vol. 3, pp. 1-19, 1980.

7. MALISKA, C.R. - A Solution Method for Three-Dimensional Parabolic Fluid Flow Problems in Nonorthogonal Coordinates, Ph.D. Thesis, University of Waterloo, Canada, 1981.

8. PEYRET, R. and VIVAND, H. - Computation of Viscous Compressible Flow Bases on the Navier-Stokes Equations, North Atlantic Treaty Organization, AGARD 212, 1975.

9. MALISKA, C.R., RAITHBY, G.D. - A Method for Computing Three-Dimensional Flows using Nonorthogonal Boundary Fitted Coordinates, to be published.

10. PATANKAR, S.V. - Numerical Heat Transfer and Fluid Flow, Hemisphere Publishing Corporation, 1980.

11. RAITHBY, G.D. and TORRANCE, K.E. - Upstream-Weighted Differencing Scheme and Their Application to Elliptic Problems Involving Fluid Flow, Computer and Fluids, vol. 2, pp. 191-206, 1974.

12. BURGGRAF, O.R. - Analytical and Numerical Studies of the Structure of Steady Separated Flows, J. Fluid Mechanics, vol. 24, pp. 113-151, 1966.

SECTION 9

NON–NEWTONIAN FLOW

THE NUMERICAL SIMULATION OF PLASTIC FLUIDS

David K. Gartling
Sandia National Laboratories
Albuquerque, New Mexico 87185

1. SUMMARY

A finite element numerical procedure is outlined for the solution of flow problems involving fluids with a yield stress. Such materials, termed plastic or viscoplastic fluids, complicate the usual numerical procedures through the introduction of a material nonlinearity. In the present work a specific form for the viscoplastic constitutive relation is described that is amenable to numerical solution. A hypermatrix technique is outlined for incorporating the deformation-dependent fluid viscosity into the finite element formulation. Two example problems are presented that illustrate the numerical method and the behavior of viscoplastic fluids in complex flows.

2. INTRODUCTION

A number of industrial and scientifically important materials can be characterized as plastic fluids, i.e. fluids that exhibit a yield stress. A partial list of such materials would include lubricating greases, slurries, blood and molten basalt (magma). The numerical simulation of such viscoplastic materials presents a challenge in that such flows often contain a type of free surface (yield surface) that must be resolved within the computational domain. Furthermore, since the fluid viscosity depends on the fluid deformation the computational problem is nonlinear even for the creeping flow or zero Reynolds number limit.

In the present paper we describe a finite element based numerical method for the solution of incompressible flows involving viscoplastic materials. The method considered is sufficiently general to treat materials of the Bingham type as well as Herschel-Buckley fluids (shear thinning fluids with a yield stress) and fluids with a temperature dependent yield stress. The remainder of this paper considers three main topics. In the next section we outline the mathematical form of the problem of interest and discuss the various fluid models to be considered. The following section briefly describes several aspects of the numerical method. Finally, in the last section a selection of example problems are presented to verify and demonstrate the computational procedure.

3. PROBLEM FORMULATION

The problems of interest concern the two-dimensional flow of an incompressible fluid. The field equations describing such a fluid motion consist of the equation of motion

$$\rho \frac{Du_i}{Dt} = \rho \frac{\partial u_i}{\partial t} + \rho u_j \frac{\partial u_i}{\partial x_j} = \frac{\partial \sigma_{ij}}{\partial x_j} + \rho g_i \tag{1}$$

and the incompressiblity constraint

$$\frac{\partial u_i}{\partial x_i} = 0 \tag{2}$$

where the i,j subscripts range from 1 to 2 with the usual summation convention. The variables in (1) and (2) are as defined in the nomenclature section of this volume.

We can specialize the above equations to a particular type of fluid through the specification of a constitutive equation that relates the stress to the fluid deformation. The viscoplastic fluids of interest here are a subclass of the generalized Newtonian fluid [1], which has a constitutive equation of the following form

$$\sigma_{ij} = -P\delta_{ij} + \tau_{ij} \tag{3}$$

where P is the pressure and τ_{ij} is the extra-stress which generally takes the form

$$\tau_{ij} = 2\mu(I_2)D_{ij} \tag{4}$$

with

$$D_{ij} = 1/2 \left(\frac{\partial u_i}{\partial x_j} + \frac{\partial u_j}{\partial x_i} \right)$$

and

$$I_2 = 1/2\ D_{i\kappa}D_{\kappa i} \quad .$$

In equation (4) the viscosity function is assumed to depend only on the second invariant (I_2) of the rate-of-deformation tensor. The type of fluid models that fall within the above description include the familiar power law fluid, the Ellis model and the Carreau model [1].

In the present work we will confine our attention to those forms of (4) that represent viscoplastic behavior. The most familiar of the viscoplastic constitutive relations is the Bingham model [2] which can be written in general form as

$$\tau_{ij} = \left(\frac{\kappa}{\sqrt{I_2}} + \eta\right) D_{ij} \quad \text{when } \frac{1}{2}\, \tau_{i\kappa}\tau_{\kappa i} \geqslant \kappa^2 \tag{6}$$

and

$$D_{ij} = 0 \quad \text{when } \frac{1}{2}\, \tau_{i\kappa}\tau_{\kappa i} < \kappa^2$$

In Equation (6) κ is the yield stress for the fluid and η is the slope of the stress, rate-of-strain curve. The inequalities in (6) describe the von Mises yield criteria for the fluid. When the extra-stress exceeds the yield stress the fluid flows with an apparent viscosity of $\{\kappa/\sqrt{I_2}+\eta\}$; for stress levels less than yield the fluid is not deformed. Note that Equation (6) can be generalized such that $\eta = \eta(I_2)$ in which case a Herschel-Buckley fluid is produced. Also, the material properties κ and η may be temperature dependent in nonisothermal flows.

For purposes of computation the constitutive equation in (6) is quite difficult to use, mainly as a result of the requirement that no deformation take place below yield. To circumvent this difficulty we have employed a modified form of Equation (6) in which

$$\tau_{ij} = \left(\frac{\kappa(1 - \eta/\eta_r)}{\sqrt{I_2}} + \eta\right) D_{ij} \quad \text{when } \frac{1}{2}\, \tau_{i\kappa}\tau_{\kappa i} \geqslant \kappa^2 \tag{7}$$

and

$$\tau_{ij} = \eta_r D_{ij} \quad \text{when } \frac{1}{2}\, \tau_{i\kappa}\tau_{\kappa i} < \kappa^2$$

where η_r is a pre-yield viscosity and $\eta/\eta_r \ll 1$. Unlike the Bingham material the fluid described by (7) can undergo deformation below the yield point though the magnitude of the deformation can be made arbitrarily small (i.e., approach the Bingham model) by increasing η_r relative to η. Also, like the Bingham model the constitutive relation in

(7) can be generalized to allow deformation and/or temperature dependent material properties.

When combined with the field equations in (1) and (2), the constitutive equations in (3) and (7) provide a description for the motion of a viscoplastic material. A set of boundary and initial conditions specialize this description to a particular flow problem. In the next section we outline a numerical method that was devised to solve the previously described boundary value problem.

4. NUMERICAL METHOD

The numerical method we have considered for the flow of viscoplastic fluids is a Galerkin based finite element procedure. As the basics of this method are common to previous work on Navier-Stokes problems [3,4], we need not be concerned here with fundamentals of the procedure. Rather, we will only outline those parts of the method that are unique to the viscoplastic constitutive model. A certain familiarity with finite element methodology will be assumed.

The concern with the use of (7) in a finite element numerical method is the appearance of a deformation dependent viscosity function. This feature impacts the numerical algorithm in two areas - the construction of the element coefficient matrices and the solution procedure. With regard to the construction of the finite element matrices the viscoplastic model requires that values of the apparent (velocity dependent) viscosity be available at various integration points within each finite element. Thus, element matrices corresponding to the viscous diffusion terms in Equation (1) must normally be reconstructed at each step of the solution process. This computationally expensive procedure can be avoided by interpolating the viscosity within an element and employing a hypermatrix representation for the variable coefficient terms. Such a technique has been used successfully for situations involving temperature dependent viscosities [5]. We have tested the hypermatrix approach in the present work and are satisfied with its ability to accurately incorporate viscosity functions that depend on derivatives of the velocity field. We have found [6] that care must be excercised in choosing points within an element at which the apparent viscosity (i.e. I_2) is to be evaluated and subsequently used in the hypermatrix. In general, attempts to evaluate the viscosity at the normally favored 2x2 Gauss integration points within the element were not successful. However, the more rudimentary procedure of using a single viscosity evaluation at the element centroid was found to be acceptable and was used in all the work reported here.

The dependence of the apparent viscosity on the velocity gradients shows the viscoplastic flow problem to be inherently nonlinear, even for cases in which inertia effects are negligible (i.e. Stokes or creeping flow). The solution of steady-state finite element problems thus requires some form of iterative solution method; transient problems also require consideration of the material nonlinearity though we do not plan to discuss this form of the problem here. The simplest steady-state algorithms are those related to Picard's method [7] (also known as successive substitution or functional iteration) in which the nonlinear coefficients in the problem are evaluated from data obtained from the previous iteration cycle. These methods have but a linear rate of convergence but work reasonably well for a wide range of problems. More complex and rapidly convergent methods, such as Newton's method, can also be applied to the problem. However, experience has shown [6,8] that in general these algorithms do not perform well for many types of generalized Newtonian fluid models. In particular, Newton's method does not work well for models with extreme shear thinning, a quality that is strongly present in the apparent viscosity of the viscoplastic model. In the present work we have used Picard's method for all material nonlinearities; Newton's method was used for the nonlinear advective terms in Equation (1). Full details on the implementation and performance of these solution algorithms are available in Reference [6].

The viscoplastic model given by Equation (7) was incorporated into an existing two-dimensional, incompressible flow code [4]. The only program particulars of interest here are the finite elements used for computation. These consisted of isoparametric, six node triangles and eight node quadrilaterals in which the velocity components were interpolated using quadratic polynomials and the pressure was approximated using linear functions. The program also allows solution of non-isothermal flow problems by appending an energy equation to the equations of motion.

5. NUMERICAL EXAMPLES

In this section we present solutions to two problems involving the flow of a viscoplastic material. These problems are intended to illustrate the behavior of fluids with a yield stress and as such will be described primarily in qualitative terms. For a more quantitative verification of the numerical procedure References [6] and [9] should be consulted.

5.1 Driven Cavity Flow

The first example is the familiar isothermal driven cavity flow. The problem consists of a unit square filled with a viscoplastic fluid. Three sides of the cavity are held fixed while the fourth side moves with a constant unit velocity in its own plane. For purposes of this example the material properties ρ, η and η_r are given the values of 1.0, 1.0 and 1000.0, respectively; the choice of η_r is based on previous experience [9] and is chosen such that the numerical solution is insensitive to its magnitude. The large value of η_r is also selected so that the constitutive relation in Equation (7) closely approximates a Bingham material. All of the results to be presented here are for a nominal Reynolds number (based on η) of 100. The finite element solutions were generated on a graded mesh of 16x16 quadrilateral elements.

Shown in Figures 1-4 are contour plots of the stream function corresponding to different values of the yield stress. The contour values for all of the plots are the same. Even for relatively small values of the yield stress it is apparent from the figures that the extent and magnitude of the fluid motion are reduced compared to the purely viscous case (Figure 1). An increasing yield stress also produces a flow field that is noticeably more symmetric with respect to the cavity centerline. Though the location of the apparent yield surface (line dividing the flow/no flow regions) is not shown on the contour plots an idea of its vertical location can be obtained from the mid-plane velocity profiles shown in Figure 5. As expected an increase in the yield stress causes an upward movement of the yield surface and an increase in the size of the no flow region.

The curves in Figure 5 are of further interest in that they show a "break" in the velocity profile, a feature that is not present in the Newtonian case. The explanation for this behavior lies in the relative magnitudes of the fluid deformation rate above and below the center of the vortex. Due to the imposed boundary condition the thin fluid layer above the vortex center is highly sheared which produces a relatively small apparent viscosity. The re-circulating fluid below the vortex center undergoes a smaller rate of deformation and therefore has a larger effective viscosity. This change in viscosity occurs across the vortex center and accounts for the different slopes in the velocity profile. Note that in Figure 5 the computed "breakpoint" in the velocity profile does not occur at zero velocity (vortex center). This small inaccuracy in location is due to the use of a constant viscosity within each element and a locally inadequate mesh refinement.

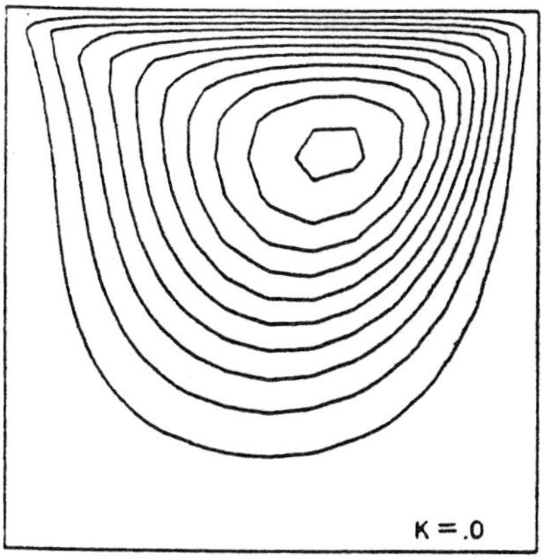

Figure 1

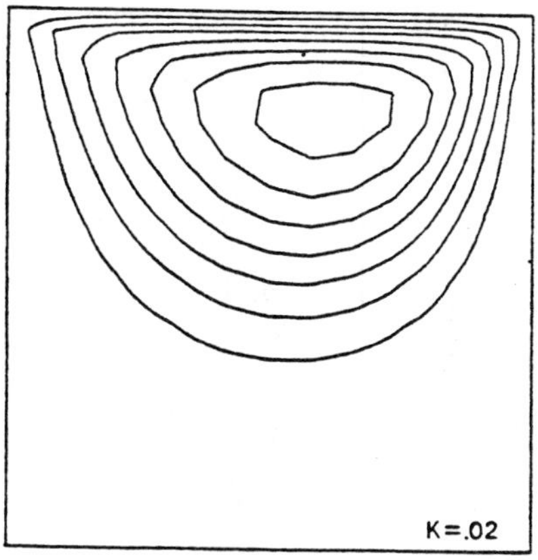

Figure 2

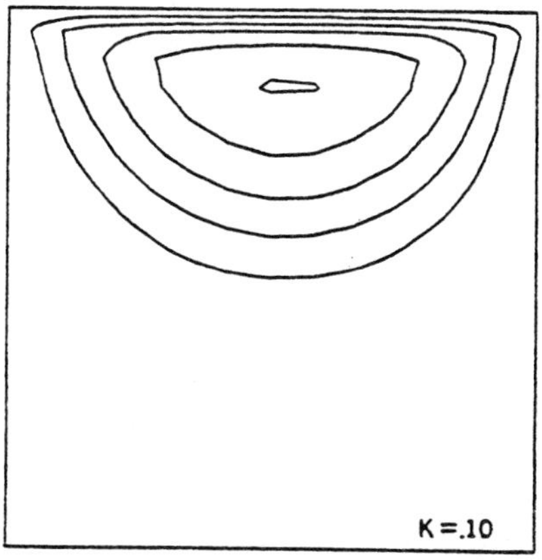

Figure 3

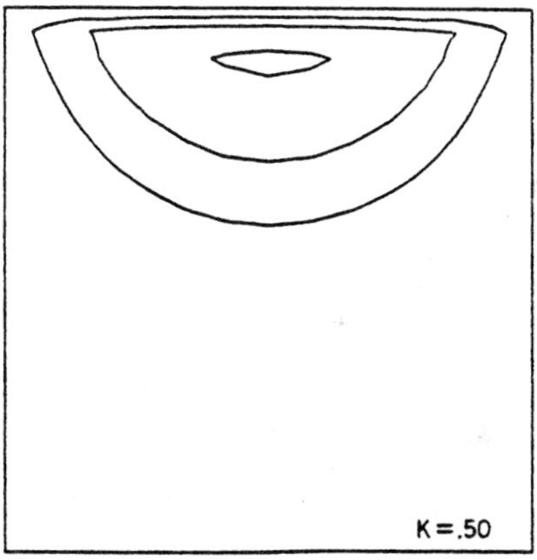

Figure 4

Figures 1-4. Streamlines for Driven Cavity Flow, Various Fluid Yield Stresses.

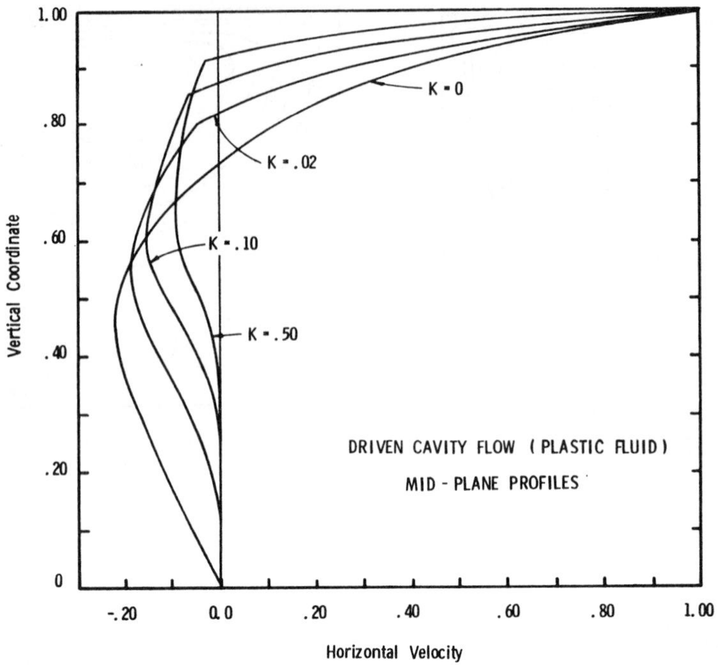

Figure 5. Mid-plane velocity profiles for the driven cavity flow.

5.2 Sudden Expansion Flow

As a second example we consider another familiar problem - the axisymmetric flow through a sudden expansion in a tube. The particular geometry of interest consists of a 2:1 expansion with the upstream tube having a unit radius. For computational purposes the tubing system is assumed to extend 2 units upstream of the expansion and 100 units downstream. The flow through the expansion is driven by a specified pressure drop that is imposed across the computational domain. The basic fluid properties are assumed to be $\rho = 1.0$, $\eta = 0.02$ and $\eta_r = 1000.0$.

Several computed solutions for this problem are shown in Figure 6 in the form of streamline plots. Each of the figures corresponds to the same imposed pressure gradient but to different assumed values of the fluid yield stress. The flow rates for each figure are different due to the varying resistance to flow; for reference the Reynolds number for the

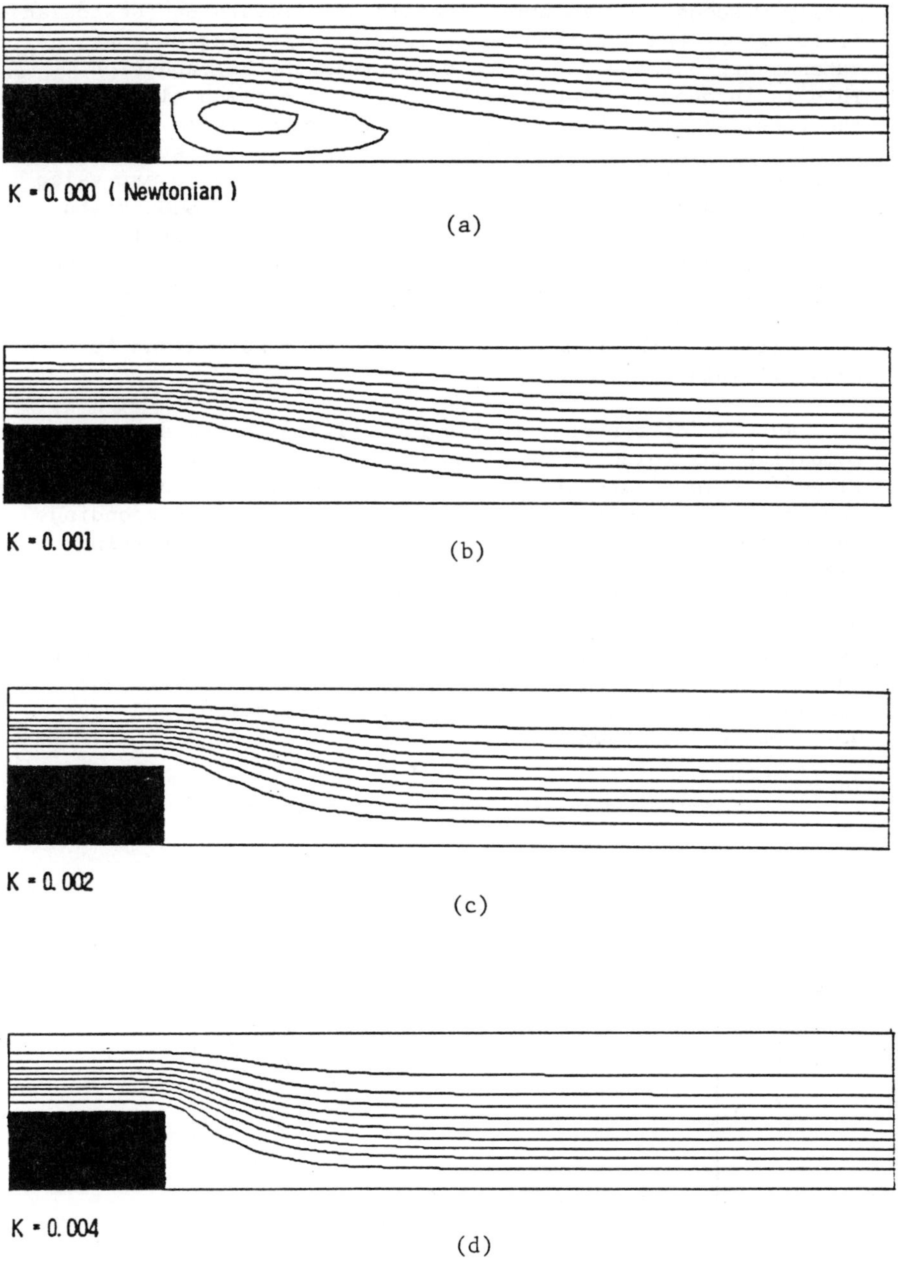

Figure 6. Streamlines for Flow Through a Sudden Expansion, Various Fluid Yield Stresses.

purely viscous flow is approximately 45. Figure 6a (Newtonian fluid) shows the existence of a recirculation region that extends approximately 4 units downstream of the expansion. This is in good agreement with previous experimental and numerical work on this problem [10]. For the cases in which a yield stress exists the flow fields show some distinct differences compared to the Newtonian flow. At the lowest value of the yield stress (Figure 6b) a recirculation zone exists but is extremely weak (30% of the Newtonian flow rate) and extends only two units downstream of the expansion. For the higher values of the yield stress (Figures 6c and 6d) the fluid in the corner region does not flow since the forces transmitted to the fluid in this area are insufficient to cause yielding.

Further differences in the flow fields can be seen by comparing the velocity profiles shown in Figure 7. The profiles in this figure are located at the expansion corner; each profile is normalized with respect to its corresponding centerline velocity. As shown by the figure an increasing yield stress produces a more plug-like flow along the tube centerline. In the present case the radial extent of the plug flow region is reduced somewhat by the complex fluid deformation field that occurs at the expansion corner. Near the tube inlet and exit planes, where the flow is very one-dimensional, the plug flow region is more distinct and covers up to 65% of the tube radius.

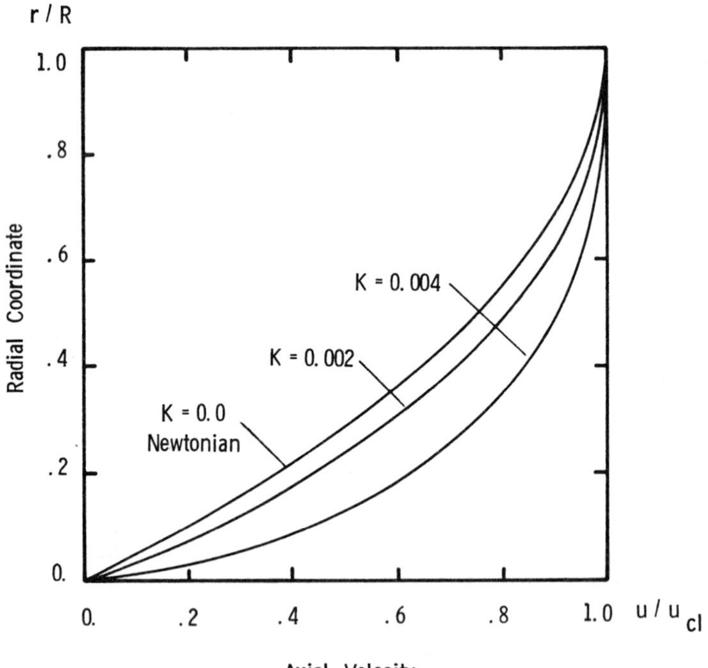

Figure 7. Velocity profiles for the flow in a sudden expansion, profiles at the expansion corner.

6. CONCLUDING REMARKS

The present note has attempted to address some of the problem areas associated with the numerical simulation of viscoplastic fluids. In particular, we have described a specific form of constitutive relation that is more suitable for computation than the classical Bingham model. In addition, procedures for implementing such a model into a finite element program were outlined. The solution of two standard flow problems served to illustrate both the capability of the numerical scheme and some of the qualitative differences between a purely viscous fluid and a viscoplastic fluid.

7. REFERENCES

1. BIRD, R., ARMSTRONG, R. and HASSAGER, D. - Dynamics of Polymeric Liquids, Wiley and Sons, New York, 1977.
2. PRAGER, W. - Introduction to the Mechanics of Continua, Ginn, Boston, 1961.
3. GARTLING, D. K. - Convective Heat Transfer by the Finite Element Method. Comp. Meth. Appl. Mech. Engr., 12, p. 365, 1977.
4. GARTLING, D. K. - NACHOS, A Finite Element Computer Program for Incompressible Flow Problems, Sandia National Laboratories Report SAND77-1333, Albuquerque, New Mexico, 1978.
5. GARTLING, D. K. - Finite Element Analysis of Convective Heat Transfer Problems with Change of Phase. Computer Methods in Fluids, Eds. K. Morgan, C. Taylor and C. Brebbia, Pentech Press, London, 1980.
6. GARTLING, D. K. - On Finite Element Procedures for Inelastic, Viscous Fluid Flow. in preparation, 1983.
7. GARTLING, D. K., NICKELL, R. E. and TANNER, R. I. - A Finite Element Convergence Study for Accelerating Flow Problems. Int. J. Num. Meth. Engr., 11, p. 1155, 1977.
8. TANNER, R. I. - private communication, 1983.
9. GARTLING, D. K. and PHAN-THIEN, N. - A Numerical Simulation of a Plastic Fluid in a Parallel Plate Plastometer. J. Non-Newtonian Fluid Mech., in press, 1983.
10. MACAGNO, E. O. and HUNG, T. K. - Computational and Experimental Study of a Captive Annular Eddy. J. Fluid Mech., 28, p. 43, 1967.

NUMERICAL SIMULATION OF THE FLOW OF FLUIDS WITH YIELD STRESSES

R.I. Tanner
Professor of Mechanical Engineering, University of Sydney

J.F. Milthorpe
Research Fellow, University of Sydney

ABSTRACT

An important class of non-Newtonian fluids possesses a yield stress which must be exceeded for deformation of the fluid to occur: for example slurries, grasses and doughs. Such fluids may be described by the Bingham model, in which the strain rate is proportional to the excess of stress over the yield stress, or the Herschel-Bulkley model, in which the strain rate is some power of the excess of stress over the yield stress. A Galerkin finite element computer program has been developed to model the deformation and flow of Bingham and Herschel-Bulkley fluids in a variety of two-dimensional and axi-symmetric boundaries including free surfaces. The program employs an iterative scheme to establish the location of the yielding surfaces and of the free surfaces. Examples are given of the extrusion of Bingham and Herschel-Bulkley fluids from slits, pipes and annular dies.

1. INTRODUCTION

A large class of materials may be described by the Bingham or Herschel-Bulkley models [1,2]: examples include greases, bread dough and slurries. The distinctive feature of these materials is a yield stress: if the absolute value of the stress at a point is less than the yield stress, the local deformation rate is zero. The Bingham model is the simplest yield-stress fluid: the deformation rate is proportional to the excess of the stress over the yield stress. The Herschel-Bulkley fluid is more general, assuming a power relationship. It would be possible to construct more elaborate models of fluids with yield stresses. Numerical modelling of Bingham flows has been examined by some authors [3,4].

The Herschel-Bulkley model is

$$\tau = \tau_y + k\dot{\gamma}^n \quad (\dot{\gamma} > 0) \tag{1}$$

$$\tau \lessgtr \tau_y \quad (\dot{\gamma} = 0) \tag{2}$$

The complete curve is an odd function of $\dot{\gamma}$. When the power-law index n = 1, and $k = \mu_p$, then we have the Bingham body; in this case μ_p is a viscosity (Pa-s).

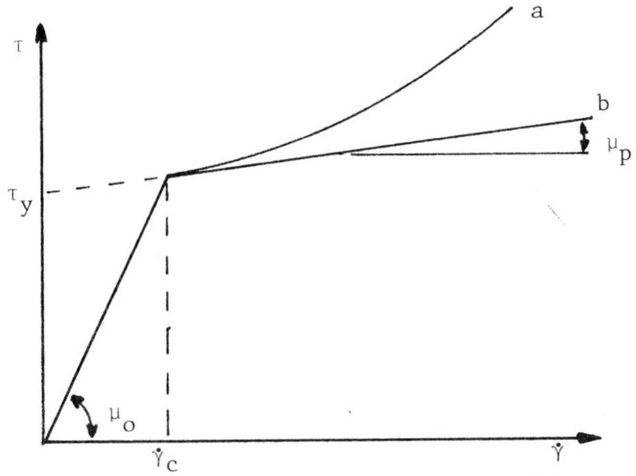

Fig. 1. Modified Herschel-Bulkley fluid (a) and Bingham fluid (b) where τ_y is the yield stress, μ_p the viscosity and μ_o the "unyielded" viscosity.

For the purposes of computation it is not convenient to distinguish between a state of shear ($\dot{\gamma} > 0$) and a state of absolute rigidity ($\dot{\gamma} = 0$). In a typical computer program we solve for the velocity components and then for the shear rates. Numerical noise will always produce some shear rate and we would not be able to locate the rigid "cores" with such models. Also, the stresses in the rigid regions would remain indeterminate. To avoid these problems the model shown in Fig. 1(a) has been adopted. Here we assume a very steep linear region around $\dot{\gamma} = 0$ so that for $\dot{\gamma} < \dot{\gamma}_c$ we have

$$\tau = \mu_o \dot{\gamma} \quad (\dot{\gamma} < \dot{\gamma}_c) \tag{3}$$

while the result (1) is used for $\dot{\gamma} > \dot{\gamma}_c$. In the Bingham case (Fig. 1(b)) there are two slopes (μ_o and μ_p) and usually we shall set in computations

$$\mu_o = 100\ \mu_p. \tag{4}$$

Then τ_y can be defined as

$$\tau_y = |\dot{\gamma}_c|(\mu_o - \mu_p) = 99\ \mu_p|\dot{\gamma}_c| \tag{5}$$

Similarly, for the Herschel-Bulkley model

$$\tau_y = \mu_o|\dot{\gamma}_c| + k|\dot{\gamma}_c|^n \tag{6}$$

Thus, to a close approximation, if $\dot{\gamma} > \dot{\gamma}_c$, we consider the material has yielded. We now need three-dimensional analogues of these laws. The law for the modified Herschel-Bulkley material may be immediately written down if we note that the viscosity for this model is

$$\eta = \mu_o (\dot{\gamma} \lessgtr |\dot{\gamma}|) \tag{7}$$

$$\eta = \tau_y |\dot{\gamma}|^{-1} + k|\dot{\gamma}|^{n-1} (|\dot{\gamma}| > \dot{\gamma}_c) \tag{8}$$

The proper generalization for three dimensions involves the use of the rate of deformation tensor d_{ij}, where

$$d_{ij} = \tfrac{1}{2}\left(\frac{\partial v_i}{\partial x_j} + \frac{\partial v_j}{\partial x_i}\right) \tag{9}$$

where v_i are the velocity components in the x_i directions. We assume that the von Mises yield criterion [5] is the relevant one in the absence of information to the contrary. (Note: it would be interesting to test this hypothesis.)

Then the material has "yielded" (i.e. $\dot{\gamma}_c$) is exceeded if

$$\dot{\gamma}_c < 2 \sum_{i=1}^{3} \sum_{j=1}^{3} d_{ij}\, d_{ij} \tag{10}$$

We shall define $\dot{\gamma}$ as the quantity on the right-hand side of (10). In simple shearing the velocity field is $v_1 = \dot{\gamma}x_2$, and the only non-zero components of d_{ij} are $d_{12} = d_{21} = \tfrac{1}{2}\dot{\gamma}$, and (11) becomes $\dot{\gamma} > \dot{\gamma}_c$ for yielding, which is consistent with the one-dimensional definition given above.)

$$\tau_{ij} = \sigma_{ij} + p\delta_{ij} = 2\mu_o\, d_{ij} \quad (\dot{\gamma} \lessgtr \dot{\gamma}_c) \tag{11}$$

$$\tau_{ij} = 2\left\{\frac{\tau_y}{|\dot{\gamma}|} + k|\dot{\gamma}|^{n-1}\right\} d_{ij} \quad (\dot{\gamma} > \dot{\gamma}_c) \tag{12}$$

where the τ_{ij} is the deviatoric stress tensor, σ_{ij} is the stress tensor, and p is a pressure; we are supposing that the materials are incompressible, so that

$$\sum_{i=1}^{3} d_{ii} = 0 \tag{13}$$

and p is the reaction to this constraint, called the pressure.

2. ANALYSIS OF EXTRUSION

The Bingham and Herschel-Bulkley models are computed with modification of the AXFINR program developed previously by us for non-Newtonian extrusion computations. For these isotropic laws, the axes of principal strain and stress coincide, and this enables us to use the effective viscosity concept of (11) and (12). One can then solve the problem in an iterative (Picard) way by

(i) assuming a (variable) viscosity at all points in the material

(ii) calculating, using the AXFINR finite element program, a new velocity field

(iii) using the new velocity field, recalculating the viscosity at all points

(iv) repeating the above steps until convergence (or divergence!) occurs.

The above algorithm is slowly convergent. Several methods have been tried to speed it up:

(i) extrapolation: if we have two estimates of the effective viscosity $\eta_{(m)}$ and $\eta_{(m-1)}$, where the subscript m represents the iteration number, then we can use

$$\eta = \alpha\, \eta_m + (1 - \alpha)\, \eta_{m-1} \tag{14}$$

When $\alpha = 1$ we have the simple Picard scheme outlined above; when $\alpha < 1$ we have under-relaxation, and when $\alpha > 1$ we have over-relaxation. Many numerical experiments on Poiseuille flow were made with power-law fluids (Eqn. (13) with $\tau_y = 0$) and with Bingham models in an effort to accelerate convergence. To be effective, α needs to be about 5 for a Bingham body and unfortunately this destabilizes the calculation and leads ultimately to divergence. Provided we take $\alpha \sim 5$ for only the first four iterations and then reset it to 1, then we have been able to obtain accurate, stable results in about 9 iterations or less. This is still relatively costly, however, and two improvements have been devised.

First, for the Bingham model, we have the option, once we are sure an element has yielded, of making use of the linearity of the relations. We can rewrite (12) for a yielded element in the form (n = 1)

$$\tau_{ij} = 2\mu_p\, d_{ij} + \frac{2(d_{ij})_m}{\dot{\gamma}_m}\, \tau_y \quad (\text{m = iteration no.}) \tag{15}$$

so that the second term on the right-hand side of (15) can be evaluated. The problem then is reduced to an ordinary Newtonian calculation with an initial stress.

The corresponding technique in elasticity theory is well-known [6]. In practice we require

$$\mu_p \dot{\gamma}_{(m)} > \tau_y \tag{16}$$

before this part of the algorithm is used. Attempts to reduce the shear rate $\dot{\gamma}$ at which this algorithm was used (by increasing μ_p to 10 μ_p and 100 μ_p in (16) led to instability. In the region between the shear rate τ_y/μ_p and the shear rate $\dot{\gamma}_c$ (Fig. 1(b)) the Picard algorithm is used. The main advantage here is that the wall shear stress is very accurately represented. We shall term this technique the "initial stress technique".

(ii) Newton-Raphson method: In the present case it can be shown that the finite element method gives rise to equations of the form

$$\sum_{i=1,N} K_{ij} \, v_j = f_i \quad (i = 1,N) \tag{17}$$

where N is the number of unknowns (v_j) in the problem, typically several hundreds to several thousands.

Now it can be shown that the stiffness matrix K_{ij} depends only on (η) in the present set of problems. Suppose we are at some stage in the computation (m^{th} iteration) and we have

$$\sum_{j=1}^{N} K_{ij} \, (\eta_{(m)}) \, v_j^{(m)} - f_i = R_i^{(m)} \tag{18}$$

where R_i is a residual vector.

Let $v_j^{(m)}$ change to $v_j^{(m)} + \Delta v_j$, η change to $\eta + \Delta\eta$. Then these better estimates of the solution obey

$$\sum_{j=1}^{n} K_{ij}(\eta + \Delta\eta) \, (v_j^{(m)} + \Delta v_j) - f_i = 0 \tag{19}$$

Expanding in a Taylor series and neglecting products of Δ^2, we find

$$\sum_{j=1}^{N} \left| \left\{ K_{ij}(\eta_m) + \frac{\partial K_{ij}}{\partial \eta} \Delta\eta \right\} v_j^{(m)} + K_{ij}(\eta) \, \Delta v_j \right| - f_i = 0 \tag{20}$$

Noting that η itself depends on v_j, so that $\frac{\partial K_{ij}}{\partial \eta}$ can be written as

$$\frac{\partial K_{ij}}{\partial \eta}\frac{\partial \eta}{\partial v_k}\Delta v_k \tag{21}$$

then (20) can be rewritten, letting $\Delta v_k = v_k - v_k^{(m)}$

$$\sum_{j=1}^{N}\left|K_{ij}\,v_j^{(m)} + K_{ij}(v_j - v_j^{(m)})\right| + \sum_{k=1}^{N}\frac{\partial K_{ij}}{\partial \eta}\frac{\partial \eta}{\partial v_k}(v_k - v_k^{m})v_j^{m}\Big| = f_i \tag{22}$$

or, collecting terms and changing indices,

$$\sum_{j=1}^{N}\left\{K_{ij}v_j + \sum_{k=1}^{N}\frac{\partial K_{ij}}{\partial \eta}\frac{\quad}{v_k}v_k^{(m)}\;v_j\right\} = f_i + \sum_{k=1}^{N}\frac{\partial K_{ij}}{\partial \eta}\frac{\partial \eta}{\partial v_k}v_k\;v_j^{(m)} \tag{23}$$

This new set of equations can be solved as before. Unfortunately this exact set of Newton-Raphson equations is highly unstable for the Bingham case and often fails to converge. A very rapidly convergent algorithm was found by taking only half of the sums over k on both sides of (23). This is half-way between the Picard and full Newton-Raphson schemes; it converges faster than any other method tried and when used with the "initial stress" it is accurate. We now give some results found with these methods.

3. COMPUTATIONAL RESULTS

The two Bingham plastic models have been tested in a number of flow geometries. The validity of the method is first demonstrated by comparison with known solutions for plans and axisymmetric Poiseuille flows under a pressure gradient. The boundary conditions used are shown in Fig. 2.

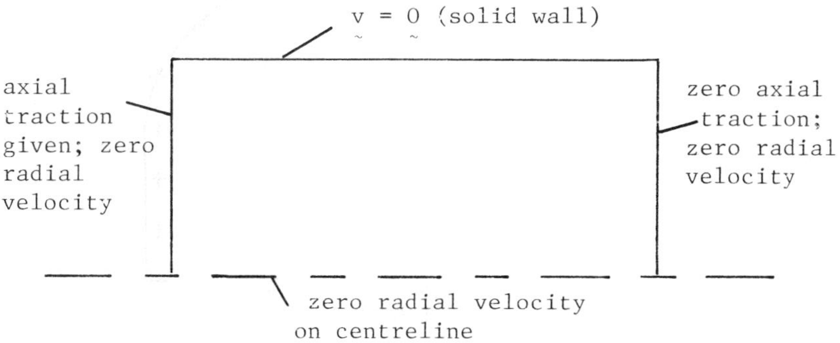

Fig. 2. Poiseuille flow problem.

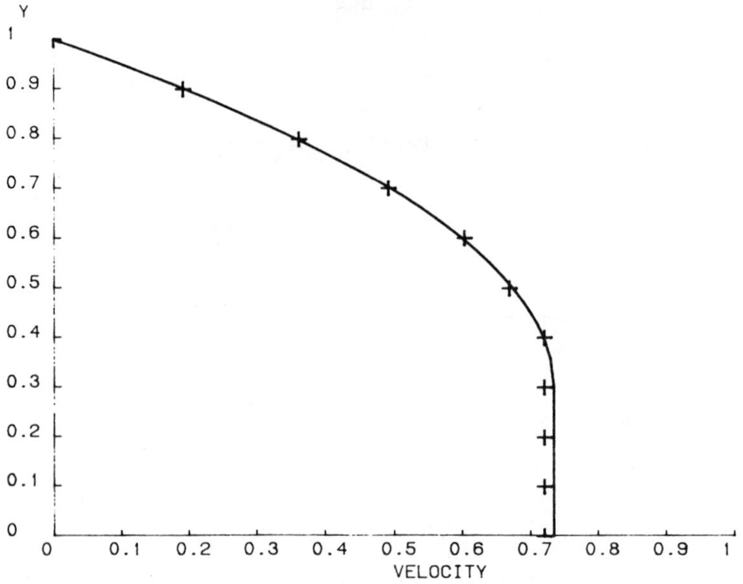

Fig. 3. Velocity profile for plane Poiseuille flow under unit pressure gradient. $\tau = .3 + .33\dot{\gamma}$: —— , exact solution; + , computational results.

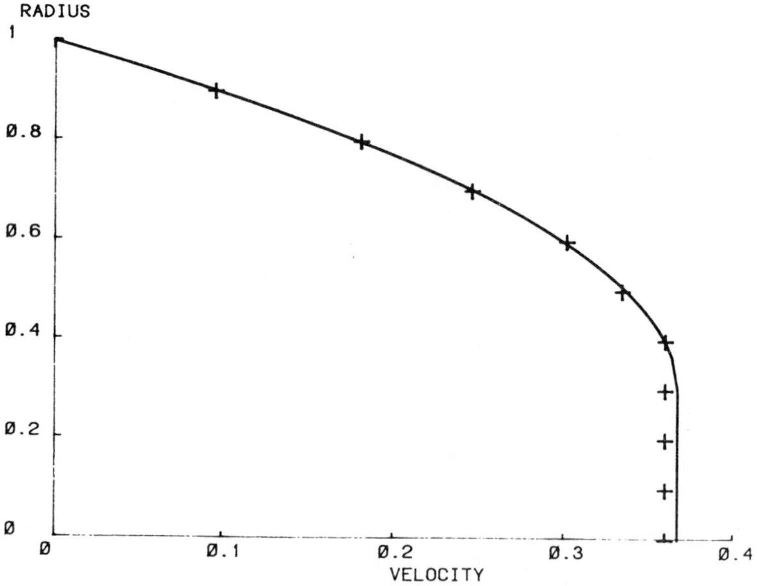

Fig. 4. Velocity profile for axisymmetric Poiseuille flow under unit pressure gradient. $\tau = .15 + .33\dot{\gamma}$: —— , exact solution; + , computational results.

Typical velocity profiles for plane and axisymmetric Poiseuille flows are shown in Fig. 3 and Fig. 4. In both cases the surface on which yielding occurs is in the centre of the elements which is the most difficult situation for the model. The comparison with the exact solution is good in both cases.

We define a dimensionless number Y

$$Y = \frac{\tau_y}{\tau_w} \tag{24}$$

where τ_y is the yield stress and τ_w is the shear stress at the wall of the tube on the solid surface. As Y increases the size of the plug expands until it occupies the entire pipe (or slit) and no flow occurs. The total rate of fluid flow through the pipe (or slit) under a given pressure gradient decreases with increasing Y. Y may be interpreted as the ratio of plug size to pipe size, or alternately as a measure of the Bingham character of the flow.

The plane Poiseuille flow under a pressure gradient (Fig. 2) has been calculated for a range of values of Y. The total flux produced by the pressure gradient is shown in Fig. 5, and compared with the analytic solution. It can be seen that the accuracy of the numerical method is very good over the whole range of Y.

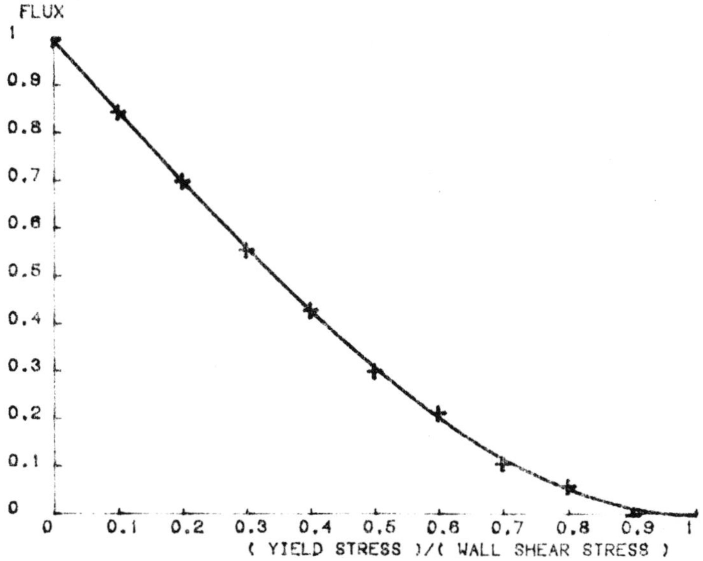

Fig. 5. Total flow rate in plane Poiseuille flow of Bingham fluid under unit pressure gradient for various Y. ——, exact solution; +, computational results.

Exact Flux		Flux .5635	Flux Error
Relaxation Factor	η_u		
1.0	1000	.5575	-1.06%
1.0	100	.5565	-1.24%
1.5	100	.5584	-0.91%

Table 1. Comparison of Flux Error in plane Poiseuille flow under unit pressure gradient, Y = 0.3, using a finite element mesh with 5 uniform elements across the flow.

The choice of the value of μ_o, the large value of viscosity used to model the unyielded plastic, is affected by considerations of accuracy and convergence rate, but neither of these is particularly sensitive to the value of μ_o. The exact value of the total flux and the calculated values are compared in Table 1. These results are not unexpected as the larger value of μ_o more closely approximates the Bingham model of zero strain rate when unyielded, and the smaller value of μ_o is closer to the initial unyielded solution. We shall usually use μ_o/μ_p = 100 henceforth. In passing, we have used an unyielded solution as an initial estimate but found that convergence was substantially slower.

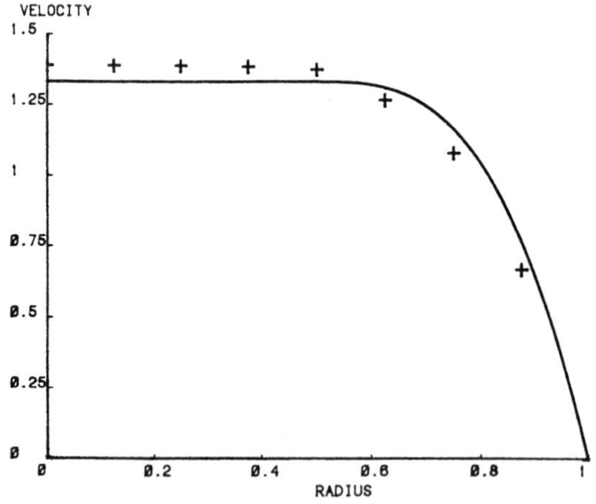

Fig. 6. Velocity profile for pipe flow of Herschel-Bulkley fluid $\tau = 1 + 0.5\dot{\gamma}^{0.5}$; —— , exact solution; + , computational results.

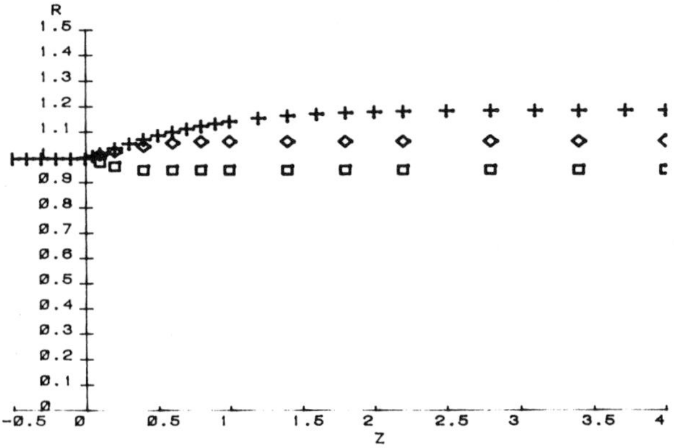

Fig.7. Plane extrusion profile for Bingham fluid at various Y. + Newtonian (Y = 0)
Y = 0.1
Y = 0.3

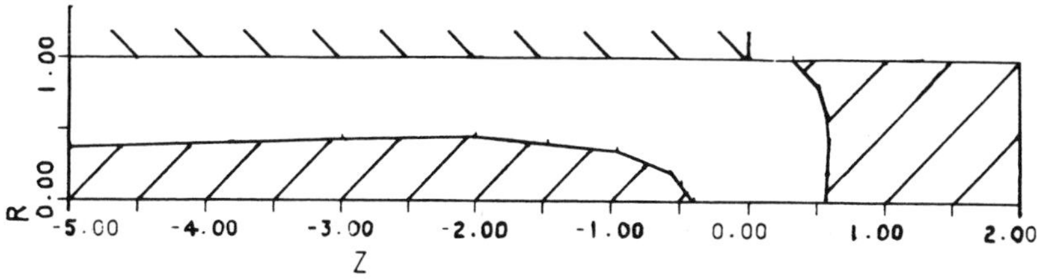

Fig. 8. Axisymmetric extrusion profile and yield surface for Herschel-Bulkley fluid $\tau = 0.5\dot{\gamma}^{0.5}$. Shaded areas are unyielded.

We have also computed accurately the Herschel-Bulkley pipe flow where $= 1.0 + 0.5\dot{\gamma}^{0.5}$ (Fig. 6).

The extrusion of both plane Bingham bodies with $\tau_y = 2$, $\mu_p = 1$, $\mu_o = 100$ and a wall shear stress (τ_w) of 6 case has been calculated. Extrusion profiles are shown for various Y in Fig. 7.

A fairly coarse (4 × 7 element) mesh was used. The plane swelling ratio is nearly 1. For the axisymmetric case the swelling ratio is about 1.036. The upstream unyielded plug is about the same in both problems (shaded area in Fig. 8) but the "freeze line" (the point after exit where the material stress state is below the yield point) is closer to the die exit in the axisymmetric case. (Fig. 8).

The small swelling ratios are in good accord with an approximate theory [7]; we have predicted negligible plane swelling and a swelling of about 5% for the axisymmetric case.

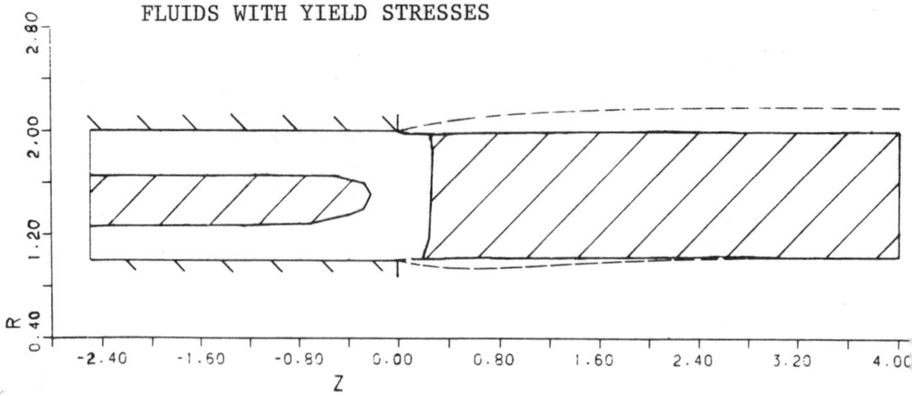

Fig. 9. Extrusion profile (solid line) and unyielded region (shaded area) for annular extrusion of Bingham material (Y = 0). The annular extrusion of a Bingham fluid has been computed with a 6 × 19 element mesh. The inner diameter of the annular is half the outer. Again we see virtually no deviation from the die diameter. The yield stress used was 0.2 and the flow was driven by a unit pressure gradient. The Newtonian profile was calculated for comparison.

ACKNOWLEDGEMENT

This work was supported by the Defence Science Research Establishment, Salisbury, South Australia.

REFERENCES

1. WILKINSON, W.L. - Non-Newtonian Fluids, Pergamon, London, 1960.

2. SKELLAND, A.H.P. - Non-Newtonian Flow and Heat Transfer, John Wiley and Sons, New York, 1967.

3. BERCOVIER, M. and ENGELMAN, M. - A Finite-Element Method for Incompressible Non-Newtonian Flow, J. Comput. Phys., Vol. 36, 313-326, 1980.

4. GLOWINSKI, R. - Numerical Methods for Non-Linear Variational Problems, Springer-Verlag, Heidelberg, 1980.

5. PRAGER, W. - Introduction to the Mechanics of Continua, Dover, New York, 1961.

6. ZIENKIEWICZ, O.C. - The Finite Element Method in Engineering Science, McGraw Hill, New York, 1971.

7. TANNER, R.I. - A New Inelastic Theory of Die-Swell, J. Non-Newt. Fluid Mech., Vol. 6, 289, 1980.

ON THE APPLICATION OF THE FINITE ELEMENT METHOD TO THE DIE-SWELL AND FIBER-SPINNING PROBLEMS INVOLVING NON-NEWTONIAN FLUIDS

V. Dakshina Murty

Assistant Professor of Mechanics, University of Portland, Mechanical Engineering

SUMMARY

Numerical solutions using Galerkin finite element technique are obtained to the die-swell and fiber-spinning problems. Two constitutive models are incorporated in this study, namely, the power-law model and the second-order fluid model. A free surface algorithm is developed, which for the class of problems studied herein, is found to be satisfactory. In the die-swell problem using second-order fluids, it is found that, for Weissenberg numbers above 0.3, computationally very expensive grids are needed. For the fiber-spinning problem, typically, it took four to six iterations to converge. For large drawing speeds, the free surface is shown to oscillate about a mean iterate.

1. INTRODUCTION

The origins of non-Newtonian fluid mechanics can be found during the period between the two world wars when chemical engineers attempted to relate the behaviour of polymeric materials to classical field mechanics. Although the equations governing the motion of non-Newtonian fluids are the same as those for Newtonian fluids except for the constitutive equation, the former are much harder to solve, the main reason being the nonlinearity of the shear stress vs. strain rate relationship. This, coupled with the advent of high speed computers has made the numerical study of non-Newtonian fluid flow phenomena essential.

Throughout the earlier development of numerical fluid mechanics, the finite difference method was the most commonly used method for solving boundary value problems in fluid mechanics. Over the past decade, however, the finite element method has gained substantial popularity.

The earliest work in the application of finite elements to

non-Newtonian fluid flow is due to Thomson et al. [1] who obtained the creeping flow solution for power-law fluids. Tanner et al. [2] solved plane and axisymmetric jet problems for a second order fluid. Crochet and his co-workers [3-5] also solved the flow of a second-order fluid through a contraction and a square cavity using both finite elements and finite differences. Kawahara and Takeuchi [6] analyzed the flow of a Maxwell fluid through an expansion. Their method introduces a large number of degrees of freedom compared to the conventional mixed finite element method where only the velocities and pressure are interpolated. Viriyayuthakorn and Caswell [7] solved the flow of a viscoelastic fluid in an expansion using the memory integral formulation of the Maxwell fluid. Chang et al. [8] solved the die-entry and stick-slip problems for Newtonian and viscoelastic liquids using both collocation and Galerkin finite elements, and obtained convergence for small Weissenberg numbers. Reddy and Tanner [9,10] solved the die-swell problem for a Newtonian fluid with surface tension and for a second order fluid for plane jets. Kiparissides and Vlachopoulous [11] solved the calendering problem for an isothermal power-law fluid using finite elements. In their formulations, however, they simplified the equations so that the problem essentially became one-dimensional. In a later investigation [12], they obtained the temperature profiles for the calendering problem using finite differences. Palit and Fenner [13,14] used finite element methods to study the flow of a power-law fluid in a rectangular channel and around cylinders. Caswell and Tanner [15] solved the problem of a wire-coating die using a power-law fluid model and obtained velocity profiles in the die.

In the present investigation, results from the numerical solution of the die-swell and fiber-spinning problems are presented. In these two problems, an additional source of nonlinearity is present, namely, the location of the free surface which is not known a priori.

2. GOVERNING EQUATIONS

Assuming that the fluids are considered to be incompressible, and the fluid motion is considered to be steady, two-dimensional and laminar, the governing equations are the continuity and linear momentum equations. They are as follows:

$$\text{div}\ \underset{\sim}{v} = o \qquad (2.1)$$

$$\rho\ \underset{\sim}{v} \cdot \nabla\ \underset{\sim}{v} = -\ \nabla p + \rho \underset{\approx}{f} + \text{div}\ \underset{\approx}{\tau} \qquad (2.2)$$

where ρ, $\underset{\sim}{v}$, $\underset{\sim}{f}$, $\underset{\sim}{\tau}$, and p represent the density, velocity, body force, stess and pressure respectively.

The constitutiive equation for the second-order fluid is given by

$$\underset{\approx}{\tau} = \eta(\underset{\approx}{L} + \underset{\approx}{L}^{\dagger}) + (\frac{\nu_1}{2} + \nu_2)(\underset{\approx}{L}^2 + (\underset{\approx}{L}^{\dagger})^2)$$

$$+ (\nu_1 + \nu_2)\, \underset{\approx}{L}\underset{\approx}{L}^{\dagger} + \nu_2\, \underset{\approx}{L}\,\underset{\approx}{L} \qquad (2.3)$$

$$- \frac{\nu_1}{2}(\underset{\sim}{v}\cdot\nabla\, \underset{\approx}{L} + \underset{\sim}{v}\cdot\nabla\underset{\approx}{L}^{\dagger})$$

where η, ν_1, and ν_2 are the viscosity, first and second normal stress coefficients respectively. If ν_1, and ν_2 are zero and η takes the form,

$$\eta = \eta_o\, (\dot{\gamma})^{n-1} \qquad (2.4)$$

then (2.3) reduces to a power-law model in which n is the power-law index, $\dot{\gamma}$ is the shear rate

$$(= \sqrt{\text{tr}\, (\nabla \underset{\sim}{v} + \nabla\, \underset{\sim}{v}^{\dagger})^2}).$$

Using Galerkin's formulation and finite element discretization (see Murty [16] for details) a set of equations of the following form is obtained:

$$[\underset{\approx}{C} + \underset{\approx}{K}]\, \{\underset{\sim}{q}\}\quad \{\underset{\sim}{F}\} \qquad (2.5)$$

where $\underset{\sim}{q}$ is the vector containing the nodal point velocities and pressures, and $\underset{\approx}{C}$, $\underset{\approx}{K}$ and $\underset{\approx}{F}$ are matrices depending upon $\underset{\approx}{q}$.

3. FREE SURFACE ALGORITHM

Consider the fluid emerging from the die into the atmosphere as shown in Figure 1. If the traction due to the surrounding air can be ignored, the free surface condition is equivalent to setting the shear traction to zero and the normal traction to the ambient pressure. Since the free surface is itself a streamline, the fluid velocity on the free surface satisfies the following condition:

$$\frac{dx}{dy} = \frac{u(x,y)}{v(x,y)} \qquad (3.1)$$

Let (x_s, y_s) be the coordinates of a point P located on the free surface at the sth iteration. Let the corresponding velocities in the x and y directions be given by u_s and v_s respectively. Note that the point A has fixed coordinates and lies on the surface during each iteration. Let (x_{s+1}, y_{s+1}) be the coordinates of the point P' at the $(s+1)^{th}$ iteration. If u_s and v_s, the velocity compenents of the point P are known, then the coordinates of the point P' can be obtained by integrating equation 3.1 as follows:

$$x^{s+1} - x_o = \int_{y_o}^{y} {}^{s+1}\frac{u_s}{v_s}\, dy \qquad (3.2)$$

Thus, using equation 3.2, an iterative updating of the free surface is done.

At the downstream end, the boundary conditions can be expressed either in terms of the velocities (zero transverse velocity and uniform axial velocity) or in terms of the tractions (zero normal and shear tractions).

4. NUMERICAL RESULTS

In this section numerical results from the finite element solution are presented.

4.1 Die-Swell Problem

The geometry and the boundary conditions for this problem are shown in Figure 2a. The shape of the jet at zero Reynolds number after five free surface iterations is shown in Figure 2b. The swell ratio for this mesh is 14.11 %. Using the same grid, the power-law index: n = 0.8 and n = 0.6. The shapes of the jet after five iterations are shown in Figures 2c and 2d. The swell ratio was found to be 10.3% and 6.4% respectively indicating that for creeping flows, the amount of swell decreases with decreasing value of n. No significant difference was observed in the axial velocity profiles compared with Newtonian fluids, except for the effect of shear thinning.

To study the effect of viscoelasticity on the behaviour of jets, the second-order fluid model is used next. The case of plane jets, that is, fluid issuing out of infinite planes is considered. A grid of 240 elements was used. The inlet boundary conditions are prescribed to be parabolic axial velocity and zero transverse velocity. For zero Reynolds number this problem was solved first for a Newtonian fluid and then for second order fluids by incrementing ν_1. The swell ratios were 18.10%, 16.6% and 16.0% for Weissenberg numbers of 0.0 (Newtonian fluid), 0.15 and 0.30, respectively.

Convergence of the numerical scheme was successfully obtained up to a Weissenberg number of 0.30 (which corresponds to ν_1 = 0.2). The axial variation of pressure for We = 0.15 and We = 0.30 are shown in Figures 3a and 3b. While the variation for lower Weissenberg number is smooth and almost identical to that of a Newtonian fluid, the variation for higher Weissenberg numbers shows oscillations. The difficulty of modelling the pressure appears to be the fundamental difficulty

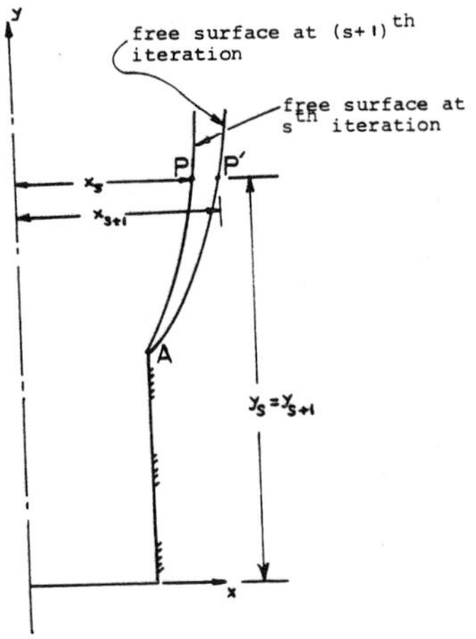

Figure 1: Location of free surface at successive iterations

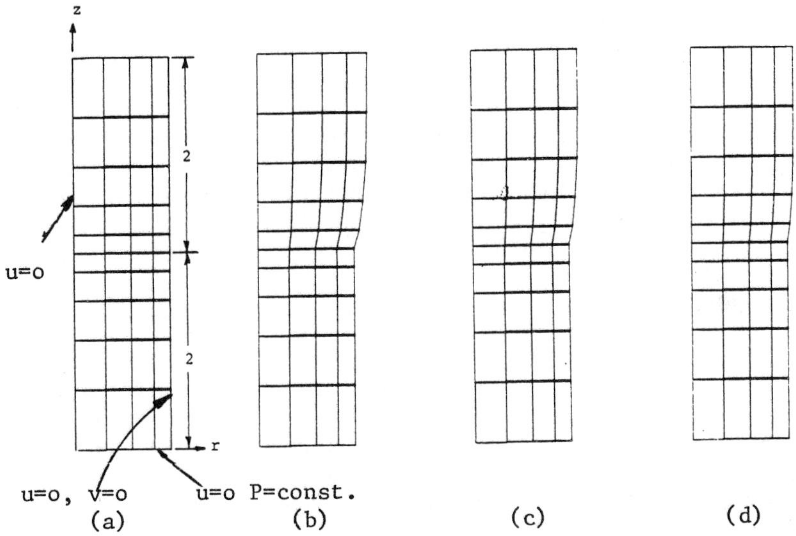

Figure 2: Die-Swell for power-law fluids
(a) grid (b) n=1.0 (c) n=0.8 (d) n=0.6

in modelling the die-swell problem of viscoelastic fluids at high Weissenberg numbers.

4.2 Fiber Spinning Problem.

A schematic diagram of the fiber spinning process is shown in Figure 4. As the extrudate emerges from the die, it swells and because of the tensile force supplied by the roll, forms a fiber before being taken up on the roll. This problem was studied for plane dies using the grid shown in Figure 5a with two different boundary conditions on the downstream boundary; prescribed axial velocity and prescribed force.

First, the effect of varying v_o on the shape of the jet was studied. In all cases, the Reynolds number was zero and Weissenberg number was 0.15. The final shape of the jet for different values of v_o is shown in Figure 5b, c, and d. Five free surface iterations were sufficient for convergence when $v_o < 1.5$. However, when $v_o = 2$ cm/s, even eight iterations were no sufficient for complete convergence. The location of the free surface at the end of each iteration for $v_o = 2$ cm/s is shown in Figure 6. Although the free surface has not converged, the iterates clearly oscillate about a mean value.

For practical considerations it is easier to control the downstream traction (by controlling the speed of the take up roll shown in Figure 4) than to control the velocity. Hence, as a second kind of boundary condition a uniform tensile force F_o is applied at the downstream end. The free surface was found to converge in five iterations when $F_o < 0.6$. The variation of F_o with the thickness of the fiber is shown in Figure 7. Results up to $F_o = 0.6$ are shown. Beyond this value it took more iterations for the free surface to converge.

5. CONCLUSIONS

A methodology using Galerkin finite element method is presented for solving the die-swell and fiber spinning problems using power-law and second-order fluids. While no significant numerical difficulties were encountered in the case of power-law fluids, the oscillation in nodal-point pressure poses a restriction in modelling viscoelastic fluids especially at high Weissenberg numbers.

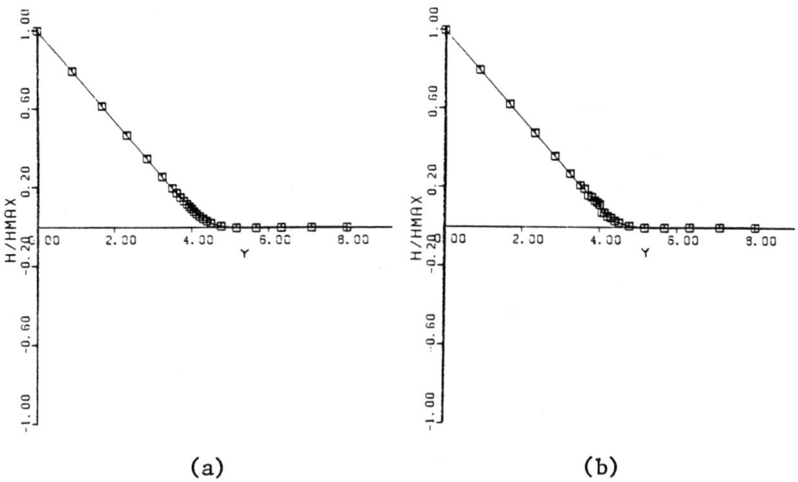

Figure 3: Variation of pressure along centerline
(a) We=0.15, Re=0.0 (b) We=0.30, Re=0.0

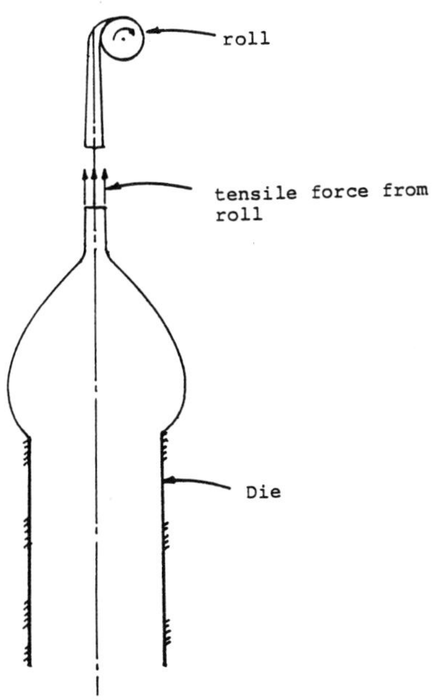

Figure 4: Fiber-spinning problem

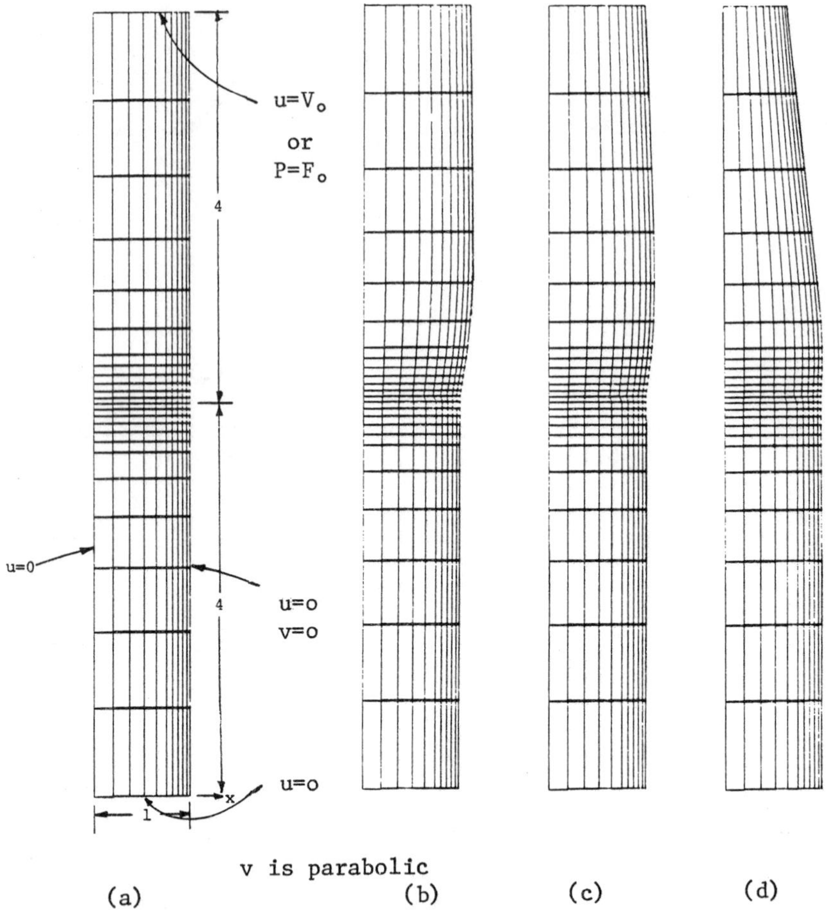

Figure 5: Fiber-spinning problem

(a) grid (b) V_o=0.9 (c) V_o=1.0 (d) V_o=1.5

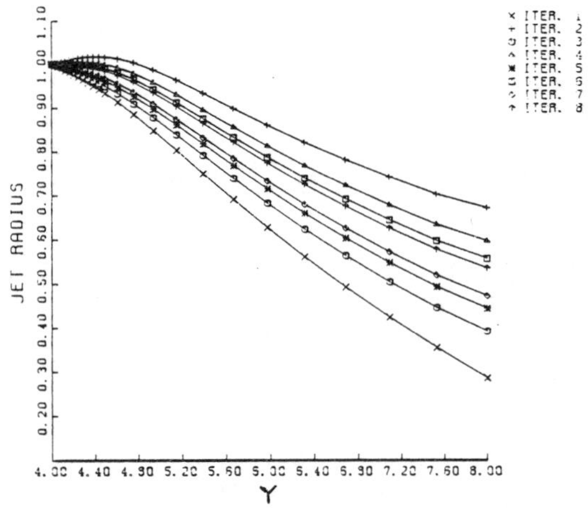

Figure 6: Location of free surface after each iteration

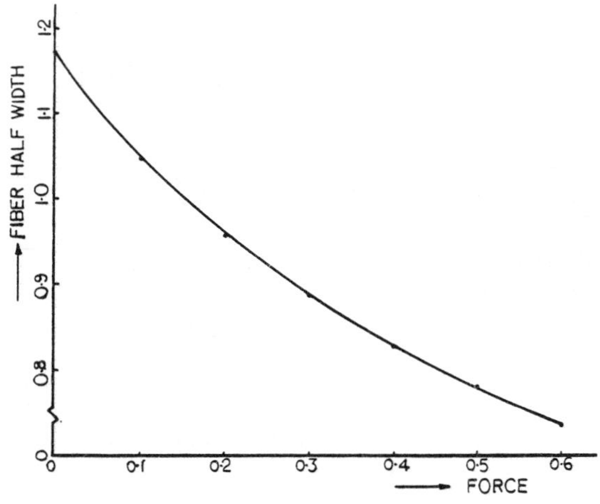

Figure 7: Variation of takeup force F_o with fiber half-width

REFERENCES

1. THOMSON, E.G., MACK, L.M., LIN, F.S., "Finite Element Method for Incompressible Slow Viscous Flow with a Free Surface," Proceedings of 11th Midwestern Mechanics Conference, Vol. 5, pp. 93-111, Iowa State University Press.

2. TANNER, R.I., NICKELL, R.E., and BILGER, R.W., "Finite Element Methods for Solution of Some Incompressible Non-Newtonian Fluid Mechanics Problems with Free Surface," Computer Methods in Applied Mechanics and Engineering, Vol. 6, 1975, pp. 247-258.

3. CROCHET, M.J. and PILATE, G., "Plane Flow of a Fluid of Second Grade Through a Contraction," Journal of Non-Newtonian Fluid Mechanics, Vol. 1, 1976, pp. 247-258.

4. CROCHET, M.J. and PILATE, G., "Numerical Solution of a Fluid of Second Grade in a Square Cavity," Computers and Fluids, Vol. 3, 1975, pp. 283-291.

5. CROCHET, M.J. and BEZY, M., "Numerical Solution for the Flow of Viscoelastic Fluids," Journal of Non-Newtonian Fluid Mechanics, Vol. 5, 1979, pp. 201-218.

6. KAWAHARA, M. and TAKEUCHI, N., "Mixed Finite Element Methods for Analysis of Viscoelastic Flow," Computers and Fluids, Vol. 5, 1977, pp. 33-45.

7. VIRIYAYUTHAKORN, M. and CASWELL, B., "Finite Element Simulation of Viscoelastic Flow," Report of Division of Engineering, Brown University, Providence, July 1979.

8. CHANG, P.W., PATTEN, T.W., and FINLAYSON, B.A., "Galerkin and Collocation Finite Element Methods for Newtonian and Viscoelastic Fluids," Computers and Fluids, Vol. 7, No. 4, pp. 267-283.

9. REDDY, K.R. and TANNER, R.I., "Finite Element Solutions of Viscous Jet Flows with Surface Tension," Computers and Fluids, Vol. 6, 1978, pp. 83-91.

10. REDDY, K.R. and TANNER, R.I., "On the Swelling of Extruded Plane Sheets," Journal of Rheology, Vol. 22, 1975, pp. 661-665.

11. KIPARISSIDES, C. and VLACHOPOULOUS, J., "Finite Element Analysis of Calendering of Power Law Fluids, " Polymer Engineering and Science, Vol. 16, 1976, pp. 712-719.

12. KIPARISSIDES, C. and VLACHOPOULOUS, J., "A Study of Viscous Dissipation in Calendering of Power Law Fluids," Polymer Engineering and Science, Vol. 18, pp.

13. PALIT, K. and FENNER, R.J., "Finite Element Analysis of Slow Non-Newtonian channel Flow," A.I.Ch.E. Journal. Vol. 18, 1972, pp. 628-633.

14. PALIT, K. and FENNER, R.J., "Finite Element Analysis of Two Dimensional Slow Non-Newtonian Flows," A.I.Ch.E. Journal, Vol. 18, 1972, pp. 1163-1170.

15. CASWELL, B. and TANNER, R.I., "Wirecoating Die Design Using Finite Element Methods," Polymer engineering and Science, Vol. 18, 1978, pp. 416-421.

16. MURTY, V.D., "Finite Element Analysis of Steady, Incompressible Non-Newtonian Fluid Flow and Heat Transfer," Ph.D. Dissertation, University of Texas, Austin, 1982.

ON FINITE ELEMENT ANALYSIS OF THE CRAVITATIONAL FLOW OF CRUSHED ORE

Kenneth Runesson(I) and Larsgunnar Nilsson(II)

SUMMARY

Crushed ore is modelled as a Newtonian-plastic fluid. Frictional yielding is assumed and is represented by Drucker-Prager´s yield criterion. The stress is found as the projection of the fictitious Newtonian stress onto the plastically admissable stress region in each point in space. A displacement Finite Element Method is obtained for the transient flow of crushed ore including the convective acceleration terms. In each time step an algorithm is deviced which is based on modified Newton iterations for the solution of the nonlinear equations arizing from the backward Euler differencing scheme. Numerical calculations for a model problem concerning the flow of crushed ore concludes the paper.

1. INTRODUCTION

In judging the behaviour of crushed ore there is a need for quantitative analysis of the gravitational flow of a dry granular material. Although our main interest (from a practical viewpoint) is in crushed ore, the theory and solution technique suggested in this paper may be applied also to other types of bulk flow, e.g. in a silo hopper.

The problem of adequately describing the motion of crushed ore, especially the constitutive properties, is indeed a difficult one. An important feature is considered to be possible existence of a "flow zone" within which the velocities may be of a order of magnitude larger than in the more stable part of the ore while the stresses are still of the same order of magnitude (or even drop to zero).

A crude method of estimating the shape (but unfortunately not the size) of the flow zone has been suggested by Bergmark and Roos, see Enstad et al [1], using elementary particulate mechanics.

(I) Dr., Chalmers Univ. of Techn., Dept. of Structural Mech., S-412 96 Göteborg, Sweden.
(II) Prof., University of Luleå, Dept. of Structural Eng., S-951 87 Luleå, Sweden.

A well-known continuum model describing the flow of a granular material was given by Jenike and Shield [2], who considered the steady creeping flow of a rigid-plastic incompressible mass with intergrain Coulomb friction. The flow was assumed to be associative in the deviatoric stress plane. As a consequence of the assumptions the velocity cannot be determined uniquely and the existence of a flowing state must be assumed a priori; the reason for this of course being that the stress state is indeterminate in the rigid region. An improved version is due to Brennen and Pearce [3], who included the convective acceleration terms into the equations of motions, which leads to uniqeness of the velocity field. However, the difficulty in defining the flow zone remains.

In this paper the crushed ore is modelled as a viscous-plastic fluid which is Newtonian with small (but non-zero) compressibility within the yield surface. Frictional yielding is assumed and is represented by Drucker-Prager´s yield criterion. Associative plastic flow is adopted for stresses on the yield surface. This model is conceptually simple but still represents the major physical characteristics of crushed ore such as the existence of a "flow zone" (the boundary of which is a part of the solution) and dilational flow within this zone. As one has to consider quite general geometry it seems inevitable to adopt a numerical technique like the Finite Element Method.

The model may be used to predict how the shape of the flow zone and the velocity pattern in general is influenced by factors such as the choice of boundary conditions (e.g. rough or smooth wall), the initial stress state within the mass, the choice of the angle of internal friction, etc. A numerical investigation of a few selected parameters is included in the paper.

2. BASIC EQUATIONS

Consider the transient flow of crushed ore occupying the region Ω with boundary Γ, which is subdivided into two parts: Γ_v with prescribed velocity $\underset{\sim}{v}$ and Γ_T with prescribed tractions $\underset{\sim}{T}$. The material is subjected to the gravitational force $\underset{\sim}{f} = (f_x, f_y, f_z) = (0,0,\rho g)$, where ρ is the bulk density and the z-axis is chosen vertical. (Cartesian coordinates are used for simplicity).

Assume that the crushed ore may be modelled as a viscous-plastic fluid, which is Newtonian with small (but non-zero) compressibility within the yield surface. Purely frictional yielding is assumed and is represented by Drucker-Prager´s yield criterion. Associative plastic flow is adopted for stress states on the yield surface so that the flow within the flow region (= the plastic zone) is effectively dilatational.

This property is true probably only at the very transient stage of the motion so that in a more realistic model a state of incompressibility representing a so called "critical state" of the granular material would be approached eventually.

To be more general, consider the convex set B of plastically admissable stresses

$$B = \{\underset{\sim}{\tau}: F(\underset{\sim}{\tau}) \leq 0\} \tag{1}$$

where $F(\underset{\sim}{\tau})=0$ defines the yield surface in stress space spanned by the Cartesian components of $\underset{\sim}{\tau}$. The subsequent arguments may be compared with those for incremental elasto-plasticity, Runesson and Booker [4].

The rate of deformation $\underset{\sim}{\varepsilon}(\underset{\sim}{v})$, where $\underset{\sim}{v}$ is the velocity, is given by the sum

$$\underset{\sim}{\varepsilon}(\underset{\sim}{v}) = \underset{\sim}{D}^{-1}\cdot\underset{\sim}{\sigma}+\underset{\sim}{\varepsilon}^{P} \tag{2}$$

where $\underset{\sim}{D}$ is a tensor representing the Newtonian viscosity and the plastic part $\underset{\sim}{\varepsilon}^{P}$ is defined by the inequality

$$(\underset{\sim}{\sigma}-\underset{\sim}{\tau})\cdot\underset{\sim}{\varepsilon}^{P} \geq 0 \qquad \forall \quad \underset{\sim}{\tau} \in B \tag{3}$$

The well-known general expression for the (compressible) Newtonian part $\underset{\sim}{\varepsilon}^{N}$ of $\underset{\sim}{\varepsilon}(\underset{\sim}{v})$ is

$$\underset{\sim}{\varepsilon}^{N} = \frac{1}{2\mu}\underset{\sim}{\sigma} - \left(\frac{1}{2\mu} - \frac{1}{3\kappa}\right)\sigma_m \underset{\sim}{I}_{(2)}, \qquad \sigma_m = \frac{1}{3}\,\mathrm{tr}\,\underset{\sim}{\sigma} \tag{4}$$

where μ and κ are the dynamic viscosities in shear and volumetric expansion respectively, and $\underset{\sim}{I}_{(2)}$ is Kronecker´s delta.

It is assumed that κ is large but finite. A particularly simple case is obtained if we assume that $\kappa/\mu = 2/3$, in which case $\underset{\sim}{\varepsilon}^{N}$ and $\underset{\sim}{\sigma}$ are colinear. In the analysis of a purely Newtonian fluid this assumption implies no advantage in particular, while it does simplify the further deductions in the case of a viscous-plastic fluid as we shall see subsequently.

The problem may now be formulated: Find $(\underset{\sim}{\sigma},\underset{\sim}{v})$ satisfying

$$\begin{aligned} &\rho(\dot{\underset{\sim}{v}}+\underset{\sim}{v}\cdot\underset{\sim}{\nabla}\underset{\sim}{v}) - \underset{\sim}{\nabla}\cdot\underset{\sim}{\sigma} = \underset{\sim}{f} && \text{in } \Omega \\ &(\underset{\sim}{\sigma} - \underset{\sim}{\tau})\cdot(\underset{\sim}{\sigma}^{N}(\underset{\sim}{\varepsilon}(\underset{\sim}{v}))-\underset{\sim}{\sigma}) \geq 0 \quad \forall\, \underset{\sim}{\tau} \in B && \text{in } \Omega \\ &\text{b.c.} && \text{on } \Gamma \end{aligned} \tag{5}$$

where (2-4) have been used, the choice $\kappa/\mu = 2/3$ has been made and the "fictitious" Newtonian stress $\underset{\sim}{\sigma}^{N}$ is defined by

$$\underset{\sim}{\sigma}^{N}(\underset{\sim}{\varepsilon}) = 2\mu\underset{\sim}{\varepsilon} \tag{6}$$

As B is convex, the constitutive inequality in (5) has the solution

$$\underset{\sim}{\sigma}(\underset{\sim}{\varepsilon}) = \Pi\underset{\sim}{\sigma}^N(\varepsilon) \tag{7}$$

where Π is the Euclidean projection operator onto B, i.e. $\underset{\sim}{\sigma}$ is the solution of

$$\min_{\underset{\sim}{\tau}\in B} |\underset{\sim}{\sigma}^N(\underset{\sim}{\varepsilon})-\underset{\sim}{\tau}|^2 \tag{8}$$

We may now formally eliminate $\underset{\sim}{\sigma}$ in (5) so that the equation of motion becomes

$$\rho(\dot{\underset{\sim}{v}}+\underset{\sim}{v}\cdot\underset{\sim}{\nabla}\underset{\sim}{v}) - \underset{\sim}{\nabla}\cdot(\Pi 2\mu\underset{\sim}{\varepsilon}(\underset{\sim}{v})) = \underset{\sim}{f} \tag{9a}$$

In the special case where no yielding occurs, i.e. where $\Pi\underset{\sim}{\sigma}^N = \underset{\sim}{\sigma}^N$, we obtain the well-known expression

$$\dot{\underset{\sim}{v}}+\underset{\sim}{v}\cdot\underset{\sim}{\nabla}\underset{\sim}{v} - \nu(\nabla^2\underset{\sim}{v}+\underset{\sim}{\nabla}\underset{\sim}{\nabla}\cdot\underset{\sim}{v}) = \frac{1}{\rho}\underset{\sim}{f} \tag{9b}$$

where ν is the kinematic viscosity.

3. STRESS PROJECTION

We may express the length of $\underset{\sim}{\sigma}$ as

$$|\underset{\sim}{\sigma}|^2 = \frac{2}{3}q(\underset{\sim}{\sigma})^2 + 3p'(\underset{\sim}{\sigma})^2 \tag{10}$$

where

$$q(\underset{\sim}{\sigma}) = \sqrt{\frac{3}{2}}\,|\underset{\sim}{\sigma}_d|, \quad \underset{\sim}{\sigma}_d = \underset{\sim}{\sigma}-\sigma_m\underset{\sim}{I}_{(2)}$$

$$p'(\underset{\sim}{\sigma}) = -\sigma_m$$

The notations q and p' are commonly used in the literature for granular materials (like soil, rock) to denote the second stress deviator invariant and the first stress invariant respectively.

In the special case when $F = F(p'(\underset{\sim}{\tau}),q(\underset{\sim}{\tau}))$ we may conclude from the fact that $\underset{\sim}{\sigma}$ is the solution of the minimization problem (8) that $\underset{\sim}{\sigma}_d$ is proportional to $\underset{\sim}{\sigma}_d^N$ (the deviatoric part of $\underset{\sim}{\sigma}^N$) and so we obtain $\underset{\sim}{\sigma} = \Pi\underset{\sim}{\sigma}^N$ as

$$\underset{\sim}{\sigma} = \frac{q}{q^N}\underset{\sim}{\sigma}_d^N - p'\underset{\sim}{I}_{(2)} \tag{11}$$

where (p',q) is the solution of

$$\min_{(\bar{p},\bar{q})} \frac{2}{q}(q^N-\bar{q})^2+(p'^N-\bar{p}')^2 \qquad (12)$$

subjected to the constrain $F(\bar{p}',\bar{q}) \leq 0$, and where $q^N = q(\underset{\sim}{\nabla}^N)$, $p'^N = p'(\underset{\sim}{\sigma}^N)$.

The yield function due to Drucker and Prager may be written in the extreme case of no cohesion

$$F(p',q) = q-mp' \qquad (13)$$

where

$$m = m_1(\Phi)\sin\Phi, \quad m_1(0) = 2$$

and where Φ is the angle of internal friction. By adopting Mohr-Coulomb´s hypothesis in compression we have, typically, $m_1(\Phi) = 6/(3-\sin\Phi)$.

Minimizing (12) under the constraint (13) gives the solution

$$\left.\begin{array}{ll} p' = \dfrac{1}{\frac{9}{2}+m^2}\left(\frac{9}{2}p'^N+mq^N\right), & \text{when } p'^N+\frac{2m}{9}q^N > 0 \\ q = mp' & \\ p' = 0, \quad q = 0 & \text{when } p'^N+\frac{2m}{9}q^N \leq 0 \end{array}\right. \qquad (14)$$

which is shown in Fig 1. The stress free state (second alternative) means that flow occurs without inter-particle contact.

Solutions:

a: $p'^N+\frac{2m}{9}q^N > 0$

b: -"- < 0

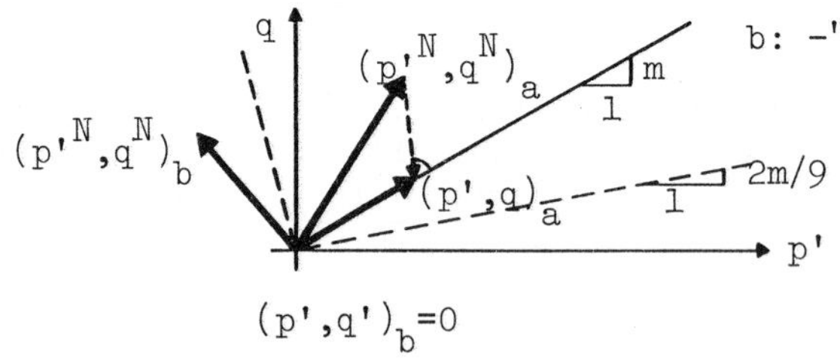

Figure 1. Stress projection for two principally different Newtonian stress states.

4. FINITE ELEMENT METHOD

We introduce the space V of admissable velocities (the regularity properties are not discussed in this paper). The variational formulation of the equation of motion in (5) is: Find $\underset{\sim}{v} \in V$ which for all $\underset{\sim}{w} \in V$ satisfies

$$\int_\Omega \rho(\dot{\underset{\sim}{v}}+\underset{\sim}{v}\cdot\underset{\sim}{\nabla}\underset{\sim}{v})\,\underset{\sim}{w}\,dx + \int_\Omega \Pi\underset{\sim}{\sigma}^N(\underset{\sim}{\varepsilon}(\underset{\sim}{v})):\underset{\sim}{\varepsilon}(\underset{\sim}{w})\,dx = \int_\Omega \underset{\sim}{f}\cdot\underset{\sim}{w}\,dx \quad (15)$$

where $\Pi\underset{\sim}{\sigma}^N(\underset{\sim}{\varepsilon})$ is given by (11) and (14) in our particular case. The formulation (15) is, of course, general with respect to any choice of yield criterion. In (15) it has also been assumed that the boundary values on Γ_T are homogeneous without loss of generality.

Let us introduce finite element approximations for the velocities by choosing $\underset{\sim}{v} \in \tilde{V} \subset V$. These approximations are constructed in the usual way (using matrix notation subsequently)

$$\underset{\sim}{v} = \Phi^T(x)v \quad (16)$$

where $\Phi(x)$ contains a set of polynomials and v is a set of nodal values. In the numerical example in this paper isoparametric eight-node elements were chosen (with biquadratic approximation in the canonical square shape).

We now introduce the notations

$$M = \int_\Omega \rho\Phi\Phi^T\,dx$$

$$N_C(v) = \int_\Omega \rho\Phi\,(\nabla\underset{\sim}{v}^T)^T\underset{\sim}{v}\,dx$$

$$N_S(v) = \int_\Omega (\tilde{\nabla}\Phi^T)^T\underset{\sim}{\sigma}(\underset{\sim}{\varepsilon}(\underset{\sim}{v}))\,dx$$

$$F = \int_\Omega \Phi\underset{\sim}{f}\,dx$$

where the operators ∇ and $\tilde{\nabla}$ are defined by $\nabla\Phi^T v = \underset{\sim}{\nabla}\underset{\sim}{v}$ and $\tilde{\nabla}\Phi^T v = \underset{\sim}{\varepsilon}(\underset{\sim}{v})$.

The finite element problem is than to find v from the discrete equivalent of (15)

$$M\dot{v}+N_C(v)+N_S(v) = F \quad (17)$$

Assuming that v_n is known at $t = t_n$ we may integrate (17) using the simple backward Euler method (which is L-stable), which leads to the non-linear system for $t = t_{n+1} = t_n+\Delta t$

$$Mv_{n+1}+\Delta tN_C(v_{n+1})+\Delta tN_S(V_{n+1}) = Mv_n+\Delta tF_n \tag{18}$$

Using a predictor-corrector algorithm, where the predictor is simply the forward Euler extrapolation and the corrector is a sequence of (modified) Newton iterations of (18) using the predicted value as starting solution, we may solve for v_{n+1}. For each time step the algorithm becomes:

Prediction:

$$v_{n+1}^{(0)} = v_n+\Delta tM^{-1}[F_n-N_C(v_n)-N_S(v_n)] \tag{19a}$$

Correction:

$$v_{n+1}^{(k+1)} = v_{n+1}^{(k)}+\alpha d^{(k)} \quad , \quad k = 1,2,\ldots$$

$$Kd^{(k)} = \Delta t[F_{n+1}-N_C(v_{n+1}^{(k)})-N_S(v_{n+1}^{(k)})] - M(v_{n+1}^{(k)}-v_n) \tag{19b}$$

where α is an acceleration factor and K is an approximation of the Jacobian of the nonlinear equation (18), c.f. Gresho et al [5]. The optimal choice of K is problem dependent but in general K should be chosen as close as possible to the Jacobian in order to achieve rapid convergence. The Jacobian J(v) is defined as

$$J(v) = M+\Delta t\left[\frac{\partial N_C}{\partial v}(v)+\frac{\partial N_S}{\partial v}(v)\right] \tag{20}$$

The term $\partial N_c/\partial v$ is disregarded in order to avoid a non-symmetric coefficient matrix K. The last matrix $\partial N_s/\partial v$ may be evaluated explicitly, as the expression $\partial \sigma/\partial \varepsilon$ is well-defined in our case. However, to simplify calculation we restrict to the choice

$$K \simeq M+\Delta tS_N \tag{21}$$

where the Newtonian matrix S_N is given by

$$S_N = \int_\Omega (\tilde{\nabla}\Phi^T)^T D\tilde{\nabla}\Phi^T \, dx \tag{22}$$

Using this iterative method in conjunction with $\alpha = 1$ for the numerical example presented subsequently, convergence was in general achieved within ten iterations. It may be noticed that no refactoring of K has to be done throughout the time-dependent process (unless Δt is changed) as K now is independent of the solution. It is also worth noticing that the elementary choice K = M leads to a formally explicit algorithm for the case of a diagonal mass matrix although the backward Euler method is implicit in character.

5. NUMERICAL EXAMPLE

Consider the model problem of plane flow of bulk material in the rectangular box shown in Fig. 2 (which also shows the finite element mesh). Results for different values of the initial stresses and for both rough and smooth walls were compared while other data were held fixed.

The values $\mu = 1.0$ GPas, $\Phi = 40^{\circ}$ were chosen. The principal stress axes in situ were assumed vertical and horisontal. The initial vertical stress $\sigma_{z,0}$ was assumed to vary linearly with depth with the rate of increase ρg, and the choice $\rho = 1350$ kg/m^3 was made. The horisontal stresses were defined as $\sigma_{x,0} = \sigma_{y,0} = K_0\sigma_{z,0}$ where K_0 is a constant to be chosen.

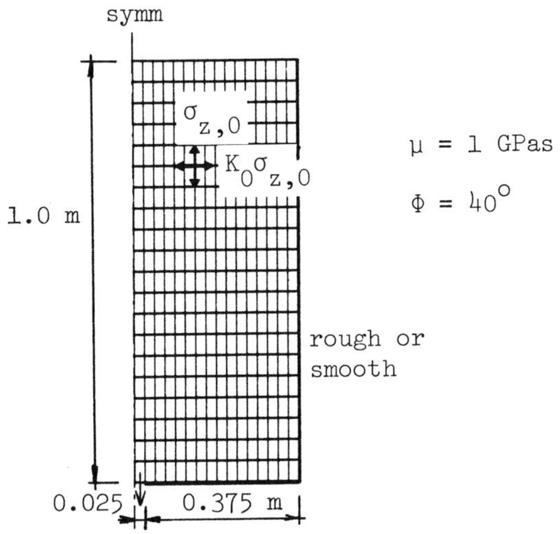

Figure 2. Granular material flowing through bottom opening of model box.

Plots of the velocity field and of the spread of the plastic zone as functions of time up to stationary conditions are given in Figs. 3 and 4 for the case of a rough wall and in Figs. 5 and 6 for the case of a smooth wall. In both cases the initial stresses were given by the choice $K_0 = 0.35$.

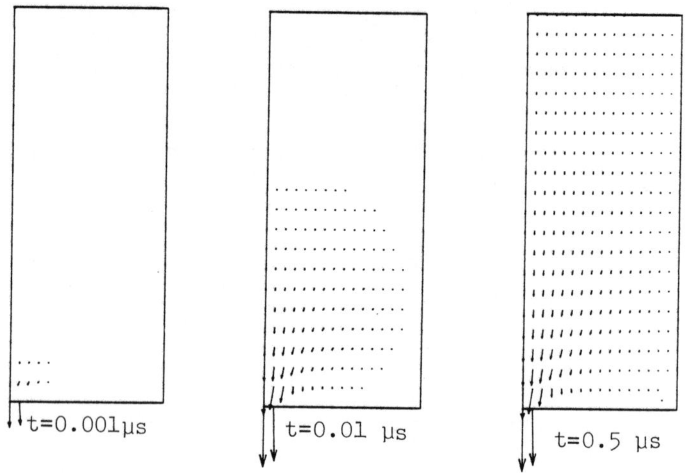

Figure 3. Development of velocity field for rough wall.

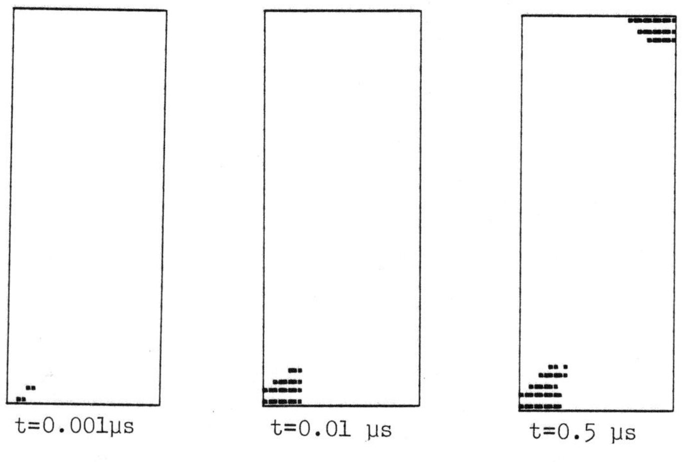

Figure 4. Development of plastic zone (flowzone) for rough wall.

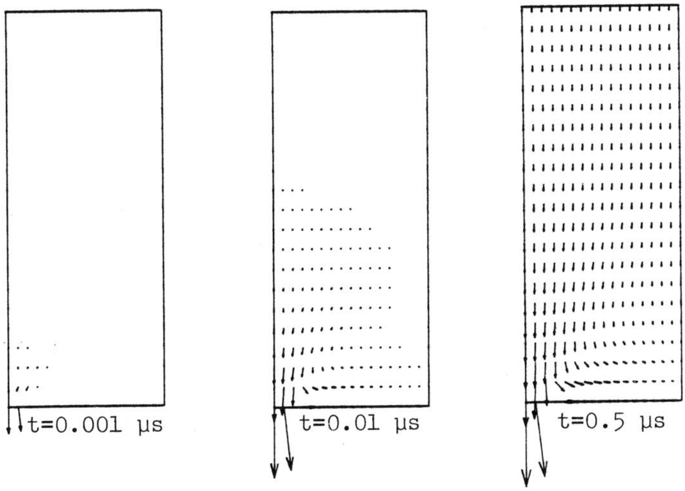

Figure 5. Development of velocity field for smooth wall.

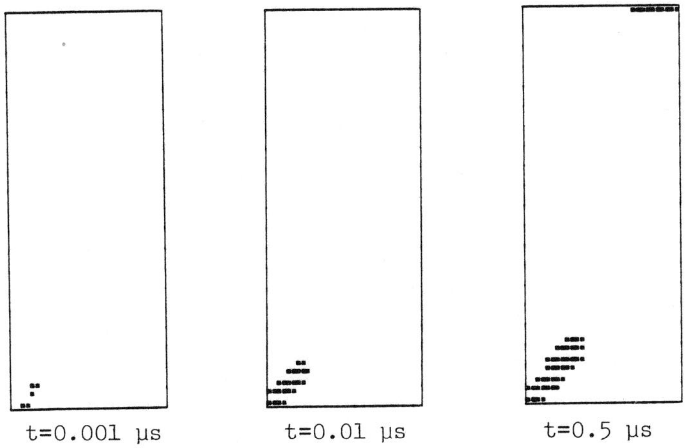

Figure 6. Development of plastic zone (flowzone) for smooth wall.

Finally, the sensibility in the results due to a change of K_0 is investigated. The development of the velocity field for a smooth wall and some selected values of K_0 are depicted in Fig. 7.

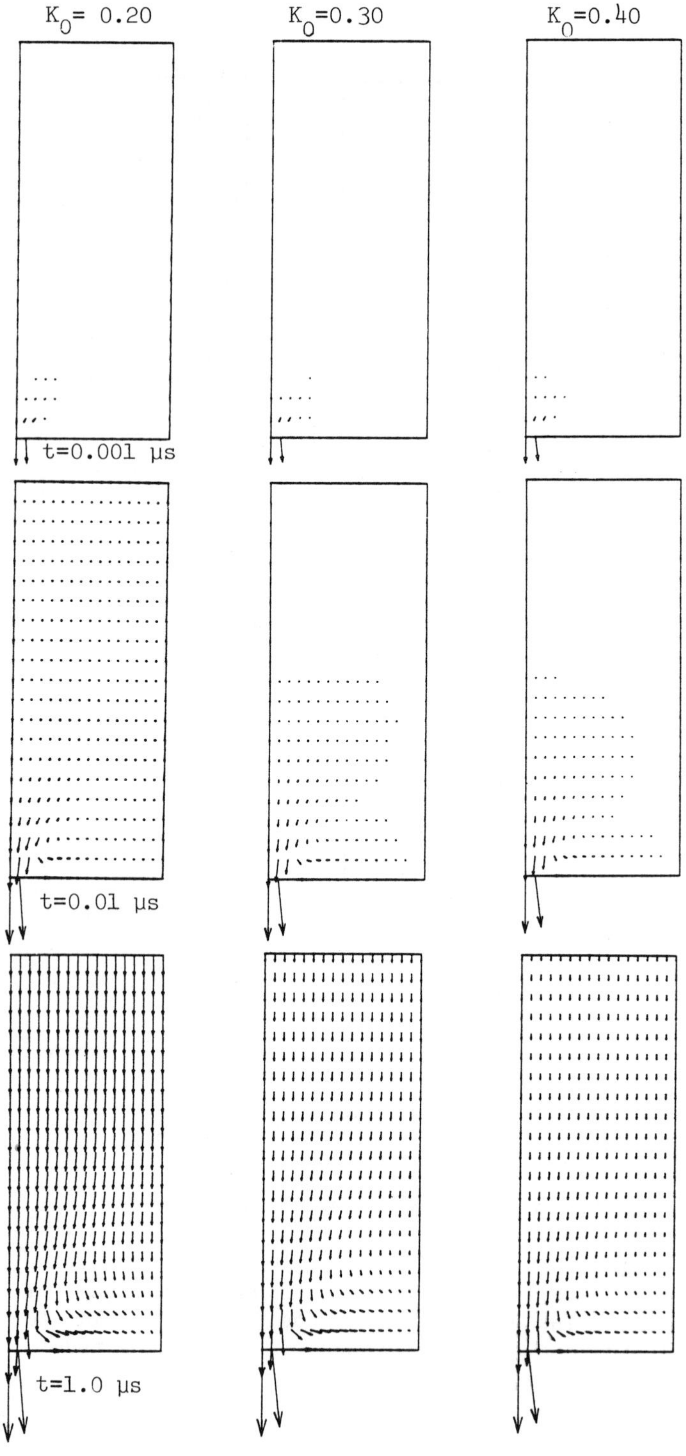

Figure 7. Development of velocity fields for smooth wall

6. CONCLUSION

Crushed ore has been modelled as a viscous-plastic fluid. The stress may be found as the projection (in this case in the Euclidean metric) of the "fictitious" Newtonian stress onto the plastically admissable stress region (in each spatial point) bounded by the yield surface. As the Newtonian stress is a linear function of the rate of deformation the complete solution is known when the velocity is known. This corresponds to a "displacement method" after finite element discretization of the velocity field.

The analyzed model problem showed that the choice of initial stress state as well as the roughness of fixed walls have significant influence on the flow pattern and the magnitude of velocities. A more thorough investigation should include the influence of the angle of internal friction and the Newtonian viscosity.

The iterative algorithm in each time step may be accelerated by a more efficient choice of iteration matrix K and by optimizing the acceleration factor.

7. REFERENCES

[1] Enstad, G.G., Heden, H. and Nilsson, L. - Gravity flow of particulate solids in mines and silos, The Chr. Michelsen Institute, Ref. No. 823105-1, Bergen 1983.

[2] JENIKE, A.W. and SHIELD, R.T. - On the Plastic Flow of Coulomb Solids beyond Original Failure, ASME, J. Appl. Mech., Vol. 26, pp. 599-602, 1959.

[3] BRENNEN, C and PEARCE, J.C. - Granular Material Flow in Two-dimensional Hoppers, ASME, J. Appl. Mech., Vol. 45, pp. 43-50, 1978.

[4] RUNESSON, K. and BOOKER, J.R. - On Mixed and Displacement Finite Element Methods in Perfect Elasto-Plasticity, Fourth Int. Conf. on FEM in Australia, Melbourne 18-20 Aug. 1982.

[5] GRESHO, P., LEE, R.L. and SANI, R.L. - On the Time Dependent Solution of the Incompressible Navier-Stokes Equations in two and three Dimensions, Recent Advances in Numerical Methods in Fluids, Ed. Taylor, C. and Morgan, K., Pineridge Press Ltd., 1980.

SECTION 10

FREE AND FORCED CONVECTION

TIME DEPENDENT DIFFUSION CONVECTION PROBLEMS USING BOUNDARY ELEMENTS

C. Brebbia and P. Skerget

Computational Mechanics Centre, Southampton and University of Maribor, Yugoslavia.

1. INTRODUCTION

In this paper the increasingly popular boundary element technique [1] is developed to solve time dependent transport problems governed by the convective-diffusion equation with particular application to heat transfer.

The solution is extended to heat non-linear material properties which are linearized using Kirchoff's transform. This method has been successfully applied by the authors to solve diffusion problems with non-linear conductivity functions [2 , 3] for which problems the technique proved to be computationally very efficient.

A time and space dependent fundamental solution for parabolic type equations is applied in this paper. This effectively reduces the dimensionality of the problem in time and space but because of the presence of convective terms domain integrals need to be computed as well. These integrals considerably increase the amount of computation required and special care has to be taken of the manner in which the numerical operations are performed. The final system of boundary integral equations is linear and can be solved using simple Gauss elimination.

Although this paper only describes the solution of the transport equation, the procedure can easily be extended to solve fluid flow problems such as those governed by the Navier-Stokes equations. These equations can be expressed in terms of a vorticity transport equation which can be interpreted as inhomogeneous parabolic equations [5,6,7,8,9]

1.2 Governing equations

Let us consider the problem governed by the diffusion-convective equation for homogeneous, isotropic media with constant conductivity k_o,

$$\frac{\partial u}{\partial t} = a\nabla^2 u - \left\{ v_x \frac{\partial u}{\partial x} + v_y \frac{\partial u}{\partial y} + v_z \frac{\partial u}{\partial z} \right\} \tag{1.1}$$

This represents an inhomogeneous parabolic partial equation for the scalar function $u(s,t)$, where c is specific heat, ρ is density and a is diffusivity ($a = k_o/c\rho$). The velocity vector is $v = (v_x, v_y, v_z)$.

The linear boundary conditions corresponding to equation (1.1) are

$$\begin{array}{lll} \text{Essential} & u = \bar{u} & \text{on } \Gamma_1 \\ \text{Natural} & k_o \dfrac{\partial u}{\partial n} = -\bar{q} & \text{on } \Gamma_2 \end{array} \tag{1.2}$$

The initial conditions needed for the solution of equation (1.1) can be written as follows,

$$u = u_i \qquad \text{in } \Omega \text{ at } t = t_i \tag{1.3}$$

When higher order time-interpolation functions are introduced, the fluxes on the boundary have also to be prescribed, i.e.

$$q = q_i \qquad \text{on } \Gamma \quad \text{at } t = t_i \tag{1.4}$$

If the conductivity k depends on the potential the governing equation (1.1) becomes non-linear, i.e.

$$\begin{aligned} c\rho \frac{\partial u}{\partial t} &= \frac{\partial}{\partial x}\left(k \frac{\partial u}{\partial x}\right) + \frac{\partial}{\partial y}\left(k \frac{\partial u}{\partial y}\right) + \frac{\partial}{\partial z}\left(k \frac{\partial u}{\partial z}\right) \\ &\quad - \left(v_x \frac{\partial u}{\partial x} + v_y \frac{\partial u}{\partial y} + v_z \frac{\partial u}{\partial z}\right) \end{aligned} \tag{1.5}$$

Kirchoff's transform

When the non-linearities in conductivity are function of the potential, the problem can be rendered into a linear one by using Kirchoff's transform. This operation results in a linear system when the problem has natural and essential boundary conditions only. The Kirchoff's transform [2,3] provides then an economic and efficient way of solving the initially non-linear problem.

The transform is defined as,

$$\psi = K[u] = \int_{u_o}^{u} k(u)du \tag{1.6}$$

$$\frac{\partial \psi}{\partial t} = k \frac{\partial u}{\partial t}$$

$$\frac{\partial \psi}{\partial x} = k \frac{\partial u}{\partial x} \; ; \quad \frac{\partial \psi}{\partial y} = k \frac{\partial u}{\partial y} \quad ; \quad \frac{\partial \psi}{\partial z} = k \frac{\partial u}{\partial z}$$

where u_o is an arbitrary reference value. Equation (1.5) can be transformed into a linear one in terms of the new variable $\psi(u)$ which produces,

$$\frac{\partial \psi}{\partial t} = a\nabla^2 \psi - \left(v_x \frac{\partial \psi}{\partial x} + v_y \frac{\partial \psi}{\partial y} + v_z \frac{\partial \psi}{\partial z}\right) \tag{1.7}$$

Thus the form of equation (1.1) is preserved.

The transformed boundary conditions (1.2) corresponding to equation (1.7) are:

$$\begin{aligned} &\text{Essential} \quad \psi = \bar{\psi} = K[\bar{u}] \quad \text{on } \Gamma_1 \\ &\text{Natural} \quad \frac{\partial \psi}{\partial n} = -\bar{q} \quad \text{on } \Gamma_2 \end{aligned} \tag{1.8}$$

The initial conditions are now,

$$\psi = \psi_i = K[u_i] \quad \text{in } \Omega \text{ at } t = t_i \tag{1.9}$$

and

$$q = q_i \quad \text{on } \Gamma \text{ at } t = t_i \tag{1.10}$$

1.3 Inhomogeneous parabolic equation

For problems described by the scalar inhomogeneous parabolic equation, i.e.

$$\frac{\partial u(s,t)}{\partial t} - a\nabla^2 (s,t) = b \quad \text{in } \Omega \tag{1.11}$$

a boundary integral equation relating boundary values for potentials and its normal derivatives can be obtained by weighting the governing equation and boundary conditions by a function $\overset{*}{u}$. As the problem is time dependent the equation has to be weighted also with respect to time. This yields the following weighted residual statement,

$$\int_{t_o}^{t_F}\int_{\Omega} [a\nabla^2 u(s,t) - \frac{\partial u(s,t)}{\partial t} + b]\overset{*}{u}(\xi,t_F;s,t)d\Omega(s)dt =$$

$$\int_{t_o}^{t_F}\int_{\Gamma_2} a[q(S,t) - \bar{q}(S,t)]\overset{*}{u}(\xi,t_F;s,t)d\Gamma(S)dt \qquad (1.12)$$

$$-\int_{t_o}^{t_F}\int_{\Gamma_1} a[u(S,t) - \bar{u}(S,t)]\overset{*}{q}(\xi,t_F;S,t)d\Gamma(S)dt$$

where t_o is the initial and t_F is the final time,

$\overset{*}{q}(\xi,t_F;s,t) = \partial\overset{*}{u}(\xi,t_F;s,t)/\partial n(s)$ and $t_o \leq t \leq t_F$.

Integrating by parts twice the Laplacian with respect to x_i gives,

$$\int_{t_o}^{t_F}\int_{\Omega} a\nabla^2\overset{*}{u}(\xi,t_F;,s,t)u(s,t)\Omega(s)dt - \int_{t_o}^{t_F}\int_{\Omega} \frac{\partial u(s,t)}{\partial t}\overset{*}{u}(\xi,t_F;s,t)d\Omega(s)d$$

$$+\int_{t_o}^{t_F}\int_{\Omega} b\overset{*}{u}(\xi,t_F;s,t)d\Omega(s)dt = \int_{t_o}^{t_F}\int_{\Gamma} a\, u(S,t)\overset{*}{q}(\xi,t_F,S,t)d\Gamma(S)dt$$

$$-\int_{t_o}^{t_F}\int_{\Gamma} a\, q(S,t)\overset{*}{u}(\xi,t_F;S,t)d\Gamma(S)dt \qquad (1.13)$$

Integrating now by parts the time derivative produces the following expression,

$$\int_{t_o}^{t_F}\int_{\Omega} [a\nabla^2 \overset{*}{u}(\xi,t_F;s,t) + \frac{\partial \overset{*}{u}(\xi,t_F;s,t)}{\partial t}] u(s,t)d\Omega(s)dt$$

$$+ \int_{t_o}^{t_F}\int_{\Omega} b\, \overset{*}{u}(\xi,t_F;s,t)d\Omega(s)dt - [\int_{\Omega} \overset{*}{u}(\xi,t_F;s,t)u(s,t)d\Omega(s)]_{t=t_o}^{t=t_F}$$

$$= \int_{t_o}^{t_F}\int_{\Gamma} a\, u(S,t)\overset{*}{q}(\xi,t_F;S,t)d\Gamma(S)dt - \int_{t_o}^{t_F}\int_{\Gamma} aq(S,t)\overset{*}{u}(\xi,t_F;S,t)d\Gamma(S)dt \qquad (1.14)$$

Let $\overset{*}{u}$ be the fundamental solution of the parabolic differential equation defined by [4].

$$a\nabla^2 \overset{*}{u}(\xi,t_F;s,t) + \frac{\partial \overset{*}{u}(\xi,t_F;s,t)}{\partial t} + \delta(\xi,s)\delta(t_F,t) = 0 \qquad (1.15)$$

where δ are Dirac delta functions given by

$$\overset{*}{u}(\xi,t_F;s,t) = \begin{cases} 0 & t > t_F \\ \dfrac{1}{(4\pi a\tau)^{d/2}} \exp\left(-\dfrac{r^2(\xi,s)}{4a\tau}\right) & t < t_F \end{cases} \qquad (1.16)$$

where $\tau = t_F - t$; and d is the spatial dimension of the problem. The fundamental solution represents the effect of the unit point source applied at the point ξ at time t_F on the reference point s at time t in an infinite region and possesses the following properties.

$$a\nabla^2 \overset{*}{u}(\xi,t_F;s,t) + \frac{\partial \overset{*}{u}(\xi,t_F;s,t)}{\partial t} = 0 \quad \text{in } \Omega \text{ and } t < t_F \qquad (1.17)$$

$$\int_{\Omega} u(s,t)\overset{*}{u}(\xi,t_F;s,t)d\Omega(s) = u(\xi,t_F) \quad \text{for } t = t_F \qquad (1.18)$$

For the two dimensional problems which will be discussed in this chapter the fundamental solution and its normal derivative are given by

$$\overset{*}{u}(\xi,t_F;s,t) = \frac{1}{4\pi a\tau}\exp\left(-\frac{r^2(\xi,s)}{4a\tau}\right)$$

$$\overset{*}{q}(\xi,t_F;s,t) = \frac{d(\xi,s)}{8\pi a^2\tau^2}\exp\left(-\frac{r^2(\xi,s)}{4a\tau}\right) \tag{1.19}$$

where $d(\xi,s) = (X(\xi)-X(s)).n_x(S) + (Y(\xi) - Y(s)).n_y(S)$ and $n_x(S)$, $n_y(S)$ are the direction cosines of the normal at the point S.

In order to investigate the singularity that occurs in the integrals of equation (1.14) at time $t = t_F$ at the peak of the Dirac delta function an arbitrarily small quantity ε can be subtracted or added to the upper limit of the integrals. The first domain integral on the left hand side of (1.14) is zero for $t_o \leq t < t_F - \varepsilon$ because of equation (1.17). Taking the limit as $\varepsilon \to 0$ and accounting for condition (1.18) equation (1.14) becomes,

$$u(\xi,t_F) + \int_{t_o}^{t_F}\int_\Gamma a\, u(S,t)\overset{*}{q}(\xi,t_F;S,t)d\Gamma(S)dt =$$

$$\int_{t_o}^{t_F}\int_\Gamma a\, q(S,t)\overset{*}{u}(\xi,t_F;S,t)d\Gamma(S)dt$$

$$+ \int_\Omega u_o(s,t=t_o)\overset{*}{u}(\xi,t_F;s,t=t_o)d\Omega(s) \tag{1.20}$$

$$+ \int_{t_o}^{t_F}\int_\Omega b\,\overset{*}{u}(\xi,t_F;s,t)d\Omega(s)dt$$

Formula (1.20) is valid for any point inside the domain, but in order to obtain a boundary integral equation, the point ξ has to be taken to the boundary yielding the following boundary equation,

$$c(\xi)u(\xi,t_F) + \int_{t_o}^{t_F}\int_\Gamma a\, u(S,t)\overset{*}{q}(\xi,t_F;S,t)d\Gamma(S)dt =$$

$$\int_{t_o}^{t_F}\int_\Gamma a\, q(S,t)\overset{*}{u}(\xi,t_F;S,t)d\Gamma(S)dt \tag{1.21}$$

$$+ \int_{\Omega} u_o(s,t=t_o)\overset{*}{u}(\xi,t_F;,s,t=t_o)d\Omega(s)$$

$$+ \int_{t_o}^{t_F}\int_{\Omega} b\overset{*}{u}(\xi,t_F;s,t)d\Omega(s)dt$$

where the coefficient $c(\xi)$ is:

$c(\xi) = 1$ ξ is in domain Ω
$c(\xi) = \frac{1}{2}$ ξ lies on a smooth boundary Γ
$c(\xi) = (\pi+\alpha_1-\alpha_2)/2\pi$ ξ lies on a non-smooth boundary Γ

α_1 and α_2 are defined in figure 1.1. It is more convenient to write equation (1.22) in incremental form, i.e. from t_{F-1} to t_F, as

$$c(\xi)u(\xi,t_F) + \int_{t_{F-1}}^{t_F}\int_{\Gamma} a\, u(S,t)\overset{*}{q}(\xi,t_F;S,t)d\Gamma(S)dt =$$

$$\int_{t_{F-1}}^{t_F}\int_{\Gamma} a\, q(S,t)\overset{*}{u}(\xi,t_F;S,t)d\Gamma(S)dt \qquad (1.22)$$

$$+ \int_{\Omega} u_{F-1}(s,t=t_{F-1})\overset{*}{u}(\xi,t_F;s,t=t_{F-1})d\Omega(s)$$

$$+ \int_{t_{F-1}}^{t_F}\int_{\Omega} b\,\overset{*}{u}(\xi,t_F;s,t)d\Omega(s)dt$$

1.4 Convective terms

Now, one can easily write the boundary integral equation for the diffusion-convective equation by taking the body force part (b) equivalent to the convective term in (1.1),

$$c(\xi)u(\xi,t_F) + \int_{t_{F-1}}^{t_F}\int_{\Gamma} a\, u(S,t)\overset{*}{q}(\xi,t_F;S,t)d\Gamma(S)dt =$$

$$\int_{t_{F-1}}^{t_F} \int_\Gamma a\ q(S,t)\overset{*}{u}(\xi,t_F;S,t)d\Gamma(S)dt \tag{1.23}$$

$$+ \int_\Omega u_{F-1}(s,t=t_{F-1})\overset{*}{u}(\xi,t_F;s,t=t_{F-1})d\Omega(s)$$

$$- \int_{t_{F-1}}^{t_F} \int_\Omega \left\{v_x(s,t)\ \frac{\partial u(s,t)}{\partial x(s)} + v_y(s,t)\ \frac{\partial u(s,t)}{\partial y(s)}\right\}\overset{*}{u}(\xi,t_F;s,t)d\Omega(s)dt$$

The last domain integral contains derivatives of potentials. Integrating by parts it can be transformed into,

$$\int_{t_{F-1}}^{t_F} \int_\Omega \left\{v_x(s,t)\ \frac{\partial u(s,t)}{\partial x(s)} + v_y(s,t)\ \frac{\partial u(s,t)}{\partial y(s)}\right\} \overset{*}{u}(\xi,t_F;s,t)d\Omega(s)dt =$$

$$\int_{t_{F-1}}^{t_F} \int_\Gamma \left\{v_x(S,t)n_x(S) + v_y(S,t)n_y(S)\right\}u(S,t)\overset{*}{u}(\xi,t_F;S,t)d\Gamma(S)dt \tag{1.24}$$

$$- \int_{t_{F-1}}^{t_F} \int_\Omega \left\{v_x(s,t)\ \frac{\partial \overset{*}{u}(\xi,t_F;s,t)}{\partial x(s)} + v_y(s,t)\ \frac{\partial \overset{*}{u}(\xi,t_F;s,t)}{\partial y(s)}\right\}u(s,t)\ \times$$

$$d\Omega(s)dt$$

Introducing (1.24) in (1.23) it follows,

$$c(\xi)u(\xi,t_F) + \int_{t_{F-1}}^{t_F} \int_\Gamma a\ u(S,t)\overset{*}{q}(\xi,t_F;S,t)d\Gamma(S)dt =$$

$$\int_{t_{F-1}}^{t_F} \int_\Gamma a\ q(S,t)\overset{*}{u}(\xi,t_F;S,t)d\Gamma(S)dt \tag{1.25}$$

$$- \int_{t_{F-1}}^{t_F} \int_\Gamma u(S,t)\{v_x(S,t)n_x(S)+v_y(S,t)n_y(S)\}\overset{*}{u}(\xi,t_F;S,t)d\Gamma(S)dt\ +$$

$$+ \int_{\Omega} u_{F-1}(s,t=t_{F-1})\overset{*}{u}(\xi,t_F;s,t=t_{F-1})d\Omega(s)$$

$$+ \int_{t_{F-1}}^{t_F} \int_{\Omega} u(s,t)\left\{v_x(s,t)\frac{\partial\overset{*}{u}(\xi,t_F;s,t)}{\partial x(s)} + v_y(s,t)\frac{\partial\overset{*}{u}(\xi,t_F;s,t)}{\partial y(s)}\right\}$$

$$\times\ d\Omega(s)dt$$

Equation (1.25) is more convenient to use in computation because it does not contain the derivatives of the potential. Notice that equation (1.25) describes the transport process in an integral form.

The first two boundary integrals represent the diffusion from the boundary and they are similar to those integrals used for the diffusion equation. The third boundary integral describes convection from the boundary. For the external flow this boundary integral may vanish because on the internal or solid boundary the velocity is zero ($\underline{v} = 0$ on Γ_S) and on the external boundary the potential is zero ($u = 0$ on Γ_∞).

The first domain integral represents the contribution of the initial conditions. The second domain integral is due to the effect of convection.

The derivatives of the fundamental solution with respect to x and y in (1.25) can be evaluated as,

$$\frac{\partial\overset{*}{u}(\xi,t_F;s,t)}{\partial x(s)} = \frac{d_x(\xi,s)}{8\pi a^2\tau^2}\exp\left\{-\frac{r^2(\xi,s)}{4a\tau}\right\}$$

$$\frac{\partial\overset{*}{u}(\xi,t_F;s,t)}{\partial y(s)} = \frac{d_y(\xi,s)}{8\pi\ a^2\tau^2}\exp\left\{-\frac{r^2(\xi,s)}{4a\tau}\right\} \qquad (1.26)$$

where $d_x(\xi,s) = x(\xi)-x(s)$ and $d_y(\xi,s) = y(\xi) - y(s)$.

1.5 Boundary element discretisation

The values of the variables u, q and v_x, v_y are assumed to vary within each element and each time step according to the space, ϕ, and time, ψ, interpolation functions i.e.;

$$
\begin{aligned}
u(\eta,t) &= \underset{\sim}{\phi}^T \underset{\sim}{\psi} \underset{\sim}{U}^n_m \\
q(\eta,t) &= \underset{\sim}{\phi}^T \underset{\sim}{\psi} \underset{\sim}{Q}^n_m \\
v_x(\eta,t) &= \underset{\sim}{\phi}^T \underset{\sim}{\psi} \underset{\sim}{V}^n_{x,m} \\
v_y(\eta,t) &= \underset{\sim}{\phi}^T \underset{\sim}{\psi} \underset{\sim}{V}^n_{y,m}
\end{aligned}
\qquad (1.27)
$$

The index n refers to the number of nodes within each element for which the nodal values of u, q, v_x and v_y are associated and the index m refers to the degree of variation of the function $\underset{\sim}{\psi}$, i.e. m = 1 if $\underset{\sim}{\psi}$ is constant, m = 1,2 if $\underset{\sim}{\psi}$ is linear etc.

Let us assume that the boundary Γ is divided into E elements with N_E boundary nodes and domain Ω discretised into C cells with N_I internal nodes so that the whole number of nodes $M = N_E + N_I$. One can write the discretised form of (1.25) as,

$$
\begin{aligned}
c(\xi)u(\xi,t_F) &+ a \sum_{e=1}^{E} \left[\int_{\Gamma_e} \underset{\sim}{\phi}^T \int_{t_{F-1}}^{t_F} \overset{*}{q}(\xi,t_F;S,t)\underset{\sim}{\psi}\, dt d\Gamma(S) \right] \underset{\sim}{U}^n_m = \\
& a \sum_{e=1}^{E} \left[\int_{\Gamma_e} \underset{\sim}{\phi}^T \int_{t_{F-1}}^{t_F} \overset{*}{u}(\xi,t_F;S,t)\underset{\sim}{\psi}\, dt d\Gamma(S) \right] \underset{\sim}{Q}^n_m \\
& - \sum_{e=1}^{E} \left\{ \int_{\Gamma_e} \underset{\sim}{\phi}^T \int_{t_{F-1}}^{t_F} \left[\underset{\sim}{\phi}^T \underset{\sim}{\psi}\, V^n_{x,m}\, n_x(S) + \underset{\sim}{\phi}^T \underset{\sim}{\psi}\, V^n_{y,m}\, n_y(S) \right] \right. \\
& \qquad \left. \times\, \overset{*}{u}(\xi,t_F;S,t)\underset{\sim}{\psi}\, dt d\Gamma(S) \right\} \underset{\sim}{U}^n_m \\
& + \sum_{c=1}^{C} \left[\int_{\Omega_c} \underset{\sim}{\phi}^T \overset{*}{u}(\xi,t_F;s,t{=}t_{F-1})\, d\Omega(s) \right] \underset{\sim}{U}^n_{t_{F-1}} \\
& + \sum_{c=1}^{C} \left\{ \int_{\Omega_c} \underset{\sim}{\phi}^T \int_{t_{F-1}}^{t_F} \left[\underset{\sim}{\phi}^T \underset{\sim}{\psi}\, V^n_{x,m} \frac{\partial \overset{*}{u}(\xi,t_F;s,t)}{\partial x(s)} + \right. \right. \\
& \qquad \left. \left. \underset{\sim}{\phi}^T \underset{\sim}{\psi}\, V^n_{y,m} \frac{\partial \overset{*}{u}(\xi,t_F;s,t)}{\partial y(s)} \right] \underset{\sim}{\psi}\, dt d\Omega(s) \right\} \underset{\sim}{U}^n_m
\end{aligned}
\qquad (1.28)
$$

The geometrical variation can be represented by expressing the coordinates in terms of the local coordinates η as

$$x(\eta) = \underset{\sim}{\phi}^T \underset{\sim}{x}^n \text{ and similarly for y and z coordinates} \tag{1.29}$$

Since the interpolation functions ϕ are expressed in terms of local coordinates, it is necessary to transform the elements of the surface differential $d\Gamma$ from the global Cartesian system to the local system of coordinates, i.e.

$$d\Gamma = |J|\ d\eta \tag{1.30}$$

and similarly for differential element $d\Omega$ as,

$$d\Omega = |J|\ d\eta_1\ d\eta_2 \tag{1.31}$$

1.6 Linear time interpolation

Let us assume a linear variation on time within each time step for u, q, v_x and v_y according to the following interpolation functions,

$$\psi_1 = \frac{t_F - t}{\Delta t_F} \quad ; \quad \psi_2 = \frac{t - t_{F-1}}{\Delta t_F} \tag{1.32}$$

where $\Delta t_F = t_F - t_{F-1}$. Applying equation (1.28) to all M boundary and internal nodes, the following matrix equations can be obtained,

$$[\underset{\sim}{H}_2 + \underset{\sim}{C}_2 - \underset{\sim}{D}_2].\underset{\sim}{U}_F - \underset{\sim}{G}_2.\underset{\sim}{Q}_F = [-\underset{\sim}{H}_1 - \underset{\sim}{C}_1 + \underset{\sim}{D}_1 + \underset{\sim}{B}].\underset{\sim}{U}_{F-1} + \underset{\sim}{G}_1.\underset{\sim}{Q}_{F-1} \tag{1.33}$$

where the elements of the matrices are given by terms such as,

$$h^n_{e1} = \frac{a}{\Delta t_F} \int_{\Gamma_e} \phi^n \int_{t_{F-1}}^{t_F} (t_F - t)\overset{*}{q}(\xi, t_F; S, t)\,dt\,d\Gamma(S)$$

$$h^n_{e2} = \frac{a}{\Delta t_F} \int_{\Gamma_e} \phi^n \int_{t_{F-1}}^{t_F} (t - t_{F-1})\overset{*}{q}(\xi, t_F; S, t)\,dt\,d\Gamma(S)$$

$$g^n_{e1} = \frac{a}{\Delta t_F} \int_{\Gamma_e} \phi^n \int_{t_{F-1}}^{t_F} (t_F - t)\overset{*}{u}(\xi, t_F; S, t)\,dt\,d\Gamma(S)$$

$$g^n_{e2} = \frac{a}{\Delta t_F} \int_{\Gamma_e} \phi^n \int_{t_{F-1}}^{t_F} (t - t_{F-1})\overset{*}{u}(\xi, t_F; S, t)\,dt\,d\Gamma(S)$$

$$b^n_c = \int_{\Omega_c} \phi^n\ \overset{*}{u}(\xi, t_F; s, t = t_{F-1})\,d\Omega(s)$$

$$c_{e1}^{n} = \frac{1}{\Delta t_F^2} \int_{\Gamma_e} \phi^n \int_{t_{F-1}}^{t_F} (t_F - t)[(t_F - t)\underset{\sim}{\phi}^T V_{x,1}^n + (t - t_{F-1})\underset{\sim}{\phi}^T V_{x,2}^n]$$

$$\times\ n_x(S)\overset{*}{u}(\xi, t_F, S, t)\,dt\,d\Gamma(S)$$

$$+ \frac{1}{\Delta t_F^2} \int_{\Gamma_e} \phi^n \int_{t_{F-1}}^{t_F} (t_F - t)[(t_F - t)\underset{\sim}{\phi}^T V_{y,1}^n + (t - t_{F-1})\underset{\sim}{\phi}^T V_{y,2}^n]$$

$$\times\ n_y(S)u^{*}(\xi, t_F; S, t)\,dt\,d\Gamma(S)$$

$$c_{e2}^{n} = \frac{1}{\Delta t_F^2} \int_{\Gamma_e} \phi^n \int_{t_{F-1}}^{t_F} (t - t_{F-1})[(t_F - t)\underset{\sim}{\phi}^T V_{x,1}^n + (t - t_{F-1})\underset{\sim}{\phi}^T V_{x,2}^n]$$

$$\times\ n_x(S)u^{*}(\xi, t_F; S, t)\,dt\,d\Gamma(S)$$

$$+ \frac{1}{\Delta t_F^2} \int_{\Gamma_e} \phi^n \int_{t_{F-1}}^{t_F} (t - t_{F-1})[(t_F - t)\underset{\sim}{\phi}^T V_{y,1}^n + (t - t_{F-1})\underset{\sim}{\phi}^T V_{y,2}^n]$$

$$\times\ n_y(S)\overset{*}{u}(\xi, t_F; S, t)\,dt\,d\Gamma(S)$$

$$d_{c,1}^{n} = \frac{1}{\Delta t_F^2} \int_{\Omega_c} \phi^n \int_{t_{F-1}}^{t_F} (t_F - t)[(t_F - t)\underset{\sim}{\phi}^T V_{x,1}^n + (t - t_{F-1})\underset{\sim}{\phi}^T V_{x,2}^n]$$

$$\times\ \frac{\partial \overset{*}{u}(\xi, t_F; s, t)}{\partial x(s)}\,dt\,d\Omega(s)$$

$$+ \frac{1}{\Delta t_F^2} \int_{\Omega_c} \phi^n \int_{t_{F-1}}^{t_F} (t_F - t)[(t_F - t)\underset{\sim}{\phi}^T V_{y,1}^n + (t - t_{F-1})\underset{\sim}{\phi}^T V_{y,2}^n] \frac{\partial \overset{*}{u}(\xi, t_F; s, t)}{\partial y(s)}$$

$$dt\,d\Omega(s)$$

$$d_{c,2}^{n} = \frac{1}{\Delta t_F^2} \int_{c} \phi^n \int_{t_{F-1}}^{t_F} (t - t_{F-1})[(t_F - t)\underset{\sim}{\phi}^T V_{x,1}^n + (t - t_{F-1})\underset{\sim}{\phi}^T V_{x,2}^n]$$

$$\times\ \frac{\partial \overset{*}{u}(\xi, t_F; s, t)}{\partial x(s)}\,dt\,d\Omega(s)$$

$$+ \frac{1}{\Delta t_F^2} \int_{\Omega_c} \phi^n \int_{t_{F-1}}^{t_F} (t - t_{F-1})[(t_F - t)\underset{\sim}{\phi}^T V_{y,1}^n + (t - t_{F-1})\underset{\sim}{\phi}^T V_{y,2}^n]$$

$$\frac{\partial \overset{*}{u}(\xi, t_F; s, t)}{\partial y(s)}\,dt\,d\Omega(s) \qquad (1.34)$$

The index n refers to the number of boundary nodes within each boundary element or to the number of domain nodes within each domain element and $H_{ij,2} = \hat{H}_{ij,2} + c_i\delta_{ij}$.

Adding all corresponding terms together, one can write

$$\underset{\sim}{A}.\underset{\sim}{U}_F - \underset{\sim}{G}_2.\underset{\sim}{Q}_F = \underset{\sim}{B}.\underset{\sim}{U}_{F-1} + \underset{\sim}{G}_1\underset{\sim}{Q}_{F-1} \qquad (1.35)$$

After applying boundary conditions the above equations can be rewritten as the following linear system

$$A.X = F \qquad (1.36)$$

where A[M×M] is system matrix and X[M] is vector of unknowns.

The time integrals in (1.34) can be computed analytically in terms of the exponential-integral function defined by

$$E_1(x) = \int_x^{\infty} \frac{e^{-t}}{t}\, dt \qquad (1.37)$$

as

$$\int_{t_{F-1}}^{t_F} (t_F - t)\overset{*}{u}(\xi,t_F;s,t)dt = \frac{\Delta t_F}{4\pi a}\,[\exp(-x_{F-1}) - x_{F-1}E_1(x_{F-1})]$$

$$\int_{t_{F-1}}^{t_F} (t - t_{F-1})\overset{*}{u}(\xi,t_F;s,t)dt = \frac{\Delta t_F}{4\pi a}\,[E_1(x_{F-1}) + x_{F-1}E_1(x_{F-1}) - \exp(-x_{F-1})]$$

$$\int_{t_{F-1}}^{t_F} (t_F - t)\overset{*}{q}(\xi,t_F;s,t)dt = \frac{d}{8\pi a^2}\,E_1(x_{F-1})$$

$$\int_{t_{F-1}}^{t_F} (t - t_{F-1})\overset{*}{q}(\xi,t_F;s,t)dt = \frac{d}{8\pi a^2}\,[\exp(-x_{F-1})/x_{F-1} - E_1(x_{F-1})]$$

$$\int_{t_{F-1}}^{t_F} (t_F^2 - 2t_Ft + t^2)\overset{*}{u}(\xi,t_F;s,t)dt = \frac{\Delta t_F^2}{8\pi a}\,[x_{F-1}^2E_1(x_{F-1}) - x_{F-1}\exp(-x_{F-1}) + \exp(-x_{F-1})]$$

$$\int_{t_{F-1}}^{t_F} (t_F t - t^2 - t_F t_{F-1} + t t_{F-1}) \overset{*}{u}(\xi, t_F; s, t)\, dt$$

$$= \frac{\Delta t_F^2}{8\pi a} [-2x_{F-1}E_1(x_{F-1}) - x_{F-1}^2 E_1(x_{F-1}) + x_{F-1}\exp(-x_{F-1}) + \exp(-x_{F-1})]$$

$$\int_{t_{F-1}}^{t_F} (t^2 - 2tt_{F-1} + t_{F-1}^2) \overset{*}{u}(\xi, t_F; s, t)\, dt =$$

$$\frac{\Delta t_F^2}{8\pi a} [2E_1(x_{F-1}) + 4x_{F-1}E_1(x_{F-1}) + x_{F-1}^2 E_1(x_{F-1})$$

$$- x_{F-1}\exp(-x_{F-1}) - 3\exp(-x_{F-1})]$$

$$\int_{t_{F-1}}^{t_F} (t_F^2 - 2t_F t + t^2) \frac{\partial \overset{*}{u}(\xi, t_F; s, t)}{\partial x(s)}\, dt = \frac{d_x \Delta t_F}{8\pi a^2} [-x_{F-1}E_1(x_{F-1})$$

$$+ \exp(-x_{F-1})]$$

$$\int_{t_{F-1}}^{t_F} (t_F t - t^2 - t_F t_{F-1} + t t_{F-1}) \frac{\partial u^*(\xi, t_F; s, t)}{\partial x(s)}\, dt = \frac{d_x \Delta t_F}{8\pi a^2} [E_1(x_{F-1})$$

$$+ x_{F-1}E_1(x_{F-1}) - \exp(-x_{F-1})]$$

$$\int_{t_{F-1}}^{t_F} (t^2 - 2tt_{F-1} + t_{F-1}^2) \frac{\partial \overset{*}{u}(\xi, t_F; s, t)}{\partial x(s)}\, dt = \frac{d_x \Delta t_F}{8\pi a^2} [\exp(-x_{F-1})/x_{F-1}$$

$$- 2E_1(x_{F-1}) - x_{F-1}E_1(x_{F-1}) + \exp(-x_{F-1})]$$

(1.38)

The argument x_{F-1} is defined as follows,

$$x_{F-1} = \frac{r^2}{4a(t_F - t_{F-1})} \tag{1.39}$$

1.6.1 Space integration

The remaining step in the numerical solution of equation (1.28) is the computation of the space integrals. Let us assume a quadratic space variation for the functions u, q, v_x, vy over each boundary element with the geometry represented by straight line segment - figure 1.2.

The interpolation functions $\underset{\sim}{\phi}$ and vectors $\underset{\sim}{U}^n$, $\underset{\sim}{Q}^n$, $\underset{\sim}{V}_x^n$ and $\underset{\sim}{V}_y^n$ are

$$\underset{\sim}{\phi}^n(\eta) = \begin{bmatrix} \frac{1}{2}\eta(\eta-1) \\ (1-\eta^2) \\ \frac{1}{2}\eta(\eta+1) \end{bmatrix} ; \quad \underset{\sim}{U}^n = \begin{bmatrix} u^1 \\ u^2 \\ u^3 \end{bmatrix} ; \quad \underset{\sim}{Q}^n = \begin{bmatrix} q^1 \\ q^2 \\ q^3 \end{bmatrix} ; \quad \underset{\sim}{V}_x^n = \begin{bmatrix} v_x^1 \\ v_x^2 \\ v_x^3 \end{bmatrix} ; \quad \underset{\sim}{V}_y^n = \begin{bmatrix} v_y^1 \\ v_y^2 \\ v_y^3 \end{bmatrix} \tag{1.40}$$

The geometrical shape functions in local coordinates are,

$$\underset{\sim}{\phi}^n = \begin{bmatrix} \frac{1}{2}(1-\eta) \\ \frac{1}{2}(1+\eta) \end{bmatrix} ; \quad \underset{\sim}{x}^n = \begin{bmatrix} x^1 \\ x^3 \end{bmatrix} ; \quad \underset{\sim}{y}^n = \begin{bmatrix} y^1 \\ y^3 \end{bmatrix} \tag{1.41}$$

and the Jacobian is simply half the boundary element length.

The domain is modelled by traingles where a quadratic variation of the function $\underset{\sim}{U}^n$, $\underset{\sim}{V}_x^n$, $\underset{\sim}{V}_y^n$ is again assumed - figure 1.3,

$$\underset{\sim}{\phi}^n(\eta) = \begin{bmatrix} \eta_1(2\eta_1-1) \\ 4\eta_1\eta_2 \\ \eta_2(2\eta_2-1) \\ 4\eta_2\eta_3 \\ \eta_3(2\eta_3-1) \\ 4\eta_3\eta_1 \end{bmatrix} ; \quad \underset{\sim}{U}^n = \begin{bmatrix} u^1 \\ u^2 \\ u^3 \\ u^4 \\ u^5 \\ u^6 \end{bmatrix} ; \quad \underset{\sim}{V}_x^n = \begin{bmatrix} v_x^1 \\ v_x^2 \\ v_x^3 \\ v_x^4 \\ v_x^5 \\ v_x^6 \end{bmatrix} ; \quad \underset{\sim}{V}_y^n = \begin{bmatrix} v_y^1 \\ v_y^2 \\ v_y^3 \\ v_y^4 \\ v_y^5 \\ v_y^6 \end{bmatrix} \tag{1.42}$$

The Cartesian coordinates of points within each domain element are expressed by the following equations

$$x(\eta) = x_1\eta_1 + x_2\eta_2 + x_3\eta_3$$
$$y(\eta) = y_1\eta_1 + y_2\eta_2 + y_3\eta_3 \qquad (1.43)$$

where η_1, η_2, $\eta_3 = 1 - \eta_1 - \eta_2$ are area coordinates - figure 1.4. The Jacobian is equal to twice the area of the triangle. Local area coordinates can be expressed in terms of the Cartesian coordinate system (x,y) [11] as,

$$\eta_i = \frac{1}{2A}(2\mathring{A}_i + b_i x + a_i y) \qquad (1.44)$$

with

$$a_i = x_k - x_j \quad ; \quad b_i = y_j - y_k \quad ; \quad 2\mathring{A}_i = x_j y_k - x_k y_j \qquad (1.45)$$

where i = 1,2,3 ; j = 2,3,1 ; k = 3,1,2 and the area of triangle is

$$A = (b_1 a_2 - b_2 a_1)/2 \qquad (1.46)$$

When the source point does not lie on the element of integration the boundary integrals h^n_{e1}, h^n_{e2}, g^n_{e1}, g^n_{e2}, c^n_{e1} and c^n_{e2} can be evaluated by standard Gauss formulas.

$$h^n_{e1} = \frac{d_{ie}}{16\pi a\Delta t_F} \sum_{k=1}^{K} E_1(x_{F-1})_k \, \phi^n_k \, w_k$$

$$h^n_{e2} = \frac{d_{ie}}{16\pi a\Delta t_F} \sum_{k=1}^{K} [\exp(-x_{F-1})/x_{F-1} - E_1(x_{F-1})]_k \, \phi^n_k \, w_k$$

$$g^n_{e1} = \frac{\ell_e}{8\pi} \sum_{k=1}^{K} [\exp(-x_{F-1}) - x_{F-1}E_1(x_{F-1})]_k \, \phi^n_k \, w_k$$

$$g^n_{e2} = \frac{\ell_e}{8\pi} \sum_{k=1}^{K} [E_1(x_{F-1}) + x_{F-1}E_1(x_{F-1}) - \exp(-x_{F-1})]_k \, \phi^n_k \, w_k$$

$$c^n_{e1} = \frac{1}{16\pi a} \sum_{k=1}^{K} \Big\{ [x^2_{F-1}E_1(x_{F-1}) + (1-x_{F-1})\exp(-x_{F-1})].[\underset{\sim}{\phi}^T \underset{\sim}{V}^n_{x,1} d_{ey}$$

$$+ \underset{\sim}{\phi}^T \underset{\sim}{V}^n_{y,1} d_{ex}] + [-x_{F-1}E_1(x_{F-1})(2+x_{F-1}) + (1+x_{F-1})].[\underset{\sim}{\phi}^T \underset{\sim}{V}^N_{x,2} d_{ey}$$

$$+ \underset{\sim}{\phi}^T \underset{\sim}{V}^n_{y,2} d_{ex}] \Big\} \phi^n_k w_k$$

$$c^n_{e2} = \frac{1}{16\pi a} \sum_{k=1}^{K} \{[-x_{F-1}E_1(x_{F-1})(2+x_{F-1})+(1+x_{F-1})\exp(-x_{F-1})].$$

$$[\underset{\sim}{\phi}^T \underset{\sim}{V}^n_{x,1} d_{ey} + \underset{\sim}{\phi}^T \underset{\sim}{V}^n_{y,1} d_{ex}] + [(2+4x_{F-1}+x^2_{F-1})E_1(x_{F-1}) -$$

$$(3+x_{F-1})\exp(-x_{F-1})][\underset{\sim}{\phi}^T \underset{\sim}{V}^n_{x,2} d_{ey} + \underset{\sim}{\phi}^T \underset{\sim}{V}^n_{y,2} d_{ex}]\}_k \; \phi^n_k \; w_k \tag{1.47}$$

ℓ_e is the boundary element length, $d_{ie} = (x^i - x^1)(y^3 - y^1) + (y^i - y^1)(x^1 - x^3)$, $d_{ey} = y^3 - y^1$, $d_{ex} = x^1 - x^3$, n = 1,2,3 and K represents the number of integration points.

For convenience during numerical computation the exponential integral function is approximated by rational and polynomials expansions [10],

$$E_1(x) = -\ \ell n\ x + a_o + a_1 x + a_2 x^2 + a_3 x^3 + a_4 x^4 + a_5 x^5 + \varepsilon(x) \tag{1.48}$$

$$0 < x \leq 1 \quad ; \quad |\varepsilon(x)| < 2.10^{-7}$$

or

$$E_1(x) = \frac{1}{xe^x} \; \frac{x^4 + b_1 x^3 + b_2 x^2 + b_3 x + b_4}{x^4 + c_1 x^3 + c_2 x^2 + c_3 x + c_4} + \varepsilon(x) \tag{1.49}$$

$$1 \leq x \leq \infty \quad ; \quad |\varepsilon(x)| < 2.10^{-8}$$

When the source point lies on the element under consideration, the integrals g^n_{e2} and c^n_{e2} contain a logarithmic singularity. In this case expanding the exponential in series

$$E_1(x) = -\ c - \ell n\ x + \sum_{n=1}^{\infty} (-1)^{n-1} \frac{x^n}{nn!} \tag{1.50}$$

where c = 0.5772156649, the integrals can be evaluated analytically for elements other than quadratic or higher order. For straight-line elements the integrals h^n_{e2} are identically zero due to the orthogonality between r and n.

Domain integrals b^n_c , d^n_{c1}, d^n_{c2} can be computed by Hammer's integration formulas for the case that source points do not lie on the element of consideration.

$$b^n_c = A_c \sum_{k=1}^{K} \frac{1}{4\pi a \Delta t_F} \exp\left(-\frac{r^2_{ik}}{4a\Delta t_F}\right) \phi^n_k \; w_k$$

$$d^n_{c,1} = \frac{A_c}{8\pi a^2 \Delta t_F} \sum_{k=1}^{K} \{[-x_{F-1} E_1(x_{F-1}) + \exp(-x_{F-1})][\underset{\sim}{\phi}^T \underset{\sim}{V}^n_{x,1} d_{ix} + \underset{\sim}{\phi}^T V^n_{y,1} d_{iy}] + [E_1(x_{F-1}) + x_{F-1} E_1(x_{F-1}) - \exp(-x_{F-1})][\underset{\sim}{\phi}^T \underset{\sim}{V}^n_{x,2} d_{ix} + \underset{\sim}{\phi}^T \underset{\sim}{V}^n_{y,2} d_{iy}]\}_k \phi^n_k w_k$$

$$d^n_{c,2} = \frac{A_c}{8\pi a^2 \Delta t_F} \sum_{k=1}^{K} \{E_1(x_{F-1}) + x_{F-1} E_1(x_{F-1}) - \exp(-x_{F-1})] [\underset{\sim}{\phi}^T \underset{\sim}{V}^n_{x,1} d_{ix} + \underset{\sim}{\phi}^T \underset{\sim}{V}^n_{y,1} d_{iy}] + [\exp(-x_{F-1})/x_{F-1} - 2E_1(x_{F-1}) - x_{F-1} E_1(x_{F-1}) + \exp(-x_{F-1})][\underset{\sim}{\phi}^T \underset{\sim}{V}^n_{x,2} d_{ix} + \underset{\sim}{\phi}^T \underset{\sim}{V}^n_{y,2} d_{iy}]\}_k \phi^n_k w_k \tag{1.51}$$

where A_c is domain element area, $d_{ix} = x_i - x(\eta)$, $d_{iy} = y_i - y(\eta)$ and $n = 1,2,3,\ldots 6$.

When the source point lies on the element of integration integrals d^n_{c2} contain singularity. In this case the integrals d^n_{c2} are divided into a regular part which is computed by Hammer's formulas and a singular part which is calculated semi-analytically. In order to perform the singular part, one can define a cylindrical coordinate system (r,ϕ) based at the singular point ξ - figure 1.5.

Local area coordinates defined by (1.44) can be expressed with polar coordinate (r,ϕ) as,

$$\bar{\eta}_i = \frac{1}{2A}(2\overset{o}{A}_i + b_i x_\xi + a_i y_\xi) + \frac{r}{2A}(b_i \cos\phi + a_i \sin\phi)$$

$$\bar{\eta}_i = \xi_{\eta_i} + \frac{r}{2A}(b_i \cos\phi + a_i \sin\phi) \tag{1.52}$$

$\xi_{\eta i}$ being the value of the area coordinate at the singular point ξ. Analytical integration with respect to r does not present any difficulties because the singularity of the integrand has been removed by introducing the polar coordinate system. The integrals can be expressed in terms of error functions defined by

$$\text{erf}(x) = \frac{2}{\sqrt{\pi}} \int_0^x e^{-t^2} dt \tag{1.53}$$

which can be approximated for convenience during numerical computation by [10],

$$\text{erf}(x) = 1-(a_1t+a_2t^2+a_3t^3+a_4t^4+a_5t^5)e^{-x^2}+\varepsilon(x) \qquad (1.54)$$

$$t = 1/(1+px) \qquad |\varepsilon(x)| \leq 1.5 \times 10^{-7}$$

The integration with respect to angle ϕ can be carried out numerically using a standard Guassian quadrature. To effect this, a new variable η is introduced such that its values, for instance, for ϕ in the range $\phi_1 \leq \eta \leq \phi_2$ is

$$\eta = \frac{2\phi - \phi_1 - \phi_2}{\phi_2 - \phi_1} \quad ; \qquad -1 \leq \eta \leq 1 \qquad (1.55)$$

1.7 Applications

To test the validity of the above formulation a simple diffusion-convection example was solved using quadratic elements in space and linear on time. The domain shown in figure 1.6 presents a constant velocity field v = 0.05 in the x direction. A discharge producing u = 1 is applied as a step function at x = 0 at t = 0. The material constants are $c = \rho = 1$ and $k_o = 0.01$.

The boundary elements mesh presented 28 boundary elements, with 56 nodes and 66 internal cells with 105 internal nodes. The quadratic finite elements mesh used for comparison purposes [13] consisted of 14 elements and 45 nodes.

The boundary elements results shown in figure were obtained using $\Delta t = 1.0$ which is a comparatively large time step. For the finite element results for instance the results were found with $\Delta t = 0.25$. Notice that the accuracy of the boundary element solution is much higher than the accuracy of the finite elements results, which appears to have a more erratic behaviour, and larger energy losses. These losses can be partly due to the relative coarseness of the finite elements mesh.

1.8 Conclusions

This paper represents the boundary element formulation for time dependent transport problem governed by diffusion-convective equation. The solution is extended to problems with nonlinear material properties by using the Kirchoff's transformation. The dimensionality of the problem is reduced in time and space by applying the fundamental solution previously used for parabolic type equations [3,4]. The procedure can be extended to solve fluid flow problems such as those governed by Navier-Stokes equations which are sometimes expressed in term of vorticity transport equation.

REFERENCES

1. BREBBIA, C. The Boundary Element Method for Engineers Pentech Press, London. Halstead Press, New York 1978. Second edition 1980.

2. SKERGET, P. and BREBBIA, C.A. Non Linear Potential Problems, Chapter 1 in Progress in Boundary Element Methods Vol. 2; Pentech Press, London, Springer-Verlag New York 1983.

3. SKERGET, P. and BREBBIA, C.A. Time Dependent Non-linear Potential Problems, Chapter 2 in Progress in Boundary Element Methods, Vol.3.

4. WROBEL, L. and BREBBIA, C.A. Time Dependent Potential Problems, Chapter 6 in Progress in Boundary Element Methods Vol. 1; Pentech Press, London, Wiley, New York 1981.

5. WROBEL, L.C. Potential and Viscous Flow Problems using Boundary Element Method. Ph.D. Thesis, Department of Civil Engineering of the University of Southampton 1981.

6. YEHIA RIZK, An Integral-representation Approach for Time-dependent Viscous Flows. Ph.D. Thesis, School of Aerospace Engineering, Georgia Institute of Technology 1980.

7. WU, J.C. and GULCAT, U. Separate Treatment of Attached and Detached Flow Regions in General Viscous Flows. AIAA Journal, Vol. 19, No.1 1981.

8. WU, J.C. Theory for Aerodynamic Force and Moment in Viscous Flows. AIAA Journal, Vol.19, No.4 1981.

9. WU, J.C. and WAHBAH, M.M. Numerical Solution of Viscous Flow Equations using Integral Representations. Lecture Notes in Physics, Vol.59, Springer-Verlag, New York 1976.

10. ABRAMOWITZ, M., STEGUN, I.A. Handbook of Mathematical Functions. Dover Publications, New York 1965.

11. BREBBIA, C.A. and WALKER, S. Boundary Element Techniques in Engineering. Butterworths 1980.

12. BREBBIA, C.A., TELLES, J. and WROBEL, L. Boundary Element Methods - Theory and Application, Springer Verlag, New York 1983.

13. ADEY, R.A. Numerical Prediction of Transient Water Quality and Tidal Motion in Estuaries and Coastal Waters. Ph.D. Thesis, University of Southampton 1974.

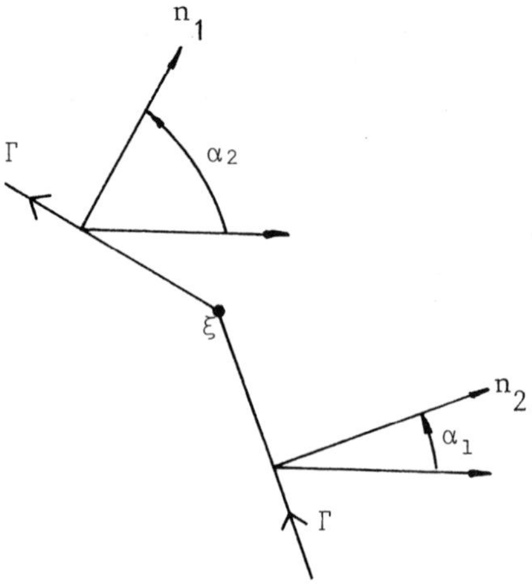

Figure 1.1 Computation of coefficient c(ξ)

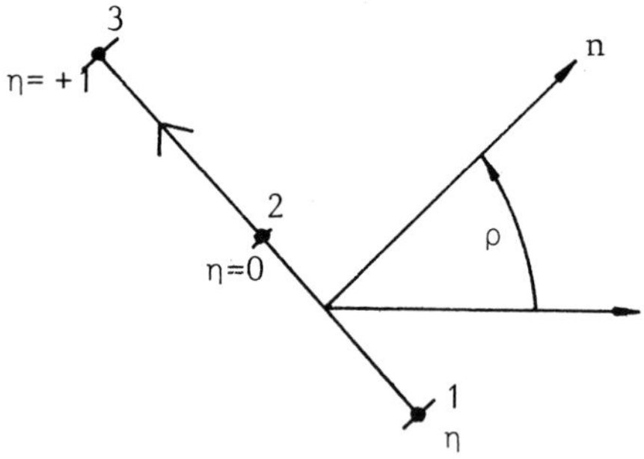

Figure 1.2 Continuous quadratic boundary element

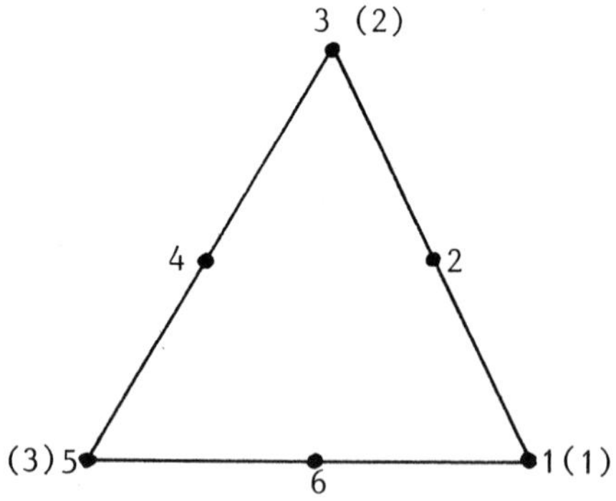

Figure 1.3 Continuous quadratic domain element

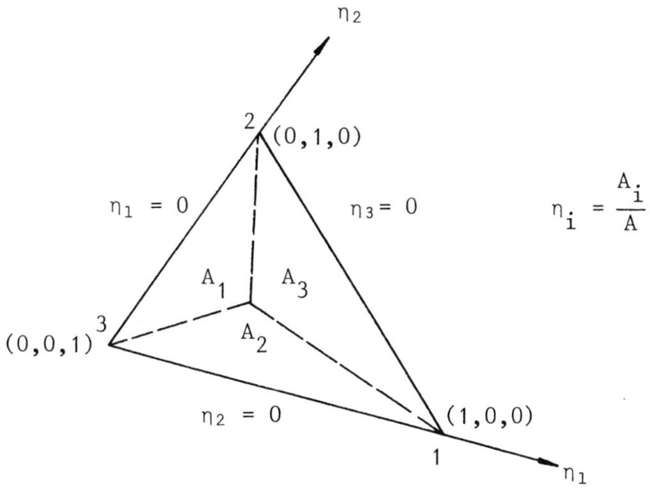

Figure 1.4 Triangular domain element and definition of area coordinate system (η_1, η_2, η_3)

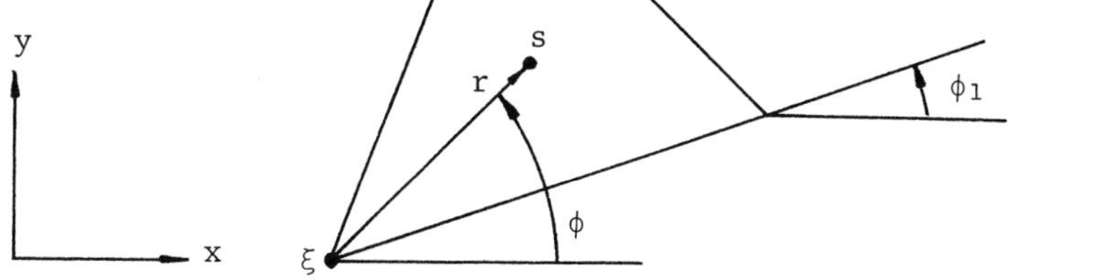

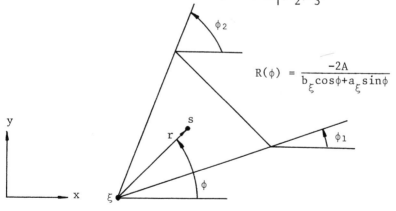

Figure 1.5 Cylindrical coordinate system based at the singular point ξ

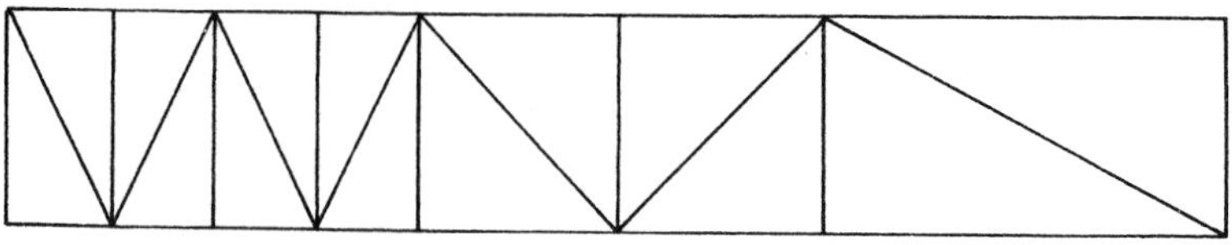

F.E. MESH

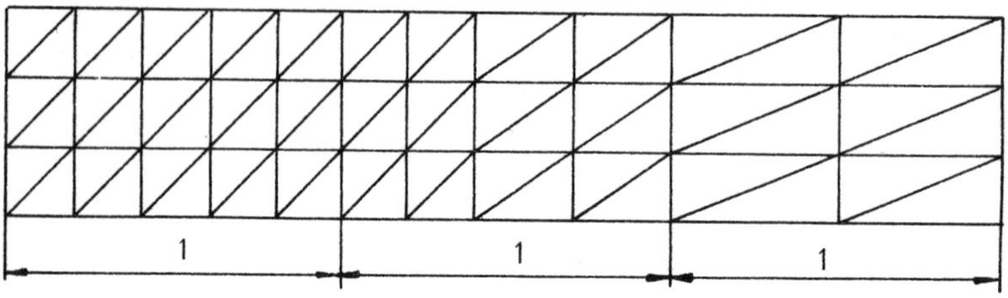

B.E. MESH

Figure 1.6 Discretisation of the problem

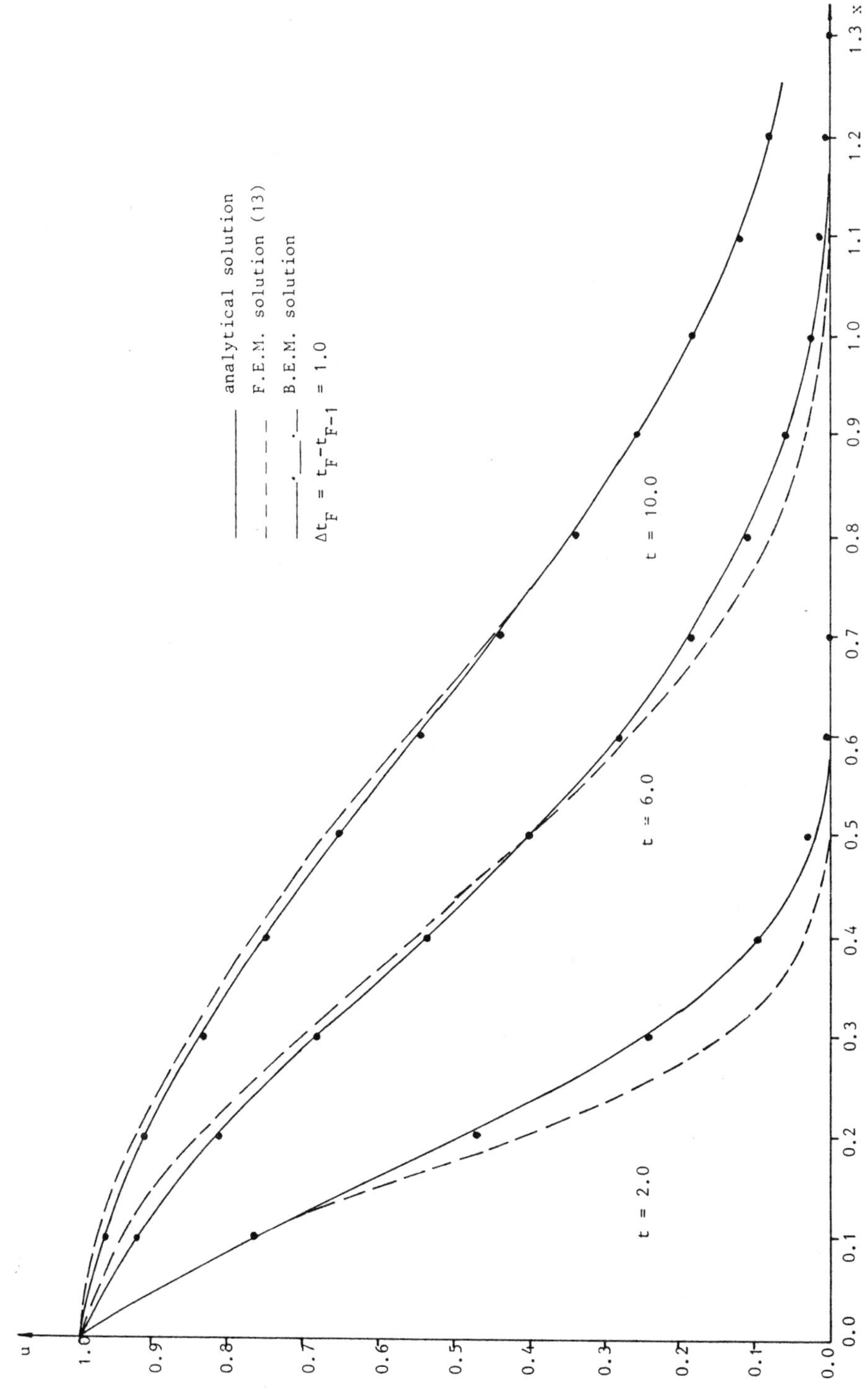

Figure 1.7 One dimensional diffusion-convection

A Method for Predicting 2-Dimensional Flows and Heat Transfer in Complex Geometries and its Application to Natural Convection in a Gas Cooled Cable

A.J. Oliver and A.J. Miller

Central Electricity Research Laboratories,
Leatherhead, Surrey, UK

SUMMARY

This paper describes a numerical method for predicting two-dimensional flows and heat transfer in general complex geometries. The particular applications for which such a model is required are elliptic problems in which buoyancy forces can play an important part. Currently the model has been developed for laminar flows, but the intention is to eventually incorporate a suitable mathematical turbulence model. Application of the method to predicting natural convection in a concentric annulus is described. A second application is also presented. This is 2-d natural convection in a complex geometry which represents a compressed gas insulated, electric power cable.

1. INTRODUCTION

There are a number of problems in the Electricity Supply Industry which involve fluid flow and heat transfer in complex geometries. The aim of the present work is to develop a computer model for general elliptic two-dimensional flows in which buoyancy forces can play an important part. At this stage only laminar flows have been considered, although a turbulence model will be included in the future. The examples discussed refer to the problem of natural convection and include buoyancy effects, although problems of forced convection will also be covered by this method.

The model uses a boundary-fitted non-orthogonal curvilinear co-ordinate system and the flow equations are solved by a finite volume method on a transformed uniform square grid. Examples of the grids used and details of the solution procedure are given. Predicted temperature profiles are presented for laminar natural convection in a horizontal concentric annulus, and these are compared with experimental measurements reported in the literature in order to test the accuracy of the model.

The model has also been applied to predicting the 2-dimensional natural convection in a complex geometry which represents a compressed gas insulated, electric power cable. This consists of three cylindrical conductors which are positioned together in a trefoil. These are contained inside a cooled cylindrical enclosure which is filled with SF_6 gas. The conductors produce heat which drives a natural convection motion in the gas. Preliminary temperature and stream function predictions are presented for this geometry.

2. THE SOLUTION PROCEDURE

2.1 The Body-Fitted Co-ordinate System

The method used to generate a body-fitted co-ordinate system is that of Thompson, Thames and Mastin [1]. This approach takes the co-ordinates as a family of curves which are solutions of a pair of elliptic partial differential equations with Dirichlet boundary conditions on all the boundaries. The two sets of curves which are the co-ordinate lines are denoted by $\xi(x,y)$ and $\eta(x,y)$.

The elliptic differential equations that are used are Poisson equations:

$$\xi_{xx} + \xi_{yy} = P(\xi,\eta) \qquad (1)$$

$$\eta_{xx} + \eta_{yy} = Q(\xi,\eta) \qquad (2)$$

These equations are transformed to the ξ-η plane and are solved on a square orthogonal grid in that plane. The transformation from the physical (x,y) plane to the transformed (ξ,η) plane is given by $\xi = \xi(x,y)$ and $\eta = \eta(x,y)$.

Interchanging dependent and independent variables for equations (1) and (2), gives:

$$\alpha x_{\xi\xi} - 2\beta x_{\xi\eta} + \gamma x_{\eta\eta} + J^2 (Px_\xi + Qx_\eta) = 0 \qquad (3)$$

$$\alpha y_{\xi\xi} - 2\beta y_{\xi\eta} + \gamma y_{\eta\eta} + J^2 (Py_\xi + Qy_\eta) = 0 \qquad (4)$$

where $\alpha = x_\eta^2 + y_\eta^2$; $\beta = x_\xi x_\eta + y_\xi y_\eta$;

$\gamma = x_\xi^2 + y_\xi^2$; $J = x_\xi y_\eta - x_\eta y_\xi$

Other transformation parameters referred to later are $\sigma = J^2P$ and $\tau = J^2Q$.

The equations (3) and (4) are solved by a program which gives the positions of the x-y grid points in terms of the corresponding ξ-η values in the transformed plane.

The Navier-Stokes equations to be solved on the curvilinear grid are solved on the transformed ξ-η plane. The new transformed system of equations will contain the transformation parameters α, β, γ, J etc. Values of these parameters are produced by a separate computer program for the particular curvilinear grid.

It can be seen that the equations in the transformed plane will be more complicated than in the physical plane, but the advantage is that for any 2-d geometry, the solutions are found for a simple square grid. Use of the boundary-fitted co-ordinate system enables the boundary conditions to be applied more readily than for other systems.

The distribution of grid points is an important consideration for the accuracy of any numerical calculation. The grid points on the body have to be specified as input, so that the spacing of these points may be chosen at will. P and Q in equations (1) and (2) are source terms which control the spacing of grid lines in the field. At present, the functions P and Q are taken as exponentials:

$$P(\xi,\eta) = -\sum_{i=1}^{n} a_i \, \mathrm{SGN}(\xi-\xi_i)\exp(-c_i|\xi-\xi_i|)$$
$$- \sum_{j=1}^{m} b_j \, \mathrm{SGN}(\xi-\xi_j)\exp\left(-d_j\sqrt{(\xi-\xi_j)^2 + (\eta-\eta_j)^2}\right)$$

where a_i, c_i, b_j, d_j are constants.

The first terms represent attraction of the co-ordinate line ξ = constant to the lines $\xi = \xi_i$, and the second terms represent attraction to the points (ξ_j, η_j). A similar expression for $Q(\xi,\eta)$ represents attraction of the co-ordinate lines η = constant. The sign of the source terms can be changed to give repulsion rather than attraction. Other forms of source terms will give different types of grid control.

An example of the type of mesh formed is shown in Fig. 1. This represents a symmetry unit of an in-line tube bank, where the curved boundaries each represent one quarter of a tube. The other boundaries represent symmetry lines. This method will be used in the study of heat transfer in boiler tube banks. Fig. 1 also shows the effect of increasing the attraction towards the line AB compared to that towards DC.

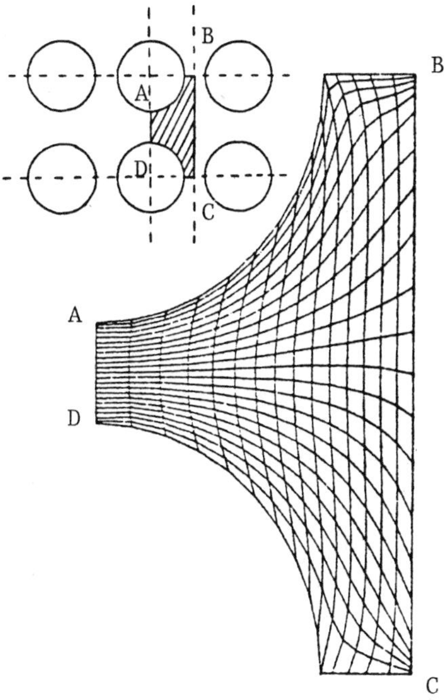

Fig. 1: A Grid for an In-line Tube Bank

2.2 The Governing Equations

The equations which govern the flow and convective heat transfer are the usual equations for the continuity of mass, conservation of energy and the Navier-Stokes equations. Introducing stream function (ψ) and vorticity (ω) give the following equations in Cartesian co-ordinates (y is vertical)

$$\frac{\partial^2\psi}{\partial x^2} + \frac{\partial^2\psi}{\partial y^2} = -\rho_a\omega \tag{5}$$

$$\frac{\partial\psi}{\partial y}\frac{\partial\omega}{\partial x} - \frac{\partial\psi}{\partial x}\frac{\partial\omega}{\partial y} = \mu\left(\frac{\partial^2\omega}{\partial x^2} + \frac{\partial^2\omega}{\partial y^2}\right) + \rho_a\hat{\beta}g\frac{\partial T}{\partial x} \tag{6}$$

$$\frac{\partial\psi}{\partial y}\frac{\partial T}{\partial x} - \frac{\partial\psi}{\partial x}\frac{\partial T}{\partial y} = \frac{K}{C_p}\left(\frac{\partial^2 T}{\partial x^2} + \frac{\partial^2 T}{\partial y^2}\right) \tag{7}$$

where the Boussinesq approximation has been used in representing the buoyancy effect, ρ_a is a mean density and $\hat{\beta}$ is the volumetric expansion coefficient. The vorticity and stream function are defined by

$$\omega = \frac{\partial v}{\partial x} - \frac{\partial u}{\partial y}$$

$$u = \frac{1}{\rho_a}\frac{\partial \psi}{\partial y} \quad , \quad v = -\frac{1}{\rho_a}\frac{\partial \psi}{\partial x}$$

The above equations were then transformed to the ξ-η co-ordinate system which consists of a square grid in ξ-η space. The equations for ψ, ω and T, equations (5)-(7) can then be expressed as

$$\frac{a_\phi}{J}\left[(\psi_\eta \phi)_\xi - (\psi_\xi \phi)_\eta\right] = b_\phi \tilde{\nabla}^2 \phi + d_\phi \quad (8)$$

where $\tilde{\nabla}^2 \phi = \frac{1}{J^2}(\alpha\phi_{\xi\xi} - 2\beta\phi_{\xi\eta} + \gamma\phi_{\eta\eta} + \sigma\phi_\eta + \tau\phi_\xi)$

and ϕ is the dependent variable. The transformation parameters $\alpha, \beta, \gamma, \sigma, \tau$ and J are defined in Section 2.1. The coefficient values in equation (8) are given in Table 1.

ϕ	a_ϕ	b_ϕ	d_ϕ
ψ	0	1	$\rho_a \omega$
ω	1	μ	$\frac{\rho_a \hat{\beta} g}{J}\left[(y_\eta \dot{T})_\xi - (y_\xi T)_\eta\right]$
T	1	$\frac{K}{C_p}$	0

Table 1. Coefficient Values for Transport Equations

The specification of the boundary conditions varies with the problem considered. However, because a boundary fitted co-ordinate system is used they are generally straightforward e.g. stream function is a constant on a wall, except for the specification of wall vorticity. When the wall is represented by a η = constant line, then the wall vorticity boundary condition becomes

$$\rho_a \omega_{wall} = -\frac{\gamma\psi_{\eta\eta}}{J^2} \quad (9)$$

and for a ξ = constant line at a wall

$$\rho_a \omega_{wall} = -\frac{\alpha\psi_{\xi\xi}}{J^2} \quad (10)$$

If a normal derivative is specified at a boundary, such as at a symmetry line or at an outlet plane having fully developed conditions, then this is represented by:

$$\frac{\partial\phi}{\partial n} = \frac{1}{J\sqrt{\alpha}}(\alpha T_\xi - \beta T_\eta) \qquad \eta = \text{constant line}$$

$$\frac{\partial\phi}{\partial n} = \frac{1}{J\sqrt{\gamma}}(\gamma T_\eta - \beta T_\xi) \qquad \xi = \text{constant line}$$

3. THE NUMERICAL EQUATIONS

The equations denoted by equation (8) were expressed in finite differences in a conservative form in ξ-η space. The diffusion term ($\tilde{\nabla}^2\phi$) was represented by second-order central differences. For the convection term $[(\psi_\eta\phi)_\xi - (\psi_\xi\phi)_\eta]$ a hybrid difference scheme was used with the upwinding part being equivalent to the second upwind-difference method as described by Roache [2] for x-y co-ordinates. The resulting finite difference equation for a node (i,j) is

$$\begin{aligned}
&[J^{-1}a_\phi(A_e + A_w + A_n + A_s) + J^{-2}b_\phi(2\alpha + 2\gamma)]\phi_{i,j} \\
&-[J^{-1}a_\phi A_e + J^{-2}b_\phi(\alpha + \tfrac{\tau}{2})]\phi_{i+1,j} - [J^{-1}a_\phi A_w + J^{-2}b_\phi(\alpha - \tfrac{\tau}{2})]\phi_{i-1,j} \\
&-[J^{-1}a_\phi A_n + J^{-2}b_\phi(\gamma + \tfrac{\sigma}{2})]\phi_{i,j+1} - [J^{-1}a_\phi A_s + J^{-2}b_\phi(\gamma - \tfrac{\sigma}{2})]\phi_{i,j-1} \\
&+ \underbrace{\frac{\beta b_\phi}{2J^2}(\phi_{i+1,j+1} - \phi_{i+1,j-1} + \phi_{i-1,j-1} - \phi_{i-1,j+1})}_{\text{cross-derivative term}} \qquad (11) \\
&= (d_\phi)_{i,j}
\end{aligned}$$

where the variable A's result from the convection terms. Equation (11) is a nine-diagonal matrix equation. The wall vorticity condition e.g. equation (9), was initially represented by a simple 1st-order formula of the form:

$$\rho_a\omega_{i,j} = -\frac{\gamma}{J^2}2(\psi_{i,j+1} - \psi_{i,j})$$

where (i,j) is a wall node. However, currently consideration is being given to replacing this by a higher order equation equivalent to Woods formula (Roache [2]). The form of this formula that represents equation (9) is

$$\rho_a\omega_{i,j} = -\frac{3\gamma}{J^2}(\psi_{i,j+1} - \psi_{i,j}) - \tfrac{1}{2}\rho_a\omega_{i,j+1}$$

The reason for using the first order formula initially is that it is expected to give better numerical stability.

4. THE METHOD OF SOLUTION

For any problem, the first stage in this solution procedure is to derive a suitable body-fitted co-ordinate system. Typically the method described in Section 2.1 could be used for this and also deriving the required transformation parameters α, β, J etc. The numerical flow equations derived in Section 3 would then be solved on the transformed grid with the appropriate boundary conditions. The grid should be chosen so that there is some grid refinement in the regions of steep gradients and because a hybrid differencing scheme is being used, the grid should be approximately aligned with the expected stream lines where possible.

The flow equations are solved iteratively using S.O.R. The sequence in an iteration is:

(i) Iterate for ω at all interior nodes

(ii) Iterate for ψ at all interior nodes

(iii) Iterate for T at all interior nodes

(iv) derive values at boundary nodes.

For some of the problems solved it has been necessary to under-relax the variations in ω and T from one iteration to the next.

Before applying the method to a problem of engineering interest, the procedure was tested against some simpler well documented flows using both orthogonal and non-orthogonal grids. Some of the flows considered, all of which have been laminar, were forced convection in a parallel plate duct, forced convection in a curved duct and natural convection in a concentric annulus. The natural convection in a concentric annulus and an engineering application are described in the next section.

5. APPLICATION AND RESULTS

5.1 Natural Convection in a Concentric Annulus

Predictions were obtained from the two-dimensional horizontal concentric annulus configuration examined experimentally by Kuehn and Goldstein [3]. The two surfaces were isothermal with the inner cylinder hotter and the annulus geometry was defined by

$$\frac{L}{D_i} = \frac{r_o - r_i}{D_i} = 0.8$$

The conditions investigated corresponded to $(R_A)_L = 4 \times 10^5$. The predictions presented were obtained using a uniform grid of concentric circles and straight radial lines. This was then transformed as usual, to a square grid in the ξ-η plane.

In order to get grid independence, 15 radial × 61 circumferential nodes were used. It was found that the top sector of 60° was more sensitive to the grid configuration than

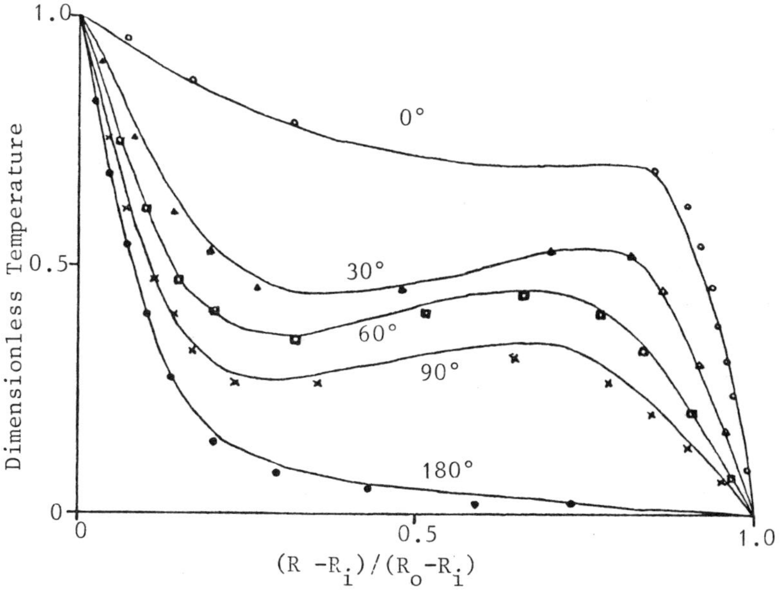

Fig. 2: Predicted and Measured Radial Temperature Profiles for an Annulus

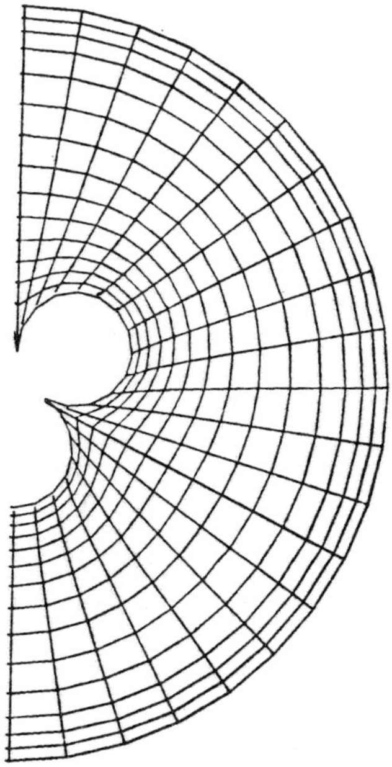

Fig. 3: The Grid for the SF_6 Cable

further round the annulus. Therefore fewer grid points could have been used by concentrating them in this sector.

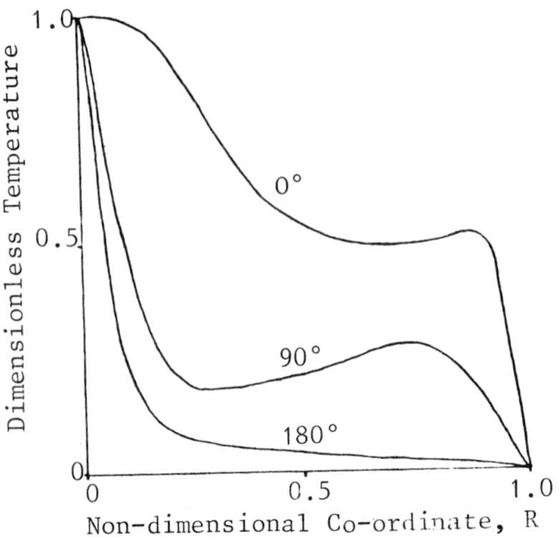

Fig. 4: Predicted temperatures for Cable (R=0, conductor Surface, R=1, casing)

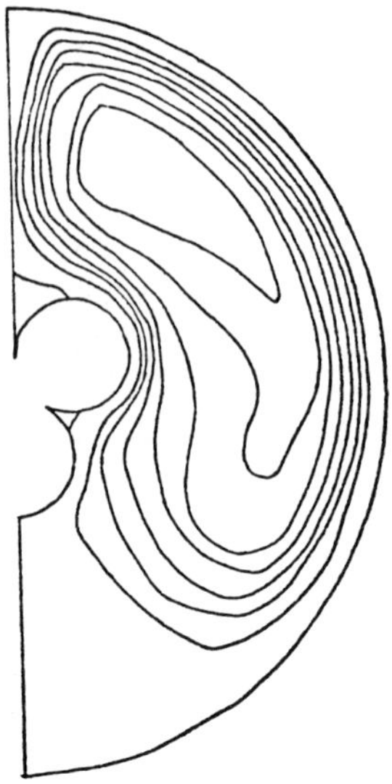

Fig. 5: Predicted Streamlines ($\Delta\psi$=0.005)

Radial temperature profiles are shown in Fig. 2 for various positions round the cylinder, measuring the angle from the top, and good agreement was obtained for all angles.

5.2 Natural Convection in a SF_6 Power Cable

One application for which the model will be used is SF_6 gas-filled power cables. The cable consists of three conductors surrounded by an outer casing, with a space filled by gas, as shown in Fig. 3. The ratio of casing radius to conductor diameter is 3.2. The inner conductors warm up when electrical current flows, and the outer casing is cooled by some means. The SF_6 circulates by natural convection and it is of interest to know details of the flow pattern, and temperature distribution so that the heat transfer coefficients from conductor to sink can be calculated. It is a reasonable assumption to make a two-dimensional calculation, since the cable will be fairly uniform along its length.

For this example, the grid was generated using the Thompson method and is shown in Fig. 3. Once again, the vertical line through the centre was assumed to be a symmetry line, which is suggested by tests on an experimental rig of the SF_6 cable. Preliminary predictions have been obtained for this geometry at R_A = 10^4 based on the conductor diameter, and a

Prandtl number of 0.7. All surfaces are regarded as isothermal with the three conductors having the same temperature. Predicted temperature profiles are shown in Fig. 4 and the predicted streamline pattern is shown in Fig. 5. The results presented are virtually grid independent.

Temperature measurements are currently being made on an experimental SF_6 cable. An initial comparison between measurements and predictions indicates that the predictions are at least qualitatively correct. For example, the predictions indicate that there is very little flow over a large area at the bottom, and this has been confirmed experimentally. However, further work on this problem is still required.

6. CONCLUSIONS

A method has been developed for predicting 2-d flows in complex geometries in which buoyancy effects may be present. The method uses a boundary-fitted non-orthogonal grid.

The grid generation scheme has been applied to two problems of engineering interest, namely a tube bank and a SF_6 power cable. The grids obtained look to be suitable for obtaining accurate predictions of flow and temperature.

The flow solution procedure has been applied to predict natural convection in a horizontal concentric annulus. The predictions agree well with available experimental data. The method has also been applied to natural convection in an SF_6 power cable. The preliminary results obtained for this look encouraging.

7. ACKNOWLEDGEMENTS

This work was carried out at the Central Electricity Research Laboratories and it is published by permission of the Central Electricity Generating Board.

8. REFERENCES

1. THOMPSON, J.F., THAMES, F.C. and MASTIN, C.W. - TOMCAT - A code for Numerical Generation of Boundary-fitted Curvilinear Co-ordinate Systems on Fields containing any number of Arbitrary two-dimensional bodies. Journal of Computational Physics, Vol. 24, pp 274-302, 1977

2. ROACHE, P.J., Computational Fluid Dynamics Hermosa Publishers, 1976

3. KUEHN, T.H., and GOLDSTEIN, R.J. - An Experimental and Theoretical Study of Natural Convection in the Annulus between Horizontal Cylinders. Journal of Heat Transfer, Vol. 74, pp 695-719, 1976

APPLICATION OF A MODIFIED FINITE ELEMENT METHOD TO THE TIME-DEPENDENT THERMAL CONVECTION OF A LIQUID METAL

Philip M. Gresho and Craig D. Upson

Lawrence Livermore National Laboratory, University of California, Livermore, CA 94550

1. SUMMARY. Starting with the Galerkin finite element method and the simplest appropriate isoparametric element for modeling the Navier-Stokes equations, the spatial approximation is modified in two ways in the interest of cost-effectiveness: the mass matrix is "lumped" and all coefficient matrices are generated via one-point quadrature. After appending a correction term to the diffusion matrices, the modified semi-discretized equations are integrated in time using the explicit Euler method in a special way to compensate for that portion of the time truncation error which is intolerable for advection-dominated flows. The scheme is completed by the introduction of a subcycling strategy which permits less frequent updates of the pressure field with little loss of accuracy. After summarizing these techniques, a simulation of free convection in a two-dimensional box at $Ra = 10^5$ and $Pr = 0.01$ is described.

2. INTRODUCTION. We have developed a numerical method for solving the time-dependent, incompressible Navier-Stokes (NS) equations and the advection-diffusion (AD) equation in two and three dimensions (2D and 3D). While the technique was originally derived via the conventional Galerkin finite element method (GFEM), we invoke two ensuing simplifying approximations that generate a scheme which is probably better described as a blend of finite elements and finite differences; i.e., an "isoparametric element, finite difference method."

The philosophy guiding the evolution of the techniques is a common one: simplicity and cost-effectiveness. Starting with the simplest GFEM approximation for spatial discretization, we invoke the simplest method for advancing the solution in time. We therefore use multilinear basis functions for the velocity (bilinear in 2D, trilinear in 3D) and (piecewise) constant approximation for the pressure. The explicit Euler method is then used to integrate the resulting oridinary differential equations (ODEs) in time.

The two a priori simplifications to the GFEM are: (1) the "mass matrix," which couples the time-derivatives in the GFEM, is replaced by a diagonal matrix via "mass lumping," thus decoupling the time derivatives and paving the way for explicit time integration, and (2) all Galerkin integrals are evaluated approximately by invoking one-point quadrature. We believe that these two ad hoc modifications, when combined with those discussed below, lead to a scheme which is generally more cost-effective than when GFEM is used (consistent mass, higher order quadrature, implicit time integration).

Two additional modifications that provide further increase in computational speed are associated with the time integration of the ODEs. The first of these is called balancing tensor diffusivity (BTD), or in particular, balancing tensor viscosity for the NS equations, and is used to permit larger time steps with no loss of accuracy — indeed, a gain in accuracy is often realized. The second is called subcycling, a procedure which permits less frequent updates of the pressure relative to the instability-limited processes of advection and diffusion.

Except for the theory behind the dynamic time step selection process, (the only new contribution to numerical methods in this paper), the techniques will only be summarized briefly; for a detailed description of these, see Gresho et al. [1].

3. GOVERNING EQUATIONS AND BASIC SPATIAL DISCRETIZATION. The governing equations of interest here are the NS equations for an incompressible, constant property fluid in the Boussinesq approximation -- in dimensionless form:

$$\partial \underline{u}/\partial t + \underline{u} \cdot \nabla \underline{u} = -\nabla P + Re^{-1}\nabla^2 \underline{u} + \underline{k}T \;, \; \nabla \cdot \underline{u} = 0 \;, \text{ and} \tag{1a,b}$$

$$\partial T/\partial t + \underline{u} \cdot \nabla T = Pe^{-1}\nabla^2 T \;, \tag{1c}$$

where, for our purposes here, $Re = \sqrt{Ra/Pr}$ and $Pe = Re\,Pr$, $Ra = \gamma g \Delta T L^3/\kappa\nu$, and $Pr = \nu/\kappa$. In the above, L is the characteristic length, $u_0 = \sqrt{\gamma \Delta T g L}$ is the characteristic (reduced free-fall) velocity, L/u_0 is the characteristic time, and ΔT is the characteristic temperature. Given an initial velocity field satisfying (1b), an initial temperature field, and appropriate boundary conditions (BCs) on $\underline{u}$ and T, (1) can be solved (in principal) for $\underline{u}$, T, and P.

If Eqs. (1) are spatially discretized using the conventional GFEM, the following differential-algebraic system obtains

$$M\dot{u} + K(u)u + CP = f \;\; ; \;\; u(0) = u_0 \;\; \text{where} \;\; C^T u_0 = 0 \;, \tag{2a}$$

$$C^T u = 0 \;, \;\; \text{and} \tag{2b}$$

$$M_S \dot{T} + K_S(u)T = f_S \;, \tag{2c}$$

where now u is a global vector containing all the nodal values of u,v (and w if 3D) and similarly for P and T. The vector f contains the buoyancy term as well as BCs on velocity. M is the mass matrix (which we henceforth consider as "lumped" via the row-sum technique so that it is diagonal), K(u) is the "advection + diffusion" matrix, C is the gradient matrix and its transpose, C^T, is the divergence matrix. Similar (and obvious) remarks apply to (2c).

In order to integrate (2) with an explicit method, we will need the consistent discretized Poisson equation for pressure, which is easily derived from (2):

$$(C^T M^{-1} C)P \equiv AP = C^T M^{-1}[f - K(u)u] \quad , \tag{3}$$

where A describes a discretized approximation to the Laplacian operator into which the appropriate BCs on velocity and temperature have been (automatically) incorporated. Henceforth we are concerned with solving (2a), (2c), and (3); the solution will always satisfy continuity; i.e., (2b). We also assume, when solving (3), that pressure modes (when present; see [1] and [2]) have been properly disposed of (typically by specifying pressures to render A nonsingular).

Since all Galerkin integrals are evaluated with one-point quadrature, an "hourglass" correction term is appended to the diffusion matrices in (2a) and (2c) in order to render them nonsingular to "2ΔX waves" and to improve the overall diffusion approximations; see [1] for details.

4. TIME INTEGRATION. Application of the explicit Euler scheme to (2a) and (2c) gives

$$u_{n+1} = u_n + \Delta t M^{-1}[f_n - K(u_n)u_n - CP_n] \quad \text{and} \tag{4a}$$

$$T_{n+1} = T_n + \Delta t M_s^{-1}[f_{sn} - K_s(u_n)T_n] \quad , \tag{4b}$$

where P_n is first obtained by solving (3) at $t = t_n$. Since the matrix A is symmetric-positive-definite and invariant with time, we have (thus far) utilized a direct method (Gaussian elimination via the skyline method) to solve (3). It is factored ($A = LDL^T$) in a preprocessor code and stored on disk (3D) or in memory (2D). Each pressure update thus consists of one forward reduction and one back substitution.

Although we have described a very simple algorithm, which is easily vectorized when one-point quadrature is employed, its cost-effectiveness warrants improvement for the following reasons:

(1) The forward Euler integration scheme exhibits (at least) two undesirable properties: (i) it (like almost any explicit method) requires the use of rather small time steps to preclude numerical instability and (ii) it reduces the effective diffusivity in the streamline direction, especially when the flow is advection-dominated (large Reynolds and/or Peclet number).

(2) For large problems (which, of course, describes most 3D simulations), the I/O cost associated with reading the factored A matrix from disk is high on the CRAY-1 computer.

The next two subsections address these items.

4.1 Balancing tensor diffusivity. In [1] it is shown that the straightforward application of the explicit Euler method to the AD equation (or to the NS equations) reduces the effective diffusivity from (say) K_{ij} (usually a scalar) to $K_{ij} - u_i u_j \Delta t/2$, where u_k is the k-th component of the continuum velocity vector. This reduction, which occurs only in the streamline direction, is especially deleterious for advection-dominated flows because it leads to the requirement that very small time steps be used to maintain stability. The simple and obvious fix, that of adding the tensor quantity $u_i u_j \Delta t/2$ to the physical diffusivity to balance the most undesirable portion of the truncation error, which we call balancing tensor diffusivity (BTD), has been found to be extremely cost-effective. It can improve both accuracy and (especially) stability. For example, the one-dimensional stability limits are increased from

$$\Delta t \leq \min(2K/u^2, \Delta x^2/2K) \quad \text{to} \tag{5}$$

$$\Delta t \leq K/u^2(\sqrt{1 + (u\Delta x/K)^2} - 1) \quad , \tag{6}$$

where K is the physical diffusivity, u is the (constant) advecting velocity, and Δx is the (constant) grid spacing. (Note that (6) gives the Courant stability limit, $\Delta t \leq \Delta x/u$, for $K \to 0$ whereas (5) is unstable in this limit.) For advection-dominated flows, the ratio between the allowable step size before and after the addition of BTD is approximately $1/P_g$, where $P_g \equiv u\Delta x/2K >> 1$ is the grid Peclet number; since often $P_g = O(10)$ to $O(100)$, BTD can permit stable integration with Δt's which are 1-2 orders of magnitude larger than otherwise. Clearly the BTD correction term is crucial to the success of explicit Euler applied to advection-dominated flows.

4.2 Subcycling. In spite of the significant gains in cost-effectiveness provided by BTD, there are still many cases wherein the stability limits imposed by (6) are too restrictive in that (barring instability) a sufficiently accurate solution to the ODEs could be obtained using a significantly larger Δt. Thus, we have devised another cost-effective strategy called subcycling, which permits a portion of the solution (basically pressure) to be based on accuracy rather than stability.

A four-step summary of the subcycling process is: (1) the minor (smaller, and fixed) time step is based on stability estimates and is used to (accurately) compute advection and diffusion processes only (these processes are subcycled to retain stability); the pressure gradient is approximated via linear extrapolation and the continuity equation is completely ignored during subcycling, (2) the approximate velocity field at the conclusion of subcycling is projected onto the subspace of (weakly) divergence-free velocity fields, (3) the compatibile pressure field is computed, and finally (4) the next major (larger) time step is dynamically computed based on the desired temporal accuracy. In more detail, these steps are:

(1) Given u_n satisfying $C^T u_n = 0$, solve the following equations between t_n and t_{n+1}:

$$M\dot{\tilde{u}} + K(\tilde{u})\tilde{u} = f - C\tilde{P} \quad ; \quad \tilde{u}(t_n) = u_n \quad , \tag{7}$$

via the explicit Euler method; i.e., for $m = 0, 1, \ldots s-1$,

$$\tilde{u}_{m+1} = \tilde{u}_m + \Delta t_s M^{-1}[f_m - K(\tilde{u}_m)\tilde{u}_m - C\tilde{P}_m] \, ; \tilde{u}_0 = u_n \text{ and } \tilde{P}_0 = P_n \, , \tag{8}$$

where Δt_s is the subcycle (minor) step size and m is the subcycle index, $\tilde{P}(t)$ is known from extrapolation, and $K(\tilde{u})$ includes the BTD correction in the viscous portion. This simple marching scheme is employed for a previously determined number, $s = \Delta t_n/\Delta t_s$, of steps where Δt_n is the current major step size (the determination of s will be discussed later) at which time we have $\tilde{u}_s$ as an approximation to the desired velocity, u_{n+1}.

(2) Given $\tilde{u}_s$, the projection onto the "solenoidal" subspace is accomplished sequentially as follows (see [1] for details):

$$A\lambda = C^T\tilde{u}_s \quad , \quad \text{and} \quad v = \tilde{u}_s - M^{-1}C\lambda \quad , \tag{9a,b}$$

where A is already available in factored form. v is taken to be the best approximation to u_{n+1}. Here λ is a vector of Lagrange multipliers and the velocity adjustment is seen to be via the gradient of a scalar.

(3) Once a mass-consistent velocity field is available (note that $C^Tv = 0$ in (9)), the compatible pressure field is obtained by solving (3) for P_{n+1} (with $v = u_{n+1}$ used to evaluate the righthand side). This second application of the A matrix shows that subcycling is only cost-effective when s is larger than 2.

(4) The last step in the process is the appropriate selection of the next major step-size, Δt_{n+1} or, equivalently, the new subcycle ratio, $s = \Delta t_{n+1}/\Delta t_s$. In the actual algorithm, the step-size calculation is the first process in the sequence rather than the last; the order of presentation was selected in the interest of greater clarity. In our previous papers discussing this scheme (e.g. [1]), we neglected the subcycle process when estimating the local truncation error. Recently we have completed the analysis of the full scheme (thanks to Dr. A. C. Hindmarsh) and, since it was employed successfully in the numerical example to be presented, we describe it in as much detail as space permits.

Introducing the matrix $B \equiv I - CA^{-1}C^TM^{-1}$, the NS system, (2a) and (3), can be formally expressed in terms of velocity only, as

$$M\dot{u} = B[f - K(u)u] \equiv Bg(u) = g(u) - CP \quad , \tag{10}$$

where $P = A^{-1}C^TM^{-1}g(u)$. Similarly, the subcycle ODE is

$$M\dot{\tilde{u}} = g(\tilde{u}) - C\tilde{P} \quad , \tag{11}$$

and the adjusted (projected) velocity, v, is given (formally) by

$$v = B^T\tilde{u} \quad . \tag{12}$$

The matrix B^T is an orthogonal projection operator; it projects vectors (such as $\tilde{u}$) onto the linear subspace of discretely divergence-free vectors: $B^Tv = v$ and $C^Tv = 0$; see also [1]. The analysis proceeds by starting at t_n where u_n (satisfying $C^Tu_n = 0$) and P_n are available. The local truncation error (between t_n and t_{n+1}) of the total scheme is required in order to estimate the next step size, Δt_{n+1}; it is $E_{n+1} \equiv v - u(t_{n+1}) = B^T\tilde{u}_s - u(t_{n+1})$, where $u(t_{n+1})$ is the exact solution of (10) between t_n and t_{n+1}. It is convenient to decompose this total error into two parts via $E_{n+1} = E^S_{n+1} + E^e_{n+1}$, the sum of a subcycle (+ projection) error and an extrapolation error. $E^S_{n+1} \equiv$

$B^T[\tilde{u}_s - \tilde{u}(t_{n+1})]$, where $\tilde{u}(t_{n+1})$ is the exact solution of (11) between t_n and t_{n+1}, given that $\tilde{u}(t_n) = u(t_n) = u_n$ and that $\tilde{P}(t_n) = P(t_n) = P_n$; also, $E^e_{n+1} \equiv B^T\tilde{u}(t_{n+1}) - u(t_{n+1})$, which is the only error if $\Delta t_s \to 0$ for Δt_n fixed. Actually, it may be clearer to first define $\tilde{E}_{n+1} \equiv \tilde{u}_s - \tilde{u}(t_{n+1})$ as the global error after s steps when solving (11), and then to use $E^s_{n+1} = B^T\tilde{E}_{n+1}$ as the "projected subcycled error"; the end result is, of course, the same. It will turn out that the extrapolation error is generally the smaller of the two and can be neglected.

To obtain E^s_{n+1}, first recall (see, e.g., [1]) that the single-step error d_{m+1} in $\tilde{u}_{m+1}$ (with $\tilde{u}_m$ assumed exact) is $-\Delta t_s^2\ddot{\tilde{u}}_m/2+O(\Delta t_s^3)$. The accumulation of the local errors during s steps can be shown to be $\sim sd_1$; i.e., invoking some global error theory for ODEs, it follows that $\tilde{E}_{n+1} = -\Delta t_n\Delta t_s\ddot{\tilde{u}}_n/2 + O(\Delta t_n^2\Delta t_s)$. Thus,

$$E^s_{n+1} = -\Delta t_n\Delta t_s B^T\ddot{\tilde{u}}_n/2 + O(\Delta t_n^2\Delta t_s) \quad . \tag{13}$$

Next we must evaluate $B^T\ddot{\tilde{u}}_n$ in terms of $\ddot{u}_n$; from (10) and (11) it is clear, given that $\tilde{u}_n = u_n$ and $\tilde{P}_n = P_n$, that $\dot{\tilde{u}}_n = \dot{u}_n$. Differentiating (10) and (11) and evaluating the result at t_n then gives $M\ddot{u}_n = g_u(u_n)\dot{u}_n - C\dot{P}_n$ and $M\ddot{\tilde{u}}_n = g_u(u_n)\dot{u}_n - C\dot{\tilde{P}}_n$, where $g_u(u) \equiv \partial g(u)/\partial u$ is the Jacobian matrix. Thus $\ddot{\tilde{u}}_n = \ddot{u}_n + M^{-1}C(\dot{P}_n - \dot{\tilde{P}}_n)$ and $B^T\ddot{\tilde{u}}_n = \ddot{u}_n$ (since $B^T\ddot{u}_n = \ddot{u}_n$ and $B^TM^{-1}C = 0$, an easily verifiable identity which is a consequence of the orthogonality of the two subspaces -- divergence-free vectors are orthogonal to vectors which are gradients of scalars). So, the subcycling error is

$$E^s_{n+1} = -\Delta t_n\Delta t_s\ddot{u}_n/2 + O(\Delta t_n^2\Delta t_s) \quad . \tag{14}$$

Before using this result, we must first dispense with the extrapolation error, E^e_{n+1}. We use Taylor series again, in the form

$$E^e_{n+1} = E^e_n + \Delta t_n\dot{E}^e_n + (\Delta t_n^2/2)\ddot{E}^e_n + (\Delta t_n^3/6)\dddot{E}^e_n + O(\Delta t_n^4) \ ,$$

or, using $E^e(t) \equiv B^T\tilde{u}(t) - u(t)$ and the previouslydetermined results, viz., $\tilde{u}_n = u_n$, $\dot{\tilde{u}}_n = \dot{u}_n$, and $B^T\ddot{\tilde{u}}_n = \ddot{u}_n$, we have $E^e_n = \dot{E}^e_n = \ddot{E}^e_n = 0$ and thus

$$E^e_{n+1} = \Delta t_n^3(B^T\dddot{\tilde{u}}_n - \dddot{u}_n)/6 + O(\Delta t_n^4) \quad . \tag{15}$$

From (10) and (11) again, we obtain $B^T\dot{\tilde{u}} - \dot{u} = B^TM^{-1}[g(\tilde{u}) - g(u)]$, where we now have used the additional identity, $M^{-1}B = B^TM^{-1}$. This leads to $B^T\ddot{\tilde{u}}-\ddot{u} = B^TM^{-1}[g_u(\tilde{u})\dot{\tilde{u}}-g_u(u)\dot{u}]$ and $B^T\dddot{\tilde{u}}-\dddot{u} = B^TM^{-1}\{\partial/\partial\tilde{u}[g_u(\tilde{u})\dot{\tilde{u}}]\dot{\tilde{u}} + g_u(\tilde{u})\ddot{\tilde{u}} - \partial/\partial u[g_u(u)\dot{u}]\dot{u} - g_u(u)\ddot{u}\}$ which, when evaluated at $t = t_n$, gives $B^T\dddot{\tilde{u}}_n - \dddot{u}_n = B^TM^{-1}g_u(u_n)[\ddot{\tilde{u}}_n - \ddot{u}_n] = B^TM^{-1}g_u(u_n)M^{-1}C(\dot{P}_n - \dot{\tilde{P}}_n)$. Thus we are left with

$$E^e_{n+1} = \Delta t_n^3\ B^TM^{-1}g_u(u_n)M^{-1}C(\dot{P}_n - \dot{\tilde{P}}_n)/6 + O(\Delta t_n^4) \tag{16}$$

as the extrapolation error. Finally, it is not hard to show that $\dot{P}_n - \dot{\tilde{P}}_n = O(\Delta t_n)$ when linear extrapolation through major time steps is used, to give $E^e_{n+1} = O(\Delta t_n^4)$. This result shows, by comparing with (14), that extrapolation introduces a higher order error and therefore we now focus on E^s_{n+1}. (Note that even a constant pressure extrapolation is viable in that E^e_{n+1} would then be $O(\Delta t_n^3)$ which is still higher order.

We briefly tested this and decided that linear extrapolation is more cost-effectve.)

In order to use (14), we first write $\ddot{u}_n = (\dot{u}_{n+1} - \dot{u}_n)/\Delta t_n + O(\Delta t_n)$ to obtain

$$E_{n+1} = E^S_{n+1} = -\Delta t_s(\dot{u}_{n+1} - \dot{u}_n)/2 \tag{17}$$

as the local error estimate, where higher-order terms have been neglected and the acceleration vectors are evaluated directly from (2a). We are finally ready to estimate the next step size. To do this, we use (14) again, in the form

$$E_{n+2} \cong -\Delta t_{n+1}\Delta t_s \ddot{u}_{n+1}/2 \cong -\Delta t_{n+1}\Delta t_s \ddot{u}_n/2 = -(\Delta t_{n+1}/\Delta t_n)E_{n+1} \quad . \tag{18}$$

To get Δt_{n+1}, we set the norm of the local error to be committed during the next time step to a user-specified tolerance, ε: $||E_{n+2}|| = (\Delta t_{n+1}/\Delta t_n)||E_{n+1}|| = \varepsilon$ to give

$$\Delta t_{n+1} = \Delta t_n(\varepsilon/||E_{n+1}||) \tag{19}$$

as the next step size, where $||E_{n+1}||$ is a relative, weighted RMS norm (see [1] for details) and is obtained from (17).

5. NUMERICAL RESULTS. The simulation to be described is a variation on the comparison problem presented in [3]; viz, thermal convection in a unit box with side-wall heating ($T = 1$ at $x = -0.5$, $T = 0$ at $x = 0.5$, top and bottom insulated). The difference is that the Prandtl number is changed from 0.71 (e.g., air) to 0.01 (e.g. liquid sodium) -- and the effects are profound; there is no longer a steady-state solution and the dynamics are rather exciting. However, we actually initiated these simulations with our steady-state GFEM code. Starting with the mesh used for the contest problem (168 elements, 745 nodes, with biquadratic velocity and temperature and discontinuous bilinear pressure) we obtained solutions at $Ra = 10^3$, 10^4, 10^5, and 2.5×10^5. Since we could not obtain a solution above $Ra = 2.5\times10^5$, and because the solution was suspiciously complex (basically a flywheel-like motion [4] with circular streamlines concentrated in a ring with radius $\bar{r} \cong 0.4$ with $\Delta r \cong 0.1$, and small-scale vortex-like structures in the corners and in the cavity center), we switched to our time-dependent code (described above) and obtained a rough map of the behavior over a Ra range of 10^4-10^7 at $Pr = 0.01$. This exploratory study, although performed with a coarse mesh (24 x 24, graded), seemed to indicate: (1) oscillatory behavior exists even as low as $Ra = 10^4$ and (2) the flow becomes exceedingly complex at high Ra. Since we then believed that a steady state does not exist and that a finer mesh was needed, we refined the mesh and narrowed our sights. It appeared as if a fine mesh was needed "almost everywhere" owing to the multitude of fine-scale structure. Our compromise was to select a uniform 70 x 70 element mesh and study just one problem: $Ra = 10^5$, $Pr = 0.01$ ($Gr = Ra/Pr = 10^7$), giving $Re = 3162$ and $Pe = 31.62$.

A (minor) time step of 0.002 was used based in part on (6), in part from an estimate of the two-dimensional diffusion limit ($\Delta t \leq$

$\Delta x^2/4K$) for the temperature equation (where $K = Pe^{-1}$) and in part by trial and error ($\Delta t \geq 0.0035$ was unstable). From (6) we get (using $u = 1$) $\Delta t \leq 0.014$ for velocity and $\Delta t \leq 0.003$ for temperature; also, $\Delta t \leq \sim 0.0016$ from $\Delta x^2/4K$. So here we have a situation in which the flow is not completely advection-dominated (the high diffusivity of the liquid metal causes the energy equation to be (nearly) diffusion-dominated), although BTD still provided additional stability for the velocity, which is advection-dominated; (5) would require $\Delta t \leq 0.0006$ if BTD were not employed. Thus, while the stability gain from BTD is modest in this case, the "theory" of balancing the truncation error still applies and we used BTD in both equations.

The simulation started from rest with an isothermal fluid ($T = 0.5$). At $t = 0$, we raised (lowered) the side wall temperatures by ± 0.5. The overall behavior exhibited skew-symmetry and was basically as follows: (1) at very early time ($t < \sim 1$), the initial flow somewhat resembled that for a higher Prandtl number ($Pr = 1$; see [5]) in that the sharp horizontal temperature gradient at the walls generated a pair of counter rotating eddies which then merged to form one large cell (of course the temperature profile developed much more quickly for the low Pr case); (2) by $t = 2$-3, when isotherm advection had just begun, the corner eddies formed, first in the upper left and lower right corners, then near the other two corners; (3) shortly thereafter, these corner vortices separated from the wall and the interesting dynamics began; (4) after a long adjustment period (50-150 time units), the system "settled down" to a complex, time-periodic behavior, which will be described in more detail below.

Figures 1 and 2 show the time histories of vertical velocity and temperature at two of the nodes indicated in Fig. 7a, as well as the dynamic Δt($\epsilon = 10^{-4}$). (The subcycle ratio reached almost 60 at

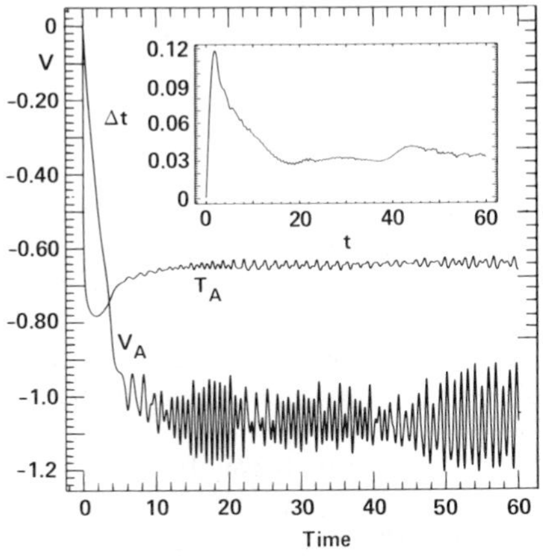

FIGURE 1. Time histories during the transient adjustment period: V_A and T_A. The inset figure shows Δt vs t during this same period of time.

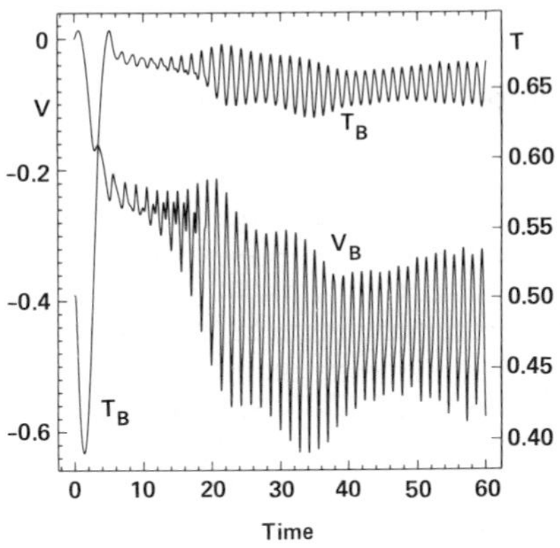

FIGURE 2. Like Fig. 1, except T_B and V_B are shown.

$t = 2$, after which it dropped to $\sim$ 15-20 to follow the action.) The first node (A) is located $2\Delta x$ from the cold wall at $y = 0$ and is in the momentum boundary layer (which thickness is $\sim 4/\sqrt{Re} = 0.07$; the thermal boundary layer is essentially nonexistent). The maximum velocity occurs 2-3 nodes further in and is ~ -1.33. Node B is located near the inside diameter of the flywheel ($x = y = 0.2$) and its time history characterizes the dynamics in the center of the cavity (viz., the revolving pair of internal rotors, to be discussed shortly). By $t = 50$-60, the major flow and temperature adjustments have occurred, although more time is required to reach a quasi-steady-state (i.e., periodic behavior); periodicity seems to have been attained by $t = 100$-150. (During quasi-SS, the dynamic Δt was $\sim 0.054 \pm 10\%$, giving $s \cong 27$). By $t = 50$-60, node A is already responding at what turned out to be the frequency of the corner vortex dynamics (with period, τ_1, $\cong 0.96$), while node B is oscillating more slowly (at $\tau_2 = 1.36 - 1.37$) and reflects the dynamics of the rotating center eddy pair.

A good portion of the overall behavior is seen quite clearly in the sequence of (nearly) two cycles ($2\tau_1$) in Fig. 3, which begins at $t \equiv t_0 \cong 190$ (during quasi-steady behavior); t_0 is chosen to coincide with the "first appearance" of a new pair of side wall eddies (one at top left, the other at bottom right; owing to the skew-symmetry exhibited by the solution, only the top half of the cavity is displayed) and the time separation is $0.125\ \tau_1$. Two cycles are shown because, while the period remains essentially constant, the amplitude does not. Thus, the first cycle shown exhibits weak corner vortices relative to the second cycle. This is also revealed in Fig. 4, which traces u_c (see Fig. 7a for the location of node c) over a period (τ_T) of 26 time units, beginning at t_0. The first cycle is weak, the second and third are strong, the next is weak, etc. The formation, strengthening, separation, weakening, and finally disappearance (on the adjacent downstream wall) of these corner vortices is interesting; and this portion of the pattern repeats (in general form, at least) with a period τ_1.

Focusing now on the center of the cavity (Fig. 3), we observe the center pair of revolving (and co-rotating) eddies. This rotor pair is apparently spun by the flywheel, but its period (τ_2) is noticeably larger than that of the corner vortices (ratio of $\sim 27/19$). Actually, τ_2 is the time required for the two rotors to exchange their positions; $2\tau_2$ is required for each rotor to complete a full revolution. Thus, in the time of $2\tau_1$ (almost shown in Fig. 3), each center eddy has rotated about $19/27 \cong 0.7$ revolution. It appears that the strength of the corner vortices is strongly related to the phase relationship between themselves and the center eddy-pair, since the latter tends to compress the streamlines in the flywheel on a line through the eddy centers, and to expand the streamlines on a line orthogonal to this. Thus, the flywheel flow is faster just outside the center eddies and, if this higher speed shear-flow exists near the walls when a corner vortex is born, this same vortex will be stronger than when it is born during a time of weaker flywheel flow. The first cycle shows weak flywheel flow during corner vortex formation while, in the second cycle, the center rotor-pair is more favorably located. This behavior is also related to the overall period, $\tau_T \cong 26$, shown in Figs. 4-6; i.e., it

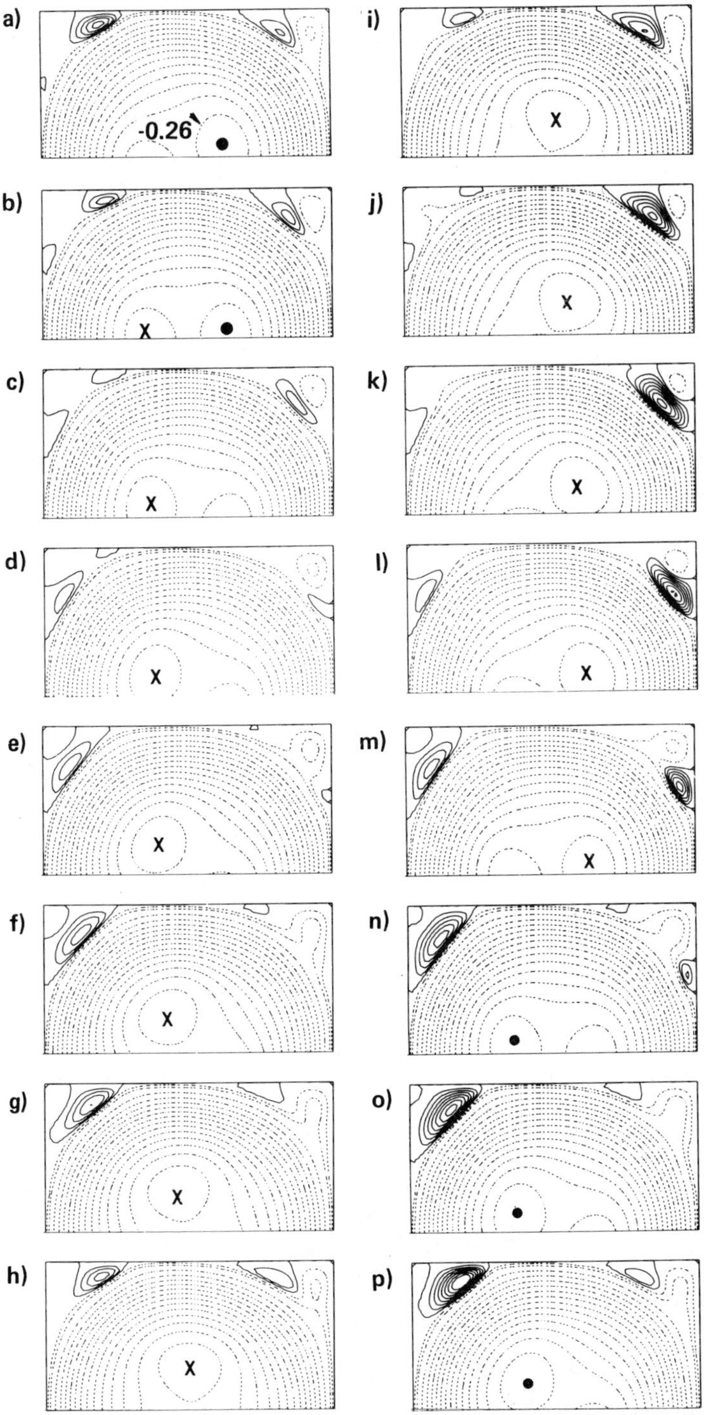

FIGURE. 3. Streamline snapshots; Δt between frames is 0.12. Dashed lines show clockwise motion ($\Delta\psi$ = 0.02 except nearest the wall, where ψ = -0.002 and -0.005 are included) solid lines ($\Delta\psi$ = 0.001) show anticlockwise motion.

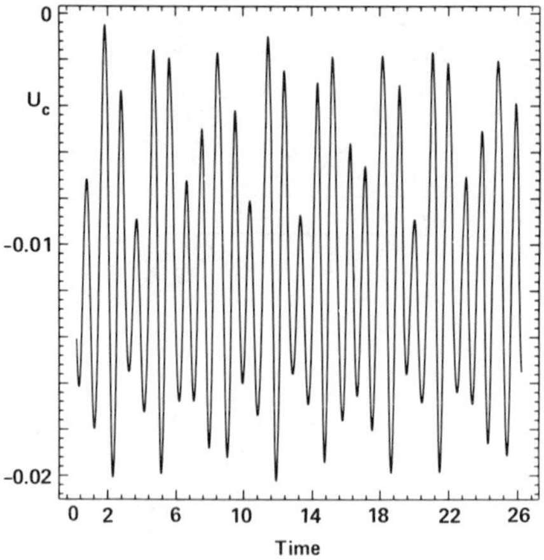

FIGURE 4. Time history of the horizontal velocity at node C (u_c) during a full period (26 time units).

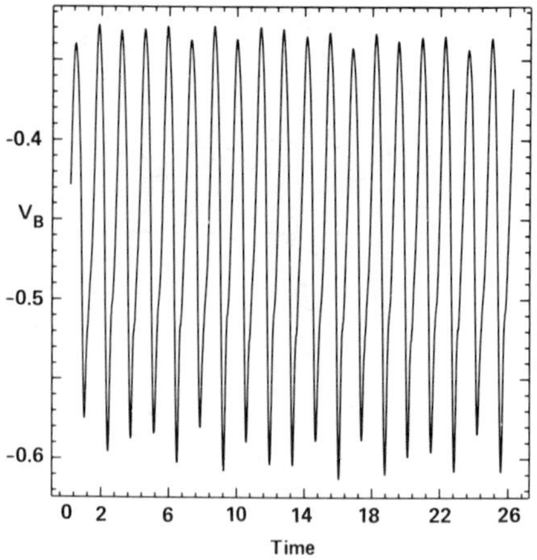

FIGURE 5. Like Fig. 4, except V_B.

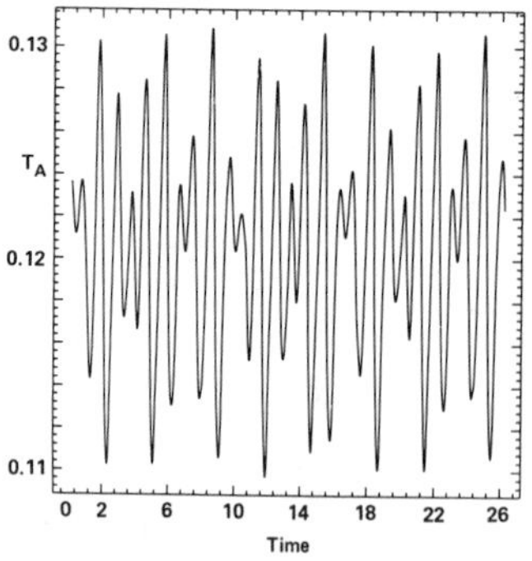

FIGURE 6. Like Fig. 4, except T_A.

requires 26 time units to obtain an exact replication of the flow (and temperature) field; the traces shown in these figures will be "precisely" duplicated during the next 26 time units and corresponds to 27 cycles of corner eddies and 19 cycles (9-1/2 full rotations) of the center eddy pair. Figure 4 shows the horizontal velocity component at node C and Fig. 5 shows the vertical component at node B during one full cycle; clearly the rotor-pair undergoes less significant amplitude changes than do the corner vortices (the interaction is more or less "one-way"). Finally, Fig. 6 shows the temperature at node A during the same period; the flow in the boundary layer sees all of the action (note too that the temperature oscillations are rather significant, considering the proximity of this node to the cold wall).

To complete the picture, Fig. 7 shows the isotherm pattern, beginning at t_0, at intervals of 0.25 τ_1 for 3/4 cycle. (Also shown are the streamlines corresponding to the rotating pair of center eddies.) These figures are to be viewed in a clockwise manner (corresponding to the principal flow direction) and supplement those shown in Fig. 3a-3g. The isotherms in the second (strong) cycle differ little from those in the first (except in details) and are not shown.

Figure 8 shows the cross-sectional velocity profiles through the cavity center (at $t = t_0$) indicating a region of nearly constant vorticity (where $u \sim y$ or $v \sim x$) in the so-called flywheel region, which here appears to extend from $r \cong 0.15$ to $r \cong 0.4$ and corresponds to the region in which the streamlines are nearly circular. Such flywheel, or inertial (rigid body) convection has also been observed in the Benard

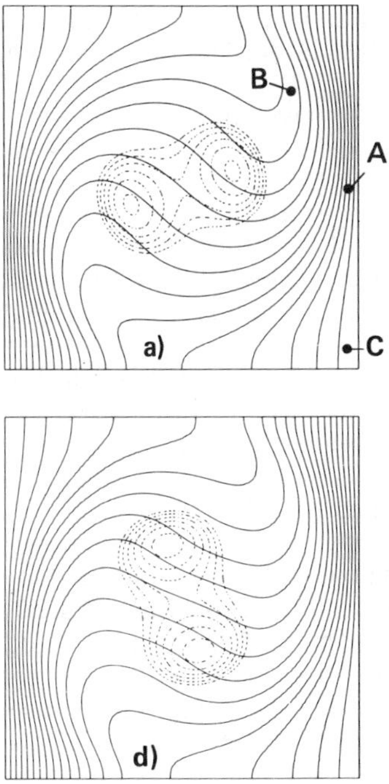

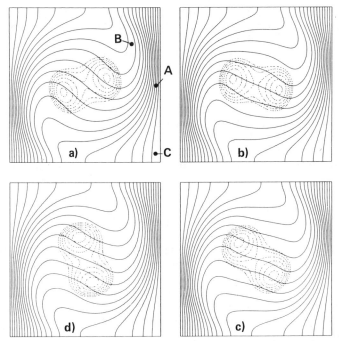

FIGURE 7. Isotherm field ($\Delta T = 0.05$) during quasi-steady state; the time interval between frames is 0.24 and the stream function is shown (dashed) in the center.

problem [4,6] at low Pr. Jones [7] has observed a similar behavior for a side-wall-heated cavity at Pr = 0.035, but points out that the mechanisms acting are fundamentally different for the two cases. He has also predicted a similar "wall vortex shedding" phenomenon.

Figure 9 shows the heat flux in terms of Nu(y) on the hot wall at two times during the time interval shown in Fig. 3; these curves

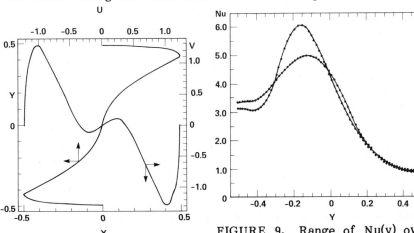

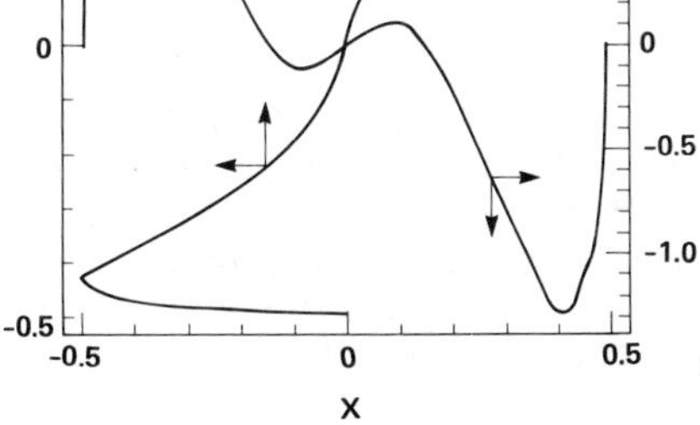

FIGURE 8. Velocity profiles through the cavity center ($t = t_0$)

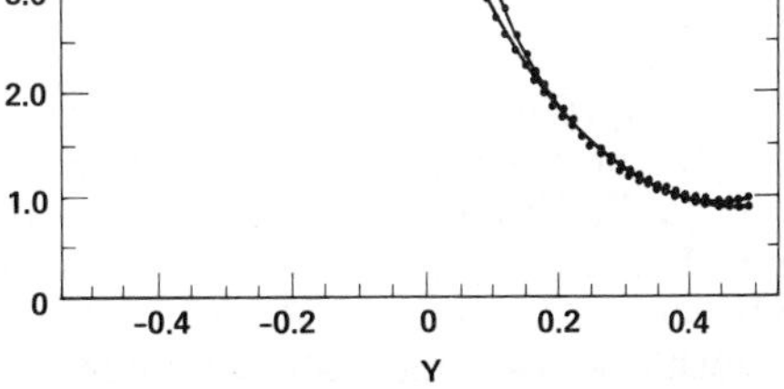

FIGURE 9. Range of Nu(y) over the two cycles shown in Fig. 3; upper curve corresponds to Fig. 3m, lower to Fig. 3p.

represent (approximately) the extremes in the heat flux behavior and, for each, the average Nusselt number is 3.16 ± 0.05 (as compared to ~ 4.52 for Pr = 0.71; [3]).

The related subject of Benard convection at low Pr has been considered in a series of papers by Busse [8,9,4,6], and the side-wall-heated case is discussed by Hart [10], Hurle et al. [11], and Gill [12]. The combined case is treated by Sweet et al. [13]. The principal result of interest here is that time-dependent (oscillatory) convection at low Pr appears to occur first as a three-dimensional instability (roll cells become wavy in the Benard case), thus tending to somewhat vitiate our current two-dimensional results. Although a full three-dimensional simulation may thus be indicated, we believe that our two-dimensional oscillatory solution is of interest in its own right, and that a similar mechanism, perhaps in conjunction with others, may also occur in the three-dimensional case. (The studies in [10-12] were concerned with aspect ratios different than unity.)

Since many of the time history traces are so complex, the signatures seem to be begging for some detailed study in the form of spectral analysis. We have had time to do only a limited study of this type, with the following results: defining $f \equiv 1/\tau$ as frequency, we detected significant energy at f_1 and f_2 (of course), and their first few harmonics, but also at $f_1 + f_2$, thus suggesting (not surprisingly) that there is some nonlinear interaction between the two basic frequencies.

The cost of the simulation was significant, but not excessive, when it is recalled that we are solving $\sim$ 15000 ODE's, an associated Poisson equation of size 4900 for pressure, and that a long time integration is required. Since $\Delta t_s = 0.002$ and $\tau_T = 26$, one overall cycle requires 13000 minor time steps (and about 500 major steps). The shortest period was covered in $\sim$ 480 minor steps and $\sim$ 15-20 major steps. The cost of one full cycle (26 time units) was about 11 minutes CPU.

6. SUMMARY AND CONCLUSIONS. We have described and demonstrated our most recent technique for solving the time-dependent, incompressible Navier-Stokes (or Boussinesq) equations. While the same method has been applied to 3D simulations (e.g. see the paper by Koseff et al. [14] in these Proceedings), the additional cost of I/O (reading the factored pressure matrix from disk) in these cases is high enough to encourage us to seek alternate methods for solving the Poisson equation. For 2D simulations, however, we are content to stay with Gaussian elimination when the entire problem can be contained in memory.

7. ACKNOWLEDGMENTS. It is our pleasure to thank the following people for their important contributions to this work: Drs. S. T. Chan and A. C. Hindmarsh (LLNL), Dr. R. L. Sani (University of Colorado), and Mr. T. Bakowsky (LLNL). This work was performed under the auspices of the U. S. Department of Energy by the Lawrence Livermore National Laboratory under contract No. W-7405-Eng-48.

8. REFERENCES

1. GRESHO, P., S. CHAN, C. UPSON, and R. LEE - "A Modified Finite Element Method for Solving the Time-Dependent, Incompressible Navier-Stokes Equations," Int. J. Num. Meth. Fluids, in press (1983); see also Lawrence Livermore National Laboratory Report UCRL-88937.
2. SANI, R., P. GRESHO, R. LEE, and D. GRIFFITHS - "The Cause and Cure(?) of the Spurious Pressures Generated by Certain FEM Solutions of the Incompressible Navier-Stokes Equations," Int. J. Num. Meth. Fluids, 1, p. 17, 1981.
3. DE VAHL DAVIS, G., and I. JONES - "Natural Convection in a Square Cavity: A Comparison Exercise," Int. J. Num. Meth. Fluids, in press (1983). Also in Numerical Methods in Thermal Problems, Vol. II, Pineridge Press, Swansea, U.K., 1981.
4. CLEVER, R. and F. BUSSE - "Low Prandtl Number Convection in a Layer Heated from Below," J. Fluid Mech., 102, p. 61, 1981.
5. GRESHO, P., R. LEE, S. CHAN, and R. SANI - "Solution of the Time-Dependent Incompressible Navier-Stokes and Boussinesq Equations Using the Galerkin Finite Element Method," in Approximation Methods for Navier-Stokes Problems, p. 203, Springer Verlag Lecture Notes in Math, No. 771, 1980.
6. BUSSE, F. and R. CLEVER - "An Asymptotic Model of Two-Dimensional Convection in the Limit of Low Prandtl Number," J. Fluid Mech., 102, p. 75, 1981.
7. JONES, I. - "Low Prandtl Number Free Convection in a Vertical Slot," AERE Harwell Report R-10416, 1982.
8. BUSSE, F. - "The Oscillatory Instability of Convection Rolls in a Low Prandtl Number Fluid," J. Fluid Mech., 52(1), p. 97, 1972.
9. CLEVER, R. and F. BUSSE - "Transition to Time-Dependent Convection," J. Fluid Mech., 65(4), p. 625, 1974.
10. HART, J. - "Stability of Thin Non-Rotating Hadley Circulations," J. Atmos. Sci, 19, p. 687, 1972.
11. HURLE, D., E. JAKEMAN, and C. JOHNSON - "Convective Temperature Oscillations in Molten Gallium," J. Fluid Mech., 64(3), p. 565, 1974.
12. GILL, A. - "A Theory of Thermal Oscillations in Liquid Metals," J. Fluid Mech., 64, p. 577, 1974.
13. SWEET, D., E. JAKEMAN, and D. HURLE - "Free Convection in the Presence of Both Vertical and Horizontal Temperature Gradients," Phys. Fluids, 20(9), p 1412, 1977.
14. KOSEFF, J., R. STREET, P. GRESHO, C. UPSON, J. HUMPHREY, and W. TO - "A Three-Dimensional Lid-Driven Cavity Flow: Experiment and Simulation," these Proceedings, 1983.

0070F/fw

A Second Order Approximation to Natural Convection in a Square Cavity

William A. Shay *
David H. Schultz **

ABSTRACT

The purpose of this paper is the application of a second order numerical technique to the problem of fluid flow within a heated closed cavity. The method is a modification of a method developed by Shay [5] and applied to the driven cavity problem. In order to test the viability of this technique, it was decided to extend the technique to the problem of natural convection in a square. Jones [2] proposed that this problem is suitable for testing techniques that may be applied to a wide range of practical problems as reactor insulation, cooling of radioactive waste containers, solar energy collection and others [1]. It is assumed both components of the velocity are zero on all boundaries and that the boundaries between the differentially heated walls are insulated. The technique proved easy to implement and has been run with Prandtl numbers as low as .0001, Rayleigh numbers as large as 100000 and mesh sizes as small as .0125. Results are displayed in the form of tables and level curves for each of the stream, vorticity, and temperature functions. Comparisons are also made with first order methods to demonstrate the superiority of this method.

THE PROBLEM

The problem is described in [4]. The equations to be satisfied in the interior of the region (Fig. 1) are as follows:

$$\nabla^2 \Psi = -\zeta \, , \tag{1}$$

$$\nabla^2 T + \Psi_x T_y - \Psi_y T_x = 0, \tag{2}$$

and $$\nabla^2 \zeta + P_R(\Psi_x \zeta_y - \Psi_y \zeta_x) + R_A T_y = 0. \tag{3}$$

* Assistant Professor, University of Wisconsin-Green Bay
** Associate Professor, University of Wisconsin-Milwaukee

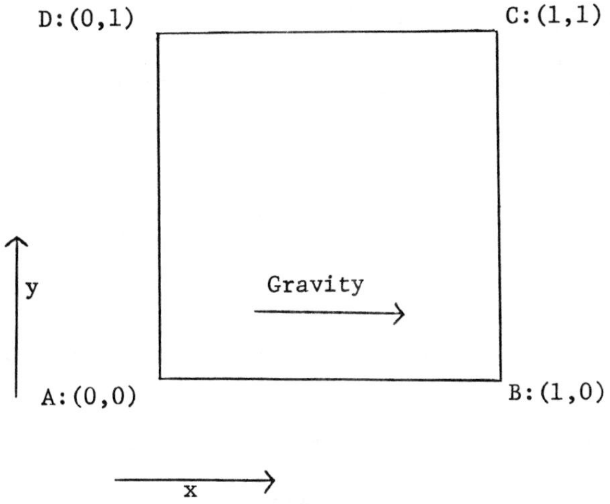

Fig. 1

The stream, vorticity, and normalized temperature functions are represented by Ψ, ζ, and T, respectively. The Rayleigh and Prandtl numbers are given by R_A and P_R, respectively.

Boundary conditions for the problem are

$$\Psi = 0 \text{ on ABCDA} \tag{4}$$

$$\partial\Psi/\partial y = 0,\ T = 0 \text{ on AB} \tag{5a}$$

$$\partial\Psi/\partial x = 0,\ \partial T/\partial x = 0 \text{ on AD and BC} \tag{5b}$$

$$\partial\Psi/\partial y = 0,\ T = 1 \text{ on CD} \tag{5c}$$

DIFFERENCE EQUATIONS

To start the method, a rectangular array of nodes is placed over the region in Fig. 1. It is assumed the vertical and horizontal spacings are equal and are described by h. Define the inner boundary as the collection of all points that lie a distance of h from the boundary. Values at these grid points are used to guarantee that the normal derivative conditions of the stream function are satisfied.

Boundary vorticities may be approximated by (Fig.2)[3].

$$\zeta_0 = -3\Psi_1/h^2 - \zeta_1/2 + O(h^2). \tag{6}$$

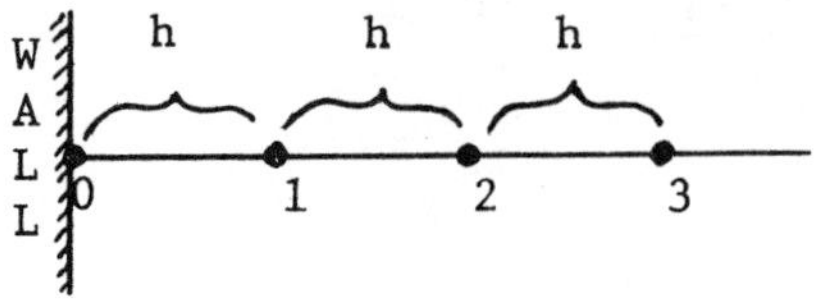

Fig. 2

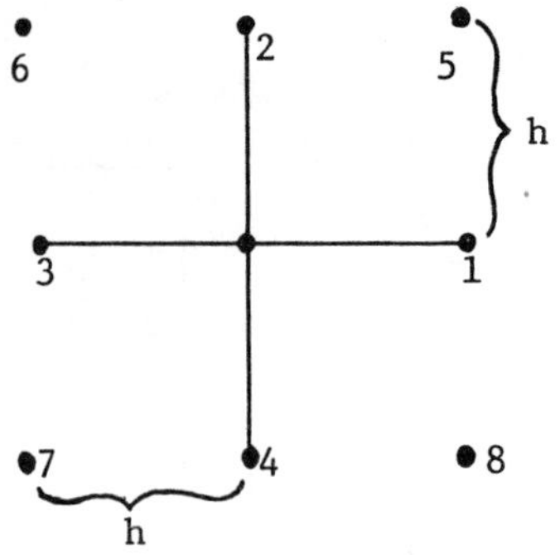

Fig. 3

Stream values on the inner boundary are determined by first writing [6].

$$\Psi_{n_0} = (-11\Psi_0+18\Psi_1-9\Psi_2+2\Psi_3)/6h + 0(h^3) \tag{7}$$

Here, Ψ_{n_0} represents the derivative of Ψ in the direction normal to the wall at the wall. Since $\Psi_{n_0}= 0$ and $\Psi_0 = 0$, Ψ_1 is expressed as

$$\Psi_1 = \Psi_2/2 - \Psi_3/9 + 0(h^4) \tag{8}$$

The higher order approximation (7) is used so that when (8) is used in (6) the $0(h^2)$ accuracy of (6) is maintained.

At any interior node (Fig. 3) $\nabla^2\Psi$ may be approximated by

$$\nabla^2\Psi = (-4\Psi_0+\Psi_1+\Psi_2+\Psi_3+\Psi_4)/h^2 + 0(h^2) \tag{9}$$

Similarly we have

$$\nabla^2\zeta = (-4\zeta_0+\zeta_1+\zeta_2+\zeta_3+\zeta_4)/h^2 + 0(h^2) \tag{10}$$

and $$\nabla^2 T = (-4T_0+T_1+T_2+T_3+T_4)/h^2 + 0(h^2) \tag{11}$$

In (3), T_y is approximated by

$$T_y = (T_2-T_4)/2h + 0(h^2) \tag{12}$$

To approximate $P_R \Psi_x \zeta_y - P_R \Psi_y \zeta_x$ in (3), write

$$P_R \Psi_x \zeta_y - P_R \Psi_y \zeta_x \Big|_{P_0} = \sum_{i=0}^{8} \alpha_i \zeta_i \tag{13}$$

where the α_i are to be determined. Expand each ζ_i in a Taylor series about the point P_0.

Next, reorganize terms of the expansion and group together the coefficients of ζ_0, ζ_{x_0}, ζ_{y_0}, etc., up through third partial derivatives. Then equate with coefficients of like terms in (13).
Therefore, in order for $\sum_{i=0}^{8} \alpha_i \zeta_i$ to approximate $P_R \Psi_x \zeta_y - P_R \Psi_y \zeta_x$ it is sufficient that the α_i satisfy

$$\sum_{i=0}^{8} \alpha_i = 0 \tag{14}$$

$$\alpha_1 - \alpha_3 + \alpha_5 - \alpha_6 - \alpha_7 + \alpha_8 = -P_R \Psi_y / h |_{P_0} \tag{15}$$

$$\alpha_2 - \alpha_4 + \alpha_5 + \alpha_6 - \alpha_7 - \alpha_8 = P_R \Psi_x / h |_{P_0} \tag{16}$$

$$\alpha_1 + \alpha_3 + \alpha_5 + \alpha_6 + \alpha_7 + \alpha_8 = 0 \tag{17}$$

$$\alpha_5 - \alpha_6 + \alpha_7 - \alpha_8 = 0 \tag{18}$$

$$\alpha_2 + \alpha_4 + \alpha_5 + \alpha_6 + \alpha_7 + \alpha_8 = 0 \tag{19}$$

$$\alpha_1 - \alpha_3 + \alpha_5 - \alpha_6 - \alpha_7 + \alpha_8 = 0 \tag{20}$$

$$\alpha_5 + \alpha_6 - \alpha_7 - \alpha_8 = 0 \tag{21}$$

$$\alpha_5 - \alpha_6 - \alpha_7 + \alpha_8 = 0 \tag{22}$$

$$\alpha_2 - \alpha_4 + \alpha_5 + \alpha_6 - \alpha_7 - \alpha_8 = 0 \tag{23}$$

However, it is generally not possible to satisfy (20) and (23) if (15) and (16) are satisfied. As a result, (20) and (23) will constitute an error term of the form

$$(-P_R \Psi_y / h) \zeta_{xxx} (h^3/6) + (P_R \Psi_x / h) \zeta_{yyy} (h^3/6) |_{P_0}$$
$$= (P_R h^2/6)(\Psi_x \zeta_{yyy} - \Psi_y \zeta_{xxx}) |_{P_0}$$

In addition, (21) and (22) are dropped since this yields more flexibility in determining the α_i and does not introduce any lower order error terms. In fact, the six remaining equations in nine unknowns result in three degrees of freedom with which to choose the α_i. If α_6, α_7, and α_8 are the independent choices then

$$\alpha_0 = 2(\alpha_6 + \alpha_8) \tag{24}$$

$$\alpha_1 = (-P_R\Psi_y/2h)|_{P_0} - \alpha_6 + \alpha_7 - 2\alpha_8 \tag{25}$$

$$\alpha_2 = (P_R\Psi_x/2h)|_{P_0} - 2\alpha_6 + \alpha_7 - \alpha_8 \tag{26}$$

$$\alpha_3 = (P_R\Psi_y/2h)|_{P_0} - \alpha_6 - \alpha_7 \tag{27}$$

$$\alpha_4 = (-P_R\Psi_x/2h)|_{P_0} - \alpha_7 - \alpha_8 \tag{28}$$

$$\alpha_5 = \alpha_6 - \alpha_7 + \alpha_8 \tag{29}$$

So for any choice of α_6, α_7, and α_8, equation (13) becomes a second order approximation with an error of

$$((h^2P_R/6)(\Psi_x\zeta_{yyy}-\Psi_y\zeta_{xxx}) + h^3(\zeta_{xxy}(\alpha_6-\alpha_7) + \zeta_{xyy}(\alpha_8-\alpha_7)))|_{P_0} \tag{30}$$

At each interior node, P_0, make the following assignments.

If $\Psi_x|_{P_0} \geq 0$ and $\Psi_y|_{P_0} \geq 0$ let

$$\alpha_6 = 0;\ \alpha_7 = (-P_R\Psi_x/2h)|_{P_0};\ \alpha_8 = (-P_R(\Psi_x+\Psi_y)/2h)|_{P_0} \tag{31}$$

If $\Psi_x|_{P_0} < 0$ and $\Psi_y|_{P_0} < 0$ let

$$\alpha_6 = (P_R(\Psi_x+\Psi_y)/2h)|_{P_0};\ \alpha_7 = (P_R\Psi_y/2h)|_{P_0};\ \alpha_8 = 0 \tag{32}$$

If $\Psi_x|_{P_0} \geq 0$ and $\Psi_y|_{P_0} < 0$ let

$$\alpha_6 = (P_R\Psi_y/2h)|_{P_0};\alpha_7 = (P_R(\Psi_y-\Psi_x)/2h)|_{P_0};\ \alpha_8 = (-P_R\Psi_x/2h)|_{P_0} \tag{33}$$

If $\Psi_x|_{P_0} < 0$ and $\Psi_y|_{P_0} \geq 0$ let

$$\alpha_6 = (P_R\Psi_x/2h)|_{P_0};\alpha_7 = 0;\ \alpha_8 = (-P_R\Psi_y/2h)|_{P_0} \tag{34}$$

Corresponding values of α_5 are

$$(-P_R\Psi_y/2h)|_{P_0};(P_R\Psi_x/2h)|_{P_0};\ 0;\ (P_R(\Psi_x-\Psi_y)/2h)|_{P_0} \tag{35}$$

respectively.

In all cases

$$\alpha_0 = (-P_R(|\Psi_x|+|\Psi_y|)/h)|_{P_0} \tag{36}$$

and $\alpha_1 = \alpha_2 = \alpha_3 = \alpha_4 = -\alpha_0/2$ (37)

The lowest order error term is

$$E=(P_R h^2/6)(\Psi_x \zeta_{yyy} + 3\Psi_x \zeta_{xxy} - 3\Psi_y \zeta_{xyy} - \Psi_y \zeta_{xxx})|_{P_0} \tag{38}$$

If $\Psi_x|_{P_0}$ and $\Psi_y|_{P_0}$ are approximated by $(\Psi_1-\Psi_3)/2h$ and $(\Psi_2-\Psi_4)/2h$, respectively, in each of the α_i, the complete approximation of (3) is

$$(-4 + \Omega_0)\zeta_0 + \sum_{i=1}^{4} (1+\Omega_i)\zeta_i + \sum_{i=5}^{8} \Omega_i \zeta_i \tag{39}$$

$$+ R_A(T_2 - T_4)h/2 = 0$$

where

$$\Omega_0 = -P_R(|\Psi_1-\Psi_3|+|\Psi_2-\Psi_4|)/2 \tag{40}$$

$$\Omega_1 = \Omega_2 = \Omega_3 = \Omega_4 = -\Omega_0/2 \tag{41}$$

and $\Omega_i = h^2\alpha_i$ for i = 5, 6, 7, 8. (42)

This method of approximation is similar to the upwind method in that coefficients are chosen based on the signs of Ψ_x and Ψ_y. As a result, contributions to diagonal elements of the coefficient matrix from the term $P_R(\Psi_x\zeta_y-\Psi_y\zeta_x)$ are guaranteed to be negative, thus adding to the magnitude of the −4 generated in the approximation of $\nabla^2\zeta$. In fact, it is not difficult to show that, in each case $\sum_{i=1}^{8} |\alpha_i|/|\alpha_0| = 3$ for any value of P_R.

To complete the description of the method, $\Psi_x T_y - \Psi_y T_x$ from (2) must be approximated. To do this, simply replace P_R with 1 and ζ with T and repeat the previous development.

Given the complete discretization of (1)-(3), the method is started by assigning initial values of Ψ, ζ, and T to each node. Initial values may consist of all 0's, results of another method, or results from the current method run with different values of h, P_R, or R_A. Then successive sweeps of the region are made, redefining the stream, vorticity, and temperature functions at each node through the use of the successive over-relaxation (S.O.R.) technique until all three functions have converged. The order in which nodes are covered within a sweep may affect stability. In an application of a similar method to another problem [6] divergence resulted when, with each sweep, the nodes were covered from left to right along each row, starting at the bottom row and proceeding to the top row. However, if in alternate sweeps the pattern were reversed, convergence resulted.

RESULTS AND CONCLUSIONS

Converged solutions have been obtained for Rayleigh numbers up to 100,000 and Prandtl numbers as small as .0001. Results have been obtained for a decreasing sequence of mesh sizes, and from the use of extrapolation on results of the two smallest mesh sizes (h = .025 and h = .0125). The extrapolated results are compared with similar results generated by DeVahl Davis [1]. As in [1], the average Nusselt number was calculated through the use of a three point approximation to $\partial T/\partial y$ at the cold wall and Simpson's rule to approximate $\int_0^1 (\partial T/\partial y)|_{y=0} dx$. Fourth order approximations were used to calculate the maximum vertical velocity on the horizontal midplane (v_{max}), maximum horizontal velocity on the vertical midplane (u_{max}), and the maximum and minimum local Nusselt numbers (Nu_{max} and Nu_{min}). Within the table, x or y indicates a coordinate of a point where the value immediately above was located. Maximum stream value is given by Ψ_{max} and Ψ_{mid} represents the stream value at the midpoint. There is excellent agreement between the results with relative differences generally less than 1%. (See Table 1)

In addition to the displayed results from [1], a collection of results from 36 outside sources are also summarized in [1]. However, none of these results were obtained with as fine a grid for R_A = 100,000 as were the current results. One source did use a similar size mesh for R_A = 1,000,000 but had difficulties preserving the symmetry of the problem. Also, none of the methods indicate success with small Prandtl numbers. It is the success of the current method with the wide range of both Rayleigh and Prandtl numbers (due primarily to the 3 degrees of freedom in choosing coefficients of approximating equations) at small mesh sizes that indicate the potential of this method as an accurate technique applicable to a wide variety of problems. Other features of the current method include the second order accuracy of all approximating equations (including boundary approximations - some authors used first order boundary approximations). The use of grid refinement and extrapolation on an already fine mesh size, easy implementation on a computer, and flexibility for adaption to other types of problems. None of the methods in [1] indicate possession of all of the above features.

Figures 4-6 contain level curves for the stream,vorticity, and temperature functions.

	R_A = 10,000		R_A = 100,000	
	[1]	Current Study	[1]	Current Study
$\overline{Nu}$	2.238	2.257	4.505	4.505
Nu_{max}	3.527	3.562	7.717	7.793
x	.143	.125	.082	.075
Nu_{min}	.586	.574	.729	.723
x	1.0	1.0	1.0	1.0
v_{max}	-16.178	-16.153	-34.77	-34.71
x	.823	.825	.854	.85
u_{max}	19.643	19.608	68.25	68.49
y	.119	.125	.066	.059
Ψ_{mid}	5.079	5.070	9.120	9.089
Ψ_{max}	n.a.	5.070	9.622	9.591
x	n.a.	.5	.399	.4
y	n.a.	.5	.713	.709

Table 1 - Comparison of results from the current method and from [1]. P_R = .71.

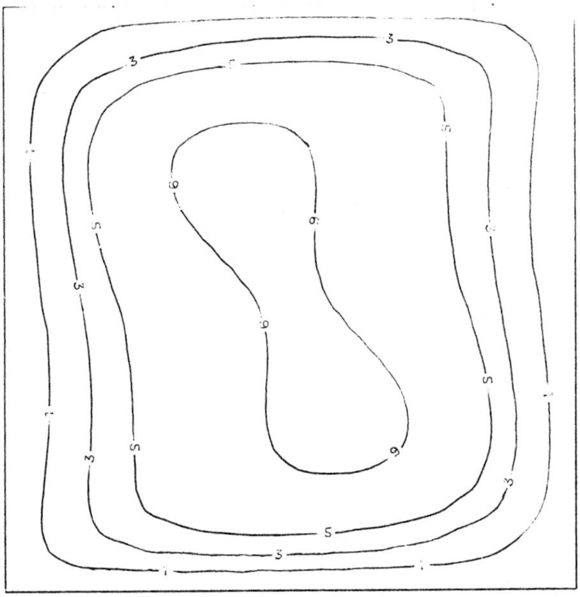

Fig. 4. Streamlines for Rayleigh Number 100,000, Prandtl Number .71, and Mesh Size .0125

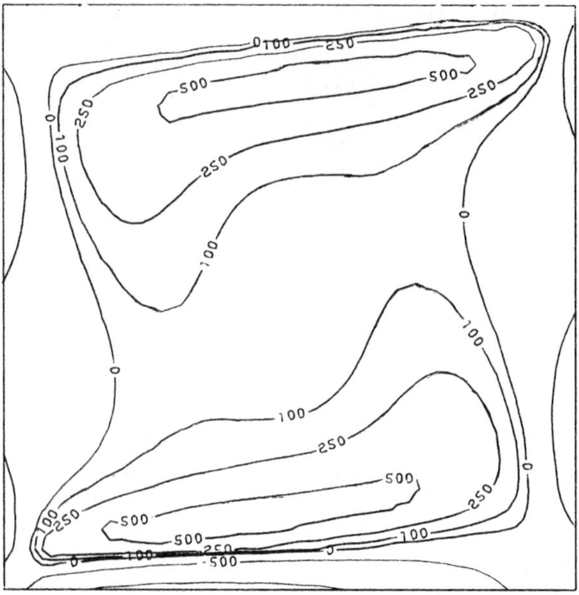

Fig. 5. Vorticities for Rayleigh Number 100,000, Prandtl Number .71, and Mesh Size .0125

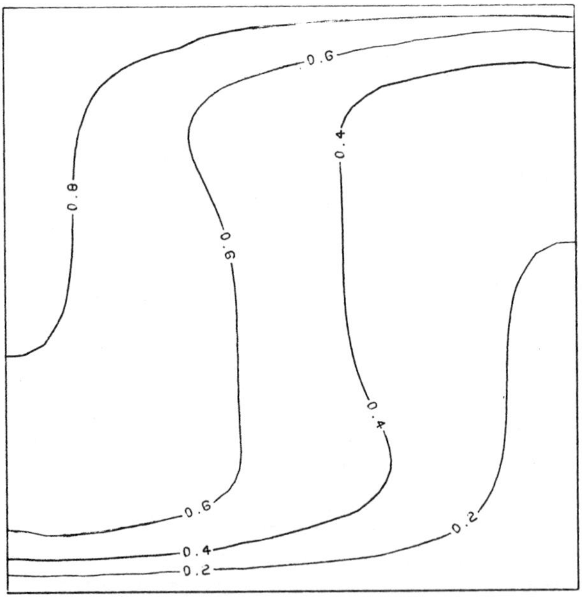

Fig. 6. Temperature for Rayleigh Number 100,000, Prandtl Number .71, and Mesh Size .0125

A comparison with a first order method on a similar problem [4] where $\partial T/\partial x = 0$ on AD and BC is replaced by T=y clearly shows the superiority of the current method over the first order technique. (Table 2)

	first order method		current method	
h	Ψ_{mid}	ζ_{mid}	Ψ_{mid}	ζ_{mid}
.1	7.962	168.0	6.764	139.4
.05	7.066	142.0	6.430	130.7
.025	6.590	135.0	6.357	129.4
.0125	6.423	132.3	-	-

Table 2 - Comparison of results from the current method and the first order method in [4]. R_A = 10,000 and P_R = .73

Table 3 contains results for R_A = 100,000 as a function of h, and a set of results extrapolated from the results of h = .025 and h = .0125.

	h = .05	h = .025	h = .0125	extrapolated
$\overline{Nu}$	4.943	4.658	4.543	4.505
Nu_{max}	7.744	7.877	7.814	7.793
x	.1	.075	.075	.075
Nu_{min}	.597	.672	.710	.723
x	1.0	1.0	1.0	1.0
v_{max}	-33.67	-34.40	-34.63	-34.71
x	.85	.85	.85	.85
u_{max}	67.27	69.28	68.69	68.49
y	.1	.075	.063	.059
Ψ_{mid}	9.31	9.24	9.127	9.089
Ψ_{max}	9.83	9.74	9.628	9.591
x	.4	.4	.4	.4
y	.7	.725	.713	.709

Table 3 - Results for R_A = 100,000, P_R = .71 as a function of h.

As previously stated the method was also run with $P_R = .0001$ and $R_A = 100$. A mesh size of .0125 was used. Values have been tabulated as follows: $\overline{Nu} = 1.00045$; $(Nu_{max}, x) = (1.0312, 0)$; $(Nu_{min}, x) = (.9688, 1)$; $(v_{max}, x) = (-.3196, .787)$; $(u_{max}, y) = (.3196, .213)$; $\Psi_{mid} = .1054$; $\Psi_{max} = .1054$

REFERENCES

1. DeVahl Davis, G. and I. P. Jones - Natural Convection in a Square Cavity, a Comparison Exercise, Numerical Methods in Thermal Problems, 2, Ed. R. W. Lewis, K. Morgan, and B. A. Schrefler, Pentech Press.

2. Jones, I. P. - A Comparison Problem for Numerical Methods in Fluid Dynamics: The 'Double-Glazing' Problem, Numerical Methods in Thermal Problems, Ed. R. W. Lewis and K. Morgan, Pineridge Press, Swansea, U. K., 1979.

3. Runchal, A. K., D. B. Spalding, and M. Wolfshtein - The Numerical Solution of the Elliptic Equations for Transport of Vorticity, Heat, and Matter in Two Dimensional Flows, Ref. No. SF/TN/14, Dept. of Mech. Eng., Imperial College, London, 1968.

4. Schultz, D. H. - Numerical Solution for the Flow of a Fluid in a Heated Closed Cavity. Q. Journal Mech. and Applied Math., Vol. XXVI, Pt. 2, 1973, pp. 173-192.

5. Shay, W. A. - Development of a Second Order Approximation for the Navier Stokes Equations, Computers & Fluids, 9, 1981, pp. 279-298.

6. Shay, W. A. - Development of a Second Order Approximation for the Navier-Stokes Equations. Ph.D. Thesis, University of Wisconsin-Milwaukee, 1978.

NUMERICAL PREDICTION OF LAMINAR FLOW WITH HEAT TRANSFER IN A TUBE WITH A TWISTED TAPE INSERT

J.P. DU PLESSIS [(i)] and D.G. KRÖGER [(ii)]

University of Stellenbosch, 7600 Stellenbosch, South Africa

SUMMARY

The numerical prediction of laminar twisted-tape flow with heat transfer is performed by means of a primitive vari= able finite difference procedure. Allowance is made for vari= able physical properties of the fluid. A parabolic, marching integration procedure is followed although iterations are necessary locally because of incomplete pressure decoupling. The numerical results are checked against experimental data and the computer programming against an independent velocity-vor= ticity procedure. Promising results are obtained although much room is still left for improvement.

1 INTRODUCTION

Twisted-tape flow, brought about by the introduction of a twisted-tape insert into a straight smooth tube, is important in present-day engineering practice. It provides a simple means for obtaining considerable gain in the overall performance of different types of heat exchangers.

Although several attempts by various authors have been launched recently in order to model twisted-tape flow, the only numerical contribution for developing flow was by Date [1] and the problem is still unresolved. Even the simpler case of flow in a smooth tube is still offering difficulties when the physical properties of the fluid are allowed to vary with fluid tempera= ture. In the case of twisted-tape flow the transversal vortex motion is complex and influences the mathematical modelling and the numerical procedures severely. In this paper the numerical work of Du Plessis [2] is discussed.

The numerical prediction of heat and momentum transfer

(i) Department of Applied Mathematics

(ii) Department of Mechanical Engineering

in the present case entails the solving of a set of non-linear, time-independent partial differential equations. These three-dimensional equations are elliptic in nature, but may be forced into a parabolic form by way of certain assumptions.

2 MATHEMATICAL MODELLING

The partial differential equations governing the laminar flow and heat transfer processes of a real fluid may be derived from basic conservation principles. If fully developed flow with isotropic fluid properties is considered, and when the body forces and the viscous dissipation are neglected, the transport equations may be written as follows:

$$\begin{aligned} &\text{Conservation of Mass} &&: \nabla\cdot(\rho\underline{v}) &&= 0 \\ &\text{Conservation of Momentum} &&: \nabla\cdot(\rho\underline{v}\underline{v}) &&= -\nabla p + \nabla\cdot\underline{\underline{\tau}} \\ &\text{Conservation of Energy} &&: \nabla\cdot(\rho\underline{v}T) &&= \frac{1}{C_p}\nabla\cdot(K\nabla T) \end{aligned} \qquad (1)$$

2.1 The Helical Coordinate System

The presence of a twisted tape in a straight tube causes a troublesome axial change in boundary conditions when used with an orthogonal cylindrical coordinate system. A helical coordi= nate system, similar to the one introduced by Date [1], is there= fore used in the present study. The tape-thickness is considered to be much smaller than the tube radius to leave effectively two semi-circular channels plaited around the axial centre-line. This assumption, however, needs to be taken into account when the real flow area is to be computed for direct comparison with experimental work.

Let (r_o,θ_o,z_o) be the usual right-handed orthogonal cy= lindrical polar coordinate system with the origin at the centre of the tape at the inlet to the tube. $\theta_o = 0$ corresponds to the tape edge at the inlet cross-section. z_o is measured downstream and the tape is assumed to possess a twist of pitch F in a di= rection opposite to that of θ_o. The helical coordinate system (r,θ,z) is then defined as follows:

$$r \equiv r_o, \ \theta \equiv \theta_o + \frac{\pi z_o}{F} \ \text{and} \ z \equiv z_o \qquad (2)$$

whence

$$\frac{\partial}{\partial r_o} = \frac{\partial}{\partial r}, \ \frac{\partial}{\partial \theta_o} = \frac{\partial}{\partial \theta} \ \text{and} \ \frac{\partial}{\partial z_o} = \frac{\partial}{\partial z} + \frac{\pi}{F}\frac{\partial}{\partial \theta} \qquad (3)$$

The incorporation of the helical coordinates into the three vector equations (1) leads to an elaborate algebraic mani= pulation in which frequent use is made of the continuity equation to obtain neater expressions for the velocity equations. In analogy with the work of Date [1] the orthogonal velocity components V_r, V_θ and V_z are retained and only the coordinates are transformed into the helical system. The expression $V_\theta + (\pi r/F)V_z$ is replaced by g_θ and $1+(\pi r/F)^2$ by G for mathematical simplicity.

2.2 Parabolic Assumptions

When only mild twist of the inserted tape is considered the curvature of the streamlines is small enough to allow the so-called parabolic assumptions to be acceptable:

(a) The pressure terms are partially uncoupled axially by defining:

$$p(r,\theta,z) \equiv \bar{p}(z) + p'(r,\theta) \quad (4)$$

(b) All stresses and diffusion fluxes on any (r,θ) plane are omitted since they would allow downstream influences to penetrate upstream. This amounts to putting all diffusive partial derivatives in z equal to zero, as well as similar terms which may have been introduced by the metric of the helical system.

Having applied all these parabolic simplifications, the following general transport equation is arrived at:

$$\frac{\partial}{\partial r}(\rho r V_r \phi) + \frac{\partial}{\partial \theta}(\rho g_\theta \phi) + \frac{\partial}{\partial z}(\rho r V_z \phi) - \frac{\partial}{\partial r}\left(r\Gamma\frac{\partial \phi}{\partial r}\right) - \frac{\partial}{\partial \theta}\left(\frac{\Gamma G}{r}\frac{\partial \phi}{\partial \theta}\right) = S + \Pi \quad (5)$$

The set of five simultaneous equations, to be solved by a numerical procedure, is obtained by taking ϕ respectively as 1, V_z, V_r, V_θ and T. The corresponding expressions are provided in Table 1 and, since the present study is aimed at highly viscous oils with only small density variations, terms concerning the latter are omitted. The first three terms of equation (5) represent the convection and the subsequent two the diffusion. On the right hand side the source terms are split into a pure source term S and another term Π which contains a pressure derivative.

2.3 Discretization of the Transport equations

The discretization of the transport equations is performed according to the work of Patankar and Spalding [3] and Spalding [4]. The differential equations are integrated over control volumes, called cells, formed by an L-wise staggered grid. Some boundary cells for V_r and V_θ are elongated to fill the entire physical flow domain. Linear interpolation is applied in order to accommodate non-uniformity of the grid. No recirculation occurs axially so that upwind differencing may be employed implicitly in the z-direction. In the radial and transverse directions hybrid differencing according to [4] is used. The complete finite difference equation is now written in the following general form where Δp is the finite difference increment of p according to the Π term:

$$\phi_P = A_N\phi_N + A_S\phi_S + A_E\phi_E + A_W\phi_W + B + E\cdot\Delta p \qquad (6)$$

3 THE PRIMITIVE VARIABLE SOLUTION PROCEDURE

The transport variables V_z, V_r, V_θ and T are solved by a marching integration procedure in the z-direction. A tridiagonal matrix algorythm (TDMA) is employed for the solving of the vari= ables at each cross-section. A schematic description of the computational procedure is given in Figure 1. When all values of variables and properties are known at the upstream cross-section the computing sequence at its downstream counterpart is as follows:

(a) Guess an approximate mean pressure $\bar{p}$ at the cross-section and compute a corresponding approximate velocity field V_z by means of the axial velocity equation.

(b) Correct V_z by means of overall continuity requirements over the cross-section and adjust $\bar{p}$ accordingly through V_z-equation.

(c) Return to (a) of corrections in $\bar{p}$ or V_z exceed some relative= ly coarse criterion (Loop IV). Without this measure the correc= tion to V_z may be large enough to upset the numerical procedure.

ϕ	Γ	Π	S	$S\{\Gamma\}$
1	0	0	0	0
V_z	μ	$-r\,\frac{d\bar{p}}{dz}$	$-\frac{\pi r}{F}\frac{\partial p'}{\partial\theta}$	$\frac{\partial\mu}{\partial r}\cdot\frac{\pi r}{F}\frac{\partial V_r}{\partial\theta}$ $-\frac{\partial\mu}{\partial\theta}\cdot\frac{\pi}{F}\frac{\partial}{\partial r}(rV_r)$
V_r	μ	$-r\,\frac{\partial p'}{\partial r}$	ρV_θ^2 $-\frac{\mu}{r}\left(V_r + 2\,\frac{\partial V_\theta}{\partial\theta}\right)$	$\frac{\partial\mu}{\partial r}\cdot r\,\frac{\partial V_r}{\partial r}$ $+\frac{\partial\mu}{\partial\theta}\cdot\left(r\,\frac{\partial}{\partial r}\left(\frac{V_\theta}{r}\right)+\frac{\pi r}{F}\frac{\partial V_z}{\partial r}\right)$
V_θ	μ	$-\frac{\partial p'}{\partial\theta}$	$-\rho V_r V_\theta$ $-\frac{\mu}{r}\left(V_\theta - 2\,\frac{\partial V_r}{\partial\theta}\right)$	$\frac{\partial\mu}{\partial\theta}\cdot\left(\frac{\partial V_r}{\partial\theta} - V_\theta\right)$ $-\frac{\partial\mu}{\partial\theta}\cdot r\,\frac{\partial}{\partial r}\left(\frac{V_r}{r}\right)$
T	K/C_p	0	0	$\frac{\partial C_p}{\partial T}\cdot\frac{rK}{C_p^2}\cdot\left[\left(\frac{\partial T}{\partial r}\right)^2 + G\left(\frac{1}{r}\frac{\partial T}{\partial\theta}\right)^2\right]$

Table 1. Coefficients for the general transport equation (5).

(d) Guess an approximate pressure distribution field p' and compute corresponding approximate velocity fields V_r and V_θ through the proper transport equations.

(e) Correct p', V_r and V_θ by means of cell-wise continuity requirements. In effect a finite difference equation for the pressure correction is iteratively solved simultaneously for all nodes at the cross-section under consideration (Loop VII).

(f) Solve the temperature equation and calculate new tempera= ture dependent properties. Return to (a) if changes in the physical properties of the fluid are liable to create intolerable errors in the obtained solutions of the flow variables (Loop II).

(g) If the end of the flow domain has not yet been reached, move downstream through a distance equalto the axial steplength and return to (a).

If fluid properties are constant the velocity field may be solved completely (Loop III) before the temperature cycle (Loop VI) is done. In cases of slow convergence the cross-stream pressure and velocities may be solved to some degree (Loop V) before returning to the T or V_z cycles. It was found to be im= possible to define a unique set of convergence criteria for the procedure due to the variation in the nature in the flow over a spectrum of tape twists and Reynolds numbers, especially with variable fluid properties. A residual source type of criterion, also used by Lilley [6] for swirl flows, is employed for con= vergence. At each node P the residual being checked for solution maturity is defined by:

$$\phi_P - A_N\phi_N - A_S\phi_S - A_E\phi_E - A_W\phi_W - B - E\cdot\Delta p \qquad (7)$$

4 THE VELOCITY-VORTICITY SOLUTION PROCEDURE

In analogy with the work of Date [1] the transport equa= tions may be transformed into a set of equations involving the axial vorticity η which is obtained by elimination of the pres= sure distribution $p'(r,\theta)$. The cross-stream velocity equations are obtained from the definition of vorticity and the continuity equation. This leaves the cell-wise mass balance requirement as an extra check on the accuracy of the solution.

A Gauss-Seidel type of solution procedure is used for each variable. The solution procedure is represented schemati= cally in Figure 2. It should be noted that the V_z equations are identical for the two procedures. In the present case the V_z source term involving $\partial p'/\partial\theta$ is retrieved from the primitive variable V_θ equation.

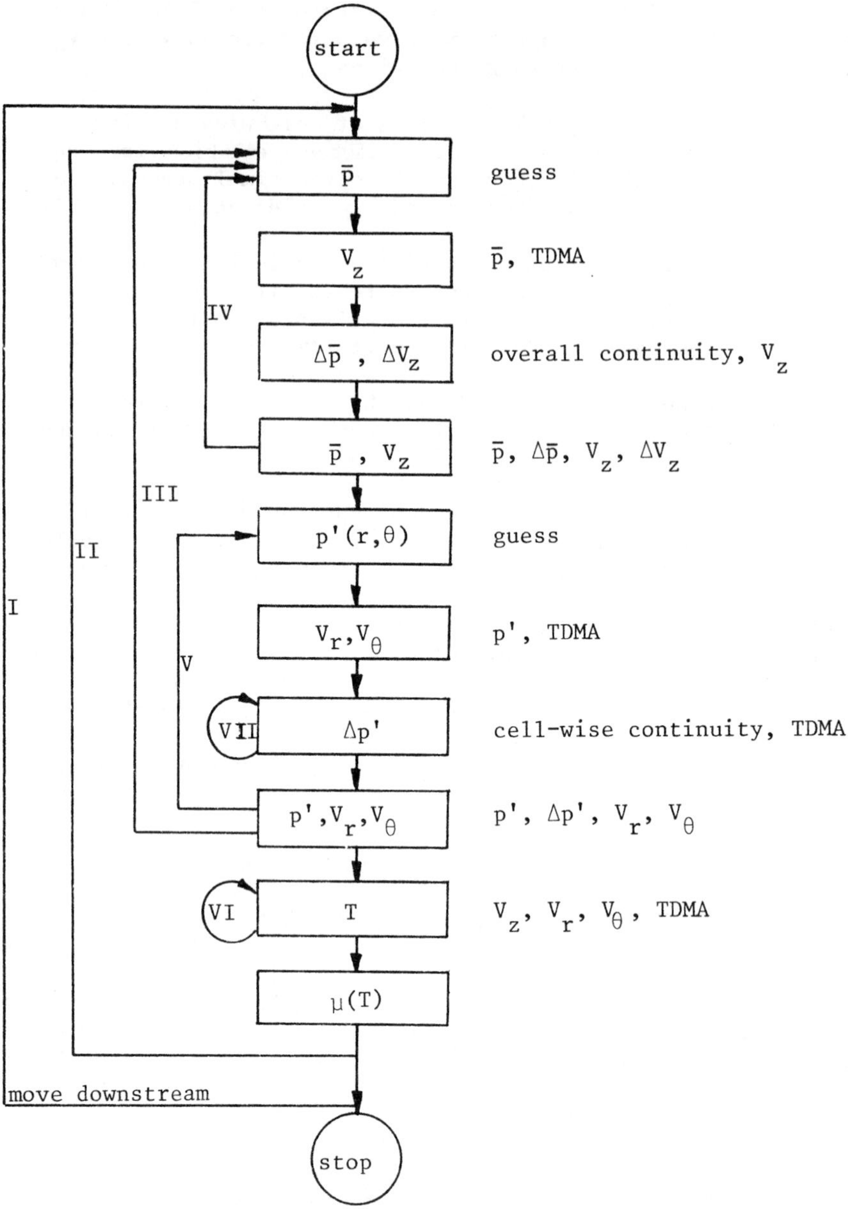

Figure 1. Schematic representation of the solution procedure. On the right hand side the principle and/or variables and/or means used to compute the variables within the specific block are indicated. The Δ's represent corrections to the respective variables.

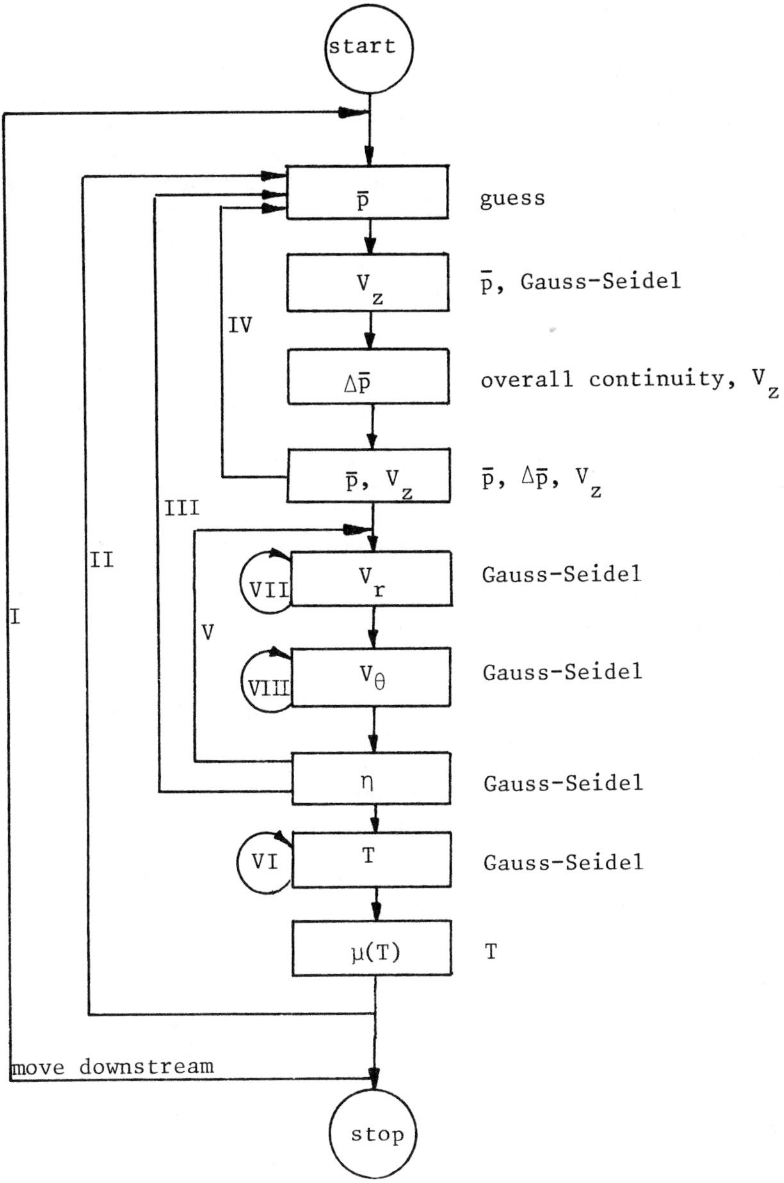

Figure 2. Schematic representation of the velocity-vorticity solution procedure. On the right hand side the principle and/or means and/or variables used to compute the variables in the respective blocks are indicated.

5 VALIDATION CHECKS

An extensive amount of computer programming is needed for the computation of such a problem and computational errors are often very difficult to detect. In lieu of this problem the velocity-vorticity procedure was used to provide solutions in=dependently to that of the primitive-variable procedure. Several test runs with identical boundary conditions were performed on both computer codes and they were corrected until the results were similar. This double check on programming did not only reveal some concealed errors, but also provided fruitful insight into both procedures.

A comprehensive set of test runs were carried out with various grid sizes and axial steplengths to ensure that the numerical solutions are independent of such factors. Cross-sectionally a non-uniform 17x17 grid was found sufficient for the purpose of the present study with an accuracy aim of $\mp$ 5%. Axially a steplength of about 0.0001 D Re_D is commenced with and this is gradually increased by a step-wise factor of 1.1 up to the fully developed region. In many cases a dynamic instability is encountered in or near the fully developed region. This is curtailed by heavy under-relaxation in conjunction with more iterations.

A parametric study with the computer program disclosed a numerical malbehaviour when the Reynolds number is high and the twist ratio is small. The present type of analysis seems applicable up to a value of about 600 for the dimensionless number $Re_D D/F$ introduced by Date and Singham [5].

As a further means of validation some carefully controlled experiments were performed on isothermal flow and both the heating and cooling of a highly viscous oil with temperature dependent physical properties. The tape twist ratio for the experimental test section was F/D = 3.289. Close agreement was obtained with the isothermal data. According to the numerical work the thermal condition of the tape has a marked influence on the results. Although the thermal condition of the tape in the present ex=periment was unknown the heat transfer data pointed towards an adiabatic behaviour of the tape. This type of behaviour is to be expected since the tape is extractable and is not in perfect thermal contact with the surrounding tube.

It is evident that the presence of strong vortex motion produces a smearing effect on the temperature. This in turn causes the persistance of the singularity in the temperature gradient at the wall for quite a distance downstream. Without vortex motion the temperature gradient normal to the wall eases off immediately to a finite value, which may be described reasonably accurately by a linear approximation between the temperatures of the wall node and its inner neighbour. This effect increases with Prandtl value.

6 RESULTS

The two solution procedures discussed in this paper have several features in common. The main point of interest is the lack of complete pressure decoupling which occurs in both procedures through the $\partial p'/\partial\theta$ term present in the V_z source term. This necessitates several iterations for the velocity fields at each cross-section (Loop III).

The primitive variable procedure was found to be more attractive since its debugging and convergence are achieved much easier and with far fewer iterations than in the vorticity case.

Some comparative results between the numerical and the experimental work are demonstrated in Figures 3 and 4. The effect of the thermal condition of the tape is very obvious and the results point to a high thermal resistance between the tape and the tube for the test section used.

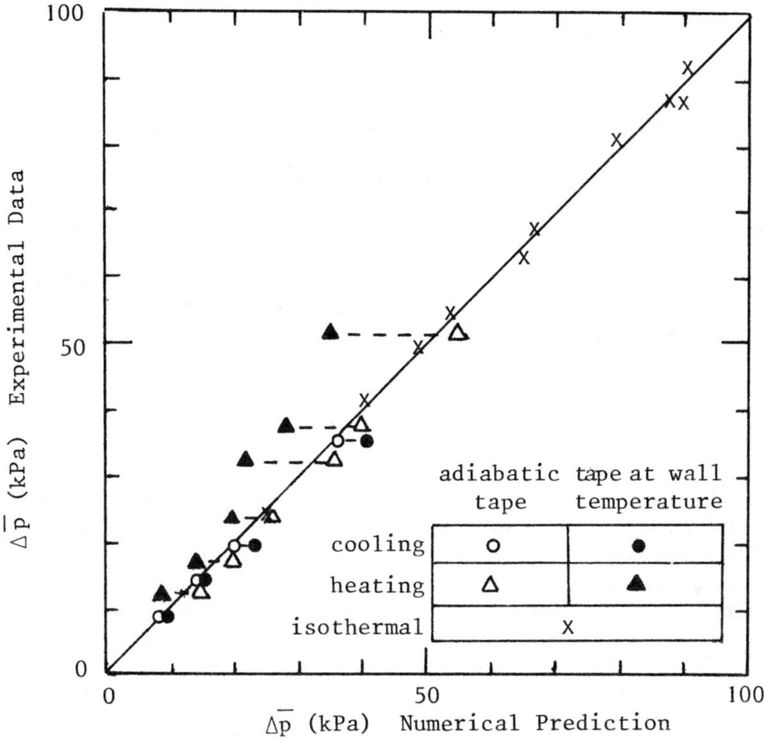

Figure 3. Comparison between the numerical predictions and the experimental results for the pressure drop over the test section. Each horizontal dashed line connects two numerical solutions which correspond to different thermal conditions of tape.

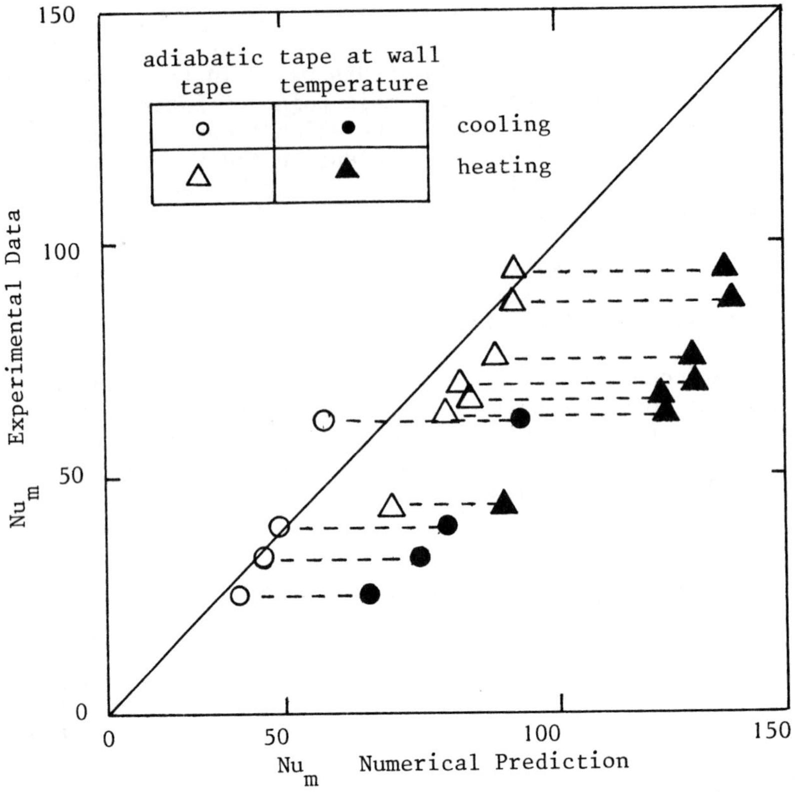

Figure 4. Comparison between the numerical predictions and the the experimental results for the mean Nusselt number for the test section. Each horizontal dashed line connects two numerical solutions which correspond to different tape thermal conditions.

7 CONCLUSIONS

An existing finite difference procedure was adapted and employed in a computer program in order to predict numerically the flow and heat transfer characteristics of twisted-tape flow. Results were tested against experimental and independent numeri= cal work by the authors. It has been demonstrated that excellent agreement between theory and experiment is possible if variable physical properties of the fluid and the thermal condition of the tape is taken into consideration.

It is suggested that a partially-parabolic procedure (Pratap [7]) should be employed in future for this type of flow to allow for the lack of complete pressure decoupling. In this way accurate results should also be obtainable for higher values of $Re_D D/F$.

REFERENCES

1. DATE, A.W. - Prediction of Friction and Heat Transfer Characteristics of Flow in a Tube Containing a Twisted Tape, Ph.D. Thesis, Dept. of Mechanical Engineering, University of London, London, 1973.

2. DU PLESSIS, J.P. - Laminar Flow and Heat Transfer in a Smooth Tube with a Twisted-Tape Insert, Ph.D. Thesis, Dept. of Applied Mathematics, University of Stellenbosch, Stellenbosch, 1982.

3. PATANKAR, S.V. and SPALDING, D.B. - A Calculation Procedure for Heat, Mass and Momentum Transfer in Three-Dimensional Parabolic Flows, Int. J. of Heat and Mass Transfer, Vol. 15, pp. 1787-1805, 1972.

4. SPALDING, D.B. - Mathematical Modelling of Fluid Mechanics, Heat Transfer and Mass Transfer Processes: a lecture course, Rep. No. HTS/80/1, Dept. of Mechanical Engineering, Imperial College of Science and Technology, London, 1980.

5. DATE, A.W. and SINGHAM, J.R. - Numerical Prediction of Friction and Heat-Transfer Characteristics of Fully Developed Laminar Flow in Tubes containing Twisted Tapes, Am. Soc. Mech. Eng., Paper 72-HT-17,1972.

6. LILLEY, D.G. - Primitive Pressure-Velocity Code for the Computation of Strongly Swirling Flows. AIAA Journal, Vol. 14, No. 6, pp 749-756, 1976.

7. PRATAP, V.S. - Flow and Heat Transfer in Curved Ducts, Ph.D. Thesis, Dept. of Mechanical Engineering, University of London, London, 1975.

A NUMERICAL METHOD FOR COMPUTING THREE-DIMENSIONAL NAVIER-STOKES EQUATIONS APPLIED TO CUBIC CAVITY FLOWS WITH HEAT TRANSFER

J.B. Cazalbou, M. Braza, H. Ha Minh.

Institut de Mécanique des Fluides de l'Institut National Polytechnique de Toulouse - L.A. CNRS 0005 -
2 rue Charles Camichel - 31077 Toulouse cédex - France.

SUMMARY

The numerical code presented here is used for solving the three-dimensional unsteady Navier-Stokes equations for either isothermal or non-isothermal flows. Its is founded on a finite-volume approach, using pressure-velocity formulation.

Cubic cavity flow configurations have been chosen as test-cases for this code which may be, in the future, applied to more complex patterns.

Different ways of setting in motion allowed to study i) forced convection, ii) free convection and iii) mixed convection. In the first case, the range of Reynolds number investigated was Re = 100-2000. Free convection was studied for Rayleigh number $R_A = 10^3$ and 10^4, and mixed convection was computed for Re = 100 and $R_A = 10^4$.

All non-isothermal problems were solved using Prandtl numbers P_R = 0.71 (air flow).

1. INTRODUCTION

The main objective of this work is to provide an efficient tool for studying the three-dimensional viscous mecanisms of instability and turbulence. With such an aim, it is vital to take great care of limiting computation times and computer storage so that the code could be applied to the more complex configurations involved in the study of instability.

This necessity explains the choice of the velocity-pressure formulation using only three transport equations for momentum and one Poisson equation for continuity condition. As a matter of fact, an extension to three-dimensional flows of the vorticity-stream function formulation intensively used for two-dimensional cases is always possible but it would need the solving of three transport equations for vector $\vec{\psi}$ (defined as $\vec{V}$ = curl $\vec{\psi}$) and three Poisson equationsfor vorticity vector $\vec{\zeta}$ = curl $\vec{V}$. The same requirements in computer storage and C.P.U. times should be necessary with the use of a $(\vec{V},\vec{\zeta})$ formulation as argued by Dennis et al. [2].

Moreover, the prescription of boundary conditions on primary variables is much easier than those concerning Helmholtz variables. At last, velocity pressure formulation needs not any hypothesis on the flow such as incompressibility or steadiness and would be therefore more suitable to be extended to compressible flows for instance.

The main features of the code performed are :

- a concept of "guess and correct" pressure field as suggested by Chorin [1] or Harlow and Amsden [5] is used in order to overcome the linking of velocity and pressure in the momentum equations ;
- all the equations are written in conservative integrated form and are solved on staggered grids, introduced for the first time in the M.A.C. method by Harlow and Welch [6] ;
- non-isothermal cases are analyzed with Boussinesq's approximation for buoyancy effects.

2. MATHEMATICAL BACKGROUND

2.1. General algorithm for isothermal cases

The Navier-Stokes equations for incompressible flows are written in conservative non-dimensional form as follows :

$$\frac{\partial u_i}{\partial t} + \frac{\partial(u_i u_j)}{\partial x_j} = -\frac{1}{\rho}\frac{\partial P}{\partial x_i} + \nu\frac{\partial^2 u_i}{\partial x_j \partial x_j} \qquad i = 1,2,3 \qquad (1)$$

$$\frac{\partial u_j}{\partial x_j} = 0 \qquad (2)$$

A semi-implicit scheme is adopted in order to linearize the convective terms in equations (1), under this assumption the forward time differencing of the momentum equations is written :

$$\frac{u_i^{n+1} - u_i^{n}}{\Delta t} + \frac{\partial}{\partial x_j}(u_j^n u_i^{n+1}) = -\frac{1}{\rho}\frac{\partial P^{n+1}}{\partial x_i} + \nu\left(\frac{\partial^2 u_i^{n+1}}{\partial x_j \partial x_j}\right) \qquad (3)$$

The linking of pressure and velocity values in this equation does not allow a straightforward solving. Hence, following a suggestion of Chorin [1], one can introduce an intermediate velocity field $\vec{V}^*$ such that :

$$\frac{u^*_i - u_i^n}{\Delta t} + \frac{\partial}{\partial x_j}(u_j^n u_i^*) = -\frac{1}{\rho}\frac{\partial P^n}{\partial x_i} + \nu \frac{\partial^2 u_i^*}{\partial x_j \partial x_j} \qquad (3^*)$$

Noticing that Curl $\vec{V}^{n+1}$ and Curl $\vec{V}^*$ satisfy the same equation, one can reasonably assume that :

$$\text{Curl } \vec{V}^{n+1} = \text{Curl } \vec{V}^* \quad \text{or} \quad \text{Curl } (\vec{V}^{n+1} - \vec{V}^*) = 0 \qquad (4)$$

Relation (4) is a sufficient condition for the existence of a function ϕ such that :

$$\vec{V}^{n+1} - \vec{V}^* = - \vec{\text{grad}}\ \phi \qquad (5)$$

A Poisson equation for ϕ is then obtained by taking the divergence of (5) and by imposing : div $\vec{V}^{n+1} = 0$:

$$\nabla^2 \phi = \text{div } \vec{V}^* \qquad (6)$$

The solution of this equation allows the evaluation of V^{n+1} from (5).

The value of P^{n+1} will be obtained from the relation :

$$P^{n+1} = P^n + \frac{\rho}{\Delta t}\phi \qquad (7)$$

that can be deduced from equations (5), (3) and (3*) under some assumptions.

Finally, the calculation of the flow field a (n+1) time step from the variables values at (n) time step is reduced to :

- the solution of equation (3*) yielding V^* ;
- the solution of equation (6) by inner iterations yielding ϕ ;
- the evaluation of the velocity and pressure fields at (n+1) time step from the relations(5) and (7).

2.2. Non-isothermal cases

Non-isothermal flow fields have been computed on the basis of the Navier-Stokes equations with Boussinesq's approximation and the classical transport equation for temperature in the form :

$$\frac{du_i}{dt} = -\frac{1}{\rho}\frac{\partial P}{\partial x_i} + \nu \nabla^2 u_i + \beta g \delta_{i3} (T - T_o)$$

$$\frac{dT}{dt} = \alpha \nabla^2 T$$

$$\text{div } \vec{V} = 0$$

Using the dimensionless variables and the classical Rayleigh and Prandtl numbers :

$$\tilde{t} = \frac{\alpha t}{d^2} \quad , \quad \tilde{u}_i = \frac{u_i d}{\alpha} \quad , \tilde{P} = \frac{Pd^2}{\rho \alpha^2} \quad , \quad \theta = \frac{T - T_o}{T_1 - T_2} \quad ,$$

The system will be solved in the form :

$$\frac{d\tilde{u}_i}{d\tilde{t}} = - \frac{\partial \tilde{P}}{\partial x_i} + P_R \nabla^2 \tilde{u}_i + P_R R_A \theta \qquad (8)$$

$$\frac{d\theta}{d\tilde{t}} = \nabla^2 \theta \qquad (9)$$

$$\operatorname{div} \tilde{V} = 0 \qquad (10)$$

The general solution procedure is basically the same as for the previous case. The velocity and pressure field at (n+1) time step are computed from equations (8) and (10), where the evaluation of the buoyancy force term is explicit, then the solution of equation (9) yields the temperature field.

2.3. Integrated form of the equation

All transport equations -and even the Poisson equations which is a degenerate transport one- can be written in the general form:

$$\frac{\partial \varphi}{\partial t} + \operatorname{div} \vec{V} \varphi = S\varphi + k\varphi \operatorname{div} \overrightarrow{\operatorname{grad}} \varphi \qquad (11)$$

where $S\varphi$ is the source term for the transportable quantity φ and $k\varphi$ its diffusion coefficient.

The integration of such an equation over an elementary cell Ω whose volume is V_Ω and surface Γ_Ω yields :

$$\iiint_{V_\Omega} \frac{\partial \varphi}{\partial t} dV_\Omega + \iiint_{V_\Omega} \operatorname{div} \vec{V}\varphi \, dV_\Omega = \iiint_{V_\Omega} S\varphi \, dV_\Omega + \iiint_{V_\Omega} k\varphi \operatorname{div} \overrightarrow{\operatorname{grad}} \varphi \, dV_\Omega \qquad (12)$$

Using Gauss theorem divergence and assuming that the temporal derivative and $S\varphi$ are constant, within Ω, equation (12) becomes :

$$\frac{\partial \varphi}{\partial t} V_\Omega + \iint_{\Gamma_\Omega} \varphi \vec{V} \, d\Gamma_\Omega = V_\Omega S\varphi + k\varphi \iint_{\Gamma_\Omega} \overrightarrow{\operatorname{grad}} \varphi \, d\Gamma_\Omega \qquad (13)$$

That is under this form that the equations will be solved.

3. NUMERICAL SOLUTION

The differencing of the equations is carried out on the now well known staggered grids (fig. 1). The use of these kinds of grids allows a better evaluation of the pressure gradient in the momentum equations and is particularly suitable to the integrated formulation.

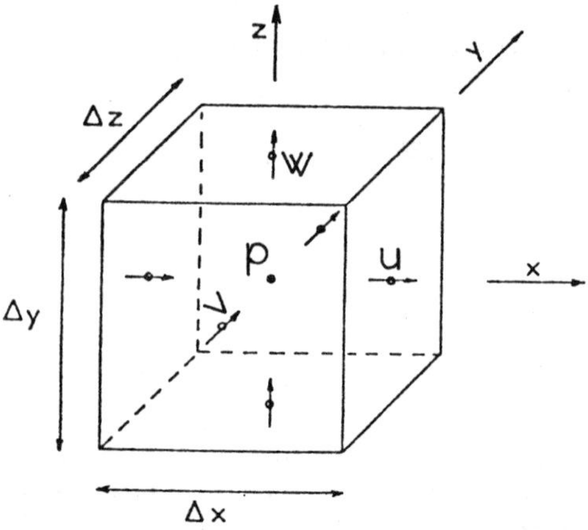

Figure 1 : The staggered grid around a pressure determination point (•)

The solution has been obtained from A.D.I. methods. Different schemes have been used whether the equation being solved was a transport equation or whether it was the Poisson equation.

Theoretically, the Peaceman-Rachford scheme [8] can give the solution of all transport equations.

For this purpose, let us consider an equation of the type :

$$\frac{\varphi^{n+1} - \varphi^{n}}{\Delta t} + X(\varphi) + Y(\varphi) + Z(\varphi) = S \tag{14}$$

where $X(\varphi)$ stands for all the x-derivatives, $Y(\varphi)$ for y-derivatives and $Z(\varphi)$ for z-derivatives.

The Peaceman-Rachford A.D.I. scheme consists in dividing the time step in three fractional steps and in computing successively the monodimensional equations :

$$\frac{\varphi^{n+1/3} - \varphi^{n}}{\Delta t / 3} + X(\varphi^{n+1/3}) = - Y(\varphi^{n}) - Z(\varphi^{n}) + S\varphi \tag{15}$$

$$\frac{\varphi^{n+2/3} - \varphi^{n+1/3}}{\Delta t / 3} + Y(\varphi^{n+2/3}) = - X(\varphi^{n+1/3}) - Z(\varphi^{n+1/3}) + S\varphi \tag{16}$$

$$\frac{\varphi^{n+1} - \varphi^{n+2/3}}{\Delta t / 3} + Z(\varphi^{n+1}) = - X(\varphi^{n+2/3}) - Y(\varphi^{n+2/3}) + S\varphi \tag{17}$$

This scheme gave good results for the transport equations, but did not lead to convergence for the Poisson equation.

It was, therefore, necessary to introduce Douglas A.D.I. scheme [3] inducing to the monodimensional expressions for equation (14) :

$$\frac{\varphi^{n+1/3} - \varphi^{n}}{\Delta t / 3} + X(\varphi^{n+1/3}) = - 2\, Y(\varphi^{n}) - 2\, Z(\varphi^{n}) - X(\varphi^{n}) + 2\, S\varphi \qquad (18)$$

$$\frac{\varphi^{n+2/3} - \varphi^{n+1/3}}{\Delta t / 3} + Y(\varphi^{n+2/3}) = Y(\varphi^{n}) \qquad (19)$$

$$\frac{\varphi^{n+1} - \varphi^{n+2/3}}{\Delta t / 3} + Z(\varphi^{n+1}) = Z(\varphi^{n}) \qquad (20)$$

Both schemes lead to the inversion of tridiagonal matrices which can be easily performed in using Gauss-Cholesky elimination algorithm.

4. RESULTS

All configurations have been studied on the same domain that is a cube with (22 x 12 x 22) grids except the case of forced convection at a high Reynolds number which needs a (32 x 22 x 32) grid. The motion of fluid inside the cavity is set either by viscosity or density effects. One can see at Fig. 2 the different sets of boundary conditions used for the study of the three types of convection processes.

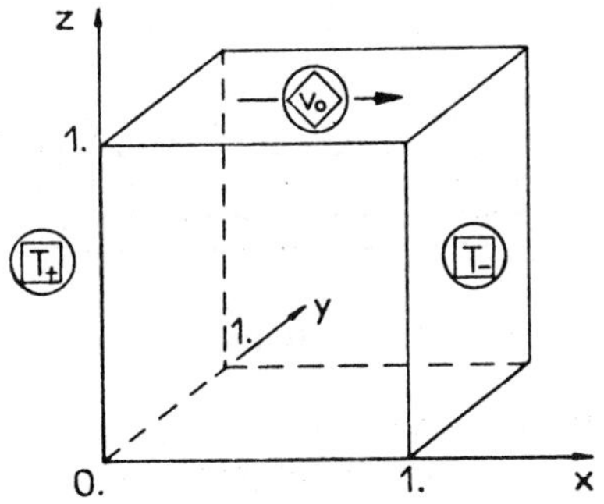

Figure 2 : The different ways of setting in the motion

◇ . Forced convection

□ . Free convection

○ . Mixed convection

4.1. Forced convection

In this configuration, the upper wall of the cavity moves along the x-axis with a unit velocity. A major vortex results from the drawing of the fluid in the vicinity of the moving wall while two secondary vortices appear at the bottom corners of the cube. A third secondary one appears in the upper downstream corner for Re = 2000 (Re=1500 , in the two dimensional case, Tuann Olson [9]). The structure of the flow is well described by the velocity fields plotted at Fig. 3(a,b,c,d).

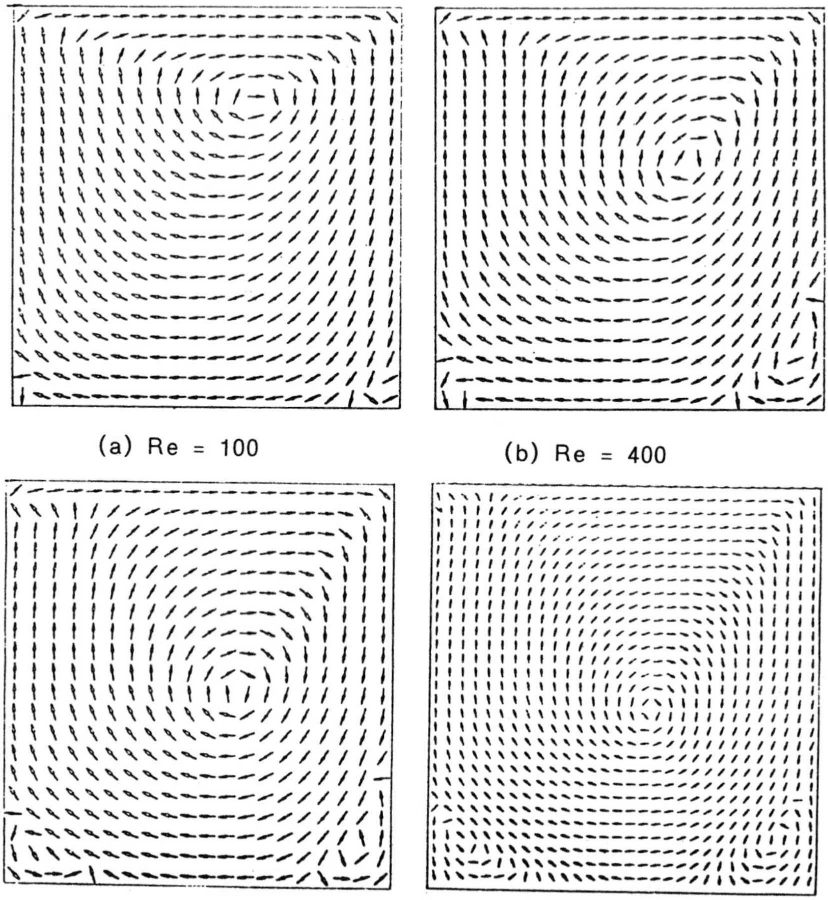

(a) Re = 100 (b) Re = 400

(c) Re = 1000 (d) Re = 2000

Figure 3 : Velocity fields in the symmetry plane for the different Reynolds numbers studied.

At Figures 4-a and 4-b, the u-velocity profiles on the vertical centerline of the cavity are compared with the results obtained by Goda [4] for Re = 100 and 400. This comparison shows a good agreement between both solutions.

For higher Reynolds numbers, no other results were available for comparison, our results for the whole range studied are presented at Figure 4(c).

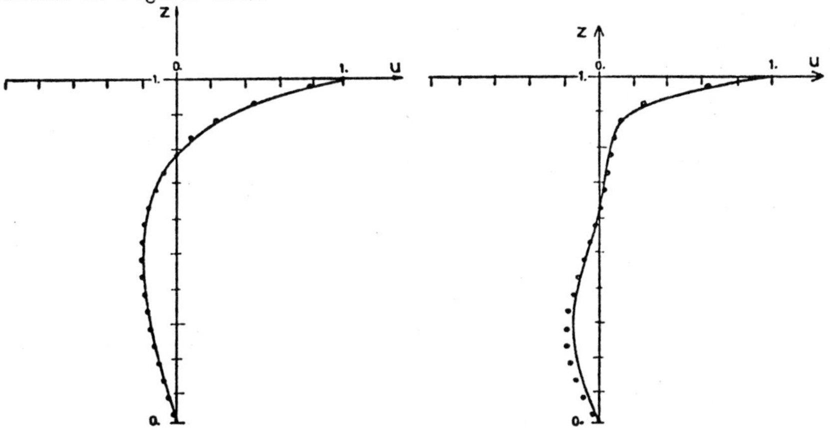

(a) Re = 100 (b) Re = 400

Figure 4 : Velocity profiles on the vertical centerline of the cavity ( ), Comparison with Goda's results (-).

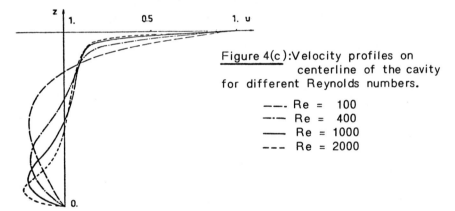

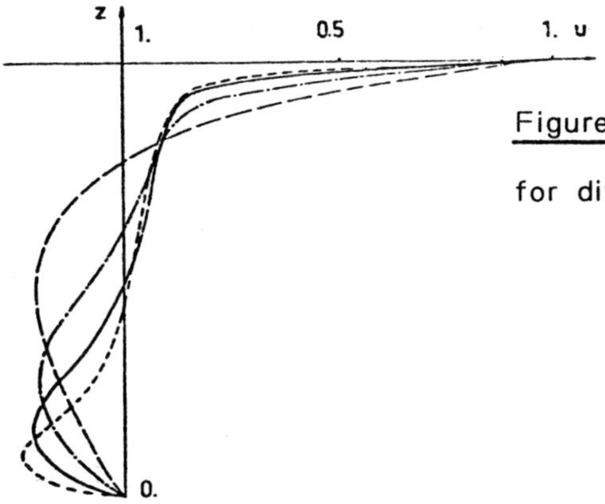

Figure 4(c):Velocity profiles on centerline of the cavity for different Reynolds numbers.

—–. Re = 100
—·— Re = 400
—— Re = 1000
- - - Re = 2000

The evolution of these profiles along the y-direction for extreme cases Re = 100 and Re = 1000 are plotted at Fig. 6 exhibiting two different behaviours,first for low Reynolds numbers there is no significant change in the shape of profiles from the center part to the vicinity of the side walls, on the contrary, for Re = 1000, the strong modification of the profile as one get closer to the side wall seems to indicate significant three dimensional effects.

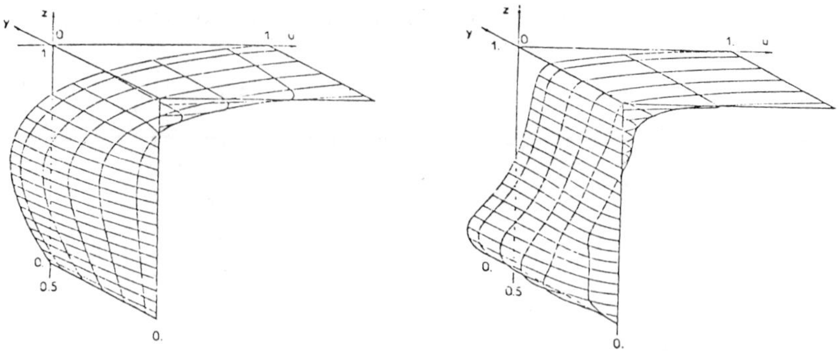

(a) Re = 100 (b) Re = 1000

Figure 6 : u-velocity profiles in the plane x = 0.5

4.2. Natural convection

In this configuration, the fluid is initially at rest at a uniform temperature $\theta=0$, while the x=0 and x=1 walls are respectively at temperature $\theta=0.5$ and $\theta=-0.5$, all the other walls being adiabatic ($\partial\theta/\partial n = 0$). A single vortex results from these conditions. Two or three vortices configurations would appear at higher Rayleigh numbers than those studied here (Fig. 6).

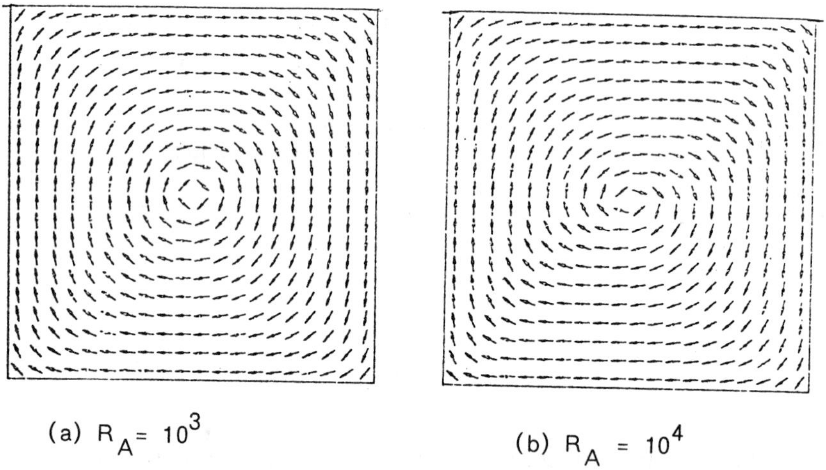

(a) $R_A = 10^3$ (b) $R_A = 10^4$

Figure 6 : Velocity fields in the symmetry plane

u(resp. w) velocity profiles on the vertical (resp. horizontal) centerline of the cavity are compared at Fig. 7 with those in 2.D of Portier et al. [10]. (•)

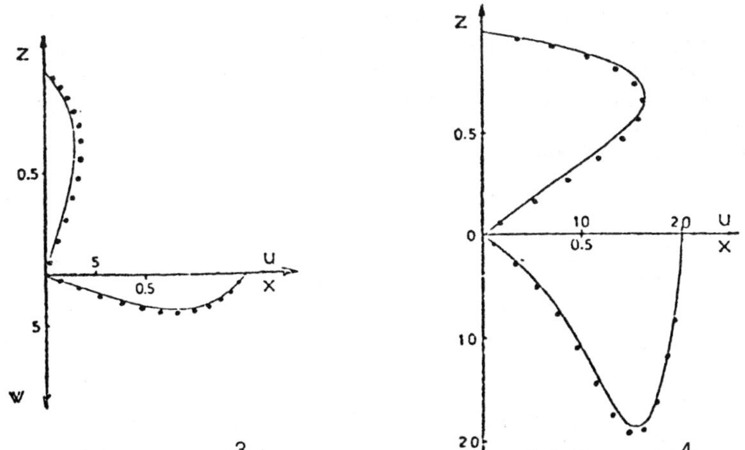

(a) $R_A = 10^3$ (b) $R_A = 10^4$

Figure 7 : Velocity profiles in the symmetry plane

Isothermal lines in the symmetry plane are plotted at Fig. 8 showing for $R_A = 10^3$ a quasi-diffusive pattern while the $R_A=10^4$ calculation shows a non negligible convective heat transfer.

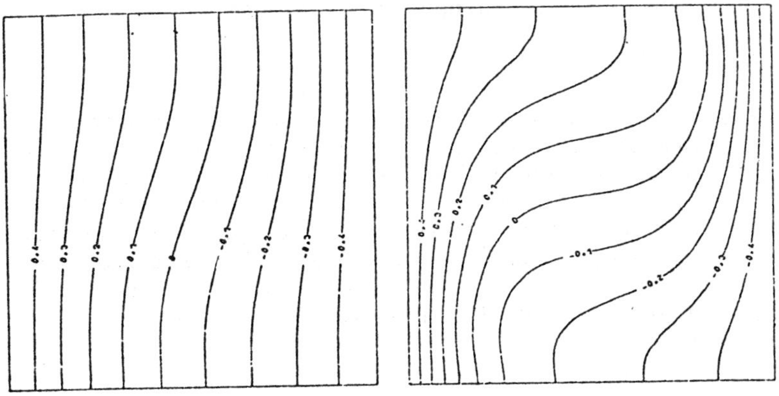

(a) $R_A = 10^3$ (b) $R_A = 10^4$

Figure 8 : Isothermal lines in the symmetry plane

4.3. Mixed convection

Mixed convection has been obtained by the simultaneous moving wall and temperature gradient effects so that both of them act in the same way to create a central vortex. The calculation was made for $R_A = 10^4$ and Re = 100. These values were chosen in order to get contributions of the same order from the two kinds of convection processes.

The resulting flow does not exhibit the secondary vortices of forced convection case, this is due to the locally counteracting buoyancy forces. Otherwise, the central symmetry observed in the natural convection case is lost. All this is clearly shown at Fig. 9.

One can see in Fig. 10, the isothermal lines distribution in the symmetry plane. The heat transfer is clearly of a convection dominated type with even reverse temperature gradients appearing in the center part of the cavity.

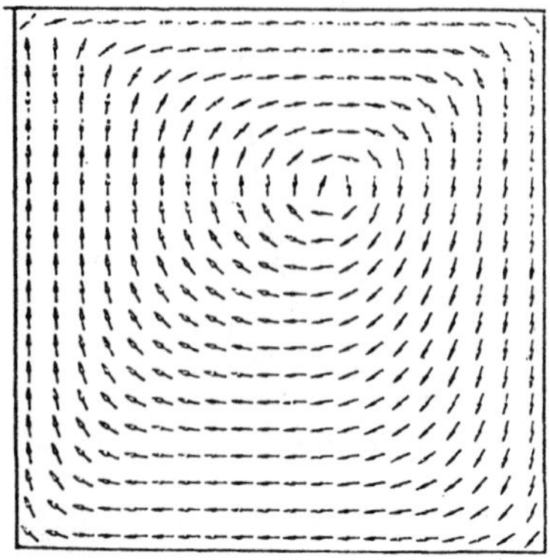

Figure 9 : Velocity field in the symmetry plane

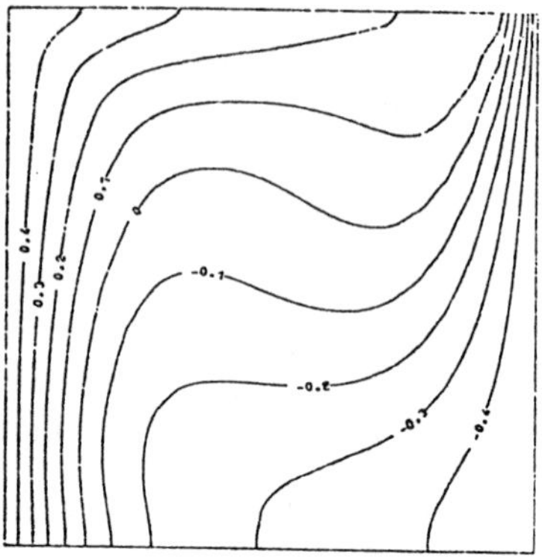

Figure 10 : Isothermal lines in in the symmetry plane

5. CONCLUSION

The numerical method presented seems to be adequate to the simulation of the three dimensional incompressible flows with reasonable computer storage and C.P.U. times. In the case of the (22 x 12 x 22) grid, the calculation never exceeds 40 mn CPU time and needs less than 280 k-bytes of storage on an IBM 3033 computer. However, a significant increase in the number of points such as that needed for the calculation of the case Re = 2000, leads to rather important CPU times (3h40mn CPU time for 420 k-bytes).

References

1. CHORIN A.J. - Numerical solution of Navier-Stokes equations Mathematics of computation, october 68(22).

2. DENNIS S.C.R., INGHAM D.B., COOK R.N. - Finite difference method for calculating steady incompressible flows in 3.D. Journal of computational Physics 33 (1979).

3. DOUGLAS J. - Alternating direction method for three space variables, Numerische Mathematik 4 (1962).

4. GODA K. - A multistep technique with implicit difference schemes for calculating two or three dimensional cavity flows. Journal of computational Physics, 30 (1979).

5. HARLOW F.H., AMSDEN A.A. - Numerical calculation of almost incompressible flow, Journal of computational Physics 3 (1968).

6. HARLOW F.H., WELCH J.E. - Numerical calculation of time dependant viscous and compressible flow of fluid with free surface, The Physics of Fluids, 8, n° 12 (1965).

7. LECOINTE Y., PIQUET J., VISONNEAU M. - "Mehrstellen" techniques for the numerical solution of unsteady incompressible viscous flow in enclosures, 2nd national Symposium on numerical method in heat transfer.

8. PEACEMAN D.W., RACHFORD H.H. - The numerical solution of parabolic and elliptic differential equations, J. Soc. Indust. Appl. Math., 3, n° 1 (1955).

9. TUANN S.Y., OLSON M.D. - Review of computing Method for recirculating flows, Journal of computational Physics, 29, n° 1 (1978).

10. Numerical solutions for a comparison problem on natural convection in an enclosed cavity, AERE Harwell HL30/3971 (C7) (1981).

A CONVECTIVE MODEL FOR A LAMINAR UNSTEADY DUCTED FLOW

R. CREFF, P. ANDRE, J. BATINA*

Laboratoire d'Energétique et de Mécanique, Université d'Orléans, 45046 Orléans Cédex, France
*Laboratoire d'Analyse Numérique, Université d'Orléans, 45046 Orléans Cédex, France

Abstract : The main lines of a numerical model especially built to analyze the dynamic and thermal developments for laminar pulsed ducted flows are presented. The energy and Navier-Stokes equations are treated by a finite-difference method using asymptotic developments for the different dynamic and thermal functions. An annular effect in temperature is presented and the evolution of its damping along the axis. Its relation to the unsteady temperature gradient at the wall and to the correlated heat transfer behaviour is shown. The importance of the unsteady velocity and temperature amplitudes in the first development steps have been pointed out. These amplitude effects could be at the origin of an added steady flux and, too, of dynamic and thermal stresses at the wall of the entry zone.

1. INTRODUCTION

The main characteristics and results of a numerical model built to treat the dynamic and thermal developments for laminar ducted flows and their related convective heat transfer are presented.

The theoretical description for combined dynamic and thermal developments in pulsed flows does not seem to have been undertaken previously. No results are available in literature for this particular problem where the developing heat transfer is initialized. In that manner, we can consider that the study of pulsation influence on the only developed dynamic and thermal flows hides a part of the physical reality. So, the processes occuring in the developed zone are a logical consequence of those appearing in the entry region with their specific entry conditions.

The proposed model allows the resolution of the four superposed developments existing since the duct entry: steady dynamic and thermal, unsteady dynamic and thermal.

Pulsed flows are treated here in terms of superposed flows : steady laminar and unsteady periodical. The different dynamic and thermal quantities are written on the form of asymptotic developments and the equation set is solved by a finite difference method. In such a manner, the discretized axial momentum and energy equations are solved by means of classical iterative processes, i.e.: predicted-corrected axial pressure gradient and axial velocity component. Two fundamental modifications are brought to the known codes [1,2,3,4] . Firstly, the continuity equation is considered as a refining "tool" for the axial velocity computing and provides also the radial velocity. Secondly, the radial momentum equation is only used to calculate the radial pressure gradient from the velocity distribution.

One of the aims of the proposed model is to point out the contribution of the different parameters on the heat transfer rates in the entry region.

2. MAIN NUMERICAL CHARACTERISTICS

The model allows the description of the four developments occuring since the entry duct : steady, unsteady, dynamic and thermal with the assumption of a constant wall temperature. Figure 1 gives the boundary conditions for the proposed study:

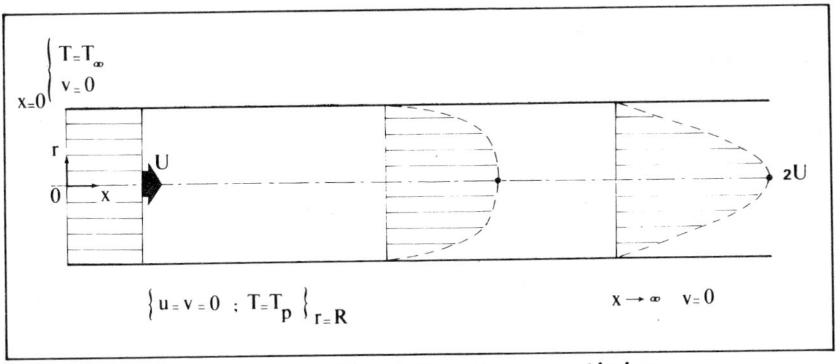

Fig.1 Boundary conditions

The equation system resolution for this physical problem is based on the fact that each function, complex or real, $f(x_1 \ldots x_p)$ admits an asymptotical development on the form :

$$f(x_1 \ldots x_p) = \sum_{n=0}^{\infty} a_n \; f_n(\varepsilon, x_1 \ldots x_p) \quad ; n \in N \qquad (1)$$

with ε : a small parameter, less than unity, called perturbation parameter

a_n : a serie of complex or real coefficients

Then, asymptotical developments for each function describing the dynamic and thermal problem can be written on the following form given here for the axial velocity u, as an example :

$$u = \sum_{n=o}^{\infty} u_n \cdot \varepsilon^n \cdot e^{jn\omega t} \tag{2}$$

The classical Navier-Stokes and energy equations are identified to the different orders.
Zeroth order:

$$[DV]_o : \frac{\partial u_o}{\partial x} + \frac{1}{r} \cdot \frac{\partial}{\partial r}(r.v_o) = 0 \tag{3}$$

$$[AX]_o : \frac{1}{r} \cdot \frac{\partial}{\partial r}(r.u_o v_o) + \frac{\partial}{\partial x}(u_o^2) = -\frac{1}{\rho} \cdot p'_{xo} + \nu\left[\frac{\partial^2 u_o}{\partial r^2} + \frac{1}{r} \cdot \frac{\partial u_o}{\partial r}\right] \tag{4}$$

$$[RD]_o : \frac{1}{r} \cdot \frac{\partial}{\partial r}(r.v_o^2) + \frac{\partial}{\partial x}(u_o v_o) = -\frac{1}{\rho} \cdot p'_{ro} +$$

$$+\nu\left[\frac{\partial^2 v_o}{\partial r^2} + \frac{1}{r} \cdot \frac{\partial v_o}{\partial r} - \frac{v_o}{r^2}\right] \tag{5}$$

$$[TH]_o : \frac{1}{r} \cdot \frac{\partial}{\partial r}(r.v_o T_o) + \frac{\partial}{\partial x}(u_o T_o) = \frac{k}{\rho C_p}\left[\frac{\partial^2 T_o}{\partial r^2} + \frac{1}{r}\frac{\partial T_o}{\partial r}\right] \tag{6}$$

First order :

$$[DV]_1 : \frac{\partial u_1}{\partial x} + \frac{1}{r} \cdot \frac{\partial}{\partial r}(r.v_1) = 0 \tag{7}$$

$$[AX]_1 : \frac{\partial u_1}{\partial t} + \frac{1}{r} \cdot \frac{\partial}{\partial r}\left[r(u_1 v_o + u_o v_1)\right] + \frac{\partial}{\partial x}(2u_o u_1) =$$

$$= -\frac{1}{\rho} \cdot p'_{x1} + \nu\left[\frac{\partial^2 u_1}{\partial r^2} + \frac{1}{r} \cdot \frac{\partial u_1}{\partial r}\right] \tag{8}$$

$$[RD]_1 : \frac{\partial v_1}{\partial t} + \frac{1}{r} \cdot \frac{\partial}{\partial r}(2r.v_o.v_1) + \frac{\partial}{\partial x}(u_o v_1 + u_1 v_o) =$$

$$= -\frac{1}{\rho} p'_{r_1} + \nu\left[\frac{\partial^2 v_1}{\partial r^2} + \frac{1}{r}\frac{\partial v_1}{\partial r} - \frac{v_1}{r^2}\right] \tag{9}$$

$$[TH]_1 : \frac{\partial T_1}{\partial t} + \frac{1}{r} \cdot \frac{\partial}{\partial r}\left[r(v_1 T_o + T_1 v_o)\right] + \frac{\partial}{\partial x}(u_o T_1 + u_1 T_o) =$$

$$= \frac{k}{\rho C_p}\left[\frac{\partial^2 T_1}{\partial r^2} + \frac{1}{r}\frac{\partial T_1}{\partial r}\right] \qquad (10)$$

Those equations have been reduced by introducing the different dimensionless quantities:

$$r^* = r/R \quad , \quad x^* = x/R$$

$$u^* = u/U \quad , \quad v^* = v/U$$

$$p^{*\prime}_x = \left(\frac{\partial p}{\partial x}\right)^* = (\partial p/\partial x) \,/\, (\partial p/\partial x)_p$$

$$p^{*\prime}_r = \left(\frac{\partial p}{\partial r}\right)^* = (\partial p/\partial r) \,/\, (\partial p/\partial x)_p$$

as well as : Ω : dimensionless pulsation = $\omega R^2/\nu$
Re = UR/ν: Reynolds number based on radius R
Pr = Re.Pr: Peclet number
with : $(\partial p/\partial x)_p$: the axial pressure gradient for the Poiseuille regime
U : the mean velocity for the Poiseuille regime
Then, the boundary conditions concerning the zeroth and the first orders are :

$$x = 0 \quad u_o = 1 \ , \quad v_o = \theta_o = 0 \qquad \forall r \ , \forall t$$

$$u_1 = v_1 = \theta_1 = 0$$

$$r = 0 \quad \frac{\partial u_o}{\partial r} = \frac{\partial \theta_o}{\partial r} = v_o = 0 \qquad \forall x \ , \forall t$$

$$\frac{\partial u_1}{\partial r} = \frac{\partial \theta_1}{\partial r} = v_1 = 0$$

$$r = 1 \quad \theta_o = 1 \ , \quad u_o = v_o = 0 \qquad \forall x \ , \forall t$$

$$u_1 = v_1 = \theta_1 = 0$$

where $*$ vanishes in order to simplify the formulation.

Numerical Scheme :

The resolution method is closed to those, classical, given in literature, but three main changes are adopted :

. AX gives an approximate value for u_i^{j+1} and p'^{j+1}_{xi} , noted $\tilde{u}_i^{j+1}$ and $\tilde{p}^{j+1}_{x_i}$, by means of the volumetric rate condition such as :

$$\dot{m} = \sum 2\pi r_i \, \Delta r_i u_i = \text{constant} \qquad (11)$$

from an arbitrary value given to the axial pressure gradient.

. DV is used as a refinement for the axial velocity and allows the computing of the radial velocity until a convergence criterion is verified on u^{j+1}:

$$\left| (u^{j+1})_n - (u^{j+1})_{n+1} \right| < \varepsilon = 10^{-3} \qquad (12)$$

(n : iteration order)

. RD, here, is only used to give the radial pressure gradient for the velocity field and so could be useless for this convective study.
In that way, this choosen scheme particularly adapted to solve this problem is faster and lighter than other known schemes.
For the radial coordinate, the finite differences for the different functions are computed for integer or fractional nodes: u, p' and θ are affected to each integer node and v to every fractional node. More details are given in reference[5]. As an example, the AX_o and AX_1 discretized equations for the zeroth (steady) and the first orders (unsteady) are given:

$$[AX_o]_i^{j+1} : \frac{1}{r_i \Delta r_i} \left[r_{i+1/2} \cdot v_{i+1/2}^j \left(\frac{u_{i+1}^{j+1} + u_i^{j+1}}{2}\right) - r_{i-1/2} \cdot v_{i-1/2}^j \times \frac{u_i^{j+1} + u_{i-1}^{j+1}}{2} \right] + \left(\frac{\ddot{u}_i^{j+1} \cdot u_i^{j+1}}{\Delta x_{j+1}} - \frac{u_i^j \; u_i^j}{\Delta x_{j+1}} \right) = \frac{8}{Re} \cdot p'^{j+1}_x +$$

$$+ \frac{1}{Re} \left[\frac{1}{\Delta r_i} \left[\frac{u_{i+1}^{j+1} - u_i^{j+1}}{\delta r_{i+1/2}} - \frac{u_i^{j+1} - u_{i-1}^{j+1}}{\delta r_{i-1/2}} \right] + \frac{1}{r_i \cdot \Delta r_i} \times (u_{i+1}^{j+1} - u_{i-1}^{j+1}) \right] \qquad (13)$$

$$[AX_1]_i^{j+1} : u_{1i}^{j+1} \left[j\frac{\Omega}{Re} + \frac{1}{2 r_i \Delta r_i} \left(r_{i+1/2} \cdot v_{o\,i+1/2}^j - r_{i-1/2} \cdot v_{o\,i-1/2}^j \right) + \frac{u_{oi}^{j+1}}{\Delta x_{j+1}} + \frac{1}{Re\, \Delta r_i} \left(\frac{1}{\delta r_{i-1/2}} + \frac{1}{\delta r_{i+1/2}} \right) \right] +$$

$$+ u_{1\,i-1}^{j+1} \left[\frac{-1}{2 r_i \Delta r_i} \left(r_{i-1/2} \cdot v_{o\,i-1/2}^j \right) - \frac{1}{Re\, \Delta r_i} \left(\frac{1}{\delta r_{i-1/2}} - \frac{1}{2 r_i} \right) \right]$$

$$+ u_{1\,i+1}^{j+1} \left[\frac{1}{2 r_i \Delta r_i} \left(r_{i+1/2} \cdot v_{o\,i+1/2}^j \right) - \frac{1}{Re \Delta r_i} \left(\frac{1}{\delta r_{i+1/2}} + \frac{1}{2 r_i} \right) \right]$$

$$= - \frac{1}{r_i \Delta r_i} \left\{ r_{i+1/2} \left[v_{1\,i+1/2}^j \; \frac{u_{o\,i+1}^{j+1} + u_{oi}^{j+1}}{2} \right] + \right.$$

$$- r_{i-1/2}\left[v^{j}_{1\,i-1/2} \frac{u^{j+1}_{oi} + u^{j+1}_{oi-1}}{2} \right]\Bigg\} +$$

$$+ \frac{8}{Re} p'^{j+1}_{x} + \frac{2}{\Delta x_{j+1}} u^{j}_{o} \cdot u^{j}_{1\,i} \qquad (14)$$

3. RESULTS

Table 1 summarizes the parameter variations used in this study :

$X^* = x/R$	$r^* = r/R$	$Re = UD/\nu$	$Pr = \mu Cp/\lambda$	$\Omega = \omega R^2/\nu$
0	29			10π
to	values	1 500	0,73 Air	25π
	from			35π
176		2 000	1,74 H_2O 100°	
	$r^* = 0$			50π
$\Delta X^* = 0,9$	to	3 000	6,76 H_2O 20°	75π
	$r^* = 1$			100π

Table 1. Ranges for the different parameters

It is well known that the steady dynamic and thermal developments since the duct entry correspond to growing boundary layer thicknesses; fully developed flows are obtained when velocity and temperature profiles keep a constant shape along the flow axis.

In figure 2, the local steady Nusselt number Nu_o defined as :

$$Nu_o = (\partial\Theta_o/\partial r^*)_{r^*=1} / (\theta_w - \overline{\theta}) \qquad (15)$$

is plotted versus the Graetz coordinate : $X^+ = 4(X/D)/RePr$, with θ_w and $\overline{\theta}$, the constant wall temperature and the mean bulk fluid temperature, respectively. For a same Reynolds number (2000) and for three different Prandtl numbers, this curve confirms that the steady developing length is shorter for smaller Pr.

It has been shown previously [5] that, for X^+ smaller than 0.01, the Nu_o values given by the model are distributed between the KAY's and ULRICHSON's data [6,7]. The last author has shown that the discrepancy between his results and those of KAYS can be devoted to the radial velocity not taken into account in the KAY's analysis. In fact, ULRICHSON himself

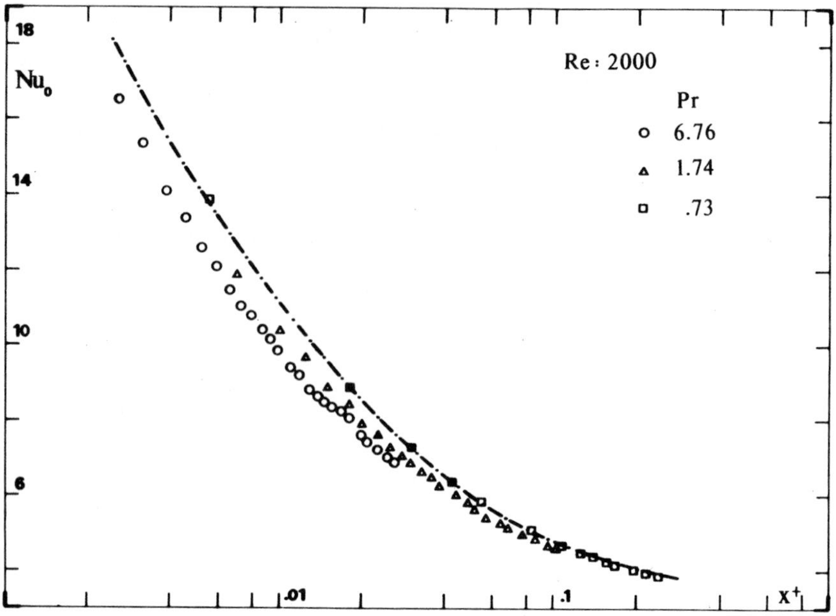

Fig.2.Local Nusselt number in the developing length

has deduced the radial velocity from the continuity equation and a linearized axial momentum equation which does not contribute to a real accuracy for the velocity field computation.

To illustrate the obtained unsteady axial velocity profiles, $u_1(r^*,x^*,t)$, this component is plotted versus r^* for different wt values and for $Re=2.10^3$, $\Omega=50\pi$, $x^*=40$. The general behaviour for the unsteady velocity field is mainly depicted by an annular effect experimentally described by RICHARDSON [8]. This particular effect, which is very intensive in the entry zone, does not exist in the fully steady developed region.

In a similar manner, figure 4 traduces the unsteady temperature profiles in the same conditions used in figure 3 but with Pr=0.73. In this section, as in other upstream and downstream sections, it appears an instantaneous temperature annular effect with a specific phase lag between the axial pressure gradient and the unsteady temperature. It indicates also the fluid temperature gradient behaviour at the wall and so the related instantaneous heat fluxes transfered to the fluid flow. A negligible amplitude is obtained on the center axis showing that pulsation effects are restricted in a concentric annular region near the wall ($0.5 \leq r^* \leq 1$).

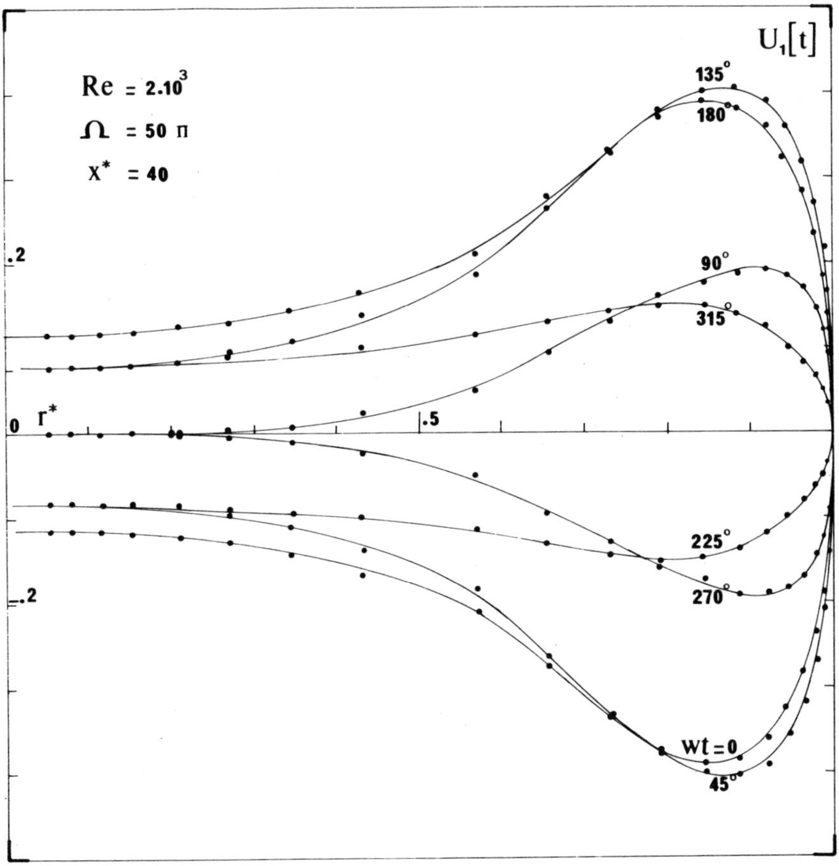

Fig.3. Unsteady-axial velocity profiles

To get a better understanding on the thermal fluid field the unsteady temperature modulus is plotted versus r* for different longitudinal sections x*: 0.2, 22, 40, 85, 175 (fig.5) The corresponding curves traduce a progressive moving for θ_{1max} and so, for the annular effect, from the wall towards the main flow axis when progressing with x*. The damping obtained for θ_{1max} with increasing x* shows that the temperature annular effect is mainly observed in the entry zone and reveals the great importance of these unsteady temperature modulus particularly in the neighbourhood of the entry section. The model has pointed out that a ratio : $|\theta_1|(x^*=25)/|\theta_1|(x^*=175)$ equal to 23 is obtained for Re=2000, Pr=1.74 and $\Omega = 100\pi$. This induces the idea of unsteady dynamic or thermal stresses in the first steps of the developing flow near the duct inlet. In a same way, thermal gradients at the wall are much more important for the first steps than farer downstream. Thus, it is possible to predict important unsteady heat exchanges between the wall and the fluid in this region. This unsteady heat transfer can

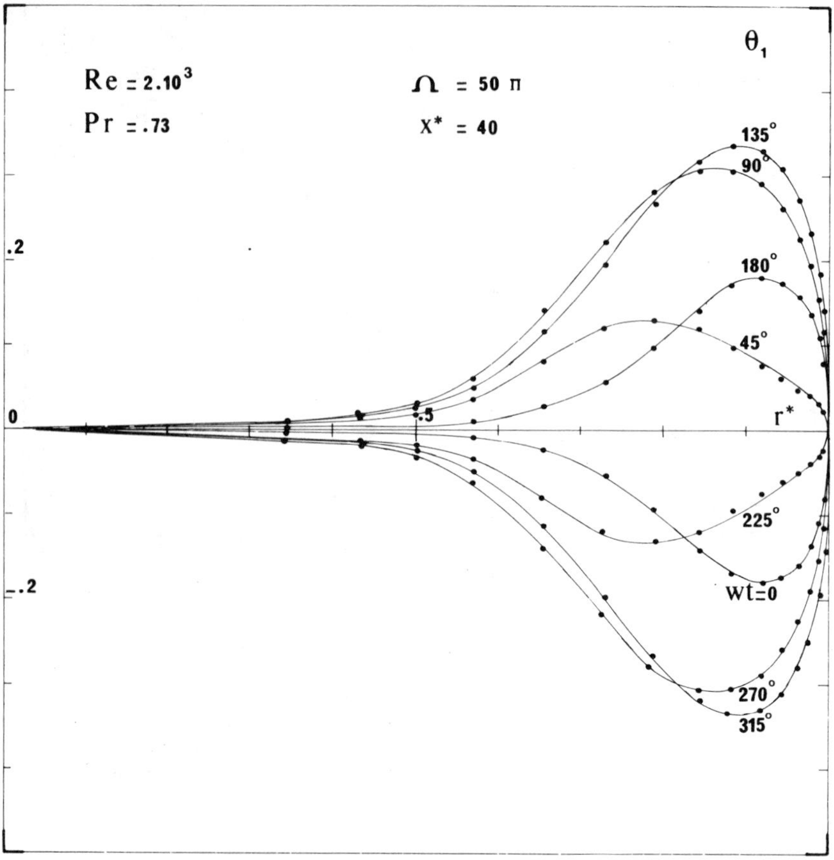

Fig.4. Unsteady temperature profiles(Pr=0.73)

be expressed by :

$$Nu_1(t) = \tau \left[\partial\theta_1(t)/\partial r^*\right]_{r^*=1} / \left[\theta_w - \tau\theta_{1M}(t)\right] \qquad (16)$$

where τ is the modulation rate and $\left[\partial\Theta_1(t)/\partial r^*\right]_{r^*=1}$ the unsteady temperature gradient at the wall. This last expression allows the study of Nu_1 along the duct wall with Ω as a parameter (10π to 100π).

These curves present a maximum close to the entry section for higher frequencies. Opposite to this, for lower frequencies, this maximum occurs far from the duct inlet with a decreasing value as the frequency decreases. The longitudinal damping for $Nu_1(t)$ is faster for "high" frequencies. These results let appear the existence of a specific frequency, here equal to 35π, giving an optimum effect averaged on the steady dynamic developing length taken as a reference length. Thus,we

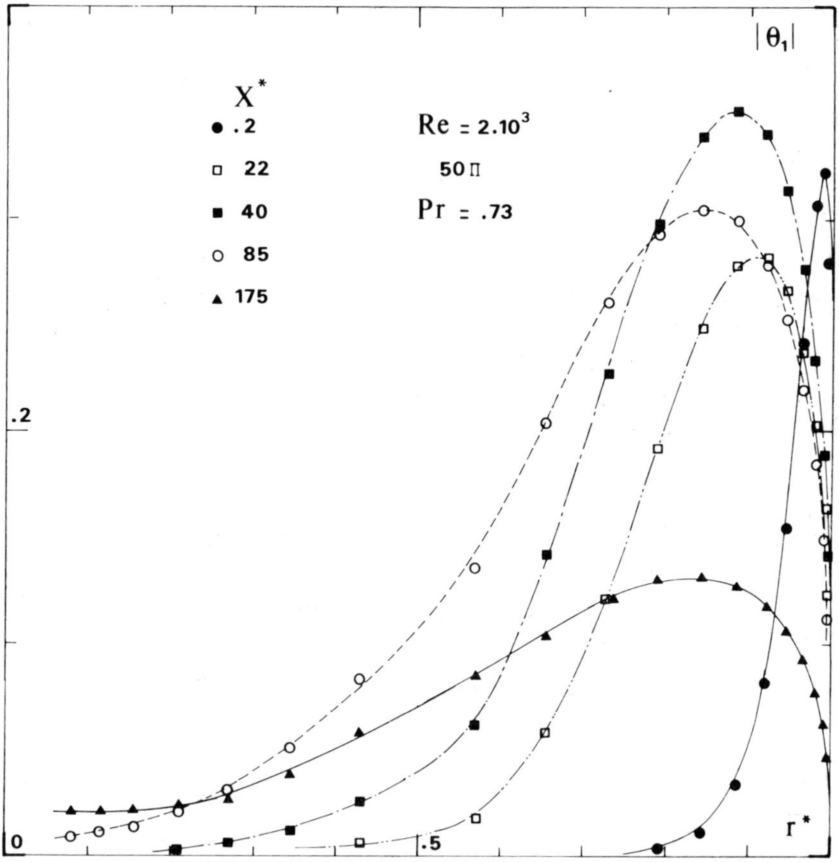

Fig.5. Profiles of the unsteady temperature amplitude

can conclude on the existence of a specific or critical combination of the main parameters such as Re and Pr leading to an optimum on heat transfer in the entry length.

REFERENCES :

1. PATANKAR S.V., SPALDING D.B. : A calculation procedure for heat, mass and momentum transfer in three-dimensional parabolic flows, Int. J. Heat Transfer, Vol. 15, 1787-1806, (1972)
2. ROACHE P.J. Computational Fluid Dynamics, Hermosa Publishers, Albuquerque, New-Mexico, (1972)
3. GOSMAN A.D., PUN W.M., RUNCHAL A.F., SPALDING D.B., WOLFSTEIN M. : Heat and Mass Transfer in Recirculating Flows, Acad. Press, London, (1969)
4. FORTIN M., PEYRET R., TEMAN R. : Résolution numérique des équations de Navier-Stokes pour un fluide incompressible, J. de Mécanique, Vol. 10, n° 3, 357-390, (1971)

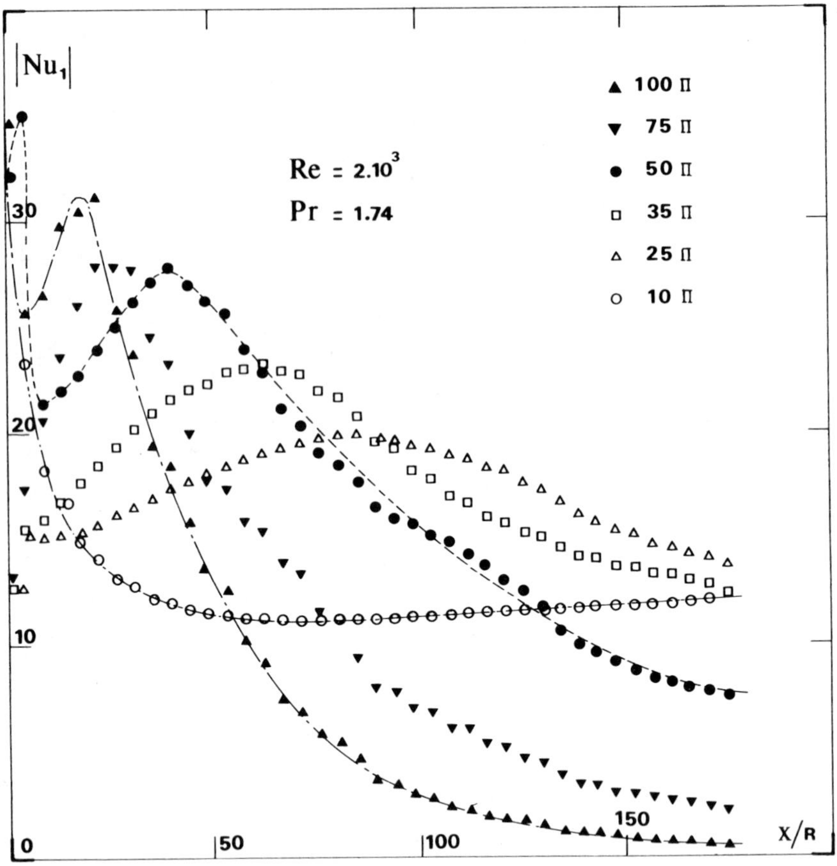

Fig.6. Unsteady Nusselt number amplitude along x*

5. CREFF R., BATINA J., ANDRE P., KARUNANITHI V.S., Numerical Model for Dynamic and Thermal Developments of a Pulsed Laminar Ducted Flow, Numerical Heat Transfer, to appear in vol.6, (1983)
7. ULRICHSON D.L., SCHMITZ R.A. : Laminar Heat Transfer in the Entrance Region of Circular Tubes; Int. J. Heat Mass Transfer, Vol. 7, p. 253, (1965)
8. RICHARDSON E.G., TYLER E. : The transverse velocity gradient near the mouths of pipes in which an alternating or continuous flow of air is established ;Proc. Roy. Soc. Vol.42, part. 1, n° 231, 1-15, (1929)

A NUMERICAL INVESTIGATION OF COUPLED CONDUCTION - MIXED CONVECTION FOR RECTANGULAR FINS

Bengt Sundén

Department of Applied Thermo and Fluid Dynamics, Chalmers University of Technology, 41296 Göteborg, Sweden

SUMMARY

The problem of coupled conduction - mixed convection for rectangular fins is considered. The thickness of the fins is small and it is assumed that laminar boundary layer flow exists around the fin surfaces. The buoyancy force is taken into account by applying Boussinesq's approximation. An iterative finite difference technique with an overall iteration loop and several sub-iteration loops has been developed. Details of the numerical method and its accuracy as well as some relevant results are presented.

1. INTRODUCTION

The heat transfer from an extended surface element such as a cooling fin is generally a combination of heat conduction within a solid and convective heat transfer in a fluid surrounding the surface element. At the interface between the solid and the fluid, the thermal conditions of continuity in temperature and heat flux have to be satisfied. There is thus a coupling between the conductive and convective heat transfer. For situations where the buoyancy force is of importance the flow field will depend on the temperature and thereby on the thermal conductivity of the fin.

In the conventional fin theory, the fin heat conduction equation is solved using a uniform value of the heat transfer coefficient and fin efficiencies can be calculated, see [1, 2]. The fin heat conduction equation can also be solved numerically for an arbitrarily varying heat transfer coefficient. However, since the temperature distribution along the fin surface is not known a priori, it is not possible to find the true distribution of the heat transfer coefficient. To overcome this problem it is necessary to solve the conductive - convective heat transfer as a coupled heat transfer problem.

The present paper concerns rectangular fins which transfer heat to or from a surrounding fluid by combined forced and natural convection. One objective of the work is to predict the influence of the buoyancy force (G_R/Re^2) and the thermal conductivity ratio (K_s/K_f, solid to fluid) on the flow field and heat transfer characteristics from rectangular fins. Another objective of the paper is to check the validity of the conventional fin theory.

2. FORMULATION OF THE PROBLEM

Consider a rectangular fin with thickness b and length L as shown in Figure 1. The fin is placed vertically and is cooled by an airflow which at far distances is described by a uniform velocity U_∞ and a uniform temperature T_∞. At the base, the fin is given the temperature T_b. The thermal conductivity of the fin material is K_s and that of the fluid is K_f. The density of the fluid is assumed to vary with temperature and the resulting buoyancy force is taken into account by applying Boussinesq's approximation. The other physical properties are assumed to be constant. The cases when the gravity force assists or counteracts the forced flow are considered. The thickness of the fin is small and it is assumed that laminar boundary layers exist around the fin surfaces. The heat conduction within the fin is taken as one-dimensional.

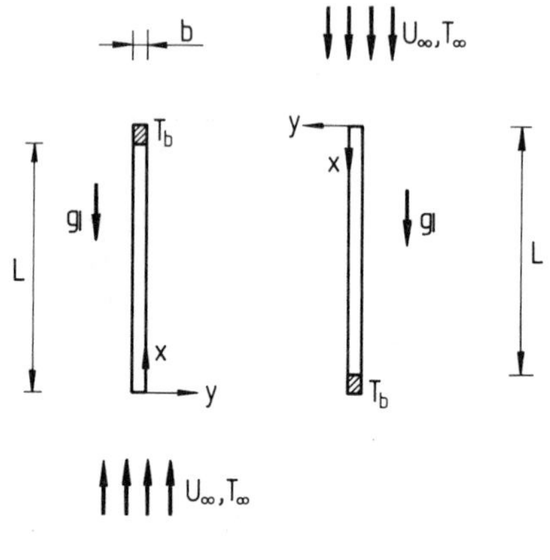

Figure 1. Problem under consideration.

Of special interest is to find the shear stress (local and total), heat flux (local and total), fin temperature distribution and the fin efficiency and how they are influenced by leading parameters such as G_R/Re^2 ($G_R = g\beta(T_b - T_\infty)L^3/\nu^2$, $Re = U_\infty L/\nu$), Re and K_s/K_f.

Papers close related to the present work have been published by Sparrow and Acharya [3], who considered conjugate pure natural convection, and Sparrow and Chyu [4], who studied conjugate forced convection. The work by Karvinen [5] is also related to the present one, but in that work a heated vertical plate fin with its base located at the side and cooled by forced or natural convection was considered.

3. GOVERNING EQUATIONS

Introduce the following non-dimensional variables:

$$x' = x/L \; ; \; y' = y\sqrt{Re}/L \; ; \; u' = u/U_\infty \; ; \; v' = v\sqrt{Re}/U_\infty \; ;$$

$$\theta = (T - T_\infty)/(T_b - T_\infty)$$

With these non-dimensional variables and the assumptions outlined, the governing equations can be written (Note: the prime used to denote non-dimensional variables is omitted from here onwards):

Flow field

$$u \frac{\partial u}{\partial x} + v \frac{\partial u}{\partial y} = \frac{G_R}{Re^2} \theta + \frac{\partial^2 u}{\partial y^2} \qquad (1)$$

$$\frac{\partial u}{\partial x} + \frac{\partial v}{\partial y} = 0 \qquad (2)$$

Convective heat transfer (fluid)

$$u \frac{\partial \theta}{\partial x} + v \frac{\partial \theta}{\partial y} = \frac{1}{P_R} \frac{\partial^2 \theta}{\partial y^2} \qquad (3)$$

Conductive heat transfer (solid)

$$\frac{d^2\theta_w}{dx^2} = -2\sqrt{Re} \frac{K_f}{K_s} \frac{L}{b} \left(\frac{\partial \theta}{\partial y}\right)_w \qquad (4)$$

From equations (1) - (4) it is obvious that the leading physical parameters are: G_R/Re^2 (ratio of buoyancy forces to inertia forces), P_R and $\sqrt{Re}\, K_f L/K_s b$. The last one may be called the conduction - convection parameter (CCP). A numerical value of this parameter thus represents several combinations of the Reynolds number, thermal conductivity ratio and the ratio of fin length to fin thickness.

Due to the buoyancy force in equation (1), the flow field cannot be solved separately.

In equation (4), θ_w means the fin temperature and $(\partial\theta/\partial y)_w$ means the temperature gradient at the wall (y = 0).

4. BOUNDARY CONDITIONS

Equations (1) - (4) have to be solved with respect to some boundary conditions. The following conditions are used:

$$y = 0: \quad u = v = 0 \tag{5}$$

$$y \to \infty: \quad u \to 1, \ \theta \to 0 \tag{6}$$

$$x = 0: \quad u = 1, \ v = 0, \ \theta = 0, \ \frac{d\theta_w}{dx} = 0 \tag{7}$$

$$x = 1: \quad \theta_w = 1 \tag{8}$$

The conditions at the leading edge (x = 0) arise since the boundary layer thicknesses are zero here. The thermal condition $d\theta_w/dx = 0$ means that the heat flux is assumed to be zero at the tip of the fin at the leading edge. This is true for a long fin with small thickness. However, as will be discussed later, the conditions at x = 0 have to be handled in a slightly different manner to enable good accuracy close to the leading edge.

5. NUMERICAL SOLUTION PROCEDURE

5.1 Finite difference equations on the fluid side

Equations (1) - (3) are solved by using second - order finite difference approximations in both the x - and y - direction. The grid notations are given by i and j, respectively. The following equations are used:

Velocity u

$$a_1 u_{i,j} = b_1 u_{i,j+1} + d_1 u_{i,j-1} + c1_{i,j} \tag{9}$$

where

$$a_1 = u_{i,j}/\Delta x + 2/\Delta y^2 \tag{10}$$

$$b_1 = -v_{i,j}/(2\Delta y) + 1/\Delta y^2 \tag{11}$$

$$d_1 = v_{i,j}/(2\Delta y) + 1/\Delta y^2 \tag{12}$$

$$c1_{i,j} = u_{i,j}(4u_{i-1,j} - u_{i-2,j} - u_{i,j})/(2\Delta x) + (G_R/Re^2)\theta_{i,j} \tag{13}$$

Velocity v

The velocity v is obtained by numerical integration of equation (2), that is

$$v_{i,j} = v_{i,j-1} - \int_{y-\Delta y}^{y} (\partial u/\partial x)dy \tag{14}$$

The integral between y-Δy and y in equation (14) is solved by using the trapetzoidal rule.

Temperature θ

$$a_2\theta_{i,j} = b_2\theta_{i,j+1} + d_2\theta_{i,j-1} + c2_{i,j} \tag{15}$$

where

$$a_2 = 3u_{i,j}/(2\Delta x) + 2/\Delta y^2/P_R \tag{16}$$
$$b_2 = -v_{i,j}/(2\Delta y) + 1/\Delta y^2/P_R \tag{17}$$
$$d_2 = v_{i,j}/(2\Delta y) + 1/\Delta y^2/P_R \tag{18}$$
$$c2_{i,j} = u_{i,j}(4\theta_{i-1,j} - \theta_{i-2,j})/(2\Delta x) \tag{19}$$

Since equations (1) and (3) are parabolic, the numerical solutions of equations (9),(14) and (15) are obtained by marching in the x - direction. At every x - position, the TDMA - algorithm is used for finding the solutions. Due to the non - linearities and coupling, an iterative procedure has to be applied at every x - position. For speeding up convergence when $G_R/Re^2 \geq 0$, relaxation parameters greater than unity are used. When $G_R/Re^2 < 0$ (adverse case), relaxation parameters less than unity have to be used.

5.1.2 Initial profiles

Since equations (1) and (3) are parabolic, initial profiles are needed at some position in the x - direction. It has been found that a direct application of the conditions in equation (7) produces results which are very much dependent on the step size Δx and of low accuracy close to the leading edge. Instead, initial profiles are created at $x = \Delta x/2$ by assuming local similarity. The following variables are then introduced:

$$\xi = x/L; \ \eta = \sqrt{U_\infty/\nu x}\ y \ ; \ \psi = f(\eta)\sqrt{\nu x U_\infty} \ ; \ u = \frac{\partial\psi}{\partial y} \ ; \ v = -\frac{\partial\psi}{\partial x}$$

The basic equations describing the transport processes in the fluid can then be written:

$$f''' + \frac{1}{2} ff'' = -\frac{G_R}{Re^2}\theta\xi \tag{20}$$

$$\theta'' + \frac{P_R}{2} f\theta' = 0 \tag{21}$$

In equations (20) and (21), the ξ - derivatives have been deleted in accordance with the assumption of local similarity.

The boundary conditions to equations (20) and(21) are:

$$\eta = 0 \ ; \ f = f' = 0, \quad \theta = \theta_w(\Delta x/2) = (\theta_w(0) + \theta_w(\Delta x))/2$$
$$\eta \to \infty \ ; \ f' \to 1, \quad \theta \to 0$$

Equations (20) and (21) can be solved by finite difference methods or Runge-Kutta methods. In the present work, a finite difference method is used.

At this point, it is important to point out that since the initial profiles are taken at $x = \Delta x/2$, equations (9) - (19) are not valid at the two gridlines closest to $x = \Delta x/2$. However, the changes, due to the varying step size in the x - direction, can easily be derived but are not given here.

5.2 Fin heat conduction equation

The finite difference equation of equation (4) is obtained by taking a heat balance of an arbitrary element situated between i-1/2 and i+1/2, where i indicates an arbitrary grid point. The following equation results:

$$\theta_w(I) = (\theta_w(I+1) + \theta_w(I-1))/2 + \frac{K_f}{K_s}\frac{L}{b}\sqrt{Re}\, q_c \Delta x^2 \qquad (22)$$

where q_c is the dimensionless convective heat flux per unit area, built up by the temperature gradients $(\partial\theta/\partial y)_w$ at positions i-1, i and i+1 .

Equation (22) is solved iteratively, thereby using a relaxation procedure with the optimal value of the relaxation parameter.

For the first element, between i=1 and i=2, also a heat balance is used. The heat conducted out from the surface at i=1 is then set to zero in accordance with the boundary condition. The resulting equation is used for finding the fin temperature at x = 0 or i=1.

5.3 Solution procedure

Due to the non - linearities and the mutual coupling, the solution must be found iteratively. The iterative procedure is rather complex, involving an overall iteration loop and a number of sub-iteration loops, and here only a very brief description of the procedure is given:

1) Set all problem parameters
2) Estimate a temperature distribution $\theta_w(x)$ in the fin
3) Solve the flow field and convective heat transfer with the estimated fin temperature $\theta_w(x)$ as a wall boundary condition
4) Calculate the convective heat flux $(\partial\theta/\partial y)_w$ at every position x
5) Solve the fin heat conduction equation with respect to these heat fluxes. This gives a new temperature distribution $\theta_w(x)$.

The sequence 3) - 5) is repeated until acceptable convergence has been reached. Then all relevant flow and heat transfer characteristics can be calculated.

In the solution procedure, several conditions have been built in. So for instance, if $G_R/Re^2 = 0$ (no buoyancy force), the flow field is solved only once. If the fin is isothermal, K_s very large, the fin heat conduction equation is not solved.

5.4 Accuracy

All calculations have been carried out on an IBM 3033N computer. Test calculations showed that a large number of grid points had to be used for obtaining high accuracy in the numerical results. Since the fin heat conduction has an elliptic behaviour and the flow field and convective heat transfer equations have an elliptic behaviour in the y - direction, it was found that calculations in double precision were necessary. However, in the x - direction the flow field and convective heat transfer equations are parabolic and thus no problem occurred with these equations when the number of grid points in this direction was increased.

When decreasing the step size in the x - direction, care must be taken so that the step size in the y - direction is small enough to give resolution in the vicinity of the leading edge, where the boundary layer thickness is quite small (zero at the leading edge). The maximum number of grid points tested, was 901 in the x - direction and 401 in the y - direction. It was found that the step size in the x - direction was most critical and less than 501 grid points should not be used if acceptable local and overall accuracy is required. In the y - direction, 71 grid points were found to be satisfactory. This was found when the cooling fluid was air. If another fluid, with a high or low Prandtl number, is considered, the proper number of grid points may be different, especially in the y - direction. For the cases with air, 35571 grid points were generally used. In what follows the results of some accuracy tests will be given.

a. Isothermal flat plate

When $CCP = (K_f L/K_s b)\sqrt{Re} = 0$ and $G_R/Re^2 = 0$, the problem treated is identical to the wellknown situation of laminar forced flow and heat transfer around an isothermal flat plate. In the standard literature, accurate solutions(called exact below) to this problem exist and thus the accuracy in the present method can be established.

The total shear stress, integrated from local values, differed by 0.68 per cent from the exact solution, while the total shear stress calculated from the momentum theorem showed no significant difference from the exact value. The total heat flux, integrated from local values, agreed with the exact value within 0.18 per cent. Over the major plate surface, the local values of the shear stress, heat flux, boundary layer thicknesses and so on agreed very well with the exact solution. The largest differences occurred close to the leading edge but were within tolerable limits. If not the initial profiles from equations (20) and (21) are used, the accuracy will be too worse.

b. Isothermal flat plate, $G_R/Re^2 \neq 0$

In [6], Lloyd and Sparrow presented results of mixed convection along a vertical flat plate by assuming local similarity over the whole plate surface. This means that equations (20) and (21) were assumed to be valid all over the plate. Under this assumption, the present numerical method produces results in excellent agreement with those of Lloyd and Sparrow.

Oosthuizen and Hart [7] reported numerical solutions of mixed convection when the wall condition was either a constant temperature or a constant heat flux. The basic equations were solved by an implicit finite difference method. As far as their results (only heat transfer data given as graphs) can be compared with the results of the present method for non-similarity cases the agreement is good.

In [8], Gryzagoridis published experimental results of combined free and forced convection from an isothermal vertical plate. Calculations against these data were also carried out with the present numerical method. The heat transfer data agreed well but the calculations predicted higher velocities than the experiments did. However, these deviations are smaller than those obtained when the experiments are compared with the results in [6] and [7].

c. Coupled conduction - convection

When CCP is greater than zero, an overall heat balance can be carried out. At x = 1 (the base of the fin), the heat enters the conduction - convection system by pure conduction. This heat flux can easily be calculated and compared with the total convective heat flux. The largest difference between these heat fluxes, 0.4 per cent, occurred for high values of the parameter CCP (of order 10). For smaller values of CCP (of order 1), these differences are quite small and not significant.

6. RESULTS

In this section some results will be presented. The cooling fluid is air for all the results presented.

Figure 2 shows the influence of the conduction - convection parameter CCP and the buoyancy force G_R/Re^2 on the total shear stress. When $G_R/Re^2 = 0$, the shear stress is independent of CCP and this value of $\tau_w = \tau_{w_o}$ is used a scaling factor for the shear stress. As is evident from this Figure , the shear stress is considerably increased by the buoyancy force, especially for small values of CCP. When CCP goes to infinity, the influence of the buoyancy force disappears and the shear stress approaches the pure forced value. This is so because an infinite value of CCP means a completely insulating fin.

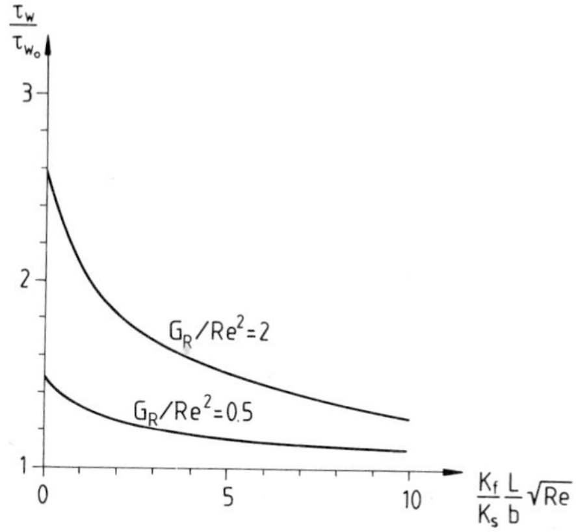

Figure 2. Total shear stress as function of G_R/Re^2 and CCP.

In Figure 3, the local heat flux distribution along the surface is shown for CCP = 2.76 and for different strength of the buoyancy force. The local heat flux is infinite at the leading edge since this is a singular point. With increasing x, the heat flux at first decreases and reaches a minimum and then starts to increase monotonically to its value at x = 1. From this Figure it is obvious that the effect of the buoyancy force is quite small for $x < 0.3$, otherwise a positive value of G_R/Re^2 increases the local heat flux. At this point it is also important to note that when CCP = 0, the local heat flux decreases monotonically with increasing x and thus CCP has a great effect on the local heat flux distribution.

Figure 4 provides distributions of the fin temperature for CCP = 2.76 and for two different values of G_R/Re^2. The full curves represent the results from the complete numerical solution while the dashed lines are calculated from the conventional fin theory (mean value of the heat transfer coefficient is calculated from the corresponding isothermal case). As is evident, the temperature distributions from the numerical solution and the conventional fin theory differ a lot especially in the vicinity of the leading edge. However, the fin efficiencies (see table in the Figure) calculated from both methods do not deviate too much. In fact, it has been found that for small values of CCP, the fin efficiency calculated from the conventional fin theory is slightly higher than the value from the numerical solution while for high values of CCP the contrary is true. This conclusion was also obtained by Sparrow and Chyu [4] for pure forced convection cases.

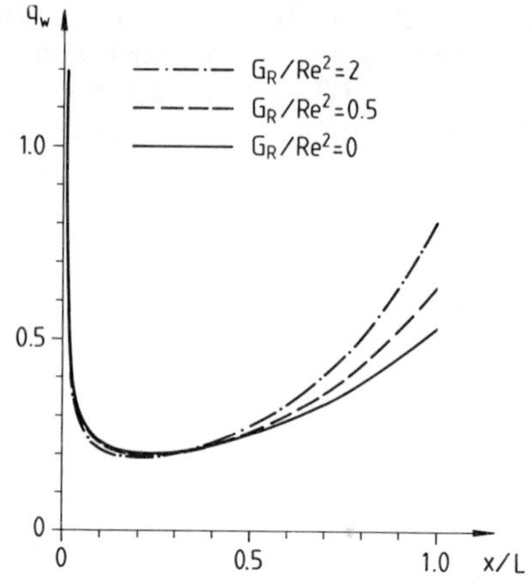

Figure 3. Local heat flux distribution for CCP = 2.76 .

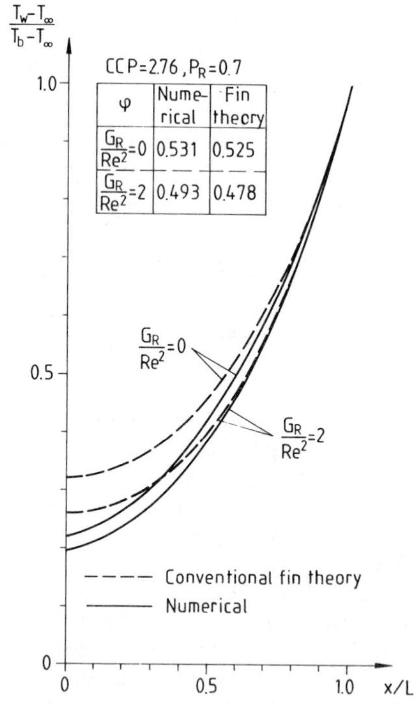

Figure 4. Fin temperature distributions for CCP = 2.76 .

In Figure 4, also the effect of the buoyancy force, G_R/Re^2, is evident. An increase in G_R/Re^2 enlarges the convective cooling and decreases the fin temperature and reduces the fin efficiency. (The fin efficiency is here defined as the ratio of the total heat flux from the actual fin and the total heat flux from an identical fin but with an infinite thermal conductivity).

When G_R/Re^2 is negative (adverse case) separation will occur when $G_R/Re^2 \lesssim -0.2$. Separation is at first detected at the base of the fin. The present numerical method can also handle such cases by way of changing the gradients with respect to x. That is, the direction of the u - velocity determines how the x - derivatives are calculated. However, if the boundary layer approximations still are applicable remains questionable.

7. ACKNOWLEDGEMENT

Financial support from the National Swedish Board for Technical Development (STU) is gratefully acknowledged.

8. REFERENCES

1. ECKERT, E.R.G. and DRAKE, R.M. JR - Analysis of heat and mass transfer, McGraw-Hill, New York, 1972.
2. KERN, D.Q. and KRAUS, A.D. - Extended surface heat transfer, McGraw-Hill, New York, 1972.
3. SPARROW, E.M. and ACHARYA, S. - A Natural Convection Fin with a Solution-Determined Nonmonotonically Varying Heat Transfer Coefficient. J. Heat Transfer, Vol. 103, pp. 218 - 225, 1981.
4. SPARROW, E.M. and CHYU, M.K. - Conjugate Forced Convection-Conduction Analysis of Heat Transfer in a Plate Fin. J. Heat Transfer, Vol. 104, pp. 204 - 206, 1982.
5. KARVINEN, R. - Natural and Forced Convection Heat Transfer from a Plate Fin. Int. J. Heat Mass Transfer, Vol. 24, pp. 881 - 885, 1981.
6. LLOYD, J.R. and SPARROW, E.M. - Combined Forced and Free Convection Flow on Vertical Surfaces. Int. J. Heat Mass Transfer, Vol. 13, pp. 434 - 438, 1970.
7. OOSTHUIZEN, P.H. and HART, R. - A Numerical Study of Laminar Combined Convective Flow over Flat Plates. J. Heat Transfer, Vol. 95, pp. 60 - 63, 1973.
8. GRYZAGORIDIS, J. - Combined Free and Forced Convection from an Isothermal Vertical Plate. Int. J. Heat Mass Transfer, Vol. 18, pp. 911 - 916, 1975.

PRACTICAL COMPUTATION OF MULTIDIMENSIONAL THERMAL FLOWS IN A GAS CENTRIFUGE*

M. H. Berger
Union Carbide Corporation, Nuclear Division
Oak Ridge, Tennessee, U.S.A.

1. SUMMARY A finite element theory is derived for Onsager's two-dimensional equation approximating the steady, viscous, gas motion in a high speed centrifuge. A new high-order tensor product element is proposed to make the computations easy. The method of weighted residuals is used to construct the stiffness matrix, associated boundary integrals and load vectors. Ekman suction conditions along horizontal surfaces are shown to be natural boundary conditions of the weak approximation. A class of pure boundary value problems are solved for the field variables of interest. We evaluate the effect of Ekman suction on the flow by computing with and without suction. Also, we compute the case of pure two-dimensional flow where the azimuthal velocity perturbation is presumed to vanish. The effect of this simplifying assumption on the end-to-end temperature difference necessary for a given circulation is discussed. Numerical results are presented graphically and we show that the so-called streamfunction must be graphed in physical coordinates for the isolines to be streamlines. Only in this form do the velocity vectors lie tangent to the contours. Also, the radial velocity is redefined for graphical purposes.

2. INTRODUCTION 2.1 Preliminaries A significant part of the U.S. Uranium Enrichment Program is based upon the gas centrifuge process with a multibillion dollar research, development, demonstration and commercialization program currently underway at Oak Ridge, Tennessee; Portsmouth, Ohio; etc. Donald Olander, University of California at Berkeley, has given a review of the history and physics of this process and the American program up to 1978 in his Scientific American article [1]. The state-of-the-art theory and process development in

*Based on work performed at Oak Ridge Gaseous Diffusion Plant, operated by Union Carbide Corporation, Nuclear Division, for the U.S. Department of Energy under U.S. Government Contract W-7405 eng 26.

the international community is discussed; however, a major omission about the fluid modeling was made regarding the linear theory. In the U.S.A., modern steady-state gas centrifuge linear flow modeling is primarily due to the late Professor Lars Onsager, Professor George F. Carrier, Harvard University, and Dr. Steven Maslen, Martin-Marrietta Corporation, who served as consultants to the U.S. Atomic Energy Commission and its descendants.

The basis of linear, compressible, centrifugal, heat conducting flow theory is Onsager's so-called "Pancake" equation [2], which involves a potential with a sixth order differential operator in x and a second order partial derivative in y. Thermodynamic effects and three components of momentum in the two coordinate directions are embodied in this formulation. A conceptual sketch of a gas centrifuge with fluid injection and removal instruments is given in Fig. 2.1.

2.2 Governing Partial Differential Equation and Boundary Conditions Linearized gas flow in a centrifuge may be described by Lars Onsager's so-called (dimensionless) Pancake equation,

$$L\chi = L_6\chi + B^2\chi_{yy} = F(x,y), \tag{2.1}$$

for the master potential χ away from the ends $y = 0$ and $y = y_o$ where,

$$L_6\chi = [e^x(e^x\chi_{xx})_{xx}]_{xx}. \tag{2.2}$$

Equation (2.1) is subject to the eight boundary conditions,

$$\chi_x(0,y) = \chi_{xx}(0,y) = 0,\ L_5\chi(0,y) = \frac{R_e}{32A^{10}}\,\overline{\theta}_y(y),\ \forall x,y \in \partial\Gamma_1 \tag{2.3a,b}$$

$$\chi_y(\infty,y) = \chi_x(\infty,y) = L_3\chi(\infty,y) = 0, \quad \forall x,y \in \partial\Gamma_2 \tag{2.3c}$$

$$\chi_y(x,0) = -\,4S^{-1/4}R_e^{-1/2}A^4[e^{x/2}\chi_x(x,0)]_x, \quad \forall x,y \in \partial\Gamma_3 \tag{2.3d}$$

$$\chi_y(x,y_o) = 4S^{-1/4}R_e^{-1/2}A^4[e^{x/2}\chi_x(x,y_o)]_x,^{I} \quad \forall x,y \in \partial\Gamma_4 \tag{2.3e}$$

where $L_5\chi = [e^x(e^x\chi_{xx})_{xx}]_x$ and $L_3\chi = (e^x\chi_{xx})_x$.

Here χ is a streamfunction potential defined by,

$$\psi = -\,2A^2\chi_x, \tag{2.4}$$

and, $$B = R_eS^{1/2}/4A^6 >> 1,\quad A = \left(\frac{MV_w^2}{2RT_o}\right)^{1/2} >> 1. \tag{2.5a,b}$$

Domain Γ and closure $\partial\Gamma$ are illustrated in Fig. 2.2.

[I] See Appendix A for derivation of the Ekman suction condition in this particular form.

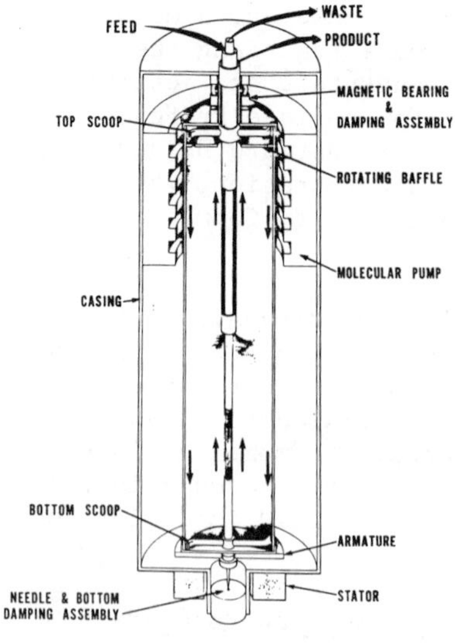

Fig. 2.1 A Gas Centrifuge

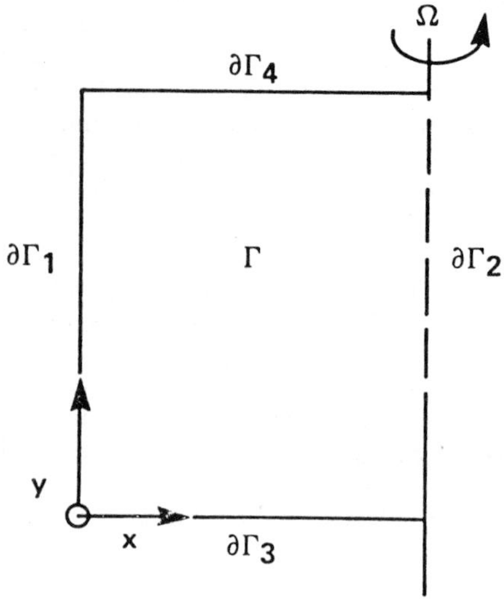

Fig. 2.2 Domain and Closure

An expression like $\lfloor\chi = F$ requires that $\chi\varepsilon C^{6,2}$ for $F\varepsilon C^0$. And since it involves eight partial derivatives there needs to be eight boundary conditions. If the eight boundary conditions are all of the generalized Neumann type, as in this case, then the solution will only be determined to within an additive constant. These differentiability requirements are relaxed in the finite element method and, in fact, this equation makes no sense if applied to the approximate solution.

3. FINITE ELEMENT THEORY 3.1 Weak Formulation Consider a restatement of the given "strong" formulation equivalently as a "weak" statement. Let,

$$\chi = \chi^e = \underline{N}^{eT}\underline{u}^e, \qquad \forall x,y \ \varepsilon \ \Gamma^e \tag{3.1}$$

where the local basis function is $\underline{N}^e$ and $\underline{u}^e$ is the discretization of χ over an elementary subdomain Γ^e involving nodal values and derivatives of χ. Presumably χ can be approximated by a finite dimensional subspace spanned by a basis that is sufficiently differentiable at interelement boundaries such that no additional contributions occur there.

The method of weighted residuals requires that one satisfy the vanishing of the weighted residual in lieu of the original differential equation where the weighted residual is an inner product over Γ of the governing differential equation and an arbitrary weight ω. Thus, over the x,y domain Γ with boundary $\partial\Gamma$,

$$-\ (\omega, \lfloor\chi - F) = 0. \tag{3.2}$$

Formally integrating by parts thrice,

$$\iint_{\Gamma} [L_3\omega\ L_3\chi + B^2\omega_y\chi_y + \omega F]\ dxdy$$

$$+ \int_{\partial\Gamma} [e^x\omega_x\ L_4\chi - e^x\omega_{xx}\ L_3\chi - \omega L_5\chi]\Big|_{x=0}^{x_T} dy - B^2 \int_{\partial\Gamma} \omega\chi_y \Big|_{y=0}^{y_o} dx = 0. \qquad (3.3)$$

In the absence of boundary integrals this reduces to,

$$(L_3\omega,\ L_3\chi) + B^2(\omega_y,\chi_y) + (\omega,F) = 0 \qquad (3.4)$$

Fortunately, for a basis with compact support it is necessary to consider only the boundary integrals for the subset of elements where $\Gamma^e \cap \partial\Gamma \neq 0$.

Discretizing χ and considering a vector of local weight functions $\underline{w}^e$, and choosing $\underline{w}^e$ the same as $\underline{N}^e$ we have the Galerkin variant of the more general method of weighted residuals. The resulting equations are,

$$\sum_e \iint_{\Gamma^e} [(\underline{L_3N^e}\ \underline{L_3N^{e^T}} + B^2\underline{N^e_y}\ \underline{N^{e^T}_y})\ \underline{u}^e + F^e\ \underline{N}^e]dxdy = \underline{0}, \qquad (3.5)$$

$$\text{or,} \quad \sum_e ([K^e]\ \underline{u}^e - \underline{f}^e) = \underline{0}, \qquad (3.6)$$

omitting for the time being those non-vanishing boundary integrals.

Also in doing the integration by parts above, numerous boundary integrals have appeared, some of which will be satisfied naturally and others which will be treated by constraining the space of admissible functions. The rule on boundary conditions [3] is that conditions which involve only derivatives below order m,n, where L is an operator of order 2m, 2n, will make sense. Those involving derivatives of order m,n or higher, will not apply; thus, the distinction between essential and natural boundary conditions. Specifically for L of order 2m = 6, 2n = 2, the appropriate boundary conditions are those of order less than m = 3 and n = 1. That is, boundary conditions on x of zeroth, first and second order, and boundary conditions on y of zeroth order are essential. All the higher order constraints are non-essential and will be satisfied naturally.

Assembly of the elemental equations into the system and correct incorporation of boundary conditions results in the usual matrix problem,

$$[K]\ \underline{u} = \underline{f}, \qquad (3.7)$$

where [K] is sparse, symmetric, positive definite and invertible.

3.2 Tensor Product Basis Functions The order of continuity of $\underline{N}^e$ required for admissibility to (3.5) is less than $C^{3,1}$. In general, if $\underline{N}^e \varepsilon C^{m-1,n-1}$ the necessary integrations can still be made and the interelement contributions vanish [3], [4]. Quite conveniently, the order of required continuity coincides with the maximum order of the essential boundary conditions. Thus for the weak form we only need $\underline{N}^e \varepsilon C^{2,0}$ for a conforming finite element approximation. Consider the family of generalized Hermite interpolating polynomials. The simplest Hermitian basis function subspaces we could choose are quintic-linear. However, we are motivated to use quintic-cubic subspaces because then the degrees of freedom coincide with the physical flow quantities of interest. Presumably something will be gained in accuracy and convergence rate as well.

Furthermore, if we restrict our attention to geometrically regular, right circular cylindrical domains we can make use of outer or tensor product basis functions defined on rectangles. For this function subspace we have nodal degrees of freedom which approximate χ, χ_x, χ_{xx}, χ_y, χ_{xy} and χ_{xxy}. Thus,

$$\underline{N}^e(\xi,\eta) = \underline{H}^e_5(\xi) \otimes \underline{H}^e_3(\eta) , \qquad (3.8)$$

where the cardinal basis functions $\underline{H}^e_5(\xi)$ [5] and $\underline{H}^e_3(\eta)$ [6] are,

$$\underline{H}^e_5(\xi) = \begin{Bmatrix} (1-\xi)^3(3\xi^2+9\xi+8)/16 \\ (1+\xi)(1-\xi)^3(3\xi+5)\ell^e_x/32 \\ (1+\xi)^2(1-\xi)^3\ell^{e^2}_x/64 \\ (1+\xi)^3(3\xi^2-9\xi+8)/16 \\ (1+\xi)^3(1-\xi)(3\xi-5)\ell^e_x/32 \\ (1+\xi)^3(1-\xi)^2\ell^{e^2}_x/64 \end{Bmatrix} , \quad \underline{H}^e_3(\eta) = \begin{Bmatrix} (2-3\eta+\eta^3)/4 \\ (1-\eta-\eta^2+\eta^3)\ell^e_y/8 \\ (2+3\eta-\eta^3)/4 \\ (1-\eta+\eta^2+\eta^3)\ell^e_y/8 \end{Bmatrix} \qquad (3.9)$$

Six of these 24 shape functions are illustrated in three-dimensional plots, Fig. 3.1a through 3.1f, on a standard square element.

3.3 Boundary Integrals and Boundary Conditions Let us discuss the boundary integrals and boundary conditions. For simplicity we consider total reflux or a fixed mass system.

Denoting the boundary integrals by I^b,

$$I^b \equiv \int_{\partial\Gamma} [e^x w_x L_4\chi - e^x w_{xx} L_3\chi - w L_5\chi] \Big|_{x=0}^{x_T} dy - B^2 \int_{\partial\Gamma} w\chi_y \Big|_{y=0}^{y_o} dx. \qquad (3.10)$$

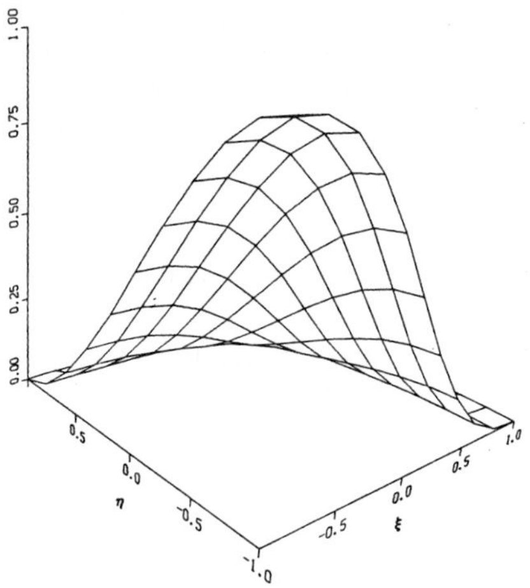

Fig. 3.1a. N_1

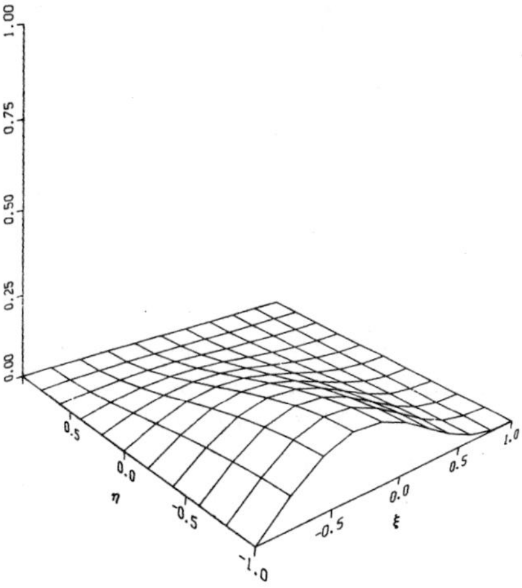

Fig. 3.1b. N_2

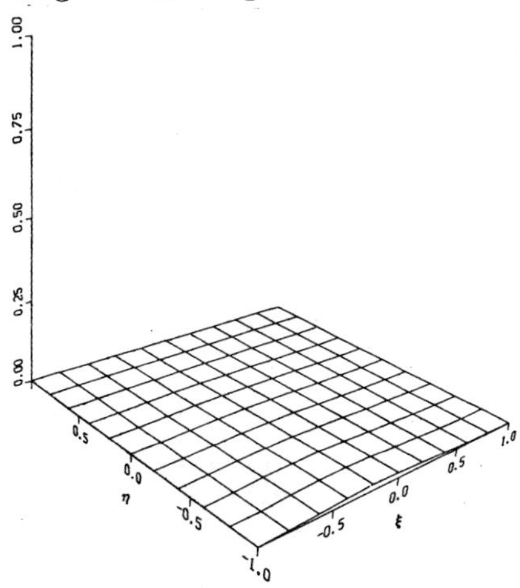

Fig. 3.1c. N_3

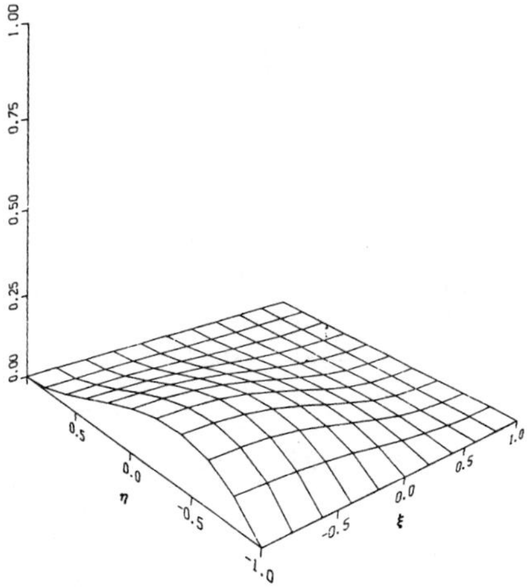

Fig. 3.1d. N_4

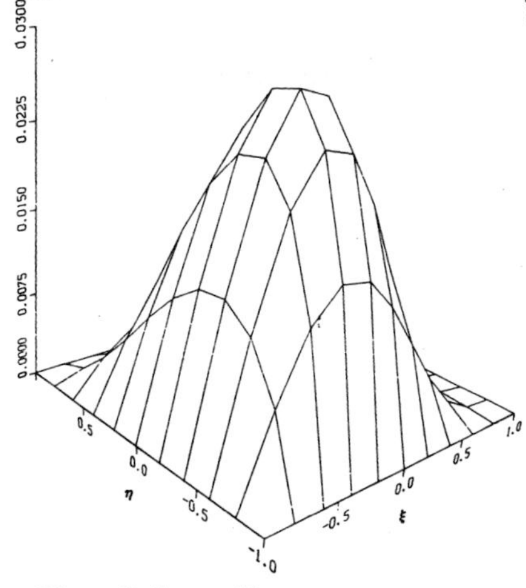

Fig. 3.1e. N_5

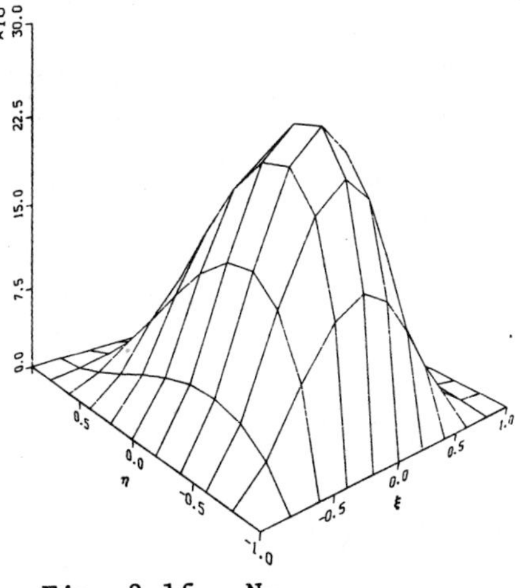

Fig. 3.1f. N_6

Fig. 3.1 Two-Dimensional Shape Functions

Incorporating essential boundary conditions, invoking Galerkin procedures, making use of the boundary conditions on $L_5\chi(0,y)$ $\chi_y(x,0)$, $\chi_y(x,y_o)$ and assuming a linear wall temperature profile, this simplifies to,

$$I^b = \frac{R_e\bar{\theta}_y}{32A^{10}} \int \omega(0,y)dy - B^2\, 4S^{-1/4}R_e^{-1/2}A^4$$

$$[\int \omega(x,y_o)[e^{x/2}\chi_x(x,y_o)]_x\, dx + \int \omega(x,0)\ [e^{x/2}\chi_x(x,0)]_x\, dx]. \quad (3.11)$$

Integrating by parts, imposing $\chi_x(x_T,y_o) = \chi_x(0,y_o) = 0$, considering a vector weight, discretizing χ and choosing $\underline{w}^e$ the same as $\underline{N}^e$ we have,

$$I^b = \sum_e \frac{R_e\bar{\theta}_y}{32A^{10}} \int \underline{N}^e(0,y)\ dy + B^2\, 4S^{-1/4}R_e^{-1/2}A^4$$

$$[\int \underline{N}^e_x(x,y_o)\ [e^{x/2}\underline{N}^e_x(x,y_o)^T]dx + \int \underline{N}^e_x(x,0)\ [e^{x/2}\underline{N}^e_x(x,0)^T]dx]\underline{u}^e. \quad (3.12)$$

This is subject to the boundary conditions,

$$\chi_x(0,y) = \chi_{xx}(0,y) = \chi(x_T,y) = \chi_x(x_T,y) = \chi_y(x_T,y) = 0, \quad (3.13)$$

$$L_3\chi(x_T,y) = 0,\ L_5\chi(0,y) = \frac{R_e}{32A^{10}}\ \bar{\theta}_y. \quad (3.14)$$

Furthermore, with the chosen tensor product basis functions it is necessary to impose additional constraints on the mixed derivatives at the wall,

$$\chi_{xy}(0,y) = \chi_{xxy}(0,y) = 0, \quad (3.15)$$

to achieve no slip between nodes at the wall. The last relation follows directly from the usual no-slip continuum assumptions. Likewise we need the additional constraint,

$$\chi_{xy}(x_T,y) = 0, \quad (3.16)$$

to achieve no flow across the zero streamline at x_T.

Thus the algorithm, after analytically integrating the boundary integral due to constant $\partial\bar{\theta}/\partial y$ is,

$$\sum_e [[\int\int_{\Gamma^e} (\underline{L_3N^e}\ \underline{L_3N^{e}}^T + B^2\underline{N}^{eT}_y)dxdy$$

$$+ 4B^2\ S^{-1/4}R_e^{-1/2}A^4 \int_{\partial\Gamma^e} (e^{x/2}\ \underline{N}^e_x(x,y_o)\ \underline{N}^e_x(x,y_o)^T$$

$$+ e^{x/2}\ \underline{N}^e_x(x,0)\ \underline{N}^e_x(x,0)^T)dx]\underline{u}^e + \frac{R_e\bar{\theta}_y}{32A^{10}}\ [\ell^e_y/2\ 0\ 0\ \ell^{e^2}_y/12\ 0\ \ldots$$

$$\ell^e_y/2\ 0\ 0\ -\ \ell^{e^2}_y/12\ 0\ 0]^T] = \underline{0}. \quad (3.17)$$

4. APPLICATIONS In all two-dimensional computations presented here we have used non-uniform grids under the assumption that these fluid flow problems are predominately thin, wall boundary layers. In accordance with the philosophy of placing nodes where they will do the most good, we have clustered elements near the rotor wall where gradients are expected to be steepest with the maximum ℓ_x^e near x_T. Also some axial grid refinement was made near the boundary y = 0. Elsewhere the axial mesh was uniform. We have used 168 elements involving 195 nodes (13 x 15) which leads to 1170 simultaneous equations for our 24 degree of freedom element. Also, F(x,y) vanishes.

The solution of these problems was obtained using the MODEL code described in detail by Akin [6]. Element matrices were formulated as described herein and the calculations employed the standard input, numerical integration, assembly, solution, and output program libraries of MODEL.

For calculations we used the May machine [7] described by the data given in Table 1. Suppose P_w = 100 torr (13330 N/m^2), V_w = 700 m/s, ΔT = 1 K and the end caps are at the temperatures of the corresponding ends of the rotor.

TABLE 1, GAS CENTRIFUGE DATA

a (radius)	0.09145 m	T_o	300 K
L (length)	3.353 m	V_w	400, 500, 700 m/s

In general, for symmetric problems one can reduce to computing on the half domain by specifying symmetry at the midplane. For symmetry $\chi_y(x,y_o/2) = 0$ and for the specified basis we add $\chi_{xy}(x,y_o/2 = \chi_{xxy}(x,y_o/2) = 0$.

4.1 Thermal Drive In Figs. 4.1 and 4.2 we have plotted 20 isolines of the finite element approximations to the potential and streamfunction in physical coordinates on half the domain. Also their magnitudes are tabulated. Potential lines are not plotted generally because their derivatives are of the most physical interest. However, we present them (once for brevity) because χ is the dependent variable occurring in our governing partial differential equation. The smoothness present in χ and similarities with ψ are apparent in the figures.

Fig. 4.2 gives lines of constant streamfunction. Only in the physical plane are these contours streamlines. For proof see Appendix B. Evidently most of the gas turns around outside the Ekman layer with only one non-zero contour intersecting the ends. This suggests that in this case about 10 percent of the flow goes into the Ekman layer. Also the locus of crossover points is approximately one scale height. These results agree very well with other solutions [8]. Both the potential and streamfunction exhibit the assumed wall boundary layer character.

We look to two-dimensional mass velocity vectors in two-dimensions to visualize the entire flow field inside the gas centrifuge. For brevity we just state here that plotting vectors in x or even $(1-\eta)$ will not give the proper result, where $\eta = r/a$. Appendix B has the details while the results are given in Fig. 4.3. A comparison shows these two-dimensional mass velocity vectors are indeed tangent to the streamlines.

Next we have a three-dimensional plot of $\rho_0 w$ in Fig. 4.4. The flow is quasi-parabolic in its axial variation with the maximum axial mass velocity occurring in the downflow stream while not much is happening high in the atmosphere. Due to Ekman suction there is a small flow evident at the end. There the streams are reduced to about 10 percent of the maximum upflow and downflow.

We note here that for the assumed operating conditions, nowhere in the centrifuge is the crossover point in good agreement with the value predicted by the asymptotic long bowl theory, $x_c = 1.256$. Even at the midplane two-dimensional effects are important.

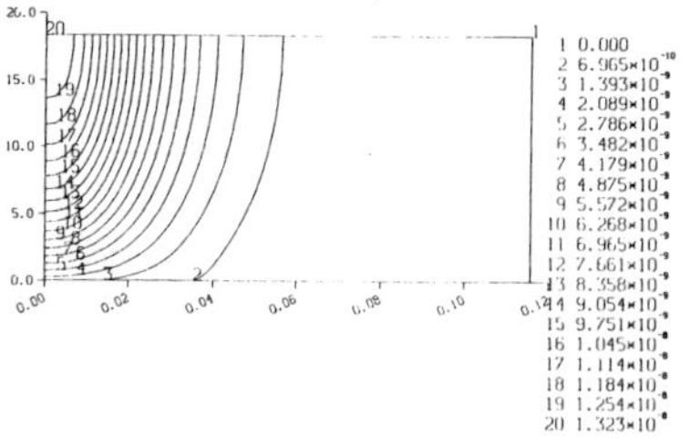

Fig. 4.1. Thermal Drive Potential Lines

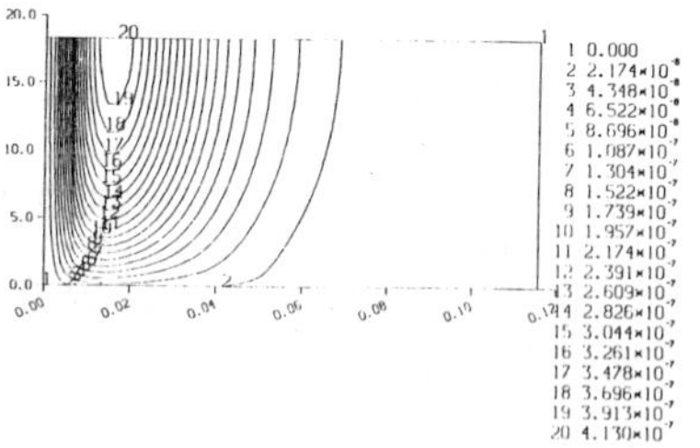

Fig. 4.2. Thermal Drive Streamlines

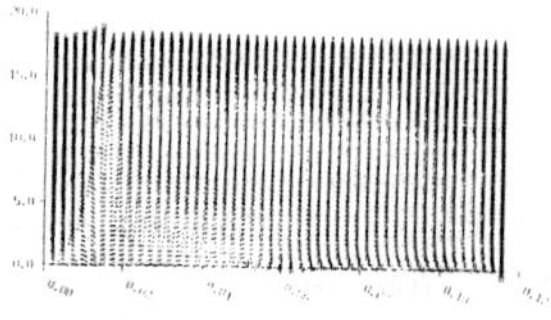

Fig. 4.3. 2-D Mass Velocity Vectors for Thermal Drive

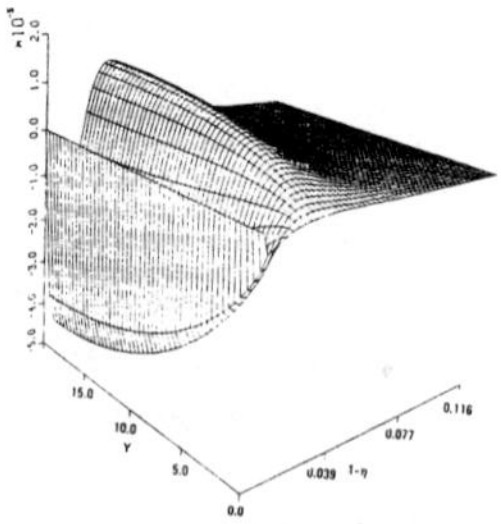

Fig. 4.4. 3-D Plot of Axial Velocity for Thermal Drive

4.2 Effects of Ekman Pumping A frequently debated question in rotating flows is how important is the Ekman pumping. As in aerodynamics, one can sometimes legitimately neglect boundary layers. Here one might neglect the end Ekman layers in a numerical model for computational efficiency since they constitute such a small part of the entire domain. Let us replace the Ekman suction conditions (2.3d), (2.3e) by the appropriate impermeable, free-slip end conditions. At $y = 0$,

$$\chi_x(x,0) = \chi_{xx}(x,0) = 0. \tag{4.1}$$

Since $\chi(x_T,y) = 0$ then we have the additional condition,

$$\chi(x,0) = 0. \tag{4.2}$$

Also, the Ekman suction boundary integral vanishes for $\omega(x,0) = 0$. Similar results apply at the other end.

The streamlines are given in Fig. 4.5, for $V_w = 700$ m/s. Comparing with our earlier results we see that ψ no longer penetrates into the ends and ψ_{max} is reduced about 10 percent. The greatest differences are evident at the ends. Also the wall gas layer is slightly thinner.

4.3 Pure Two-Dimensional Flow As another interesting variation let us consider purely two-dimensional centrifugal flows. That is where the azimuthal velocity perturbation, ω, is identically zero everywhere. It can be shown that this amounts to replacing S by S-1 everywhere. In the long bowl limit only axial flow exists, i.e., the azimuthal flow is necessarily zero. Results for the same operating conditions are presented in Fig. 4.6. The overall flow pattern looks primarily unchanged, however, ψ is uniformly increased by about 50 percent, which is a simple scaling. In other words, approximately the same circulation is obtained at about two-thirds the temperature difference since energy is not transferred into rotational motion. The streamlines are not exactly similar for purely two-dimensional flow; the wall boundary layer thickness is slightly thicker than for non-zero ω.

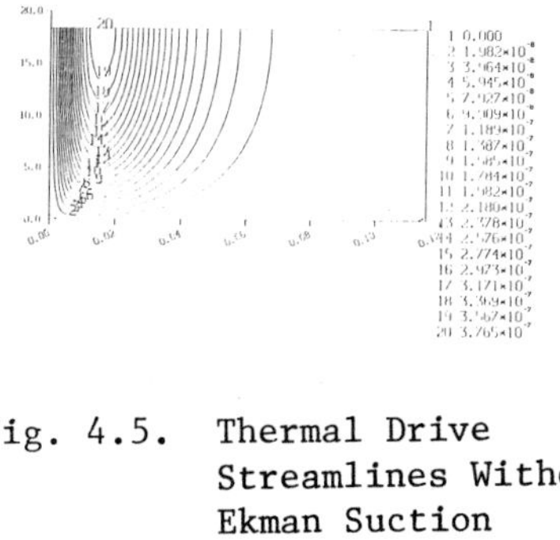

Fig. 4.5. Thermal Drive Streamlines Without Ekman Suction

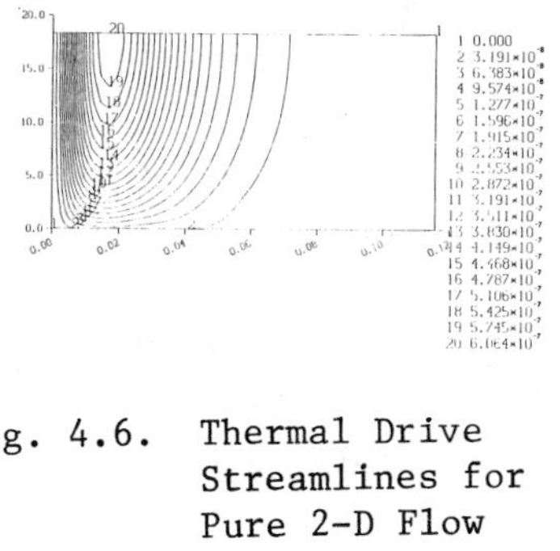

Fig. 4.6. Thermal Drive Streamlines for Pure 2-D Flow

5. REFERENCES

1. OLANDER, D.-The Gas Centrifuge, Scientific American, Vol. 2, 1978.

2. ONSAGER, L.-Approximate Solutions of the Linearized Flow Equations, unpublished manuscript, 1965.

3. STRANG, G. and G. FIX-An Analysis of the Finite Element Method, Prentice Hall, Inc., 1973.

4. ZIENKIEWICZ, O. C.-The Finite Element Method, McGraw-Hill Book Company, Third Edition, 1977.

5. KAWAI, T. and M. WATANABE-Analysis of a Solitary Water Wave Problem by the Method of Weighted Residuals, Second International Symposium on Finite Element Methods in Flow Problems, S. Margherita Liguere, Italy, June 14-18, 1976.

6. AKIN, J. E.-Finite Element Analysis-Its Implementation and Application, TICOM (Texas Institute for Computational Mechanics) Report 77-13, December 1977.

7. MAY, W. G.-Separation Parameters of Gas Centrifuges, AIChE Symposium Series, Vol. 73, No. 169, American Institute of Chemical Engineers, New York, 1977.

8. WOOD, H. G. III and J. B. MORTON-Onsager's Pancake Approximation for the Fluid Dynamics of a Gas Centrifuge, Journal of Fluid Mechanics, Vol. 101, Part 1, pp. 1-31, 1980.

APPENDIX A, DERIVATION OF EKMAN MATCHING CONDITION From Wood and Morton [8] the connection between the Ekman layers and the solution away from the ends is given by,

$$-4S^{3/4}R_e^{1/2}\,[-2A^2\chi_x(x,0) + \frac{\dot{m}_o}{2\pi}(x)]$$

$$= e^{-x/2}\,[\phi(x,0)] - \bar{\phi}(x,0)] + 2S^{1/2}e^{-x/2}\,\bar{U}(x,0), \qquad (A.1)$$

near the end $y = 0$, since $\psi \equiv -2A^2\chi_x$. Multiplying by $e^{x/2}$, and differentiating with respect to x,

$$-4S^{3/4}R_e^{1/2}\,\{-2A^2[e^{x/2}\chi_x(x,0)]_x + \frac{1}{2\pi}[e^{x/2}\dot{m}_o(x)]_x\}$$

$$= \phi_x(x,0) - \bar{\phi}_x(x,0) + 2S^{1/2}\,\bar{U}_x(x,0). \qquad (A.2)$$

In the absence of sources and sinks on the boundary we have [8],

$$\phi_x = \frac{-2R_eS}{A^2}\,[\chi_y(x,0) - \chi_y(x_T,0)]. \qquad (A.3)$$

After multiplying by $\frac{-A^2}{2R_eS}$,

$$\chi_y(x,0) = -4S^{-1/4}R_e^{-1/2}A^4[e^{x/2}\chi_x(x,0)]_x$$

$$+\frac{S^{-1/4}}{\pi} R_e^{-1/2}A^2 [e^{x/2}\dot{m}_o(x)]_x - \frac{A^2}{2R_eS}[\overline{\phi}_x(x,0) - 2S^{1/2}\overline{U}_x(x,0)]. \quad (A.4)$$

The term $\chi_y(x_T,0)$ may be taken to vanish since $\chi_y(\infty,y) = 0$ is a boundary condition on Onsager's equation for cases with no feed from the rotor centerline. This equation can be considerably simplified in the case of thermal drive at total reflux.

Setting $\dot{m}_o(x) = 0$ (solid end) and $\overline{U}(x,0) = 0$ (no endcap radial motion) and $\overline{\phi}_x(x,0) = 0$ (uniform end temperature) this reduces to,

$$\chi_y(x,0) = -4S^{-1/4}R_e^{-1/2}A^4 [e^{x/2}\chi_x(x,0)]_x, \quad (A.5)$$

and likewise for the other end $y = y_o$ except the sign changes.

APPENDIX B, FLOW VISUALIZATION Let us look at contours of the streamfunction and mass velocity vectors in both scale heights and physical coordinates. Recall our definition of x is,

$$x = 2A^2 (1-\eta). \quad (B.1)$$

From Wood and Morton [8],

$$\rho_o u = -\psi_y, \quad \rho_o w = -2A^2\psi_x \quad (B.2)$$

so, $\rho_o w/\rho_o u = 2A^2\, dy/dx$, (B.3)

where dy/dx is a tangent to a contour.

Now consider the physical plane (η,y). Using the chain rule,

$$\psi_x = -(1/2A^2)\psi_\eta, \quad (B.4)$$

then,

$$\rho_o w/\rho_o u = \psi_\eta/\psi_y = -dy/d\eta. \quad (B.5)$$

We simply redefine the radial velocity such that,

$$U = -u. \quad (B.6)$$

So finally we have,

$$\rho_o w/\rho_o U = dy/d\eta. \quad (B.7)$$

The fluid dynamic notation used herein mostly follows reference 8.

A NUMERICAL STUDY OF HEAT TRANSFER FOR A TURBULENT FLOW OF SUPERCRITICAL HELIUM.

M.C.M. Cornelissen '
C.J. Hoogendoorn ''

' Graduate research assistent
'' Professor
Heat Transfer Section
Dept. of Applied Physics
Delft University of Technology
The Netherlands.

SUMMARY

The heat transfer to turbulent supercritical helium can be studied numerically. The supercritical fluid is characterised by large variations in fluid properties. A reasonable agreement between numerical results and experimental data can be obtained. Three models of turbulence were considered the mixing length model, the k-model and the k-ε model. In the variabel property environment the mixing length model and k-model lead to reasonable results. Performance of the k-ε model is worse. The best agreement can be found for low heat fluxes. For higher heat fluxes agreement with experimental data is worse.

The numerical method we employed is a finite difference formulation of the problem. The parabolic character of the equations, due to the large velocities enables us to use a marching solution procedure. The near critical effects can be observed in this study. For low heat fluxes the heat transfer coefficient shows a large peak when the bulk temperature reaches the pseduocritical value. A sudden drop in the wall temperature can be observed for larger heat fluxes when the bulk temperature reaches the pseudocritical value. It results in a sudden increase of the heat transfer coefficient.

INTRODUCTION

The heat transfer to a turbulent flow of supercritical helium was studied numerically. This kind of heat transfer plays an important role in superconducting magnet design, and in other areas where a forced colling at such low temperatures is required. The critical point of helium is at a pressure of 2.26 atm. and a temperature of 5.04 K. The area we study is above the critical point.

Boiling phenomena do not play a role However fluid properties do vary importantly in this area (McCarty [11], Arp [1]. The specific heat Cp shows a sharp peak near the critical point. This peak deminishes for higher pressures (it can still be distinguished for pressures up to 20 atm.). The precise temperature ere this peak occurs for a specific supercritical pressure is known as the pseudocritical temperature. The density and viscosity show an important drop near this pseudocritical point. Thermal conductivity shows a maximum near the pseudocritical point. These variations of fluid properties lead to important deteriorations in the heat transfer characteristics.

FLOW EQUATIONS

The coordinate system employed in this model will be a circular one, since the pipe shows a circular symmetry. The momentum balance leads to two equations, one for the axial velocity and one for the radial velocity:

$$\rho(Vz \frac{\partial Vz}{\partial z} + Vr \frac{\partial Vz}{\partial r}) = \frac{1\partial}{r\partial r} \mu r \frac{\partial Vz}{\partial r} + \frac{\partial}{\partial z} \mu \frac{\partial Vz}{\partial z} - \frac{\partial p}{\partial z} \qquad (1)$$

$$\rho(Vz \frac{\partial Vr}{\partial z} + Vr \frac{\partial Vr}{\partial r}) = \frac{1\partial}{r\partial r} \mu r \frac{\partial Vr}{\partial r} + \frac{\partial}{\partial z} \mu \frac{\partial Vr}{\partial z} - \frac{\partial p}{\partial r} \qquad (2)$$

The mass balance leads to the continuity equation:

$$\frac{\partial \rho vz}{\partial z} - \frac{1}{r}\frac{\partial}{\partial r} \rho r\, Vr = 0 \qquad (3)$$

A Newtonian fluid is assumed, and the properties may be variable in this formulation. The energy balance is represented by the enthalpy equation:

$$\rho(Vz \frac{\partial H}{\partial z} + Vr \frac{\partial H}{\partial r}) = \frac{1}{r}\frac{\partial}{\partial r}\frac{\lambda}{Cp} r \frac{\partial H}{\partial r} + \frac{\partial}{\partial z}\frac{\lambda}{Cp}\frac{\partial H}{\partial z} + Vz \frac{\partial P}{\partial z} + Vr \frac{\partial P}{\partial r} \qquad (4)$$

In case of a turbulent flow λ and μ should be modified, by adding a turbulent λ_t and μ_t The turbulent Prandtl number is assumed to be 1. We will use a turbulence model for this task.

An extensive number of models is available (Launder and Spalding [10]. The complexity of the models can be characterised by the number of additional differential equations that the model employs. The most simple model (mixing length model) occupies no additional differential equations. Also models with one additional differential equation (k-model), or two additional differential equations (k-ε model) are considered. An important problem are the large variations in fluid properties. The turbulence models are all developed for constant property flow. However, we will use them for this variable property flow.

MIXING LENGTH MODEL

The Prandtl mixing length model describes the eddy diffusivity as the product of a length scale and e velocity scale:

$$\mu_t = \rho \, l_m \, l_m \left|\frac{\partial U}{\partial y}\right| \tag{5}$$

U is the velocity parallel to the wall, y is the coordinate perpendicular to the wall. For pipe flow the expression for the mixing length can be (Schlichting [17]):

$$l_m/R = 0.14 - 0.08\,(1-y/R)^2 - 0.06\,(1-y/R)^4 \tag{6}$$

where y is the distance from the wall.

A correction on this mixing length for the transition region is developed by Van Driest [2]. It describes the damping effects on the turbulence due to viscosity effects near the wall. The resulting correction is:

$$l'_m = l_m(1 - \exp(y^+/A) \tag{7}$$

where $y^+ = (\tau_w \rho)^{1/2}\, y/\mu$ with τ_w the wall shear stress.

$A^+ = 26$ is an emperical constant.

For fluids with variable properties the same equations are used, with the fluid properties evaluated locally (Pletcher [13], Patankar and Spalding [12]). According to these references the local shear stress shold be used rather than the wall shear stress. However in our situation of a circular pipe this leads to erronous results for the special case of a constant property flow. So we will not employ this practice.

k-MODEL.

The k-model, also known as the Prandtl-kolmogorov energy model [14,8], introduces a better guess of the velocity in the turbulent viscosity (eq. 5). In this model the velocity scale is replaced by the square root of the kinetic energy of turbulence k. It is the squared averaged of the velocity fluctuations. Now the equation describing the turbulent viscosity (eq. 5) is:

$$\mu_t = \rho \, l_m C_\mu \, k^{1/2} \tag{8}$$

where C_μ is a constant

The length scale is equal to the length scale in the mixing length model, multiplied by a factor 2.5. This factor is accounted for in the value of C_μ. The Van Driest correction for damping effects near the wall can still be applied.

An equation describing k can be derived from the Navier-Stokes equations after multiplication by u'. The resulting equation is:

$$\rho Vz \frac{\partial k}{\partial z} = \frac{1}{r}\frac{\partial}{\partial r} r \; (\mu+\mu_t/\sigma_k) \frac{\partial k}{\partial r} + \mu_t \; (\frac{\partial Vz}{\partial r})^2 - \rho\varepsilon \qquad (9)$$

Hassid and Poreh [5] introduce different values for the damping factor in the turbulent diffusivity and the dissipation terms:

$$\mu_t = \rho C_\mu l \; k^{1/2} \; (1-\exp\;(-A_\mu Re_k)) \qquad (10)$$

$$\varepsilon = (C_D k^{3/2} / l) \; (1-\exp\;(-A_\mu Re_k)) + 2\mu k/\rho l^2$$

where $Re_k = \rho y k^{1/2}/\mu$

The relations are similar to the Van Driest hypothesis. They proposed the set of constants:

$C_D = 0.416$, $C_\mu = 0.22$, $A_\mu = 0.012$, $\sigma_k = 1.0$

k-ε MODEL

In the preceeding models an emperical mixing lenght is introduced. It is possible to calculate this quantity,or a related one from the Navier Stokes equation. We will use the dissipation rate for this task.

Jones and Launder [6,7] described it as:

$$\rho\, Vz \frac{\partial \varepsilon}{\partial z} = \frac{1}{r}\frac{\partial}{\partial r}((\mu+\mu_t/\sigma_\varepsilon)r\frac{\partial \varepsilon}{\partial r}) + \qquad (11)$$

$$C_1 \frac{\varepsilon}{k} \mu_t (\frac{\partial Vz}{\partial r})^2 - (C_2 \rho \varepsilon^2/k) + 2(\mu\mu_t/\rho)\;(\partial^2 Vz/\partial r^2)^2$$

The k-equation is the same as presented in the preceeding paragraph, but the dissipation is evaluated from eq. 11. The turbulent viscosity is now written as:

$$\mu_t = \rho \; C_\mu \; k^2/\varepsilon \qquad (12)$$

This set of equations should describe turbulence in the full turbulent region adequately, however they are not able to describe the influences of a nearby wall.

This influence can either be modelled by means of wall functions or by adapting the equations. Lam and Bremhorst [9] modelled the influence of the wall by adapting the set of constants. Their set of constants is:

$$C_\mu = .09 \; (1-\exp(-.0165 \; Re_t))^2 \; (1+20.5/Re_t) \qquad (13)$$

$$C_1 = 1.44 \;\; (1+(.0045/C_\mu)^3)$$

$$C_2 = 1.92 \; (1-\exp\;(-Re_t^2)), \; \sigma_k = 1.0, \; \sigma_\varepsilon = 1.3$$

NUMERICAL FORMULATION

Boundary conditions are a no slip condition for velocity and a constant wall heat flux for the energy equation.

Free convective (buoyancy) effects are not included in this model. The numerical method we employ is a finite difference formulation. The difference equations are derived by a control volume approach, which balances sources and diffusive and convective flows of a quantity (velocity, enthalpy) for a small volume. A staggered grid is employed. The pressure is used to fulfil the continuity equation by correcting the velocities. The Reynolds number of the flow is high ($Re \sim 100{,}000$). This leads to a parabolic character of the equations (Pratap and Spalding [15] ,[16]). The pressure drop in axial and radial direction are treated independently. The first is used to correct the axial velocities in order to fulfil the continuity equation on a pipe cross section. The latter guarantees cell by cell continuity in a cross section. The same pressure drop is assumed for all axial velocities in a cross section. Thus a marching solution can be obtained.

Pressure is fixed at the inlet, as is the velocity profile. A constant heat flux is assumed at the wall. The heating is started some distance from the pipe entrance to allow for inlet effects in the velocity profiles. All simulations were performed using a down stream stepsize between .01 cm and .005 cm dependent on eventual instabilities that may occur when too large steps are employed. The radial distribution of grid points was chosen to be very dense near the wall and rather coarse in the bulk of the fluid. Of the total of 20 points in a cross section 3 to 5 points were located in the laminar boundary layer. No special treatment of the wall by means of wall functions was employed. All difference equations are solved right to the wall. The pipe diameter was 0.213 cm. Convergence was assumed when no significant changes in velocity and enthalpy occurred (changes less than 0.01%).

NUMERICAL RESULTS COMPARED WITH EXPERIMENTAL DATA.

The experiments of Giarratano and Jones [3] are simulated using the described procedure. All results are given by a heat transfer coefficient which is represented by h. All experiments are simulated using the mixing length model. A number of experiments is also simulated with a k-model or k-ε model of turbulence. Three different inflow conditions are covered (massflow G and inlet temperature T):

- $G = 72\ kg/m^2s$, $T = 5.04$ K
- $G = 72\ kg/m^2s$, $T = 4.06$ K
- $G = 118\ kg/m^2s$, $T = 4.08$ K

The pressure is for all experiments 2.5 atm. Heat flux varies between 280 and 7130 W/m^2. The results presented in this section are obtained using the mixing length model.

The results for a mass flow of 72 kg/m^2s are presented in fig. 1.
The entrance temperature was 5.04 K (corresponding to an enthalpy of 15 kJ/kg). This is rather near the pseudocritical temperature of 5.37 K.

For the two experiments with the higher heat fluxes, the bulk temperature alsmost reaches the pseudocritical temperature, although this temperature is not passed. All experimental data tend to drop for a distance along the pipe over 8 cm. For all other experiments not presented in this figure this can also be observed. Giarratano and Jones [3] do not mention this. It could be contributed to experimental errors. The numerical results do cover the experiments quite well. Except at the end of the pipe all experimental and numerical results agree within 10%. This suggests that the experimental errors are less than the 55% that is expected for high flow velocity and low heat flux. The experiments for low heat fluxes agree as good as the high heat flux experiments. This will be shown later.

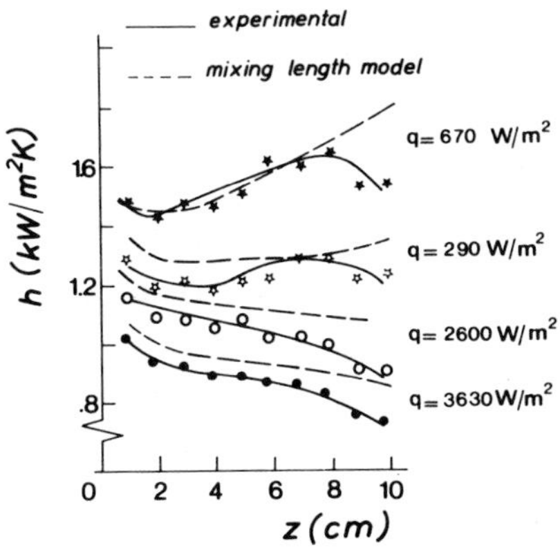

Figure 1. Experimental and numerical heat transfer coefficient for G=72 kg/m^2s and inlet temperature of 5.04K.

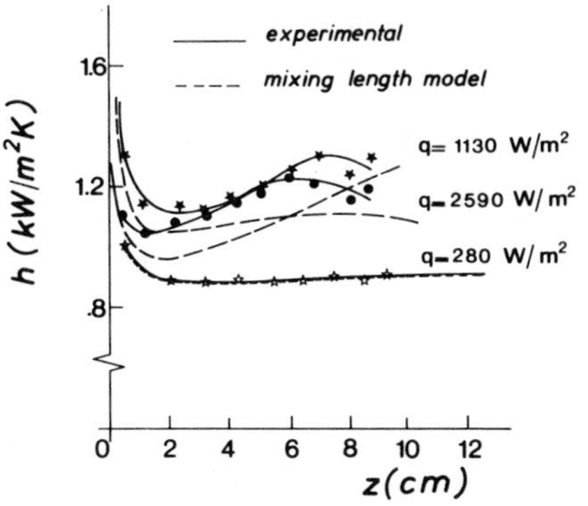

Fig. 2. Experimental and numerical heat transfer coefficient for G=72 kg/m^2s and inlet temperature of 4.06 K.

For the same mass flow but lower entrance temperature the results are shown in fig. 2. Here the entrance temperature is 4.06 K, which is well below the pseudo critical temperature. The agreement between experiment and numerical simulation is quite good again. The differences do not exceed the 10% again, except for the doubtful region at the end of the pipe.

The results presented in fig. 3 are for a higher mass flow of 118 kg/m^2s.

In these results the differences between numerical study and experimental work are larger than in the preceeding simulations. The differences are 15-20% and for the high heat flux ($q=7130$ W/m^2) even larger. Except for the latter the qualitative beahviour is still correct.

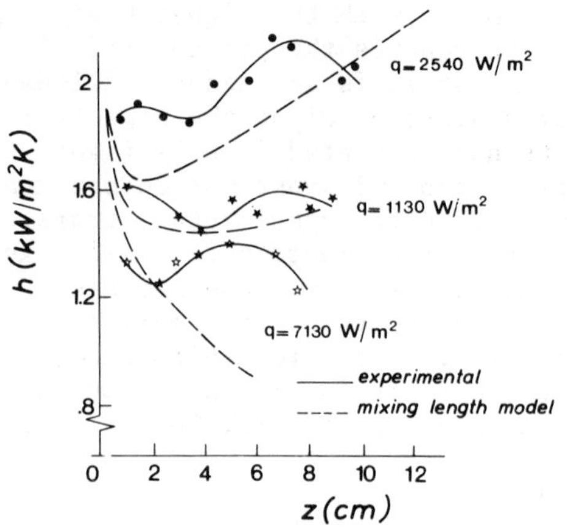

Figure 3. Experimental and numerical heat transfer coefficient for G=118 kg/m^2s and inlet temperature of 4.08 K.

COMPARISON OF THE MODELS OF TURBULENCE.

A number of experiments are simulated with different models of turbulence. In fig. 4 the experiments of fig. 2 are presented for the three models of turbulence: mixing length, k-and k-ε model.

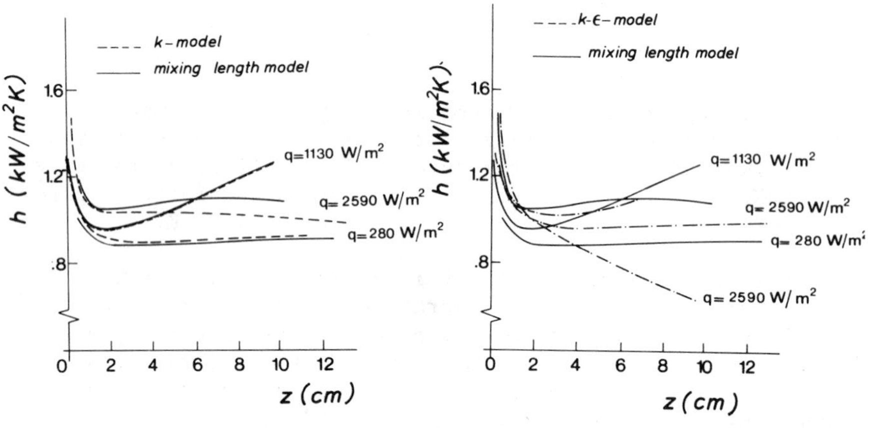

Figure 4. Heat transfer coefficients for the three models of turbulence, G=72 kg/m^2s and inlet temperature 4.06 K.

Only for the high heat flux of 2590 W/m^2 a significant difference between mixing length and k-model can be observed. In that simulation the k-model results are further away from the experimental data than the mixing length results.

In all simulations the k-ε model leads to results that show important differences from the mixing length and k-model results. When comparing to the experimental results the k-model results are far worse than the others.

The same tendency can be observed in fig. 5 where the experiments of fig. 3 are simulated for all turbulence models. The mixing length and k-model differ significantly for a heat flux of 2540 W/m^2 only. This difference is an improvement of the k-model when comparing to the experimental results. The k-ε model again leads to erronous results.

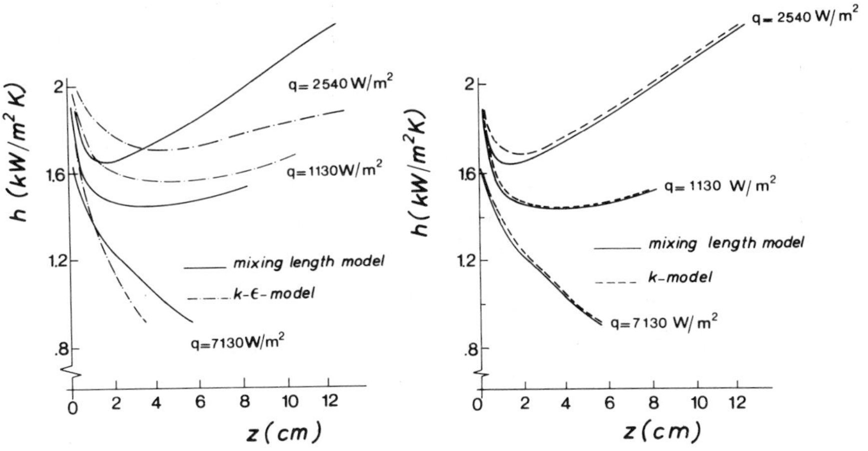

Figure 5. Heat transfer coefficients for the three models of turbulence, G=118 kg/m^2s and inlet temperature 4.08 K.

From this it appears that the k-ε model can't be used in this variable property flow. Since the k-equation is the same as in the k-model the ε-equation can be assumed to cause the difficulties. The profiles in the bulk of the fluid did not differ significantly. The derivatives of velocity and enthalpy are slightly different neat the wall, what has important consequences for the heat transfer coefficient. The k-model and mixing length model do not differ significantly. By observing the heat transfer coefficient for temperatures both under and above the supercritical temperature the effects of the supercritical fluid properties on heat transfer can be studied. We simulated situations where the inlet temperature is well below the pseudocritical value (4.2 K, enthalpy 10 kJ/kg), and the outlet temperature 9 K or higher (enthalpy 55 kJ/kg). A mixing length model is employed. The pseudocritical temperature is 5.37K (enthalpy 23 kJ/kg). Mass flows of 10.5, 53.6 and 111.2 kg/m^2s are simulated.

Schnurr [18] suggests a correlating factor:

$$\phi = q\, D^{0.2}/G^{0.8} \tag{14}$$

It is defined by Goldman [4]. In fig. 6 the heat transfer coefficients are plotted as a function of the dimensionless position for $\phi=10$ and $\phi=30$.
For a specific value of ϕ all profiles have the same shape. The absolute value of the heat transfer coefficients differ, as the mass flow rate and heat flux do. For $\phi=30$ the heat transfer coefficeint decreases until the minimum value is reached at the point where the bulk temperature equals the pseudocritical value. The heat transfer coefficient then increases sharply to a small peak and levels out.

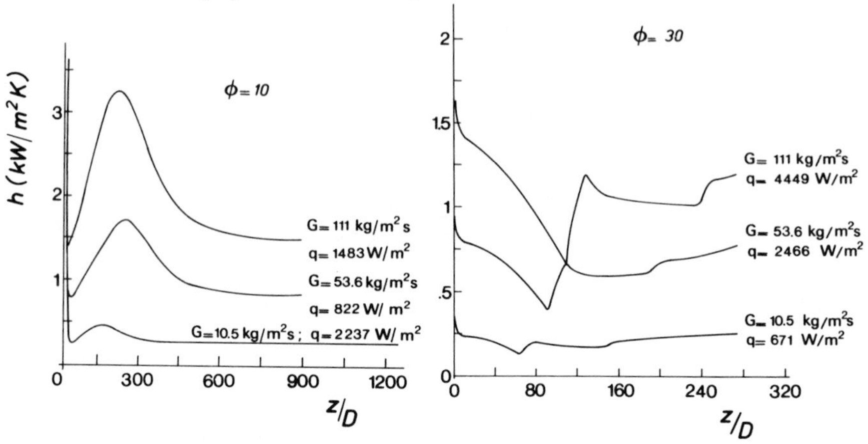

Fig. 6. Heat transfer coefficient as function of dimensionless distance for $\phi=10$ (a) and $\phi=30$ (b).

At this place the wall temperature shows a sharp decrease. It appears that the enthalpy of the fluid in a cross section is redistributed. The little step further on along the pipe also corresponds to a small decrease in temperature. For $\phi=10$ no minimum is observed, but a pronounced maximum occurs. The position of the maximum is near the place where the wall temperature reaches the pseudocritical value. For $\phi=20$ the same applies however the effects are not very pronounced.

REFERENCES

1. Arp V., New forms of state equations for helium, Cryogenics, 1974, pg 593-598.
2. Driest E.R. van, On turbulent flow near a wall, J. of the Aeronautical Sciences, 1956, pg 1007-1011.
3. Giarratano P.J., Jones M.C., Deterioration of heat transfer to supercritical helium at 2.5 atmosphere, Int. J. Heat Mass Transfer, 18, 1975, pg 649-653.
4. Goldmann, K., Heat Transfer to supercriticla water and other fluids with temperature dependent properties, Nuclear Engineering, C.E.P. symposium series, 50, no 11, pg 105-113.

5. Hassid S., Poreh M., A turbulent energy model for flows with drag reduction, J. of Fluid Mechanics, ASME, 1975, pg 234-241.
6. Jones W.P., Launder B.E., The prediction of laminarization with a two-equation model of turbulence, Int. J. Heat Mass Tr., 15, 1972, pg 301-314.
7. Jones W.P., Launder B.E., The calculation of low-Reynolds number phenomena with two-equation model of turbulence, Int. J. Heat Mass Tr., 16, 1973, pg 1119-1130.
8. Kolmogorov A.N., Equations of turbulent motion of an incompressible turbulent flow, Izv. Akad. Nauk SSSR Ser. Phys. VI, 1-2, pg 56, 1942.
9. Lam C.K.G., Bremhorst K., A modified form of the k-eps model for predicting wall turbulence, Journal of Fluids Engineering, Trans. ASME, 1981, vol 103, pg 456-460.
10. Launder B.E., Spalding D.B., Mathematical models of turbulence, Academic Press Inc., 1972.
11. McCarty R.D. Thermophysical properties of helium-4 from 4 to 3000 R wity pressures to 15000 PSIA, NBS technical note 622, 1972.
12. Patankar S.V., Spalding D.B., Heat and mass transfer in boundary layers, intertext Books, London, 1970.
13. Pletcher R.H., On a solution for turbulent boundary layer flow with heat transfer, pressure gradients, and wall blowing or suction, Proc. 4th. International heat transfer conference, Paris, Vol 2, paper FC 2.9.
14. Prandtl L. Ueber ein neues Formelsystem fur die ausgebildete Turbulenz, Nachrichten von der Akad der Wissenshcaft in Goettingen, 1945.
15. Pratap V.S., Spalding D.B., Numerical computations of the flow in curved ducts, Aeronautical quarterly, 1975.
16. Pratap V.S., Spalding D.B., Fluid flow and heat transfer in three dimensional duct flows, Int. J. Heat Mass Tr., 19, 1976, pg 1183-1188.
17. Schlichting H., Grenzschicht Theorie, Verlag H. Braun, 1951.
18. Schnurr N.M., Numerical predictions of heat transfer to supercritical helium in turbulent flow through circular tubes, J. of Heat Tr., 99, 1977, pg 580-585.

Multi-Element Collocation Solution For Convective Dominated Transport

H. J. Migliore and E. G. McReynolds
Portland State University
Portland, Oregon, USA

SUMMARY

Global collocation was successfully applied to a two-dimensional hyperbolic PDE containing nonlinear dissipative terms and was attempted on the Navier-Stokes equations, in penalty function form for the driven-cavity problem. Results from the latter indicated the need to develop a multi-element approach to better capture sharp gradients. The one-dimensional convection-diffusion transport equation was selected as the vehicle for developing multi-element collocation. In order to more fully explore the advantages of using collocation in future work in structure-fluid interaction, traditional finite element shape functions and assembly procedures were avoided. Interelement points were treated as boundary conditions for evaluating residuals within each element. Two continuity schemes were investigated, C^1 and mixed C^0, C^1 across element boundaries. Results favorably compared to exact solutions. The mixed continuity appears to offer advantages in terms of less numerical ringing without inducing dispersion, but increased computational effort was required.

1. INTRODUCTION

A global collocation approach was successfully applied to the problem of ocean cables subjected to hydrodynamic loading, surface excitation, and change of length with time [1]. Fluid effects were modeled as forces applied to the cable and were functions of added mass and velocity-squared drag. The governing equation for the cable was basically a wave equation with nonlinear dissipative drag terms, and with specified and differential boundary conditions. The collocation formulation was cast in terms of an approximate expression for derivatives which were directly substituted into the differential equation. This approach suggests a generality for application of collocation which is not particularly sensitive to the order of differential equation or nonlinearities. Furthermore, boundary conditions are explicity expressed, along with the system of residual equations at interior collocation points. Thus, a variety of boundary functions can be accomodated including the possibility of treating the boundary as an approximate quantity in a mixed residual fashion. Future work will attempt to explore these advantages as applied to structure-fluid (Navier-Stokes) interaction problems.

Global collocation suffered from significant ringing whenever sharp gradients were encountered and required small time integration steps when collocation points were closely packed, as was found when global collocation was applied to the driven-cavity problem with a Navier-Stokes fluid.As a consequence, a multi-element collocation solution technique was developed as first applied to one-D convective-dominated transport, with two different sets of boundary conditions. The form of transport equation is

$$\frac{\partial Z}{\partial t} = \propto \frac{\partial Z}{\partial X} + \delta \frac{\partial^2 Z}{\partial X^2} \qquad (1)$$

where Z is the transport field variable, X is the spatial variable, t is time, $\propto$ is the convection (advection) coefficient, and δ is the diffusion coefficient.

From the onset the developed collocation solution was not tied to conventional finite element (FEM) shape functions and assembly procedures; the commitment to a new multi-element approach is justified as follows. Finite element has already been applied to the same problem with a high degree of success [2]. In addition, collocation has been used in a finite element sense [3,4] by using FEM type shape functions, but utilizing collocation for shortcutting quadrature evaluation and element assembly. Both FEM and Finite Element Collocation use the same Galerkin principle and shape functions. The authors feel that these approaches do not take full advantage of collocation as a formulation principle in terms of using higher order functions and nodal degrees of freedom which are not tied apriori to a FEM assembly procedure.

The following section will describe two multielement collocation approaches. Both are based on higher order approximation functions within an element with element end points acting as boundaries. Function values at element boundary points are determined from interelement continuity requirements. The chief distinction between the two approaches was the role of continuity. In the first approach, that of Villadsen and Finlayson [5,6] C^1 interelement continuity is maintained. In the developed technique, C^1 continuity was tied to the second order derivatives while C^0 continuity was associated with first order derivatives.

2. MULTIELEMENT COLLOCATION

For simplicity the spatial domain is divided into m equal length elements each of which have n+1 points (or nodes) including the two element boundary points.

The discretization yields the vectors:

$$X_{COL} = (X_0^1, X_1^1, X_2^1, \ldots, X_{n-1}^1, X_1^2, \ldots\ldots X_{n-1}^m, X_n^m)$$

$$X_{IE} = (X_n^1, X_n^2, \ldots, X_n^{m-1})$$

where X_k^e represents the kth sample point of the eth element.

X_{COL} represents the so-called collocation points, while X_{IE} is composed of interelement boundary points.

The above points are chosen so that for each element e, the set of points $\{X_o^e, X_1^e, \ldots X_n^e\}$ are Lobatto quadrature points for unit weight on the interval $[X_o^e, X_n^e]$, and at the intersection of elements $X_0^{e+1} = X_n^e$ for e = 1, m - 1.

Corresponding to the vectors of sample points are the values of the dependent variable at those points. These are denoted by:

$$Z_{COL} = (Z_0^1, Z_1^1, \ldots, Z_{n-1}^1, Z_1^2, \ldots Z_{n-1}^m, Z_n^m)$$

and

$$Z_{IE} = (Z_n^1, Z_n^2, \ldots, Z_n^{m-1})$$

For the above discretization X_{COL} and Z_{COL} will have m(n-1) + 2 components.

The requirement that the dependent variable be continuous will make,

$$Z_0^{e+1} = Z_n^e \qquad \text{for } e = 1, m-1.$$

2.1 First Approach

By the selection of polynomials as approximants to Z, matrices $D^1 = [d^1_{ik}]$ and $D^2 = [d^2_{ik}]$ may be constructed [1] from X_{COL} and X_{IE} such that

$$\frac{\partial Z^e_i}{\partial X} = \sum_{k=0}^{n} d^1_{ik} Z^e_k \quad \text{and} \quad \frac{\partial^2 Z^e_i}{\partial X^2} = \sum_{k=0}^{n} d^2_{ik} Z^e_k \tag{2}$$

where

$\frac{\partial Z^e_i}{\partial X}$ and $\frac{\partial^2 Z^e_i}{\partial X^2}$ represent the first and second spatial derivatives of Z at the ith sample point of the eth element. These expressions may be used directly to specify the discretized form of the transport equation provided that values for Z_{IE} are given.

In order to specify the values making up the vector Z_{IE}, C^1 continuity between elements is assumed [5,6]. C^1 continuity is satisfied whenever,

$$\frac{\partial Z_0^{e+1}}{\partial X} = \frac{\partial Z_n^{e}}{\partial X} \qquad e = 1, m - 1$$

or

$$\sum_{k=0}^{n} d^1_{0k} Z^{e+1}_k = \sum_{K=0}^{n} d^1_{nk} Z^e_k, \qquad e = 1, m - 1$$

These m - 1 relations contain function values at both interelement and collocation points, Z_{IE} and Z_{COL}, and can be solved, [5] yielding

$$Z_{IE} = Q Z_{COL}$$

where Q is an m - 1 by m(n - 1) + 2 matrix.

C^1 continuity provides boundary conditions for each element which are in effect incorporated into elemental trial solutions. Those in turn are used to formulate a discrete Galerkin approximate solution to the given PDE over each element and hence over the entire domain. This procedure results in the following set of ODE's to be solved in time:

$$\frac{\partial Z_i^e}{\partial t} = \propto \frac{\partial Z_i^e}{\partial X} + \delta \frac{\partial^2 Z_i^e}{\partial X^2} \qquad e = 1, m \qquad i = 1, n - 1$$

$$Z_n^e = \sum_{i=1}^{m} \sum_{j=1}^{n-1} q_{ij}^e Z_j^i + q_{(1)(0)}^e Z_0^1 + q_{m0}^e Z_n^m \qquad e=1, m-1 \tag{3}$$

Boundary conditions are specified relative to Z_0^1 and Z_n^m

Two sets of boundary conditions were examined:

1. For the case of an unconstrained out flow boundary, i = 1, n in the governing equation, first of Equations (3) for e=m, and Z_0^1 is given.

2. Natural boundary approximating outflow.

$$\frac{\partial Z_n^m}{\partial X} = 0 \text{ and } Z_0^1 \text{ is given}$$

2.2 Second Approach

For the problem of pure convection (δ=0.0) a C^0 continuous collocation model was constructed in which the residual at interelement points was required to be zero. This was accomplished by defining the material derivative at such points as the average of the derivatives from the right and left where those derivatives were given by the relation:

$$\frac{\partial Z_n^e}{\partial X} = \frac{1}{2} \left(\sum_{k=0}^{n} d_{nk}^1 Z_k^e + \sum_{k=0}^{n} d_{0k}^1 Z_k^{e+1} \right) \qquad e=1, m-1 \tag{4}$$

on the left and right elements in the conventional collocation method.

When these results were compared to those obtained from a C^1 continuous collocation solution of the same problem a significant reduction of solution dispersion was observed to have occurred. An attempt was then made to construct a solution method for highly convective transport problems which would take advantage of the C^0 solution properties.

For such a formulation it was necessary to add an additional degree of freedom, V, at each interelement boundary node.

$$V_{IE} = (V_n^1, V_n^2, \ldots, V_n^{m-1})$$

$$V_{IE} = Q\, Z_{COL}$$

These terms were then utilized to determine the values of second material derivatives by their employment as function values in place of Z_{IE}.

$$\frac{\partial^2 Z_n^e}{\partial X^2} = \frac{1}{2}\Bigg[d_{n0}^2 V_n^{e-1} + d_{nn}^2 V_n^e + \sum_{k=1}^{n-1} d_{nk}^2 Z_k^e \tag{5}$$

$$+ d_{00}^2 V_n^e + d_{0n}^2 V_n^{e+1} + \sum_{k=1}^{n-1} d_{nk}^2 Z_k^{(e+1)}\Bigg]$$

Where, in the same manner as for first derivatives, the second material derivative was defined as the average of left and right second derivatives. First and second spatial derivatives at the collocation points are evaluated as in the first approach, Equations (2), with V_{IE} substituted for Z_{IE} in the expression for second derivatives. With both derivatives defined at interelement points, the residual can be equated to zero as in the C^0 treatment. This yielded the system of ODE's:

$$\frac{\partial Z_i^e}{\partial t} = \propto \frac{\partial Z_i^e}{\partial X} + \delta \frac{\partial^2 Z_i^e}{\partial X^2} \qquad e=1,m; \quad i = 1,n \tag{6}$$

$$V_n^e = \sum_{i=1}^{m} \sum_{j=1}^{n-1} q_{ij}^e Z_j^i + q_{(1)(0)}^e Z_0^1 + q_{m0}^e Z_n^m \qquad e=1, \quad m-1$$

Z_0^1 is given

for an unconstrained outflow problem, noting that $Z_n^e = Z_0^{e+1}$.

The developed formulation does not necessarily yield a discrete Galerkin solution since the second derivatives as obtained through Equations [2] are not in all cases derivatives of the trial solution. The definition of V_{IE} and the method by which second derivatives were computed emulated C^1 continuity for the approximate solution, i.e., V's are computed by means of Q. No schemes for second order differentiating which abanboned the C^1 property were found to be convergent.

Both the conventional and developed collocation ODE systems were integrated in time by the Lax-Wendroff technique.

3. NUMERICAL RESULTS

The problems that were solved are:

$$\frac{\partial Z}{\partial t} = \propto \frac{\partial Z}{\partial X} + \delta \frac{\partial^2 Z}{\partial X^2}, \qquad X\varepsilon[0,\infty)$$

$$Z(0,X) = e^{\frac{-(X-3.75)^2}{2}} \tag{7}$$

$$Z(t,0) = e^{\frac{-(3.75)^2}{2}}, \qquad Z(t,\infty) = 0$$

We only solve over the finite domain where $X\varepsilon[0,25]$ hence we replace the B.C. at infinity with an approximating B.C. at X = 25. Two such B.C.'s were employed:

Unconstrained outflow:

$$\frac{\partial Z}{\partial t}(t,25) + \propto \frac{\partial Z}{\partial X}(t,25) = \delta \frac{\partial^2 Z}{\partial X^2}(t,25) \tag{8}$$

Natural boundary approximation as employed in [7]:

$$\frac{\partial Z}{\partial X}(t,25) = 0 \tag{9}$$

Results for unconstrained outflow boundary are shown in Figures 1 and 2, for $\propto = -1.0$, $\delta = 0.1$ and $\delta = 0.001$ respectively. For the natural boundary condition, which simulated outflow, results are shown in Figures 3 and 4, for $\propto = -1.0$, $\delta = 0.1$ and $\delta = 0.001$ respectively. Three curves are shown on all plots: exact solution, conventional collocation (C^1 continuity), and the developed collocation (mixed C^0 and C^1 continuity), all at t = 20 seconds. The exact solutions at this time for $\propto = -1.0$ are [7]:

$$Z(X) = \frac{e^{-(X-23.75)^2/10}}{5} \qquad \delta = 0.1$$

$$Z(X) = \frac{e^{-(X-23.75)^2/2.08}}{1.04} \qquad \delta = 0.001$$

Figures 1 and 3 show that both collocation techniques were essentially identical and correlated with the exact solution, for the ratio $\propto/\delta = 10$.

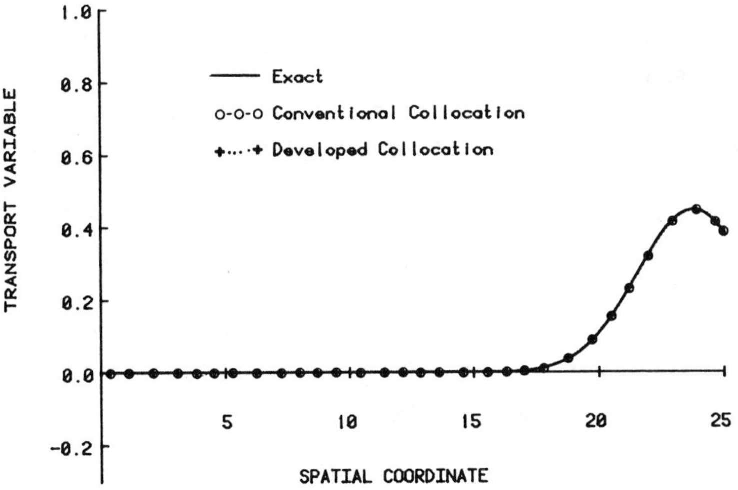

Figure 1 OUTFLOW BOUNDARY $a/\delta=10$

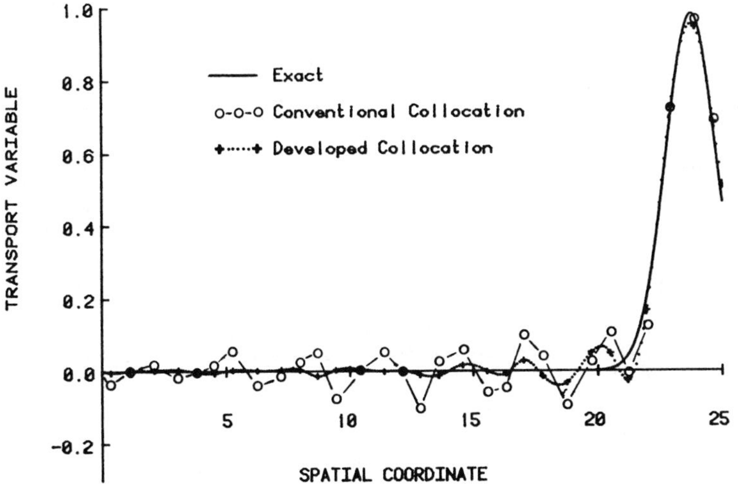

Figure 2 OUTFLOW BOUNDARY $a/\delta=1000$

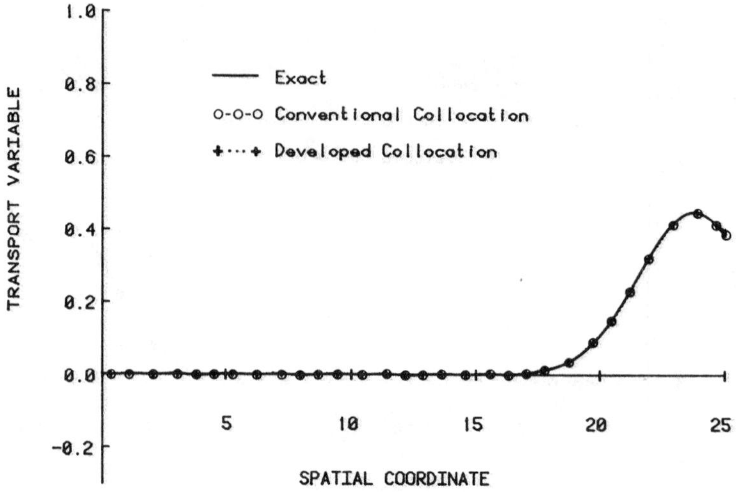

Figure 3 NATURAL BOUNDARY $\alpha/\delta=10$

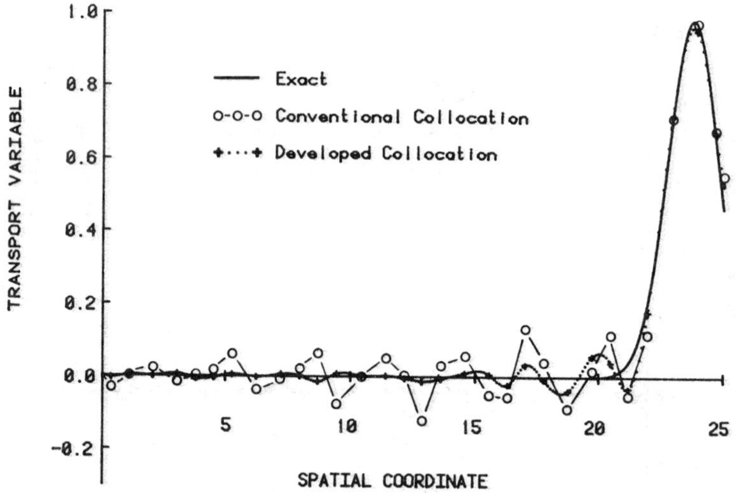

Figure 4 NATURAL BOUNDARY $\alpha/\delta=1000$

As the ratio α/δ increases, numerical oscillations increase which is characteristic of advective dominated flow, as can be seen by comparing Figures 1 and 3 to Figures 2 and 4. However, the developed collocation techniques displays significant reduction of these oscillations for both boundaries, as can be seen by comparing Figure 2 with Figure 4. Comparison to exact solution for the greater ratio, $\alpha/\delta = 1000$, shows reasonable correlation to the exact solution, e.g. little numerical diffusion, in the wave pulse region $20 < X < 25$. For the trailing region $X < 20$, the conventional technique displayed significant ringing relative to the exact solution, whereas the developed technique has to a large extent reduced this ringing. Note also that the developed technique did not introduce noticeable numerical diffusion for both cases. In other words, control of numerical ringing was not achieved at the cost of significant added diffusion.

Both sets of boundary conditions, outflow and natural, were selected to test the sensitivity of the techniques for simulating an outflow boundary, but neither technique offered an advantage over the other for either version. It should also be pointed out that no attempt was made for optimum time integration and that the time increment was 0.1, in all cases. The developed technique, however, required more cpu time in the ratio of 37/32.

4. CONCLUSIONS AND RECOMMENDATIONS

The developed collocation technique differs from previous multi-element collocation in that the newer technique effectively enforces C^0 interelement continuity for first order derivatives while retaining C^1 for second order derivatives. It appears that this newer multi-element collocation approach inherently reduces numerical oscillation or dispersion without an apparent link to conventional upwinding techniques and without numerical diffusion. This advantage would indicate the need for further study in nonconventional multi-element collocation, especially in three areas: underlying theoretical basis, i.e. Galerkin, least squares, etc. and possible links to upwind collocation (3) and cost effectiveness.

Future work will be directed toward applying multi-element collocation to two-dimensional Navier-Stokes, with interactive boundary conditions.

5. ACKNOWLEDGEMENT

The development of multi-element collocation was sponsored by the Ocean Technology Program of the Office of

Naval Research under contract N00014-78-C-0631. The participants in the 1982 Summer Research Apprentice Program are also acknowledged for their valuable assistance.

6. REFERENCES

1. MIGLIORE, H.J., and MCREYNOLDS, E.G. - Ocean Cable Dynamics Using An Orthogonal Collocation Solution. AIAA Journal, Vol. 20, No. 8, August 1982.

2. KERAMIDAS, G.A. - Finite Element Modeling for Convection-Diffusion Problems. NRL Memorandum Report 4255, Naval Research Lab, Washington, D.C., 1980.

3. PINDER, G.E., and SHAPIRO, A. - A New Collocation Method for the Solution of the Convection-Dominated Transport Equation. Water Resources Research, Vol. 15, No. 5, October 1979.

4. FINLAYSON, B.A. - Weighted Residual Methods and Their Relation to Finite Element Methods In Flow Problems. Finite Elements in Fluids, Volume 2, John Wiley & Sons, New York, 1975.

5. VILLADSEN, J. and MICHELSEN, M.L. -Solution of Differential Equation Models by Polynomial Approximation, Prentice-Hall, Englewood Cliffs, New Jersey, 1978.

6. CAREY, G. F. and FINLAYSON, B.A. - Orthogonal Collocation on Finite Element. Chemical Engineering Science, Vol. 30, 1975.

7. GRESHO, P.M., et. al. - On the Time-Dependent Solution of the Incompressible Navier-Stokes Equations in Two and Three Dimensions. prepared for Recent Advances in Numerical Methods in Fluids, Pineridge Press, Ltd., Swansen, U.K.

NUMERICAL COMPUTATION OF NATURAL CONVECTION IN VERTICAL DUCTS

Ole Melhus* and Bjørn F. Magnussen**

Division of Thermodynamics
Norwegian Institute of Technology, N-7034 TRONDHEIM-NTH

SUMMARY

The present paper describes a numerical procedure for steady, two-dimensional, elliptic computation of natural convection in vertical ducts. The flow field is directly calculated even though the velocity profile at the inlet of the calculating domain, and the mass flowrate, initially are unknown. Static pressure must be specified where fluid crosses the boundaries, either directly or as a function of the velocity at the boundaries. The procedure is well suited when solving complicated problems involving both conjugate heat transfer and recirculation. The method used is based on finite differences and the SIMPLE algorithm.

Calculations are done for ducts heated in three different ways. Some of the results are compared to experimental data, whilst others are compared to results obtained from other computational methods.

It is concluded that the elliptic method for calculation of natural convection in vertical ducts used in this paper, is better suited than the earlier used parabolic methods.

1. INTRODUCTION

1.1 Statement of the problem

Optimum design of heat transfer equipment consisting of vertical, parallel plates or vertical tubes involves computation of natural convection. The flow field and the temperature field must be simultaneously calculated due to the strong coupling between these fields. Unfortunately, the boundary conditions for the fluid flow does not seem to be sufficiently known. Both the velocity level and profile at the inlet (and also at the outlet) of the duct are unknown. Static pressure at the entrance is unknown too, but there the

*Associate Professor and **Professor

relationship between velocity and static pressure can be modelled by means of Bernoulli´s equation. At the outlet of the calculating domain static pressure is equalised to that of the surroundings.

Natural convection between parallel plates in heat transfer equipment usually involves conjugate heat transfer, i.e. computation of the heat transfer in both the fluid and the solid walls must be carried out simultaneously. Due to the conduction in the duct walls and radiation between the walls, the problem is elliptic. Sometimes the flow will be recirculating too. This implies that such problems have to be solved by an elliptic method.

1.2 Previous work

Numerical computation of natural convection in vertical ducts consisting of two parallel plates have traditionally been performed by means of methods based on parabolic equations. The velocity profile at the inlet has been prescribed, either as flat (rectangular) or parabolic. In earlier works, the pressure drop due to acceleration of the fluid was assumed to be zero both at the entrance and at the exit region, i.e. static pressure at inlet and outlet was equalized to that of the surrondings. In later works, the pressure drop due to acceleration is taken into account at the entrance region. The velocity level is found by a trial and error method.

It is not possible to solve complicated problems involving conjugate heat transfer and recirculation with a parabolic method. Therefore the method is applied to simple convection systems with either given wall temperature or given heat flux into the fluid flow. To the author´s knowledge, no works deal with complicated problems involving computation of the developing flow field and the temperature field in both the fluid and the walls.

1.3 Scope of present work

The scope of this work is to present an elliptic computation procedure for direct calculation (i.e. no trial and error) of natural convection in vertical ducts. In contrast to the use of parabolic methods, the present procedure avoids any specification of velocity profile and velocity level at the inlet prior to calculation. However, static pressure must be specified in one or another way where fluid crosses the boundaries.

Since the computational method used is elliptic, the present procedure is well suited when solving complicated problems involving both conjugate heat transfer and recirculation. When the domain of interest consists of both a fluid and a solid, the heat transfer for the whole domain can be calculated simultaneously.

The method used is based on finite differences and the SIMPLE algorithm, presented by D. B. Spalding and S.V. Patankar [19].

1.4 Contents of paper

Section 2 describes the mathematical model, section 3 the numerical solution procedure while section 4 presents results obtained with the present procedure. The various cases shown include only laminar flow. Section 5 deals with the conclusion.

2. MATHEMATICAL MODEL

2.1 Basic equations

The 2-dimensional, steady fluid flow in the duct consisting of two vertical, parallel plates is described by Navier-Stokes equations an the continuity equation in the following form:

$$\frac{\partial}{\partial x}(\rho u) + \frac{\partial}{\partial y}(\rho v) = 0 \qquad (1)$$

$$\frac{\partial}{\partial x}(\rho uu) + \frac{\partial}{\partial y}(\rho vu) = \frac{\partial}{\partial x}(\mu\frac{\partial u}{\partial x}) + \frac{\partial}{\partial y}(\mu\frac{\partial u}{\partial y}) - \frac{\partial p_d}{\partial x} + (\rho_\infty - \rho)g \qquad (2)$$

$$\frac{\partial}{\partial x}(\rho uv) + \frac{\partial}{\partial y}(\rho vv) = \frac{\partial}{\partial x}(\mu\frac{\partial v}{\partial x}) + \frac{\partial}{\partial y}(\mu\frac{\partial v}{\partial y}) - \frac{\partial p_d}{\partial y} \qquad (3)$$

p_d is the reduced local pressure, i.e. the difference between the actual local static pressure, and the local hydrostatic pressure in the surroundings, p_∞:

$$p_d = p - p_\infty \qquad (4)$$

$$\frac{\partial p}{\partial x} = \frac{\partial}{\partial x}(p_\infty + p_d) = \frac{\partial}{\partial x}(p_{ref} - \rho_\infty \cdot g \cdot x + p_d) = \frac{\partial p_d}{\partial x} - \rho_\infty \cdot g \qquad (5)$$

The heat transfer is described by the energy equation. When compressibility and dissipation effects are neglected, the energy equation can be written in a form similar to the N-S-equations:

$$\frac{\partial}{\partial x}(\rho uT) + \frac{\partial}{\partial y}(\rho vT) = \frac{\partial}{\partial x}(\frac{K}{c_p} \cdot \frac{\partial T}{\partial x}) + \frac{\partial}{\partial y}(\frac{K}{c_p} \cdot \frac{\partial T}{\partial y}) + \frac{S}{c_p} \qquad (6)$$

In (6), S is a general source term. Thermal properties like K, C_p and μ will be a function of temperature as long as laminar flow is considered. Density ρ is found from the equation of state.

2.2 Boundary conditions

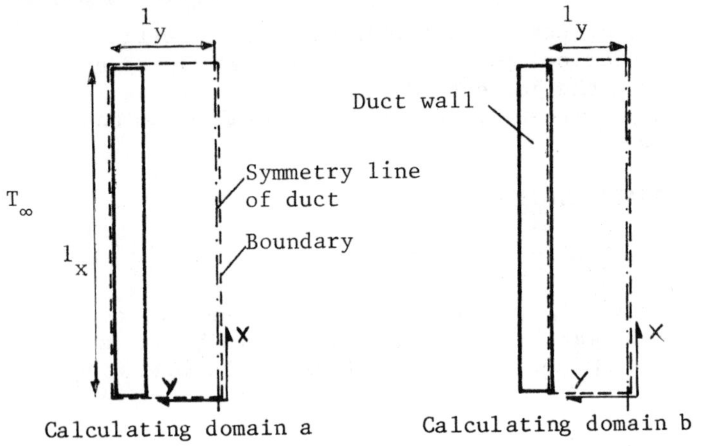

Fig. 1 Two actual calculating domains

At the inlet of the duct (x = 0):

$$\frac{\partial u}{\partial x} = 0 \qquad \frac{\partial v}{\partial x} = 0 \qquad T = T_\infty$$

$$p_d = -\rho \frac{u^2+v^2}{2} \tag{7}$$

At the outlet of the duct ($x = l_x$):

$$\frac{\partial u}{\partial x} = 0 \qquad \frac{\partial v}{\partial x} = 0 \qquad \frac{\partial T}{\partial x} = 0 \qquad p_d = 0 \tag{8}$$

At the symmetry line (y = 0):

$$\frac{\partial u}{\partial y} = 0 \qquad v = 0 \qquad \frac{\partial T}{\partial y} = 0 \tag{9}$$

Boundary value for static pressure is only necessary where fluid crosses the boundary.

At the wall ($y = l_y$):

$$u=0 \qquad v = 0 \tag{10a}$$

The boundary value for temperature will depend on the problem considered. Three different cases are shown:

1. Insulated wall: $\frac{\partial T}{\partial y} = 0$ (outer side, domain a) (10b)

2. Given temp. : $T = T_{wall}$ (inner side, domain b) (10c)

3. Given heat transfer coefficient : h_c (outer side, domain a) (10d)

3. NUMERICAL SOLUTION PROCEDURE

Finite-difference methods are used to solve the set of partial differential equations together with the boundary conditions. The calculation domain is divided into a finite number of control volumes each surrounding a grid point (fig. 2). For the velocity components u and v, staggered grids are used. For the velocity component u there is a "half control volume" at the inlet and outlet boundaries (fig. 3).

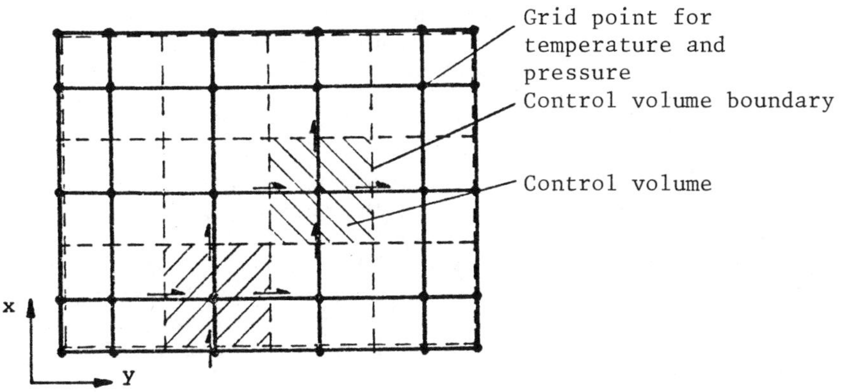

Fig. 2. Grid including control volumes for temperature and pressure.

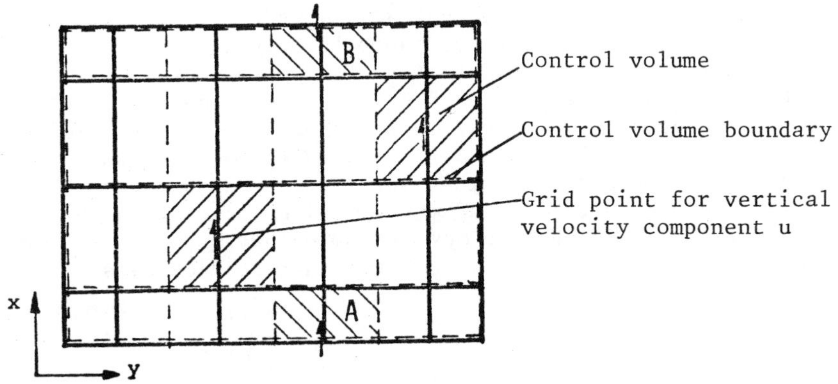

Fig. 3. Grid including control volumes for vertical velocity component u.

Each differential equation is integrated over control volumes by application of assumptions about the variations of the variables between grid points. This yields a set of non-linear algebraic equations which are solved by iteration.

Because of the half control volumes for velocity component u at inlet and outlet, the whole flow field including u at these boundaries is calculated, under the assumption of given static pressure and $\frac{\partial u}{\partial x} = 0$ at the boundaries. At the entrance, static pressure is calculated from eq. (7), where the values of u and v result from the preceding iteration.

The special problem of coupling between momentum and mass conservation equations is solved by use of the SIMPLE method given by Patankar and Spalding [19] . A more extensive description of the method used can be found in the thesis by O. Melhus [16].

4. COMPUTATIONS

The present elliptic procedure for computation of natural convection in vertical parallel plate channels is tested for ducts heated in three different ways.

4.1 Case 1

For a duct with uniformly distributed heat sources in the fluid, and with walls insulated on the outside, results are compared to results obtained by another elliptic computer program, written by S.V. Patankar [20]. However, that program is only suited for problems involving given velocity at the boundaries. The results are very similar when Patankar's program is given the velocity profile at inlet estimated by the present computer program.

4.2 Case 2

For a duct with isothermal walls the computed results are compared with experimental data [6,14]. In the case of ducts with large height/width ratios, there is very good agreement. The comparison is still satisfactory in the case of ducts with a moderate to small height/width ratio. The computed velocity level is however somewhat less than the measured value. This is because the flow in low and wide ducts is more influenced by uncertain boundary conditions for static pressure than the flow in high and narrow ducts, where wall friction is much more important. The results may indicate that static pressure at the outlet really is somewhat lower than static pressure of the surroundings.

4.3 Case 3

For a duct with a simple heating element, fig. 4, results are

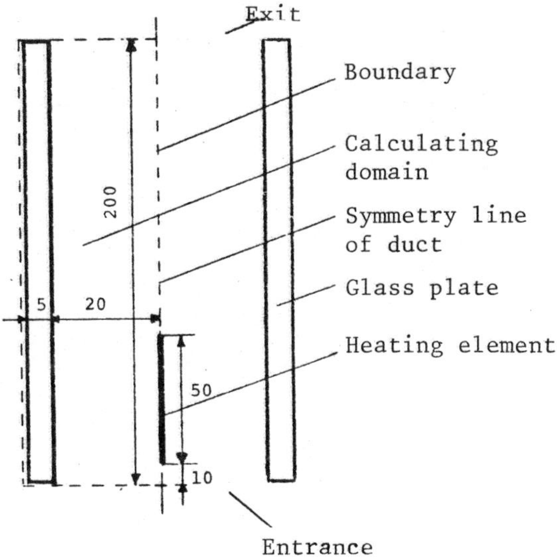

Fig. 4   Computational model for a duct heating element. Dimensions in mm.

compared to both experimental data and theoretical results obtained by a parabolic boundary layer computer program 11 .

Because of heat conduction in the walls, radiation between solid surfaces and possibly recirculation, the conditions in the duct can not be found by use of the parabolic procedure alone. The temperature on all surfaces inside the duct and the velocity profile at inlet are first computed by the present procedure, then this information is given as input to the parabolic program. The reuslts obtained in this way by the two computer programs, agree.

The velocity field was experimentaly investigated by use of a laser-doppler-anemometer. Comparison with computed values is shown in fig. 5. Just as in Case 2 the computed velocity level is somewhat less than the measured values.

The measured temperature profile on the outside of the duct wall is compared to calculations in fig. 6. Qualitatively there is agreement, but it is reason to believe that the measured temperature level is too high.

Fig. 7 and fig. 8 show the complete vertical velocity and temperature fields inside the duct (and in the glass wall too). These gives at a glance an impression of the flow condition in the duct. It is very interesting to discover that the resultant velocity profile at the entrance is almost rectangular, even though no assumption was made regarding the shape prior to calculation.

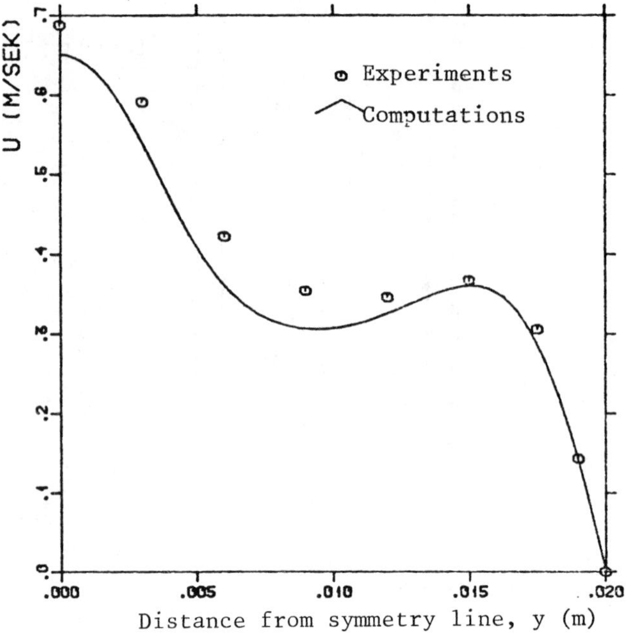

Fig. 5 Vertical velocity inside duct 120 mm from entrance.

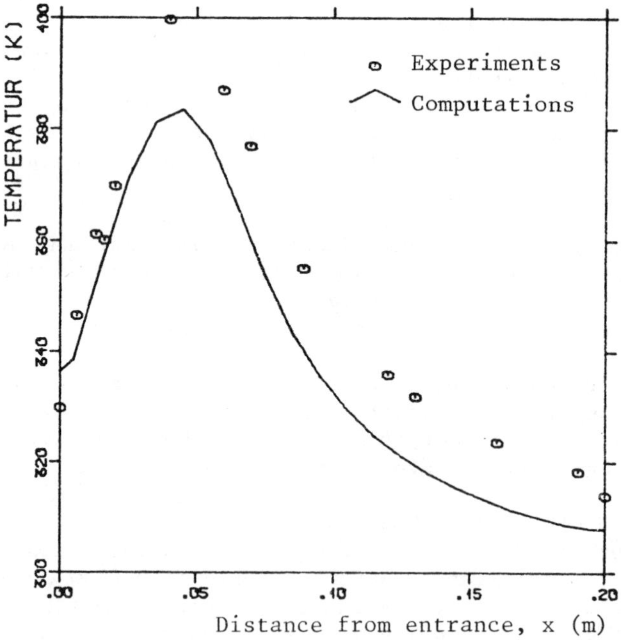

Fig. 6 Temperature profile on the outer side of the duct wall.

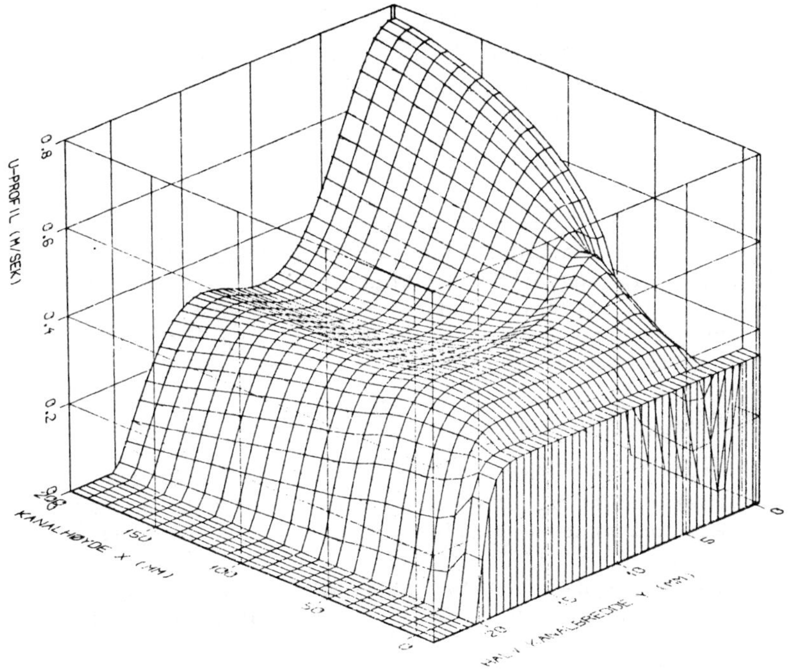

Fig. 7 Vertical velocity field inside duct with a flat heating element.

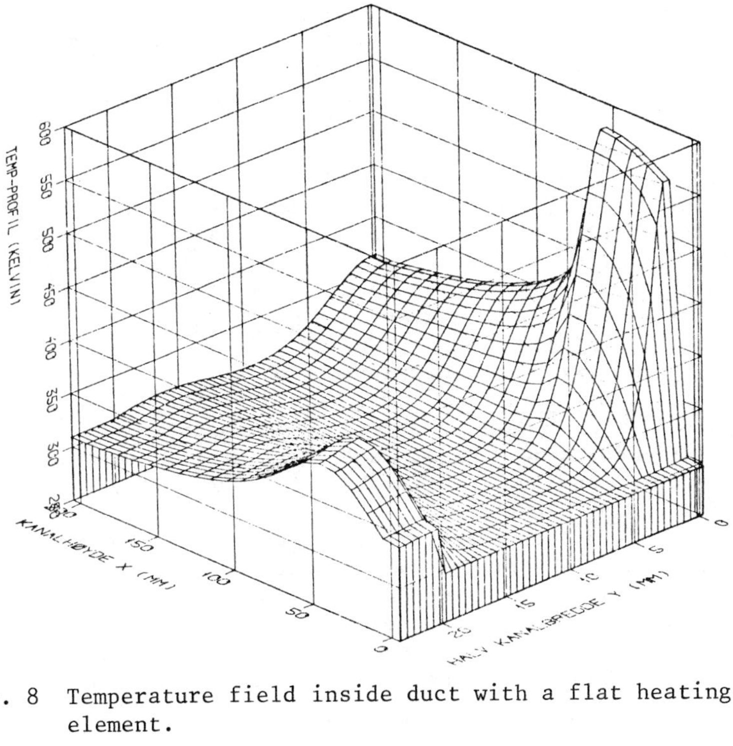

Fig. 8 Temperature field inside duct with a flat heating element.

4.4 Computational details

The computations for case 1 and case 3 were performed on a 16-bits computer, NORD 10/S (Norsk Data), while computations for case 2 were performed on a 32-bits VAX 11/780.

5. CONCLUSION

The previous sections have shoved that the elliptic method for calculation of natural convection in vertical ducts used in this paper, is better suited than the earlier used parabolic methods. The elliptic method is simpler, and thus faster and less expensive to use. Above all it is applicable for much more complicated problems.

REFERENCES

1. AIHARA,T. - Effects on inlet boundary conditions on numerical solution of free convection between vertical parallel plates. Rep.Inst.High Speed Mech. Tohoku Univ. 28 Ser. B, s. 1/27 (1973)
2. AUNG,W., FLETCHER,L.S., SERNAS,V. - Developing laminar free convention between vertical flat plates with asymmetric heating. Int. J. Heat Mass Transf. 15, s. 2293/2308 (1972)
3. BAYLEY, F.J., OWEN, J.M., TURNER,A.B., - Heat Transfer Nelson, England (1972)
4. BODOIA, J.R., J.F. OSTERLE, - The development of free convection between heated vertical plates. Trans. ASME 84, Ser. nr. 1, s.40/44 (1962)
5. CARPENTER,J.R., BRIGGS, D.G., SERENAS,V., - Combined radiation and developing laminar free convection between vertical flat plates with asymmetric heating. Trans ASME 98, Ser.C, nr.1., s.95/100 (1976)
6. CURRIE, I.G., NEWMAN, W.A. - Natural convection between isothermal vertical surfaces. Heat Transfer 4. Int. Konf. in Paris, bd. 4, Dusseldorf VDI-Verlag NC 2.7 (1970)
7. DYER, J.R. - Natural-convective flow through a vertical duct with a restricted entry. Int. J. Heat Mass Transfer, vol. 21, s. 1341/1354 (1978)
8. ELENBAAS, W. - Heat dissipation of heat by free convection. Physica 9, nr.1, s. 1/28 (1942)
9. GEBHART, B. - Heat transfer, 2. ed. McGraw-Hill (1971)
10. HJERTAGER, B.H. - Flow, heat transfer and combustion in three-dimensional rectangular enclosures. Thesis, Trondheim, NTH (1979)
11. HOVE, S. - Measurements and computation of natural convection in a vertical duct. Thesis,M.Sc., Division of Thermodynamics,NTH,Trondheim (1980)
12. HAALAND,S. - Private communication.
13. KETTLEBOROUGH, C.F. - Transient laminar free convection between heated vertical plates including entrance effects. Int. J. Heat Mass Transfer 15, nr. 5, s. 883/96

(1972)
14. KLAN, H. - Uber den Warmeubergang bei freier Konvection von Luft in beheizten senkrechten Kanalen. Diss. TH Darmstadt (1971)
15. KU, A. - A numerical model for the prediction of unsteady buoyant flows with application to a heated room corridor. Dissertation, Univ. Notre Dame, Indiana (1976)
16. MELHUS, O. - Numerical analysis of natural convection in vertical ducts. Thesis, Trondheim, NTH (1981)
17. MIYATAKE, O., FUJII, T. - Natural convection heat transfer between vertical parallel plates at unequal uniform temperatures. Heat Transfer-Jap. Res. 2, nr. 4, s. 79/88 (1973)
18. MORI, S., KAWAMURA, Y., TANIMOTO, A. - Conjugated heat transfer to laminar flow with internal heat source in a parallel plate channel. Canadian J. Chemical Eng., vol. 57, s. 698/703 (1979)
19. PATANKAR, S.V., SPALDING, D.B., - A calculation procedure for heat, mass and momentum transfer in three-dimensional parabolic flows. Int. J. Heat Mass Transfer, vol. 15, s. 1787/1806 (1972)
20. PATANKAR, S.V. - Computational heat transfer and fluid flow. Lectures at Division of Thermodynamics, NTH, (1977)
21. PATANKAR, S.V. - Numerical heat transfer and fluid flow. McGraw-Hill, USA (1980)
22. RAGNARSON, A. - Private communication.
23. SPARROW, E.M., PRAKASH, C. - Interaction between internal natural convection in an enclosure and an external natural convection boundarylayer flow. Int. J. Heat Mass Transfer, vol. 24, no. 5, s. 895/907 (1981)

FINITE ELEMENT ANALYSIS FOR LAMINAR FLOW WITH HEAT TRANSFER IN A SQUARE DUCT

M. Rahman
Department of Applied Mathematics
Technical University of Nova Scotia
P.O. Box 1000, Halifax, N.S.
Canada B3J 2X4

ABSTRACT

A numerical analysis is developed to predict the axial velocity profiles, the temperature distributions and the secondary flow behaviour in a square duct. The isoparametric eight noded finite element method is employed to solve the conservation equations describing the fully developed parabolic laminar flow with heat transfer. The numerical solutions substantially agree with both experimental results and analytic solutions.

1. INTRODUCTION

Temperature distribution within the rod bundle of a nuclear reactor is of major importance in nuclear fuel design. However, a knowledge of the hydrodynamic behaviour of the coolant is a prerequisite to the determination of the temperature distribution. The nuclear fuel elements composed of uranium oxide pallets, generally consist of rod bundles with the coolant (heavy water) flowing axially through the bundles in the space between the rods, thereby transporting thermal energy of the fuel to the heat exchangers and turbines which rotate to produce electricity. Therefore, a thorough understanding of the hydrodynamic behaviour of the coolant under normal conditions is extremely essential for the design of a fuel bundle. Heat transfer calculations form an important part in the design of such fuel elements, which can be carried out only if sufficient information about the velocity field is available.

Although in real situations, the flows in the reactor pressure tubes are highly turbulent, a theoretical study of the laminar flow case is the first step towards the evaluation of the effects of turbulence. In this study, we have considered the fully developed laminar parabolic flow with heat transfer in a square duct without the introduction of any fuel rods in the flow channel. This problem forms a basis in formulating the finite element technique to be applied to the more complicated pressure tube-fuel bundle geometry.

Sparrow and Loeffler [1] published the results of a theoretical study of laminar coolant flow through a parallel array of rods arranged in an equilateral triangular pattern. Siegel and Sparrow [2] have studied laminar forced-convection heat transfer in a flat duct with uniform heat flow at the walls. They have obtained empirical expressions for velocity and temperature distributions for fully developed flow situation. Then Sparrow, Loeffler and Hubbard [3] used the result as the starting point for a similar derivation of fully developed temperature distributions corresponding to the boundary conditions of uniform wall temperature in the circumferential direction and uniform wall heat flux along the axial direction. Later, Dwyer and Berry [4] repeated these calculations, obtaining the same results, and also extended the investigation to include the thermal boundary condition of an entirely uniform wall heat flux. These theoretical studies were concerned with laminar flow with heat transfer between the cylinders. Ramachandra and Spalding [5] studied the fully developed laminar flow and heat transfer behaviour in rod bundles having equilateral triangular arrangements by employign a non-orthogonal finite difference grid.

For laminar flow heat transfer in a horizontal square duct, the body forces due to the density differences are characterized by the Grashof number or more correctly by its product with the Prandtl number, the Rayleigh number. In real situations, when flowing fluid is heated in a horizontal duct, the fluid near the heated surfaces is warmer and, therefore lighter, than the rest of the fluid. This lighter fluid, thus, tends to flow upward along the heated surfaces. As a result, a secondary fluid motion is established that is, ideally, symmetrical to a vertical plane. This study takes into account the effect of the buoyancy forces on the secondary flow. Taylor and Hood [6] have made a remarkable comparative study of the use of the velocity and pressure variables versus use of the stream function and vorticity variables in simple flow problems. It was concluded that, to achieve a good accuracy, the velocity-pressure variable formulation required much less computational time than the stream function and vorticity variable formulation.

In the present investigation, the fully developed parabolic flow with heat transfer in a square duct without regard to rod bundles has been considered as a first step toward evaluating flow behaviour in the pressure tube-fuel bundle geometry. Analytic solutions of the axial velocity distribution and the temperature distribution have been obtained together with the volumetric flow rate and wall shear stresses.

The finite element method with eight noded isoparametric quadrilateral element has been employed in the present study. By using the Galerkin's weighted residual concept, the governing differential equations were discretized, and a set of matrix equations has been obtained. Computer results were obtained by FEM and analytical technique. The predicted values are compared with that of the experimental data and show excellent agreement.

2. MATHEMATICAL MODEL

We consider the steady, hydrodynamically and thermally developed incompressible laminar flow of a viscous Newtonian fluid in a square duct. If the boundary layer approximations are applied for this particular situation, the conservation equations may be written as follows:

Continuity Equation: $$\frac{\partial V}{\partial Y} + \frac{\partial W}{\partial Z} = 0 \qquad (2.1)$$

Momentum Equation in the axial, x-direction

$$V \frac{\partial U}{\partial Y} + W \frac{\partial U}{\partial Z} = -\frac{1}{\rho}\frac{\partial \bar{P}}{\partial X} + \nu\left[\frac{\partial^2 U}{\partial Y^2} + \frac{\partial^2 U}{\partial Z^2}\right] \qquad (2.2)$$

Momentum Equation in the y-direction

$$V \frac{\partial V}{\partial Y} + W \frac{\partial V}{\partial Z} = -\frac{1}{\rho}\frac{\partial P}{\partial Y} - g + \nu\left[\frac{\partial^2 V}{\partial Y^2} + \frac{\partial^2 V}{\partial Z^2}\right] \qquad (2.3)$$

Momentum Equation in the z-direction

$$V \frac{\partial W}{\partial Y} + W \frac{\partial W}{\partial Z} = -\frac{1}{\rho}\frac{\partial P}{\partial Z} + \nu\left[\frac{\partial^2 W}{\partial Y^2} + \frac{\partial^2 W}{\partial Z^2}\right] \qquad (2.4)$$

Energy Equation

$$U \frac{\partial T}{\partial X} + V \frac{\partial T}{\partial Y} + W \frac{\partial T}{\partial Z} = \alpha\left[\frac{\partial^2 T}{\partial Y^2} + \frac{\partial^2 T}{\partial Z^2}\right] \qquad (2.5)$$

where, U is the streamwise velocity, V and W are the transverse velocity components, X is the streamwise coordinate, Y, and Z transverse coordinates, T is the temperature, α is the thermal diffusivity, ρ is the density, μ is the dynamic viscosity, ν is the kinematic viscosity and P is the pressure.

The pressure field is assumed to be of the form:

$$P(X,Y,Z) = \bar{P}(X) + P(Y,Z) - \rho_b gY \qquad (2.6)$$

where g is the acceleration due to gravity. It is to be noted here that

$$\frac{\partial U}{\partial X} = 0, \quad \frac{\partial V}{\partial X} = 0, \quad \frac{\partial W}{\partial X} = 0, \quad \text{and} \quad \frac{\partial^2 T}{\partial X^2} = 0$$

follow from the assumption of a steady, fully developed velocity and temperature profiles.

The initial conditions at X = 0 are:

$$U(0,Y,Z) = U_0 \qquad T(0,Y,Z) = T_0$$
$$V(0,Y,Z) = 0 \qquad W(0,Y,Z) = 0 \quad . \tag{2.7}$$

The boundary conditions at the walls of the duct are:

$$U = V = W = 0, \text{ and } \quad T = T_w \quad . \tag{2.8}$$

The above governing equations are made dimensionless by introducing the following characteristic variables:

$$\theta = \frac{T - T_w}{T_0 - T_w} \qquad u = \frac{U}{b^2(-\frac{1}{\mu}\frac{dP}{dX})}$$
$$x = \frac{X}{L}, \quad y = \frac{Y}{b}, \quad z = \frac{Z}{b} \tag{2.9}$$
$$v = \frac{V}{V_c}, \quad w = \frac{W}{V_c}, \quad p = \frac{P}{P_c}$$

where, L is the characteristic axial length of the duct, 2b is the width of the duct, V_c is the characteristic velocity $(=\alpha/b)$, P_c is the characteristic pressure, T_w is the wall temperature and T_0 is the temperature at the entrance of the duct.

Using these dimensionless variables, the conservation equations (2.1) to (2.5) may be written as:

$$\frac{\partial v}{\partial y} + \frac{\partial w}{\partial z} = 0 \tag{2.10}$$

$$\frac{\partial^2 u}{\partial y^2} + \frac{\partial^2 u}{\partial z^2} = -1 \tag{2.11}$$

$$\frac{\partial^2 v}{\partial y^2} + \frac{\partial^2 v}{\partial z^2} = \lambda \frac{\partial p}{\partial y} - R_A\theta \tag{2.12}$$

$$\frac{\partial^2 w}{\partial y^2} + \frac{\partial^2 w}{\partial z^2} = \lambda \frac{\partial p}{\partial z} \tag{2.13}$$

$$\frac{\partial^2 \theta}{\partial y^2} + \frac{\partial^2 \theta}{\partial z^2} = u\,(\gamma \frac{\partial \theta}{\partial x}) \tag{2.14}$$

where,

$$\gamma = \frac{b^4(-\frac{1}{\mu}\frac{d\bar{P}}{dX})}{L\alpha}$$ is a dimensionless parameter in which

$\alpha = \frac{k}{\rho C_p}$ is the thermal diffusivity.

$$\lambda = \frac{P_c b}{\mu V_c} \qquad V_c = \frac{k}{\rho\, b\, C_p}$$

$$R_A = \text{Rayleigh number} = G_R \cdot P_R$$

$$G_R = \frac{g\, \beta\, \rho^2\, b^3\, (T_o - T_w)}{\mu} = \text{Grashof number}$$

$$P_R = \frac{\mu\, C_p}{k} = \text{Prandtl number}$$

$\beta = \frac{-1}{\rho} \left(\frac{\partial \rho}{\partial T}\right)_p$ = coefficient of thermal expansion at constant pressure

C_p = specific heat at constant pressure

k = thermal conductivity.

In terms of the stream function ψ and the vorticity ζ, the momentum equations (2.12) and (2.13) of the secondary flow may be written in nondimensional variables as

$$\nabla^2 \zeta = R_A \frac{\partial \theta}{\partial z} \tag{2.15}$$

$$\nabla^2 \psi = \zeta \tag{2.16}$$

where,

$$\nabla^2 = \frac{\partial^2}{\partial y^2} + \frac{\partial^2}{\partial z^2} \quad . \tag{2.17}$$

The velocity components (v,w) in the (y,z) directions are defined as

$$w = \frac{\partial \psi}{\partial z} , \qquad v = - \frac{\partial \psi}{\partial y} \tag{2.18}$$

and the vorticity is defined as

$$\zeta = \frac{\partial w}{\partial y} - \frac{\partial v}{\partial z} \quad . \tag{2.19}$$

The boundary conditions can be restated as follows:

$$\theta = u = v = w = \psi = 0 \qquad \text{at the duct walls} \tag{2.20}$$

$$\frac{\partial \psi}{\partial n} = 0 \qquad \text{at the duct walls} \tag{2.21}$$

$$\zeta = 0 \qquad \text{at the centerlines} \tag{2.22}$$

where,

$\frac{\partial}{\partial n}$ is the normal gradient with respect to the wall.

Initial conditions are:

$$u = u_0, \quad \theta = 1, \quad v = w = 0 \qquad \text{at } x = 0 . \tag{2.23}$$

3. ANALYTIC SOLUTIONS

The analytic solutions of the equations (2.11) and (2.14) may be obtained by the method of separation of variables and the concept of Fourier series and integrals. Mathematical developments in details can be found in the work of Rahman [7,8]. These solutions may be written as follows:

Axial Velocity

$$u = \frac{1}{2}(1-y^2) - \frac{16}{\pi^3}\sum_{n=0}^{\infty}\frac{(-1)^n}{(2n+1)^3} \cdot \frac{\cosh[(2n+1)\frac{\pi z}{2}]}{\cosh[(2n+1)\frac{\pi}{2}]} \cdot \cos[(2n+1)\frac{\pi y}{2}] \quad . \tag{3.1}$$

Temperature Solution

$$\theta = \theta_p - \frac{5B}{24} + \frac{B}{6\pi}(P)\int_{-\infty}^{\infty}[\sin\lambda\,(\frac{1}{\lambda} + \frac{3}{\lambda^3} - \frac{6}{\lambda^5}) - \cos\lambda(\frac{1}{\lambda^2} - \frac{6}{\lambda^4})] \cdot \frac{\cosh\lambda z}{\cosh\lambda} \cdot \cos\lambda y \; d\lambda \tag{3.2}$$

in which (P) stands for the Cauchy Principle Value of the integral, where

$$\theta_p = \frac{B}{4}(y^2 - \frac{y^4}{6}) - \frac{16\;By}{\pi^4}\sum_{n=0}^{\infty}\frac{(-1)^n}{(2n+1)^4} \cdot \frac{\cosh[(2n+1)\frac{\pi}{2}z]}{\cosh[(2n+1)\frac{\pi}{2}]} \cdot \sin[(2n+1)\frac{\pi y}{2}] \tag{3.3}$$

and

$$\gamma\frac{\partial\theta}{\partial x} = \text{constant} = B \quad .$$

4. GALERKIN'S FINITE ELEMENT METHOD

The finite element matrix equations corresponding to equations (2.11), (2.14), (2.15) and (2.16) are derived by using Galerkin's method and the resulting equations may, respectively, be written as

$$[K]\;\{u\} = \{R\} \tag{4.1}$$

$$[K]\;\{\theta\} = \{F\} - [S]\;\{u\} \tag{4.2}$$

$$[K]\;\{\zeta\} = -\,R_A\;[T]\;\{\theta\} \tag{4.3}$$

$$[K]\;\{\psi\} = -\,[G]\;\{\zeta\} \tag{4.4}$$

where

[K] = influence matrix

$$= \iint_D \left[\frac{\partial}{\partial y}[N]^T \frac{\partial [N]}{\partial y} + \frac{\partial [N]^T}{\partial z} \cdot \frac{\partial [N]}{\partial z}\right] dy\, dz \qquad (4.5)$$

{R} = column vector accounting for the driving force

$$= \int_s [N]^T X^* \, ds - \iint_D [N]^T c \, dy\, dz \qquad (4.6)$$

$$\{F\} = \int_s [N]^T y^* \, ds \qquad (4.7)$$

$$[S] = B \iint_D [N]^T [N] \, dy\, dz \qquad (4.8)$$

$$[T] = \iint_D [N]^T \frac{\partial [N]}{\partial z} \, dy\, dz \qquad (4.9)$$

$$[G] = \iint_D [N]^T [N] \, dy\, dz \qquad (4.10)$$

in which

$$X^* = \left[\frac{\partial u}{\partial y} \ell y + \frac{\partial u}{\partial z} \ell z\right] \qquad y^* = \frac{\partial \theta}{\partial y} \ell y + \frac{\partial \theta}{\partial z} \ell z \quad . \qquad (4.11)$$

The above equations may be assembled into a set of global equations of the form

$$[K^{(g)}] \{u^{(g)}\} = \{R^{(g)}\} \qquad (4.12)$$

$$[K^{(g)}] \{\theta^{(g)}\} = \{F^{(g)}\} - [S^{(g)}] \{u^{(g)}\} \qquad (4.13)$$

$$[K^{(g)}] \{\zeta^{(g)}\} = -R_A \, [T^{(g)}] \{\theta^{(g)}\} \qquad (4.14)$$

$$[K^{(g)}] \{\psi^{(g)}\} = - [G^{(g)}] \{\zeta^{(g)}\} \quad . \qquad (4.15)$$

This set of simultaneous equations is solved sequencially after introducing the appropriate boundary conditions. The computer results are depicted in Figures 1, 2 and 3.

5. RESULTS AND CONCLUSIONS

The computer codes have been developed using the iso-parametric finite element method to obtain the axial velocity profiles, the temperature distributions and the secondary flow behaviour in a square duct. Assuming the symmetric property of the flow, the positive quarter of the square duct has been considered and discretized into 100 elements with 341 nodal points. Figure 1 compares the predicted isotach with those of the experimental data of Leutheusser [9]. The finite element calculations of the temperature distributions are compared with the analytic solutions in Figure 2. The vorticity distributions along the wall y = 1 are compared in Figure 3 when the Rayleigh numbers, $R_A = 10^2, 10^3, 10^4, 10^5$ and 10^6. Favourable agreements between the FEM predictions with the analytic and experimental results have been found. The computations were performed on a CDC Cyber 170 computer at the Technical University of Nova Scotia in Halifax.

6. REFERENCES

1. SPARROW, E.M. and LOEFFLER, A.L. Jr., "Longitudinal Laminar Flow Between Cylinders Arranged in Regular Array" A.I.Ch.E. Journal, Vol. 5, No. 3, pp. 325-330, 1959.

2. SIEGEL, R. and SPARROW, E.M.," Simultaneous Development of Velocity and Temperature Distributions in a Flat Duct with Uniform Wall Heating",A.I.Ch.E. Journal, Vol. 5, No. 1, pp. 73-75, 1959.

3. SPARROW, E.M., LOEFFLER, A.L. Jr. and HUBBARD, H.A., "Heat Transfer to Longitudinal Laminar Flow Between Cylinders", J. Heat Trans. (ASME), Vol. 83, pp. 451-422, 1961.

4. DWYER, O.E. and BERRY, H.C., "Laminar Flow Heat Transfer for In-Line Flow Through Unbaffled Rod Bundles", Nucl. Sc. and Eng., Vol. 42, pp. 81-88, 1970.

5. RAMACHANDRA, V. and SPALDING, D.B., "The Numerical Prediction of Laminar Heat Transfer in Rod Bundle Geometries", Mech. Eng. Dept., Imperial College Report, HTS/78/4, May 1978.

6. TAYLOR, C. and HOOD, P., "A Numerical Solution of the Navier-Stokes Equations Using the Finite Element Technique", Computers Fluids, Vol. 1, No. 1, 1973.

7. RAHMAN, M., "Finite Element Analysis of Laminar Flow in a Square Duct", A Technical Report, Dept. of Applied Mathematics, Technical University of Nova Scotia, Halifax, March 1982.

8. RAHMAN, M., "Finite Element Analysis of Heat Transfer to Fully Developed Laminar Flow in a Square Duct", A Technical Report, Dept. of Applied Mathematics, Technical University of Nova Scotia, Halifax, June 1982.

9. LEUTHEUSSER, H.J., "Turbulent Flow in Rectangular Ducts", Proc. Am. Soc. Civ. Engg., Vol. 89, HY-3, pp. 1-19, 1963.

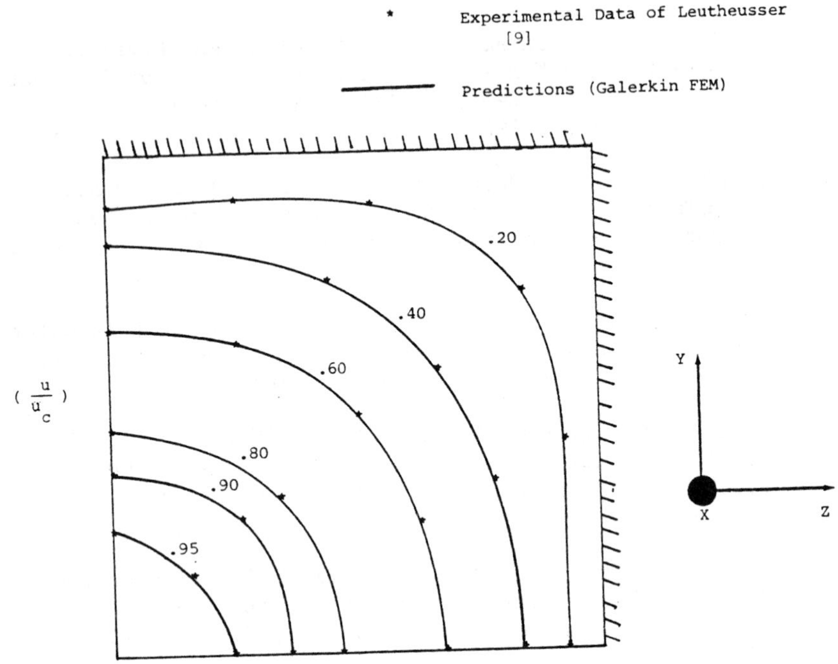

Figure 1: Laminar Veclocity Distribution ($\frac{u}{u_c}$) in a Square Duct (ISOTACH)

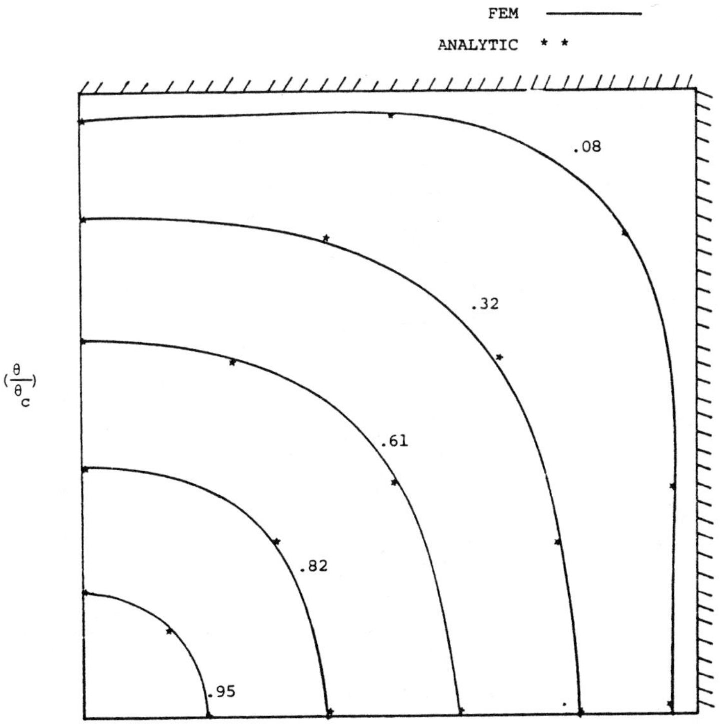

Figure 2. Temperature Distribution $(\frac{\theta}{\theta_c})$ in a Square Duct (Isotherms) for B = - 1.0.

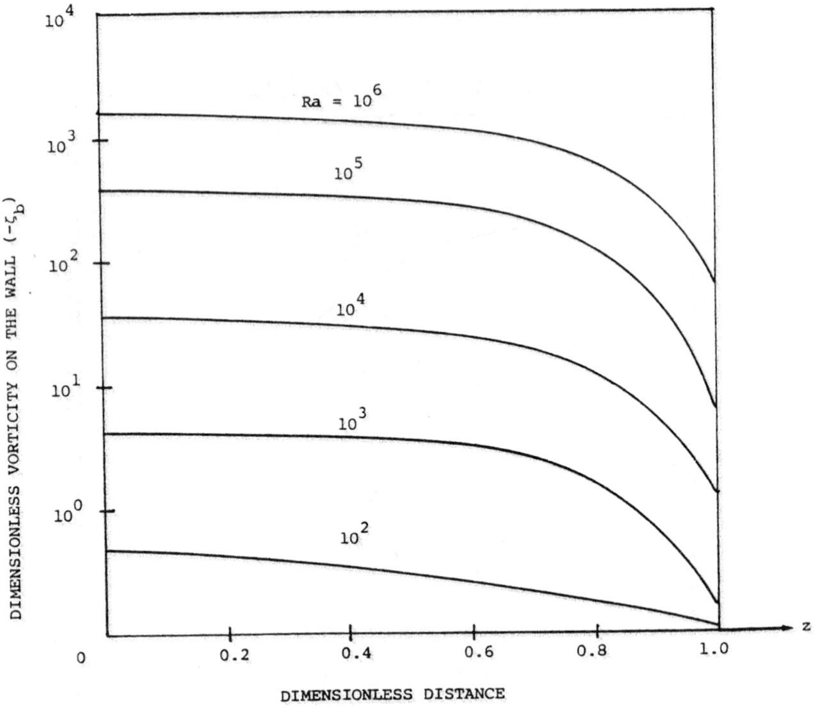

FIGURE 3: DISTRIBUTION OF WALL VORTICITY ALONG THE WALL y = 1.

A STUDY OF THERMAL INSTABILITY IN AN ATMOSPHERE WITH BASIC FLOW AND ROTATION

N. RAMANATHAN*

SUMMARY

In this investigation the constant heat flux hypothesis with respect to the formation of organized convection pattern with basic flow is studied. The basic flow is taken to be an antisymmetric polynomial with maximum shear at the ground with negative curvature in the entire convective layer. Viscous and coriolis effects are included. The hydrodynamic boundary conditions correspond to that of a free upper surface with a rigid lower boundary. The characteristics of thermal modes are determined in terms of the properties of the basic state and wave numbers of the superimposed perturbations. It is concluded that constant heat flux hypothesis is not suitable in atmospheric convection studies with a zonal flow. Its effect on the developed modes is stabilizing in nature. Basic flows with shear and curvature stabilizes the thermal modes more than flows with constant curvature. Longitudinal modes are the preferred mode of convection irrespective of thermal boundary conditions.

1. INTRODUCTION

Satellite observations have revealed that frequent occurrence of three dimensional cellular cloud patterns over oceans and cloud bands over land areas. Generally they are capped by an inversion layer and occur at a height of 1-2 km from the earth's surface. This study is concerned with the formation of such organized patterns with a basic flow in the convective layer. The studies for organised convection with basic flows possessing either linear or variable shear [1], [2], [5] have shown their inhibiting influence on the thermal unstable modes with preference to formation of convective streeting. However, in the above investigation either the shear or curvature of the basic flow is retained in the governing equations. An investigation including both the shear and curvature of the

* Senior Scientific Officer , Indian Institute of Technology New Delhi 110 016, India.

basic flow has not been attempted. By including coriolis force, momentum, energy and vorticity equations are coupled and this facilitates inclusion of shear and curvature simultaneously into the system of equations solved. This study adopts this approach. The inclusion of coriolis force is justified as in the atmosphere the organized convection patterns which consists of numerous cloud elements are maintained for a considerable period of time in contrast to single cumulus cloud whose life may be of few minutes only.

1.1 Outline of the present study and objectives

In this study the basic state velocity profile is taken to be a cubic polynomial an approximated form of the u component of baroclinic Ekmen profile. No unqiue formulation has been devised to describe the wind flow in an unstable baroclinic boundary layer due to the difficulty of defining boundary layer thickness. An assumption of infinitely deep boundary layer would be physically unrealistic. Hence, in this study the boundary layer thickness is taken to be a constant defined by the height of the inversion layer which caps off the organized convection. In the deduced basic flow profile the essential features are retained, namely, the shear decreases with height having its maximum at the ground as observed in the atmosphere. Moreover, the shear gradient is a function of height and is negative in the convective layer. Thus, the profile differs from those considered in previous investigations and appears to be in better agreement with observations. The neglect of the meridional velocity ($\bar{V}$) in the basic state is of no significance in this study as we are not concerned with the orientation of organized convection pattern with respect to basic flow. As Kuttener [5] has observed, 'convective cloud bands in the earth atmosphere tend to form in strong flows heated from below with curved velocity profile of rather uniform direction.' That is, the shear direction can be uniform. Thus, a study with plane parallel flow in the basic state is appropriate.

Further, in this study we have neglected diabatic heating effects such as latent heat release which is not a serious limitation as thermoconvective eddies arranged in regular patterns occur even in clear air [4].

Observations show that these patterns which occur with different aspect ratio (ratio of width to height) is a function of heat flux through the boundaries. Thus, attention to heat transfer on the bounding surfaces (known as thermal boundary conditions) is called for since through these surfaces heat exchange takes place to and from the convective layer. Sparrow et al. [6] have shown that in initially motionless liquids with constant flux heat boundaries convection can occur at lower Rayleigh numbers (i.e. at a weaker gravitational instability and the aspect ratio is a function of heat flux through the boundaries. The controlling influence of constant heat flux boundary conditions on the growth of perturbations in a flowing medium has not been explored. Thus, the objectives of this investigation are (1) to determine the influence of constant heat flux boundary conditions on the stability of the organized convection and (2) to study the stability of thermal modes under the combined influence of shear and curvature of the basic flow.

In the present investigation, perturbation technique is used to linearize the governing equations. Assuming the perturbations are periodic in the zonal and meridional directions the problem is reduced to an eigenvalue problem. Solutions are obtained using finite difference method.

2. MATHEMATICAL FORMULATION

As is customary in the study of convection, it is assumed the convective layer is bounded by horizontal flat surfaces separated by a distance H, the lower surface taken to be earth's surface with the upper surface coinciding with the base of inversion layer where the organized convection patterns appear. Further, it is assumed the fluid (atmosphere) in the convective layer is incompressible, viscous and infinite horizontal extension in x direction over a flat ground (topography neglected). The basic flow speed varies with height and not its direction. In the basic state we assume gravitationally unstable configuration and the temperature profile for the basic state is given by $T = T_o - \Gamma z$, where T_o is the temperature at the lower boundary and Γ is constant lapse rate. The processes which establish the necessary constant lapse rate in the convective layer are not modelled here. It is further assumed the coefficients of eddy viscosity and conductivity remain invariant throughout the medium. With the traditional linearization technique assuming $\bar{V} = \bar{W} = 0, U$ with its derivatives not equal to zero, we obtain the governing equations for perturbations as below,

$$\frac{\partial u}{\partial t} + \bar{U}\frac{\partial u}{\partial x} + W\frac{d\bar{U}}{dz} - fv = -\frac{1}{\rho_o}\frac{\partial p}{\partial x} + K_M \nabla^2 u \qquad (1)$$

$$\frac{\partial v}{\partial t} + \bar{U}\frac{\partial v}{\partial x} + fu = -\frac{1}{\rho_o}\frac{\partial p}{\partial y} + K_M \nabla^2 v \qquad (2)$$

$$\frac{\partial w}{\partial t} + \bar{U}\frac{\partial w}{\partial x} = -\frac{1}{\rho_o}\frac{\partial p}{\partial z} + g\alpha T + K_M \nabla^2 w \qquad (3)$$

$$\frac{\partial u}{\partial x} + \frac{\partial v}{\partial y} + \frac{\partial w}{\partial z} = 0 \qquad (4)$$

$$\frac{\partial T}{\partial t} + \bar{U}\frac{\partial T}{\partial x} + w\frac{d\bar{T}}{dz} = K_H \nabla^2 T \quad (5)$$

where u, v, w, are the velocities in x,y,z directions, f is coriolis parameter, ρ_o basic state density, K_H and K_M are the coefficients of eddy conductivity for heat and momentum, α is coefficient of thermal expansion, p pressure, T is temperature and barred variables represent the basic state. We have applied Boussinesq approximation in the third equation. Eliminating u,v,p the above set of equations are non-dimensionalized choosing the depth of convective layer (H) as characteristic length, the free stream velocity (U_*) as characteristic velocity and imposed temperature difference ($T^* = \Gamma H$) as characteristic temperature. Assuming the perturbations (u,v,w,ζ,T) proportional to (U,V,W,Z,θ) exp $(i(k_x x + k_y y) + \sigma t)$ the non-dimensional linearized perturbation equations can be written as

$$[\sigma + i K_x \bar{U} - Re^{-1}(D^2-k^2)](D^2-k^2)W - \frac{d^2\bar{U}}{dz^2} i k_x W + F\frac{dZ}{dz} + Ri\, k^2\theta = 0 \quad (6)$$

$$[\sigma + i k_x \bar{U} - Re^{-1} P_R^{-1}(D^2-k^2)]\theta - W = 0 \quad (7)$$

$$[\sigma + i k_x \bar{U} - Re^{-1}(D^2-k^2)]Z - i k_y W\frac{d\bar{U}}{dz} - F\frac{dW}{dz} = 0 \quad (8)$$

where $F = \frac{fH}{U_*}$, $P_R = \frac{K_M}{K_H}$, $Re = \frac{HU_*}{K_M}$, $Ri = \frac{g\alpha H T^*}{U_*^2}$,

$R_A = \frac{g\alpha H^3 T^*}{K_M K_H}$, $D^2 = \frac{d^2}{dz^2}$, $\sigma = \sigma_r + i\sigma_i$ (complex frequency)

$k^2 = k_x^2 + k_y^2$ k_x and k_y are the wave numbers in x and y directions. Since a constant lapse rate is assumed $-\frac{D\bar{T}}{dz}$ is taken to be unity in eqn. (7). The eqn. (8) is the vorticity equation obtained from the momentum equations. U,V,W,Z,θ are the complex amplitude functions of the perturbations. It is necessary to note that a relation $R_A = P_R \cdot Ri\, Re^2$ exists among the four dimensionless parameters. Since we are concerned here with unstable stratification the temperature at the lower boundary is higher than that at the upper boundary and Ri defined here has a sign opposite from the conventional one. With the coriolis force taken into account the above set of equations are coupled and form three simultaneous equations in W, θ and Z. Thus, to obtain the desired solution these equations must be solved simultaneously with relevant boundary conditions discussed below.

2.1 Hydrodynamic boundary conditions

Studies of this type differ in the nature of bounding surfaces (e.g.) smooth, rigid etc. The critical values for the onset of stationary convection are a function of hydrodynamic boundary conditions [3]. The assumption of symmetrical (both free or rigid) hydrodynamic boundary conditions at the bounding surfaces does not appear to correspond to any realistic physical situation in the atmosphere. The base of the inversion layer which forms the upper boundary in this investigation is taken to be that of free surface (i.e.) tangential viscous stress vanish. The lower boundary is taken to be rigid and coincides with the earth's surface. By this specification no slip occurs and all the velocity components, including perturbation quantities vanish.

2.2 Thermal boundary conditions

These depend on whether or not the boundaries are conducting, insulated or allow constant heat flux. The assumption of insulating surfaces on either one of the boundaries is not realistic since the inversion layer gets destabilized in course of time and the earth's surface is neither a good conductor or perfect insulator. Furthermore, this assumption would alter the temperature of the insulated surface drastically. Since such alterations are not desirable in this study the assumption of insulated surfaces need not be considered. The assumption of fixed temperatures on perfectly conducting boundaries would mean they have infinite thermal conductivity. Because the heat transfer coefficients are finite, the boundary temperature will not remain constant during convection process. Thus, a boundary condition of constant heat flux at both boundaries would be more realistic. For comparison purposes we have adopted perfectly conducting boundaries at both surfaces.

2.3 Empirical deduction of the velocity profile

With the above hydrodynamical boundary conditions, the basic state velocity profile in convection layer is deduced from the u component of baroclinic Ekman profile, since we essentially consider an unidirectional flow in the basic state. The u component of baroclinic Ekman profile neglecting higher order terms more than third power we obtain

$$\bar{U} = U_g \, [az - \frac{a^3 z^3}{2}] + bz$$

where a,b are constants. Our guess profile corresponding to this can be written as $\bar{U} = U_* \, (A X^3 + BX + C)$, where A,B,C are constants, $X = z/H$ and U_* is free stream velocity. Applying the assumed hydrodynamical boundary conditions we obtain

$$\bar{U} = U_* \left(\frac{3}{2} X - \frac{X^3}{2}\right) \tag{9}$$

This profile is taken to specify the basic state in our investigation and is a generalization of Couette flow. However, we do not propose any mechanism for the maintenance of the basic profile in the atmosphere.

The hydrodynamical boundary conditions for this study may be written as

$$W = 0, \quad \frac{dW}{dz} = 0, \quad Z = 0 \text{ at } z = 0 \tag{10}$$

$$W = 0, \quad \frac{d^2W}{dz^2} = 0, \quad \frac{dZ}{dz} = 0 \text{ at } z = 1 \tag{11}$$

First condition $W = 0$ arises due to assumption that there is no overshooting at the top and bottom of the convective layer (i.e.) at $z = 0$ and $z = 1$. The thermal boundary condition for constant heat flux becomes,

$$\frac{d\theta}{dz} = 0 \text{ at } z = 0 \text{ and } z = 1 \tag{12}$$

For perfectly conducting boundaries

$$\theta = 0 \text{ at } z = 0 \text{ and } z = 1 \tag{13}$$

The set of differential equations (6) - (8) are solved with the respective boundary conditions eqns. (10) - (12) or (13) using finite difference technique dividing the convective layer in n equal intervals. With constant temperature boundary conditions we obtain 3n-2 equations in 3n-2 unknown variables of W, θ and Z. With constant heat flux boundary conditions we obtain 3n linear equations. The additional two equations arise as θ does not vanish at the boundaries. The stability characteristics of the flow is determined from the eigenvalues obtained by the solution of the above equations cast in matrix form. Since convergence is obtained at $n = 12$ we have selected $n = 12$ in this study.

3. RESULTS AND DISCUSSION

The parameters which control the growth rate of the imposed perturbations are the depth of convective layer (H), wave numbers of the imposed perturbations (k_x, k_y) coefficients of eddy conduction and momentum (K_H,K_M), coriolis parameter (f) and the basic state velocity ($\bar{U}$). Coriolis parameter (f) is specified as 10^{-4} sec^{-1}. The depth of convective layer (H) is assumed to be 1 km in agreement with observations. In the absence of any definite conclusion regarding the values of K_H and K_M they are treated as constant, isotropic and equal.

Thus, the Prandtl number P_R is taken to be unity which is strictly valid in neutral atmosphere. Since the lower region of the atmosphere remains turbulent and as we have chosen a lapse rate quite close to the adiabatic value choosing a Prandtl number with the value of unity is justified.

The Rayleigh number which denotes the strength of convection or temperature difference between the bounding surfaces is taken to be 10^{+4} which yields a super adiabatic lapse rate quite close to the adiabatic value in the range of values of parameters taken. The parameters which are varied during the investigation are the wave numbers of imposed perturbations (k_x and k_y) and vertical shear (Ri) of the basic flow. The variation of k_x, k_y or both specifies the geometry of the cells and wavelength of imposed perturbations. In the presented results $k_x = k_y$ is assumed. The wavelength of imposed perturbation $2\pi/k$ ranges from 1 - 20 km. The Richardson number is varied between 0.2 to 2. The influence of the thermal boundary conditions are determined by comparison of results obtained with conducting boundaries.

(a) Comparison of stability characteristics under different thermal boundary conditions

The amplification rates (σ_r) of the developed unstable modes with horizontal wave number k for different shears (Ri) in the basic flow are shown in figures 1 and 2. From the diagrams, it can be seen that there exist two cut off limits for the onset of instability one at a small wave number and the other at a large wave number (k). There is a threshold limit on Richardson number Ri for the onset of instability below which the current is rendered stable. With constant heat flux boundary conditions this limit is close to Ri = 0.5 (Fig. 2). However, with conduction boundaries the current is rendered stable only at Ri = 0.25 (i.e.) at a higher shear. In other words, the wave number of maximum instability does not decrease without limit with the decrease of Ri or increase in the baroclinicity of the flow. The shortest wavelength, below which the current is stable, corresponds to cumulus wavelength (1 km) while the largest corresponds to mesoscale range (20 km). Table 1 shows a comparison of growth rates (σ_r) and wave speeds ($C_r = \sigma_i/k_x$) of the unstable modes with different thermal boundary conditions for a typical value of Ri. It is clear that the amplification rate of unstable modes with conducting boundaries is always more than with constant heat flux conditions results. The results of previous investigations have shown that conducting boundaries provide a stronger constraint against perturbations of the temperature profile than does the conditions on the temperature deviation at the surface.

k	(a) σ_r	(a) C_r	(b) σ_r	(b) C_r
0.5	0.017	0.671	0.014	0.714
1.0	0.103	0.674	0.100	0.718
2.0	0.235	0.683	0.213	0.708
3.0	0.232	0.700	0.189	0.718
5.0	0.023	0.803	stable	

Table 1 : A typical comparison of stability characteristics of the unstable modes with (a) constant temperature (b) constant heat flux boundary conditions for different values of wave number (k) with Ri = 1.0.

The above results apply only to initially motionless media. However, the results of this study shows that in the presence of an initial zonal flow the constant heat flux hypothesis stabilizes the thermal modes but with increased phase speed as shown in Table 1. Further, in both cases the unstable modes travel with a speed greater than the velocity of the basic flow averaged over the entire depth of convective layer ($C_r > 0.5$). Viscous induced unstable waves can arise with the assumed boundary conditions (i.e.) rigid lower surface. However, it is verified, in the range of parameters considered, the unstable modes obtained are of thermal origin modified by shear/shear gradient and rotation.

From Fig.2 we notice, the highest wavelength, that develops instability decreases as shear increases, in the basic current (see Ri = 0.5 curve). This is in contrast with the results obtained with conducting boundaries. However, from Fig.2 one can note that the highest wavelength an unstable mode can attain is of the order of 20 km while with conducting boundaries we see that the highest unstable wavelength is 12 km (see L.H.S. of the figures). Moreover, the preferred perturbation which has maximum growth rate is of the order of 6 km with conducting boundaries which decreases to 4.5 km with constant heat flux hypothesis. Thus, results of this study show the conducting boundaries are more suitable for atmospheric convection studies than with constant heat flux hypothesis assumption. The slight increase in the growth rates obtained at higher wavelengths at lower shears is nullified by a decrease in wavelength as shear increases. Since in the atmosphere there is an ever-present zonal flow and associated vertical shear with its gradient, it appears certain that the mesoscale convection patterns attain maximum growth due to conditions other than constant heat flux at the boundaries. Thus, the theories which explain the aspect ratio is meso-scale convection patterns with constant heat flux hypothesis become of an academic interest only, while considering the zonal motion in the atmosphere. An increase in Rayleigh number does not essentially alter the above results. Thus an increase in degree of stratification alone may not be sufficient. However, in media without the basic flow the constant heat flux hypothesis at the boundaries allow convection to occur at weaker gravitational stability. As shown in Fig. 2 at very little Ri unstable modes appear in higher wavelengths with considerable growth rates rather than with conducting boundaries case. With zonal motion and constant heat flux boundary conditions the maximum unstable wavelength is of the order of 4 km. In case of conducting boundaries the preferred perturbation is 6 km wavelength. It is certain in the presence of zonal flow with shear/shear gradient gravitational convection can occur in meso-scale wavelength irrespective of thermal boundary conditions. However, the growth rates remain quite small.

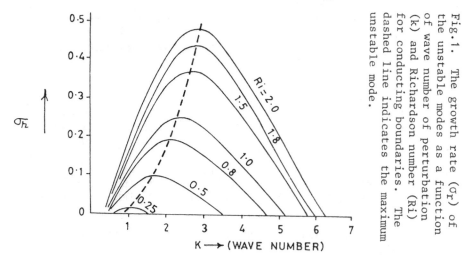

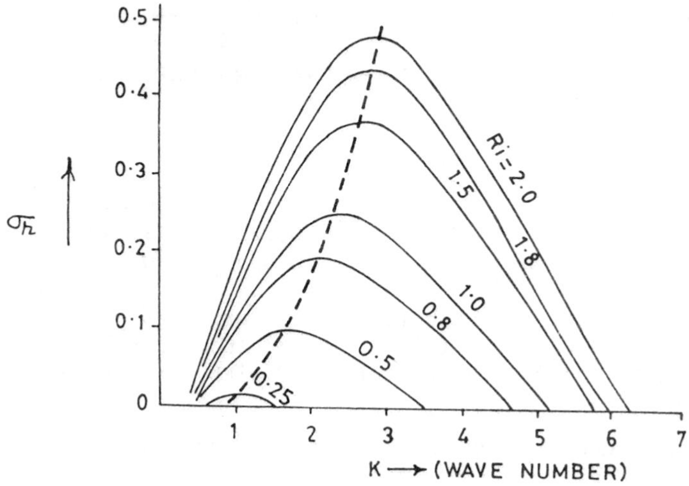

Fig.1. The growth rate (σ_r) of the unstable modes as a function of wave number of perturbation (k) and Richardson number (Ri) for conducting boundaries. The dashed line indicates the maximum unstable mode.

(b) Comparison of constant shear (Couette) flow with power law (cubic) profile flow

As mentioned before, neglecting cubic power term in our deduced basic state profile we arrive at the profile for Couette flow. For comparison purposes, the constant (3/2) in the first term is also dropped out. The stability characteristics of the unstable modes are compared in Table 2, given below, for Ri = 1.0. Due to asymmetric hydrodynamic boundary condition we obtain a single unstable mode, taken to be stationery unstable perturbation, for the basic flow with constant shear.

From the above table we find that the unstable mode with constant shear in the basic state travels with a velocity close to the velocity found midway between the two boundaries while the phase speed of unstable mode obtained with the cubical profile moves with a velocity greater than the velocity averaged over the entire layer. The amplification rates ($_r$) are larger than that obtained with a Couette flow.

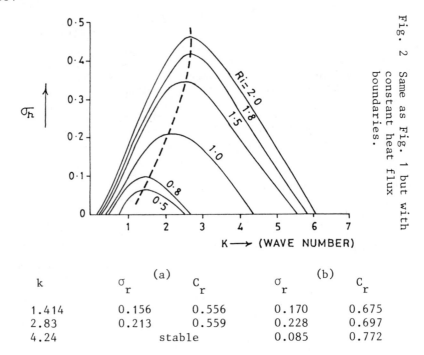

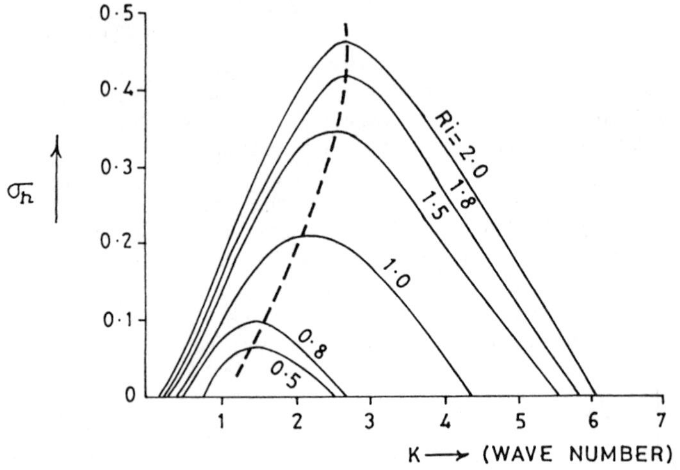

Fig. 2 Same as Fig. 1 but with constant heat flux boundaries.

k	(a) σ_r	(a) C_r	(b) σ_r	(b) C_r
1.414	0.156	0.556	0.170	0.675
2.83	0.213	0.559	0.228	0.697
4.24	stable		0.085	0.772

Table 2: Comparison of the stability characteristics of constant shear flow (a) with cubic (power law) velocity profile (b) for Ri = 1.0 with conduction boundaries

(c) Comparison of constant curvature flow with cubical profile flow

To consider the stability characteristics of thermal modes due to curvature of the flow alone we have chosen the following velocity profile for the basic flow

$$\bar{U} = U_* \, [X - X^2] \text{ where } X = z/H \tag{14}$$

It is necessary to note the averaged or net shear of this basic flow in the entire conductive layer is zero while the curvature is constant and negative in the entire convective layer. The growth characteristics of the unstable modes is compared in Fig. 3, with the cubical profile flow taken in this study with conducting boundaries. It is clear from the Fig. 3 that constant curvature flow has less inhibiting influence on the developed modes than the flow with varying curvature and shear in the entire convective layer.

(d) Changes with geometry of modes

The growth rate (σ_r) of the unstable thermal modes for different values of Ri and k_x/k_y are shown in Fig.4 with constant heat flux boundaries. $k_x/k_y \gg 1$ denotes a longitudinal mode while $k_x/k_y \ll 1$ specifies a transverse mode. From the Fig.4 it is clear irrespective of the value of Ri the growth rate of longitudinal modes are higher than transverse modes. From this we conclude irrespective of thermal boundary conditions longitudinal modes or convection bands oriented in the direction of basic flow are the preferred type of convection. A similar result is obtained (not shown here) with conducting boundaries.

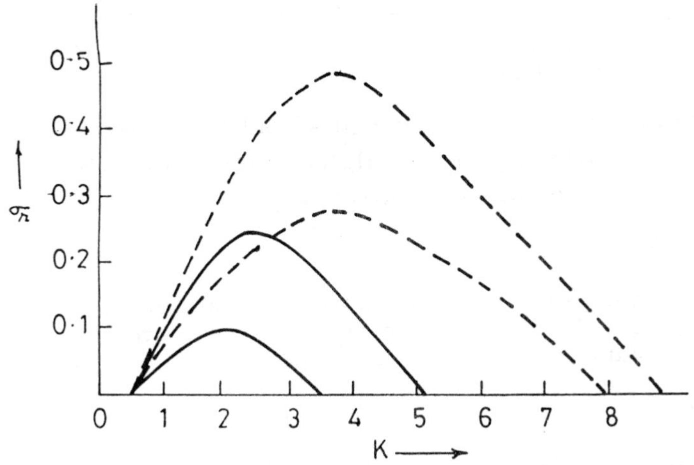

Fig.3. Growth rates (σ_r) comparison of a constant curvature flow (Dashed line) with cubical profile (Thickline) flow used in this study for different Ri with conducting boundaries.

4. CONCLUSIONS

Constant heat flux boundary conditions with basic flow in the convective layer stabilizes the thermal modes. This is in contrast with the results obtained in an initially motionless media. Basic flow with shear and curvature stabilizes the thermal modes more in comparison with a basic flow with a constant curvature and zero average shear in the convective layer. Irrespective of thermal boundary conditions longitudinal perturbations are the preferred mode of convection.

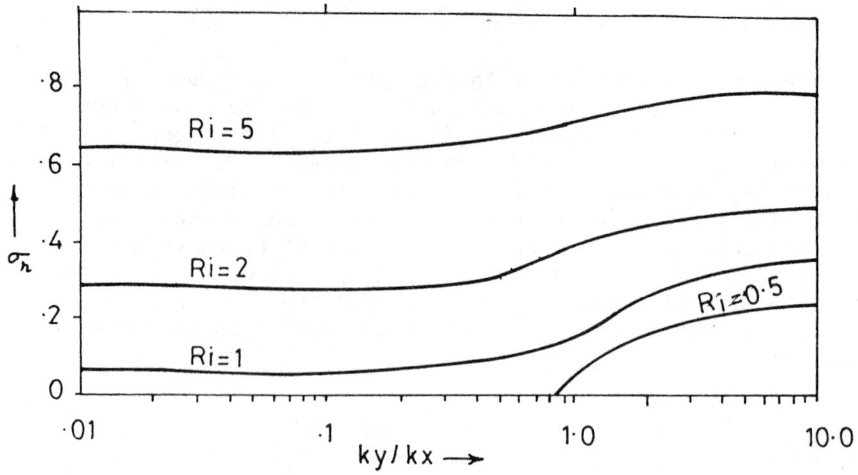

Fig.4. Variation of amplification rates (σ_r) for different ratios of ky/kx. ky/kx ⪌ 1. Refer longitudinal, square and transverse modes respectively.

5. ACKNOWLEDGEMENTS

The author expresses his deep gratitude to his Ph.D. thesis adviser Prof. R.L. Inman, for his guidance, encouragement and support during his studies in U.S.A.

6. REFERENCES

[1] ASAI, T. - Three dimensional features of thermal convection in a plane Couette flow. J. Met. Soc. Japan, vol. 48, pp. 18-29, 1970a.

[2] ASAI, T. - Stability of a plane parallel flow with variable shear and unstable stratification, Ibid, Vol. 48, pp. 128-139, 1970b.

[3] JEFFREYS, H. - Some cases of instability in fluid motion Proc. Roy. Soc. A. Vol.118, pp. 195-208, 1928.

[4] KONRAD, T.G. - The alignment of clear air convective cells, Proc. Intl. Conf. Cloud Physics, Toronto, Canada pp. 539-543, 1968.

[5] KUTTENER, J.P. - Cloud bands in the earth's atmosphere, Tellus, Vol. 23, pp. 404-425, 1971.

[6] SPARROW, E.M. et al - Thermal instability in a horizontal fluid layer; Effect of boundary conditions and non-linear temperature-profile, J. Fluid Mech., Vol.28, pp. 513-538, 1964.

NATURAL CONVECTION BOUNDARY LAYER OF A RADIATING GAS ALONG A VERTICAL PLATE

A. R. BESTMAN

Mathematics Programme, Federal University of Technology, Bauchi, Nigeria.

SUMMARY

This paper studies the boundary layer development along a vertical porous flat plate maintained at a constant temperature. It is assumed that the temperatures involved are large enough for both convection and radiation effects to be significant. Two problems are analysed. First the asymptotic suction profile valid at large distances from the tip of the plate is discussed. Here an example is given on the dominance of viscous dissipation on natural convection flow fields of large size. Next, the developing boundary layer flow from the tip of the plate is studied. In both cases solutions are effected either by straightforward numerical integration or by asymptotic expansion followed by numerical integration. The temperature characteristic of the flows is discussed quantitatively.

1. INTRODUCTION

It is now a known fact that the effect of viscous dissipation play a significant role in natural convection flow fields prevalent in a variety of physical and technological situations. A model study of this problem was undertaken by Gebhart[1], and much of the earlier works done in this field are enumerated in this paper.

The above studies do not take into account heat transfer by radiation, which is significant when we are concerned with space applications and high operating temperatures. Such applications form the primary motivation of this paper. The model deployed is a vertical semi-infinite porous hot flat plate in fluid of constant properties except for the density which varies only in the buoyancy force term. This is the celebrated Boussinesq approximation.

Two problems are analysed. The first is the asymptotic suction profile, far away from the edge of the plate, where flow variables vary with only the transverse coordinate. For a hot plate in the presence of a hot gas, the Cogley et al[2] differential approximation for a radiating optically thin non-grey gas near equilibrium may be adopted. This example highlights the importance of viscous dissipation heat in natural convection flow fields of large size. In the case of arbitrary wall and freestream temperatures the Rosseland's differential approximation (see Vincenti and Kruger[3]) for an optically thick gas is convenient in representing the radiation term.

The second problem studies the developing boundary layer from the tip of the plate. Similarity solution is possible for an optically thick gas. The proceedure adopted below is as follows.

In section 2 the boundary layer equations and the asymptotic suction solutions are presented for both optically thin and thick gases. The similarity problem of the developing boundary layer for an optically thick gas is discussed in section 3. Finally in section 4 the temperature profiles are discussed quantitatively.

2. EQUATIONS OF MOTION AND ASYMPTOTIC SUCTION PROFILE

The equations of motion, in the light of boundary layer approximations, are

$$\frac{\partial u}{\partial x} + \frac{\partial v}{\partial y} = 0, \tag{1}$$

$$u\frac{\partial u}{\partial x} + v\frac{\partial u}{\partial y} = \nu\frac{\partial^2 u}{\partial y^2} + g\beta(T-T_\infty) \tag{2}$$

$$u\frac{\partial T}{\partial x} + v\frac{\partial T}{\partial y} = \kappa\frac{\partial^2 T}{\partial y^2} + \frac{\nu}{C_P}\left(\frac{\partial u}{\partial y}\right)^2 - \frac{1}{\rho_\infty C_P}\frac{\partial q_y}{\partial y}\,. \tag{3}$$

Here y is perpendicular to the plate and x parallel to it. Subscript ∞ refers to condition at infinite where the fluid is at rest, $\kappa = k/\rho_\infty C_P$ is the diffusivity and q_y is the component of radiative flux. In this paper we consider two differential approximations for the radiative flux of the form

$$\frac{\partial q_y}{\partial y} = 4(T-Tw) \int_0^\infty \left(\alpha_\lambda \frac{\partial B_\lambda}{\partial T}\right)_w d\lambda \tag{4}$$

for an optically thin non-grey gas near equilibrium (see Cogley et al[2]), and

$$q_y = -\frac{16\sigma}{3\alpha} T^3 \frac{\partial T}{\partial y} \tag{5}$$

which is the Rosseland's approximation for an optically thick grey gas. B_λ is the Planck's function where λ is the frequency, α_λ is the absorption coefficient and σ is the Stefan-Boltzmann constant. When α_λ is independent of λ, we designate it by α and it will be assumed to be constant in equation (5). Helliwell and Mosa[4] adopted a power law dependence of the absorption coefficient upon the temperature and comparison with constant absorption coefficient show very little difference in results. The boundary conditions for the problem are

$$u=o,\ v=-V(x),\ T=Tw \text{ on } y=o, \tag{6}$$

$$u=o,\ T=T_\infty \text{ as } y\to\infty,$$

and subscript 'w' stands for conditions at the wall.

Far away from the leading edge of the plate, the flow is fully developed and all dependence on x disappears. Thus $v=-V_o \equiv$ constant from (1), and introducing the non-dimensional quantities

$$Y=\frac{V_o y}{\nu},\quad U=u/V_o,\quad P_R=\frac{\mu C_P}{k},$$

$$\Theta=\begin{cases}\dfrac{T-T_\infty}{Tw-T_\infty}\\[2ex] \dfrac{T}{T_\infty}\end{cases},\quad E=\begin{cases}\dfrac{V_o^2}{C_P(Tw-T_\infty)}\\[2ex] \dfrac{V_o^2}{C_P T_\infty}\end{cases},\quad GR=\begin{cases}\dfrac{\beta g\nu(Tw-T_\infty)}{V_o^3}\\[2ex] \dfrac{\beta g\nu T_\infty}{V_o^3}\end{cases},\quad N=\begin{cases}\dfrac{4\nu^2}{kV^2}\displaystyle\int_0^\infty \left(\alpha_\lambda\frac{dB_\lambda}{dT}\right)_w d\lambda, & \text{thin gas}\\[2ex] \dfrac{16\sigma T_\infty^3}{3\alpha k}, & \text{thick gas}\end{cases}$$

we have

$$\frac{d^2U}{dY^2} + \frac{dU}{dY} + G_R\Theta = o,$$

$$\frac{d^2\Theta}{dY^2} + P_R\frac{d\Theta}{dY} + PrEc\left(\frac{dU}{dY}\right)^2 - N(\Theta-1) = o, \qquad (7)$$

$U=o$, $\Theta=1$ on $Y=o$, $U=o$, $\Theta=o$ as $Y\to\infty$,
for the thin gas, and

$$\frac{d^2U}{dY^2} + \frac{dU}{dY} + G_R(\Theta-1) = o,$$

$$\frac{d^2\Theta}{dY^2} + P_R\frac{d\Theta}{dY} + P_REc\left(\frac{dU}{dY}\right)^2 + N\frac{d}{dY}\left(\Theta^3\frac{d\Theta}{dY}\right) = o, \qquad (8)$$

$U=o$, $\Theta=\Theta w$ ($\equiv Tw/T_\infty$) on $Y=o$,
$U=o$, $\Theta=1$ as $Y\to\infty$,
for the thick gas.

2.1 Solution for thin gas limit

Equation (7) is valid for the optically thin non-grey gas near equilibrium if the difference between Tw and T_∞ is small. Now for all incompressible fluids $Ec<<1$; but if Ec is set equal to zero in the second equation of system (7), no solution to these system of equations exist that satisfy the boundary conditions of the system. However the full nonlinear system can be studied by considering the iterated solutions

$$U^{(i+1)''} + U^{(i+1)'} + G_R\Theta^{(i+1)} = o,$$

$$\Theta^{(i+1)''} + P_R\Theta^{(i+1)'} - N\Theta^{(i+1)} + 2P_REcU^{(i)'}U^{(i+1)} = \Theta^{(i)''}$$

$$+ P_R\Theta^{(i)'} - N\Theta^{(i+1)} + 2P_RE_CU^{(i)}U^{(i)'}, \qquad (9)$$

$i=0,1,2,\ldots\ldots,$
and a dash denote differentiation with respect to the argument.

To solve (9), we put
$y_o=o;\ y_j=y_{j+1}+l,\ j=1,2,3,\ldots J;\ y_J=\infty.$

If we write the function $\Theta(y_j)$, say, as
$\Theta(y_j) = \Theta(j)$,
equations (9) can be expressed in the finite difference form

$$(2+l)U^{(i+1)}(j+1) - 4U^{(i+1)}(j) + (2-l)U^{(i+1)}(j-1) + 2G_R\cdot l^2\Theta^{(i+1)}(j) = o,$$

$$(2+P_R l)\Theta^{(i+1)}(j+1)-4\Theta^{(i+1)}(j)+(2-Prl)\Theta^{(i+1)}(j-1)$$
$$-2Nl^2\Theta^{(i+1)}(j)+2P_RE_Cl[U^{(i)}(j+1)-U^{(i)}(j-1)] \qquad (10)$$
$$U^{(j+1)}(j)=(2+Prl)\Theta^{(i)}(j+1)-4\Theta^{(i)}(j)+(2-Prl)\Theta^{(i)}(j-1)$$
$$-2Nl^2\Theta^{(i)}(j)+2P_RE_Cl[U^{(i)}(j+1)-U^{(i)}(j-1)]U^{(i)}(j),$$

subject to the conditions

$$U^{(i+1)}(o)=o,\ \Theta^{(i+1)}(o) = 1, \qquad (11)$$
$$U^{(i+1)}(J)=o,\ \Theta^{(i+1)}(J) = o.$$

Equations (10) and (11) reduce to 2J+2 equations in the 2J+2 unknowns, namely,

$(U^{(i+1)}(-1),\ \Theta^{(i+1)}(-1);(U^{(i+1)}(j),\Theta^{(i+1)}(j)),$
$j=1,2,3,..,J-1;\ (U^{(i+1)}(J+1),\ \Theta^{(i+1)}(J+1),$

once we choose

$$\Theta^{(o)}=e^{-\gamma Y},\gamma=\tfrac{1}{2}\{P_R+(P_R^2+4N)^{\frac{1}{2}}\};\Theta^{(o)}(-1)=o \qquad (12)$$
$$U^{(o)}=\frac{G_R}{\gamma(\gamma-1)}(e^{-Y}-e^{-\gamma Y});\ U^{(o)}(-1)=o.$$

The solution is easily effected by writing the coefficient matrix as a tridiagonal matrix whose elements are 2x2 matrices. The choice of equation (12) ensures a rapidly converging iterated solutions.

2.2 Solution for the thick gas limit

The solution to equation(8) can be effected by assuming a perturbation expansion of the form

$U=U^{(o)}+Re\ U^{(1)}$ etc.

We obtain the system of equations

$$U^{(o)''}+U^{(o)'}+G_R(\Theta^{(o)}-1) = o,$$
$$\Theta^{(o)''}+P_R\Theta^{(o)'}+N(\Theta^{(o)3}\Theta^{(o)'})'=o, \qquad (13)$$
$$U^{(o)}=o,\ \Theta=\Theta_W \text{ on } Y = o,$$
$$U^{(o)}=o,\ \Theta=1 \text{ as } Y \to \infty,$$

and

$$U^{(1)''}+U^{(1)'}+G_R\Theta^{(1)}=o,$$
$$(1+N\Theta^{(o)3})\Theta^{(1)''}+(P_R+6N\Theta^{(o)'})\Theta^{(1)'}+3\Theta^{(o)}(2\Theta^{(o)'2}$$
$$+\Theta^{(o)}\Theta^{(o)''})\Theta^{(1)}=o, \qquad (14)$$
$$U^{(1)}=o=\Theta^{(1)} \text{ on } Y=o,$$
$$U^{(1)}=o=\Theta^{(1)} \text{ as } Y\to\infty.$$

The solutions to equation (13) can be obtained by straightforward integration. The results are

$$(1+N)\ln\left[\frac{\Theta^{(o)}-1}{\Theta_W - 1}\right]+N\{\frac{1}{3}(\Theta_W^3-\Theta^{(o)3})+\tfrac{1}{2}(\Theta_W^2-\Theta^{(o)2})+\Theta_W-\Theta^{(o)}\},$$

$$=-P_R Y \tag{15}$$

$$U^{(o)}=G_R[e^{-Y}\int_o^Y(e^x-1)\{\Theta^{(o)}(z)-1\}dz+(1-e^{-Y})\int_Y^\infty\{\Theta^{(o)}(z)-1\}dz].$$

With the help of (15), equation (14) can be solved by finite difference techniques in the same fashion as for (9), again on the assumption that

$\Theta^{(o)}(-1) = o.$

3. SIMILAR BOUNDARY LAYER PROBLEM

If we introduce the stream function Ψ, such that $\frac{\partial\Psi}{\partial y}$, $v = -\frac{\partial\Psi}{\partial x}$,

equations (1) - (3) for an optically thick gas given by (5) can be expressed as

$$\Psi_y\Psi_{yx} - \Psi_x\Psi_{yy} = \nu\Psi_{yyy} + g\beta T_\infty(\Theta-1),$$

$$\Psi_y\Theta_x - \Psi_x\Theta_y = \kappa\Theta_{yy}+ \frac{\nu}{C_P T_\infty}\Psi_{yy}^2+\frac{16\sigma T_\infty^3}{3\alpha\rho_\infty C_P}(\Theta^3\Theta_y)_y. \tag{16}$$

We put

$$\Psi=4c\nu x^{3/4}[f_o(\eta)+4\varepsilon(x)f_1(\eta)+16\varepsilon^2(x)f_2(\eta)+\ldots],$$

$$\Theta=g_o(\eta)+4\varepsilon(x)g_1(\eta)+16\varepsilon^2(x)g_2(\eta)+\ldots, \tag{17}$$

$$V{:}-\nu c x^{-\frac{1}{4}}[1+28\varepsilon(x)+176\varepsilon^2(x)+\ldots].$$

Following Pohlhausen (see Schlichting[5]) we choose η and c as

$$\eta=c.\frac{y}{x^{\frac{1}{4}}},\quad c = \left(\frac{g\beta T_\infty}{4\nu^2}\right)^{\frac{1}{4}},$$

and following Gebhart[1] we set

$\varepsilon(x)=g\beta x/C_P$

where $\varepsilon(x)$ is the dissipation number which is known to be very small for a large variety of fluids.

If we substitute equation (17) into equations (6) and (16) we can deduce that

$$f_o''' + 3f_o f_o'' - 3f_o'^2 + g_o = 1$$

$$g_o'' + 3P_R f_o g_o' + N(g_o^3 g_o')' = o \qquad (18)$$

$$f_o(o) = 1/3, \; f_o'(o) = o, \; g_o(o) = \Theta_W$$

$$f_o'(\infty) = o, \; g_o(\infty) = 1,$$

which is the basic nonlinear system, and

$$f_1''' + 3f_o f_1'' - 8f_o' f_1' + 7f_o'' f_1 + g_1 = o$$

$$(1+Ng_o^3)g_1'' + P_R(3f_o g_1' + 7g_o' f_1 - 4f_o g_1 + f_o^{''2} + 6Ng_o^2 g_o' g_1' +$$

$$3Ng_o(2g_o'^2 + g_o g_o'')g_1 = o, \qquad (19)$$

$$f_1(o) = 1, f_1'(o) = o, g_1(o) = o,$$

$$f_1'(\infty) = o, \; g_1(\infty) = o,$$

which is linear so that all subsequent higher order linear approximations can be treated as this one.

No similarity solution exists for an optically thin non-grey gas near equilibrium.

3.1 Method of solution

To solve the nonlinear system (18), we replace it by the iterated set of linear equations

$$f_o^{(i+1)'} = h_o^{(i+1)}, \quad h_o^{(i+1)''} + 3f_o^{(i)} h_o^{(i+1)'} - 4h_o^{(i)} h_o^{(i+1)}$$

$$+ 3h_o^{(i)'} f_o^{(i+1)} + g_o^{(i+1)} = h_o^{(i)''} + 6f_o^{(i)} h_o^{(i)'} - 4h_o^{(i)2} + g_o^{(i)},$$

$$(1+Ng_o^{(i)3}\; g_o^{(i+1)''} + 3(P_R f_o^{(i)} + 2Ng_o^{(i)2} g_o^{(i)'})g_o^{(i+1)'}$$

$$+ 3Ng_o^{(i)}(g_o^{(i)} g_o^{(i)''} + 2g_o^{(i)'2})g_o^{(i+1)} + 3P_R g_o^{(i)'} f_o^{(i+1)} =$$

$$(1+Ng_o^{(i)3})g_o^{(i)''} + 3(P_R f_o^{(i)} + 2Ng_o^{(i)2} g_o^{(i)'})g_o^{(i)'} + 3N(g_o^{(i)} \cdot$$

$$g_o^{(i)''} + Ng_o^{(i)2})g_o^{(i)2}, \quad f_o^{(i+1)}(o) = \frac{1}{3}, h_o^{(i+1)}(o) \qquad (20)$$

$$= o = h_o^{(i+1)}(J), g_o^{(i+1)}(o) = \Theta_w, g_o^{(i+1)}(J) = 1.$$

In replacing the set of equations (20) by a finite difference system, we replace the first equation in the set by a centre-difference; thus

$$\{f_o^{(i+1)}(j)-f^{(i+1)}(j-1)\}/l=\tfrac{1}{2}\{h_o^{(i+1)}(j)+h^{(i+1)}(j-1)\}.$$

The remaining two equations in the set are replaced by conventional central differences. The end result is

$$2f_o^{(i+1)}(j)-2f_o^{(i+1)}(j-1)-lh^{(i+1)}(j)-lh_o^{(i+1)}(j-1)=o,$$

$$3l[h_o^{(i)}(j+1)-h_o^{(i)}(j-1)]\,f_o^{(i+1)}(j)+2l^2g_o^{(i+1)}(j)+[2+$$

$$3lf_o^{(i)}(j)]\,h_o^{(i+1)}(j+1)-4[1+2l^2h_o^{(i)}(j)]\,h_o^{(i+1)}(j)-$$

$$[2-3lf_o^{(i)}(j)]\,h_o^{(i+1)}(j-1)=2l^2g_o^{(i)}(j)+\;2[1+3f_o^{(i)}(j)]\cdot$$

$$h_o^{(i)}(j+1)-4[1+2l^2h_o^{(i)}(j)]\,h_o^{(i)}(j)+2[1-3lf_o^{(i)}(j)]\cdot$$

$$h_o^{(i)}(j-1),\;3P_Rl[g_o^{(i)}(j+1)-g_o^{(i)}(j-1)]f_o^{(i+1)}(j)+\{2.$$

$$[1+Ng_o^{(i)3}(j)]+3[P_Rlf_o^{(i)}(j)+Ng_o^{(i)2}(j)g_o^{(i)}(i+1)-Ng_o^{(i)2}(j)$$

$$g_o^{(i)}(j-1)]\,g_o^{(i+1)}(j+1)+3Ng_o^{(i)}(j)\{2g_o^{(i)}(j)[g_o^{(i)}(j+1)$$

$$-2g_o^{(i)}(j)+g_o^{(i)}(j-1)]+[g_o^{(i)}(j+1)-g_o^{(i)}(j-1)]^2$$

$$-4[1+Ng_o^{(i)3}(j)]\}g_o^{(i+1)}(j)+\{2[1+Ng_o^{(i)3}(j)]-3[P_Rlf_o^{(i)}(i)$$

$$+Ng_o^{(i)2}(j)g_o^{(i)}(j+1)-Ng_o^{(i)2}g_o^{(i)}(j-1)]\}g_o^{(i+1)}(j-1)$$

$$=2[1+Ng_o^{(i)3}(j)]\,[g_o^{(i)}(j+1)-2g_o^{(i)}(j)+g_o^{(i)}(j-1)]+3.$$

$$[P_Rlf_o^{(i)}(j)+Ng_o^{(i)2}(j)g_o^{(i)}(j+1)-N^2g_o^{(i)2}(j-1)][g_o^{(i)}(j+\nu)$$

$$-g_o^{(i)}(j-1)]+\;3Ng_o^{(i)2}(j)\{2g_o^{(i)}(j)[g_o^{(i)}(j+1)-$$

$$2g_o^{(i)}(j)+g_o^{(i)}(j-1)]+4Nl^2g_o^{(i)2}(j)\}. \qquad (21)$$

If we choose the initial solutions as

$$f_o^{(o)}(\eta)=\frac{1}{3},\;\; h_o^{(o)}(\eta)=o,$$

$$(1+N)\ln\left[\frac{g_o^{(o)}-1}{\Theta_w-1}\right] + N\{\frac{1}{3}(\Theta_w^3-g_o^{(o)3})+\frac{1}{2}(\Theta_w^2-g^{(o)2})+$$

$$\Theta_w-g_o^{(o)}\}=-P_R\eta\ ,\ f_o^{(o)}(-1)=o=h_o^{(o)}(-1)=g_o^{(o)}(-1),\quad (21)$$

the equation (20) reduces to the solution of 3J+3 equations in 3J+3 unknowns

$$\{f_o^{(i+1)}(-1),h_o^{(i+1)}(-1);g_o^{(i+1)}(-1)\},\{f_o^{(i+1)}(j),$$
$$h_o^{(i+1)}(j),g_o^{(i+1)}\},j=1,2,..,J-1;\{f_o^{(i+1)}(J),h_o^{(i+1)}$$
$$(J+1),g_o^{(i+1)}(J+1)\},$$

which unknowns can be solved economically on computer by writing the coefficient matrix as a tridiagonal one whose elements are 3x3 matrices. The LU decomposition could then be effectively applied. The choice in (21) ensures rapid convergence; and the third equation in set (21) follows by substituting for the first two in this set into the second equation in (20). The problem is then the same as that of the temperature field in equation (13).

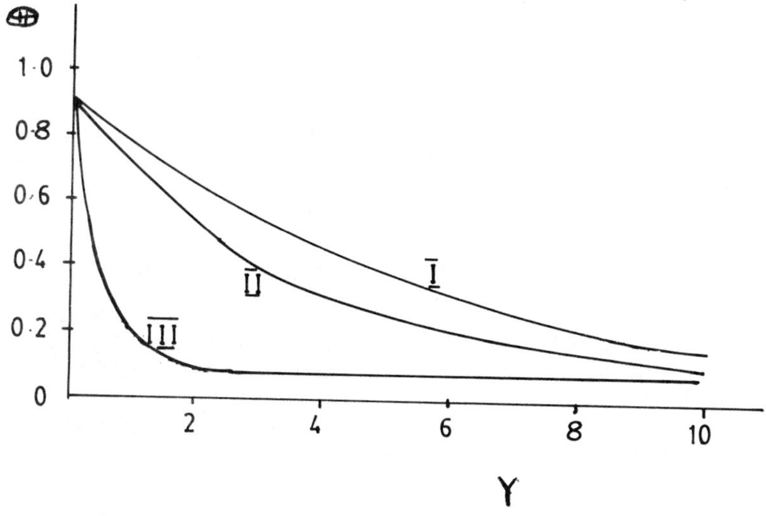

Figure 1: Asymptotic temperature distribution for thin gas (for legend see Fig.2)

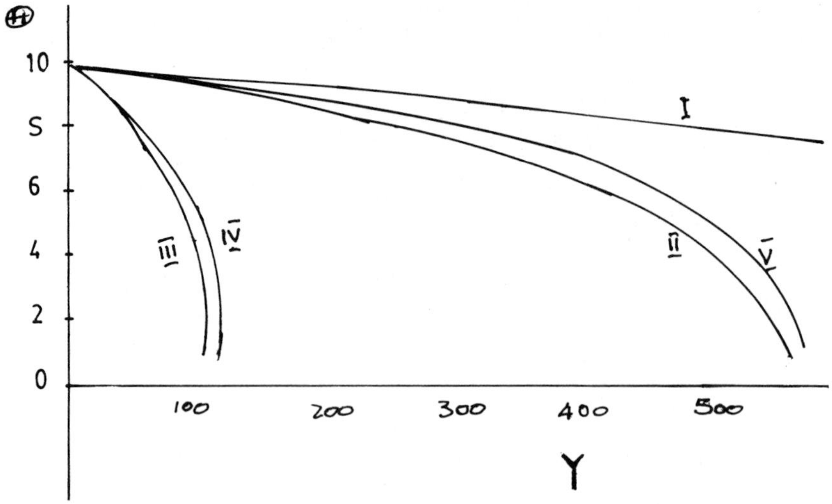

Figure.2. Assymptotic temperature distribution for thick gas

N	P_R	E_C	
0.5	0.71	0.01	I
1.0	0.71	0.01	II
0.5	7.0	0.01	III
0.5	7.0	0.02	IV

4. DISCUSSION

The asymptotic temperature profiles are plotted in figs. 1 and 2 for E_C=0.01 and G_R=5.0. The common features of the two profiles are decrease in temperature with increase in N and P_R and an increase in temperature with increase in E_C. The thin gas is characterized by fast exponential type decay and as was noticed earlier, it is very sensitive to E_C. For bounded natural convection flows, asymptotic expansion for E_C is possible (see Bestman 6). When the gas is thick, the profile is now of the slowly varying logarithmic decay type.

Figure 3 depicts the boundary layer temperature distribution for an optically thick gas. For air and water (P_R=0.71 and P_R=7.0 respectively) the thermal boundary thickness is of order hundred times the value of the corresponding non-radiating fluid. This thickness decreases rapidly with increase in P_R. The heat transfer characteristic at the wall can also be shown to be smaller for the radiating case.

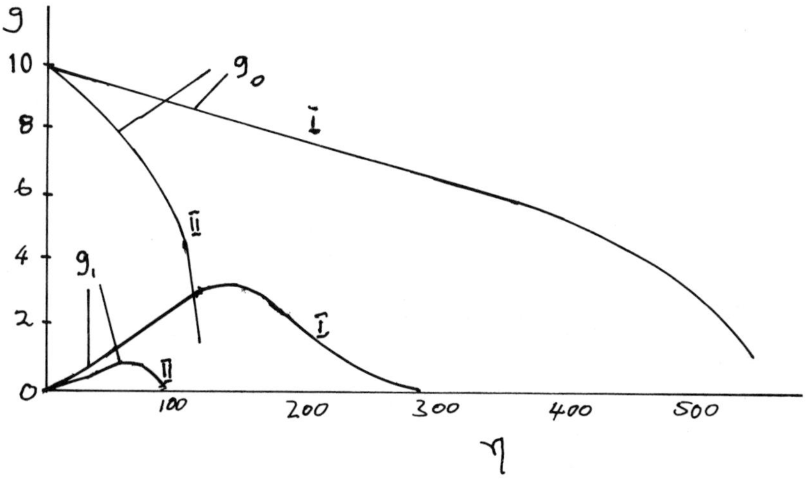

FIGURE 3. Similar boundary layer temperature distribution

N	Pr	
1.0	0.71	I
0.5	7.0	II

ACKNOWLEDGEMENT

The author gratefully acknowledges a financial support from the Federal University of Technology.

REFERENCES

1. GEBHART,B. Effects of Viscous Dissipation in Natural Convection, J.fluid Mech.,Vol.14,pp.225-232, 1962
2. COGLEY,A.C,VINCENTI,W.G.and GILLES,S.E.-Differential Approximation to Radiative Transfer in a Non-Grey Gas Near Equilibrium,A.I.A.A Jl., Vol.6,pp.551-553, 1968
3. VINCENTI,W.G.and KRUGER,C.S. Physical Gasdynamics, Krieger Publishing Coy., New York,1965
4. HELLIWELL,J.B.and MOSA.M.R-Radiative Heat Transfer in a Horizontal Magnetohydrodynamic Channel Flow with Buoyancy Effects and an Axial Temperature Gradient,Int.J.Heat Mass Transfer,Vol.22,pp. 657-668, 1979
5. SCHLICHTING,H.-Boundary Layer Theory,McGraw. Hill, New York, 1951
6. BESTMAN, A.R.-Numerical Studies of Laminar Convection of a Radiating Non-Newtonian fluid in a Vertical Porous Channel,Numerical Methods in Thermal Problems, Vol3,pp....,1983

APPLICATION OF THE ALTERNATING DIRECTION IMPLICIT METHOD TO THE COMPUTATION OF TIME-DEPENDENT COMPRESSIBLE CONVECTIVE FLOWS

Kwing L. Chan

Applied Research Corporation
Landover, MD 20785 USA

In this paper, we shall discuss the time accuracy of the ADI method for the computation of time-dependent compressible convective flows. Numerical experiments indicate that the time accuracy is most sensitive to the maximum CFL number and insensitive to other parameters.

1. INTRODUCTION

The Alternating Direction Implicit (ADI) method is a very efficient technique whose CPU time per step is almost competitive with explicit methods - this is true even in vector machines (see APPENDIX). At the same time, its implicitness allows larger time steps to be taken.

However, most applications of the ADI method are in finding steady state solutions [1-3]. Very few literatures touch upon the subject of time accuracy of the ADI method for computing transients [1,4-6], and most of the discussions are very brief. It is of great interest to know the accuracy of the ADI method for application to time dependent problems. In this short paper, we shall report the results of a systematic study of the time accuracy of the ADI method as applied to problems of internal convection. Even though this is only a special application of the ADI method, it may shed some light on the general temporal behavior of the ADI approach.

In an earlier paper by Chan and Wolff [7] (CW hereafter), a consistent scheme based on the two-level ADI method [5] with staggered mesh was discussed and applied to the study of compressible convection of a stratified fluid. Using the same code and the same physical problem, we studied the behavior of transients for different time steps and parameters.

2. THE TESTING GROUND

Details of the numerical scheme and the formulation of the convection problem can be found in CW; we only summarise the key features here. For the present study, the fluid is an ideal gas with a constant ratio of specific heats equal to 5/3; the kinematic viscosity (ν) and the specific conductivity are taken to be constant and the Prandtl number is taken to be 1. The gas is confined in a square box with physical depth equal to d. The temperatures at the top and the bottom are fixed and the side walls are adiabatic. The velocities has stress free boundary conditions. The fluid originally has a polytropic distribution with index 1.4 (see CW), and the depth of the stratification is specified by the dimensionless parameter $Z = (T_1 - T_2)/T_2$ where T stands for temperature, and the subscripts 1 and 2 denote quantities at the bottom and the top of the box, respectively. Given Z and the Rayleigh number Ra (see CW for definition), the convection problem is fixed. We shall express velocities in units of $(P_2/\rho_2)^{\frac{1}{2}}$ and time in units of $d\,(P_2/\rho_2)^{-\frac{1}{2}}$ where P and ρ stand for pressure and density respectively.

In our treatment of the Navier-Stokes equations, only the part which can be split spatially is treated implicitly. The fractional degree of implicitness for this part is described by the parameter β (see CW). The spatial operators with cross-derivatives are treated purely explicitly. Therefore, the time discretization is always first-ordered, independent of β.

There are two important numbers which describe approximately how much the usual stability conditions for explicit methods are violated. They are

$$N_{CFL} = \mathrm{Max}\left(\frac{\Delta t C_s}{\Delta x} \right) \tag{1}$$

$$N_{\nu} = \mathrm{Max}\left(\frac{\Delta t \nu}{\Delta x^2} \right) \tag{2}$$

where C_s is the sound speed, Δx is the grid spacing, and Δt is the size of the time step. Notice that the ratio N_ν/N_{CFL} is independent of Δt.

To obtain different ranges of N_{CFL}, N_ν, and other parameters, we use three different physical configurations for testing. The first configuration has Z equal to 1, Ra equal to 5000, and the mesh is uniform (equal spacing in each direction) with 9 X 37 grids. N_ν/N_{CFL} is equal to 0.086. We made most detailed study in this case. The effects of using different values of β is tested. The second configuration has Z equal to 0.1 and Ra equal to 5000, and the mesh is uniform with 22 X 27 grids. The minimum grid spacing is wider in this case so that the values of N_ν are very small compared to N_{CFL} (N_ν/N_{CFL} = 0.0084). All the runs use β = 0.54 for this configuration. The third configuration has Z equal to 20 and Ra equal to 50, and the 9 X 37 mesh is non-uniform in the vertical direction with about 5 grids per pressure scale height (the distance that p changes e-fold). The value of β is fixed to be 0.54. The very small grid spacing near the top makes N_ν very large (N_ν/N_{CFL} = 250) and the time steps are restricted to be very small. The runs for this case were not carried to the stationary state; only early portions of the transients were followed. A summary of all the computed cases is given in Table 1.

We use the maximum speed in the fluid, V_{max}, as the indicator for the evolution of the numerical system. All cases in each configuration start with a V_{max} equal to 0.001. The asymptotically stationary values of V_{max} for the configurations 1 and 2 are 0.1483 and 0.01551, respectively. In each configuration, the case with minimum Δt is used as the standard for comparison. Since the difference between the result of this case and that with the next smallest Δt is very small, it should be sufficiently accurate to represent the exact transient.

3. RESULTS FOR TRANSIENTS

In Fig. 1, curves a-e plot V_{max} versus time, t, for the cases with β= 1, N_{CFL} = 0.72, 5.79, 11.6, 23.2, 46.3 in configuration 1. It is obvious that the time accuracy deteriates very fast as N_{CFL} increases, but the most interesting thing is that these curves are almost identical to those produced by cases with Δt two times larger and β one-half smaller, namely 0.5. Therefore, with the same requirement of time accuracy, almost two times faster computational speed can be obtained if β is

Configuration 1		Configuration 2		Configuration 3	
N_{CFL}	N_{ν}	N_{CFL}	N_{ν}	N_{CFL}	N_{ν}
0.72	2^{-4}	0.93	2^{-7}	0.50	125
1.45	2^{-3}	3.71	2^{-5}	1.00	250
2.90	2^{-2}	7.43	2^{-4}	2.00	500
5.79	2^{-1}	14.9	2^{-3}	4.00	1000
11.6*	1	29.7	2^{-2}		
23.2*	2	59.4	2^{-1}		
46.3*	2^2				
92.6*	2^3				
185.	2^4				

Table 1 N_{CFL} and N_{ν} of computed cases.
(*) $\beta = 0.5$ also

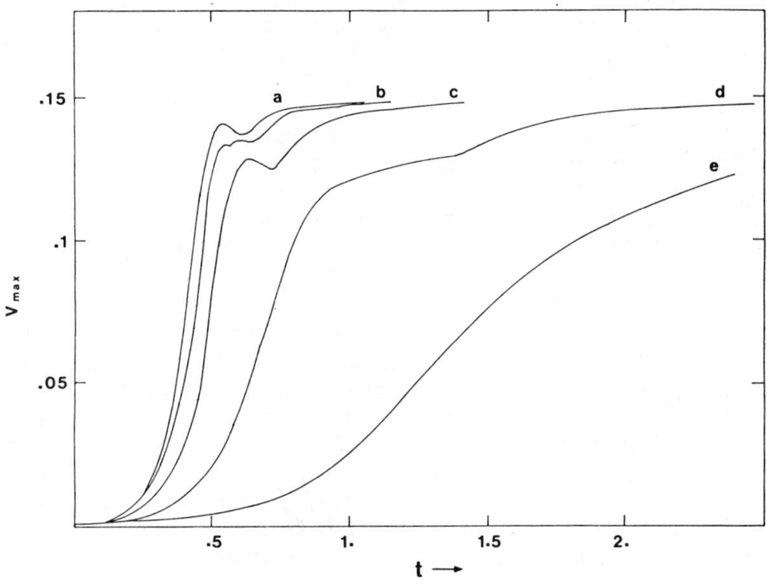

Figure 1 The transients of V_{max} for configuration 1.

chosen to be closed to 0.5 instead of 1.

Fig. 2 summarizes the time accuracies of the different cases versus N_{CFL}. For configurations 1 and 2, the time accuracies are determined by comparing the times at which the maximum speeds reach one-half their stationary values. For configuration 3, the cases were computed only until V_{max} grows to 0.007. The time spans for this portion of the transients are then compared. The ordinate of Fig. 2 is the percentage of error in time measured with respect to the standard cases. The solid line shows the percentages of error for the $\beta = 1$ cases in configuration 1; it is higher than all other curves with smaller β. The dotted curve is for the $\beta = 0.5$ cases in the same configuration. The dashed curve shows the errors for configuration 2 and the dot-dashed curve shows the errors for configuration 3. An important observation is that all these curves lie in a narrow band along the line

$$\% \text{ of error} \simeq 0.1\, N_{CFL}^2 \quad (3)$$

Since the three configurations represent very wide distribution of N_ν/N_{CFL} (over a range differed by a factor of 30000) and all other parameters (e.g. Z, Ra, grid spacing), this figure indicates that the time accuracy is essentially only dependent on N_{CFL}. Another point to observe is that in the region under consideration ($N_{CFL} > 1$), the errors grow nonlinearly with N_{CFL} (or Δt). It is actually worse than what a first-order time scheme can do. Therefore, the particular implicit form of this scheme accelerates the destruction of time accuracy. It is suspected that the extra term proportional to $(\beta \Delta t)^2$ introduced by the spatial operator splitting may be the cause of such problem.

We have tested the efficiency of the explicit Euler method for configuration 1 and found that it needs N_{CFL} to be $\leq 1/4$ for stability, even though it is three times faster in each step. Therefore, for the ADI method to be relatively more economical, it is necessary that $N_{CFL} > 1$. If the demand for the time accuracy is not too high, say 3%, Eq (3) shows that $N_{CFL} \leq 5$ can satisfy the requirement, and the ADI scheme can be quite economical.

The ADI method is more efficient in the search of stationary solutions for which time accuracy is not a matter of concern. Certainly, efficiency does not increase indefinitely by increasing Δt. For different sizes of N_{CFL}, Fig. 3 plots the

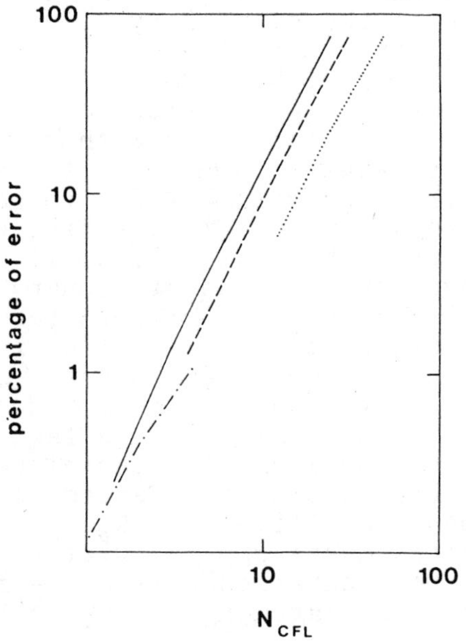

Fig. 2 The percentage of error in time VS. N_{CFL} for different configurations.

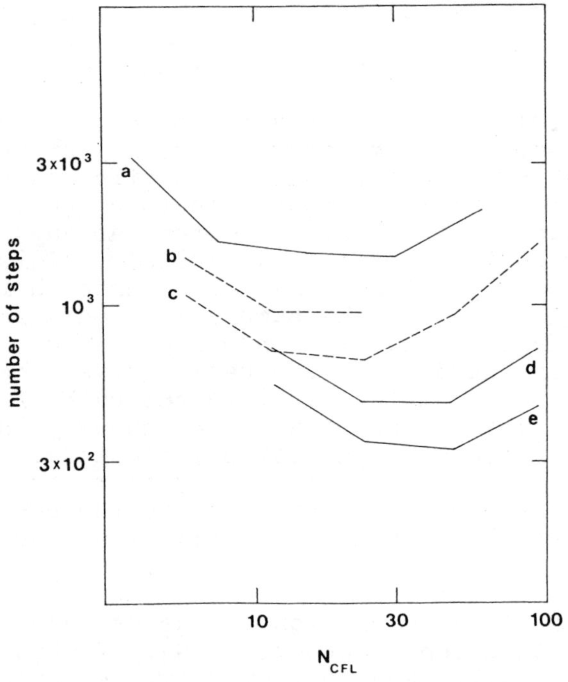

Fig. 3 Number of steps for V_{max} to reach a certain percentage of the stationary value.

number of steps required for V_{max} to reach a certain percentage of the stationary value. Curve a is for V_{max} to reach 99% of the stationary value of configuration 2 ($\beta = 0.54$). Curves b, c, d, and e are for configuration 1. b is for cases with $\beta = 1$ to reach 99.9%; c is for cases with $\beta = 1$ to reach 99%; d is for cases with $\beta = 0.5$ to reach 99.9%; and e is for cases with $\beta = 0.5$ to reach 99%. All these curves show minimum at some finite values of N_{CFL}. By comparing b, c with d, e, it can be seen that the convergence to the stationary value can be much expediated by using $\beta = 0.5$ (about a factor of 2 more efficient). Now, let us focus our attention on curves a, d and e which have β closed to 0.5. All these curves have minimum around $N_{CFL} = 30$, even though N_ν/N_{CFL} differs by a factor of 10 between configuration 1 and 2. therefore, it seems that the optimal choice for fastest convergence does not depend sensitively on N_ν. By choosing a time step which makes $N_{CFL} = 30$, the number of steps required for convergence is at least 10 times less than that of marching an explicit method.

Before we leave the discussion of transients, we remind the reader that the results obtained here are derived only from the convection problem; they vary for other physical problems.

4. EXPERIMENTS FOR A TURBULENT SYSTEM

For turbulent flows which never become stationary, the question of time accuracy turns out to be less well-defined. Due to the stochastic nature of turbulence, it is almost impossible to compute an exact evolution of a turbulent system, as any small perturbation on the system grows quickly. This phenomenon is called 'sensitive dependence on initial conditions'[8]. In the numerical simulation of a turbulent system, any small numerical error (truncation or round-off) eventually drives the model system to a completely different path of evolution. As numerical errors are continuously introduced at every step of computation, the model system is constantly being shifted away from its previous path of evolution. Here, we use the results of some numerical experiments to illustrate such behavior.

We use a slight variation of the CW code to simulate the turbulent convection [9]. Instead of a constant viscosity and conductivity, we employed a shear dependent formula to represent subgrid scale turbulent viscosity and diffusivity [10].

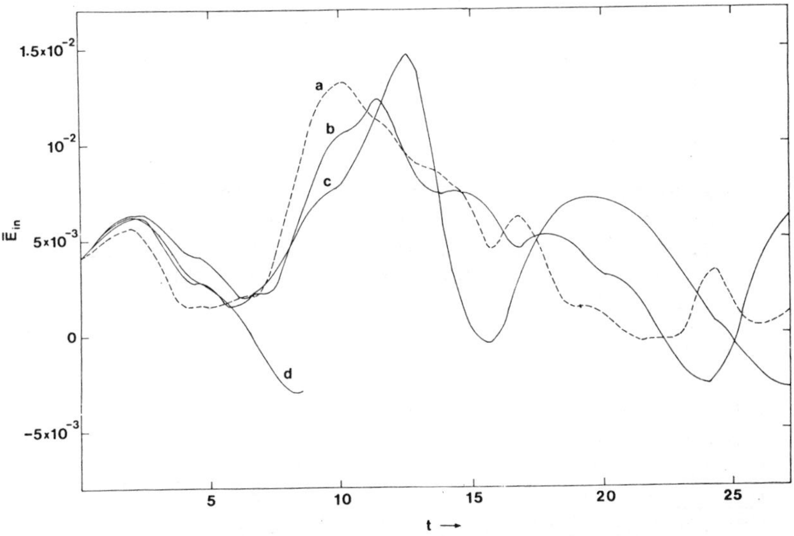

Fig. 4 The evolution of $\bar{E}_{in}$ for cases with different time steps.

The size of the box is now 2d in width and d in depth. The dimensionless depth parameter Z has a value of 20 and the polytropic index is 0.999 X 1.5. The mesh size is 37 X 42 with non-uniform vertical grid spacing smaller at the top. We shall look at the time accuracy of the ADI method for computing the horizontally averaged energy flux input from the bottom ($\bar{E}_{in}$) of a gas undergoing turbulent convection.

Fig. 4 compares the evolutions of $\bar{E}_{in}$ for cases with different sizes of time steps. The value of β is fixed at 0.55 and the initial condition is the same for all cases. Curves a, b, c, and d are for N_{CFL} = 1, 2, 4, and 8 respectively. Using the N_{CFL} = 1 case as the reference, other cases with larger N_{CFL} can be seen to diverge from it at different times. The larger N_{CFL} is, the earlier the corresponding curve diverges away from the reference. Another observation is that the divergences occur quite abruptly. For example, the N_{CFL} = 8 case agrees with the reference case quite well before t =5.7 (over 600 steps for the reference case) after which it suddenly follows another path of evolution. The divergences are not due to a numerical instability. Statistically speaking, the different paths are valid solutions. However, different time steps do introduce perturbations to the turbulent system at different rates which eventually knock the model system away from its previous paths. For a turbulent system, numerical simulation is most useful in providing statistical statements concerning the behavior of the system.

ACKNOWLEDGEMENTS

The author would like to thank Drs. C.L. Wolff and S. Sofia for helpful discussions, and Dr. S.P.S. Anand for his encouragement and support. This research is partially supported by a grant from NASA (RTOP # 672-40-04).

REFERENCES

1. BRILEY, W.R., and MCDONALD, H. - Solution of the Three-Dimensional Compressible Navier-Stokes Equations by an Implicit Technique, Proceedings of Fourth International Conference on Numerical Methods in Fluid Dynamics, Springer-Verlag, 1975.

2. BEAM, R.M., and WARMING, R.F. - An Implicit Finite-Difference Algorithm for Hyperbolic Systems in Conservation-Law Form. J. Computational Phys., Vol. 22, p.87, 1976.

3. STEGER, J.L., and KUTLER, P. - Implicit Finite-Difference Procedures for the Computation of Vortex Wakes. AIAA Journal, Vol. 15, p.581, 1977.

4. PEACEMAN, D.W., and RACHFORD, H.H., Jr. - The Numerical Solution of Parabolic and Elliptic Differential Equations. J. Soc. Ind. Appl. Math., Vol. 3, p.28, 1955.

5. BRILEY, W.R., and MCDONALD, H. - Solution of the Multidimensional Compressible Navier-Stokes Equations by a Generalized Implicit Method. J. Computational Phys., Vol. 24, p.372, 1977.

6. MAKHVILADZE, G.M., and SHCHERBAK, S.B. - Numerical Method of Investigating Nonstationary Spatial Motions of a Compressible Gas. J. of Engineering Phys., Vol. 38, p.317, 1980.

7. CHAN, K.L., and WOLFF, C.L. - ADI on Staggered Mesh - A Method for the Calculation of Compressible Convection. J. Computational Phys., Vol. 47, p.109, 1982.

8. ECKMANN, J.D. - Road to Turbulence in Dissipative Systems. Rev. Mod. Phys., Vol. 53, p.643, 1981.

9. CHAN, K.L., SOFIA, S., and WOLFF, C.L. - Turbulent Compressible Convection in a Deep Atmosphere. I. Preliminary Two-Dimensional Results. Astrophys. J., Vol. 263, p.935, 1982.

10. DEARDORFF, J.W. - On the Magnitude of the Subgrid Scale Eddy Coefficient. J. Computational Phys., Vol. 7, p.120, 1971.

APPENDIX: VECTORIZATION OF THE ADI SCHEME

The ADI scheme has the nice property that in each direction, the split differential operator (and the associated matrix) only mixes spatial indices in that particular direction, leaving indices in all other directions untouched. Therefore, an operation along one direction can be treated as a vector operation with vector length equal to the total number of grid points in the other directions.

In Table 2, the grind times of a few grid configurations are shown. The ADI scheme discretizes the Navier-Stokes equations in spherical coordinates and is vectorized in a CYBER 205 (2 pipes) machine. It can be seen that the vectorization for three-dimenaional cases is more effective as the vector lengths are longer.

Grid configuration	Explicit part only	Total
22 X 22		51
12 X 12 X 12	13	31
22 X 22 X 12	10	26

Table 2 Grind times (CPU time/ grid/ step; in 10^{-6} sec.) for the vectorized ADI method in CYBER 205.

ON THE EFFECT OF FLOW DIRECTION ON MIXED CONVECTION FROM A HORIZONTAL CYLINDER

H.M.Badr
Mechanical Engineering Department
University of Petroleum & Minerals
UPM Box # 322
Dhahran, Saudi Arabia

ABSTRACT

The influence of free stream direction on mixed (natural and forced) convection heat transfer from a circular cylinder is investigated. The cylinder, which has an isothermal surface, is placed with its axis horizontal. The direction of the oncoming flow, which is always normal to the cylinder axis, vary from vertically upward (parallel flow) to vertically downward (contra flow). The investigation is based on the time integration of the unsteady, two-dimensional equations of motion and energy until reaching steady conditions. The study is limited to Reynold's numbers up to Re=40 and Grashof numbers of $G_r = Re^2$. The results are compared with the available experimental data and the agreement is satisfactory.

1. INTRODUCTION

The problem of laminar mixed convection heat transfer from a circular cylinder is one of the fundamental problems which has received an extensive attention because of its many engineering applications. Many experimental studies were carried out to investigate the effect of different factors on the heat transfer process. Some of these studies resulted in experimental correlations, however, no correlation could successfully predict the overall heat transfer coefficient and take into consideration all the parameters involved in the process.

The first experimental investigation on the influence of free stream direction on the rate of heat transfer from a horizontal cylinder was carried out by Hatton, et al [1] who studied the problem up to Reynold's number Re = 45 and Grashof number Gr = 10. Oosthuizen and Madan [2] studied the same problem when the forced flow direction makes an angle of 0, 90°, 135° and 180° with the direction of natural convection. The study was conducted at relatively high Re and Gr values compared to the range in [1]. Other experimental studies are found in references [3-6].

On the other hand, most of the theoretical studies found in the literature dealt with the case of parallel flow. For example the work by Acrivos [7] who studied combined convection in laminar boundary layer flow to obtain the Nusselt number distribution near a stagnation point when $Pr \to 0$ and $Pr \to \infty$. The approach requires the existence of boundary layer flow and is limited to the region surrounding the stagnation point. Sparrow and Lee [8] obtained an approximate solution for the Nusselt number distribution in the region between the forward stagnation point and the point of separation for the case of parallel flow. Joshi and Sukhatme [9] and Merkin [10] solved the same problem, however, their solution could not exceed the point of separation since it is based on boundary layer flow assumption. It appears from the literature that there is no complete solution to the problem of laminar mixed convection from a horizontal cylinder even for the relatively simple case of parallel flow.

The main objective of this work is to conduct a theoretical investigation on the effect of flow direction on the velocity and thermal fields and consequently on the heat transfer process. Of special interest here is the case when the effect of natural convection is of the same order of magnitude as that of forced convection. The study is based on the solution of the full conservation equations of mass, momentum and energy.

2. PROBLEM STATEMENT AND ASSUMPTIONS

The problem considered here is that of mixed convection heat transfer from a horizontal circular cylinder of radius, a. The cylinder surface has a constant temperature, T_s, and is placed in a uniform stream of velocity, u_∞, and temperature, T_∞. The direction of the free stream, which is always normal to the cylinder axis, vary from vertically upward ($\gamma = 0$) to vertically downward ($\gamma = 180°$), where γ is the angle between the oncoming flow and the vertical upward directions. The cylinder is assumed to be long enough so that the end effects can be neglected and the flow is considered two-dimensional. The fluid properties are also assumed to behave according to the Boussinisq approximations. Consider the line $\theta = 0$ to represent the radius through the rearmost point on the cylinder surface (see Fig. 1). Using the modified polar coordinates (ξ, θ), where $\xi = \ln \frac{r}{a}$, the governing equations of motion and energy can be written as

$$e^{2\xi} \frac{\partial \zeta}{\partial t} = - \frac{\partial \psi}{\partial \theta} \frac{\partial \zeta}{\partial \xi} + \frac{\partial \psi}{\partial \xi} \frac{\partial \zeta}{\partial \theta} + \frac{2}{Re} \left(\frac{\partial^2 \zeta}{\partial \xi^2} + \frac{\partial^2 \zeta}{\partial \theta^2} \right) - e^{\xi} \frac{Gr}{2Re^2}$$

$$\left[\frac{\partial \phi}{\partial \xi} \sin (\gamma - \theta) - \frac{\partial \phi}{\partial \theta} \cos (\gamma - \theta) \right] \qquad (1)$$

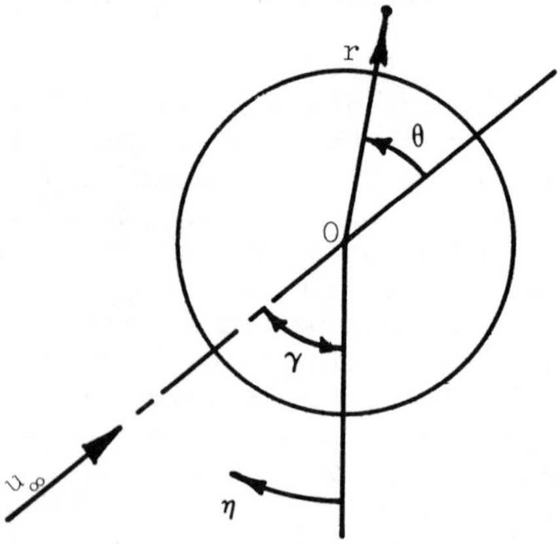

Figure 1. Coordinate System

$$e^{2\xi}\ \zeta = \frac{\partial^2\psi}{\partial\xi^2} + \frac{\partial^2\psi}{\partial\theta^2} \tag{2}$$

$$e^{2\xi}\ \frac{\partial\phi}{\partial t} = -\frac{\partial\psi}{\partial\theta}\frac{\partial\phi}{\partial\xi} + \frac{\partial\psi}{\partial\xi}\frac{\partial\phi}{\partial\theta} + \frac{2}{Pe}\left(\frac{\partial^2\phi}{\partial\xi^2} + \frac{\partial^2\phi}{\partial\theta^2}\right) \tag{3}$$

where t, ψ, ζ, ϕ, Re, Gr and Pe are all dimensionless quantities defined as

$$t = t'\ u_\infty/a, \quad \psi = \psi'/au_\infty, \quad \zeta = \zeta' a/u_\infty,$$

$$\phi = (T - T_\infty)/(T_s - T_\infty), \quad Re = 2au_\infty/\nu, \quad Gr = g\beta(2a)^3(T_s - T_\infty)/\nu^2$$

and $Pe = Pr\ Re$

where Re is the Reynolds number, Gr is the Grashof number, Pe is the Pecelet number, Pr is the Prandtl number, β is the coefficient of thermal expansion and ν is the kinematic viscosity. All quantities with primes are dimensionless quantities.

The dimensionless velocity components $v_r \left(= \frac{v_r'}{u_\infty}\right)$ and $v_\theta \left(= \frac{v_\theta'}{u_\infty}\right)$ are related to ψ by

$$v_r = e^{-\xi}\frac{\partial\psi}{\partial\theta} \quad \text{and} \quad v_\theta = -e^{-\xi}\frac{\partial\psi}{\partial\xi} \tag{4}$$

The boundary conditions for the velocity and thermal fields are,

$$\left.\begin{aligned} &\psi = \frac{\partial\psi}{\partial\theta} = \frac{\partial\psi}{\partial\xi} = 0 \quad \text{and} \quad \phi = 1 \quad \text{at} \quad \xi = 0 \\ &e^{-\xi}\frac{\partial\psi}{\partial\theta} \to \cos\theta, \quad e^{-\xi}\frac{\partial\psi}{\partial\xi} \to \sin\theta, \\ &\zeta \to 0 \quad \text{and} \quad \phi \to 0 \qquad \text{as} \quad \xi \to \infty \end{aligned}\right\} \tag{5}$$

The conditions in Eq.(5) are based on the no-slip and isothermal conditions on the cylinder surface and the free stream conditions away from it.

3. METHOD OF SOLUTION

The method used for solving the governing equations (1-3) to obtain the steady velocity and temperature distributions is based on studying the time development of the velocity and thermal fields until reaching steady conditions. In this method the uniform stream is assumed to start suddenly from rest at $t = 0$ with no temperature difference between the cylinder surface and the oncoming flow. Following this start, the velocity boundary layer develops with time while no body forces present. At a latter time, when the boundary layer thickens, the cylinder surface temperature is assumed to be suddenly raised to T_s. Following this temperature rise the velocity and thermal fields develop simultaneously with time until the steady conditions are achieved. The approach is similar to that used by Badr [11].

In the cases of $\gamma = 0$ and $\gamma = 180$ the streamlines and isotherms are symmetrical about a vertical plane passing by the cylinder axis. However, for any other value of γ the velocity and temperature fields are asymetric. Accordingly we express the functions ψ, ζ and ϕ in terms of Fourier series expansions as

$$\left.\begin{aligned}
\psi &= \frac{1}{2} F_o(\xi,t) + \sum_{n=1}^{N} f_n(\xi,t)\sin n\theta + F_n(\xi,t)\cos n\theta \\
\zeta &= \frac{1}{2} G_o(\xi,t) + \sum_{n=1}^{N} g_n(\xi,t)\sin n\theta + G_n(\xi,t)\cos n\theta \\
\phi &= \frac{1}{2} H_o(\xi,t) + \sum_{n=1}^{N} h_n(\xi,t)\sin n\theta + H_n(\xi,t)\cos n\theta
\end{aligned}\right\} \qquad (6)$$

Using Eq.(6) together with Eq.(1) results in the following set of partial differential equations (pde)

$$e^{2\xi} \frac{\partial G_o}{\partial t} - \frac{2}{Re} \frac{\partial^2 G_o}{\partial \xi^2} = X_o \qquad (7a)$$

$$2e^{2\xi} \frac{\partial g_n}{\partial t} - \frac{4}{Re}\left(\frac{\partial^2 g_n}{\partial \xi^2} - n^2 g_n\right) - n F_n \frac{\partial G_o}{\partial \xi} + n G_n \frac{\partial F_o}{\partial \xi} = X_{n1} \qquad (7b)$$

$$2e^{2\xi} \frac{\partial G_n}{\partial t} - \frac{4}{Re}\left(\frac{\partial^2 G_n}{\partial \xi^2} - n^2 G_n\right) + n f_n \frac{\partial G_o}{\partial \xi} - n g_n \frac{\partial F_o}{\partial \xi} = X_{n2} \qquad (7c)$$

where $f_n = f_n(\xi,t)$, $g_n = g_n(\xi,t)$, ..., etc., and X_o, X_{n1} and X_{n2} each represent a known function of γ, ξ and t. The function X_o takes the form,

$$X_o(\gamma,\xi,t) = e^{\xi} \frac{Gr}{2Re^2}\left[\sin\gamma\left(\frac{\partial H_1}{\partial\xi} + H_1\right) - \cos\gamma\left(\frac{\partial h_1}{\partial\xi} + h_1\right)\right]$$

$$+ \sum_{n=1}^{N} n\left(F_n \frac{\partial g_n}{\partial\xi} - f_n \frac{\partial G_n}{\partial\xi} + g_n \frac{\partial F_n}{\partial\xi} - G_n \frac{\partial f_n}{\partial\xi}\right) \quad (8)$$

Similar expressions are obtained for X_{n1} and X_{n2}. If we again substitute from Eq.(6) in Eq.(2) and use simple mathematical analysis we obtain

$$\frac{\partial^2 F_o}{\partial\xi^2} = e^{2\xi} G_o \quad (9a)$$

$$\frac{\partial^2 f_n}{\partial\xi^2} - n^2 f_n = e^{2\xi} g_n \quad (9b)$$

$$\frac{\partial^2 F_n}{\partial\xi^2} - n^2 F_n = e^{2\xi} G_n \quad (9c)$$

Finally by using Eq.(6) together with Eq.(3) we obtain the following set of pde for the functions H_o, h_n and H_n

$$e^{2\xi}\frac{\partial H_o}{\partial t} - \frac{2}{Pe}\frac{\partial^2 H_o}{\partial\xi^2} = Z_o \quad (10a)$$

$$2e^{2\xi}\frac{\partial h_n}{\partial t} - \frac{4}{Pe}\left(\frac{\partial^2 h_n}{\partial\xi^2} - n^2 h_n\right) - n F_n \frac{\partial H_o}{\partial\xi} + n H_n \frac{\partial F_o}{\partial\xi} = Z_{n1} \quad (10b)$$

$$2e^{2\xi}\frac{\partial H_n}{\partial t} - \frac{4}{Pe}\left(\frac{\partial^2 H_n}{\partial\xi^2} - n^2 H_n\right) + n f_n \frac{\partial H_o}{\partial\xi} - n h_n \frac{\partial F_o}{\partial\xi} = Z_{n2} \quad (10c)$$

where Z_o, Z_{n1} and Z_{n2} are known functions of ξ and t. The boundary conditions for all the functions given in Eqs.(7), (9) and (10) can be easily deduced from Eqs.(5) and (6).

In the first part of the motion (following $t = 0$) the streamlines are symmetric about the line $\theta = 0$ since the buoyancy force is zero. In this part the boundary layer coordinate x, where $x = \xi/2(2t/Re)^{\frac{1}{2}}$, is used instead of ξ since the boundary layer thickness is very small. The method of solution in this part is exactly the same as that used by Badr and Dennis [12] for the special case of no cylinder rotation. The integration process in this part is terminated at a time $t = \frac{Re}{8}$, at which $\xi = x$, and the boundary layer becomes thick enough to use the original coordinate ξ. At the start of the second part of the motion $\left(t = \frac{Re}{8}\right)$ the cylinder surface temperature is suddenly increased to T_s and allowing both velocity and thermal boundary layers to develop with time. In this part the three sets of pde [Eqs.(7), (9) and (10)] are integrated to advance the solution of ψ, ζ and

ϕ in time. The integration process is based on a Crank-Nicolson finite-difference scheme for Eqs.(7) and (10) to obtain the functions G_o, g_n, G_n, H_o, h_n and H_n at the new time step. To solve Eqs.(9) for the functions F_o, f_n and F_n, a direct solution for Eq.(9a) is obtained using central differences while a step-by-step integration scheme, similar to that used by Dennis and Chang [13] is used to solve Eqs.(9b) and (9c). The solution procedure of Eqs.(7), (9) and (10) is similar to that used by Badr [11] for tackling the problem of cross mixed convection.

4. RESULTS AND DISCUSSION

The effect of flow direction on the rate of heat, transfer from a horizontal cylinder is studied for Reynold's numbers of Re = 5, 10, 20 and 40 and Grashof numbers of Gr = Re . The flow direction varied from $\gamma = 0$ (parallel flow) to $\gamma = 180°$ (contra flow) with steps of 30°. To compare and discuss results let us define the local and average coefficients of heat transfer h and $\bar{h}$ as

$$h = \dot{q}/(T_s - T_\infty) \quad \text{and} \quad \bar{h} = \frac{1}{2\pi}\int_0^{2\pi} h \, d\theta \qquad (11)$$

where $\dot{q} = k\left(\frac{\partial T}{\partial r'}\right)_{r'=a}$ is the rate of heat transfer per unit area. Define also the local and average Nusselt numbers Nu and $\overline{Nu}$ such that

$$Nu = 2ah/k \quad \text{and} \quad \overline{Nu} = 2a\bar{h}/k \qquad (12)$$

The relationship between each of Nu and $\overline{Nu}$ and the functions H_o, h_n and H_n can be easily deduced from Eqs.(6), (11) and (12).

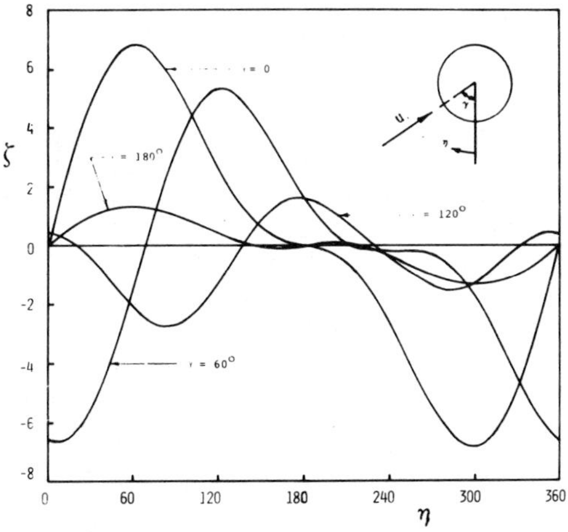

Figure 2. The variation of vorticity on the cylinder surface for different free stream directions. (Case of R_e = 20, G_r = 400 and P_r = 0.7)

It is found from the results that the velocity field is highly influenced by the free stream direction. Figure 2 shows the vorticity distribution on the cylinder surface at different flow directions ($\gamma = 0$, 60°, 120° and 180°). It can be seen from the figure that $|\zeta|$ decreases over most of the cylinder surface as γ increases. This is mainly due to the fact that when $\gamma = 0$ the buoyant forces are aiding the flow and accordingly causing higher velocities near the cylinder surface. This effect decreases as γ increases until reaching $\gamma = 180^\circ$ at which the buoyant forces are directly opposite to the forced flow direction.

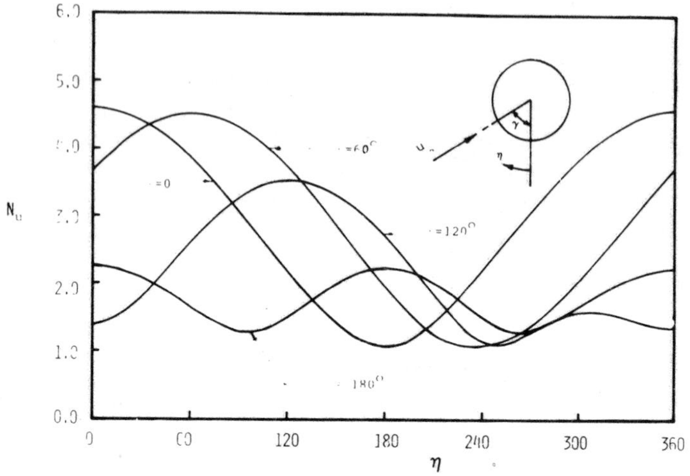

Figure 3. The Nusselt number distribution on the cylinder surface for different free stream directions. (Case of R_e=20, G_r=400 and P_r = 0.7)

It is also found that the local Nusselt number distribution is greatly affected by the direction of the main stream. The variation of Nu around the cylinder surface at different values of γ can be seen in Fig. 3. The figure shows that Nu decreases with the increase of γ over most of the cylinder surface. This is expected to occur as a natural result of the decrease of the surface vorticity. Figures 4 and 5 show the streamline and isotherm patterns for the case of Re = 10 and different values of γ. It can be seen from Fig. 4 that the size and orientation of the wake region is completely dependent on γ. When $\gamma = 0$ the wake disappears and flow separation occurs at the backward stagnation point ($\eta = 180^\circ$). On the other hand, when $\gamma = 180^\circ$ the flow field is divided into two distinct zones. The first one is surrounding the cylinder with the enclosed circulating flow driven by the buoyant forces. The heat transfer regime in this zone is dominated by natural convection. The second zone is outside the first one and the buoyant forces there have smaller effect. The heat transfer from the first zone to the second one is mainly due to conduction at the boundaries since the fluid in the first zone never gets entrained to the main stream.

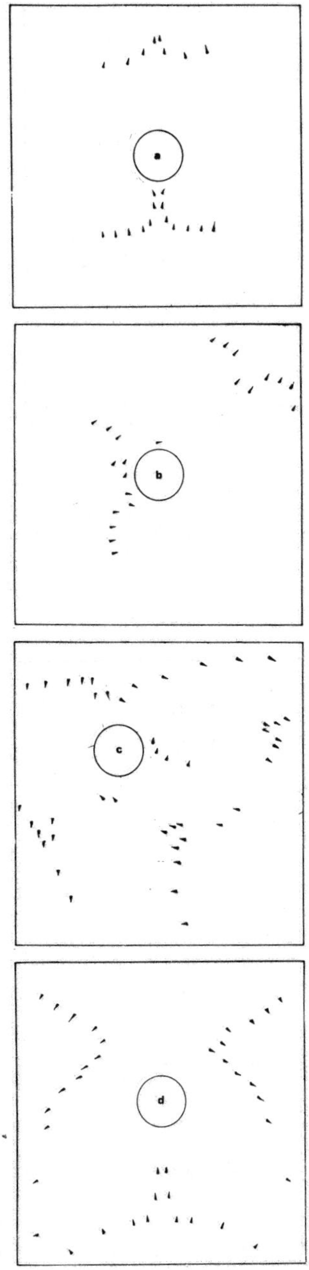

Figure 4. Streamline patterns for the case of Re = 10, Pr = 0.7, Gr = 100 and a) $\gamma = 0$, b) $\gamma = 60°$, c) $\gamma = 150°$, d) $\gamma = 180°$.

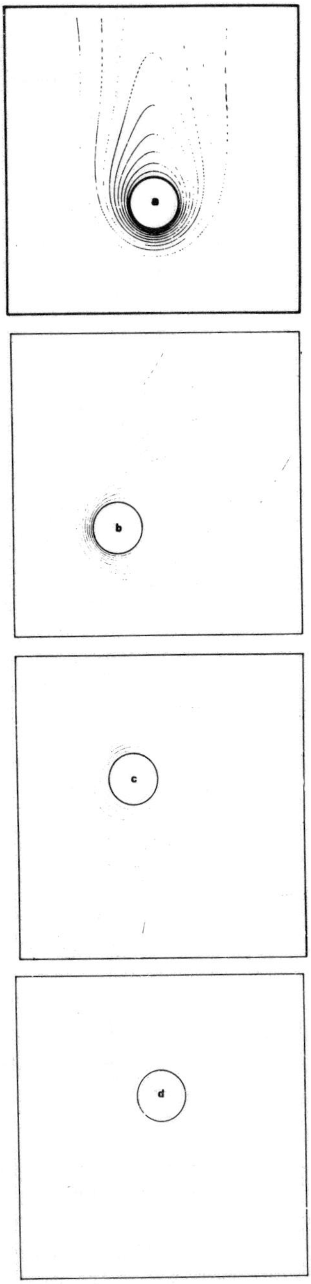

Figure 5. Isotherm patterns for the case of Re = 10, Pr = 0.7, Gr = 100 and a) $\gamma = 0$, b) $\gamma = 60°$, c) $\gamma = 150°$, d) $\gamma = 180°$.

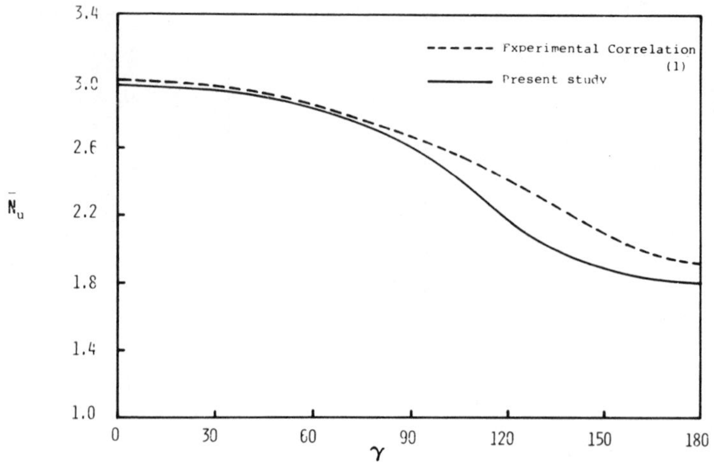

Figure 6. Comparison between the values of $\bar{N}u$ obtained from the present work and Hatton's experimental correlation [1] for the case of Re = 20, Pr = 0.7 and Gr = 400.

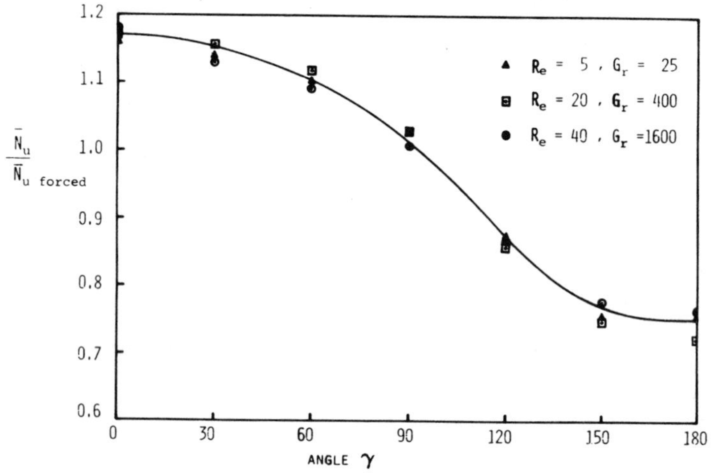

Figure 7. Variation of the ratio $\bar{N}u/\bar{N}u_{forced}$ with the free stream direction.

A comparison between the predsent results of the $\bar{N}u$ and the experimental correlation obtained by Hatton, et al [1] can be seen in Fig. 6. The agreement between the two is satisfactory up to $\gamma = 90°$, however as γ approaches 180° Hatton's correlation becomes inapplicable. Figure 7 shows the variation of $\bar{N}u/\bar{N}u_{forced}$, where $\bar{N}u_{forced}$ is the average Nusselt number in case of forced convection, with the angle γ for all the cases considered. The figure shows that the maximum value of $\bar{N}u$ occurs when $\gamma = 0$ and the minimum value occurs when $\gamma = 180°$ provided that the flow direction is the only variable.

ACKNOWLEDGEMENTS

The author wishes to acknowledge the support received from the University of Petroleum & Minerals during this study.

REFERENCES

1. Hatton, A.P., James, D.D., and Swire, H.W., 1970, J.Fluid Mech. 42, 17.

2. Oosthuizen, P.H., and Madan, S., 1971, J.Heat Transfer, 93, 240.

3. Sharma, G.K., and Sukhatme, S.P., 1969, J.Heat Transfer, 91, 457.

4. Gebhart, B., and Pera, L., 1970, J.Fluid Mech., 45, 49.

5. Oosthuizen, P.H., and Madan, S., 1970, J.Heat Transfer, 92, 194.

6. Fand, R.M., and Keswani, K.K., 1973, Int.J. Heat Mass Transfer, 16, 1175.

7. Acrivos, A., 1966, Chem. Eng. Sci., 21, 343.

8. Sparrow, E.M., and Lee, L., 1976, Int. J. Heat Mass Transfer, 19, 229.

9. Joshi, N.D., and Sukhatme, S.D., 1971, J. Heat Transfer, 93, 441.

10. Merkin, J.H., 1976, Int. J. Heat Mass Transfer, 20, 73.

11. Badr, H.M., "A Theoretical Study of Laminar Mixed Convection From a Horizontal Cylinder in a Cross Stream", in press, Int. J. Heat Mass Transfer.

12. Badr, H.M., and Dennis, S.C.R., Proceedings of the Eighth Canadian Congress of Applied Mechanics, 1981, p.659.

13. Dennis, S.C.R., and Chang, G., "Numerical Integration of the Navier-Stokes Equations in Two-Dimensions", Math. Research Center, U.S.Army, Tech. Rep. No.859, 1969.

NUMERICAL SOLUTIONS OF LINEARIZED SECONDARY GAS FLOW WITHIN ROTATING CYLINDER

Jerzy T.Sielawa and Elia R.C.San Miguel

Instituto de Estudos Avançados, CTA, São José dos Campos, SP, Brazil.

Summary

The linearized, steady state, axisymmetric equations of motion of a secondary flow of a gas within a rotating cylinder are presented. The boundary conditions at solid surfaces are decomposed into symmetric and antisymmetric parts.

Due to certain invariance properties, new boundary conditions for both cases are formulated at the middle plane of the cylinder. In consequence, only a fourth of the cross-sectional area of the cylinder has to be considered in the numerical solution of either case. The final solution is obtained as the superposition of both solutions.

1.- Equations of motion

As is well known, [1], [2], the full set of steady state gasdynamic equations in cylindrical coordinates, with axial symmetry is:

$$\begin{cases} \frac{1}{r}\frac{\partial}{\partial r}(\rho r u)+\frac{\partial}{\partial z}(\rho w) = 0 \\ \rho\left(u\frac{\partial u}{\partial r}+w\frac{\partial u}{\partial z} - \frac{v^2}{r}\right) = -\frac{\partial p}{\partial r} + \mu\left[\left(\nabla^2-\frac{1}{r^2}\right)u+\frac{1}{3}\frac{\partial}{\partial r}\operatorname{div}\mathbf{V}\right] \\ \rho\left(u\frac{\partial v}{\partial r}+w\frac{\partial v}{\partial z}+\frac{uv}{r}\right) = \mu\left(\nabla^2-\frac{1}{r^2}\right)v \\ \rho\left(u\frac{\partial w}{\partial r}+w\frac{\partial w}{\partial z}\right) = -\frac{\partial p}{\partial z}+\mu\left(\nabla^2 w+\frac{1}{3}\frac{\partial}{\partial z}\operatorname{div}\mathbf{V}\right) \\ \rho C_v\left(u\frac{\partial T}{\partial r}+w\frac{\partial T}{\partial z}\right) + p\operatorname{div}\mathbf{V} = \kappa\nabla^2 T + \Phi_{visc} \end{cases} \quad (1)$$

where

$$\Phi_{visc} = 2\mu\left\{\left(\frac{\partial u}{\partial r}\right)^2 + \left(\frac{\partial w}{\partial z}\right)^2 + \frac{u^2}{r^2} + \frac{1}{2}\left[\left(\frac{\partial v}{\partial z}\right)^2 + \left(\frac{\partial u}{\partial z}+\frac{\partial w}{\partial r}\right)^2 + \left(\frac{\partial v}{\partial r}-\frac{v}{r}\right)^2\right] - \frac{1}{3}(\mathrm{div}\mathbf{V})^2\right\} \tag{2}$$

and

$$\mathrm{div}\mathbf{V} = \frac{1}{r}\frac{\partial}{\partial r}(ru) + \frac{\partial w}{\partial z} \tag{3a}$$

$$\nabla^2 = \frac{\partial^2}{\partial r^2} + \frac{1}{r}\frac{\partial}{\partial r} + \frac{\partial^2}{\partial z^2} \tag{3b}$$

Assuming perfect gas, one obtains

$$p = R\varrho T \tag{4}$$

The cylindrical coordinates are r, θ, z, the corresponding velocity components are u, v, w ($\mathbf{V}$ being the velocity vector). The thermodynamic variables are pressure p, density ϱ and temperature T. The viscosity μ, the thermal conductivity κ and the specific heat at constant volume C_v are assumed constants. The term Φ_{visc} represents the viscous dissipation and R is the gas constant.

If the gas is contained in a vertical cylinder (whose axis coincides with the z axis), bounded by two horizontal end caps (disks), rotating as a whole with an angular velocity Ω and a wall temperature T_o, one obtains the so called rigid body solution (denoted by asterisks) of the above system, namely:

$$\begin{cases} u^* = 0 \\ v^* = \Omega r \\ w^* = 0 \\ \varrho^*/\varrho_w = p^*/p_w = \exp\left\{-A\left[1-(r/r_w)^2\right]\right\} \\ T^* = T_o \end{cases} \tag{5}$$

where the subscript "w" indicates "wall". The dimensionless quantity A is defined by

$$A = \frac{\Omega^2 r_w^2}{2RT_o} \tag{6}$$

which corresponds roughly to the squared Mach number.

2.- Linearization

In order to obtain the linearized equations of motion governing a small perturbation about the rigid body rotation defined by (5), the following scaled nondimensional perturbation quantities (denoted by tildes) are introduced:

$$\begin{cases} u = \Omega r_w \epsilon \tilde{u} \\ v = \Omega r_w \tilde{r} - \Omega r_w \epsilon \tilde{v} \\ w = \Omega r_w \tilde{w} \\ p = p^*(1 + \epsilon \tilde{p}) \\ \rho = \rho^*(1 + \epsilon \tilde{\rho}) \\ T = T^*(1 + \epsilon \tilde{T}) \end{cases} \quad (7)$$

where

$$r = r_w \tilde{r} \; ; \quad z = r_w \tilde{z} \; , \quad (8)$$

ϵ being a Rossby number that measures the magnitude of the perturbation.

Introducing (7) and (8) into (1), neglecting the second-order terms in ϵ and omitting the tildes, one can obtain the following, [2]:

$$\begin{cases} r\frac{\partial u}{\partial r}+(2Ar^2+1)u+r\frac{\partial w}{\partial z} = 0 \\ r^2\frac{\partial^2 u}{\partial r^2}+r(1-\frac{2}{3}Ar^2)\frac{\partial u}{\partial r}+r^2\frac{\partial^2 u}{\partial z^2}-(1+\frac{2}{3}Ar^2)u+2r^2Hv-r^3HT+ \\ \qquad -\frac{1}{2A}r^2H\frac{\partial p}{\partial r} = 0 \\ -2r^2Hu+r^2\frac{\partial^2 v}{\partial r^2}+r\frac{\partial v}{\partial r}+r^2\frac{\partial^2 v}{\partial z^2} - v = 0 \\ -\frac{2}{3}Ar^2\frac{\partial u}{\partial z}+r\frac{\partial^2 w}{\partial r^2}+\frac{\partial w}{\partial r}+r\frac{\partial^2 w}{\partial z^2}-\frac{1}{2A}rH\frac{\partial p}{\partial z} = 0 \\ 4r^2BHu+r\frac{\partial^2 T}{\partial r^2}+\frac{\partial T}{\partial r}+r\frac{\partial^2 T}{\partial z^2} = 0 \end{cases} \quad (9)$$

with

$$H = E^{-1}\exp\{-A(1-r^2)\} \quad (10)$$

where E and B are two characteristic nondimensional quantities (Ekman number and Brinkman number, respectively), given by:

$$\begin{cases} E = \dfrac{\mu}{\rho_w \Omega r_w^2} & (11a) \\ B = \dfrac{\gamma-1}{2\gamma} P_R A & (11b) \end{cases}$$

where γ is the specific heats ratio and P_R is the Prandtl number,

$$P_R = \frac{\gamma \mu C_v}{\kappa}$$

Observe that E measures the magnitude of viscosity effects relative to the Coriolis force effects, while B represents the coupling between thermal and dynamical effects.

3.- Boundary conditions at solid surfaces.

If the angular velocity of the end caps is not exactly Ω and/or if the temperature at solid surfacesis not exactly T_o, the rigid body rotation of the gas will be perturbed. If the perturbation is small, it will generate a secondary flow, governed by equations (9).

Consider the following boundary conditions (with z axis along the cylinder symmetry axis and with the end caps at $z = \pm\lambda/2$)

$$\begin{cases} z = -\lambda/2 \Rightarrow v = \epsilon_1 r \quad , \quad T = \tau_1 \\ z = \lambda/2 \Rightarrow v = \epsilon_2 r \quad , \quad T = \tau_2 \\ r = 1 \Rightarrow v = 0 \quad , \quad T = (\tau_2 - \tau_1)\frac{z}{\lambda} + \frac{\tau_1 + \tau_2}{2} \end{cases} \tag{13}$$

where all variables are nondimensional.

Observe that at r=1 the temperature distribution is linear with $T=\tau_1$ at $z=-\lambda/2$ and $T=\tau_2$ at $z=\lambda/2$.

Assuming that

$$\epsilon_1, \epsilon_2, \tau_1, \tau_2 \ll 1,$$

the perturbation of the rigid body rotation is small and, thus, is governed by equations (9).

The rest of the boundary conditions is assumed to remain the same as in the case of the rigid body rotation, i.e.:

$$\begin{cases} u = v = 0 \quad \text{at } z = \pm\lambda/2 \text{ and at } r = 1 \\ v = 0 \quad \text{at } r = 1 \end{cases} \tag{14}$$

4.- Symmetric and atisymmetric parts of the perturbation.

Following [2], the boundary conditions (12) can be written as:

$$v = \begin{cases} \epsilon_s r + \epsilon_a r & \text{at } z = -\lambda/2 \\ \epsilon_s r - \epsilon_a r & \text{at } z = \lambda/2 \end{cases}$$

$$T = \begin{cases} \tau_s + \tau_a & \text{at } z = -\lambda/2 \\ \tau_s - \tau_a & \text{at } z = \lambda/2 \\ \tau_s - 2\tau_a z/\lambda & \text{at } r = 1 \end{cases} \tag{15}$$

where

$$\epsilon_s = \frac{\epsilon_1+\epsilon_2}{2} , \quad \tau_s = \frac{\tau_1+\tau_2}{2} \tag{16a}$$

and

$$\epsilon_a = \frac{\epsilon_1-\epsilon_2}{2} , \quad \tau_a = \frac{\tau_1-\tau_2}{2} \tag{16b}$$

and where "s" stands for "symmetric" and "a" for "antisymmetric".

5.- Symmetric case.

Consider the boundary conditions:

$$\begin{cases} v = \epsilon_s r \ , \quad T = \tau_s & \text{at } z = \pm \lambda/2 \\ T = \tau_s & \text{at } r = 1 \end{cases} \tag{17}$$

together with conditions (14).

It is easy to verify that the linearized system (9) and the above boundary conditions are invariant under the following transformation:

$$\begin{cases} r = \bar{r} \ , \quad z = -\bar{z} \\ u = \bar{u} \ , \quad v = \bar{v} \ , \quad w = -\bar{w} \ , \quad p = \bar{p} \ , \quad T = \bar{T} \end{cases} \tag{18}$$

This indicates that in the symmetric case the variables u, v, p and T are even functions and w is an odd function with respest to z. Thus, new boundary conditions can be formulated at $z = 0$ (the middle plane of the cylinder) as:

$$\frac{\partial u}{\partial z} = \frac{\partial v}{\partial z} = w = \frac{\partial p}{\partial z} = \frac{\partial T}{\partial z} = 0 \ . \tag{19}$$

6.- Antisymmetric case.

Consider now the boundary conditions:

$$\begin{cases} v = \epsilon_a r \ , \quad T = \tau_a & \text{at } z = -\lambda/2 \\ v = -\epsilon_a r \ , \quad T = -\tau_a & \text{at } z = \lambda/2 \\ T = -2\tau_a z/\lambda & \text{at } r = 1 \ , \end{cases} \tag{20}$$

together with conditions (14).

The system (9) and the above boundary conditions are now invariant under the following transformation:

$$\begin{cases} r = \bar{r} \ , \quad z = -\bar{z} \\ u = -\bar{u} \ , \quad v = -\bar{v} \ , \quad w = \bar{w} \ , \quad p = -\bar{p} \ , \quad T = -\bar{T} \end{cases} \tag{21}$$

This indicates that in the antisymmetric case, the variables u, v, p and T are odd functions and w is an even function with respect to z.

Thus, in the middle plane (z=0), new boundary conditions can be formulated in the following manner:

$$u = v = \frac{\partial w}{\partial z} = p = T = 0 \tag{22}$$

7.- Superposition.

Let equations (9) be represented in operational form by:

$$L[u] = 0 \tag{23}$$

where u is a vector composed by the dependent variables and let the boundary conditions (17) and (14) be represented by:

$$M[u] = g_s \tag{24a}$$

and (20) & (14) by:

$$M[u] = g_a \tag{24b}$$

Then, the original conditions (13) & (14) are

$$M[u] = g_s + g_a \tag{24c}$$

Due to the linearity of the operators L and M, it is easy to see that if $u=u_s$ is the solution of (23) with (24a) and $u=u_a$ is the solution of (23) with (24b), then

$$u = u_s + u_a$$

is the solution of the original problem, namely the system (23) with the boundary conditions (24c).

Thus, in order to numerically solve the original problem, it can be decomposed into two uncoupled parts. The numerical solution of either part is much easier than that of the original problem, since, by formulation of the boundary conditions at the middle plane (z=0), only half of the original area has to be spanned by grid points, increasing, thus, the accuracy and/or saving the computational time.

8.- Boundary conditions on the axis of rotation.

Since the flow is assumed axisymmetric, it is easy to verify that the system (9) with the boundary conditions (13) & (14) remains invariant under the transformation:

$$\begin{cases} r = -\bar{r} \ , \ z = \bar{z} \\ u = -\bar{u} \ , \ v = -\bar{v} \ , \ w = \bar{w} \ , \ p = \bar{p} \ , \ T = \bar{T} \end{cases}$$

This indicates that u and v are odd functions and w, p and T are even functions with respect to r. Thus, at the axis of rotation (r=0),

$$u = v = \frac{\partial w}{\partial r} = \frac{\partial p}{\partial r} = \frac{\partial T}{\partial r} = 0 \ . \tag{25}$$

This way, only a fourth of the cross-sectional area of the cylinder has to be considered in the numerical solution.

9.- Numerical procedures.

As an example of the numerical procedure, the antisymmetric case is considered.

Let the grid points be introduced in the area of interest as shown in Fig.1. The points (i,j) with $0<i<m$ and $0<j<n$ will be called "interior points". Let $h=1/m$ and $k=\lambda/2n$ be the horizontal and vertical distances between the points.

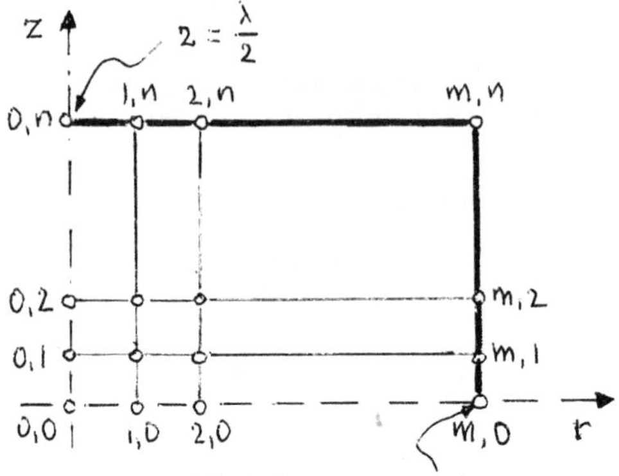

Fig.1

Introducing the second-order differences, equations (9) are discretized in the interior points as follows:

$$-\frac{i}{2}u_{i-1,j} + (2Ai^2h^2+1)u_{i,j} + \frac{i}{2}u_{i+1,j} + \frac{i}{2}\frac{h}{k}w_{i,j+1} + $$
$$-\frac{i}{2}\frac{h}{k}w_{i,j-1} = 0$$

$$(i^2-\frac{i}{2}+\frac{A}{3}i^3h^2)u_{i-1,j} - (2i^2+2i^2\frac{h^2}{k^2}+1+\frac{2}{3}Ai^2h^2)u_{i,j} + $$
$$+(i^2+\frac{i}{2}-\frac{A}{3}i^3h^2)u_{i+1,j} + \frac{i^2h^2}{k^2}u_{i,j-1} + \frac{i^2h^2}{k^2}u_{i,j+1} + $$
$$+2i^2h^2H(r_i)v_{i,j} - i^3h^3H(r_i)T_{i,j} + $$
$$-\frac{1}{4A}i^2hH(r_i)p_{i-1,j} + \frac{1}{4A}i^2hH(r_i)p_{i+1,j} = 0$$

$$-2i^2h^2H(r_i)u_{i,j} + (i^2-\frac{i}{2})v_{i-1,j} - (2i^2+2\frac{i^2h^2}{k^2}+1)v_{i,j} + $$
$$+(i^2+\frac{i}{2})v_{i+1,j} + \frac{i^2h^2}{k^2}v_{i,j-1} + \frac{i^2h^2}{k^2}v_{i,j+1} = 0$$

$$\frac{Ai^2h^2}{3k}u_{i,j-1} - \frac{Ai^2h^2}{3k}u_{i,j+1} + (\frac{i}{h}-\frac{1}{2h})w_{i-1,j} + $$

$$-(2\frac{i}{h}+2\frac{ih}{k^2})w_{i,j}+(\frac{i}{h}+\frac{1}{2h})w_{i+1,j}+\frac{ih}{k^2}w_{i,j-1}+$$

$$+\frac{ih}{k^2}w_{i,j+1}+\frac{i}{4Ak}H(r_i)p_{i,j-1}-\frac{i}{4Ak}p_{i,j+1} = 0$$

$$4iBhH(r_i)u_{i,j}+(\frac{i}{h}-\frac{1}{2h})T_{i-1,j}-(2\frac{i}{h}-2\frac{ih}{k^2})T_{i,j}+ \tag{26}$$

$$+(\frac{i}{h}+\frac{1}{2h})T_{i+1,j}+\frac{ih}{k^2}T_{i,j-1}+\frac{ih}{k^2}T_{i,j+1} = 0$$

These are 5(m-1)(n-1) equations. Since the total number of unknowns is 5(m+1)(n+1), one has to formulate 10(m+n) additional equations - which cannot be obtained solely from boundary conditions, since there are no expressions for pressure, neither at $z=\lambda/2$ nor at r=1.

The experience has shown that probably the best way to overcome this problem is to replace the pointwise continuity equations, i.e. the first of equations (26) by its integral form within (mn) mesh rectangles. To do so, observe that:

$$\int_{r_i}^{r_{i+1}}\rho^* w\cdot 2\pi r\,dr = \frac{\pi}{3}\rho_w e^{-A(1-ih)}h^2\{(3i+1)w_{i,j}+ +(3i+2)e^{Ah}w_{i+1,j}\} \tag{27}$$

$$\int_{z_j}^{z_{j+1}}\rho^* u\cdot 2\pi r_i dz = \pi\rho_w e^{-A(1-ih)}ihk(u_{i,j}+u_{i,j+1})$$

Observe that m-1 additional equations can be obtained from the fourth of equations (26), at z=0. i.e. substituting j=0. Recalling that $u_{i,-1}=-u_{i,1}$, etc, one gets:

$$-\frac{2}{3}A\frac{i^2h^2}{k}u_{i,1}+(\frac{i}{h}-\frac{1}{2h})w_{i-i,0}-(2\frac{i}{h}+2\frac{ih}{k^2})w_{i,0}+$$

$$+(\frac{i}{h}+\frac{1}{2h})w_{i+1,0}+2\frac{ih}{k^2}w_{i,1}-\frac{i}{2Ak}H(r_i)p_{i,1} = 0 \tag{28}$$

Now one has already

$$4(m-1)(n-1)+mn+m = 5mn-3m-4n+4$$

equations. Still needed are 8m+9n+1 ones. These (with only one remaining to be established) can be obtained directly from the boundary conditions:

$$u_{i,0} = v_{i,0} = p_{i,0} = T_{i,0}, \quad 0\leq i\leq m$$

$$u_{i,n} = w_{i,n} = 0, \ T_{i,n} \text{ and } v_{in} \text{ - prescribed} \quad 1\leq i\leq m \tag{29}$$

$$u_{0,j} = v_{0,j} = \left(\frac{\partial w}{\partial r}\right)_{0,j} = \left(\frac{\partial p}{\partial r}\right)_{0,j} = \left(\frac{\partial T}{\partial r}\right)_{0,j} = 0$$

$$1 \leqslant j \leqslant n$$

$$u_{m,j} = v_{m,j} = w_{m,j} = 0 \ , \ T_{m,j} \text{ - prescribed}$$

$$1 \leqslant j \leqslant n-1$$

One has to be careful not to include redundant informations like, for instance $T_{i,0}=0$ $(0 \leqslant i \leqslant m)$ and $(\partial T/\partial r)_{0,0}=0$.

The number of the above equations is 8m+9n; the last remaining equation should contain $p_{m,n}$, since it is not present in any of the equations so far treated. It can be obtained from the fourth of equations (9), applied to the (mn) corner:

$$\left(\frac{\partial p}{\partial z}\right)_{m,n} = 0 \qquad (30)$$

Finally, the discretization of expressions in (29) and (30), containing first derivatives could be achieved by forward or backward second-order differences. It appears from experience that the first-order is more desirable in this case.

10.- Numerical results.

By the application of the above equations to the symmetric and to the antisymmetric case, the typical countercurrent flow patterns, represented schematicly in Fig.2 and 3, respectively, are obtained.

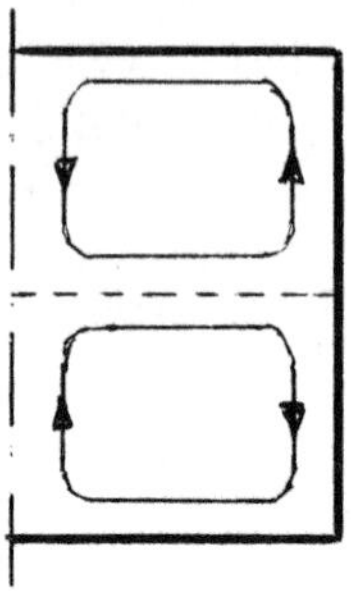

Fig.2

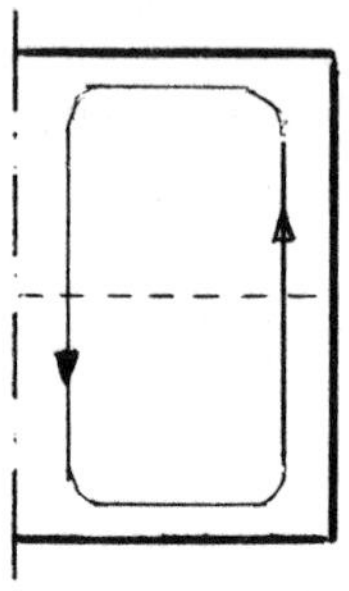

Fig.3

The calculations were performed for moderate values of A and E, and for $B \ll 1$. For $A \gg 1$ and $E \ll 1$, rapid variations in flow parameters appear near solid surfaces. In these cases it is not practical to maintain the mesh points equidistant. Consequently, slight reformulation of equations (26) in this case is required.

References.

1 - VILLANI, S. (ed.) - Uranium Enrichment - Topics in Applied Physics, Vol.35, Springer-Verlag, New York, 1979

2 - SIELAWA, M.M.F. - Secondary Flow in a Rotating Fluid - Analytic Solution Based on Properties of Symmetry and Antisymmetry, (in Portuguese) - PhD Thesis, Instituto Tecnológico de Aeronáutica, São José dos Campos, Brazil, 1982.

FINITE ELEMENT ANALYSIS OF THREE-DIMENSIONAL FLOW FIELD AROUND A PAIR OF BUBBLES IN SUBCOOLED NUCLEATE BOILING

Dr.A.L.Chandraker(1) and Dr.A.Gopalakrishnan(2)

SUMMARY

The most effective heat transfer processes occur in the fully-developed subcooled nucleate boiling domain which is characterized by reasonably low surface-to-fluid temperature differentials and by the vigorous growth and collapse of a large number of bubbles from discrete nucleation centres on the heated surface [1]. An understanding of heat transfer mechanism, in this case, requires a knowledge of the flow field as one of the input variables. In the present work, a three-dimensional quasi-steady hydrodynamic model is proposed to study the flow field surrounding a pair of bubbles, hemispherical in shape and each of them generated at different times. The mathematical model, which is solved here by the finite element method is subjected to zero Neumann conditions on the solid surfaces of a large vessel containing stagnant fluid and vapour bubbles. Available semi-emperical solution of energy equation for bubble growth/collapse rate is incorporated in the form of a non-zero Neumann condition imposed on the bubble-liquid interface. Dirichlet condition is prescribed on the free surface separating liquid and ambient medium.

The computer code is first applied to an isolated bubble and the corresponding numerical result is compared with an analytical solution for a volume source (representing a stationary bubble) placed in an unbounded space. The comparison appears to be interesting since it brings out the wall

---

(1) Deputy Manager } Energy Systems Group, BHEL
(2) General Manager } Vikas Nagar, Hyderabad-500593, INDIA

effect. Subsequently, the code is applied to the present problem of a pair of bubbles. Isopotential lines on the vertical central plane are plotted for various growth and decay phases. As expected, the isopotential lines are concentrated around the bubbles; more densely near the bubble with higher growth or collapse rates. The lines flatten out as the distance from the bubbles increases.

1. INTRODUCTION

As in any other convective process the heat exchange between bubbles and the surrounding fluid depends on the flow conditions of the fluid. An understanding of the flow and flow regime is, therefore a prerequisite for an analysis of the heat transport mechanism. Many semi-empirical models have been proposed for the extremely large heat transfer coefficients observed in the nucleate boiling regime [1,2,3]. A systematic study of bubble dynamics and the associated flow field needs solutions of mass, momentum and energy equations at three distinct regions; viz., liquid region, bubble vapour region and bubble-liquid interface. The third region, i.e., bubble-liquid interface introduces Neumann type of boundary condition for the other two regions. In the present work, the first region, i.e. liquid region is analysed from the hydrodynamic point of view.

2. MATHEMATICAL MODEL

The fluid is considered here as incompressible, inviscid, isentropic and predominantly stagnant (some unsteadiness is introduced by the motion of bubble-liquid interface which is accounted here by suitable time-dependent boundary condition). Under the above assumptions, the flow may be considered as a potential flow, as shown in this section. The appropriate form of governing equations in the liquid region may be written as (neglecting body force).

Mass: $$\frac{1}{\rho}\frac{D\rho}{Dt} + \underline{\nabla}\cdot\underline{u} = 0 \quad (1)$$

Momentum: $$\frac{\partial \underline{u}}{\partial t} + (\underline{u}\cdot\underline{\nabla})\,\underline{u} = -\frac{1}{\rho}\,\underline{\nabla} p \quad (2)$$

Isentropy: $$dp = c^2\, d\rho \quad (3)$$

The symbols p, ρ, $\underline{u}$, c and t denote fluid pressure, density, instantaneous velocity, sound speed and time, respectively. Taking the scalar product of (2) with $\underline{u}$ and making use of (1) and (3), one obtains

$$\frac{\partial \underline{u}^2}{\partial t} + \frac{1}{2}\,\underline{u}\cdot\underline{\nabla}(\underline{u}^2) - c^2\,\underline{\nabla}\cdot\underline{u} = -\frac{1}{\rho}\frac{\partial p}{\partial t} \quad (4)$$

where use is made of the following vector identity

$$(\underline{u}\cdot\underline{\nabla})\,\underline{u} \quad = \quad \underline{\nabla}\,(\underline{u}^2/2) \quad - \underline{u}\times\underline{\Omega} \tag{5}$$

$\underline{\Omega}$ is the vorticity vector $\underline{\nabla}\times\underline{u}$. Further on taking the time derivative of (2) and making use of the definition of a potential function ϕ as

$$\underline{u} \quad = \quad \underline{\nabla}\,\phi \tag{6}$$

one obtains

$$\frac{\partial^2\phi}{\partial t^2} + \frac{1}{2}\,\frac{\partial}{\partial t}\,(\underline{\nabla}\,\phi)^2 \quad = \quad -\frac{1}{\rho}\,\frac{\partial p}{\partial t} \tag{7}$$

Substitution of (6) and (7) in (4) yields after some rearrangement [4]

$$\frac{1}{c^2}\left[\frac{\partial^2\phi}{\partial t^2} + \frac{\partial}{\partial t}(\underline{\nabla}\,\phi)^2 + \frac{1}{2}\,\underline{\nabla}\phi\cdot\underline{\nabla}\,(\underline{\nabla}\,\phi)^2\right] - \underline{\nabla}^2\phi = 0 \tag{8}$$

For an incompressible fluid $c \rightarrow \infty$; thereby eq.(8) reduces to the well-known Laplace equation

$$\underline{\nabla}^2\phi \;=\; 0 \tag{9}$$

The idealized physical description of the flow domain is as follows: a large container of square cross section, filled with fluid which is essentially incompressible and stagnant, is heated from the bottom to attain a wall temperature which

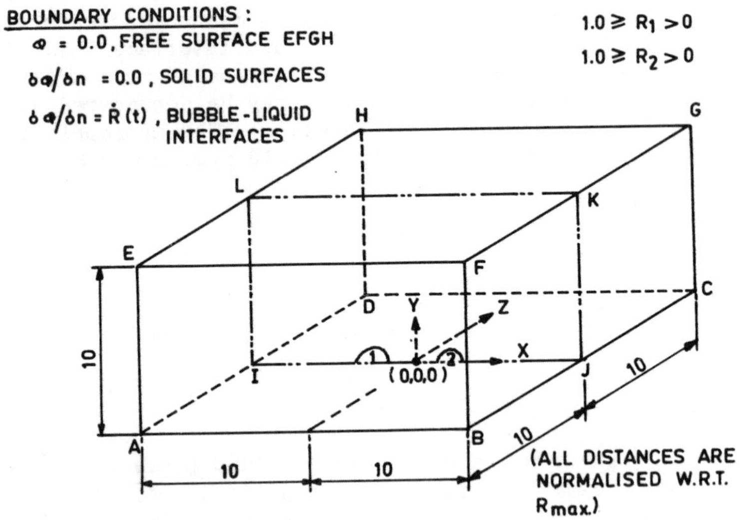

FIG.1. THREE DIMENSIONAL FLOW DOMAIN

can sustain nucleate boiling. The time-dependent growth of the bubble introduces a corresponding displacement and growth of the bubble-liquid interface. The bubbles are assumed to remain hemispherical and attached to the bottom surface throughout their life; thereby neglecting the effect of buoyancy. The appropriate boundary conditions are as shown in Fig.1.

The bubble radius R and its growth rate $\dot{R} = dR/dt$ may be derived by writing down energy-balance equation at the interface. It is assumed here that the diffusion mode of heat transfer is predominant throughout the bubble life, thus neglecting the effects of surface tension and inertia which are significant, however, only at the initial stage of bubble growth.

For the sake of generality, the governing equation $\nabla^2\phi = 0$ and variables R and $\dot{R}$ are normalized with respect to maximum bubble radius R_{max} and time t_{max} (defined as the time at which $R = R_{max}$ or $\dot{R} = 0$). Thus

$$1 \geq (\hat{R} = R/R_{max}) > 0$$

during its life $t/t_{max} = 2$ time-units. Laplace equation remains of the same form even after normalizing (with $\hat{\phi} = \phi / R^2_{max}\, t_{max}$). The maximum bubble radius R_{max} follows $R_{max} \simeq \sqrt{\frac{12}{\pi}}\; J_a \sqrt{\alpha_L t_{max}}$ where t_{max} depends on experimental set up. Symbols J_a and α_L denote Jacob number and thermal diffusivity respectively.

Two of the suitable growth and implosion laws in non-dimensional form chosen in the present work are the followings [5].

Growth: $\hat{R} = \sqrt{\hat{t}}\;(2 - \sqrt{\hat{t}})$; $1 \geq \hat{t} > 0$

Implosion: $\hat{R} = \sqrt{2 - \hat{t}}\;(2 - \sqrt{2 - \hat{t}})$; $2 > \hat{t} \geq 1$

3. FINITE ELEMENT FORMULATION

An appropriate integral equation corresponding to (9) may be derived by Galerkin's residual technique by multiplying (9) by a weighting function W and applying Gauss theorem while integrating over an elemental volume $V^{(e)}$. The resulting equation for $\hat{\phi}$ in normalized Cartesian system (x,y,z) turns out to be (omitting the symbol ' ' henceforth)

$$\int_{V^{(e)}} \left(\frac{\partial W}{\partial x} \frac{\partial \phi^{(e)}}{\partial x} + \frac{\partial W}{\partial y} \frac{\partial \phi^{(e)}}{\partial y} + \frac{\partial W}{\partial z} \frac{\partial \phi^{(e)}}{\partial z} \right) dV^{(e)} = \oint_{S^{(e)}} W \frac{\partial \phi^{(e)}}{\partial n} ds^{(e)} \quad (10)$$

where

$$dV^{(e)} = dx^{(e)}\, dy^{(e)}\, dz^{(e)} \quad (11)$$

$\underline{n}$ is the positive normal on a boundary surface element $S^{(e)}$, pointing outward. A superscript (e) denotes element identity.

The flow field is discretized into a set of 20-node isoparametric three-dimensional elements. The potential over an element (e) is approximated in terms of 'shape function' N_j [6].

$$\phi^{(e)} = \Sigma\, N_j \cdot \phi_j^{(e)} \quad (12)$$

The summation is over all the nodes of an element. With (12) and using the shape function as weight function [6]

$$W = N_i \quad (13)$$

eq.(10) takes the matrix form, for each element

$$[K]^{(e)} \{\phi\}^{(e)} = \{f\}^{(e)} \quad (14)$$

where co-efficient matrix $[K]^{(e)}$ and load vector $\{f\}^{(e)}$ are

$$K_{ij}^{(e)} = \int_{V^{(e)}} \left(\frac{\partial N_i}{\partial x} \cdot \frac{\partial N_j}{\partial x} + \frac{\partial N_i}{\partial y} \cdot \frac{\partial N_j}{\partial y} + \frac{\partial N_i}{\partial z} \cdot \frac{\partial N_j}{\partial z} \right) dV^{(e)} \quad (15)$$

$$f_i^{(e)} = \int_{S^{(e)}} N_i \frac{\partial \phi^{(e)}}{\partial n} ds^{(e)} \quad (16)$$

Eq.(14), on assembly over all the elements yields a "global system" of simultaneous algebraic equations.

A finite element program is developed and applied to an isolated hemispherical bubble placed at the centre of the bottom plate of a vessel which is of circular section. Both the radius of the circular section and height of vessel are chosen to be 15 length-units (with maximum bubble radius R_{max} = 1 unit). Analytically, one may expect that the potential function for a volumetric source in an unbounded space is to follow

$$\phi = A/r$$

with average value of constant A on the bubble surface is equal to $A = \phi R = -0.55$. r denotes the distance from the bubble centre. The following table shows an interesting comparison between potential fields of a volumetric source in an unbounded space and a hemispherical bubble placed in a circular vessel.

Table 1

r	FEM solution for the bottom plane	Analytical solution
2	- 0.265	- 0.277
4	-0.131	- 0.138
6	- 0.089	- 0.092
10.5	- 0.053	- 0.053

Note that FEM solution has departed from the relation $\phi = A/r$ due to wall effects.

4. NUMERICAL RESULT

The vessel is chosen of square cross-section, each side 20 length units long and of height 10 units. The two bubbles are placed at equi-distance from the origin located at the centre of square bottom plate (Fig.1). The bubble ① is assumed to start growing at 0.3 time-unit later than the bubble ② . The fluid domain is discretized into 72 elements involving 400 nodes. Nodal positions on the two planes ABCD and IJKL are shown in Figs. 2 and 3.

The isopotential lines are shown for the vertical central plane IJKL for two growth instants (Figs. 4 & 5) and two decay instants (Figs. 6 & 7). The lines are, as expected physically, concentrated around the bubble having higher growth/implosion rates; and flattens out at large distances from the bubble centres. Since the streamlines are ortho-normal to potential lines, one may conclude that the resultant velocity vector in the central plane is at some angle, rather than being normal, near the inner face of the bubble with lesser growth/implosion rate; due to presence of another stronger bubble.

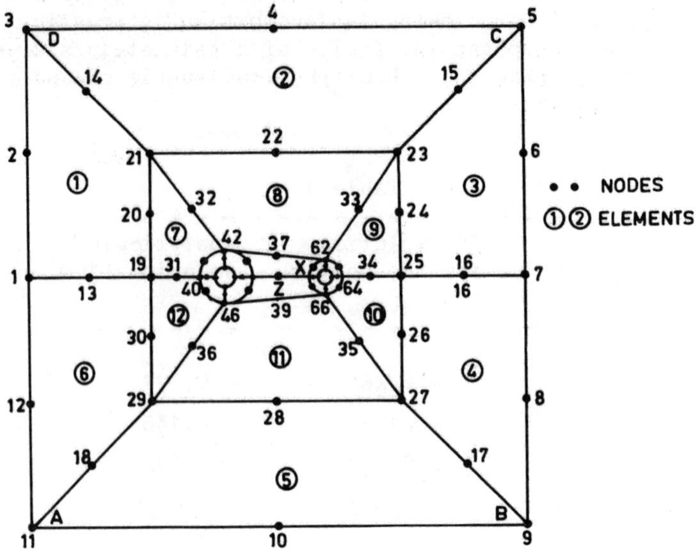

FIG.2. NODES ON BOTTOM PLANE ABCD

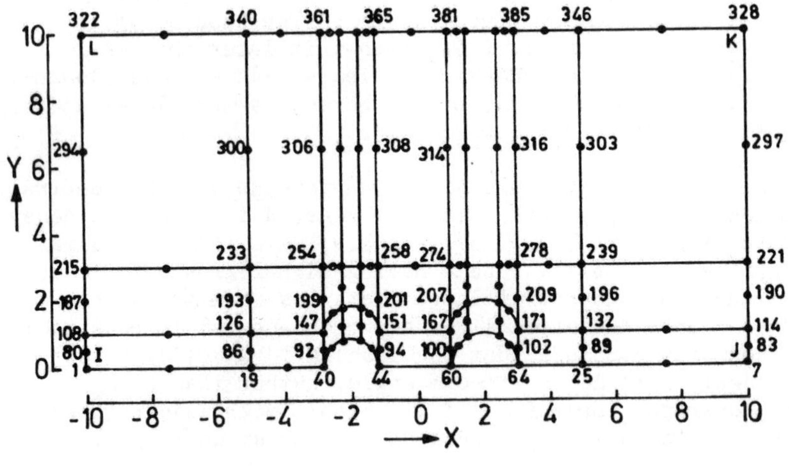

FIG.3. NODES ON CENTRAL SECTION IJKL (Z=0, PLANE)

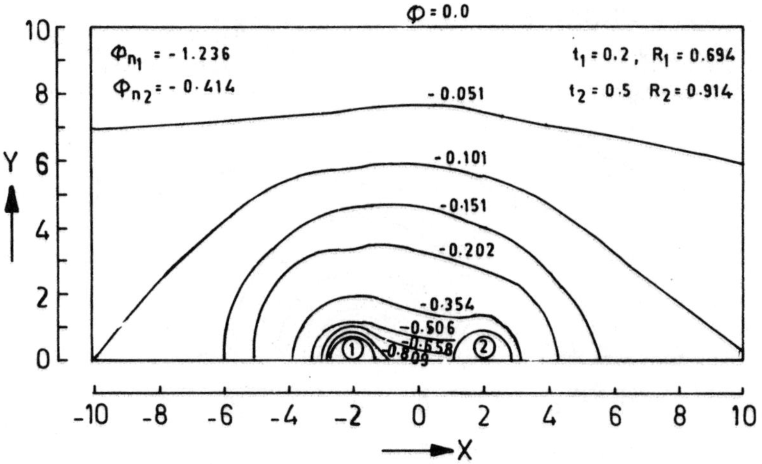

FIG.4. ISOPOTENTIAL LINES - GROWTH PHASE I

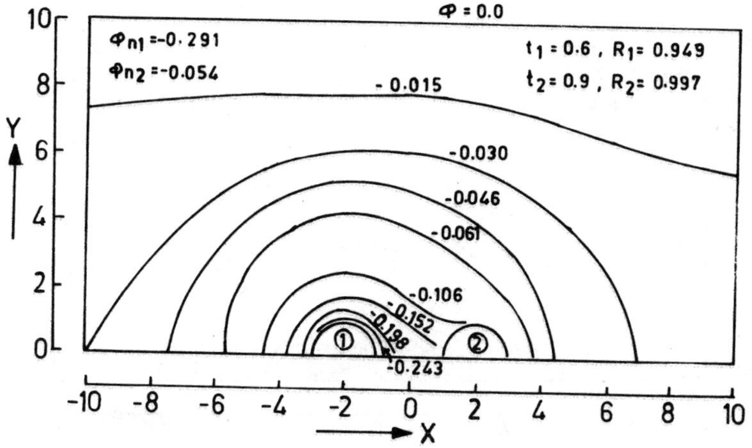

FIG.5. ISOPOTENTIAL LINES - GROWTH PHASE II

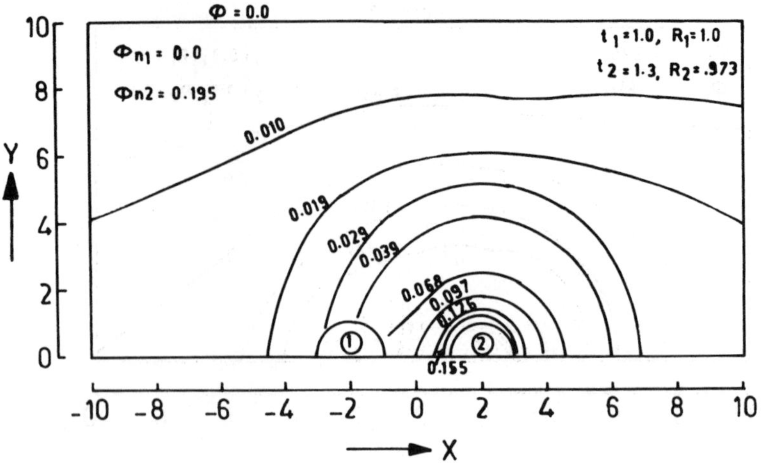

FIG.6. ISOPOTENTIAL LINES - DECAY PHASE I

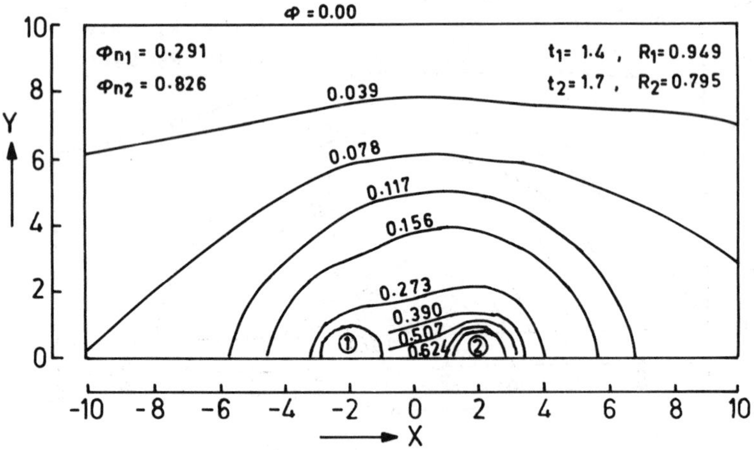

FIG.7. ISOPOTENTIAL LINES - DECAY PHASE II

As a concluding remark it is worthwhile to mention that the present work is only a part of general study for the investigation of the bubble dynamics and nucleate boiling phenomena. The next work being taken up concerns with the thermal modelling of the field surrounding the bubbles; which in turn, requires flow pattern obtained from the present study as a necessary input.

5. REFERENCES

1. VAN STRALEN, S. and COLE, R. - Boiling Phenomena Vol.1., McGraw-Hill, New York, 1979.

2. PLESSET, M.S. and ZWICK, S.A. - The growth of Vapour Bubbles in Superheated Liquids, J. Appl. Phys., Vol.25, pp.493-500, 1954.

3. FORSTER, H.K. and ZUBER, N. - Growth of a Vapour Bubble in Superheated Liquid, J.Appl.Phys., Vol.25, pp.474-478, 1954.

4. THOMPSON, P.A. - Compressible Fluid Dynamics, McGraw-Hill, New York, 1972.

5. LEPPERT, G. and PITTS, C.C. - Boiling, Advances in Heat Transfer, Vol.1, pp.185-263, 1964.

6. ZIENKIEWICZ, O.C. - The Finite Element Method, McGraw-Hill, New York, 1977.

SECTION 11

CONVECTION/DIFFUSION

A TAYLOR-GALERKIN METHOD FOR CONVECTIVE TRANSPORT PROBLEMS

Jean Donea

Commission of the European Communities
Joint Research Centre - Ispra Establishment, 21020 Ispra (Va) - Italy

ABSTRACT

The recently proposed Taylor-Galerkin method is briefly illustrated and applied to derive time-accurate finite-element schemes for pure advection and mixed advection-diffusion problems governed by the non-linear Burgers' equation. By contrast with the Petrov-Galerkin formulations of upwind-type, the Taylor-Galerkin approach provides a rational basis to improve the order of accuracy of the time-discretised equations without introducing special weighting functions, nor free parameters. To produce an accurate time discretisation, the method employs Taylor series expansions in the time-step including second-order and, when possible, third-order time derivatives. These are then evaluated from the governing partial differential equation, so as to obtain simple two-level time-stepping schemes. The resulting time-discretised equation is successively discretised in space using the standard Bubnov-Galerkin finite-element formulation.

1. INTRODUCTION

The numerical solution of initial-boundary-value problems generally involves two distinct discretisation processes: temporal discretisation and spatial discretisation. In recent years, considerable interest has developed in the potential advantages of the finite-element method for the spatial discretisation of fluid dynamics problems and, in particular, those governed by hyperbolic partial differential equations. Although it was soon established that, for hyperbolic problems, a high spatial accuracy can be obtained with a Galerkin semidiscrete formulation, rather severe difficulties were encountered in coupling the spatial approximation provided by the finite-element method to the time discretisation. In fact, owing to the role of characteristics in the solution of hyperbolic partial differential equations, disadvantages of the Galerkin formulation do appear as soon as a standard, low-order, time discretisation is introduced.

To illustrate the problem, consider the hyperbolic evolutionary equation

$$u_t + Lu = 0 \tag{1.1}$$

where L is a first-order spatial operator. In the semidiscrete Galerkin formulation, the approximate solution $U(\underline{x},t)$ to eq.(1.1) is represented as a product of shape functions $N_I(\underline{x})$ which are independent of time and nodal values $U_I(t)$ which are independent of $\underline{x}$ and incorporate the time dependence

$$U(\underline{x},t) = \sum_I N_I(\underline{x})\, U_I(t) \tag{1.2}$$

The Galerkin finite-element approximation to eq.(1.1) is then obtained by setting

$$< U_t + LU, N_I > = 0, \text{ for all I,} \tag{1.3}$$

where $< u,v >$ denotes the L_2 inner product $\int uv d\underline{x}$ over the domain of the problem. If the N_I's are linear shape functions on a uniform mesh of size h, the semidiscrete nodal equations resulting from (1.3) are accurate to $0(h^4)$ when the operator L represents a first-order hyperbolic system, either linear or quasi-linear. This high spatial accuracy is a direct consequence of the presence of a consistent "mass" matrix in the finite-element formulation. However, the system of ordinary differential equations emanating from (1.3) still has to be integrated forward in time to produce the transient response, and the fourth-order accuracy can be quickly eroded as the time-step Δt is increased, if the finite-difference approximation to the time-derivative term U_t is not taken of comparable accuracy. This is indeed the case for the usual time-stepping methods, which should therefore be operated with small time-steps in order to preserve the attractive properties of the Galerkin formulation. In these conditions, the implicit algorithms (e.g. Crank-Nicolson) do not appear to be cost-effective, and one would prefer using explicit time-stepping methods. Unfortunately, a reduced stability range of the standard explicit schemes is generally obtained when they are coupled to the Galerkin formulation. For example, the stability range for the leap-frog method is reduced by a factor $\sqrt{3}$, while Euler's first-order method is unstable for the advection equation unless $\Delta t = 0(h^2)$.

Many investigators have sought means to remedy the above defects and several of them are based on a Petrov-Galerkin approach. Of particular relevance for convective transport problems are the works of Hughes and collaborators [1-3] and of Morton and co-workers [4-6].

An alternative approach was taken by the present author [7] with the introduction of the Taylor-Galerkin method as a means of generating high-order accurate time-stepping schemes to be coupled to the high spatial resolution attainable with the Galerkin/finite element formulation. The essence of the Taylor-Galerkin method is briefly recalled in Section 2. It consists of developing improved versions of the usual time-stepping methods (Euler, leap-frog, Crank-Nicolson) on the basis of Taylor series expansions including second and third order terms in the time-step. Then, to preserve the simplicity and ease of implementation of the usual two-level time-stepping methods, the second and third order time derivatives are evaluated from the original partial differential equation. This process yields a generalised governing equation which is discretised in time only, the spatial variables being left continuous. Such equation is successively discretised in space using the conventional Bubnov-Galerkin finite-element method.

The effectiveness of the Taylor-Galerkin method has been illustrated in [7] for the case of pure advection problems in one and two space dimensions. In [8] the method was further extended to deal with mixed advection-diffusion problems. Herein, it will be illustrated for the solution of the non-linear Burgers' equation (Section 3). We first consider the solution of the inviscid form of Burgers' equation by a generalised Euler-Taylor-Galerkin method. Then, we discuss Burgers' equation with dissipation and introduce a splitting-up method of solution in which advection and diffusion are treated separately by appropriate Taylor-Galerkin methods.

2. THE TAYLOR-GALERKIN METHOD

As a model equation for illustrating the Taylor-Galerkin method, consider the scalar conservation law in one dimension

$$u_t + \partial_x f(u) = 0 \tag{2.1a}$$

or

$$u_t + a(u)u_x = 0 \tag{2.1b}$$

where

$$a(u) = \partial f(u) \mid \partial u \tag{2.1c}$$

To produce high-order accurate methods for the time discretisation of eq. (2.1b), consider the following Taylor series expansions in the time-step Δt:

$$u^n = u^{n+1} - \Delta t\, u_t^{n+1} + \tfrac{1}{2}\Delta t^2 u_{tt}^{n+1} - \tfrac{1}{6}\Delta t^3 u_{ttt}^{n+1} + \ldots \tag{2.2}$$

$$u^{n+1} = u^n + \Delta t\, u_t^n + \tfrac{1}{2}\Delta t^2 u_{tt}^n + \tfrac{1}{6}\Delta t^3 u_{ttt}^n + \ldots \tag{2.3}$$

$$u^{n-1} = u^n - \Delta t\, u_t^n + \tfrac{1}{2}\Delta t^2 u_{tt}^n - \tfrac{1}{6}\Delta t^3 u_{ttt}^n + \ldots \tag{2.4}$$

where superscript n indicates the time level, so that $t^n = n\Delta t$.

2.1 Generalised Forward-Time, or Euler, Time-Stepping

To generalise the usual forward-time, or Euler, time-stepping method, we make use of expansion (2.3) to write

$$u_t^n = \frac{u^{n+1} - u^n}{\Delta t} - \tfrac{1}{2}\Delta t\, u_{tt}^n - \tfrac{1}{6}\Delta t^2 u_{ttt}^n - 0(\Delta t^3) \tag{2.5}$$

By time-differentiation of the governing equation (2.1b) we obtain

$$u_{tt} = - a_t u_x - a\, u_{xt} \tag{2.6}$$

which, noting that in view of (2.1c) and (2.1b)

$$a_t = a_u u_t = - a\, a_u u_x = - a\, a_x \tag{2.7}$$

and

$$u_{xt} = - a_x u_x - a\, u_{xx} \tag{2.8}$$

may be transformed into

$$u_{tt} = 2a\, a_x u_x + a^2 u_{xx} = \partial_x (a^2 u_x) \tag{2.9}$$

Similarly, by time-differentiation of (2.9) and use of (2.7) we find

$$u_{ttt} = - \partial_x (2a^2 a_x u_x) + \partial_x (a^2 u_{tx}) \tag{2.10}$$

We can now replace the first, second and third time derivatives in (2.5) by expressions (2.1b), (2.9) and (2.10), respectively, and obtain

$$\frac{u^{n+1} - u^n}{\Delta t} = -a u_x^n + \tfrac{1}{2}\Delta t \partial_x (a^2 u_x)^n - \tfrac{1}{3}\Delta t^2 \partial_x (a^2 a_x u_x)^n + \tfrac{1}{6}\Delta t^2 \partial_x (a^2 u_{tx})^n \tag{2.11}$$

or, approximating u_{tx}^n by $\partial_x (\frac{u^{n+1} - u^n}{\Delta t})$,

$$\{1 - \partial_x (\frac{a^2 \Delta t^2}{6} \partial_x)\} (\frac{u^{n+1} - u^n}{\Delta t}) =$$

$$- a u_x^n + \tfrac{1}{2}\Delta t \partial_x \{a^2 (1 - \tfrac{2}{3}\Delta t\, a_x) u_x\}^n \tag{2.12}$$

Eq.(2.12) represents a generalised, third-order accurate, Euler time-stepping method. In spite of the appearance of eq.(2.12), it should be clear that the terms involving second-order spatial derivatives are not to be thought of as additional artificial diffusion terms, but rather as part of the difference approximation to u_t^n. Notice also that the third time-derivative in Taylor series expansion (2.5) has purposely been expressed in (2.10) in a mixed spatial-temporal form to avoid the introduction of third-order spatial derivatives, which would prevent us from using C° finite elements for the spatial discretisation.

2.2 Generalised Leap-Frog Time-Stepping

A generalised leap-frog time-stepping may be derived by subtracting (2.4) from (2.3). The result is:

$$\frac{u^{n+1} - u^{n-1}}{2\Delta t} = u_t^n + \tfrac{1}{6}\Delta t^2 u_{ttt}^n + 0(\Delta t^4) \tag{2.13}$$

or, using (2.1b) and (2.10) with u^n_{tx} replaced by $\partial_x(\frac{u^{n+1}-u^{n-1}}{2\Delta t})$:

$$\{1-\partial_x(\frac{a^2\Delta t^2}{6}\partial_x)\}(\frac{u^{n+1}-u^{n-1}}{2\Delta t}) =$$

$$-a u_x^n - \partial_x(\frac{a^2\Delta t^2}{3}a_x u_x)^n \qquad (2.14)$$

This scheme represents a generalised, fourth-order accurate, leap-frog method.

2.3 Generalised Crank-Nicolson Time-Stepping

A generalisation of the implicit Crank-Nicolson method is obtained by subtracting (2.2) from (2.3) which gives

$$\frac{u^{n+1}-u^n}{\Delta t} = \frac{1}{2}(u_t^n + u_t^{n+1}) + \frac{1}{4}\Delta t(u_{tt}^n - u_{tt}^{n+1}) + \frac{1}{12}\Delta t^2(u_{ttt}^n + u_{ttt}^{n+1}) \qquad (2.15)$$

Substituting the time derivatives by their expressions (2.1b), (2.9), (2.10) and letting

$$\frac{1}{2}(u_{tx}^n + u_{tx}^{n+1}) = \partial_x(\frac{u^{n+1}-u^n}{\Delta t}) \qquad (2.16)$$

we obtain

$$\frac{u^{n+1}-u^n}{\Delta t} = -\frac{1}{2}\{(a u_x)^n + (a u_x)^{n+1}\} +$$

$$\frac{\Delta t}{12}\partial_x\{(a^2 u_x)^n - (a^2 u_x)^{n+1}\} -$$

$$\frac{\Delta t^2}{6}\partial_x\{(a^2 a_x u_x)^n + (a^2 a_x u_x)^{n+1}\} \qquad (2.17)$$

Scheme (2.17) is a fourth-order accurate generalisation of the Crank-Nicolson time-stepping method.

2.4 Finite-element Spatial Discretisation

To obtain a fully discrete equation, we apply the Galerkin formulation to the above time-discretised generalised equations, with local approximations of the form (1.2). This gives a system of nodal equations which for Euler's scheme (2.12) takes the following form

$$M_{IJ}(U_J^{n+1} - U_J^n) = F_I \qquad (2.18)$$

where

$$M_{IJ} = \sum_e \int_{\Omega^e} \{N_I N_J + \frac{\partial N_I}{\partial x}\frac{a^2\Delta t^2}{6}\frac{\partial N_J}{\partial x}\} d\Omega^e \qquad (2.19)$$

is a generalised, but symmetric, "mass" matrix, and

$$F_I = \sum_e - \int_{\Omega^e} \{N_I + \frac{a\Delta t}{2}(1-\tfrac{2}{3}\Delta t a_x)\frac{\partial N_I}{\partial x}\}\, a\Delta t U_x^n\, d\Omega^e \qquad (2.20)$$

represents generalised nodal "loads". Eqs.(2.18) – (2.20) are clearly valid for arbitrary shape functions N_I and form the basis for developing Euler-Taylor-Galerkin finite-element schemes for the scalar conservation law in eq.(2.1). The Galerkin equations associated with the generalised leap-frog and Crank-Nicolson methods are derived in a similar manner.

3. APPLICATION TO BURGERS' EQUATION

Let us first consider the solution by the Euler-Taylor-Galerkin method of the inviscid form of Burgers' equation in one dimension:

$$u_t = -\partial_x (\tfrac{1}{2} u^2) \qquad (3.1)$$

with initial data $u(x,0) = 1$ for $x \leqslant 0$, $u(x, 0) = 0$ for $x > 0$. Eq. (3.1) is a particular form of (2.1b) in which $a(u) = u$. Therefore, according to (2.12), the generalised Euler time-stepping method for the inviscid Burgers' equation has the form

$$\{1 - \partial_x(\frac{u^2 \Delta t^2}{6}\partial_x)\}\,(\frac{u^{n+1} - u^n}{\Delta t}) =$$

$$- \partial_x(\tfrac{1}{2}u^2) + \partial_x\,\{\frac{u^2\Delta t}{2}(1 - \tfrac{2}{3}\Delta t\, u_x)\, u_x\}^n \qquad (3.2)$$

To obtain a fully discrete equation, the Galerkin formulation was applied to (3.2) assuming a uniform mesh of piecewise linear elements with size h=0.01. The numerical results obtained at time $t = 1$ with $\Delta t = 0.5h$ and $\Delta t = h$ are shown in Fig. 1.

By comparison with the analytical solution, it is noted that the shock speed is well predicted with both time steps and that the shock remains very sharp, with virtually no oscillations in the case $\Delta t = h$, i.e. when the generalised Euler scheme is operated at its stability limit.

As a second test problem, we consider Burgers' equation with dissipation

$$u_t = -\partial_x (\tfrac{1}{2}u^2) + \nu u_{xx} \qquad (3.3)$$

The initial data at time $t = 1$ are chosen in the form [9]

$$u(x,1) = x/[1 + t_0^{-1/2} \exp(x^2/4\nu)] \qquad (3.4)$$

where $t_0 = \exp(1/8\nu)$. Assuming that $u(0,t) = 0$, the problem has the following analytical solution:

$$u(x,t) = \frac{x/t}{1 + \exp(x^2/4\nu t)\,(t/t_0)^{1/2}} \qquad (3.5)$$

As explained in Section 2, the presence of a diffusion operator in eq.(3.3) prevents us from using a third-order accurate Euler scheme in conjunction with C^{o} finite elements. To partially circumvent this difficulty, we have proposed in [8] to use a splitting-up method in which advection and diffusion are treated separately in two distinct phases of the time integration procedure. The advection phase is treated first by the third-order Euler method in (3.2) and yields intermediate values $u^{n+1/2}$. These are then used for the diffusion phase which is based on the following second-order Euler method:

$$(1 - \frac{\nu\Delta t}{2}\partial_x^2)(\frac{u^{n+1} - u^{n+1/2}}{\Delta t}) = \nu u_{xx}^{n+1/2} \qquad (3.6)$$

The above splitting-up method was applied to solve eqs. (3.3) - (3.4) on a uniform mesh of 100 linear elements (h = 0.01) with a time-step $\Delta t = h$. Fig. 2.a shows the computed propagation of the initial shock for $\nu = 5 \cdot 10^{-3}$, and Fig. 2.b that for the more advective case, $\nu = 5 \cdot 10^{-4}$. In both cases the numerical predictions are seen to be in excellent agreement with the analytical solutions. For the purpose of comparison, we have reported in Fig. 2.c the solution obtained for $\nu = 5 \cdot 10^{-4}$ with the standard Euler-Galerkin formulation. Here, the computed shock speed is slightly in error and rather severe oscillations do appear at the shock front.

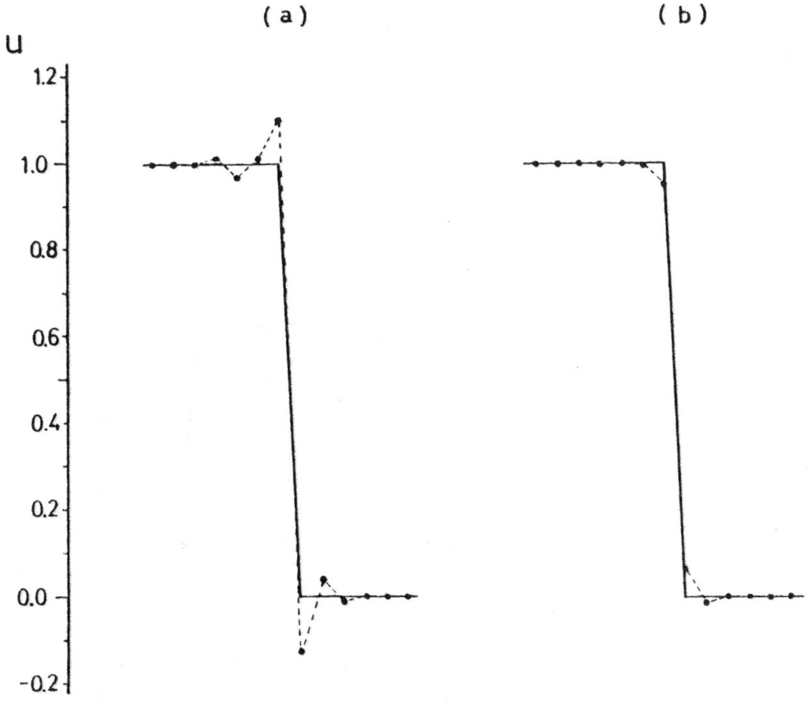

Fig. 1 : Solution of inviscid Burgers' equation by Euler-Taylor-Galerkin method : (a) $\Delta t/h = 0.5$; (b) $\Delta t/h = 1.0$.

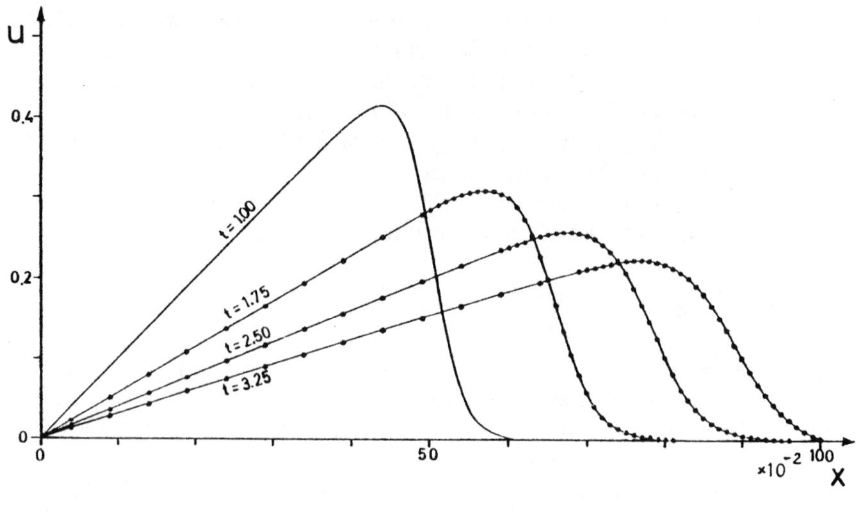

Fig. 2a : Solution of Burgers' equation with dissipation (ν = 5.10^{-3}) by splitting-up Taylor-Galerkin method .

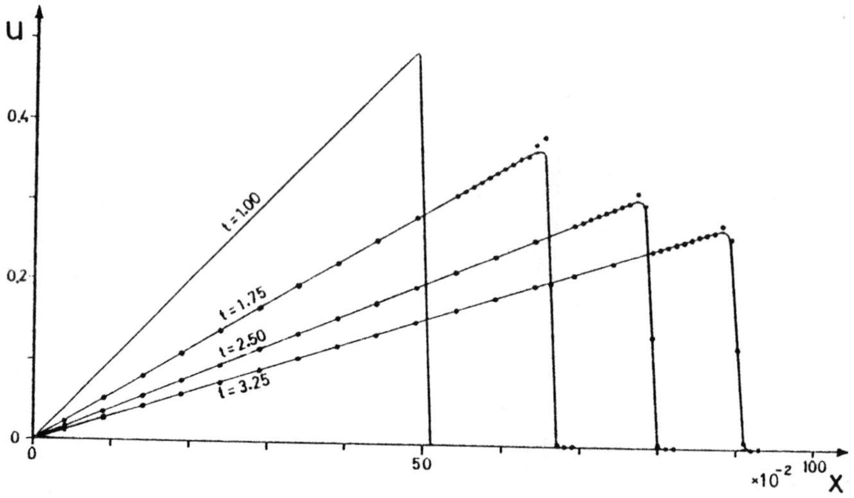

Fig. 2b : Solution of Burgers' equation with dissipation (ν = 5.10^{-4}) by splitting-up Taylor-Galerkin method.

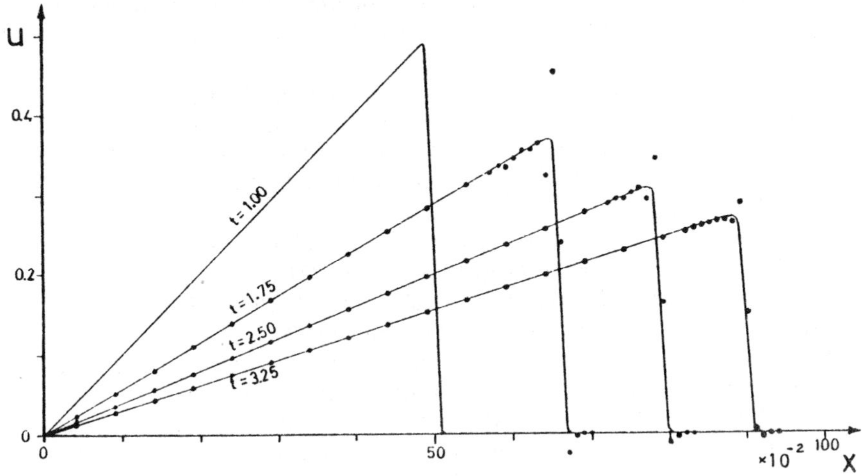

Fig. 2c : Solution of Burgers' equation with dissipation ($\nu = 5.10^{-4}$) by standard Euler-Galerkin method.

4. CONCLUSIONS

The Taylor-Galerkin method appears to be capable of producing numerical schemes of high accuracy for the solution of time-dependent convective and convective-diffusive transport problems, including non-linear situations. As shown in [7,8], the method is easily applied to multi-dimensional problems and, in contrast to the Petrov-Galerkin weighted residual formulations of upwind-type, it does not require the use of special weighting functions, nor the determination of free parameters to maximize the accuracy.

ACKNOWLEDGEMENTS

The author is indebted to S. Giuliani for the computations in Section 3.

REFERENCES

1. HUGHES, T.J.R. and BROOKS, A.N. - A multi-dimensional upwind scheme with no crosswind diffusion, **Finite Element Methods for Convection Dominated Flows**, Ed. T.J.R. Hughes, AMD Vol.34, ASME (New York), pp.19-35, 1979.
2. HUGHES, T.J.R. and BROOKS, A.N. - A theoretical framework for Petrov-Galerkin methods with discontinuous weighting functions: application to the streamline-upwind procedure. To appear in **Finite Elements in Fluids**, Vol.4, Ed. R.H. Gallagher, J. Wiley & Sons (New York).
3. HUGHES, T.J.R., TEZDUYAR, T.E. and BROOKS, A.N. - A Petrov-Galerkin finite element formulation for systems of conservation laws with

special reference to the compressible Euler equations. Prepared for the **Proceedings of the IMA Conference on Numerical Methods in Fluid Dynamics**, March 29th - 31st, 1982, University of Reading, England. To be published by Academic Press.

4. MORTON, K.W. and PARROTT, A.K. - Generalised Galerkin methods for first-order hyperbolic equations. J. Comp. Phys. Vol.36, pp.249-270,1980.
5. MORTON, K.W. and STOKES, A. - Generalised Galerkin methods for hyperbolic equations. To appear in **Proc. MAFELAP 1981 Conf.**, Ed. J.R. Whiteman.
6. MORTON, K.W. - Generalised Galerkin methods for steady and unsteady problems. Prepared for the **Proceedings of the IMA Conference on Numerical Methods in Fluid Dynamics**, March 29th - 31st, 1982, University of Reading, England. To be published by Academic Press.
7. DONEA, J. - A Taylor-Galerkin method for convective transport problems. To appear in Int. J. for Numerical Methods in Engineering, 1983.
8. DONEA, J., GIULIANI, S., LAVAL, H. and QUARTAPELLE, L. - Time-accurate solution of advection-diffusion problems by finite elements. To appear in a special issue of Comp. Meths. in Appl. Mech. and Engng. on **"Optimal FEM's for Convection Dominated Phenomena"**, Ed. T.J.R. Hughes, 1983.
9. LOHAR, B.L. and JAIN, P.C. - Variable mesh cubic spline technique for N-wave solution of Burgers' equation, J. Comp. Phys., Vol.39, pp.433-442, 1981.

NEW EXPLICIT METHODS FOR THE NUMERICAL SOLUTION OF THE DIFFUSION-CONVECTION EQUATION

A.R. Abdullah[I] and D.J. Evans[II]

Department of Computer Studies
Loughborough University of Technology
Loughborough, Leics. U.K.

SUMMARY

We consider the generalized finite difference approximation to the diffusion-convection equation, from which new explicit formulae are obtained which are asymmetric. These explicit schemes can then be used to develop a new class of methods called Group Explicit as introduced in [2].

Theoretical aspects of the stability, consistency, convergence and truncation errors of this new class of methods is briefly discussed and numerical evidence presented to confirm our recommendations.

1. INTRODUCTION

Recently numerical methods involving both *explicit* and *implicit* schemes for the solution of the diffusion-convection equation, i.e.,

$$\frac{\partial u}{\partial t} = \varepsilon \frac{\partial^2 u}{\partial x^2} - k\frac{\partial u}{\partial x} , \tag{1.1}$$

have been studied extensively. It is necessary for this equation to be treated separately from the ordinary diffusion equation because of the presence of spatial derivatives of first order.

Briefly, the implicit methods, i.e. Crank Nicolson, etc. normally offer unconditionally stable schemes but require the solution of system of equations at each time step. Meanwhile the explicit schemes as usual suffer from a restrictive stability condition.

[I] On leave from the Department of Computer Science, National University of Malaysia, Kuala Lumpur, Malaysia.

[II] Professor

However, it can be shown that by using different combinations and types of approximation for the terms $\frac{\partial^2 u}{\partial x^2}$ and $\frac{\partial u}{\partial x}$ in (1.1), a generalized finite difference approximation for (1.1) is obtained from which almost all the earlier two time-level schemes can be derived.

Furthermore we can obtain from this generalised form; a class of stable semi-explicit schemes which is of similar structure to the semi-explicit scheme introduced by [3],[4],[5] and [6].

The simplicity of explicit methods of solution prompts us to seek such a method with increased stability characteristics and with the capability that the solution can be obtained at many points concurrently on 'next generation' array/parallel computers. The introduction of this new class of explicit method called the Group Explicit method will enable the explicit methods to compete with implicit methods on level terms again.

2. FINITE-DIFFERENCE APPROXIMATION

We now consider equation (1.1) in the domain $(0,1)\times(0,\infty)$ with the initial condition

$$u(x,0) = f(x) \ , \quad 0 \leqslant x \leqslant 1 \ , \tag{2.1a}$$

and boundary conditions

$$\begin{aligned} u(0,t) &= g_0(t) \ , \quad t>0, \\ u(1,t) &= g_1(t) \ , \quad t>0. \end{aligned} \tag{2.1b}$$

As usual, the open-rectangular domain is covered by a rectangular grid, with spacing $\Delta x, \Delta t$ in the x,t directions respectively. The values of Δx and Δt are assumed uniform throughout the region and the grid points (x,t) denoted by $x=x_i=i\Delta x$, $i=0,1,2...,m$, $m=1/\Delta x$ and $t=t_j=j\Delta t$, $j=0,1,2,...$.

In [1], the diffusion-convection equation (1.1) is approximated in a general way at the point $(i,j+\frac{1}{2})$ by the two time-level finite difference representation

$$\begin{aligned} \frac{u_{i,j+1}-u_{i,j}}{\Delta t} = {} & \frac{\varepsilon}{(\Delta x)^2}\left[(\theta_1\delta_x u_{i+\frac{1}{2},j+1}-\theta_2\delta_x u_{i-\frac{1}{2},j+1})\right. \\ & \left.+(\theta_1'\delta_x u_{i+\frac{1}{2},j}-\theta_2'\delta_x u_{i-\frac{1}{2},j})\right] - \frac{k}{2\Delta x}\left[\alpha_1\Delta_x u_{i,j+1}+\alpha_2\nabla_x u_{i,j}\right] \\ & - \frac{k}{2\Delta x}\left[\alpha_1'\nabla_x u_{i,j+1}+\alpha_2'\Delta_x u_{i,j}\right] , \end{aligned} \tag{2.2}$$

where δ_x, Δ_x and ∇_x are the central-, forward- and backward differences operators with respect to the x variable respectively. The parameters $\theta_1, \theta_2 \theta_1', \theta_2', \alpha_1, \alpha_2, \alpha_1'$ and α_2' which have been introduced are related to each other by certain conditions which will determine the accuracy and consistency of the approximation.

After the insertion of the values for the difference operators, equation (2.2) will lead to

$$\begin{aligned}-(E\theta_2+K\alpha_1')u_{i-1,j+1}&+[1+E(\theta_1+\theta_2)-K(\alpha_1-\alpha_1')]u_{i,j+1}\\ -(E\theta_2-K\alpha_1)u_{i+1,j+1} &= (E\theta_2'+K\alpha_2)u_{i-1,j}+[1-E(\theta_1'+\theta_2')+K(\alpha_2'-K\alpha_1)]\\ &\quad u_{i,j}+(E\theta_1'-K\theta_2')\,u_{i+1,j}\ , \qquad (2.3)\end{aligned}$$

where $E=\varepsilon r$, $K=k\Delta t/2\Delta x=kr\Delta x/2$ and as usual $r=\Delta t/(\Delta x)^2$.

The approximation (2.3), due to several reasons [1], is found to be a generalized two time-level finite difference representation of the equation (1.1) with the generalised principal part of the local truncation error given by

$$\begin{aligned}T_{i,j+\frac{1}{2}} &= [\tfrac{\varepsilon}{2}(\tfrac{\Delta t}{\Delta x})(\theta_2-\theta_1-\theta_2'+\theta_1')+\tfrac{k\Delta t}{4}(\alpha_1'+\alpha_1-\alpha_2-\alpha_2')]\frac{\partial^2 u}{\partial x\partial t}+\frac{\Delta t^2}{24}\ \frac{\partial^3 u}{\partial t^3}\\ &+\frac{k}{6}(\Delta x)^2\ \frac{\partial^3 u}{\partial x^3}+\frac{k}{24}(\Delta t)^2\ \frac{\partial^3 u}{\partial x\partial t^2}-\frac{k\Delta x}{4}(-\alpha_1'+\alpha_1-\alpha_2+\alpha_2')\frac{\partial^2 u}{\partial x^2}\\ &+[\tfrac{\varepsilon}{12}\Delta t(-\theta_2-\theta_1+\theta_2'+\theta_1')+\frac{k\Delta x\Delta t}{24}(-\alpha_1'+\alpha_1+\alpha_2-\alpha_2')]\frac{\partial^3 u}{\partial x^2\partial t}\ .\end{aligned}\qquad (2.4)$$

For stability, the generalised form has to fulfill the inequality,

$$\begin{aligned}&(A_1-A_2)^2+4A_1A_2\sin^2\frac{\beta\Delta x}{2}-A_1-A_2\\ &\qquad\leqslant (B_1-B_2)^2+4B_1B_2\sin^2\frac{\beta\Delta x}{2}+B_1+B_2\ ,\end{aligned}\qquad (2.5)$$

where

$$A_1 = E\theta_2'+K\alpha_2\ ,\qquad A_2 = E\theta_1'-K\alpha_2'\ ,$$

$$B_1 = E\theta_2+K\alpha_1'\ ,\qquad B_2 = E\theta_1-K\alpha_1\ .$$

From this generalized form, it will be seen later that for some explicit algorithms it is possible to relax the stability restriction without incurring a considerable loss of accuracy.

3. SOME STABLE EXPLICIT SCHEMES

From the generalized finite difference approximation (2.3) some interesting explicit schemes can be obtained, with the local truncation error (L.T.E) and stability conditions derived from equations (2.4) and (2.5) respectively. They are:

(i) $\theta_1=\theta_2'=1$, $\theta_2=\theta_1'=0$, $\alpha_1=\alpha_2=1$ and $\alpha_1'=\alpha_2'=0$ to give (Fig.3.1),

$$[1+(\varepsilon r-\frac{kr\Delta x}{2})]u_{i,j+1}-(\varepsilon r-\frac{kr\Delta x}{2})u_{i+1,j+1} = (\varepsilon r+\frac{kr\Delta x}{2})u_{i-1,j} +[1-(\varepsilon r+\frac{kr\Delta x}{2})]u_{i,j}\ , \tag{3.1}$$

with L.T.E. given by,

$$T_{3.1} = -\varepsilon(\frac{\Delta t}{\Delta x})\frac{\partial^2 u}{\partial x\partial t} + \frac{\Delta t^2}{24}\frac{\partial^3 u}{\partial t^3} + \frac{k}{24}(\Delta t)^2\frac{\partial^3 u}{\partial x\partial t^2} + \frac{k}{6}(\Delta x)^2\frac{\partial^3 u}{\partial x^3} + \frac{k\Delta x\Delta t}{12}\frac{\partial^3 u}{\partial x\partial t}\ , \tag{3.2}$$

and requires for stability the condition

$$0 < r \leqslant \frac{1}{k\Delta x}\ , \tag{3.3}$$

to be satisfied.

This condition for stability is always rather favourable since with the values k=1.0, Δx=0.1, $k\Delta x$=0.1 this is always less restrictive than the condition for the classical explicit formula.

(ii) $\theta_2=\theta_1'=1$, $\theta_1=\theta_2'=0$, $\alpha_1'=\alpha_2'=1$ and $\alpha_1=\alpha_2=0$ will result in (Fig. 3.2),

$$-(\varepsilon r+\frac{kr\Delta x}{2})u_{i-1,j+1}+[1+(\varepsilon r+\frac{kr\Delta x}{2})]u_{i,j+1} = [1-(\varepsilon r-\frac{kr\Delta x}{2})]u_{i,j} +(\varepsilon r-\frac{kr\Delta x}{2})u_{i+1,j}\ , \tag{3.4}$$

with the L.T.E. given by,

$$T_{3.4} = \varepsilon(\frac{\Delta t}{\Delta x})\frac{\partial^2 u}{\partial x\partial t} + \frac{\Delta t^2}{24}[\frac{\partial^2 u}{\partial t^3} + k\frac{\partial^3 u}{\partial x\partial t^2}] + \frac{k}{6}(\Delta x)^2\frac{\partial^3 u}{\partial x^3} - \frac{k\Delta x\Delta t}{12}\frac{\partial^3 u}{\partial x^2\partial t}\ . \tag{3.5}$$

For stability it requires the condition,

$$2\varepsilon r(kr\Delta x+1)\geqslant 0\ , \tag{3.6}$$

to be satisfied, which is fulfilled by all values of $r>0$.

Due to the opposite signs of the truncation errors in (3.2) and (3.5), the following algorithms which are similar to those suggested in [4] can be obtained by:

1. Use of equation (3.1) in a right-to-left direction (UNE).

2. Use of equation (3.4) in a left-to-right direction (UPOS).

3. Use of equation (3.1) at the jth time-level in a right-to-left direction and alternatively use of equation (3.4) at the (j+1)th time-level in a left-to-right direction (ALDC).

4. Use of equation (3.1) as in (1) and equation (3.4) as in (2) at each time-level and then average the results (UAV).

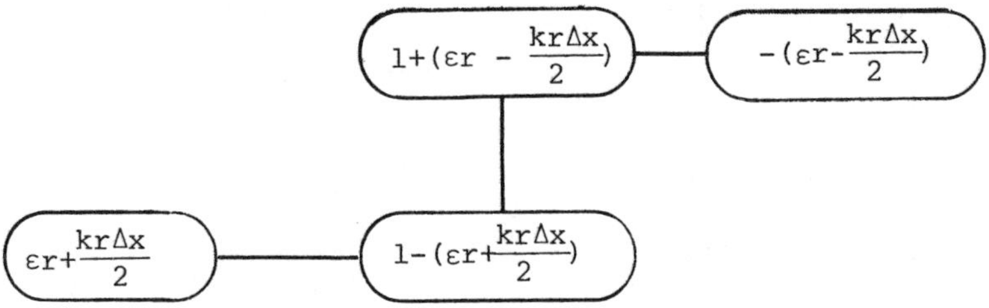

FIGURE 3.1

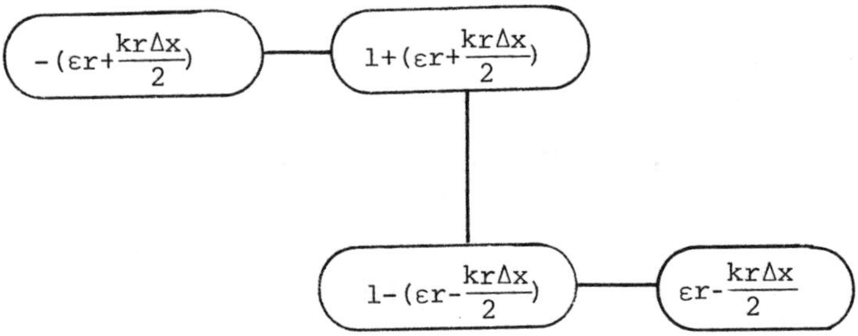

FIGURE 3.2

Recently in [2] a new method of using the asymmetric equations (3.1) and (3.4) was presented. The method is developed by applying the equations (3.1) and (3.4) to a group of adjacent nodal points in a coupled manner as illustrated in the next section.

4. GE FORMULATION AND ALGORITHMS

Consider now any two points $(i,j+1)$ and $(i+1,j+1)$ and use equation (3.1) at point $(i,j+1)$ and use equation (3.4) at point $(i+1,j+1)$ to give,

$$[1+(\varepsilon r - \frac{kr\Delta x}{2})]u_{i,j+1} - (\varepsilon r - \frac{kr\Delta x}{2})u_{i+1,j+1} = (\varepsilon r + \frac{kr\Delta x}{2})u_{i-1,j} + [1-(\varepsilon r + \frac{kr\Delta x}{2})]u_{i,j}, \quad (4.1)$$

and

$$-(\varepsilon r + \frac{kr\Delta x}{2})u_{i,j+1} + [1+(\varepsilon r + \frac{kr\Delta x}{2})]u_{i+1,j+1} = [1-(\varepsilon r - \frac{kr\Delta x}{2})]u_{i+1,j} + (\varepsilon r - \frac{kr\Delta x}{2})u_{i+2,j}, \quad (4.2)$$

respectively. Equations (4.1) and (4.2) will then form a small system of 2×2 linear equations, i.e.,

$$\begin{bmatrix} 1+(\varepsilon r - \frac{kr\Delta x}{2}) & -(\varepsilon r - \frac{kr\Delta x}{2}) \\ -(\varepsilon r + \frac{kr\Delta x}{2}) & 1+(\varepsilon r + \frac{kr\Delta x}{2}) \end{bmatrix} \begin{bmatrix} u_{i,j+1} \\ u_{i+1,j+1} \end{bmatrix} = \begin{bmatrix} 1-(\varepsilon r + \frac{kr\Delta x}{2}) & 0 \\ 0 & 1-(\varepsilon r - \frac{kr\Delta x}{2}) \end{bmatrix}$$

$$\begin{bmatrix} u_{i,j} \\ u_{i+1,j} \end{bmatrix} + \begin{bmatrix} (\varepsilon r+\frac{kr\Delta x}{2})u_{i-1,j} \\ (\varepsilon r-\frac{kr\Delta x}{2})u_{i+2,j} \end{bmatrix} \tag{4.3}$$

Since,

$$\begin{bmatrix} 1+(\varepsilon r-\frac{kr\Delta x}{2}) & -(\varepsilon r-\frac{kr\Delta x}{2}) \\ -(\varepsilon r+\frac{kr\Delta x}{2}) & 1+(\varepsilon r+\frac{kr\Delta x}{2}) \end{bmatrix}^{-1} = \frac{1}{1+2\varepsilon r}\begin{bmatrix} 1+(\varepsilon r+\frac{kr\Delta x}{2}) & r\varepsilon-\frac{kr\Delta x}{2} \\ \varepsilon r+\frac{kr\Delta x}{2} & 1+(\varepsilon r-\frac{kr\Delta x}{2}) \end{bmatrix},$$

then (4.3) can be explicitly represented by,

$$\begin{bmatrix} u_{i,j+1} \\ u_{i+1,j+1} \end{bmatrix} = \beta\left\{\begin{bmatrix} 1-(\varepsilon r+\frac{kr\Delta x}{2})^2 & (\varepsilon r-\frac{kr\Delta x}{2})[1-(\varepsilon r-\frac{kr\Delta x}{2})] \\ (\varepsilon r+\frac{kr\Delta x}{2})[1-(\varepsilon r+\frac{kr\Delta x}{2})] & 1-(\varepsilon r-\frac{kr\Delta x}{2})^2 \end{bmatrix}\right.$$

$$\left.\begin{bmatrix} u_{i,j} \\ u_{i+1,j} \end{bmatrix} + \begin{bmatrix} (\varepsilon r+\frac{kr\Delta x}{2})[1+(\varepsilon r+\frac{kr\Delta x}{2})]u_{i-1,j}+(\varepsilon r-\frac{kr\Delta x}{2})^2 u_{i+2,j} \\ (\varepsilon r+\frac{kr\Delta x}{2})^2 u_{i-1,j}+(\varepsilon r-\frac{kr\Delta x}{2})[1+(\varepsilon r-\frac{kr\Delta x}{2})]u_{i+2,j} \end{bmatrix}\right\},$$

where $\beta=1/(1+2\varepsilon r)$. (4.4)

In the case where there is any ungrouped point near either boundary, we use equation (3.4), i.e.,

$$u_{1,j+1} = \frac{1}{[1+(\varepsilon r+\frac{kr\Delta x}{2})]}\{(\varepsilon r+\frac{kr\Delta x}{2})u_{0,j+1}+[1-(\varepsilon r-\frac{kr\Delta x}{2})]u_{1,j}+ (\varepsilon r-\frac{kr\Delta x}{2})u_{2,j}\}, \tag{4.5}$$

for the left ungrouped point and equation (3.1), i.e.,

$$u_{m-1,j+1} = \frac{1}{[1+(\varepsilon r-\frac{kr\Delta x}{2})]}\{(\varepsilon r-\frac{kr\Delta x}{2})u_{m,j+1}+[1-(\varepsilon r+\frac{kr\Delta x}{2})]u_{m-1,j}+ (\varepsilon r+\frac{kr\Delta x}{2})u_{m-2,j}\}, \tag{4.6}$$

for the right ungrouped point.

To derive the algorithms which form the class of Group Explicit methods, we use the implicit form (4.3). Also we assume that the space interval x is divided into an even number of sub-intervals which implies that the (m-1) points are odd. With the notations,

$$\alpha_1 = \varepsilon - \frac{k\Delta x}{2},$$

and

$$\alpha_2 = \varepsilon + \frac{k\Delta x}{2}, \tag{4.7}$$

the following algorithms can be established:-

1. <u>Group Explicit with Right-Ungrouped Point (GER)</u>: Use

equation (4.3) for the first (m-2) points and equation (4.6) for the last unknown point. This will give the system,

$$(I+rG_1)\underline{u}_{j+1} = (I-rG_2)\underline{u}_j + \underline{b}_{1,j} , \qquad (4.8)$$

where

$$G_1 = \begin{bmatrix} \begin{matrix} \alpha_1 & -\alpha_1 \\ -\alpha_2 & \alpha_2 \end{matrix} & & & 0 & \\ & \ddots & & & \\ & & \begin{matrix} \alpha_1 & -\alpha_1 \\ -\alpha_2 & \alpha_2 \end{matrix} & & \\ 0 & & & \alpha_1 & \end{bmatrix} , \qquad (4.9)$$

$$G_2 = \begin{bmatrix} \alpha_2 & & & & \\ & \begin{matrix} \alpha_1 & -\alpha_1 \\ -\alpha_2 & \alpha_2 \end{matrix} & & 0 & \\ & & \ddots & & \\ & 0 & & \begin{matrix} \alpha_1 & -\alpha_1 \\ -\alpha_2 & \alpha_2 \end{matrix} \end{bmatrix} , \qquad (4.10)$$

$$\underline{b}_{1,j}^T = [r\alpha_2 u_{0,j}, 0, \ldots, 0, r\alpha_1 u_{m,j+1}] \text{ and}$$

$$\underline{u}_j^T = [u_{1,j}, u_{2,j}, \ldots, u_{m-1,j}].$$

2. <u>Group Explicit with Left-Ungrouped Point (GEL)</u>: Use equation (4.5) for the first unknown point from the left of the boundary and equation (4.3) for the remaining $(\frac{m-2}{2})$ pairs of points. This will result in the system,

$$(I+rG_2)\underline{u}_{j+1} = (I-rG_1)\underline{u}_j + \underline{b}_{2,j} , \qquad (4.11)$$

with $\underline{b}_{2,j}^T = [r\alpha_2 u_{0,j+1}, 0, \ldots, 0, r\alpha_1 u_{m,j}]$.

3. <u>Alternating Group Explicit (S)AGE</u>: Use equation (4.8) at the (j+1)th time level and equation (4.11) at the (j+2)th time level, i.e.,

$$\left.\begin{aligned} (I+rG_1)\underline{u}_{j+1} &= (I-r(G_2)\underline{u}_j + \underline{b}_{1,j} , \\ (I+rG_2)\underline{u}_{j+2} &= (I-rG_1)\underline{u}_{j+1} + \underline{b}_{2,j+1} , \end{aligned}\right\} \qquad (4.12)$$

4. <u>Alternating Group Explicit (D)AGE</u>: In this algorithm, the group explicit formulae are incorporated alternately within the four time-levels with the direction reversed at the third time level. This will give the equations,

$$\left.\begin{aligned}(I+rG_1)\underline{u}_{j+1} &= (I-rG_2)\underline{u}_j+\underline{b}_{1,j}\\(I+rG_2)\underline{u}_{j+2} &= (I-rG_1)\underline{u}_{j+1}+\underline{b}_{2,j+1}\\(I+rG_2)\underline{u}_{j+3} &= (I-rG_1)\underline{u}_{j+2}+\underline{b}_{2,j+2}\\(I+rG_1)\underline{u}_{j+4} &= (I-rG_2)\underline{u}_{j+3}+\underline{b}_{1,j+3}\end{aligned}\right\} . \tag{4.13}$$

These are only a few of the examples of the algorithms which can be established from the original formulae (4.4)-(4.6). There are a few more algorithms which the authors have omitted for brevity.

The estimate of the truncation errors of all the schemes mentioned is given in [1] and can be shown to be of order $O(\Delta t+(\Delta x)^2+\frac{\Delta t}{\Delta x})$. However for the (S)AGE and (D)AGE schemes, they are of the order approximately $O(\Delta t+(\Delta x)^2)$, with the consistency condition $(\Delta t/\Delta x)\to 0$ for $\Delta t\to 0, \Delta x\to 0$ applicable.

The stability analysis of this class of method can be obtained by using the matrix method [1]. It was proved that the GER and GEL schemes are stable provided

$$r \leqslant \frac{1}{\max\{|\varepsilon-\frac{k\Delta x}{2}|, |\varepsilon+\frac{k\Delta x}{2}|\}},$$

and for the (S)AGE and (D)AGE schemes they are unconditionally stable for all $r>0$ provided $\Delta x\leqslant 2\varepsilon/k$.

5. NUMERICAL EXAMPLE

In this example the equation (1.1) together with the initial condition $f(x)=0$ and boundary conditions $g_0(t)=0$ and $g_1(t)=1$ is used as a model problem. This problem can be shown by the method of separation of variables to have the exact solution

$$u(x,t) = \frac{e^{\frac{kx}{\varepsilon}}-1}{e^{\frac{k}{\varepsilon}}-1} + \sum_{n=1}^{\infty} \frac{(-1)^n n\pi}{(n\pi)^2+(\frac{k}{2\varepsilon})^2} e^{\frac{k}{2\varepsilon}(x-1)} \sin(n\pi x)\, e^{-[(n\pi)^2\varepsilon+k^2/4\varepsilon]t} \tag{5.1}$$

The solution of some of the numerical schemes presented earlier are then compared with this exact solution in terms of their absolute errors. A comparison is also made with the Crank-Nicolson upwinding (CNU) scheme (in the case of (2.3) when $\theta_i=\theta_i'=1/2$, $\alpha_2=\alpha_1'=1, \alpha_1=\alpha_2'=0$). The results are given in Tables (5.1)-(5.2) and graphically in Fig.(5.1).

From the table and graph, it can be seen that the scheme in this class of method are much more accurate than the CNU method. For r=0.5 the (D)AGE scheme appears to be better than any other scheme.

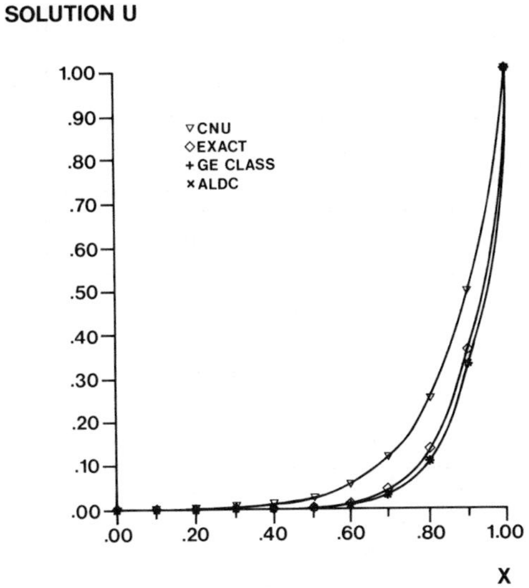

$\epsilon=0.1$, $k=1.0$, $r=1.0$, $t=1.0$, $\Delta t=0.01$, $\Delta x=0.1$

FIGURE (5.1)

$k=1,0,\varepsilon=1,0,\Delta t=0.005,\Delta x=0.1,r=0.5,t=0.5$

Method \ x	0.1	0.2	0.3	0.4
CNU	1.7×10^{-3}	3.2×10^{-3}	4.5×10^{-3}	5.4×10^{-3}
Eqn.3.1(UNE)	1.9×10^{-4}	3.5×10^{-4}	4.7×10^{-4}	5.2×10^{-4}
Eqn.3.4(UPOS)	3.3×10^{-4}	6.1×10^{-4}	8.1×10^{-4}	9.1×10^{-4}
ALDC Eqn.3.1 & 3.4	1.1×10^{-4}	2.4×10^{-4}	3.6×10^{-4}	4.6×10^{-4}
AVERAGE Eqn.3.1 & 3.4	0.7×10^{-4}	1.3×10^{-4}	1.7×10^{-4}	2.0×10^{-4}
(D)AGE	0.2×10^{-4}	0.6×10^{-4}	0.5×10^{-4}	0.8×10^{-4}
EXACT SOLUTION	0.06043	0.12730	0.20136	0.28345

cont.....

0.5	0.6	0.7	0.8	0.9
6.1×10^{-3}	6.2×10^{-3}	5.8×10^{-3}	4.7×10^{-3}	2.8×10^{-3}
5.1×10^{-4}	4.4×10^{-4}	3.3×10^{-4}	2.0×10^{-4}	0.8×10^{-4}
9.1×10^{-4}	8.2×10^{-4}	6.5×10^{-4}	4.4×10^{-4}	2.2×10^{-4}
5.2×10^{-4}	5.3×10^{-4}	4.8×10^{-4}	3.8×10^{-4}	2.1×10^{-4}
2.0×10^{-4}	1.9×10^{-4}	1.6×10^{-4}	1.2×10^{-4}	0.7×10^{-4}
0.8×10^{-4}	0.8×10^{-4}	0.9×10^{-4}	0.6×10^{-4}	0.6×10^{-4}
0.37447	0.47539	0.58724	0.71114	0.84830

TABLE 5.1

$k=1,0, \varepsilon=1,0, \Delta t=0,01, \Delta x=0,1, r=1,0, t=1,0$

Method \ x	0.1	0.2	0.3	0.4
CNU	1.5×10^{-3}	2.9×10^{-3}	4.1×10^{-3}	4.9×10^{-3}
Eqn.3.1 (UNE)	2.4×10^{-5}	4.5×10^{-5}	6.3×10^{-5}	7.7×10^{-5}
Eqn.3.4(UPOS)	3.3×10^{-5}	6.2×10^{-5}	8.5×10^{-5}	10.3×10^{-5}
ALDC Eqn.3.1 & 3.4	3.2×10^{-5}	6.2×10^{-5}	8.8×10^{-5}	10.8×10^{-5}
AVERAGE Eqn.3.1 & 3.4	2.8×10^{-5}	5.3×10^{-5}	7.4×10^{-5}	9.0×10^{-5}
(D)AGE	2.7×10^{-5}	5.2×10^{-5}	7.2×10^{-5}	8.8×10^{-5}
EXACT SOLUTION	0.06120	0.12884	0.20360	0.28621

cont. ....

0.5	0.6	0.7	0.8	0.9
5.5×10^{-3}	5.7×10^{-3}	5.3×10^{-3}	4.3×10^{-5}	2.6×10^{-5}
8.6×10^{-5}	8.9×10^{-5}	8.4×10^{-5}	7.0×10^{-5}	4.3×10^{-5}
11.3×10^{-5}	11.4×10^{-5}	10.4×10^{-5}	8.4×10^{-5}	5.0×10^{-5}
12.1×10^{-5}	12.3×10^{-5}	11.5×10^{-5}	9.3×10^{-5}	5.5×10^{-5}
9.9×10^{-5}	10.1×10^{-5}	9.4×10^{-5}	7.7×10^{-5}	4.6×10^{-5}
9.8×10^{-5}	10.0×10^{-5}	9.4×10^{-5}	7.6×10^{-5}	4.6×10^{-5}
0.37752	0.47843	0.58996	0.71322	0.84945

TABLE 5.2

6. CONCLUSIONS

The explicit schemes (3.1) and (3.4) obtained from the generalised approximation are both very easy and economical to implement. As they are unconditionally stable in a practical sense and also accurate, therefore they are strongly recommended.

The GE schemes derived are also comparably accurate and strongly stable. For $r \lesssim 2.0$, the GE schemes ((D)AGE in particular) are still to be recommended against the CNU schemes.

One point worth noting here is that this class of method which is made up of approximations to $\partial u/\partial x$ by both forward and backward differences at different time levels is always superior than the CNU schemes where $\partial u/\partial x$ is always approximated by the backward difference.

The scheme discussed in this chapter can be easily extended and adapted for multi-dimensional problems.

Finally, we can establish that since the method is explicit and highly stable, it can be recommended as an alternative competitive method for solving the diffusion-convection equation.

REFERENCES

1. ABDULLAH, A.R.: The Study of Some Numerical Methods for Solving Parabolic Partial Differential Equations, Ph.D. Thesis, Loughborough University of Technology, 1983.

2. EVANS, D.J. and ABDULLAH, A.R., A New Explicit Method for the Diffusion Equation, presented at the International Conference on Numerical Methods in Thermal Problems, The University of Washington, Seattle, U.S.A., 1983.

3. SAUL'YEV, V.K.: Integration of Equations of Parabolic Type by the Method of Nets, Macmillan, New York, 1964.

4. LARKIN, B.K., Some Stable Explicit Difference Approximations to the Diffusion Equation, Math.Comp., 18, 196-202, 1964.

5. BARAKAT, H.Z. and CLARK, J.A.: On the Solution of The Diffusion Equation by Numerical Methods, Jour. of Heat Transfer, Trans. A.S.M.E., Series C, 88, 421-427, 1966.

6. EVANS, D.J., Simplified Implicit Methods for the Finite Difference Solutions of Parabolic and Hyperbolic Partial Differential Equations, J.I.M.A., 2, 1-13, 1966.

THE CODE DISCO-2 FOR PREDICTING THE BEHAVIOUR OF DISCRETE PARTICLES IN TURBULENT FLOWS, AND ITS COMPARISONS AGAINST THE CODE DISCO-1 AND EXPERIMENTS.

A. BERLEMONT, A. PICART, G. GOUESBET

Laboratoire de Thermodynamique - LA CNRS N° 230
Faculté des Sciences et Techniques de Rouen
BP 67 76130 MONT-SAINT-AIGNAN (France)

Abstract :

That paper describes the code DISCO-2 (in the framework of the Non-Discrete Dispersive Approach) to predict the behaviour of discrete particles embedded in turbulent flows, and compares its structure with respect to the code DISCO-1. Numerical results obtained from DISCO-2 are compared against DISCO-1 results, and experiments. The effect of the crossing-trajectory effects are included when necessary. Agreements are very satisfactory.

I - THE STRUCTURE OF THE CODE DISCO-2

Previous published works describe the code DISCO-1 to compute the behaviour of discrete particles embedded in turbulent flows [1,2]. The structure of the code DISCO-2 is similar to the one for DISCO-1 and contains mainly two parts.

Part I : turbulence

The turbulence is predicted from a (k-ε)-model supplemented with algebraic relations deduced from a second-order closure scheme. See [1,2] and [3,4].

Part II : particles

In the code DISCO-1, an approximate Eulerian spectra $E_{fE}(\omega)$ was first computed, then a simple rule of Eulerian $\rightarrow$ Lagrangian transformation was applied before computing the cœfficients of dispersion of the discrete particles. Although rather satisfactory for engineering purposes, the process was found time-consumming and costfull in terms of computations. It is now given up.

In the code DISCO-2, the computations of the coefficients of dispersion are based on the theory described in [5], starting from a Frenkiel's family of Lagrangian correlation functions :

$$R_{fL}(\tau) = \exp\left[\frac{-\tau}{(m^2+1)\tau_L}\right] \cos\left[\frac{m\tau}{(m^2+1)\tau_L}\right] \qquad (1)$$

where τ is the time shift, m the loop parameter and τ_L is the Lagrangian time macroscale. When the Basset's term is negligible [5,6], the coefficient of dispersion then reads [5] :

$$\varepsilon_p = \overline{u_{fL}^2}\ \tau_L\{1+\psi_o e^{-at}+[\psi_1 \cos(mt/f)+\psi_2 \text{Sin}(mt/f)]\, e^{-t/f}\} \qquad (2)$$

where all the terms are defined in the references [5]. The scheme is closed when m and τ_L are known. m will be taken as a model constant, and the usual recommanded value is $\simeq$ 1. But τ_L must be more precisely evaluated.

II - EVALUATION OF THE LAGRANGIAN TIME MACROSCALE.

Following Tennekes and Lumley [7], in homogeneous isotropic turbulence :

$$\nu_T \simeq \overline{u_{fL}^2}\ \tau_L \qquad (3)$$

where ν_T is the turbulent kinematic viscosity of the fluid. On the other hand, the closure equation of the (k-ε)-model reads :

$$\mu_T = C_\mu\, \rho\, \frac{K^2}{\varepsilon} \qquad (4)$$

where C_μ is usually taken equal to 0.09, and K is the turbulent energy. Identifying the mean square of the Lagrangian fluid fluctuating velocity $\overline{u_{fL}^2}$ in a given direction and the mean square of the Eulerian fluid fluctuating velocity $\overline{u^2}$ in the same direction (which is rigorous in homogeneous turbulences), we obtain :

$$\tau_L \simeq 0.20\ \frac{\overline{u^2}}{\varepsilon} \qquad (5)$$

That relation will be used to evaluate τ_L in the code DISCO-2. The constant 0.2 is possibly to be adjusted but it has been found satisfactory in the presently given comparisons. τ_L will be made depending on the direction :

$$\tau_{L,i} = 0.20\ \frac{\overline{u^2_{(i)}}}{\varepsilon} \qquad (6)$$

where the anisotropy of the flow is reintroduced through the variances. It could be argued that τ_L is a scalar and should be written :

$$\tau_L = 0.20 \frac{(2K/3)}{\varepsilon} \qquad (7)$$

In that case, the anisotropy would be only reintroduced in the formula used to compute the coefficient ε_p (relation 2). That way to work has been tested but does not give the same agreement as Relation (6).

III - COMPARISON OF THE DISPERSION COEFFICIENTS PREDICTED WITH DISCO-1 AND DISCO-2.

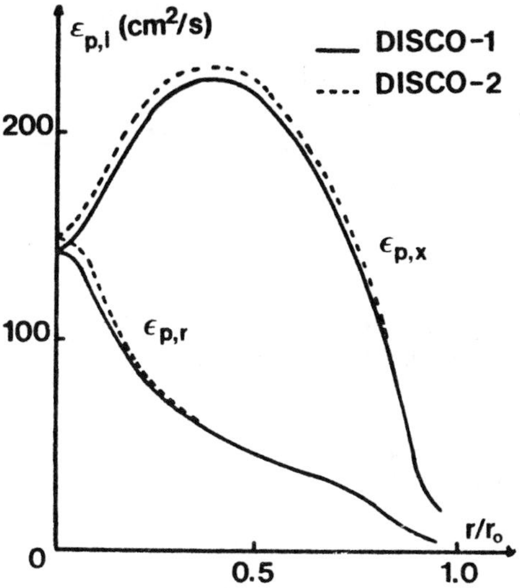

Fig. 1

The DISCO-2 predictions are now compared against the DISCO-1 predictions [8] in the case of water droplets embedded in a turbulent pipe flow of air. Asymptotic values $\varepsilon_{p,x}(\infty)$ and $\varepsilon_{p,r}(\infty)$ in all the grid are shown on the figure 1. The comparison between the two codes is very satisfactory. The figures 2 and 3 give the ratios $\varepsilon_{p,x}(t)/\varepsilon_{f,x}(\infty)$ and $\varepsilon_{p,r}(t)/\varepsilon_{f,r}(\infty)$ as a function of the dispersion time, for two diameters of the particles, 1 µm and 100 µm respectively, and three values of m(0,1,2). The agreement is satisfactory between DISCO-1 and DISCO-2, when m ranges between 0 and 1 and confirms the validity of the Eulerian → Lagrangian transform given in DISCO-1 [8]. As well known, the value m = 0 is to be rejected on a theoritical point of view [9], while the value m = 2 appears much too high, as far as we are discussing DISCO-1 comparisons. So the value m = 1 will be kept from now on. Note that the ratios $\varepsilon_{p,i}(t)/\varepsilon_{f,i}(\infty)$ can be higher than 1, a fact produced both by DISCO-1 and DISCO-2 and which has been explained in [10].

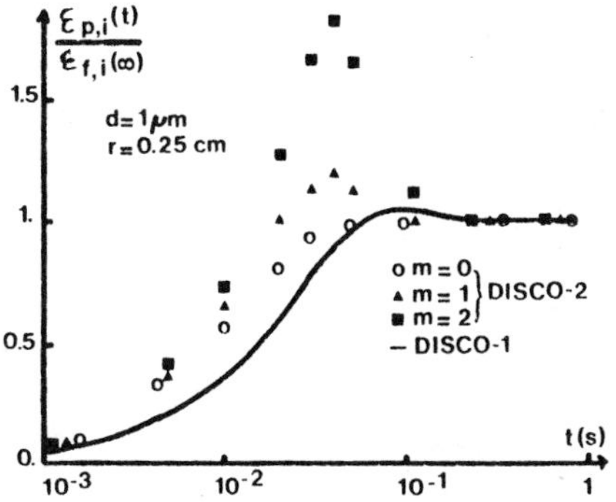

Fig. 2

DISCO-2 will be now used, since it produces sensibly the same results but much more easily, less costfull and less time-consuming.

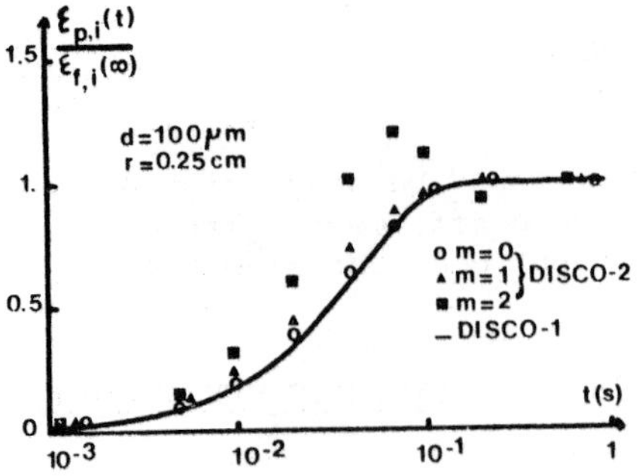

Fig. 3

IV - NON-DIAGONAL TERMS OF THE DISPERSION TENSOR

In the code DISCO-1, only the diagonal terms of the dispersion tensor were considered. They are now reintroduced in the code DISCO-2, and express the coupling of the dispersion between two directions.

The correlation tensor, leading to the dispersion tensor, is symmetric and thus a coordinates system can be found where it is diagonal. In that system, the diagonal terms of the dispersion tensor, which is also diagonal, are linked to the spectral tensor by [8] :

$$\varepsilon'_{p,i}(t) = \overline{u^2_{fL,i}}\ \tau_{L,i} \int_0^\infty \eta^2(\omega)\ E_{fL,i}(\omega)\ \frac{\sin \omega t}{\omega}\ d\omega \qquad (8)$$

The change of coordinate system is done using a rotation angle θ on the third axis, leading to the matrix of rotation :

$$a_{ij} = \begin{pmatrix} \cos\theta & \sin\theta \\ -\sin\theta & \cos\theta \end{pmatrix} \qquad (9)$$

After some straightforward calculations, we find :

$$\left.\begin{aligned} u'_{xx} &= u_{xx} + u_{xy}\ \mathrm{tg}\ \theta \\ u'_{xy} &= 0 \\ u'_{yy} &= u_{yy} - u_{xy}\ \mathrm{tg}\ \theta \end{aligned}\right\} \qquad (10)$$

$$\mathrm{tg}2\theta = \frac{2\ u_{xy}}{u_{xx} - u_{yy}} \qquad (11)$$

where u_{ij} and u'_{ij} are the correlation tensors in the old and the new systems respectively.

The dispersion coefficients are then evaluated. We get [11] :

$$\left.\begin{aligned} \varepsilon_{xx} &= \varepsilon'_{xx} - \sin^2\theta\ (\varepsilon'_{xx} - \varepsilon'_{yy}) \\ \varepsilon_{xy} &= \frac{\sin 2\theta}{2}\ (\varepsilon'_{xx} - \varepsilon'_{yy}) \\ \varepsilon_{yy} &= \varepsilon'_{yy} + \sin^2\theta\ (\varepsilon'_{xx} - \varepsilon'_{yy}) \end{aligned}\right\} \qquad (12)$$

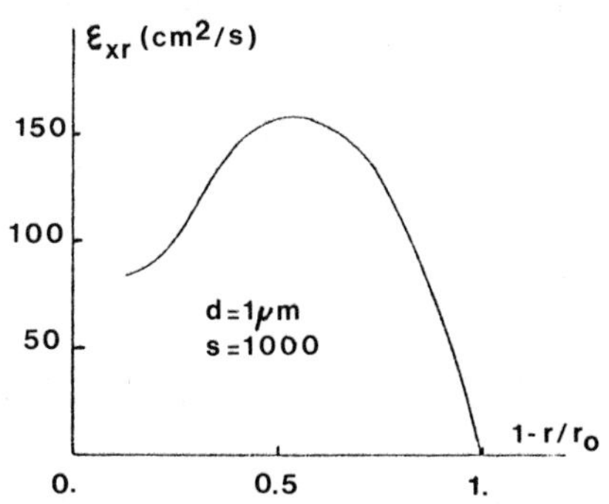

Fig. 4

To illustrate this purpose, the figure 4 shows the variations of ε_{xr} versus r in the case of the fully developped pipe flow [2], for a long diffusion time (assymptotic values). Note that the maximum value of ε_{xr} is 0,65 times the maximum value of ε_{xr} and equal to the maximum value of ε_{rr}.

Once the dispersion coefficients are known, the transport equation for the mean number of particles per unit of volume can be solved, which reads in a cartesian coordinates system :

$$\frac{\partial \bar{n}}{\partial t} + U_i \frac{\partial \bar{n}}{\partial x_i} = \frac{\partial}{\partial x_i} \varepsilon_{p,ij} \frac{\partial \bar{n}}{\partial x_j} \qquad (13)$$

The main advantage of such an equation is that it can be solved using the algorithm of the TEACH-T [12] code which is employed for the turbulence predictions, the terms of the r.h.s. for $i \neq j$ being introduced in the source term.

V - CROSSING TRAJECTORIES EFFECTS.

One of the main assumption in the Tchen's theory is that the discrete particle remains in the same fluid particle, during its motion. There is generally a lack of coïncidence between the discrete particles and the fluid particles trajectories, for instance when the particles are submitted to gravity forces, or when there is an external force producing a drift of the particles.

Yudine [13] and Csanady [14,15] considered the case of the turbulent diffusion of heavy particles in the atmosphere, falling fastly enough to cut the fluid particles trajectories. This is called since Yudine the crossing trajectories effect (C.T.E). Analyzing that phenomenon, Csanady derives an expression for the dispersion in the vertical direction, that means parallely to the free fall velocity :

$$\varepsilon_{p//}(\infty) = \varepsilon^{T}_{p//}(\infty) \left[1+\beta^2 \frac{\overline{f^2}}{\overline{w^2}}\right]^{-1/2} \qquad (14)$$

where $\varepsilon^{T}_{p//}(\infty)$ is the coefficient of dispersion in the Tchen's theory at an infinite time of dispersion, parallely to the free fall velocity $\vec{f}$, β is the ratio τ_L/τ_E of Lagrangian and

Eulerian integral time scales, and w^2 the mean square of the fluctuating velocity in the direction of $\vec{f}$.

Similarly, when the direction perpendicular to $\vec{f}$ is considered :

$$\varepsilon_{p\perp}(\infty) = \varepsilon_{p\perp}^{T}(\infty)\left[1+4\beta^2 \frac{f^2}{\overline{w^2}}\right]^{-1/2} \qquad (15)$$

the relation being slightly different because of the continuity of the flow.

Nevertheless these relations cannot be rigorously valid, for instance when the drift is not constant, or when an infinite time of dispersion in not considered, and a quite important uncertainty remains on the coefficient β. Rather to complicate the scheme, the philosophy will be to simplify it, as the aim of the code is essentially concerned by engineering purposes.

In a first step, assuming that the correction on the dispersion coefficient depends on the turbulence energy K and the square (for dimensional reasons) of the velocities difference $(U_{p,i} - U_{f,i})^2$, and following Csanady's analysis, we suggest :

$$\varepsilon_p = \varepsilon_p^T\left(1+C_\beta \frac{(U_{p,i} - U_{f,i})^2}{2K/3}\right)^{-1/2} \qquad (16)$$

where C_β is expected to have the order of magnitude of 1. For particles only submitted to gravity forces, we get :

$$\varepsilon_p = \varepsilon_p^T \left(1+C_\beta \frac{f^2}{2K/3}\right)^{-1/2} \qquad (17)$$

which is to be compared with (14) and (15).

The simplification in that first approach is mainly due to the facts that no difference is made between $\varepsilon_{p//}$ and $\varepsilon_{p\perp}$, the correction is introduced whatever the time, and C_β is an adjustable constant whose value will be deduced from experiments.

As it is going to be showed, the semi-empirical relation (17) will prove quite successful when discussing comparisons between predictions and experiments.

VI - COMPARISONS WITH GRID TURBULENCE EXPERIMENTS.

VI - 1 : Snyder and Lumley's experiments

Snyder and Lumley [16] studied the dispersion of different

particles in a turbulent grid flow. The main direction of the air flow is vertical, the Reynolds number about 10^4 and the influence of the walls is negligible. Turbulence measurements were made and energy decay laws on the centerline of the test section, established. Four kinds of particles are used by Snyder and Lumley, namely hollow glass, solid glass, corn and copper. The main relevant parameters are listed on the table I.

	hollow glass	solid glass	corn	copper
d (μm)	46.5	87.0	87.0	46.5
ρ (g/cm^3)	0.26	2.5	1.0	8.9
f (cm/s)	1.67	44.2	19.8	48.3

Table I

where d is the diameter of the particles, ρ their density and f their terminal velocity, i.e the Stokes velocity corrected by a method recommended by Fuchs [17].

The particles are injected on the centerline of the flow, 51 cm after the grid, with a mean velocity equal to the mean velocity of the flow. Measurements of the mean square displacement $\overline{Y^2}$ are made at different locations, from 173 cm up to 427 cm.

In order to be as close as possible from Snyder and Lumley's experiments, the decay laws for turbulence given analitically are introduced in the code.

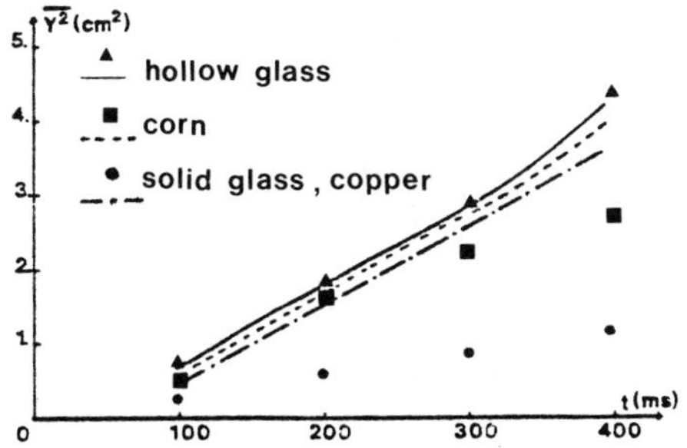

Fig. 5 Lines : DISCO-2-Pts: [16]

The figure 5 gives the results of $\overline{Y^2}$ versus the time t, for the four particles, without the C.T.E. As expected, a good agreement is obtained for the hollow glass particles, wich follow quite well the fluid and are not concerned by the C.T.E since their terminal velocity is small compared to the turbulence energy (f = 1.67, 2K/3 = 200). On the other hand, a large difference is observed for the three other particles.

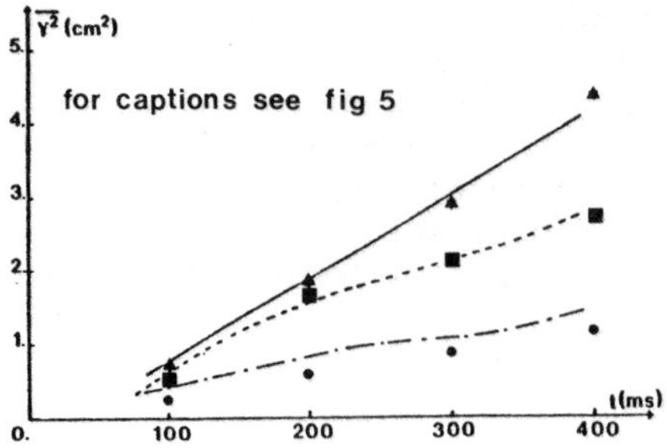

Fig. 6

The figure 6 presents the same results, but taking into account for the C.T.E, using the relation (17) with C_β = 0.85. The comparison is satisfactory for all particles and confirm that C_β has the order of magnitude of 1.

VI - 2 : Well's experiments

The turbulence situation considered by Wells [18] is similar to Snyder and Lumley's experiments, but the main direction of the flow is horizontal. Turbulence measurements are made and energy decay laws characterizing the grid flow are given, quite identical to Snyder and Lumley's results. The particles are glass spheres, with a diameter of 5 and 57 µm, and a density equal to 2.45 g/cm^3. They are injected on the centerline of the flow, and submitted to an electrical field which allows to control the crossing trajectory effects by adjusting the terminal velocity. Measurements of the mean square displacement $\overline{Y^2}$ are made, from 50 cm up to 178 cm.

Similarly to the case of Snyder and Lumley, the decay laws for the turbulence given by Wells are introduced in the program.

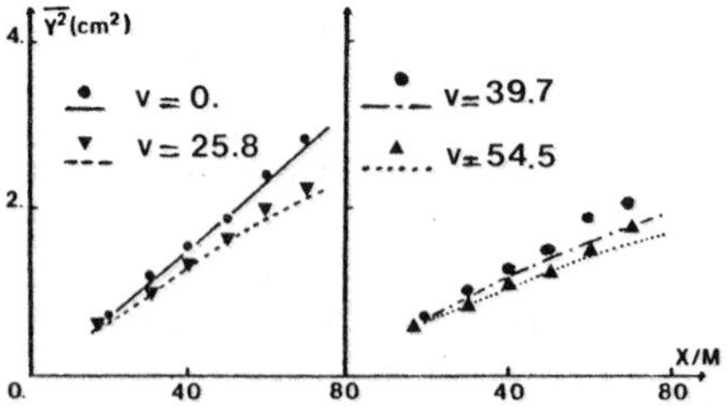

Fig.7:Pts:Wells Lines DISCO-2

The figure 7 shows the mean square displacement $\overline{Y^2}$ versus X/M(M = 2.54 cm) for 57 µm particles, with four different terminal velocities, namely 0, 25.8, 39.7 and 54.5 cm/s. The same value (0.85) is taken for C_β. We can observe that the predictions are smaller than the experimental results, but the agreement remains satisfactory.

VII - COMPARISON WITH PARTICLES DISPERSION IN A TURBULENT PIPE FLOW.

Arnason's experiments [19] deals with an air flow in a pipe, aligned vertically to prevent gravity from changing the axial symmetry of the flow which carries particles. The diameter of the pipe is 9 cm and the Reynolds number 50 000.

The particles are the same as those used by Wells and are injected on the centerline of the flow, isokinetically, 50 diameters from the beginning of the pipe in order to reach the fully developed part of the flow.

The code DISCO-2 has been used to predict all the experimental situation, both for turbulence and particles dispersion. Good agreements has been found between predictions and experimental results for the turbulence field, and are available from [20], but are not repeated here as they are quite similar to the results obtained when comparing predictions and Laufer's results [21], extensively discussed elsewhere [4,8].

The figures 8,9 and 10 show the mean concentration reduced profiles $\bar{n}/\bar{n}_c$, where $\bar{n}_c$ is the mean number of particles on the axis, versus r/R, where R is the pipe radius, for three different locations, namely 31.8, 50.2 and 67.9 cm from the seeding point. The considered particles are the 5 μm diameter ones. The predictions are obtained using the last version of the code, and are compared with Arnason's experimental results. A good agreement is observed in the three cases, and confirms the validity of the code DISCO-2.

VIII - CONCLUSION

The code DISCO-2 for the prediction of particles dispersion in turbulent flows has been described particularly the Lagrangian integral time scale evaluation and the non-diagonal terms of the dispersion tensor. Results of the dispersion coefficients using the code DISCO-2 have been compared against the predictions of the code DISCO-1 from which it is derived. The agreement is very satisfactory. Comparisons have also been made against experimental results, in two different turbulence fields, a grid flow and a pipe flow respectively, and a good agreement has been found when the crossing trajectory effects are taken into account for, using a correction term which has been defined.

The results leads us to the conclusion that the code DISCO-2 should be further developed, since this approach appears to be very promising.

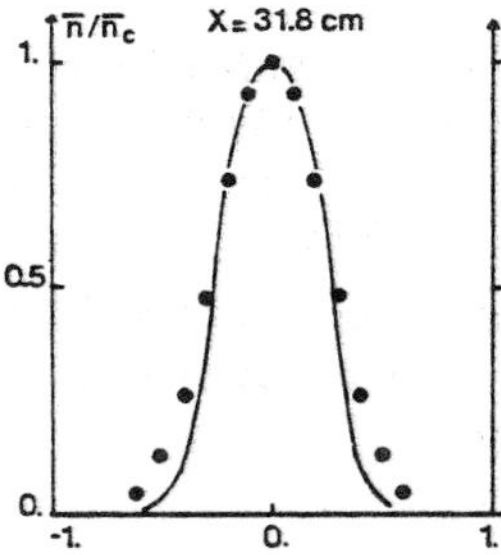

Fig. 8

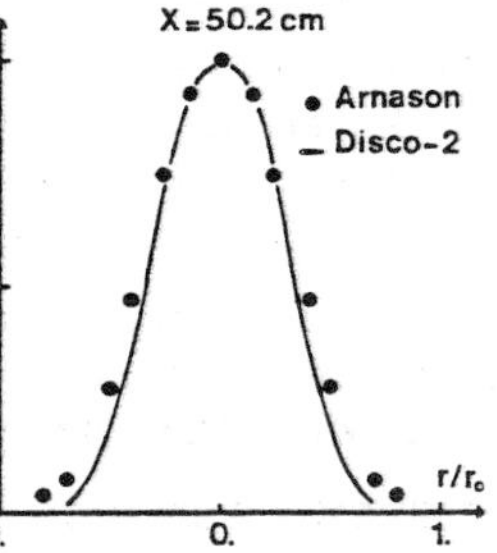

Fig. 9

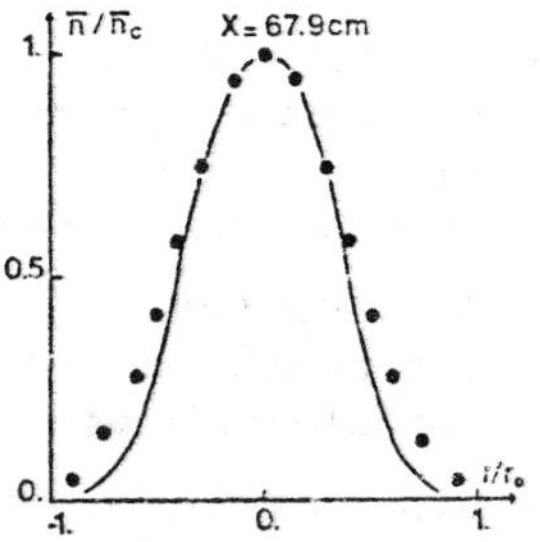

Fig. 10

1 - A. BERLEMONT and G. GOUESBET. The dispersion of a cloud of particles released in a fully developed pipe flow, using the code DISCO-1. Letters in Heat and Mass Trans. Vol 9, 11-19, 1982.

2 - A. BERLEMONT and G. GOUESBET. Prediction of the behaviour of a cloud of discrete particles released in a fully developed turbulent pipe flow, using a dispersive approach. Numerical methods in laminar and turbulent flows conferences. Venizia, July 13-16th, 1981, Proceedings.

3 - A. BERLEMONT and G. GOUESBET. Prediction of the turbulent round free jet, including fluctuating velocities correlations, by means of a simplified second-order closure scheme. Letters to Heat and Mass Trans., 8, 207-217, 1981.

4 - G. GOUESBET and A. BERLEMONT. Prediction of turbulent fields, including fluctuating velocities correlations and approximate spectra by means of a simplified second-order closure scheme : the round free jet and developing pipe flow. Numerical methods in laminar and turbulent flows conference. Venizia, July 13-16th, 1981, Proceedings.

5 - G. GOUESBET, A. PICART, A. BERLEMONT. Dispersion of discrete particles by turbulent continuous motions using a Frenkiel's family of Lagrangian correlation functions, in the non-discrete dispersive approach. Numerical methods in laminar and turbulent flow conferences. August 8-11th 1983, The University of Washington, Seattle, USA.

6 - A. PICART, A. BERLEMONT, G. GOUESBET. De l'influence du terme de Basset sur la dispersion de particules discrètes dans le cadre de la théorie de Tchen. CRAS, Série II, tome 295, 305, 1982.

7 - H. TENNEKES, J.L. LUMLEY. A first course in turbulence. MIT Press, Cambridge, Mass, 1972.

8 - A. BERLEMONT, G. GOUESBET. Modelling and predicting the behaviour of discrete particles embedded in turbulent flows using the Non-Discrete Dispersive Approach and the code DISCO-1. Int. Rep. MADO/81/11/XI.

9 - J.O. HINZE. Turbulence. Mac Graw Hill book Co, 1959.

10 - G. GOUESBET, A. BERLEMONT, A. PICART. On the Tchen's theory of discrete particles dispersion : can dense discrete particles disperse faster than fluid particles ? Letters in Heat and Mass Transfer, 9, 5, Sept-Oct., 1982.

11 - A. BERLEMONT, G. GOUESBET, A. PICART. The code DISCO-2 for the prediction of the dispersion of discrete particles in turbulent flows. Int. Report MADO/82/9/V.

12 - A.D. GOSMAN, F.J.K. IDERIAH. TEACH T. Dept. Mech. Eng. Imperial College London 1976.

13 - M.I. YUDINE. Physical considerations on heavy particle diffusion. Adv. in Geophysics, 6, 185-191, New York, Academic Press 1959.

14 - G.T. CSANADY. Turbulent Diffusion in the environment. Reidel Publishing Company 1973.

15 - G.T. CSANADY. Turbulent diffusion of heavy particles in the atmosphere. Journal of the Atm. Sci. 20, 201, 1963.

16 - W.H. SNYDER, J.L. LUMLEY. Some measurements of particle velocity autocorrelation functions in a turbulent flow. J. Fluid. Mech., 1971.

17 - N.A. FUCHS. The mechanics of Aerosols. MacMilan 1964.

18 - M.R. WELLS. The effects of crossing trajectories on the diffusion of particles in a turbulent fluid. Ph. D. Thesis Washington State University, 1982.

19 - G. ARNASON. Measurements of particle dispersion in turbulent pipe flow. Ph. D. Thesis Washington State University 1982.

20 - G. GOUESBET, A. PICART, A. BERLEMONT. Miscellaneous problems and results connected with the improvements and applications of the code DISCO-2 for the prediction of the behaviour of discrete particles in turbulent flows. Int. Report MADO/GPB/1/83/I 1983.

21 - LAUFER J. The structure of turbulence in fully developed pipe flow. Nat. Ad. Com. Aero. Report 1/74, 1954.

NUMERICAL PROBLEMS ASSOCIATED WITH SIMULATION OF CONTAMINANT TRANSPORT UNDER CONFINED CONDITIONS

by

A. Pandit[1] and S.C. Anand[2]

1. SUMMARY

The Galerkin finite element technique is used to simulate the transport of contaminants in a two dimensional confined aquifer. The mathematical model is formulated in terms of a flow equation and a transient transport equation. As the two equations are coupled, the velocity vectors become a function of the contaminant concentration. For extreme values of some nondimensional field parameters the spatial locations of concentration contours at a given time vary significantly with the value of the selected time step size, Δt. This phenomenon is attributed to the fact that nodal concentrations near the contaminant source oscillate with time. These oscillations remain despite the usage of extremely small spatial and temporal discretizations. Calculations show that, for some extreme parameters, concentrations affect the flow field to yield relatively large negative values of the cross dispersion tensors, D_{xy} and D_{yx}, near the contaminant source. However, D_{xy} and D_{yx} are positive for all stable cases. Further analysis indicates that stable solutions are possible even for extreme parameter values if the governing equations are uncoupled.

2. INTRODUCTION

Due to the versatile nature of its technique, the Galerkin finite element method is becoming increasingly popular with engineers involved in solving flow problems in porous media. However, with the progressive use of this technique, users are

(1) Asst. Prof., Dept. of Civil. Engr., Florida Inst. of Technology, Melbourne, FL., 32901, USA.

(2) Prof., Dept. of Civil Engr., Clemson University, Clemson, S.C., 29631, USA.

also becoming aware of the numerical problems associated with the application of the finite element method. Some of the numerical instabilities that appear in the solution of flow equations in porous media have been extensively documented by Mercer and Faust [1].

In this paper, an effort is made to document a peculiar form of instability observed by the authors while simulating the movement of contaminants in a confined aquifer. The instability exists only for some typical values of the field parameters. Attempts made to remove this instability are described in detail.

3. MATHEMATICAL FORMULATION

The mathematical model is similar to the h-c model proposed by Huyakorn and Taylor [2] and is formulated in terms of a flow equation and a transport equation which are coupled by the velocity and density terms. These equations may be written as

$$V_i = - K_{ij} \left[\frac{\partial h}{\partial x_j} + \frac{\varepsilon C}{C_1} n_j \right] , \qquad (1)$$

and

$$\frac{\partial}{\partial x_j} \left[D_{ij} \frac{\partial C}{\partial x_j} \right] - V_i \frac{\partial C}{\partial x_i} = \frac{\partial C}{\partial t} \qquad (2)$$

in which K_{ij} = hydraulic conductivity tensor, h = hydraulic head, ε = normalized density difference between contaminant and fresh water, C_1 = contaminant concentration at the source, V_i = interstitial pore velocities and t = time. D_{ij} in Eq. (2) is the dispersion tensor, and its terms for a planer condition can be expressed as (Bear [3])

$$D_{xx} = \alpha_L \frac{V_x^2}{|V|} + \alpha_T \frac{V_y^2}{|V|} + D_d \tau_{xx} , \qquad (3)$$

$$D_{yy} = \alpha_T \frac{V_x^2}{|V|} + \alpha_L \frac{V_y^2}{|V|} + D_d \tau_{yy} , \qquad (4)$$

and

$$D_{xy} = D_{yx} = (\alpha_L - \alpha_T) \frac{V_x V_y}{|V|} , \qquad (5)$$

in which α_L and α_T = longitudinal and transverse dispersivities, D_d = molecular diffusion coefficient, τ_{xx} and τ_{yy} = principal components of the tortuosity tensor, and n_j = 1 for j = 2 (vertical direction). Introducing the nondimensional variables

$$x' = x/d \ , \ y' = y/d \ , \ h' = h/d \ , \ C' = C/C_1 \ ,$$

$$K'_{yy} = K_{yy}/K_{xx} \ , \ K'_{xx} = K_{xx}/K_{xx} = 1 \ , \ V'_x = V_x/U \ , \qquad (6)$$

$$V'_y = V_y/U \ , \ \alpha'_T = \alpha_T/\alpha_L \ , \ \alpha'_L = \alpha_L/\alpha_L = 1 \ , \ t' = tU/d,$$

in which U and d = characteristic velocity and depth of aquifer, respectively, Eqs. (1) and (2) can be nondimensionalized (neglecting molecular diffusion) by substituting Eqs. (3) to (5) in these which after rearrangement, and neglecting primes for convinience, yields

$$K_{xx} \frac{\partial^2 h}{\partial x^2} + K_{yy} \frac{\partial^2 h}{\partial y^2} = - K_{yy} \varepsilon \frac{\partial C}{\partial y} \quad (7)$$

and

$$\left(\frac{V_x^2}{|V|} + \alpha_T \frac{V_y^2}{|V|}\right)\frac{\partial^2 C}{\partial_x^2} + \left(\alpha_T \frac{V_x^2}{|V|} + \frac{V_y^2}{|V|}\right)\frac{\partial^2 C}{\partial y^2} + (1-\alpha_T)\frac{V_x V_y}{|V|}\frac{\partial^2 C}{\partial x \partial y}$$

$$+ (1-\alpha_T)\frac{V_x V_y}{|V|}\frac{\partial^2 C}{\partial y \partial x} - \left(V_x \frac{\partial C}{\partial x} + V_y \frac{\partial C}{\partial y} + \frac{\partial C}{\partial t} R_d\right) d_L = 0, \quad (8)$$

respectively, in which $d_L = d/\alpha_L$.

3.1 Finite Element Technique

Using a triangular element with no internal or midside nodes, and assuming the shape functions to be linear, the senior author [4] has shown that Eqs. (7) and (8) can be written symbolically in the matrix form as

$$[K] \{h\} = \{P\} \quad (9)$$

and

$$[R] \{C\} + [S] \{\partial C/\partial t\} = \{Q\} , \quad (10)$$

respectively, where [K], [R] and [S] are coefficient matrices and [P] and [Q] are force vectors containing the boundary fluxes. Specific expressions for these matrices are given in Reference [4].

3.2 Temporal Approximation

Approximating the time derivative term in Eq. (10) using the implicit scheme (Pinder and Gray [5]) leads to

$$([R] + 1/\Delta t\ [S])\{C\}_{t+\Delta t} = 1/\Delta t\ [S]\{C\}_t + \{Q\}_{t+\Delta t} \quad (11)$$

in which Δt is the size of the time step. Equations (9) and (11) represent a coupled, nonlinear, transient system and are solved by an iterative algorithm developed by the authors [6].

4. PROBLEM DEFINITION

The phenomenon of contaminant transport in porous media is investigated for a two-dimensional, saturated, rectangular, confined aquifer. The pollutant is assumed to discharge continuously from the top boundary of the confined aquifer due to leaky conditions. The rate of discharge into the aquifer, however, is not necessarily constant. Furthermore, it is assumed that the leakage occurs over a length of d/5 where d is the depth of the aquifer. Mass transport in the unsaturated zone is considered to be negligible. The dimensions and boundary conditions utilized in this problem are shown in Fig. 1.

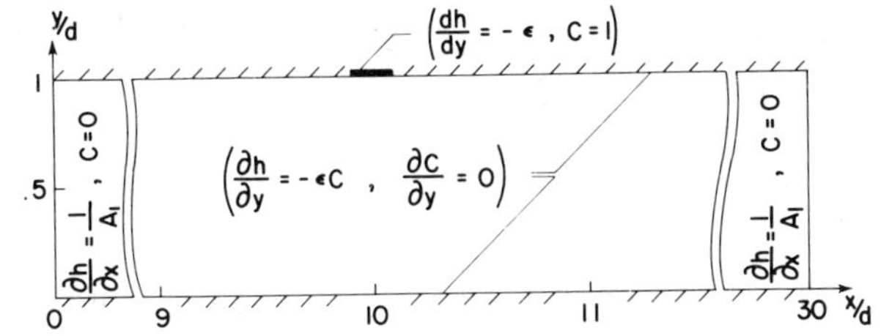

Fig. 1 Boundary Conditions and Region of Rectangular Aquifer.

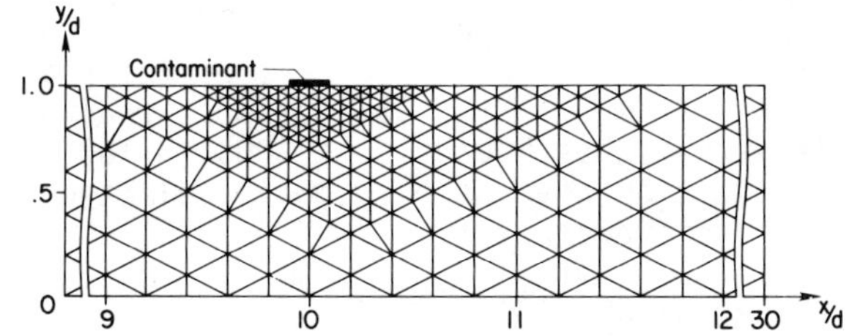

Fig. 2 Finite Element Mesh

Results are simulated for a wide range of parameters using a region size of 30:1. The location of the center of the source, for convenience, is kept at a nondimensional distance of 10 from the left-hand boundary.

4.1 Boundary Conditions

For Flow Equation	For Transport Equation
$\frac{\partial h}{\partial x}(0,y,t) = -\frac{1}{A_1}$	$C(0,y,t) = 0$,
$\frac{\partial h}{\partial y}(x,1,t) = -\varepsilon C$,	$\frac{\partial C}{\partial y}(x,1,t) = 0$, $0 \leq x \leq 9.9$
$\frac{\partial h}{\partial y}(x,1,t) = -\varepsilon$,	$C(x,1,t) = 1$, $9.9 \leq x \leq 10.1$
$\frac{\partial h}{\partial y}(x,1,t) = -\varepsilon C$,	$\frac{\partial C}{\partial y}(x,1,t) = 0$, $10.1 \leq x \leq 30$
$\frac{\partial h}{\partial x}(30,y,t) = -\frac{1}{A_1}$	$C(30,y,t) = 0$,
$\frac{\partial h}{\partial y}(x,0,t) = -\varepsilon C$	$\frac{\partial C}{\partial y}(x,0,t) = 0$.

In these equations, A_1 is the inverse of the undisturbed hydraulic gradient.

4.2 Initial Conditions

$$C\ (x,\ 1,\ 0) = 1\ ,\quad 9.9 \leq x \leq 10.1$$

$$C\ (x,\ y,\ 0) = 0\ .$$

5. VALUES OF VARIOUS FIELD PARAMETERS

Dispersivity α_L in the field is typically much larger than its values obtained from laboratory experiments and is reported to be in the range of 0.1 m to 100 m. In view of the fact that the depth, d, of an aquifer may vary anywhere from 1 m to 500 m, it is not unreasonable to assign a range of 1 to 20 for d_L ($d_L = d/\alpha_L$), keeping in mind that larger aquifers should also have larger dispersivities. Bounds from 0.5 to 1.0 for the nondimensional parameters α_T and K_{yy} are considered quite reasonable. The gradient of the piezometric surface can vary from 0.1 % to 1.0 % for most realistic situations. This establishes bounds from 1000 to 100 for the parameter A_1. In addition a value of 0.05 is selected for the normalized density difference, ε.

6. RESULTS AND DISCUSSION

For the purpose of this investigation, 12 combinations of the nondimensional parameters, d_L, α_T and A_1 are considered and are defined as specific cases shown in Table 1. In this

Table 1. Various Cases Analyzed for Total Time T = 1.60

Case	d_L	α_T	A_1	L_1	L_2
1	1	1	100	3.35	3.45
2			1000	4.40	4.80
3		.05	100	2.55	-
4			1000	2.60	3.85
5	10	1	100	1.80	1.65
6			1000	2.60	4.85
7		.05	100	1.85	-
8			1000	No Acceptable Solutions	
9	20	1	100	1.70	-
10			1000	2.10	5.20
11		.05	100	1.80	-
12			1000	No Acceptable Solutions	

table, L_1 and L_2 are the distances (in the downstream direction) from the contaminant source to the points of intersection of the 0.1 Concentration Contour with the top and bottom boundaries of the aquifer, respectively. The finite element mesh utilized for this investigation is shown in Fig. 2 and consists of 1938 elements and 1137 nodal points. Four time step schemes with gradually increasing Δt are utilized to conduct various analyses and are shown in Table 2. The

Table 2. Time Step Schemes

Scheme	Time Step #	Δt	Total Elapsed Time, T	Scheme	Time Step #	Δt	Total Elapsed Time, T
J	1	0.02	0.02	L	1-2	0.005	0.01
	2	0.04	0.06		3-4	0.01	0.03
	3	0.08	0.14		5-6	0.02	0.07
	4-46	0.16	6.38		7-36	0.04	1.27
K	1-2	0.005	0.01	M	1-2	0.005	0.01
	3-4	0.01	0.03		3-21	0.01	0.20
	5-6	0.02	0.07				
	7-8	0.04	0.15				
	9-46	0.08	3.19				

results shown in Table 1 are obtained utilizing Scheme J. It may be noted that the solutions for cases 8 and 12 have been termed unacceptable because certain oscillatory trends in concentration values were observed for these analyses. An attempt is made here to describe this oscillatory phenomenon, and the efforts undertaken to remove the oscillations. Analysis is also conducted to find correlations between the occurance of these oscillations and the values of the field parameters, d_L α_T, A_1 and ε

6.1 Results With Time Step Scheme J

Concentration contours for Case 8 at nondimensional times T = 1.58, 3.38, and 6.38 are shown in Fig. 3. Two observations, based on the location of the 0.3 Contour, create some doubt regarding the validity of the solutions. First, it can be seen that at T = 1.58 this concentration isopleth touches the bottom boundary. However, instead of continuing to spread along this boundary, as one would expect, it moves in an upward direction and at a later time, T = 3.18, it does not touch the bottom boundary at all. Second, at T = 6.38, the location of the 0.3 Contour indicates that the spread of the pollutant has shrunk with time, and that the bulk of the contaminant exists near the top of the aquifer close to the pollutant source. Since the source is discharging continuously into the confined aquifer, although not necessarily at a constant rate, these observed phenomena do not appear to be physically possible. As no negative concentrations are observed even near the pollutant source, one is inclined to believe that the contaminant shrinkage occurs because the temporal discretizations are too large and the spatial discretizations have no appreciable influence.

It is also of interest to note that the rate at which the contaminant is introduced into the confined aquifer due to dispersion may be given by

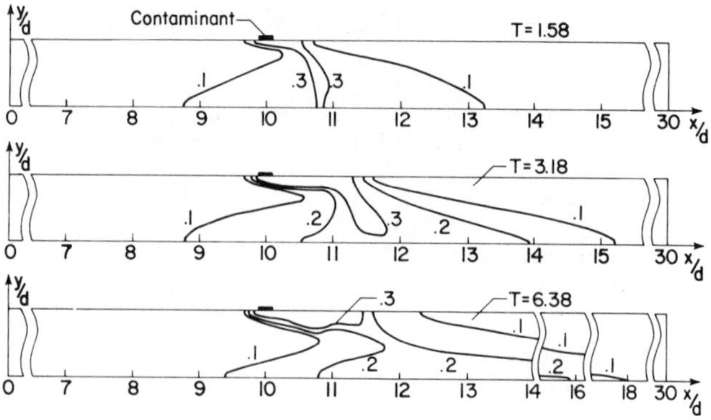

Fig. 3 Concentration Contours for $d_L = 10$, $\alpha_T = .05$ and $A_1 = 1000$ at Various Times Using Scheme J.

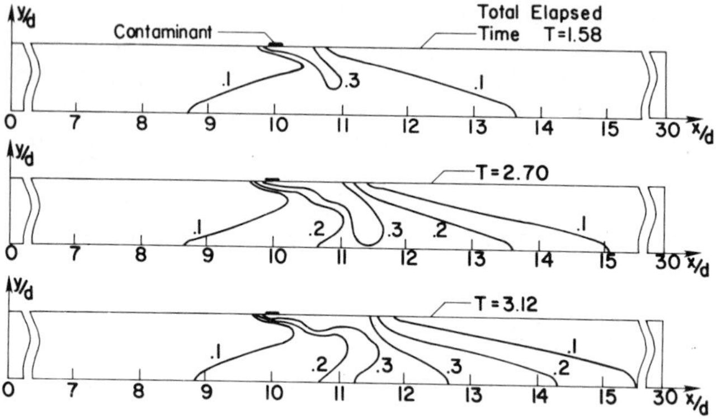

Fig. 4 Concentration Contours for $d_L = 10$, $\alpha_T = .05$ and $A_1 = 1000$ at Various Times Using Scheme K.

$$Q_c = D_{yy}\ dC/dy \tag{12}$$

where dC/dy represents the concentration gradient near the pollutant source. However, the boundary condition C = 1, applied at the leaky boundary, causes this gradient to vary with time. Since the contaminant concentration C is initially equal to zero near the source, a sharp concentration gradient exists in this region for the first few time steps. As a result, the value of Q_c is also large during this period, further reinforcing the hypothesis that a relatively large value of Δt at the beginning of a simulation can lead to erroneous results. Consequently, smaller time step sizes are more desirable at the initial stages of the pollutant transport. Therefore, a new time step scheme, Scheme K, shown in Table 2, with a smaller initial time step of size 0.005 is utilized.

6.2 Results With Time Step Scheme K

Concentration isopleths for Case 8 at various levels, utilizing time step Scheme K, are plotted at different intervals in time and are shown in Fig. 4. It can be seen that the 0.3 Concentration Contour does not touch the bottom boundary of the aquifer at T = 1.58, contrary to what was observed in Fig. 3. This indicates that a change in the size of the time step, Δt, does indeed alter the results. Moreover, the 0.3 Concentration Isopleth, once it touches the lower boundary, continues to spread along this boundary and does not tend to rise as had been observed previously. However, an inspection of the 0.2 and 0.3 Contours near the pollutant source at T = 3.12 shows that a waviness exists in this region. A similar observation can be made by inspecting these same isopleths at T = 2.70.

To investigate the reasons for this waviness, concentrations at three of the five closest nodes to the pollutant source versus time T are shown in Figure 5 for Case 8. These nodes are numbered 1 through 5 for convenience. The nondimensional distance of Nodes 1, 3, and 5 from the source is 0.05, whereas, Nodes 2 and 4 are at a distance of 0.025. It can be seen from this figure that concentrations at all three nodes reach a peak and then decrease just as sharply within a period of T = 0.2. This is followed by a transition period during which concentration reaches an equilibrium position. This is especially true for Nodes 1 and 2. However, after T = 1.5, these values tend to oscillate around some equilibrium position. Similar observations were also made for Nodes 3 and 4. On the other hand, a plot of concentration vs. time for a node at a nondimensional vertical distance of 0.20 from the source indicated no oscillatory behavior.

At this point, two observations are made. First, since the oscillations at every node begin at approximately T = 1.50, one can hypothesize that the value of Δt = 0.08 is too large and is the cause for the oscillations. Second, the sharp peak that occurs within the nondimensional time 0.5 could be due to a relatively large initial time step size and may be the cause for oscillations at a later time. Hence, Time Step Schemes L and M are devised to further investigate this phenomenon. The maximum values of Δt are 0.04 and 0.01 in Schemes L and M, and the concentrations are determined up to a total nondimensional time of 1.27 and 0.20, respectively. The corresponding results are shown in Figs. 6 and 7, which indicate that the oscillations and initial peaks still remain. In view of these results it appears that although the size of the time step in the later stages of the simulation is not the cause of these oscillations, it does alter the frequency. Moreover, the use of smaller time increments at the beginning of the simulation does alter the results.

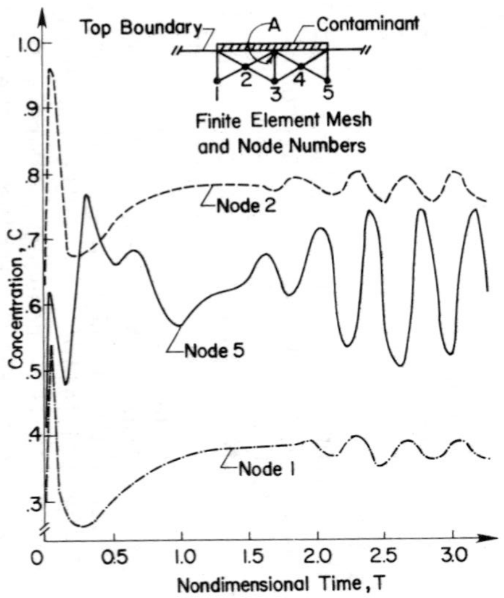

Fig. 5 Nodal Concentrations vs. Time for Case 8 Using Scheme K.

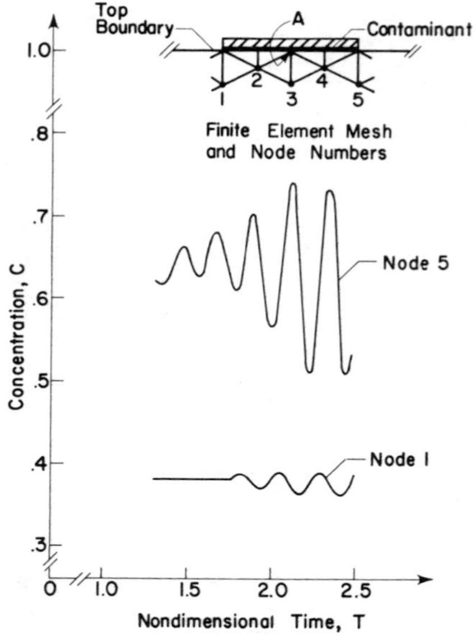

Fig. 6 Nodal Concentrations vs. Time for Case 8 Using Scheme L.

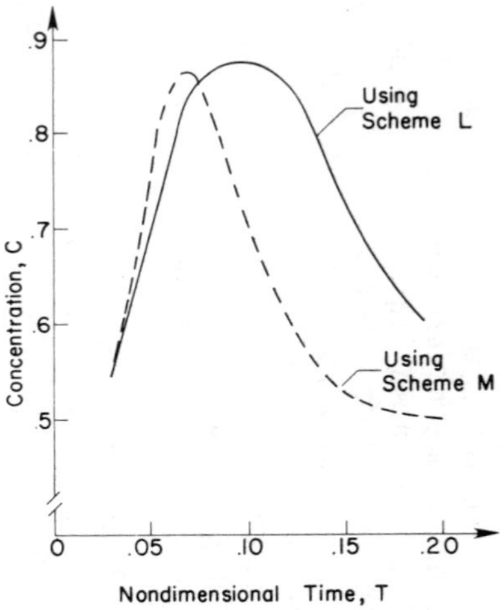

Fig. 7 Concentrations vs. Time at Node 3 Using Schemes L and M.

To further understand the causes for the presence of the initial peak and subsequent oscillations, the flow and transport equations are solved for three new cases. These cases, numbered 13 through 15 in Table 3, are designed to study the effects of various field parameters on the stability of the solution.

Table 3. Typical Values of Coefficients of Hydrodynamic Dispersion at the Center of the Leaky Boundary for $d_L = 10$

Case	α_T	A_1	ε	T	D_{xx}	D_{yy}	$D_{xy} = D_{yx}$
				.005	2.11	2.11	0
6	1.0	1000	.05	.03	3.06	3.06	0
				.15	4.95	4.95	0
				.005	0.07	0.09	-0.07
7	.05	100	.05	.03	0.02	0.10	-0.03
				.15	0.04	0.12	-0.06
				.005	0.50	1.10	-0.66
*8	.05	1000	.05	.03	0.40	1.41	-0.65
				.15	9.76	1.90	-3.52
				.005	0.82	0.26	-0.40
*13	.05	1000	.02	.03	0.13	0.67	-0.24
				.15	5.08	0.41	-0.84
				.005	1.00	0.05	0.00
14	.05	1000	.00	.03	1.00	0.05	0.00
				.15	1.00	0.05	0.00
				.005	0.11	0.14	-0.04
15	.5	100	.05	.03	0.10	0.15	-0.03
				.15	0.12	0.20	-0.03

* Solutions are considered unstable.

6.3 Transient Solutions for Uncoupled Equations

The equations can be uncoupled by assigning a value of zero to the normalized density difference, ε. This means that the contaminant concentrations no longer affect the flow field. Thus, a new case, Case 14 is introduced in which d_L=10, α_T = 0.05, A_1 = 1000 and ε = 0. Concentrations plotted for Nodes 1 through 5, shown in Fig. 8 reveal that no oscillations occur for this case. Since the values of all parameters in Cases 8 and 13 are the same as in Case 14 except for the value of ε, one can argue that for these cases large concentrations near the leaky boundary cause abrupt changes in the flow field leading to oscillations. Results of Case 13, in which ε = 0.02 is rather small, were also found to be unacceptable due to the presence of oscillations. Therefore, it appears, that at the extreme parametric values, oscillations will occur even if the degree of coupling is rather small.

6.4 Solutions for Some Stable Cases

It should be of interest to look at the concentration vs. time curves for those cases in which the flow and transport equations are coupled and the solutions are stable. These plots are shown in Fig. 9 for Cases 6, 7 and 15 and exhibit no oscillatory trend. Except for the value of α_T, which is equal to 0.5, all other parametric values in Case 15 are identical to Case 8 that yields unstable results. This indicates that the solutions tend to become unstable as the value of α_T decreases. Again, since the only difference between Cases 7 and 8 is in the value of A_1, it appears that an increase in the value of A_1 also leads to unstable results.

6.5 Dispersion Coefficients at Middle of Pollutant Source

Some typical values of the coefficients of hydrodynamic dispersion are calculated for various cases at the middle of the leaky boundary and are shown in Table 3. A comparison of these values indicates that they are a magnitude larger for Cases 6, 8 and 13. Since Case 6 is considered to yield stable solutions there does not appear to be any correlation between these values and the stability of the solution.

On the other hand, a comparison of the values of the cross-dispersivity tensors, D_{xy} and D_{yx}, for the various cases leads to an interesting observation. It can be seen that for the unstable cases the values of D_{xy} and D_{yx} are not only

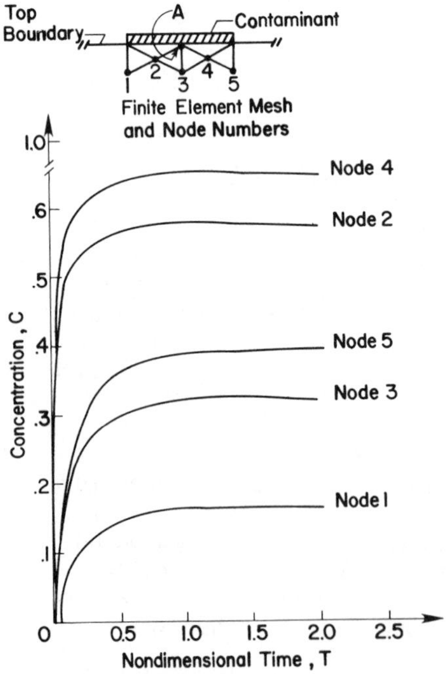

Fig. 8 Nodal Concentratios vs. Time for Uncoupled Case (14) Using Scheme K.

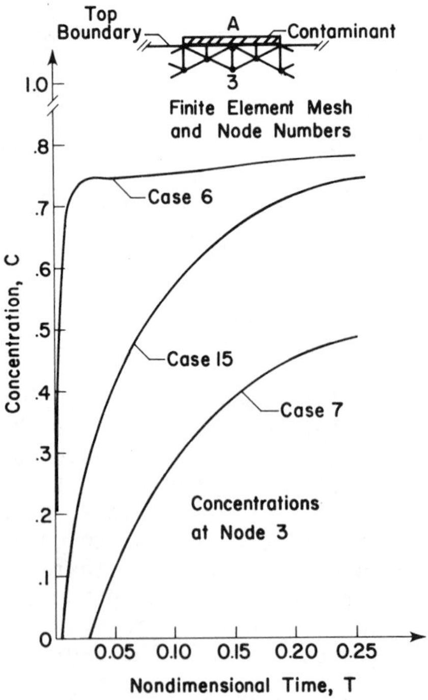

Fig. 9 Concentration vs. Time at Node 3 for Various Cases Using Scheme K.

negative but have a relatively larger magnitude than those calculated for the other cases. For example, all D_{xy} and D_{yx} in Cases 7 and 15 are of an order less than 0.1, whereas, they are equal to zero for Cases 6 and 14. For the unstable Cases 8 and 13, on the contrary, these values are of the order of 1 or more. It appears that the large negative values of the cross-dispersivity tensor D_{xy} and D_{yx} can lead to unstable results

7. CONCLUSIONS

It is not easy and obvious to exactly define the range of values for the field parameters for which the solutions, that predict contaminant transport in a saturated confined aquifer by the Galerkin finite element method, become unstable. However, the results of various solutions indicate that instability occurs if the values of the cross-dispersion tensors, D_{xy} and D_{yx}, near the contaminant source become negative with magnitudes equal to or greater than one. Further research is needed to investigate the influence on the stability of solutions using higher order elements.

8. ACKNOWLEDGMENTS

The research reported in this paper was supported by a research grant No. A-051-SC from the Office of Water Research and Technology Department of the Interior, to Clemson University. The Financial support of OWRT is gratefully acknowledged.

9. REFERENCES

1. MERCER J.W. and FAUST, C.R. - An Application of Finite Element Techniques to Immicible Flow in Porous Media, Finite Elements in Water Resources, Ed. by Gray, W.G., Pinder, G.F., and Brebbia, C.A., Pentech Press, 1977.

2. HUYAKORN, P. and C. TAYLOR. - Finite Element Models for Coupled Groundwater Flow and Convective Dispersion, Finite Elements in Water Resources, Ed. by Gray, W.G., Pinder, G.F., and Brebbia, C.A., Pentech Press, 1977.

3. BEAR, J. - On The Tensor Form of Dispersion, Jour. Geophys. Res., Vol. 66, pp. 1185-1197, 1961.

4. PANDIT, A. - Numerical Simulation of Contaminant Transport Problems in Groundwater Using the Finite Element Method, Ph.D. Diss., Dept. of Civil Eng., Clemson Univ., August 1982.

5. PINDER, G.F. and GRAY, W.G. - Finite Element Simulation in Surface and Subsurface Hydrology, New York Academic Press, 1977.

6. ANAND, S.C. and PANDIT, A. - Finite Element Solutions of Coupled Groundwater Flow and Transport Equations Under Transient Conditions Including the Effect of Selected Time Step Sizes, Finite Elements in Water Resources, Ed. by Holtz et al., Springer Verlag, 1982.

MATHEMATICAL MODEL OF RESERVOIR FLUSHING

R.Cavor and M. Slavic
ENERGOPROJEKT
Beograd, Yugoslavia

ABSTRACT

One of the methods of recovering lost storage is the flushing of deposited sediments with the water available in the reservoir, by opening fully the bottom outlets and lowering the water levels abruptly. A mathematical model is presented in the paper describes the unsteady flow of water and the entrainment of sediments during the flushing. In addition to the Saint-Venant´s equations for the flow of water, the sediment balance is expressed by turbulent diffusion equation, in which the entrainment of sediment is defined by the Velikanov transport capacity formula. The system of equations is solved by an variant of the explicite weighted method of finite differences. The model has been calibrated by making use of two flushing operations. By its subsequent use operational rules are being developed in order to make the flushing more efficient.

1. INTRODUCTION

Deposition of sediments upstream from dams built on sediment carrying rivers in an unavoidable process, the consequence of which is a gradual loss of live storage of the reservoir and a reduction of its useful life. The supply of sediments into the reservoir can, in principle, be reduced by soil conservation and erosion control in the catchement; in case of large river basins, which are partly inaccessible, effects of such measures can hardly be expected. The only measures which can successfully be applied consists in preventing the deposition of sediments in the live storage and removing, if possible, the already deposited sediments from the reservoir.

If the reservoir is partly full in the period of low river discharges and if a certain volume of stored water can be used for the removal of sediment deposits, the water level in the reservoir can rapidly be lowered by releasing comparatively high discharges through the bottom outlets (Fig.1).

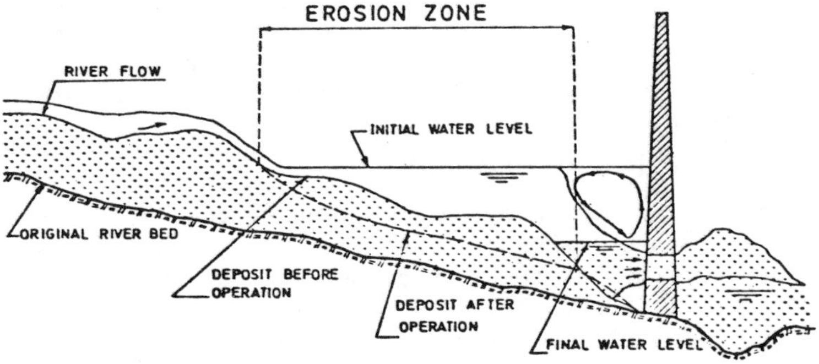

Fig. 1: Sediment flushing

This method is regularily applied in the some reservoirs in the last years (and called by the French word "chasse"), thanks to the circumstance that after the end of the irrigation season, some water remains stored in the reservoir and can be used without risk of shortage in the next year. The efficiency of the above method can be improved by analysing the operational conditions by means of mathematical models, which simulate the phenomena and by which the effects of the operations can be predicted, depending upon the hydraulic conditions. Such a model can be successfully used only if properly calibrated, which means the adjustments of relevant parameters in such a way that the results of calculations fit prototype observations to a desired level of accuracy. Thus, the existance of appropriate field observations in a pre-requisite of successful model development and a pre-condition of reliable predications by the model.

2. DESCRIPTION OF THE PROBLEM

By opening the bottom outlets at a certain stage of the water in the reservoir, the water levels decrease rapidly and an unsteady flow towards the otulets begins. The velocities are increasing and the highest rate of increase takes places at the upstream end of the reservoir. The increase of velocities leads to an increased transport capacity of the flow and sediment particles are entrained from the deposits and carried away in suspension. The water-sediment mixture approaches the dam, the released discharges being practically free of sediments until the sediment carrying water wave reaches the dam. The flowing water cuts its way through the thick deposits of sediment and a channel is formed expending gradually its dimensions as the scour of the deposits progresses.

When the water level in the reservoir reaches a lowest permissable stage, the bottom outlets are shut again, reducing the released discharge to a prescribed minimum value; the progression of the sediment in suspension continues for a while, due to the unsteady flow of the water, bringing high concentrations near to the dam. The inflow in the reservoir being higher than the outflow, the level in the reservoir begins to raise again up to a certain level, and the chasse operation can be repeated.

3. BASIC ASSUMPTIONS AND EQUATIONS

In view of the above description of the reservoir during the "chasse", a model simulating the unsteady flow of water and sediment had to be developed. The suspending of sediment particles and their transport is linked to turbulent diffusion phenomena so that equations describing the latter had to be developed. The interaction of the flow with the solid boundary (the reservoir bed) could be represented by introducing a source-and-sink term into the equations. Except in the vicinity of the dam, one-dimensional approximation was used (Fig. 2) by making use of the following equations:

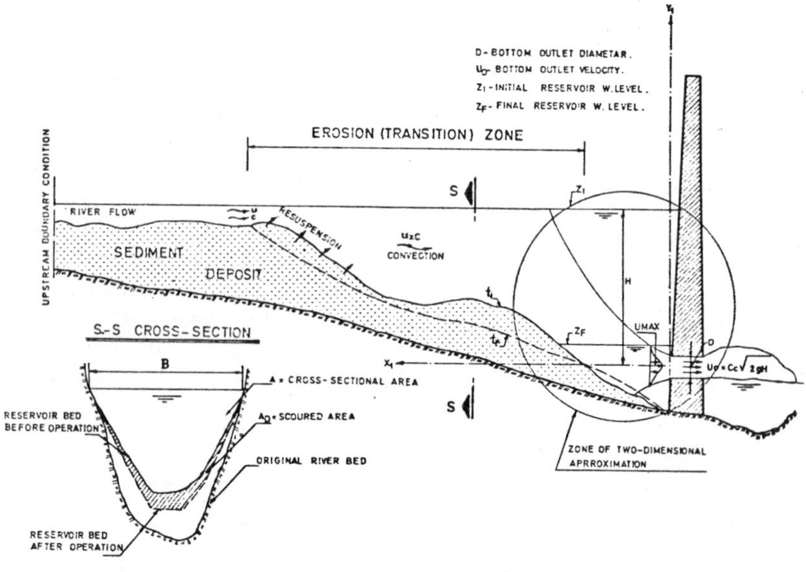

Fig. 2: Definition sketch

For the flow of water:

- the dynamic equation:

$$\frac{\partial Q}{\partial t} + \frac{\partial (Qu)}{\partial x} + \frac{g}{\gamma m} \left(A \frac{\partial z}{\partial x} + \frac{Qun^2}{R^{4/3}} \right) = 0 \qquad (1)$$

- the equation of continuity:

$$\frac{\partial z}{\partial t} + \frac{1}{B} \cdot \frac{\partial Q}{\partial x} = 0 \qquad (2)$$

Close to the dam, the velocity field was approximated by a two-dimensional distribution of velocities as in a jet

$$u(x_1, y_1) = u_{max}\, e^{\frac{1}{2}\left(\frac{y_1}{0.081 x_1}\right)^2} \qquad (3)$$

$$u_{max} = 6{,}2\, \frac{D}{X_D} \cdot u_o \qquad (4)$$

In the above equations the following notations are used:

Q	- water discharge	m3/s
z	- water level elevation	m
n	- the Manning roughness coefficient	$s/m^{1/3}$
γ_m	- specific gravity of suspension	t/m3
R	- hydraulic radius	m
B	- width of water surface	m
X_D	- distance from the dam	m

For the transport of sediments:
- the equation for turbulent diffusion of two phase flow:

$$\frac{\partial c}{\partial t} + \frac{\partial (uc)}{\partial x} = \frac{\partial}{\partial x}\left(K_T \frac{\partial c}{\partial x}\right) + Q_{ss} \qquad (5)$$

where Q_{ss} is a sink-source term, defining the exchange with the solid boundary, assumed in the following form:

$$Q_{ss} = \psi_{1,2} \cdot \frac{\partial C_T}{\partial t} \qquad (6)$$

In this equation C_T is the transport capacity of the flow, expressed by the Velikanov formula:

$$C_T = \eta \cdot \frac{\gamma m}{\gamma_s - \gamma} \cdot \frac{u}{W} J_e \qquad (7)$$

The coefficient of turbulent diffusion in eq. (5) is defined as follows:

$$K_T = 7{,}25 \cdot d\, u_* \cdot \left(\frac{u}{u_*}\right)^{1/4} \qquad (8)$$

The modification of the channel is then obtained from the continuity equation of sediment transport:

$$\frac{\partial Ad}{\partial t} + \frac{\partial (cA)}{\partial t} + \frac{\partial (Qc)}{\partial x} = 0 \qquad (9)$$

The following notations were used in the former equations:

c - volume concentration
C_T - transport capacity

K_T - coefficient of turbulent diffusion m2/s
Q_{ss} - exchange with the solid boundary (resuspension or sedimentation)
$\psi 1$ - coefficient of resuspension
$\psi 2$ - coefficient of sedimentation
d - depth of water m
J_e - slope of the energy line
u_* - shear velocity m/s
η - ratio between the energy needed to suspend solid particles and the total energy of flow
W - settling velocity of sediment particles m/s
γ_s - specific gravity of sediment t/m3
γ - specific gravity of water t/m3
Ad - increment of cross sectional area m2

Equations (1) to (9) define the mathematical model of unsteady flow of water and suspended sediments. They have to be completed with assumptions about the modification of the cross sections of the reservoir. The simplest assumption would suppose a uniform deformation of the cross section due to scour, but was rejected as remote from reality. Instead, it was assumed that the changes of cross sections are proportional to the local depth, which is much more realistic than a uniform distribution of the scour across the width of the reservoir.

By the above assumption for the calculation of channel deformations, the mathematical model is completed. It contains several assumptions and numerical constants, which have to be calibrated by comparing the results of calculations with the available observations.

4. NUMERICAL SOLUTION

The numerical integration of the basic equations (1,2,5,9) was made by means of a variant of the explicit weighted method of finite differences using special procedure in order of solution steps. The derivatives of the variables (Q, c, z, A_d) are written as below. For the sake of simplicity, an arbitrary function f(x,t) is used here, instead of the actual variables.

- weighted time derivative in Equations 1 and 5:

$$\frac{\partial f_i}{\partial t} = \{f_i^{t+\Delta t} - [\omega_i f_i^t + (1-\omega_i) \cdot \left(\frac{\Delta x_u}{\Delta x} f_{i+1}^t + \frac{\Delta x_d}{\Delta x} f_{i-1}^t \right)]\} \frac{1}{\Delta t} \tag{10}$$

- time derivative in Equations 2 and 9:

$$\frac{\partial f_i}{\partial t} = \frac{f_i^{t+\Delta t} - f_i^t}{\Delta t} \qquad (11)$$

- derivative along the axis x in Equations 1 and 5:

$$\frac{\partial f_i}{\partial x} = \frac{f_i^t - f_{i-1}^t}{2\Delta x_u} + \frac{f_{i-1}^t - f_i^t}{2\Delta x_d} \qquad (12)$$

- derivative along the axis x in Equations 2 and 9:

$$\frac{\partial f_i}{\partial x} = \frac{f_i^{t+\Delta t} + f_i^t - f_{i-1}^{t+\Delta t} - f_{i-1}^t}{4\Delta x_u} + \frac{f_{i+1}^{t+\Delta t} + f_{i+1}^t - f_i^{t+\Delta t} - f_i^t}{4\Delta x_d} \qquad (13)$$

Figure 3 illustrated applied method and solution procedure.

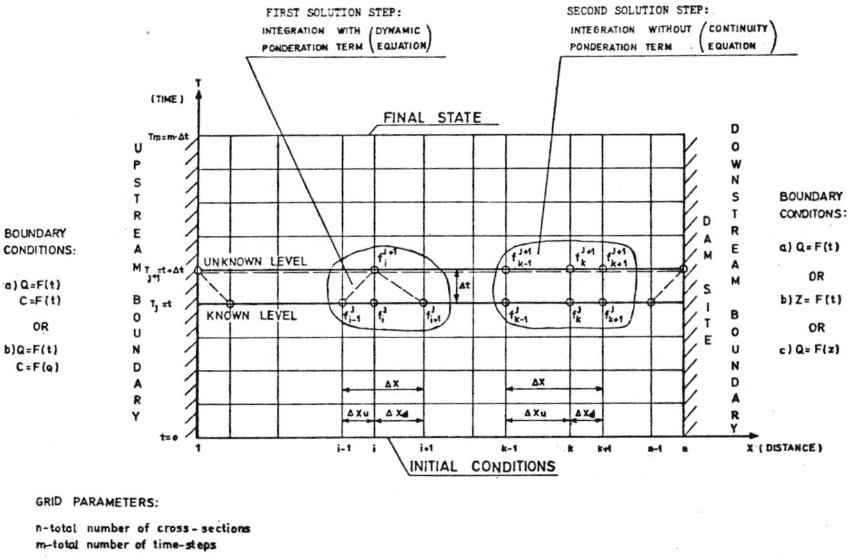

Fig. 3: Computational grid and discretization scheme

By means of the above expressions, the set of equations can be easyly written in a form suited for numerical calculation.

Solution procedure. The integration of the basic equations for a time interval Δt beginning from an initial state, proceeds in several steps:

1. From the dynamic equation for water (1), the discharges are calculated in all cross sections for the next time interval t + Δt.
2. From the continuity equation for water (2), the water levels are determined in all cross sections, by making use of the discharges determined before.
3. The velocities and cross sectional areas of the "jet" are determined close to the dam (Eq. 3-4) and the transport capacities calculated (Eq. 7).
4. By means of the diffusion equation (5), the sediment concentrations are determined in all cross sections.
5. From the sediment continuity equation (9), the cross sectional area increments are calculated, taking into account the previously determined concentrations.
6. The modification of the cross sections is calculated according to the adopted assumption and the new geometry of the cross sections determined.

The equations describing the flow of water and the transport of sediments are solved within the same time interval independetly, assuming that the channel geometry of the previous interval remains unchanged. Thus, the interaction between the two phases (the fluid and the solid phase) is neglected withing a time interval Δt, assuming that the modifications within one phase do not affect the solution of the other phase. The choice of the length of the time interval in the integration must respect the above constrains, and also the conditions of stability.

Stability conditions. For the above described numerical procedure (Eq.10-13), different stability requirements are obtained, so that equations (1,2) and (5,9) should be integrated in different (x, t) planes; this would cause great problems in solving the equations. Therefore, an inverse way of obtaining the stability conditions was applied: the time step was selected first, and by means of the stability conditions, the weighting coefficients were determined next.

- for the dynamic equation for water:

$$\omega_{iw} = \sqrt{1-\left(2u_i \cdot \frac{\Delta t}{\Delta x'}\right)^2 - \frac{g \cdot \Delta t}{\gamma m} \cdot \left(\frac{1}{u_i} \cdot \frac{\Delta z}{\Delta x'} + \frac{u_i n_i^2}{R_i^{4/3}}\right)} \tag{14}$$

- for the diffusion equation:

$$\omega_{ic} = \sqrt{1 - \left(u_i \frac{\Delta t}{\Delta x'}\right)^2 + 2K_T \cdot \frac{\Delta t}{(\Delta x')^2}} \tag{15}$$

$$\Delta x' = \frac{1}{2}\left(\Delta x_u + \Delta x_d\right) \tag{16}$$

The above conditions are obtained by the standard method of representing the solutions by means of Fourrier series and examining the n-th term of the solution.

5. CALIBRATION OF THE MODEL

The main parameters of the model which had to be adjusted in the process of calibration were the values of ψ_1 (resuspension) and ψ_2(sedimentation), aiming at a satisfactory similarity between model performance and prototype observations. The principal indicator of the simulation was the exit concentration, measured downstream from the dam, and compared to concentrations of the same place obtained by the calculations. In order to calibrate the main parameters and to examine the behaviour of the model, a large number of runs was made (about 30), varying the model parameters (the ψ_1,ψ_2 values, and the grainsize characteristics of bed sediments, as well as their distribution along the reservoir).

Some typical results of the calculations are shown on Figures 4a, 4b, 4c, 4d, simulating the "chasse" of 1981. The process of improving the choice of parameters can be followed on these diagrams, till the achievement of a satisfactory simulation (Fig. 4d).

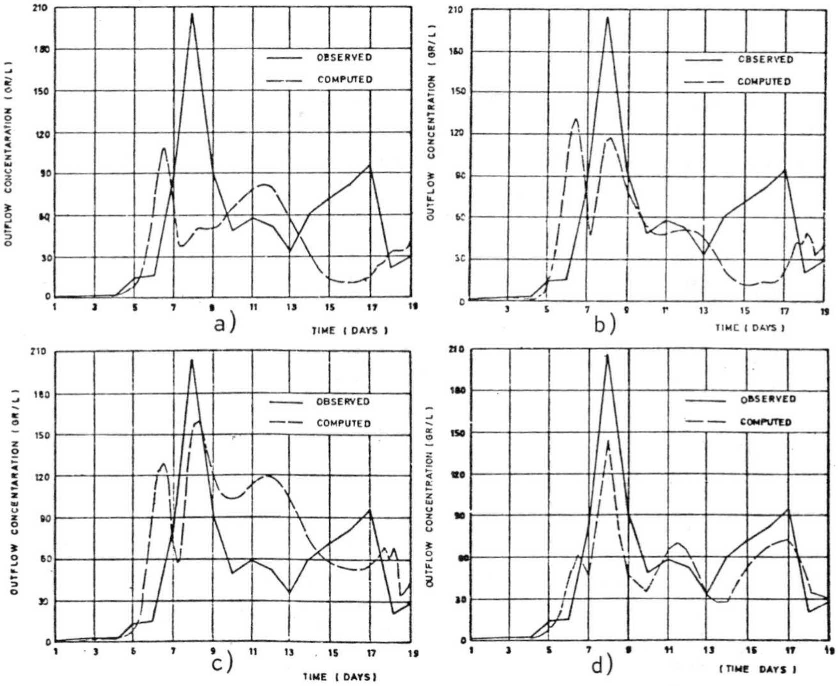

Fig. 4: Calibration of the model - year 1981

After having completed the first phase of the calibration, by making use of the observed values during the 1981 flushing in the second phase the previous flushing 1980 was simulated (Fig. 5). A satisfactory simulation could be obtained much easier than in the first phase, although an additional adjustment of the parameters was nevertheless necessary.

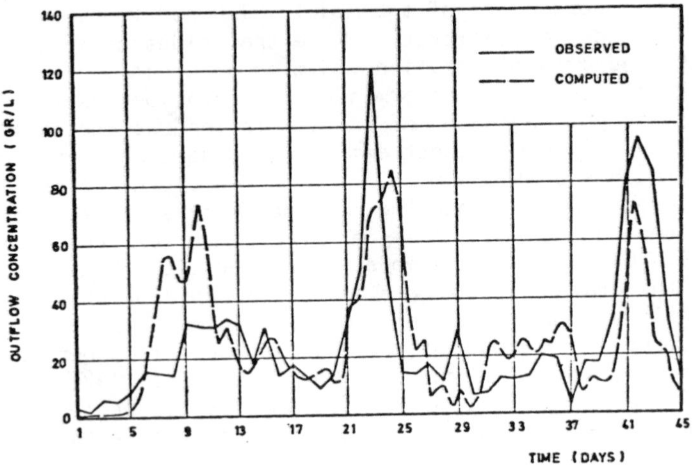

Fig. 5: Simulation of the sediment washing out-year 1980

The change of values of ψ_1, ψ_2 parameters from one year to the other is quite logical and in accordance with field observations. This could also be expected, since the most easily movable parts of the deposits were washed away during the first year chasse. Higher values of the coefficient ψ_1 (second year) indicate that more energy is needed to erode and suspend the deposits.

6. CONCLUSIONS

The mathematical model developed by Energoprojekt - Energodata is based on theoretical concepts and past experience of investigations in several countries, such as China, Yugoslavia, Japan, France and others. The model simulates the physical phenomena which occur in the course of the flushing operations considering the most relevant factors which determine the effects of the flushing.

The calibration of the model is achieved through the comparison of the results of calculation with actual observations during flushing operations in two years, and is reflected in the adjustment of the most important parameters on which the effects of the flushing depends (the transport coefficients ψ_1 and ψ_2).

The model, however, may well be used for analysing possible improvements of the current flushing operations, in order to develop operational methods by which the maximum amount of sediment can be flushed with the available amount of water in reservoir.

7. REFERENCES

1. ASCE - Sedimentation Engineering, American Society of Civil Engineers, New York, 1975.

2. K.Mahmood and V.Yevjevich - Unsteady Flow in Open Channels, Water Resources Publications, Fort Collins, Colorado 1975.

3. A.J. Raudkivi and R.A. Callander - Advanced Fluid Mechanics, Edvard Arnold, London, 1975.

4. Fan Jiahua and Jiang Rujin - On Methods for the Desiltation of Reservoirs, International Seminar of Experts on Reservoir Desiltation, Tunis July 1-4, 1980.

5. Fan Jiahua and Jiang Rujin - Expiriences on Reservoir desiltation in China, Research Institute of Water Conservancy and Hydroelectric Power, Academia Sinica and Ministry of Water Conservancy and Electric Power, China, 1981.

6. B.Geza and K.Bogic - Some observations on Density Currents in the Laboratory and in the Field, Serbian Academy of Sciences, Yugoslavia, 1954.

7. B.Geza, M.Boreli, S.Jovanovic - On Silting and on Washing out of the Storage Reservoir on the River Treska - Water Resources Institute "Jaroslav Cerni", Beograd, Yugoslavia, 1954.

8. Kerssens, Prins and Rijn - Model for Suspended Sediment Transport, ASCE Journal of the Hydraulics Division, May 1979, pp.461-475

9. Takamatsu, Naito, Shiba, Ueda - Effects of Deposit Resuspension On Settling Basin, ASCE Journal of Enviromental Engineering Division, August 1974, pp. 883-903.

10. Itakuro and Kishi - Open Channel Flow with Suspended Sediments, ASCE Journal of the Hydraulics Division, August 1980, pp. 1325-1343.

11. S.Takatsu - Hydraulics Design and Model Tests on a Sediment Release Facility of Unazuki Dam, XIV International Congress on Large Dams, vol. 2, pp. 21-39, Rio De Janeiro, 1982.

DISPERSION OF DISCRETE PARTICLES BY TURBULENT CONTINUOUS MOTIONS USING A FRENKIEL'S FAMILY OF LAGRANGIAN CORRELATION FUNCTIONS, IN THE NON-DISCRETE DISPERSIVE APPROACH.

G. GOUESBET, A. PICART, A. BERLEMONT

Laboratoire de Thermodynamique - LA CNRS N° 230
Faculté des Sciences et Techniques de Rouen
BP 67 76130 MONT-SAINT-AIGNAN (France)

Abstract :

The dispersion of discrete particles embedded in turbulent flows is first described in a three-dimensional way as a synthesis between the monodimensional Tchen's dispersion theory and the three-dimensional Batchelor's diffusion theory of fluid particles. The results are then specified assuming a two parameters Frenkiel's family of Lagrangian correlation functions. The situations in which the so-called Basset's term can be neglected are discussed. It is also shown that even dense discrete particles may disperse faster than fluid particles. That paper is linked to a second one in the same conferences where predictions of particles dispersions are achieved in various cases and compare favourably with experiments.

I - INTRODUCTION

The dispersion of discrete particles by continuous motions in turbulent flows has become a major field of research with the aim of improving understanding of multi-phases flows processes connected with the present crisis of energy, plants control and environmental care. A second paper [1] will concern this applied point of view. But the present topic is also of great fundamental importance due to the fact that the dispersion of discrete particles is a more general concept that the one of diffusion of fluid particles. The present paper is concerned with new results concerning the theory of dispersion in homogeneous, "steady", non-isotropic turbulence fields.

II - BASIC FORMALISM

The theory of the three-dimensional diffusion of fluid particles has been given by Batchelor [2]. To adapt that theory to the dispersion of discrete particles, the basic work consists mainly in rewriting Batchelor's formalism with appropriate changes of notations to emphasize the more general topic under discussion. Since proofs are available from elsewhere [2,3], the basic formalism is as a whole recalled without demonstration. The key-mathematical entity is a dispersion tensor for the discrete particles. It is always possible to proceed to a rotation of a local Cartesian coordinates system in order to make the symmetric dispersion tensor diagonal. Furthermore, in numerous situations such as two-dimensional parabolic dispersions, the non-diagonal terms of the dispersion tensor are without any influence so that they can be neglected. Thus, in all cases, the complete three-dimensional problem can be simplified to the juxtaposition of two (or three) monodimensional problems where we are only concerned with dispersion coefficients. For sake of shortness, the basic formalism will be mostly recalled in that monodimensional context. The reader can help itself by going back to other works previously published by the authors [4,5,6] and also to [7].

In the homogeneous and steady turbulence under study, the coefficient of dispersion of discrete particles at time t of dispersion reads :

$$\varepsilon_p(t) = \overline{u_{pL}^2} \int_o^t R_{pL}(\tau)\, d\tau \qquad (1)$$

where $\overline{u_{pL}^2}$ is the mean value of the square of a (fluctuating) Lagrangian particle velocity, and $R_{pL}(\tau)$ the Lagrangian correlation for particles velocities defined by :

$$\overline{u_{pL}^2}\; R_{pL}(\tau) = \overline{u_{pL}(o)\, u_{pL}(\tau)} \qquad (2)$$

The Fourier transform of $R_{pL}(\tau)$ is the monodimensional spectrum for particles $E_{pL}(\omega)$. Normalizing the Fourier transform such as :

$$\overline{u_{pL}^2} = \int_o^\infty E_{pL}(\omega)\, d\omega \qquad (3)$$

the relation (1) also reads :

$$\varepsilon_p(t) = \int_o^\infty E_{pL}(\omega)\, \frac{\sin \omega t}{\omega}\, d\omega \qquad (4)$$

In the framework of the Tchen's theory, the spectrum $E_{pL}(\omega)$ is linked to the equivalent spectrum $E_{fL}(\omega)$ for fluid particles through the simple relation :

$$E_{pL}(\omega) = \eta^2(\omega)\; E_{fL}(\omega) \qquad (5)$$

where $E_{fL}(\omega)$ is the Fourier transform of $R_{fL}(\tau)$, the Lagrangian correlation function for fluid particles velocities defined similarly as relation (2). And $\eta(\omega)$ is the Tchen's amplitude ratio given by :

$$\eta = [(1+f_1)^2 + f_2^2]^{1/2} \tag{6}$$

$$f_1 = \omega(\omega+c(\pi\omega/2)^{1/2})\ (b-1)/D \tag{7}$$

$$f_2 = \omega(a+c(\pi\omega/2)^{1/2})\ (b-1)/D \tag{8}$$

$$D = (a+c(\pi\omega/2)^{1/2})^2 + (\omega+c(\pi\omega/2)^{1/2})^2 \tag{9}$$

$$a = 18\nu/[(s+\tfrac{1}{2})d^2] \tag{10}$$

$$b = 3/[2(s+\tfrac{1}{2})] \tag{11}$$

$$c = 9(\nu/\pi)^{1/2}/[(s+\tfrac{1}{2})d] \tag{12}$$

where ω is an angular frequency, ν the fluid kinematic viscosity, d the diameter of the (monodispersed) particles, $s = \rho_p/\rho_f$ the ratio of particles material and fluid densities respectively.

Rather than computing the coefficient of dispersion, it is often convenient to compute the ratios :

$$S = \frac{\varepsilon_p(t)}{\varepsilon_f(t)} = \frac{\int_0^\infty \eta^2(\omega)\ E_{fL}(\omega)\ \frac{\sin \omega t}{\omega}\ d\omega}{\int_0^\infty E_{fL}(\omega)\ \frac{\sin \omega t}{\omega}\ d\omega} \tag{13}$$

and :

$$S_\infty = \frac{\varepsilon_p(t)}{\varepsilon_f(t \to \infty)} = \frac{2}{\pi E_{fL}(\omega=0)} \int_0^\infty \eta^2(\omega)\ E_{fL}(\omega)\ \frac{\sin \omega t}{\omega}\ d\omega \tag{14}$$

The adimensional number S(13) compares the coefficient of dispersion at time t to the coefficient of diffusion at the same time, and permits to determine how efficiently the particles disperse relatively to the fluid particles. The second adimensional number S_∞(14) compares $\varepsilon_p(t)$ to its asymptotic value and indicates whether the process of dispersion occurs at short times ($S_\infty \ll 1$), at large times ($S_\infty \to 1$), or in a transitory regime.

III - THE FORMALISM USING A FRENKIEL'S FAMILY OF CORRELATION FUNCTIONS.

In order to compute the coefficients of dispersion $\varepsilon_p(t)$, the fluid Lagrangian spectrum $E_{fL}(\omega)$ (relations 4 and 5) must be known. This is a very hard problem without complete answer up to now. In our previous works, (5,6), an approximate Eulerian spectra $E_{fE}(\omega)$ was first computed, then we applied a simple rule of Eulerian → Lagrangian transformation. The process was found rather satisfactory for engineering purposes but required costful and time-consuming computations. In the present work, we shall start from a fluid Lagrangian correlation function $R_{fL}(\tau)$ with two parameters, used by Frenkiel, and having empirical support (8,9,10) :

$$R_{fL}(\tau) = \exp\left[\frac{-\tau}{(m^2+1)\tau_L}\right] \cos\left[\frac{m\tau}{(m^2+1)\tau_L}\right] \qquad \tau \geqslant o \qquad (15)$$

where τ_L is the Lagrangian time macroscale, and the parameter m is linked to the occurence and the importance of negative loops in the function.

Fourier transforming, we obtain after rather lengthy but straightforward calculus :

$$E_{fL}(\omega) = \frac{2}{\pi}\,\overline{u^2_{fL}}\; f\,\frac{1+m^2+\omega^2 f^2}{[1-m^2+\omega^2 f^2]^2+4m^2} \qquad (16)$$

where :

$$f = (m^2+1)\tau_L \qquad (17)$$

and $\overline{u^2_{fL}}$ is $\overline{u^2_{pL}}$ specified for fluid particles (for instance, by making the diameter d tending to zero).

Thus (relations 4,5,16) :

$$\varepsilon_p(t) = \int_o^\infty \eta^2(\omega)\,.\frac{2}{\pi}\,\overline{u^2_{fL}}\; f.\frac{1+m^2+\omega^2 f^2}{[1-m^2+\omega^2 f^2]^2+4m^2}\;.\;\frac{\sin \omega t}{\omega}\,d\omega \qquad (18)$$

Relation (13) becomes :

$$S = \frac{\displaystyle\int_o^\infty \eta^2(\omega)\;\frac{1+m^2+\omega^2 f^2}{[1-m^2+\omega^2 f^2]^2+4m^2}\;.\;\frac{\sin \omega t}{\omega}\,d\omega}{\displaystyle\int_o^\infty \frac{1+m^2+\omega^2 f^2}{[1-m^2+\omega^2 f^2]^2+4m^2}\;.\;\frac{\sin \omega t}{\omega}\,d\omega} \qquad (19)$$

and relation (14) becomes :

$$S_\infty = \frac{2(m^2+1)}{\pi} \int_0^\infty \eta^2(\omega) \frac{1+m^2+\omega^2 f^2}{[1-m^2+\omega^2 f^2]^2+4m^2} \cdot \frac{\sin \omega t}{\omega} d\omega \tag{20}$$

Due to the complexity of the integrals involved in the above relations, the computations of the dispersion coefficients again need numerical quadratures which are costful and time-consuming. The next section discusses a case where all that simplifies drastically.

IV - THE FORMALISM WHEN THE BASSET'S TERM IS NEGLIGIBLE.

One of the terms appearing in the equation of motion of the particle is the so-called Basset's term or history integral. From a mathematical point of view, the Basset's term has no influence when the parameter c (relation 12) is equal to zero [7]. The expression for the amplitude ratio η (relation 6) then simplifies dramatically. After some algebra, we obtain :

$$\eta_o^2(\omega) = \frac{a^2+b^2\omega^2}{a^2+\omega^2} \tag{21}$$

From relation (18), the coefficient of dispersion becomes :

$$\varepsilon_p = \frac{2}{\pi} \overline{u_{fL}^2} f \int_0^\infty \frac{a^2+b^2\omega^2}{a^2+\omega^2} \cdot \frac{1+m^2+\omega^2 f^2}{[1-m^2+\omega^2 f^2]^2+4m^2} \cdot \frac{\sin \omega t}{\omega} d\omega \tag{22}$$

which gives, after some lenghty but straightforward computations :

$$\varepsilon_p = \overline{u_{fL}^2} \tau_L \{1+\psi_o e^{-at} + [\psi_1 \cos(mt/f) + \psi_2 \sin(mt/f)] e^{-t/f}\} \tag{23}$$

where :

$$\psi_o = \frac{(m^2+1)(b^2-1)(1+m^2-a^2 f^2)}{[a^2 f^2+m^2-1]^2 + 4m^2} \tag{24}$$

$$\psi_1 = \frac{-[a^4 f^4 - a^2 f^2 [(1+b^2)(1+m^2)-4m^2] + b^2 (1+m^2)^2]}{[a^2 f^2+m^2-1]^2 + 4m^2} \tag{25}$$

$$\psi_2 = m\,\psi_1 + 2m\,\frac{a^4f^4-2a^2f^2(1-m^2)+b^2(1+m^2)^2}{[a^2f^2+m^2-1]^2 + 4\,m^2} \qquad (26)$$

The coefficient of diffusion can be obtained from relation (23) as a special case where $b = 1(\rho_p = \rho_f)$.

We have :

$$\left.\begin{array}{l} \psi_0(b = 1) = 0 \\ \psi_1(b = 1) = -1 \\ \psi_2(b = 1) = +m \end{array}\right\} \qquad (27)$$

and thus :

$$\varepsilon_f(t) = \overline{u_{fL}^2}\,\tau_L\{1+[m\,\sin(mt/f)-\cos(mt/f)]\,e^{-t/f}\} \qquad (28)$$

The relations (19) and (20) become :

$$S = \frac{1+\psi_o e^{-at}+[\psi_1\cos(mt/f)+\psi_2\sin(mt/f)]\,e^{-t/f}}{1+[m\,\sin(mt/f)-\cos(mt/f)]\,e^{-t/f}} \qquad (29)$$

$$S_\infty = 1+\psi_o e^{-at}+[\psi_1\cos(mt/f)+\psi_2\sin(mt/f)]\,e^{-t/f} \qquad (30)$$

V - CASES WHERE THE BASSET'S TERM IS NEGLIGIBLE.

Comparison between section III and IV shows that, for practical purposes, the situation is much simpler when the Basset's term can be neglected. So, it is worthwhile to get criteria or laws to decide in which situations the aforementioned approximation is valid.

Such a problem has been previously discussed by Hjelmfelt and Mockros [11] not only in terms of the Tchen's amplitude ratio $\eta(\omega)$ but also in terms of the phase angle $\beta(\omega)$ which appears in the Tchen's theory as expressed by most authors. Nevertheless, from the dispersion point of view, the discussion was both overprecise and underprecise. It was overprecise since the phase angle does not appear in the computations of the dispersion coefficients. But it was underprecise because no systematic discussion of the influence of turbulence parameters on the importance of the Basset term was given. In the present case, the basic relevant turbulent parameter is the time macroscale τ_L. The discussion would be different according to the ratio between τ_L and a particulate time $\tau_p = 1/a$. It also exists a general procedure which consists to compare the values of the coefficients of dispersion given respectively by the relations (18) and (23). Such a complete discussion has been published elsewhere [12] and we shall only repeat here some examples of computations.

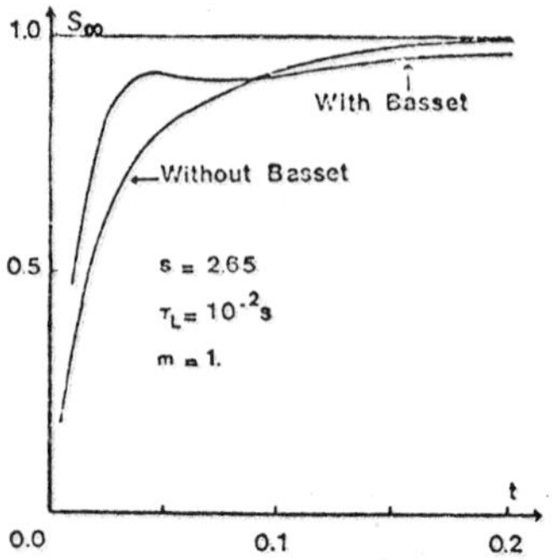

Fig. 1 : Influence of the Basset's term

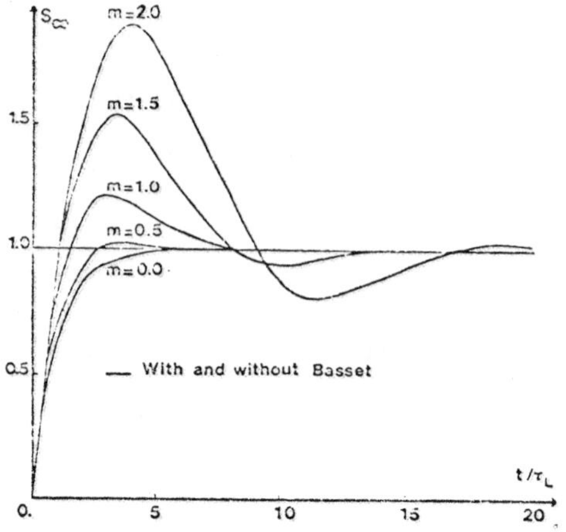

Fig. 2 : Reduced dispersion coefficient

The case is particles of sand in water ($s = \rho_p/\rho_f = 2.65$). The diameter of the particles is d = 500 μm and the Lagrangian time macroscale is $\tau_L = 10^{-2}$s. The loop parameter is 1, an usually recommanded value. The ratio S_∞ has been computed either with the relation (18) (with Basset) and with the relation (23) (without Basset). Clearly, the Basset's term cannot be neglected (figure 1). The particulate time τ_p is here 0.044s which has the same order of magnitude as τ_L.

Similar results are given on the figure 2 for a different turbulence where τ_L = 1s for five loop parameters m ranging from 0 up to 2. Here, the Basset's term can be negligible. Note that $\tau_p \ll \tau_L$ in the present case . When $\tau_p \ll \tau_L$, the particle follows strictly the fluid since, for each harmonic of frequency ω present in the stochastic turbulent process, the particle reacts instantaneously. No notion of particle motion remains in the description of the dispersion phenomenon. Furthermore, dispersion identifies with diffusion, meaning that the particles exactly diffuse as fluid particles.

This difference between the cases $\tau_L = 10^{-2}$s and τ_L = 1s emphasizes the advantage of our analysis compared to Hjelmfelt and Mockros' one since we state different conclusions on the influence of the Basset's term, depending on the time τ_L.

VI - CAN DENSE DISCRETE PARTICLES DISPERSE FASTER THAN FLUID PARTICLES ?

Looking at relation (1), it appears that, when $R_{pL}(\tau)$ is given, the properties of $\varepsilon_p(t)$ are determined. When the correlation $R_{pL}(\tau)$ has no negative loop, the quantity $\varepsilon_p(t)/\overline{u_{pL}^2}$ will increase steadily from 0 at time 0 up to a particulate Lagrangian time macroscale τ_{pL}, at time infinite. Then, $\varepsilon_p(t)$ is always smaller than (or equal to) $\varepsilon_f(\infty) = \varepsilon_p(\infty)$.

The situation is different when $R_{pL}(\tau)$ has zero values at times t_n where its sign changes, that means the correlation exhibits negative loops. $\varepsilon_p(t)/\overline{u_{pL}^2}$ then increases first steadily from 0 up to a first maximal value $\int_o^{t_1} R_{pL}(\tau)\, d\tau$, then decreases down to $\int_o^{t_2} R_{pL}(\tau)\, d\tau$ due to the negative contribution of the negative loop to the integral, increases again, and so on, until convergence to the τ_{pL} value. Thus the coefficients of dispersion $\varepsilon_p(t)$ can become larger than $\varepsilon_p(\infty)$ during intervals of time. Furthermore, it is easy to prove that maximal and minimal values of the ratio $\varepsilon_p(t)/\varepsilon_p(\infty)$ correspond to the zeros of $R_{pL}(\tau)$ (just derive relation 1).

Let us limit the following to the case where $\eta(\omega) < 1$ (except at $\omega = 0$) that is to say to the case of dense particles. Unexpectedly, not only the ratio $\varepsilon_p(t)/\varepsilon_p(\infty)$ but also the ratio $\varepsilon_p(t)/\varepsilon_f(t)$ can become higher than 1 at certain times t. That means that the dense discrete particles at time t can disperse faster than the fluid particles at the same time.

This can be understood by noting that :

$$S = \frac{\varepsilon_p(t)}{\varepsilon_f(t)} = \frac{\overline{u_{pL}^2}}{\overline{u_{fL}^2}} \cdot \frac{\int_o^t R_{pL}(\tau)\, d\tau}{\int_o^t R_{fL}(\tau)\, d\tau} \tag{31}$$

At small times, in the limit $t \to o$, S is smaller than 1 since it is well-known that :

$$S(o) = \frac{\varepsilon_p(o)}{\varepsilon_f(o)} = \frac{\overline{u_{pL}^2}}{\overline{u_{fL}^2}} \tag{32}$$

which is smaller than 1 when $\eta(\omega) < 1$ according to relation (3):

$$\frac{\overline{u_{pL}^2}}{\overline{u_{fL}^2}} = \frac{\int_o^\infty \eta^2(\omega)\, E_{fL}(\omega)\, d\omega}{\int_o^\infty E_{fL}(\omega)\, d\omega} < 1 \tag{33}$$

Just remember that the spectra are positive.

The ratio S will then remain smaller than 1 when t increases up to a certain critical time t_c. In the range $o \leq t < t_c$, it is possible to say that the respective values of $\overline{u_{pL}^2}$ and $\overline{u_{fL}^2}$ control the sign of [1-S] (relation 31).

Let us now discuss the second ratio $\int_o^t R_{pL}(\tau)\,d\tau / \int_o^t R_{fL}(\tau)\,d\tau$ in the r.h.s of (31). The spectrum $E_{pL}(\omega) = \eta^2 E_{fL}(\omega)$ is always smaller than the spectrum $E_{fL}(\omega)$ since $\eta(\omega) < 1$. Due to the fact that ω and τ are conjugate variables, the correlations $R_{pL}(\tau)$ are conversely (as a whole) larger than $R_{fL}(\tau)$. That means physically that, due to inertia, the discrete particles tend to keep their previous motions longer than fluid particles. When $t > t_c$, the fact that $\int_o^t R_{pL}(\tau)\,d\tau$ is larger than $\int_o^t R_{fL}(\tau)\,d\tau$ will overcome the fact that $\overline{u_{pL}^2} < \overline{u_{fL}^2}$, and the discrete particles will disperse faster than fluid particles at the same time t. Other details on this problem are available from the reference [13].

VII - CONCLUSION

The dispersion of discrete particles has been discussed in the framework of the Tchen's theory. The results have been specified assuming a two-parameters Frenkiel's family of Lagrangian correlation functions. Emphasis is put on the case where the Basset's term can be neglected, leading to algebraic calculations of the dispersion coefficients. It is also shown that dense discrete particles may disperse faster than fluid particles. That paper is followed by a second one in the same conferences where predictions of particles dispersions are achieved in various cases and compare favourably with experiments.

REFERENCES

1 - A. BERLEMONT, A. PICART, G. GOUESBET. The code DISCO-2 for predicting the behaviour of discrete particles in turbulent flows, and its comparisons against the code DISCO-1 and experiments. Third International Conference on numerical methods in laminar and turbulent flow.
August 8th-11th 1983, The University of Washington, Seattle, USA.

2 - G.K. BATCHELOR. Diffusion in a field of homogeneous turbulence.
Aust. Journal Sci. Res., 2, 437, 1949.

3 - G. GOUESBET, A. BERLEMONT. A three dimensional dispersion model for the behaviour of vaporizing particles in a turbulent fluid.
Sixth Biennal Symposium on turbulence. Oct. 8-10, 1979, University of Missouri, Rolla, USA.

4 - G. GOUESBET et A. BERLEMONT. Une approche dispersive pour la modélisation du comportement de particules dans un champ turbulent. C.R. Acad. Sci. A, 2420, 21 Mai 1979.

5 - A. BERLEMONT and G. GOUESBET. The dispersion of a cloud of particles released in a fully developed pipe flow, using the code DISCO-1.
Letters in Heat and Mass Transfer, vol. 9, 11-19, 1982

6 - A. BERLEMONT and G. GOUESBET. Prediction of the behaviour of a cloud of discrete particles released in a fully developed turbulent pipe flow, using a dispersive approach. Numerical methods in laminar and turbulent flows conferences.
Venizia, July 13-16th, 1981. Proceedings.

7 - J.O. HINZE. Turbulence, Mc Graw-Hill, 1959.

8 - FRENKIEL F.N. - ONERA - Etude statistique de la turbulence. Fonctions spectrales et coefficients de correlation. Rapport Technique n° 34, 1948.

9 - CALABRESE R.V., MIDDLEMAN S. The dispersion of discrete particles in a turbulent fluid field.
AICHE Journal, 25, 6, 1979.

10 - SNYDER W.H. and LUMLEY J.L. Some measurements of particle velocity autocorrelation functions in a turbulent flow. Journal of Fluid Mechanics, 48, 1, 1971.

11 - HJELMFELT and MOCKROS. Motion of discrete particles in a turbulent field.
App. Sci. Res., 16, 149, 1966.

12 - A. PICART, A. BERLEMONT et G. GOUESBET.
De l'influence du terme de Basset sur la dispersion de particules discrètes dans le cadre de la théorie de Tchen. C.R. Acad. Sci., Paris, Série II, 305-308, 295, 27 Sept. 1982.

13 - G. GOUESBET, A. BERLEMONT and A. PICART.
On the Tchen's theory of discrete particles dispersion : can dense discrete particles disperse faster than fluid particles ?
Letters in Heat and Mass Transfer, Vol. 9, 407-419, 1982.

NUMERICAL INVESTIGATION OF JET-INDUCED CONFINED FLOW AND MIXING

Åge S. Ulserød* and Bjørn F. Magnussen**

Division of Thermodynamics
Norwegian Institute of Technology
N-7034 TRONDHEIM-NTH NORWAY

SUMMARY

A confined, axisymmetric jet system is investigated for isothermal flow. The purpose is to develope a numerical computational tool for practical engineering use for entrainment and mixing calculations in a gas disperser.

An experimental investigation is briefly described and presented.
The jet is described by the conservation equations for momentum and mass in a two-dimensional form. A two-equation model, the k-ε-model for turbulence, has been applied. The differential equations are transformed by finite difference technique to algebraic equations and solved numerically on a computer.

The calculations are compared with own experimental data and data from the literature. The agreement is very good downstream of the nozzle, near the nozzle the computed profiles for velocity and turbulence intensity deviates from the experiments. Mass entrainment seems to be well predicted, and concentration profiles at the outlet of the tube are in good agreement with experimental data.

1. INTRODUCTION

1.1 Statment of the problem

Confined jet induced flow play and important role in many technical appliances like combustion chambers, gas dispersers etc. It is of great engineering importance to analyze and predict the behaviour of such flows to design and operate technical equipment in an optimum way. Till recently this design and operation has been based on simple empirical relations. These empirical relations are in general related to certain design

*Associate professor **Professor

and operation characteristics. Their applicability is therefore limited.

The problem considered here is to calculate the behaviour of a confined jet flow and the entrained secondary stream by numerical simulation of the flow. Turbulent transport, generation and dissipation are considered and the transport of momentum and mass are studied.

1.2 Related work

Of the most referred reports regarding confined jet flow are the papers of Thring and Newby [1] in -53, Craya and Curtet [2] in -55, Becker [3] in -63 and the book of Abramovich [4] in -63. In -71 Razinsky and Brighton [5] did an experimental investigation of confined jet flow without recirculation. The investigation is very good, describing velocity and distribution of turbulence intencity both radially and axially along the symmetry axis. Pressure distribution along the wall is also presented.

One of the first numerical calculations of the flow pattern in confined jet flow is presented in a thesis of Hendricks [6] from -75. He applies the boundary layer equations e.g. no recirculation is possible. A two equation model for turbulence is applied and he studies the influence of swirl velocity on flow and mixing. Later Elgobashi, Pun and Spalding [7] did a calculation of velocity and concentration profiles in confined jet flows in -77. They applied the fundamental equations on elliptic form and a two-equation model for turbulence. The so-called law of the wall was used for taking care of the influence of the wall on the flow field. The results for concentration profiles and recirculation zone was very promising. In -78 Kang and Suzuki [8] did a numerical calculation and compared with the experiments of Razinsky and Brighton [5]. The agreement was fairly good.

Common for all these cases is that the secondary stream is given. No paper is found presenting numerical calculations where the secondary ambient stream is completely induced by the primary jet.

1.3 Scope of the present work

The present work is concerned with numerical calculation of confined jet-induced flow in open tubes for different diameters and velocities. The entrained mass in the secondary stream and concentration profiles at the outlet of the tube is calculated. The calculations are compared with experiments, and also tested against confined jets where the secondary stream is controlled. The purpose is to contribute to the development of a numerical computational tool for practical engineering use.

1.4 Contents of paper

Section 2 gives a short presentation of the experimental investigation. Section 3 describes the flow mathematically and section 4 gives a brief presentation of the numerical solution method. The results are presented and discussed in section 5 and finally conclusive remarks are dealt with in section 6.

2. EXPERIMENTAL INVESTIGATION

2.1 Equipment

An experimental investigation was done for measuring axial velocity, axial turbulent intensity, concentration profiles at the outlet of the confining tube and the pressure along the tube wall, see fig. 1 for geometry. The confining tube is of plexiglass.

Laser Doppler Technique was applied for velocity and turbulence intensity measurements. Frequency shift was used and varied between 2 and 5 MHz to pick up possible negative velocities, and to get stronger signals at very low velocities. The laser was an argon ion type, Lexel 95, and the frequency counter was an TSI 1096. In addition a multi channel analyzer was used for analyzing the signals, which were finally computerized.

The concentration measurements were done by adding a very little controlled quantity of CO_2 to the jet flow. The concentration at the outlet of the tube was measured by an infrared gas analyzer sucking small gas quantities from the stream.

The wall pressure was measured simply by making holes with 1 mm diameter in the tube wall and measure the static pressure by water manometer.

2.1 Experiments

The experiments were done on a geometry shown in figure 1
The different cases are shown in table 1 as UL1 to UL4 for the velocity and turbulence intencity measurements. Seven different radial profiles and the axial profile were measured for each case. For concentration just UL1 and UL3 was measured radially at the outlet. Wall pressure measurement was done for case UL1 only. For details see Ulserød (9).

3. MATHEMATICAL ANALYSIS

3.1 Basic equations

The flow situation regarded is axisymmetric and isothermal, and

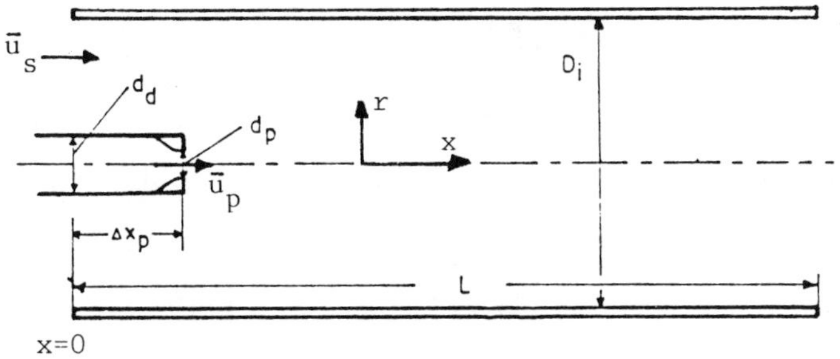

Fig. 1 Some important parameters for confined jet flow.

Cases	D_i m	d_p m	d_d m	L m	Δx_p m	$\bar{u}_p$ m/s	$\bar{u}_s$ m/s	Ct	Re_p
UL1	0,092	0,006	0,024	0,776	0,1L	135	-	0,90	54000
UL2	0,194	0,006	0,024	0,776	0,1L	45	-	0,88	18000
UL3	0,194	0,006	0,024	0,776	0,1L	135	-	0,88	54000
UL4	0,194	0,006	0,024	0,776	0,1L	325	-	0,88	130000
RB	0,1524	0,0254	0,0254	1,372	0	45,7	4,57	0,62	78000

Table 1 Different cases.

there is assumed no tangential velocity. The flow is than compeletely described by the general two dimensional steady state equations for continuity and conservation of momentum. If concentration profiles are wanted, the conservation equations for different components in the primary and secondary steam are necessary.

The flow is turbulent and therefore the well-known Reynolds equations combined with the Boussinesq analogy give the conservation equations for the time mean quantities ϕ in the form

$$\frac{\partial}{\partial x}(\rho\bar{u}\phi) + \frac{1}{r}\frac{\partial}{\partial r}(r\rho\bar{v}\phi) = \frac{\partial}{\partial x}(\Gamma_\phi \frac{\partial\phi}{\partial x}) + \frac{1}{r}\frac{\partial}{\partial r}(r\Gamma_\phi \frac{\partial\phi}{\partial r}) + S_\phi \qquad (1)$$

The meaning of each ϕ, Γ_ϕ and S_ϕ are given i table 2.

3.2 Turbulence model

The turbulent shear stresses are treated similar to the viscous shear stress according to the Boussinesq eddy viscosity model. The turbulent viscosity arising from this assumption is related to laminar viscosity as:

$$\mu_e = \mu + \mu_T \tag{2}$$

The turbulent viscosity μ_T is discribed by the well known k-ε-model

$$\mu_T = c_D \frac{k^2}{\varepsilon} \rho \tag{3}$$

This turbulence model is well tested for various two-dimensional situations for example in a paper of Launder and Spalding [10]. Each of the quantities k and ε are found from solving conservation equations for these quatities, see table 2 and eq. (1). The constants in the equations and the model are found in table 3.

Transport-equation for	ϕ	Γ_ϕ	S_ϕ
Continuity	1	0	0
Velocity in x-direction	$\bar{u}$	μ_e	$-\frac{\partial p}{\partial x}$
Velocity in r-direction	$\bar{v}$	μ_e	$-\frac{\partial p}{\partial r} - 2\mu_e \frac{\bar{v}}{r^2}$
Turbulent kinetic energi	k	μ_T/σ_k	$G - \rho\varepsilon$
Dissipations-velocity for k	ε	μ_T/σ_ε	$\frac{\varepsilon}{k}(c_1 G - c_2\varepsilon)$
Mass fraction for component i	ω	μ_e/σ_ω	0

$$G = \mu_T \left\{ 2\left[\left(\frac{\partial \bar{u}}{\partial x}\right)^2 + \left(\frac{\partial \bar{v}}{\partial r}\right)^2 + \left(\frac{\bar{v}}{r}\right)^2 \right] + \left[\frac{\partial \bar{u}}{\partial r} + \frac{\partial \bar{v}}{\partial x} \right]^2 \right\}$$

Table 2 Conservation equations

σ_k	σ_ε	σ_ω	c_D	c_1	c_2	E	κ
1,0	1,3	0,7	0,09	1,44	1,92	9,0	0,42

Table 3 Model constants

3.3 Boundary conditions

a) Inlet

For the secondary stream u is calculated from continuity, v=0 and ω and p is given. For k and ε:

$$k = \text{const.} \cdot \bar{u}^2 \tag{4}$$

$$\varepsilon = \frac{c_D\, k^{1.5}}{l} \tag{5}$$

For the jet flow u is also given and all other parameters are as far the secondary stream.

b) Wall

Large gradients appears close to the wall. A lot of grid points in the numerical calculations would be nescessary to be able to simulate that layer. One much applied method is the socalled wall function method given in [10]. This method is valid for fully developed flow, but gives good results for other flow situtations also. The wall shear stress are then given by

$$\tau_w = \frac{\rho (c_D^{.5}\, k)^{.5}\, \kappa\, \bar{u}}{\ln(E\, y^+)} \tag{6}$$

ε is given in the point next to the wall by

$$\varepsilon = \frac{c_D^{.75}\, k^{1.5}}{\kappa\, y} \tag{7}$$

For turbulent energy and mass fraction the gradient is zero at the wall.

c) Outlet/symmetry axis:

All gradients are equal to zero. At symmetry axis, v=0.

d) Dimensionless parameters

A dimensionless parameter introduced by Becker [3] is much applied for discribing confined jet flows. If uniform velocity profiles are assumed at outlet of nozzle and inlet of the secondary stream this parameter may be expressed as

$$Ct = \left(\frac{\dot{m}_s\, \bar{u}_s}{\dot{m}_p (\bar{u}_p - \bar{u}_s)}\right) \tag{8}$$

The value of Ct determines if recirculation or not.

4. METHOD OF SOLUTION

Finite-difference methods are used to solve the set of partial differential equations together with the boundary conditions. The calculation domain is divided into a finite number of control volumes, each surrounding a grid point. Each differential equation is integrated over the control volumes to bring the equations over to the finite difference form. The non-linear algebraic equations are solved by a tri-diagonal-matrix-algorithm and iteration. The special problem of coupling between momentum and mass conservation equations is solved by the SIMPLE method given by Patankar and Spalding [11]. A more extensive description of the method can be found in the book of Patankar [12].

5. RESULTS AND DISCUSSION

5.1 Cases

The calculation cases are shown in table 1. In addition to the cases UL1 - UL4, the case RB, Razinsky and Brighton [5], is calculated for testing the program against a case with other conditions. Some typical results are presented and discussed below.

5.2 Results and discussion

Figure 2 shows the axial velocity and axial turbulence intensity along the centerline for the case UL3. It can be observed that the calculated velocity after the nozzle outlet is lower than measured. In the same area the calculated turbulence intensity is too high, so that the physical coupling between turbulence intensity and velocity is correct. The main problem is probably that there are too few gridpoints available (see section 5.3) for describing the potensial core after the nozzle satisfactory. Further downstream the calculations are fairly good.

In figures 3 and 4 three radial profiles showing axial velocity and turbulence intensity at the same locations are presented. The profiles at the inlet of the secondary stream are very satisfactory both for velocity and turbulence intensity. In the middle of the tube the velocity calculation is still in good agreement with the experiments, but turbulence intensity is calculated too low both near the centerline of the tube and closer to the wall.

An anisotropic turbulence on the centerline is probably the most important reason for the deviation there. The potensial core in the secondary stream has disappeared according to experiments. Due to calculations the disturbance from the inlet of the secondary stream is not taken care of, and the potential core in the secondary stream will exist further downstream.

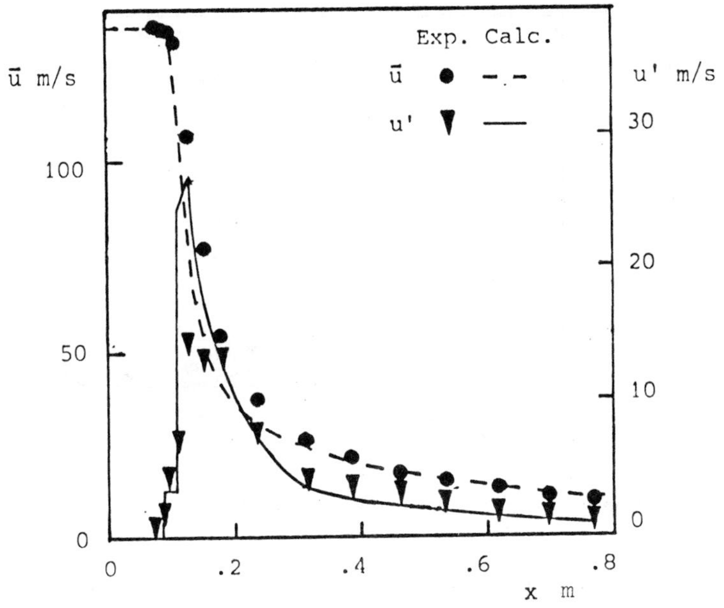

Fig. 2 Axial velocity and turbulence intensity along centerline, case UL3.

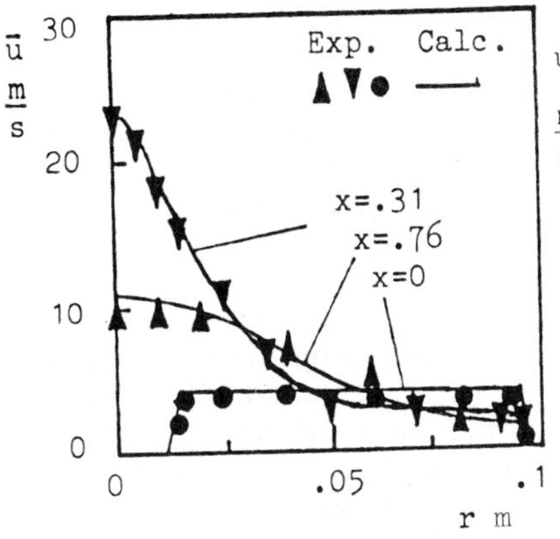

Fig. 3 Radial distribution for axial velocity, case UL3.

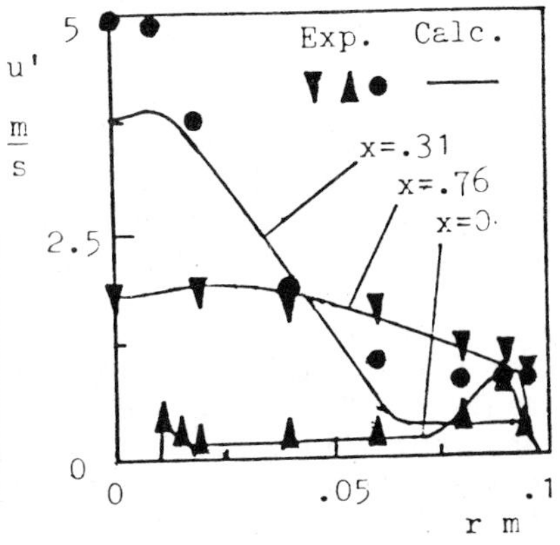

Fig. 4 Radial distribution for turbulence intensity, case UL3.

At the outlet the velocity profiles show that the flow is more fully developed than calculated. This is probably due to the very strong turbulence generated between the jet and the secondary stream where the gradients are very steep. This area cannot be good enough simulated with that few gridpoints.

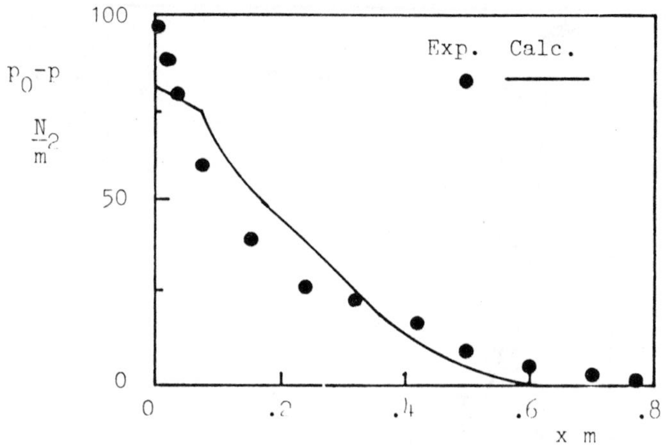

Fig. 5 Wall pressure distribution, case UL1.

Generally the calculated results may be done considerably better by changing the constants in the k-ε model. c_2 in the ε-equation is tested for other values, but the changing is not recommended because of loss of generality.

The profiles for the cases UL2 and UL4 are very similar with UL3 and are not shown here.

In figure 5 the wall pressure distribution for case UL1 is shown. The calculated pressure is for the point next to the wall, since the wall pressure is not calculated. Downstream of the nozzle the agreement between calculated and experimental data is seen to be fairly satisfactory.

Calculations are also compared with the very extensive experimental investigation done by Razinsky and Brighton [5], case RB in table 1. Here inlet velocity is given also for the secondary stream. Figure 6 shows the static wall pressure, axial velocity and turbulence intensity along the center line. p_s is the inlet pressure in the secondary stream. The pressure is too high after the potential core of the jet, and the turbulence intensity too low. This is again probably due to the too few grid- points in the potential core even if it is more than for the cases UL. Further downstream the gradients are less, the turbulence more isotrop and the agreement with experiments is very good.

Fig. 7 shows the entrained mass $\dot{m}_s$ as a function of jet mass $\dot{m}_p$ for the cases UL2, UL3 and UL4 both experimentally and calculated. The experimental values are determined by integrating the measured velocity profiles for the secondary stream. For the primary jet the mass flowrate is measured by flowmeter.

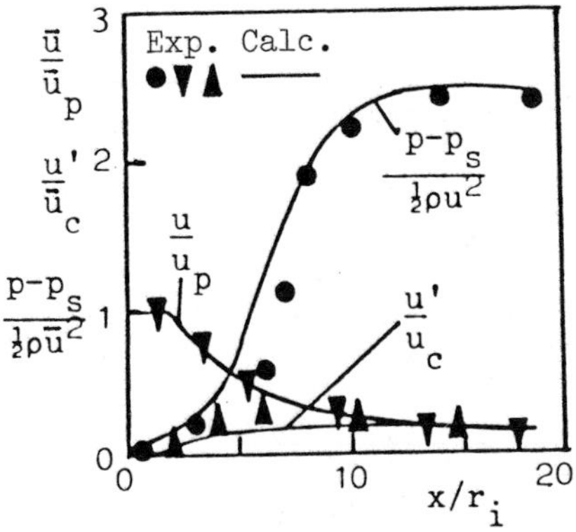

Fig.6. Dimensionless pressure velocity and turbulence intencity for case RB

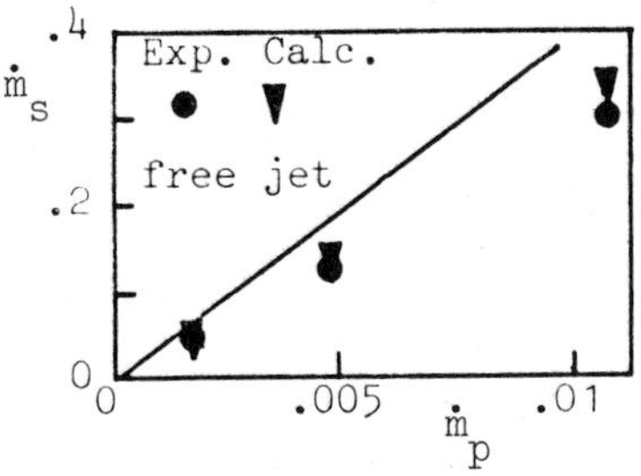

Fig.7. Mass entrainment for cases UL2-UL4

The ratio $\dot{m}_s/\dot{m}_p$ is slightly bigger for UL2 than for the other cases. The equation for the linear part is

$$\dot{m}_s = 28 \cdot \dot{m}_p \qquad (9)$$

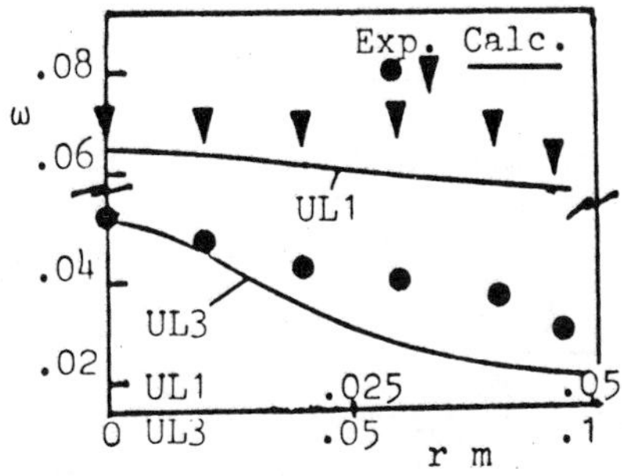

Fig.8. Mass fractions for the primary jet

Figure 8 shows the mass fraction profiles at the outlet of the tube for UL1 and UL3. The total mass for both cases are calculated less than measured.

5.3 Some computational details

The calculations were performed on a NORD-10S computer with 48 bits wordlength. All the variables including density and turbulent viscosity had a relaxation factor of 0.5 except pressure which had 0.8. Approximately 300 iterations were needed to obtain converged solution for the flow calculations and 500 iterations for the concentration calculations. It was applied a nonuniform grid with 25 x 25 gridpoints for all the calculations and that was the maximum capacity for the computer. Further details are given in [9].

6. CONCLUSIVE REMARKS

It is concluded that the numerical procedure developed is well suited for practical engineering use for scaling, optimizing and parametric study of such equipment as discussed here. The computed results could be more exact with finer grid, but finer grid gives a higher computational cost, and this has to be judged against the engineering benefit. It is also difficult to say for some areas if it is the turbulence model or the lack of grid points which is the origin of the inaccurate deviation from experimental data.

REFERENCES

1. THRING, M.W. and NEWBY, M.P. Fourth Symposium (International) on Combustion, p. 789, 1953
2. CRAYA, A., CURTET, R. On the Spreading of a Confined Jet. Compte-rendus Académi des Sciences, Paris, vol. 241, pp. 621 - 622, 1955
3. BECKER, H.A., Hottel, H.C. and WILLIAMS, G.C. 9th Symposium (International) on Combustion, Academic Press, p. 7, 1963
4. ABRAMOVICH, G.N. The Theory of Turbulent Jets, MIT, 1963
5. RAZINSKY, E. and BRIGHTON, J.A. Confined Jet Mixing for Nonseparating Conditions. Trans. ASME J. of Basic Eng., pp. 333 - 349, Sept. 1971
6. HENDRICKS, C.J. The Prediction of Turbulent Confined Jet Mixing with and without Swirl. Thesis, Pennsylv. State University, March 1974
7. ELGOBASHI, S.E., PUN, W.M., SPALDING, D.B. Concentration Fluctuations in Isothermal Turbulent Confined Coaxial Jets. Ch. Eng. Sc. Vol. 32, pp. 161 - 166. 1977
8. KANG, V.M. and SUZUKI, K. Numerical Study of Confined Jets I. Prediction of Flow Pattern and Turbulence Quantities with Two-equation Model of Turbulence. Mem. Fac. Eng. Kyoto Univ., v. 40, p. 41 - 61, 1978
9. ULSERØD, AA. S. Numerical and experimental investigation of jet-induced flow, mixing and combustion in an open tube. Thesis, Norwegian Institute of Technology, Norway, 1981
10. LAUNDER, B.E. and SPALDING, D.B. The numerical computation of turbulent flows. Comput. Meth. Appl. Mech. Engng. No. 3, 269-289, 1974
11. PATANKAR, S.V., SPALDING, D.B. A Calculation Procedure for Heat, Mass and Momentum Transfer in Three-Dimensional Parabolic Flows, Int. J. Heat and Mass Transfer, Vol. 15, pp. 1787 - 1806, 1972
12. PATANKAR, S.V. Numerical Heat Transfer and Fluid Flow. Hemisphere Publ. Corp., 1980

Additional notation:

D,d	- diameters
c_D, c_1, c_2	- constants in turbulence model
Ct	- dimensionless parameter
E	- constant in wallfunction
G	- generation term in turbulence model
L	- length of confining tube
S	- general source term
$\bar{u}$	- axil velocity
u'	- turbulence intencity
x	- axial coordinate
y	- normal distance from wall
y^+	- dimensionless normal distance

greek symbols:

Φ	- general dependent variabel
Γ	- general transport coefficient
ε	- dissipation of turbulent kinetic energy
σ	- general Schmidt-Prandtl number
ω	- mass fraction

subscripts:

0	- surroundings
c	- centerline
i	- internal
p	- primary stream (nozzle)
s	- secondary stream
w	- wall

SECTION 12

TWO/MULTI PHASE FLOW

PREDICTION OF TURBULENT TWO-PHASE FLOW IN ANNULAR VERTICAL TUBES

M.M.M. Abou-ELLAIL (i) and H.A. WARDA (ii)

SUMMARY

A numerical procedure for the compution of single and two-phase turbulent flow, in the entrance region of an annular gap, with a non rotating inner cylinder, is presented with particular reference to oil production wells.

In the method employed, separate conservation equations are formulated for the liquid phase, in conjunction with similar equations for the gas phase. The interaction between the phases is accounted for by including the interfacial drag terms in the corresponding momentum equations.

The numerical procedure is based on the solution of the finite-difference form of the governing partial differential equations. The governing equations involve a number of turbulent correlation terms for each phase. These terms, modelled through turbulent exchange coefficients and mean gradients, are obtained from the solution of a two-equation model of turbulence, for kinetic energy of turbulence and its dissipation rate.

The computation is applied to the flow in a vertical annulus of a mixture of oil and gas bubbles with an emphasis on the effect of the dispersed

---

(i) Mech. Engg. Dept.,Cairo University,Cairo,EGYPT.

(ii) Mech. Engg. Dept., Alex. University,Alex.EGYPT.

gas phase, the bubble size and void fraction on the flow pattern, pressure gradient and turbulence characteristics.

1. INTRODUCTION

The prediction of pressure drop and liquid holdup occuring during two-phase gas-liquid flow in pipes is of particular interest to the petroleum, chemical and nuclear design engineers. In nuclear reactors, two-phase flow occurs and liquid hold up greatly affects heat transfer. Two-phase flow occurs frequently in chemical processing, and the design of processing equipment and piping systems requires knowledge of pressure drop, liquid hold-up, and often flow pattern.

In the petroleum industry, two-phase flow occurs in pipelines and in oil and gas wells. As the natural oil reservoir energy is depleted, many wells have to be equipped with artificial lift systems such as gas lift. In such a system a continuous small volume of high pressure gas may be injected down the tubing string in order to aerate or lighten the fluid column until reduction of the bottom hole pressure of the fluid column will allow a sufficient differential across the sand face, causing the well to produce the desired rate of flow up the tubing-casing annular space. To design these systems, a method of predicting two-phase flow pressure gradient is required.

Although extensive research in two-phase flow has been conducted, most of it has been concentrated on either horizontal or vertical pipe flow (1,2). Meanwhile, various models for the analysis of annular two-phase flow have been built to predict various overall quantities such as pressure gradient, void fraction, critical heat flux etc. (3,4). All these models are one-dimensional and are based on combination of emperical correlations.

In order to reduce the dependence on emperical correlations, a numerical prediction of turbulent gas-liquid flow in the entrance region of an annulus is presented. In the method employed, separate time-averaged conservation equations are formulated for both phases after modelling the Reynolds stresses and other correlation terms by Boussinesq hypothesis. The coupling between the two phases is accounted for through the exchange of momentum.

The computed results demonstrate the role of interphase friction and buoyancy force, in affecting the process of settlement of the fully developed flow. A detailed analysis of the results is presented which demonstrates their plausability.

Details of the differential equations and other aspects of the mathematical formulation are given in section 2; the finite-difference forms of the equations and an outline of the solution procedure can be found in section 3, and the results of the illustrative calculation in section 4.

2. THE MATHEMATICAL MODEL

The mathematical model used for the present study solves the finite-difference analogues of the partial differential equations that govern the two-dimensional flow of two distinct fluid phases, a carrier fluid and a dispersed gas phase. Effect of inertia, convection, gravity and interface momentum transfer are all included. Interphase mass transfer is neglected. Special features arising in two-phase flows, notably the effects of fluctuations in void fraction and the interaction between the fluctuating velocities of the two phases are included.

As it is usually the case in turbulent flow problems the instantaneous value of any dependant variable is decomposed into a time-mean value and a fluctuating component. For compactness, all the time-averaged conservation equations for mass and momentum may be expressed in the following compact tensor notation although they are actually solved in their cylindrical polar form.

2.1 Governing equations

The field equations for the two-fluid model are:

a) Mass conservation :

$$\frac{\partial}{\partial x_i}\left(\rho_1 r_1 u_i\right) + \frac{\partial}{\partial x_i}\left(\rho_1 \overline{r_1' u_i'}\right) = 0 \qquad (1)$$

for the liquid phase, and

$$\frac{\partial}{\partial x_i}\left(\rho_g r_g v_i\right) + \frac{\partial}{\partial x_i}\left(\rho_g \overline{r_g' v_i'}\right) = 0 \tag{2}$$

for the gas phase

b) Momentum conservation in the j-direction :

$$\frac{\partial}{\partial x_i}\left(\rho_1 r_1 u_i u_j - r_1 \sigma_{1_{ij}} + \rho_1 r_1 \overline{u_i' u_j'} + \rho_1 u_j \overline{r_1' u_i'} + \rho_1 u_i \overline{r_1' u_j'}\right)$$

$$= - r_1 \frac{\partial P}{\partial x_j} + f\, r_g (v_j - u_j) + f\, \overline{r_g' (v_j' - u_j')} \tag{3}$$

for the liquid phase, and

$$\frac{\partial}{\partial x_i}\left(\rho_g r_g v_i v_j - r_g \sigma_{g_{ij}} + \rho_g r_g \overline{v_i' v_j'} + \rho_g v_j \overline{r_g' v_i'} + \rho_g v_i \overline{r_g' v_j'}\right)$$

$$= -r_g \frac{\partial P}{\partial x_j} - f\, r_g (v_j - u_j) - f\, \overline{r_g' (v_j' - u_j')} + r_g (\rho_g - \rho_1) g_j \tag{4}$$

for the gas phase.

Where u_i and v_i are the time-averaged liquid carrier and dispersed gas velocities in the x_i direction respectively, ρ is the density, r is the volume fraction, σ_{ij} is the laminar stress tensor, g_j is the component of gravitational acceleration in the x_j direction and f is the drag function; f is dependant on the flow regime, bubbles mean diameter, liquid density, and bubble Reynolds' number. The over-bars and-dashes indicate time average and fluctuating values respectively.

In the above equations P stands for the reduced pressure which is defined as $P = p+\rho_l g_j x_j$, where p is the static pressure. It is because of the use of P that there is no explicit term for the gravity effects in the liquid momentum equation. The practice of using reduced pressure has been found to be advantageous for numerical stability (5) particularly in flows with significant body forces such as buoyancy forces.

Finally, the summation of volume fractions must be equal to unity, i.e.

$$r_l + r_g = 1 \tag{5}$$

Equation (5) indicates that $r_l' = - r_g'$ and also that $\frac{\partial r_l}{\partial x_i} = - \frac{\partial r_g}{\partial x_i}$

The interphase momentum exchange, represented by the term f in equation 3 and 4 is assumed to arise soley from drag. Further, as the gas phase may be considered as a swarm of spherical bubbles interspersed in the liquid, the total drag force per unit volume can be evaluated as the sum of the drag force exerted by the liquid on each individual bubble contained in that volume. The drag exerted on an individual bubble is obtained emperically from the consideration of the flow around a spherical bluff body representing a bubble.
It follows that (6) ,

$$f = \frac{3}{8} \frac{1}{R} C_D \; |v_j - u_j| \tag{6}$$

where R is the aggregate radius of the bubbles and C_D is the drag coefficient. For low Reynolds number, as in the present case, Stocks drag is assumed valid and hence,

$$C_D = \frac{24}{RN} \; .$$

2.2 MODELLING TASK

In the above equations, triple correlations (e.g. $\overline{r_l' u_i' u_j'}$) and strain rate-volume fraction correlations are neglected due to their relative small contribution. It can easily be shown that the double correlation terms appearing in the governing equations must be modelled before an attempt is made to solve the above equations.

A second order turbulence model was proposed (8) for predicting incompressible two-phase flows. The model is based on rigorously-derived equations for the kinetic energy of turbulence and its dissipation rate for two-phase flows, with several turbulent correlations appearing in the equations. However, for simplicity, only the following terms were retained:

1- Reynolds stress terms ; $\overline{u_i' u_j'}$ and $\overline{v_i' v_j'}$

2- Turbulent diffusion of each phase; $\overline{r_l' u_i'}$ & $\overline{r_g' v_i'}$

3- Inter-phase turbulent diffusion; $\overline{r_g' u_i'}$ & $\overline{r_l' v_i'}$

2.3 TURBULENCE MODEL

In the case of single phase flow, only the Reynolds stresses ($\overline{u_i' u_j'}$) need to be modelled. In general the Reynolds stresses are computed either by introducing the eddy diffusivity concept or solving transport equations for these stresses. In the present work, the first approach is adopted and hence $\overline{u_i' u_j'}$ and $\overline{v_i' v_j'}$ are defined as follows :

$$- \rho_l \overline{u_i' u_j'} = \mu_t \left(\frac{\partial u_i}{\partial x_j} + \frac{\partial u_j}{\partial x_i} - \frac{2}{3} \delta_{ij} \frac{\partial u_m}{\partial x_m}\right) - \frac{2}{3} \rho_l k \delta_{ij} \qquad (7)$$

Similar relation exists for $\overline{v_i' v_j'}$

where k ($= \frac{1}{2} \overline{u'_m u'_m}$) is the kinetic energy of turbulence and μ_t is a turbulent viscosity which may be related to the kinetic energy of turbulence (k) and its dissipation rate (ϵ) through the following equation,

$$\mu_t = C_\mu \rho_1 k^2/\epsilon \tag{8}$$

where C_μ is a constant of the model, k and ϵ are to be determined from the solution of their transport equations which are respectively:

$$\frac{\partial}{\partial x_i}(\rho_1 r_1 u_i k) = \frac{\partial}{\partial x_i}\left(r_1 \frac{\mu_t}{Pr_k} \frac{\partial k}{\partial x_i}\right) + r_1 (G - \rho_1 \epsilon) \tag{9}$$

$$\frac{\partial}{\partial x_i}(\rho_1 r_1 u_i \epsilon) = \frac{\partial}{\partial x_i}\left(r_1 \frac{\mu_t}{Pr_\epsilon} \frac{\partial \epsilon}{\partial x_i}\right) + r_1 \left(C_1 G \frac{\epsilon}{k} - C_2 \rho_1 \frac{\epsilon^2}{k}\right) \tag{10}$$

where Pr_k , Pr_ϵ represents turbulent Prandtl numbers for the parameter in question and C's are the constants of the model.

G is a generation term defined by

$$G = -\rho_1 \overline{u'_i u'_j} \frac{\partial u_i}{\partial x_j} \tag{11}$$

The turbulent diffusion terms of the carrier fluid ($\overline{r'_1 u'_i}$) and the dispersed phase ($\overline{r'_g v'_i}$) are modelled by the following two equations:

$$-\rho_1 \overline{r'_1 u'_i} = \frac{\mu_t}{Pr_1} \frac{\partial r_1}{\partial x_i} \tag{12}$$

$$-\rho_g \overline{r'_g v'_i} = \frac{\mu_t}{Pr_g} \frac{\partial r_g}{\partial x_i} \tag{13}$$

where Pr_l and Pr_g are turbulent Schmidt numbers of the carrier and dispersed phases respectively of order unity. Using equation (5) and assuming equal Schmidt numbers, it can be proved that the turbulent diffusion of one phase is counter balanced by the diffusion of the other, i.e.

$$\rho_l \overline{r_l' u_i'} = - \rho_g \overline{r_g' v_i'}$$

Finally, the turbulent inter-phase diffusion terms are modelled as follows :

$$\overline{r_g' u_i'} = - \overline{r_l' v_i'} = \left(\frac{\mu_t}{\rho_l Pr_l}\right) \frac{\partial r_l}{\delta x_i}$$

$$= - \left(\frac{\mu_t}{\rho_l Pr_l}\right) \frac{\partial r_g}{\delta x_i} \qquad (14)$$

The above-mentioned constants C_μ, C_1, C_2, Pr_k, Pr_ϵ, Pr_l and Pr_g are given values 0.09, 1.44, 1.92, 0.7, 1.22, 1.0 and 1.0 respectively. Equations (7-8) and (11-14) are substituted in equations (1-5) and (9-10) to yield a closed set of transport equations for the dependant variables u_i, v_i, p, r_l, r_g, k and ϵ.

2.4 BOUNDARY CONDITIONS

The boundary conditions to be satisfied by the governing equations are as follows:

a) Inlet Section : The axial velocity profiles for both phases are assumed to be uniform at the inlet section. The inlet profiles of k and ϵ were assumed to conform to turbulent pipe flows,

$$k_{in} = C_k \, W^2_{in} \qquad (15)$$

$$\epsilon_{in} = C_\mu k_{in}^{3/2} / \left(\frac{D_h}{2} C_\epsilon\right) \qquad (16)$$

where $D_h = D_o - D_i$ is the hydraulic diameter, D_o and D_i are outer and inner diameters of the annulus respectively, $C_k = 0.003$, $C_\epsilon = 0.03$ and subscript 'in' refers to inlet conditions.

b) Exit Section: A fully developed flow conditions with zero gradients are assumed at the exit section, i.e,

$\frac{\partial \emptyset}{\partial x_j} = 0$, where $\emptyset$ is any of the depandant variables and x_j is the axial direction.

c) ALONG THE WALL

All the velocity components were presumed zero along the wall. The resultant shear stress τ_w at the near wall node was obtained from the wall function given by (7), with modification to account for the existance of the dispersed phase:

$$\tau_w = r_1 \rho_1 (v_w - v_p) \, C_\mu^{\frac{1}{4}} \, k_p^{\frac{1}{2}} \frac{k}{l_n(Ey_p^+)}$$

$$y_p^+ = \rho_1 y_p \, C_\mu^{\frac{1}{4}} \, k_p^{\frac{1}{2}} / C_\mu$$

Similar relations exists for the gas phase, where $k = 0.4187$, $E = 9.79$, and subscript p and w denote node p and wall condition respectively.

The generation term G in equation is modified in order to take into consideration the shear stress τ_w calculated from the wall function. Also, the dissipation term ϵ is modified at the point adjoining the wall in the following manner.

$$\epsilon_p = \frac{1}{k} C_\mu^{\frac{3}{4}} k_p^{3/2} / y_p$$

3. FINITE-DIFFERENCE SOLUTION OF THE EQUATIONS:

3.1 THE FINITE DIFFERENCE EQUATIONS

The differential equations are solved by a finite difference technique employing a staggered grid (7) ; where the solution domain is subdivided into a number of finite volumes or 'cells' each of which encloses an imaginary grid node at which all scallar variables (p, r_1, r_g, k and ϵ) are stored, while the velocity components u_i and v_i are chosen to lie on the cell boundaries where they are used for mass flux computations. The general form of the equations to be integrated can be written, for a variable $\emptyset$, as:

$$\frac{\partial}{\partial x_i}(\rho r w_i \emptyset - \Gamma_\emptyset r \frac{\partial \emptyset}{\partial x_i}) = S_\emptyset , \qquad (19)$$

where $\Gamma_\emptyset$ is the corresponding effective diffusivity, $S_\emptyset$ is a source term, ρ and r are density and volume fraction of the carrier or dispersed phases depending on the variable in question. When $\emptyset$ is replaced by either r_l or r_g (mass continuity), r and $S_\emptyset$ should be set equal to unity and zero respectively. The finite-difference equations are derived by integrating the differential equations over the volume enclosing a grid node, with appropriate assumptions for the variation of variables between the nodes. These equations relate the value of a variable at a grid node, p, to those at the four nodes, N,E,Wands, surrounding it; the general form of these equations can be written as:

$$a_p \emptyset_p = \sum_{n=N,E,W,S} a_n \emptyset_n + S_p \qquad (20)$$

where $\sum$ signifies the summation over the four neighbouring nodes. The a's are influence coefficients, expressing the effects of convection and, where appropriate, diffusion between the nodes; there values are rendered positive by the use of "upwind" differencing. S_p is the integrated source term.

There is one equation like (20) for each dependant variable at each grid node. The difference equations of boundary nodes must be modified to incounter the boundary conditions imposed there.

3.2 SOLUTION OF THE DIFFERENCE EQUATIONS

The steps involved in a typical computation can be summarised as follows: The velocity components of the two phases (u_i^*, v_i^*) are first computed by solving the difference equations assuming an approximate pressure field p^*; the dispersed phase volume fraction difference equations are solved for the 'r_g' field, and carrier fluid volume fraction field 'r_l' is then obtained ($r_l = 1 - r_g$); momentum based velocity components u_i^* and v_i^* are then substituted in the 'overall' mass conservation

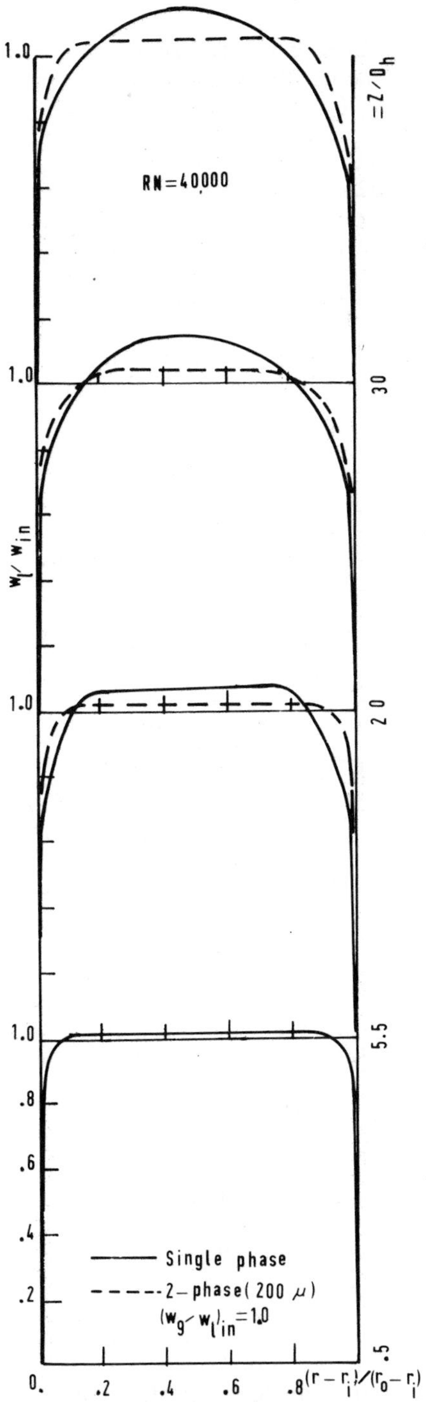

Fig.1 Radial distribution of axial velocity along the annulus.

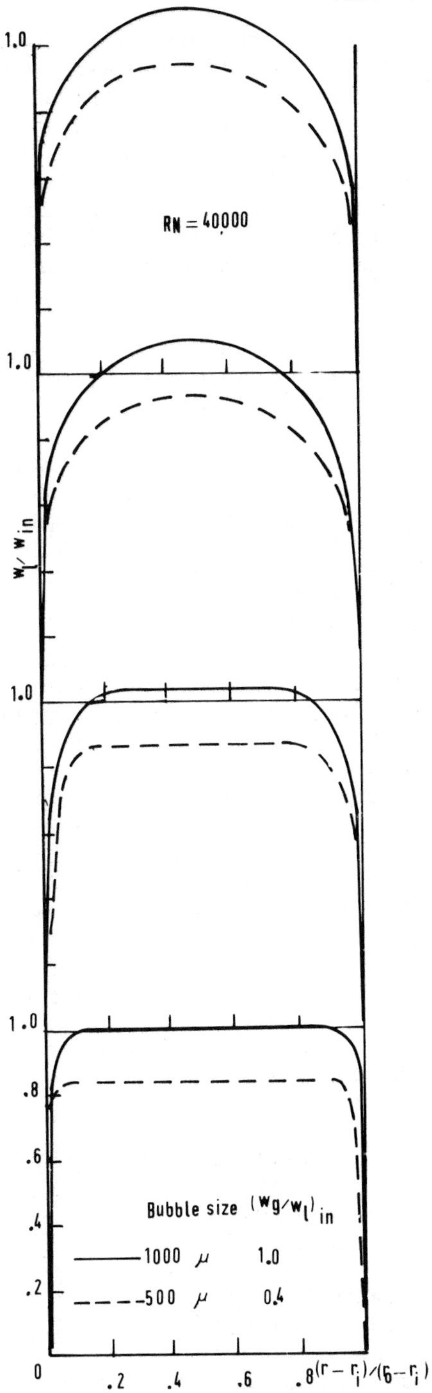

Fig.2 Effect of bubble size & slip velocity on axial liquid velocity.

equation to obtain mass imbalance source at each grid node; these mass sources are then used to compute pressure perturbations (p') which yield an updated values of p, u_i and v_i simultaneously, to satisfy the overall continuity; the kinetic energy of turbulence and its dissipation rate are then computed from their difference equations and hence the turbulent viscosity is obtained at each node; the above computation loop is repeated until satisfactory convergence has been achieved.

The solution of equation (20) is obtained by a Gaus-elimination line-by-line alternating direction algorithm.

4. DISCUSSION AND PRESENTATION OF RESULTS

The numerical computations in the present study are carried out for an upward flow of oil with dispersed gas bubbles in an annulus of 18 cm. outer diameter (casing) and 7.5 cm. inner diameter (tubing). The problem is formulated to simulate the actual two-phase gas lift process, in oil production wells, in which gas is injected down the tubing string and producing up the tubing-casing annular space, at an inlet pressure, liquid and gas phase densities of 100 bar, 10^3 kg/m^3 and 10^2 kg/m^3 respectively. In the present study the densities of the two phases were taken as constants. This simplificaction is adopted for convenience and is not imposed by the present computation method. Three values for the bubble diameter were investigated (200,500,1000 microns). In the computations, a non-uniform grid was employed in the radial direction, while a uniform grid was used in the axial direction. For the flow domain considered, a grid having 16 nodes in the radial direction and 70 nodes in the axial direction, to cover a total length of 3.5 m., is found to give a grid-independent solution.

The radial distribution of the mean axial velocity of oil (W_l), at various axial locations (Z/D_h), for both single and two-phase flows are presented in fig.1. Both velocities were normalized with respect to the inlet velocity(W_{in}), for an inlet RN=40000. The two-phase velocity profile was obtained for a voild fraction (r_g) of 0.3 and bubble size 200 micron. This figure shows that the presence of gas bubbles resulted in an increased velocity gradients near the walls and a more flat velocity distribution in the middle of the annular space. The effect of the bubble size and slip velocity at inlet on the velocity profile is shown in fig.2.

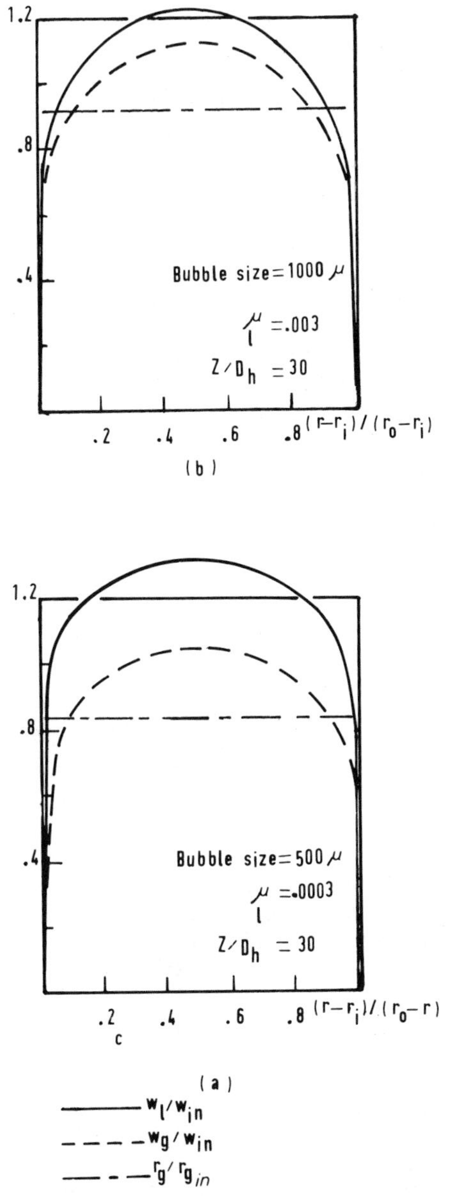

Fig.3 Radial distribution of axial gas & liquid velocities & void fraction.

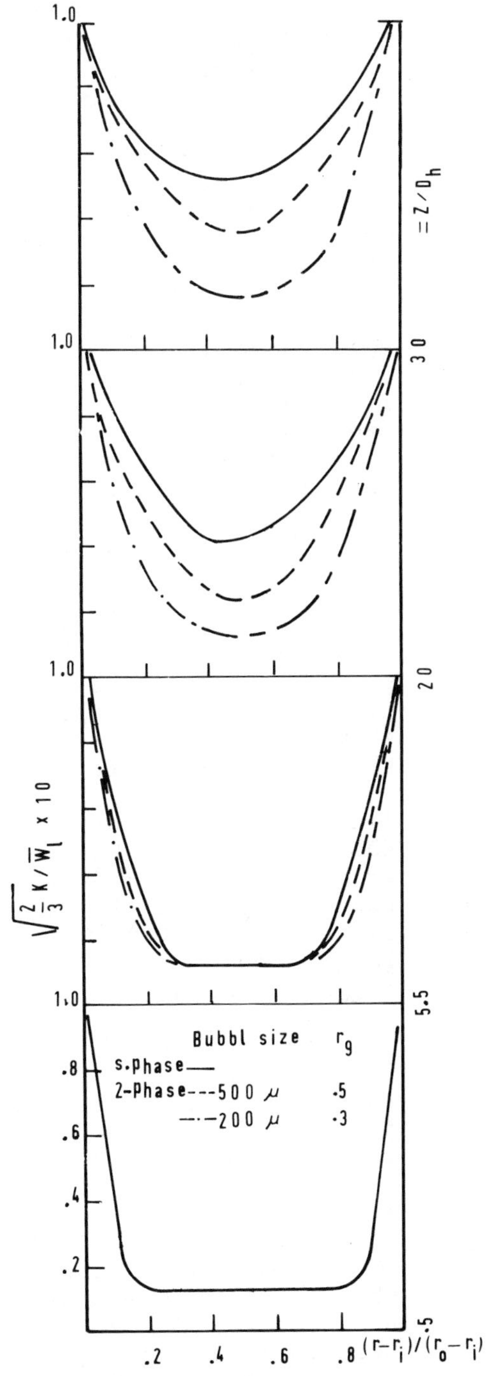

Fig.4 Radial distribution of turbulence intencity at various sections.

By comparing fig.1 and 2 it becomes evident that the influence of the smaller sized bubbles(200μ) on the development of the turbulent flow is more pronounced than the effect of larger bubbles (1000μ) which may be attributed to the increased effect of the interfacial drag. Fig.2 also shows that the large slip velocity at inlet has no effect on the velocity profiles, however the mean axial velocity of oil decreases as a result of an increase of void fraction. Fig.3 (a) and (b) demonstrates the significance of the buoyancy force and viscous drag on the radial velocity profiles of both phases, and on the void fraction. Fig.3a is obtained for a bubble size 1000μ while fig.3b is obtained for smaller viscosity (viscosity of fresh water). For larger bubbles, the buoyancy force increases and the velocity of oil approaches that of the gas resulting in a reduced slip velocity and consequently lower void fraction. On the other hand, as the liquid viscosity decreases, the slip velocity and void fraction increases.

The radial distribution of the turbulence intensity along the passage for single and two-phase flows is depicted in fig.4. For high void fraction and small sized bubbles,it is clearly shown that the presence of bubbles tends to reduce the turbulence intensity as a result of the reduced velocity gradients. The axial variation of turbulent kinetic energy at various radial positions is shown in fig.5. The figure shows a remarkable effect of smaller sized bubbles, whose presence tends to retard the development rate of the turbulent two-phase flow. Meanwhile, the turbulence length scale is reduced as a result of the presence of gas bubbles as shown in fig.6.

The effect of void fraction on the friction factor or pressure drop along the annulus is shown in fig.7. The predicted behaviour indicates a pronounced decrease in the pressure drop as the void fraction increases.

5. CONCLUSION :

A method for the prediction of turbulent two-phase flow in the entrance region of annular tubes has been described. Modelling of turbulence has been investigated by introducing a number of correlation terms. The results indicate that friction loss in two-phase bubbly flows is greatly affected by the void fraction; a phenomena which is used in

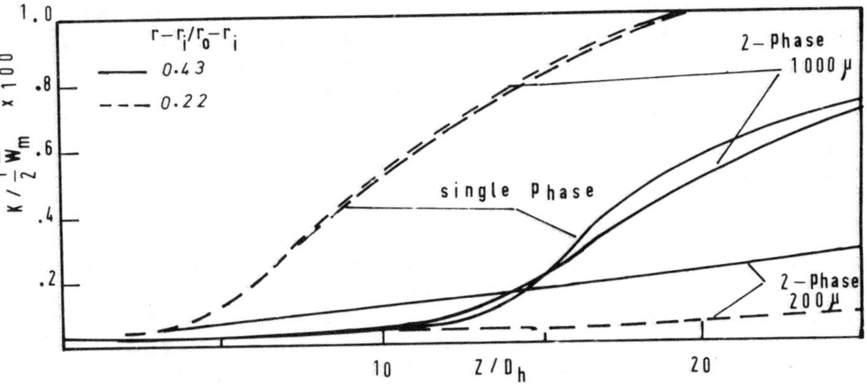

Fig.5 Axial variation of turbulence kinetic energy at different axial positions:

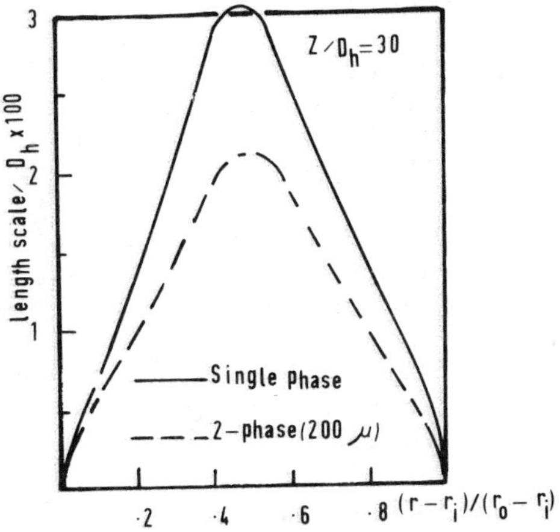

Fig.6 Radial distribution of the turbulence length scale at Z/D_h=30, for single & two-phase flows.

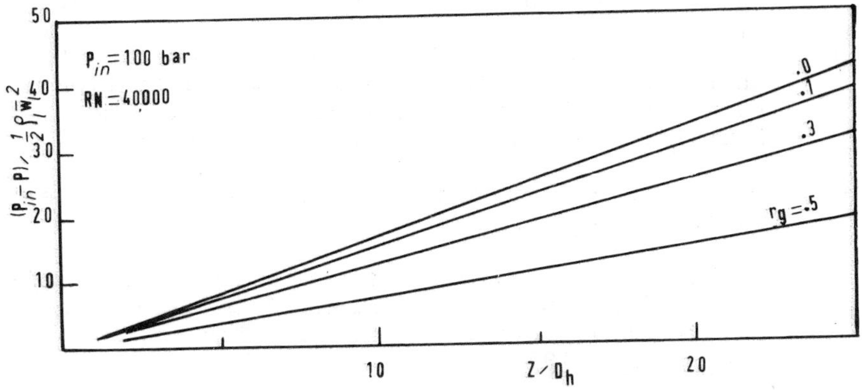

Fig.7 Effect of void fraction on the axial pressure dtop.

oil production wells to improve the well deliverability of depleted wells by introducing gas lift process. Meanwhile the presence of gas bubbles, was found to have a pronounced effect on the turbulence characteristics, namely in increasing appreciably the development length, particulary for small sized bubbles.

In order to apply the present procedure to actual gas lift application, mass transfer between the phases must be considered. Therefore more attention must be paid to the phase change process due to gas liberation under reduced pressures. However, further extension to incounter the above phenomena in the solution is being undertaken by the authors.

REFERENCES:

1. Poettmann, F.H. and Carpenter, P.C. - The multiphase flow of gas, oil, and water through vertical flow strings, Drill. and Prod. Prac., API, pp. 257-261, 1952

2. ORKISZEWSKI- Predicting two-phase pressure drops in vertical pipe, Journal of Petroleum Technology, pp. 829-834, 1967.

3. SAitc, T. et al. - Multifluid modelling of annular two-phase flow, Nucl. Engg. Design, vol.50, pp. 225-271, 1978.

4. Graver, R.B. - Two dimensional modelling of annular two-phase flow, Proceeding of the International Heat Transfer Conference, 1982

5. MARKatos, N.C. and SINGHAL, A.K.- Numerical analysis of one- dimensional, two-phase flow in a vertical cylindrical passage, Ad. Eng. Software, vol.4, No.3, pp. 99-104, 1982.

6. Soo, S.L. - Fluid dynamics of multi-phase systems, Blaisdell, 1967

7. Spalding, O.B.- Numerical computation of multiphase flow and heat transfer, Recent advances in numerical methods in fluids, Taylor, C. & Morgan, K., 1980.

8. Elghobashi, S.E., and Abou-Arab, T.W.- A second order turbulence model for two-phase flows, Proceeding of International Heat Transfer Conference, TF4, pp.219-223, 1982.

NUMERICAL MODELING OF UNSTEADY GAS-PARTICLE FLOWS AROUND RECTANGLES INSIDE CHANNELS

R. W. Davis[(I)], E. F. Moore[(I)] and C. T. Crowe[(II)]

This paper presents numerical solutions for gas-particle flows around rectangles inside two-dimensional channels. Vortex shedding frequencies are seen to compare well with the results of a wind tunnel experiment. Trajectories of individual physical particles through this highly unsteady flow are presented for varying combinations of Stokes number and gravitational force. The numerical scheme utilizes an explicit Leith-type of temporal differencing and quadratic upwind differencing for convection. A fast direct method is used to solve the Poisson equation for pressure at each time step. An infinite-to-finite mapping is employed near the downstream boundary of the computational mesh. This enables a fully-developed parabolic velocity profile to be prescribed at the mesh exit. The computational results indicate the presence of possible surface erosion problems for some combinations of the tested parameters.

1. INTRODUCTION

The motion of a particle-laden flow around a bluff body is of interest in engineering problems involving, for instance, particle collection devices and erosion of flow sensors. As it is difficult to perform experimental measurements in these types of flows, computer simulations assume great importance. The purpose of this paper is to present numerical results for

---

(I) Center for Chemical Engineering, FM 105, National Bureau of Standards, Washington, DC 20234, U.S.A.

(II) Department of Mechanical Engineering, Washington State University, Pullman, Washington 99164, U.S.A.

two-dimensional incompressible gas-particle flows around rectangles inside channels. The basic parameters involved here are Reynolds number based on rectangle width, blockage ratio, rectangle aspect ratio, upstream velocity profile shape, particle Stokes number, and gravitational force. Multishaped passive marker particles are used to illuminate the structure of the vortices in the wake of the rectangle. Trajectories of physical particles are presented for both small and large Stokes number.

The numerical method employed here is of recent origin [1,2]. It utilizes quadratic upwind differencing for convection and an explicit Leith-type of temporal differencing. This scheme has been successful in modeling the vortex shedding from rectangles in infinite domains at Reynolds numbers up to approximately 1000 using only moderate computer resources [3,4]. As the Navier-Stokes equations are solved in primitive form, a Poisson equation for pressure must also be solved at each time step. This is accomplished by means of a fast direct method. The difficult problem of an exit boundary condition in the unsteady wake region downstream of the rectangle is solved by use of an infinite-to-finite mapping. This allows a steady, fully-developed parabolic velocity profile to be prescribed at the exit from the computational mesh. The shape of the upstream velocity profile ahead of the rectangle is arbitrary. No-slip boundary conditions are imposed on the surface of the rectangle and on the channel walls. Rebounding of physical particles from surfaces is modeled based on experimental data [5]. The numerical method will be described in the next section.

2. NUMERICAL METHOD

The two-dimensional Navier-Stokes and continuity equations for a viscous incompressible fluid are

$$\frac{\partial \underset{\sim}{q}}{\partial t} + (\underset{\sim}{q} \cdot \nabla)\, \underset{\sim}{q} = -\nabla p + \nu \nabla^2 \underset{\sim}{q} \tag{1}$$

$$\nabla \cdot \underset{\sim}{q} = 0. \tag{2}$$

Here $q = (u,v)$, where u and v are velocity components in the x- and y- directions, respectively, in a cartesian reference frame; p is the ratio of pressure to constant density; ν is kinematic viscosity; and t is time. A solution of these equations is required for flow around a rectangle between semi-infinite parallel walls, as illustrated in Fig. 1. The rectangle has length A and width B and is situated symmetrically between two walls a distance H apart. The inlet velocity profile is arbitrary with centerline value U_o, while the exit velocity profile at an infinite distance downstream from the inlet is parabolic. The Reynolds number for this flow is defined as

$Re = \frac{U_o B}{\nu}$. Henceforth, all lengths are nondimensionalized with respect to B, all velocities with respect to U_o, time with respect to B/U_o, and p with respect to U_o^2. Thus, in Fig. 1, $B = U_o = 1$ by definition, A becomes the aspect ratio of the rectangle, and H^{-1} becomes the blockage ratio. The distance from the inlet to the front face of the rectangle is 8.5.

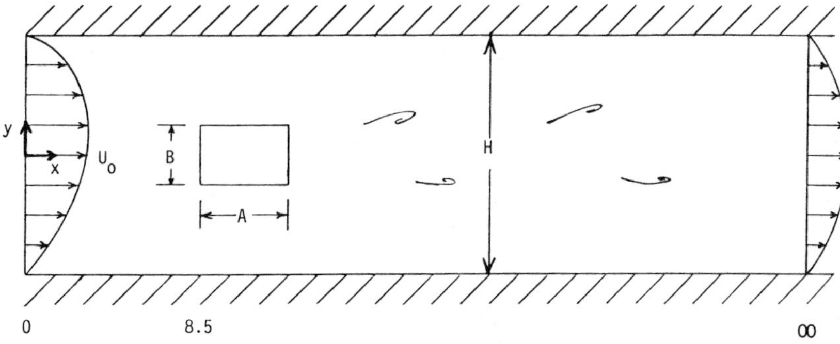

Fig. 1. Configuration Definition

The numerical solution of Eqs. (1) and (2) is accomplished on a variably-spaced staggered mesh in which pressures are defined at cell centers and normal velocities at cell faces. Mesh cells are concentrated in the areas around the rectangle and along the walls. As noted previously, the basic numerical scheme employed here is the same as that used in [3] and [4], i.e., a multidimensional version of the QUICKEST scheme [2]. This scheme employs quadratic upwind differencing for convection, an explicit Leith-type of temporal differencing, and requires a Courant number less than one. Unlike [3] and [4], the Poisson equation for pressure is solved at each time step by a fast direct method. This method utilizes cyclic reduction and is implemented in subroutine BLKTRI of the FISHPAK package of FORTRAN subprograms for the solution of separable elliptic partial differential equations developed at the National Center for Atmospheric Research [6]. Its implementation via the capacitance matrix technique on irregular regions, as occurs here due to the presence of the rectangle inside the channel, is described in [7].

The boundary conditions employed in this study are no-slip along all solid surfaces (with normal pressure gradients consistent with the full momentum equations), inlet velocity profile specified at x=0, and fully-developed flow as $x \to \infty$. Although the inlet velocity profile is arbitrary, all results presented in this paper are for the parabolic fully-developed case unless otherwise noted. The exit boundary condition is implemented by use of an infinite-to-finite mapping. The mapping used is $Z=K-K_1/(x-4)$ for $x>24$, where the constant K_1 is chosen so that $\frac{\partial}{\partial Z} = \frac{\partial}{\partial x}$ at $x=24$, making for a smooth convective transition into the transformed region. The parabolic exit velocity profile is prescribed at $Z \simeq K$.

The equation of motion for the physical particles is

$$\frac{d\underset{\sim}{q}_p}{dt} = \frac{f}{St} (\underset{\sim}{q} - \underset{\sim}{q}_p) - g, \quad (3)$$

where $\underset{\sim}{q}_p = (u_p, v_p)$ is the particle velocity vector, g is gravity (acting in the -y direction), St is Stokes number, and f is the ratio of the particle drag coefficient to Stokes drag [8]. Note that the nondimensional gravity is an inverse Froude number squared. The Stokes number is the ratio of the aerodynamic response time of a particle [8] to the characteristic flow time (B/U_o). After each time step of the flow computation, Eq.(3) is integrated twice for each physical particle in order to obtain its new location. The rebounding of particles from solid surfaces is based on experimental data for coefficient of restitution and rebound angle of steel spheres striking an aluminum surface [5]. Note that the results to be presented here are applicable only for lightly loaded systems as particle-particle collisions are not allowed and only one-way coupling is employed, i.e., the particles are affected by the fluid but not the reverse.

The two nonuniform computational meshes employed in this study were 76 x 42 for $H^{-1}=1/4$ and 76 x 52 for $H^{-1}=1/6$, with the latter being shown in Fig. 2 for $x<24$. The time step, Δt, was set between 0.04 and 0.055 in order to maintain the maximum Courant number in the flow at less than unity. A typical run time on the NBS UNIVAC 1100/82 required to compute trajectories for 20 physical particles was approximately 3 hours.

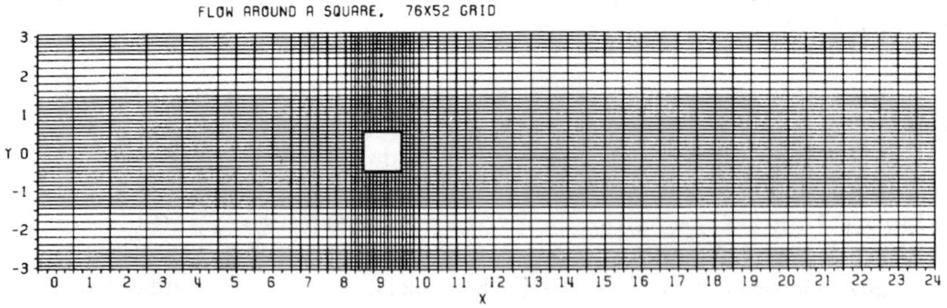

Fig. 2. The 76x52 Nonuniform Mesh

3. NUMERICAL RESULTS

A comparison between computed Strouhal numbers (nondimensional vortex shedding frequencies) and those obtained experimentally in the NBS Low Velocity Airflow Facility is shown in Fig. 3. These results are for a square (A = 1) with experimentally-determined inlet velocity profiles fit by error functions. It can be seen that the comparison appears satisfactory for the three blockage ratios shown up to Reynolds numbers of about 1000.

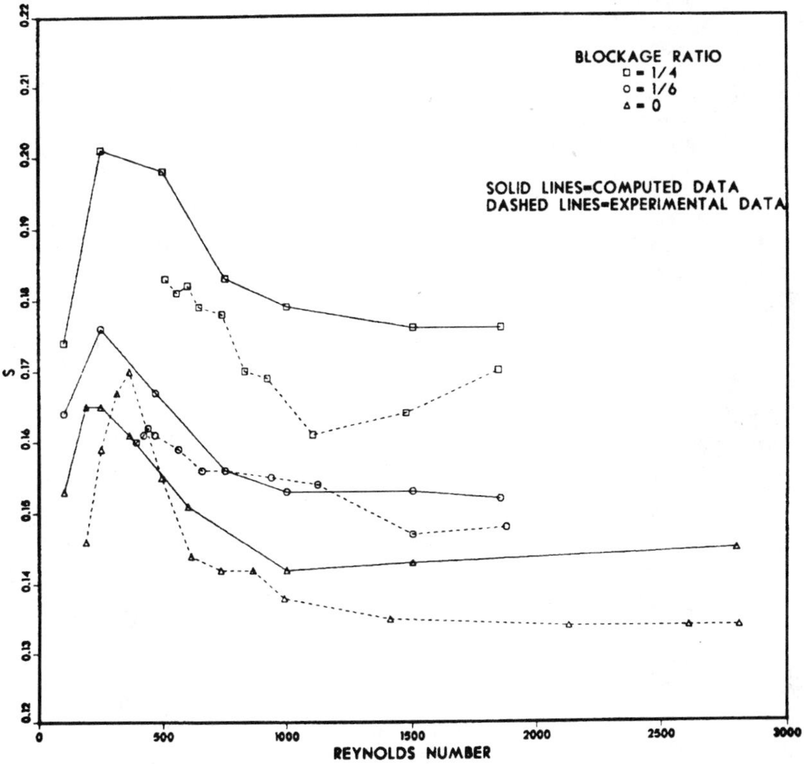

Fig. 3. Numerical-Experimental Strouhal Number Comparison

A streakline plot composed of passive multishaped marker particles [4] is shown in Fig. 4 for the case Re = 1000, A = 1, and $H^{-1}=1/6$. The structure of the wake can be clearly seen here, with its organization decreasing markedly as x increases (the mapped region here is for x>44). Trajectories of individual physical particles through this same flow are shown in Figs. 5 and 6 for zero gravity and St = 0.10 and 10.0, respectively. In each figure ten particles of radius 0.01 are introduced simultaneously upstream of the square at x = 7 and then tracked using Eq.(3). It can be seen that the particles

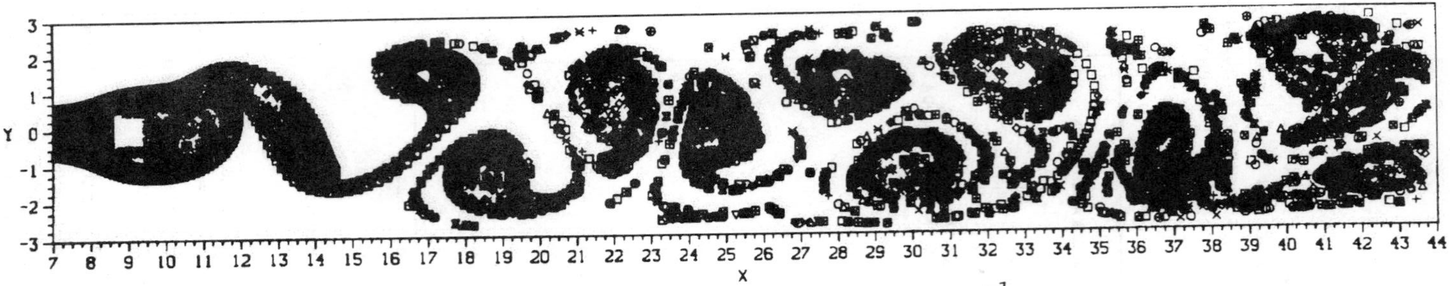

Fig. 4. Streakline Plot: Re = 1000, A = 1.0, H^{-1} = 1/6

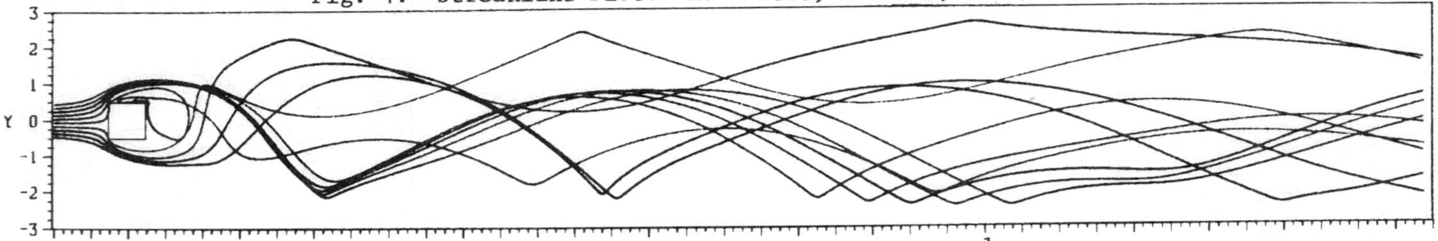

Fig. 5. Physical Particle Trajectories: Re = 1000, A = 1.0, H^{-1} = 1/6, St = 0.10, g = 0

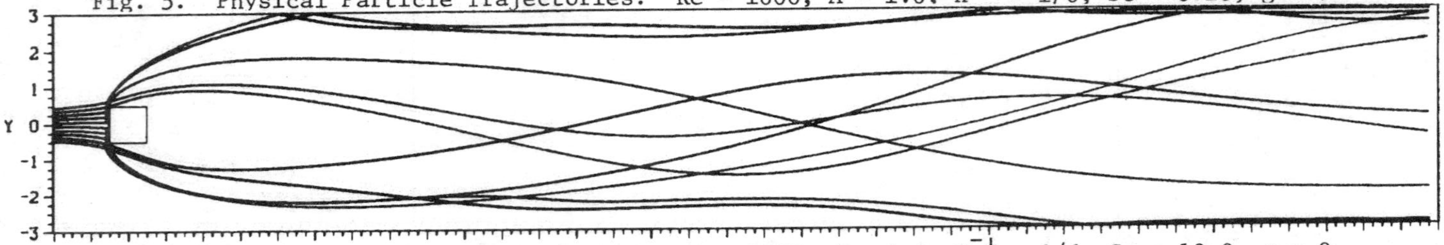

Fig. 6. Physical Particle Trajectories: Re = 1000, A = 1.0, H^{-1} = 1/6, St = 10.0, g = 0

respond much more rapidly to changes in the flow at the lower Stokes number. The capture of the particles by the vortices downstream of the square is clear from Fig. 5. For St = 10, erosion of the front face of the square as well as the channel walls appears likely to be a problem. Erosion of the channel walls would seem to be an even larger problem for the case shown in Fig. 7. Here Re = 1000, A = 0.6, H^{-1}=1/4, St = 10.0 and g = 0. The case of Re = 250, A = 1.0, H^{-1}=1/6, St = 0.10 and g=0 is shown in Fig. 8. It can be seen that, when compared with Fig. 5, the transverse transport of the particles is less at this lower value of Re. The effect of increasing gravity is shown in Fig. 9 for Re = 1000, A = 1, H^{-1}=1/6, and St = 10.0. It can be seen that, as g varies from 0.01 to 0.10, the particles' angle of descent increases until they avoid even colliding with the square. Changing the Stokes number in Fig. 9c from 10.0 to 0.10 leads to Fig. 10, where the effect of gravity is seen to be rather small.

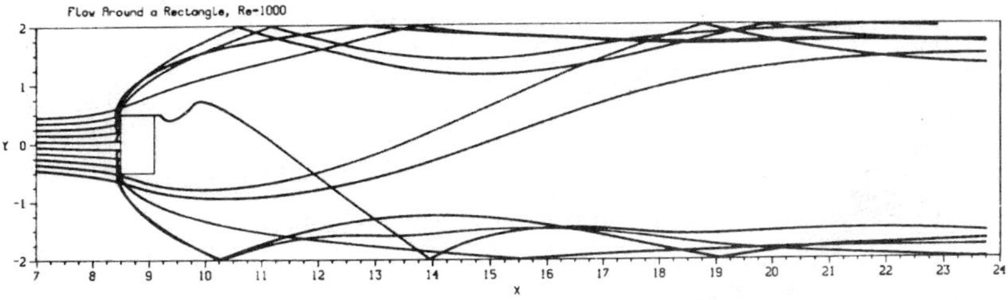

Fig. 7. Physical Particle Trajectories: Re = 1000, A = 0.6, H^{-1} = 1/4, St = 10.0, g = 0

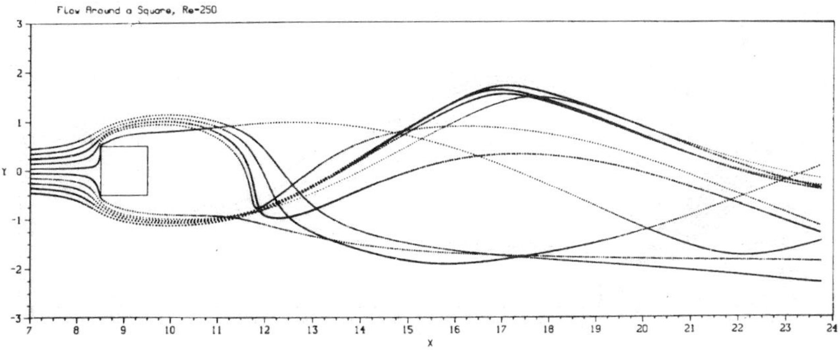

Fig. 8. Physical Particle Trajectories: Re = 250, A = 1.0, H^{-1} = 1/6, St = 0.10, g = 0

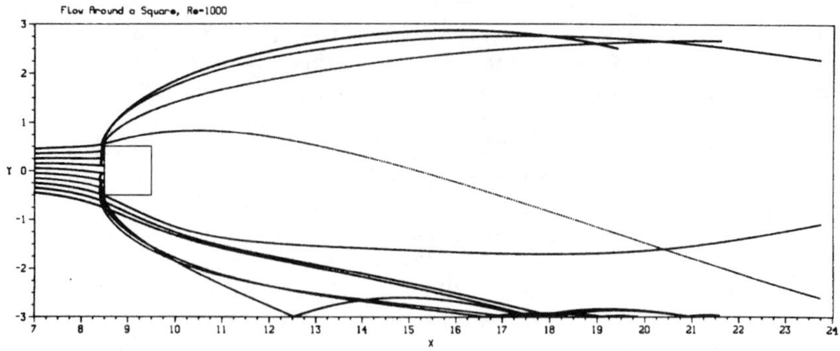

a. g = 0.01

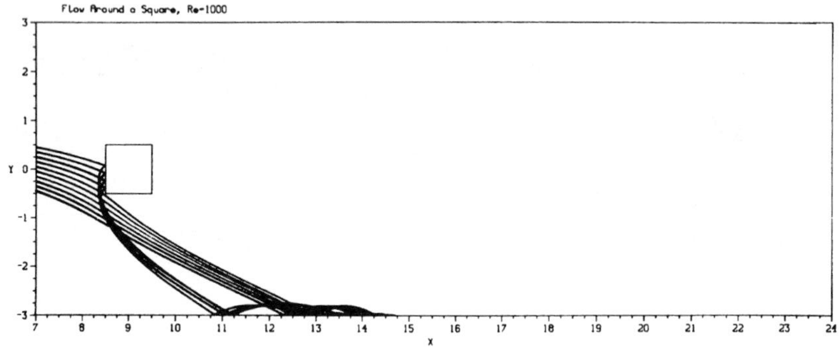

b. g = 0.05

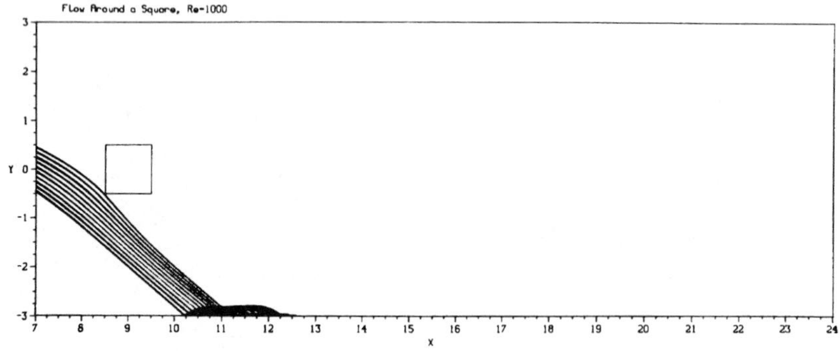

c. g = 0.10

Fig. 9. Physical Particle Trajectories: Re = 1000, $A = 1.0$, $H^{-1} = 1/6$, $St = 10.0$

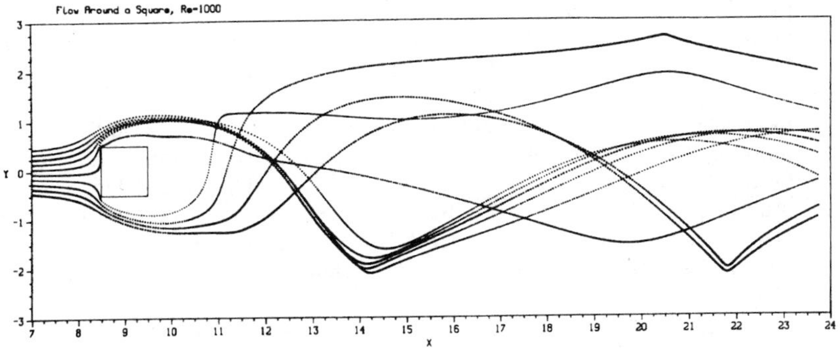

Fig. 10. Physical Particle Trajectories: Re = 1000, A = 1.0, H^{-1} = 1/6, St = 0.10, g = 0.10

4. CONCLUSIONS

A two-dimensional computer code has been developed which can simulate the flow around a bluff body inside a channel. The vortex shedding frequency has been shown to be reasonably reproduced as a function of blockage ratio and Reynolds number. The trajectories of physical particles through this type of flow have been obtained and shown to provide useful information on erosion effects. The importance of not relying on a steady flow computation to compute particle trajectories is clear, as then the all-important vortical structures in the wake region would be replaced by a simple recirculation zone. The great value of quadratic upwind differencing for convection is once again illustrated. It is anticipated that use of this type of differencing scheme will become more frequent during the next several years.

5. REFERENCES

1. LEONARD, B. P., LESCHZINER, M. A. and MCGUIRK, J. - Third-Order Finite Difference Method for Steady Two-Dimensional Convection, Numerical Methods in Laminar and Turbulent Flow, Ed. C. Taylor, K. Morgan and C. A. Brebbia, John Wiley, New York, pp. 807-819, 1978.
2. LEONARD, B. P. - A Stable and Accurate Convective Modelling Procedure Based on Quadratic Upstream Interpolation, Comp. Meth. Appl. Mech. and Eng., Vol. 19, pp. 59-98, 1979.
3. DAVIS, R. W. and MOORE, E. F. - The Numerical Simulation of Flow Around Squares, Proceedings of the Second International Conference on Numerical Methods in Laminar and Turbulent Flow, Ed. C. Taylor and B. A. Schrefler, Pineridge Press, Swansea, pp. 279-290, 1981.

4. DAVIS, R. W. and MOORE, E. F. - A Numerical Study of Vortex Shedding From Rectangles, J. Fluid Mech., Vol. 116, pp. 475-506, 1982.
5. SHELDON, G. L., MAJI, J. and CROWE, C. T. - Erosion of a Tube by Gas-Particle Flow, ASME J. Eng. Matl. and Tech., Vol. 2, pp. 138-142, 1977.
6. SWARZTRAUBER, P. and SWEET, R. - Efficient FORTRAN Subprograms for the Solution of Elliptic Equations, Nat. Center for Atmos. Research Tech. Note IA-109, 1975.
7. BUZBEE, B. L., DORR, F. W., GEORGE, J. A. and GOLUB, G. H.- The Direct Solution of the Discrete Poisson Equation on Irregular Regions, SIAM J. Num. Anal., Vol. 8, pp. 722-736, 1971.
8. SHARMA, M. P. and CROWE, C. T. - A Novel Physico-Computational Model for Quasi One-Dimensional Gas-Particle Flows, J. Fluids Eng., Vol. 100, pp. 343-349, 1978.

MIXED FINITE ELEMENTS APPLIED TO A NON-LINEAR DIFFUSION-CONVECTION EQUATION MODELLING A PROBLEM OF DIPHASIC FLOW IN POROUS MEDIA

Gary COHEN.

I.N.R.I.A., B.P. 105. 78153

LE CHESNAY CEDEX FRANCE

ABSTRACT

We study the numerical resolution of a non-linear convection diffusion equation coming out of a problem of incompressible diphasic flow in porous media rewritten by CHAVENT by using a mixed finite elements method. We discretize the non-monotonous terme of convection by adaptation of the GODUNOV scheme which was studied by LEROUX in the case of finite differences to these finite elements. After definition of different boundary conditions used for this problem, numerical results are given.

O - INTRODUCTION

The modelling of flow in porous media and, particularly, of polyphasic flow is a difficult problem arising from secondary recuperation in an oil reservoir by injection of water, and its numerical resolution by finite difference schemes (frequently used) often does not provide precise enough solutions because of the diffusive character of such schemes.

This paper is devoted to the study of the numerical resolution by a mixed finite element method of the problem of incompressible, immiscible diphasic flow in a superficial porous medium representing (part of) an oil reservoir (case 1) or in a dihedra section around a production well (case 2). In both cases, the problem is reduced to a two-dimensional problem, the thickness of the reservoir or the angle of the dihedron being supposed "small" enough to be neglected.

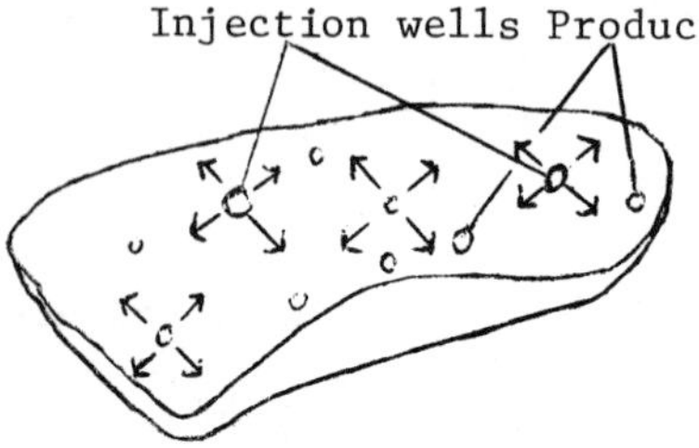

Case 1 : (x,y)

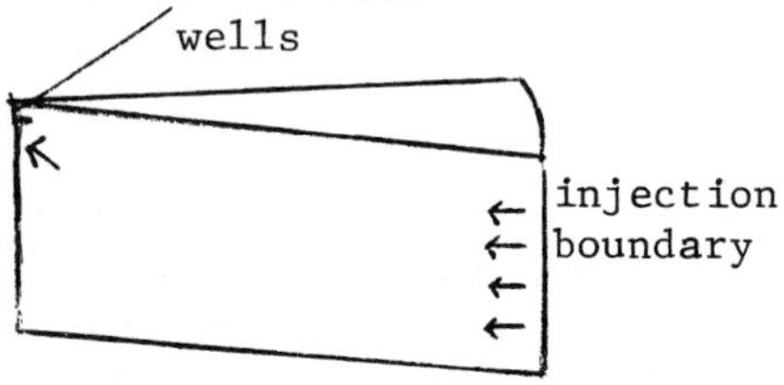

Case 2 : (ρ,z)
(water coning)

This problem leads to a diffusion-convection equation (for the saturation) coupled with an elliptic equation (for the pressure) derived from a new formulation of the nonstationary Darcy's law introduced by CHAVENT. The diffusion and convection terms are both non-linear and non-monotonous.

This work was done in the frame of a cooperation with the Société Nationale ELF-AQUITAINE and the Institut Français des Pétroles (IFP).

1 - THE CONTINUOUS EQUATIONS OF THE PROBLEM

Let Ω be open set of R^2, Γ its boundary, $\vec{\nu}$ the exterior normal on Γ and $]0,T[$ an interval of R. Γ is divided into three parts :

- Γ_i : injection boundary
- Γ_p : production boundary
- Γ_ℓ : impermeable boundary

By using the notations of Appendix 1 (at the end of the paper), the equations of the problem are $(x = (x1,x2) \in \Omega)$:

-<u>Saturation equations</u>

(1) $\lambda \phi \frac{\partial u}{\partial t} + \mathrm{div}(\vec{r} + \vec{w}) = 0$

(2) $w = \sum_{j=0}^{1} b_j(u) \vec{q}_j$

(3) $r = - \lambda [K_i \frac{\partial \alpha(u)}{\partial x_i}] \quad i = 1,2$

b) Pressure equations

(4) $\vec{q}_o = - \lambda[K_i d(u) \frac{\partial P}{\partial x_i}] \quad i = 1,2 \quad + d_1(u)\vec{q}_1$

(5) $\mathrm{div}\ \vec{q}_o = 0$

c) Gravity field

(6) $\vec{q}_1 = - \lambda[K_i \frac{\partial P}{\partial x_i}] \quad i = 1,2$

(7) $\lambda = 1$ in the first case and $\lambda = x1$ in the second case.

Besides this, we define the continuous water and oil velocities as follows :

(8) $\vec{\phi}_1 = \vec{q}_o + \vec{w} + \vec{r} \qquad ; \vec{\phi}_2 = \vec{q}_o - \vec{w} - \vec{r}$

so that $\vec{q}_o$ will represent half of the global velocity of the two fluids.

For the boundary conditions many possibilities occur, and one can find an exhaustive list in COHEN ; but in this paper, we shall describe the most "physical" ones and their actual way of discretization.

- on Γ_ℓ, the condition of impermeability will be expressed by :

(9) $\vec{\phi}_1 . \vec{\nu} = 0$ and $\vec{q}_o . \vec{\nu} = 0$

- Γ_i is generally composed of many connex parts (denoted Γ_{ij} j=i,Np) which represent a part of the reservoir around the injection wells. On each part, we shall impose the global velocity of the fluids and the local velocity of the water, i.e :

(10) $\int_{\Gamma ij} \vec{q}_o . \vec{\nu} = Q_j$ and $\vec{\phi}_1 . \vec{\nu} = f_1(x,t)$

- Γ_p is generally composed like Γ_i. We shall impose the pressure at (the bottom of) the well P_{bj} and no diffusion term.

(11) $P_{bj} = P_j$ and $\vec{r}.\vec{\nu} = 0$

In order to have enough boundary conditions on Γ_i and Γ_p, we must add to (10) and (11) the following Fourier relations between q_o and P on the boundary :

(12) $\vec{q}_o . \vec{\nu} = \alpha_j (P_{bj} - P)$

where α_j is a given scalar coefficient depending on x.

Remarks -1- Since each component of the boundary is a curve, the wells cannot be exactly represented and their dimensions are generally too small to be accurately taken into account in the discretization. So, conditions (10) and (11) where chosen to enable us to connect the general model of the reservoir with a local model around the well. In this case the coefficients α_j could be provided by the local model.

-2- In the case of water coning, we have generally j=1.

-3- The conditions for $\vec{q}_o$ and P could be permuted on Γ_i and Γ_p.

After this description of the continuous problem, we shall now define its discretization by a mixed finite element method.

2 - THE FUNCTIONAL SPACES OF THE METHOD

2.1 Local spaces on the unit square

Let Λ be an open set of $\mathbb{R}^2$. One defines the following family of spaces of polynomials :

$$P_{k,\ell}(\Lambda) = \{\Pi : \Lambda \longrightarrow \mathbb{R} \mid \Pi(\rho,\sigma) = \sum_{i=0}^{k} \sum_{j=0}^{\ell} c_{ij}\, \rho^i \sigma^j,$$

$$c_{ij} \in \mathbb{R}\ ;\ (i,j) \subset N^2\}$$

Let $\hat{Q}$ denote the unit square. Then one can define the following spaces :

(13) $V^k_{\hat{Q}} = P_{k,k}(\hat{Q})$

(14) $H^k_{\hat{Q}} = P_{k+1,k}(\hat{Q}) \times P_{k,k+1}(\hat{Q})$

2.2 Extension to any quadrilateral

The open set Ω will be approximated by a grid of quadrilaterals Ω_h. Let Q be any quadrilateral of the set K_h of quadrilaterals composing Ω. We know that one can determine a unique canonical mapping $\vec{F}_Q$ so that $\vec{F}_Q(\hat{Q}) = Q$.

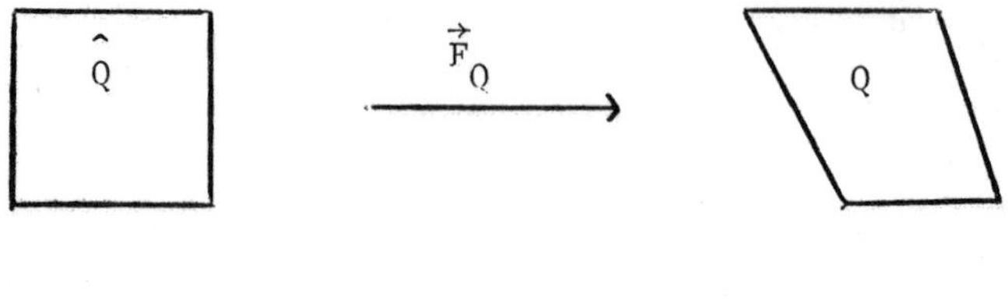

With the aid of this mapping and with (14) one defines the following space :

(15) $H_Q^k = \{\vec{s} : Q \longrightarrow \mathbb{R}^2 \mid \vec{s} = \frac{1}{J_Q} D\vec{F}_Q\ \hat{s} o \vec{F}_Q^{-1}\ ;\ \hat{\vec{s}} \in H_Q^k\}$

where $D\vec{F}_Q$ and J_Q respectively denote the jacobian matrix and its determinant defined as usual. This complicated definition of H_Q^k ensures us the continuity of the flows through the edges of the grid.

2.3 Functional spaces on Ω

In order to define the general discrete spaces, we must first define the following space of continuous functions :

$$H(div,\Omega) = \{\vec{q} : \Omega \longrightarrow R \mid \vec{q} \in [L^2(\Omega)],\ div q \in L^2(\vec{\Omega})\}$$

We can now define the functional spaces of mixed finite elements on the whole grid Ω_h.

(16) $V = \{v_h \in L^2(\Omega) \mid \forall Q \in K_h,\ v_h|_Q o\vec{F}_Q \in V_Q^k\}$

(17) $H = \{s_h \in H(div,\Omega) \mid \forall Q \in K_h, \vec{s}_h|_Q \in H_Q^k\}$

These two spaces are respectively approximations of the spaces $L^2(\Omega)$ and $H(div,\Omega)$ and their elements are polynomials in the quadrilaterals of K_h and are discontinuous at the edges of Ω_h. The discontinuous character of the functions of V_h^k enables us to have a good approximation of the stiff solutions, frequently encountered in convection problems. On the other hand, the continuity of the normal component of functions of H_h^k ensures us the conservativity of the scheme. More details may be found in RAVIART-THOMAS for theorical properties of these spaces, and in COHEN, for their practical construction.

3 - DISCRETE FORMULATION OF THE PROBLEM

3.1 Approximation of the convection term : the Godunov scheme

Since the function b_1 is not monoton, the approximation of the convection term $\vec{w}$ cannot be done by a classical "upwind scheme" The Godunov method, studied by LEROUX in a finite difference scheme, takes into account this non-monotonicity as follows :

Let us give the convection equation :

(18) $\frac{\partial u}{\partial t} + \frac{\partial f(u)}{\partial x} = 0\ x \in]a,b[\ ;\ \ t \in]0,T[$

and a discretization of]a,b[x]0,T[in N intervals of space (lower index) and M intervals of time (upper index). The Godunov scheme applied to this problem gives the following algorithm of resolution :

(19) $u_i^o = u_{oi} \quad \forall i = 1,N+1$

(20) $u_i^{n+1} = u_i^n + \frac{\tau}{h}[f(\xi_{i+1}^n) - f(\xi_i^n)]$

with ξ_i^n solution of

(21) $\mathrm{sgn}(u_{i+1}^n - u_i^n) f(\xi_i^n) = \min \mathrm{sgn}(u_{i+1}^n - u_i^n) f(\mu) \ ; \ \mu\in \mathcal{J}(u_i^n, u_{i+1}^n)$

where $\tau = \frac{T}{M}$; $h = \frac{b-a}{N}$; $\mathcal{J}(\lambda_1, \mu_1) = [\min(\lambda_1,\mu_1), \max(\lambda_1,\mu_1)]$

In the (generic) case when (21) has a unique solution, ξ_i^n may be interpreted as the value at the point i at the instant u+0 of the exact solution of the Riemann problem the initial value of which at the instant n is the discontinuity at the point i.

Let us now apply this scheme to our discrete problem.

We discretize first the interval]0,T[into M equal intervals and τ will denote the time-step as above. For any function σ, σ_h^n will denote its approximed value at the instant t_n.

Then, multiplying the convection term by $v\in V$ and applying the Green's formula, we obtain :

(22) $$\int_Q \mathrm{div}\, \vec{w}(u_h^n)\, v\, dx = -\int_Q \vec{w}(u_h^n).\overrightarrow{\mathrm{grad} v}\, dx + \int_{\partial Q} \vec{w}(u_h^n).\vec{\nu}_Q v d\Gamma$$

The discontinuity of the discrete solution u_h at the edges of the grid appears by the boundary term of (22). We shall apply the above method to this term as following :

Let A be an edge of the boundary ∂Q of Q and $Q'\in \mathcal{K}_h$ the quadrilateral so that $A = Q\cap Q'$. We choose a normal $\vec{\nu}_A$, for instance, going from Q to Q' and we denote by u^- and u^+ the upwind and downwind values of u_h^n respectively.

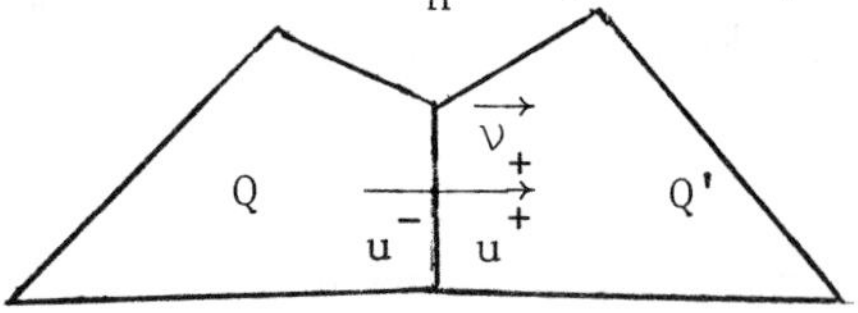

We can now define the Godunov value ξ_A^n of u_h^n along the edge A at the instant n :

(23) $$\begin{cases} \xi_A^n G \mathcal{J}(u^-,u^+) \\ \mathrm{sgn}(u^+ - u^-)\vec{w}(\xi_A^n).\vec{\nu}_A = \underset{\mu\in\mathcal{J}(u^+,u^-)}{\mathrm{Min}} \mathrm{sgn}(u^+ - u^-)\vec{w}(\mu).\vec{\nu}_A \end{cases}$$

One can easily check that ξ_A^n does not depend on the choice of $\vec{\nu}_A$.

Finally (22) becomes :

$$(24) \quad \int_Q \mathrm{div}\vec{w}(u_h^n) v \, dx = - \int_Q \vec{w}(u_h^n).\overrightarrow{\mathrm{grad}} v \, dx + \sum_{A \in \partial Q} \int_A \vec{w}(\xi_A^n).\vec{\nu}_Q v \, d\Gamma$$

In order to ensure the conservativity of the scheme, we shall approximate the water and oil velocities by :

$$(25) \quad \overrightarrow{\phi}_{1h}^{n} = \overrightarrow{q}_{oh}^{n} + \overrightarrow{r}_{h}^{n} + \overrightarrow{W}_n(\xi_A^n)$$

$$(26) \quad \overrightarrow{\phi}_{2h}^{n} = \overrightarrow{q}_{oh}^{n} - \overrightarrow{r}_{h}^{n} - \overrightarrow{W}_n(\xi_A^n)$$

3.2 <u>The approximed problem</u>

Using the notations of paragraph III-1, the values of u_h^n and P_h^n will be sought in spaces V_h^k and $V_h^{k'}$ wich are approximations of $L^2(\Omega)$. Therefore we cannot define traces of these functions on the boundary because of their lack of regularity. So the boundary values of the solutions will be replaced by "exterior" values without any condition of continuity with the approximed solution. These values will be denoted u_b and P_b.

With these notations, the approximed problem is :

Find $u_h \in [V_h^k]^{M+1}$, $\vec{r}_n \in [H_h^{k'}]^{M+1}$, $P_n \in [V_h^{k'}]^{M+1}$, $\vec{q}_{oh} \in [H_h^{k'}]$, $\vec{q}_{1h} \in H_h^{k'}$, $u_b \in [L^2(\Gamma_n)]^{M+1}$, $P_b \in [L^2(\Gamma_h)]^{M+1}$ so that :

$$(27) \quad \int_Q \lambda\phi \frac{u_h^{n+1} - u_h^n}{\tau} v \, dx + \int_Q \mathrm{div}\, \vec{\nu}_h^n \, v \, dx - \int_Q \vec{w}_h^n(u_h^n) \, v \, dx + \int_Q \vec{w}_h^n(\xi^n).\vec{\nu} \, v \, d\Gamma = 0$$

$\forall v \in V_h^k$; $\forall Q \in K_h$

$$(28) \quad \vec{W}_h^n(\mu) = b_o(\mu)\, \vec{q}_{oh}^{\,n} + b_1(\mu)\, \vec{q}_{1h}$$

ξ^n defined as in (23)

$$(29) \quad \int_{\Omega_h} \frac{\vec{r}_h^n.\vec{s}}{\lambda\psi} dx = \int_{\Omega_h} \alpha(u_h^n) \, \mathrm{div}\vec{s} \, dx - \int_{\Gamma_h} \alpha(u_h^n)\vec{s}.\vec{\nu} \, d\Gamma \quad \forall s \in H_h^{k'}$$

$$(30) \quad \int_{\Omega_h} \frac{\overrightarrow{q_{oh}}.\vec{s}}{\lambda\psi d(u_h^n)} dx = \int_{\Omega_h} P_h^n \, \mathrm{div}\vec{s} \, dx - \int_{\Gamma_h} P_b^n \, \vec{s}.\vec{\nu} \, d\Gamma + \int_{\Omega_h} \frac{d_1(u_h^n)}{\psi d(u_h^n)} \overrightarrow{q_{1h}}.\vec{s} \, dx \quad \forall \vec{s} \in H_h^{k'}$$

(31) $\int_{\Omega_h} \operatorname{div} \overrightarrow{q_{oh}}\, v\, dx = 0 \quad \forall v \in V_h^k$

(32) $\int_{\Omega_h} \frac{\overrightarrow{q_{1h}} \cdot \vec{s}}{\lambda\psi}\, dx = \int_{\Omega_h} \pi_1(x) \operatorname{div} \vec{s}\, dx - \int_{\Gamma_h} \pi_1(x) \vec{s} \cdot \vec{\nu}\, d\Gamma \quad \forall s \in H_h^{k'}$

One can find in JAFFRE that the best value of k' is k-1.

3.3 The discrete boundary conditions

Generally, the discrete boundary conditions in pressure are immediately derived from the continuous conditions defined in I. It is not the same for the conditions in saturation that are expressed by means of the oil and water velocities in a non-linear way. Therefore, their discretization will depend on the presence of the term of diffusion. So we shall study these conditions in two cases.

a) Case without diffusion.

Let us first suppose $\overrightarrow{r_n} \equiv \vec{0}$ in the saturation equation.

In this case, the exterior value u_b^n of u_h^n appears only through the Godunov values along the edges of the boundary. Let A be an edge of the boundary and the water velocity on this edge. The relation (25) enables us to have $\vec{w}_h^n \cdot \vec{\nu}$ without computing u_b^n explicitely :

(33) $\vec{w}_h^n \cdot \vec{\nu} = f_1^n(x) - \vec{q}_{oh}^n \cdot \vec{\nu}$

In particular, (9) is turned into the condition :

$$\vec{w}_h^n \cdot \vec{\nu} = 0$$

b) Case with diffusion

When $\overrightarrow{r_h} \not\equiv 0$, one must have the explicit value of u_b^n which appears in (29). Now, on each edge, this value will be coupled with the value of r by means of (25) and (29). This coupling makes non-linear the saturation equations and the actual computation of the solution is too difficult to use these equations just as they are. Therefore, one must find out a means to take off the coupling of these equations without making too much error on the solution.

On Γ_i, we can easily suppose that $\overrightarrow{r_h} \cdot \vec{\nu} = 0$, since the flow $\overrightarrow{q_{oh}}$ is big enough to neglect the effect of the diffusion term. Then, we come back to the situation of a).

On the other hand, we cannot suppose the same thing on Γ_p, since the value of $\overrightarrow{q_{oh}}$ is, on the contrary, generally very small. Therefore we shall uncouple the two equations by the following algorithm :

- One chooses an initial value of u_b^o.
- One deduces $\vec{r}_h^n$ from $\vec{\phi}_{1h}^n . \vec{\nu} = 0$, i.e. $\vec{r}_h^n . \vec{\nu} = - \overrightarrow{q_{1h}} b(\xi^n) . \vec{\nu}$
- One computes u_b^n with (29).

This algorithm holds as long as the gravity term is not prevailing (that is generally true in practice).

3.4 Actual computation

a) The used spaces.

The principal functional spaces used are the following :

- V_h^o : constants on the unit square. ($\dim V_h^o = 4$)
- V_h^1 : functions of first degree in each variable. ($\dim V_h^1 = 4$)
- H_h^o : vectorial functions with linear components. ($\dim H_h^o = 4$)

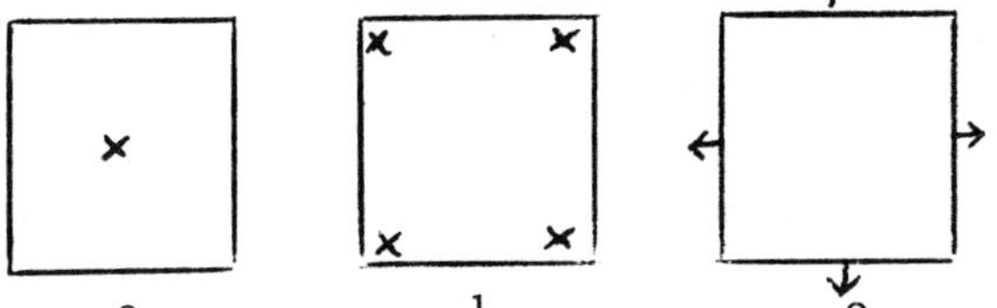

V_h^o $\quad$ V_h^1 $\quad$ H_h^o

Différent degrees of freedom

We sought u_h^n in V_h^1, p_k^n in V_h^o and $\vec{q}_{oh}^n, \vec{r}_h^n, \overrightarrow{q_{1h}}$ in H_h^o.

b) Computation of the pressure.

In order to simplify the matrices occuring in the discrete pressure equations, we have used an augmented lagrangien method the lagrangian of which is the following :

$$(34) \qquad (\vec{q},P) = \frac{1}{2} \int_{\Omega_h} \frac{|\vec{q}|^2}{\lambda \psi d(u_h^n)} dx - \int_{\Omega_h} P \operatorname{div}\vec{q}\, dx -$$

$$- \int_{\Omega_h} \frac{d_1(u_h^n)}{\psi d(u_h^n)} \overrightarrow{q_{1h}} . \vec{q}\, dx + \frac{1}{2\varepsilon_1} \int_{\Omega_h} (\operatorname{div}\vec{q})^2 dx +$$

$$+ \frac{1}{2\varepsilon_2} \sum_{k=1}^{np} \theta_k \lambda_k \{ \int_{\Gamma_k} \frac{\vec{q}.\vec{\nu}}{\lambda} - \frac{1}{2} Q d_i \}^2 + \sum_{k=1}^{np} \lambda_k \sum_{j=1}^{mp} P_{kj} \int_{\Gamma_{kj}} \vec{q}.\vec{\nu} d\Gamma +$$

$$+ \frac{1}{2\varepsilon_3} \sum_{k=1}^{np} \theta_k \sum_{j=1}^{mk} [\int_{\Gamma_{kj}} \frac{\vec{q}.\vec{\nu} d\Gamma}{\lambda} - \tilde{\alpha}_{ij} (P_{fi} - P_{ij})]^2 +$$

$$+ \sum_{k=1}^{np} (1 - \lambda_k) \sum_{j=1}^{mp} (P_{fi} - \frac{1}{\alpha_{ij}} \int \frac{\vec{q}.\vec{\nu})}{\lambda}$$

The integration of the boundary conditions to the lagran- enables us to obtain all the possible classical boundary conditions on flow and pressure only by modifying the coefficients θ_k and λ_k.

4 - NUMERICAL RESULTS

4.1 Superficial experiments (X,Y)

Two kinds of experiments were made in the superficial case :

a-a quarter of five spot models.

b-a complete reservoir simulation (with 3 injection wells and 2 production wells).

The physical data are given below :

Fiel size : a-210x210m ; b-600x600m, Densities $\rho_1=1,\rho_2=0,75$

Thickness : a-5m ; b-$15\leq\sigma(x)\leq 20$m

Permeability : 100mD

Porosity : 0,3

Viscosities : $\mu_2=1,43$cP
$\mu_1=$a-0.33cP, b-1cP.

Discretization : a-64 elements ;
b-95 elements.

Mobility ratio : a-3; b-1

Dipping : a-30% towards the east, : b-10% towards the west.

Injection rate : a-5m^3/day b-20m^3/day /well

Number of unknows for saturations : a-256 ; b-380.
Numer of unknows for the flows : a-144 ; b-220.

The saturation contours go from 0.1 to 0.9 with a step of 0.1. Now, let us give the physical and auxiliary functions of the problem.

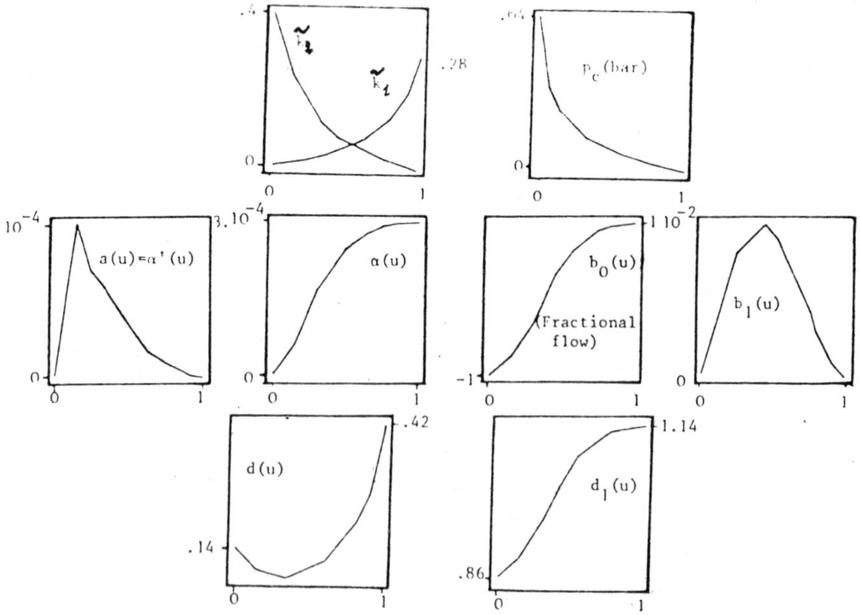

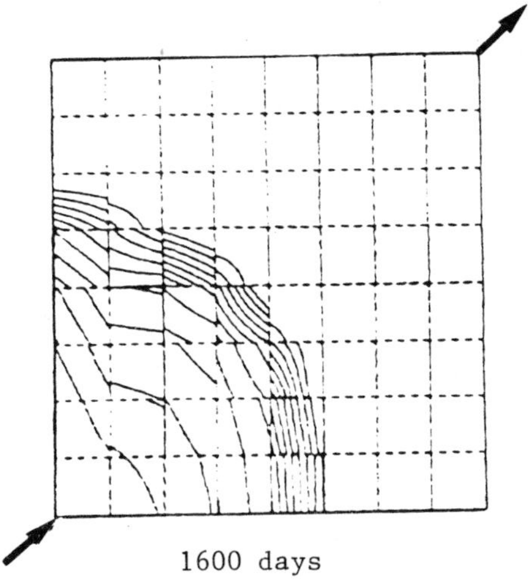

1600 days

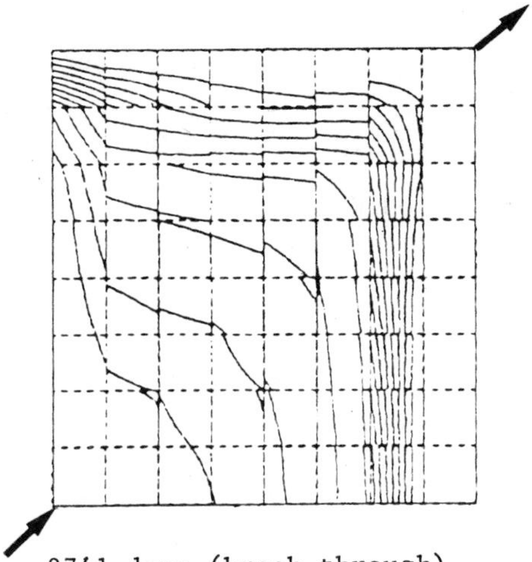

3741 days (breah-through)

Quarter of five-spot.

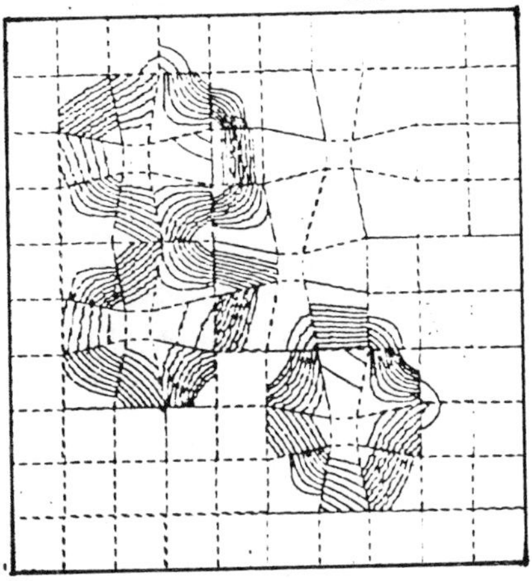

3000 days

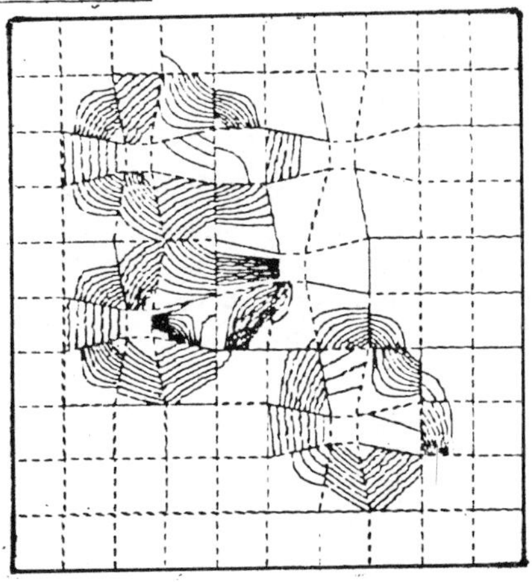

3000 days

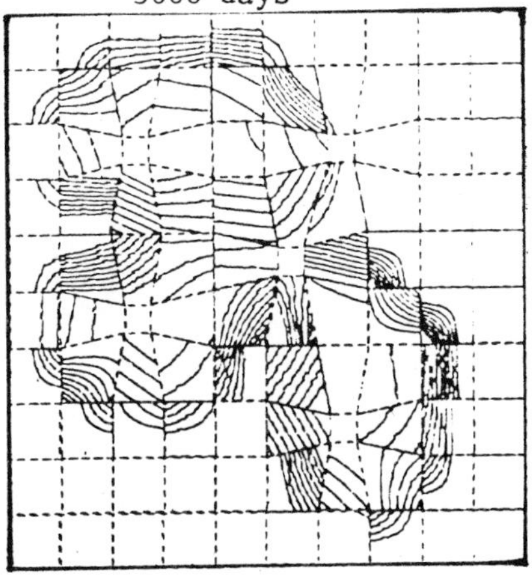

6000 days
without gravity

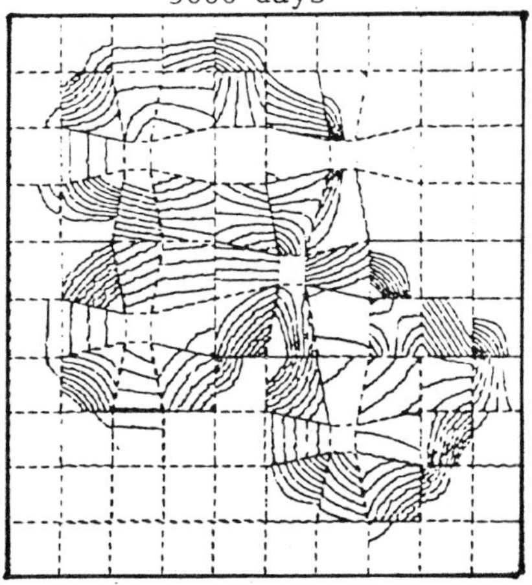

6000 days
with gravity

Complete reservoir simulation

4.2 Water coning experiments (ϕ,z)

We give now the contour saturations and flow lines for a coning experiment with the same data except :

Size 300x60m Permeability : K_ϕ=250mD, K_z=500mD

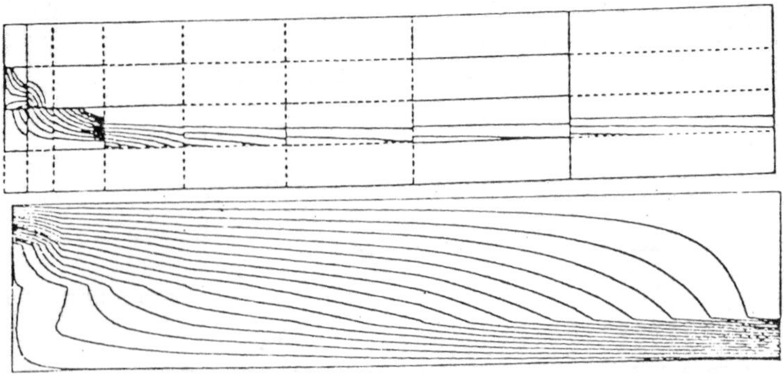

Saturations contours and flow lines for a water coning experiment at 175 days (injection rate : 50m^3/day in the aquifer).

The time of computation given on IBM 370 is for 6000 days of the complete reservoir simulation, 10 minutes with gravity and 2 minutes without gravity. This difference of time comes from the constraint on the time step arising from the gravity field which creates physical (digitation) the numerical instabilities.

APPENDIX 1

SYMBOLS AND NOTATIONS

1 — index for water

2 — index for oil

$\sigma(x)$ — thickness of the reservoir (case 1) or angle of the dihedron (case 2)

$Q(x)$ — porosity

$K(x)$ — absolute permeability

$\phi = Q \times \sigma \times (S_M - S_m)$ (cf $S(x,t)$)

$\psi = K \times \sigma$

g — gravitational acceleration

$S(x,t)$ — water saturation ($S_m \leq S \leq S_M$)

$u(x,t) = \dfrac{S(x,t)-S_m}{S_M - S_m}$ reduced water saturation

P — global pressure

$\tilde{k}_1(u), \tilde{k}_2(u)$ relative permeabilities

ρ_1, ρ_2 — densities

μ_1, μ_2 — viscosities

$P_c(u)$ — capillary pressure

$k_i(u) = \tilde{k}_i(u)/\mu_i = 1, 2$ mobilities

$$\frac{d\alpha}{du} = \frac{k_1 \; k_2}{k_1+k_2} \; \frac{dp_c}{du} \; (\alpha(0) = 0)$$

$$b_0 = \frac{k_1-k_2}{k_1+k_2}$$

$$b_1 = \frac{k_1 \; k_2}{k_1+k_2}$$

$$d = \frac{1}{2} (k_1+k_2)$$

$$d_1 = \frac{k_1\rho_1+k_2\rho_2}{k_1 + k_2} \; \frac{2}{\rho_1+\rho_2}$$

$$\Pi_1(x) = (\rho_1-\rho_2)gZ(x)$$

REFERENCES

G.CHAVENT.
"A new formulation of diphasic incompressible flows in porous media"- Lectures notes in Mathematics - Vol.503 - Springer 1976.

G.COHEN.
"Eléments finis mixtes appliqués à un problème d'écoulement diphasique incompressible bidimensionnel en milieu poreux en présence de gravité"-
Rapport I.N.R.I.A. n° 138., 1982.

J.JAFFRE.
"Décentrage et éléments finis mixtes pour les équations de diffusion-convection" - To appear in Calcolo.

A.Y.LEROUX.
"A numerical conception of entropy for quasi-linear equations" - Mathematics of computation - Vol. 31 - n° 140 - Pages 848-872 - Octobre 1977.

P.A.RAVIART., J.M.THOMAS
"A mixed finite elements method for order elliptic problems" - Lectures notes in Mathematics - Vol. 606 Springer, 1977.

A BOUNDARY INTEGRAL APPROACH TO A MOVING BOUNDARY PROBLEM IN POROUS MEDIA

Christine D Gnanathurai[(i)] and David C Wilson[(ii)]

SUMMARY

Laminar flow conditions prevail in a petroleum reservoir when oil is produced from a well. Close to the well, the pressure gradients that are set up can raise the oil water contact from its initial horizontal level to form a water cone. Coning phenomena are moving boundary problems and have been tackled as such in this study.

A boundary integral approach has been adopted for the model development. At each time level the model reduces to the solution of two integral equations on the oil water interface. The model therefore, has the outstanding advantage that the only variables which need to be calculated and stored are those at the oil water contact, which is precisely where the flow information is needed.

The accuracy of the solution technique is demonstrated and numerical solutions obtained from the model are compared where possible with existing analytical solutions.

(i) Research student, Petroleum Engineering Section, Imperial College, London.

(ii) Reader in Petroleum Engineering, Imperial College, London.

1. THE CONING PROBLEM

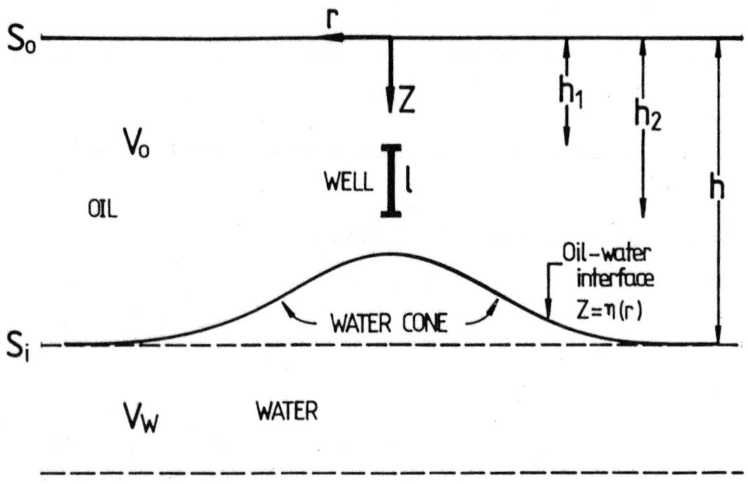

Figure 1 : Water Coning

Once production from an oil reservoir commences, the pressure gradients that are set up around the well raise the oil water contact from its initial horizontal level to form a cone. For small well flow-rates, this interface reaches a new, stable, equilibrium position. However, for flow rates exceeding a certain critical rate, the cone becomes unstable and water breaks through into the well.

A diagramatic representation of the reservoir model, we shall use, is shown in Figure 1. The reservoir is assumed to be cylindrical in shape, and cylinderical co-ordinates (r,z) are used in the model. S_0 is an upper impermeable surface and S_i represents the oil water interface. The region V_0 represents the oil domain and V_w the water domain. The well is placed at the centre ($h_1 \leq z \leq h_2$) and is of length ℓ.

In this paper, a model will be developed to trace the evolution of the water cone until its critical height or steady state equilibrium position is reached. The critical height represents an altitude between the original interface and the bottom of the well, beyond which the water cone becomes unstable and water breakthrough occurs almost instantaneously [5, 6].

2. THE MATHEMATICAL MODEL

Saturated flow will be assumed in both V_0 and V_w, i.e. it is assumed that only one phase is the flowing phase - oil in V_0, and water in V_w. We shall also assume that the system is incompressible.

Then, velocity potentials Φ_i (i = 0,w for the oil and water phase respectively) can be defined in accordance with Darcy's law.

$$\Phi_i = \frac{\kappa_i}{\mu_i}(p_i - \rho_i gz) \tag{2.1}$$

Governing equations:

By conservation of mass,

$$\nabla^2\Phi_0 = 0 \quad \forall\ \underline{r} \in V_0, \text{ except at the well} \tag{2.2}$$

$$\nabla^2\Phi_w = 0 \tag{2.3}$$

Interface conditions:

Let the interface S_i be represented by $z = \eta(r,t)$. On the interface the following conditions hold.

- continuity of pressure, neglecting capillary pressure

$$P_0 = P_w \quad \text{on } z = \eta(r,t).$$

Therefore, using equation (2.1)

$$\eta(r,t) = a_0\Phi_0 + a_w\Phi_w \tag{2.4}$$

where $a_0 = \dfrac{\mu_0}{\kappa_0(\rho_w-\rho_0)g}$, $a_w = \dfrac{-\mu_w}{\kappa_w(\rho_w-\rho_0)g}$

- continuity of velocity normal to the interface

$$\frac{\partial\Phi_0}{\partial n} = \frac{\partial\Phi_w}{\partial n} \tag{2.5}$$

- material surface condition, i.e. any fluids at the interface move in the direction and at the velocity of the interface.

$$\frac{DF}{Dt} = 0, \quad F \equiv z-\eta(r,t) = 0$$

where $\dfrac{D}{Dt}$ in the operator $(\dfrac{\partial}{\partial t} - \nabla\Phi_i\cdot\nabla)$.

Noting that, $\frac{\partial \Phi_i}{\partial n} = \nabla\Phi_i \cdot \underline{n} = \nabla\Phi_i \cdot \frac{\nabla(z-\eta)}{|\nabla(z-\eta)|}$

$$\frac{\partial \eta}{\partial t} = -\frac{\partial \Phi_i}{\partial n}\left(1 + \left(\frac{\partial \eta}{\partial r}\right)^2\right)^{\frac{1}{2}} .$$

Writing $\frac{\partial \eta}{\partial r} = \tan \beta$, where β is the angle the interface makes with the horizontal,

$$\frac{\partial \eta}{\partial t} = -\frac{\partial \Phi_i}{\partial n} \frac{1}{(\cos\beta)} \tag{2.6}$$

Well boundary condition:

Approximating the well by a line sink,

$$\lim_{r\to 0} r \frac{\partial \Phi_0}{\partial r} = \frac{q}{2\pi \ell} \qquad (h_1 \leq z \leq h_2) \tag{2.7}$$

Boundary condition on S_0:

$$\frac{\partial \Phi_0}{\partial n} = 0 \qquad \text{on } z = 0 \tag{2.8}$$

Far-field condition:

At large distances away from the well, the oil water contact is undisturbed.

Therefore,

$$\lim_{r\to\infty} \eta(r,t) = h \tag{2.9}$$

Initial conditions:

$$\left.\begin{array}{l} \eta(r,0) = h \\ \Phi_0, \Phi_w \text{ specified} \end{array}\right\} \text{, for all } r \tag{2.10}$$

3. THE PERTURBATION PROBLEM

As in our earlier work on the steady state problem [2], the concept of a basic-flow is introduced to remove the well singularity.

Let the basic flow be taken to be the flow towards the line sink $(h_1 \leq z \leq h_2)$ in the presence of the no-flow boundary S_0, but with no water cone present. It will also be made to satisfy the flow behaviour at infinity.

Then, using the classical theory of sources and sinks and the method of images, the basic flow potential, Φ_1 is calculated to be

$$\Phi_1 = \frac{q}{4\pi\ell}\left(\ell n\left|[(z-h_2)^2+r^2]^{\frac{1}{2}}+(z-h_2)\right|\left|[(z+h_1)^2+r^2]^{\frac{1}{2}}+(z+h_1)\right| - \ell n\left|[(z-h_1)^2+r^2]^{\frac{1}{2}}+(z-h_1)\right|\left|[(z+h_2)^2+r^2]^{\frac{1}{2}}+(z+h_2)\right|\right) + C \quad (3.1)$$

where $C = \lim_{r\to\infty} \Phi_1(r,z)$

The total flow for the coning problem can now be considered as the basic flow, together with the flow disturbance created by superimposing the water cone on the basic flow.

$$\Phi_0 = \Phi_1 + \tilde{\Phi}_0$$

$$\Phi_w = \Phi_1 + \tilde{\Phi}_w$$

$\tilde{\Phi}_0$ and $\tilde{\Phi}_w$ are perturbation potentials that are harmonic and non-singular throughout the flow domain.

The governing equations, boundary conditions and initial conditions (2.2) to (2.10) now become,

$$\nabla^2\tilde{\Phi}_0 = 0 \quad (3.2)$$

$$\nabla^2\tilde{\Phi}_w = 0 \quad (3.3)$$

$$\eta(r,t) = (a_0+a_w)\Phi_1+a_0\tilde{\Phi}_0+a_w\tilde{\Phi}_w \quad (3.4)$$

$$\frac{\partial\tilde{\Phi}_0}{\partial n} = \frac{\partial\tilde{\Phi}_w}{\partial n} \quad , \text{ on } z = \eta(r,t) \quad (3.5)$$

$$\frac{\partial\eta}{\partial t} = -\frac{\partial}{\partial n}(\Phi_1+\tilde{\Phi}_i)\cos\beta^{-1}, \quad \text{on } z = \eta(r,t) \quad (3.6)$$

$$\frac{\partial\tilde{\Phi}_0}{\partial n} = 0, \text{ on } z = 0 \quad (3.7)$$

$$\eta(r,0) = h \quad . \quad (3.8)$$

Also $\tilde{\phi}_0$, $\tilde{\phi}_w \to 0$ as $r \to \infty$.

4. THE BOUNDARY INTEGRAL EQUATIONS

Green's 2nd identity is used to obtain the boundary equations. If two functions U and V are continuous and have continuous 1st and 2nd partial derivatives in the closure of a domain D, bounded by a surface S, then,

$$\int_D (u\nabla^2 v - v\nabla^2 u)\, dV = \int_S (U \frac{\partial V}{\partial n} - V \frac{\partial U}{\partial n})\, dS. \tag{4.1}$$

It is apparent from equation (4.1) that if two functions U and V are chosen such that they are harmonic in the domain D, then then the volume integral on the Left Hand side of (4.1) vanishes altogether. This property will be exploited to reduce the dimensionality of the problem by one.

For the domain V_0, U is chosen to be $\tilde{\phi}_0$ and V to be the Green's function g.

$$g = \left| (r\cos\theta - r_i \cos\theta_i)^2 + (r\sin\theta - r_i \sin\theta_i)^2 + (z-z_i)^2 \right|^{-\frac{1}{2}}$$

$$+ \left| (r\cos\theta - r_i \cos\theta_i)^2 + (r\sin\theta - r_i \sin\theta_i)^2 + (z+z_i)^2 \right|^{-\frac{1}{2}}$$

(g is a solution to Laplace's equation and it also satisfies the zero flux condition at z = 0).

Similarly for V_w, U is chosen to be $\tilde{\phi}_w$ and V to be g. Then, after introduction of an axisymmetric Green's function G, [3].

Where, $G = \int_0^{2\pi} g d\theta$, the surface integrals of equation (4.1) reduce to the contour integral equations

$$- \alpha_{0i}\tilde{\phi}_{0i} = \int_{\Gamma i} r(\tilde{\phi}_0 \frac{\partial G}{\partial n} - G \frac{\partial \tilde{\phi}_0}{\partial n})\, ds \tag{4.2}$$

$$\alpha_{wi}\tilde{\phi}_{wi} = \int_{\Gamma i} r(\tilde{\phi}_w \frac{\partial G}{\partial n} - G \frac{\partial \tilde{\phi}_0}{\partial n})\, ds. \tag{4.3}$$

α_{0i} and α_{wi} are constants that are calculated from the solid angles of the interface geometry, Γi is the generating contour of the surface S_i and $\underline{n}$ is the unit normal in Γi directed away from the volume V_o.

5. THE TIME-STEPPING EQUATIONS

The time stepping equations are derived from the material

surface conditions (3.6). N nodal points P_j (j=1,...N) are placed on the interface, and on each nodal point P_j, equation (3.6) is replaced by its finite difference analogue,

$$\eta_j^{(k+1)} = \eta_j^{(k)} - \frac{\Delta t}{\cos\beta^{(K)}}\left\{\left(\frac{\partial\Phi_1}{\partial n} + \frac{\partial\tilde{\phi}_0}{\partial n}^{(k+1)}\right)_j \theta + \left(\frac{\partial\Phi_1}{\partial n}+\frac{\partial\tilde{\phi}_0}{\partial n}\right)^k (1-\theta)\right\}$$

for all j = 1, ... N.

Using equation (3.4) to eliminate η and after some simplification, the time-stepping equations become,

$$a_0\tilde{\phi}_{0_j}^{(k+1)} + a_w\tilde{\phi}_{w_j}^{(k+1)} + c_j^{(k)} \frac{\partial\tilde{\phi}_0}{\partial n}_j^{(k+1)} = d_j^{(k)} \qquad (5.1)$$

$$j = 1,...N$$
$$k = 0,1,2,...$$

where $c_j^{(k)} = \dfrac{\Delta t\theta}{(\cos\beta_j)^{(k)}}$

$$d_j^{(k)} = -\frac{\Delta t}{(\cos\beta_j)^{(k)}}\left[\frac{\partial\Phi_1}{\partial n}^{(k)} + (1-\theta)\left(\frac{\partial\Phi_0}{\partial n}\right)^{(k)}\right]_j$$

$$+ a_0\tilde{\phi}_{0_j}^{(k)} + a_w\tilde{\phi}_{w_j}^{(k)} .$$

(θ is chosen to be 0.5 for all computations).

6. NUMERICAL SOLUTION

The following steps are followed:

(a) A finite number of points P_j (j=1,...N) are selected to represent the initial interface.

(b) The independent variables $\tilde{\phi}_0, \tilde{\phi}_w$ $\frac{\partial\tilde{\phi}_0}{\partial n}$ are represented on the interface by linear interpolation formula. (higher order formulae may be used). For example, over the element Γ_j between a pair of nodal points P_j and P_{j+1}, $\tilde{\phi}$ may be approximated as

$$\tilde{\phi} = \left(\frac{\xi_{j+1}-\xi}{\xi_{j+1}-\xi_j} \quad \frac{\xi-\xi_j}{\xi_{j+1}-\xi_j}\right)\begin{pmatrix}\tilde{\phi}_j \\ \tilde{\phi}_{j+1}\end{pmatrix}$$

(c) The points P_j (j= 1,...N) are used successively as origins for the Green's function and 2N integral equations are generated. After substituting the interpolation formula in the integral equations 2N algebraic equations are derived.

$$\left.\begin{aligned} [H^0]\{\phi_0\} + [G]\left\{\frac{\partial\tilde{\phi}_0}{\partial n}\right\} &= 0 \\ [H^w]\{\phi\ \} + [G]\left\{\frac{\partial\tilde{\phi}_w}{\partial n}\right\} &= 0 \end{aligned}\right\} \qquad (6.1)$$

The co-efficients of the matrices $[H^0]$, $[H^w]$ and $[G]$ are geometry dependent and may be evaluated easily by using a Gaussian quadrature formula.

(d) Conditions at time level $t^{(k)}$, $k = 0, 1,2,...$ ($k=0$ refers to initial conditions) are used to evaluate the co-efficients $c_j^{(k)}$, $d_j^{(k)}$ of equations (5.1). Assembling the time-stepping equations together with equations (6.1), we obtain a matrix equation of the form

$$[A]\{x\}^{(k+1)} = \{b\}\ , \qquad (6.2)$$

where $\{x\}^T = \left\{\phi_{01} \cdots \phi_{0N}, \phi_{w_1} \cdots, \phi_{w_N}, \frac{\partial\phi_0}{\partial n}, \ldots \frac{\partial\phi_{0N}}{\partial n}\right\}^{(k+1)}$.

(e) The matrix equation (6.2) is a complete system of equations for the unknown vector $\{x\}$ at time level $t^{(k+1)}$. The matrix A is invariably a full matrix and therefore a Gaussian elimination algorithm is suitable for the solution of equation (6.2).

(f) The pressure continuity condition (3.4) can now be used to update the interface position, by shifting each nodal point forward in time.

(g) If steady state (i.e. if $|\eta^{(k+1)} - \eta^{(k)}| < \varepsilon_d$, where ε_d is small) or critical height is reached the algorithm is terminated. If not, the procedure is repeated starting at (b). The critical height is evaluated by using the steady state model [2] to determine the height of the water cone for a flow-rate that is just below the critical

7. NUMERICAL EXAMPLES AND ACCURACY

The numerical model was tested on a model reservoir whose characteristics are given below. The steady state model [2] was used to compute steady state interface profiles, the critical flow-rate and critical height.

Reservoir characteristics:

oil-formation thickness	100ft
well completion interval (h_1, h_2)	40 - 50ft
oil water contact	100ft below datum

ρ_w		1.05g/cc
ρ_0		0.65g/cc
K_w		80 mD
K_0		100 mD
μ_w		0.4 cp
μ_0		0.25 cp
critical flow rate	for this	850 cc/sec
critical cone height	reservoir	28ft.

For convenience all computations were performed in dimensionless variables. Graphical displays (Figure 2-7) are also in dimensionless variables. To evaluate actual reservoir distances (in feet) multiply dimensionless distances by 100, and to evaluate time in seconds multiply dimensionless time by 1.195×10^7.

7.1 Interface profiles and comparison with the linearized solution

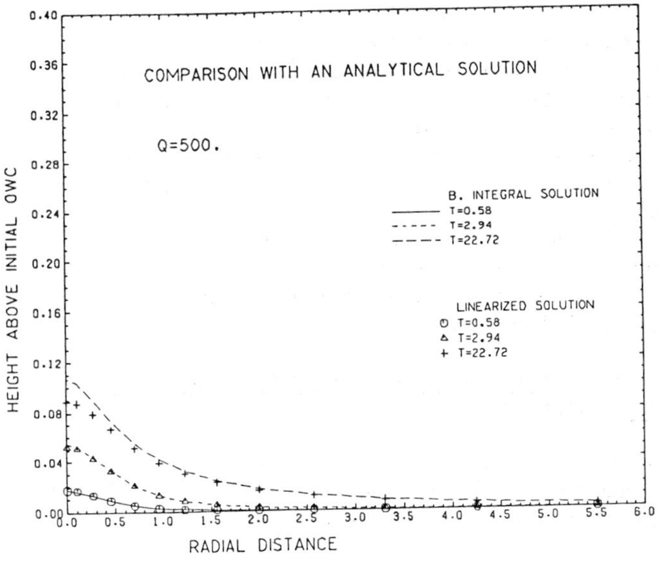

Figure 2: Interface profiles - sub-critical case

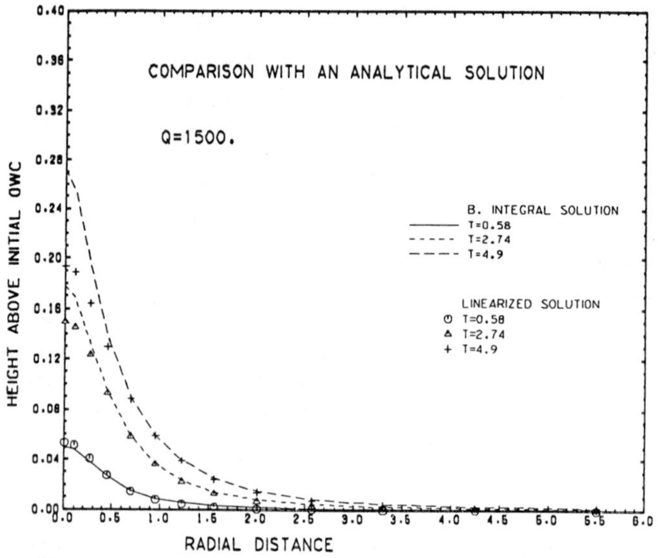

Figure 3: Interface profiles - super-critical case

Figure 2 traces cone evolution up to the steady state position for a sub-critical flow-rate, whilst Figure 3 traces cone evolution up to breakthrough position for a super-critical rate. The broken line profiles are obtained by the boundary integral model. The symbols corresponds to the solution obtained by the linearized model due to Camy [1]. The linearized solution is for the particular case where there is no upper impermeable surface S_0 present. The comparison between the two solutions is excellent for the sub-critical case. As to be expected, the linearized solution breaks down when the cone becomes large and non-linearities become important. Therefore, there is substantial disagreement at larger times for the super-critical case.

7.2 Influence of the number of boundary points on the numerical solution

The influence of boundary points on the numerical model was investigated by deriving solutions using N, 2N, 4N, 8N boundary points for a range of flow rates at different time levels

In Figure 4 we plot - log (error) against the number of boundary points for a typical flow-rate at three time levels. From the graph we deduce that the relative error e_N, over the whole interface behaves approximately as Aexp(-BN), where A and B are constants. The relative error, e_N, is defined as

$$e_M = \left\{ \sum_{i=1}^{M} \left(\frac{z_{8N}(r_i) - z_M(r_i)}{z_{8N}(r_i)} \right)^2 \right\}^{\frac{1}{2}} \frac{1}{M}$$

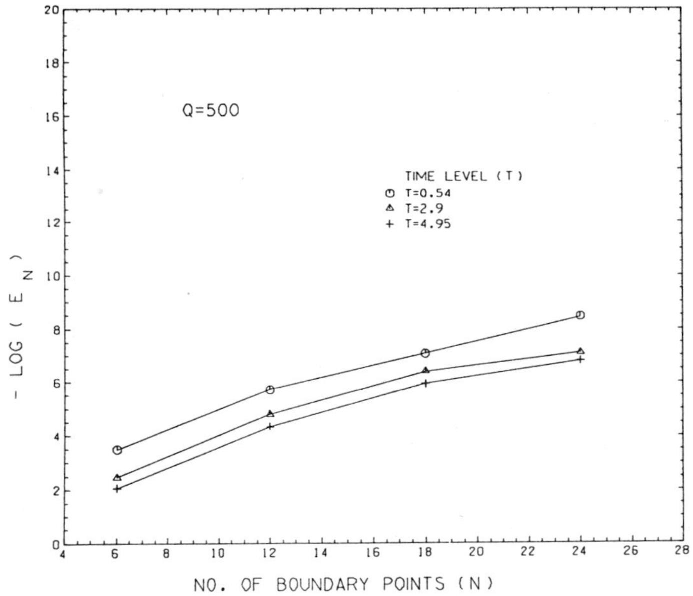

Figure 4: Influence of boundary points on solution

7.3 Influence of time step-size on numerical solution

[a] Constant time step model

A major difficulty with finite difference time marching processes is that the solution obtained for the first few time steps is likely to be in error by around 7 to 10%. This is partly due to the forward difference approximation of using initial data at t=0 to calculate the solution at the next time step and so on, and partly due to approximating a step change of input in the physical system by a mathematical description [4]. In view of this difficulty, it is necessary to use very small time steps (of the order of ½ a day in field data and .005 in dimensionless units) for the first few time levels. After that a much larger time step may be used.

In figure 5 we compare interface profiles obtained by using a constant dimensionless time step size (Δt) of 0.1 and 0.2. The time level, t=3.2 is very nearly at breakthrough position. Agreement between the two sets of profiles is good indicating that the solution converges to the true numerical solution asymptotically.

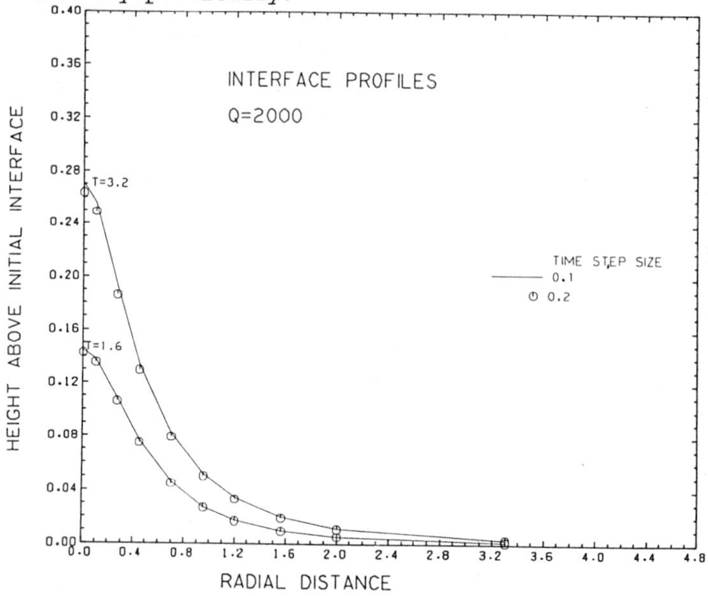

Figure 5: Comparison of interface profiles for Δt=0.1,0.2

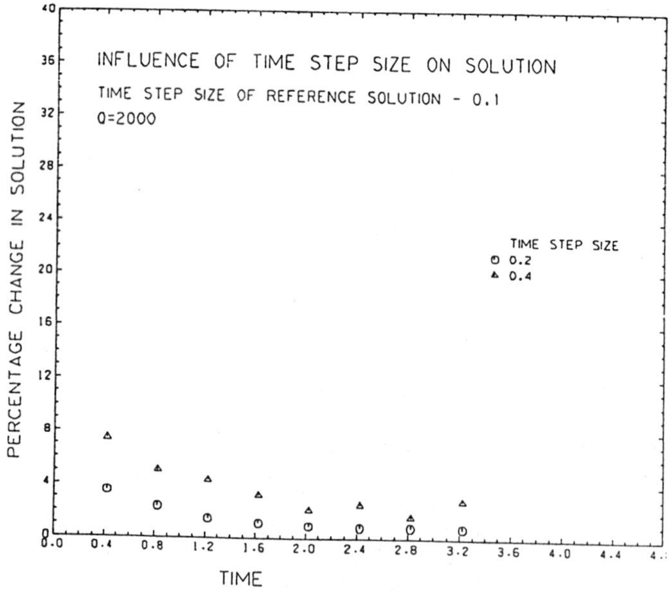

Figure 6: Percentage change in solution

Figure 6 examines the percentage relative change in the numerical solution from Δt=0.1 (the reference case) to Δt = 0.2,0.4. We see that the change in solution is at worst 7% and therefore conclude that the constant time step size solution is stable.

[b] Variable time step model

As it is the change in potential or pressure that determines how far the interface progresses, the criteria for the selection of time step size was now made to depend on the amount of change in potential over the previous time step. So at each time level the time step size was chosen such that the maximum potential change over the previous time step was less than some prescribed value. We took this value to be 1% of the initial potential distribution.

As for the case of constant time step size, the first few time steps taken to be very small. After that the time step size, Δt, was calculated as follows:

$$\Delta t^{(n+1)} = \Delta t^{(n)} \times \frac{k}{[\Delta\Phi_{max}]^{(n)}}$$

$$k = 1\% \min [\Phi_0^{(0)}, \Phi_w^{(0)}].$$

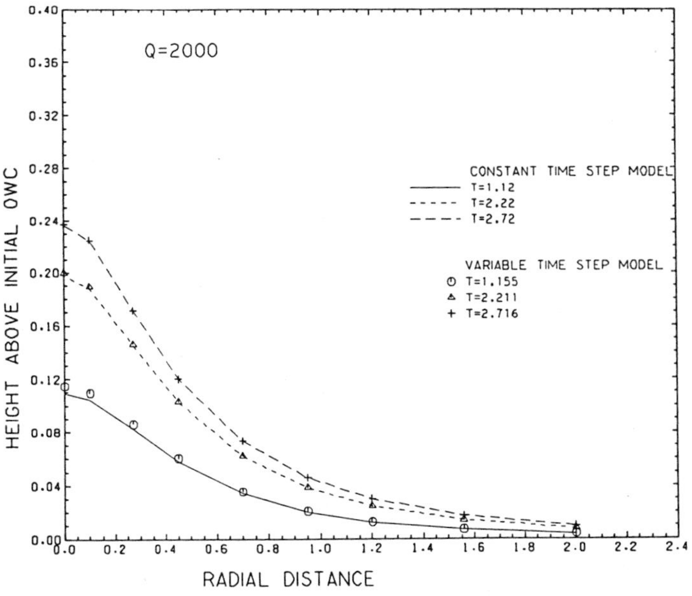

Figure 7: Comparison of solutions using constant and variable time step size

In Figure 7 we compare the interface profiles obtained using variable time step and constant time step. For this particular case, the constant time step model reached breakthrough after 35 time steps, whilst the variable time step model reached breakthrough in 16 steps.

The agreement between the two solutions is excellent. We therefore conclude that the same degree of accuracy can be obtained with far fewer time steps using the variable time step-potential control scheme.

8. CONCLUSIONS

A mathematical model has been developed to study coning behaviour. Numerical calculations are performed on the oil water interface only. A variable time step-potential control feature has also been introduced so that the maximum potential change over each time level is checked. This additional feature minimizes the number of time steps that need to be used.

The accuracy of the model has been demonstrated and it may be used as a suitable model for predicting water front positions, water breakthrough times and other details of engineering importance.

Acknowledgements

The authors would like to thank Societe Nationale Elf Aquitaine (Production) and Elf UK for financial support of this work. We would also like to express our appreciation to Professor Colin Wall and Dr Richard Dawe for their interest, and to Mr Sezgin Daltaban for helpful comments in the development of the variable time stepping scheme.

References

1. CAMY, J. - Prediction of coning breakthrough time and limiting water oil ratio for lateral drive systems, Ph.D. Thesis, Stanford University, 1975.

2. GNANATHURAI, C.D., and WILSON, D.C. - An integral equation approach to a free surface problem occurring in petroleum reservoirs, Computational and Asymptotic Methods for Boundary and Interior Layers, Boole Press, Dublin, 1982.

3. LENNON, G.P., LIU, P.L.F., and LIGGETT, A. - Boundary integral solutions to axisymmetric potential folows, Water Resources Research, V15, No.5, pp.1102-1106,1979.

4. MORSE, P., and FESHBACH, H. - Methods of Theoretical Physics, VII, McGraw-Hill, New York, 1953.

5. MUSKAT, M. - The flow of Homogeneous fluids through porous media, McGraw-Hill, 1937.

6. SOBOCINSKY, D.P., and CORNELIUS, A.J. - A correlation for predicting water coning time, J Petroleum Tech.,1965.

SECTION 13

COMBUSTION

INVESTIGATION OF PARTICLE-LADEN TURBULENT FLOW IN FREE SHEAR TURBULENT COMBUSTION[1]

Alfred C. Buckingham[2], Wigbert J. Siekhaus[3], Janet Ellzey[4], and John W. Daily[5]

Explicit numerical mixed phase simulations are described which couple random gasdynamic motions to inertially interactive gas borne particles. These simulations are numerical experiments intended to provide data for investigating the interaction between a developing turbulent free shear layer and gas borne solid particles it entrains. The simulations predict most probable distributions of dispersed phase trajectories, standard deviations, and gas phase mixing dynamics which include the concomitant back-influences of the particle phase on the carrier gas flow. Data for refinement of the computational scheme and physical verification is provided by experiment. The experimental evidence is developed in a splitter plate divided, two-channel free shear mixing combustion tube. A variety of particle concentrations and particle size distributions are admitted into non-combusting or combusting flows with selected heat release levels. The computations, in turn, provide guidance on design and selection of new experiments.

1. BACKGROUND

Observations reveal that non-combusting (or nearly inert) finely divided particles significantly influence carrier gas combustion. The interaction process is of particular impor-

---

[1]Work performed under the auspices of the U.S. Department of Energy by Lawrence Livermore National Laboratory under contract #W-7405-Eng-48, and supported by U.S. Army Research Office (DRXRO-EG) and ARRADCOM, Large Caliber Weapons Systems Laboratory.
[2]Physicist, Theoretical & Applied Mechanics, H Divison/ Physics, [3]Surface Scientist, Materials Science, Chemistry & Materials Sciences, Lawrence Livermore National Laboratory.
[4]Doctoral Candidate, Mechanical Engineering, [5]Professor, Mechanical Engineering, University of California at Berkeley.

tance in solid (coal) combustion, burner flow field analysis, and in solid propellant ballistic flow field studies. The influence of almost thermo-chemically inert particles appears to reduce the combustion mixing zones by enhancing dissipation of the turbulent gas field. The influence appears to be associated with a breakdown in the large scale, persistent and most energetic eddy structure, which drives the production and transfer of energy to smaller dissipative-diffusional turbulent scale motions. The kinematics of the breakdown and transfer process appear to result, on a temporally averaged scale, in a reduction of the vigor and extent of the turbulent mixing and mixing-enhanced combustion regions.

The previous numerical-experimental investigations of Buckingham and Siekhaus[1] are herein extended by experimental evidence of particle-turbulence structure influences as revealed by a two channel, two-color LDA/LDV diagnostic technique. This procedure provides us with spatially and temporally coincident recordings of instantaneous particle and gas motions. Our theoretical and computational collaboration between the Livermore and Berkeley groups,[2] provides data for describing particle additive influences on reducing wall erosion[3,4]. Here we extend our investigations with two and three dimensional numerical simulations and turbulent free shear layer experimental observations which give us a more penetrating view of the evolving fine-scale turbulent structures when undergoing modulation by the entrained particles.

Our study includes numerical simulations of particle laden turbulent flows and experiments in free shear layers and within the boundary layer adjacent to solid walls. Specific attention is placed on the dissipative influence of particles on the carrier gas turbulent field, as well as the predicted dispersal of particles subjected to the random influence of the turbulent gas field and consequent mean and fluctuating components of gas-particle forces. Near the wall, the influence of particle-to-gas and particle-to-solid surface boundary effects is tested in the opposing limits of perfect surface-particle momentum and energy accommodation and of pure elastic scattering[4]. In this paper our attention is focused on the free shear layer mixing region near the centerline of a two-dimensional two channel combustion tube. We have and will publish separately our discussions of the erosive boundary layer modeling in the two phase flow at a reactive wall[3,4].

2. DISCUSSION

2.1 Formulation

This discussion presents a substantial change to our previous numerical method which was based on use of more or less traditional Reynolds'-Favre' averages for the mean, compressible Navier-Stokes solutions applied to transitional and developing turbulent flow. In our applications, however, compressibility must be considered in relation to combustion energy release. As a first approximation we calculate the velocity fluctuations and mean flow components satisfying a locally incompressible Poisson equation based on a pressure potential, $\pi = P/\bar{\rho}$. Here $\bar{\rho}$ is the mesh averaged local mean density at the previous time step. Particle fractional volume effects will be described subsequently. The random density statistical fluctuations are assumed small enough to neglect with respect to the velocity fluctuations. The mean flow density is then corrected for compressibility by relaxing to a new mean state with an iterative correction to the pressure and simultaneously the internal energy and mean velocity fields in the spirit of the almost incompressible all Mach number schemes introduced by the Los Alamos group[10,11]. The present hybrid scheme is adopted for expediency in implementing both pseudo-spectral and LES procedures. As such it is an intermediate step to developing a self-consistent approximation to both the mean continuous phase density field and any significant density fluctuations (and later, necessary joint density-velocity correlations).

For simplicity we will ignore the reaction-species production phase of our computations in this description. For reference to the reactive phase numerical procedures, particularly near the eroding surface, the reader is invited to review our previous discussions[3,4]. We consider the continuous phase (unsubscripted) and dispersed particle (subscript d) phases as continuum fields at the mixture stage. We write the conservation equations on a geometrically regular three-dimensional cartesian mesh with position x_j, and velocity, u_j vectors, $j = 1,2,3$.

$$\frac{\mathcal{D}\rho}{\mathcal{D}t} = 0, \quad \frac{\mathcal{D}(\ )}{\mathcal{D}t} \equiv \frac{\partial(\ )}{\partial t} + u_j \frac{\partial(\ )}{\partial x_j}\delta_{ij}, \text{ (gas)} \tag{1}$$

$$\frac{\mathcal{D}\rho_p}{\mathcal{D}t} = 0, \tag{2}$$

automatically, under the Lagrangean particle description for thermochemically inert, monodisperse, size invariant particles. The vapor void fraction is implicit in Eq. (2),

$$\rho = \alpha\rho^*, \quad \rho\ = (1-\alpha)\,\overset{*}{\rho}_p\ ,$$

where superscript (*) refers to the uncorrected density terms and the void fraction,

$$\alpha \equiv \text{volume of gas}/\Sigma\ \text{(volumes of gas + particles)}. \tag{3}$$

Momentum conservation,

$$\frac{\mathscr{D}(\rho u_i)}{\mathscr{D} t} + \frac{\partial}{\partial x_j}(\alpha p \delta_{ij}) - \frac{\partial}{\partial x_j}(\alpha\mu \frac{\partial u_i}{\partial x_j}) + F_i = 0 \tag{4}$$

$$\frac{\mathscr{D}(\rho_p u_{p,i})}{\mathscr{D} t} - F_i = 0 \ , \tag{5}$$

Note that the momentum exchange forces, F_i, between dispersed and continuous phases appear explicitly and that the dispersed particle phase momentum, for thermochemically inert particles transfers no stress particle-to-particle in the absence of inelastic collisions. Also we note that the continuous phase stress tensor is modified by the void fraction to account for the change to the proportion of the local surface area over which the stress acts which is obscured by the particles. The local viscosity coefficient μ is replaced by the kinematic viscosity coefficient, $\nu \equiv \mu/\rho$ in our almost incompressible formulation step.

Energy Conservation

$$\frac{\mathscr{D}(\rho H)}{\mathscr{D} t} - \frac{\partial \alpha p}{\partial t} - \frac{\partial}{\partial x_j}(\alpha\mu \frac{\partial u_i}{\partial x_j}) u_j + \alpha \frac{\partial q_j}{\partial x_j} \delta_{ij} + u_{p,i} F_i = 0 \tag{6}$$

$$\frac{\mathscr{D}(\rho_p \, u_p^2/2)}{\mathscr{D} t} = \frac{d}{dt}(\frac{Nm}{V} \frac{u_p^2}{2}) \tag{7}$$

Note that the divergence of the heat flux vector, q_j is in the total enthalpy (H) expression, Eq. (6) (sum of specific enthalpy plus kinetic energy per unit mass), is also "shielded" from the area occupied by the particles. Also note that the continuous phase releases energy to the particle phase (due to interphase coupling force, F_i) and transports it at a rate given by the product, $u_{p,i}F_i$. A self-consistent set of coupling forces, including velocity (Reynolds number) modified drag, Basset viscous force, and Saffman lift for the curvelinear eddy trajectory followed by an entrained particles, is adapted from the mixed phase system laws analyzed by Crowe[12]. In the particle system, Eq. (7) describes the spontaneous rate of change of the kinetic energy of N particles, each containing m (mass) in a volume, V.

Closure and averaging relationships are not imposed in the conventional sense. The fluctuating velocity field is initiated from an instability in the initial viscous profile[7] or from imposed initial averaged velocity profiles derived from experimental analysis for particle shearing flows[5].

2.2 General Numerical Procedure

As an introduction, we outline the procedure. The first step involves modeling the continuous vapor phase including the effects of an assigned initial particle distribution by means of an averaged void fraction and pseudo-spectral or spectrally averaged LES methods. Particles are considered inert with no interphase mass or thermal transfer and the particle sizes are assumed invariant with no agglomeration or breakup.

Statistical decomposition by random selection of the velocity field produces the fluctuational vapor components and forces on the particles, randomly distributed at selected grid mesh point sites. The probability distribution integration is approximated by ensemble averaging over a finite summation with "importance sampling" techniques invoked to reduce the number of samples required for statistical significance[13]. Sampling of the fluctuations must be repeated for a statistically significant number of gas phase trajectories in phase space. Each trajectory represents an independent random probability density function distribution for a particle "cluster" centered at some fixed position and time in the computation mesh. Gas phase fluctuations about the mean flow speed are thereby simulated as a weighted sum of previous increments plus a currently evaluated random fluctuation sampled from a one-component probability density function (pdf) distribution. The incremental sampling procedure and summation over a finite number of events, replaces the exact integral formation. This converges to the integral value provided a sufficiently large number of samples is taken. The base distribution is taken to be the Gaussian normal distribution for weighted phase space spectrum samples of the random variables.

Following determination of the most-probable cluster trajectories, particle aggregate cloud concentrations and spatial distributions are reevaluated for the next time interval and the void fraction distribution modified gas phase Navier-Stokes solution is updated.

2.3 Boundary Conditions

We impose law-of-the-wall slip conditions at interior solid contours and continuous flow or periodic boundary conditions on streamwise entrance and wake. Unsteady particle field streamwise conditions are imposed in the streamwise particle current from initially selected population size and momentum distributions at inlet. These represent the prescribed particle injection, emission or entrainment situation under study. Exhaust of particles is incrementally accounted for so that the finite array of particles selected for a computation can be replenished for re-entrance at a later time interval, if necessary. A major assumption is associated with the particle flux at the wall. Tangential flow is assumed free slip with no surface accommodation. At other then tangent incidence, either specular elastic reflection or perfect accommodation of momentum and energy are maintained as the two limiting options for the computation.

2.4 Experimental

We study pre-mixed two-channel separated flows, as seen in Fig. 1. These flows are dominated by large scale structures which grow by entrainment and coalescence. In combusting flows it is the entrainment process which governs the rate of combustion. For combustion to take place, the fuel and oxidizer must be brought into molecular contact by the mixing process. The rate at which this is done generally controls the rate at which products are formed, thus the name diffusion flame. In premixed flames, one stream is premixed oxidizer and fuel, the other is hot combustion products, generally at zero axial velocity. In this case the burning rate is controlled by the rate at which the already hot products heat premixed reactants to the ignition temperature.

The splitter plate divided, two-channel flow experiments are conducted in the two-dimensional, two-stream combustion facility at the University of California at Berkeley. The combustor is operated at fixed entrance velocities and densities and the amount of heat release within the mixing layer is varied, as the injected particle flux and size distribution. High speed Schlieren movies are taken of the combusting flow. Structure passing frequencies are determined from

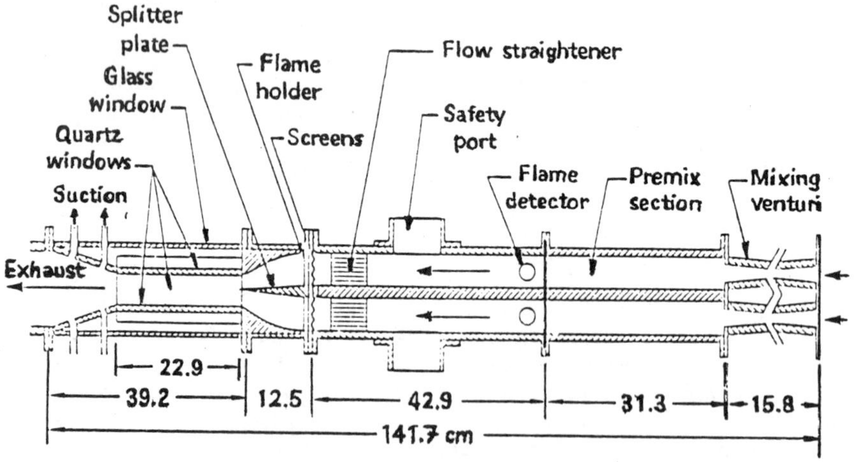

Fig. 1 Two channel, rectangular combustion tube free shear layer mixing chamber at the University of California, Berkeley experimental facility.

the movies at various downstream positions, and information about the vortex structure pairing process is deduced. Supplemental LDA/LDV, hot film and hot wire measurements are made to obtain the fine scale turbulent features and statistical measurements. Further discussions on the experimental set-up and diagnostic procedures appear in current publications[2,4].

2.5 Observations and Simulations

Even the finely ground, sub-micron dust particles consistently influence the breakdown of the largest eddy mixing structures in the shear layer. The significance of this breakdown is associated with an increase in the population of the smaller structures out to the dissipation (high frequency) range where the viscous energy transfer is accelerated by the increased population of small high frequency motions. Figure 2 illustrates this breakdown and population of higher frequency structures by recording the measured normalized probability density distribution of passing frequencies at a fixed station, x, downstream of the splitter plate trailing edge. The upper channel is a lean unburned mixture of air and propane at an equivalence ratio of ϕ_1 = 0.6. The lower channel is premixed, ignited and combusting prior to entering the two channel mixing zone. The first three positions reveal the breakdown. The final station, however, shows some

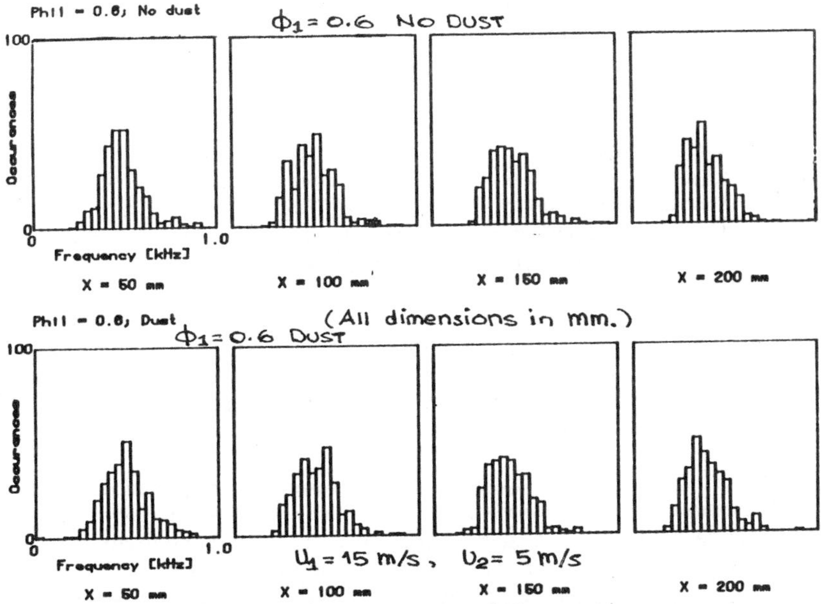

Fig. 2 Dusty (upper row) vs dust free (lower row) passing frequencies from Hy-Cam (5000 s^{-1}) framing motion picture images of cross channel optical Schlieren patterns at indicated stations.

of the effects of two dimensional vortex pairing and growth in the larger scales which appears to shift the pdf distribution back to the lower frequencies.

Figure 3 shows the LDA measured Reynolds stress in the combustion tube for a cold, non-combusting flow situation with channel velocities of 15 m/s (upper) and 5 m/s (lower). The measurements and digital processing provide cross and auto-correlations and higher moments: skewness and Kurtosis of the velocity field. For comparison, Fig. 4 shows numerically simulated profiles of the normalized Reynolds stress (upper figure) and normalized turbulence energy (lower figure). These computations were made without and with particles at the initial mass loading (k_m) indicated.

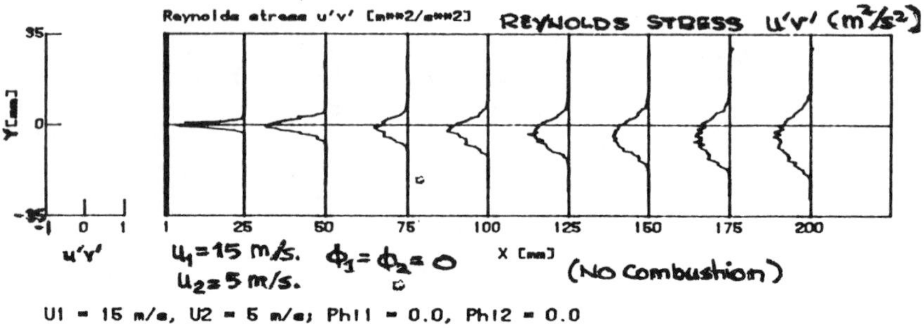

Fig. 3 Measured profiles of Reynolds stress in the two channel combustion chamber at the listed flow conditions. The y = 0 axis is on the splitter plate centerline.

In other numerical tests we have predicted comparable comparisons of the higher moments of the unsteady velocity field and both auto and joint correlations to compare with the experiments. Publication space limits preclude showing more of these results in this description.

To summarize, sufficient concentrations of finely divided particles interact with a turbulent mixing layer so as to reduce the intensity and extent of the mixing zones. This is of great importance in combustion situations where sidewall erosion is a problem.

Pseudo-spectral or spectrally averaged procedures appear to be necessary for numerically simulating these interactive flows since they are able to provide sufficient resolution and accuracy to trace the fluctuating motions over a usefully broad range of scales, without excessive computer capacity.

The major problems encountered in our current simulations are the proper, efficient and accurate treatment of the particle flow boundary conditions at the wall and the treatment of compressibility effects. These considerations are the focus of our current and on-going research and method development.

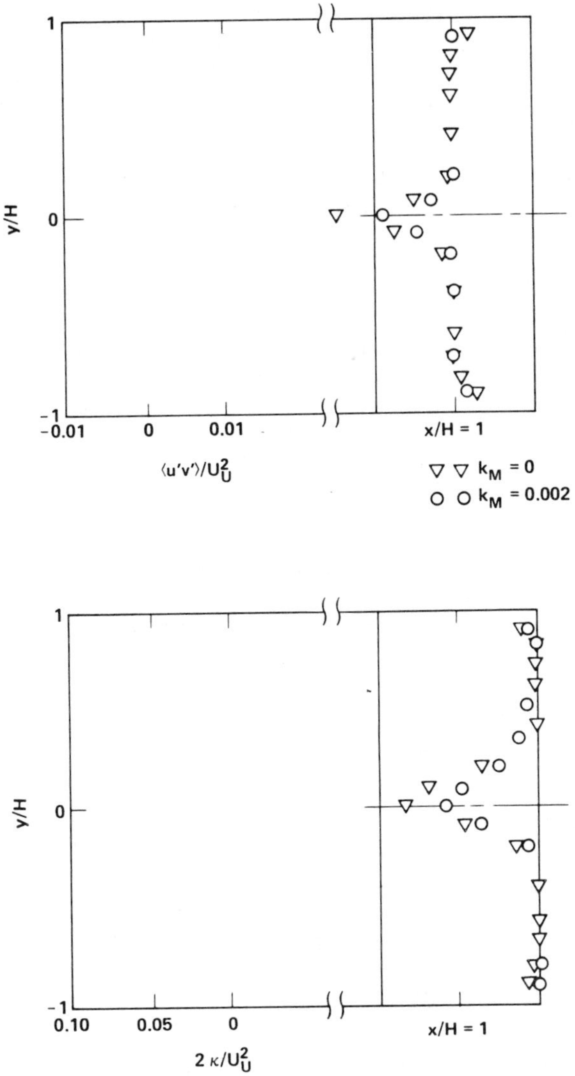

Fig. 4 Results of numerical simulations variations in Reynolds stress (upper figure) and turbulence kinetic energy (lower figure) with listed particle mass loading.

REFERENCES

1. BUCKINGHAM, A. C. - Simulating Interactions between Turbulence and Particles in Erosive Flow and Transport, Numerical Methods in Laminar and Turbulent Flow, Taylor,

C. and Schrefler, B. A., eds., Pineridge Press, Swansea, U.K. (1981)

2. BUCKINGHAM, A. C., SIEKHAUS, W. J., KELLER, J. O., ELLZEY, J., HUBBARD, G., DAILY, J. W. - Computed and Experimental Interactions between Eddy Structure and Dispersed Particles in Developing Free Shear Layers, AIAA/ASME 3rd Joint Thermophysics, Fluids, Plasma, and Heat Transfer Conference, St. Louis, Mo., June 7-11, 1982.

3. BUCKINGHAM, A. C. - Dusty Gas Influences on Transport in Erosive Propellant Flow, AIAA Journ. 19, No. 1, pp. 501-510, 1981.

4. BUCKINGHAM, A. C., SIEKHAUS, W. J., PRICE, C. W. - Erosion Mechanisms, Proc. of the 29th Sagamore Army Materials Research Conference, Weiss, V. and Mescall, J., Lake Placid, NY, June 19-23, 1982.

5. KWAK, D., REYNOLDS, W. C., FERZIGER, J. H. - Three-Dimensional Time-Dependent Computation of Turbulent Flow, Stanford University Thermosciences Division Report TF-5, May 1975.

6. MANSOUR, N. N., FERZIGER, J., REYNOLDS, W. C. - Large Eddy Simulation of a Turbulent Mixing Layer, Stanford University Thermosciences Division Report TF-11, April 1978.

7. ORSZAG, S. A. - Numerical Simulation of Turbulent Flows, Handbook of Turbulence, Frost, W. and Moulden, T. H., Plenum Press, NY, pp. 281-313, 1977.

8. GOTTLIEB, D., ORSZAG, S. A. - Numerical Analysis of Spectral Methods, SIAM Monograph, Philadelphia, 1977.

9. PATTERSON, G. S., ORSZAG, S. A. - Numerical Simulation of Turbulence, Atmos. Technol. 3, pp. 71-78, 1973.

10. HARLOW, F. H., AMSDEN, A. - Journ. of Comp. Phys. 3, pp. 197-213, 1971.

11. HARLOW, F. H., AMSDEN, A. - Journ. of Comp. Phys. 16, pp. 1-19, 1974.

12. CROWE, C. T. - Vapor-Droplet Flow Equations, Lawrence Livermore National Laboratory Report UCRL-51877, August 18, 1975.

13. FAIST, M. B., MUCKERMAN, J. T. - Importance Sampling and Histogrammic Representations of Reactivity Functions and Product Distributions in Monte Carlo Quasiclassical Trajectory Calculations, J. Chem. Phys., 69, p. 9, 1978.

NUMERICAL SIMULATION OF COMBUSTING FLOWS IN TWO DIMENSIONS

Julius Brandeis(I)

ABSTRACT

Use is made of the two-dimensional combustion (TDC) computer code to simulate various problems involving combustion of gaseous fuels. The code solves the full hydrodynamic conservation equations together with the equations for chemical species conservation and diffusion, heat conduction, and chemical kinetics. The problems addressed in this study range in physical dimensions from laboratory apparatus ($\ell \sim 1$ m) to unconfined combustion of large quantities of gaseous fuels ($\ell \sim 1000$ m). Through a number of such case studies, the flexibility as well as limitations of the method are examined. The usefulness of numerical simulations as a means of supplementing experimental data is demonstrated.

1. INTRODUCTION

With the advent of the current generation of computers, specifically machines belonging to the class of Cray 1, the routine two-dimensional numerical simulations of combustion systems became a reality. Although three-dimensional numerical models for combustion computation are now becoming feasible, their practical use is still severely restricted by computer speed and storage limitations. Even in two-dimensions, the current computer capabilities put constraints on model sophistication as well as on the size and complexity of the problems to be addressed. Thus, it is not yet practical in two dimensions to include a detailed, multi-step chemical kinetics model, such as the one developed by Westbrook and Dryer [1] and used for one-dimensional computation. For many purposes, however, simpler global or one-step chemistry description is sufficient for combustion modeling. The problem-related limitations made necessary by the present computer constraints are associated

(I) Physicist, Liquefied Gaseous Fuels Program
Lawrence Livermore National Laboratory

with the following factors identified by Oran and Boris [2] which commonly arise in combustion modeling. 1) Multiple time scales (stiffness): temporal resolution constrained by the shortest physical time scale in the problem. 2) Multiple space scales: spatial resolution constrained by the steep gradients (i.e., flame fronts, shocks). 3) Geometric complexity: many practical problems transcend the two-dimensionality limitations of the model. In addition there are limitation dictated by the state of the art, most notably in the area of turbulence modeling.

In order to overcome the problems enumerated above, compromises must be accepted both in model capabilities and in the quantity and distribution of grid points making up the computational mesh. These necessarily affect the quality of the results obtained from a model. In the present study, the numerical model, TDC, developed by Haselman [3], is used to simulate several cases involving combusting flow. The code, which briefly will be described later, is subject to assumptions and limitations consistent with the problems discussed above. As will be shown by way of examples, the model effectively reproduces the features observed in combustion experiments of various physical dimensions ranging over three orders of magnitude in characteristic length. The numerical results are often more abundant and more detailed than those obtained experimentally. This is especially true when such combustion tests are too large to be carried out in the laboratory environment and must be done at a remote, sparsely instrumented site. The prime benefit of a predictive technique is, however, derived from applying the model to a problem for which no experimental data can be obtained for the reasons of safety or economy. For this purpose the reliability of the method must be established and its limitations must be understood. The examples presented in this work can be categorized as follows: 1) simulations of two laboratory-scale combustion experiments ($\ell \sim 1$ m) for which detailed data exist. These serve primarily to establish confidence in the code; 2) simulation of some aspects of a gasoline fire in a highway tunnel ($\ell \sim 100$ m). An attempt was made here to address the actual accident which took place in the Caldecott tunnel, California, in April 1982; 3) simulation of combustion in large clouds of gaseous fuel ($\ell \sim 1000$ m). Only very general experimental observations are available for this case.

2. DESCRIPTION OF THE COMPUTATIONAL METHOD

The numerical combustion model, TDC, (see Ref. 3 for full description) was used in this study with minor changes and additions. TDC is a finite difference, two-dimensional code which solves the full hydrodynamic conservation equations (compressible Navier-Stokes) by an explicit method of second-order accuracy. In addition to the viscous hydrodynamic equations, equations for chemical species conservation and diffusion, heat conduction and chemical kinetics are solved in a coupled manner. The chemistry of combustion is modeled by the following one-step reaction equation (written here for reaction in air):

$$\text{Fuel} + \text{Oxygen} + \text{Nitrogen} \rightarrow \text{Products} + \text{Nitrogen} \qquad (1)$$

Specific products must be entered on the right-hand side of Eq. (1) in order to carry out stoichiometric balance, but for the purpose of computation one lump atomic weight is used, equal to the sum of the atomic weights of all products multiplied by their respective stoichiometric coefficients. The reaction-rate relation incorporating an Arrhenius term is written as

$$k(T) = AT^{\alpha} \exp(-\beta/T) [C_F]^{\gamma} [C_O]^{\delta} \quad (2)$$

where the following definitions apply. A is the pre-exponential factor, influencing the flame speed; T is the temperature in °K; C_F is the fuel mass fraction; C_O is the oxidizer mass fraction; β is the activation energy divided by the gas constant ($\beta = E_a/R$); and α, γ, δ are empirical constants. These constants and factors have been compiled by Westbrook and Dryer [4] for various reaction systems. The turbulence model included in TDC is of the simple gradient diffusion type. This was found adequate for some cases presented, but insufficient for others.

The finite difference equations derived from the governing equations are solved first in the Lagrangian coordinate system and are periodically mapped back onto the Eulerian grid. Since explicit differencing is used, the time step in the finite difference hydrodynamic equations is limited (for stability) by the sound speed. In order to alleviate the stiffness problem (especially when the flow is subsonic), the Lagrangian hydrodynamics are run a number of subcycles for each cycle of the rest of computation. The numerical diffusion, occurring mainly in the remapping phase, is minimized by taking a number of Lagrangian cycles (normally up to 20) for each Eulerian cycle.

The solution to the finite difference equations is carried out on a non-uniform, stationary grid specified by the user. Since about five mesh points are necessary by TDC to define a flame, relatively high grid density is needed in the region where combustion (and especially ignition) takes place. This proved to be the main constraint in defining a grid. Since only a limited number of points could economically be included in the mesh system, and since the system was static, the flame fronts were smeared over an unphysically large region. This was acceptable considering that the flame speed was not significantly affected. Ignition in this study was accomplished by specifying a volume of preburned material.

The boundary conditions applied at the rigid walls require impermeability and tangency of the flow. At the open boundaries extrapolation conditions are imposed by setting to zero the derivatives of the flow quantities. These conditions are adequate, but they are reflective and the solution away from the boundaries may be artificially affected. For this reason an attempt was made to remove these boundaries as far as practically possible from the region of interest. A more serious problem arose concerning the upstream boundary when a prescribed constant-velocity wind-driven inflow had to be maintained. The imposition of this boundary condition essentially made the inflow conditions insensitive to any events occurring downstream, effectively ruling out any combustion-driven inflow. No satisfactory remedy for this problem was found within the present structure of the code, even with zero inflow specified.

3. APPLICATION AND DISCUSSION

The first two examples of the use of the numerical model consist of simulation of two laboratory experiments. Figures 1 and 2 address the problem of flame propagation in a channel 90 x 30 x 15 cm, filled with a combustible propane/air mixture [5]. Figure 1, which compares the flame propagation in a closed and an open channel without obstacles, is a good illustration of the method's use in a quantitative mode. The set of results corresponding to the "lid on" case serves as a calibration run, in which the reaction rate is set to give good agreement with the experimental results. No further changes are made. It is seen that for the open channel the computed results lie within the experimental bars, denoting distribution of results from several runs, indicating that the effects of confinement on the flame propagation are reproduced by the method. In Fig. 2 results are presented for the open channel with an array of raised obstacles in the form of temperature and velocity plots. The rather complex flame geometry is clearly seen. The spread in the flame profile towards the top of the frame is caused by the expanding mesh which was used to remove the upper (open) boundary as far as possible while utilizing a manageable number of grid points. Also noteworthy is the jetting forward of the flame underneath the obstacles, caused by the interaction with the flow field. The vortices downstream of each obstacle act to stimulate the burning. Unfortunately, the computed

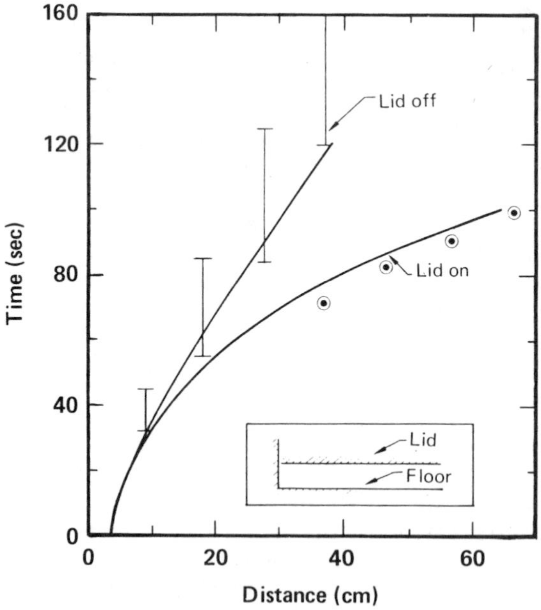

Fig. 1. Time-distance plot for flame propagation in closed and open channels. Symbols and error bars denote experimental results (premixed propane/air).

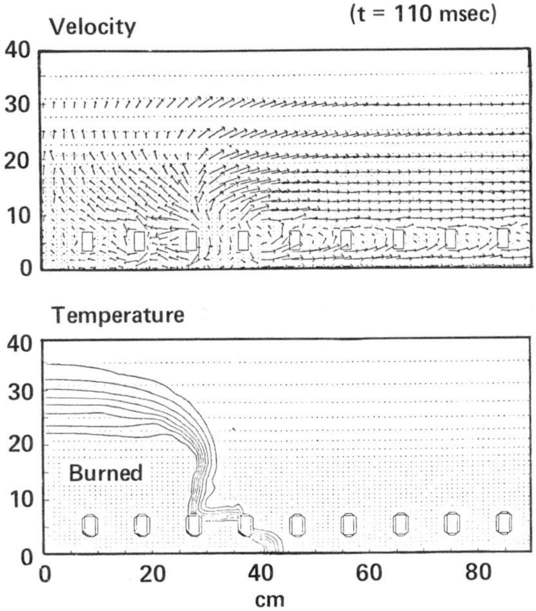

Fig. 2. Computed results for flame propagation in an open channel with raised obstacles (premixed propane/air).

results showed only a small increase in flame speed due to the presence of the obstacles, though the shape of the flame front was greatly effected. The experimental results, using various obstacle configurations indicated quite a significant flame acceleration. This points towards the inadequacy of the simple turbulence model for this purpose.

The example shown in Fig. 3 is taken from the simulation of the experiment by Keller et al. [6], in which combustible propane/air mixture was driven through a test chamber approximately 40-cm long containing a rearward-facing step. The Schlieren record shown is one frame of a sequence showing an unsteady, turbulent burning mode called humming in Ref. 6. It is seen from Fig. 3 that the numerical model reproduces the flame characteristics very well. The period of the burning layer was not precisely matched, but then there was no attempt made to precisely duplicate the experimental conditions in these preliminary simulations. The small dimensions of the test chamber made it possible to use a uniform computational net. The outflow boundary, however, proved to be a limiting factor in this case, as the computation ceased when the flame reached that boundary. The problem ran long enough (t > 8 msec), however, to allow several cycles of the burning layer.

The next application (see Figs. 4 and 5) is inspired by the Caldecott tunnel fire mentioned in the introduction. The actual dimensions of the tunnel are used, but only the middle 130 m of the tunnel have been modeled (only a segment of this is shown in the figures). The ventilation system has been simulated by allowing an

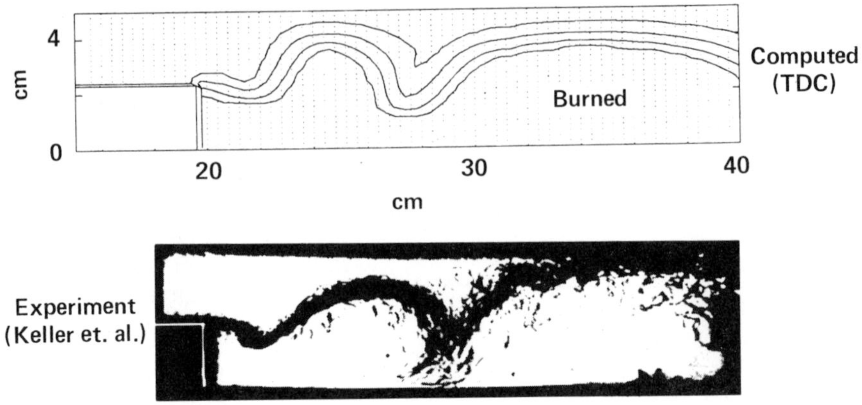

Fig. 3. Computed and experimental [6] results for combusting flow over rearward-facing step. (Premixed propane air flowing left to right.) (Schlieren courtesy of Prof. Oppenheim.)

independent air inflow and outflow for the section of the tunnel above the slotted partition. Needless to say, no physical measurements are available for comparison. Two cases, representing possible burn scenarios, are presented in Fig. 4 for comparison. In one case the tunnel is filled with a premixed gas/air cloud prior to ignition, while in the other a continuous addition of combustible material through a source was used. Low inflow velocities were used for both the vent duct and the tunnel. In both of these examples there is a strong interaction between the flow field and combustion, emphasizing the importance of the coupled approach to modeling such problems. Both flows were quite unsteady. In Fig. 5 low speed venting is compared with high speed venting in its effect on the flow. The effectiveness of the high speed venting in dispersing both the fuel and the products is seen. Moreover, the flame is extinguished in the latter case. The limitation due to the static nature of the computational grid is apparent in this example . The grid has been highly stretched towards the inflow and the outflow boundaries, and relatively dense (0.2 x 0.3 m) in the region surrounding the ignition location. This was satisfactory for the continuous source cases, as the flame was always within the dense grid region. In the premixed case, however, the flame burning through the cloud quickly entered the coarse grid region, and consequently its thickness reached tens of meters. Here the need for a flame-following grid is quite obvious.

Finally, attention is focused on prediction of the combustion characteristics in large clouds of heavier-than-air (cold) natural gas. This example is motivated by accident scenarios involving releases of large quantities of liquefied natural gas (LNG) from storage or transit

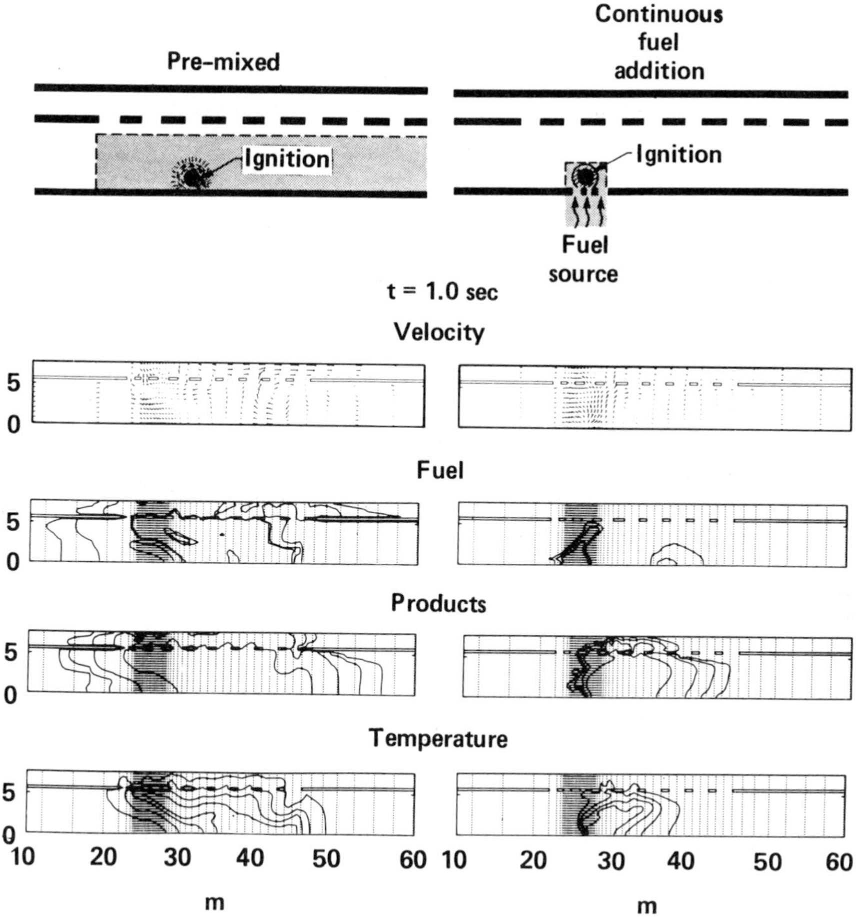

Fig. 4. Gasoline fire in a tunnel--comparison of premixed case to case with continuous fuel addition.

facilities. Figure 6 presents results for two such cases. In one of these a premixed, combustible methane/air cloud is specified. In the second case, the methane was continuously injected at the source and allowed to spread by gravity and diffusion, simulating a real accident situation. All other parameters were identical for the two cases in order to facilitate a comparison of results. The common feature in both cases, frequently observed in this type of situation, is the buoyant plume of hot gases (often referred to as a fireball) rising hundreds of meters into the air. A vortex structure develops as the result of this upward motion, and is correctly reproduced by the model as shown in Fig. 7. It is apparent from Fig. 6, that the premixed combustion produces the larger and faster-rising fireball of the two cases. It is sustained by large amounts of combustible material sucked

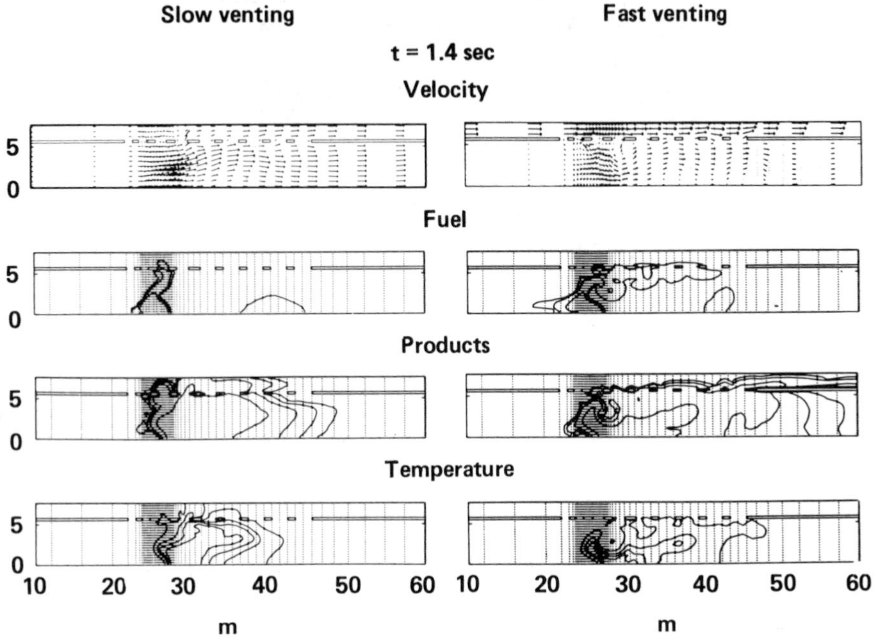

Fig. 5. Gasoline fire in a tunnel--effect of venting velocity.

into the stem and burning there. By contrast, at t = 10 seconds, the diffusion case sustains combustion only at the very base of the stem. The present application is another good example of the utilization of the numerical method for qualitative, comparitive study of hazards and risk-assessment.

4. CONCLUSIONS

Using a number of examples illustrating a broad spectrum of combustion problems, TDC was shown to be a useful research tool. The primary value of the complex numerical code lies in its ability to provide valid qualitative results for configurations for which experimental data is either sparse or nonexistent. The code is then used as a predictive tool in hazard assessment and configuration evaluation. It is also useful in exploring the effect and origin of phenomena observed in experiments. As a quantitative tool, TDC is hindered by the need to fix various empirical parameters necessary because of inability to model various aspects of the problem from first principles. These parameters must often be fixed by trial and error, using experimental results for calibration. Furthermore, some

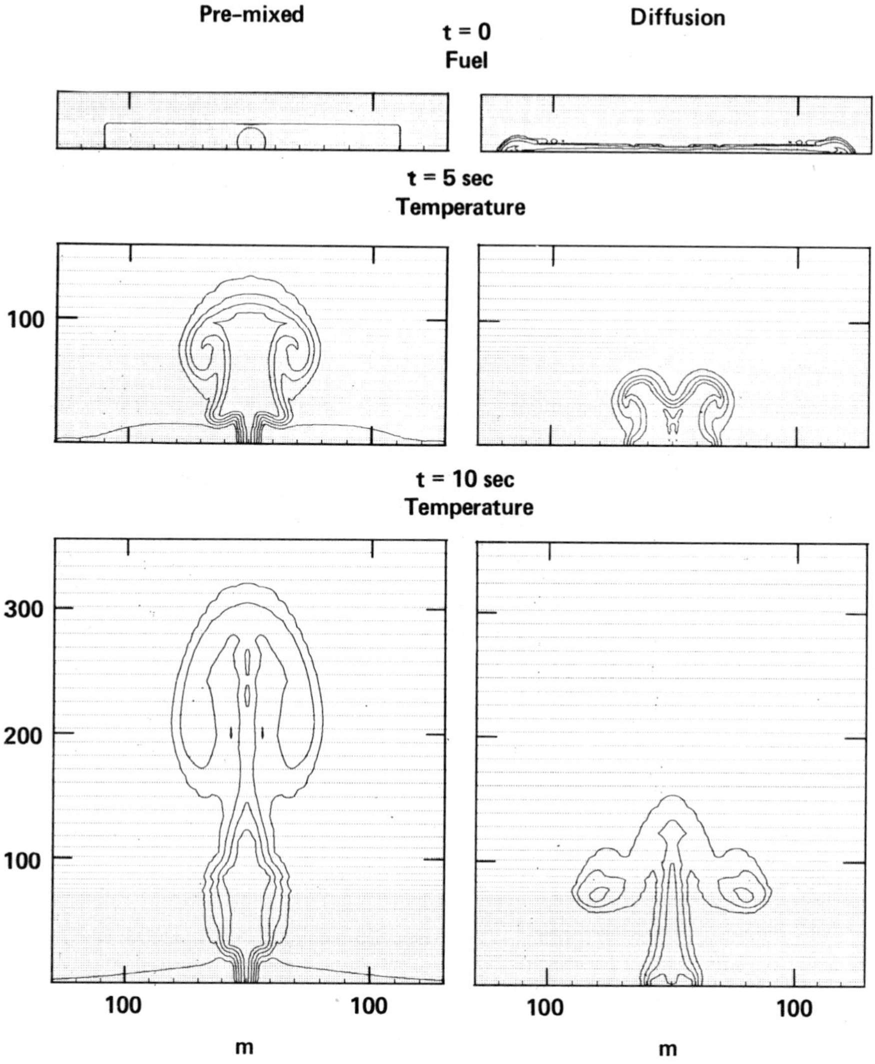

Fig. 6. Fireball developing during combustion in premixed and diffusion methane clouds.

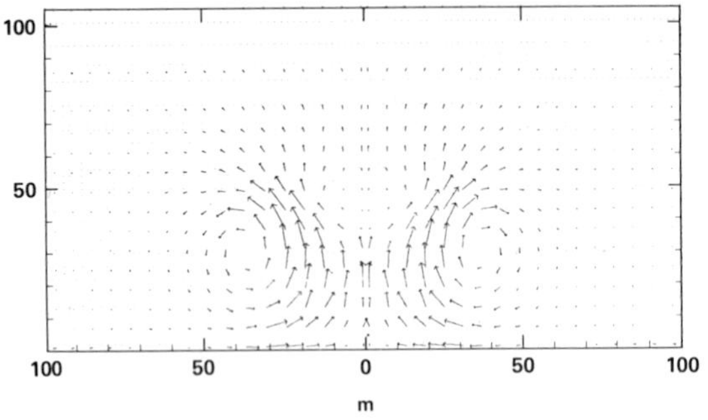

Fig. 7. Velocity plot showing the vortex structure in one frame of Fig. 6 corresponding to the diffusion case at t = 5 sec.

of these parameters may be affected by grid size. The computation costs associated with the model are quite high. The execution time per grid point per major cycle is on the order of 10^{-4} sec. on Cray 1. The execution time for a problem is directly proportional to the number of points in the grid and is inversely proportional to the minimum grid spacing. Therefore, reducing the number of grid points will economize the computation. It is recognized that improvements in boundary condition formulation (i.e., non-reflective conditions) and use of adaptive grids to resolve moving gradients (i.e., flames, shocks) could reduce the grid point requirement and enhance the numerical model.

ACKNOWLEDGMENTS

The author is grateful to Mr. D.J. Bergmann for his assistance with part of the computations and to Dr. L.C. Haselman for his help in understanding some of the more subtle features of TDC.

Work performed under the auspices of the U.S. Department of Energy by the Lawrence Livermore National Laboratory under contract number W-7405-ENG-48.

REFERENCES

1. WESTBROOK, C.K. and DRYER, F.L., - Chemical Kinetics and Modeling of Combusion Processes, Eighteenth Symposium (International) on Combustion, The Combustion Institute, Pittsburgh, pp 749-767, 1980.

2. ORAN, E.S. and BORIS, J.P., - Detailed Modeling of Combustion Systems, Prog. Energy Combust. Sci., Vol 7, pp 1-72, 1981.

3. HASELMAN, L.C., TDC - Computer Code for Calculating Chemically Reacting Flows in Two Dimensions, Lawrence Livermore National Laboratory, Livermore, Ca., UCRL-52931.

4. WESTBROOK, C.K. and DRYER, F.L., - Simplified Reaction Mechanisms for the Oxidation of Hydrocarbon Fuels in Flames, Combustion Science and Technology, Vol 27, pp 31-43, 1981.

5. URTIEW, P.A., BRANDEIS, J. and HOGAN, W.J., - Experimental Study of Flame Propagation in Semiconfined Geometries with Obstacles, Lawrence Livermore National Laboratory, Livermore, Ca., UCRL-87088, 1982 (to appear in Combustion Science and Technology).

6. KELLER, J.O., VANEVELD, L., KORSCHELT, D., HUBBARD, G.L., GHONIEM, A.F., DAILY, J.W. and OPPENHEIM, A.K., - Mechanism of Instabilities in Turbulent Combustion Leading to Flashback, AIAA J., Vol 20, No. 2, pp 254-262, 1982.

MIXED-DIMENSIONALITY FLUX-CORRECTED TRANSPORT CALCULATIONS WITH ADAPTIVE GRIDDING

Mark A. Fry[1]
David L. Book[2]

I. SUMMARY

A numerical technique has been developed for capturing nonsteady shocks that maintains accurate results throughout a very distorted finite difference grid. This method is a natural extension of the flexibility inherent in coordinate-split Flux-Corrected Transport (FCT) algorithms. The FCT module used in the present calculations solves one-dimensional (1-D) fluid equations in Cartesian, cylindrical, or spherical geometry. It provides a finite difference approximation to the conservation laws in the form:

$$\frac{\partial}{\partial t} \int_{\delta V(t)} \phi \, dV = - \int_{\delta A(t)} \phi \, (u-u_g) \cdot dA + \int_{\delta A(t)} \tau \, dA \, , \qquad (1)$$

where ϕ is the density of either mass, momentum, energy, or species number in a cell of volume $\delta V(t)$, u and u_g denote the fluid and grid velocities, and τ represents the pressure terms. Such a formulation allows the grid to slide with respect to the fluid without introducing any additional numerical diffusion. One can then concentrate zones to resolve surfaces of discontinuity. Coordinate-splitting allows extension to multidimensional problems. However, on a two dimensional (2-D) mesh this "sliding grid" technique leaves many zones with abnormal aspect ratios and thus yields

(1) Mark A. Fry
Science Applications, Inc.
McLean, Va 22102

(2) David L. Book
Naval Research Laboratory
Washington, D.C. 20375

inaccurate solutions to the conservation laws. By laying down an accurate 1-D solution (which also utilizes the sliding grid) in the part of the mesh where the free-field solution is applicable, one can overcome these deficiencies. Then evolution of shock phenomena over long simulation times and with enhanced resolution becomes possible. The method is illustrated by calculating 2-D Mach shock reflection from airblasts. Comparison is made with available experimental data.

II. NUMERICAL APPROACH

The description of blast and shock phenomena include the flows from two phase motion of chemical explosion products. In addition, shock propagation, rarefaction waves, and contact discontinuities through nonideal media (thermally stratified atmosphere) and the interaction with structural surfaces must be simulated. Flux-Corrected Transport (FCT) algorithms employed in two-dimensional (2-D) computer codes have been shown to be an accurate and flexible tool for solving such nonsteady compressible flow problems [1,2]. The sharp discontinuities present in the gas dynamics of blast waves have been successfully simulated with the addition of a nondiffusive adaptive gridding schedule [3]. Unfortunately, the adaptive gridding scheme introduces errors outside the area of interest; moreover, in time the solution further deteriorates because of fluid moving from coarse to fine gridded regions. This undesirable side-effect has been removed by coupling a one-dimensional (1-D) solution to the numerical experiment.

In general the FCT algorithm can be described as a set of operations on a conserved quantity such as density, momentum or energy. Consider the following three-point transport scheme:

$$\hat{\rho}_j = \rho_j^o - \eta(\rho_{j+1}^o - \rho_{j-1}^o) + \kappa(\rho_{j+1}^o - 2\rho_j^o + \rho_{j-1}^o);$$

$$\bar{\rho}_j = \hat{\rho}_j - \theta(\rho_{j+1}^o - \rho_{j-1}^o) + \lambda(\rho_{j+1}^o - 2\rho_j^o + \rho_{j-1}^o); \qquad (2)$$

$$\rho_j^n = \bar{\rho}_j - \mu(\phi_{j+1/2} - \phi_{j-1/2}),$$

where $$\phi_{j+1/2} = \hat{\rho}_{j+1} - \hat{\rho}_j.$$

The arrays $\{\rho_j^o\}$ and $\{\rho_j^n\}$ are the old and new densities, $\hat{\rho}_j$ and $\bar{\rho}_j$ are temporary intermediate densities, and η, θ, κ, and μ are velocity-dependent coefficients. Here κ and λ are diffusion coefficients, and μ is the antidiffusion coefficient. In the actual algorithm, $\phi_{j+1/2}$ is corrected (hence the name FCT) to a value $\phi_{j+1/2}^c$ chosen so no extrema in $\bar{\rho}_j$ can be enhanced or new ones introduced in ρ_j^n.

Older FCT algorithms had $\theta = 0$; the widely used ETBFCT and related algorithms [4] have in addition $\kappa = 0$. If we define ρ_j to be sinusoidal with wave number k on a mesh with uniform spacing δx so that $\rho_j^o = \exp(ij\beta)$ where $\beta = k\delta x$, then the new density array satisfies

$$\rho_j^n/\rho_j^o \equiv A = 1-2i(\eta+\theta)\sin\beta + 2(\kappa+\lambda)(\cos\beta-1) \tag{3}$$

$$- 2\mu(\cos\beta-1)\left[1-2i\eta\sin\beta+2\kappa(\cos\beta-1)\right].$$

From A we can determine the amplification $\alpha = A$ and relative phase error $R = (1/\varepsilon\beta)\tan^{-1}(-\text{Im}A/\text{Re}A)-1$, where $\varepsilon = v\delta t/\delta x$ is the Courant number. Expanding in powers of β we find

$$\alpha = 1 + a_2\beta^2 + \alpha_4\beta^4 + \alpha_6\beta^6 + \ldots ;$$

$$R = R_0 + R_2\beta^2 + R_4\beta^4 + R_6\beta^6 + \ldots . \tag{4}$$

First-order accuracy entails making R_0 vanish, which requires that $\eta + \theta = \varepsilon/2$. Second-order accuracy ($\alpha_2 = 0$) implies that $\mu = \kappa + \lambda - \varepsilon^2/2$. Analogously, the "reduced-phase-error" property $R_2 = 0$ determines $\mu = (1-\varepsilon^2)/6$, thus leaving two free parameters. One of these can be used to make R_4 vanish also. The resulting phase error $R(\beta)$ is small not only as $\beta \to 0$, but also for larger values of β corresponding to the short wavelengths. The remaining parameter η can be chosen to relax the Courant number restriction needed to ensure positivity from $\varepsilon \leqslant 1/2$ to $\varepsilon \leqslant 1$.

In summary, FCT is a finite-difference technique for solving the fluid equations in problems where sharp discontinuities arise, e.g., shocks, contact surfaces, and steep concentration gradients. It modifies the linear properties of a second- (or higher) order algorithm by adding a diffusion term, then subtracting it out "almost everywhere." The residual diffusion is just large enough to prevent dispersive ripples from arising at the discontinuity, thus ensuring that quantities like density remain positive. FCT captures shocks accurately over a wide range of parameters. This means that no information about the number or nature of the surfaces of discontinuity need be provided prior to initiating the calculation.

The FCT routine used in the present calculation, called JPBFCT," consists of a flexible, general transport module which solves 1-D fluid equations in Cartesian, cylindrical, or spherical variables. It incorporates a sliding rezone, so that the fluid quantities being transported can be deposited nondiffusively onto a new mesh at each timestep. Thus, knowing where the features of greatest interest are located, one can concentrate fine zones where they will resolve these features most effectively as the system evolves (Fig. 1).

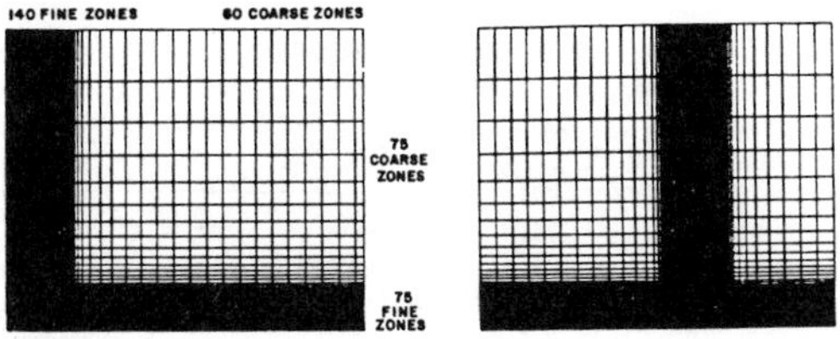

Fig. 1. Adaptive grid for height-of-burst problem shown (a) initially and (b) at time when transition to Mach reflection occurs.

The fact that FCT algorithm is 1-D allows us to easily combine a 1-D and 2-D numerical calculation to "everywhere" use the 1-D solution where the flow field remains 1-D. More difficult is the algorithm that decides where the transition from 1-D to 2-D occurs. In practice one can search from the outside inward to find a discontinuity in pressure and density which defines a shock. Generally, for a blast wave solution in one-dimension only one such shock is present. Subsequent shocks are indicative of reflections or 2-D interactions.

III. APPLICATION

A particular problem that this technique is applied to is shown in Fig. 2. Shown are contour plots of pressure and density, velocity vectors and the adaptive grid after 3000 cycles. The fine-grid region has captured the shock and the double mach structure near the ground. The calculation is a simulation of one kiloton ($4.2x10^{19}$ ergs) of energy deposited in the air 104 feet above the ground surface. Because it is a simulation of a nuclear explosion the temperatures and sound speeds are very high behind the shock front. In the chemical explosive case the temperatures are many times less. The reflected shock from the ground which makes the problem 2-D passes through the region behind the spherically expanding shock. Since the reflected shock velocity depends on the local temperature, the nuclear case becomes 2-D sooner than the chemical explosive case. Nonetheless, the improved definition of the flow field in the large zones results in a better overall solution.

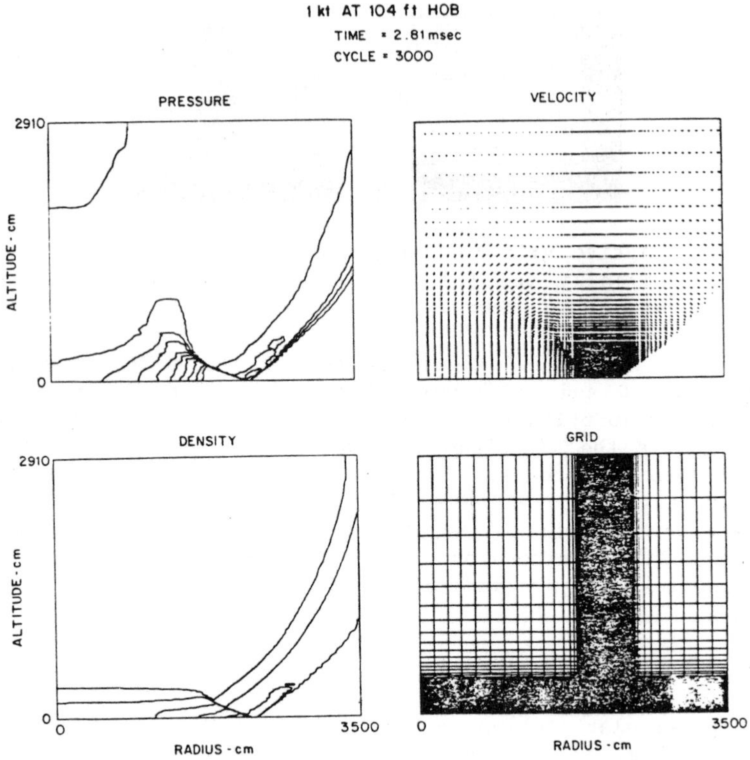

Fig. 2. Adaptive grid FCT solution for nuclear case.

The above methodology has also been applied to a series of test problems initiated by a spherical high-explosive (HE) detonation in air. An ideal Chapman-Jouguet detonation was used to specify the initial conditions; afterburning was neglected. In the absence of reflecting surfaces, spherical symmetry is maintained and the calculation remains one-dimensional, (Fig. 3). A nonuniform radial grid was used with extremely fine zoning near the shock front. The grid was moved so that the shock remained approximately fixed with respect to the mesh.

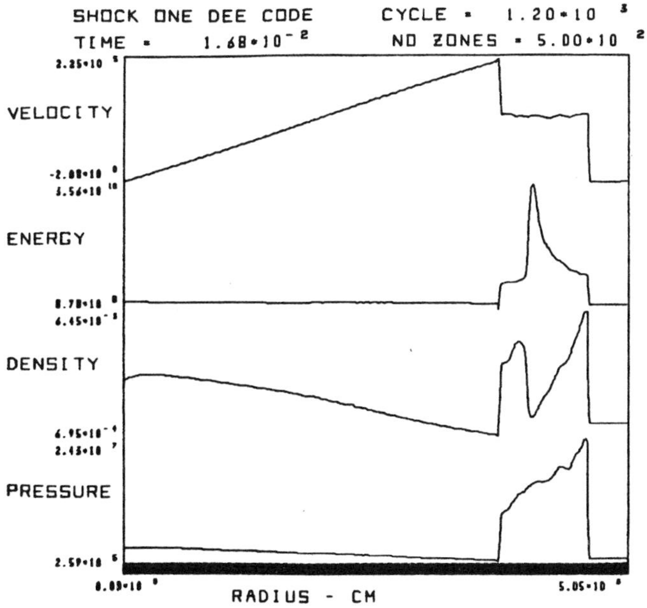

Fig. 3. One-dimensional solution of expanding HE products and air calculated with the new algorithm using 500 equally spaced zones. Note contact surface separating He products from air.

Next, 2-D numerical calculations were performed to simulate height-of-burst experiments. The previous fine-zoned 1-D calculation was used to initialize the problem. It was mapped onto the 2-D grid just prior to the onset of reflection. The solution was then advanced in time, with pressure being calculated from a real-air equation of state and a JWL equation of state for the combustion products. The front of the blast wave was captured in a finely gridded region which moved outward horizontally. Special care was taken to ensure that the grid moved smoothly. Fig. 4a and 4b are comparisons of the 2-D simulations. Fig. 4a shows adaptive gridding without the 1-D, 2-D combination. Fig. 4b includes the new methodology. Results are sharper and better defined in all parts of the computational grid.

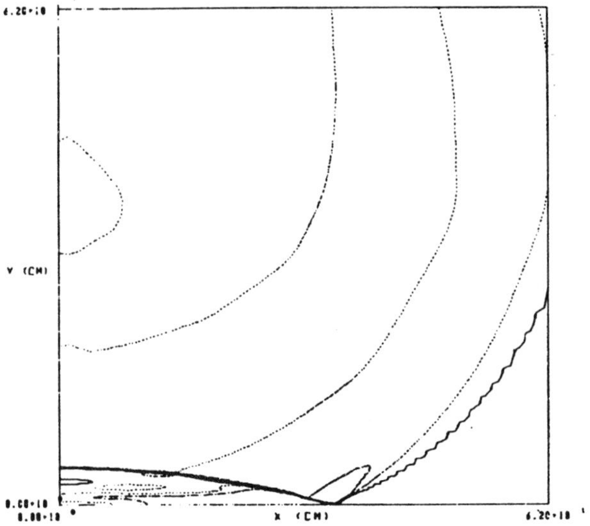

Fig. 4a. Sample Chemical Explosion Pressure Contours Note resolution at upper part of grid.

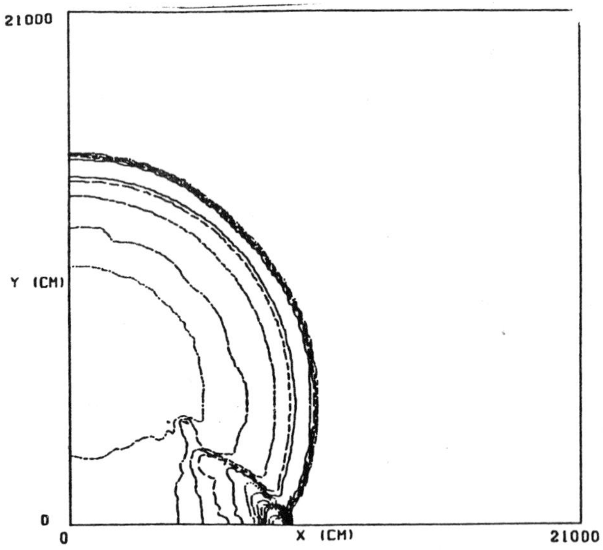

Fig. 4b. Sample Chemical Explosive Pressure Contours Note improved resolution.

1. BORIS, J., and BOOK D. - Methods in Computational Physics, J. Killeen, Ed., Academic Press, New York, Vol. 16, p. 85, 1976.

2. FRY, M., et. al. - Shock-Capturing Using FCT Algorithms with Adaptive Gridding, NRL Memo Report 4629, 1981.

3. BOOK, et. al. - Seventh International Conference on Numerical Methods in Fluid Dynamics, Stanford, 1980.

4. BORIS, J. - Flux-Corrected Transport Modules for Generalized Continuity Equations, NRL Memo Report 3327, 1976.

NUMERICAL PREDICTION OF FLOW CHARACTERISTICS OF CONFINED TURBULENT PREMIXED FLAME

G.DEVAPAUL* V.GANESAN@ T.L.SITHARAMA RAO*

* Department of Mechanical Engineering, Regional Engineering College, Warangal-506004. INDIA.

@ Department of Mechanical Engineering, Indian Institute of Technology, Madras-600036. INDIA.

ABSTRACT:

An understanding of the flow field in turbulent premixed combustion system leads to a rational design of combustion chambers for gas turbines and jet engines. This paper presents a numerical prediction of the flow field primarily of axial velocity and concentration of species in a confined premixed turbulent flame stabilized by a hot auxiliary flow. The prediction makes use of a finite difference technique using the marching integration method. The eddy-break-up model with modified constants and Prandtl mixing length hypothesis have been incorporated for the reaction rate and hydrodynamics respectively.

The results are compared with the available experimental data and it is found that there is a fairly good agreement between the predicted and the experimental velocity field. However the concentration field shows some discrepancies. A complex reaction mechanism instead of the single step reaction assumed in this analysis may improve the predictions.

1. INTRODUCTION:

Confined turbulent premixed flames fall under two categories: flames stabilized by bluff bodies and flames stabilized by a parallel hot pilot flow. Till about 1975 experimental and theoritical investigations centred around flames of the first category. Williams et al [1], Wright and Zukoski [2], Howe et al [3] and Cushing et al [4] have studied experimentally the flow field of rod stabilized flames. Fabri et al [5] and Iida [6] have given theoritical predictions based on equations of flow. Moreau and Boutier [7] and Moreau [8,9] have reported experimental investigations on turbulent flames of the second category.

With the advancement of numerical techniques for solving partial differential equations that govern flow and combustion process Spalding [10,11] predicted numerically the flow field of a rod stabilized flame investigated by Howe et al [3]. Similarly Borghi and Moreau [12] predicted the flow field of flames stabilized by hot pilot flow based on the expected shape of probability density functions from velocity measurements.

An attempt is made in this work to predict the transverse velocity, concentration of reactants and carbondioxide and reaction rate for confined turbulent premixed flames of the second category by making use of the method of numerical solutions proposed by Spalding [13] for two dimensional parabolic phenomena.

The results are compared with published experimental data [7,8,15].

2. THE COMBUSTION SYSTEM CONSIDERED:

The combustion system considered for the theorėtical analysis is the same as experimentally investigated by Moreau [8], the basic features of which are given in Fig. 1

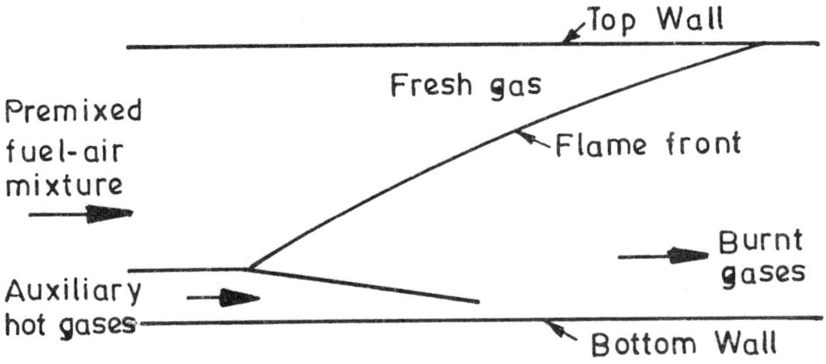

FIG.1 SYSTEM CONSIDERED

Premixed methane-air mixture at a temperature of about $600^{o}K$ flows through a suitable device to create the desired level of turbulence and enters the combustion chamber of cross section 100 x 100 mm and length 1300 mm. A divider plate located about 19 mm from the bottom wall of the combustion chamber at its entrance helps to separate the two streams before they mix at the trailing edge. Completely burnt products of chemically correct methane-air mixture enter the lower part of the combustion chamber at a temperature of about $2000^{o}K$ and a velocity of about twice that of the main flow over it. In the test section the two streams mix due to the intense shear at the interphase and the turbulent flame spreads into the fresh mixture as the fl(w proceeds along the combustion chamber.

3. GOVERNING EQUATIONS:

Making use of the $x \sim \omega$ co-ordinate system proposed by Spalding and Patankar [10] the partial differential equations for mean velocity, u, mass fraction of fuel, m_{fu}, mass fraction of oxygen, m_{ox}, mass fraction of products, m_{pr}, the conserved properties $(m_{ox} - m_{fu}S)$ and $\{m_{pr} + m_{fu}(1 + S)\}$, enthalpy, H, etc. can be expressed in the general form;

$$\frac{\partial \phi}{\partial \omega} + (a + b\omega)\frac{\partial \phi}{\partial \omega} = \frac{\partial}{\partial \omega}\left(c\frac{\partial \phi}{\partial \omega}\right) + d \qquad (1)$$

where ϕ stands for any of the above independent or dependent quantities

$$\omega = \frac{\psi - \psi_I}{\psi_E - \psi_I} \qquad (2)$$

$$a = r_I \dot{m}''_I / (\psi_E - \psi_I) \qquad (3)$$

$$b = (r_E \dot{m}''_E - r_I \dot{m}''_I) / (\psi_E - \psi_I) \qquad (4)$$

$$c = r^2 \rho u \Gamma_{\phi,e} / (\psi_E - \psi_I)^2 \qquad (5)$$

$$\text{and } d = (1/\rho u)\, S_\phi \qquad (6)$$

In these relationships ψ_I and ψ_E are functions of x, the whole of the region of interest being contained within the internal and external boundaries indicated by I and E (which in this case refer to the lower and upper walls of the combustion chamber respectively), $\dot{m}''_I$ and $\dot{m}''_E$ are the mass fluxes at the I and E boundaries respectively, r_I and r_E are the distances from the symmetry axis, $\Gamma_{\phi,e}$ is the effective exchange coefficient of the variable ϕ, and S_ϕ is the source term which takes the appropriate equation for the variable under consideration.

The finite difference equations of the implicit formula for the equations of the form (1) can be written as:

$$D_i \phi_{i,D} = A_i \phi_{i+1,D} + B_i \phi_{i-1,D} + C_i \qquad (7)$$

where A_i, B_i and C_i are constants, i represents the ith grid number and D stands for downstream.

The above constants can be evaluated by integrating the partial differential equations over the control volume. Finite difference equations for each grid point other than those on the internal and external boundaries and each independent variable, $u, H, m_{fu}, (m_{ox}-m_{fu}S)$ etc. are solved simultaneously by the tri-diagonal matrix algorithm.

4. EDDY-BREAK-UP MODEL:

Spalding [11] has expressed that the Arrhenius reaction rate fails to enlarge the influence of the local turbulence level on the rate of reaction in turbulent premixed flames and hence proposed the eddy-break-up model[14]. According to this model the reaction zone of turbulent flames is supposed to consist of interspersed sheets and filaments of fully reacted and completely unreacted material and the rate of reaction depends upon the rate of stretching of the sheets and filaments and the latter is taken simply proportional to $\partial u/\partial y$. Thus the reaction rate is given by:

$$R_{fu} = - C_{EBU} \frac{(m_{fu,u} - m_{fu})(m_{fu} - m_{fu,b})}{(m_{fu,u} - m_{fu,b})} \rho \left|\frac{\partial u}{\partial y}\right| \qquad (8)$$

where $m_{fu,u}$ stands for the mass fraction of fuel in the completely unburnt mixture having the locally present value of $(m_{ox}-m_{fu}S)$ and $m_{fu,b}$ stands for the mass fraction of the fuel that would be present in the fully burnt mixture and C_{EBU} is known as the eddy-break-up constant.

5. MIXING LENGTH HYPOTHESIS:

According to Prandtl mixing length hypothesis

$$\mu_e = \mu_l + \mu_T \qquad (9)$$

$$\text{and } \mu_T = \rho l_p^2 \left|\frac{\partial u}{\partial y}\right| \qquad (10)$$

The mixing length constant is assumed to be 0.14. Thus the mixing length is taken as 0.14 times the width of the shear layer or 0.435 times the distance from the wall whichever is smaller

6. NUMERICAL PREDICTION:

Computations were performed in I.B.M. 370/155 system with a grid number of 40. The step size was kept around 5 mm. The total time taken was less than two and half minutes. Since a single value for the eddy-break-up constant could not predict the temperature profiles properly it has been modified so that the value could be varied with respect to x,y and equivalence ratio, ϕ. Single step reacton was assumed so that the fuel is oxidised to carbondioxide and water vapour directly.

7. DISCUSSION:

7.1 Velocity Profiles:

Fig. 2 gives the cross stream velocity profiles of the flow at two stations in the test section for inlet gas velocities of 130 and 65 m/s and methane-air equivalence ratio 0.80.

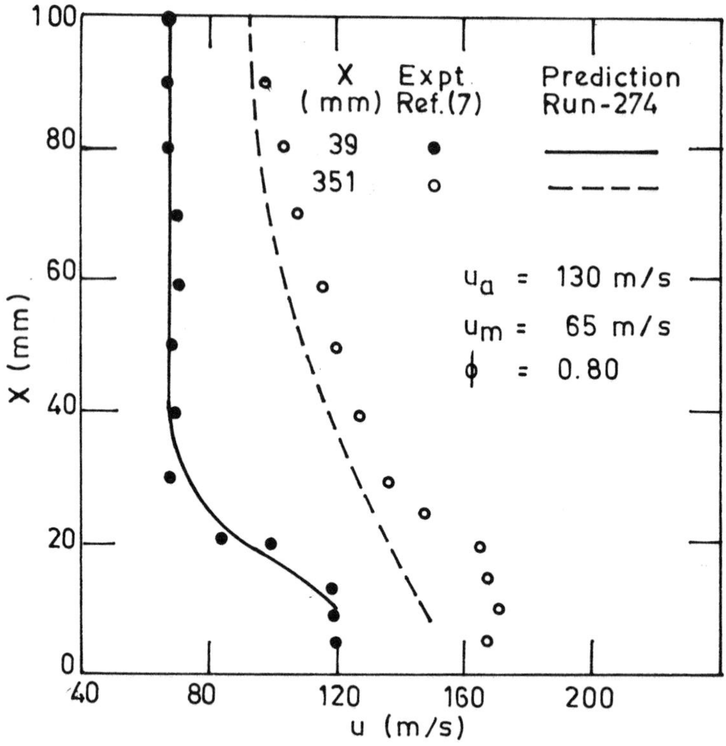

FIG.2 VELOCITY PROFILES

The predicted values agree well with the measured values [7] at x=39 mm. However at x = 351 mm the predicted values are lower than the experimental values. Improper matching of the temperature profile at this station may be the cause for this discrepancy.

7.2 Concentration Profiles:

Fig. 3 (a) and (b) give the mass concentration profiles of fuel and oxygen respectively. The agreement between the predicted and the experimental values [15] is good.

Fig. 4 gives the molar concentration profiles of methane related to the molar concentration of nitrogen for two stations. At station x = 322 mm there is a satisfactory agreement between the predicted and the experimental values [8] as observed in

Fig. 3 (a). At station x = 522 mm the predicted values are larger than those measured by experiment upto a duct height of about 60 mm beyond which there is a satisfactory agreement.

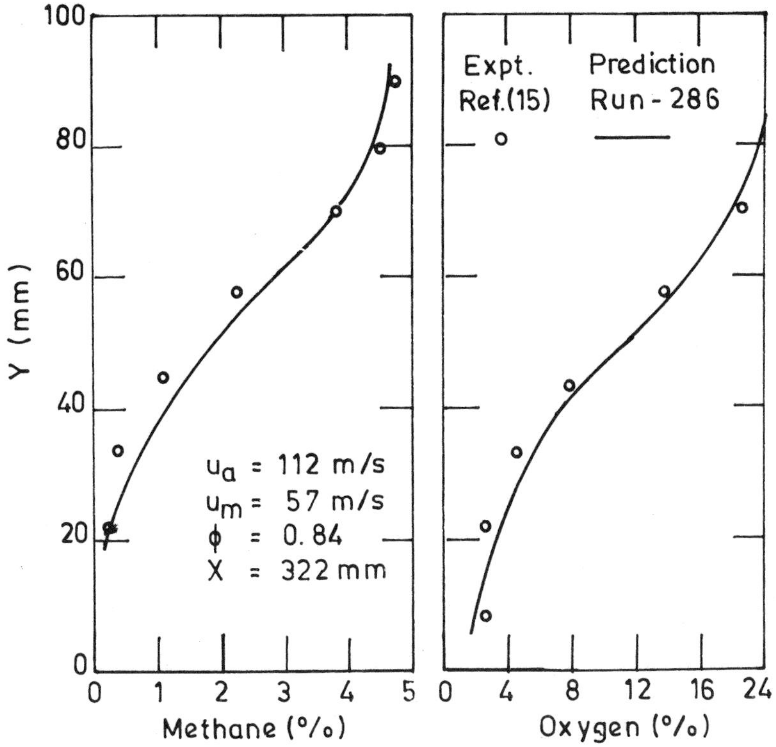

FIG.3 MASS FRACTION PROFILES

Fig. 5 gives the concentration profiles of oxygen and the trends are similar to those observed in Fig. 3 (b) at x=322 mm. At x = 522 mm the predicted values are larger than experimentally obtained values upto a duct height of about 60 mm.beyond which there is a satisfactory agreement. The general trend is similar to what is observed in Fig. 4 for methane concentration.

Fig. 6 gives the profiles of carbondioxide concentration. At x = 322 mm the measured values are lower than the predicted values at all points across the duct. At x = 522 mm there seems to be a reasonable agreement between the two values except in the hot gas zone near the bottom wall where the measured concentration is lower than the predicted vlaues. This discrepancy may be partly due to the dissociation of carbondioxide.

7.3 Reaction Rate Profiles:

Fig. 7 gives the variation of the predicted reaction rate across the combustion chamber at station x = 318 mm for two equivalence ratios. The reaction rate in the hot gas zone appears to be about 100 to 1000 times greater than the rate at the region near the top wall. Further it appears to be independent of the distance from the bottom wall in the hot gas zone for an equivalence ratio of 0.61. However for nearly chemically correct mixture the reaction rate reaches a maximum and then decreases rapidly towards the top wall.

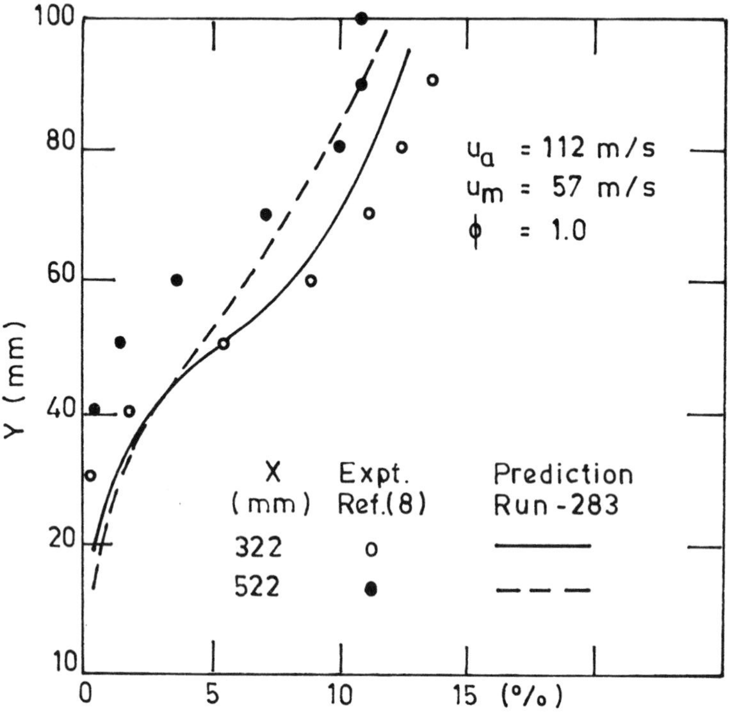

FIG.4 CONCENTRATION PROFILES FOR CH_4/N_2

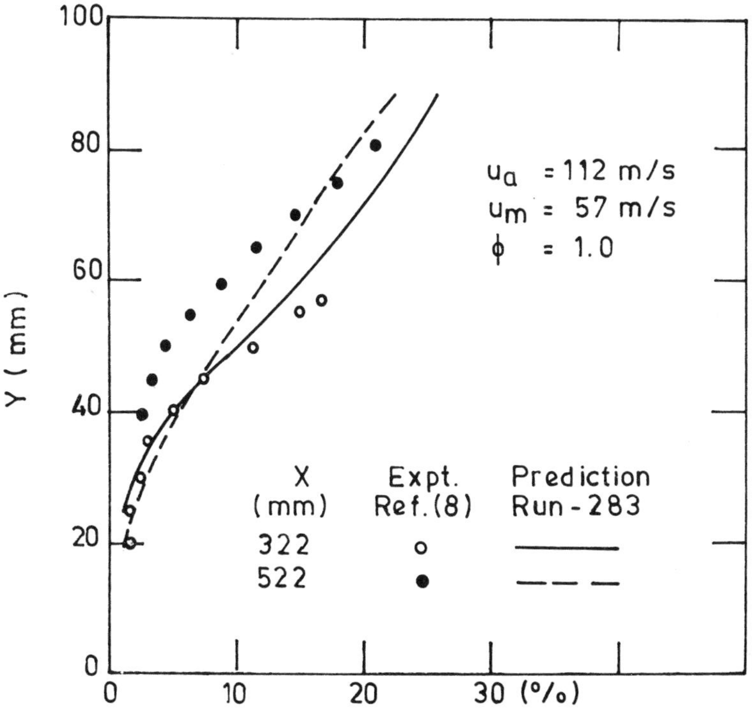

FIG. 5 CONCENTRATION PROFILE FOR O_2 / N_2

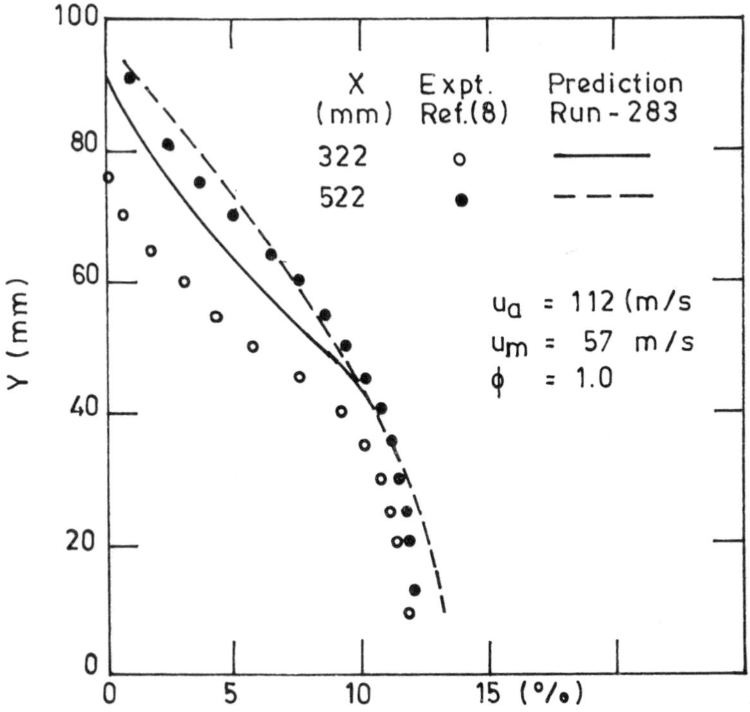

FIG.6 CONCENTRATION PROFILES FOR CO_2/N_2

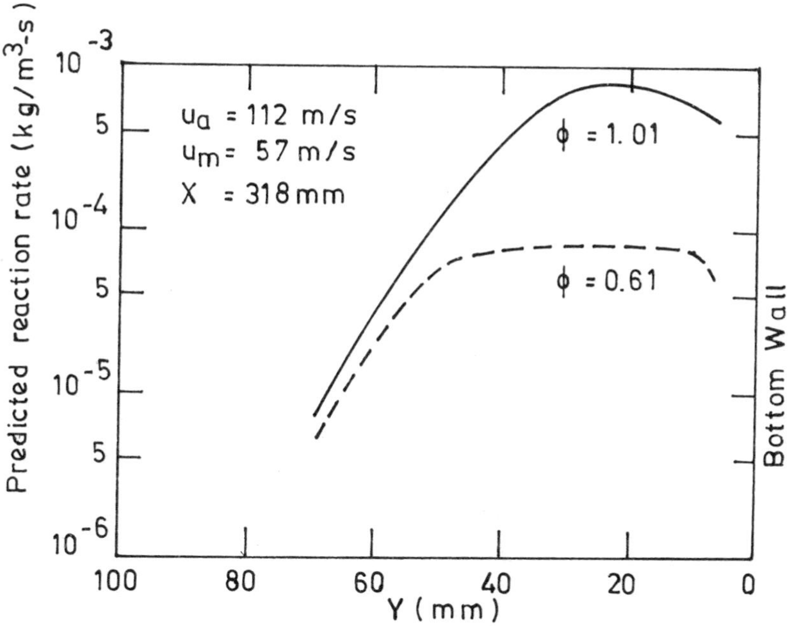

FIG.7 REACTION RATE PROFILE

8. CONCLUSION:

The eddy-break-up model with modified constant coupled with the Prandtl mixing length hypothesis is capable of predicting the flow characteristics in the case of a turbulent premixed flame stabilized by an auxiliary hot gas flow. However, with a complex reaction mechanism in place of the single step reaction assumed in this work the flow field may be predicted more acurately.

REFERENCES:

1. WILLIAMS, G.C., HOTTEL,H.C. and SCURLOCK A.C.- Flame stabilization and propagation in high velocity gas streams, Third Symposium (International) on Combustion, Williams and Wilkins, Baltimore, pp. 21-39, 1949.

2. WRIGHT, G.H. and ZUKOSKI, E.E.- Flame spreading from bluff-body flame holders, Eighth Symposium (International) on Combusion, Williams and Wilkins, Baltimore, pp.933-943, 1962.

3. HOWE, N.M., JR., SHIPMAN, C.W. and VRANOS, A.-Turbulent mass transfer and rates of combustion in confined turbulent flames, Ninth Symposium (International) on Combustion, The Combustion Institute, Pittsburgh, pp.36-47, 1963.

4. CUSHING, B.S., FAUCHER, J.E.,GANDBIR, S., and SHIPMAN,C.W.- Turbulent mass transfer rates of combustion in confined turbulent flames II, Eleventh Symposium (International) on Combustion, The Combustion Insittute, Pittsburgh, pp. 817-824, 1967.

5. FABRI, J., SIESTRUNCK, R., AND FOURE, C.- On the aerodynamic field of stabilized flames, Fourth Symposium (International) on Combustion, Williams and Wilkins, Baltimore, pp. 443-450, 1953

6. Iida, H.- Combustion in turbulent gas streams, Sixth Symposium (International) on Combustion, Reinhold, pp.341-350, 1958.

7. MOREAU, P. and BOUTIER, A.- Laser velocimeter measurement in a turbulent flame, Sixteenth Symposium (International) on Combustion, The combustion Institute, Pittsburgh, pp. 1746-1756, 1976.

8. MOREAU, P.- Turbulent flame development in a high velocity pre-mixed flow, AIAA Paper No. 77-49

9. MOREAU, P.- Experimental determination of probability density functions with a turbulent high velocity premixed flame, Eighteenth Symposium (International) on Combustion, The Combustion Institute, Pittsburgh, pp. 993-1000, 1981.

10. SPALDING, D.B. and PATANKAR, S.V.- Heat and mass transfer in boundary layers, Morgan-Grampian, London, 1967.

11. SPALDING,D.B.- Mixing and chemical reation in steady confined turbulent flames, Thirteenth Symposium (International on combustion, pp. 649-657, 1971.

12. BORGHI, R. and MOREAU, P.- Turbulent combustion in premixed flow, Acta Astronautica, Vol. 4,Pergamon Press, pp. 321-341, 1977.

13. SPALDING, D.B.- Genmix: A general computer programme for two dimensional parabolic phenomena, Pergamon Press,Oxford, 1977.

14. SPALDING, D.B.-Mathematical models of turbulent flames: A Review, Combustion Science and Technology, Vol. 13, pp.3-25, 1976.

15. SINGH,V.P., BORGHI, R. and MOREAU,P.-Reactions chimiques dans les flammes turbulentes, Second Int. Symposium on dynamics of chemical reations, Padoue, 15-17, Dec. 1975.

THE APPLICATION OF FINITE DIFFERENCE AND FINITE ELEMENT METHODS TO A REACTION-DIFFUSION SYSTEM IN COMBUSTION

J.I. RAMOS*

SUMMARY

The ozone-decomposition flame is studied by means of fixed-node finite difference, and adaptive and moving finite element methods. The finite difference schemes employed in this study include a fourth-order accurate method of lines, a fourth-order accurate majorant operator-splitting technique and three time linearization algorithms. These algorithms have different temporal and spatial resolutions and are aimed at assessing the effects of the truncation errors on the ozone laminar flame speed. The finite element methods use a Galerkin approximation and linear basis, and are based on a projection technique. These methods move the grid according to the computed flame speed or to the maximum temperature gradient, and keep a number of fixed elements within the flame front. Calculations are presented for 121 grid points and 36 elements and show that the finite difference and finite element methods predict as accurate answers as other adaptive finite difference procedures. It is shown that the wave speed accuracy improves with the order of the truncation error, and that adaptive and moving finite element methods may yield accurate results.

1. INTRODUCTION

The ozone-decomposition flame has been the subject of numerous numerical studies [1-4]. Bledjian [1] employed a second-order accurate method of lines technique to solve the equations which govern the propagation of one-dimensional flames in cartesian coordinates; he computed a wave speed of 54.3 cm/s. Margolis [2] studied a premixed combustible mixture emerging from a flat plate burner, introduced lagrangian coordinates and reduced the problem to a system of reaction-diffusion equations. He then introduced appropriate spline basis for the spatial variations into the lagrangian equations and, after imposing collocation and boundary conditions, obtained a stiff initial value problem. Margolis [2] used sixth-order B-splines and 272 collocation points, compared with the 100 grid points used by Bledjian; the computed wave speed was 49.7 cm/s. Meintjes [3] employed an explicit predictor-

*Assistant Professor, Department of Mechanical Engineering, Carnegie-Mellon University, Pittsburgh, PA 15213, U.S.A.

corrector method algorithm to compute the ozone-decomposition flame speed and showed that the wave speed approaches a constant value of 48 cm/s. Meintjes estimates that this value may be in error by 2 cm/s because of the integrations required to calculate the coordinates and the velocity; he used a grid consisting of 121 points. Reitz [4] employed an adaptive finite-difference explicit method and showed that his second-order accurate scheme computes an average wave speed of 49.8 ±0.1 cm/s. This result is very much the same as that calculated by Margolis [2] using a fourth-order accurate finite element method. The wave speed computed by Reitz [4] was obtained with 30 grid points whose location depended on the minimization of the mesh spacing changes. The mesh was searched for the point at which the largest temperature gradient occurs, and the minimum value of the mesh spacing was placed at this point.

In this paper we study the ozone-decomposition flame by means of finite-difference and finite element methods. The finite-difference schemes employed in this study include a fourth-order accurate method of lines, a fourth-order accurate majorant operator-splitting method and three time-linearization algorithms. All of these finite-difference procedures use fixed spatial grids and are aimed at assessing the influence of different spatial and temporal approximations on the computed flame speeds.

The propagation of flames through gaseous mixtures is characterized by the presence of a very steep front; away from the flame front, the species and temperature profiles are almost uniform in unconfined adiabatic flames. The presence of sharp moving boundaries within the computational domain can be resolved by using adaptive and moving finite-difference and finite element techniques. These techniques concentrate the grid points in regions where rapid changes occur or where the local truncation errors exceed a certain specified tolerance. In addition, these grids may move with the sharp boundary thus reducing the computational cost. Finite element methods offer the advantage that errors can be estimated when the mesh is adapted to the solution procedure. Lee and Ramos [5], for example, employed a semi-discrete Galerkin method to study the propagation of laminar flames through confined gaseous mixtures and developed a moving finite element method (FEM) whose grid points are located at the flame front. Miller [6] developed a moving FEM method whose nodal locations and nodal amplitudes change in time; penalty functions were included to limit the minimum separation between nodes. Miller's method was employed to compute Bürgers' equation; the results were very accurate.

In this paper we present two FEMs; the first method moves the elements with the previously computed flame speed which has to be accurately known at the beginning of the calculations. The second method uses an adaptive grid which is concentrated around the maximum temperature gradient at the flame front. This grid moves in time; the calculation of the nodal amplitudes in the new grid is based on a projection method and the minimization of a certain functional. Comparisons between the fixed-node finite-difference and adaptive finite element method results are presented; the methods are also compared with other numerical techniques and moving FEMs.

2. FORMULATION OF THE PROBLEM

We consider the propagation of a one-dimensional adiabatic flame through a premixed combustible mixture of O_2 and O_3. Neglecting the body forces and assuming that the Mach number is small and that the species diffuse according to Fick's law with equal mass diffusivities for all of them, the conservation equations of mass, momentum, energy and species mass fractions are

$$\partial \rho/\partial t + \partial(\rho u)/\partial x = 0, \tag{1}$$

$$p = \text{constant}, \tag{2}$$

$$\rho\, C_p\, [\partial T/\partial t + u\partial T/\partial x] = \partial(\lambda\partial T/\partial x)/\partial x - \sum_{i=1}^{3} h^o_i\, w_i, \tag{3}$$

$$\rho[\partial Y_i/\partial t + u\partial Y_i/\partial x] = \partial \rho D\partial Y_i/\partial x)/\partial x + w_i,\ i = 1, 2, \tag{4}$$

$$Y_3 = 1 - Y_1 - Y_2, \tag{5}$$

$$\rho = p/RT \sum_{i=1}^{3} Y_i/W_i), \tag{6}$$

where ρ is the density, u the velocity, t the time, x the spatial coordinate ($-\infty<x<\infty$), p the pressure, T the temperature, C_p the specific heat at constant pressure, λ the thermal conductivity, h^o the enthalpy of formation, w_i the reaction rate of species i, Y the species mass fraction, D the mass diffusivity, R the universal gas constant, and W the molecular weight. In the above equations, the viscous dissipation terms have been neglected, and the specific heats at constant pressure for the different species have been considered equal and constant. Equations (1), (3) and (4) can be reduced to a system of reaction-diffusion equations by means of the mapping $(t,x) \rightarrow (t*,\Psi)$ defined by

$$t* = t, \qquad (t \geq 0), \tag{7}$$

$$\partial\Psi/\partial t = -\rho u \quad \text{and} \quad \partial\Psi/\partial x = \rho. \tag{8}$$

Introducing Equations (7) and (8) into Equations (3) and (4) we obtain

$$\partial T/\partial t* = \alpha\partial^2 T/\partial\Psi^2 - \sum_{i=1}^{3} h^o_i\, w_i/\rho, \tag{9}$$

$$\partial Y_i/\partial t* = \alpha\partial^2 Y_i/\partial\Psi^2 + w_i/\rho,\ i = 1, 2, \tag{10}$$

where $\alpha = \rho^2 D$ is assumed constant and the Lewis number was taken equal to one.

Equations (9), (10), (5) and (6) yield the values of the temperature, species mass fractions and density, respectively. Once these variables are known the x-coordinate can be calculated from Equation (8) as

$$\Delta x = -\int_{\Psi}^{\Psi+\Delta\Psi} d\Psi/\rho, \tag{11}$$

whereas the velocity can be calculated from Equation (1) to yield

$$\Delta(\rho u) = -\int_{x}^{x+\Delta x} (\partial \rho/\partial t)\, dx. \tag{12}$$

In order to fully specify the problem the reaction rate terms, w_i, need to be considered. In the calculations reported here, we have employed the seven-reaction mechanism for the combustion of O_2 and O_3 [2]. This mechanism can be written as

$$O_3 + X \Longleftrightarrow O_2 + O + X, \tag{13}$$

$$O + O_3 \Longleftrightarrow 2\, O_2, \tag{14}$$

$$O_2 + X \Longleftrightarrow 2\, O + X, \tag{15}$$

where X represents any of the three species, i.e., O, O_2 and O_3. The values of the reaction rates, activation energies, pre-exponential factors and other constants are given by Margolis [2] and are not repeated here. It is sufficient to say that the reaction rates are very nonlinear functions of the temperature and that in the present study Equation (10) was employed to calculate the mass fractions of O_2 and O_3, whereas Equation (5) was used to compute the mass fraction of O which is much smaller than the other species mass fractions. Equations (9) and (10) were nondimensionalized (cf. Margolis [2]) and solved subject to the appropriate initial and boundary conditions. Initially the velocity was assumed equal to zero, and temperature and species mass fraction profiles were used to simulate the fresh and burned mixture [2]. The velocity upstream of the flame was taken equal to zero, and adiabatic conditions were used for both, upstream and downstream, locations. The physical domain, $-\infty<\Psi<\infty$, was truncated to reduce the computational time; the truncated domain was selected so that the locations of the upstream and downstream boundaries do not affect the computational results.

The steady state flame speeds, U_i, were defined as

$$\partial Y_i/\partial t^* + U_i\, \partial Y_i/\partial \Psi = 0,\ i = 1,2. \tag{16}$$

Equation (16) is used to calculate $\partial Y_i/\partial t^*$, which is then substituted into Equation (10); the resulting equations are then integrated to yield the following flame speeds based on O_2 and O_3

$$U_i = \int_{-\infty}^{\infty} d\Psi w_i/\rho/\left(Y_i(-\infty,t^*)-Y_i(\infty,t^*)\right),\ i = 1,\ 2. \tag{17}$$

Equations (9) and (10) can be written in vector notation as

$$\partial A/\partial t^* = \alpha\, \partial^2 A/\partial \Psi^2 + S, \tag{18}$$

where $A = (Y_1, Y_2, T)$ and $S = (w_1/\rho,\ w_2/\rho,\ -\sum_{i=1}^{3} h^\circ_i\, w_i/\rho)$. The reaction rate w_3 is given by $w_3 = -w_1 - w_2$.

Equation (18) represents a system of three reaction-diffusion equations for the species mass fractions of O_2 and O_3, and the temperature. The source terms, S, in Equation (18) were written in terms of Y_1, Y_2 and T by employing Equations (5) and (6); thus the only unknowns are Y_1 (O_2), Y_2 (O_3) and T.

3. THE NUMERICAL SCHEMES

Equation (18) was solved by means of finite-difference and finite element methods. The finite-difference procedures include a fourth-order-accurate method of lines, a fourth-order-accurate majorant operator-splitting method, and three time-linearization techniques. These techniques use different spatial and temporal approximations and are aimed at assessing the influence of truncation errors and time linearization on the numerical calculations. The finite element methods use piecewise linear basis and a Galerkin approximation. The methods move (or adapt) the grid according to the computed wave speed (moving FEM) or the maximum temperature gradient at the flame front (adaptive FEM). The interpolation of the dependent variable in the new grid is based on a projection method and the minimization of a certain functional. The moving FEM has been previously used by Lee and Ramos [5] in the computation of a nonlinear reaction-diffusion equation which has an exact traveling wave solution.

3.1 A FOURTH-ORDER-ACCURATE METHOD OF LINES

In this technique the following fourth-order-correct central difference approximation

$$(\partial^2 A/\partial\Psi^2)_i = (-A_{i-2} + 16A_{i-1} - 30A_i + 16A_{i+1} - A_{i+2})/12\Delta\Psi^2, \qquad (19)$$

was introduced into Equation (18) to yield a set of ordinary differential equations for A_i which was solved by means of a fourth-order-accurate Runge-Kutta method. The truncation error of this scheme is $O(\Delta t^{*4}, \Delta\Psi^4)$.

3.2 A FOURTH-ORDER-ACCURATE MAJORANT OPERATOR-SPLITTING METHOD

Equation (18) has also been solved by means of a majorant operator-splitting method. This method allows one term to be considered alone during each fractional step, while the remaining terms are ignored. Equation (18) is split into a reaction and diffusion operator as follows

$$\text{(Reaction Operator) } L_R\text{: } dA_i/dt^* = S_i, \qquad (20)$$

$$\text{(Diffusion Operator) } L_D\text{: } \partial A_i/\partial t^* = \alpha\partial^2 A_i/\partial\Psi^2. \qquad (21)$$

Thus, the solution is advanced one time step by the sequence

$$A^{n+1}_i = L_D\, L_R\, A^n_i. \qquad (22)$$

In our calculations Equation (20) was first integrated by means of a fourth-order-accurate Runge-Kutta method to yield the solution A_i. During the integration of Equation (20) the time step was varied as necessary to maintain a specified accuracy and was in general much smaller than $\Delta t^* = t^{*n+1} - t^{*n}$. Once the solution of the reaction operator is known, the diffusion equation (21) can be solved to yield A^{n+1}_i. This equation was solved by means of the compact three-point scheme for the diffusion terms given by

$$\left(\partial^2 A/\partial\Psi^2\right)_i = \left[\delta^2/(1 + \delta^2/12)\, A_i\right]/\Delta\Psi^2 + O(\Delta\Psi^4), \qquad (23)$$

where δ^2 is the second-order-accurate central difference operator defined by $\delta^2 A_i = A_{i+1} - 2A_i + A_{i-1}$.

Equation (21) is then solved by means of the following Crank-Nicolson-like algorithm

$$\frac{A^{n+1}_i - \bar{A}_i}{\Delta t} = \frac{a}{\Delta\Psi^2}\frac{\delta^2}{1 + \delta^2/12}\left(\frac{A^{n+1}_i + \bar{A}_i}{2}\right), \qquad (24)$$

which can be written as

$$(-\gamma+1/12)A^{n+1}_{i-1} + (1+2\gamma-2/12)A^{n+1}_i + (-\gamma+1/12)A^{n+1}_{i+1} =$$
$$(\gamma+1/12)\,(\bar{A}_{i+1}-2\bar{A}_i + \bar{A}_{i-1}) + \bar{A}_i, \qquad (25)$$

where $\gamma = a\Delta t/2\Delta\Psi^2$. Equation (25) represents a tridiagonal matrix which can be solved by the standard tridiagonal matrix algorithm.

3.3 TIME-LINEARIZATION ALGORITHMS

Equation (18) represents a system of highly nonlinear and coupled reaction-diffusion equations whose solutions may be accelerated by linearization algorithms which use different spatial and temporal approximations. These algorithms use time-linearization as opposed to iteration-linearization in which the nonlinear terms are linearized with respect to the previous iterated value. Both, time-linearization and iteration-linearization schemes give rise to a system of linear and uncoupled reaction-diffusion equations which can be solved by means of the standard tridiagonal matrix algorithm. Although both schemes result in tridiagonal matrices, it should be noted that the accuracy of the time-linearization deteriorates as longer time steps are taken in the computations. This problem does not occur in the iteration-linearization because one has to iterate until a specified convergence criterion is reached. The algorithms just describe are to be distinguished from the linear-block-tridiagonal algorithm in which the nonlinear terms are linearized with respect to all of the dependent variables; this leads to a system of linear, but coupled, reaction-diffusion equations which can be solved, for example, by using LU decomposition.

We consider the following time linearization schemes

$$A^{n+1}_i - A^n_i = \frac{a\Delta t^*}{\Delta\Psi^2}\,\delta^2 A^{n+1}_i + \left[S^n_i + \left(\frac{\partial S}{\partial A}\right)^n_i I\,(A^{n+1}_i - A^n_i)\right]\Delta t^*, \qquad (26)$$

$$\frac{1}{2}\left(\frac{\partial S}{\partial A}\right)^n_i I\,(A^{n+1}_i - A^n_i)\Big]\,\Delta t^*, \frac{A^{n+1}_i - A^n_i}{\Delta t^*} = \frac{a}{\Delta\Psi^2}\frac{\delta^2}{1 + \delta^2/12} \qquad (27)$$

$$\left[\frac{A^n_i + A^{n+1}_i}{2}\right] + \frac{1}{2}\left[S^n_i + S^{n+1}_i\right], \qquad (28)$$

where I denotes the unit matrix.

Equation (26) corresponds to the standard implicit method expression except that the reaction terms have been linearized, whereas Equations (27) and (28) use the Crank–Nicolson scheme and linearization of the reaction terms. Equation (28) also uses a compact, fourth-order-accurate, three-point scheme for the diffusion terms, which can be written as

$$\left[I/12 - \gamma I/2 - \Delta t^* P^n_{i-1}/24\right] A^{n+1}_{i-1} + \left[I - 2I/12 + \gamma I - \Delta t^* P^n_i/2 + \Delta t^* P^n_i/12\right] A^{n+1}_i + \left[I/12 - \gamma I/2 - \Delta t^* P^n_{i+1}/24\right] \bullet A^{n+1}_{i+1} = \gamma \delta^2 A^n_i + \Delta t^* S^n_i + \Delta t^* \delta^2 S^n_i/12 + \left[I/12 - \gamma I/2 - \Delta t^* P^n_{i-1}/24\right] A^n_{i-1} + \left[I - 2I/12 - \gamma I - \Delta t^* P^n_i/2 + \Delta t^* P^n_i/12\right] A^{n+1}_i + \left[I/12 - \gamma I/2 - \Delta t^* P^n_{i+1}/24\right] A^{n+1}_{i+1}, \quad (29)$$

where $P^n_i = (\partial S/\partial A)^n_i$ and $\gamma = \alpha \Delta t^*/\Delta\Psi^2$.

The truncation errors of Equations (26), (27) and (28) are $O(\Delta t^*, \Delta\Psi^2)$, $O(\Delta t^{*2}, \Delta\Psi^2)$ and $O(\Delta t^{*2}, \Delta\Psi^4)$, respectively.

3.4 FINITE-ELEMENT METHODS

The finite element methods described in this section employ moving nodes and adaptive mesh refinement. Two methods are developed and employed in the calculation of the ozone-decomposition flame speed. The first method moves the grid with the previously computed wave speed, while the second method calculates the location of the maximum temperature gradient and places the finite elements around it. A brief description of the methods follows.

3.4.1 MOVING FINITE ELEMENT METHOD

The moving finite element methods developed in this paper differ from the moving finite element method proposed by Miller [6] in several respects. First, Miller's method does not presume any knowledge of the solution; ours does. Second, Miller's method employs elements whose location, as well as the nodal amplitudes, changes (moves) with the solution; our method, however, uses a grid whose elements are moved with the previously computed wave speed and are concentrated, in a specified manner, at the flame front. In addition, our method uses a projection method to interpolate the nodal amplitudes into the new grid.

At time t = 0 we calculate the location of $(\partial T/\partial\Psi)_{max}$, which corresponds to the location of the temperature profile inflection point. We also calculate the location of $(\partial T/\partial\Psi \sim 0)$ where we use a prescribed tolerance, e.g., $T_u \sim 0.99\ T_{unburned}$ and $T_b \sim 0.99\ T_{burned}$, where $T_{unburned}$ and T_{burned} denote the temperatures of the unburned and burned gases, i.e., 300°K and 1250°, respectively. The locations of T_u, T_b and $(\partial T/\partial\Psi)_{max}$ are denoted by Ψ_u, Ψ_b and Ψ_{max}, respectively. For $\Psi_b \leq \Psi \leq \Psi_u$ a very refined grid is employed; in the present calculations 20 grid points (19 equally-sized elements) were used at the flame front. In the regions $-\infty < \Psi \leq \Psi_b$ and $\Psi_u \leq \Psi < \infty$ the species and temperature profiles are almost uniform and will remain so except in the region $\Psi_u \leq \Psi < \infty$ due to the flame propagation. We have used 5 elements in the region $-\infty < \Psi \leq \Psi_b$, while the number of elements employed in the unburned mixture region depended on the frequency of mesh adaption. We know that if the flame is U (based on the O_2 and O_3 profiles), the flame will advance a distance $\Delta x \sim U\Delta t$ in a time step Δt^*, which corresponds to $\Delta\Psi \sim \rho U \Delta t$ [Equation (8)], where ρ is the density at Ψ_{max}, which, in steady state, is a constant. If the grid adaption takes place every time step and if M denotes the number of elements in $\Psi_b \leq \Psi \leq \Psi_u$, we use $M(1 + \Delta\Psi/(\Psi_u - \Psi_b))$ elements of equal size in $\Psi_b \leq \Psi \leq \Psi_u + \Delta\Psi$ and 11 elements in $\Psi_u + \Delta\Psi \leq \Psi < \infty$. If mesh adaption is not performed every time step we use the same expressions to calculate the number of elements in the region ahead of the flame, i.e., $M\Delta\Psi/(\Psi_u - \Psi_b)$, and 11 elements in the unburned mixture region. In the calculations reported here mesh adaption was performed every twenty time steps; calculations performed with mesh adaptions every more than twenty time steps yielded inaccurate results because the distance, $\Delta\Psi \sim \rho U \Delta t$, ahead of the flame front was underestimated. This occurs because the number of elements placed ahead of the flame front is proportional to the density at the flame inflection point, which is about 800°K; this density is about 2.7 times smaller than the fresh mixture density. The accuracy of the numerical calculations could have been improved by using the fresh mixture density in $\Delta\Psi \sim \rho U \Delta t$ but this required more elements.

Once the initial number of elements has been adequately placed at the flame front, we reduce the reaction-diffusion equations [Equations (18)] to a system of ordinary differential equations by means of a Galerkin method [5]. The reaction terms were linearized and the resulting set of ordinary differential equations was discretized by using a trapezoidal rule. Thus, the solution vector $A(\Psi,t^*)$ was assumed to be

$$A(\Psi,t^*) = \sum_{l=1}^{N} a_l(t^*)\, \Phi_l(\Psi), \tag{30}$$

where $\Phi_l(\Psi)$ are the linear roof basis, l the grid point, and a_l the nodal amplitude. Equation (18) was multiplied by Φ_i and integrated to yield

$$\int_{-\infty}^{\infty} \frac{\partial A}{\partial t^*} \Phi_i d\Psi = -a \int_{-\infty}^{\infty} \frac{\partial A}{\partial \Psi}\frac{\partial \Phi_i}{\partial \Psi} d\Psi + \int_{-\infty}^{\infty} S\Phi_i d\Psi. \tag{31}$$

In this equation we substitute the value of A given by Equation (30) and obtain

$$\sum_{l=1}^{N} (M_{il} a_l + a\, K_{il}\, a_l) = \int_{-\infty}^{\infty} S\Phi_i \partial\Psi. \tag{32}$$

where

$$M_{il} = \int_{-\infty}^{\infty} \Phi_i\, \Phi_l\, d\Psi \quad \text{and} \quad K_{il} = \int_{-\infty}^{\infty} \frac{\partial \Phi_i}{\partial \Psi}\frac{\partial \Phi_l}{\partial \Psi} d\Psi. \tag{33}$$

Equation (32) represents a system of ordinary differential equations for the nodal amplitudes, a_l; this system is discretized using a midpoint rule to yield

$$\sum_{l=1}^{N}\left[M_{il} \frac{a^{n+1}_l - a^n_l}{\Delta t^*} + \frac{a}{2} K_{il}\, (a^n_l + a^{n+1}_l)\right] =$$

$$\frac{1}{2}\left[\int_{-\infty}^{\infty} S^n \Phi_i \, d\Psi + \int_{-\infty}^{\infty} S^{n+1} \Phi_i \, d\Psi\right], \tag{34}$$

where S^{n+1} was linearized as $S^{n+1} = S^n + (\partial S/\partial A)^n (A^{n+1} - A^n)$ (35)

Convergence within the time step was established when the L^2 norm defined by

$$L^2 = \sqrt{\sum_{i=1}^{N} |(A_i^2)^{k+1} - (A_i^2)^k| / N} \tag{36}$$

was less than 10^{-3}. In Equation (36), k stands for the k-th iteration required to solve the system of algebraic equations given by Equations (34) and (35), and N is the number of grid points.

Suppose that at time $(t^*)^n$ the solution is known and can be written as

$$A^n = \sum_{l=1}^{N} a_l^n \Phi_l^n, \tag{37}$$

where Φ_l^n are the linear basis at $(t^*)^n$. Suppose that at $(t^*)^n$ the grid is to be moved to its new location at $(t^*)^{n+1}$; this location is given by Ψ_{max}^{new}, Ψ_u^{new} and Ψ_b^{new}. In order to calculate the initial conditions in the new grid which has the same number of elements in $\Psi_b^{new} \leq \Psi \leq \Psi_u^{new}$ as in $\Psi_b^n \leq \Psi \leq \Psi_u^n$, and whose new set of linear basis is given by Φ_l^{n+1} we use a projection method as follows. The initial conditions in the new grid can be written as

$$A^{new} = \sum_{l=1}^{N} a_l^{new} \Phi_l^{n+1}, \tag{38}$$

where a_l^{new} are determined by minimizing the following functional

$$I = \frac{1}{2} (A^{new}, A^{new}) - (A^n, A^{new}), \tag{39}$$

where

$$(A^n, A^{new}) = \int_{-\infty}^{\infty} A^n A^{new} \, d\Psi. \tag{40}$$

3.4.2 ADAPTIVE FINITE ELEMENT METHOD

The moving finite element method described before requires a knowledge of the previously computed wave speed. This speed has to be known very precisely to minimize the risk of obtaining inaccurate answers. This can be accomplished by defining an adaptive grid which places the elements at the flame front; this front can be characterized by the location of the maximum temperature gradient. The method and equations are similar to those described in Section 3.4.1 except for the way in which the new grid is placed in the computational domain. Thus, suppose that at $(t^*)^n$ the mesh is to be redefined. At this time we locate Ψ_{max}^{new}, Ψ_u^{new}, Ψ_b^{new} and place the same number of elements in $\Psi_b^{new} \leq \Psi \leq \Psi_u^{new}$ as we had in $\Psi_b^n \leq \Psi \leq \Psi_u^n$. This procedure does not require to know the flame speed and uses the location of the flame front as the adaption criterion; the method can be applied to other problems characterized by the presence of only one sharp boundary. In contrast, Miller's method can be used in problems characterized by the presence of multiple sharp boundaries or shocks [6].

4. PRESENTATION AND DISCUSSION OF RESULTS

The finite-difference and finite element methods described in Section 3 were used to compute the ozone-decomposition flame speed in a gaseous mixture (at 300°, and 0.821 atm) composed of 25% (by volume) O_3 and 75% O_2. The finite-difference calculations were performed with 121 grid points, while those based on the finite elements methods employed 36 elements; five elements were placed on the burned gas region, twenty elements at the flame front, and the remaining were placed ahead of the flame front in the manner indicated in Section 3.4.1. The computations were carried out up to the same time in all the methods employed; this time corresponds to 2 milliseconds. The flame speeds at 2 milliseconds were 49.57 cm/s, and 49.51 cm/s for the fourth-order accurate method of lines and the majorant operator-splitting technique, respectively; the wave speeds predicted by the time linearization procedures were 48.91 cm/s, 48.97 cm/s and 49.38 cm/s. While those predicted by the moving and adaptive methods were 49.26 cm/s and 49.41 cm/s.

The wave speeds computed with the time linearization schemes show that the temporal accuracy of a scheme is not very important in computing the ozone-decomposition flame; for example, there is very little difference between first- and second-order temporal discretizations, i.e., 48.91 cm/s and 48.97 cm/s. There is, however, substantial accuracy improvement when the fourth-order accurate time linearization scheme is used; in this case the computed wave speed, 49.38 cm/s, is comparable to that obtained using the fourth-order accurate method of lines and the majorant operator-splitting technique. The flame speed calculated by these methods is in very good agreement with those calculated by using the moving and adaptive finite element methods, and with those obtained by Margolis [2] and Reitz [4].

Although the flame speeds predicted by the time linearization schemes, particularly the fourth-order accurate method, are in very good agreement with those computed by using the method of lines, it cannot, in general, be concluded that time linearization schemes are as accurate as other schemes. In the particular case considered here the linearization of the highly nonlinear terms seems to be appropriate due to the small time steps used in the computations. However, the accuracy of the calculations deteriorates when large time steps are considered; in some cases, the linearization of the reaction terms may not yield a diagonally dominant matrix.

The accuracy of the moving and adaptive finite element methods is slightly lower than that of the fourth-order accurate method of lines and operator-splitting technique. The accuracy of the moving finite element method deteriorates with time; initially, the wave speed used to move the elements was that computed by the method of lines, i.e., 49.57 cm/s. The small deterioration can be observed at 2 milliseconds; at this time, the wave speed has decreased by 0.31 cm/s. Although this inaccuracy is small, it should be noted that it propagates in time in such a way that the computed wave speed keeps on decreasing. This occurs because the grid is adapted in terms of the wave speed computed in previous time steps.

The adaptive finite-element method is more accurate than the moving technique because it concentrates the elements where the solution is changing rapidly rather than in a region whose velocity is approximately known. The accuracy of both, moving and adaptive, finite element methods can be improved by increasing the number of elements at, and ahead of, the flame front. If the number of points at the flame front is increased more accurate results can be obtained; however, the system of o.d.e.'s becomes stiffer and smaller time steps need to be used in the integration.

5. CONCLUSIONS

Fixed-node finite-difference, a moving FEM and an adaptive FEM have been used to study the ozone-decomposition flame speed. It is shown that the accuracy of time linearization schemes depends very much on the spatial approximations, but it is almost independent of the temporal truncation error. A fourth-order accurate method of lines and a majorant operator-splitting technique predict a wave speed comparable to that obtained with adaptive finite-difference schemes and methods using B-splines.

It is also shown that moving and adaptive finite element procedures which concentrate most of the grid points at the flame front can give accurate results. In particular, an adaptive method which places the elements within the flame front and around the maximum temperature gradient gives more accurate results than a method which moves the elements with the previously computed wave speed. In the moving method, the flame speed has to be accurately known at the beginning of the computation; this problem does not arise when the adaptive FEM method is used.

REFERENCES

1. BLEDJIAN, L. - Computation of Time-Dependent Laminar Flame Structure, Combustion and Flame, Volume 20, pp.5-17, 1973.

2. MARGOLIS, S.B. - Time-Dependent Solution of a Premixed Laminar Flame, Journal of Computational Physics, Volume 27, pp.410-427, 1978.

3. MEINTJES, K. - Predictor-Corrector Methods for Time Dependent Compressible Flows, Ph.D. Thesis, Princeton University, Princeton, New Jersey, 1979.

4. REITZ, R.D. - The Application of an Explicit Numerical Method to a Reaction-Diffusion System in Combustion, The Courant Institute of Mathematical Sciences, New York University, New York, 1979.

5. LEE, D.N. and RAMOS, J.I. - Application of the Finite Element Method to One-Dimensional Flame Propagation Problems, AIAA Journal, Volume 21, pp.262-269, 1983.

6. MILLER, K. - Moving Finite Element Methods. II, SIAM Journal of Numerical Analysis, Volume 18, pp.1033-1057, 1981.

NUMERICAL MODELLING OF TWO DIMENSIONAL TURBULENT COMBUSTION INITIATED BY SURFACE IGNITION

Vasanth Kumar Victor*, Shankar Balakrishnan** & V.Ganesan***

*Graduate Student, Dept. of Ind. Engg.,University of Wisconsin, Madison, U.S.A.

**Graduate Student, Dept. of Ind. Engg.,University of California, Berkeley, U.S.A.

***Asst. Professor, Dept. of Mech. Engg., I.I.T., Madras-36 India.

ABSTRACT:

This paper reports the results of numerical modelling of surface ignition phenomena. The flow is assumed to be two dimensional, steady and turbulent. Hydrodynamics and combustion phenomena have been modelled using Prandtl's mixing length hypothesis and eddy break up model respectively. A finite difference procedure using marching integration technique has been used for prediction. Velocity profiles, temperature profiles and concentration profiles of fuel and oxygen have been predicted when a fuel air mixture is ignited by a hot surface.

1. INTRODUCTION:

The tendency of certain fuels to ignite easily from hot spots is called the surface ignition phenomenon. When a liquid comes into contact with a hot surface, there is rapid vaporisation and a fuel air mixture is formed. Ignition occurs when the surface of the material activates the fuel-air mixture. Different factors affect the ignition process to different extend. The most important ones are i)the surface temperature, ii) the material of the surface, iii)the pressure of the fuel-air mixture, iv) the area of the hot surface.

The surface of the material has a catalytic action on the combustion. At high temperature all materials exhibit catalytic properties. Certain materials, however, have a catalytic action at low temperatures also. They make ideal 'hot spot' materials.

It has been found as a result of certain bench tests that alcohols are particularly well suited to take part in the surface ignition phenomenon. Among the fuels tested it was found that methanol was the best fuel for surface ignition ethanol was next best; octane is not very well suited. Methane is the least suitable fuel for surface ignition among the above four fuels. The temperatures required to ignite the above fuels range from 600 to 800 degree centigrade.

The main advantage of a simulated experiment is the tremendous economy possible. Once a reliable model that yields plausible results has been developed, it is possible to vary the independant variables for almost no additional cost, and study the effect on the particular dependant variable of interest. Thus the value of the dependant variable can be optimised and the corresponding values of the independant variables determined.

2. THE PROBLEM CONSIDERED:

A semi infinite long plate is kept at a fixed temperature. Methanol-air mixture is assumed to flow over the hot surface. Ignition is initiated at the surface and proceeds down streams. Velocity, temperature, fuel concentration and oxygen concentration profiles have been predicted. Flame spread along the longitudanal direction have been calculated from the temperature profiles. Three air-fuel ratios viz., one rich, one lean and chemically correct have been considered.

Four partial differential equations have been solved in their finite difference form using marching integration technique. Hydrodynamics and combustion processes have been modelled using the Prandtl's mixing length hypothesis and eddy break up model respectively.

3. THE GOVERNING DIFFERENTIAL EQUATION:

All the differential equations viz., for mass, momentum, enthalpy, mass fraction of fuel and mixture have been brought to a general form [1]* by suitable simplifications. The general form is given by:

$$\frac{\partial \Phi}{\partial x} + (a+b\omega)\frac{\partial \Phi}{\partial \omega} = \frac{\partial}{\partial \omega}\left(c\,\frac{\partial \Phi}{\partial \omega}\right) + d \qquad (1)$$

---

* The numbers within the parantheses indicate the references at the end of the paper.

The various terms in the equation are:

i) $\frac{\partial \Phi}{\partial x}$: effect of longitudanal convection

ii) $(a+b\omega)\frac{\partial \Phi}{\partial \omega}$: effect of lateral convection

iii) $\frac{\partial}{\partial \omega}\left(c\frac{\partial \Phi}{\partial \omega}\right)$: effect of viscous action, heat conduction or diffusion

iv) d : effect of generation or destruction.

4.INTEGRATION PROCEDURE:

Integration, in our context, is the establishment of the solution of the differential equations which describe the physical processes; so it entails finding out what values of temperature, velocity etc. prevailing at each point in the domain of interest.

'Marching' integration is that kind of integration which starts by determining the values at one end of the domain, then determines the values over a front displaced just a little from that end, and so gradually moves the integration front towards the other end of the domain until the required values have been determined everywhere. Iteration is not required because the later determined quantities cannot influence those determined earlier; this is the characteristic of the parabolic phenomenon. The ability to use marching integration is useful in practice because the confinement of the integration to a single sweep diminishes the necessary computer time; further, the necessity to visit each point in the fiels only once reduces the dimensionality of computer storage.

The solution procedure employs six-node formulae, which cannect the values of fluid variables of six nearby points, three upstream and three down stream. Since each formula will have more than one unknown (downstream point), the equations for the downstream values at the nodes all along the line must be solved simultaneously; thus the equations are implicit. The implicit system of equation is obviously more troublesome to solve than an explicit system. Hence, it must have some strongly countervailing advantage - and it does have. It allows the magnitude of the 'forward step' i.e., the distance between the upstream and the next downstream lines of the grid to be freely chosen without the incidence of numerical instability which assails the explicit shemes when the forward step size is large. It is therefore in the interest of economy of number of steps, and hence the computer time that the implicit method is employed.

It is possible to solve the system of euuations only if the following values are specified: values of the variables for all grid nodes lying on the upstream edge of the grid where the integration starts; and values of the variables at the edges of the domain of integraion.

The differential equations have to be transferred into their finite difference equivalents, before their numerical solutions can be obtained. The justification for the transformation lies in the fact that properly formed finite-difference equations have solutions which approach those of the differential equations when the number of grid nodes tends to infinity; in practice, the smallest number of grid points is employed that will provide acceptable accuracy. It may be remarked here that the transformation is effected by means of integration over the controlvolume.

To obtain the solutions of the finite difference equations one needs an algorithm which will solve the simultaneous equations economically. For this, a particular form of Gaussian elimination technique, known as the tridiagonal matrix algorithm (TDMA) is used. The finite-difference equations, after suitable manipulation, are written in a matrix form where the coefficients are non-zero only along the diagonal and along the two adjacent lines on either side. These are then converted, by algebraic substitution, into the set $\Phi_i = P_i \Phi_{i+1} + Q_i$. The TDMA is then employed. In conclusion, it is mentioned that the computational procedure employed by the TDMA is a highly efficient one, both in time and storage.

5.PHYSICAL MODELLING:

The hydrodynamic model consists of calculation of effective exchange coefficeents which involves the calculation of effective viscosity (μ_{eff})

$$\mu_{eff} = \mu_l + \mu_T \tag{2}$$

where μ_T is calculated using Prandtl's mixing length hypothesis.

$$\mu_T = \rho\, l^2 \left|\frac{\partial u}{\partial y}\right| \tag{3}$$

where l is the mixing length. The magnitude of the mixing length is taken as 0.14 times width of the shear layer or 0.435 times the distance from the wall whichever is smaller.

The combustion process is modelled using the eddy break up model proposed by spalding [2]. The reaction rate, R_{fu}, of fuel air mixture is given by the formula:

$$R_{fu} = -C_{EBU} \frac{(m_{fu,u} - m_{fu})(m_{fu} - m_{fu,b})}{(m_{fu,u} - m_{fu,b})} \rho \left|\frac{\partial u}{\partial y}\right|$$

where $m_{fu,u}$ stands for the mass fraction of fuel in completely unburned mixture and $m_{fu,b}$ stands for mass fraction of fuel in completely burned mixture. C_{EBU} is known as eddy break up constant.

6. RESULTS AND DISCUSSION:

Figure 1 shows a typical velocity distribution at x=500 mm from the leading edge of the plate. As expected the velocity shows the boundary layer profile near the wall. Higher velocity is seen near the wall due to combustion and velocity dips to free stream value after the reaction zone.

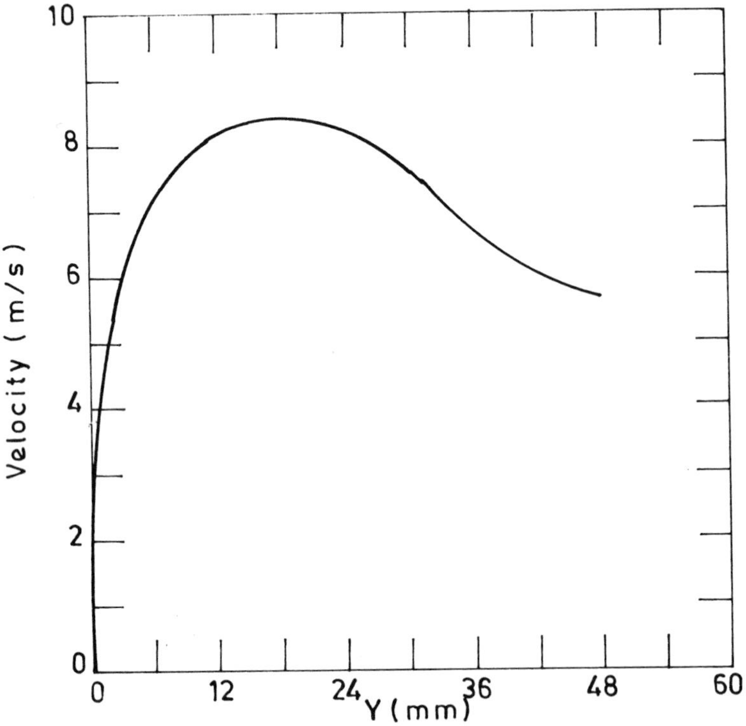

FIG.1 VELOCITY PROFILE

Figure 2 shows the temperature profile for rich, chemically correct and lean mixtures of methanol and air. As expected chemically correct mixtures produce higher temperature than the rich and lean mixtures. Because reaction takes place due to surface ignition temperature shoots up near the wall.

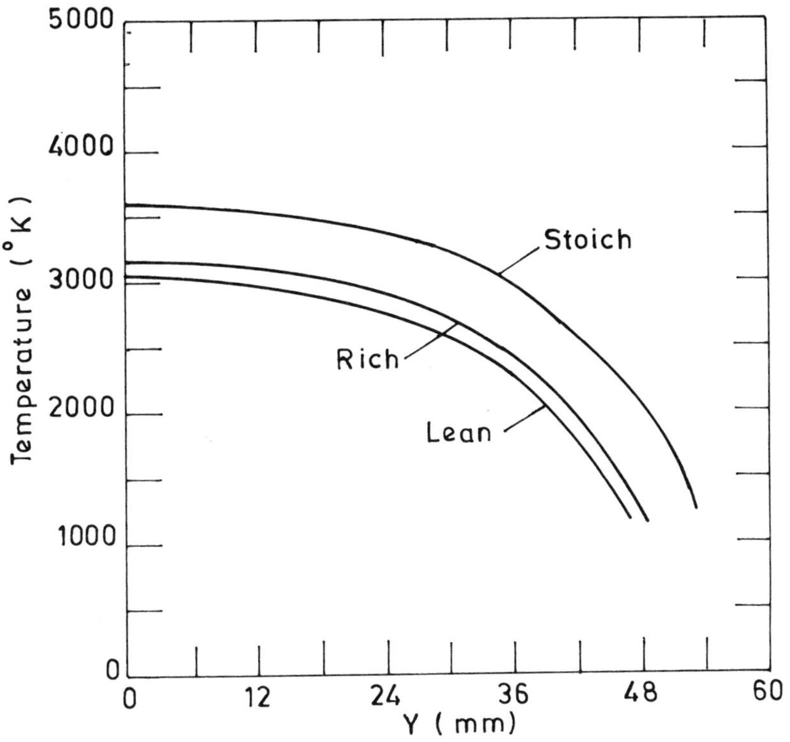

FIG.2 TEMPERATURE PROFILE

Figure 3 shows that fuel concentration profiles for chemically correct, rich and lean mixtures. The unburned fuel concentration is more for rich mixture indicating incomplete combustion followed by lean and stoichiometric mixture. This is in confirmity with the temperature profile

Figure 4 gives the details of oxygen concentration profiles for the 3 mixtures studied. The left out oxygen concentration is more for lean mixtures. Since the incoming oxygen concentration is more lean mixtures this is to be expected. Compared to chemically correct mixture the oxygen concentration is more for the rich mixture due to the incomplete combustion.

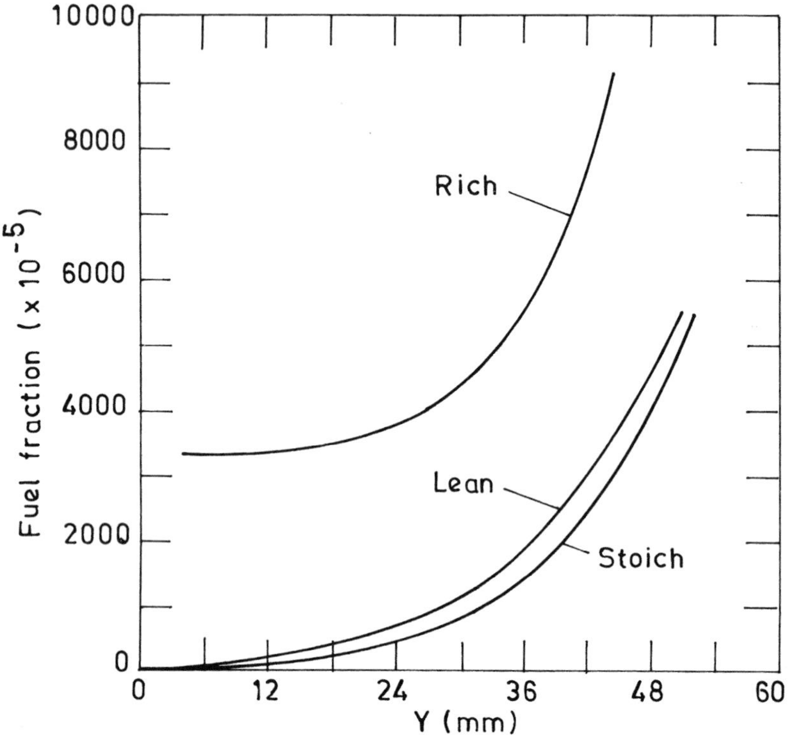

FIG. 3 FUEL CONCENTRATION PROFILE

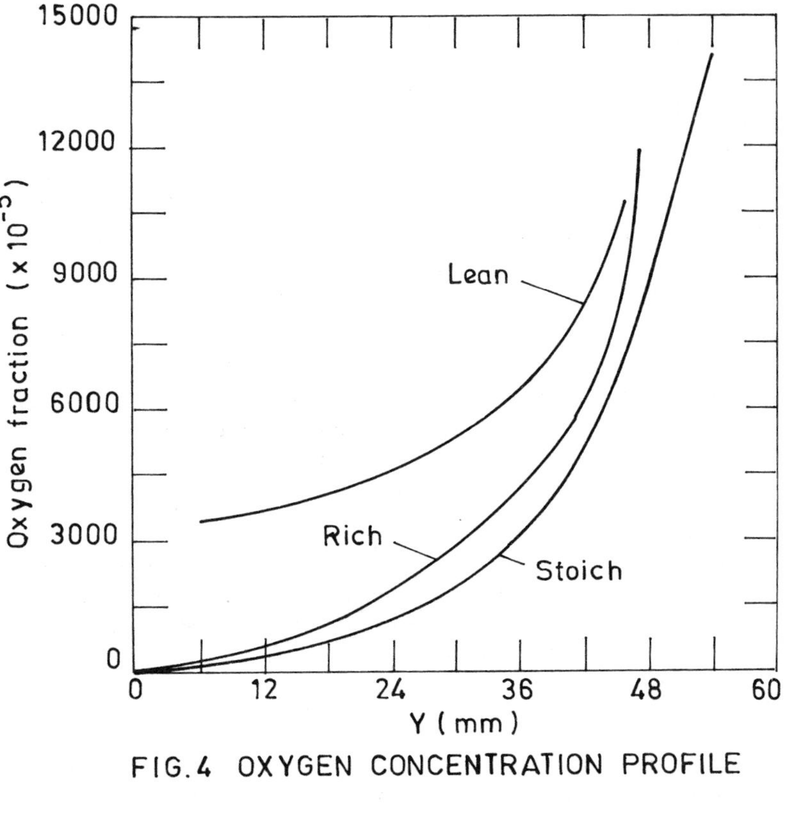

FIG.4 OXYGEN CONCENTRATION PROFILE

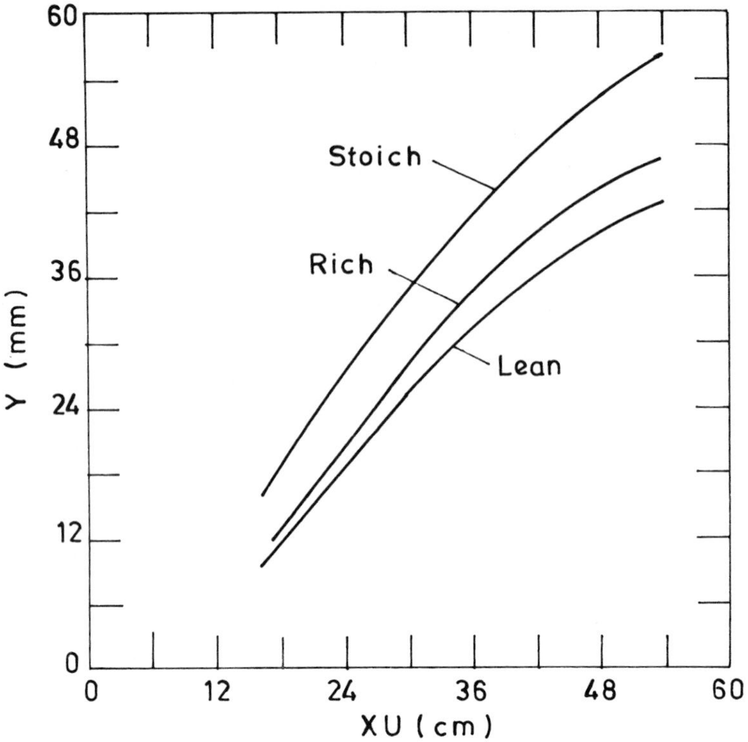

FIG. 5 FLAME SPREAD

Figure 5 illustrates the flame spread for the different mixtures and is seen that flame spreads faster in chemically correct mixtures followed by rich and then lean.

7. CONCLUSION:

The present method predicts the surface ignition phenomena satisfactorily. It is found that the method is quite fast and cheap. However, experimental details are necessary to validate the predictions.

8. REFERENCES:

[1] Spalding, D.B., GENMIX: A general computer programme for two dimensional parabolic phenomena, Pergamon Press, 1977.

[2] Spalding, D.B.., Development of the eddy break up model of turbulent flow, Sixteenth Symposium (International) of combusion Institute, Pittsburg, pp. 1653-1663,1976.

TURBULENT FLOW FIELD CALCULATIONS IN AN INTERNAL COMBUSTION ENGINE EQUIPPED WITH TWO VALVES

M.H. CARPENTER* and J.I. RAMOS**

SUMMARY

The turbulent flow field within an axisymmetric motored internal combustion engine equipped with two values has been studied by using a finite-difference scheme. The numerical algorithm uses the conservation form of the governing equations, which are discretized by employing a control-volume approach. The inlet valve has been simulated as an annular orifice located at the cylinder head, while the exhaust valve consists of an infinitesimally thin disk which penetrates into the engine cylinder. Calculations have been performed with a two-equation model of turbulence in order to study the effects of the intake and exhaust valves on the mean velocity field and turbulence levels within the cylinder. The results show that in the intake stroke an air jet is drawn into the cylinder. This jet strikes on the piston and creates two toroidal vortices. One of these vortices is located near the cylinder wall and persists well into the expansion stroke. The other vortex is located between the exhaust valve and the piston, and disappears in the compression stroke. The turbulence levels at top dead center of the compression stroke are almost spatially uniform and are controlled by the rpm and the inlet valve geometry.

1. INTRODUCTION

The design of an internal combustion (IC) engine embraces the areas of fluid mechanics, convective heat and mass transfer, turbulence, thermodynamics, chemical kinetics, radiation and solid mechanics. The present-day IC engines have reached a high level of performance and efficiency, but they are still considered far from ideal. The design of an IC engine combustion chamber, for example,

---

*Graduate Student, Department of Mechanical Engineering, Carnegie-Mellon University, Pittsburgh, PA 15213, U.S.A.

**Assistant Professor, Department of Mechanical Engineering, Carnegie-Mellon University, Pittsburgh, PA 15213, U.S.A.

has been developed as more of an *art* than a science, guided by a myriad of empirical, and largely algebraic, correlations each of which being the result of a very detailed, but restricted, range of experimental conditions.

The availability of very large, high speed digital computers and nonintrusive diagnostics techniques such as the Laser Doppler Velocimeter have encouraged efforts to study the flow field within an internal combustion engine. These theoretical and experimental studies have been performed mostly in *idealized piston-cylinder configurations* and have been aimed at understanding the characteristics of the flow under *motored* conditions. Some of the numerical predictions have been compared with the available experimental data and the results have been encouraging. However, more theoretical and experimental work is needed to assess the numerical models.

This paper describes a numerical model which has been developed to analyze the flow field within an *idealized motored* piston-cylinder configuration equipped with two valves. The inlet valve has been simulated as an annular orifice located at the cylinder head. This orifice opens and closes suddenly at specified crank angles in such a way that the 36° overlap between the inlet and exhaust valves corresponds to that of a production IC engine. The exhaust valve has been simulated as a infinitesimally thin disk which penetrates into the engine cylinder.

The numerical scheme used in the calculations reported in this paper has been employed in other, more idealized, piston-cylinder configurations. The results of these calculations have compared favorably with the available experimental data [1-2]. Ramos *et al* [1] studied the flow field in a motored IC engine provided with a centrally located orifice at the cylinder head. This orifice was employed for both the intake and exhaust strokes. In another set of experiments, Ramos *et al* [1] used a 45°-beveled valve in order to assess its influence on the flow field and turbulence levels within the cylinder. This valve was simulated as a thin disk located at a fixed distance from the cylinder head. The comparisons between the theoretical model predictions and the Laser Doppler Velocimeter (LDV) data showed that the turbulence levels in both configurations are similar near the intake orifice. The numerical results indicate that the valve bends the intake air jet. This jet strikes on the cylinder wall and creates additional turbulence there. In the absence of a valve, the turbulence levels were higher at the intake air jet shear layer. Similar results were found by Gosman *et al* [3] in a piston-cylinder configuration not equipped with an intake-exhaust valve.

Ramos *et al* [2] studied the flow field in a motored reciprocating engine equipped with *one* valve which was used for the intake and exhaust strokes. The intake stroke results were similar to those obtained without compression in Reference 1. The numerical predictions compared favorably with the LDV data except near the cylinder head where the details of the valve geometry are important. This result was expected since the theoretical model used an infinitesimally thin valve, whereas in the experiments a 45°-beveled valve was employed. The results showed that the flow field in the four-stroke engine cycle was controlled by the intake stroke. In this stroke, turbulence was created at the intake air jet shear layers and at the cylinder wall where the jet struck. The air jet created two toroidal vortices: one of them was located at the corner between the cylinder head and cylinder wall and referred to as the *head vortex*; the other was located between the valve and the piston and named the *valve vortex*. The dynamic interaction of these vortices depended on the bore-to-stroke ratio, compression ratio, rpm and intake air jet angle. This angle was used to represent the valve seat angle in a real engine. The results of the calculations reported in Reference 2 showed that the turbulence levels at top-dead-center (TDC) of the compression stroke, i.e., at 360° after top-dead-center (ATDC), were uniform throughout the engine cylinder. This was not so when the intake air jet angle was about 45° [4].

Schock *et al* [4] showed that the flow field within a motored engine was very sensitive to the valve seat angle. For intake angles of 0° and 22.5° the flow consisted only of the head and valve vortices. The latter merged with the head vortex in such a way that by the end of the compression stroke only a vortex was observed at the corner between the cylinder head and the cylinder wall. However, for a 45° intake angle the valve vortex was stretched and broke into two new vortices, which merged with the head vortex. This vortex was also stretched and by the end of the intake stroke was located along the cylinder wall.

In order to understand the dynamic interactions between the vortices created in the intake stroke of a four-stroke IC engine and their influence on the turbulence levels at TDC of the compression stroke, an axisymmetric piston-cylinder configuration equipped with *two* valves is numerically simulated. Mean axial velocity profiles and turbulence levels are presented throughout the four-stroke engine cycle in order to identify the physical mechanisms which control the flow. The periodicity of the flow is also studied.

2. MATHEMATICAL FORMULATION

The axisymmetric piston-cylinder configuration studied in this paper is shown in Figure 1. The intake port is an annular orifice of radii R_i = 1.375 cm and R_o = 1.875 cm and is located at the cylinder head. The intake air jet angle is 0°. The exhaust valve consists of an infinitesimally thin disk of radius R_v = 0.785 cm. This valve penetrates into the cylinder in such a way that its profile and the 36° overlap between the intake and exhaust valves correspond to those of a production engine. The rpm is 160, the compression ratio is 7 and the bore is 7.77 cm. The connecting rod length is 20.32 cm and the clearance is 1.27 cm.

Due to the piston motion the problem has a moving boundary. In addition, the exhaust valve penetration into the cylinder provides the problem with an internal moving boundary. These moving boundaries are transformed into a fixed and a moving boundary in such a way that in the transformed coordinates the piston is fixed while the valve moves discretely. This is accomplished by defining a fictitious plane perpendicular to the engine cylinder axis. The only restriction on this plane is that its distance to the cylinder head must be smaller than the clearance. From this plane to the piston, the axial coordinate is normalized by the time-dependent distance between the fictitious plane and the piston; thus, this domain nondimensional distance is one. Between the cylinder head and the fictitious plane the axial coordinate is nondimensionalized by the axial distance between them. This domain is then fixed in both the physical and transformed coordinates so that the valve can move within it in an arbitrary, but prescribed, manner independently of the piston motion. The distance between the fictitious plane and the cylinder head must be smaller than the clearance so that the grid points located between the piston and the plane do not merge into a line when the piston is at TDC of the compression stroke [5].

This two-domain transformation permits studying the flow field between the cylinder head and valve where the flow field is affected by the geometry and intake port conditions. Gosman *et al* [6] did not study the flow between the valve and cylinder head, but rather they provided boundary conditions on the surface of a cylinder of radius equal to that of the valve. In this approach there is no need for a two-domain technique but still the axial coordinate is normalized by the time-dependent distance between the cylinder head and piston, i.e., the distance between grid points stretches and compresses like in an accordion. The problem with this approach is that the valve has to move with the grid which is completely stretched at bottom-dead-center of the intake stroke. In order to simulate a realistic profile, the valve has to jump towards the cylinder head. This is the case regardless of the number of grid points employed in the calculations. The two-domain technique does not exhibit this problem because the valve motion is uncoupled from the grid through the two coordinate transformations. However, the two-domain technique is discontinuous at the fictitious plane.

2.1 THE EQUATIONS

The equations describing the unsteady axisymmetric flow field in a piston cylinder configuration are given below. In these equations the laminar and turbulent Prandtl numbers are equal to 0.73 and 1.00, respectively. The flow is assumed to be adiabatic, i.e., p/ρ^{γ} = constant, where p, ρ and γ are the pressure, density and ratio of specific heats, respectively. In the calculations reported here γ is equal to 1.4 and the flow is considered isothermal in the intake and exhaust strokes. The equations is nontransformed coordinates are

$$\frac{\partial}{\partial t}(\rho\Phi) + \frac{1}{r}\frac{\partial}{\partial r}(\rho v r\Phi) + \frac{\partial}{\partial z}(\rho u\Phi) = \frac{1}{r}\frac{\partial}{\partial r}\left(r\mu_e \frac{\partial\Phi}{\partial r}\right) +$$

$$\frac{\partial}{\partial z}\left(\rho\mu_e \frac{\partial\Phi}{\partial z}\right) + S_\Phi \quad (1)$$

where t is time, v the radial velocity, u the axial velocity, μ_e the effective viscosity, i.e., the sum of the laminar and turbulent viscosities, Φ is any scalar variable, and S_Φ the source term. In this equation Φ stands for 1 (continuity equation), u (axial momentum equation), v (radial momentum equation), k (turbulent kinetic energy) and ϵ (dissipation rate of turbulent kinetic energy). The ideal gas law is used in the calculations with a molecular weight of 28.91 g/mole.

2.2 THE SOLUTION PROCEDURE

Equation 1 is written in terms of the two-domain coordinates defined above and integrated using a control-volume approach [7]. The resulting finite difference equations are solved line-by-line along the cylinder axis using an underrelaxation procedure and a hybrid scheme for the convection terms. A grid of 22 x 22 points is used in the computations; the time step is the reciprocal of the rpm. The local pressure variations are accounted for through the continuity equation, while the global pressure variations, i.e., those associated with the thermodynamics of the compression and expansion strokes, are obtained from the global conservation of mass in the piston-cylinder configuration. In previous papers [7-8] a global temperature correction was introduced to account for the temperature increase (decrease) during the compression (expansion) stroke. This perturbation was introduced to smear out the temperature discontinuity at the fictitious plane. In the present calculations no global temperature correction is employed.

2.3 BOUNDARY CONDITIONS

At the cylinder centerline the radial gradients of the flow variables, except for the radial velocity, are zero. The radial velocity at the centerline is equal to zero. At a solid wall the normal component of the velocity at the boundary is equal to the normal component of the boundary velocity, and the logarithmic law of the wall is applied for the tangential component of the velocity at the boundary. The normal component of the turbulent kinetic energy gradient at the boundary is set to zero, while the dissipation rate of turbulent kinetic energy is calculated by assuming an equilibrium boundary layer, where production of turbulent kinetic energy is set equal to its dissipation rate. The valve is considered as a two-sided solid wall and boundary conditions are applied on both sides. At the intake port, the values of the flow variables are specified during the intake stroke; for example, the turbulent intensity at the inlet port is 0.3% in the present calculations. In the exhaust stroke, the axial derivatives of the flow variables at the exhaust port are set to zero except for the axial velocity which is calculated in such a way the outgoing air mass flow rate is equal to the mass flow rate displaced by the piston plus the mass flow rate coming into the cylinder through the inlet port.

3. PRESENTATION AND DISCUSSION OF RESULTS

In Figure 1 we show the mean axial velocity and turbulent kinetic energy profiles at four axial locations within the cylinder and at selected crank angles. The figure illustrates the main characteristics of the flow throughout the intake, compression, expansion and exhaust strokes. In addition Figure 1 shows the velocity and turbulent kinetic energy profiles at 780°, i.e., in the second cycle, 60° ATDC of the intake stroke.

The intake stroke is characterized by the presence of an air jet which impinges on the piston and returns along the cylinder wall and cylinder axis. Two toroidal vortices are created: the *cylinder* vortex is located near the cylinder wall, while the *valve* vortex is located near the cylinder axis. Turbulence is generated at the shear layers of the intake air jet and is greater near the cylinder axis. Additional turbulence is created at the piston where the air jet strikes. The piston stretches the cylinder and valve vortices. The cylinder vortex is elongated during the intake stroke, but its size is about the distance between the cylinder head and piston. The valve vortex has broken up at 120° ATDC and is concentrated near the cylinder head; its intensity decreases throughout the intake stroke. In the compression stroke, cf. 240° ATDC, the cylinder vortex has almost completely merged with the valve vortex and is being compressed; its axial dimension decreases, while its diameter increases. There is still some vorticity near the cylinder head and cylinder axis; this vorticity is due to the valve vortex. By 300° ATDC, the valve vortex has completely merged with the cylinder vortex which has a diameter equal to half the cylinder bore. At TDC of the compression stroke, the *cylinder vortex is still present*, but its intensity has substantially diminished. In the expansion stroke, the piston drives the flow; the axial velocity profile vary almost linearly with their distance from the cylinder head and the turbulence levels are lower than those of the intake and exhaust strokes. Of course, turbulence is generated at the solid walls and at the corner between the piston and the cylinder wall; however, this generation is not properly accounted for in these numerical calculations due to the coarseness of the grid employed in the calculations.

When the exhaust valve opens, a vortex is formed between the piston and the valve; this vortex is clearly seen at 540° ATDC. Another small vortex is created at the corner between the cylinder wall and cylinder head (cf. 660° ATDC). At the end of exhaust stroke these two vortices are very elongated along the cylinder axis and the cylinder wall respectively. The latter merges with the cylinder vortex created in the intake stroke, while the valve vortex is reinforced. The second cycle is almost identical to the first; this can be observed by comparing the axial velocity and turbulent kinetic energy profiles at 60° ATDC and 780° ATDC and suggests that only one cycle may be adequate to obtain quantitative information on the velocity and turbulence fields within the engine.

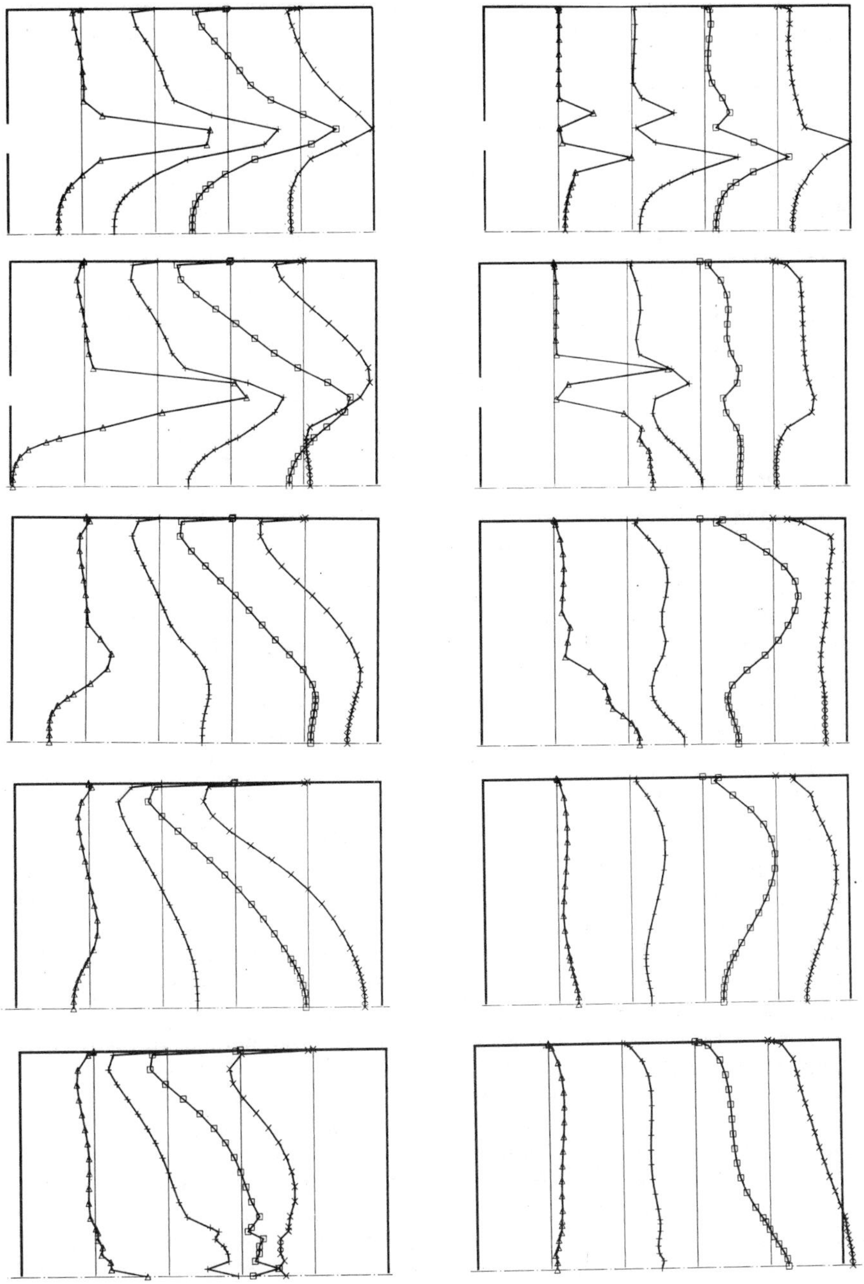

Figure 1. Mean axial velocity (left) and turbulent kinetic energy (right) profiles at 60°, 120°, 180°, 240°, and 300°.

Figure 1 also shows that the levels of turbulence within the engine cylinder are controlled by the intake stroke; turbulent kinetic energy is created at the shear layers of the intake air jet. Additional turbulence is created at the piston where the air jet strikes. However, the turbulence levels decrease very rapidly in the deceleration part of the intake stroke and in the compression stroke. Turbulence is diffused and dissipated in the compression and expansion strokes and is almost spatially uniform at 360° ATDC. The turbulent kinetic energy levels decrease in the expansion stroke, when the piston drives the flow; however, turbulence is created in the exhaust stroke near the exhaust valve. This generation is due to the local flow acceleration and is greater than in the intake stroke. The local increase of turbulence energy near the exhaust valve has also been observed in calculations performed in piston-cylinder configurations equipped with *only* one valve [4-5]. The LDV data show a much smaller increase [2] and, in general, a much more uniform turbulent kinetic level than the numerical predictions. The differences between the theoretical and experimental data may be due to the boundary conditions employed at the exhaust port; however, if the turbulent kinetic energy boundary condition were fixed at the exhaust port, the turbulence levels would still be larger than those observed experimentally. This suggests that farther experimental data are needed to validate the boundary conditions employed in the numerical model.

It is worthwhile to compare the velocity fields in piston-cylinder configurations equipped with *one* [4-6] and *two* (the present study) valves. In an axisymmetric motored engine equipped with *one* valve the flow field in the intake stroke is characterized by the presence of a valve vortex and a head vortex. If the valve seat angle is less than 22.5°, the valve vortex merges with the head vortex; the latter disappears in the compression. If the valve seat angle is 45°, the valve vortex breaks up into two new vortices which then merge with the head vortex. The merging takes place near the cylinder wall and reinforces the head vortex which is very elongated. The calculations reported in this paper show similar vortex dynamic interactions to those in an engine provided with *only one* 45°-beveled valve. In both configurations, the head vortex is still present at 360° ATDC but disappears in the expansion stroke. Despite this *similarity in the mean velocity fields*, the turbulence levels and characteristics are *quite* different, particularly in the intake stroke. When the engine is equipped with an annular intake valve, turbulence kinetic energy is generated at the shear layers of the intake air jet; some turbulence is also created at the piston. However, if the engine is equipped with *only one* valve, turbulence is generated at the shear layers of the inclined air jet and at the cylinder wall. In the last part of the intake stroke the vortex merging and interaction create turbulence along a cylinder concentric with the cylinder axis. The creation of turbulence along this cylindrical surface is similar to that observed in the present study; however, the physical mechanisms are different: turbulence in an engine equipped with *two* valves is generated along the *shear layers* of the intake air jet whereas in an engine equipped with *one* valve turbulence is created by *vortex merging and interaction.*

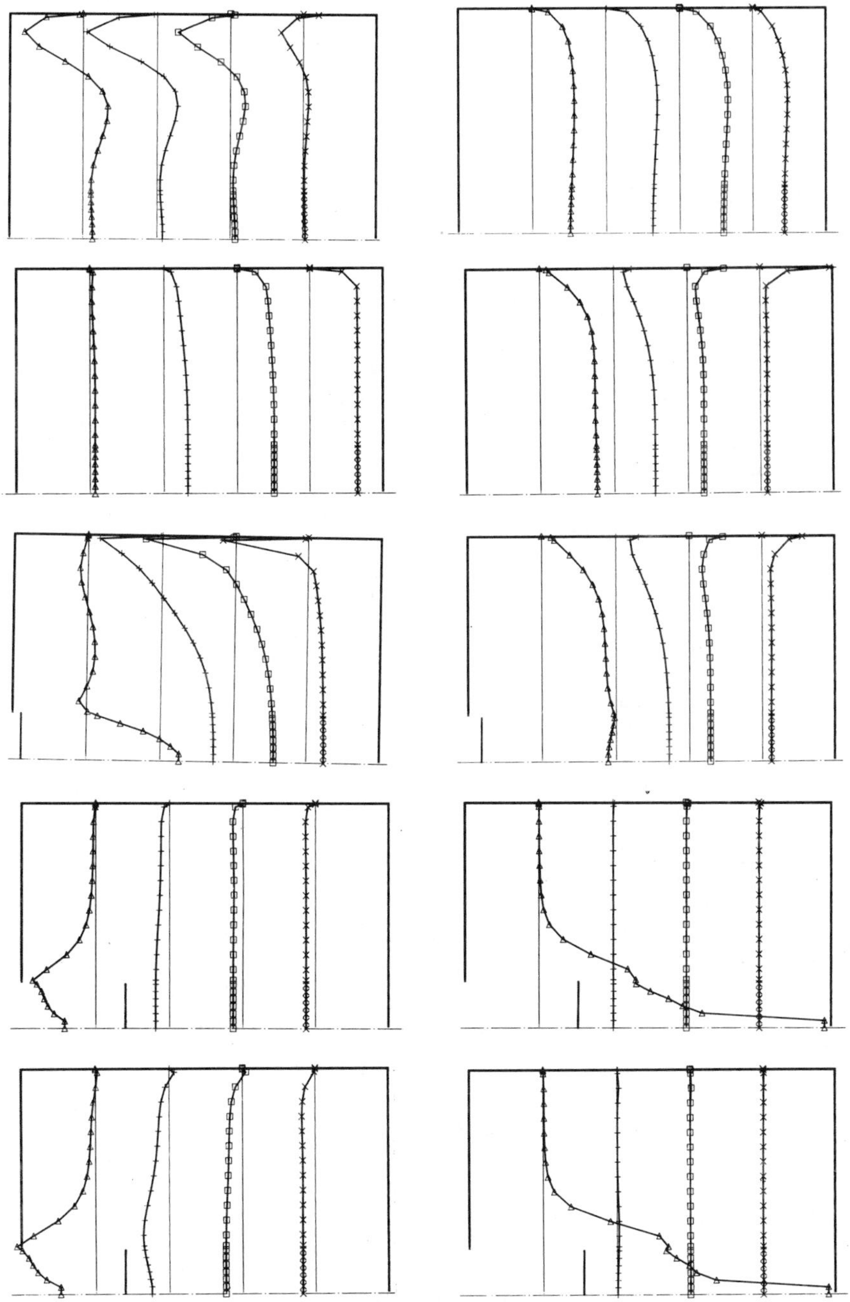

Figure 1 (cont'd). Mean axial velocity (left) and turbulent kinetic energy (right) profiles at 360°, 480°, 540°, 600° and 660°.

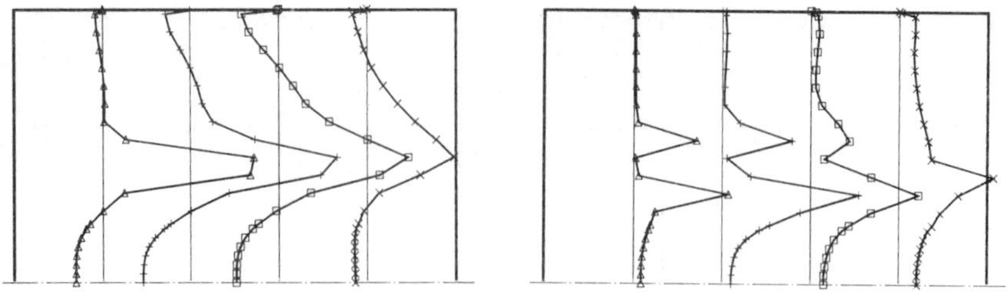

Figure 1 (cont'd). Mean axial velocity (left) and turbulent kinetic energy (right) profiles at 780°.

4. CONCLUSIONS

The flow field in a four-stroke motored internal combustion engine has been studied by means of an implicit algorithm which uses hybrid differences for the convection terms. The numerical scheme solves the axisymmetric form of the conservation equations of mass, momentum and two equations for the turbulent kinetic energy and its dissipation rate. The algorithm uses the conservation form of the governing equations, a global pressure correction technique to account for the thermodynamic pressure variations and two coordinate transformations which map the moving boundary associated with the piston into a fixed boundary. Calculations are reported for the case of an engine equipped with an annular intake valve located at the cylinder head and a moving exhaust valve. The numerical results show that the flow field within the cylinder is very similar to that observed in engines equipped with only one valve when the valve seat angle is 45°. The mean velocity field is characterized by the formation of a valve vortex and a cylinder vortex in the intake stroke. The valve vortex merges with the cylinder vortex in the compression stroke and persists into the expansion stroke. Small vortices are created in the exhaust stroke at the corner between the cylinder head and cylinder wall, and between the valve and the piston. The turbulence levels within the engine cylinder are controlled by the generation of turbulence at the shear layers of the intake air jet. The turbulent kinetic energy decreases during the compression and expansion strokes, and is almost uniform at top-dead-center of the compression stroke. Turbulence is generated at the exhaust port due to the local flow acceleration. The interaction between vortices in the intake stroke has been compared with that observed in motored engines equipped with only one valve. The results of this comparison indicate that vortex interactions play an important role in determining the velocity field and turbulence levels within an internal combustion engine.

ACKNOWLEDGEMENTS

This research was supported by the NASA Lewis Research Center under Grant NAG 3-21. The help and advice provided by Dr. H.J. Schock, the technical monitor of the Grant, is deeply appreciated.

REFERENCES

1. RAMOS, J.I., GANY, A., and SIRIGNANO, W.A. - The Recirculating Flow Field in a Two-Stroke, Motored Engine: Comparisons Between Theory and Experiments, Momentum and Heat Transfer Processes in Recirculating Flows, HTD-Vol. 13, Eds., Launder, B.E. and Humphrey, J.A.C., ASME, 1980.

2. RAMOS, J.I., GANY, A. and SIRIGNANO, W.A. - Study of Turbulence in a Motored Four-Stroke I.C. Engine, AIAA Journal, Vol. 19, No. 5, pp. 595-600, 1981.

3. GOSMAN, A.D., MELLING, A., WHITELAW, J.H. and WATKINS, P. - Axisymmetric Flow in a Motored Reciprocating Engine, Proceedings of the Institution of Mechanical Engineers, Vol. 192, pp. 213-223, 1978.

4. SCHOCK, H.J., SOSOKA, D.J. and RAMOS, J.I. - Numerical Studies of the Formation and Destruction of Vortices in a Motored Four-Stroke Piston-Cylinder Configuration, AIAA Paper No. 83-0497, AIAA 21st Aerospace Sciences Meeting, Reno, Nevada, 1983.

5. RAMOS, J.I. and SIRIGNANO, W.A. - Axisymmetric Flow Model in a Piston-Cylinder Arrangement with Detailed Analysis of the Valve Region, SAE Paper No. 800286, SAE Congress and Exposition, Detroit, Michigan, 1980.

6. GOSMAN, A.D., JOHNS, R.J.R. and WATKINS, P.A. - Development of Prediction Methods for In-Cylinder Processes in Reciprocating Engines, Combustion Modeling in Reciprocating Engines, Eds. Mattavi, J.N. and Amann, C.A., pp. 69-129, Plenum Press, New York, 1980.

7. RAMOS, J.I., HUMPHREY, J.A.C. and SIRIGNANO, W.A. - Numerical Prediction of Axisymmetric, Laminar and Turbulent Flows in Motored, Reciprocating Internal Combustion Engines, SAE Transactions, Vol. 88, pp. 1217-1242, 1980.

8. RAMOS, J.I. and SIRIGNANO, W.A. - The Calculation of the Pressure in Unsteady Flows Using a Control-Volume Approach, Journal of Computational Physics, Vol. 14, No. 1, pp. 211-216, 1981.

SECTION 14

GENERAL APPLICATIONS

AND

MATHEMATICAL CONCEPTS

ON THE CONVERGENCE OF THE LAD, NOS, AND SPLIT NOS METHODS FOR SOLVING THE STEADY-STATE NAVIER-STOKES EQUATIONS

Markku Lindroos *

SUMMARY

This paper is devoted to the study of the convergence properties of the methods which P.J. Roache has proposed for solving the Navier-Stokes equations for steady planar flow of an incompressible viscous fluid. The methods analyzed are in fact generalizations of those devised by Roache, in the sense that relaxation parameters are employed to hasten the convergence. A limitation of practical importance is that Dirichlet boundary conditions are imposed for both dependent variables, which are the stream function and the vorticity. It is shown that in certain special cases the consideration of the vorticity and stream-function iterations involved can be separated. This inspires consideration of the vorticity iteration with fixed values of the stream function. Mathematical analysis of the range of parameters which yield convergence for this vorticity iteration is simplified by a restriction to constant-coefficient model problems.

1. INTRODUCTION

The purpose of this paper is to study the convergence characteristics of the methods which Roache [4] has proposed for solving the nondimensional Navier-Stokes equations

$$\Delta\zeta - \mathrm{Re}\nabla\cdot(\bar{V}\zeta) = 0 \tag{1}$$

$$\Delta\psi + \zeta = 0, \tag{2}$$

which govern the steady two-dimensional flow of an incompressible ($\nabla\cdot\bar{V} = 0$) viscous fluid. Here, ψ is the stream function, ζ the vorticity, Re the Reynolds number, and $\bar{V} = (u,v)^T$ the velocity vector with components defined by $u = \partial\psi/\partial y$ and $v = -\partial\psi/\partial x$, respectively; Δ and ∇ denote the Laplacian and gradient operators in the two-dimensional space. Equation (1) is referred to as the vorticity equation, and equation (2) as the stream-function equation.

* Computing Centre, Helsinki University of Technology, SF-02150 Espoo 15, Finland

The shape of the region considered and the boundary conditions to be specified depend on the problem in question. In this paper we assume that the region is a rectangle with sides parallel to the coordinate axes, and that the values of ζ and ψ are known along the perimeter of the rectangle. This limits the usefulness of the results derived, as in practice the equations (1) and (2) are usually supplemented by other boundary conditions.

The rectangular region is subdivided by introducing a uniform mesh of grid lines parallel to the coordinate axes. Let M and N denote the numbers of mesh intervals which span the sides of the rectangle in the x and y directions, and let Δx and Δy be the mesh intervals themselves. The mesh points are identified by a pair of indices (i,j), with i running from 0 to M and j running from 0 to N; the extreme values refer to the points on the boundaries.

At each interior grid point, equations (1) and (2) are replaced by their finite-difference approximations. This results in a set of simultaneous algebraic equations

$$\Delta_h\zeta_{ij} - \mathrm{Re}\nabla_h\cdot(\bar{V}_{ij}\zeta_{ij}) = 0\,, \tag{3}$$

$$\Delta_h\psi_{ij} + \zeta_{ij} = 0, \quad i = 1, \ldots, M-1, \quad j = 1, \ldots, N-1, \tag{4}$$

which relate the nodal values of the stream function to each other and to the values of the vorticity. The finite differences to be used are those of the usual second-order central-difference approximations ($\Delta_h\psi_{ij}$ is defined analogously with $\Delta_h\zeta_{ij}$):

$$\Delta_h\zeta_{ij} = \frac{\zeta_{i+1,j} - 2\zeta_{ij} + \zeta_{i-1,j}}{(\Delta x)^2} + \frac{\zeta_{i,j+1} - 2\zeta_{ij} + \zeta_{i,j-1}}{(\Delta y)^2}\,, \tag{5}$$

$$\nabla_h\cdot(\bar{V}_{ij}\zeta_{ij}) = \frac{u_{i+1,j}\zeta_{i+1,j} - u_{i-1,j}\zeta_{i-1,j}}{2\Delta x} + \frac{v_{i,j+1}\zeta_{i,j+1} - v_{i,j-1}\zeta_{i,j-1}}{2\Delta y} \tag{6}$$

with

$$u_{ij} = \frac{\psi_{i,j+1} - \psi_{i,j-1}}{2\Delta y}\,, \quad v_{ij} = -\frac{\psi_{i+1,j} - \psi_{i-1,j}}{2\Delta x}\,. \tag{7}$$

In the sequel the subscripts i and j are dropped from the contexts in which they are not necessary for the presentation.

The Laplacian driver (LAD) method, in the form suggested by Roache, is defined by the iterative scheme

$$\Delta_h\zeta^k - \mathrm{Re}\nabla_h\cdot(\bar{V}^{k-1}\zeta^{k-1}) = 0 \tag{8}$$

$$\Delta_h\psi^k + \zeta^k = 0, \tag{9}$$

where k is the iteration index; an initial guess ζ^o, ψ^o is, of course, required to start the iteration. At each iteration step, a fast Poisson solver may first be employed to solve (8) for ζ^k, and then to solve (9) for ψ^k. For our considerations, however, it is immaterial how the equations are solved for the new values ζ^k and ψ^k.

The numerical Oseen (NOS) method is defined by

$$\Delta_h \zeta^k - \mathrm{Re}\nabla_h \cdot (\bar{V}^{k-1} \zeta^k) = 0 \tag{10}$$

$$\Delta_h \psi^k + \zeta^k = 0 . \tag{11}$$

Efficient use of this method requires a direct (non-iterative) linear solver which is more general than the Poisson solver. The EVP method [4] is one possibility, although its applicability is limited by the mesh restrictions. Since the coefficient matrix of the discrete vorticity equation varies with k, a new inversion is required at each step. However, for the considerations to follow, it is again immaterial how the system is solved for ζ^k and ψ^k.

The Split NOS method was introduced by Roache to combine some virtues of both the LAD and NOS methods. The velocity vector $\bar{V}$ is written as $\bar{V} = \bar{V}^o + \bar{V}'$, where $\bar{V}^o$ is an initial guess at the velocity field. The Split NOS method is then given by

$$\Delta_h \zeta^k - \mathrm{Re}\nabla_h \cdot (\bar{V}^o \zeta^k) - \mathrm{Re}\nabla_h \cdot (\bar{V}'^{k-1} \zeta^{k-1}) = 0 \tag{12}$$

$$\Delta_h \psi^k + \zeta^k = 0. \tag{13}$$

If $\bar{V}^o$ is held fixed during the iteration, the linear solver employed in the vorticity equation must be initialized only once.

Roache presents some numerical results which demonstrate the convergence properties of the three methods discussed above. Our aim is to shed some more light on the understanding of the convergence of the iterations by invoking a general theory of the iterative solution of nonlinear equations in several variables. Other iterative methods for solving the steady-state Navier-Stokes equations have been analyzed in a similar way in papers [1] and [2].

The methods to be analyzed are generalizations of those proposed by Roache, in the sense that parameters r_ζ and r_ψ are employed to hasten the convergence of the iterations. If the vector of the vorticity values solved from the vorticity equation is for a moment denoted by ζ^{*k}, the accepted value at step k is $\zeta^k = r_\zeta(\zeta^{*k} - \zeta^{k-1}) + \zeta^{k-1}$. This will be employed in the stream-function equation which yields ψ^{*k}, and the ultimate accepted value is $\psi^k = r_\psi(\psi^{*k} - \psi^{k-1}) + \psi^{k-1}$. In the case of the Split NOS method, the accelerated scheme is as follows:

$$\frac{1}{r_\zeta}[\Delta_h \zeta^k - \mathrm{Re}\nabla_h \cdot (\bar{V}^o \zeta^k)] + (1 - \frac{1}{r_\zeta})[\Delta_h \zeta^{k-1} - \mathrm{Re}\nabla_h \cdot (\bar{V}^o \zeta^{k-1})] - \mathrm{Re}\nabla_h \cdot (\bar{V}'^{k-1} \zeta^{k-1}) = 0 \tag{14}$$

$$\zeta^k + \frac{1}{r_\psi}\Delta_h\psi^k + (1 - \frac{1}{r_\psi})\Delta_h\psi^{k-1} = 0. \tag{15}$$

The accelerated LAD method is obtained from this scheme by choosing $\bar{V}^o = 0$, $\bar{V}'^{k-1} = \bar{V}^{k-1}$, and the accelerated NOS method by choosing $\bar{V}^o = \bar{V}^{k-1}$, $\bar{V}'^{k-1} = 0$ (in the latter case $\bar{V}^o$ is not constant with respect to k but is chosen to be the most recent approximation to the velocity vector). The unaccelerated methods proposed by Roache are obtained with $r_\zeta = r_\psi = 1$. For the brevity of discussion, the accelerated methods are still simply referred to as the LAD, NOS, and Split NOS methods.

2. ANALYSIS OF CONVERGENCE FOR THE NONLINEAR SYSTEM

We introduce a vector α, the components of which are the nodal unknowns, and rename these components:

$$\begin{aligned}\alpha = {} & (\zeta_{11},\zeta_{12}, \ldots,\zeta_{1,N-1},\zeta_{21}, \ldots,\zeta_{M-1,N-1}, \\ & \psi_{11},\psi_{12}, \ldots,\psi_{1,N-1},\psi_{21}, \ldots,\psi_{M-1,N-1})^T \\ = {} & (\alpha_1,\alpha_2, \ldots, \alpha_{2(M-1)(N-1)})^T.\end{aligned} \tag{16}$$

The nonlinear system of equations (3) and (4) can then be written in the vector form

$$F(\alpha) = 0 . \tag{17}$$

The first half of the components of F is made up of the left-hand sides of the discrete vorticity equation, and the second half embraces the discrete representations of $\Delta\psi + \zeta$ at the inner nodal points.

The Jacobian matrix of $F(\alpha)$ is

$$\frac{\partial F}{\partial \alpha} = \begin{pmatrix} A + B & C \\ I & A \end{pmatrix} \tag{18}$$

where A, B, C, and I (the unit matrix) are square matrices of order (M-1)(N-1). The two A's are formed of the partial derivatives of the discrete representations of $\Delta\zeta$ and $\Delta\psi$ with respect to ζ and ψ, respectively, B and C are composed of the partial derivatives of the nonlinear (convection) terms in the vorticity equation with respect to ζ and ψ, respectively, and I corresponds to the vector ζ in the stream-function equation. The matrix A is independent of ζ and ψ, B is a function of ψ, and C is a function of ζ.

Each of the three methods, accelerated with r_ζ and r_ψ, can be seen to be of the form

$$G(\alpha^k,\alpha^{k-1}) = 0, \quad k = 1,2, \ldots, \tag{19}$$

where G is a mapping from $R^{2(M-1)(N-1)} \times R^{2(M-1)(N-1)}$ to $R^{2(M-1)(N-1)}$. Let $\partial_1 G(\alpha,\beta)$ and $\partial_2 G(\alpha,\beta)$ denote the partial derivatives of $G(\alpha,\beta)$ with respect to α and β, respectively.

There exists a general theory for the convergence of iterations of the form (19); see e.g. Ortega and Rheinboldt [3, p. 325].

To apply this theory, we have to assume that our discrete system has a solution α^*, i.e. $G(\alpha^*,\alpha^*) = 0$. If $\partial_1 G(\alpha^*,\alpha^*)$ is nonsingular and

$$\rho(-\partial_1 G(\alpha^*,\alpha^*)^{-1}\partial_2 G(\alpha^*,\alpha^*)) < 1\,, \tag{20}$$

where ρ stands for the spectral radius of the matrix within the parentheses, and if the initial guess is sufficiently close to α^*, then the iteration converges to α^*.

For the iterative scheme of equations (14) and (15), the definition of G is straightforward, and it is most naturally made in such a way that the derivatives $\partial_1 G$ and $\partial_2 G$ are as follows:

$$\partial_1 G = \begin{pmatrix} \frac{1}{r_\zeta}(A + B^o) & 0 \\ I & \frac{1}{r_\psi}A \end{pmatrix}, \quad \partial_2 G = \begin{pmatrix} (1 - \frac{1}{r_\zeta})(A + B^o) + B' & C \\ 0 & (1 - \frac{1}{r_\psi})A \end{pmatrix} \tag{21}$$

where the matrix B of equation (18) has been split into the sum $B = B^o + B'$; the parts B^o and B' correspond to $\bar{V}^o$ and $\bar{V}'$, respectively, in the sum $\bar{V} = \bar{V}^o + \bar{V}'$. For the LAD method $B^o = 0$, $B' = B$, and for the NOS method $B^o = B$, $B' = 0$.

The eigenvalues, λ, of $-\partial_1 G(\alpha^*,\alpha^*)^{-1}\partial_2 G(\alpha^*,\alpha^*)$ are determinable from $\det(\lambda\partial_1 G(\alpha^*,\alpha^*) + \partial_2 G(\alpha^*,\alpha^*)) = 0$. In our case this yields

$$\begin{vmatrix} [1 + r_\zeta^{-1}(\lambda - 1)](A + B^o) + B' & C \\ \lambda I & [1 + r_\psi^{-1}(\lambda - 1)]A \end{vmatrix} = 0\,, \tag{22}$$

or

$$\det(([1 + r_\zeta^{-1}(\lambda - 1)](A + B^o) + B')[1 + r_\psi^{-1}(\lambda - 1)]A - \lambda C) = 0\,. \tag{23}$$

It is quite easy to see that if the vorticity values are all equal in the solution α^*, then $C = C(\alpha^*)$ is the null matrix, and equation (23) decomposes into the two separate equations

$$\det([1 + r_\zeta^{-1}(\lambda - 1)](A + B^o) + B') = 0\,, \tag{24}$$

$$\det([1 + r_\psi^{-1}(\lambda - 1)]A) = 0. \tag{25}$$

These are the characteristic equations for the iteration matrices of the ζ and ψ iterations, respectively, when the values of the other dependent variable are frozen at those included in α^*. Since the determinant of A is nonzero (as it is easy to see, for example, by means of the identity (30) below), the roots of the latter equation are all equal to $\lambda = 1 + r_\psi$, and for convergence it is required that $0 < r_\psi < 2$. The condition to be imposed for r_ζ is more involved. For the NOS method, however, $B' = 0$ and equation (24) leads to the condition $0 < r_\zeta < 2$ (in this special case in which $C(\alpha^*) = 0$). As a simple example of a case for which $C(\alpha^*) = 0$, we can mention the uniform flow.

The above results can clearly be generalized to regions other than rectangles. In the rest of this paper, however, it is essential that the region is a rectangle, as in other regions an analytical treatment of the problem seems to be prohibitively difficult.

3. RESULTS FOR A LINEAR MODEL PROBLEM

In the previous section it was shown that in certain special cases the convergence question of the iteration reduces to the convergence questions of the ζ and ψ iterations. This motivates the study of the vorticity iteration when the values of the stream function are held fixed. (The ψ iteration with fixed values of ζ converges for $0 < r_\psi < 2$, and its characteristics can be inferred from those of the ζ iteration by passing to the limit $u = 0$, $v = 0$.) To further simplify the problem, and to enlarge the possibilities of our analytical treatment, restriction is made to the model problem

$$-u\frac{\partial\zeta}{\partial x} - v\frac{\partial\zeta}{\partial y} + \frac{1}{Re}\left(\frac{\partial^2\zeta}{\partial x^2} + \frac{\partial^2\zeta}{\partial y^2}\right) = 0, \qquad (26)$$

in which the velocity components u and v are constant throughout the rectangular calculation domain. By introducing the abbreviations

$$p = u\Delta x Re,\quad q = v\Delta y Re,\quad \gamma = (\Delta x/\Delta y)^2 \qquad (27)$$

($|p|$ and $|q|$ may be called the grid Reynolds numbers), the discrete equations can be written as

$$a\zeta_{ij} + b\zeta_{i+1,j} + c\zeta_{i-1,j} + d\zeta_{i,j+1} + e\zeta_{i,j-1} = 0, \qquad (28)$$

where

$$a = -2(1+\gamma),\quad b = 1-p/2,\quad c = 1+p/2,\quad d = \gamma(1-q/2),\quad e = \gamma(1+q/2). \qquad (29)$$

If these grid equations are taken column by column from left to right and each column is treated from the bottom to the top, in accordance with equation (16), the $(M-1)(N-1) \times (M-1)(N-1)$ coefficient matrix of the linear system has the block-tridiagonal structure

$$\begin{pmatrix} A & B & & & \\ C & A & B & & \\ & \ddots & \ddots & \ddots & \\ & & C & A & B \\ & & & C & A \end{pmatrix}, \text{ where } A = \begin{pmatrix} a & d & & & \\ e & a & d & & \\ & \ddots & \ddots & \ddots & \\ & & e & a & d \\ & & & e & a \end{pmatrix}, \ B = bI, \text{ and } C = cI.$$

The square matrices A, B, C, and the unit matrix I are of order N-1.

In paper [2] we can find the identity

$$\begin{vmatrix} A & B & & \\ C & \ddots & \ddots & \\ & \ddots & \ddots & B \\ & & C & A \end{vmatrix} = \prod_{m=1}^{M-1}\prod_{n=1}^{N-1}\left(a - 2\sqrt{bc}\cos(m\pi/M) - 2\sqrt{de}\cos(n\pi/N)\right), \qquad (30)$$

which is valid for any real or complex numbers a, b, c, d, e and for any integers $M \geqq 2$, $N \geqq 2$. Substitution of the values given in (29) into the right-hand side of (30) shows that the determinant of the coefficient matrix is nonzero, so that the equation system has a unique solution. The identity (30) will be very useful in the determination of the eigenvalues of the iteration matrix, as will be seen later.

We choose $\bar{V}^o = \alpha\bar{V} = \alpha(u,v)^T$ and $\bar{V}' = (1-\alpha)(u,v)^T$, where α is a scalar parameter, and consider the Split NOS method

$$\frac{1}{r_\zeta}[\Delta_h\zeta^k - \mathrm{Re}\alpha\nabla_h\cdot(\bar{V}\zeta^k)] + (1 - \frac{1}{r_\zeta})[\Delta_h\zeta^{k-1} - \mathrm{Re}\alpha\nabla_h\cdot(\bar{V}\zeta^{k-1})] - \mathrm{Re}(1-\alpha)\nabla_h\cdot(\bar{V}\zeta^{k-1}) = 0. \quad (31)$$

The LAD method is obtained with $\alpha = 0$, and the NOS method with $\alpha = 1$. The solution of the model problem could, of course, be obtained in one iteration step by choosing $\alpha = 1$ and $r_\zeta = 1$. As our purpose is to contribute to the understanding of the iteration in general, the use of other values of α and r_ζ is also meaningful.

Equation (31) can be written as

$$(1+p\alpha/2)\zeta^k_{i-1,j} - 2(1+\gamma)\zeta^k_{ij} + (1-p\alpha/2)\zeta^k_{i+1,j} + \gamma(1+q\alpha/2)\zeta^k_{i,j-1} + \gamma(1-q\alpha/2)\zeta^k_{i,j+1} + (r_\zeta-1+p(r_\zeta-\alpha)/2)\zeta^{k-1}_{i-1,j} + 2(1+\gamma)(1-r_\zeta)\zeta^{k-1}_{ij} + (r_\zeta-1-p(r_\zeta-\alpha)/2)\zeta^{k-1}_{i+1,j} + \gamma(r_\zeta-1+q(r_\zeta-\alpha)/2)\zeta^{k-1}_{i,j-1} + \gamma(r_\zeta-1-q(r_\zeta-\alpha)/2)\zeta^{k-1}_{i,j+1} = 0. \quad (32)$$

If λ is an eigenvalue of the iteration matrix and e_{ij}, $i = 1, \ldots,$ $M-1$, $j = 1, \ldots, N-1$, are the components of the respective eigenvector, and if we agree that $e_{i0} = e_{iN} = 0$, $i = 1, \ldots, M-1$, and $e_{0j} = e_{Mj} = 0$, $j = 1, \ldots, N-1$, then

$$(\lambda+r_\zeta-1+p(\alpha\lambda+r_\zeta-\alpha)/2)e_{i-1,j} - 2(1+\gamma)(\lambda+r_\zeta-1)e_{ij} + (\lambda+r_\zeta-1-p(\alpha\lambda+r_\zeta-\alpha)/2)e_{i+1,j} + \gamma(\lambda+r_\zeta-1+q(\alpha\lambda+r_\zeta-\alpha)/2)e_{i,j-1} + \gamma(\lambda+r_\zeta-1-q(\alpha\lambda+r_\zeta-\alpha)/2)e_{i,j+1} = 0 \quad (33)$$

for $i = 1, \ldots, M-1$, $j = 1, \ldots, N-1$. Nontrivial solutions $\{e_{ij}\}$ exist for this linear homogeneous system if and only if the determinant of the system vanishes. By making use of the partitioning (30), it is found that the eigenvalues are determinable from the equations

$$2(\lambda+r_\zeta-1) - \sqrt{4(\lambda+r_\zeta-1)^2 - p^2(r_\zeta-\alpha+\alpha\lambda)^2}\cos(m\pi/M) + \gamma\{2(\lambda+r_\zeta-1) - \sqrt{4(\lambda+r_\zeta-1)^2 - q^2(r_\zeta-\alpha+\alpha\lambda)^2}\cos(n\pi/N)\} = 0, \quad (34)$$

$$m = 1, \ldots, M-1, \quad n = 1, \ldots, N-1.$$

Removing of the square-root signs leads into a set of polynomial equations which are of order 4 or less in λ. The author has developed a computer program which solves these equations for λ and then determines the spectral radius of the iteration matrix. By means of this program one may determine, for example, the region of the p,q plane in which convergence is achieved for given values of r_ζ, α, γ, M, and N. Figure 1 shows this region for some values of r_ζ and α when $\gamma = 1$, $M = 10$, and $N = 10$ (only half of the region, which is symmetrical with respect to the line $|p| = |q|$, has been drawn).

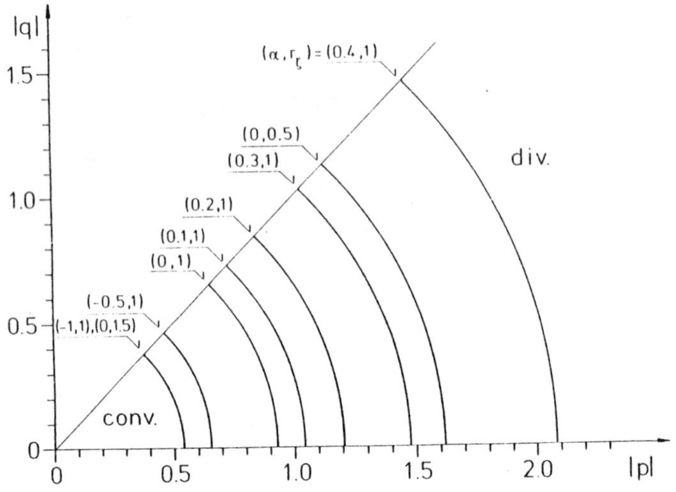

Figure 1. Variation of the convergence region with α and r_ζ when $\Delta x/\Delta y = 1$, $M = 10$, and $N = 10$.

One may want to know the values of α and r_ζ which yield convergence for any p, q, γ, M, and N. Figure 2 answers the question; the convergence region is shown by the hatched area. This can be proved by reasoning along lines quite similar to those laid down in our previous paper [2], where other iterative methods were studied. Here we can only give an outline of the proof.

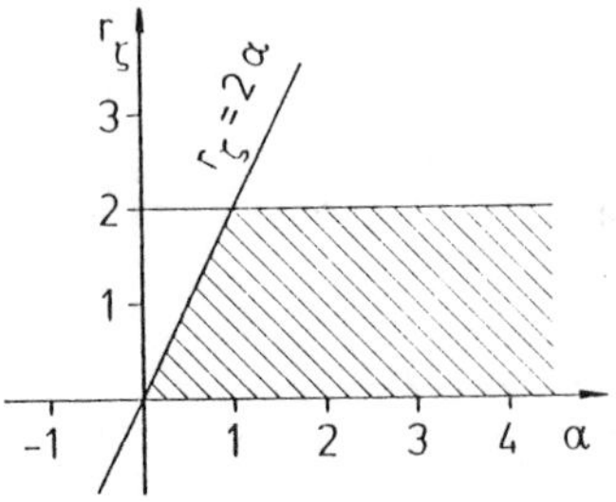

Figure 2. The values of (α, r_ζ) resulting in convergence for any p, q, γ, M, and N.

Equation (34) can be written as

$$f(\lambda, r_\zeta, \alpha, p, \cos(m\pi/M)) + \gamma f(\lambda, r_\zeta, \alpha, q, \cos(n\pi/N)) = 0 \tag{35}$$

if the function f is defined as follows:

$$f(z, r, \alpha, p, \phi) = 2(z+r-1) - \sqrt{4(z+r-1)^2 - p^2(r-\alpha+\alpha z)^2}\ \phi . \tag{36}$$

In the polynomial equation into which equation (35) can be cast, the coefficient of the highest power of λ is nonzero, and all coefficients are continuous in p and q. This implies that the eigenvalues λ are continuous in p and q. For $p = q = 0$, all of the eigenvalues are equal to $\lambda = 1 + r_\zeta$. In the hatched area, these are less than one in absolute value. If $\lambda = e^{i\theta}$, $\theta \in [0,\pi]$, cannot be a root of equation (35) for any pair (p,q), the eigenvalues are less than unity in modulus for all (p,q).

Equation (35) is satisfied if and only if the two (generally complex-valued) terms on the left are equal in modulus and of the opposite sign. This leads us to consider, for fixed values of r_ζ, α, and $z = e^{i\theta}$, the image of the set

$$S = \{(p,\phi) \mid -\infty < p < \infty,\ -1 \leqq \phi \leqq 1\} \tag{37}$$

in the mapping defined by f. When ϕ varies between -1 and 1, $f(e^{i\theta},r_\zeta,\alpha,p,\phi)$ traverses the line segment connecting the points $f(e^{i\theta},r_\zeta,\alpha,p,-1)$ and $f(e^{i\theta},r_\zeta,\alpha,p,1)$. So it remains to determine the loci of the end points of these line segments as p varies from 0 to ∞ (f is an even function of p). An elaborate study reveals that if α and r_ζ are chosen from the hatched region, there cannot be any two image points which would locate on opposite sides of the origin of the complex plane. As a consequence, equation (35) cannot have a root $\lambda = e^{i\theta}$ for any (p,q), and so the spectral radius is less than unity for any p, q, γ, M, and N (i.e., the iteration converges for any u, v, Re, Δx, Δy, M, and N).

4. RESULTS FOR A ONE-DIMENSIONAL MODEL PROBLEM

In this section, we consider the numerical solution of the one-dimensional model problem

$$-u\frac{\partial\zeta}{\partial x} + \frac{1}{Re}\frac{\partial^2\zeta}{\partial x^2} = 0\,. \tag{38}$$

This allows for an analytical determination of the spectral radius as a function of the pertinent parameters. The iterative scheme corresponding to the Split NOS method can be obtained from (32) by setting $\gamma = 0$ and dropping the subscript j. The eigenvalues of the iteration matrix can be determined from the equations

$$2(\lambda+r_\zeta-1) - \sqrt{4(\lambda+r_\zeta-1)^2 - p^2(r_\zeta-\alpha+\alpha\lambda)^2}\ \cos(m\pi/M) = 0, \tag{39}$$
$$m = 1, \ldots, M-1.$$

By comparing the structures of equations (34) and (39), it is easy to see that if $M = N$ and $|p| = |q|$, then the spectral radius of the two-dimensional problem with parameters r_ζ, α, p, q, γ, M, and N is greater than or equal to the spectral radius of the one-dimensional problem with parameters r_ζ, α, p, and M. As a consequence, consideration of the one-dimensional problem gives valuable information about the two-dimensional problem.

The eigenvalues for the one-dimensional problem can be calculated from quadratic equations. The expression for the spectral radius depends on whether M is odd or even and on the location of α and r_ζ in the α, r_ζ plane. In the hatched areas of figure 3, the square of the spectral radius is given by

$$\rho^2 = \begin{cases} \dfrac{4(r_\zeta-1)^2\cos^2(\pi/(2M))+p^2(r_\zeta-\alpha)^2\sin^2(\pi/(2M))}{4\cos^2(\pi/(2M))+p^2\alpha^2\sin^2(\pi/(2M))}, & M \text{ odd}, \\ (r_\zeta-1)^2, & M \text{ even}, \end{cases} \tag{40}$$

while in other areas it is given by

$$\rho^2 = \frac{4(r_\zeta-1)^2\sin^2(\pi/M)+p^2(r_\zeta-\alpha)^2\cos^2(\pi/M)}{4\sin^2(\pi/M)+p^2\alpha^2\cos^2(\pi/M)} . \tag{41}$$

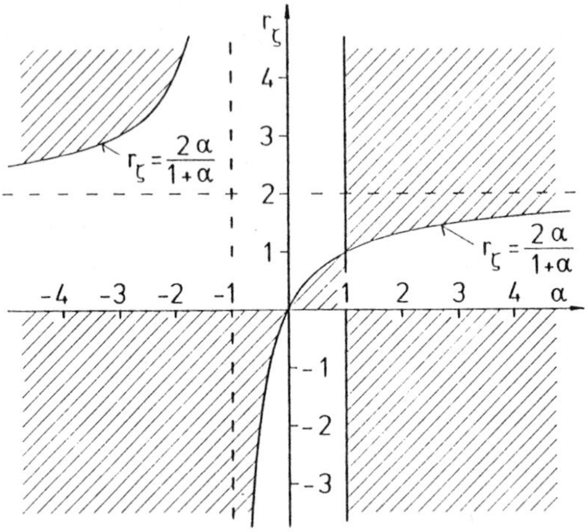

Figure 3. Chart for the choice of the right expression for the spectral radius: equation (40) is valid in the hatched areas, equation (41) elsewhere.

By requiring that $\rho^2 < 1$, we can derive the following conditions for convergence:

$$0 < r_\zeta < \bar{r}_\zeta = \begin{cases} \dfrac{8\cos^2(\pi/(2M))+2\alpha p^2\sin^2(\pi/(2M))}{4\cos^2(\pi/(2M))+p^2\sin^2(\pi/(2M))} & \text{for } \alpha \geqq 1,\ M \text{ odd}, \\ 2 \quad \text{for } \alpha \geqq 1,\ M \text{ even}, \\ \dfrac{8\sin^2(\pi/M)+2\alpha p^2\cos^2(\pi/M)}{4\sin^2(\pi/M)+p^2\cos^2(\pi/M)} & \text{for } -\dfrac{4}{p^2}\tan^2(\pi/M) < \alpha < 1, \end{cases} \tag{42}$$

or

$$\frac{8\cos^2(\pi/(2M))+2\alpha p^2\sin^2(\pi/(2M))}{4\cos^2(\pi/(2M))+p^2\sin^2(\pi/(2M))} < r_\zeta < 0 \text{ for } \alpha < -\frac{4}{p^2}\cot^2(\pi/(2M)),\ M \text{ odd.} \tag{43}$$

If M is odd, there exist two separate regions from which (α, r_ζ) can be chosen to achieve convergence. In figure 4 these regions are illustrated for $|p| = 3$, $M = 5$ by the hatched areas. If M is even, the left region is absent and the upper boundary of the right region coincides with the horizontal line $r_\zeta = 2$. (One may imagine that the more gently sloping straight line has turned around the point (1,2) and become horizontal. Compare also with figure 2.)

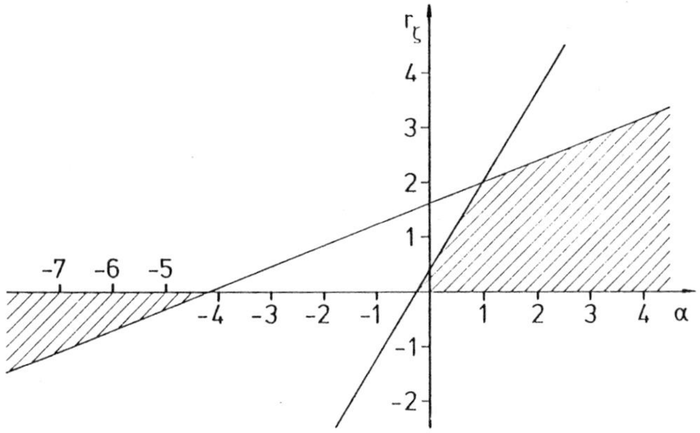

Figure 4. The values of (α, r_ζ) resulting in convergence for $|p| = 3$, $M = 5$.

Figure 5 presents $\bar{r}_\zeta$, the right-hand side of inequality (42), as a function of the grid Reynolds number $|p| = |u|\Delta x Re$ for some values of α and M. It can be seen, for example, that if $|p|$ is large, strong under-relaxation is required in the LAD method $(\alpha = 0)$ to attain convergence.

If we choose $\alpha = 0$ and $r_\zeta = 1$, which corresponds to the LAD method in the unaccelerated form suggested by Roache, the spectral radius is

$$\rho = \frac{|p|}{2\tan(\pi/M)} \tag{44}$$

and the condition for convergence can be written as

$$|p| = |u|\Delta x Re < 2\tan(\pi/M) \quad . \tag{45}$$

This is a very restrictive condition indeed. If M is large, the condition (45) can be further reduced to $|p| \lesssim 2\pi/M$.

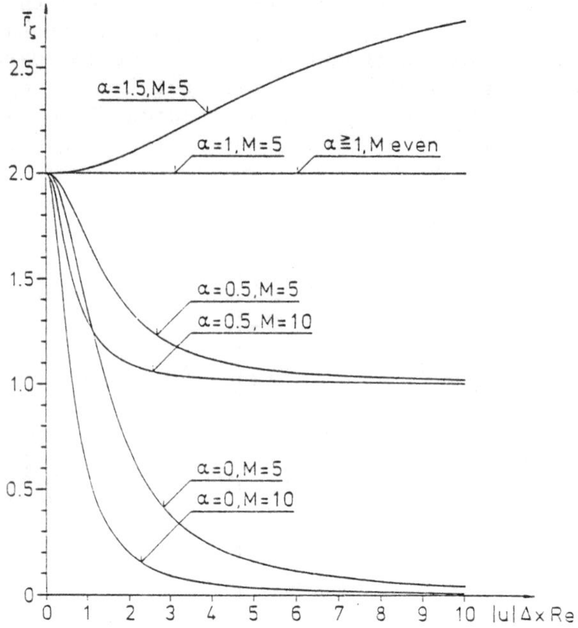

Figure 5. Variation of $\bar{r}_\zeta$ with $|p| = |u|\Delta x Re$, α, and M.

REFERENCES

1. LINDROOS, M. — On the convergence of iterative methods for solving the steady-state Navier-Stokes equations by finite differences. Proceedings of the Seventh International Conference on Numerical Methods in Fluid Dynamics (held at the Stanford University and the NASA/Ames Research Center, June 23-27, 1980), Lecture Notes in Physics, Vol. 141, pp. 272-278, Springer-Verlag, Berlin, 1981.

2. LINDROOS, M. — On a deferred-correction procedure for determination of central-difference solutions to the Navier-Stokes equations. Proceedings of the Second International Conference on Numerical Methods in Laminar and Turbulent Flow (held in Venice, Italy, July 13-16, 1981), pp. 1129-1142, Pineridge Press, Swansea, United Kingdom, 1981.

3. ORTEGA, J.M., and RHEINBOLDT, W.C. — Iterative Solution of Nonlinear Equations in Several Variables. Academic Press, New York, 1970.

4. ROACHE, P.J. — The LAD, NOS and Split NOS methods for the steady-state Navier-Stokes equations. Computers and Fluids, Vol. 3, pp. 179-195, 1975.

ON THE USE OF HIGHER ORDER WEIGHTING FUNCTIONS IN THE BOUNDARY INTEGRAL METHOD FOR FLUID FLOW

C. Patterson and M. A. Sheikh

Department of Mechanical Engineering, University of Sheffield, Sheffield, Sheffield, S1 3JD, U.K.

SUMMARY

A second order harmonic is employed as kernel function for the solution of Laplacian problems in the Boundary Element Method. This is achieved in conjunction with the Regular Boundary Element Method where the singular point of the fundamental solution is located outside the domain of a given problem. This approach leads to improved diagonal dominance in the resulting system of algebraic equations and makes it possible to carry out numerical integration only upto a certain distance from the singularity and not over the entire domain. In this paper, the method is applied to two-dimensional, steady-state, inviscid, laminar flow. Two simple test problems are examined having regular and singular solutions respectively, and the results are compared with those obtained by using Singular Boundary Element Method. For each case the best position of the singularity outside the domain is determined.

1. INTRODUCTION

The abundant success of the Finite Element Method has led to progressively more exacting demands upon it. Whereas previously designers were content with a simplified two dimensional approach there is now an increasing demand for more detailed three dimensional analyses. However, the increased computing overhead in going from two to three dimensions is considerable so that there is urgent need to explore methods which may be more efficient than the Finite Element Method.

Because of the square-cube relation of degrees of freedom, any boundary domain method is attractive in this light. One such method is the Boundary Element Method which is now well known as a valid numerical technique for the solution of field problems, performing equally efficient as the Finite Element Method in some cases [1].

Basic to the method is the replacement of the classical field problem by a problem on the boundary surface expressed as an infinite system of boundary integral equations. These are discretized by the introduction of boundary elements (constant, linear, quadratic, etc.) after the manner of finite elements. In the conventional approach singular boundary integral equations are encountered because the fundamental solution with singularity on the boundary is used in generating the boundary problem. This requires numerical evaluation of singular integrals in the discretized problem. A Regular Boundary Integral Method has been proposed by the authors [2] and applied to the problems of fluid flow. Here the location of the singular point is taken outside the domain of the problem thereby giving regular equations. When the singular point of the kernel function is taken outside the domain of the problem kernels with higher order singularities can be employed in deriving the boundary integral equations. The use of higher order singularities is attractive because of their more rapid spatial decay implying improved diagonal dominance in the algebraic equations [3].

In this paper the second order harmonic (the second spatial derivative of the fundamental solution) is used as kernel function with discretization achieved as usual using constant and linear boundary elements. Two test problems are examined, having regular and singular solutions and for each case a systematic study has been carried out to determine the best position of singularity outside the domain. It is found that the singular point located at about one and a half lengths of the element away from the boundary with 4-point integration order generates valid results which compare favourably, in quality and cost, with solutions obtained by other boundary element methods.

2. THEORY

The fundamental statement of the Boundary Element Method formulation of potential problems (governed by Laplace's equation) is the integral equation [4],

$$c_i\phi_i + \int_S \phi^*_{,n} \ \phi \ dS = \int_S \phi^* \ \phi_{,n} \ dS \qquad (1)$$

relating potentials and normal fluxes ($\phi_{,n} = \frac{\partial\phi}{\partial n}$) over the boundary 'S' of domain Ω. The boundary conditions of the problem can be of the Neumann, Dirichlet or Cauchy (mixed) types [1]. The value of c_i depends upon the type of boundary under consideration. In the Regular Boundary Element Method

[5], [6], [7], where the singular point of the fundamental solution is located outside the domain of the problem (Figure 1); $c_i = 0$, and the resulting Regular Boundary Integral Equation is:

$$\int_S \phi^*_{,n} \, \phi \, dS = \int_S \phi^* \, \phi_{,n} \, dS \tag{2}$$

In this case, no special attention is required to evaluate the singular integrands and additionally, higher order functions can be employed as kernels.

Fig. 1

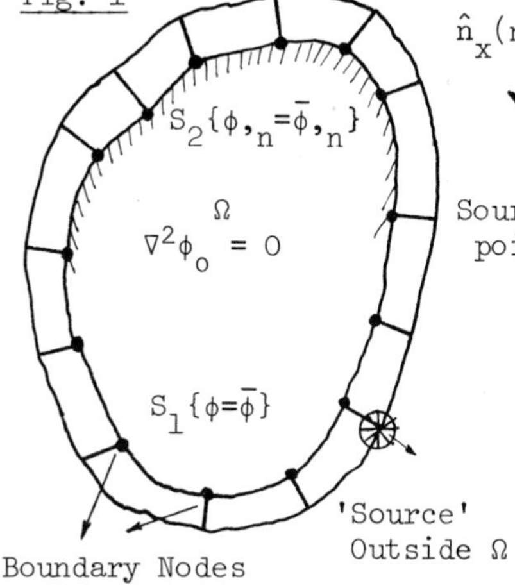

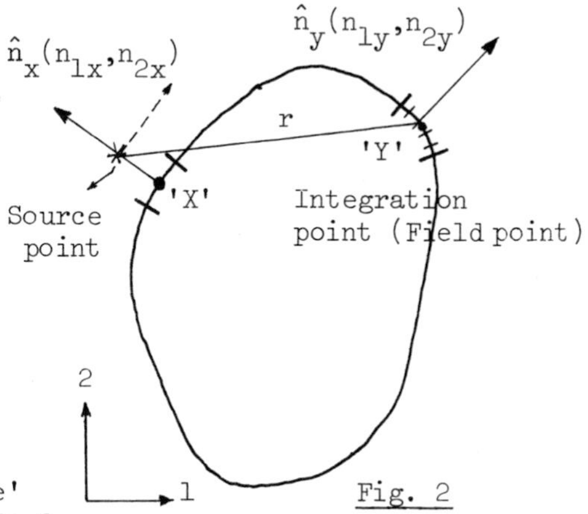

Fig. 2

The fundamental solution ϕ^*_o for the two dimensional case is given as:

$$\phi^*_o \, (x,y) = \frac{1}{2\pi} \, \ell n \, \frac{1}{r(x,y)} \tag{3}$$

where 'r' is the distance between the 'source' and 'observation' points and $\phi^*_{o,n}$ is the normal derivative given by:

$$\frac{\partial \phi^*_o}{\partial n} (x,y) = \frac{1}{2\pi r} \sum_{\ell=1}^{2} n_{\ell y} \left(\frac{x_\ell - y_\ell}{r} \right) \tag{4}$$

where $x \equiv (x_1,x_2)$ is the singular (source) point

$y \equiv (y_1,y_2)$ is the field (observation) point

$n \equiv n_y$ is the normal at point 'y' (see Fig. 2)

In the present study, higher order functions are obtained by taking successive spatial derivatives with respect to the normal at the freedom point corresponding to the source. The kernels thus obtained are of the form $\sim \frac{1}{r^k}$ where $k \in \{1,2,3,4,..\}$ is the order of the kernel.

Referring to Figure 2, the first order kernel is:

$$\left.\begin{aligned}\phi^*_I &= \hat{n}_x \cdot \nabla\phi^*_o = n_{1x} \cdot \frac{\partial\phi^*_o}{\partial x_1} + n_{2x} \cdot \frac{\partial\phi^*_o}{\partial x_2} \\ &= \frac{1}{2\pi r} \sum_{\ell=1}^{2} n_{\ell x} \frac{x_\ell - y_\ell}{r} = \frac{1}{2\pi r} \sum_x \end{aligned}\right\} \qquad (5)$$

where ∇ is the gradient operator. The normal derivative associated with this new kernel is given by:

$$\left.\begin{aligned}\phi^*_{I,n} &= \frac{\partial\phi^*_I}{\partial n_y} = \hat{n}_y \cdot \nabla\phi^*_I \\ &= \frac{1}{2\pi r^2} \left(2 \sum_x \sum_y - \hat{n}_x \cdot \hat{n}_y \right) \end{aligned}\right\} \qquad (6)$$

Similarly, the 2nd order kernel is:

$$\left.\begin{aligned}\phi^*_{II} &= \hat{n}_x \cdot \nabla\phi^*_I = \hat{n}_x \cdot \nabla \left(\frac{1}{2\pi r} \sum_x \right) \\ &= \frac{1}{2\pi r^2} \left[2\left(\sum_x\right)^2 - 1 \right] \end{aligned}\right\} \qquad (7)$$

and the normal derivative:

$$\phi^*_{II,n} = \frac{\partial\phi^*_{II}}{\partial n_y} = \hat{n}_y \cdot \nabla\phi^*_{II} \qquad (8)$$

These new kernels which still satisfy the governing Laplace's equation, can now be used in the Regular Boundary Element Method to generate a system of higher order boundary integral equations. It should be noted that the first derivative with respect to 'n_x' actually, is a linear combination of the first derivatives of ϕ^* with respect to the global coordinatate axes, (x_1,x_2) where there is no obligation to use 'n_x' components (n_{1x},n_{2x}). Any linear combination, for instance $(1/\sqrt{2},1/\sqrt{2})$, can be used to generate the same solution on the boundary provided that all the components exist [8]. Moreover, these higher order kernels are used to generate the solution only on the boundary. Once that is determined, conventional kernels can be employed to find the internal solution of the problem using Green's third identity.

3. APPLICATIONS

Two 2-dimensional, inviscid, laminar fluid flow problems are analysed using the second order kernels given by equation (7) in section 2. A critical comparison of the results is made with the conventional Boundary Element Method solutions. For both methods, constant elements are employed for the discretization of the problems and all integrals are evaluated using a 4-point Gauss integration scheme. Computed values at some selected boundary points are also listed for different singularity locations.

3.1 Flow past a circular obstacle in a channel

The problem has been analysed using the Finite Element Method with 55 nodes and 80 elements [9] and 72 nodes and 110 elements [10]. The governing equation of the problem in stream function formulation is:

$$\frac{\partial^2 \psi}{\partial x_1{}^2} + \frac{\partial^2 \psi}{\partial x_2{}^2} = 0 \qquad (9)$$

where ψ is the stream function and its normal derivative gives velocities along the boundary. The problem is defined in Fig. 3a and its discretization using a quarter domain with 32 constant elements is shown in Fig. 3b. Boundary solutions for the Conventional (Singular) and Higher Order (2nd order kernels) Boundary Element Methods are compared in Fig. 4. The solution obtained for the higher order kernels corresponds to the best position of the singularity outside the domain. For a 4-point Gauss numerical integration scheme, this position is found to be one and a half lengths of the elements away from the freedom along the positive normal, as indicated in Table 1. It should be noted that if the singularity is brought closer to the boundary from this optimum location, then higher order integration schemes should be employed to sustain the accuracy of the solution.

3.2 Flow past a disc in a channel

The problem is defined in Fig. 5a. The flow in the channel is considered to be at normal incidence to the disc and is represented by a stream function ψ. Only a quarter of the domain is considered for analysis due to the symmetry and is discretized using 37 constant elements (Fig. 5b). An infinite speed is acquired by the stream at point 'O', the edge of the disc [11], thereby giving a singularity in the mathematical solution. Consequently, a refined mesh is employed around this point for better representation of the singular behaviour of the solution. Results obtained for both conventional and Higher Order (second order kernels) Boundary Element Methods, for the same discretization and using a 4-point Gauss integration scheme, are shown in Fig. 6. Again, the solution given by the Higher Order Boundary Element Method is obtained for

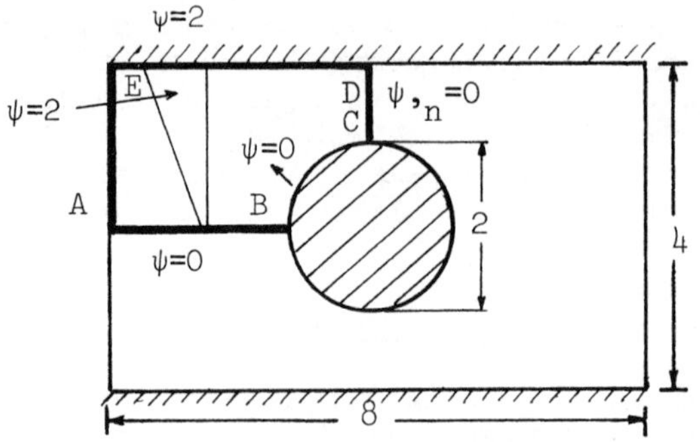

Fig.3a. Problem Definition.

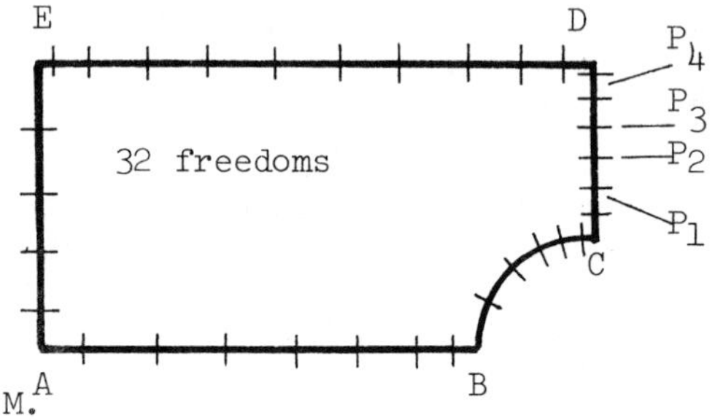

Fig.3b. Discretization.

——— Singular B.E.M.

- - - - Higher Order B.E.M.

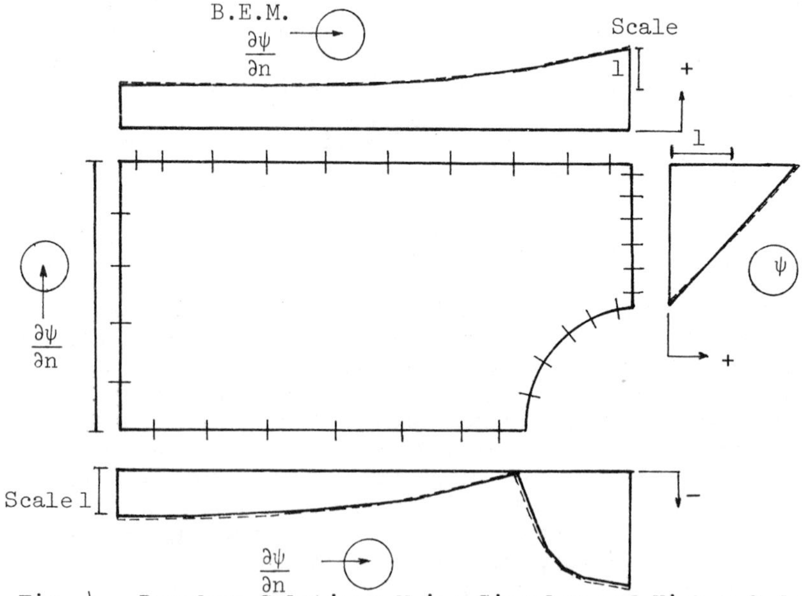

Fig. 4. Boundary Solutions Using Singular and Higher Order Boundary Element Methods.

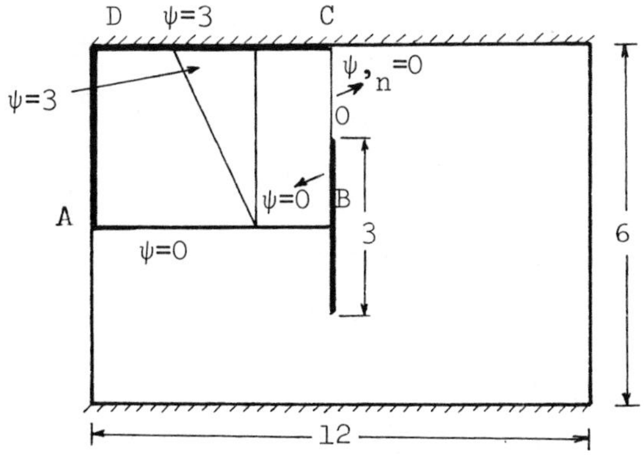

Fig. 5a. Problem Definition

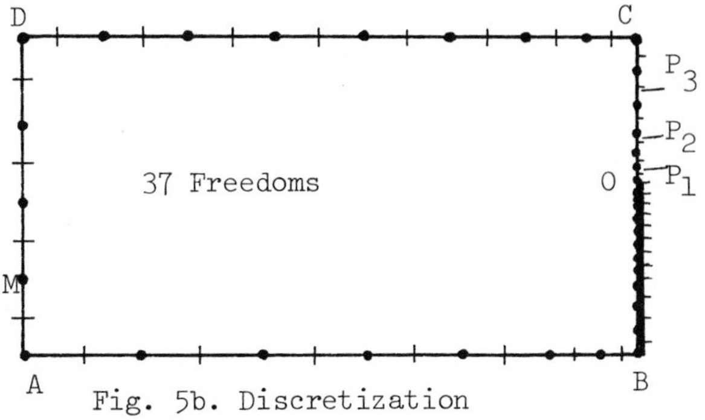

Fig. 5b. Discretization

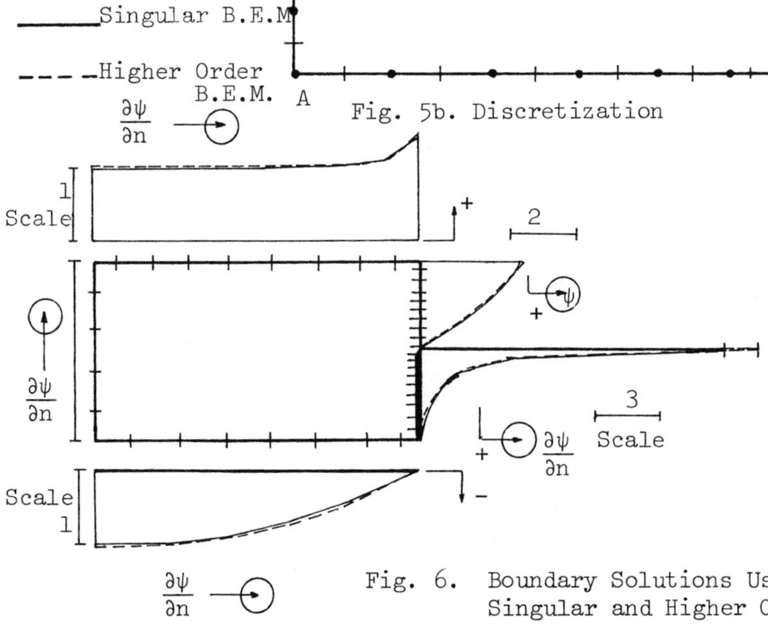

Fig. 6. Boundary Solutions Using Singular and Higher Order Boundary Element Methods

Freedom Point	Coordinate	Conventional B.E.M	Higher Order B.E.M(ϕ^*_{II})		
			Singularity Location		
			L/2	L	3/2L**
P_1	1.16	0.368	0.381	0.372	0.369
P_2	1.33	0.752	0.762	0.756	0.753
P_3	1.54	1.169	1.210	1.191	1.172
P_4	1.76	1.575	1.629	1.591	1.571

Table 1. Flow past a cylinder. Computed ψ values at 4 points along CD for 3 different singularity locations.

Freedom Point	Coordinate	Conventional B.E.M	Higher Order B.E.M(ϕ^*_{II})		
			Singularity Location		
			L	3/2L**	2L
P_1	1.8	1.10	1.16	1.09	1.07
P_2	2.3	2.28	2.25	2.24	2.16
P_3	2.8	2.72	2.87	2.85	2.85

Table 2. Flow past a disc. Computed ψ values at 3 different points along OC for 3 different singularity locations.

the best position of the singularity outside the domain which is indicated in Table 2.

4. DISCUSSION AND CONCLUSIONS

In this paper a Higher Order Boundary Element Method has been presented for two dimensional Laplacian fluid flow problems. Here the second spatial derivative of the fundamental solution with respect to the normal at the 'source point' has been employed as kernel function. The singularity (source) is located outside the domain of the problem, giving an infinite system of higher order regular integral equations. On discretizing the system in the usual manner, the resulting higher order kernels are everywhere regular over the boundary.

The case study results presented here, for harmonic fluid flow problems, show that the Conventional and Higher Order methods have similar convergence characteristics provided for the latter case the singularity is placed in its optimum position outside the domain. This position hinges on the order of kernels employed, and numerical integration. For the second order kernels and 4-point Gauss integration scheme, used for the analysis presented in this paper, best results are obtained if the singularity is located at one and a half lengths of the element away from the element along the outward normal to a node.

The discretization of the two simple problems analysed here, was achieved using constant elements; other higher order kernels could equally be employed. In the Boundary Solutions, using conventional and higher order methods, variations of around 2% were found. The solution for the problem of flow past a disc is singular, nevertheless by choosing a suitably refined mesh, both methods perform well in the neighbourhood of the singularity.

In conclusion: Higher order kernels can validly be employed in the Regular Boundary Element Method to generate Higher Order Regular Boundary Integral Equations for both regular and singular problems. These kernels are of potentially great interest because they decay more quickly with distance than does the Fundamental Solution and thus improve diagonal dominance in the algebraic system. Moreover they offer the possibility that the equations need only be evaluated upto a certain distance from the singularity, the remaining contributions effectively vanishing. If this is so, then the labour required to set up the algebraic equations can be considerably reduced. Early results are promising.

REFERENCES

1. WROBEL, L. C. and BREBBIA, C. A. - The Boundary Element Method for Steady State and Transient Heat Conduction, Numerical Methods in Thermal Problems, Eds, Lewis, R. W. and Morgan, K., Pineridge Press, 1980.

2. PATTERSON, C. and SHEIKH, M. A. - Regular Boundary Integral Equations for Fluid Flow, Numerical Methods in Laminar and Turbulent Flow, Eds. Taylor, C. and Schrefler, B.A., Pineridge Press, 1981.

3. PATTERSON, C. and SHEIKH, M. A. - A Higher Order Boundary Element Method for Fluid Flow, Finite Element Flow Analysis, Ed. Kawai, T., Univ. of Tokyo Press, 1982.

4. BREBBIA, C. A. - The Boundary Element Method for Engineers, Pentech Press, London, 1978.

5. PATTERSON, C. and SHEIKH, M. A. - Regular Boundary Integral Equations for Stress Analysis, Boundary Element Methods, Ed. Brebbia, C. A., Springer Verlag, 1981.

6. PATTERSON, C. and EL-SEBAI, N. A. S. - A Regular Boundary Method Using Non-Conforming Elements for Potential Problems in Three Dimensions, Boundary Element Methods in Engineering, Ed. Brebbia, C. A., Springer Verlag, 1982.

7. PATTERSON, C. and SHEIKH, M. A. - A Regular Boundary Element Method for Coupled Subdomains, Numerical Methods for Coupled Problems, Eds. Hinton, E., Bettess, P and Lewis, R. W., Pineridge Press, 1981.

8. EL-SEBAI, N. A. S. - An Investigation of the Regular Boundary Element Method in Three Dimensions, Ph.D. Thesis, University of Sheffield, 1982.

9. SEGERLIND, L. J. - Applied Finite Element Analysis, John Wiley & Sons, New York, 1976.

10. MARTIN, H. C. - Finite Element Analysis of Fluid Flow, Matrix Methods in Structural Mechanics, AFFDL TR 68-150, Patterson Air Force Base, 1969.

11. MILNE-THOMPSON, L. M. - Theoretical Hydrodynamics, Macmillan, London, 1968.

STRUCTURING COMPUTATIONAL FLUID DYNAMIC ANALYSIS SYSTEMS

Robert G. Hopcroft(I)

SUMMARY

This paper describes an integrated computational fluid dynamic system being developed to effectively use CFD in the design process. The structure of this system is the basis of a new generation of computer software developed to facilitate flow analysis. It will allow intelligent use of automation in choosing solution procedures based on the simultaneous generation of numerical error information. It is characterized by two distinct levels, a flexible top level or macrostructure to accommodate the infinite variations that arise in fluid flow analysis and a precise bottom level or microstructure to exactly replicate the highly structured concept of space. The initial implementation of this structure has allowed the selection of advantageous solution procedures. The goal is to achieve good results at a reasonable cost in a timely manner for design purposes.

1. INTRODUCTION

An information revolution touching all phases of human activity is presently taking place. Its beginning is rooted in the rapidly expanding capability of computer hardware. To bring this revolution to full fruition in CFD will require a similar development in computer software. As robotics automates the manufacturing process, the information provided by this new generation software will automate the use of CFD in the design process.

(I) Specialist Engineer, Boeing Computer Services

Many problems of practical interest in the design process involve the analysis of complex three-dimensional laminar and turbulent flows over or through three-dimensional geometries. To solve these flow analysis problems requires the use of a strategy which couples or interfaces various flow analysis algorithms for subregions of the flow. The necessity for coupling or interfacing may be due to geometry, flow or length scale problems. A complex geometry may require a body fitted mesh or involve the use of overlapping grids. A complex flow may involve regions of very different flows such as regions of separated flow. A complex flow may also involve regions of flow characterized by very different length scales and requiring different turbulence models.

Interfacing a variety of fluid flow codes poses difficulties in solving these problems. Interfacing codes and designing a data flow consistent with an overall solution strategy is difficult if the data structure and I/O formats vary from code to code. Such a system may also fail because the available codes lack the generality required for a particular problem; for example, if the code utilizes an inappropriate data structure, solver or boundary condition for the problem at hand.

To overcome these deficiences, a structured approach is being implemented to develop an effective computational fluid dynamic workstation. This approach features an integrated approach from the geometry definition through the grid generation and solution procedure to the interactive display of the analytic predictions. In addition this structure allows for the future automation of solution procedures based on numerical error norm assessment.

The details of the structured approach for finite difference methods are contained in the following sections titled macrostructure and microstructure. Due to space limitations they can be only briefly described. Examples of the use of this system will be given in a future paper. Although details of software design have been described in Reference 1 and details of many fluid codes, including several systems, have been described, such as Reference 2; to the best of this author's knowledge, no comparable CFD system design has been described in the literature.

2. MACROSTRUCTURE

The top level or macrostructure involves a functional overview. Its purpose is to clearly define the various elements and functions required and to organize flexible relationships among them. It answers the following questions.

What types of fluid flow are involved? What types of analysis are required? How do the analyses interrelate and how do they function together? Finally what hardware and software is required? How does the hardware and software interrelate and how do they function together?

2.1 Elements, Functions and Relationships

To answer the above questions involves categorizing the extensive variety of fluid flows involved in design. The flow may consist of a combination of subsonic, transonic, supersonic and hypersonic regions. It may be both viscid and inviscid, rotational and irrotational. The flow may be internal, external or both with arbitrarily complex geometries. Such cases include lobed mixers, inlet/duct, nacelle/strut/wing and duct/nozzle with jet mixing. In addition, some flows also involve multiple phases, heat transfer, chemical reactions, infrared or noise predictions and radar cross-section calculations.

To solve the preceding flows requires a complete range of analysis algorithms from potential and Euler with boundary layer to parabolized, partially parabolic and full Navier Stokes. To address the most general case requires that these algorithms be multi-dimensional, orientatable in any coordinate direction, and compatible with a variety of boundary conditions and solvers. In addition these algorithms must be multigridded to obtain high efficiency solutions, be amenable for extracting error norms, and have available turbulence modeling for the flows to be analyzed. Finally these algorithms must function together so that the most appropriate algorithm can be selected and applied in each region of the flow field.

A study of the functional structure required for the preceding analyses revealed that 75% of the software could be common to all the analyses. This dictated that a set of modular elements be developed which are independent of the particular equations being solved and the particular solution procedure used. The software functional structure is shown schematically in Figure 1. From the workstation the user defines the control functions to be exercised for a given flow problem creating a data control file. This data file is then routed to the computer selected. The control data determines which boundary conditions are to be selected for each variable, how metric and source terms are to be calculated, which algorithm and solver is to be used, and which data will be returned to the workstation for the graphical display. After viewing the data the user can restart the problem changing the parameters and functions desired. Thus the flow problem is at all times under the control of the user.

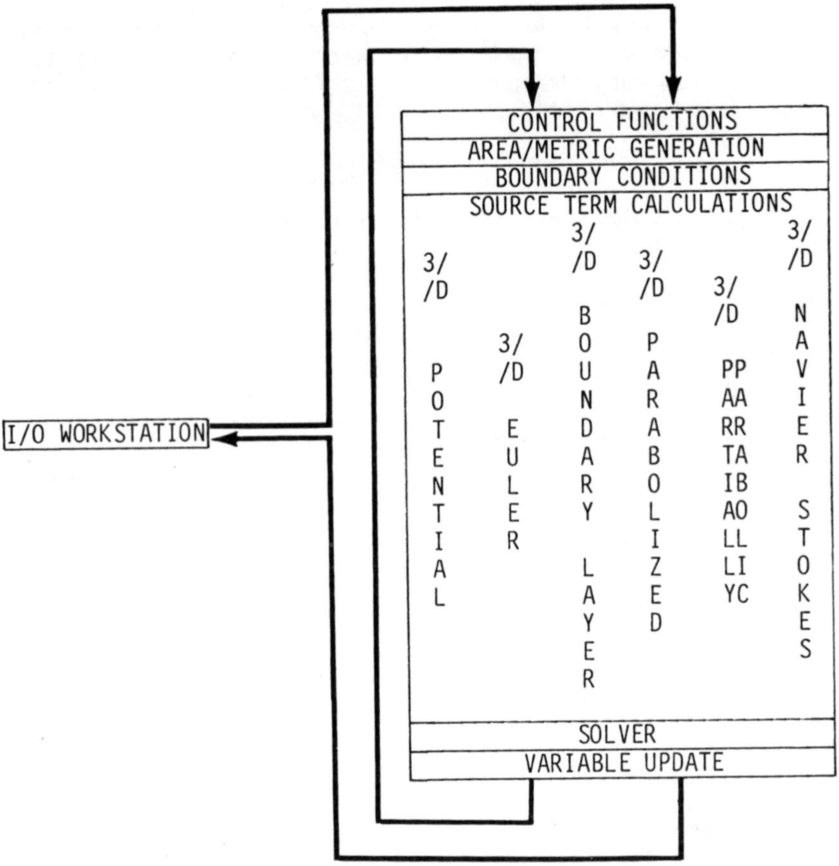

Figure 1. Software Functional Structure

2.2 Hardware/Software Relationships

Special considerations are required to handle the data flow for large problems involving a number of zones or regions requiring different types of flow analysis involving a varying number of multigrid levels. This involves hardware efficiency considerations since the data base is usually far larger than the main memory of the computer.

In the case of the CRAY I, data stored on disk is addressable in blocks of 512 words. This dictates that the data should be cycled in multiples of 512 words. This is accomplished by structuring the variable arrays so that any subset of the variables can be arranged in any order and blocked into multiples of 512 words. This structure is also useful in the case of a virtual memory system since it will minimize paging.

3. MICROSTRUCTURE

The bottom level or microstructure involves maintaining a precise relationship among problem variables by the use of symmetry operations and principles of group theory. This section first reviews the concept of both the permutation group and the binary group. This is followed by a review of the concept of a coordinate system and its relationship to the product of a permutation group and a binary group. Finally the relationship between the location of the problem variables with respect to the computational cell and the group concept of a coordinate system is discussed. For completeness this section is developed for zero, one, two and three dimensional spaces. At first a zero dimensional space consisting of a single point or cell may seem strange or absurd. However, it can, for example, be used in conjunction with a point solver as a consistency check on boundary conditions.

3.1 Groups

A group consists of a set and a binary operation between any of the members of the set such that the result is a member of the set. For details of group theory see Reference 3. Given an ordered set $\{(a,b,c)\}$, a permutation is a reordering of the set. In cycle notation the permutation 123 means interchanging the first and second elements followed by interchanging the new first element with the third element. Thus the permutation 123 on the set $\{(a,b,c)\}$ gives the set $\{(c,a,b)\}$. The multiplication tables for the permutation or symmetry group S_n are given in Table 1, for n=0,1,2,3. The subgroups consisting of the even permutations and the identity element 1 which is always a subgroup are shown by partitions.

S_0

1

S_1

1

S_2

1	12
12	1

S_3

1	123	132	12	13	23
123	132	1	13	23	12
132	1	123	23	12	13
12	23	13	1	132	123
13	12	23	123	1	132
23	13	12	132	123	1

Table 1. Permutation or Symmetry Groups

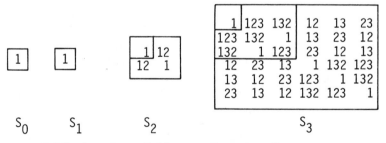

To obtain the binary group, start with the binary set $B = \{0,1\}$. Take the Cartesian product of B with itself n times. This is shown by Equation (1). Equations (2-5) give the explicit elements for B^n, n = 0,1,2,3.

$$B^n = \{(b_1, b_2, \ldots, b_n) : b_i \in B\} \quad (1)$$

$$B^0 = \{0\} \quad (2)$$

$$B^1 = \{0,1\} \quad (3)$$

$$B^2 = B \times B = \{(0,0),(1,0),(0,1),(1,1)\} \quad (4)$$

$$B^3 = B \times (B \times B) = B \times B^2 = \{(0,0,0),(1,0,0),(0,1,0),(1,1,0),(0,0,1),(1,0,1),(0,1,1),(1,1,1)\} \quad (5)$$

A group formed by the set B under the operation of addition modulo 2 is shown in Table 2. Remember that 0 is the identity element for addition.

0 + 0 = 0	0 + 1 = 1
1 + 0 = 1	1 + 1 = 0

=

0	1
1	0

Table 2. Formation of a Binary Group

Proceding in a similar fashion, all groups B_n can be formed from the corresponding sets B^n by addition modulo 2 performed element by element in the ordered doublets, triplets, or n-lets. This operation can be carried out in FORTRAN 77 by the logical nonequivalence operator (.NEQV.). The multiplication tables for the resulting group B_n are given in Table 3, for n = 0,1,2,3.

0

0	1
1	0

00	10	01	11
10	00	11	01
01	11	00	10
11	01	10	00

000	100	010	110	001	101	011	111
100	000	110	010	101	001	111	011
010	110	000	100	011	111	001	101
110	010	100	000	111	011	101	001
001	101	011	111	000	100	010	110
101	001	111	011	100	000	110	010
011	111	001	101	010	110	000	100
111	011	101	001	110	010	100	000

Table 3. Binary Groups

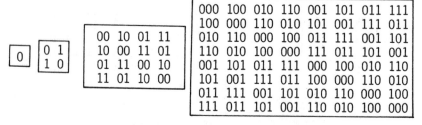

3.2 Coordinate System

Within a Euclidean vector space an orthonormal reference frame or coordinate system can be constructed. Once the mutually perpendicular lines of the reference orthonormal coordinate system have been selected there still remains two choices to be made. Both the order for labeling the lines and the positive/negative sense of the lines must be selected. The number of possible choices is made up of the product of two parts, namely the number of permutations of the labels times the sum of the appropriate binomial coefficients. The number of choices for the orthonormal coordinates is expressed by Equations (6) and (7) for a space of n dimensions and listed in Table 4 for spaces of zero, one, two and three dimensions. This clearly illustrates the rapid rate of increase in complexity with space dimension.

$$N_n = P_n^n \sum_{r=0}^{n} (P_r^n / P_r^r) \qquad (6)$$

$$P_r^n = n! / (n - r) ! \qquad (7)$$

Number of space dimensions	0	1	2	3
Number of label permutations	1	1	2	6
Sum of binomial coefficients	(1)	(1+1)	(1+2+1)	(1+3+3+1)
Number of coordinate choices	1	2	8	48

Table 4. Number of Orthonormal coordinates

Each of the orthonormal coordinate systems is relative to all the rest. That is, anyone can be arbitrarily chosen as primary and all the others are related by the transformations shown in Table 5 for one, two and three dimensions or the permutation or interchange of the rows or columns.

Number of Space Dimensions	1	2	3
Coordinate Transformation Matrix	$\begin{vmatrix} \pm 1 \end{vmatrix}$	$\begin{vmatrix} \pm 1 & 0 \\ 0 & \pm 1 \end{vmatrix}$	$\begin{vmatrix} \pm 1 & 0 & 0 \\ 0 & \pm 1 & 0 \\ 0 & 0 & \pm 1 \end{vmatrix}$

Table 5. Orthonormal Coordinate Transformation Matrices

Three-dimensional space contains one-dimensional space and two-dimensional space each in three ways. This can easily be seen in Table 5. The three-dimensional transformation matrix contains the one-dimensional transformation matrix as each of the three elements of the diagonal. The three-dimensional transformation matrix also contains the two-dimensional transformation matrix in three ways, namely, the upper left two by two elements, the lower right two by two elements and the elements of the four corners. Carrying the discussion one step further suggests identifying the first binomial coefficient with the one zero-dimensional space and labeling it zero, identifying the second binomial coefficient with the three one-dimensional spaces and labeling them one, two and four, identifying the third binomial coefficient with the three two-dimensional spaces and labeling them three, five and six, and finally identifying the fourth binomial coefficient with the one three-dimensional space and labeling it seven. Table 6 shows the logic point, line, square and cube with the labels expressed in terms of their binary equivalent.

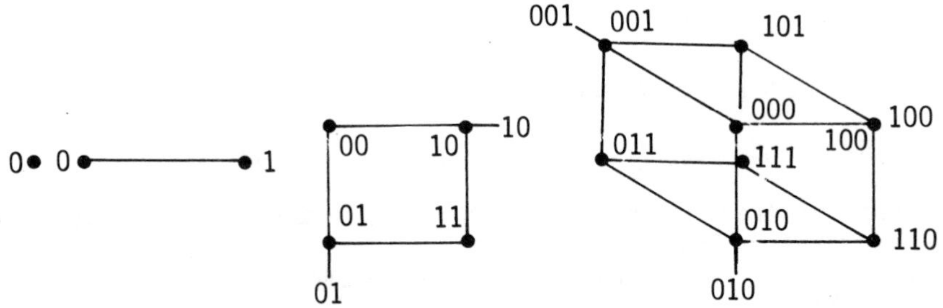

Table 6. Logic Point, Line, Square and Cube

Table 7 shows the logic cube composition containing one zero-dimensional space, three one-dimensional spaces and three two-dimensional spaces.

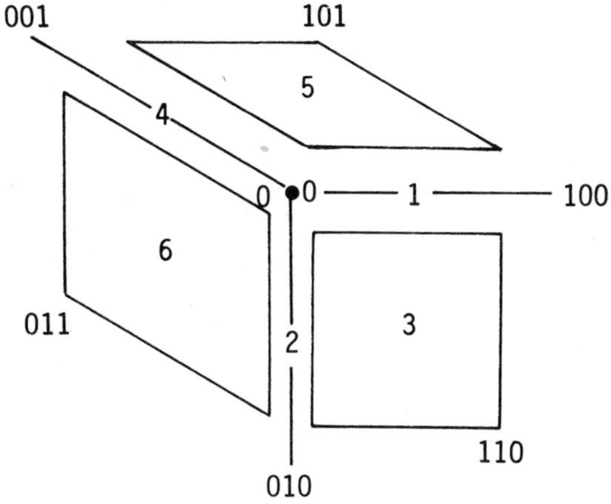

Table 7. Logic Cube Composition

At this point a comparison of the logic point, line, square and cube Table 6 with the binary group Table 3, should suggest a relationship. If the binary group nonequivalence operator is thought of as a move in the appropriate binary direction, then the multiplication tables for the binary group gives the outcome or final position for all moves from all points. In addition permuting the bits within the labels is equivalent to permuting the orthonormal coordinate directions. Thus the product of the permutation group and the binary group generates a group which gives the outcome or final position for all moves from all points between all orthonormal coordinate systems.

The source term and solver software is generalized for the general three-dimensional case but structured such that the three one-dimensional and three two-dimensional cases use only the appropriate sections determined by the location of the one bits in their labels. For a point solver there exists 48 paths through the source term and solver software, whereas for a line solver there exists 24 paths since one degree of freedom or motion is removed by simultaneously solving for a complete line. Again the appropriate path is determined by the location of the one bits in the appropriate labels.

The volume of the computational cell can be computed by tetrahedral decomposition. To compute an averaged symmetric cell volume, a triple scalar product loop is cycled through ten times. An index counter starts at zero and increments by one each cycle. The location of the one bits in the index counter increment the appropriate variable indices while the bit count of the index counter determines whether the permutation is odd or even.

3.3 Variable Relationships

In various fluid flow analyses, the problem variables are located at a variety of positions with respect to each other and with respect to the logic space. To handle this problem, generate a logic subspace by contracting the logic space by one half in each dimension maintaining a bijective mapping (the correspondence is one-to-one and onto) between the labels of the logic space points and the labels of the logic subspace points. Each variable is now labeled according to its position with respect to the logic subspace using the corresponding logic subspace label. Thus the orientation relationships among the problem variables can be maintained by an isomorphism with the relationships among the points of the logic spaces or subspaces. In the case of three dimensions, various variables may be located at the corner (C0), at the edge centers (E1,E2,E4), at the face centers (F3,F5,F7) or at the volume center (V7) of the computational cell. The isomorphism between the problem variable labels and the logic space labels is shown in Table 8.

C0	E1	E2	F3	E4	F5	F6	V7
E1	C0	F3	E2	F5	E4	V7	F6
E2	F3	C0	E1	F6	V7	E4	F5
F3	E2	E1	C0	V7	F6	F5	E4
E4	F5	F6	V7	C0	E1	E2	F3
F5	E4	V7	F6	E1	C0	F3	E2
F6	V7	E4	F5	E2	F3	C0	E1
V7	F6	F5	E4	F3	E2	E1	C0

000	100	010	110	001	101	011	111
100	000	110	010	101	001	111	011
010	110	000	100	011	111	001	101
110	010	100	000	111	011	101	001
001	101	011	111	000	100	010	110
101	001	111	011	100	000	110	010
011	111	001	101	010	110	000	100
111	011	101	001	110	010	100	000

Table 8. Label Isomorphism

If all variable arrays are of the same dimensions as the logic space, then only the one variable location to which the logic subspace contracts will completely fill the variable array. This leads to the consideration of upstoring or downstoring a variable in the variable array. This problem can be handled by labeling the computer storage arrays to coincide with the label of the variable location to which the logic subspace contracts. Thus the relationship established among the labels and the points of the logic space is valid whether the labels refer to the orthonormal coordinates themselves, the problem variable locations or the relative computer storage locations. In another field, quantum chromodynamics, GEORGI has pointed out a similar relationship among the elementary particles and quarks and the corners of a cube Reference 4.

4. CONCLUSION

The initial implementation of the preceding structure into a computational fluid dynamic system, designed and optimized for the CRAY I computer, has provided the user with an arbitrary choice in the selection of problem solution for one, two or three dimensional problems. Data can be arbitrarily ordered in any of the 24 right-handed or 24 left-handed three-dimensional coordinate systems. The 24 orientations for a line solver can be arbitrarily selected in any sequence to provide an optimal path to convergence. In addition, solution variables, boundary conditions, solvers and acceleration procedures can be arbitrarily selected. This generalized program structure has facilitated the patching or interfacing of various regions in the flow field. In many cases, by being able to readily select the most advantageous solution procedure, good results are being achieved at a favorable cost reduction in a timely manner for design purposes. The continued implementation of the preceding structure and the generation of information will be effective in expanding the analysis capability and in increasing its versatility. In addition as automation criteria are developed, they can be quickly implemented into the analysis system.

5. REFERENCES

1. PRATHER, R. E. - Discrete Mathematical Structures for Computer Science, HOUGHTON MIFFLIN COMPANY, BOSTON, 1976.

2. SPALDING, D. B. - A General Computer Program for Fluid-Flow, Heat-Transfer and Chemical-Reaction Processes. International FEM Congress, Bohen-Baden, Germany, 17-18 November, 1980.

3. ROTMAN, J. J. - The Theory of Groups, An Introduction, Second Edition, ALLYN and BACON, 1973.

4. GEORGI, H. - a Unified Theory of Elementary Particles and Forces, Scientific American, pp. 48-63, April 1981.

A CONSERVATIVE FINITE DIFFERENCE SCHEME FOR PRESSURE CALCULATION FROM VELOCITY FIELD*

M. BOURIOT

Assistant-Professor
University of Poitiers, France
C.E.A.T.-E.N.S.M.A.

Abstract

When the pressure field is computed afterwards the velocity field by solving a Poisson equation, the conservation of momentum and kinetic energy has to be imposed through the finite difference scheme for pressure calculation. The two-dimensionnal pressure problem to solve here, is a Neuman problem and it is shown that it is essential that the boundary data satisfied an integral relationship. A finite difference scheme designed for this purpose is validated in the case of a Laplace back-step flow and of a viscous time dependent shear flow.

1. INTRODUCTION

In order to keep some physical meaning to the pressure field computations of an incompressible flow, the conservation of momentum and kinetic energy in a bounded volume of fluid, without participation of external pressure variations, should be hold by the numerical scheme used. These two former properties can be straight enforced in the finite difference scheme of the primitive velocity-pressure equations, because the local evolution of momentum and kinetic energy is directly controlled by these equations. In the vorticity-stream function formulation, the pressure field is computed afterwards the velocity field by solving a Poisson's equation, so the conservation of these two quantities has to be imposed through the numerical pressure scheme. The Poisson's equation $\Delta p = S$, to solve over the region of interest $\mathcal{D}$, proceeds from the application of the divergence operator, and continuity property to the 2D-momentum equations. With the normal gradient pressure $\partial p/\partial n$ specified on the boundary $\mathcal{D}^*$, this equation forms a Neuman problem. By taking the surface integral of the pressure equation over $\mathcal{D}$, and using Green's theorem, an important integral property can be derived :

$$(1) \quad \oint_{D^*} \frac{\partial p}{\partial n} \vec{n} \, ds = \iint_D S \, d\sigma$$

having to be satisfied both, by the boundary data and the source term S. By successive differenciation of products and using the continuity equation div $\vec{u} = 0$, the source term S can be expressed in three forms :

$$(2) \quad S = -\frac{\partial}{\partial x}\left(\frac{\partial u^2}{\partial x} + \frac{\partial uv}{\partial y}\right) - \frac{\partial}{\partial y}\left(\frac{\partial v^2}{\partial y} + \frac{\partial uv}{\partial x}\right)$$

$$(3) \quad S = -\frac{\partial}{\partial x}\left(u \frac{\partial u}{\partial x} + v \frac{\partial u}{\partial y}\right) - \frac{\partial}{\partial y}\left(u \frac{\partial v}{\partial x} + v \frac{\partial v}{\partial y}\right)$$

$$(4) \quad S = -2\left(\frac{\partial v}{\partial x}\frac{\partial u}{\partial y} - \frac{\partial u}{\partial x}\frac{\partial v}{\partial y}\right) = -2\left[\frac{\partial v}{\partial x}\frac{\partial u}{\partial y} + \left(\frac{\partial u}{\partial x}\right)^2\right]$$

$$= -2\left[\frac{\partial v}{\partial x}\frac{\partial u}{\partial y} + \left(\frac{\partial v}{\partial y}\right)^2\right]$$

Only the first two expressions (2) (3) have a divergence form, and the corresponding momentum equations are :

$$(5) \quad -\frac{\partial p}{\partial x} = \frac{\partial u}{\partial t} - R_e^{-1} \Delta u + \frac{\partial u^2}{\partial x} + \frac{\partial uv}{\partial y}$$

$$-\frac{\partial p}{\partial y} = \frac{\partial v}{\partial t} - R_e^{-1} \Delta v + \frac{\partial v^2}{\partial y} + \frac{\partial uv}{\partial x}$$

$$(6) \quad -\frac{\partial p}{\partial x} = \frac{\partial u}{\partial t} - R_e^{-1} \Delta u + u \frac{\partial u}{\partial x} + v \frac{\partial u}{\partial y}$$

$$-\frac{\partial p}{\partial y} = \frac{\partial v}{\partial t} - R_e^{-1} \Delta v + u \frac{\partial v}{\partial x} + v \frac{\partial v}{\partial y}$$

It is easy to see that the integral constraint (1) is respected by the source terms and boundary data (2)(5) and (3)(6), then reported to be conservative, but not with the source terms (4). We build up the corresponding finite difference scheme by successive applications of the simple centered difference operator at each derivative steps of the former expressions. The source terms are evaluated at grid points (i,j). The gradients are computed at mid-points (i±1/2,j), (i,j±1/2), (i±1/2,j±1/2) and lead back by simple average at grid points (i,j) of the boundary. The values of the stream function needed at the mid-points are defined by simple average over the closest adjacent grid points. With these rules the finite difference scheme fulfill the integral property (1). The numerical pressure solutions are post-checked directly for momentum conservation, and for the conservation of kinetic energy through the total pressure $p+\vec{u}^2/2$.

2. LAPLACE TEST PROBLEM

The test flow is a two-dimensional Laplace back-step flow

of a non viscous fluid, where the total pressure should be constant all over the computed field. In such a flow the stream function is solution of a Laplace equation $\Delta\psi=0$, with slip-conditions $\vec{u}\times\vec{n}=0$ on boundaries EA, AB, CD, a uniform flow $u=u_o$ at upstream boundary DE, and a simple parallel flow v=0 on down-stream boundary CD, all conditions translated in terms of stream function ψ. Two pressure solutions of $\Delta P=S$ are computed with conservative (2) and non-conservative (4) forms of S, respectively in stream function ψ formulation :

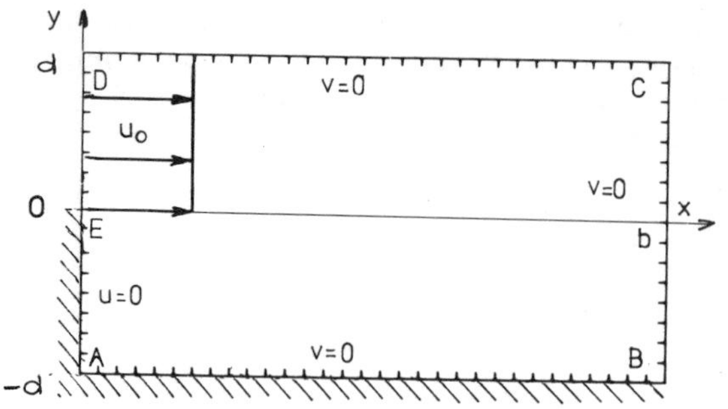

FIG.1 - Test geometry

$$(7)\quad S = -\left\{\frac{\partial}{\partial x}\left[\frac{\partial}{\partial x}\left(\frac{\partial\psi}{\partial y}\right)^2 + \frac{\partial}{\partial y}\left(-\frac{\partial\psi}{\partial x}\frac{\partial\psi}{\partial y}\right)\right] + \frac{\partial}{\partial y}\left[\frac{\partial}{\partial y}\left(-\frac{\partial\psi}{\partial x}\right)^2 + \frac{\partial}{\partial x}\left(-\frac{\partial\psi}{\partial x}\frac{\partial\psi}{\partial y}\right)\right]\right\}$$

$$(8)\quad S = 2\left[\frac{\partial^2\psi}{\partial x^2}\frac{\partial^2\psi}{\partial y^2} - \left(\frac{\partial^2\psi}{\partial x\partial y}\right)^2\right]$$

The associated momentum equations are (5), and in this sta-tionnary Laplace flow can be reduced to a single expression for normal gradient at upstream boundary :

$$(9)\quad \frac{\partial p}{\partial x}_{AD} = -\left[\frac{\partial}{\partial x}\left(\frac{\partial\psi}{\partial y}\right)^2 + \frac{\partial}{\partial y}\left(-\frac{\partial\psi}{\partial x}\frac{\partial\psi}{\partial y}\right)\right]$$

and anywhere else a zero normal gradient.

2.1. Discretization

The discretization of the domain ABCD, Fig.1, is done by a square mesh (x_i,y_j), $x_i = (i-1)\delta x$; $i = 1,n$; $y_j = (j-1)\delta x$; $j = 1,m$. In order to allow the use of centered schemes for $\partial p/\partial n$ and S on the boundary $\mathcal{D}^*$, a supplementary row : i=0 ; n+1 and j = 0 ; m+1 extend the integration domain. The numeri-cal solution of Laplace's equation, for the stream function $\psi_{i,j}$, use the classical 5 points, second order accurate, finite difference scheme, and a direct Fast Fourier Transform method {1}.

2.1.a. Source terms

The common part of conservative and non-conservative pressure finite difference scheme is the Laplacian operator :

$$(10)\quad p_{i+1,j} + p_{i-1,j} + p_{i,j+1} + p_{i,j-1} - 4p_{i,j} = S_{i,j}\times\delta x^2$$

$$i=1,n\ ;\ j=1,m$$

In the non-conservative form (8), the derivatives are simply replaced by second order centered schemes :

$$(11)\quad S_{ij} = \frac{2}{\delta x^4}\left[(\psi_{i+1,j} + \psi_{i-1,j} - 2\psi_{i,j})(\psi_{i,j+1} + \psi_{i,j-1} - 2\psi_{i,j}) - \frac{1}{16}(\psi_{i+1,j+1} - \psi_{i+1,j-1} - \psi_{i-1,j+1} + \psi_{i-1,j-1})^2\right]$$

$$i=1,n\ ;\ j=1,m$$

The discretization procedure described at the beginning is applied in order to keep the conservative property (1) in the finite difference scheme of expression (7). This finite difference equivalent should be evaluated at grid points (i,j) by successive terms decomposition. For example, the first term can be split into :

$$\frac{\partial}{\partial x}_{i,j}\left[\frac{\partial}{\partial x}\left(\frac{\partial\psi}{\partial y}\right)^2\right] = \frac{1}{\delta x}\left[\frac{\partial}{\partial x}_{i+\frac{1}{2},j}\left(\frac{\partial\psi}{\partial y}\right)^2 - \frac{\partial}{\partial x}_{i-\frac{1}{2},j}\left(\frac{\partial\psi}{\partial y}\right)^2\right]$$

$$\frac{\partial}{\partial x}_{i+\frac{1}{2},j}\left(\frac{\partial\psi}{\partial y}\right)^2 = \frac{1}{\delta x}\left[\left(\frac{\partial\psi}{\partial y}\right)^2_{i+1,j} - \left(\frac{\partial\psi}{\partial y}\right)^2_{i,j}\right]$$

$$\left(\frac{\partial\psi}{\partial y}\right)^2_{i+1,j} = \frac{1}{\delta x^2}\left(\psi_{i+1,j+\frac{1}{2}} - \psi_{i+1,j-\frac{1}{2}}\right)^2$$

where the values $\psi_{i+1,j+1/2}$ and $\psi_{i+1,j-1/2}$ at mid-points are obtained by simple average of the closest grid-point values, and in the same way for the second term :

$$\frac{\partial}{\partial x}_{i,j}\left[\frac{\partial}{\partial y}\left(-\frac{\partial\psi}{\partial x}\frac{\partial\psi}{\partial y}\right)\right] = \frac{1}{\delta x}\left[\frac{\partial}{\partial y}_{i+\frac{1}{2},j}\left(-\frac{\partial\psi}{\partial x}\frac{\partial\psi}{\partial y}\right) - \frac{\partial}{\partial y}_{i-\frac{1}{2},j}\left(-\frac{\partial\psi}{\partial x}\frac{\partial\psi}{\partial y}\right)\right]$$

$$\frac{\partial}{\partial y}_{i+\frac{1}{2},j}\left(-\frac{\partial\psi}{\partial x}\frac{\partial\psi}{\partial y}\right) = \frac{1}{\delta x}\left[\left(-\frac{\partial\psi}{\partial x}\frac{\partial\psi}{\partial y}\right)_{i+\frac{1}{2},j+\frac{1}{2}} - \left(-\frac{\partial\psi}{\partial x}\frac{\partial\psi}{\partial y}\right)_{i+\frac{1}{2},j-\frac{1}{2}}\right]$$

$$\left(-\frac{\partial\psi}{\partial x}\frac{\partial\psi}{\partial y}\right)_{i+\frac{1}{2},j+\frac{1}{2}} = \left(-\frac{\partial\psi}{\partial x}\right)_{i+\frac{1}{2},j+\frac{1}{2}} \times \left(\frac{\partial\psi}{\partial y}\right)_{i+\frac{1}{2},j+\frac{1}{2}}$$

$$\left(-\frac{\partial\psi}{\partial x}\right)_{i+\frac{1}{2},j+\frac{1}{2}} = -\frac{1}{\delta x}\left(\psi_{i+1,j+\frac{1}{2}} - \psi_{i,j+\frac{1}{2}}\right)$$

These procedures repeated until working out of derivative terms of the expression (7), and using symmetry property, lead to a 9 points, second order finite difference scheme for S at grid point (i,j) :

$$S_{i,j} = -\frac{1}{4\delta x^4}\Big[(\psi_{i+1,j+1}-\psi_{i+1,j-1})^2 - 2(\psi_{i,j+1}-\psi_{i,j-1})^2$$
$$+(\psi_{i-1,j+1}-\psi_{i-1,j-1})^2+(\psi_{i-1,j+1}-\psi_{i+1,j+1})^2-2(\psi_{i-1,j}-\psi_{i+1,j})^2$$
$$+(\psi_{i-1,j-1}-\psi_{i+1,j-1})^2+2\Big[-(\psi_{i+1,j+1}+\psi_{i+1,j}-\psi_{i,j+1}-\psi_{i,j})$$
$$(\psi_{i+1,j+1}+\psi_{i,j+1}-\psi_{i+1,j}-\psi_{i,j})+(\psi_{i,j+1}+\psi_{i,j}-\psi_{i-1,j+1}-\psi_{i-1,j})$$
$$(\psi_{i,j+1}+\psi_{i-1,j+1}-\psi_{i,j}-\psi_{i-1,j})+(\psi_{i+1,j}+\psi_{i+1,j-1}-\psi_{i,j}-\psi_{i,j-1})$$
$$(\psi_{i+1,j}+\psi_{i,j}-\psi_{i+1,j-1}-\psi_{i,j-1})-(\psi_{i,j}+\psi_{i,j-1}-\psi_{i-1,j}-\psi_{i-1,j-1})$$
$$(\psi_{i,j}+\psi_{i-1,j}-\psi_{i,j-1}-\psi_{i-1,j-1})\Big]\Big] \qquad (12)$$

2.1.b. Neuman boundaries conditions

On sides AB, BC, CD the zero normal gradient pressure condition are simply :

(13) $p_{i,0} = p_{i,2}$; $p_{i,m+1} = p_{i,m-1}$ $\quad i = 1,n$

(14) $p_{n+1,j} = p_{n-1,j}$ $\quad j = 1,m$

On AD, the expression (9) is computed with similar successive terms decomposition. The gradient values are evaluated at mid-points then lead back to the boundary grid point (i,j) by simple average :

$$\frac{\partial p}{\partial x}_{i,j} = \frac{1}{2}\left(\frac{\partial p}{\partial x}_{i+\frac{1}{2},j} + \frac{\partial p}{\partial x}_{i-\frac{1}{2},j}\right)$$

$$\frac{\partial p}{\partial x}_{i+\frac{1}{2},j} = -\frac{\partial}{\partial x}_{i+\frac{1}{2},j}\left(\frac{\partial\psi}{\partial y}\right)^2 - \frac{\partial}{\partial y}_{i+\frac{1}{2},j}\left(-\frac{\partial\psi}{\partial x}\frac{\partial\psi}{\partial y}\right) .$$

The two ψ partial derivatives at mid-points i+1/2,j has been previously computed as intermediary steps in the design of S scheme (12). With these expressions, a similar 9 points scheme is obtained for the gradient expression (9) :

(15) $$\frac{\partial p}{\partial x}_{i,j} = \frac{1}{8\delta x^3}\Big[-(\psi_{i+1,j+1}-\psi_{i+1,j-1})^2+(\psi_{i-1,j+1}-\psi_{i-1,j-1})^2$$
$$+(\psi_{i+1,j+1}+\psi_{i+1,j}-\psi_{i,j+1}-\psi_{i,j})(\psi_{i+1,j+1}+\psi_{i,j+1}-\psi_{i+1,j}-\psi_{i,j})$$
$$+(\psi_{i,j+1}+\psi_{i,j}-\psi_{i-1,j+1}-\psi_{i-1,j})(\psi_{i,j+1}+\psi_{i-1,j+1}-\psi_{i,j}-\psi_{i-1,j})$$

$$-(\psi_{i+1,j}+\psi_{i+1,j-1}-\psi_{i,j}-\psi_{i,j-1})(\psi_{i+1,j}+\psi_{i,j}-\psi_{i+1,j-1}-\psi_{i,j-1})$$
$$-(\psi_{i,j}+\psi_{i,j-1}-\psi_{i-1,j}-\psi_{i-1,j-1})(\psi_{i,j}+\psi_{i-1,j}-\psi_{i,j-1}-\psi_{i-1,j-1})\Big)$$

The particular expression, at corners A, B, is :

$$(16)\quad \frac{\partial p}{\partial x}\Big|_{1,j} = \frac{1}{2\delta x}(p_{2,j} - p_{0,j}) \qquad j = 1 \text{ and } m$$

The two sets of finite difference expressions, conservative (10, 12 to 16) and non-conservative (10,11,13 to 16) are solved by a direct Fast Fourier Transform method described in {2}.

2.2. Run tests

All the tests are performed on a grid of n=32 by m=17 nodes. A misconservation indicator of momentum is defined by the difference ε_m, between the momentum of inflow and outflow, normalized by u_o^2/d :

$$(17)\quad \varepsilon_m = \frac{1}{u_o^2 d} \times \left(\int_{x=b} (p+u^2)\, dy - \int_{x=0} (p+u^2)\, dy \right)$$

The value $\varepsilon_k(x,y)$ of the total pressure variation between a given point (x,y) and a reference point $\mathcal{D}$, normalized by the dynamic pressure at the inflow $u_o^2/2$, indicates the degree of conservation of kinetic pressure :

$$(15)\quad \varepsilon_k = \frac{2}{u_o^2}\left(p_t(x,y) - p_t(0,d)\right)$$

Numerical solutions are displayed on pressure maps, with 20 isobars regulary spaced between minimum and maximum values.

In test T_4' the standard non conservative scheme was used, the solution shows a relative error $\varepsilon = 0.0525$ on the integral relation (1). Consequently there are large values $\varepsilon_m = -1.68$ and $\varepsilon_k(b-\delta x,0) = -1.41$ which indicate the poor quality of this pressure solution, the corresponding map, figure 2, exhibits an important longitudinal parasitic gradient. In an attempt to improve this solution, the values of $\partial p/\partial n$ on $\mathcal{D}^*$ (test T'5), or S values over $\mathcal{D}$ (T'6) was revised prorata to its values to ensure the respect

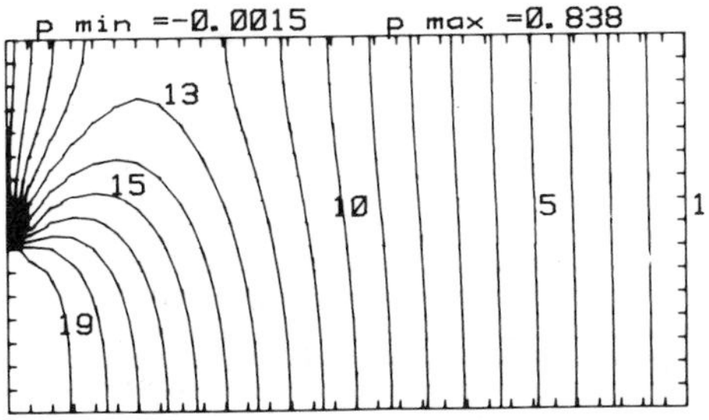

FIG.2 - Test T'_4

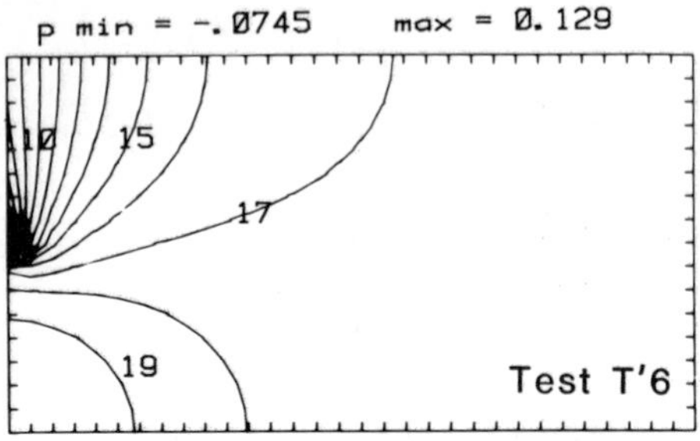

FIG.3 - (No discernible difference for T'5 map)

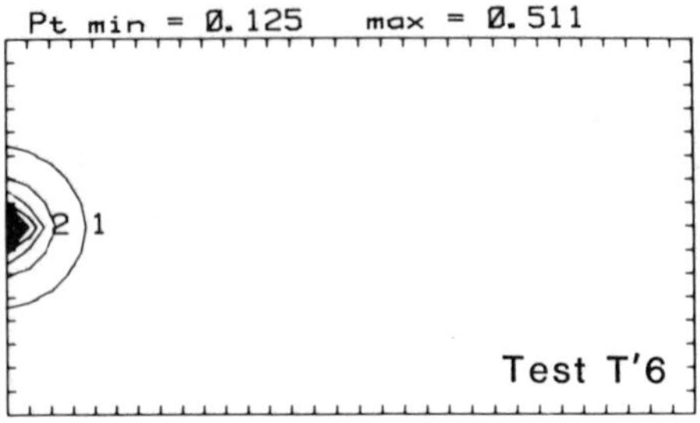

FIG.4 - Total pressure plot

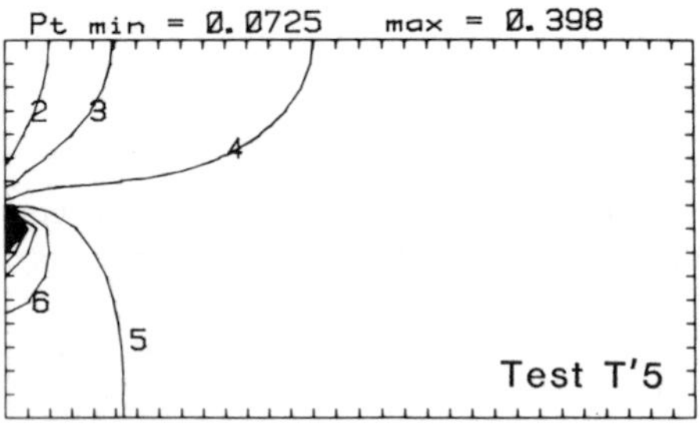

FIG.5 - Total pressure plot

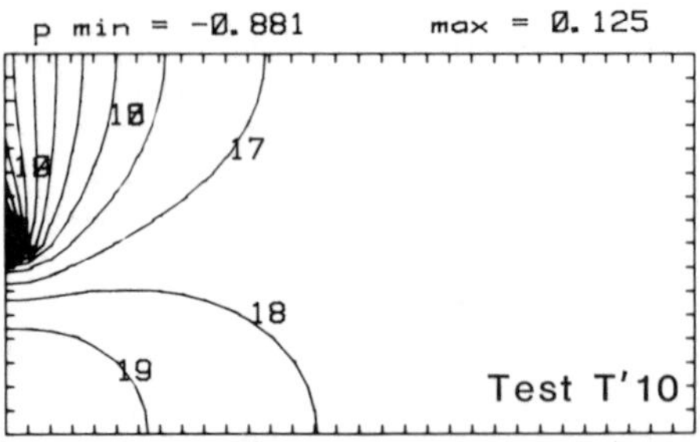

FIG.6 - Conservative scheme

of the finite difference equivalent of the integral relation (1). An important reduction of the parasitic gradient occuring in test T'4, is obtained on these two pressure solutions, figure 3. The use of this stratagem seems to be justified for the correction of discretization errors, in particular in the nearness of discontinuities, but its weight on the resulting pressure field is suspicious upon the total pressure plots of figure 4 and figure 5.

In test T'10 the use of the conservative scheme, divergence form for S and corresponding expression of pressure gradients (10,12 to 16), allows the respect of the integral relation (1) with a relative error $\varepsilon = -0.000025$. The pressure solution shows a correct conservation of momentum $\varepsilon_m = -0.0156$ and negligible variations $\varepsilon_k(b-\delta x,0) = 9\ 10^{-5}$ $\varepsilon_k(\delta x,0) = 6\ 10^{-5}$, of the total pressure between the upstream and downstream limits. The plot of the pressure field figure 6 is relatively close to the one of the figure 3 (rectification of S distribution) except for minimum values in the vicinity of sharp edge E. The main result of these tests is to show off the importance of the keeping of an integral property (1) by the finite difference scheme, in order to hold the conservation properties of momentum and kinetic energy in the pressure solution.

3. TWO-DIMENSIONAL SHEAR FLOW

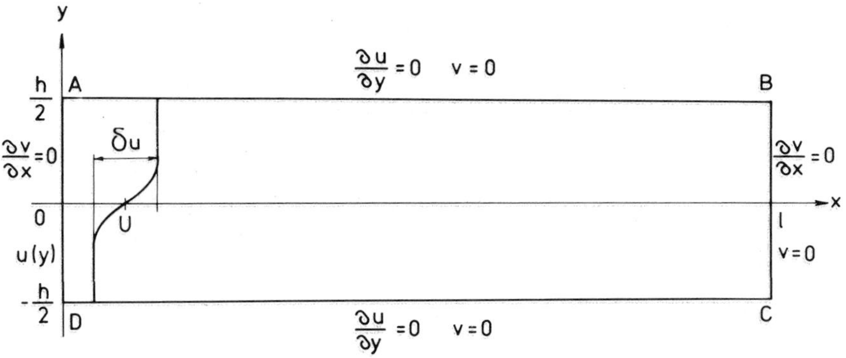

FIG.7 - Simulation domain geometry

A finite difference velocity solution of the two-dimensionnal time dependent Navier-Stokes equations, in a mixing layer extending from an initial tangent hyperbolic velocity profile was previously computed {2} and stored on a magnetic tape. The momentum thickness θ_0 and the Reynolds number $R_{e\theta_0} = \theta_0 \delta U/\nu = 100$ characterize this realization. From this velocity solution extended in the same way than Laplace test velocity field, the time dependent pressure field can be computed at each time step by the solution of a Neuman problem. The conservative scheme (10,12 to 16 - Test T'10) should be adjusted in the present viscous, and time dependent case, in which the entire expressions for gradient pressure could be used :

$$(19)\quad \frac{\partial p}{\partial x} = -\frac{\partial}{\partial t}\left(\frac{\partial \psi}{\partial y}\right) + R_e^{-1}\Delta\left(\frac{\partial \psi}{\partial y}\right) - \left[\frac{\partial}{\partial x}\left(\frac{\partial \psi}{\partial y}\right)^2 + \frac{\partial}{\partial y}\left(-\frac{\partial \psi}{\partial x}\frac{\partial \psi}{\partial y}\right)\right]$$

$$(20)\quad \frac{\partial p}{\partial y} = -\frac{\partial}{\partial t}\left(-\frac{\partial \psi}{\partial x}\right) \quad R_e^{-1}\Delta\left(-\frac{\partial \psi}{\partial x}\right) - \left[\frac{\partial}{\partial y}\left(-\frac{\partial \psi}{\partial x}\right)^2 + \frac{\partial}{\partial x}\left(-\frac{\partial \psi}{\partial x}\frac{\partial \psi}{\partial y}\right)\right]$$

3.1. Discretization

The finite difference scheme previously designed for equation (9) is used here for the square brackets of (19) and a symmetrical form is made up for the bracket of equation (20). Using space derivatives of ψ at mid-points, the time derivatives are approximated at the middle of the time step δt

$\frac{\partial}{\partial t}^{k+1/2}\left(\frac{\partial \psi}{\partial y}\right)_{i,j+1/2}$, then leads back to grid point i,j by simple averaging with the adjacent mid point expression

$$\frac{\partial}{\partial t}^{k+1/2} \left(\frac{\partial\psi}{\partial y}\right)_{i,j-1/2} \quad .$$

$$(21) \quad -\frac{\partial}{\partial t}^{k+1/2}\left(\frac{\partial\psi}{\partial y}\right)_{i,j} = -\frac{1}{2\delta x\delta t}(\psi^{k+1}_{i,j+1}-\psi^{k+1}_{i,j-1}-\psi^{k}_{i,j+1}+\psi^{k}_{i,j-1})$$

$$(22) \quad -\frac{\partial}{\partial t}^{k+1/2}\left(-\frac{\partial\psi}{\partial x}\right)_{i,j} = -\frac{1}{2\delta x\delta t}(-\psi^{k+1}_{i+1,j}+\psi^{k+1}_{i-1,j}+\psi^{k}_{i+1,j}-\psi^{k}_{i-1,j})$$

In the following, the time superscript k are understood. The transposition of the discretization rules, previously used for the estimation of velocity time derivatives scheme, to velocity Laplacian scheme, leads to an eight points scheme, collecting 2 points on each side of the application grid point (i,j) :

$$(23) \quad R_e^{-1}\Delta_{i,j}\left(\frac{\partial\psi}{\partial y}\right) = \frac{R_e^{-1}}{2\delta x^3}(\psi_{i+1,j+1}-\psi_{i+1,j-1}+\psi_{i-1,j+1}-\psi_{i-1,j-1}$$
$$+\psi_{i,j+2}-\psi_{i,j-2}-4\psi_{i,j+1}+4\psi_{i,j-1})$$

$$(24) \quad R_e^{-1}\Delta_{i,j}\left(-\frac{\partial\psi}{\partial x}\right) = \frac{R_e^{-1}}{2\delta x^3}(-\psi_{i+2,j}+\psi_{i-2,j}+\psi_{i-1,j+1}-\psi_{i+1,j+1}$$
$$+\psi_{i-1,j-1}-\psi_{i+1,j-1}-4\psi_{i-1,j}+4\psi_{i+1,j})$$

With these revised gradient pressure schemes, the respect of integral constraint (1) is preserved, and we can test the pressure solution for our 2-D shear flow. At the integration domain corners A,B,C,D respectively : (1,1), (n,1), (1,m), (n,m), schemes (23) and (24) catch points outside the extended grid (0,n+1), (0,m+1) calling for a special treatment. On the x=0, and x=ℓ vertical boundaries, the contribution of the Laplacian term $\Delta(\partial\psi/\partial y)$ to the gradient (19) can be devided in two equal parts : $\Delta_{i,j+1/2}$ and $\Delta_{i,j-1/2}$. So it is sufficient to choose the part corresponding to the inside domain. For example, the special expression of these Laplacian terms at points (1,1) and (n,1) is :

$$(25) \quad R_e^{-1}\Delta\left(\frac{\partial\psi}{\partial y}\right)_{i,1} = \frac{R_e^{-1}}{2\delta x^3}(\psi_{i+1,2}-\psi_{i+1,1}+\psi_{i-1,2}-\psi_{i-1,1}+\psi_{i,0}+\psi_{i,3}$$
$$-5\psi_{i,2}+5\psi_{i,1}) \qquad i = 1 \text{ and } n \quad .$$

The same decomposition can be used for time derivative terms, since these terms are also calculated at mid-points first, then leads back to grid points by simple average :

$$(26) \quad \frac{\partial}{\partial t}\left(\frac{\partial\psi}{\partial y}\right)_{i,1} = \frac{1}{\delta x\delta t}(\psi^{k+1}_{i,2}-\psi^{k+1}_{i,1}+\psi^{k}_{i,1}-\psi^{k}_{i,1}) \qquad i = 1 \text{ and } n.$$

The set of equations (10,12,19,20) solved by a direct FFT method {2}, gives a conservative way for pressure calculation from a velocity field of a viscous unsteady flow.

3.2. Run test

On a 118×32 square mesh, the unsteady pressure solution has been computed until a dimensionless time $t\delta u/\theta_0 = 1000$. For the whole pressure computation the relative error on the integral relation (1) is always contained inside a range from 10^{-7} to 10^{-4}, of the order of magnitude of computer error. The small relative difference ε_m between the momentum integral on the inflow and the outflow boundaries : $-0.109\ 10^{-3} \leqslant \varepsilon_m \leqslant -0.430\ 10^{-7}$ means a correct conservation of momentum by this scheme. The total pressure conservation is not checked in this complex flow beacuse stream-lines are strongly disturbed by the moving vortex row. Finally, the shape of the pressure field computed in a shear layer seems to be right on the pressure map, of the figure 8, where 25 isobars are drawn between the maximum and minimum values of the pressure field at the intermediate dimensionless time of 500.

p min = -0.314   p max = 0.098

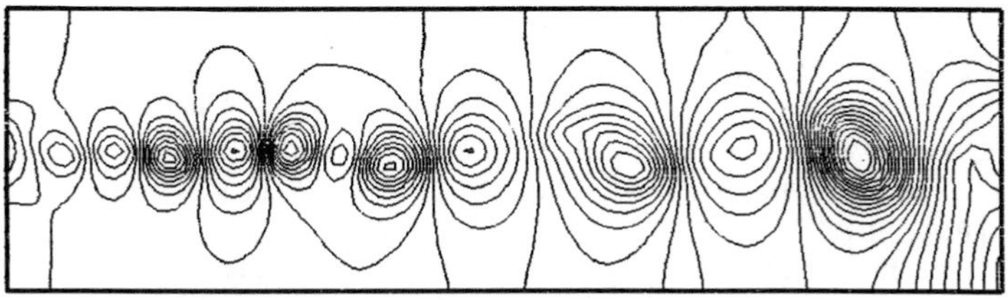

FIG.8 - Pressure map at t = 500

REFERENCES

1. HOCKNEY, R.W.- A fast direct solution of Poisson's equation using Fourier analysis, J. of Ass.Comp. Mech., Vol.12, n°1, pp.95-113 (1965)

2. BOURIOT, M. - Contribution à l'étude de la simulation numérique de la turbulence, Thèse d'Etat, Université de Poitiers, à paraître (1983)

* - *This research was supported by French Ministry of Defence, DRET contract 77/544 under Tsen L.F. Scientific responsibility, and DRET convention 82/169.*

COORDINATE GENERATION METHOD FOR FLOWS THROUGH PERIODIC CHANNELS

J.M. FLORYAN*

Assistant Professor

SUMMARY

A coordinate generation method for channel flows has been developed. An arbitrary channel, with boundaries consisting of straight elements in the physical plane, is mapped into a straight channel in the computational plane. Transformation is given separately for the periodic and non-periodic configurations. The parameters of the transformations have to be determined through successive approximations. A simple iteration scheme, that converges quite rapidly even with a poor initial guess, is presented. Solutions for different configurations are displayed to illustrate the capabilities of the method. The extension of the method, that includes channels consisting of curved elements, is also available.

1. INTRODUCTION

Numerical grid generation is a fairly common tool in computational fluid dynamics, however, its applications extend to solutions of partial differential equations arising from all physical problems involving field properties. The common goal of different grid generation procedures is to construct boundary fitted coordinates to permit writing of general codes that may treat field equations in domains of arbitrary shapes. Grid generation for boundaries of arbitrary shapes has to be done numerically, since analytical solutions exist for only a limited class of geometrical configurations. The generated coordinates are, in general, curvilinear. The form of the field equations is considerable simplified if the coordinates satisfy condition of orthogonality. When the coordinates are also conformal, additional simplifications result from the application of the Cauchy-Riemann relations.

---

*Faculty of Engineering Science, The University of Western Ontario, London, Ontario, Canada, N6A 5B9

This paper describes a method for the generation of conformal coordinates for channel flows. The basic idea is to map the channel of an arbitrary shape in the physical plane into a straight channel in the computational plane. Thus, the solution of the flow problems may be carried in the computational plane with the help of a simple rectangular grid. Transformations being presented are of the Schwarz-Christoffel type and represent extension of ideas introduced in Ref. [1]. Transformation dealing with a non-periodic channel is described in Section 2, while Section 3 deals with periodic configurations. Both transformations involve parameters defined in the computational plane and thus are not known apriori for the configuration of interest in the physical plane. Section 4 describes the numerical procedure that permits determination of the required parameters for a specified configuration. The method is limited to two dimensions due to the nature of the complex variables. An extension to the case of an axisymmetric channel can be made by rotating two-dimensional systems described in this paper.

2. TRANSFORMATION FOR A NON-PERIODIC CHANNEL

An arbitrary non-periodic channel, consisting of straight elements in the physical plane, is mapped into a straight channel in the computational plane. The transformation is illustrated in Fig. 1. The individual mappings have the form

$$w = \frac{h}{\pi}\,\ell n t; \qquad \frac{dz}{dt} = M\,t^{-1-\delta} \prod_{j=1}^{j=n} (t - c_j)^{\alpha_j} \tag{1}$$

where M is a complex constant, n determines number of corners and $\alpha\pi$ denotes corner turning angles. Angles $\alpha\pi$ are taken to be positive for the clockwise rotation when the channel is circled in the counterclockwise sense. Elimination of the t-plane results in an equation directly relating physical and computational planes

$$\frac{dz}{dw} = R\,\exp\left[\frac{w\pi}{2h}(\phi-\delta)\right] \prod_{j=1}^{j=n} \left[\sinh \frac{\pi}{2h}(w-a_j)\right]^{\alpha_j} \tag{2}$$

The reader should note that the a_j's in the above equation are complex. The transformation is complete provided locations of the points a_j in the computational plane, corresponding to the corners in the physical plane, are known and the complex constant R is determined. This is achieved by applying the numerical procedure described in Section 4.

3. TRANSFORMATION FOR A PERIODIC CHANNEL

A description of transformation mapping a channel, consisting of an infinite number of segments of the same geometry

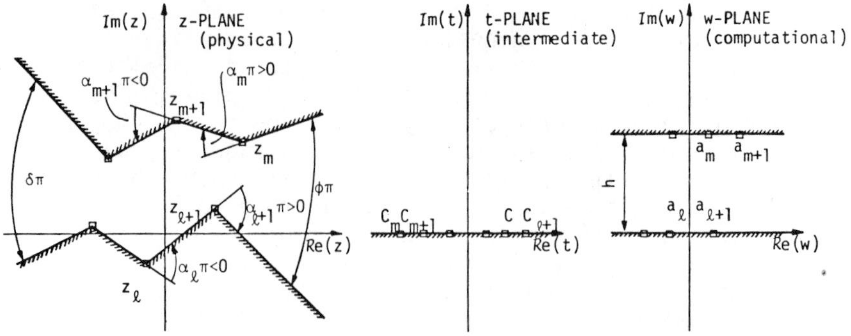

Fig. 1 Mapping of an Arbitrary Channel into a Straight Channel

in the physical plane, into a straight channel in the computational plane, is carried on with the help of configuration shown in Fig. 2. The channel shown has a straight top; its

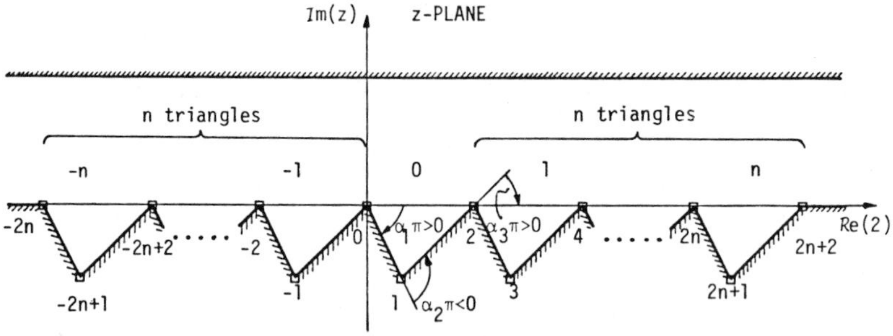

Fig. 2 Channel With (2n + 1) Triangular Indentations on the Bottom

bottom consists of (2n + 1) triangular indentations of the same form and size. The transformation for this particular channel is obtained from Eq. (2)

$$\frac{dz}{dw} = R \prod_{\ell=-n}^{\ell=n} \left[\sinh \frac{w - w_{2\ell}}{2h} \pi\right]^{\alpha_1} \left[\sinh \frac{w - w_{2\ell + 1}}{2h} \pi\right]^{\alpha_2} \left[\sinh \frac{w - w_{2\ell + 2}}{2h} \pi\right]^{\alpha_3} \tag{3}$$

where subscript ℓ, which denotes a particular triangle, assumes all integer values between $\ell = -n$ and $\ell = n$. The reader should note that the turning angles, corresponding to the corners located on the borders between different segments, are defined as those between the appropriate wall element and the real axis. The total turning angle associated with each segment (triangle) is $\alpha_1 + \alpha_2 + \alpha_3 = 0$. The configuration in the physical plane becomes periodic when the number of triangle pairs increases indefinitely $n \to \infty$. It is obvious that if the right hand side of Eq. (3) in the limit $n \to \infty$ is periodic (ie. it describes a certain periodic configuration in the w-plane), the corresponding configuration in the physical plane is also periodic. It remains to be shown that the reverse statement is true, and this is a relatively simple matter. The question of periodicity will not be pursued any further due to the lack of space. The important conclusion is that a periodic configuration in the z-plane corresponds to a periodic configuration in the w-plane and vice-versa.

The generalization of the transformation (3) to the case of a channel consisting of a repeatable segment of an arbitrary shape, when a particular shape is constructed out of straight elements, is straightfoward.

$$\frac{dz}{dw} = R \prod_{\ell=-\infty}^{\ell=+\infty} \prod_{j=1}^{j=n} \left[\sinh \frac{\pi}{2h}(w - \ell D - a_j)\right]^{\alpha_j} \tag{4}$$

In the above, D denotes length of the segment in the w-plane, n stands for the number of corners in the segment, a_j's denote relative locations of the corners in the segment, and α_j's are the turning angles. The symbols are illustrated in Fig. 3, where locations of the corners inside the segment are defined relative to the left bottom and top corners. The reader should

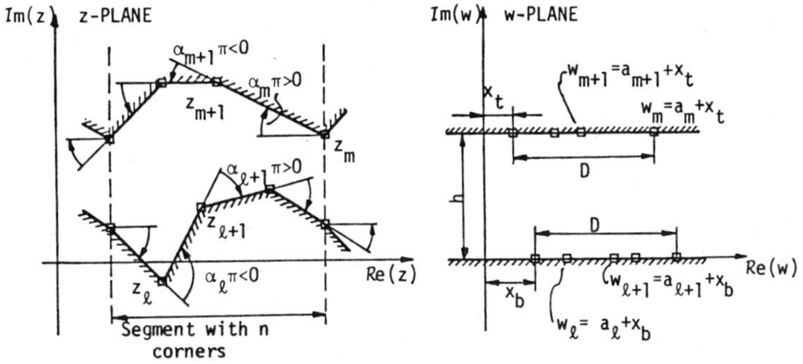

Fig. 3 Mapping of a Channel Consisting of an Infinite Number of Segments of the Same Form and Size

note that the turning angles at the edges of the segment are those between the appropriate wall element and the real axis. The sum of the turning angles for the segment is zero. Transformation (4) is complete provided that the complex constant R is known and that the locations of corners a_j and the length D of the segment, corresponding to a particular channel, have been determined. This is done with the help of the numerical procedure described in Section 4.

4. DETERMINATION OF THE PARAMETERS OF THE TRANSFORMATION

The difficulties in applying transformations (2) and (4) are due to the fact that the relation between a particular channel shape, and the values of the parameters of the appropriate transformation, is not explicit. Therefore, the required parameters have to be determined through a method of successive approximations, as described in Ref. [1].

4.1 Iteration Procedure

Transformations being considered in this paper are of the Schwarz-Christoffel type. In general, they have three constants that can be chosen arbitrarily. The details of the numerical procedure and selection of constants are given for the case of a non-periodic symmetric channel shown in Fig. 4.

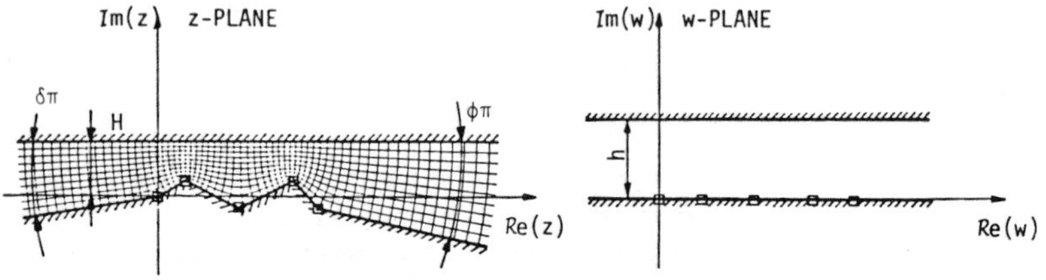

Fig. 4 Mapping of a Symmetric Non-Periodic Channel

Only half of the channel needs to be considered due to symmetry. Location of a point in the z-plane is given as

$$z = \int_0^w f(w)\, dw + N \tag{5}$$

where f(w) stands for the right hand side of Eq. (2). The complex constant N is chosen to be zero and thus the origins of the z and w planes coincide. The reader should note that one of the corners (first corner in the present case) may be placed at the origin of both planes. The third constant may be chosen by selecting the location in the computational plane of any point belonging to the top of the channel. This constant has been, in the present case, selected implicitly during the iteration procedure. All the remaining parameters, ie. the complex constant R and the corner locations a_j, have to be determined from Eq. (2). Constant R can be written as $R = |R| \exp(-i\pi\phi)$ where $\phi\pi$ is the angle shown in Fig. 4 and $|R|$ is not known. An initial guess is made for $|R|$ and for locations of corners a_j. The a_j's are picked along the real axis such that $Re(a_{j-1}) < Re(a_j) < Re(a_{j+1})$ where Re denotes a real part. The new value of $|R|$ is adopted by imposing condition

$$Im\left[\int_{(o,o)}^{(o,h)} f(w)\, dw\right] = H \tag{6}$$

where *Im* denotes the imaginary part. Integration is carried subsequently along the bottom of the channel to determine locations of the corners in the physical plane. The computed locations do not, in general, coincide with the specified locations. The new guess is made for the a_j's. It is assumed that the a_j's should be rescaled according to the scaling indicated by the errors in the distances between the corners in the physical plane.

$$\frac{|a_{cj} - a_{c(j-1)}|}{|a_{gj} - a_{g(j-1)}|} = \frac{|z_{cj} - z_{c(j-1)}|}{|z_{gj} - z_{g(j-1)}|} \tag{7}$$

Here subscript c denotes correct values, g stands for guessed values, and j denotes corner number. The above procedure coverges quite rapidly, even with a poor initial guess. Results for one of the cases considered are displayed in Fig. 4.

The iteration procedure described above assumes that the integration of Eq. (2) does not pose any problems. In fact, the required integration can rarely be done analytically and, therefore, a numerical integration has to be introduced to permit applications of transformations (2) and (4) to arbitrary shapes.

4.2 Numerical Integration

The numerical integration of Eq. (2) is complicated due to singularities present when $\alpha_j < 0$. An attempt to integrate this equation across the singularity, by using say a trapezoidal rule, will result in errors due to the non-analytic nature of the integrand there. The difficulties are avoided by properly adopted integration method described in Ref. [1].

A modified trapezoidal rule which integrates exactly any non-analytic term occuring at a corner is described by examining two adjacent corners, say k and k + 1. Eq. (2) may be viewed as being made up of three factors on the right hand side

$$\frac{dz}{dw} = [F_1(w)G_1(w)][F_2(w)G_2(w)]\; f(w) \tag{8}$$

where

$$G_1(w)=[\tfrac{\pi}{2h}(w-a_k)]^{\alpha_k}; \quad F_1(w)=[\sinh\tfrac{\pi}{2h}(w-a_k)]^{\alpha_k}/G_1(w)$$

$$G_2(w)=[\frac{\pi}{2h}(w-a_{k+1})]^{\alpha_{k+1}}; \quad G_2(w)=[\sinh\frac{\pi}{2h}(w-a_{k+1})]^{\alpha_{k+1}}/G_2(w)$$

and f(w) is a well behaved function near corners k and k + 1 resulting from all the remaining corners in the problem. The reader should note that functions F_1 and F_2 are non-singular, and that functions G_1 and G_2 may be singular depending on the sign of α. Functions F_1 and F_2 are calculated for small values of arguments with the help of the appropriate Taylor expansions. The integration near k can be carried as

$$z_{m+1} - z_m = \bar{F}_1\bar{G}_2\bar{F}_2\bar{f}\ [(\frac{\pi}{2h})^{\alpha_k} \frac{(w - a_k)^{\alpha_k+1}}{\alpha_k + 1}]_{w_m}^{w_{m+1}} \tag{9}$$

and near k + 1

$$z_{m+1} - z_m = \bar{G}_1\bar{F}_1\bar{F}_2\bar{f}\ [(\frac{\pi}{2h})^{\alpha_{k+1}} \frac{(w - a_{k+1})^{\alpha_{k+1}+1}}{\alpha_{k+1} + 1}]_{w_m}^{w_{m+1}} \tag{10}$$

where $\bar{F}_1$, $\bar{F}_2$, $\bar{G}_1$, $\bar{G}_2$, $\bar{f}$ are values of F_1, F_2, G_1, G_2 and f evaluated at $(w_m + w_{m+1})/2$, which is proper for well behaved functions. Subscript m denotes the integration step. Far away from corners k and k + 1 both expressions become

$$z_{m+1} - z_m = \bar{G}_1\ \bar{F}_1\ \bar{G}_2\ \bar{F}_2\ \bar{f}\ \ \Delta w \tag{11}$$

The multiplicative composite formula of the type given in Refs. [1] and [2] and valid through the whole integration domain, is developed by multiplying the right hand sides of Eqs. (9) and (10) and dividing by Eq. (11). The general formula, that includes all corners, has the form

$$z_{m+1} + z_m = R\frac{\exp[\frac{\pi}{2h}(\phi-\delta)]}{(\Delta w)^{n-1}} \prod_{j=1}^{n} \bar{F}_j \left[\frac{(w-a_j)^{\alpha_j+1}}{\alpha_j + 1}\right]_{w_m}^{w_{m+1}} \tag{12}$$

where terms of the type $(\pi/2h)^{\alpha_k}$ were incorporated into the constant R. Eq. (12) correctly accounts for singularities and is second-order accurate.

4.3 Non-Periodic Channel

The iteration procedure for a symmetric channel has been described in Section 4.1. The case of a non-symmetric channel is sketched in Fig. 5. The complex constant R in Eq. (2) is rewritten as $R = |R|\ \exp[i\pi(-\phi_B + 1/2\ \phi_T + 1/2\ \delta_T)]$, where the

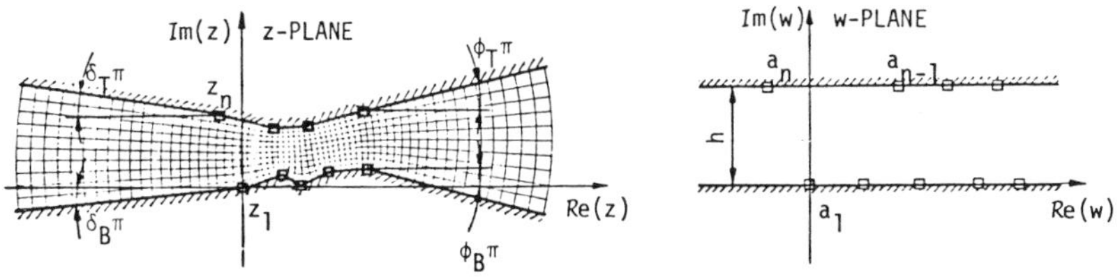

Fig. 5 Mapping of a General Non-Periodic Channel

appropriate angles are shown in Fig. 5. An initial guess is made for $|R|$ and for locations of corners a_j. The a_j's are numbered in the counterclockwise direction starting with a_1 being located at the origin. The initial guess has to be such that $Re(a_{j-1}) < Re(a_j) < Re(a_{j+1})$ along the bottom and $Re(a_{k-1}) > Re(a_k) > Re(a_{k+1})$ along the top, where e denotes the real part. Bottom corners are located along the real axis and top corners along the axis $w = ih$ (Fig. 5). The new value of $|R|$ is chosen by imposing condition

$$Im\left[\int_{(o,o)}^{a_n} f(w)\, dw\right] = Im(z_n) \tag{13}$$

where f(w) stands for the right hand side of Eq. (2), subscript n corresponds to the top left corner, z_n stands for location of the top left corner in the physical plane, a_n denotes location of the same corner in the transformed plane, and *Im* stands for the imaginary part. The integration is subsequently carried with the new value of $|R|$ along the bottom corners in the physical plane. Since the computed locations do not, in general, coincide with the specified locations, a new guess is made for a_j's in exactly the same way as described in Section 4.1. The evaluation of the new improved a_j's corresponding to the top of the channel is done by noting that the location of one of the top corners in the w-plane can be selected freely (Section 4.1). In the present case the location of the top left corner is kept constant during the iteration. The integration is carried from a_n along the top of the channel to determine the location of the remaining top corners in the physical plane. The new improved a_j's are determined by utilizing the same procedure as applied to the bottom of the channel. The above procedure converges quite rapidly, even with a poor initial guess. The results are illustrated in Fig. 5.

4.4 Periodic Channel

The mapping of a periodic channel of an arbitrary shape into a straight channel is described by Eq. (4), where the infinite product may be truncated after n terms. The estimate of the error of approximation is described with the help of configuration shown in Fig. (2). Equation (4) is rewritten as

$$\frac{dz}{dw} = R\, M(w) \prod_{j=1}^{\infty} R_j(w)\, L_j(w) \tag{14}$$

where M(w) describes terms arising due to the middle segment (segment 0 in Fig. 2), $R_j(w)$ denotes terms due to the j-th segment on the right side of the middle segment and $L_j(w)$ stands for terms due to the j-th segment on the left side (see Fig. 2). Behaviour of a single term in the product, say n-th term, may be estimated for large n by considering its logarithm. The result is given as

$$L_n(w)\, R_n(w) = 1 + P \exp(- \frac{nD}{h} \pi) + O[\exp(- \frac{2nD}{h} \pi)] \tag{15}$$

where P is a complex constant, and D is the length of the segment in the computational plane. The truncated part T_n of the product in Eq. (4) can be estimated by considering its logarithm. The result is

$$T_n = 1 + \hat{P} \exp(- \frac{nD}{h} \pi) + O[\exp(- \frac{2nD}{h} \pi)] \tag{16}$$

where $\hat{P}$ is a complex constant. The magnitude of the truncated part approaches unity exponentially with n, which suggests that the infinite product in Eq. (4) may be replaced with a very few initial terms.

The iteration procedure used to determine parameters a_j and D in Eq. (4) is analogous to the case of a non-periodic channel. The actual iteration procedure, as described by Eqs. (7) and (14), is carried out only for the middle segment (Fig. 2). The locations of corners, corresponding to all the remaining segments, are determined at each iteration step based on the condition of periodicity. In the case of a non-symmetric channel, the new locations of corners along the upper wall are normalized at each iteration step to assure the same periodicity along the top and bottom of the channel. The procedure described above converges quite rapidly, even with a poor initial guess. The results are illustrated in Figs. 6a and 6b. In this particular case the parameters of the transformation (4) were determined with sufficient accuracy for n = 2.

5. CONCLUDING REMARKS

A coordinate generation method for channel flows has been developed. The basic idea of the method is to map the channel of an arbitrary shape in the physical plane into a straight

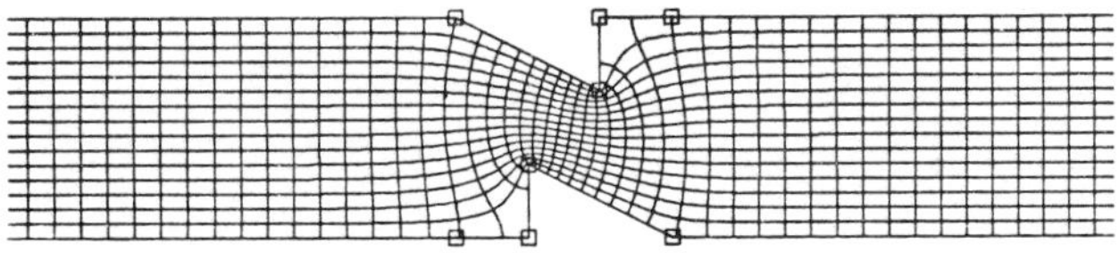

Fig. 6a Coordinate System for a Single Segment Fitted into a Straight Channel

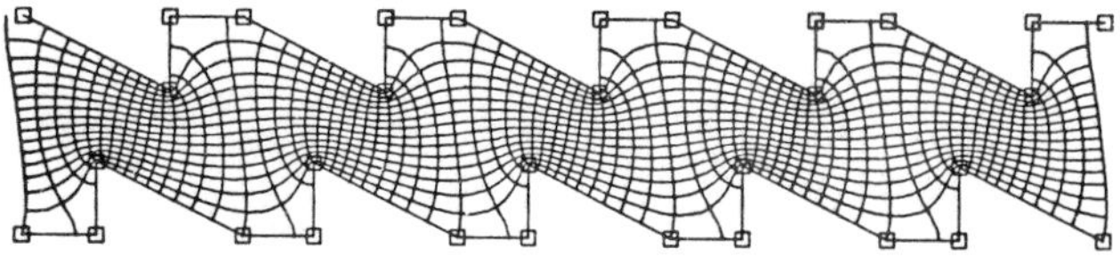

Fig. 6b Coordinate System for a Channel Consisting of an Infinite Number of Similar Segments (Periodic Channel)

channel in the computational plane. The mapping, which is given explicitly, involves parameters defined in the computational plane and thus not known apriori for a specified configuration in the physical plane. A simple numerical procedure, permitting determination of the required parameters, is described. The procedure has a very simple logic, is easy to program and converges quite rapidly.

6. REFERENCES

DAVIS, R.T. - Numerical Methods for Coordinate Generation Based on the Schwarz-Christoffel Transformations, AIAA Paper No. 79-1463, 4th Computational Fluid Dynamics Conference, 1979.

VAN DYKE, M. - Perturbation Methods in Fluid Mechanics, The Parabolic Press, 1975.

FLUID DYNAMIC SIMULATIONS IN GENERAL COORDINATES

Peter R. Voke* and Michael W. Collins*

SUMMARY

We discuss the advantages of spectral methods for the simulation of turbulence, and the use of Chebyshev polynomials for channel flow. A brief outline is given of a method for the simulation of the incompressible Navier-Stokes equations using the primitive variables, in a plane channel. We then develop the theory for the simulation of the equations in general coordinates, discussing in detail the steps taken to compute the geometric terms. A form of the generalised Navier-Stokes equations is given which does not explicitly involve the affine connection coefficients, resulting in considerable savings of computation time and storage space needed for simulations in three dimensional geometries. We conclude that the use of general coordinate transformations in two dimensions is viable, and that the extension to three dimensions is a practical possibility.

---

*Department of Mechanical Engineering, The City University, London, U.K.

1. ADDITIONAL NOTATION

1.1 Roman

$c^i{}_j$	transformation matrix
d_i	covariant derivative gradient operator
d^2	covariant Laplacian operator
E_i	explicitly computed inertial acceleration
F_i	driving force
g^{ij}	metric tensor (also g_{ij})
J	Jacobian of transformation
k_i	wavenumber for direction i
$k_h{}^2$	$k_1{}^2 + k_2{}^2$
N_i	number of mesh points in direction i
S_{ij}	stress tensor
T_k	Chebyshev polynomial of degree k
u_i	velocity components (Cartesian)
u^i	physical contravariant velocity components
v^i	contravariant velocity components
w^i	Cartesian velocity components
x_i	Cartesian coordinates
x^i	general coordinates
y^i	Cartesian coordinates

1.2 Greek

$\Gamma_{ij}{}^k$	affine connection coefficients
∂_t	time derivative
∂_i	vector gradient operator (grad)
∂^2	Laplacian $\partial_i\ \partial_i$
Δt	time step

1.3 Sub- and Superscripts

i, j, k, ℓ, m	vector indices (= 1, 2, 3)
^	Fourier/Chebyshev transformed quantity
+	time advanced quantity
-	time retarded quantity
(1), (2)	Chebyshev expansion of first or second derivative.

2. INTRODUCTION

The spectral-collocation method is an accurate and attractive approach to the three-dimensional numerical simulation of turbulent flows. Typically finite Fourier expansions are employed in homogeneous directions, with the size of the periodic box significantly greater than the expected correlation length of the turbulence. In inhomogeneous directions, with one or two walls bounding the flow, Chebyshev polynomial expansions offer an almost ideal representation, since the natural collocation points become clustered near the walls, and thus the resolution of the wall layers is good. The finite Fourier and Chebyshev summations and expansions necessary for the explicit evaluation of the nonlinear terms in the Navier-Stokes equations may both be performed efficiently by fast Fourier transform algorithms.

This combination of spectral expansions has been used successfully to simulate transitional flows in the plane channel [10], and is expected to be useful in large-eddy simulation [8, 11] for higher Reynolds numbers. The spectral expansions are described as infinite-order accurate, meaning that the errors decrease faster than any power of the number of degrees of freedom N, as N increases [6]. Typically the accuracy of a second-order finite difference method is matched at a fairly low value of N (16 or 32) [11], and for higher N the spectral method has an increasing advantage.

The main drawback of Chebyshev and Fourier series methods is that they demand particular boundary conditions; Fourier expansions are appropriate where periodic boundary conditions apply, Chebyshev expansions where conditions of any type are imposed at two fixed terminal points ($x = \pm 1$). We wish to use Chebyshev expansions in situations where the boundary conditions are imposed at arbitrary points (the walls of a distorted duct). We therefore use coordinate transformation to map the interior of the distorted duct onto a rectangular region, in such a way that the boundary walls are at $x = \pm 1$ in the transformed coordinates. We trade simple equations with complex boundary conditions for more complex equations with simple boundary conditions [12].

3. CARTESIAN COORDINATES

3.1 Spectral Method

We simulate the Navier-Stokes equations in three dimensions by a collocation-spectral method, with semi-implicit time stepping of the primitive variables u_i and p. In Cartesian tensor notation, x_i are the coordinates; x_1 is the streamwise coordinate, x_2 spanwise and x_3 is perpendicular to the walls. The summation convention for indices repeated in any term is used everywhere: thus $\partial_i u_i$ is the divergence of the velocity. The Laplacian $\partial_i \partial_i$ is abbreviated as ∂^2. With this notation, the continuity and Navier-Stokes equations (in vorticity form) may be written:

$$\partial_i u_i = 0 \tag{1}$$

$$\partial_t u_i = -\partial_i p + 2 u_j \omega_{ij} + \nu\partial^2 u_i \quad (i=1,2,3) \tag{2}$$

where

$$\omega_{ij} = \frac{1}{2} (\partial_i u_j - \partial_j u_i) \tag{3}$$

The variable p is the total pressure, per unit mass. The dependence on time and space of u_i, p and the vorticity ω_{ij} is understood.

To simulate these equations, u_i and p are Fourier transformed with respect to x_1 and x_2, and "Chebyshev transformed" with respect fo x_3, thus:

$$\hat{u}_i (t; k_1, k_2, k_3) =$$

$$= \sum_{x_1=1}^{N_1} \sum_{x_2=1}^{N_2} \sum_{x_3=0}^{N_3-1} e^{-2\pi i (x_1 k_1/N_1 + x_2 k_2/N_2)}$$

$$\times T_{k_3} u_i (t; x_1, x_2, x_3) \tag{4}$$

In this wavenumber or k-space representation, derivative operators are very simple. For example

$$\partial_1 p \to i k_1 \hat{p} \tag{5}$$

$$\partial_3 p \to \hat{p}^{(1)} \tag{6}$$

$$\partial_3 \partial_3 \to \hat{p}^{(2)} \tag{7}$$

The superscripted $\hat{p}^{(1)}$ and $\hat{p}^{(2)}$ represent sequences of Chebyshev coefficients which are recursively related in a simple way to the sequence p [6, 3]. Thus equations (1) and (2) become matrix equations for the k-space modes $\hat{u}_i(t; k_j)$ and $\hat{p}(t; k_j)$:

$$\partial_t \hat{u}_i = i k_i \hat{p} + 2 \widehat{(u_j \omega_{ij})} - \nu k_h^2 \hat{u}_i + \nu \hat{u}_i^{(2)} + F_i \quad (i=1,2) \tag{8}$$

$$\partial_t \hat{u}_3 = -\hat{p}^{(1)} + 2 \widehat{(u_j \omega_{3j})} - \nu k_h^2 \hat{u}_3 + \nu \hat{u}_3^{(2)} \tag{9}$$

$$i k_1 \hat{u}_1 + i k_2 \hat{u}_2 + \hat{u}_3^{(1)} = 0 \tag{10}$$

A driving force F_i has been included; in our simulations this is a mean pressure gradient in the x_1 direction.

The non-linear terms $(u_j\omega_{ij})$ must be evaluated in x-space by the spectral-collocation or pseudospectral method. The velocity components and the three non-zero vorticity components are transformed to x-space, multiplied together at each mesh point, and the product vector $u_i\omega_{ij}$ transformed back to k-space to obtain $(u_i\omega_{ij})$. Altogether nine three-dimensional transforms are involved; we find this to be the most computationally demanding part of our simulations.

3.2 Time Advance

The non-linear or term is thus treated explicitly, using the second order Adams method for time advance. The remaining terms on the right hand side of equations (8) and (9) are treated implicitly using the second order Crank-Nicholson scheme. The time-stepping equations derived from (8), (9) and (10) are (omitting the $\wedge$ accents):

$$2u_i^+/\Delta t + ik_i p^+ + \nu k_h^2 u_i^+ - \nu u_i^{+(2)} =$$

$$= 2u_i/\Delta t - ik_i p - \nu k_h^2 u_i + \nu u_i^{(2)} + 3E_i - E_i^- + 2F_i \qquad (i = 1,2\ ) \qquad (11)$$

$$E_i = (2u_j\omega_{ij}) \qquad (i = 1,\ 2,\ 3) \qquad (12)$$

$$2u_3^+/\Delta t + p^{+(1)} + \nu k_h^2 u_3^+ - \nu u_3^{+(2)} =$$

$$= 2u_3/\Delta t - p^{(1)} - \nu k_h^2 u_3 + \nu u_3^{(2)} + 3E_3 - E_3^- \qquad (13)$$

$$ik_1 u_1^+ + ik_2 u_2^+ + u_3^{+(1)} = 0 \qquad (14)$$

To solve these equations, (11) and (13) are recast into a recursive form [6] by integrating with respect to x_3 twice; this makes the matrices banded in k_3 (they are block diagonal in k_1 and k_2) with u_1, u_2, u_3 and p coupled. Six equations are found to be lacking; these of course are the physical boundary conditions $u_i^+(x_3 = \pm 1) = 0$, which in k-space read:

$$\sum_{k_3 \text{odd}} u_i^+(k_1,\ k_2,\ k_3) = \sum_{k_3 \text{even}}{}' \ u_i^+(k_1,\ k_2,\ k_3) = 0 \qquad (15)$$

These conditions make the block matrices bordered [14] . The resulting equations agree with those given by Moin and Kim[8] apart from a few factors of 2 which depend on the type of Chebyshev expansion used.

It should be noted that for the mode $k_1 = k_2 = 0$, equation (14) implies that $u_3^+ = 0$ (all Chebyshev coefficients), and thus (15) degenerates into a simple integral for p^+. This special treatment also applies if $k_1 = N_1/2$ (or 0) and $k_2 = N_2/2$ (or 0).

4. GENERAL CO-ORDINATES

4.1 Transformation

For the simulation of flow in a distorted channel, we employ a simple co-ordinate transformation which maps the distorted region in the Cartesian $y^1 - y^3$ plane to a rectangular channel in the $x^1 - x^3$ plane. The transformation is generated numerically, but not by any sophisticated technique; essentially it is a programmed version of a hand mapping transformation [13].

The transformation is not orthogonal. The choice of a general co-ordinate transformation is motivated by the observation that orthogonal transformations have difficulty in obtaining sufficient resolution at concave corners, and by our long-term aim of producing codes capable of dealing with three-dimensional geometries. As we shall show, this may be a practical possibility.

4.2 The Generalised Equations

The generalised Navier-Stokes equations were first given by Gal-Chen and Somerville [4, 5], but have not found extensive application. Spectral methods have been used with co-ordinate transformation in a simple way[7, 9] .

The contravariant components v^i of velocity in the transformed system are related to the Cartesian components w^i through the transformation matrix $c^i_{\cdot j}$

$$w^i = c^i_{\cdot j} \, v^j$$

$$c^i_{\cdot j} = \partial y^i / \partial x^j \qquad (16)$$

y^i and x^j are related by the transformation function. In practice it is convenient to use the "physical" velocity components which are scaled by the Jacobian J:

$$u^i = J \, v^i \qquad (17)$$

$$J = \det(c^i_{\cdot j}) \qquad (18)$$

It can be shown that the use of u^i preserves the continuity equation in an explicitly conservative form:

$$\partial_i u^i = 0 \qquad (19)$$

In terms of u^i, the generalised Navier-Stokes equations read:

$$\partial_t u^i = -Jg^{ij}\partial_j p + 2u^j\omega^i_{\ j} + \nu J d^2 (J^{-1}u^i) \qquad (20)$$

The tensor g^{ij} is the contravariant metric, defined by

$$g^{ij} = (g_{ij})^{-1} \qquad (21)$$

$$g_{ij} = \sum_{k=1}^{3} c^k_{\ i} c^k_{\ j} \qquad (22)$$

The other geometric objects required are the affine connection coefficients, defined by

$$\Gamma_{ij}^{\ \ k} = \tfrac{1}{2}\, g^{k\ell}(\partial_i g_{j\ell} + \partial_j g_{i\ell} - \partial_\ell g_{ij}) \qquad (23)$$

The operator d is our notation for the covariant derivative; the usual semicolor notation is somewhat confusing in this context, and obscures the formal similarity between equations (20) and (2). For the meaning of the covariant derivative, the role played by Γ, and the use of g in raising and lowering indices, the reader is referred to standard texts [1, 2].

4.3 Computation

In order to simulate equations (19) and (20), the geometric terms contained in (20) must be evaluated explicitly. We separate the pressure term thus:

$$Jg^{ij}\partial_j p = \delta^{ij}\partial_j p + (Jg^{ij} - \delta^{ij})p \qquad (24)$$

The first term is treated implicitly to find a solution p^+; the second is treated explicitly, requiring three extra triple transforms.

The non-linear term in ω is also computed explicitly. However, the expression for $\omega^i_{\ j}$ in terms of u^i is complicated and would require nine extra triple transforms to evaluate directly. The simplest method is to transform u^i to x space (it is needed to find $u^j\omega^i_{\ j}$ eventually), to form v_i from

$$v_i = g_{ij}\, u^j/J \qquad (25)$$

and then to transform this back to k space to find ω_{ij}:

$$2\,\omega_{ij} = d_i v_j - d_j v_i$$

$$= \partial_i v_j - \partial_j v_i \qquad (26)$$

Finally ω_{ij} is transformed back to x space to form the contraction with u. Twelve triple transforms are needed, instead of nine in the Cartesian calculation, but this is an acceptable extra burden.

The final term in (21) is also computed in a special way. The full expression for the covariant Laplacian of a vector field has six terms, five of which involve the connection coefficients. We use the fact that

$$d_i v^i = \partial_i u^i = 0 \qquad (27)$$

to rewrite the term as follows:

$$d^2 v^i = 2 d_j s^{ij} = -2 d_j \omega^{ij} \qquad (28)$$

The stress form is familiar (and fundamental), the vorticity form less so. Nevertheless, we propose to use the latter. We observe that

$$\nu J d^2 v^i = -2\nu J d_j \omega^{ij}$$

$$= -2\nu \left[\partial_j (J\omega^{ij}) + \Gamma_{jk}{}^{i}\, {}^{kj} \right]$$

$$= -2\nu \partial_j (J\omega^{ij}) \qquad (29)$$

The final step follows since $\Gamma_{jk}{}^{i}$ is symmetric but ω^{kj} is antisymmetric in j and k. This is the reason for using ω rather than s. The computation of (29) requires three further triple transforms to obtain $J\omega^{ij}$ in k-space. The part $\nu\partial_j\partial_j u^i$ of (29) is treated implicitly, the remainder explicitly.

The major result presented here is that the components of the affine connection (or Christoffel symbols) do <u>not</u> need to be computed or used. Combining the manipulations of the previous section, the equations read:

$$\partial_i u^i = 0 \qquad (30)$$

$$\partial_t u^i = -J_g{}^{ij} \partial_j p + 2u^j \omega^i{}_j - 2\nu\partial_j (J\omega^{ij}) \qquad (31)$$

The vorticity is computed using (26), without involving the connection.

4.4 Computational Requirements

We find that the computation in general co-ordinates requires thirteen working fields (by a field we mean an array containing $N_1N_2N_3$ real values) and eighteen triple transforms, compared to nine fields and nine triple transforms for the Cartesian simulation. These are comparable to the extra requirements of a large-eddy simulation using a subgrid model based on stress rather than on vorticity.

In a two-dimensional geometry, we store all of y^i(2 values), $c^i{}_j$(4 values), g^{ij}(3 values), g_{ij} (3 values) and J(1 value), since they need only be stored for N_1N_3 distinct positions. c^{ij} is computed from y^i by spectral methods in k-space using the same routines employed in the simulation code. The $c^i{}_j$ are then transformed to x-space, where J is found as the determinant, g_{ij} by multiplication, and g^{ij} by inversion of g^{ij}.

In a three-dimensional geometry, we suggest that the same spectral methods may be used to find g^{ij}, but that only the six independent components of this tensor should be stored. Our computational scheme requires the use of g_{ij} only once, and J three times. These should be found from g^{ij} as required, using equation (21) and

$$J^{-2} = \det\,(g^{ij}) \qquad (32)$$

This would require nineteen working fields, and some additional numerical work.

REFERENCES

1. ARIS, R. -"Vectors, Tensors and the Basic Equations of Fluid Mechanics". Prectice-Hall, London,1962

2. COBURN, N. - "Vector and Tensor Analysis". Dover, New York, 1970; Constable & Co., London, 1970.

3. FOX, L. and PARKER, I.B. - "Chebyshev Polynomials in Numberical Analysis". Oxford University Press, London, 1968.

4. GAL-CHEN, T. and SOMERVILLE, R.C.J. - "On the Use of Co-ordinate Transformation for the Solution of the Navier-Stokes Equations". J.Comput.Phys. 17, 209-228, 1975.

5. GAL-CHEN, T. and SOMERVILLE, R.C.J. - "Numerical Solution of the Navier-Stokes Equations with Topography". J.Comput.Phys. 17, 276, 1975.

6. GOTTLIEB, D. and ORSZAG, S.A. - "Numerical Analysis of Spectral Methods: Theory & Applications". SIAM, Philadelphia, 1977.

7. McCRORY, R.L. and ORSZAG, S.A. - "Spectral Methods for Multi-dimensional Diffusion Problems". J.Comput.Phys. 37, 93-112, 1980.

8. MOIN, P. and KIM, J. - "On the Numerical Solution of Time-Dependent Viscous Incompressible Flows Involving Solid Boundaries". J.Comput.Phys. 35, 381-392, 1980.

9. ORSZAG, S.A. - "Spectral Methods for Problems in Complex Geometries". J.Comput.Phys. 37,70-92, 1980.

10. ORSZAG, S.A. and KELLS, L.C. - "Transition to Turbulence in Plane Poiseuille & Plane Couette Flow". J.Fluid Mech. 96, Part 1, 159-205, 1980.

11. SCHUMANN, U., GROTZBACH, G. and KLEISER, L. - "Direct Numerical Simulation of Turbulence". PSB-Bericht 680 (Kl.I), Kernforschungszentrum Karlsruche, 1979.

12. THOMPSON, J.F., THAMES, F.C. and MASTIN, C.W. - "Automatic Numerical Generation of Body-Fitted Curvilinear Co-ordinate System for Field Containing Any Number of Arbitrary Two-Dimensional Bodies". J.Comput.Phys. 15, 299-319, 1974.

13. VOKE, P.R. - "Fluid Dynamic Simulations in Curvilinear Co-ordinates". AERE-R10548, United Kingdom Atomic Energy Authority, Harwell, U.K.

14. VOKE, P.R., and COLLINS, M.W. - "Large-eddy Simulation: Retrospect and Prospect". AERE-R10716, UKAEA, Harwell, U.K.

CLASSES OF SPLITTING-UP SCHEMES FOR SOLVING TWO AND THREE-DIMENSIONAL NAVIER-STOKES EQUATIONS

by Pierre LAVAL+

Office National d'Etudes et de Recherches Aérospatiales
29 avenue de la Division Leclerc, 92320 CHATILLON-SOUS-BAGNEUX
Ad. Postale : BP 72 92322 CHATILLON Cédex (France)

1. SUMMARY

New classes of explicit second order splitting-up schemes are introduced for solving the full time-dependent, three-dimensional compressible viscous Navier-Stokes equations of fluid mechanics. The flux vectors are first split in two or three parts : a purely hyperbolic and either a parabolic one or a purely parabolic and a part with mixed derivatives. Second order schemes are used at fractional steps (such as Lerat-Peyret's schemes for the hyperbolic part and extensions of them for the parabolic one). The hyperbolic part is then split. A new class of schemes, which allows optimal time steps, is introduced in order to solve this part. The stability analysis of the newly-derived (resulting) schemes, which is carried out analytically, shows that the present method should be quicker than the other existing splitting-up methods. Implicit or hybrid splitting-up schemes are also in current development and applications of these different schemes to transonic flow are planned.

2. INTRODUCTION

Various numerical methods have been developed, during the past decade, to solve the full time-dependent compressible viscous Navier-Stokes equations of fluid mechanics. We can distinguish three approaches : the splitting-up methods (such as explicit MacCormack-Baldwin's [1] , and Abarbanel-Gottlieb's [2] or hybrid MacCormack's [3]), the implicit methods (such as alternating direction implicit (ADI) Beam-Warwing's [4] and Briley-McDonald's [5] or bi-diagonal MacCormack's [6]) and the hopscotch-type methods (such as Greenberg's [7]), which combine the positive features of the explicit and implicit methods.

+Senior Scientist, ONERA

Among the three main splitting-up methods : stabilization (generalization of the ADI method), predictor-corrector and disintegration, the latter has the widest field of application since it can be applied to non-linear parabolic [8] [9] and non-linear hyperbolic [10] [11] problems. The disintegration method is also well adapted to computations on new vector and array processors. Thus we introduce new general classes of explicit or implicit or hybrid (explicit and implicit) disintegration schemes for solving two and three-dimensional Navier-Stokes equations. As it is not possible to present all these classes in this paper, it will only be devoted to the classes of explicit schemes in the framework of the three-dimensional Navier-Stokes equations (the two-dimensional case can be easily deduced).

We first present two general classes of disintegration schemes. In both classes, second order schemes are used at fractional steps (Lerat-Peyret's schemes [12] for the hyperbolic part, extensions of them for the parabolic one). In one class, the mixed derivative part is discretized as the non-homogeneous part of parabolic equations through the disintegration method [8]. These classes are extensions of the general class of disintegration schemes which has been introduced for solving non linear hyperbolic systems [10] [11] and applied to solve Euler equations in the framework of two-dimensional steady and unsteady transonic flows with shocks [10] and three-dimensional unsteady transonic flows in a compressor with distorted inflow [13].

We then introduce an "optimal" class of disintegration schemes (it is optimal in the sense that it allows optimal time steps). It is obtained from one of the first classes by means of a splitting of the hyperbolic part and by using a new class of "optimal" one-dimensional schemes to solve this part.

3. CLASSES OF EXPLICIT SECOND ORDER DISINTEGRATION SCHEMES

Let us consider the system of time-dependent, compressible three-dimensional Navier-Stokes equations written in conservation form :

$$(1) \qquad U_t + \sum_{l=1}^{3} F^l_x \; (U, U_{x_1}, U_{x_2}, U_{x_3}) = 0 \, ,$$

where vector $U \in R^5$ has the following components : $U_1 = \rho$, $U_{m+1} = \rho\, u_m$ $(m = 1, 2, 3)$, $U_5 = E$; ρ is the density, u_m is the component of the velocity V in the x_m direction and E is the total energy per unit volume which can be expressed for a perfect gas (γ = cte) as $E = (\gamma - 1)^{-1} p + 2^{-1} \rho V^2$, where p is the static pressure. U_t means $\partial U/\partial t$ (similarly $F^l_{x_l} \equiv \partial F^l/\partial x_l$). Each vector F^l $(l = 1, 2, 3)$ is first split in three parts :

$$(2) \; F^l = F^l_H \, (U) + F^l_P \, (U, U_{x_l}) + F^l_M \, (U, U_{x_r}, U_{x_s}) \; (r \neq s \neq l),$$

where $(F_H^\ell)_{x_\ell} = A^\ell(U)U_{x_\ell}$, $F_P^\ell = -B^\ell(U)U_{x_\ell}$, $F_M^\ell = -(C_r^\ell(U)U_{x_r} + C_s^\ell(U)U_{x_s})$.

We note a_k^ℓ, b_k^ℓ and c_k^ℓ ($k = 1,\ldots, 5$) the respective eigenvalues of matrices A^ℓ (jacobian of vector F_H^ℓ, which appears in the hyperbolic part), B^ℓ (triangular, which appears in the purely parabolic part), $C_r^\ell + C_s^\ell$ (part with mixed derivatives). These eigenvalues are given by : $a_k^\ell = u_\ell$ ($k = 1, 2, 3$), $a_4^\ell = u_\ell + a$, $a_5^\ell = u_\ell - a$; $b_1^\ell = 0$, $b_2^\ell = b_3^\ell = \rho^{-1}\mu$, $b_4^\ell = \rho^{-1}(\lambda + 2\mu)$, $b_5^\ell = (\rho P_r)^{-1}\gamma\mu$; $c_k^\ell = 0$ ($k = 1, 2, 3$), $c_4^\ell = \rho^{-1}(\lambda + \mu)$, $c_5^\ell = -c_4^\ell$ (a is the speed of sound, μ and λ are the viscosity coefficients, P_r is the Prandtl Number : $P_r = K^{-1}\mu C_p$, where C_p and K are respectively the specific heat and the thermal conductivity coefficient). We introduce the following notations : U_i^n approximation of solution U $((x_1)_i, (x_2)_j, (x_3)_k, t^n)$ at mesh points $(x_1)_i = i\,\Delta x_1$, $(x_2)_j = j\,\Delta x_2$, $(x_3)_k = k\,\Delta x_3$ and at time $t^n = n\Delta t$, $(F^\ell)_i^n = F^\ell(U_i^n)$ (similarly $(B^\ell)_i^n = B^\ell(U_i^n)$, ..., etc), $\sigma_{x_\ell} = \Delta t/\Delta x_\ell$; $\mathcal{L}_{x_\ell}^H(\Delta t, \alpha_m, \beta_m) \equiv (S_{\beta_m}^{\alpha_m})_{x_\ell}(\Delta t)$ ($l = 1, 2, 3$; $m = l$ and $l + 3$) one-dimensional purely hyperbolic finite difference operator in the x_l direction ; $\mathcal{L}_{x_\ell}^P(\Delta t, \alpha_m, \beta_m) \equiv (S'^{\alpha_m}_{\beta_m})_{x_\ell}(\Delta t)$ ($l = 1, 2, 3$; $m = l + 6$ and $l + 9$) one-dimensional purely parabolic operator in the x_l direction ; $\mathcal{L}_\ell^{xyz}(\Delta t, \alpha_m, \beta_m) \equiv (S''^{\alpha_m}_{\beta_m})_{xyz}(\Delta t)$ ($l = 1, 2, 3$; $m = l + 6$ and $l + 9$) and $\mathcal{L}^M(\Delta t) \equiv S_{xyz}^M(\Delta t)$ three-dimensional parabolic operators. $S_{\beta_m}^{\alpha_m}$, $S'^{\alpha_m}_{\beta_m}$ and $S''^{\alpha_m}_{\beta_m}$ are second order schemes. Each of them depends on two parameters $\alpha_m \neq 0$ and β_m. They are respectively the hyperbolic one-dimensional Lerat-Peyret's scheme [12], the parabolic one-dimensional Peyret's scheme [14] (see relation (5)) and an extension of the latter. In particular S_1^1, S_0^1, S'^1_1, S'^1_0 are MacCormack's schemes, $S_{1/2}^{1/2}$ and $S'^{1/2}_{1/2}$ are Richtmyer's scheme. Scheme S_{xyz}^M is also second order and is given by relation (6).

Let us now consider system (1) in the two following forms: systems (1') and (1"). In system (1'), the flux vectors are split in three parts (relation (2)) : a purely hyperbolic, a purely parabolic and a part with mixed derivatives. In system (1"), the flux vectors are split in two parts : a purely hyperbolic and a parabolic one which includes mixed derivatives. Applying the disintegration method, the three-dimensional finite difference operator is split as a product of the already mentioned operators. In order to retain the second order accuracy, a commutative product must be applied. The product is made of thirteen or twelve operators according as the first (system 1') or the second (system 1") flux vectors splitting is used. So, if we assume that solution U_i^n is known, solution U_i^{n+2} can be obtained through the disintegration method by the two following classes of schemes :

(3) $$U_i^{n+2} = \mathcal{L}_{12}(6\Delta t/7)\,\mathcal{L}^M(2\Delta t/7)\,\mathcal{L}_{11}(6\Delta t/7)\,U_i^n,$$

with $$\mathcal{L}_{11} = \prod_{\ell=3}^{\ell=1} \mathcal{L}_{x_\ell}^P\left(\frac{\Delta t}{7}, \alpha_{\ell+6}, \beta_{\ell+6}\right)\mathcal{L}_{x_\ell}^H\left(\frac{\Delta t}{7}, \alpha_\ell, \beta_\ell\right),$$

$$\mathcal{L}_{12} = \prod_{\ell=1}^{\ell=3} \mathcal{L}^{H}_{x_\ell}(\Delta t/7, \alpha_{\ell+3}, \beta_{\ell+3}) \mathcal{L}^{P}_{x_\ell}(\Delta t/7, \alpha_{\ell+9}, \beta_{\ell+9}).$$

(4) $$U_i^{n+2} = \mathcal{L}_{22}(\Delta t)\mathcal{L}_{21}(\Delta t) U_i^n ,$$

with $\mathcal{L}_{21} = \prod_{\ell=3}^{\ell=1} \mathcal{L}^{xyz}_{\ell}(\Delta t/6, \alpha_{\ell+6}, \beta_{\ell+6}) \mathcal{L}^{H}_{x_\ell}(\Delta t/6, \alpha_\ell, \beta_\ell)$,

$$\mathcal{L}_{22} = \prod_{\ell=1}^{\ell=3} \mathcal{L}^{H}_{x_\ell}(\Delta t/6, \alpha_{\ell+3}, \beta_{\ell+3}) \mathcal{L}^{xyz}_{\ell}(\Delta t/6, \alpha_{\ell+9}, \beta_{\ell+9}).$$

These schemes depend on two, at least ($\alpha_i = \alpha_1$, $\beta_i = \beta_1$, $i = 2,.., 12$) and twenty four, at most (α_i , β_i ; $i = 1,.., 12$) parameters. The other schemes are obtained from schemes (3) and (4) by permutating the split operators. These two classes are, as aforesaid, extensions of the general class of hyperbolic disintegration schemes [10]. So, schemes $S^{\alpha}_{\beta_m}$($m = 1,.., 6$) are used at fractional steps to solve the purely hyperbolic part. They discretize pseudo-unsteady one-dimensional systems $M\, U_t + F^{\ell}_{x_\ell}(U) = 0\ (\Delta t^2, \Delta x_\ell^2)$, with $M = 6^{-1}$ or 7^{-1} and $\ell = 1$, 2, 3. Among the fractional step purely parabolic schemes, we consider for example, the scheme $U_i^{N+1/7} = \mathcal{L}^{P}_{x_1}(\Delta t/7, \alpha, \beta)\, U_i^N$ ($\alpha = \alpha_7$, $\beta = \beta_7$ and $N = n + 3/7$). This scheme discretizes the pseudo-unsteady system $7^{-1} U_t = [B^1 U_{x_1}]_{x_1} + 0\ (\Delta t^2, \Delta x_1^2)$. It is carried out in the following way (S'^{α}_{β} scheme [14]) :

(5) $$\begin{cases} \tilde{U}_i = (1-\beta)\, U_i^N + \beta U_{i+1}^N - \alpha \sigma_{x_1} (V_{i+1}^N - V_i^N) \\ U_i^{N+1/7} = U_i^N - \dfrac{\sigma_{x_1}}{2\alpha} \Big\{ (\alpha-\beta) V_{i+1}^N + (2\beta - 1) V_i^N \\ \quad + (1-\alpha-\beta) V_{i-1}^N + \tilde{V}_i - \tilde{V}_{i-1} \Big\} , \end{cases}$$

where $V_i^N = -(B^1)_i^N D_i U_i^N$, $\tilde{V}_i = -B^1(\tilde{U}_i) D_i \tilde{U}_i$,

with $D_i U_i^N = \Delta x_1^{-1}\left[(\beta-1) U_{i-1}^N + (1-2\beta) U_i^N + \beta U_{i+1}^N\right]$ and

$$D_i \tilde{U}_i = \Delta x_1^{-1}\left[(\alpha(2\beta-1)+1-3\beta)\tilde{U}_{i-1} + (1-2\beta)(2\alpha-3)\tilde{U}_i + (\alpha(2\beta-1)+2-3\beta)\tilde{U}_{i+1}\right]$$

Among the fractional step parabolic schemes, the scheme $U_i^{N+1/6} = \mathcal{L}^{xyz}_1(\Delta t/6, \alpha, \beta)\, U_i^N$ ($N = n + 1/2$), for example, which discretizes the system $6^{-1} U_t = [B^1 U_{x_1} + C_2^1 U_{x_2} + C_3^1 U_{x_3}]_{x_1} + 0\ (\Delta t^2, \Delta x_1^2, \Delta x_1 \Delta x_2, \Delta x_1 \Delta x_3)$, has a similar expression, in which V_i^N and $\tilde{V}_i$ are given by : $V_i^N = -(B^1)_i^N D_i U_i^N - (C_2^1)_i^N D_j U_i^N - (C_3^1)_i^N D_k U_i^N$ and $\tilde{V}_i = -B^1(\tilde{U}_i) D_i \tilde{U}_i - C_2^1(\tilde{U}_i) D_j \tilde{U}_i - C_3^1(\tilde{U}_i) D_k \tilde{U}_i$. $D_j U_i^N$ and $D_k U_i^N$, approximations of $(U_i^N)_{x_2}$ and $(U_i^N)_{x_3}$, are easily deduced from $D_i U_i$ which approximates $(U_i^N)_{x_1}$, ($D_j \tilde{U}_i$ and $D_k \tilde{U}_i$ are deduced from $D_i \tilde{U}_i$).

The scheme $U_i^{n+8/7} = \mathcal{L}^M (2\Delta t/7)\, U_i^{n+6/7}$ discretizes the mixed derivative part (as the non-homogeneous part of parabolic equations through the disintegration method [8]) and is given by :

(6) $$U_i^{n+8/7} = U_i^{n+6/7} + 2\Delta t \sum_{\ell=1}^{3} \left[F_M^\ell\right]_{x_\ell}^{n+1} ,$$

where $(F^{\ell}_{M})^{n+1} = C^{\ell}_{r}(U^{n+1})U^{n+1}_{x_r} + C^{\ell}_{s}(U^{n+1})U^{n+1}_{x_s}$, with $U^{n+1} = U^{n} + \Delta t\, U^{n}_{t} = U^{n} - \Delta t \sum_{\ell=1}^{3} F^{n}_{x_\ell}$ $(r \neq s \neq \ell)$.

All the fractional step schemes discretize pseudo-unsteady systems, but we have shown that the newly-derived (resulting) schemes (3) and (4) are consistent with system (1) with a second order accuracy [15]. These schemes satisfy the condition of stability :

$$\Delta t \leq \mathrm{Min}\left[\frac{\Delta x_\ell}{|u_\ell| + a}, \frac{\rho P_r \Delta x_\ell^2}{2\mu\gamma}, R^M_\ell\right] \quad (\ell = 1,2,3), \tag{7}$$

where $R^M_\ell = R^{(3)} = 2^{-1}\rho(\lambda+\mu)^{-1}\left(6^{-1}\sum_{\substack{r,s=1\\ r\neq s}}^{3} \Delta x_r^{-2}\Delta x_s^{-2}\right)^{-\frac{1}{2}}$ (schemes (3))
and $R^M_\ell = R^{(4)}_\ell = \Delta x_\ell^2/\delta_\ell$, with $\delta_\ell = [\rho(\lambda+3\mu)]^{-1}\{(\lambda+2\mu)\mu - 2^{-2}\Delta x_\ell^2(\Delta x_r^{-2} + \Delta x_s^{-2})\lambda\mu\}$ (schemes (4)), $r \neq s \neq \ell$).

This (linear) condition is obtained analytically considering the L_2 norm of schemes (3) and (4). They are stable if the norm of each split operator is bounded by unity. The details of this stability analysis being given in [15], we just note here that the fractional step scheme $U_i^{n+8/7} = \mathcal{L}^M U_i^{n+6/7}$ is not always stable (its amplification matrix has two eigenvalues greater than 1) even when the condition $\Delta t \leq \mathrm{Min}[(2\mu\gamma)^{-1}\rho P_r \Delta x_\ell^2 , R^{(3)}]$ is respected. However, we can show [15] that the resulting schemes (3) are stable under a condition which is more restrictive than the latter (the new condition relates to a split operator such as $\mathcal{L}^P_{x_1}\mathcal{L}^M\mathcal{L}^P_{x_1}$ instead of $\mathcal{L}^M$). Thus, we point out that it will be better to use the class of schemes (4). Effectively, in this case, the norm of each split operator is always bounded by unity (under condition (7)) and schemes (4) lead to an optimal time step (for the parabolic part). Before introducing an optimal class of disintegration schemes, we briefly compare the present method, based on these two classes of schemes, with the other existing explicit splitting-up methods. Scheme (3), obtained by a splitting of the flux vectors in three parts, can be considered as a generalization of the Abarbanel-Gottlieb's scheme [2], although the mixed derivative operators are different. Actually, when the parameters, appearing in scheme (3), are defined by $\alpha_i = 1$, $\beta_i = 0$ (i = 1,.., 12), the purely hyperbolic and purely parabolic fractional step schemes are MacCormack's schemes S_0^1 and S'^1_0 (which are used to discretize the hyperbolic and parabolic parts in [2]).

Schemes (3) and (4) allow larger time steps than the MacCormack-Baldwin's method [3], in which an usual split in the respective spatial-coordinate direction is used. Such a splitting leads (in the case of Navier-Stokes equations) to an estimated criterion, which is obviously more restrictive than the (exact) condition (7) obtained by means of a flux vectors splitting.

4. OPTIMAL CLASS OF EXPLICIT SECOND ORDER DISINTEGRATION SCHEMES

It is well known that the difficulty to solve system (1), at high Reynolds numbers, is due to the fact that the magnitude of the inertial forces described by the hyperbolic terms is much larger than that of the viscous forces described by the parabolic terms. Thus, the last condition (7), in which the smallest value of Δt will be generally given by $\Delta t_H = \mathrm{Min}\,[(|u_\ell| + a)^{-1} \Delta x_\ell]$ ($\ell = 1, 2, 3$), may be very restrictive in some cases. Such a case occurs, in the calculation of inviscid-viscous interacting flow, in the thin viscous layer near the wall when the velocity transversal component u_2 is very small ($u_2 \ll a$) and where it is necessary to use a very fine mesh : $\Delta x_2 \ll \Delta x_E$ (where Δx_E is the space step away from the wall). It is clear that this difficulty will disappear if we can build fractional step schemes for the hyperbolic part in such a way that Δt_H becomes $\Delta t'_H \simeq \mathrm{Min}\,[|u_\ell|^{-1} \Delta x_\ell]$ ($\ell = 1, 2, 3$). These schemes are thus carried out : we first investigate the system of Euler equations in non conservation form : $U'_t + \sum_{\ell=1}^{3} A'^{\ell}(U')\, U'_{x_\ell} = 0$. Each matrix A'^{ℓ} is split in two matrices : $A'^{\ell} = A'^{\ell}_1 + A'^{\ell}_2$ in such a way that the eigenvalues of A'^{ℓ}_1 and A'^{ℓ}_2 (triangular, respectively upper and lower) are equal to $2^{-1} u_\ell$ (property P_1). The hyperbolic part of system (1) : $U_t + \sum_{\ell=1}^{3} A^{\ell}(U)\, U_{x_\ell} = 0$ is then split in the same way : $A^{\ell}(U) = A^{\ell}_1(U) + A^{\ell}_2(U)$. Matrices A^{ℓ}_1 and A^{ℓ}_2, which share property P_1 with matrices A'^{ℓ}_1 and A'^{ℓ}_2, are obtained by using the similarity rule between the two last systems ($A^{\ell}(U) = P(U')\, A'^{\ell}(U')\, P^{-1}(U')$). So, we are able to define <u>a new class of explicit second order schemes (4'), which are deduced from schemes (4) by replacing each one-dimensional hyperbolic operator $\mathcal{L}^{H}_{x_\ell}(\Delta t/6)$ by a product of two operators</u> :

$$\mathcal{L}^{H}_{x_\ell}\left(\tfrac{\Delta t}{6}, \alpha_\ell, \beta_\ell\right) = \mathcal{L}^{(2)}_{x_\ell}\left(\tfrac{\Delta t}{12}\right) \mathcal{L}^{(1)}_{x_\ell}\left(\tfrac{\Delta t}{12}\right); \quad \mathcal{L}^{H}_{x_\ell}\left(\tfrac{\Delta t}{6}, \alpha_{\ell+3}, \beta_{\ell+3}\right) = \mathcal{L}^{(1)}_{x_\ell}\left(\tfrac{\Delta t}{12}\right) \mathcal{L}^{(2)}_{x_\ell}\left(\tfrac{\Delta t}{12}\right)$$

The new second order fractional step schemes, which discretize pseudo-unsteady systems $12^{-1} U_t + A^{*}(U)\, U_{x_\ell} = 0$ (where $A^{*}(U) = A^{\ell}_1(U)$ or $A^{\ell}_2(U)$), depend on two (at least) and seven (at most) parameters . A supplementary condition is imposed on them to obtain the schemes with two parameters : to have the same equivalent systems (same dispersive and dissipative error terms) as the Lerat-Peyret's schemes [12]. We give the expression of the scheme $U_i^{n+1/12} = \mathcal{L}^{(1)}_{x_1}(\Delta t/12, \alpha, \beta)\, U_i^{n}$ ($\alpha_1 = \alpha$, $\beta_1 = \beta$) in this case :

$$(8)\quad \begin{cases} \tilde{U}_i = (1-\beta)\, U_i^n + \beta\, U_{i+1}^n - \alpha\, \sigma_{x_1} A_i^{(1)} \left(U_{i+1}^n - U_i^n\right) \\ U_i^{n+\frac{1}{12}} = U_i^n - \dfrac{\sigma_{x_1}}{2\alpha} \Big\{ \alpha A_i^{(2)} \left(U_{i+1}^n - U_{i-1}^n\right) + A_i^{(3)} \big[(2\beta-1)\, U_i^n \\ \qquad + (1-\beta)\, U_{i-1}^n - \beta\, U_{i+1}^n \big] + \tilde{A}_i \left(\tilde{U}_i - \tilde{U}_{i-1}\right) \Big\} \end{cases}$$

where $A_i^{(1)} = 2^{-1}\left(A_1^1(U_i^n) + A_1^1(U_{i+1}^n)\right)$, $A_i^{(2)} = 2^{-1}\left(A_1^1(U_{i-1}^n) + A_1^1(U_{i+1}^n)\right)$

$$A_i^{(3)} = \left[3^{-1} - \beta(1-2\beta)\right] A_1^1(U_i^n) + \left(2.3^{-1} - \beta^2\right) A_1^1(U_{i-1}^n) + \beta(1-\beta)\, A_1^1(U_{i+1}^n),$$

$$\tilde{A}_i = 3^{-1}\left[A_1^1(\tilde{U}_i) + 2\, A_1^1(\tilde{U}_{i-1})\right].$$

It is important to note that if the hyperbolic fractional step schemes (as (8)) discretize one-dimensional systems in quasi-linear form, the newly-derived schemes (4') are, as schemes (3) and (4), consistent with system (1), in conservation form, with a second order accuracy [15]. Moreover, they satisfy a condition of stability which is obtained from condition (7) (for schemes (4)) by replacing $1/(|u_\ell| + a)$ by $2/|u_\ell|$. This new condition is generally less restrictive than (7) and much less restrictive in a flow region where $u_\ell << a$ and $\Delta x_\ell << \Delta x_E$ (where Δx_E is the space step away from it). Thus, the new class of schemes (4') must be considered as an "optimal" class (in the sense that it allows optimal time steps) and the present explicit method, based on the use of this class, should be quicker than the other existing explicit splitting-up methods [1] [2] (since they satisfy a condition of stability really more restrictive).

We also point out that a similar condition : $\Delta t_H \leq \Delta x_2/|u_2|$ was already used (for the hyperbolic part) through the hybrid splitting-up MacCormack's method [3] to compute the two-dimensional inviscid-viscous interacting flow. The assumption that u_2 is negligible compared to a (speed of sound) and an explicit numerical method, based on the characteristics theory, were utilized in the thin viscous layer near the wall. But there is an important difference between this method and the present one. The first requires a particular treatment in the viscous layer since it is found upon the afore-mentioned assumption which is valid only in this layer. On the other hand, the newly-derived fractional step schemes, which satisfy the C.F.L. condition $\Delta t \leq 2\Delta x_\ell / |u_\ell|$, are available in all the flow because they are built without any hypothesis.

These schemes (as (8)) can obviously also be applied to solve two or three-dimensional Euler equations, since they are well adapted to compute flows or flow regions where the velocity (or one component) is small and where it is necessary to use a fine mesh (like the neighbourhood of a stagnation point).

5. CONCLUSION AND RECOMMANDATIONS

New classes of explicit second order disintegration schemes have been introduced for solving three-dimensional unsteady Navier-Stokes equations (the two-dimensional case may easily be deduced). An "optimal" (which allows optimal time steps) class has been, in particular, obtained by means of a splitting of the hyperbolic part and by using a new class of "optimal" one-dimensional schemes to solve this part. Thus, the present method, based on the use of this class, should be quicker than the other existing explicit splitting-up methods [1] [2]. The applications of these newly-derived schemes to two-dimensional transonic flows with shocks and three-dimensional unsteady flows in a compressor are planned. Such applications have already been carried out assuming that the flow was inviscid and using disintegration schemes [10] [13]. The treatment of boundary conditions is more delicate than with an alternating-direction method, because of the lack of consistency at fractional steps. Nevertheless, it is possible to overcome these difficulties by a careful investigation of this treatment [10] [13] and, on the other hand, the disintegration method is very well adapted to stability analysis. Effectively, this analysis is more complex with other methods, especially in the three-dimensional case [8]. The results obtained, either for flows with shocks [10] or for unsteady flows in a compressor [13], show that these disintegration schemes, which easily lend themselves to a splitting of the flow field in different sub-domains, are well adapted to compute such flows. Thus, the existing computer programs will only be modified in the regions where the viscous effects must be taken into account. Parabolic operators will be added to the one-dimensional hyperbolic operators.

Though these explicit schemes are well adapted to vector and array processors, it is however very useful to also develop classes of implicit or hybrid disintegration schemes. Such implicit schemes are in current developments : they use standard second order implicit schemes such as Hollanders-Peyret's [16] (in particular, bi-diagonal MacCormack's [6]) or Lerat's [17] for the hyperbolic part and extensions of these schemes or Crank-Nicholson's for the parabolic part. Another approach may also be followed, which seems to be fruitful. It consists of connecting disintegration and hopscotch methods in order to allow large time steps as in implicit methods but without any matrix inversion.

6 . REFERENCES

1. MacCORMACK, R.W. and BALDWIN, B.S. - A Numerical Method for Solving the Navier-Stokes equations with application to Shock-Boundary Layer Interactions, AIAA paper, 75-1, January 1975.

2. ABARBANEL, S. and GOTTLIEB, D. - Optimal Time Splitting for Two and Three-Dimensional Navier-Stokes Equations with Mixed Derivatives, Jour. of Comp. Physics, Vol. 41, N° 1, pp. 1-33, 1981.

3. MacCORMACK, R.W. - An Efficient Numerical Method for Solving the Time-Dependent Compressible Navier-Stokes Equations at High Reynolds Number, NASA TM X-73 129, July 1976.

4. BEAM, R.M. and WARMING, R.F. - An Implicit Factored Scheme for the Compressible Navier-Stokes Equations, AIAA paper 77-645, 1977.

5. BRILEY, W.R. and McDONALD, H. - Solution of the Multidimensional Compressible Navier-Stokes Equations by a Generalized Implicit Method, Jour. of Comp. Physics, Vol. 24, N° 4, pp. 372-397, Aug. 1977.

6. MacCORMACK, R.W. - A Numerical Method for Solving the Equations of Compressible Viscous Flow, AIAA paper 81-0110, January 1981.

7. GREENBERG, J.B. - Semi-Implicit Hopscotch - Type Methods for the Time-Dependent Navier-Stokes Equations, AIAA Journal, Vol. 20, N° 8, pp. 1064-1070, 1981.

8. LAVAL, P. - Theory and Numerical Methods for Solving Partial Differential Equations, ONERA Technical Note N° 1979-2. English translation ESA-TT 606, 1980.

9. LAVAL, P. - Survey of Numerical Methods for Three-Dimensional, Unsteady Thermal Problems, 2nd Int. Conf. on Numerical Methods in Thermal Problems, LEWIS, R.W., MORGAN, K., SCHREFLER, B.A., PINERIDGE PRESS. VENISE, 1981.

10. LAVAL, P. - Explicit Second Order Splitting-up Schemes for Solving Hyperbolic Non linear Problems : Theory and Applications to Transonic Flows, ONERA Technical Note N° 1981-10. English translation ESA-TT 768, 1982.

11. LAVAL, P. -New Explicit Second Order Splitting-up Schemes Yielding Shock Profiles Without Oscillations, BAIL II, 2nd Int. Conf. on Computational and Asymptotic Methods for Boundary and Interior Layers, MILLER, J.J.M. BOOLE press, Dublin, 1982.

12. LERAT, A. and PEYRET, R. - Dispersive and Dissipative Properties of a Class of Difference Schemes for Non-Linear Hyperbolic Systems, La Recherche Aérospatiale, 1975-2, pp. 61-79, 1975.

13. BRY, P., LAVAL, P. and BILLET, G. - Distorted Flow Field in Compressor Inlet Channels, ASME paper 82 GT-125, 1982.

14. PEYRET,R. and TAYLOR, T.D. - Computational Methods for Fluid Flow, Springer-Verlag, 1983.

15. LAVAL, P. and BILLET, G. - New Splitting-up Schemes for Solving Hyperbolic and Parabolic Non Linear Problems, La Recherche Aérospatiale, 1983-3, 1983.

16. HOLLANDERS, H. and PEYRET, R. - Two Step Implicit Schemes for the Solution of a Conservation Equation, La Recherche Aérospatiale, 1981-4, pp. 287-294, 1981.

17. LERAT, A. - On the Computation of Weak Solutions of Hyperbolic Systems of Conservation Laws by Difference Schemes, ONERA Publication 1981-1. English translation ESA-TT-762, 1982.

DESCRIPTION AND VALIDATION OF A PREDICTION PROCEDURE FOR DIESEL ENGINE SWIRL CHAMBER TWO-PHASE FLOWS

M.M.M. ABOU-ELLAIL AND M.M. ELKOTB
MECHANICAL ENGINEERING DEPARTMENT, CAIRO UNIVERSITY, CAIRO, EGYPT.

SUMMARY

A numerical procedure, capable of predicting unsteady two-phase turbulent flow and heat transfer in diesel engine swirl chambers, is described. Seperate conservation equations are formulated for the gas phase and for the liquid fuel phase ; gas phase equations are formulated for a fixed co-ordinate system, while the liquid equations are formulated for a moving system of co-ordinates. The interaction between the fuel spray and the gas phase is accounted for by including the interficial drag and heat exchange terms in the corresponding momentum and energy equations.

The numerical procedure is based on the solution of the finite-difference form of the gas phase partial differential equations which are coupled to a Runga-Kutta-4 method for solving the liquid phase ordinary differential equations.

The computed results for the gas and liquid phases, inside a swirl chamber, during compression and expansion strokes, are compared with experimental data showing good agreement.

1. INTRODUCTION

The perfection of mixture formation and combustion in multifuel engines is the main concern of Diesel engine design engineers. These processes are, to a great extent, controlled by two-phase flow motion. Recently, several investigations have been carried out to achieve fast combustion by increasing the turbulence level in the engine swirl chamber. The increased swirl motion and turbulence intensity will decrease the delay period, accelerate the fuel mixing and ensure fast evaporation of the fuel spray.

Perhaps the combustion process in Diesel engine swirl chambers, using multifuel, is the most complicated process in industrial combustion chambers. During the compression stroke, air flows from the cylinder (main chamber) to the swirl chamber through a tangential port, creating rotating air flow inside it. During compression the air inside the swirl chamber is accelerated and gains high enthalpy. Fuel is injected, approximately 20 to 30 degrees crank angle before the end of the compression stroke. Injection continues, during expansion, to some 10 degrees after TDC. During the injection period, Diesel fuel (or Diesel - gasoline mixture in the case of multifuel engines) droplets penetrate the swirl chamber until finally they hit the walls. A thin film of fuel is thus formed where the final part of evaporation takes place.

The present paper describes a two-phase simulation of the above unsteady gas-droplets flow and heat transfer processes occuring in a cylindrical swirl chamber prior to combustion.

Details of the differential equations and other aspects of the mathematical formulation are given in section 2 ; the finite-difference forms of the governing equations and an outline of the solution procedure can be found in section 3 ; the computed results and the corresponding experimental data are given in section 4.

2. THE MATHEMATICAL MODEL

The mathematical model, used for the present study, is based on a Lagrangian-Eulerian approach. While the gas phase governing equations are solved in an Eulerian fashion, the droplets (liquid phase) equations are solved for a Lagrangian system of co-ordinates. Thus, the gas phase equations are solved in a grid fixed in space while droplets equations are solved in a system of co-ordinates which follow the droplet trajectory.

As it is usually the case in turbulent flow problems, the instantaneous value of any dependent variable is decomposed into an ensamble-averaged value and a fluctuating component. For compactness, all the governing equations for momentum, energy and mass may be expressed in the following compact tensor notation ; however, they are actually solved in their cylindrical polar form, to adequately describe the axisymmetric geometry of the swirl chamber.

2.1. THE GOVERNING EQUATIONS

In the following governing equations mass exchange between the phases and radiation heat transfer are neglected, for simplicity.

2.1.1.GAS PHASE EQUATIONS

The gas conservation equations governing the transport of momentum, energy and mass for an unsteady two-phase flow in a cylindrical swirl chamber of a Diesel engine can be expressed in tensor form as (1,2)

a) Momentum conservation in the j-direction :

$$(\rho u_j)_{,t} + (\rho u_i u_j - \sigma_{ij} + \rho \overline{u_i' u_j'})_{,i} = -p_{,j} + f_m r_m (v_{mj} - u_j) \quad (1)$$

b) Energy conservation :

$$(\rho h)_{,t} + (\rho u_i h - \Gamma_h h_{,i} + \rho \overline{u_i' h'})_{,i} = p_{,t} + u_i p_{,i} + g_m r_m (T_m - T) \quad (2)$$

c) Mass conservation

$$\rho_{,t} + (\rho u_i)_{,i} = 0 \quad (3)$$

Where u_i and v_{mi} are the mean gas and droplet velocities in the x_i direction, ρ is the gas density, σ_{ij} is the laminar stress tensor, p is pressure, h is the total enthalpy of gas, Γ_h is thermal diffusivity, r_m is the volume fraction of droplets size group 'm' and f_m and g_m are the corresponding drag and interphase heat transfer functions ; f_m and g_m are dependent on the flow regime, droplet diameter, liquid fuel density and droplet Raynolds number (3). The over-bars and over-dashes indicate ensamble-averaged and fluctuating values respectively. Index m indicates droplet size group m having a diameter range D_m to D_m + dD. Subscripts ',t' and

',i' indicate derivatives with respect to time and x_i-coordinate respectively. Repeated index indicates summation over that index which also includes droplet size index 'm'.

2.1.2. LIQUID PHASE EQUATIONS

The fuel spray may be considered as a swarm of spherical droplets with different sizes. The total drag force per unit volume can be evaluated as the sum of the drag force exerted by the gas on each individual droplet contained in that volume ; other forces acting on the droplet, e.g. pressure and apparent mass forces, are neglected. The drag exerted on an individual droplet is obtained emperically from the consideration of the flow around a spherical bluff body representing a droplet. It follows that the momentum and energy equations for a single droplet of diameter D_m may be written as (4)

$$v_{mj,t} = f_m \, (u_j - v_{mj}) \, / \rho_L \qquad (4)$$

and

$$T_{m,t} = g_m \, (T - T_m) \, / \, (\rho_L c_{PL}) \qquad (5)$$

Where T and T_m are temperatures of gas and droplet size group m respectively, and ρ_L and c_{PL} are the liquid fuel density and constant-pressure specific heat respectively.

2.2. THE TURBULENCE MODEL

It can be easily shown that the double correlation terms appearing in the governing equations must be modelled before an attempt is made to solve the above equations. In general the Reynolds stresses $(\overline{u_i' u_j'})$ are computed either by introducing the eddy diffusivity concept or solving transport equations for these stresses. In the present work, the first approach is adopted and hence $\overline{u_i' u_j'}$ is defined as :

$$-\rho \overline{u_i' u_j'} = \mu_t \, (u_{i,j} + u_{j,i} - (2/3)\delta_{ij} u_{n,n}) - (2/3)\rho k \delta_{ij} \qquad (6)$$

and $\overline{u_i' h'}$ as

$$-\rho\overline{u_i'h'} = (\mu_t/Pr_h).h_{,i} \quad (7)$$

Here Pr_h is a turbulent Prandtl number, k is the kinetic energy of turbulence and μ_t is a turbulent viscosity which may be related to k and its dissipation rate (ϵ) through the following equation (5),

$$\mu_t = c_\mu \rho \, k^2/\epsilon \quad (8)$$

Where c_μ is a constant of the model, k and ϵ are to be determined from their transient transport equations which are respectively (5) :

$$(\rho k)_{,t}+(\rho u_i k-(\mu_t/Pr_k)k_{,i})_{,i}= G-\rho\epsilon \quad (9)$$

$$(\rho\epsilon)_{,t}+(\rho u_i\epsilon-(\mu_t/Pr_\epsilon)\epsilon_{,i})_{,i}=(c_1 G-c_2\rho\epsilon)\epsilon/k \quad (10)$$

where Pr_k, Pr_ϵ , c_1 and c_2 are further constants of the model ; G is a generation term defined by :

$$G=\mu_t(u_{i,j}(u_{i,j}+u_{j,i})-(2/3){u^2}_{i,i})-(2/3)\rho k u_{i,i} \quad (11)$$

Here $u_{i,i}$ is the divergence of the gas velocity which reflects the effect of compressibility at high pressure variations occuring in Diesel engine swirl chambers.

The constants of the model of turbulence c_μ, c_1, c_2,Pr_k, Pr_ϵ, and Pr_h are given values 0.09, 1.44, 1.92, 0.7, 1.22, 0.9 respectively (5,6,7).

2.3. AUXILIARY EQUATIONS

In addition to the above governing differential equations, auxiliary equations are needed to define f_m, g_m, r_m, T, ρ in terms of the main variables u_i, v_i, p, h, T_m, k and ϵ.

The drag function f_m and the interphase heat transfer function g_m are defined as (4) :

$$f_m = \frac{3}{4}\frac{\rho}{D_m} C_D |v_{mj} - u_j| \tag{13}$$

and

$$g_m = 6NuK/D_m^2 \tag{14}$$

Here C_D and Nu are drag coefficient and droplet Nusselt number respectively ; C_D and Nu are functions of the droplet Reynolds number $Re(=\rho D_m|V_{mj} - U_j|/\mu_l)$ and are given by the following two expressions (4)

$$C_D = \frac{24}{Re} C_r \tag{15}$$

and

$$Nu = 2 + 0.459(Pr_l^{0.33})(Re^{0.55}) \tag{16}$$

Here subscript 'l' indicates laminar properties, and C_r is a function of Re which approaches unity for low Reynolds number (4).

For a droplet size group 'm', r_m may be related to the number of droplets per unit gas volume dN_m by the following equation :

$$r_m = D_m^3 (dN_m)/6 \tag{17}$$

Moreover, the number of droplet dN_m is related to D_m by a size distribution function of the form (8).

$$dN_m = a\, D_m^{\alpha} \exp(-bD^{\beta})dD \tag{18}$$

where a, b, α and β are constants dependent on the type of injector and operating conditions (8).

Finally T and P are related to the main variables through thermodynamic relations, namely :

$$h = k + u_i u_i/2 + \int_0^T C_p\, dT \tag{19}$$

and $\rho = P/RT$ (20)

Here R and C_p are the gas constant and the constant-pressure specific heat of gas, respectively.

2.4. INITIAL AND BOUNDARY CONDITIONS

Still air at atmospheric pressure and temperature is assumed as the condition prevailing initially inside the swirl chamber and the fuel is assumed to be injected radially during the injection period as shown in Fig. 1.

The imposition of the boundary conditions for the gas phase is not as straight forward as it may seem. At chamber wall : the velocity components are equal to zero to conform to the no slip condition ; the walls are assumed to be insulated and hence the wall heat flux is taken equal to zero ; the turbulence fluctuations and their dissipation rate are also zero. Wall functions are used for grid nodes adjoining the walls, in order to account economically for wall shear stresses in the boundary sub-layer(5). At the inlet plane - i.e. the tangential port which connects the swirl chamber to the engine cylinder - the inlet velocity, density and temperature vary with crank angle. They are specified by solving an ordinary differential equation for the inflow/outflow from the engine cylinder to the swirl chamber (9).

3. NUMERICAL SOLUTION PROCEDURE

The gas governing equations (1-3) and (9-10) may be completely represented by a single general equation for an arbitrary dependent variable $\emptyset$ (i.e. $\emptyset$ stands for u_i,h,k & ϵ), namely

$$(\rho\emptyset)_{,t} + (\rho u_i\emptyset - \Gamma_\emptyset \emptyset_{,i})_{,i} = S_\emptyset \qquad (21)$$

where $S_\emptyset$ is a source/sink term which also includes the interphase source terms.

3.1. THE FINITE-DIFFERENCE EQUATIONS

For the purpose of derving the finite-difference equations of the gas, the swirl chamber is overlied with a two-dimensional staggered grid of nodes, formed by the intersections of meridunal lines and circles. Equation (21) is formally integrated over imaginary control volume surrounding a typical grid node p and for a time interval δt to give the following equation (1) :

$$(A_p^n - S_p)\emptyset_p^n = \sum_m A_m^n \emptyset_m^n + A_p^o \emptyset_p^o + S_u \qquad (22)$$

where $\sum_m$ denotes summation over the four neighbouring nodes of a typical node p, the A's are influence

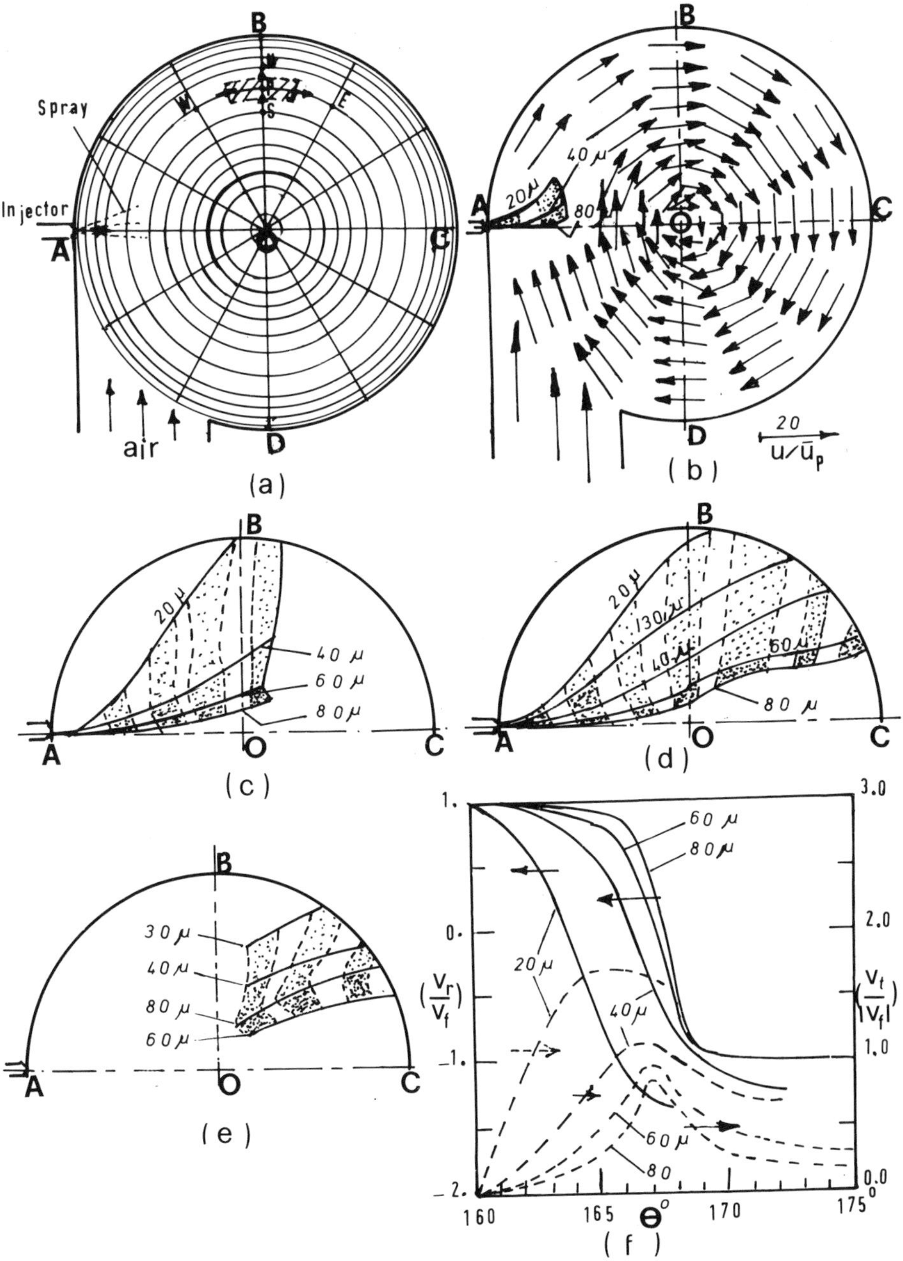

Fig.1.(a) Computational grid ; (b-e) predicted trajectories of droplets inside the swirl chamber at crank angles(θ^o) : (b) 163°, (c) 169°, (d) 190°, (e) 197°, (f); (————), predicted v_r ; (-----), predicted v_t ; (N=1500 RPM, $|v_f|$ = 30m/s).

coefficients representing the net flux across each face of the control volume caused by convection and diffusion and '0' and 'n' denote 'old' and 'new' values at times t and t+δt respectively ; S_p and S_u are coefficients of a linearized source term obtained by integrating $S_\emptyset$.

A difference equation, similar to equation (22), for the so-called pressure correction p' may be obtained from the gas mass continuity equation as explained in details in ref. (1) ; for unsteady flow, p' is used to correct simultaneously u_i, p and ρ so as to satisfy the local mass continuty (1) ; in addition to p', a global pressure correction is also considered (6).

3.2. SOLUTION ALGORITHM

At any instant of time t, before injection of fuel, the $\emptyset^o$ fields of the gas variables are known, predictions for a time increment δt are then obtained by solving interatively the difference equations(22) for the $\emptyset^n$ fields (1,7) ; in this way the solution is marched forward in time upto begining of injection.

During and after the injection period, a Runga-Kutta-4 solution algorithm for the liquid fuel equations is imbedded in the overall interative solution procedure of the gas difference-equations. For this purpose the fuel spray is discretized both in time and droplet size.

During the injection period, for each time step, a 'new' portion of the fuel is injected having a mass dependent on the actual fuel injection curve obtained experimentally (9). The 'new' portion of the fuel spray is approximated by a finite number of droplet size groups upto 100 Micron with incremental increase of 10 Micron and the corresponding number of droplets are obtained from Eq.(18). The liquid fuel equations are solved not only for the 'new' portion of spray but also for all the 'old' portions; thus, the contribution of all spray portions, and droplet size groups, to the droplet-gas interaction source terms, appearing in equations (1-2), is included. After the injection period no 'new' spray portions are considered and only the 'old' portions are tracked inside the swirl chamber.

The computed interaction source terms are substituted in equation (22) and a new field of the

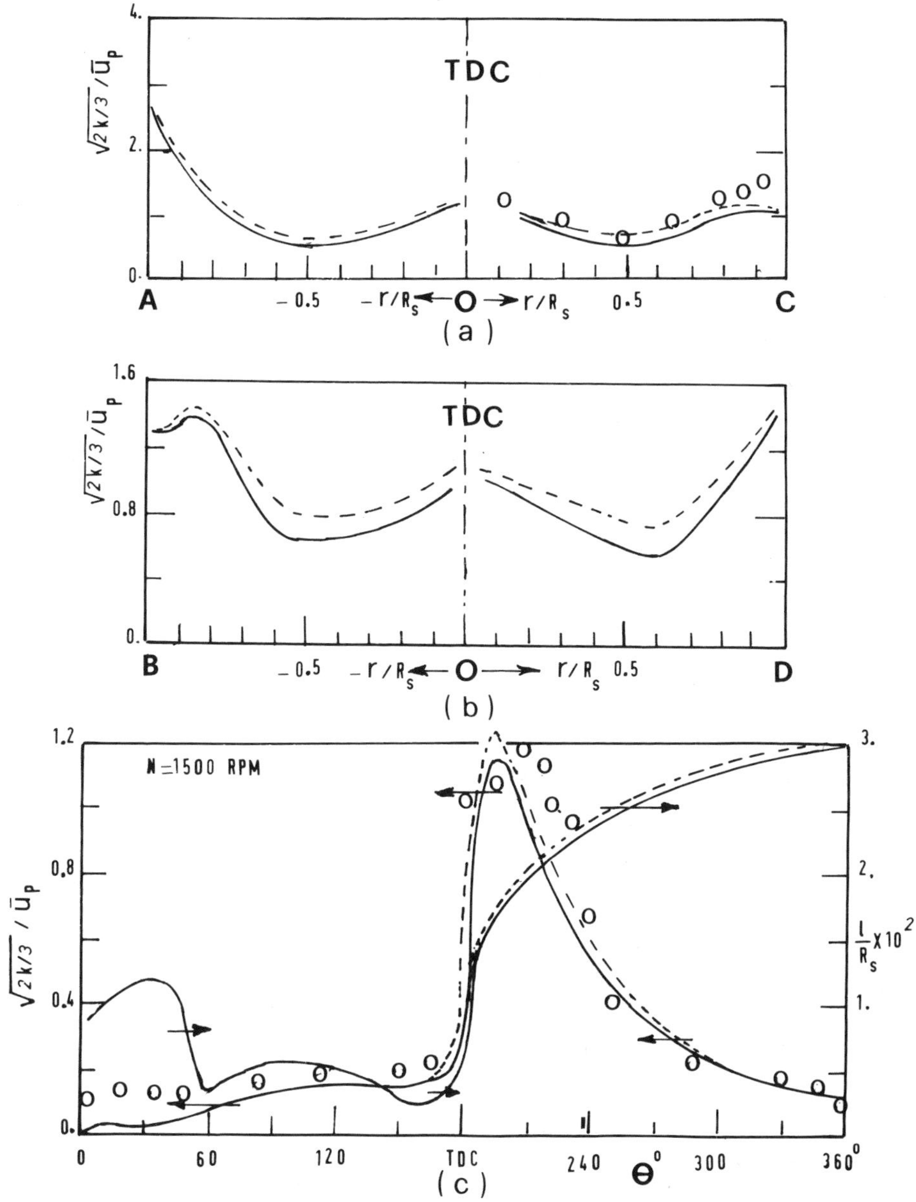

Fig.3. Turbulent intensity and eddy length scale variations inside the swirl chamber. Radial profiles of turbulent intensity at TDC in (a) AOC plance & (b) BOD plane. (c) Variations of turbulent intensity and length scale, at $0.87R_s$, in OC plane, versus crank angle ; predictions, (———) without injection, (-----) with injection ; (o o), measurements without injection (7).

gas dependent variables are obtained by an interative Gauss-elimination technique. A solution is considered converged when it satisfies the difference equation (22) to within 1% of a reference value ; otherwise, the above solution steps are repeated, for the time step in question, a number of times until convergence is obtained.

4. PRESENTATION AND DISCUSSION OF RESULTS

The numerical computations in the present study are carried out for a Diesel engine swirl chamber of 48.5 mm diameter, tangential port of diameter 15 mm and inclination of 45° to cylinder axis; engine speed, stroke, bore, and compression ratio are 1500 RPM, 140 mm, 125 mm, and 17 respectively. The injection period is 30 DCA (starting 20° BTDC) and the fuel injected per cycle is 3.5 x 10^{-5} Kg. A computational grid of 14 x 14 was used (Fig.1). To study the effect of fuel injection, the gas field dependent variables were computed for two cases : (a) with fuel injection and (b) without fuel injection. A 1° crank angle was used as a time interval. Subscripts 'r' and 't' will be used, throughout the rest of the paper, to indicate radial and tangential velocities respectively. Figure 1.a shows the swirl chamber and the computational grid. The trajectories of droplets, having diameters 20, 40 and 80 Microns, are depicted in Fig. 1.b, for crank angle 163°; on the same plane, air velocity vectors are plotted indicating a high swirling motion. As the fuel injection continues, the spray starts to develop inside the swirl chamber as shown in Fig.1.(b-e). At a crank angle of 169°, a small amount of fuel, from the first two spray portions, comes in contact with the chamber wall (Fig.1.c). At 190 DCA, all the fuel has been injected with a big amount hitting the walls, as shown in Fig.1.d. The variation of the droplets radial and tangential velocities (v_r & v_t) in the chamber with respect to crank angle are depicted in Fig.1.f. The velocities are normalized with respect to the initial fuel velocity v_f. It can be seen that the smaller sized droplets gain very quickly angular momentum (v_t) and at the same time quickly lose radial momentum(Fig.1.f).

Non-dimensional tangential gas velocity ($u_t/\bar{u}_p$) and radial velocity ($u_r/\bar{u}_p$) variations inside the swirl chamber are shown in Fig.2.(a-c), where $\bar{u}_p$ (= 7m/s) is the mean piston speed. Figure 2.a shows the radial distribution of predicted u_t, in AOC plane

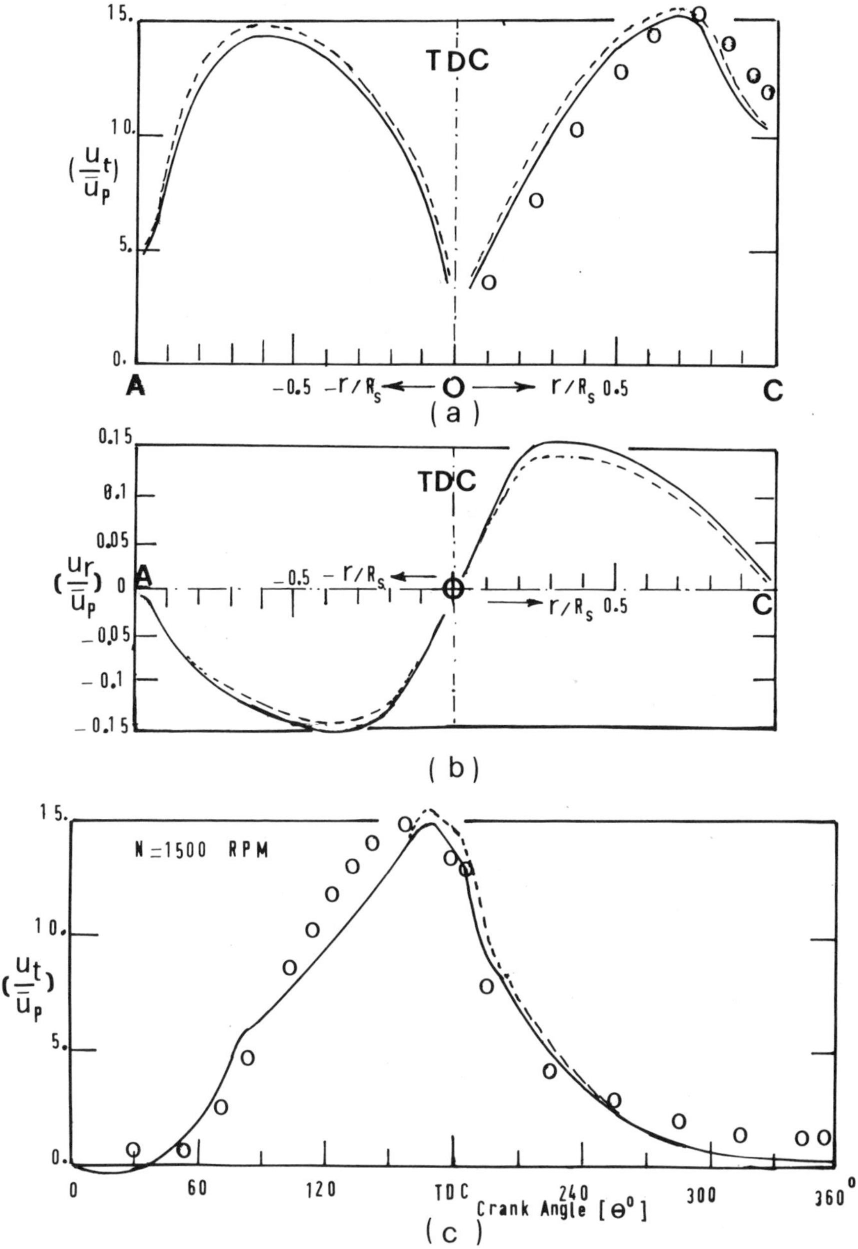

Fig.2. Tangential and radial air velocities(u_t & u_r) variations in the AOC plane. (a) Radial profiles of u_t at TDC, (b) radial profiles of u_r at TDC, (c) variation of u_t, at $0.87R_s$, versus crank angle; predictions,(———) without fuel injection,(-----) with fuel injection ; (o o), measurements without injection (7).

at TDC, for cases of pure air (solid line) and air-droplet flow (dashed line). The corresponding variation, with respect to crank angle (or time), of the tangential gas velocity at a radial location of $0.87R_s$ is depicted in Fig.2.c, R_s being the swirl chamber radius. The corresponding experimental data, for air only, of ref.(1,7) are also shown in Fig.2.a and Fig.2.c, showing good agreement between predictions and measurements. It can be concluded from Fig.2 that the presence of the fuel droplets has little effect on air velocity inside the swirl chamber.

The non-dimensional turbulence intensity ($\sqrt{(2k/3)}/\bar{u}_p$) radial distribution, at TDC, in two perpendicular planes (AOC and BOD), is depicted in Fig.3.(a-b). Although the fuel is injected radially at point A, the effect of the presence of the droplets on the turbulence intensity is more pronounced in the BOD plane; the presence of the fuel droplets increases the level of turbulence. The variations of the turbulence intensity and the eddy length scale (l) with respect to crank angle ($r/R_S = 0.87$) are shown, together with the experimental data (7), in Fig.2.c. During the expansion stroke the length scale steadily increases, as expected, while during compression l increases with crank angle upto 45 DCA where the value of l sharply falls.

5. CONCLUSIONS

A Lagrandian-Eulerian numerical method for the prediction of unsteady two-phase flow and heat transfer in Diesel engine swirl chambers has been described. The method enables the design engineer to obtain a fairly good picture of the shape of the fuel spray in addition to the air velocity pattern. The obtained agreement between predictions and experimental data is fairly good. It may be concluded, from the predictions, that the presence of the fuel droplets increases the level of turbulence inside the swirl chamber.

5.1. ACKNOWLEDGEMENT

This work was supported by the European Research Office, U.S. Army, through Grant No. DA-ERO-79-G-0017.

6. REFERENCES

1. ABOU-ELLAIL, M.M.M. and ELKOTB, M.M. - Prediction and Measurement of Flow and Heat Transfer in Motored Diesel Engine Swirl Chambers. Proc.

3rd Symposium on Turbulent Shear Flows, Davis, pp.5.1 -5.8, 1981.

2. ABOU-ELLAIL,M.M.M. and WARDA,H.A. - Prediction of Turbulent Two-Phase Flow in Annular Vertical Tubes. To be presented at the 3rd International Conference on Numerical Methods in Laminar and Turbulent Flow, Seattle, August, 1983.

3. ABOU-ELLAIL,M.M.M. and KHALIL,E.E. - A Mathematical Model of Gas-Liquid Interaction in the Turbulent Reacting Flows. J. Applied Mathematical Modelling, Vol.4, pp. 136-138, April, 1980.

4. SOO, S.L. - Fluid Dynamics of Multi-Phase System, Blaidell, 1967.

5. LAUNDER, B.E.and SPALDING, D.B. - Mathematical Models of Turbulence, Academic Press, 1972.

6. GOSMAN, A.D. and WATKINS, A.P. - A Computer Prediction Method for Turbulent Flow and Heat Transfer in Piston/Cylinder Assemblies. Proc. 1st Symposium on Turbulent Shear Flows, Pennsylvania, April, 1977.

7. ELKOTB, M.M., ABOU-ELLAIL, M.M.M. and SALAH SALEM, I. - Influence of Swirl Chamber Geometry on Turbulence in Multifuel Engines. Paper presented at the International Conference on Combustion, Oxford University, April, 1983

8. ABOU-ELLAIL, M.M.M. et al - Effect of Fuel Pressure, Air Pressure and Air Temperature on Droplet Size Distribution in Hollow-Cone Kerosene Sprays. 1st International Conference on Liquid Atomization and Spray Systems, No. 4-2, pp.85-92, Tokyo, 1978.

9. ELKOTB, M.M. et al - Spray Behaviour inside a Swirl Chamber of a Diesel Engine. Proc. 1st Conference of Mechanical Power Engineering, Cairo, 1977.

Unsteady Boundary Layers with and without Analogy between Momentum and Energy Transfer

Dr. R. Yalamanchili

Technology Branch, Armament Division
Fire Control & Small Caliber Weapon Systems Laboratory
US Army Armament Research & Development Command
Dover, NJ 07801

ABSTRACT

The physical problem under investigation involves boundary layers not only highly unsteady in nature but also possessing arbitrarily varying gradients of pressure, density, and temperature. A combination of the method of weighted residuals, in particular the method of moments, method of characteristics, and Euler method are utilized to solve the governing equations. The velocity boundary layer thickness increases not only along the axial direction but also decreases near the upstream end. First, the boundary layer thickness increases with respect to time and decreases afterwards. The convective heat transfer increases as Prandtl number is increased. The convective heat transfer coefficient is also computed with the use of continuity, momentum, and analogy between momentum and energy transfer. Significant differences are noticed between the two different results for convective heat transfer coefficient especially for small and large times. The convective heat transfer is maximum near the projectile and first increases and later decreases with respect to time.

1. INTRODUCTION

The state of the art in unsteady boundary layers [1] and turbulent models [2] is quite limited. One of the earliest to study two-dimensional unsteady boundary layers is Moore [3] who obtained an approximate solution with quasi-steady state assumptions with no pressure gradients. Stewartson [4] studied the leading edge effects on the flow due to an impulsive start of a semi-infinite flat plate in its own plane. He utilized a certain type of similarity transformation. Lam and Crocco [5] used Crocco's transformation in their analysis of this problem. These and other earlier works on unsteady compressible boundary layers, mostly with no pressure gradients were reviewed in detail by Stewartson [6, 7].

Mirels [8, 9] investigated the boundary layer behind a shock or thin expansion wave advancing into a stationary fluid as it happens in a shock tube. This again has no pressure gradients and this problem is reduced to solving equations similar to Blasius equation. There were other approximate methods of solving unsteady boundary layers such as the one by Yang and Huang [10] who obtained the small time solution for flow on a flat plate moving with time varying velocity. Panton and Berger [11] also used approximate techniques for the analysis of boundary layers for unsteady flow of gas in a duct. These methods, however, are of limited range of applicability.

Nash and Patel [12] used an explicit finite difference scheme to obtain the solutions of two-dimensional, unsteady, turbulent boundary layers for two different types of problems: One is the flat plate boundary layer with sinusoidal variation with time in the free stream velocity. Another one is a boundary layer under oscillatory adverse pressure gradient. How well this scheme would work for more general problems is not known. Piquet [13] utilized Crocco transformation and a finite difference scheme for solving flat plate unsteady boundary layers behind a shock in a shock tube. Of course, this does not include flows with pressure gradients.

Koob and Abbott [14] solved an unsteady incompressible boundary layer problem on a semi-infinite flat plate with zero pressure gradients. However, no criteria is given either in the selection of approximate solution form or in the selection of weighting function due in the application of method of weighted residuals. Hall [15] developed a finite difference scheme to compute the development of unsteady two-dimensional boundary layers in laminar incompressible flow and computed the flow development from Rayleigh flow conditions to Blasius flow conditions over a segment of a semi-infinite flat plate.

Yalamanchili [16] developed the method of weighted residuals and the method of lines for the solution of unsteady compressible boundary layers with arbitrary pressure gradients. This is applied not only to Rayleigh-Blasius flow but also to shock-induced boundary layer flow problems [17] which may be considered similar to a boundary layer flow in a gun barrel. A second-order accurate numerical scheme is utilized [18,19] for more general problems, i.e., the unsteady compressible boundary layers with arbitrary pressure gradients and arbitrary free stream conditions. The results indicate drastic increase in computational times over analytical counterparts, and also extreme sensitivity (i.e., severe oscillations) to the input of initial and boundary conditions. Undoubtedly, there are other investigators who utilized superior mechanics of turbulence models and other numerical methods. Since these methods demand extensive and sophisticated input data (which is not readily available for many practical problems of interest), large scale computers, and excessive computational times, it is desirable to pursue further the development of alternative techniques.

Rapid-fire automatic weapons permit discharge of hundreds of projectiles per minute. As a result, the surface temperatures of the tube easily reach 800°C at the inner surface, and 500°C at the outer surface in a short period of time, even though both surfaces were initially at ambient temperature. As the projectile moves ahead because of the high pressure (300 m/m^2) and high temperature (3000 °K) gases, the propellant gas will be set into motion starting from rest and reaches about 1500 m/sec. in a time frame of the order of 1 millisecond. The time is the most predominant factor and therefore should be considered in any modeling and in seeking a solution, thereafter.

Since the governing equations of fluid dynamics for many problems of interest are a system of nonlinear partial differential equations and are also dominated by real gas and nonequilibrium effects, no general solutions exist that allow arbitrary initial and boundary conditions. Instead of attempting to solve the coupled and complicated continuity, momentum and energy equations simultaneously, one may be able to either develop empirical/semiempirical expressions for unsteady turbulent skin friction or formulate more accurate analytical/numerical procedures for the solution of only continuity and momentum equations and there predict unsteady convective heat transfer by analogy, if one exists, without any loss in the generality of unsteady compressible turbulent boundary layers with viscous dissipation and arbitrarily varying pressure gradients.

Numerous empirical or semiempirical methods which provide reliable estimates of turbulent skin friction for zero pressure gradient and steady compressible turbulent boundary layer flows are available. These methods can be used to predict heat transfer provided that a valid analogy between skin friction and heat transfer is available. There are various relations for analogy between momentum and energy transfer: Reynolds, Colburn, Karman, Kozlov, Chi and Spalding; Martinelli, etc. Most of them do not differ significantly and vary within about 25 percent [20]. Since not much literature is available and the state-of-art is limited in unsteady boundary layers and turbulence models, even though the boundary layer research was initiated about 80 years ago, various investigators in the past solved only momentum and continuity equations for turbulent skin friction and utilized these analogies for predicting convective heat transfer [21].

There are several objectives for performing this research. One objective is based on a need for development of analytical-numerical procedure for calculation of skin-friction and convective heat transfer without the use of any analogy between momentum and energy transfer. The technical approach, presented in the next section, is well suited for such an objective. Another objective is to calculate only the skin friction coefficient by the use of only continuity and momentum equations and to calculate the convective heat transfer, later on, by the use of any one of the analogies mentioned above. The technical approach of the next section may not be optimal in this respect. However, this approach is chosen to study the development of unsteady boundary layers in both ways, i.e., with and without analogy between momentum and energy transfer. If the analogy fares reasonably well, one may be able to either develop empirical/semiempirical expressions for unsteady turbulent skin friction or formulate more accurate analytical/numerical procedures for the solution of only continuity and momentum equations and then predict unsteady convective heat transfer by analogy without any loss in the generality of unsteady compressible turbulent boundary layers with viscous dissipation and arbitrarily varying pressure gradients.

2. TECHNICAL APPROACH

There are two techniques to evaluate the objective, i.e., does or does not the analogy hold good for unsteady turbulent boundary layers? The experimental approach involves the simultaneous measurement of skin friction and heat transfer for unsteady cases. Even though this is desirable, it is, indeed, very difficult to achieve, in practice, especially for the flow in a gun barrel. Therefore, no further consideration is given to it. The analytical approach involves the development of a solution for simultaneous equations of continuity, momentum, and energy and also a separate solution for continuity, momentum, and analogy equations and a comparison between the two solutions in the end.

An integral method is chosen due to lack of general solution and shear stress and convective heat transfer are only of interest here. This is a particular variation of the method of weighted residuals, i.e., the method of moments with a weighting function of unity. The appropriate governing equations can be obtained by application of Prandtl's boundary layer assumptions to the Navier-Stokes equations across the boundary layer. Otherwise, one can also derive these equations by consideration of control volume and setting up continuity (mass), momentum, and energy balances for the same:

Momentum

$$\tau_w = \frac{\partial}{\partial t}\int_0^\delta \rho(u_1-u)\,dy + \frac{\partial u}{\partial x}\int_0^\delta \rho(u_1-u)\,dy + \frac{\partial}{\partial x}\int_0^\delta \rho u(u_1-u)dy \qquad (1)$$

$$-\frac{q_w}{c_p} + \frac{\mu}{c_p}\int_0^{\zeta} (\frac{\partial u}{\partial y})^2\, dy = \frac{\partial}{\partial t}\int_0^{\zeta} \rho T\, dy + \frac{\partial}{\partial x}\int_0^{\zeta} \rho u T\, dy$$

$$- T_1 [\frac{\partial}{\partial t}\int_0^{\zeta} \rho\, dy + \frac{\partial}{\partial x}\int_0^{\zeta} \rho u\, dy] - \frac{1}{c_p}\int_0^{\zeta} [\frac{\partial p}{\partial t} + u\frac{\partial p}{\partial x}]\, dy \qquad (2)$$

The continuity is utilized to arrive at the above equations. The pressure gradient term in the momentum equation is expressed in terms of fre stream velocity gradients. The major decisions, in application of a method of weighted residuals, are selection of a function to represent the unknown dependent variable and selection of a weighting function. No way presently seems to be available to select the functions systamatically for all problems. Since the power law profiles for steady turbulent boundary layers are well-established, the following classical velocity and temperature profiles for the boundary layer are introduced:

$$\frac{u}{u_1} = (\frac{y}{\delta})^{1/n} \quad , \quad \frac{\theta}{\theta_1} = (\frac{y}{\zeta})^{1/m} \quad , \quad \theta = T - T_w \qquad (3)$$

Basically, five criteria are available for weighting functions. To name a particular criteron as a best one is impossible. The choice may depend upon the problem to be solved, the assumed solution form, and also the parameter of interest in that problem. The method of moments yields least error than any other criteria for boundary layer thickness in Blasius flow. Since the boundary layer thickness is an important parameter which is a function of x and t in the assumed solution form, the method of moments is chosen. The resulting equations are as follows:

$$\frac{n}{n+2}\frac{\partial\delta}{\partial x} + \frac{1}{u_1}\frac{\partial\delta}{\partial t} = R_1 \qquad (4)$$

$$(1-\beta)\frac{\partial\zeta}{\partial x} + \frac{1}{u_1}\frac{\partial\zeta}{\delta t} = R_2 \qquad (5)$$

$$\beta = \frac{m}{n+m+mn}(\frac{\delta}{\zeta})^{\frac{m+1}{m}} \quad ; \quad u_1 = \frac{x}{X} V$$

$$R_1 = \frac{n+1}{\rho u_1^2}\tau_w - \frac{\delta}{u_1}[\frac{1}{V}\frac{\partial V}{\partial t} + \frac{2n}{n+2}\frac{V}{X} + \frac{1}{\rho}\frac{\partial\rho}{\partial t} + \frac{n}{n+2}\frac{u_1}{\rho}\frac{\partial\rho}{\partial x}]$$

$$R_2 = \frac{(m+1)q_w}{\rho c_p \theta_1 u_1} + \frac{\zeta}{u_1}[\frac{m}{\theta_1}\frac{\partial\theta_1}{\partial t} - \frac{1}{\rho}\frac{\partial\rho}{\partial t}] + (m+1)[\frac{1}{n+1} - \frac{m}{n+m+mn}(\frac{\delta}{\zeta})^{1/m}]\frac{\partial\delta}{\partial x}$$

$$+ [-\zeta+\delta(\frac{m+1}{n+1} - \frac{m}{n+m+mn}(\frac{\delta}{\zeta})^{1/m})](\frac{1}{\rho}\frac{\partial\rho}{\partial x} + \frac{1}{u_1}\frac{\partial u_1}{\partial x}) + \frac{m}{\theta_1}[\zeta$$

$$- \frac{m\delta}{n+m+mn}(\frac{\delta}{\zeta})^{1/m}]\frac{\partial\theta_1}{\partial x} - \frac{m+1}{\rho c_p\theta_1}[(\zeta - \frac{\delta}{n+1})\frac{\partial p}{\partial x} + \frac{\zeta}{u_1}\frac{\partial p}{\partial t}] \qquad (6)$$

The following relations are assumed for closure purposes:

$$\tau_w = C_n \rho u_1^2 \left(\frac{\nu}{u_1 \delta}\right)^{\phi}, \qquad q_w = \frac{K\theta_\ell}{y_\ell}, \qquad \phi = \frac{2}{1+n}$$

$$y_\ell = 14.3\,\nu \left(\frac{\rho}{\tau_w}\right)^{1/2}, \qquad \theta_\ell = \theta_1 \left(\frac{y_\ell}{\zeta}\right)^{1/m} \tag{7}$$

C_n = .0228, .0174, .0143, .0117 for n= 7,8,9,10 respectively.

The transformed governing equations, namely (4) and (5), are still partial differential equations, but with only two independent variables, x and t instead of the original three independent variables x,y, and t. These equations can be reduced to the following ordinary differential equations by a procedure sometimes referred to as method of characteristics:

$$\frac{d\delta}{dt} = u_1 R_1, \qquad \frac{d\zeta}{dt} = u_1 R_2 \tag{8}$$

along the respective characteristic directions:

$$\frac{dx}{dt} = \frac{nu_1}{n+2}, \qquad \frac{dx}{dt} = (1-\beta)\, u_1 \tag{9}$$

The above procedure permits one to solve four ordinary differential equations instead of two original partial differential equations.

3. NUMERICAL ANALYSIS

The equation (8) is applicable along specific curves called the velocity characteristic curves and the temperature characteristic curves. These are obtained by the use of equation (9). It is to be noted that the velocity characteristic equation is independent of the temperature characteristic equation. Therefore, the velocity characteristic curve can be obtained by integration:

$$\frac{x}{x_r} = \left(\frac{X}{X_r}\right)^{\frac{n}{n+2}} \tag{10}$$

Now, the velocity boundary layer thickness can be obtained by numerical integration of equation (8). The Euler method of integration is chosen due to lack of analytical integration. This is an iterative method. For the first iteration, the initial and final velocity boundary layer thickness is assumed to be the same for that time step. The resulting velocity boundary layer thickness is used in further iterations until the desired degree of convergence is reached. It is found that no more than three iterations are required for the type of nonlinearities present in the physical example, shown in Fig. 1.

The physical problem is similar to flow in a tube with a closed end and a moving boundary at the other end. This is somewhat similar to the flow in a shock tube. The characteristics of a projectile with zero mass in a barrel can approach the characteristics of a shock in a shock tube for identical driving gas generating conditions. The typical velocity and temperature characteristic curves are shown in Fig. 2.

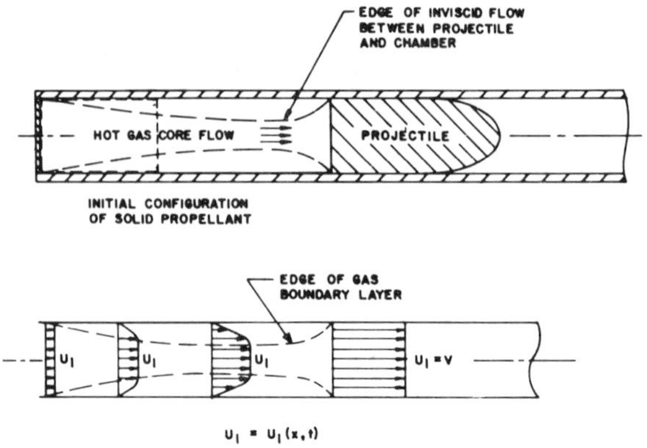

FIGURE 1. SCHEMATIC OF GUN BARREL AND ASSOCIATED VELOCITY PROFILES

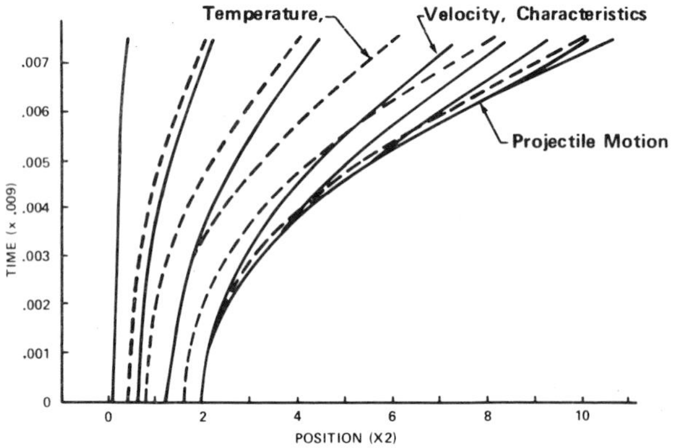

FIGURE 2. TEMPERATURE AND VELOCITY CHARACTERISTICS

Once the location of velocity characteristic curve and the velocity boundary layer thicknesses are known for a number of velocity characteristics at various times, the necessary information is available to integrate the temperature characteristic curve equation (9) and the temperature boundary layer thickness equation (8). However, the integration procedure is different from the above velocity counterparts. The temperature characteristic curve equation and the temperature boundary layer thickness equation are coupled. In general, the two equations have to be solved simultaneously. In such circumstances, the usual procedure is to construct the characteristic curve one step at a time and then obtain the temperature boundary layer thickness for that time step before the characteristic curve for the next time step can be constructed. Sometimes, this amounts to an iterative procedure between the construction of characteristic curves and the solution of temperature boundary layer thickness. Otherwise, small time steps are necessary.

The initial positions of the characteristic curves are chosen so that the region of interest is covered. Since the velocity characteristic curves have a steeper slope as shown, it is necessary to use characteristic curves originating both in the original chamber and at the projectile or shock base during the motion. Since the temperature characteristic curves do not have a steep slope, it is not necessary to generate additional temperature characteristic curves originating at the projectile base at later times. It is important to note that the characteristic curves and boundary layer thickness parameters can be calculated without reference to the adjoining characteristics. Therefore, the accuracy does not depend on number of characteristic curves chosen. A large number of characteristic curves means the final results are given at a large number of axial locations. Very small time steps are not necessary either due to the use of an iterative technique.

4. NUMERICAL RESULTS

The flow is highly unsteady due to the event starting from rest and coming to rest, occuring in a tube in less than one second. This process is repeated hundreds of times in a minute in some automatic weapon systems. Extensive evidence indicates that the pressure varies nonlinearly between the closed end and moving projectile. Similar variations are anticipated for density and temperature too. Of course, the free stream velocity could vary from zero to 1000 meters/second along the barrel. The chosen technical approach has provisions to take into account all these variations.

A problem, similar to the physical example, is set up to study the unsteady boundary layer characteristics. The input data is shown in Table 1. Axial variations are not taken into account in this case. The dimensionless wall temperature is 0.082 (389°K).

The typical velocity boundary layer thickness is shown in Fig. 3. The boundary layer grows not only with time but also with increase in axial location except near the base of the projectile. The growth is steeper at the projectile base than near the closed end. This approach predicts the growth and also the decay continuously without any separate treatment. The boundary layers are much thicker than predicted by Nordheim [22]. Once the velocity boundary layer thickness parameter is known, it is straightforward to compute any other boundary layer parameter, such as, displacement thickness, momentum thickness, velocity thickness, energy dissipation thickness etc.

TABLE 1. TYPICAL DIMENSIONLESS INPUT DATA

Time	Projectile Location	Projectile Velocity	Pressure	Density	Temperature
0.000	1.000	0.000	0.375	0.310	1.000
0.056	1.015	0.046			
0.111	1.062	0.092			
0.166	1.141	0.138			
0.222	1.250	0.184	0.575	0.555	0.893
0.277	1.397	0.250			
0.333	1.587	0.315			
0.388	1.822	0.380			
0.444	2.100	0.444	1.000	1.000	0.859
0.500	2.425	0.519			
0.556	2.800	0.593			
0.611	3.225	0.667			
0.667	3.700	0.741	0.450	0.621	0.732
0.722	4.215	0.784			
0.778	4.758	0.827			
0.833	5.331	0.870			
1.000		1.000	0.150	0.276	0.579

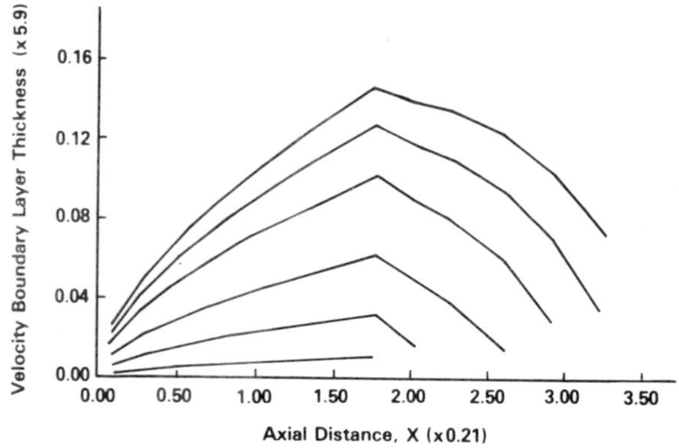

FIGURE 3. VARIATION OF BOUNDARY LAYER THICKNESS

The peak dimensionless heat transfer obtained from a typical output data is shown in Fig. 4. The exact peak value is unknown due to lack of information on when (what time) and where (location) it will occur and which characteristic will sweep through the point of interest. The heat transfer is maximum near the moving boundary (projectile) and minimum near the closed end. The peak rate of heat transfer happens at a time slightly lower than one-half of the interior ballistic cycle time. The effect of Prandtl number is also shown in the same figure. As Prandtl number increases, the rate of heat transfer also increases. The Prandtl number is 0.88 for typical propellant gases and is also found to be insensitive to temperature variations.

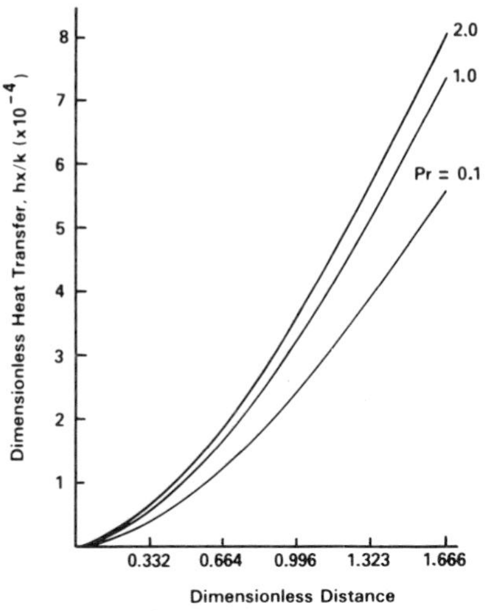

FIGURE 4. FORCED CONVECTIVE HEAT TRANSFER

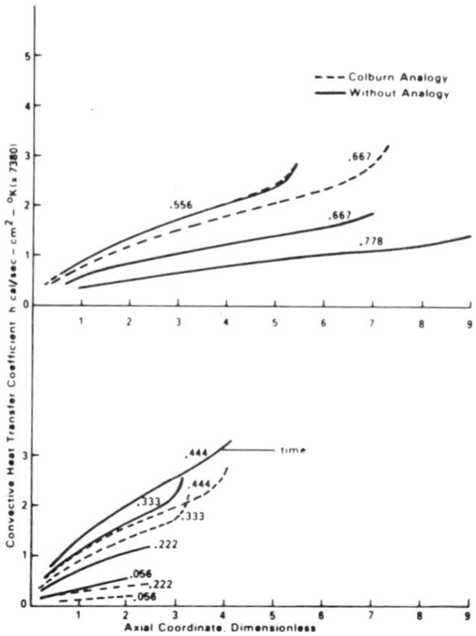

FIGURE 5 CONVECTIVE HEAT TRANSFER COEFFICIENTS

The convective heat transfer coefficient is defined as the ratio of heat transfer at the wall and the difference in temperature between the free stream and the wall. The heat transfer at the wall is determined by the use of the temperature profile and laminar sublayer thickness relationship. The convective heat transfer coefficient distribution at various times is shown in Fig. 5. The heat transfer coefficients are maximum near the projectile or shock base and decrease as the closed end or chamber is approached. The drastic decrease near the projectile is due to the requirement of zero boundary layer thickness at the base because of continuity at the interface between the projectile and the gas stream. The large heat flux which is directly proportional to the convective heat transfer coefficient is due to the presence of large velocities and dense flows in addition to large temperature differences between the free stream and the wall.

The skin friction coefficient is obtained by the use of only continuity and momentum equations, i.e., without the use of the energy equation. Then, the convective heat transfer coefficient is calculated by the use of this skin friction coefficient and the Colburn analogy. The results are shown in Fig. 5 by dotted lines. The trends are similar. However, the difference between the results obtained by analogy and the simultaneous solution of complete governing partial differential equations is enormous especially for small and large times. The results based on Colburn analogy underestimate about 100 percent for small times, however, the same analogy overestimate the convective heat transfer coefficients up to 50 percent for large times. It is predicted that the Colburn analogy yields almost identical convective heat transfer coefficients as obtained without the use of analogy at a specific time which is slightly less than one-half of the projectile travel time inside the tube. A time-dependent factor may patch the differences between the two different results. This functional relationship is unknown and requires further analytical and experimental research. For the time being, it is suggested to solve complete governing equations, even if necessary with more restrictive assumptions, rather than solve by utilization of analogy.

5. REFERENCES

1. PIQUET, J., and RADYADOUR KH. ZEYTOUNIAN, "Recent Research in the Field of Unsteady Boundary Layers," Proc. of IUTAM, Laval Press, Laval University, Quebec, Canada, May 1971.

2. MARKOVIN, M. V., and S. J. KLINE, "Calculation of Incompressible Turbulent Boundary Layers - A Review of the AFOSR-IFD-Stanford 1968 Conference," Symposium on Compressible Turbulent Boundary Layers, Langley Research Center, NASA-SP-216, Hampton, VA (1968).

3. MOORE, F. K., NACA TN 2471 (1951).

4. STEWARTSON, K., Quarterly Journal of Mechanics and Applied Mathematics, Vol. 4, pp. 182 (1951).

5. LAM, H., and CROCCO, L., Princeton University Technical Report #428 (1958).

6. STEWARTSON, K., Advances in Applied Mechanics, Vol. 6, Academic Press (1960).

7. STEWARTSON, K., "The Theory of Laminar Boundary Layers in Compressible Fluids," Oxford University Press, 1964.

8. MIRELS, H., NACA TN 3401 (1955).

9. MIRELS, H., NACA TN 3712, 1955.

10. YANG, W., and HUANG, H., "Unsteady Compressible Laminar Boundary Layer Flow Over a Flat Plate," AIAA Journal, Vol. 7, #1, pp. 100-105 (1969).

11. PANTON, R., and S. A. BERGER, "Unsteady Compressible Boundary Layer For Flow of Gas In A Tube," AIAA Journal, Vol. 5, #1, pp. 2121-2128 (1967).

12. NASH, J. F., AND PATEL, V. C., "Recent Research in the Field of Unsteady Boundary Layers," Proc. of IUTAM, Laval University, Quebec, Canada, (1971).

13. PIQUET, J., "Numerical Calculation of Laminar Compressible Boundary Layers in Unsteady Flow," NASA TT F 14, 410 (Aug 72).

14. KOOB, S. J., and ABBOT, D. E., "Investigation of a Method for the General Analysis of Time Dependent Two-Dimensional Laminar Boundary Layers," Journal of Basic Engineering, Trans. of the ASME, pp. 563-571 (1968).

15. HALL, M. G., "A Numerical Method for Calculating Unsteady Two-Dimensional Laminar Boundary Layers," Ingenieur-Archiv, Vol. 38, No. 32 (1969).

16. YALAMANCHILI, R., AND P. D. BENZKOFER, "Unsteady Compressible Boundary Layers With Arbitrary Pressure Gradients," 11th Aerospace Sci Mtg, Wash. D.C. 1973, AIAA Paper #73-132, (AD746235).

17. YALAMANCHILI, R., "Shock Induced Boundary Layers By Weighted Residuals and Method of Lines," Proc. of 1st International Conference on Computational Methods in Non-linear Mechanics, University of Texas, Austin, TX (1974).

18. REDDY, K. C., W. L. SICKLES, and R. YALAMANCHILI, "Computation of Unsteady Boundary Layers," Proc. of Unsteady Aerodynamics Symposium, University of Arizona, Tucson, Arizona, (1975).

19. YALAMANCHILI, R., AND K. C. REDDY, "Numerical Simulation of General Purpose Boundary Layers," Numerical/Laboratory Computer Methods in Fluid Mechanics, The American Society of Mechanical Engineers (Editors A.A. Pouring and V.L. Shah), (1976).

20. YALAMANCHILI, R., "Convective Heat Transfer in Unsteady Turbulent Boundary Layers," Proc. of Third Int. Conf. on Numerical Methods in Thermal Problems, Seattle, WA, (Aug 1983).

21. DAHM, T.J., and L.W. ANDERSON, "Propellant Gas Convective Heat Transfer in Gun Barrels," Aerotherm Report #70-18, Mt. View, CA, (Aug 1970).

22. NORDHEIM, L.W., SOODAK, H., and G. NORDHEIM, "Thermal Effects of Propellant Gases in Erosion Vents and in Guns," National Defense Research Committee, Armor and Ordinance Report # A-262, Division 1, (May 1944).

ANALYSIS OF FLUID FLOWS AT MODERATE REYNOLDS NUMBERS THROUGH TUBE WITH CONSTRICTION

Q. Hasan*
R. Natarajan** and V.Seshadri**

SUMMARY:

Analysis of Fluid Flows through locally constricted tubes has applications in many areas, specially blood flow through volved vessels. The distribution of pressure and shear stress through the constriction which has been discussed by many authors using different idealizations is one of the factors that has potential medical significance.

In the present work the flow through the axisymmetric bell shaped constrictions has been analysed at different Reynolds numbers, by using Finite element method. Velocity-pressure formulation is used in the analysis. The system of non-linear simultaneous equations have been solved by applying two methods, namely, Crank Nichlson method and perturbation method. Here only later is described. The results for velocity field, distribution of shear stresses and pressure in the region of constrictions have been computed.

It has been observed that perturbation method requires less computer time for convergence and the methodology can be applied for higher Reynolds numbers. Further the flow with different geometries of constrictions can also be analysed by the present methodology.

---

* Assistant Professor, Department of Mechanical Engineering, Z.H. College of Engineering and Tech. Aligarh Muslim University, Aligarh 202 001. INDIA

** Professors, Department of Applied Mechanics, Indian Institute of Technology, Hauzkhas, New Delhi 110 029.INDIA

1. INTRODUCTION:

The knowledge of the effect of a constriction on the velocity and shear field is important in order to understand the mechanism of endothelial cell damage or the growth of the stenosis. Lee and Fund [1] have analysed the blood flow in a circular cylindrical tube with a local constriction by using a finite difference scheme. They presented stream-line patterns and distribution of velocity, pressure and shear stress for Reynolds numbers in the range of 0-25. Deshpande et al.[2] have solved numerically the steady flow through axisymmetric constrictions in a rigid tube considering the full Navier-Stokes equations. Morgan et al.[3] have employed momentum and energy integral equations to obtain approximate solutions to the axial velocity component of flow in the region of the stenosis. The method did not give the radial velocity component and treated the axial momentum equation as a parabolic partial differential equation by neglecing the second derivative of velocity in the axial direction. Young and Tsai [4] have described a series of steady flow in-vitro experiments in which important hydrodynamic factors, including pressure drop, separation and turbulance have been considered. Azume et al.[5] have a performed experiments for blood flow through stenotic blood vessels. In physiological applications, complicated boundary conditions and irregular flow geometries are invariably encountered and since F.E.M. has got the advantage of incorporating these effects, the present study has been done by using finite element method based on Galerkin Weighted residual technique. The analysis has been made for three diameter ratios, (d/D),0.4,0.5 and 0.6 (d - diameter of the narrowest section, D - diameter of the vessel) for Re in the range of 0-80. The results for velocity field, distribution of shear stress and additional pressure drop in the region of stenosis have been computed and compared with the available data in the literature.

2. GOVERNING EQUATIONS:

The present study is restricted to a steady, axisymmetric flow of an incompressible, Newtonian fluid. The governing equations for such a flow are the familiar Navier-Stokes equations and the continuity equation. They are given by :

Momentum equation along the radial direction:

$$\rho \left[u \frac{\partial u}{\partial r} + w \frac{\partial u}{\partial z} \right] = \mu \left[\frac{\partial^2 u}{\partial r^2} + \frac{1}{r}\frac{\partial u}{\partial r} + \frac{\partial^2 u}{\partial z^2} - \frac{u}{r^2} \right] - \frac{\partial p}{\partial r} \quad (1)$$

Momentum equation along the axial direction:

$$\rho \left[u \frac{\partial w}{\partial r} + w \frac{\partial w}{\partial z} \right] = \mu \left[\frac{\partial^2 w}{\partial z^2} + \frac{1}{r}\frac{\partial w}{\partial r} + \frac{\partial^2 w}{\partial r^2} \right] - \frac{\partial p}{\partial z} \quad (2)$$

and the continuity equation,

$$\frac{\partial u}{\partial r} + \frac{u}{r} + \frac{\partial w}{\partial z} = 0 \quad (3)$$

where u, and w are the velocity components in r and z directions respectively in the cylindrical co-ordinate system, p is pressure, ρ is fluid density and μ dynamic viscosity of the fluid.

The Reynolds number for a flow through a circular vessel of diameter 'D' is given by,

$$Re = \rho \bar{w} D/\mu \quad (4)$$

where $\bar{w}$ is the average fluid velocity. In the analysis, $\bar{w}$ and D are taken as constant and different values of Re are incorporated by changing the value of μ.

3. FINITE ELEMENT FORMULATIONS:

In the analysis isoparametric quadratic ring elements with cross-sections as quadrilateral have been employed. [8] The interpolating equations for velocity and pressure in an element are expressed in the following forms:

$$u^e = N_\alpha u_\alpha$$
$$w^e = N_\alpha w_\alpha , \quad p^e = N^p_\lambda p_\lambda \quad (5)$$

where N_α and N^p_λ are the interpolating functions for velocity and pressure, respectively, and u_α, w_α and p_λ represent the values of field variables at the αth Node of an element.[6] The velocities are taken as nodal variables at all the nodes while pressure is defined only at the corner nodes to ensure the same degree of approximation for velocity and pressure. Now applying the Galerkin weighted residual method to governing equations 1-3 and substituting for u,w and p in terms of nodal variables, we have,

$$\int_{V^e} [\rho N_\alpha (N_\beta u_\beta \frac{\partial N_\gamma}{\partial r} u_\gamma + N_\beta w_\beta \frac{\partial N_\gamma}{\partial z} u_\gamma) + \mu (\frac{\partial N_\alpha}{\partial r} \frac{\partial N_\beta}{\partial r} u_\beta + \frac{\partial N_\alpha}{\partial z} \frac{\partial N_\beta}{\partial z} u_\beta + \frac{N_\alpha N_\beta}{r^2} u_\beta) - (N_\alpha \frac{N_\lambda^p}{r} + N_\lambda^p \frac{\partial N_\alpha}{\partial r}) p_\lambda]\, dV$$

$$= \int_{S^e} N_\alpha [\mu(\frac{\partial N_\beta}{\partial r} u_\beta n_r + \frac{\partial N_\beta}{\partial z} u_\beta n_z) - n_r N_\lambda^p p_\lambda] dS \qquad (6)$$

$$\int_{V^e} [\rho N_\alpha (N_\beta u_\beta \frac{\partial N_\gamma}{\partial r} w_\gamma + N_\beta w_\beta \frac{\partial N_\gamma}{\partial z} w_\gamma) + \mu(\frac{\partial N_\alpha}{\partial z} \frac{\partial N_\beta}{\partial z} w_\beta + \frac{\partial N_\alpha}{\partial r} \frac{\partial N_\beta}{\partial r} w_\beta) - N_\lambda^p \frac{\partial N_\alpha}{\partial z} p_\lambda] dV = \int_{S^e} N_\alpha [\mu(\frac{\partial N_\beta}{\partial z} w_\beta n_z + \frac{\partial N_\beta}{\partial r} w_\beta n_r) - N_\alpha N_\lambda^p p_\lambda n_z] dS \qquad (7)$$

$$\text{and } \int_{V^e} N_\lambda^p [\frac{\partial N_\alpha}{\partial r} u_\alpha + \frac{N_\alpha u_\alpha}{r} + \frac{\partial N_\alpha}{\partial z} w_\alpha]\, dV = 0 \qquad (8)$$

where n_r and n_z represent the direction cosines of the normal $\bar{n}$ to the surface of an element. Equations (6-8) are recast in the following forms by making substitution like

$$B^1_{\alpha\beta\gamma} = \int_{V^e} N_\alpha N_\beta \frac{\partial N_\gamma}{\partial r} dV \text{ so on.}$$

They are as follows:

$$B^1_{\alpha\beta\gamma} u_\beta u_\gamma + B^2_{\alpha\beta\gamma} w_\beta u_\gamma + B^7_{\alpha\beta} u_\beta + B^6_{\alpha\lambda} p_\lambda = R_1 \qquad (9)$$

$$B^1_{\alpha\beta\gamma} u_\beta w_\gamma + B^2_{\alpha\beta\gamma} w_\beta w_\gamma + D^7_{\alpha\beta} w_\beta + D^1_{\alpha\lambda} p_\lambda = R_2 \qquad (10)$$

$$\text{and } T^7_{\alpha\lambda} u_\alpha + T^3_{\alpha\lambda} w_\alpha = 0 \qquad (11)$$

where R_1 and R_2 correspond to surface integral terms. In the present analysis, R_1 and R_2 become zero when the components of velocity are specified on all the boundaries. Equations (9-11) form a non-linear simultaneous equation system. In order to solve these equations, a perturbation method has been applied [6]. This method has been described in detail by Kawabara et al.[6] and, hence, only the salient points are mentioned here for the sake of completeness. The velocity and pressure terms are assumed to be expanded in Taylor series in terms of a small perturbation parameter, ϵ, as follows:

$$u_\beta = u_\beta^{(0)} + \varepsilon u_\beta^{(1)} + \varepsilon^2 u_\beta^{(2)} + \ldots$$

$$p_\lambda = p_\lambda^{(0)} + \varepsilon p_\lambda^{(1)} + \varepsilon^2 p_\lambda^{(2)} + \ldots \qquad (12)$$

Introducing these values in equations (9),(10) and (11) and equating the coefficients of the same order terms in ε, the following linear simultaneous equation system can be obtained. For the zeroth order term, we get,

$$B^1_{\alpha\beta\gamma}\, u_\beta^{(0)} u_\gamma^{(0)} + B^2_{\alpha\beta\gamma}\, w_\beta^{(0)} u_\gamma^{(0)} + B^7_{\alpha\beta} u_\beta^{(0)} + B^6_{\alpha\lambda}\, p_\lambda^{(0)} = 0 \qquad (13)$$

$$B^1_{\alpha\beta\gamma}\, u_\beta^{(0)} w_\gamma^{(0)} + B^2_{\alpha\beta\gamma}\, w_\beta^{(0)} w_\gamma^{(0)} + D^7_{\alpha\beta}\, w_\beta^{(0)} + D^1_{\alpha\lambda}\, p_\lambda^{(0)} = 0 \qquad (14)$$

and $T^7_{\beta\lambda}\, u_\beta^{(0)} + T^3_{\beta\lambda}\, w_\beta^{(0)} = 0 \qquad (15)$

In the similar way, the nth order terms can be written as follows:

$$B^1_{\alpha\beta\gamma}(u_\beta^{(0)} u_\gamma^{(n)} + u_\gamma^{(0)} u_\beta^{(n)}) + B^2_{\alpha\beta\gamma}(w_\beta^{(0)} u_\gamma^{(n)} + u_\gamma^{(0)} w_\beta^{(n)}) + B^7_{\alpha\beta}\, u_\beta^{(n)} + B^6_{\alpha\lambda}\, p_\lambda^{(n)} = -B^1_{\alpha\beta\gamma} \sum_{r=1}^{n-1} u_\beta^{(r)} u_\gamma^{(n-r)} - B^2_{\alpha\beta\gamma} \sum_{r=1}^{n-1} w_\beta^{(r)} u_\gamma^{(n-r)} \qquad (16)$$

$$B^1_{\alpha\beta\gamma}(u_\beta^{(0)} w_\gamma^{(n)} + w_\gamma^{(0)} u_\beta^{(n)}) + B^2_{\alpha\beta\gamma}(w_\beta^{(0)} w_\gamma^{(n)} + w_\gamma^{(0)} w_\beta^{(n)}) + D^7_{\alpha\beta}\, w_\beta^{(n)} + D^1_{\alpha\lambda}\, p_\lambda^{(n)} = -B^1_{\alpha\beta\gamma} \sum_{r=1}^{n-1} u_\beta^{(r)} w_\gamma^{(n-r)} - B^2_{\alpha\beta\gamma} \sum_{r=1}^{n-1} w_\beta^{(r)} w_\gamma^{(n-r)} \qquad (17)$$

and $T^7_{\beta\lambda}\, u_\beta^{(n)} + T^3_{\beta\lambda}\, w_\beta^{(n)} = 0 \qquad (18)$

For computing the nth order increment, equations (16),(17) and (18) are written in the matrix form [6]. In the simplified form, the matrix equation is

$$[K]\ (\phi) = (F) \qquad (19)$$

where [K] is the characteristic matrix.
(ϕ) is the matrix of the unknown variables.
and (F) is the matrix corresponding to the R.H.S. terms(Load matrix).

The characteristic equations for different elements are assembled to obtain a set of albegraic equations for nodal variables and they are solved using the front solution technique [7]. In the first iteration, the terms $u_\gamma^{(0)}$ and $w_\gamma^{(0)}$ are assumed to be zero which reduces the matrix equations to creeping flow equations. After this, the coefficient matrix is modified by taking the creeping flow values for velocity and pressure and the subsequent iteration is performed. After this, without changing the coefficient matrix, the equations are solved by updating the load matrix during subsequent iterations. The increments of -

$$u_\beta^{(k)},\ w_\beta^{(k)} \text{ and } p_\lambda^{(k)} \quad (k = 1,2,\ldots, n)$$

are substituted in equations (15) and the components of velocity and pressure at each nodal point of the flow field are computed by taking ε equal to $Re^{-1/2}$.

4. FLOW FIELD DISCRETIZATION AND BOUNDARY CONDITIONS

Fig.1 shows the flow domain along with the boundary conditions. The flow domain has been discretized by using isoparametric ring elements whose cross sections are quadrilateral. The total length of the flow domain is taken long enough to ensure the Poiseuille profile at the downstream side of the stenosis at the chosen Reynolds Number. At the inlet as well as at the outlet the fully developed Poiseuille profile is specified. No-slip as well as symmetry conditions are imposed at the wall and the axis, respectively. An arbitrary value of pressure is specified at the inlet section.

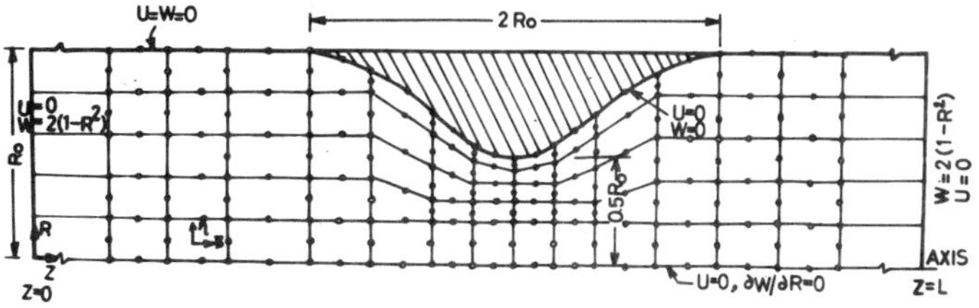

FIG. 1. FLOW DOMAIN DISCRETIZATION AND BOUNDARY CONDITIONS

5. RESULTS AND DISCUSSION:

5.1 Velocity Profiles:

The different cross-sections at which the velocity profiles have been plotted are shown in Fig.2. Only for d/D = 0.5 and Re in the range of

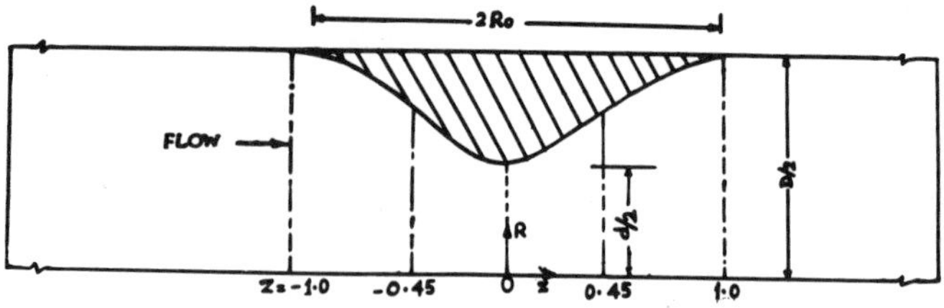

FIG.2 FLOW FIELD WITH SECTIONS IN STENOSIS REGION

0-80, the velocity profiles have been presented here. It is observed from Fig.3(a) that at $R_e = 0$, (creeping flow) the upstream velocity profiles are the mirror image of the downstream profiles. For higher Reynolds numbers (Re > 0) at all d/D ratios on the down stream side near the stenosis wall, negative velocity is observed which increases with a decrease in d/D and with an increase in Re. The velocity profiles, for Re > 0 are shown in Fig.3(b) to (d). Further, it is observed that the velocity profiles at Z = 0 become more blunt as Re increases. The results shown in the present study are in qualitative agreement with that given by Lee and Fung [1].

It has been also noticed that this bluntness in velocity reduces as the flow proceeds and the velocity profiles become Poiseuille - like at a distance of about three times the radius measured from the middle section of the stenosis (Z = 0) with d/D = 0.5 for Reynolds number of 25. With the decrease in d/D and increase in Re, this length increases.

5.2 Distribution of shear stresses along the stenosis wall:

Figures 4,5 and 6 show the variation of non-dimensional shear stress along the wall of the stenosis for different diameter ratios (d/D = 0.6, 0.5 and 0.4). The shear stress is calculated by determining the velocity gradients from the computed velocity profiles. The shear stress distribution on the tube wall is symmetric at zero Re for all diameter ratios but it is asymmetric when the Re is finite. There is sudden fall in shear stress values on the downstream side for Re > 0. The sign of shear stress changes on the downstream side for

Re > 0 and the magnitude of this negative shear stress increases with increasing Re. It is also observed that the peak value of shear stress increases with increase in Re. The negative shear stress on the down stream side of the stenosis indicates the presence of recirculatory flow down stream of the stenosis. Lee and Fung [1] have analysed the flow through a stenotic vessel at d/D = 0.5 for Re in the range 0 to 25.

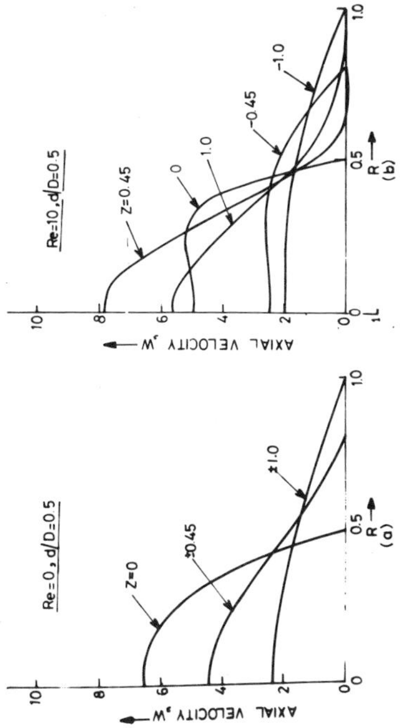

FIG 3. COMPUTED VELOCITY PROFILES AT DIFFERENT SECTIONS IN A TUBE WITH STENOSIS

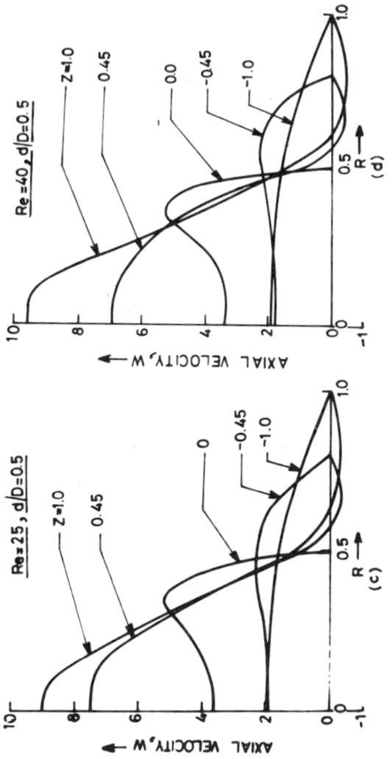

FIG. 3. COMPUTED VELOCITY PROFILES AT DIFFERENT SECTIONS IN A TUBE WITH STENOSIS

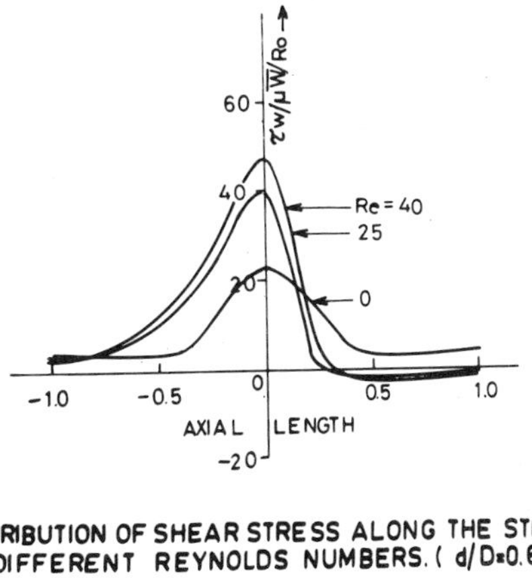

FIG. 4. DISTRIBUTION OF SHEAR STRESS ALONG THE STENOSIS WALL AT DIFFERENT REYNOLDS NUMBERS. (d/D=0.6)

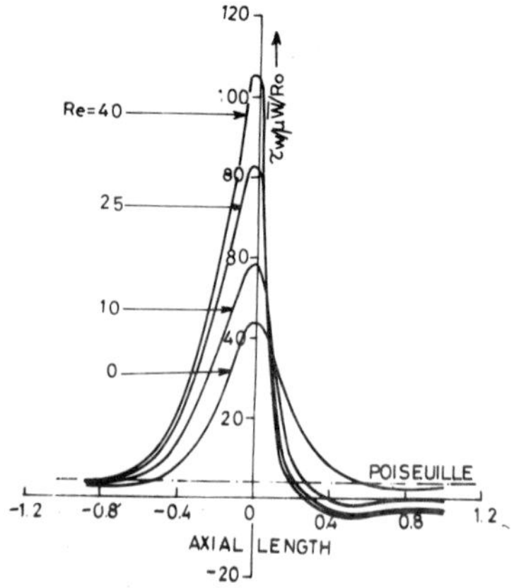

FIG. 5. DISTRIBUTION OF SHEAR STRESS ALONG THE STENOSIS WALL

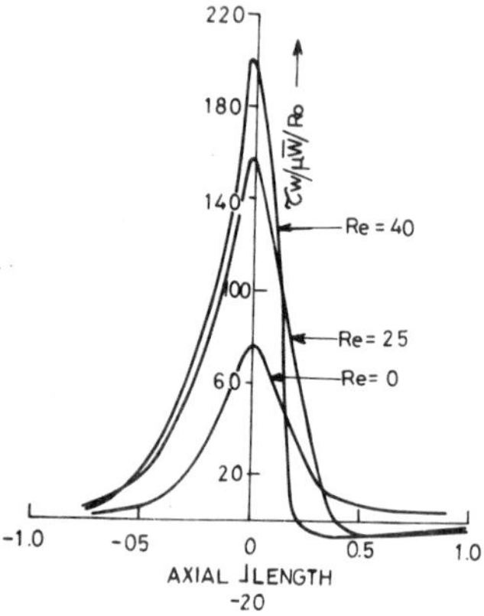

FIG. 6. DISTRIBUTION OF SHEAR STRESS ALONG THE STENOSIS WALL AT DIFFERENT REYNOLDS NUMBERS. (d/D=0.4)

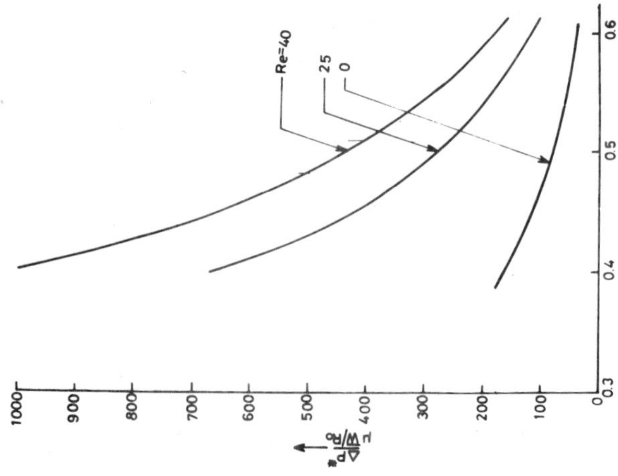

FIG. 7. ADDITIONAL PRESSURE DROP AT DIFFERENT STAGES OF STENOSIS.

5.3 Additional Pressure Drop:

When the pressure drop corresponding to Poiseuille flow is substracted from the total pressure difference, the resulting additional pressure drop represents the loss due to the geometric constriction. It is observed that at Re = 0 and for all d/D, half of the pressure head is lost in the converging part of the constriction and the other half in the diverging part of the constriction. At non-zero Reynolds numbers a greater portion of the pressure head is lost in the converging part of the stenosis. Further, it is observed that there is positive pressure gradient on the downstream side of the stenosis at the higher Re and this effect-increases with increase in Re. The positive pressure gradient on the down steam side of the stenosis is attributed to the fact that there is reversed flow in the downsteam side and the size of this region of recirculatory flow increases with increase in Re. The non-dimensional additional pressure drops for different stages of stenosis (d/D = 0.6,0.5 and 0.4) are shown in Fig.7. The figure shows the variation in additional pressure drop with Re. It is seen that the additional pressure drop increases very rapidly with increase in the blockage. It is also seen that as the Re increase the magnitude of P* also increases.

6. CONCLUSIONS:

The present analysis has been carried out by using a finite element method. The non-linear simultaneous system of equations have been solved using perturbation methodfor different Re. The methodology with some modifications can be applied for higher Re. The computed results are in good agreement with data available in literature. The additional advantage of the present methodology is that any arbitrary geometry of the stenosis can be easily incorporated in the analysis and the flow field can be analysed accurately, while the accuracy with other analytical and numerical methods is comparatively less.

REFERENCES:

1. Lee, J.S. and Fung, Y.C. - Flow in locally constricted tubes at Low Reynolds numbers. J. of Applied Mechanics, Vol.37, pp.9-16, 1970.

2. Deshpande, M.D., Giddens, D.P. and Malon, R.F.- Steady Laminar Flow through Modeled Vascular Stenosis. J. Bio-mechanics, Vol.9, pp.165-174, 1976.

3. Morgan, B.E. and Young, D.F.-An Integral method for the analysis of flow in Arterial stenosis. Bull. Maths. Biol., Vol.36, pp.39-53, 1974.

4. Young, D.F. and Tsai, F.Y. - Flow characteristics in Models of artèrial stenosis, 1 - steady flow. J. Biomechanics, Vol.6, pp.395-410, 1973.

5. Azuma T. and Fukushima, T. - Flow Patterns in stenotic blood vessel models. Biorheology, Vol. 13, pp.337-355, 1976.

6. Kawahara, M., Yoshimura, N., Nakagawa, K. and Ohsaka, H. - Steady and unsteady finite element analysis of in compressible Viscous Fluid. Int. J. Num. Methods Engg. Vol.10, pp.437-456, 1976.

7. Natarajan, R. - Front Solution Program for transmission tower Analysis. J. of Computers and Structures, Vol.5, pp.59-64, 1975.

8. Hubner,K.H. - The Finite element method for Engineers. John Wiley and Sons, New York. 1975.

CONVECTIVE HEAT TRANSFER BY AXIAL FLOW IN AN ANNULUS WITH POROUS LINING

M.N. Channabasappa, K.G. Umapathy and I.V. Nayak

Karnataka Regional Engineering College, Surathkal, P.O. Srinivasnagar - 574157, INDIA.

SUMMARY

The paper presents a theoretical model for the study of convective heat transfer by axial flow in an annulus bounded by two long concentric circular cylinders, the outer surface of the inner cylinder being provided with a non-erodible porous lining and the two bounding surfaces being maintained at constant temperatures. The analysis makes use of the 'velocity slip' and the 'temperature slip' boundary conditions at the interface and employs Brinkman model to obtain the velocity field in the porous zone. While the Navier-Stokes equation (for the free flow zone) and Brinkman's equation (for the porous zone) admit exact solutions, the energy equations in both the zones present difficulties in their exact integration. We, therefore, solve the second order boundary value problem concerning the temperature field on the computer by using a quasinumerical technique which we will call the Nested Numerical Quadrature Technique (NNQT). The influence of various parameters on the velocity and temperature fields is studied.

1. INTRODUCTION

The study of flow through and past porous media has attracted the attention of many research workers because of its application potential in industrial, biophysical and hydrological problems. Most of the available studies involve impermeable boundaries. The object of the present paper is to investigate the change in the velocity and temperature on account of the porous lining provided at one of the boundaries.

Heat transfer in an annulus

The study makes use of Brinkman model [1] for analysing the flow within the porous lining. The interaction between the velocity fields in the porous lining and outside it is brought about through the use of velocity slip boundary condition at the interface due to Beavers and Joseph [2]. Similarly the interaction between the temperature fields in the two zones is brought about by the use of temperature slip boundary condition at the interface developed by Rudraiah et al [3,4]. The velocity field in both the zones has been obtained exactly. The temperature field is obtained by using a quasi numerical approach.

2 MATHEMATICAL FORMULATION OF THE PROBLEM

The physical model (see Fig.1) consists of two long concentric circular cylinders of radii R_1 and R_2 ($>R_1$). The outer surface of the inner cylinder is lined with a non-erodible porous material of thickness b and is maintained at a constant temperature T_1 where as the outer impermeable cylinder is maintained at a constant temperature T_2 ($> T_1$).

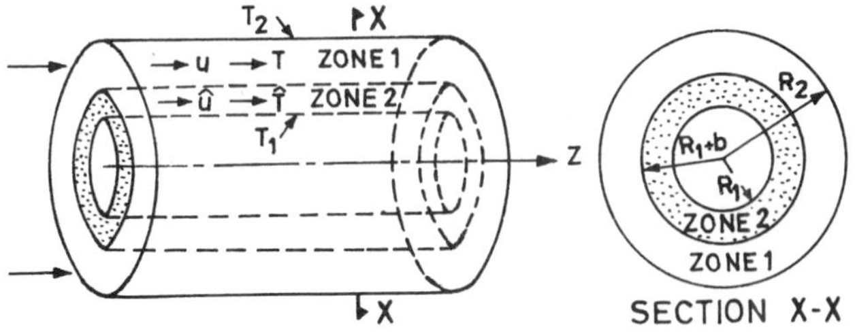

Fig.1. Physical Model

Thus the flow regime is divided into two zones: Zone 1 from the impermeable outer cylinder to the surface of the porous lining (called the nominal surface), zone 2 from the nominal surface to the impermeable inner cylinder.

2.1 Basic Equations

Assuming that (i) the flow in both the zones which is driven by a common pressure gradient dp/dz is steady and fully developed with constant fluid properties and (ii) the porous medium is homogeneous and isotropic, the equations for velocity and temperature fields are:

For zone 1:
$$\frac{d^2u}{dR^2} + \frac{1}{R}\frac{du}{dR} = \frac{1}{\mu}\frac{dp}{dz} \qquad (2.1)$$

$$K\frac{1}{R}\frac{d}{dR}\left(R\frac{dT}{dR}\right) + \mu\left(\frac{du}{dR}\right)^2 = 0 \qquad (2.2)$$

For zone 2:
$$\frac{d^2\hat{u}}{dR^2} + \frac{1}{R}\frac{d\hat{u}}{dR} - \frac{\hat{u}}{s} = \frac{1}{\mu}\frac{dp}{dz} \qquad (2.3)$$

$$K\frac{1}{R}\frac{d}{dR}\left(R\frac{d\hat{T}}{dR}\right) + \mu\left(\frac{d\hat{u}}{dR}\right)^2 + \frac{\mu}{s}\hat{u}^2 = 0 \qquad (2.4)$$

The equations (2.1) and (2.2) are the usual Navier-Stokes equation and the energy equation for the viscous flow. The equation (2.3) is the Brinkman[1] equation for flows in the porous medium and is a modification of the usual Darcy law. Similarly (2.4) is the modified form of the energy equation, s being the permeability of the porous medium.

2.2 The Boundary Conditions

(a) Velocity fields: Beavers and Joseph [2], hereafter called BJ, have postulated the existence of a streamwise slip velocity at the nominal surface in a parallel plate channel flow, which has been supported by their own experiments and by those of several others. The BJ condition is of the form

$$\left.\frac{du}{dR}\right|_{R=R_1+b} = \frac{\alpha}{\sqrt{s}}(u_B - Q), \qquad (2.5)$$

Q being the Darcy velocity in the porous medium existing away from the nominal surface and α is the slip parameter, a property of the porous material, which is taken to be known from experiments. It may be noted that the thickness of the porous medium in the BJ condition is infinite.

In order to suit our problem involving finite thickness of the porous lining, we propose to use (2.5) with a modification. In the BJ condition the slip velocity u_B at the nominal surface changes to the constant Darcy velocity Q through a thin layer called the velocity slip layer, whose thickness was given by BJ as of order $\sqrt{s}$. This thickness subsequently has been shown to be equal to $\sqrt{s}$ (see Rudraiah et al [3,4]). We identify Q in (2.5) with the value of $\hat{u}$ (velocity in zone 2) at a distance $\sqrt{s}$ below the nominal surface and thus get the boundary condition

$$u = u_B \quad \text{at} \quad R = R_1 + b \tag{2.6}$$

where u_B is given by

$$\left.\frac{du}{dR}\right|_{R=R_1+b} = \frac{\alpha}{\sqrt{s}}\left(u_B - \hat{u}\big|_{R=R_1+b-\sqrt{s}}\right) \tag{2.7}$$

The boundary condition at the outer impermeable cylinder is the usual no slip condition

$$u = 0 \quad \text{at} \quad R = R_2 \tag{2.8}$$

The boundary conditions on $\hat{u}$ are

$$\hat{u} = u_B \quad \text{at} \quad R = R_1 + b \tag{2.9}$$

$$\hat{u} = 0 \quad \text{at} \quad R = R_1 \tag{2.10}$$

(b) Temperature fields: The first boundary condition on the temperature field T is

$$T = T_2 \quad \text{at} \quad R = R_2 \tag{2.11}$$

In analogy with the BJ condition, as proposed by Rudraiah et al [3,4], and to suit the criterion of the finite thickness of the porous lining, we propose the boundary condition on the temperature at the nominal surface in the form

$$T = T_B \quad \text{at} \quad R = R_1 + b \tag{2.12}$$

where T_B is given by

$$\left.\frac{dT}{dR}\right|_{R=R_1+b} = \frac{\alpha}{\sqrt{s}}\left(T_B - \hat{T}\big|_{R=R_1+b-\sqrt{s}}\right), \tag{2.13}$$

$\hat{T}$ being the temperature in zone 2. It may be pointed out that the physical basis for the existence of the temperature slip layer is the existence of the velocity slip layer and the dependence of the temperature field on the velocity field. The boundary conditions on $\hat{T}$ are

$$\hat{T} = T_B \quad \text{at} \quad R = R_1 + b \tag{2.14}$$

$$\hat{T} = T_1 \quad \text{at} \quad R = R_1 \tag{2.15}$$

3 VELOCITY DISTRIBUTION

On using the nondimensionalisation

$$[v,\hat{v},r,\varepsilon,S,\lambda,\pi,\xi,Re,P]=[\frac{u}{\bar{u}},\frac{\hat{u}}{\bar{u}},\frac{R}{R_2},\frac{b}{R_2},$$

$$\frac{R_2}{\sqrt{s}},\frac{R_1}{R_2},\frac{p}{\frac{1}{2}\rho\bar{u}^2},\frac{z}{R_2},\frac{\rho\bar{u}R_2}{\mu},\frac{Re}{2}\frac{d\pi}{d\xi}] \qquad (3.1)$$

with $\bar{u}$ denoting the average velocity in zone 1 of Fig.1, the solution of (2.1) under the boundary conditions (2.6) and (2.8) is

$$v=\frac{P}{4}r^2+\frac{\log r}{\log(\lambda+\varepsilon)}[v_B+\frac{P}{4}\{1-(\lambda+\varepsilon)^2\}]-\frac{P}{4} \qquad (3.2)$$

To obtain v_B we proceed as follows:
The solution of (2.3) satisfying (2.9) and (2.10)is

$$\hat{v}=-\frac{P}{S^2}+C\,I_o(S\,r)+D\,K_o(S\,r) \qquad (3.3)$$

where

$$C=\frac{1}{G}[(v_B+\frac{P}{S^2})K_o(S\lambda)-\frac{P}{S^2}K_o(S\,\overline{\lambda+\varepsilon})] \qquad (3.4)$$

$$D=\frac{1}{G}[\frac{P}{S^2}I_o(S\,\overline{\lambda+\varepsilon})-(v_B+\frac{P}{S^2})I_o(S\lambda)] \qquad (3.5)$$

in which

$$G=I_o(S\,\overline{\lambda+\varepsilon})K_o(S\lambda)-K_o(S\,\overline{\lambda+\varepsilon})I_o(S\lambda), \qquad (3.6)$$

I_o, K_o being the modified Bessel functions of first and second kind of order zero.
On using (3.2) and (3.3) in (2.7) we get, after some simplififation,

$$v_B=\frac{-\frac{P}{2}(\lambda+\varepsilon)-\frac{P}{4}[\frac{1-(\lambda+\varepsilon)^2}{(\lambda+\varepsilon)\log(\lambda+\varepsilon)}]+\frac{\alpha P}{S}-\frac{\alpha P}{SG}[\{K_o(S\lambda)-K_o(S\,\overline{\lambda+\varepsilon})\}I_o(S\,\overline{\lambda+\varepsilon-\frac{1}{S}})+\{I_o(S\,\overline{\lambda+\varepsilon})-I_o(S\lambda)\}K_o(S\,\overline{\lambda+\varepsilon-\frac{1}{S}})]}{\frac{1}{(\lambda+\varepsilon)\log(\lambda+\varepsilon)}+\alpha S[-1+\frac{1}{G}\{K_o(S\lambda)I_o(S\,\overline{\lambda+\varepsilon-\frac{1}{S}})-I_o(S\lambda)K_o(S\,\overline{\lambda+\varepsilon-\frac{1}{S}})\}]} \qquad (3.7)$$

It is important to note that in view of the non-dimensional form of (2.7), $\varepsilon>1/S$. Also the physical configuration is such that $0<\varepsilon<1$.

In order to bring out the effect of porous lining on the mass flow rate in the annulus, we compare m(mass flow rate with porous lining) with m*(mass flow rate without porous lining), where

$$m = 2\pi \int_{\lambda+\varepsilon}^{1} vr\, dr \tag{3.8}$$

$$m^* = 2\pi \int_{\lambda}^{1} v^* r\, dr, \tag{3.9}$$

v^* being the velocity distribution in the annulus without porous lining and is the solution of (2.1) under the boundary conditions $v^* = 0$ at $r = 1$ and at $r = \lambda$.

4 TEMPERATURE DISTRIBUTION

On using the nondimensionalisation

$$[\theta, \hat{\theta}, P_R E] = [\frac{T-T_1}{T_2-T_1}, \frac{\hat{T}-T_1}{T_2-T_1}, \frac{\mu\, \bar{u}^2}{K(T_2-T_1)}], \tag{4.1}$$

with $P_R E$ denoting the product of Prandtl number and Eckert number, the solution of (2.2) under the boundary conditions (2.11) and (2.12) is

$$\theta = -P_R E[\frac{P^2}{64} r^4 + \frac{A_1^2}{2}(\log r)^2 + \frac{PA_1}{4} r^2] + C_1 \log r + C_2$$

where (4.2)

$$C_2 = 1 + P_R E(\frac{P^2}{64} + \frac{PA_1}{4}) \tag{4.3}$$

$$C_1 = \frac{1}{\log(\lambda+\varepsilon)}[\theta_B - C_2 + P_R E \{ \frac{P^2}{64}(\lambda+\varepsilon)^4 + \frac{A_1^2}{2}(\log \overline{\lambda+\varepsilon})^2 + \frac{PA_1}{4}(\lambda+\varepsilon)^2 \}] \tag{4.4}$$

To obtain θ_B we proceed as follows:
Using (4.1) and (3.1) in (2.4) we get

$$\frac{1}{r}\frac{d}{dr}(r \frac{d\hat{\theta}}{dr}) = - P_R E[(\frac{d\hat{v}}{dr})^2 + S^2 \hat{v}^2] \tag{4.5}$$

On using (3.3) this becomes

$$\frac{d}{dr}(r \frac{d\hat{\theta}}{dr}) = B(r) \tag{4.6}$$

where

$$B(r) = -r P_R E[\{CS\, I_1(Sr) - DS\, K_1(Sr)\}^2 + \{- \frac{P}{S} + CS\, I_o(Sr) + DS\, K_o(Sr)\}^2], \tag{4.7}$$

I_1, K_1 being the modified Bessel functions of first and second kind of order one.
Integrating (4.6) w.r.t.r twice in $[\lambda, r]$ and using (2.15) we get

$$\hat{\theta} = \lambda\eta(\log r - \log\lambda) + f(r) \tag{4.8}$$

where $f(r) \equiv \int_\lambda^r \frac{A(r)}{r}\,dr$, $A(r) \equiv \int_\lambda^r B(r)\,dr$

and
$$\eta = \left.\frac{d\hat{\theta}}{dr}\right|_{r=\lambda} = \frac{\theta_B - f(\lambda+\varepsilon)}{\lambda\log(1+\frac{\varepsilon}{\lambda})}, \tag{4.9}$$

which is obtained by using (2.14) in (4.8).
On using (4.2) and (4.8) in (2.13) we get, after some simplification,

$$\theta_B = \frac{\alpha S C_5 + \frac{C_3}{\lambda+\varepsilon} - P_R E\left[\frac{P^2}{16}(\lambda+\varepsilon)^3 + \frac{A_1^2\log(\lambda+\varepsilon)}{\lambda+\varepsilon} + \frac{PA_1}{2}(\lambda+\varepsilon)\right]}{\alpha S(1-C_4) - \frac{1}{(\lambda+\varepsilon)\log(\lambda+\varepsilon)}} \tag{4.10}$$

where

$$\left.\begin{aligned} C_3 &= \frac{1}{\log(\lambda+\varepsilon)}\left[-C_2 + P_R E\left\{\frac{P^2}{64}(\lambda+\varepsilon)^4 + \frac{A_1^2}{2}(\log\overline{\lambda+\varepsilon})^2 + \frac{PA_1}{4}(\lambda+\varepsilon)^2\right\}\right] \\ C_4 &= \frac{\log(\lambda+\varepsilon-\frac{1}{S}) - \log\lambda}{\log(1+\frac{\varepsilon}{\lambda})} \\ C_5 &= -C_4\, f(\lambda+\varepsilon) + f(\lambda+\varepsilon-\tfrac{1}{S}). \end{aligned}\right\} \tag{4.11}$$

The Nusselt numbers at the outer and inner cylinders respectively are given by

$$(Nu)_o = \left.\frac{d\theta}{dr}\right|_{r=1} = -P_R E\left(\frac{P^2}{16} + \frac{PA_1}{2}\right) + C_1 \tag{4.12}$$

$$(Nu)_i = \left.\frac{d\hat{\theta}}{dr}\right|_{r=\lambda} = \eta \tag{4.13}$$

5 NUMERICAL TECHNIQUE FOR COMPUTING THE TEMPERATURE FIELDS

The temperature fields θ and $\hat{\theta}$ in zones 1 and 2 respectively are known if $f(r)$ is known. As the integrand $B(r)$ given by (4.7) is a complicated expression, obtaining (in closed form) its integral

A(r) and again that of A(r) as f(r) is a formidable problem. So, we propose to evaluate A(r) and f(r) for any specified r by using a Numerical Quadrature formula in a nested fashion. The process involves nesting in the sense that the several values of A(r) required in the evaluation of f(r) with the help of a chosen numerical quadrature formula are themselves got through the use of the same quadrature formula as applied to B(r). This technique, we call, The Nested Numerical Quadrature Technique (NNQT). The numerical quadrature formula used in the present computations is Simpson's one-third rule. We prefer this for its simplicity and efficiency.

The computation of f(r) is achieved by writing three subprograms (see Appendix) - A function subprogram defining B(r) which is known, a subroutine subprogram for the evaluation of A(r) and a second subroutine subprogram for the evaluation of f(r). The second subroutine employs the first subroutine and the function subprogram by using them as its arguments. The subroutine for the evaluation of two integrals contain ACQ (the accuracy) as an argument which when specified will give the values of the integrals correct to that accuracy. The subroutines are designed to improve the values of integrals through the use of interval halving till the absolute value of the difference between two successive values become less than or equal to ACQ. For a specified set of parameters the computation of the velocity and temperature fields at ten points in each of the zones 1 and 2 took about two seconds with ACQ = 0.000001 and with Double precision arithmetic on the DEC 1090 system.

6 DISCUSSION OF RESULTS AND CONCLUSIONS

Expressions have been obtained for the velocity and temperature in zones 1 and 2, mass flow rate in the annulus and Nusselt numbers at the walls. Numerical values of these quantities have been obtained for different combinations of S and ε from which the following conclusions can be drawn:

(i) Inspite of the reduction of the region of free flow in the annulus on account of the porous lining, the mass flow rate increases (in comparison with the mass flow rate in the annulus without porous lining) upto a certain thickness of the porous lining. It is also seen that for a given thick-

ness, the mass flow rate increases with increasing permeability.
(ii) The velocity and temperature in the annulus are enhanced on account of the porous lining. The present results reduce to the corresponding ones in the annulus without porous lining in the limit of $S \longrightarrow \infty$ and $\varepsilon \longrightarrow 0$.
(iii) The Nusselt number at the inner wall decreases with increasing ε, and for a given thickness increases with increasing permeability while it exhibits the opposite behaviour at the outer wall (Fig. 2).

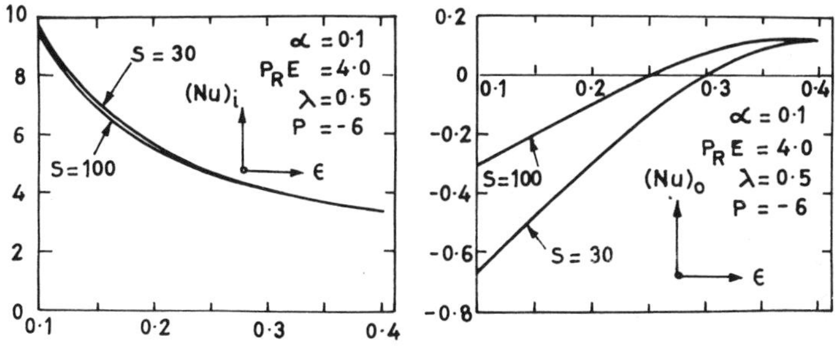

Fig.2. Nusselt numbers at the walls.

The numerical technique developed in this paper is of general utility in the sense that it can be used to solve the energy equations in many regions of flow of practical importance like the parallel plate channel, the circular pipe and the annular region. It would be particularly useful in heat transfer problems involving Non-Newtonian fluids whose velocity fields are described by highly complicated expressions. Also the technique can be used to solve similar higher order two-point boundary-value problems using any Quadrature formula in a nested fashion an appropriate number of times. The only difficulty that may be encountered is the rounding error, which can be minimised with the help of Double or Triple precision arithmetic.

REFERENCES

1 BRINKMAN, H.C. - A Calculation of the Viscous Force Exerted by a Flowing Fluid on a Dense Swarm of Particles. Appl.Sci. Res.Al, pp. 27-34, 1947.
2 BEAVERS, G.S. and JOSEPH, D.D. - Boundary Conditions at a Naturally Permeable Wall.

J.Fluid Mech, Vol.30, pp.197 - 207, 1967.
3 RUDRAIAH, N. and VEERABHADRAIAH, R. - Effect of Buoyancy on the Free Surface Flow Past a Permeable Bed. Wärme-und Stoffübertragung, 11, pp. 265 - 275, 1978.
4 RUDRAIAH, N. and VEERABHADRAIAH, R. - Temperature Distribution in Couette Flow Past a Permeable Bed. Proc. Indian Acad. Sci. Vol.86A, No.6, pp. 537 - 547, Dec.1977.

APPENDIX

```
C     SUBPRØGRAMS FØR NNQT
C     SUBRØUTINE F FØR SECØND INTEGRATIØN ØF ENERGY
C     EQUATIØN BY SIMPSØNS ØNE-THIRD RULE.
C     ARGUMENTS: (AL,Y,ACQ,G,A,B)=(LØWER LIMIT,
C     UPPER LIMIT,ACCURACY,VALUE ØF THE INTEGRAL,
C     SUBRØUTINE A, FUNCTIØN B).
      SUBRØUTINE F (AL,Y,ACQ,G,A,B)
      IMPLICIT REAL*8(A-H,Ø-Z)
      EXTERNALB
      H=(Y-AL)/2.
      C=0.0
      CALLA(AL,Y,ACQ,D,B)
      D=D/Y
      SUM1=C+D
      SUMA=0.
      X=AL+H
      SUM2=0.
    3 CØNTINUE
      CALLA(AL,X,ACQ,E,B)
      E=E/X
      SUM2=SUM2+E
      X=X+H
      IF(X+H/2.-Y)3,50,4
    4 SUM4=0.
      X=AL+H/2.
   10 CØNTINUE
      CALLA(AL,X,ACQ,AA,B)
      AA=AA/X
      SUM4=SUM4+AA
      X=X+H
      IF(X-Y)10,50,20
   20 SIMP=H/6. *(SUM1+2.*SUM2+4. *SUM4)
   50 CØNTINUE
      G=SIMP
      IF(DABS(SIMP-SUMA)-ACQ)1,1,30
   30 SUM2=SUM2+SUM4
      SUMA=SIMP
      H=H/2.
      GØTØ4
```

Heat transfer in an annulus

```
    1 RETURN
      END
C     SUBR0UTINE A F0R FIRST INTEGRATI0N 0F ENERGY
C     EQUATI0N BY SIMPS0NS 0NE-THIRD RULE -
C     ARGUMENTS: (AL,Y,ACQ,A1,B)= (L0WER LIMIT,
C     UPPER LIMIT,ACCURACY,VALUE 0F THE INTEGRAL,
C     FUNCTI0N B)
      SUBR0UTINE A(AL,Y,ACQ,A1,B)
      IMPLICIT REAL*8(A-H,0-Z)
      H=(Y-AL)/2.
      SUM1=B(AL)+B(Y)
      SUMA=0.
      X=AL+H
      SUM2=0.
    3 SUM2=SUM2+B(X)
      X=X+H
      IF(X+H/2.-Y)3,50,4
    4 SUM4=0.
      X=AL+H/2.
   10 SUM4=SUM4+B(X)
      X=X+H
      IF(X-Y)10,50,20
   20 SIMP=H/6.*(SUM1+2.*SUM2+4.*SUM4)
   50 C0NTINUE
      A1=SIMP
      IF(DABS(SIMP-SUMA)-ACQ)1,1,30
   30 SUM2=SUM2+SUM4
      SUMA=SIMP
      H=H/2.
      G0T04
    1 RETURN
      END
C     FUNCTI0N SUBPR0GRAM B(R)
      FUNCTI0NB(R)
      IMPLICIT REAL*8(A-H,0-Z)
      SR=SIG*R
      CALL I0(SR,SI0)
      CALL INUE(SR,1,SI0,SI)
      CALL BESK(SR,0,BK6,KER)
      CALL BESK(SR,1,BK7,NER)
      Z1=-P/SIG+C*SIG*SI0+D*SIG*BK6
      Z2=Z1*Z1
      Z3=C*SIG*SI-D*SIG*BK7
      Z4=Z3*Z3
      B=-PRE*R*(Z2+Z4)
      RETURN
      END
```